Methods in Enzymology

Volume XXVIII
COMPLEX CARBOHYDRATES
Part B

METHODS IN ENZYMOLOGY

EDITORS-IN-CHIEF

Sidney P. Colowick   Nathan O. Kaplan

Methods in Enzymology

Volume XXVIII

Complex Carbohydrates

Part B

EDITED BY

Victor Ginsburg

NATIONAL INSTITUTE OF ARTHRITIS AND METABOLIC DISEASES
NATIONAL INSTITUTES OF HEALTH
BETHESDA, MARYLAND

1972

ACADEMIC PRESS   New York and London

ACADEMIC PRESS, INC.
111 Fifth Avenue, New York, New York 10003

United Kingdom Edition published by
ACADEMIC PRESS, INC. (LONDON) LTD.
24/28 Oval Road, London NW1

LIBRARY OF CONGRESS CATALOG CARD NUMBER: 54-9110

PRINTED IN THE UNITED STATES OF AMERICA

Table of Contents

Section I. Analytical Methods

Section II. Preparations

Section III. Purification and Properties of Carbohydrate-Binding Proteins

Section IV. Enzymes of Sugar Activation

Section V. Synthesis of Complex Carbohydrates

Section VI. Degradation of Complex Carbohydrates

Contributors to Volume XXVIII

Article numbers are in parentheses following the names of contributors.
Affiliations listed are current.

ROSALIE J. ACKERMAN (127), *Department of Biochemistry and Biophysics, Iowa State University, Ames, Iowa*

B. B. L. AGRAWAL (34), *Department of Physiology, Wayne State University, Detroit, Michigan*

K. M. L. AGRAWAL (92, 93), *Department of Biochemistry, State University of New York at Buffalo, Buffalo, New York*

JAMES K. ALEXANDER (130, 131), *Department of Biological Chemistry, Hahneman Medical College, Philadelphia, Pennsylvania*

DAVID AMINOFF (97), *Henry Simpson Memorial Institute for Medical Research, University of Michigan, Ann Arbor, Michigan*

HELMUT ANKEL (53, 76, 77), *Department of Biochemistry, Medical College of Wisconsin, Milwaukee, Wisconsin*

GILBERT ASHWELL (14, 15, 17), *National Institute of Arthritis and Metabolic Diseases, National Institutes of Health, Bethesda, Maryland*

OM P. BAHL (92, 93, 93a, 93b, 94, 95), *Department of Biochemistry, State University of New York at Buffalo, Buffalo, New York*

JOHN R. BAKER (7, 85), *School of Dentistry and Department of Medicine, Institute of Dental Research, University of Alabama in Birmingham, Birmingham, Alabama*

CLINTON E. BALLOU (66), *Department of Biochemistry, University of California at Berkeley, Berkeley, California*

DIETHARD BARON (57), *Biological Institute II, Freiburg University, Freiburg, West Germany*

ROBERT W. BARTON (121), *Department of Medicine, Washington University School of Medicine, St. Louis, Missouri*

SHMARYAHU BLUMBERG (43, 44 Add.), *Department of Biophysics, Weizmann Institute of Science, Rehovoth, Israel*

OLGA O. BLUMENFELD (21), *Department of Biochemistry, Albert Einstein College of Medicine, Bronx, New York*

ANNA MARIE BOWSER (54), *Department of Microbiology, University of Pittsburgh School of Medicine, Pittsburgh, Pennsylvania*

ROSCOE O. BRADY (109, 110, 115, 116), *National Institute of Neurological Diseases and Stroke, National Institutes of Health, Bethesda, Maryland*

J. L. BRESLOW (120), *National Heart and Lung Institute, National Institutes of Health, Bethesda, Maryland*

BARBARA ILLINGWORTH BROWN (106), *Department of Biological Chemistry, Washington University School of Medicine, St. Louis, Missouri*

DAVID H. BROWN (106), *Department of Biological Chemistry, Washington University School of Medicine, St. Louis, Missouri*

NORMAN E. BROWN (73), *Hoffmann-La Roche Medical Laboratories, Nutley, New Jersey*

ENRICO CABIB (80, 137), *National Institute of Arthritis and Metabolic Diseases, National Institutes of Health, Bethesda, Maryland*

MICHAEL CANTZ (121), *University of Kiel Kinderklinik, Kiel, Germany*

DON M. CARLSON (2, 3, 26), *Department of Biochemistry, Case Western Re-*

serve University School of Medicine, Cleveland, Ohio

HÉCTOR CARMINATTI (137), Instituto de Investigaciones Bioquimicas, Buenos Aires, Argentina

J. P. CHAMBERS (62), Department of Biochemistry, University of Texas Medical School at San Antonio, San Antonio, Texas

STEPHEN CHIPOWSKY (96), Department of Biology, Johns Hopkins University, Baltimore, Maryland

J. ANTHONY CIFONELLI (7), Department of Pediatrics, University of Chicago, Chicago, Illinois

PHILIP COHEN (134), Department of Biochemistry, University of Washington, Seattle, Washington

B. COLVIN (67, 68), Department of Biochemistry, Oklahoma State University, Stillwater, Oklahoma

PEDRO CUATRECASAS (122), Department of Pharmacology, Johns Hopkins University School of Medicine, Baltimore, Maryland

M. A. CYNKIN (82), Department of Biochemistry and Pharmacology, Tufts University School of Medicine, Boston, Massachusetts

GLYN DAWSON (114), Departments of Pediatrics and Biochemistry, University of Chicago School of Medicine, Chicago, Illinois

PARIMAL R. DESAI (46), Department of Immunology, Northwestern University, Evanston, Illinois

JACK DISTLER (63), Rackham Arthritis Research Unit, University of Michigan, Ann Arbor, Michigan

K. E. EBNER (67, 68), Department of Biochemistry, Oklahoma State University, Stillwater, Oklahoma

FRANK EISENBERG, JR. (11), National Institute of Arthritis and Metabolic Diseases, National Institutes of Health, Bethesda, Maryland

ALAN D. ELBEIN (62, 70, 71, 78, 140), Department of Biochemistry, The University of Texas Medical School at San Antonio, San Antonio, Texas

WALTER J. ESSELMAN (8, 10), Department of Biochemistry, Michigan State University, East Lansing, Michigan

MARILYNN ETZLER (39), Department of Biochemistry and Biophysics, University of California at Davis, Davis, California

DER-FONG FAN (52, 55, 56), Department of Microbiology, University of Pittsburgh School of Medicine, Pittsburgh, Pennsylvania

DAVID SIDNEY FEINGOLD (52, 54, 55, 56), Department of Microbiology, University of Pittsburgh School of Medicine, Pittsburgh, Pennsylvania

ANN HEY FERGUSON (140), Department of Biochemistry, University of Texas Medical School at San Antonio, San Antonio, Texas

EDMOND H. FISCHER (133, 134), Department of Biochemistry, University of Washington, Seattle, Washington

D. K. FITZGERALD (67, 68), Department of Biochemistry, Oklahoma State University, Stillwater, Oklahoma

W. T. FORSEE (62), Department of Biochemistry, University of Texas Medical School at San Antonio, San Antonio, Texas

MICHAEL FOSSET (133), Department of Biochemistry, University of Washington, Seattle, Washington

JEFFREY FOX (74), Department of Biochemistry and Biophysics, University of California at Davis, Davis, California

FRANK E. FRERMAN (83), Department of Microbiology, Medical College of Wisconsin, Milwaukee, Wisconsin

OTHMAR GABRIEL (30, 59), Department of Biochemistry, Georgetown University Schools of Medicine and Dentistry, Washington, D. C.

WILLIAM GALBRAITH (35), Riker Laboratories, Minnesota Mining and Manufacturing Center, St. Paul, Minnesota

MARY A. GAUNT (53), Department of Biochemistry, Medical College of Wisconsin, Milwaukee, Wisconsin

MOHAMMED ALI GHALAMBOR (139), Department of Biochemistry, Pahlavi University of Medicine, Shiraz, Iran

J. M. GILBERT (82), National Heart and Lung Institute, National Institutes of Health, Bethesda, Maryland

VICTOR GINSBURG (13), National Institute of Arthritis and Metabolic Diseases, National Institutes of Health, Bethesda, Maryland

LUIS GLASER (58, 136), Department of Biological Chemistry, Washington University School of Medicine, St. Louis, Missouri

C. P. J. GLAUDEMANS (47), National Institute of Arthritis and Metabolic Diseases, National Institutes of Health, Bethesda, Maryland

SARA H. GOLDEMBERG (132), Instituto de Investigaciones Bioquimicas, Buenos Aires, Argentina

I. J. GOLDSTEIN (16, 34, 35), Department of Biological Chemistry, University of Michigan, Ann Arbor, Michigan

PEDRO GONZALEZ-PORQUÉ (60), Department of Biochemistry, Harvard University, Boston, Massachusetts

J. A. GORDON (44 Add.), Department of Biophysics, Weizmann Institute of Science, Rehovoth, Israel

SYDNEY GOVONS (74), Department of Biochemistry and Biophysics, University of California at Davis, Davis, California

ROBERT W. GREEN (38), Biochemistry Department, Duke University Medical Center, Durham, North Carolina

ELAINE GREENBERG (27, 28), Department of Biochemistry and Biophysics, University of California at Davis, Davis, California

WALTER T. GREGORY (40), Department of Medicine, Washington University School of Medicine, St. Louis, Missouri

HANS GRISEBACH (57, 61), Biological Institute II, Freiburg University, Freiburg, West Germany

SEN-ITIROH HAKOMORI (9, 19), Department of Microbiology, University of Washington School of Medicine, Seattle, Washington

STEN HAMMARSTRÖM (45), Department of Immunology, Wenner Gren Institute, Stockholm, Sweden

J. S. HAWKER (75), Department of Biochemistry and Biophysics, University of California at Davis, Davis, California

EDWARD C. HEATH (29, 48, 49, 83, 139), Department of Biochemistry, University of Pittsburgh School of Medicine, Pittsburgh, Pennsylvania

TORSTEN HELTING (85), Behringwerke A. G., Marburg/Lahn, West Germany

KARL HIMMELSPACH (18), Max-Planck-Institüt für Immunbiologie, Freiburg/Br., West Germany

PETER HOVINGH (123), Department of Biological Chemistry, University of Utah College of Medicine, Salt Lake City, Utah

CALDERON HOWE (20), Department of Microbiology, Louisiana State University Medical Center, New Orleans, Louisiana

HANAKO ISHIHARA (29, 48, 49), Department of Internal Medicine, Nagoya University School of Medicine, Nagoya, Japan

RICHARD L. JACKSON (5), Department of Medicine, Baylor College of Medicine and the Methodist Hospital, Houston, Texas

PETER L. JEFFREY (106), Monash University, Department of Biochemistry, Clayton, Australia

WILLIAM G. JOHNSON (115, 116), National Institute of Neurological Diseases and Stroke, National Institutes of Health, Bethesda, Maryland

GEORGE W. JOURDIAN (63), Rackham Arthritis Research Unit, University of Michigan, Ann Arbor, Michigan

MICHAEL M. KABACK (117), Department of Pediatrics, The Johns Hopkins University School of Medicine, Baltimore, Maryland

A. JOSEPH KALB (43), Department of

Biophysics, Weizmann Institute of Science, Rehovoth, Israel

EDWARD L. KEAN (51, 138), Departments of Ophthalmology and Biochemistry, Case Western Reserve University School of Medicine, Cleveland, Ohio

S. KIRKWOOD (31), Department of Biochemistry, University of Minnesota, St. Paul, Minnesota

GERD KLEINHAMMER (18), Max-Planck-Institüt für Immunbiologie, Freiburg/Br., West Germany

AKIRA KOBATA (24, 69), Department of Biochemistry, Kobe University School of Medicine, Kobe, Japan

MICHIAKI KOHNO (102), Department of Biochemistry, Kyoto University, Kyoto, Japan

ROSALIND KORNFELD (40), Department of Medicine, Washington University School of Medicine, St. Louis, Missouri

STUART A. KORNFELD (40), Department of Medicine, Washington University School of Medicine, St. Louis, Missouri

HANS KRESSE (121), Physiologisch-Chemisches Institute, University of Münster, Münster, West Germany

ROGER A. LAINE (8, 10, 62), Department of Biochemistry, University of Texas Medical School at San Antonio, San Antonio, Texas

DAVID F. LAPP (70), Department of Biochemistry, University of Texas Medical School at San Antonio, San Antonio, Texas

JOSEPH LARNER (73), Department of Pharmacology, University of Virginia School of Medicine, Charlottesville, Virginia

LUCILLE T. LEE (20), Department of Microbiology, Columbia University College of Physicians and Surgeons, New York, New York

Y. C. LEE (6, 89), Department of Biology, The Johns Hopkins University, Baltimore, Maryland

L. LEHLE (72), Biology Department, Regensburg University, Regensburg, West Germany

LORETTA LEIVE (23), National Institute

of Arthritis and Metabolic Diseases, National Institutes of Health, Bethesda, Maryland

W. J. LENNARZ (79), Department of Physiological Chemistry, The Johns Hopkins University School of Medicine, Baltimore, Maryland

SU-CHEN LI (90, 91), Department of Biochemistry, Tulane University School of Medicine, New Orleans, Louisiana

YU-TEH LI (90, 91), Department of Biochemistry, Tulane University of Medicine, New Orleans, Louisiana

ULF LINDAHL (85a), Institute of Medical Chemistry, University of Uppsala, Uppsala, Sweden

BENGT LINDBERG (12), Institution für Organisk Kemi, Stockholms Universitet, Stockholm, Sweden

ALFRED LINKER (123), Department of Biological Chemistry, University of Utah College of Medicine, Salt Lake City, Utah

HALINA LIS (44, 44 Add.), Department of Biophysics, Weizmann Institute of Science, Rehovoth, Israel

TEH-YUNG LIU (4), Biology Department, Brookhaven National Laboratory, Upton, New York

KENNETH O. LLOYD (20), Department of Dermatology, Columbia University College of Physicians and Surgeons, New York, New York

CHERYL R. McBROOM (16), Department of Biological Chemistry, University of Michigan, Ann Arbor, Michigan

M. J. McDONALD (93a), Department of Biochemistry, State University of New York at Buffalo, Buffalo, New York

EDWARD J. McGUIRE (96), Department of Biology, The Johns Hopkins University, Baltimore, Maryland

WILLIAM L. McLELLAN (126), Department of Pathology, Columbia University, College of Physicians and Surgeons, New York, New York

FRANK MALEY (25, 99, 100, 101), Developmental Biochemistry Laboratories, State of New York Department of Health, Albany, New York

V. T. Marchesi (22, 42), *National Institute of Arthritis and Metabolic Diseases, National Institutes of Health, Bethesda, Maryland*

Donald M. Marcus (135), *Departments of Medicine, Microbiology and Immunology, Albert Einstein College of Medicine, Bronx, New York*

Luis R. Maréchal (132), *Instituto de Investigaciones Bioquimicas, Buenos Aires, Argentina*

Martin B. Mathews (7), *Department of Biochemistry, University of Chicago, Chicago, Illinois*

Isamu Matsumoto (36), *Faculty of Pharmaceutical Sciences, University of Tokyo, Tokyo, Japan*

K. L. Matta (94, 95), *Department of Biochemistry, State University of New York at Buffalo, Buffalo, New York*

Mike Matula (71, 140), *Department of Biochemistry, University of Texas Medical School at San Antonio, San Antonio, Texas*

John Mauck (136), *Department of Biological Chemistry, Washington University School of Medicine, St. Louis, Missouri*

R. Mawal (67, 68), *Department of Biochemistry, Oklahoma State University, Stillwater, Oklahoma*

Miriam Meisler (108), *Roswell Park Memorial Institute, State of New York Department of Health, Buffalo, New York*

John J. Mieyal (129), *Department of Pharmacology, Northwestern University Medical School, Chicago, Illinois*

Mike Mitchell (71), *Department of Biochemistry, University of Texas Medical School at San Antonio, San Antonio, Texas*

George Mook (116), *National Institute of Neurological Diseases and Stroke, National Institutes of Health, Bethesda, Maryland*

Anatol G. Morell (14), *Albert Einstein College of Medicine, Department of Medicine, Bronx, New York*

David C. Morrison (23), *Department of Experimental Pathology, Scripps Clinic and Research Foundation, La Jolla, California*

Larry W. Muir (133), *Department of Biochemistry, University of Washington, Seattle, Washington*

L. Müller (82), *Department of Microbiology, University of Connecticut Health Center, Farmington, Connecticut*

Elizabeth F. Neufeld (121), *National Institute of Arthritis and Metabolic Diseases, National Institutes of Health, Bethesda, Maryland*

Larry Nielson (133), *Department of Biochemistry, University of Washington, Seattle, Washington*

Tadayoshi Okumura (103), *Department of Biochemistry, Kyoto University, Kyoto, Japan*

R. Ortmann (61), *Biological Institute II, Freiburg University, Freiburg, West Germany*

Toshiaki Osawa (36, 37), *Faculty of Pharmaceutical Sciences, University of Tokyo, Tokyo, Japan*

M. J. Osborn (82), *Department of Microbiology, University of Connecticut Health Center, Farmington, Connecticut*

Hachiro Ozaki (50), *Department of Biochemistry and Biophysics, University of California at Davis, Davis, California*

J. L. Ozbun (75), *Department of Biochemistry and Biophysics, University of California at Davis, Davis, California*

Betty W. Patterson (70, 140), *Department of Biochemistry, University of Texas Medical School at San Antonio, San Antonio, Texas*

John H. Pazur (86, 128), *Department of Biochemistry, Pennsylvania State University, University Park, Pennsylvania*

James J. Plantner (3), *Department of Biochemistry, Case Western Reserve University School of Medicine, Cleveland, Ohio*

T. H. PLUMMER, JR. (100), *Departmental Biochemistry Laboratories, State of New York Department of Health, Albany, New York*

R. D. PORETZ (41), *Department of Biochemistry and Molecular Biology, University of Kansas Medical Center, Kansas City, Kansas*

JANET L. POTTER (81), *Department of Biochemistry, University of Texas Medical School at San Antonio, San Antonio, Texas*

MICHAEL POTTER (47), *National Cancer Institute, National Institutes of Health, Bethesda, Maryland*

JACK PREISS (27, 28, 50, 74, 75), *Department of Biochemistry and Biophysics, University of California at Davis, Davis, California*

NORMAN S. RADIN (32, 64, 65, 111, 113), *Neuroscience Laboratory, University of Michigan, Ann Arbor, Michigan*

JOHN F. ROBYT (127), *Department of Biochemistry and Biophysics, Iowa State University, Ames, Iowa*

LENNART RODÉN (7, 85), *Department of Pediatrics, University of Chicago School of Medicine, Chicago, Illinois*

SAUL ROSEMAN (26, 96), *Department of Biology, The Johns Hopkins University, Baltimore, Maryland*

ADDISON M. ROSENKRANS (73), *Department of Pharmacology, University of Virginia School of Medicine, Charlottesville, Virginia*

HARVEY J. SAGE (38), *Biochemistry Department, Duke University Medical Center, Durham, North Carolina*

COLLEEN H. SAMANEN (16), *Department of Biological Chemistry, University of Michigan, Ann Arbor, Michigan*

HEINRICH SANDERMAN (57), *Biological Institute II, Freiburg University, Freiburg, West Germany*

HARRY SCHACHTER (29, 48), *Department of Biochemistry, University of Toronto, Toronto, Canada*

MALKA SCHER (79), *Department of Physiological Chemistry, The Johns Hopkins University School of Medicine, Baltimore, Maryland*

KEITH K. SCHLENDER (73), *Department of Pharmacology and Therapeutics, Medical College of Ohio, Toledo, Ohio*

JOHN S. SCHUTZBACH (53, 76, 77), *Department of Biochemistry, Medical College of Wisconsin, Milwaukee, Wisconsin*

NANCY B. SCHWARTZ (85), *Department of Pediatrics, University of Chicago, Chicago, Illinois*

JERE P. SEGREST (5), *National Institute of Arthritis and Metabolic Diseases, National Institutes of Health, Bethesda, Maryland*

NATHAN SHARON (44, 44 Add.), *Department of Biophysics, Weizmann Institute of Science, Rehovoth, Israel*

LAURA C. SHEN (13), *Department of Pharmacology, University of Virginia School of Medicine, Charlottesville, Virginia*

BADER SIDDIQUI (9), *Department of Microbiology, University of Washington School of Medicine, Seattle, Washington*

M. SINGH (82), *Department of Microbiology, University of Connecticut Health Center, Farmington, Connecticut*

HOWARD R. SLOAN (118, 119, 120), *National Heart and Lung Institute, National Institutes of Health, Bethesda, Maryland*

CARL H. SMITH (73), *Department of Pathology, Washington University School of Medicine, St. Louis, Missouri*

SIMONETTA SONNINO (137), *Centro de Investigaciones Medicas Albert Einstein, Buenos Aires, Argentina*

MARY JANE SPIRO (84), *Department of Medicine, Harvard Medical School and The Elliott P. Joslin Research Laboratory, Boston, Massachusetts*

ROBERT G. SPIRO (1, 84), *Department of Biological Chemistry, Harvard University Medical School and The Elliott P. Joslin Research Laboratory, Boston, Massachusetts*

GEORG F. SPRINGER (46), *Department of Microbiology, Northwestern University, Evanston, Illinois*

PHILIP D. STAHL (107), *Department of Molecular Biology, Vanderbilt University, Nashville, Tennessee*

K. J. STONE (33), *Department of Biochemistry, Harvard University, Boston, Massachusetts*

ALLEN C. STOOLMILLER (85), *Department of Pediatrics, University of Chicago, Chicago, Illinois*

JACK L. STROMINGER (33, 60, 87, 88), *Biological Laboratories, Harvard University, Cambridge, Massachusetts*

KAZUYUKI SUGAHARA (98), *Department of Biochemistry, Kyoto University, Kyoto, Japan*

T. SUKENO (100), *Developmental Biochemistry Laboratories, State of New York Department of Health, Albany, New York*

KUNIHIKO SUZUKI (112), *Rose F. Kennedy Center for Research in Mental Retardation and Human Development, Albert Einstein College of Medicine, Bronx, New York*

SAKARU SUZUKI (124, 125), *Department of Chemistry, Nagoya University, Nagoya, Japan*

N. SWAMINATHAN (94), *Department of Biochemistry, State University of New York at Buffalo, Buffalo, New York*

CHARLES C. SWEELEY (8, 10), *Department of Biochemistry, Michigan State University, East Lansing, Michigan*

JOHN F. TALLMAN (109), *National Institute of Neurological Diseases and Stroke, National Institutes of Health, Bethesda, Maryland*

W. TANNER (72), *Biology Department, Regensburg University, Regensburg, West Germany*

ANTHONY L. TARENTINO (99, 100, 101), *Developmental Biochemistry Laboratories, State of New York Department of Health, Albany, New York*

OSCAR TOUSTER (107), *Department of Molecular Biology, Vanderbilt University, Nashville, Tennessee*

SATOSHI TOYOSHIMA (37), *Faculty of Pharmaceutical Sciences, University of Tokyo, Tokyo, Japan*

FREDERIC TROY (83), *Department of Biochemistry, University of California School of Medicine at Davis, Davis, California*

JAY UMBREIT (88), *Biological Laboratories, Harvard University, Cambridge, Massachusetts*

MARC URAM (54), *Department of Microbiology, University of Pittsburgh School of Medicine, Pittsburgh, Pennsylvania*

LEE VAN LENTEN (15), *National Institute of Arthritis and Metabolic Diseases, National Institutes of Health, Bethesda, Maryland*

CARLOS VILLAR-PALASI (73), *Department of Pharmacology, University of Virginia School of Medicine, Charlottesville, Virginia*

L. WARD (58), *Department of Biological Chemistry, Washington University School of Medicine, St. Louis, Missouri*

NEAL J. WEINREB (110), *National Institute of Neurological Diseases and Stroke, National Institutes of Health, Bethesda, Maryland*

ROBERT A. WEISMAN (81), *Department of Biochemistry, University of Texas Medical School at San Antonio, San Antonio, Texas*

BERNARD WEISSMANN (104, 105), *Department of Biological Chemistry, University of Illinois College of Medicine, Chicago, Illinois*

ECKARD WELLMANN (57), *Biological Institute II, Freiburg University, Freiburg, West Germany*

GARY G. WICKUS (87), *Biological Laboratories, Harvard University, Cambridge, Massachusetts*

SADAKO YAMAGATA (85), *Aïchi Cancer Research Institute, Nagoya, Japan*

TATSUYA YAMAGATA (85), *Department of Chemistry, Nagoya University, Nagoya, Japan*

IKUO YAMASHINA (98, 102, 103), *Depart-*

ment of Biochemistry, Kyoto University, Kyoto, Japan

JOSEPH YARIV (43), Department of Biophysics, Weizmann Institute of Science, Rehovoth, Israel

EDWARD C. YUREWICZ (139), Department of Biochemistry and Biophysics, University of California at Davis, Davis, California

JURIS ZALITIS (54), School of Bio-

chemistry, University of New South Wales, Kensington, Australia

HAROLD ZARKOWSKY (58), Department of Biological Chemistry, Washington University School of Medicine, St. Louis, Missouri

BINA ZVILICHOVSKY (21), Unit for Research in Aging, Albert Einstein College of Medicine, Bronx, New York

Preface

This volume covers material on complex carbohydrates that has appeared in the literature since the publication in 1966 of Volume VIII on this subject. It has the same subdivisions as Volume VIII plus a section on the purification and properties of carbohydrate-binding proteins.

It is my pleasure to thank all the authors for their contributions that made this volume possible. I would also like to thank Miss Juanita Harris for her secretarial assistance, Miss Donna-Beth Howe for expertly preparing the Subject Index, and the staff of Academic Press for their friendly cooperation.

VICTOR GINSBURG

METHODS IN ENZYMOLOGY

EDITED BY

Sidney P. Colowick and Nathan O. Kaplan

VANDERBILT UNIVERSITY
SCHOOL OF MEDICINE
NASHVILLE, TENNESSEE

DEPARTMENT OF CHEMISTRY
UNIVERSITY OF CALIFORNIA
AT SAN DIEGO
LA JOLLA, CALIFORNIA

METHODS IN ENZYMOLOGY

EDITORS-IN-CHIEF

Sidney P. Colowick   Nathan O. Kaplan

VOLUME XX. Nucleic Acids and Protein Synthesis (Part C)
Edited by KIVIE MOLDAVE AND LAWRENCE GROSSMAN

VOLUME XXI. Nucleic Acids (Part D)
Edited by LAWRENCE GROSSMAN AND KIVIE MOLDAVE

VOLUME XXII. Enzyme Purification and Related Techniques
Edited by WILLIAM B. JAKOBY

VOLUME XXIII. Photosynthesis (Part A)
Edited by ANTHONY SAN PIETRO

VOLUME XXIV. Photosynthesis and Nitrogen Fixation (Part B)
Edited by ANTHONY SAN PIETRO

VOLUME XXV. Enzyme Structure (Part B)
Edited by C. H. W. HIRS AND SERGE N. TIMASHEFF

VOLUME XXVI. Enzyme Structure (Part C)
Edited by C. H. W. HIRS AND SERGE N. TIMASHEFF

VOLUME XXVII. Enzyme Structure (Part D)
Edited by C. H. W. HIRS AND SERGE N. TIMASHEFF

VOLUME XXVIII. Complex Carbohydrates (Part B)
Edited by VICTOR GINSBURG

Section I

Analytical Methods

[1] Study of the Carbohydrates of Glycoproteins

By Robert G. Spiro

In recent years there has been an increase in the study of glyco-proteins because of the recognition of their great biological importance in the form of transport proteins of plasma, enzymes of many types, gonadotrophins, immunoglobulins, phytohemagglutinins, collagens, and extracellular membranes.[1] Particular attention has been focused on the carbohydrate-containing proteins as components of plasma membranes, on which they act as antigenic determinants, serve as hormone and virus receptors, and play a role in cell surface interactions.

Analysis of these diverse glycoproteins requires reliable and sensitive methods for measuring their sugar components, which include D-galactose, D-mannose, D-glucose, L-fucose, D-xylose, L-arabinose, N-acetyl-D-gluco-samine, and N-acetyl-D-galactosamine, as well as the sialic and uronic acids. Furthermore, methods are necessary for elucidating the structure of the carbohydrate units and determining their peptide attachment. The techniques used in structural analysis can also serve to modify the car-bohydrate units in the preparation of acceptors for the purpose of studying their enzymatic assembly and evaluating the role which the carbohydrate plays in determining the biological properties of the glycoproteins.

Many procedures useful for accomplishing these purposes have been described in an earlier volume of this series.[2] It is our purpose to describe more recently developed methods that expand the tools available for the analysis and structural characterization of the carbohydrate units of glycoproteins.

Analysis of Monosaccharides

Various hydrolytic, chromatographic, and colorimetric techniques for the release, identification, and estimation of the sugar components of glycoproteins have been previously described (Vol. 8 [1]). Since that time there have come into wide use methods for the separation and esti-mation of neutral sugars and hexosamines that rely on automated ion exchange chromatography and gas–liquid chromatography and yield improved sensitivity, resolution, and accuracy.

---

[1] R. G. Spiro, *Annu. Rev. Biochem.* **39**, 599 (1970).
[2] R. G. Spiro, Vol. 8 [1] and [2].

Determination of Neutral Sugars

Analysis of the neutral sugars of glycoproteins can be carried out satisfactorily by borate-complex anion exchange chromatography or by gas–liquid chromatography. Either procedure is preferable to the quantitative paper chromatographic procedure (Vol. 8 [1]), which, however, is sufficiently reliable to be employed if the specialized equipment required for the other methods is not available.

For the release of neutral sugars, dilute solutions of glycoproteins (1–5 mg/ml) should be hydrolyzed in 1–2 N HCl or H_2SO_4 for 4–6 hours in sealed tubes at 100°. Conditions for optimal release will vary with the protein under study depending on the stability of the glycosidic linkages to be split and the resistance to acid degradation of the released sugars. Hydrolysis in 1 N HCl for 4–6 hours at 100° has proved to be satisfactory for a number of glycoproteins. Higher recovery of monosaccharides is obtained if the acid is diluted from glass-distilled, constant-boiling HCl. Since fucose is particularly labile under the conditions necessary to liberate the more internal sugars, milder conditions of acid hydrolysis are required, or direct determination of the sugar by the Dische-Shettles cysteine-sulfuric acid method (Vol. 8 [1]) may be employed.

After acid hydrolysis, a neutral sugar fraction can be obtained by passage of the diluted hydrolyzate through coupled columns of Dowex 50 (H^+ form) and Dowex 1 (formate form) (Vol. 8 [1]). The neutral sugars in this fraction can be analyzed directly by borate-complex ion exchange chromatography or after derivatization by gas–liquid chromatography. It is recommended that prior to the use of these techniques paper chromatography in several solvent systems be carried out to ensure proper identification of the neutral sugars present in the sample.

Automated Borate-Complex Anion Exchange Chromatography

The demonstration that the borate complexes of neutral sugars can be separated by anion exchange chromatography[3] has formed a basis for their determination in glycoproteins, and a number of automated systems have been reported.[4-8]

The Technicon system, to be described here with some modifications,

[3] J. X. Khym and L. P. Zill, J. Amer. Chem. Soc. 74, 2090 (1952).
[4] J. G. Green, Nat. Cancer Inst. Monogr. 21, 447 (1966).
[5] Technicon Sugar Chromatography Brochure (Technicon Chromatogr. Corp., Ardsley, New York, 1966) and Technicon Development Bull. No. 124, (1968).
[6] J. I. Ohms, J. Zec, J. V. Benson, Jr., and J. A. Patterson, Anal. Biochem. 20, 51 (1967).
[7] Y. C. Lee, J. F. McKelvy, and D. Lang, Anal. Biochem. 27, 567 (1969).
[8] E. F. Walborg, Jr., D. B. Ray, and L. E. Öhrberg, Anal. Biochem. 29, 433 (1969).

has proved well suited for the separation of the sugars present in glycoproteins and has given values in close agreement with those obtained by
quantitative paper chromatography.

Reagents and Buffers

Orcinol reagent. For 1 liter of this reagent, 1 g of orcinol (K & K
Laboratories) is dissolved in 70% H_2SO_4. Fresh reagent should
be made up prior to each analysis.
Internal standard, rhamnose, 0.25 mM
Buffer 1, 0.1 M boric acid, pH 8.0
Buffer 2, 0.1 M boric acid, pH 8.0, containing 0.02 M sodium sulfate
Buffer 3, 0.1 M boric acid, pH 8.0, containing 0.04 M sodium sulfate
Buffer 4, 0.2 M boric acid, pH 8.0
Buffer 5, 0.2 M boric acid, pH 9.5, containing 0.08 M sodium sulfate
Potassium tetraborate ($K_2B_4O_7 \cdot 4H_2O$), 10%

The buffers are made by dissolving the appropriate amount of boric
acid in distilled water and titrating to the indicated pH with 2 N NaOH.
The sodium sulfate is conveniently added as a 1.0 M solution prior to
titration and adjustment to volume.

Column and Resin. Type-S Chromo-Beads, low pressure (Technicon),
anion exchange resin is packed into a water-jacketed glass column (6 ×
750 mm) to a height of 650 mm. The column temperature is maintained
at 45° ± 0.5° with a circulating water bath (Haake).

Instruments. Components of the Technicon sugar chromatography
system, which include a 9-chamber autograd, high pressure micropump,
pressure gauge, mixing coil, proportioning pump, heating bath (95°)
with two reaction coils (40 feet long each, 1.6 mm internal diameter)
connected in series, water-jacketed cooling coil, fluorescent sensitizing
lamp and coil, colorimeter with 420 nm filter and 15 mm flow cell, and
logarithmic recorder. Acidflex tubing (Technicon) is used for all solutions
containing H_2SO_4.

Procedure. The resin is regenerated into the borate form in the morning by passage of the 10% potassium tetraborate solution (0.75 ml per
minute) for 4 hours. It is then equilibrated with Buffer 1 at the same rate
for 1.5 hours. The sample, which should contain 0.05–0.5 micromole of
each neutral sugar to be determined is taken to dryness in *vacuo* (Evapomix, Buchler) at 40° after the addition of 1 ml of the internal standard
(0.25 μmole of rhamnose). It is then dissolved in 0.5 ml of Buffer 1
and applied to the column under nitrogen pressure and washed in with
0.2 ml of the same buffer.

Elution is achieved with the following gradient in the nine-chambered autograd: chamber 1, 25 ml of Buffer 1 and 20 ml of Buffer 2; chambers 2 and 3, 15 ml of Buffer 3 and 30 ml of Buffer 4; chambers 4 through 7, 22 ml of Buffer 4 and 23 ml of Buffer 5; chambers 8 and 9, 45 ml of Buffer 5. The flow rate during elution is 0.48 ml per minute.

Since the entire gradient is completed at 14 hours (glucose emerges at about 9 hours), a reservoir containing Buffer 5 equipped with a time-activated valve is set to empty into the autograd at 12 hours. This enables liquid to move through the heating coils throughout the night and prevents formation of deposits on the interior of the coils from the reagents. In the morning, the coils should be extensively washed with distilled water (2–3 hours), 20% methanol (0.5 hour), and again with distilled water.

Good separation of all the sugars occurring in glycoproteins can be achieved with this gradient (Fig. 1), with the exception of arabinose and fucose. Although these two sugars have not as yet been reported to occur together in a glycoprotein, this possibility can be assessed by paper chromatography in 1-butanol–ethanol–water (10:1:2) for 4 days, in which the R_{Glc} of arabinose is 1.33 while that of fucose is 1.86.

The quantity of each sugar in a sample analyzed by the borate-complex chromatography is determined by comparing the area under its peak with that of the internal standard, rhamnose. This is best accomplished by means of the Technicon Integrator-Calculator (logarithmic), but can also be done by manual measurement of net height (in absorbancy units) and the width at half height (in cm). The molor color yield of each sugar varies in this system, and color adjustment factors relative to rhamnose must be determined from the ratio of the area under the rhamnose peak (0.25 μmole) to the area under the peak of an equi-

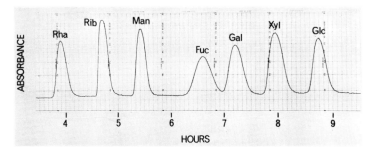

FIG. 1. Separation of a standard mixture of neutral sugars (0.25 μmole of each) by borate-complex anion exchange chromatography. The elution of rhamnose, ribose, mannose, fucose, galactose, xylose, and glucose is shown. The conditions for the chromatography are given in the text.

molar amount of the other sugar. These factors are used to calculate the actual amount of each component according to the following equation:

$$\text{Micromoles of sugar} = \frac{\text{area under sugar peak}}{\text{area under rhamnose peak}}$$

$$\times\ 0.25 \times \text{color adjustment factor}$$

The color adjustment factors vary significantly with different gradients and should be determined for each system. For the gradient described above, the values listed in Table I have been obtained.

A more gradual gradient in which the sodium sulfate concentration of buffer 5 is 37 mM and the column jacket is kept at 60° has also resulted in an effective separation of the neutral sugar components.[9]

Borate-complex anion exchange chromatography can also be employed to separate oligosaccharides containing neutral sugars and/or sugar acids obtained during graded acid hydrolysis or alkaline borohydride treatment of glycoproteins. It has been used to separate the reduced di- and trisaccharides of galactose obtained after alkaline borohydride treatment of earthworm cuticle collagen as well as the reduced glucuronic acid-mannose disaccharide formed after such treatment of *Nereis* cuticle collagen.[10]

TABLE I

RELATIVE ELUTION TIMES AND COLOR ADJUSTMENT FACTORS OF NEUTRAL
SUGARS DURING AUTOMATED BORATE-COMPLEX ANION
EXCHANGE CHROMATOGRAPHY[a]

Sugar	Relative elution time[b]	Color adjustment factor[c]
L-Rhamnose	1.00	1.00
D-Ribose	1.22	0.51
D-Mannose	1.40	0.74
L-Fucose	1.72	1.05
D-Galactose	1.88	0.87
D-Xylose	2.06	0.63
D-Glucose[d]	2.27	0.84

[a] Chromatography at 45° under conditions described in text.

[b] Elution time of rhamnose is 3 hours, 52 minutes.

[c] Ratio of area under rhamnose peak to that of area under peak of equimolar amount of the sugar.

[d] D-Glucuronic acid has a relative elution time of 2.57.

[9] T. Arima, M. J. Spiro, and R. G. Spiro, *J. Biol. Chem.* **247**, 1825 (1972).

[10] R. G. Spiro and V. D. Bhoyroo, *Fed. Proc., Fed. Amer. Soc. Exp. Biol.* **30**, 1223 (1971).

Gas–Liquid Chromatography of Alditol Acetates

Since it was shown that alditol acetates of monosaccharides can be well separated by gas–liquid chromatography,[11] these derivatives have been used to analyze the neutral sugar components present in glycoproteins.[12,13] The alditol acetates have an advantage over other derivatives commonly employed, such as the trimethylsilyl ethers, in that the reduction of the carbonyl group performed during their preparation eliminates the possibility of anomer formation or ring isomerization and therefore results in only one peak for each sugar. The details of the method as employed in our laboratory are here given.

Reagents

Anhydrous pyridine. Prepared by refluxing for 1 hour with ninhydrin (10 g/liter) followed by distillation. Stored in an open bottle in a desiccator containing NaOH pellets.

Acetic anhydride

Sodium borohydride, $0.25 M$, freshly made up in cold distilled water

Dowex 50-X4, 200–400 mesh (H^+ form) (analytical grade)

Acetic acid, $1 N$

Methanol

Internal standard, 2-deoxy-D-glucose, 0.15 mM

Column and Column Packing

Two glass coiled columns, 2 mm I.D. × 1.83 m, packed with 3% (w/w) ECNSS-M on Gas Chrom Q, 100–120 mesh, one to serve as the analytical and the other the reference column.

Instrument

Perkin-Elmer Model 900 or equivalent instrument equipped with dual column oven, dual flame ionization detector, linear temperature programmer, linear recorder, nitrogen carrier gas.

Procedure. To samples (neutral sugar fraction from acid hydrolyzate) containing about 0.05–0.30 μmoles of each neutral sugar is added 1.0 ml of the internal standard (0.15 μmoles of 2-deoxyglucose) in 18 × 150 nm tubes. The solution is taken to dryness in vacuo at 40° (Evapomix). Re-

[11] J. S. Sawardeker, J. H. Sloneker, and A. Jeanes, Anal. Chem. 37, 1602 (1965).

[12] J. H. Kim, B. Shome, T.-H. Liao, and J. G. Pierce, Anal. Biochem. 20, 258 (1967).

[13] W. F. Lehnhardt and R. J. Winzler, J. Chromatogr. 34, 471 (1968).

duction is accomplished by adding to the dry samples 2.0 ml of the fresh sodium borohydride solution. After the samples have stood at 2° overnight, 3.0 ml of water is added and the excess borohydride is destroyed by acidification to pH 4.5 with acetic acid. The samples are then passed through columns (8.5 mm internal diameter) containing 2.0 g of the Dowex 50 resin. The columns are washed with 25 ml of water, and the effluent plus wash is taken by dryness by lyophilization or on a vacuum rotator. The boric acid is then volatilized in a vacuum rotator as the methyl borate at 37° by several additions of methanol. The samples are transferred with water to 18 × 150 mm borosilicate tubes, taken to dryness *in vacuo* (Evapomix) and then further dried overnight in a vacuum desiccator over Drierite.

Acetylation is carried out by the addition of 0.25 ml of the pyridine and 0.25 ml of acetic anhydride to the dry sample. The tubes are quickly sealed with a flame and heated in a 100° bath for 1 hour. The samples are then rapidly dried in the opened tubes with a stream of dry nitrogen at 37° in an electric heating block or an oil bath. They are then transferred to small capped polyethylene tubes with a minimal amount of the pyridine and again dried with a stream of dry nitrogen. They are stored in the capped tubes in a jar with Drierite at −20° until use.

Prior to analysis the sample is redissolved in an appropriate amount of the pyridine (about 20–30 μl), and 1–2 μl of this solution is injected into the analytical column. For the analysis of such samples, the instrument can usually be set at an amplifier range of 10 and an attenuation of ×4 to ×16. The frequent occurrence of artifactual components (contributed to by the lack of specificity of the detecting system compared, for example, to the much more specific colorimetric reaction which is employed after borate column chromatography) necessitates the use of relatively high attenuations and prevents utilization of the full sensitivity of the instrument.

For optimal separation with this instrument and column, the following conditions for chromatography should be employed: carrier gas (nitrogen) flow, 20 ml/min; column temperature, initial, 160° for 4 minutes; final, 205°; program rate, 1° per minute; injector temperature, 250°; manifold temperature, 225°; chart speed, 60 inches per hour. To compensate for the shifting baseline due to the "bleeding" of the column, it is important to employ a well balanced reference column. All the neutral sugars in glycoproteins separate under these conditions (Fig. 2) and their relative retention times are given in Table II.

The quantity of each sugar in a sample is determined by comparing the area under its peak to that of the internal 2-deoxyglucose standard.

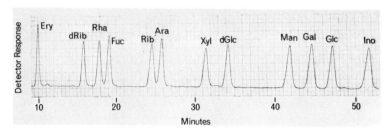

Fig. 2. Gas–liquid chromatography of alditol acetates of a neutral monosaccharide mixture showing the emergence of the acetates of erythritol, 2-deoxyribitol, rhamnitol, fucitol, ribitol, arabinitol, xylitol, 2-deoxyglucitol, mannitol, galactitol, glucitol, and myoinositol. The conditions for the chromatography are given in the text.

This area can be calculated by planimetry, by measurement of net height and width at half height, or with the Technicon Integrator-Calculator (linear).

The molar response relative to 2-deoxyglucose varies with each sugar and response adjustment factors for this system are given in Table II. These factors are calculated from the ratio of the area under the 2-deoxyglucose peak to the area under the peak of an equimolar

TABLE II

RELATIVE RETENTION TIMES AND RESPONSE ADJUSTMENT FACTORS OF ALDITOL
ACETATES DURING GAS–LIQUID CHROMATOGRAPHY[a]

Alditol	Relative retention time[b]	Response adjustment factor[c]
Erythritol	0.29	1.18
2-Deoxy-D-ribitol	0.46	1.22
L-Rhamnitol	0.52	1.02
L-Fucitol	0.56	0.95
Ribitol	0.72	0.93
L-Arabinitol[d]	0.76	0.92
Xylitol	0.92	1.02
2-Deoxy-D-glucitol	1.00	1.00
D-Mannitol	1.22	0.83
Galactitol	1.29	0.85
D-Glucitol	1.37	0.91
Myoinositol	1.50	0.78

[a] Conditions of chromatography described in text.

[b] Retention for 2-deoxy-D-glucitol (pentaacetate) is 35 minutes.

[c] Ratio of area under 2-deoxy-D-glucitol peak to that of area under peak of an equimolar amount of the alditol.

[d] L-Arabinose and L-lyxose produce the same alcohol.

amount of the sugar. To calculate the amount of each component the following equation is used:

$$\text{Micromoles of sugar} = \frac{\text{area under sugar peak}}{\text{area under 2-deoxyglucose peak}}$$
$$\times\ 0.15 \times \text{response adjustment factor}$$

Gas–liquid chromatography on ECNSS-M columns, such as described above, can be effectively used for the separation of the alditol acetates of methylated neutral monosaccharides.[14] For this purpose, the neutral sugars obtained after acid hydrolysis of methylated glycoproteins or glycopeptides are reduced with sodium borohydride and acetylated with acetic anhydride in pyridine as described for the unsubstituted monosaccharides. We have found that with an initial column temperature of 150° for 6 minutes and a program rate of 1° per minute to 205° good resolution of many of the tetra-, tri-, di-, and mono-O-methyl derivatives of galactose, mannose, and glucose can be achieved.

This gas–liquid chromatography system is also useful for the separation as their acetates of polyols resulting from the reduction of the products of periodate oxidation. Erythritol emerges ahead of the pentose and hexose components (Fig. 2). It has been our experience that the triacetate of glycerol can best be separated using an initial column temperature of 150° for 8 minutes, under which circumstances it has an elution time of 3.5 minutes.

Determination of Hexosamines and Hexosaminitols

The hydrolysis of glycoproteins for the determination of hexosamines is best accomplished in sealed tubes at 100° in 4 N HCl for 6 hours. In most cases it is convenient to first separate the amino sugars from neutral sugars and some of the amino acids and peptides by adsorption on and elution from a small column of Dowex 50 (H⁺ form) and to survey this fraction by paper chromatography (Vol. 8 [1]).

If both glucosamine and galactosamine are present, these two sugars can best be separated and determined by cation exchange chromatography using the amino acid analyzer. In most cases this can be accomplished directly with the Technicon system employing the gradient used for amino acid separation. However, in the presence of large amounts of amino acids it is necessary to perform the analysis after equilibrating the column with a buffer at pH 5.0. At this pH the amino sugars are well separated from the neutral amino acids, which emerge close to the

---

[14] H. Björndal, B. Lindberg, and S. Svensson, *Acta Chem. Scand.* **21**, 1801 (1967); see also this volume [12].

void volume, from each other, and from the basic amino acids, which are retarded. Good separation is achieved when the column is equilibrated with 0.1 M (Na$^+$) citrate buffer at pH 5.0, and elution is performed with a gradient detailed in a later section of this chapter dealing with the separation of hydroxylysine-linked carbohydrate units.

Alternatively, glucosamine and galactosamine can be separated on the amino acid analyzer in citrate–borate buffers, which can also be employed for the separation of the alcohols of these two sugars. Hexosaminitols may result from studies involving borohydride reduction of oligosaccharides during reducing sugar end-group determinations and from investigations employing alkaline borohydride for the cleavage by β-elimination of glycopeptide bonds involving serine and/or threonine. While glucosamine and glucosaminitol separate well on the amino acid analyzer (Technicon) with the usual gradient or in the pH 5.0 gradient, glucosaminitol, galactosamine, and galactosaminitol do not resolve from each other in these systems. These three components also fail to separate by paper chromatography, although glucosamine and glucosaminitol can be resolved in the Fischer–Nebel system[15] in which pyridine–ethyl acetate–water–acetic acid (5:5:3:1) is used as the solvent in the trough and pyridine–ethyl acetate–water (11:40:6) is placed on the bottom of the chromatographic cabinet (R_{GlcN} of glucosaminitol is 0.87).

Chromatography in Citrate–Borate Buffers

Separation of glucosamine and galactosamine and their respective alcohols has been achieved by employing citrate–borate buffers.[16,17] We have found the procedure here described to be effective employing the Technicon amino acid analyzer.

Reagents and Buffer

Ninhydrin reagent. For 1 liter of this reagent, 6.55 g of ninhydrin (Pierce) and 365 mg of hydrindantin (Pierce) are dissolved in 575 ml of methyl Cellosolve and then bubbled with nitrogen for 15 minutes, after which 240 ml of water followed by 185 ml of 4 M sodium acetate buffer, pH 5.5, are added, and the mixture is bubbled an additional 15 minutes with nitrogen. Fresh reagent should be made up prior to each analysis. This reagent is used in all the methods employing the amino acid analyzer described in this article.

Citrate-borate buffer, pH 5.28, containing 0.2 M Na$^+$, 0.35 M boric

[15] F. G. Fischer and H. J. Nebel, *Hoppe-Seyler's Z. Physiol. Chem.* **302**, 10 (1955).
[16] A. S. R. Donald, *J. Chromatogr.* **35**, 106 (1968).
[17] A. M. Bella, Jr., and Y. S. Kim, *J. Chromatogr.* **51**, 314 (1970).

acid. For 10 liters of this buffer, 0.67 mole of trisodium citrate dihydrate (197 g) and 3.5 moles of boric acid (216 g) are dissolved in 8000 ml of deionized water to which 100 ml of a 33% (w/v) aqueous solution of BRIJ 35 is added. The pH is titrated to 5.28 with 6 N HCl, and the solution is made to volume with deionized water.

Internal standard, phenylalanine, 0.25 mM

Column and Resin

Type A Chromo-Beads (Technicon) packed into a water-jacketed glass column (0.6 × 140 cm) to a height of 130 cm. This type of column is used in all the methods employing the amino acid analyzer described in this chapter.

Procedure. The Technicon analyzer system is used with the column jacket temperature at 60° and a flow rate of 0.48 ml per minute. After regeneration with 0.2 N NaOH, the column is equilibrated with the above buffer until a constant pH is attained.

The amino sugar sample, which has been adsorbed on and eluted from a small Dowex 50 column, and contains 0.05–0.75 μmole of each component is taken to dryness *in vacuo* (Evapomix) after the addition of 1.0 ml (0.25 μmole) of phenylalanine. Phenylalanine can be used as an internal standard as this amino acid does not elute with HCl from the small Dowex 50 columns under the conditions employed.

The sample is placed on the column in the pH 5.28 buffer, and elution is performed with the same buffer. The elution times relative to phenylalanine, which emerges at 1 hour 46 minutes, are as follows: galactosaminitol, 2.69 glucosaminitol, 2.79 glucosamine, 3.90 galactosamine, 4.48

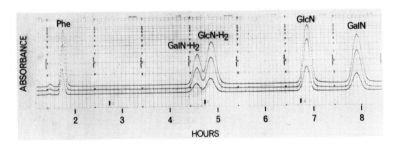

Fig. 3. Chromatography of hexosamines and hexosaminitols on the amino acid analyzer (Technicon) with a pH 5.28 citrate-borate buffer. The position of elution of phenylalanine (0.25 μmole) and of galactosaminitol, glucosaminitol, glucosamine, and galactosamine (0.50 μmole of each) is shown. The conditions for chromatography are given in the text.

(Fig. 3). The neutral amino acids and tyrosine emerge before phenylalanine while the basic amino acids elute considerably later than the galactosamine. The color yield of the hexosaminitols with ninhydrin is lower than that of the hexosamines and color adjustment factors for each should be determined under the conditions employed.

Mannosamine, which has so far not been reported to occur in glycoproteins, does not separate adequately from galactosamine in either the regular or citrate–borate buffer system. However, this sugar can be distinguished from galactosamine by ninhydrin degradation (Vol. 8 [1]), in which it yields arabinose rather than lyxose.

Preparation and Separation of Glycopeptides

As a first step in the elucidation of the structure of the carbohydrate units of glycoproteins and the nature of the glycopeptide bond, it is useful to digest the protein extensively with a protease of broad specificity. After such digestion, the carbohydrate of the glycoprotein may usually be obtained with a minimum number of amino acids attached. Fractionation of such glycopeptides permits the separation of carbohydrate units of different types, and also makes possible the resolution of a given type of unit into its variants which almost always exist owing to the microheterogeneity imparted by the nature of the biosynthetic mechanism involved. A study of the intact protein generally fails to reveal the differences which may occur in its carbohydrate units.

The investigation of glycopeptides facilitates the identification of the amino acid and sugar involved in the glycopeptide bond. Glycopeptides moreover are more accessible to degradation by glycosidases than the intact protein and studies involving periodate oxidation, methylation, and graded acid hydrolysis are more easily performed and interpreted than on the undigested protein.

Proteolytic Digestion

A variety of proteases have been successfully employed for the preparation of glycopeptides, but the enzyme Pronase from *Streptomyces griseus* has been used most effectively. To achieve maximal proteolysis, prolonged digestion (see Vol. 8 [2]) is usually carried out at 37° in the presence of toluene to prevent bacterial growth at a concentration of 25 mg of the glycoprotein per milliliter with several additions of enzyme totaling up to 2% of the substrate weight. Pronase contains less than 2% carbohydrate[9] and therefore does not contribute significant amounts of glycopeptides to the digest. On occasion Pronase has been reported to split the *N*-acetylglucosaminylasparagine bond linking neutral carbohydrate units to the peptide chain.[9] If this occurs, the resulting oligosaccharides can readily be separated from glycopeptides by ion exchange chromatography and can be identified by the finding of 1 mole of glucosaminitol per carbohydrate unit after sodium borohydride reduction.

Fractionation of Glycopeptides

Gel Filtration

In order to separate glycopeptides in a proteolytic digest from peptide material not containing carbohydrate, gel filtration in volatile buffers is the procedure of choice.

Columns of Sephadex G-25 (fine) equilibrated with 0.1 M pyridine acetate buffer, pH 5.0, and having a dimension of about 2.0 $\times$ 80 cm can fractionate digests of up to 1 g of protein, while columns of 3.0 $\times$ 80 cm have been used for digests for 5–6 g of glycoprotein.[9] The glycopeptides can usually be resolved from the peptides and amino acids, as the latter are of smaller size and penetrate more deeply into the gel. If the glycoprotein contains carbohydrate units which differ significantly in molecular weight, their glycopeptides can often be separated in the same filtration from each other as well as from the peptide material, as in the case of a digest of glomerular basement membrane (Fig. 4), where the peptides containing the disaccharide unit (peak 2) penetrated the gel, while those with the heteropolysaccharide unit appeared in the void volume.[18] The glycopeptides of porcine ribonuclease containing the mannose-N-acetylglucosamine unit have been separated from those containing the larger units made up of mannose, N-acetylglucosamine, galactose, fucose, and sialic acid, by filtration on a very long (0.9 $\times$ 400 cm) Sephadex G-25 column.[19]

When pyridine acetate buffers are used in the filtration they can readily be removed from pooled fractions by lyophilization.

Ion Exchange Chromatography

Various forms of ion exchange chromatography can be used effectively to resolve glycopeptides. DEAE-cellulose has been widely used to separate sialic acid-containing glycopeptides from each other and from those with neutral carbohydrate units (Vol. 8 [2]). Dowex 1-X2, 200–400 mesh (acetate form) has been employed at pH 5.0 to resolve acidic glucuronic acid-containing glycopeptides from *Neresis* cuticle collagen,[10] whereas chromatography on Dowex 50-X2, 200–400 mesh

[18] R. G. Spiro, *J. Biol. Chem.* **242**, 1923 (1967).
[19] R. L. Jackson and C. H. W. Hirs, *J. Biol. Chem.* **245**, 624 (1970).

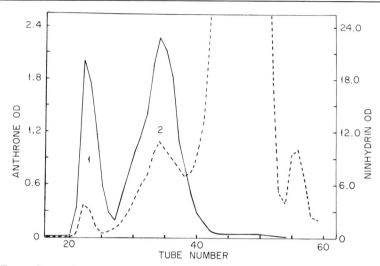

FIG. 4. Separation of glycopeptides by gel filtration on Sephadex G-25. The collagenase-Pronase digest of 400 mg of bovine glomerular basement membrane was placed on a column 2.1 × 82 cm, and elution was performed with 0.1 M pyridine acetate buffer, pH 5.0, while fractions of 5.2 ml were collected. The two glycopeptide peaks emerge prior to the large amount of carbohydrate-free peptide material. ——, Anthrone; - - -, ninhydrin. From R. G. Spiro, *J. Biol. Chem.* **242**, 1923 (1967).

(pyridine form) at pH 3.4 has resulted in the fractionation of the basic hydroxylysine-linked glycopeptides containing the glucosylgalactose units of the glomerular basement membrane.[18]

Dowex 50 chromatography at an acidic pH and low ionic strength has also been used to advantage for resolving glycopeptides containing neutral mannose-*N*-acetylglucosamine carbohydrate units of oval-bumin[20,21] and thyroglobulin.[9] Utilizing this approach, glycopeptides from a Pronase digest of thyroglobulin obtained after filtration on Sephadex G-25 have been chromatographed at room temperature on columns (1.9 × 120 cm) of Dowex 50-X2, 200–400 mesh (pyridine form) in the following manner.[9]

The resin (previously washed with 2 N NaOH, water, 3 N HCl, and water, in that order) is put into the pyridine form with a 2 M pyridine solution. It is then equilibrated with 0.10 M pyridine formate at pH 3.0, followed by 1 mM pyridine formate at pH 3.0 until the effluent reaches this pH value (large volumes of buffer are required to achieve this). The sample containing 80–800 mg of hexose is titrated to pH 3.0 with

[20] L. W. Cunningham, *in* "Biochemistry of Glycoproteins and Related Substances, Cystic Fibrosis" (E. Rossi and E. Stoll, eds.), Part II, p. 141. Karger, Basel, 1968.
[21] C.-C. Huang, H. E. Mayer, Jr., and R. Montgomery, *Carbohyd. Res.* **13**, 127 (1970).

formic acid and is placed on the column in a volume of about 25 ml. After application of the sample the column is washed with a large volume of the initial buffer (2500 ml) and chromatography is continued with a linear concentration gradient consisting of 1800 ml of 1 mM pyridine formate, pH 3.0, in the mixing chamber and 1800 ml of 0.25 M pyridine formate, pH 3.0 in the reservoir. A flow rate of 45 ml per hour is maintained and fractions of 15 ml are collected.

Such a column results in the resolution of glycopeptides containing the mannose-N-acetylglucosamine unit (Fig. 5) and demonstrates the extensive microheterogeneity which exists in the mannose content. Much of the resolution is achieved by affinity chromatography, as the amino acid composition and net charge of these compounds are very similar. The glycopeptides with the larger carbohydrate content emerge from

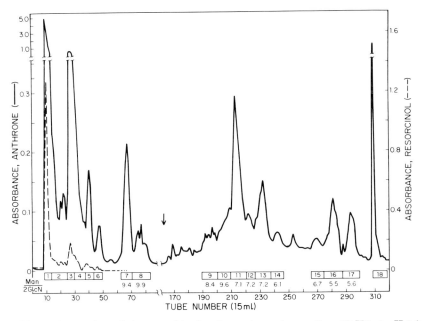

Fig. 5. Separation of glycopeptides by chromatography on Dow 50-X2 at pH 3.0. The glycopeptides (105 mg of hexose) from a Pronase digest of human thyroglobulin were applied to a column (1.9 × 120 cm) in 1 mM pyridine formate buffer, pH 3.0, followed by elution with this buffer. A linear gradient of pyridine formate buffer at this pH was started (arrow) as described in text from 1 mM (tube 171) to 0.155 M (tube 320). The molar ratio of mannose to 2 glucosamine residues is given for peaks 7–17 which contain the mannose-N-acetylglucosamine unit (unit A) of thyroglobulin. Peaks 1 to 6 contain primarily the larger carbohydrate unit made up of sialic acid, fucose, galactose, N-acetylglucosamine and mannose (unit B). From T. Arima, M. J. Spiro, and R. G. Spiro, *J. Biol. Chem.* **247**, 1825 (1972).

the column earlier than those with the smaller number of sugar residues. The peptide portion, on the other hand, has a retarding influence on the emergence of these compounds. The sialic acid-containing glycopeptides emerge with the void volume or early in the low ionic strength wash by virtue of their negative charge and large size. It is essential to include a prolonged wash with the initial buffer prior to the start of the gradient as glycopeptides which come off too rapidly to be resolved with the gradient can often be well separated during the buffer wash.

Structure of Carbohydrate Units

Studies on the structure of the carbohydrate of glycoproteins are best carried out on purified glycopeptides which contain only one type of carbohydrate unit. Optimally, variants of a unit representing different degrees of its completion should also be resolved. While it is almost always possible to separate glycopeptides containing different carbohydrate units of a given glycoprotein on the basis of charge and size, it is occasionally not possible to separate the glycopeptides containing all the variants of a unit, and heterogeneity may only become evident upon structural investigation of the carbohydrate portion. In such cases, further resolution can often be achieved by a fractionation of oligosaccharides derived from the glycopeptides.

Since some of the major tools useful in the study of the sugar sequences and linkages of glycoprotein carbohydrate units have been described (Vol. 8 [2]), only newer techniques and applications will be here set forth.

Glycosidases

One of the most effective ways to determine the sequence and anomeric configuration of the sugars in the carbohydrate units of glycoproteins is with the use of glycosidases, and in conjunction with periodate oxidation and methylation, the complete structure of a unit can often be determined. A large number of such enzymes have now been purified and characterized from a wide variety of sources (Table III) and have been used successfully to remove the various sugars occurring in glycoproteins. The preparation of some of these glycosidases are given in Section VI of this volume.

These enzymes are exoglycosidases and act by removing only the terminal nonreducing sugar from the polymer. Therefore, if the complete structure of a carbohydrate unit is to be established it is necessary to submit it to sequential digestion by a number of enzymes. After each digestion a new sugar becomes uncovered and can then be released with the appropriate glycosidase. In addition to releasing monosaccharides from

TABLE III

Enzyme	Enzyme source	Examples of glycoproteins studied[a]
Neuraminidase	*Vibrio cholerae*	Submaxillary glycoprotein,[b] fetuin,[c] red blood cell membrane glycoprotein,[d] thyroglobulin,[e] glomerular basement membrane[f]
	Clostridium perfringens	α_1-Acid glycoprotein, submaxillary glycoprotein, red blood cell membrane glycoprotein[g]
	Diplococcus pneumoniae	α_1-Acid glycoprotein[h]
β-Galactosidase	*Escherichia coli*	Fetuin,[c] glomerular basement membrane[i]
	Jack bean	Immunoglobulins,[j] red blood cell membrane glycoprotein,[k] fetuin[l]
	Phaseolus vulgaris	Fetuin,[m] chorionic gonadotrophin,[n] α_1-acid glycoprotein[o]
	Aspergillus niger	Fetuin, chorionic gonadotrophin, α_1-acid glycoprotein[p]
	Diplococcus pneumoniae	α_1-Acid glycoprotein[h]
	Clostridium perfringens	α_1-Acid glycoprotein[q]
α-Galactosidase	*Aspergillus niger*	Earthworm cuticle collagen[r]
	Coffee bean	Earthworm cuticle collagen[s]
β-N-acetylglucos-aminidase	Jack bean	Ovalbumin,[t] α_1-acid glycoprotein,[t] immunoglobulins,[j] red blood cell membrane glycoprotein,[k] thyroglobulin,[u] fetuin[l]
	Pig epididymis	Fetuin,[v] immunoglobulin,[w] transferrin[x]
	Phaseolus vulgaris	Fetuin,[m] α_1-acid glycoprotein,[m] chorionic gonadotrophin[n]
	Aspergillus niger	Fetuin, α_1-acid glycoprotein, chorionic gonadotrophin, ovalbumin[p]
	Diplococcus pneumoniae	α_1-Acid glycoprotein[y]
	Clostridium perfringens	α_1-Acid glycoprotein[q]
α-Mannosidase	Jack bean	Ovalbumin,[z] α_1-acid glycoprotein,[o] ribonuclease B,[aa] immunoglobulins,[j] thyroglobulin,[u] stem bromelain,[bb] fetuin[l]
	Phaseolus vulgaris	Chorionic gonadotrophin[n]
	Almond emulsin	Fetuin,[v] stem bromelain[bb]
	Turbo cornutus	Ovalbumin,[cc] Taka-amylase,[cc] stem bromelain[dd]
β-Mannosidase	Hen oviduct	Ribonuclease B, ovalbumin[ee]
α-N-acetylgalactos-aminidase	Porcine liver	Submaxillary glycoproteins, stomach blood group A glycoprotein[ff]
	Lumbricus terrestris	Submaxillary glycoprotein[gg]
	Clostridium perfringens	Submaxillary glycoprotein[q]
α-Fucosidase	*Clostridium perfringens*	Porcine submaxillary glycoprotein[hh]
	Rat epididymis	Immunoglobulin G, luteinizing hormone[ii]

(Continued)

TABLE III (*Continued*)

Enzyme	Enzyme source	Examples of glycoproteins studied[a]
	Aspergillus niger	Submaxillary glycoproteins[ii]
α-Glucuronidase	*Helix pomatia*	*Nereis* cuticle collagen[kk]
β-Xylosidase	*Charonia lampas*	Xylosyl protein,[ll] stem bromelain[dd]

[a] Either glycoproteins or glycopeptides derived from indicated glycoprotein were used; see references for details of preparation of substrates and enzymes. Preparation of many enzymes also given in this volume Section VI and in Vol. 8.

[b] E. R. B. Graham and A. Gottschalk, *Biochim. Biophys. Acta* **38**, 513 (1960).

[c] R. G. Spiro, *J. Biol. Chem.* **237**, 646 (1962).

[d] R. H. Kathan and R. J. Winzler, *J. Biol. Chem.* **238**, 21 (1963).

[e] R. G. Spiro and M. J. Spiro, *J. Biol. Chem.* **240**, 997 (1965).

[f] R. G. Spiro, *J. Biol. Chem.* **242**, 1915 (1967).

[g] J. T. Cassidy, G. W. Jourdian, and S. Roseman, *J. Biol. Chem.* **240**, 3501 (1965).

[h] R. C. Hughes and R. W. Jeanloz, *Biochemistry* **3**, 1535 (1964).

[i] R. G. Spiro, *J. Biol. Chem.* **242**, 4813 (1967).

[j] R. Kornfeld, J. Keller, J. Baenziger, and S. Kornfeld, *J. Biol. Chem.* **246**, 3259 (1971).

[k] R. Kornfeld and S. Kornfeld, *J. Biol. Chem.* **245**, 2536 (1970).

[l] R. G. Spiro and V. D. Bhoyroo, unpublished observations (1971).

[m] K. M. L. Agrawal and O. P. Bahl, *J. Biol. Chem.* **243**, 103 (1968).

[n] O. P. Bahl, *J. Biol. Chem.* **244**, 575 (1969).

[o] P. V. Wagh, I. Bornstein, and R. J. Winzler, *J. Biol. Chem.* **244**, 658 (1969).

[p] O. P. Bahl and K. M. L. Agrawal, *J. Biol. Chem.* **244**, 2970 (1969).

[q] S. Chipowsky and E. J. McGuire, *Fed. Proc., Fed. Amer. Soc. Exp. Biol.* **28**, 606 (1969).

[r] L. Muir and Y. C. Lee, *J. Biol. Chem.* **245**, 502 (1970).

[s] R. G. Spiro and V. D. Bhoyroo, unpublished observations (1969).

[t] S.-C. Li and Y.-T. Li, *J. Biol. Chem.* **245**, 5153 (1970).

[u] T. Arima and R. G. Spiro, *J. Biol. Chem.* **247**, 1836 (1972).

[v] R. G. Spiro, see Vol. 8 [2].

[w] J. R. Clamp and F. W. Putnam, *J. Biol. Chem.* **239**, 3233 (1964).

[x] G. A. Jamieson, M. Jett, and S. L. DeBernardo, *J. Biol. Chem.* **246**, 3686 (1971).

[y] R. C. Hughes and R. W. Jeanloz, *Biochemistry* **3**, 1543 (1964).

[z] Y.-T. Li, *J. Biol. Chem.* **241**, 1010 (1966).

[aa] A. Tarentino, T. H. Plummer, Jr., and F. Maley, *J. Biol. Chem.* **245**, 4150 (1970).

[bb] J. Scocca and Y. C. Lee, *J. Biol. Chem.* **244**, 4852 (1969).

[cc] T. Muramatsu and F. Egami, *J. Biochem.* **62**, 700 (1967).

[dd] Y. Yasuda, N. Takahashi, and T. Murachi, *Biochemistry* **9**, 25 (1970).

[ee] T. Sukeno, A. L. Tarentino, T. H. Plummer, Jr., and F. Maley, *Biochem. Biophys. Res. Commun.* **45**, 219 (1971).

[ff] B. Weissmann and D. F. Hinrichsen, *Biochemistry* **8**, 2034 (1969).

[gg] E. Buddecke, H. Schauer, E. Werries, and A. Gottschalk, *Biochem. Biophys. Res. Commun.* **34**, 517 (1969).

[hh] D. Aminoff and K. Furukawa, *J. Biol. Chem.* **245**, 1659 (1970).

[ii] R. B. Carlsen and J. G. Pierce, *J. Biol. Chem.* **247**, 23 (1972).

[jj] O. P. Bahl, *J. Biol. Chem.* **245**, 299 (1970).

[kk] R. G. Spiro and V. D. Bhoyroo, *Fed. Proc., Fed. Amer. Soc. Exp. Biol.* **30**, 1223 (1971).

[ll] M. Fukuda, T. Muramatsu, and F. Egami, *J. Biochem.* **65**, 191 (1969).

their linkages to more internal sugars, these enzymes can split O-glycosidic glycopeptide bonds, such as the N-acetylgalactosaminyl-serine (threonine), galactosyl-hydroxylysine, and xylosyl-serine linkages.

Generally the enzymes act optimally on glycopeptides where steric hindrances imposed by the peptide chain of the whole glycoprotein are not present. Modification of the glycopeptides may be necessary to permit enzymatic action, such as the N-acetylation of hydroxylysine which is required for the splitting of the galactosylhydroxylysine bond.[22] Release of a sugar can sometimes be effected by an enzyme of given specificity from one source but not from another. Therefore, from the failure to release a sugar by the action of a given glycosidase, it cannot be concluded that this component is not present in the terminal position or anomeric configuration for which the enzyme is specific. Such resistance of a terminal sugar to enzyme action is exemplified by the glucose residue of the hydroxylysine-linked glucosylgalactose disaccharide unit, which is not split by α-glucosidase (*A. niger*) treatment of its glycopeptides (native or N-acetylated) but can be readily cleaved by this enzyme from the free disaccharide obtained after partial acid hydrolysis.[22]

While the activity of glycosidases is often determined with the use of synthetic substrates, such as p-nitrophenyl compounds, these assays may bear little quantitative relevance to the action of the enzymes on glycopeptides. For maximal release, glycopeptides should be incubated at high concentrations (10–20 mg/ml), in the presence of a large amount of enzyme for extended periods of time (up to 7 days), at the pH optimum of the enzyme and at 37° in the presence of toluene. Under these conditions, other glycosidase activities which show up only as trace amounts by the brief p-nitrophenol assay, may express themselves with the release of substantial amounts of sugars other than anticipated. It is therefore essential to perform a time study of the release of the sugars. If the enzyme preparation has more than one glycosidase activity this will become evident from such a time-release study and advantage can be taken of this information in establishing sequence (Vol. 8 [2]). For example, incubation of thyroglobulin glycopeptides containing the mannose-N-acetylglucosamine unit with a jack bean α-mannosidase preparation which had some β-N-acetylglucosaminidase activity indicated that all the mannose residues were released before one N-acetylglucosamine residue was liberated indicating the internal location of the latter sugar[23] (Fig. 6).

The release of sialic acid and N-acetylhexosamines can be followed colorimetrically by performing the thiobarbituric acid reaction (Vol.

[22] R. G. Spiro, *J. Biol. Chem.* **242**, 4813 (1967).
[23] T. Arima and R. G. Spiro, *J. Biol. Chem.* **247**, 1836 (1972).

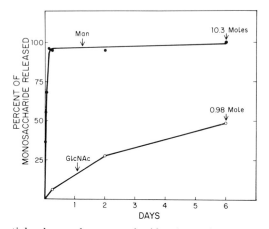

Fig. 6. Sequential release of monosaccharides from glycopeptides by jack bean glycosidases. Glycopeptides containing the mannose-N-acetylglucosamine unit (unit A) of calf thyroglobulin were incubated with a glycosidase preparation containing α-mannosidase and β-N-acetylglucosaminidase activity in a ratio of 30 to 1. The numbers given above the arrow refer to the moles of mannose and N-acetylglucosamine per mole of glycopeptide released after 6 days of digestion. All the mannose could be released prior to the release on 1 mole of N-acetylglucosamine. The second residue of N-acetylglucosamine (which is linked to asparagine) could not be removed even upon further treatment with β-N-acetylglucosaminidase. From T. Arima and R. G. Spiro, *J. Biol. Chem.* **247,** 1836 (1972).

8 [1]) and Morgan–Elson reaction[24] (after pH adjustment of sample), respectively, on aliquots of the incubation. Total hexose release can be determined after separating the released sugars from enzyme and remaining substrate by passage of aliquots of the incubation mixture through small columns of charcoal-Celite (Darco G-60, Celite 535, 1:1, w/w) Vol. 8 [2]) and performing the anthrone reaction (Vol. 8 [1]) on the effluent and water wash (30–40 ml/ml of column). Alternatively, the glycopeptides, if small enough, can be adsorbed on columns of Dowex 50-X2, 200–400 mesh (H⁺ form) and the effluent and wash from these columns containing the released monosaccharides analyzed.[23]

To determine whether more than one sugar has been released, paper chromatography of the desalted (Dowex 50 and Dowex 1) charcoal–Celite effluent and wash or Dowex 50-X2 effluent and wash should be performed. If acetate buffers are employed for the incubation, only Dowex 50 is necessary to desalt the charcoal–Celite treated material and the Dowex 50-X2 effluent and wash can be chromatographed directly.

[24] J. L. Reissig, J. L. Strominger, and L. F. Leloir, *J. Biol. Chem.* **217,** 959 (1955).

For quantitation of released neutral sugars, borate-complex chromatography or gas–liquid chromatography may be employed.

To recover the degraded glycopeptide after glycosidase digestion, it is best to employ gel filtration in pyridine acetate buffer, pH 5.0, choosing a gel (Sephadex G-25 or G-50 or Bio-Gel P-4 or P-6) which will separate it from the enzyme and the released monosaccharides. Alternatively, with small glycopeptides, the enzyme can be precipitated with trichloroacetic acid and the glycopeptides in the ether-extracted supernatant separated from released monosaccharides by adsorption on Dowex 50-X2, 200–400 mesh (H^+ form) followed by elution with cold $1.5\,N$ NH_4OH or $1.0\,M$ pyridine acetate buffer, pH 5.0.

Graded Acid Hydrolysis

Measurement of the timed release of monosaccharides during mild acid hydrolysis provides information in regard to sugar sequences (Vol. 8 [2]). Such an approach is particularly helpful if sequential release by glycosidases cannot be effected and works best if the stabilities of the glycosidic bonds of the sugars in the carbohydrate unit vary substantially, as in the case of the hydroxylysine-linked disaccharide unit of the basement membrane (Fig. 7). During hydrolysis of glycopeptides from this membrane in $0.1\,N$ H_2SO_4 at $100°$, glucose is released at a much more rapid rate than the galactose, owing to the stabilization of the galactosylhydroxylysine linkage by the positive charge on the ϵ-amino group of the hydroxylysine adjacent to the glycosidic bond.[22]

Oligosaccharides

Oligosaccharides can be isolated from partial acid hydrolyzates of glycoproteins (Vol. 8 [2]). Moreover, when the carbohydrate units are linked to the peptide by O-glycosidic bonds involving serine and/or

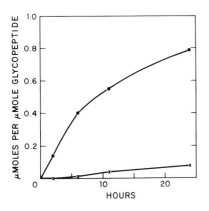

Fig. 7. Differential release of monosaccharides from glycopeptides by graded acid hydrolysis. Glomerular basement membrane glycopeptides containing glucose, galactose, and hydroxylysine in a ratio of 1:1:1 were hydrolyzed for various periods of time in $0.1\,N$ H_2SO_4 at $100°$. ●——●, Glucose; ×——×, galactose. From R. G. Spiro, *J. Biol. Chem.* **242**, 4813 (1967).

threonine, they can be obtained in the form of reduced oligosaccharides after β-elimination by mild alkaline treatment in the presence of sodium borohydride. Information in regard to the carbohydrate units of a number of glycoproteins has been obtained through a study of oligosaccharides obtained in this manner.[10,25-27]

For the formation of these reduced oligosaccharides, the concentration of alkali, temperature, and length of incubation should be sufficient to result in maximal β-elimination, and the concentration of the borohydride must be high enough to prevent alkaline degradation by a "peeling" reaction which is especially likely to take place if the terminal reducing sugar is substituted at C-3. The use of 1.0 M sodium borohydride will ensure rapid reduction and therefore stabilization of the terminal sugar before alkaline degradation can proceed.[25,28] For each protein the conditions giving the highest yield of oligosaccharides should be determined. This can often conveniently be done by measuring, after acid hydrolysis of the treated protein, the amount of alditol formed from the sugar involved in the glycopeptide bond (see also [2]).

Conditions which in our experience have proved satisfactory for the preparation of reduced oligosaccharides by β-elimination have employed 0.3 M sodium borohydride in 0.1 N sodium hydroxide (earthworm cuticle collagen) and 1.0 M sodium borohydride in 0.1 N sodium hydroxide (fetuin, alkali-labile units) at 37° for 48 hours at a protein concentration of 5–10 mg/ml. The earthworm collagen oligosaccharides are made up of $1 \rightarrow 2$ linked sugar residues and therefore do not require the high borohydride concentration to prevent degradation.

At the end of the alkaline borohydride treatment, the sample is cooled by immersion in ice, titrated with acetic acid to pH 4.0 (considerable foaming may occur at this step), diluted with 0.01 N formic acid to a sodium ion concentration of 0.1 M and passed through a column of Dowex 50-X2, 200–400 mesh (H+) containing 4 times the number of equivalents of sodium ions present, and washed with 5 to 6 column volumes of 0.01 N formic acid. After lyophilization of the effluent and wash from the columns, the boric acid is volatilized as methyl borate by several additions of methanol.

The released oligosaccharides can then be fractionated by gel filtration or charcoal–Celite, anion, or preparative paper chromatography. If the glycoprotein under study also contains carbohydrate units linked to the peptide chain by alkali-stable bonds, glycopeptides of these units

[25] D. M. Carlson, J. Biol. Chem. 243, 616 (1968).
[26] L. Muir and Y. C. Lee, J. Biol. Chem. 244, 2343 (1969).
[27] D. B. Thomas and R. J. Winzler, J. Biol. Chem. 244, 5943 (1969).
[28] J. M. Mayo and D. M. Carlson, Carbohyd. Res. 15, 300 (1970).

may be present in the Dowex 50 effluent and wash and must be separated from the oligosaccharides by these fractionation techniques.

Methylation

Application of the methylation method of Hakamori[29] to glycopeptides can provide important structural information about their carbohydrate units in regard to linkages between sugar residues and the extent of branching.

The following procedure is based on our experience with glycopeptides from the glomerular basement membrane[22] and thyroglobulin.[23]

Reagents

Sodium acetate, $4.5\,M$

Acetic anhydride

Dimethyl sulfoxide

Methylsulfinyl carbanion, $1\,M$, freshly prepared in the following manner. Sodium hydride (obtained as a 50% dispersion in mineral oil from Alfa Inorganics, Beverly, Massachusetts) is added rapidly to dimethyl sulfoxide (48 mg of the dispersion per milliliter) in an 18×150 mm sidearm tube containing a small magnet and fitted at the top with a rubber stopper through which passes a glass tube attached to a Drierite trap. The reaction is performed under dry nitrogen with stirring at 70° for 1 hour with the tube immersed in an oil bath placed on a stirring hot plate. This reagent is cooled to room temperature before it is added to the glycopeptides.

Methyl iodide

Chloroform

NaCl, $5.5\,M$

Dowex 50-X4, 200–400 mesh (H⁺ form)

Procedure. Prior to methylation the glycopeptides are N-acetylated. In this form more complete reaction takes place, and ready extraction of the glycopeptides into chloroform is made possible.

For acetylation the glycopeptides are dissolved in the sodium acetate at a concentration of about 3 μmoles/ml. Acetic anhydride in 25-fold molar excess over the available amino and hydroxyl groups is added to the sample at room temperature in five equal portions over a period of 1 hour. The reaction is terminated by diluting the sample with 20 volumes of water and heating in a boiling water bath for 10 minutes. The samples are freed of sodium ions by passage through columns of the

[29] S. Hakomori, *J. Biochem. (Tokyo)* **55**, 205 (1964).

Dowex 50 used in 4-fold excess. The columns are washed with 8 volumes of water and the effluent and wash are lyophilized.

For methylation, the acetylated glycopeptides containing a total of about 15 μmoles of monosaccharide residues are pipetted into 18×150 mm sidearm tubes and titrated to pH 8.5. The sample is then dried with a stream of nitrogen and kept in a vacuum oven at 50° overnight. The dry sample is dissolved in 0.5 ml of dimethyl sulfoxide, followed by the addition of 2.5 ml of fresh methylsulfinyl carbanion, and the mixture is stirred magnetically at room temperature for 3 hours under nitrogen and employing anhydrous conditions. After that period of time, 2.5 ml of methyl iodide, preceded by 2.5 ml of dimethyl sulfoxide, are added and the stirring continued under nitrogen for another 24 hours. The reaction is terminated by the addition of 15 ml of the NaCl solution, the sample is brought to pH 1.5 with HCl, and the methylated glyco-peptides extracted into chloroform (5 times 8 ml). The chloroform ex-tracts are washed with small amounts of $0.05 N$ HCl (4 times 3 ml) and then taken to dryness.

Methylated glycopeptides can be hydrolyzed ($2 N$ H_2SO_4 or $1 N$ HCl, 4 hours, 100°) to obtain the O-methylmonosaccharides. The hydrolyzate is passed through coupled columns of Dowex 50-X4, 200–400 mesh (H^+ form) and Dowex 1-X8, 200–400 mesh (formate form) and the columns are washed with 10 column volumes of 20% (v/v) methanol. The effluent and wash containing the neutral sugars can be dried in a vacuum rotator or lyophilized after dilution with water. Methylated hexosamines are eluted from the Dowex 50 at 4° with 20% methanol containing $1.5 N$ NH_4OH and dried by lyophilization after dilution with water.

While complete methylation of neutral glycopeptides takes place under these conditions, less favorable yields are obtained with sialic acid-containing glycopeptides.

Identification of Methylated Monosaccharides

Neutral Sugars. The identification of the O-methyl derivatives of the neutral sugars gives information in regard to their position in the oligo-saccharide chains. Tetra-O-methyl ethers indicate a terminal nonreducing position; tri-O-methyl ethers are obtained from singly substituted in-ternally located sugar residues; while di-O-methyl derivatives are formed from monosaccharides serving as branch points.

For the identification of the methyl ethers of the neutral sugars of glycoproteins, gas–liquid chromatography of their alditol acetates may be employed.[14] Paper chromatography in 1-butanol saturated with water accomplishes group separation of tetra-, tri-, di-, and mono-O-methyl derivatives of a given sugar. Separation of most of the tri-O-methyl derivatives of galactose and mannose can be achieved by thin-layer chromatography on silica gel plates in the system of Stoffyn[30] using acetone–water–concentrated NH_4OH (250:3:1.5). The migrations of the tri-O-methyl ethers of galactose relative to 2,3,4,6-tetra-O-methylgalactose are as follows[22]: 3,4,6-, 0.46; 2,3,4,-, 0.51; 2,4,6-, 0.74; and 2,3,6-, 0.84. The migration of tri-O-methyl ethers of mannose relative to 2,3,4,6-tetra-O-methylmannose are as follows[23]: 3,4,6-, 0.68; 2,3,4-, 0.78; and 2,3,6-, 0.85.

We have observed that 3,4,6-tri-O-methyl ethers of both galactose and mannose can be distinguished from their other tri-O-methyl isomers by their borate complex formation. On paper electrophoresis in $0.2\,M$ borate buffer, pH 10, they move faster than the other derivatives[22]; on borate-complex anion exchange chromatography, they emerge later from the column[23]; and on chromatography on borate-impregnated paper in 1-butanol saturated with $0.1\,M$ sodium borate[31] they move slower.

Hexosamines. The method of Hakamori results in N-methylation of hexosamines and therefore yields after acid hydrolysis the O-methyl ethers of N-methylamino sugars. These derivatives can be identified by paper chromatography of the hexosamine fraction and detected by the aniline hydrogen phthalate reagent.[23] They stain only very weakly with ninhydrin. The migration on paper chromatography of the di-O-methyl ethers of N-methylglucosamine in pyridine–ethyl acetate–water–acetic acid (5:5:3:1) with pyridine–ethyl acetate–water (11:40:6) in the bottom of the chromatocab has been determined.[23] In relation to 3,4,6- tri-O-methyl-N-methylglucosamine, the migration of the 3,6-derivative is 0.84, the 4,6-, 0.90, and the 3,4-, 0.80. The 6-O-methyl-N-methylglucosamine migrates 0.71 the distance of the tri-O-methyl-N-methylglucosamine.

Clear separation of these O-methyl derivatives of N-methyl-D-glucosamine can be obtained on the Technicon amino acid analyzer with sodium citrate buffer at pH 5.00 with the following gradient in the 9-chamber autograd[23]: chambers 1 and 2, 75 ml of $0.1\,M$ (Na^+); chamber 3, 50 ml of $0.1\,M$ (Na^+) and 25 ml of $0.18\,M$ (Na^+); chambers 4 through 9, 75 ml of $0.18\,M$ (Na^+) buffer. With a column temperature of 60° and a flow rate of 30 ml per hour, the order of elution is as given in Table IV. The components are detected with the ninhydrin reagent, but give only 10–15% of the color of free glucosamine with this reagent.

The elution times of the O-methyl derivatives of N-methyl-D-gluco-

[30] P. J. Stoffyn, *J. Amer. Oil Chem. Soc.* **43**, 69 (1966).
[31] Y. C. Lee and C. E. Ballou, *J. Biol. Chem.* **239**, 1316 (1964).

TABLE IV

RELATIVE ELUTION TIMES OF SOME O-METHYL DERIVATIVES OF
N-METHYL-D-GLUCOSAMINE ON AMINO ACID ANALYZER[a]

Component	Relative elution time[b]
Phenylalanine	1.00
3,6-Di-	4.83
6-Mono	5.28
4,6-Di-	5.79
3,4-Di-	6.91
3,4,6-Tri-	7.37

[a] Chromatography at 60° under conditions described in text.
[b] Elution time of phenylalanine is 2 hours, 16 minutes [T. Arima and R. G. Spiro, J. Biol. Chem. 247, 1836 (1972)].

samine and N-methyl-D-galactosamine, with the Spinco amino acid analyzer (Model 120C), have been published.[32,33]

Periodate Oxidation

A study of the products of periodate oxidation of glycoproteins can give valuable information in regard to the structure of their carbohydrate units, and the general methodology used for this purpose has been described (Vol. 8 [2]). Moreover serial periodate oxidation in the form of Smith degradations has proved to be a powerful tool in the stepwise degradation of carbohydrate units (Vol. 8 [2]).

Recently periodate oxidation has been employed to determine the position on which the N-acetylhexosamines involved in various glycopeptide bonds are substituted.[23,27,34] In this approach, reduced oligosaccharides obtained through β-elimination by alkaline borohydride treatment of glycoproteins (glycopeptides) containing the N-acetylgalactosamine-serine (threonine) bond or oligosaccharides produced by enzymatic cleavage of the N-acetylglucosamine-asparagine bond, followed by borohydride reduction, are submitted to oxidation with periodate. After reduction and acid hydrolysis of the products of such periodate oxidations, an amino alcohol can be obtained that will be characteristic of the position on the terminal hexosamine to which the penultimate sugar is linked[35] (Table V).

This approach is illustrated by the procedure employed for the de-

---

[32] P. A. J. Gorin and A. J. Finlayson, Carbohyd. Res. 18, 269 (1971).
[33] P. A. J. Gorin, Carbohyd. Res. 18, 281 (1971).
[34] A. Tarentino, T. H. Plummer, Jr., and F. Maley, J. Biol. Chem. 245, 4150 (1970).
[35] A. B. Foster, D. Horton, N. Salim, M. Stacey, and J. M. Webber, J. Chem. Soc. 2587 (1960).

TABLE V

AMINO POLYOLS OBTAINED FROM OLIGOSACCHARIDES TERMINATING IN
N-ACETYLHEXOSAMINITOL AFTER PERIODATE OXIDATION AND
BOROHYDRIDE REDUCTION[a]

	Terminal hexosaminitol	
Linkage to hexosaminitol	N-Acetyl-D-glucosaminitol	N-Acetyl-D-galactosaminitol
$1 \rightarrow 3$	L-Threosaminitol	L-Threosaminitol
$1 \rightarrow 4$	D-Xylosaminitol	L-Arabinosaminitol
$1 \rightarrow 6$	Serinol	Serinol
$1 \rightarrow 3$ plus $1 \rightarrow 6$	L-Threosaminitol	L-Threosaminitol
$1 \rightarrow 3$ plus $1 \rightarrow 4$	D-Xylosaminitol	L-Arabinosaminitol
$1 \rightarrow 4$ plus $1 \rightarrow 6$	D-Glucosaminitol	D-Galactosaminitol

[a] The reduced oxidation products are hydrolyzed in acid to obtain the amino polyols (see text).

termination of the linkage to the terminal N-acetylglucosamine of the mannose-N-acetylglucosamine unit of thyroglobulin[23] (Fig. 8). The neutral reducing oligosaccharide obtained after enzymatic cleavage of the N-acetylglucosaminylasparagine bond, is treated with sodium borohydride by the addition of a 400-fold molar excess as $0.2 M$ sodium borohydride in $0.2 M$ sodium borate, pH 8.0. The reaction is allowed to proceed in an ice bath for 16 hours and is then terminated by lowering the pH to 5 with acetic acid and passing the sample through a column of Dowex 50 (H+ form) to remove sodium ions. The effluent and water wash from the column is lyophilized and the boric acid volatilized as methyl borate.

Oxidation of the reduced oligosaccharide is then carried out in the dark for 8 hours at 4° by the addition of a 10-fold molar excess of sodium metaperiodate to total sugar residues in the form of a $0.035 M$ solution in 35 mM sodium acetate buffer, pH 4.5. The oxidation is terminated by the addition of an 8-fold molar excess of ethylene glycol. The oxidized oligosaccharide is then reduced for 16 hours in an ice bath at pH 8.0 by the addition of a 30-fold excess of sodium borohydride over the periodate used. The reduced sample is acidified and passed through coupled columns of Dowex 50 (H+) and Dowex 1 (formate) and the effluent and water wash lyophilized. After removal of the boric acid as the methyl borate, hydrolysis in 4 N HCl for 4 hours at 100° is performed. The amino polyols can then be identified and quantitated by analysis on the amino acid analyzer.

Excellent separation of all possible products resulting from an oligosaccharide containing terminal glucosamine, namely, glucosamine, gluco-

Fɪɢ. 8. Scheme for determining the position of linkage to a reducing terminal *N*-acetylhexosamine residue. The reactions depicted were performed on a mannose-*N*-acetylglucosamine oligosaccharide of thyroglobulin obtained after enzymatic cleavage of the glycopeptide bond [T. Arima and R. G. Spiro, *J. Biol. Chem.* **247,** 1836 (1972)]. The mannose (Man) residue shown linked to di-*N*-acetylchitobiose serves as a branch point for more periperal mannose residues and is therefore not destroyed during the periodate oxidation step. After step 4 the oligosaccharide is submitted to acid hydrolysis. The formation of 1 mole of D-xylosaminitol is consistent with the occurrence of a $1 \rightarrow 4$ linkage to the terminal glucosamine (Table V).

saminitol, xylosaminitol, threosaminitol, and serinol can be achieved on the Technicon amino acid analyzer using a conventional gradient at a flow rate of 0.48 ml per minute and a jacket temperature of 60° (Fig. 9). Alanine can be conveniently used as an internal standard. The elution times relative to glucosamine, which emerges at 7 hours, 22 minutes are as follows: glucosaminitol, 1.09: arabinosaminitol, 1.15; xylosaminitol, 1.23; threosaminitol, 1.38; and serinol, 1.52. The molar color yields of the amino polyols are considerably lower than that of glucosamine and should be determined for a given analytical system.

Paper chromatographic separation of amino polyols can be achieved in 1-butanol–acetic acid–water (4:1:5) in which glucosamine, galactosamine, glucosaminitol, and galactosaminitol migrate together but are separated from the shorter chain compounds which have the following migrations relative to glucosamine: xylosaminitol, 1.20; threosaminitol,

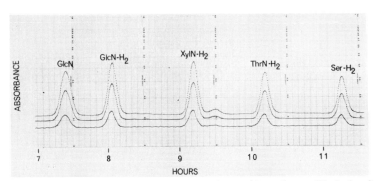

FIG. 9. Separation of aminopolyols on amino acid analyzer (Technicon). The position of elution of glucosaminitol (GlcN-H₂) 0.30 μmole; xylosaminitol (XylN-H₂) 0.30 μmole; threosaminitol (ThrN-H₂) 0.45 μmole; and serinol (Ser-H₂) 0.50 μmole are shown in reference to glucosamine (GlcN) 0.15 μmole. The conditions for the chromatography are given in the text. In this system arabinosaminitol appears just prior to xylosaminitol, from which it is completely separated.

1.33; and serinol, 1.67. These compounds are best detected with the silver reagent.

Identification and Quantitation of Glycopeptide Bonds

One of the most important aspects in the study of a glycoprotein is the characterization of the bonds which attach its carbohydrate units to the peptide chain. Several quite distinct types of glycopeptide bonds have been described,[1] and in recent years methodology for their identification has been developed.

All bonds involve C-1 of the most internal sugar residue of the carbohydrate unit and a functional group on an amino acid in the peptide chain. The linkages can be placed into three groups: (1) the glycosylamine bond which always involves N-acetylglucosamine and the amide group of asparagine; (2) the alkali-labile O-glycosidic bond to serine or threonine which can involve N-acetylgalactosamine, galactose, xylose, or mannose as the sugar component; and (3) the alkali-stable O-glycosidic bond of galactose to hydroxylysine. An O-glycosidic linkage involving arabinose and hydroxyproline which is stable to alkali has also been described.[36]

In identifying the glycopeptide bonds, advantage should be taken of their difference in stability to alkali and their susceptibility to enzymatic cleavage. Optimally, a characteristic sugar-amino acid linkage compound should be isolated, or if that is not feasible, derivatives of the

[36] D. T. A. Lamport, *Biochemistry* **8**, 1155 (1969).

sugar and amino acid resulting from chemical cleavage of the glyco-peptide bond should be obtained. Since a number of proteins have been shown to contain more than one type of glycopeptide bond it is necessary to account quantitatively for the peptide linkage of all of the carbo-hydrate. If evidence for the presence of two or more bonds is obtained, glycopeptides containing these should be resolved and studied separately.

Glycosylamine Bond

In order to prove the existence of a glycosylamine bond, it is neces-sary to isolate the linkage compound, namely GlcNAc-Asn (2-acetamido-1-N-β-L-aspartyl-2-deoxy-β-D-glucopyranosylamine) and compare its properties to that of the synthetic standard.[37,38] The experience of several laboratories with ovalbumin,[37-42] ribonuclease B,[34,43] and the mannose-N-acetylglucosamine unit of thyroglobulin[23] has provided a general ap-proach to accomplish this.

The glycoprotein under study should first be extensively digested with Pronase to obtain glycopeptides which contain only asparagine or asparagine plus a minimal number of other amino acids. To obtain max-imal proteolysis high concentration of substrate (40–50 mg/ml), high temperature (47°–50°) in the presence of toluene and large amounts of enzyme (up to 25% of the substrate weight) have been used. Moreover, repeated digestions of the glycopeptide fraction (separated each time by gel filtration on Sephadex G-25) may be required to obtain the carbo-hydrate with only asparagine attached. The best results have been ob-tained with neutral carbohydrate units; if a sialic acid-containing unit is to be studied, it is helpful to remove this sugar from the glycoprotein or glycopeptides before performing these extensive proteolytic digestions.

Upon completion of the Pronase digestion, the glycopeptides should be separated from released amino acids and enzyme (most of which is autodigested to amino acids and short peptides) by filtration on Sephadex G-25 or a polyacrylamide gel (P-4 or P-6). Further fractionation by chromatography on Dowex 50-X2 at an acidic pH (see previous section of this chapter) will separate glycopeptides containing only asparagine from those containing more amino acids and will also resolve asparagine-linked oligosaccharides from each other on the basis of differences in

[37] G. S. Marks, R. D. Marshall, and A. Neuberger, *Biochem. J.* **87**, 274 (1963).
[38] R. D. Marshall and A. Neuberger, *Biochemistry* **3**, 1596 (1964).
[39] I. Yamashina, K. Ban-I, and M. Makino, *Biochim. Biophys. Acta* **78**, 382 (1963).
[40] I. Yamashina and M. Makino, *J. Biochem. (Tokyo)* **51**, 359 (1962).
[41] M. Makino and I. Yamashina, *J. Biochem. (Tokyo)* **60**, 262 (1966).
[42] R. Montgomery, Y. C. Lee, and Y.-C. Wu, *Biochemistry* **4**, 566 (1965).
[43] T. H. Plummer, Jr., A. Tarentino, and F. Maley, *J. Biol. Chem.* **243**, 5158 (1968).

their carbohydrate content. In a study of the glycopeptide bond, it is not necessary to first separate the variants of the carbohydrate unit, as subsequent steps will serve to remove the outer sugar residues and thereby eliminate this polydispersity.

Release of the external sugars to yield GlcNAc-Asn can be accomplished by means of partial acid hydrolysis, serial application of the Smith periodate oxidation procedure, use of glycosidases or a combination of these techniques. Treatment of the glycopeptides with $2 N$ HCl for 12–20 minutes[37–39] at 100° results in the appearance of GlcNAc-Asn (28% yield in 12 minutes).[38] Partial hydrolysis ($1 N$ HCl, 20–30 minutes) of dansylated glycopeptides may be performed if only very limited amounts of glycopeptides are available to yield the dansyl derivative of GlcNAc-Asn.[43]

Application of three Smith periodate oxidations to glycopeptides from ovalbumin has resulted in about a 60% yield of GlcNAc-Asn.[41] Treatment of thyroglobulin glycopeptides with α-mannosidase and β-N-acetylglucosaminidase has been used to obtain GlcNAc-Asn.[23] GlcNAc-Asn was formed from ribonuclease B glycopeptides by removal of all but one mannose residue with α-mannosidase to give Man-GlcNAc-GlcNAc-Asn, followed by release of the mannose residue by Smith periodate degradation and subsequent cleavage of the outer N-acetylglucosamine residue by the action of β-N-acetylglucosaminidase.[34]

The GlcNAc-Asn obtained after the enzymatic treatment or acid hydrolysis can be conveniently separated from released neutral sugars by adsorption on Dowex 50-X4, 200–400 mesh (H+), and elution with cold $1.5 N$ NH$_4$OH or $1.0 M$ pyridine acetate, pH 5.0.[23] GlcNAc-Asn can be determined on the amino acid analyzer and has an elution time relative to aspartic acid of 0.27 (in the Technicon system). It has a low molar color yield with the ninhydrin reagent (25% of that of aspartic acid) and the color produced has a high 440 nm/570 nm absorbancy ratio (0.82 compared to 0.21 for aspartic acid). Large asparagine-linked oligosaccharides emerge even earlier from the amino acid analyzer and can be partly resolved by elution with the starting buffer at low ionic strength.

Chromatography[34] of such asparagine-linked oligosaccharides on a column of Aminex-H (Bio-Rad) (0.9 × 57 cm) at 53° by elution with $0.1 M$ sodium citrate buffer, pH 2.80, containing 3% propanol (after application of the sample in a pH 2.2 buffer) results in the following order of elution relative to aspartic acid: (Man)$_6$GlcNAc-GlcNAc-Asn, 0.14; (Man)$_1$GlcNAc-GlcNAc-Asn; 0.17; GlcNAc-GlcNAc-Asn, 0.21; and GlcNAc-Asn, 0.33.

GlcNAc-Asn can be identified by paper chromatography in 1-butanol–

acetic acid–water (4:1:5). In this system it migrates[23] with an R_{Asp} of 0.54, while GlcNAc-GlcNAc-Asn (2-acetamido-4-O-(2-acetamido-2-deoxy-β-D-glucopyranosyl)-1-N-β-L-aspartyl-2-deoxy-β-D-glucopyranosylamine) moves with an R_{Asp} of 0.35. The latter compound is obtained by enzymatic or periodate degradation of asparagine-linked oligosaccharides.[23,34] Both compounds have been isolated by preparative paper chromatography employing this solvent system.[23] GlcNAc-Asn and GlcNAc-GlcNAc-Asn give a brown color when the paper chromatographs are stained with a 1% (w/v) solution of ninhydrin in acetone containing 2% (v/v) pyridine (heated at 80°)[37] or with ninhydrin (0.5% w/v) in acetone containing 2% (v/v) of a mixture of collidine–lutidine (1:3, v/v) and 2% (v/v) acetic acid (at room temperature).[23]

The isolated GlcNAc-Asn should yield 1 mole of aspartic acid, glucosamine, and ammonia upon acid hydrolysis. It should readily be split into N-acetylglucosamine and aspartic acid by glycosyl asparaginase[44,45] (see also this volume [101, 102]) and should be resistant to the action of β-N-acetylglucosaminidase. Larger asparagine-linked oligosaccharides are also split by the glycosyl asparaginase, with the release of aspartic acid and the oligosaccharide with N-acetylglucosamine as the terminal reducing residue. A neutral oligosaccharide can be separated from such a digest by its failure to adsorb on Dowex 50 (H⁺). Its reduction with sodium borohydride followed by acid hydrolysis should yield 1 mole of glucosaminitol.[23]

GlcNAc-Asn is stable to mild alkali treatment such as can be used to split the O-glycosidic bond to serine or threonine. However, the bond can be split by stronger alkaline conditions[37] and in the presence of sodium borohydride yields glucosaminitol or an oligosaccharide terminating in glucosaminitol.[46] We have found that hydrolysis in 2 N NaOH containing 2 M sodium borohydride at 80° for 16 hours in polypropylene tubes gives about 75% yield of glucosaminitol.[47] Under these conditions essentially complete deacetylation of the hexosamine takes place and the released amino sugar or oligosaccharide can be adsorbed on Dowex 50. After elution with 1.5 N NH₄OH, the released carbohydrate can be restored to neutrality by acetylation with acetic anhydride (for conditions see section on methylation) and separated from amino acids by passage through Dowex 50 resin.

[44] M. Makino, T. Kojima, T. Ohgushi, and I. Yamashina, *J. Biochem.* (*Tokyo*) **63**, 186 (1968).
[45] A. L. Tarentino and F. Maley, *Arch. Biochem. Biophys.* **130**, 295 (1969).
[46] Y. C. Lee, *Fed. Proc., Fed. Amer. Soc. Exp. Biol.* **30**, 1223 (1971).
[47] R. G. Spiro and V. D. Bhoyroo, unpublished observations (1971).

O-Glycosidic Linkage to Serine and Threonine

When glycoproteins containing carbohydrate units linked by O-glyco-sidic bonds to the α-amino-β-hydroxy acids are treated under relatively mild alkaline conditions, the glycopeptide bond is split through the process of β-elimination with the release of a reducing oligosaccharide and the formation in the peptide chain of an unsaturated amino acid.[48] The linkage sugar can be identified by performing the alkaline treatment in the presence of sodium borohydride and measuring after acid hydrolysis the sugar alcohol formed (see previous section on oligosaccharides), while the amino acid involved in the glycopeptide bond can be determined after reduction of its unsaturated product or by its conversion through a sulfite addition reaction to its sulfonyl form.

Preliminary evidence for the existence of an O-glycosidic bond in a glycoprotein exists if mild alkaline treatment results in a significant decrease in its serine and or threonine content. While β-elimination studies can be effectively carried out on intact glycoproteins, final proof of the existence of such a bond should include the isolation of glyco-peptides containing the linkage and submission of these to alkaline treatment. The β-elimination reaction, however, will not proceed satisfactorily if the glycosidated amino acid residue is in terminal position on the peptide chain, and in such a case the blockage of the amino or carboxyl group by substituents would be required.

In order to achieve β-elimination, various strengths of alkali (0.05–0.5 N NaOH), temperatures (0–45°) and length of incubation (15–216 hours) have been used.[25-28,48-50] In our experience incubation in 0.1 NaOH at 37° for 48–72 hours has given the best results,[10] but the conditions should be established separately for each glycoprotein or glycopeptide.

If the β-elimination is carried out in the absence of a reducing agent, identification of the amino acid believed to be involved in the glyco-peptide linkage can be done indirectly only through the loss of serine and threonine, while the sugar involved in the bond cannot at all be designated, as it itself and also more internally located sugars are likely to undergo alkaline degradation. For the identification of the sugar component, it is essential to perform the alkaline treatment in the presence of sodium borohydride and concentrations of 0.15 M to 1.0 M of this reagent have been effectively used (see the section on oligosac-

[48] B. Anderson, N. Seno, P. Sampson, J. G. Riley, P. Hoffman, and K. Meyer, *J. Biol. Chem.* **239**, PC 2716 (1964).

[49] V. P. Bhavanandan, E. Buddecke, R. Carubelli, and A. Gottschalk, *Biochem. Biophys. Res. Commun.* **16**, 333 (1964).

[50] K. Tanaka and W. Pigman, *J. Biol. Chem.* **240**, PC 1487 (1965).

charides). The reduced oligosaccharides are then hydrolyzed in acid to their sugar constituents to identify and determine the amount of sugar alditol formed (e.g., galactitol, galactosaminitol, xylitol, or mannitol). The amino sugar alcohol can be identified on the amino acid analyzer with the citrate–borate system. Neutral sugar alcohols are readily identified by paper chromatography in butanol–ethanol–water (10:1:2) where the R_{Gal} of galactitol is 1.18, the R_{Man} of mannitol is 0.81, and the R_{Xyl} of xylitol, 0.87, as well as by gas–liquid chromatography as their acetates.

Reduction with sodium borohydride and palladium chloride of the unsaturated amino acid formed during β-elimination makes possible the identification of the amino acid involved in the glycopeptide bond not only by its decrease after the alkaline treatment, but also by the increase in alanine from serine through the reduction of the α-aminoacrylic acid and the appearance of α-aminobutyric acid (which elutes between alanine and glucosamine on the Technicon amino acid analyzer) from threonine through reduction of α-aminocrotonic acid (see also [3]). While it has been reported that α-aminoacrylic acid can be readily converted to alanine even by sodium borohydride alone, the conversion of α-aminocrotonic acid to α-aminobutyric acid requires the additional presence of palladium chloride.[50] Even under these conditions the formation of α-aminobutyric acid is not quantitative, reaching at the best an 85% conversion.[50] In our experience the extent of formation of this amino acid is considerably less and even the conversion to alanine is substantially smaller than expected from the serine destroyed.

Alkaline-Sulfite Treatment

When β-elimination is carried out in the presence of sodium sulfite the unsaturated product of the hydroxyamino acids involved in the O-glycosidic glycopeptide bond are converted to their sulfonyl derivatives, namely, cysteic acid from serine and α-amino-β-sulfonylbutyric acid from threonine.[51] The appearance of these two components can serve as evidence for the involvement of the α-amino-β-hydroxy acids in glycopeptide bonds and has in our experience proved to be a much more reliable and specific technique than alkaline-borohydride palladium chloride treatment.

Since cysteic acid and α-amino-β-sulfonylbutyric acid cannot be resolved on the amino acid analyzer, both appearing in the void volume, we have developed a chromatographic procedure employing Dowex 1 which clearly separates these two components from an acid hydrolyzate.

[51] S. Harbon, G. Herman, and H. Clauser, *Eur. J. Biochem.* **4**, 265 (1968).

While the conversion of serine through α-aminoacrylic acid to cysteic acid during the alkaline-sulfite treatment can be made quantitative, only about a 75% conversion of threonine through α-aminocrotonic acid to α-amino-β-sulfonylbutyric acid is effected. Since variable conversion of half-cystine to cysteic acid occurs during alkaline sulfite treatment, it is best to perform this reaction on glycopeptides or glycoproteins free of half-cystine or to use performic acid-oxidized material in which all the half-cystine has been converted to cysteic acid prior to alkaline-sulfite treatment. Any additional cysteic acid in such material would then be the result of the β-elimination.

The procedure which we have employed is as follows. The glyco-peptide (1.0 μmole/ml) or glycoprotein (8 mg/ml) is incubated with shaking at 37° for 48–72 hours in 0.1 N NaOH containing 0.5 M sodium sulfite. The reaction is terminated by acidification with constant-boiling HCl, and the sample is taken to dryness *in vacuo* (Evapomix) at 40°. Subsequently it is hydrolyzed with glass-distilled constant-boiling HCl in a sealed tube under nitrogen for 28 hours. A control sample which is acidified prior to the addition of the alkaline sulfite is hydrolyzed in the same manner.

An aliquot of the hydrolyzate can be placed on the amino acid analyzer to determine the decrease in serine and threonine caused by the alkaline treatment. This analysis will also provide the sum of cysteic acid and α-amino-β-sulfonylbutyric acid formed by the reaction. If the sugar involved in the glycopeptide bond is a hexosamine and is itself substituted at C-3, an additional component equimolar to the cysteic acid and α-amino-β-sulfonylbutyric acid will appear shortly after the cysteic acid. This component is believed to be a 3-sulfonylhexosamine formed by an addition reaction involving the unsaturated terminal N-acetylhexosamine exposed by the alkaline treatment.[52]

Dowex 1 Chromatography for Separation of Cysteic Acid and α-Amino-β-sulfonylbutyric Acid

  Buffer 1

  Sodium acetate, 0.1 M, pH 3.80. For 10 liters of this buffer 82.0 g of anhydrous sodium acetate are dissolved in 8500 ml deionized water to which 100 ml of a 33% (w/v) aqueous solution of BRIJ 35 is added. The pH is titrated to 3.80 with glacial acetic acid and the solution is made to volume with deionized water.

[52] P. Weber and R. J. Winzler, *Arch. Biochem. Biophys.* **137,** 421 (1970).

Buffer 2

Sodium acetate, 0.5 M, pH 3.80. For 10 liters 410.2 g of anhydrous sodium acetate are dissolved in 7000 ml deionized water to which 100 ml of the BRIJ solution are added. The pH is titrated to 3.80 with glacial acetic acid and the solution is made to volume with deionized water.

Resin and Column

Dowex 1-X8, minus 400 mesh (Bio-Rad) anion exchange resin is put into the acetate form by successive washing with 3 N sodium acetate (until the eluate is negative for chloride ions), 5 N acetic acid, and deionized water. The resin is packed at room temperature into a column 6 mm internal diameter (Technicon) to a height of 125 cm. It is equilibrated with the 0.1 N pH 3.80 sodium acetate buffer at a rate of 0.48 ml per minute.

Internal Standard

O-Phospho-L-serine, 0.25 mM

Instruments

Technicon amino acid analyzer system

Procedure. The sample containing 0.03 to 0.75 μmole of each sulfonyl-amino acid is titrated to pH 3.8 with 0.05 N NaOH after the addition of 1.0 ml of the internal standard (0.25 μmole of O-phospho-L-serine). It is then taken to dryness *in vacuo* (Evapomix) and subsequently dissolved in 0.5 ml of Buffer 1 in which it is applied to the column under nitrogen pressure, followed by a wash with this buffer.

Elution is achieved at 0.48 ml per minute at room temperature with the following gradient in a nine-chambered autograd: chamber 1, 73 ml of Buffer 1 and 2 ml of methanol; chamber 2, 70 ml of Buffer 1 and 5 ml of methanol; chambers 3 and 4, 48 ml of Buffer 1, 20 ml of Buffer 2 and 7 ml of methanol; chambers 5 and 6, 72 ml of Buffer 2 and 3 ml of methanol; chambers 7, 8, and 9, 75 ml of Buffer 2. The column is regenerated for 45 minutes with Buffer 2 followed by 120 minutes with Buffer 1.

This gradient achieves good separation of cysteic acid and α-amino-β-sulfonylbutyric acid (Fig. 10). Other acidic amino acids including glutamic acid, aspartic acid, O-phosphothreonine, and O-phosphoserine emerge prior to these components. The elution times relative to O-phosphoserine which emerges at 7 hours are as follows: glutamic acid, 0.24;

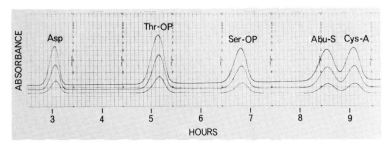

Fig. 10. Separation of O-phospho- and sulfonyl amino acids by chromatography on Dowex 1-X8 at pH 3.8. The elution of the products of alkaline sulfite treatment, namely α-amino-β-sulfonylbutyric acid and cysteic acid (CysA) are shown in reference to aspartic acid (Asp), O-phosphothreonine (Thr-OP), and O-phosphoserine (Ser-OP). The latter two components can be used as internal standards for a hydrolyzate obtained after alkaline sulfite treatment of a glycoprotein or glycopeptide.

aspartic acid, 0.45; O-phosphothreonine, 0.75; α-amino-β-sulfonylbutyric acid, 1.26; and cysteic acid, 1.33.

Cysteic acid and α-amino-β-sulfonylbutyric acid can also be identified by paper chromatography of an acidic amino acid fraction. For this purpose the acid hydrolyzate of the alkaline sulfite treated glycoprotein or glycopeptide, after removal of the HCl, is dissolved in sufficient water to make a sodium ion concentration of less than 50 mM and titrated to pH 4.0. It is then placed on a small column (10-fold excess over anions present) of Dowex 1-X8, 200–400 mesh (formate form) previously equilibrated with 5 mM pyridine formate buffer, pH 4.0. After the sample has passed through, the column is washed with 8 volumes of the 5 mM buffer and the acidic amino acids are then eluted with 5 volumes of 4 N formic acid. After dilution with water, the formic acid is removed by lyophilization and the acidic amino acids including aspartic acid, glutamic acid, α-amino-β-sulfonylbutyric acid, and cysteic acid are chromatographed in 1-butanol–acetic acid–water (4:1:5) for 5 days. The R_{Asp} in this system for cysteic acid is 0.42 and for α-amino-β-sulfonylbutyric acid, 0.55.

O-Glycosidic Bond to Hydroxylysine

The presence of a galactosylhydroxylysine linkage (5-O-β-D-galactopyranosylhydroxylysine) can be considered only in those glycoproteins (primarily belonging to the collagen family) which contain hydroxylysine as one of their amino acid constituents.[1,53] Because of the stability of this bond even to strong alkali, the presence of carbohydrate units linked

---

[53] R. G. Spiro, in "Glycoproteins" (A. Gottschalk, ed.), 2nd ed. Part B, p. 964, Elsevier, Amsterdam (1972).

to hydroxylysine (which occur either as single galactose residues or as the 2-O-α-D-glucosylgalactose disaccharide) can readily be established. After alkaline hydrolysis of sufficient strength to split all the peptide bonds of a protein these carbohydrate units linked to hydroxylysine, namely, glucosylgalactosylhydroxylysine (Glc-Gal-Hyl) and galactosyl-hydroxylysine (Gal-Hyl), can be identified and quantitated on the amino acid analyzer. Glycopeptides containing the hydroxylysine-linked carbohydrate units can be separated from other glycopeptides after collagenase and Pronase digestion of the protein by gel filtration on Sephadex G-25[18,54] (see also this volume [84]).

The following procedure has been employed in our laboratory for the quantitation and identification of the hydroxylysine-linked carbohydrate units.[22,54] The glycoprotein (15–20 mg/ml) or glycopeptides (about 5 μmoles/ml) are hydrolyzed in 2 N NaOH at 105° for 24 hours in tightly capped polypropylene tubes. The hydrolyzate is then neutralized with HCl and diluted to a fixed volume. An aliquot is taken for direct analysis on the amino acid analyzer to determine the Glc-Gal-Hyl, Gal-Hyl, and free hydroxylysine content. Analysis on the Technicon amino acid analyzer employing a regular gradient results in the elution of Glc-Gal-Hyl either just ahead of methionine, between the methionine and alloisoleucine, or between the alloisoleucine and isoleucine, depending on the lot of Technicon resin A used, while Gal-Hyl elutes immediately after phenylalanine.

Because of the variability of elution of these components in the regular radient, we have devised a simplified elution scheme starting at pH 5.0 which gives clear separation of Glc-Gal-Hyl, Gal-Hyl, as well as free hydroxylysine, without the possibility of interference with other amino acids.

Separation of Hydroxylysine-Linked Carbohydrate Units on the Amino Acid Analyzer at pH 5.0

Buffer 1

Sodium citrate, 0.1 M Na⁺, pH 5.00. For 10 liters of this buffer, 69.7 g of trisodium citrate dihydrate and 16.9 g of sodium chloride are dissolved in 8500 ml of deionized water to which 100 ml of a 33% (w/v) aqueous solution of BRIJ 35 is added. The solution is titrated to pH 5.00 with 6 N HCl and made to volume with deionized water.

[54] R. G. Spiro, *J. Biol. Chem.* **244**, 602 (1969).

Buffer 2

Sodium citrate, 0.18 *M* Na⁺, pH 5.00. For 10 liters of this buffer, 127.3 g of trisodium citrate dihydrate and 29.2 g sodium chloride are dissolved as above and titrated to pH 5.00.

Buffer 3

Sodium citrate, 0.35 *M* Na⁺, pH 4.50. For 10 liters of this buffer 127.3 g of trisodium citrate dihydrate and 128.6 g of sodium chloride are dissolved as above and titrated to pH 4.50 with HCl.

Procedure. The column of the Technicon animo acid analyzer is equilibrated with Buffer 1 at a rate of 0.48 ml per minute at a jacket temperature of 60° after the usual regeneration with 0.2 *N* NaOH. The sample containing 0.025–0.75 μmole of hydroxylysine-linked carbohydrate units is titrated to pH 4.5 and taken to dryness *in vacuo* after the addition of 0.25 μmole of glucosamine hydrochloride as an internal standard (1.0 ml of a 0.25 m*M* solution). Glucosamine may be employed as a standard as any amount of this sugar present in the sample is completely destroyed during alkaline hydrolysis.

Elution is achieved at 0.48 ml per minute at 60° with the following gradient in a nine-chambered autograd: chamber 1, 75 ml of Buffer 1; chamber 2, 60 ml of Buffer 1 plus 15 ml of Buffer 2; chamber 3, 35 ml of Buffer 2 and 40 ml of Buffer 3; chambers 4 through 9, 75 ml of Buffer 3.

Neutral and acidic amino acids emerge early in this elution scheme (Fig. 11), and the order of elution of the components which resolve in this gradient is given in Table VI. The diastereoisomers of Gal-Hyl and hydroxylysine are separated while those of Glc-Gal-Hyl merge into one peak. Since the amount of destruction of Glc-Gal-Hyl, Gal-Hyl, hydroxylysine, and phenylalanine under these conditions of alkaline hydrolysis is similar (with a recovery of approximately 85%) the values for Glc-Gal-Hyl, Gal-Hyl, and free hydroxylysine can be corrected for such destruction on the basis of the phenylalanine content of the sample compared to the amount of phenylalanine obtained after acid hydrolysis. The sum of Glc-Gal-Hyl, Gal-Hyl, and free hydroxylysine should equal the total hydroxylysine content of the protein (as determined after acid hydrolysis).

The hydroxylysine-linked carbohydrate units can also be identified by paper chromatography or electrophoresis of the desalted alkaline hydrolyzate. For this purpose the hydrolyzate is diluted with water to 0.1 *N* NaOH and acidified to pH 3.0 with HCl. It is then passed through a column of Dowex 50-X4, 200–400 mesh (H⁺ form) containing 5 times

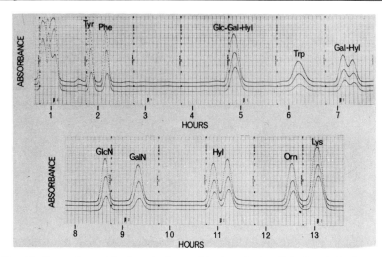

FIG. 11. Separation of hydroxylysine and hydroxylysine-linked carbohydrate units on the amino acid analyzer (Technicon) with a pH 5.0 citrate buffer gradient. The neutral and acidic amino acids emerge before tyrosine (Tyr) and phenylalanine (Phe). The position of elution of glucosylgalactosylhydroxylysine (Glc-Gal-Hyl), galactosylhydroxylysine (Gal-Hyl) and hydroxylysine (Hyl) in reference to these amino acids as well as tryptophan (Trp), ornithine (Orn), and lysine (Lys), which are also present in alkaline hydrolyzates, is shown. The position of glucosamine (GlcN) and galactosamine (GalN) is shown, as these two components, which are not present in alkaline hydrolyzates, may be used as internal standards. Glc-Gal-Hyl and Hyl are resolved into their diastereoisomers. The conditions for the chromatography are given in the text.

TABLE VI

RELATIVE ELUTION TIMES OF COMPONENTS FROM AMINO ACID ANALYZER WITH THE "pH 5.0" GRADIENT[a]

Component	Relative elution time[b]
Tyrosine	0.23
Phenylalanine	0.26
Glc-Gal-Hyl	0.55
Tryptophan	0.73
Gal-Hyl	0.83, 0.85[c]
Glucosamine	1.00
Galactosamine	1.09
Hydroxylysine	1.26, 1.30[c]
Ornithine	1.46
Lysine	1.53

[a] Chromatography at 60° under conditions described in text.

[b] Elution time of glucosamine is 8 hours, 40 minutes.

[c] Gal-Hyl and hydroxylysine are resolved into their diastereoisomers.

the equivalents of NaOH used in the hydrolysis, and after extensive washing of the column with water (10 column volumes) is eluted with about 8 column volumes of 1.5 N NH$_4$OH.[22] The ammonia is removed by lyophilization. On paper chromatography in 1-butanol–acetic acid–water (4:1:5) for 5 days Glc-Gal-Hyl (R_{Hyl} = 0.52) and Gal-Hyl (R_{Hyl} = 0.69) are well separated from all the other amino acids.[22] In pyridine–ethyl acetate–water–acetic acid (5:5:3:1) the R_{Hyl} of Glc-Gal-Hyl is 0.65 while that of Gal-Hyl is 0.72.[22]

Upon electrophoresis at pH 3.5, Glc-Gal-Hyl migrates to the cathode at a rate 0.60 that of hydroxylysine while Gal-Hyl moves at 0.73 the speed of this amino acid.[22]

Glc-Gal-Hyl can be isolated from desalted alkaline hydrolyzates of glycopeptides or glycoproteins by gel filtration on a Sephadex G-15 (fine) column (2.1 × 80 cm) equilibrated and eluted with 0.1 M pyridine acetate buffer at pH 5.0. The Glc-Gal-Hyl emerges at an elution volume of about 150–165 ml from such a column.[22]

Glc-Gal-Hyl can be converted to Gal-Hyl by hydrolysis in 0.1 N H$_2$SO$_4$ at 100° for 28 hours.[22]

[2] Assay for N-Acetylgalactosaminitol[1]

By DON M. CARLSON

N-Acetylgalactosamine is linked O-glycosidically to serine or threonine in several glycoproteins. Alkali treatment releases N-acetylgalactosamine by β-elimination with concomitant formation of dehydroalanine from serine and α-aminocrotonic acid from threonine. Degradation of the N-acetylgalactosamine is controlled by the reduction of its reducing group with sodium borohydride,[2] forming N-acetylgalactosaminitol, which is then measured as described below.

Assay Method[1]

Principle. Acid hydrolysis converts N-acetylgalactosaminitol and N-acetylhexosamines into galactosaminitol and amino sugars. The resulting free amino groups are N-acetylated with [^{14}C]acetic anhydride. After removal of excess acetic anhydride, the [^{14}C]N-acetylated derivatives

[1] D. M. Carlson, *Anal. Biochem.* **20**, 195 (1967).
[2] D. M. Carlson, R. N. Iyer, and J. Mayo, "Blood and Tissue Antigens" (D. Aminoff, ed.), p. 229. Academic Press, New York, 1970.

are separated by paper electrophoresis in sodium tetraborate buffer and quantitated by liquid scintillation counting.

Reagents

Galactosamine·HCl, 0.1 mM (standard solution)
Hydrochloric acid, 4 N
Sodium bicarbonate, 1 M
Sodium tetraborate, 1%
[14C]Acetic anhydride, specific activity about 50,000 cpm/μmole
Methanol
N-Acetylgalactosaminitol. Prepare by reducing N-acetylgalactosamine with sodium borohydride, and crystallize as described by Crimmin.[3]
Dowex 50-X8 H+ form 200–400 mesh
Toluene scintillation fluid

Procedure. The compounds to be assayed and standard amounts of galactosamine are hydrolyzed for 5 hours in 1 ml of 4 N HCl at 100° in a sealed tube (screw-cap tubes with Teflon liners are permissible). A blank without hexosamine is included with each experiment as a control. HCl is removed *in vacuo*, and the samples are dissolved in 0.2 ml of water. N-Acetylation is performed essentially as described by Distler *et al.*[4] The following amounts of reagents are added to the samples in an ice bath: 0.025 ml of methanol, 0.025 ml of 1 M NaHCO$_3$ (pH must be maintained at 7 or above), and 0.005 ml of [14C]acetic anhydride. The mixtures are kept in ice for at least 60 minutes with frequent mixing. Excess acetic anhydride is destroyed by addition of about 0.1 ml of dry Dowex 50 H+. Each mixture is then passed through a small column prepared from a disposable pipette plugged with glass wool and containing about 0.1 ml of Dowex 50 H+.

The reaction tubes and column are washed with a total of 1 ml of water as four equal aliquots, and the combined eluates are evaporated to dryness. Each sample is dissolved in 0.15 ml of water, and 0.10 ml is applied as a 1-inch streak on Whatman 3 MM paper. Electrophoresis is carried out in 1% sodium tetraborate buffer at 50 V/cm for the time necessary (about 45 minutes) for the N-acetylgalactosaminitol to migrate about 20 cm toward the anode. One-inch segments of the electrophoresis strips are counted in the toluene system. N-Acetylmannosaminitol and N-acetylglucosaminitol, prepared by sodium borohydride reduction of the N-acetylamino sugars, are not separated from

[3] W. R. C. Crimmin, *J. Chem. Soc.* **1957**, 2838.
[4] J. J. Distler, J. M. Merrick, and S. Roseman, *J. Biol. Chem.* **230**, 497 (1958).

ASSAY OF N-ACETYLGALACTOSAMINE AND N-ACETYLGALACTOSAMINITOL[a]

| | Yield (μmoles)[b] | | | |
| | N-Acetylgalactosamine | | N-Acetylgalactosaminitol | |
Compound assayed	Theory	Found	Theory	Found
Galactosamine	0.100	0.100	0	<0.001
N-Acetylgalactosaminitol	0	<0.001	0.064[c]	0.058
	0	<0.001	0.032[c]	0.031
Oligosaccharide V[d]	0	<0.001	0.100	0.104
	0	<0.001	0.200	0.210
Oligosaccharide IV[d]	0.080	0.092	0.080	0.086
	0.160	0.181	0.160	0.156

[a] D. M. Carlson, $Anal. Biochem.$ **20**, 195 (1967).
[b] Calculated from galactosamine standard curve.
[c] Based on nitrogen determination and dry weight.
[d] Oligosaccharide V contains fucose, galactose, and N-acetylgalactosaminitol; oligosaccharide IV contains fucose, N-acetylgalactosamine, galactose, and N-acetylgalactosaminitol. Theoretical concentrations are based on fucose and N-acetylgalactosamine values, respectively.

N-acetylgalactosaminitol. The standard of [^{14}C]N-acetylgalactosamine migrates about 7.5 cm.

A linear increase of radioactivity in N-acetylgalactosamine is found with increasing galactosamine concentrations.[1] Values obtained for N-acetylgalactosaminitol and oligosaccharides isolated from pig submaxillary mucins,[5] based on the galactosamine standard curve, are presented in the table.

Comments. Although this method is adequate for determining N-acetylgalactosamine and N-acetylgalactosaminitol in purified oligosaccharides, caution should be used in applying the method to crude preparations. The N-acetylation reaction is not specific and will acetylate other free amino groups. If samples contain protein, it is necessary to isolate the hexosamine fraction by the method of Boas[6] prior to N-acetylation. Treatment of the N-acetylated samples with a mixed-bed ion-exchange resin should remove any remaining amino acids.

The quantitative determination of glucosamine, mannosamine, and galactosamine is possible with this procedure, but electrophoresis for a longer period of time is required for adequate separation.[1] The relative distances of migration (cm) in 90 minutes are N-acetylglucosamine, 2; N-acetylgalactosamine, 9; N-acetylmannosamine, 16.

[5] D. M. Carlson, $J. Biol. Chem.$ **243**, 616 (1968).
[6] N. F. Boas, $J. Biol. Chem.$ **204**, 553 (1953).

[3] Assay for Olefinic Amino Acids: Products of the β-Elimination Reaction in Glycoproteins

By James J. Plantner and Don M. Carlson

An alkali-catalyzed β-elimination reaction releases carbohydrate chains linked O-glycosidically to serine and threonine in glycoproteins.[1] Amino acid derivatives that absorb at 240 nm are the unsaturated products of the β-elimination reaction.[2] In studies on the protein structure of pig submaxillary mucin (PSM), we found it necessary to investigate the relative rates of β-elimination of O-substituted serine and threonine. An assay procedure which simplifies the quantitation of the olefinic amino acids formed by alkali treatment is described.

Assay Method

Principle. The carbohydrate side chains of PSM are released by treating the glycoprotein with 0.5 N NaOH at 50°. The extent of β-elimination is determined by the increase in absorbance at 240 nm.[2] The concentration of both pyruvate and α-ketobutyrate, products of acid hydrolysis of the olefinic amino acid residues,[2,3] is measured by lactic acid dehydrogenase (LDH). LDH, coupled with NADH, catalyzes the reduction of both pyruvate and α-ketobutyrate.[4] Pyruvate is rapidly reduced by a low (1 unit) level of LDH, whereas α-ketobutyrate requires a much higher (40 units) level of enzyme.

Reagents

Pig submaxillary mucin, isolated as described previously[5]
NaOH, 0.5 N
Sodium phosphate buffer, 0.2 M, pH 7.5
Sodium pyruvate, 0.001 M
Sodium α-ketobutyrate, 0.001 M
NADH, 1.4 mg/ml, in 0.2 M sodium phosphate buffer
Lactic acid dehydrogenase, 100 units/ml and 4000 units/ml in sodium phosphate buffer

[1] D. M. Carlson, R. N. Iyer, and J. Mayo, *in* "Blood and Tissue Antigens" (D. Aminoff, ed.), p. 229. Academic Press, New York, 1970.

[2] D. H. Neiderhiser, J. J. Plantner, and D. M. Carlson, *Arch. Biochem. Biophys.* **145**, 155 (1971).

[3] S. Harbon, G. Herman, and H. Clauser, *Eur. J. Biochem.* **4**, 265 (1968).

[4] A. Meister, *J. Biol. Chem.* **184**, 117 (1950).

[5] D. M. Carlson, *J. Biol. Chem.* **243**, 616 (1968).

Procedure. Carbohydrate chains are released from PSM by treatment of the glycoprotein (2 mg/ml) with 0.5 N NaOH at 50°.[6] A Gilford multiple absorbance recording spectrophotometer, equipped with a temperature probe, is used to follow the β-elimination reaction at 240 nm. The molar extinction coefficient for 2-aminopropenoic acid is 6050.[7]

The olefinic amino acids produced by the β-elimination reaction are converted into pyruvate and α-ketobutyrate by hydrolysis in 3 N HCl for 4 hours at 100°, conditions which give maximal yields of the α-keto acids. The hydrolyzate is adjusted to between pH 7 and 8 with NaOH.[8] The assay mixture contains the following components in a final volume of 1.0 ml: 0.3 ml of sodium phosphate buffer, 0.1 ml of NADH, and 0.1–0.6 ml of the neutralized hydrolyzate containing 0.02–0.10 μmole of α-keto acids. Absorbance at 340 nm is recorded, and the reaction is initiated by the addition of 1 unit of LDH (0.01 ml of 100 units/ml). The reduction of pyruvate is essentially complete in 3–5 minutes; this decrease in absorbance represents the amount of pyruvate in the sample. A larger amount of LDH is then added (0.01 ml of the 4000 units/ml LDH); the subsequent decrease in absorbance is a measure of the amount of α-ketobutyrate in the sample.[2] Recoveries of

OLEFINIC AMINO ACID, α-KETO ACID, AND GALACTOSAMINE CONTENT OF
PIG SUBMAXILLARY MUCIN (PSM) BEFORE AND AFTER ALKALINE
ELIMINATION

	Content (μmoles/mg PSM)		
Component	Before	After	Difference
Pyruvate	0	0.37	0.37
α-Ketobutyrate	0	0.27	0.27
Galactosamine	0.88	0.14	0.74
Olefinic amino acids	0	0.77[a]	0.77

[a] This value is obtained from the A_{240} using an extinction coefficient of 6050. [J. Greenstein and M. Winitz, "Chemistry of the Amino Acids," p. 859. Wiley, New York, 1961.] The molar extinction coefficient for 2-amino-2-butenoic acid has not been determined. An extinction coefficient similar to that for 2-aminopropenoic acid is assumed.

[6] J. J. Plantner, unpublished observations, 1970. The duration of alkali treatment should be determined for different mucins. For PSM this period is 60 minutes.
[7] J. Greenstein and M. Winitz, "Chemistry of the Amino Acids," p. 859. Wiley, New York, 1961.
[8] Loss of α-keto acids will occur if the volume is reduced by vacuum techniques after the acid hydrolysis.

pyruvate and α-ketobutyrate are determined by adding known quantities of the α-keto acids to a duplicate sample before hydrolysis.

Comments. As determined by this assay procedure, close to stoichiometric amounts of pyruvate and α-ketobutyrate are recovered, based on the loss of galactosamine and increase in absorbance at 240 nm. Typical results are presented in the table. Furthermore, the ratio of pyruvate to α-ketobutyrate produced (1.4:1) is the same as the ratio of the hydroxyamino acids, serine to threonine, destroyed.

Pyruvate and α-ketobutyrate from the acid hydrolyzate were characterized as the 2,4-dinitrophenylhydrazone derivative,[9] prepared by a modification of the method of Kun and Garcia-Hernandez.[10] In addition, alanine and α-aminobutyrate are formed from the respective 2,4-dinitrophenylhydrazone derivatives by catalytic hydrogenolysis.[11,12]

[9] H. Katsuki, T. Yoshida, C. Tanegashima, and S. Tanaka, *Anal. Biochem.* **24**, 112 (1968).
[10] E. Kun and M. Garcia-Hernandez, *Biochim. Biophys. Acta* **23**, 181 (1957).
[11] H. C. Brown and C. Brown, *J. Amer. Chem. Soc.* **84**, 1495, 2829 (1962).
[12] F. Downs and W. Pigman, *Biochemistry* **8**, 1760 (1969).

[4] Determination of Sialic Acid Using an Amino Acid Analyzer[1]

By Teh-Yung Liu

Sialic acids are acylated derivatives of an aminodeoxynonulosonic acid called neuraminic acid, 5-amino-3,5-dideoxy-D-glycero-D-galacto-nonulosonic acid. The free amino compound does not occur in nature and has not yet been synthesized. In sialic acid isolated from biological material, the amino group is always substituted by acetyl or glycoyl radicals. Some natural sialic acids contain O-acetyl groups in addition. Sialic acids are present in erythrocytes, in various serum proteins, in glycoproteins, mucoproteins, and in various bacteria,[2] notably some strains of *Escherichia coli*[3] and meningococcus.[4]

[1] Research carried out at Brookhaven National Laboratory under the auspices of the U.S. Atomic Energy Commission and by a contract with the U.S. Army Medical Research and Development Command, Office of the Surgeon General (MIPR 9959).
[2] G. F. Springer (Ed.), *Fifth Macy Conference on Polysaccharides in Biology*, Josiah Macy Jr. Foundation, New York, June 1–3 (1959).
[3] G. T. Barry and W. F. Goebel, *Nature (London)* **179**, 206 (1957).

Numerous colorimetric reactions for the sialic acid have been developed, including the Bial orcinol,[5] resorcinol, diphenylamine,[6] direct Ehrlich,[7] and tryptophan–perchloric acid reactions, and a method involving periodate oxidation and coupling with 2-thiobarbituric acid which has been used for the detection of 2-deoxyribose and 2-keto-3-deoxysugar acid.[8-11] Most of the reactions are used for the determination of other carbohydrates as well and cannot be applied directly to tissues or mixtures containing other carbohydrates. The Bial orcinol reaction gives identical colors with ketohexoses and sialic acid and can be used only when these substances are present in minimal quantity. More recently, Reinhold et al.[12] utilized the gas–liquid chromatographic procedure of Sweeley et al.[13] for the determination of sialic acid in porcine ribonuclease as its trimethylsilylated methyl glycoside.

The method to be described utilizes $2 N$ methanesulfonic acid as the catalyst for methanolysis to cleave glycosidic linkages and at the same time removes the amino acyl group from sialic acid. The amino acid analyzer is used for the quantitative estimation of sialic acid as methoxyneuraminic acid. The method in its present form has been applied to several polymers of sialic acids, glycoproteins, and bird's nest and was found to be successful in the determination of sialic acid in these samples.[4,14]

Methods

Reagents. Methanesulfonic acid $(2 N)$ in anhydrous methanol is prepared by pipetting 0.65 ml of methanesulfonic acid (reagent grade, Eastman Kodak) into a 5.0-ml volumetric flask, it is brought to volume with anhydrous methanol (reagent grade absolute methanol which has been soaked with Molecular Sieves, Fisher M-514, 10 g/100 ml of liquid).

Sialic Acid Standard. A standard solution containing 2.5 μmoles sialic acid (purchased from Pierce Chemical Co., Rockford, Illinois and fur-

[4] T. Y. Liu, E. C. Gotschlich, F. T. Dunne, and E. K. Jonsson, *J. Biol. Chem.* **246,** 4703 (1971).

[5] L. Svennerholm, *Ark. Kemi* **10,** 577 (1957).

[6] A. Saifer and H. A. Siegel, *J. Lab. Clin. Med.* **53,** 474 (1959).

[7] I. Werner and L. Odin, *Acta Soc. Med. Upsal.* **57,** 230 (1952).

[8] V. S. Waravdekar and L. D. Saslaw, *Biochim. Biophys. Acta* **24,** 439 (1957).

[9] A. Weissbach and J. Hurwitz, *J. Biol. Chem.* **234,** 705 (1959).

[10] D. Aminoff, *Virology* **7,** 355 (1959).

[11] L. Warren, *J. Biol. Chem.* **234,** 1971 (1959).

[12] V. N. Reinhold, F. T. Dunne, J. C. Wriston, M. Schwarz, L. Sarda, and C. H. W. Hirs, *J. Biol. Chem.* **243,** 6482 (1968).

[13] C. C. Sweeley, R. Bentley, M. Makita, and W. W. Well, *J. Amer. Chem. Soc.* **85,** 2497 (1963).

[14] T. Y. Liu and Y. H. Chang, manuscript in preparation.

ther recrystallized by the method of McGuire and Binkley[15]) is lyophilized. This sample is used to check the procedure and to establish the color factor for methoxyneuraminic acid on a Beckman-Spinco Model 120C amino acid analyzer. Crystalline sample of methoxyneuraminic acid prepared according to the procedure of Klenk and Faillard[16] can also be used for this purpose. The color values obtained for both samples should agree within ±3%.

Methanolysis and Analysis. Samples containing 0.1–2.5 μmoles of sialic acids are placed in glass tubes (Kimble 45066A, 12 × 150 mm) equipped with Teflon-lined screw caps. Methanesulfonic acid (2 N) in anhydrous methanol, 0.5 ml, is added, and the tubes are flashed with nitrogen and sealed with the cap. The tubes are immersed into a heating block (Exacta-Heat, Model 218, Techni Laboratory Instruments, Pequannock, New Jersey; hole depth 50 mm) maintained at 65 ± 1° for various lengths of time. At the end of each incubation period, the tubes are attached with a short section of Tygon to the condensor of a rotary evaporator which can be operated with the condensor axis at a downward tilt of about 30°. The methanol is removed in about 20 minutes at 40°. Alternatively, the solvent can be removed by evaporation in a stream of nitrogen at 40°. The latter procedure is more time consuming.

The product of methanolysis, methyl (methyl D-neuraminid)ate [compound (II), Fig. 1], is converted to methoxyneuraminic acid [compound (III) Fig. 1] by saponification in the following manner. The methanolysate is treated with 1.10 ml of a 1.0 N NaOH for 60 minutes at 25° (pH should be 12–13). The solution is transferred quantitatively to a 2.0-ml or a 5.0-ml volumetric flask and made up to volume with water. An aliquot (0.5–2.0 ml) of the sample is used for analysis on the 60-cm column of the amino acid analyzer with the pH 3.25 buffer as eluent. The amino acid analyzer constant for methoxyneuraminic acid determined with an authentic crystalline sample was 5.12 for an instrument for which the aspartic acid constant is 8.84. The elution volume of methoxyneuraminic acid and aspartic acid are 40 and 65 ml, respectively, on this instrument.

Calculations. It is important to note that the rate of cleavage of glycosidic bonds involving sialic acids differs from one sample to the other and that some destruction of sialic acid is unavoidable during methanolysis; the degree of destruction being dependent upon the composition and the concentration of the sample used and the time of meth-

[15] E. McGuire and S. B. Binkley, *Biochemistry* 3, 247 (1964).
[16] E. Klenk and H. Faillard, *Hoppe-Seyler's Z. Physiol. Chem.* 298, 230 (1954).

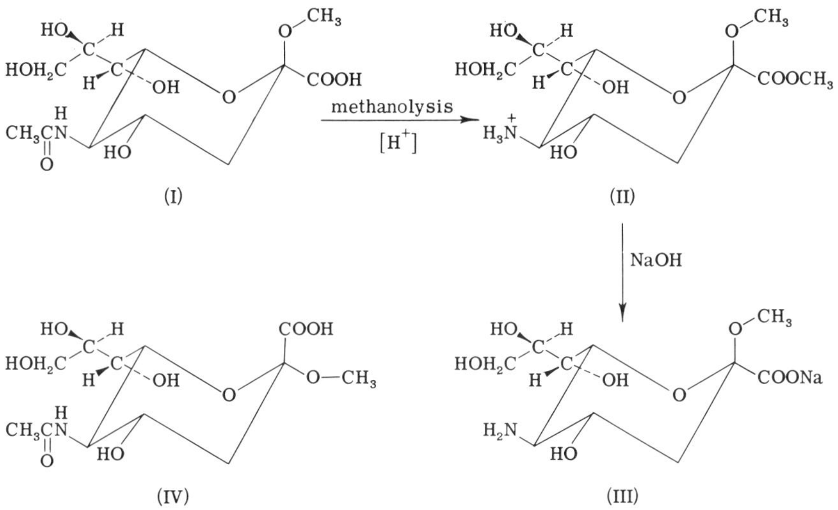

(I) β-Methoxy-N-acetylneuraminic acid
(II) β-Methyl(methyl D-neuraminid)ate
(III) β-Methoxyneuraminic acid
(IV) α-Methoxy-N-acetylneuraminic acid

Fig. 1. Structure of N-acetylneuraminic acid and its derivatives.

anolysis. A more accurate evaluation of the content of sialic acid is obtained from the study of multiple analyses with samples heated for varying periods of time. The values obtained at the two times of methanolysis, for instance, at 16 and 32 hours, are extrapolated to zero time assuming first-order kinetics.

Comments

Successful analyses of amino sugars in glycoproteins or polysaccharides depends upon: (a) the complete cleavage of every glycosidic bond; (b) prevention of destruction of amino sugars during hydrolysis; and (c) quantitative procedure suitable for the characterization and determination of the amino sugar released. Such a condition seems to have been achieved by the combined use of methanolysis in $2\,N$ methanesulfonic acid at 65° for the cleavage of glycosidic bonds and the amino acid analyzer column for the quantitative estimation of the released sialic acid. This conclusion is justified by the results of analyses of a number of sialic acid-containing glycoproteins and polysaccharides as shown in the table.

Sialic acids can be liberated from glycoproteins or polysaccharides by three different procedures. These include: (1) mild acid hydrolysis

ANALYSES OF SIALIC ACIDS IN GLYCOPROTEINS AND POLYSACCHARIDES[a]

Samples	Method of analyses		
	Neuraminidases: Warren test[b]	0.1 N H_2SO_4: Warren test[c]	Methanolic methanesulfonic acid: Amino acid analyzer[d]
1. α_1-Glycoprotein (human)[e]	37.7	37.0	39.9
2. Submax mucin (sheep)[f]	103.0	92.1	98.1
3. Desialized submax mucin (sheep)[f]	<1.0	<1.0	<1.0
4. Bird's nest (swallow)	31.2	30.1	40.1
5. *Escherichia coli* colominic acid[g]	245.0	75.0	273.1
6. Meningococcal B-polysaccharide[h]	263.0	84.0	260.0
7. Meningococcal C-polysaccharide[h]	<2.5	125.0	269.0
8. Group B streptococcus[i]	<1.0	52.9	98.7

[a] Results are expressed as micromoles of sialic acid per 100 mg of samples.

[b] Samples were digested with neuraminidase (*Clostridium perfringens*, obtained from Pierce Chemical Co.) at 37° for 24 hours essentially according to the method of Cassidy *et al.* [J. T. Cassidy, G. W. Jourdian, and S. Roseman, *J. Biol. Chem.* **240**, 3501 (1965)].

[c] The conditions used were 0.1 N H_2SO_4, 80°, 1 hour (R. G. Spiro, Vol. 8, p. 14).

[d] See text for detail. The values reported were obtained by extrapolation of 16- and 32-hour values to zero time or infinite time, assuming first-order kinetics.

[e] Obtained as a gift from Dr. E. A. Popenoe of Brookhaven National Laboratory.

[f] Sheep submax mucin R-2 samples and its desialized product were obtained from Dr. S. Roseman of the Johns Hopkins University in 1962.

[g] *E. coli* colominic acid was prepared from strain K235 by the method of Gotschlich. [E. C. Gotschlich, T. Y. Liu, and M. S. Artenstein, *J. Exp. Med.* **129**, 1349 (1969)].

[h] T. Y. Liu, E. C. Gotschlich, F. T. Dunne, and E. K. Jonsson, *J. Biol. Chem.* **246**, 4703 (1971).

[i] Obtained from Drs. Rebecca Lancefield and Emil C. Gotschlich of The Rockefeller University.

with 0.1 N H_2SO_4 at 80° for 1 hour[17]; (2) the action of neuraminidases; and (3) acid-catalyzed methanolysis with anhydrous acid such as 2 N methanesulfonic acid.

Sialic acid is in general unstable in aqueous acidic conditions. The standard mild conditions (0.1 N H_2SO_4, 80°, 1 hour) commonly used are good perhaps only for the liberation of sialic acid from glycoproteins because of its terminal position in these molecules. However, stronger conditions in aqueous media will rapidly cause complete destruction of sialic acid. When this method was applied to polymer of sialic acid

[17] R. G. Spiro, Vol. 8 [1].

with molecular weight in excess of 100,000 such as the meningococcal
B- and C-polysaccharides and the *E. coli* colominic acid (see the table),
the method failed to yield more than 20% of the sialic acid content in
60 minutes. Prolonged incubation (3–4 hours) resulted in higher recovery
of sialic acid (30–50%), but the yield never exceeded 55% from both
B- and C-polysaccharides of meningococcus.[4] The failure of this procedure to yield more than 55% of the sialic acid from these polysaccharides is most likely caused by compensating factors; continuous release
and destruction of the released sialic acid during the hydrolysis in
0.1 N H_2SO_4 at 80°.

All or a large part of the sialic acid present in glycoproteins and
polysaccharides can also usually be released by the action of neuraminidase (see the table). However, in some instances enzymatic hydrolysis
of the sialic acid containing polysaccharide did not result in the release
of any appreciable amount of sialic acid as are shown in the case of
the C-polysaccharide from meningococcus[4] and the polysaccharide isolated from a strain of group B streptococcus. Evidently, these neuraminidases are not capable of hydrolyzing certain glycosidic linkages of
sialic acid or its O-acetylated derivatives.

When 2 N methanesulfonic acid is used as a catalyst in methanolysis,
the release of sialic acid from the polysaccharides or the glycoproteins
has been consistently higher, as is shown in the table.

For the estimation of sialic acid content in polysaccharide and glycoproteins, the advantages of using methanesulfonic acid as a catalyst for
methanolysis, and the amino acid analyzer for the quantitative estimation of this amino sugar, are 3-fold. First, the reagent is effective in
causing more complete cleavage of sialic acid from the polysaccharides
and the glycoproteins. Second, the product of methanolysis after removal of solvent and saponification can be analyzed without further
derivatization such as is required for the gas chromatographic procedure. Third, the product, methoxyneuraminic acid, is the most stable
derivative of sialic acid and it is eluted at a unique position on the
chromatogram of the amino acid analyzer column, which serves to
identify this amino sugar.

For the analysis of amino sugars in glycoproteins or polysaccharides,
examination of the data as a function of hydrolysis time is important.
It provides an opportunity to ascertain whether the particular amino
sugar in question is stable under the condition of hydrolysis and if it
has been completely liberated from the glycoproteins or the polysaccharides. It permits an extrapolation of the values, either to zero
time or to infinite time to correct for the hydrolysis time on the
destruction or release of amino sugars. This method of calculation prob-

ably comes as close to giving accurate results as is possible at present. For this purpose at least two companion hydrolyzates, heated for 16 and 32 hours, are required. The values obtained at the two times of hydrolysis are extrapolated to give best figures.

[5] Molecular Weight Determination of Glycoproteins by Polyacrylamide Gel Electrophoresis in Sodium Dodecyl Sulfate

By JERE P. SEGREST and RICHARD L. JACKSON

Electrophoresis on polyacrylamide gels in the detergent sodium dodecyl sulfate $[CH_3(CH_2)_{10}CH_2OSO_3Na]$, abbreviated SDS, is a rapid and often employed technique for the determination of the molecular weights of proteins.[1-3] The usefulness of this procedure for accurate molecular weight determinations depends upon two factors. (1) Proteins in general bind constant amounts of SDS per gram when saturated.[4,5] The protein then has an overall negative charge that masks its intrinsic charge,[2,3] resulting in a constant charge to mass ratio for proteins.[4,5] (2) Proteins saturated with SDS take on a rodlike configuration, the length of the structure being proportional to its polypeptide chain length, and thus its molecular weight.[5]

Principle

This procedure is not directly applicable to molecular weight determinations of glycoproteins. Glycoproteins containing more than 10% carbohydrate behave anomalously during SDS polyacrylamide gel electrophoresis when compared to standard proteins.[6,7] The cause of this anomalous behavior is a decreased binding of SDS per gram of glycoprotein as compared with standard proteins.[7] The lower SDS binding results in a decreased charge to mass ratio for glycoproteins versus

[1] A. L. Shapiro, E. Viñuela, and J. V. Maizel, Biochem. Biophys. Res. Commun. 28, 815 (1967).
[2] K. Weber and M. Osborn, J. Biol. Chem. 244, 4406 (1969).
[3] A. K. Danker and R. R. Rueckert, J. Biol. Chem. 244, 5074 (1969).
[4] J. A. Reynolds and C. Tanford, Proc. Nat. Acad. Sci. U.S. 66, 1002 (1970).
[5] J. A. Reynolds and C. Tanford, J. Biol. Chem. 245, 5161 (1970).
[6] M. S. Bretscher, Nature (London) New Biol. 231, 229 (1971).
[7] J. P. Segrest, R. L. Jackson, E. P. Andrews, and V. T. Marchesi, Biochem. Biophys. Res. Commun. 44, 390 (1971).

standard proteins, a decreased mobility during SDS gel electrophoresis, and thus a higher apparent molecular weight. However, with increasing polyacrylamide gel cross-linking, of the two factors (charge and molecular sieving) involved in electrophoresis in SDS gels, molecular sieving predominates and the anomalously high apparent molecular weights of glycoproteins decrease, approaching, in an asymptotic manner, values close to their real molecular weights.[7]

Based upon this phenomenon a largely empirical technique for the estimation of the molecular weights of glycoproteins by SDS gel electrophoresis has been developed. The procedure involves determination of apparent molecular weights on glycoproteins over a range of polyacrylamide gel concentrations (usually 5, 7.5, 10, and 12.5%). From the curves obtained by plotting apparent molecular weight versus percent acrylamide, one can estimate an asymptotic minimal molecular weight. The degree of the resultant anomaly for each glycoprotein is some direct function of its percent carbohydrate and can be used as an approximate indicator of how close the asymptotic minimum is to the real molecular weight. For glycoproteins or glycopeptides with very large carbohydrate contents, 15% gels may be required for confident estimation of an asymptotic minimum.

Procedure

Protein Standards

Protein standards covering a sufficiently broad molecular weight range are required to construct standard plots of log molecular weight versus mobility for each of the polyacrylamide gel concentrations used. The following four protein standards generally provide an acceptable molecular weight range: Bovine serum albumin (BSA) (MW 68,000); aldolase (MW 40,000); α-chymotrypsinogen A (MW 25,700); myoglobin (MW 17,200). BSA is convenient because the dimer and tetramer which appear on SDS gels can be used as markers for higher molecular weights. Other standards that have been used are ovalbumin (MW 45,000), cytochrome c (MW 12,400) and insulin (MW 6000). Insulin is useful only in 10% and higher gels; with 5 and 7.5% gels the protein migrates with the tracking dye.

Gels

Reagents and Buffers

Solution A: N grams of acrylamide (where N = percent gel desired, and $N/25$ grams of N,N'-methylenebisacrylamide (Bis) (e.g., for 5% gels, 5 g of acrylamide and 200 mg of Bis) is made to a volume of 100 ml with stock phosphate buffer. Do not refrigerate.

Stock phosphate buffer: 0.1% SDS, 0.1 M sodium phosphate (pH 7.1), 0.012% sodium azide

Tetramethylethylenediamine (TEMED): 5 ml of TEMED to total volume of 100 ml with distilled water. Refrigerate.

Ammonium persulfate (1.5%): 1.5 g of ammonium persulfate per 100 ml of distilled water. Refrigerate; make a fresh solution every 5 days.

Preparation. To prepare gels with acrylamide concentrations of 5–15%, 19 ml of solution A of desired percentage of acrylamide (the higher percentage solutions should be deaerated), 1.0 ml of ammonium persulfate solution and 0.2 ml of TEMED are stirred for 2 minutes and added to gel tubes to within 1.5 cm of top. Water is carefully added to the top of the gel. The gels are allowed to polymerize for 1 hour at room temperature, the water is poured off, and the top is blotted with tissue paper. The gels are now ready for immediate sample application.

Preparation and Electrophoresis of Samples

The standard proteins and glycoproteins in solution are added to 12 × 75-mm test tubes (10 μg of each standard protein and 50–100 μg of each glycoprotein per gel) and lyophilized. The lyophilized samples are dissolved in 1% SDS containing 10^{-2} M dithiothreitol (for sulfhydryl bond reduction) and incubated for 1 hour at 37°. Before electrophoresis an equal volume of 0.002 M sodium phosphate buffer, pH 7.1 in 8 M urea, is added for a final SDS concentration of 0.5%; 10 μl of sample is added to the top of each gel, and the gels are placed in the reservoir tank for electrophoresis at room temperature. The tracking dye is electrophoresed to approximately 1 cm of the bottom of the gel.

Fixing and Staining of Gels

Reagents and Buffers

Fixative solution for periodic acid-Schiff (PAS): 40% ethanol, 5% glacial acetic acid, and 55% distilled water. Store at room temperature.

Fixative solution for Coomassie Blue: 50% methanol, 5% glacial acetic acid, and 45% distilled water. Store at room temperature.

Schiff reagent: Dissolve 10 g of basic fuchsin in 2 l of distilled water with heating. After cooling, add 200 ml of 1 N HCl and

17 g of sodium metabisulfite; mix the solution until it is de-colorized. Stir with HCl washed charcoal and centrifuge charcoal to avoid contact with filter paper. Filter the supernatant through glass wool to remove remaining charcoal; the filtrate should be clear and colorless. The solution is stored in a brown bottle at 4°.

Coomassie Brilliant Blue stain: Add 1.25 g of Coomassie Blue and 46 ml of glacial acetic acid to 454 ml of 5% methanol and stir for 30 minutes. The solution can be filtered if desired, but this is usually not necessary. Can be stored at room temperature for several days.

Periodic acid solution (0.7%): Dissolve 1.4 g of periodic acid in 200 ml of 5% acetic acid.

Sodium metabisulfite (0.2%): Dissolve 0.4 mg of sodium meta-bisulfite in 200 ml of 5% acetic acid.

Destaining solution: 875 ml of water, 50 ml of methanol, 75 ml of glacial acetic acid.

PAS Staining. For PAS staining of glycoproteins, each gel is fixed overnight in 100–200 ml of PAS fixative solution. The gels are then treated with the 0.7% periodic acid solution (covering the gels) for 2–3 hours, followed by treatment with 0.2% sodium metabisulfite for 2–3 hours with one solution change after 30 minutes. The gels after clearing are put in tubes and the tubes filled with Schiff reagent. Only plastic tops should be used to seal the tubes, not corks. Color develops in 12–18 hours at room temperature. Thereafter the gels should be stored at 4°.

Coomassie Blue Staining. For Coomassie Blue staining of proteins, each gel is fixed overnight in 100–200 ml of Coomassie Blue fixative and stained for 30–60 minutes, longest for the higher percentage gels, in sufficient Coomassie Brilliant Blue solution to cover the gels. The gels are then destained with 3–4 changes of 100–200 ml of destaining solu-tion and stored in tubes filled with destain solution at room temperature.

Interpretation of Results

Standard Proteins

The mobilities of the four protein standards—BSA, aldolase, α-chymotrypsinogen A, and myoglobin—when plotted against the log of their molecular weights are linear in each of four acrylamide gel con-centrations studied (5, 7.5, 10, 12.5%; Fig. 1). Ideally the higher molec-ular weight standards should always be within the linear portion of the curve for a particular gel concentration, but this will not always be

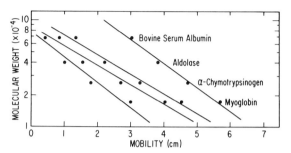

FIG. 1. Log of molecular weight versus mobility by sodium dodecyl sulfate gel electrophoresis of four standard proteins run in increasing concentrations of acrylamide. The plots from right to left represent 5, 7.5, 10, and 12.5% gels.

possible. As an example, BSA in the 12.5% gels is probably slightly beyond the point of linearity for this gel (Fig. 1). Nonlinearity at the low molecular weight end of standard curves can also be a problem. Figure 2 shows that the mobility of a protein standard, such as insulin, with a molecular weight of less than 10,000 fails to lie on the linear portion of a standard curve of higher molecular weight proteins even in 15% gels, suggesting this nonlinearity to be a property of the SDS complex itself rather than of gel concentration.[5] Therefore molecular weights of glycoproteins or glycopeptides below approximately 10,000 can only be approximated, in lieu of the use of peptide standards with a wide range of values below this level.

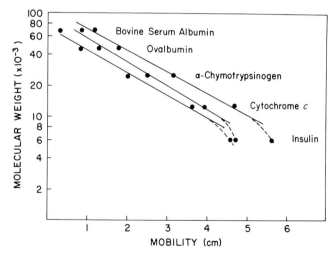

FIG. 2. Log of molecular weight versus mobility by sodium dodecyl sulfate gel electrophoresis of five standard proteins, including insulin. The plots from right to left represent 10, 12.5, and 15% gels.

Glycoproteins and Glycopeptides

When the apparent molecular weights of two glycoproteins and two glycopeptides (human RBC membrane glycoprotein, tryptic RBC membrane glycopeptide, and porcine ribonuclease, obtained as noted elsewhere,[7] and a discrete fragment of the RBC membrane glycoprotein produced by incomplete tryptic digestion[8]) are extrapolated from the appropriate standard curves of Fig. 1 and plotted against gel concentration (Fig. 3), a distinctly nonlinear relationship is found in each case. Noncarbohydrate containing polypeptide chains plot as horizontal straight lines intersecting their molecular weights (Fig. 1). In each example of Fig. 3 the apparent molecular weight decreases with increasing gel concentration, leveling off asymptotically at some characteristic value (asymptotic minimal molecular weight). This is what one would expect if decreased mobility due to low SDS binding is compensated for by increasing gel sieving (i.e., increasing gel concentration).

An asymptotic minimal molecular weight can be readily estimated for each curve in Fig. 3 except for the tryptic RBC glycopeptide. Here the apparent molecular weight is still falling precipitously between the 10% and 12.5% gels; for a more accurate estimate of the asymptotic minimum an additional point represented by data from 15% gels would be helpful.

Two reasonable assumptions would seem to follow. (1) The greater

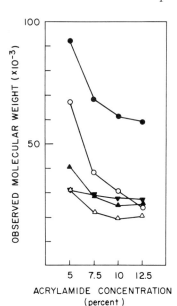

FIG. 3. Observed molecular weights of four glycoproteins calculated by electrophoretic mobility relative to the standard curves in Fig. 1 versus acrylamide gel concentration. ●——●, Human erythrocyte membrane glycoprotein; ○——○, human erythrocyte membrane tryptic glycopeptide; ▼——▼, fragmented human erythrocyte membrane glycopeptide; ▲——▲, porcine ribonuclease, higher molecular weight species; △——△, porcine ribonuclease, lower molecular weight species.

[8] R. L. Jackson, J. P. Segrest, and V. T. Marchesi, manuscript in preparation.

the percentage of carbohydrate for a given glycoprotein, the greater the anomalous effect. (2) The asymptotic minimal molecular weight, for a given glycoprotein, equals its real molecular weight as a first approximation.

When the approximate percentage of carbohydrate for each glycoprotein is known, the curves in Fig. 3 support the first assumption. When the difference between the apparent molecular weight of each glycoprotein on 5% and 7.5% gels is plotted against its approximate percentage of carbohydrate (Fig. 4), the resultant points fit a straight line to the limit of the data. This suggests that the degree of molecular weight anomaly for a given glycoprotein is a direct linear function of its percentage of carbohydrate. It should be noted, however, that this curve applies only to sialoglycoproteins. Glycoproteins without sialic acid are another matter and will be discussed in the next section.

The second assumption, that the asymptotic minimum approximates a true molecular weight, is supported by the curves in Fig. 3 for porcine ribonuclease, the best characterized of the glycoproteins included in this figure. This glycoprotein has been shown to be heterogeneous with a molecular weight range of 21,000–17,000[9]; the estimated asymptotic minimal molecular weights from Fig. 3 are 24,000 and 19,000, respectively.

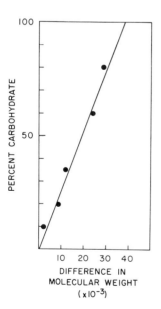

FIG. 4. Percent carbohydrate for each of the glycoproteins in Fig. 3 versus difference in the apparent molecular weight of each on 5 and 7.5% gels.

[9] V. N. Reinhold, F. T. Dunne, J. C. Wriston, M. Schwartz, L. Sarda, and C. H. W. Hirs, J. Biol. Chem. 243, 6482 (1968).

As a rough approximation the asymptotic minimal molecular weight for a given glycoprotein appears to be about 1000 daltons above its real molecular weight for every 10% of the glycoprotein represented by carbohydrate. The human RBC membrane glycoprotein which is 60% carbohydrate has an asymptotic minimum of 57,000 and thus an estimated real molecular weight of 51,000, a value in excellent agreement with chemical data based upon carbohydrate composition and cyanogen bromide peptides.[10] Similarly the tryptic RBC glycopeptide with an estimated minimum of 22,000 and 80% carbohydrate has an estimated real molecular weight of 14,000, near the actual value of 16,000.[8] No independent evidence exists for the real molecular weight of the partial trypsin fragment but the curve in Fig. 3 suggests it to be approximately 25,000.

Until more data are available, molecular weights determined for glycoproteins by the techniques described here should be considered to be only approximations. Sedimentation equilibrium should still be the method of choice for accurate molecular weights on most glycoproteins. However, when convenience is a main consideration or for glycoproteins that form aggregates or micellar structures in aqueous solutions (e.g., membrane glycoproteins[11]), the techniques described here can be useful.

Nonsialic Acid-Containing Glycoproteins and Glycopeptides

Removal of the sialic acid residues from the human erythrocyte glycoprotein and tryptic glycopeptide by neuraminidase[7] further accentuates the anomalous migratory patterns of these molecules (Fig. 5). Allowing for the loss of molecular weight represented by sialic acid from the human erythrocyte membrane glycoprotein and glycopeptide (25% and 50%, respectively), the desialized glycoprotein has a 20% increase in molecular weight and the desialized tryptic glycopeptide more than a 100% increase for each gel concentration examined.

The low mobility of the desialized glycopeptide even in 5% gels suggests a low binding of SDS to this molecule; the "native" tryptic glycopeptide is shown in the table to bind only 4% of the SDS per gram that standard proteins bind. These results indicate that the negatively charged sialic acid residues of glycoproteins partially compensate for low SDS binding. It is unlikely that the low mobility of the desialized glycopeptide represents aggregation since this molecule forms sharp bands well away from the top of glycine-buffered acrylamide gels.

The results of Fig. 5 suggest that the standard curve in Fig. 4 of

[10] J. P. Segrest, R. L. Jackson, and V. T. Marchesi, manuscript in preparation.
[11] V. T. Marchesi, R. L. Jackson, and J. P. Segrest, manuscript in preparation.

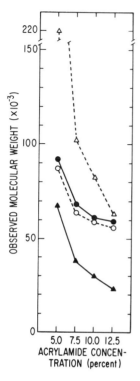

Fig. 5. Observed molecular weights of native and desialized glycoproteins on sodium dodecyl sulfate gel electrophoresis versus acrylamide gel concentration. ●——●, Native human erythrocyte membrane glycoprotein; ○——○, desialized human erythrocyte membrane glycoprotein; ▲——▲, "native" human erythrocyte membrane tryptic glycopeptide; △——△, desialized human erythrocyte membrane glycopeptide.

percentage of carbohydrate versus molecular weight anomaly might not apply to nonsialic acid containing glycoproteins; the desialized tryptic glycopeptide certainly does not fit this curve. Until more data are available on the degree of molecular weight anomaly for glycoproteins with

BINDING OF [³⁵S]SODIUM DODECYL SULFATE TO PROTEINS[a]

Protein	Total cpm bound × 10⁻³	SDS:protein (g:g)
Bovine serum albumin	1009	0.73
Ovalbumin	1213	0.88
Orosomucoid	560	0.41
Erythrocyte glycoprotein	531	0.38
Erythrocyte glycopeptide	32	0.023

[a] Each protein (4 mg) in 4 ml of 0.1 M potassium phosphate buffer, pH 7.2, containing 10^{-3} M dithiothreitol was dialyzed against 400 ml of the same buffer which contained 2×10^{-3} M [³⁵S]sodium dodecyl sulfate (Amersham, 343,545 cpm/mg). After 72 hours the samples were removed and radioactivity was determined with a Beckman scintillation spectrometer.

no sialic acid, the methods for molecular weight determination described here should be used with caution on these glycoproteins.

Other Possible Anomalies

A third factor which could influence molecular weight determinations on glycoproteins in SDS polyacrylamide gels is the distribution of a major portion of the mass of these molecules as multiple side chains branching out from the main polypeptide chain. There is as yet no evidence to suggest what effect this might have on the mobility of such a molecule when it is complexed with SDS. Some effect might be expected, however, since glycoprotein–SDS complexes would be sterically unlike standard protein–SDS complexes that are in rodlike configurations with lengths proportional to the molecular weight of their polypeptide chains.[5]

A final word of caution is in order concerning the use of SDS polyacrylamide gels for molecular weight determinations on certain noncarbohydrate-containing proteins, especially those associated with membranes. We have observed several cyanogen bromide fragments of membrane proteins that have falsely high molecular weights calculated on the basis of SDS gel mobility.[12] Curiously enough this anomalous mobility is not changed by varying the acrylamide concentration. The cause for this phenomenon is not presently understood, but may involve unusual amino acid compositions.

[12] Unpublished observations.

[6] Analysis of Sugars by Automated Liquid Chromatography

By Y. C. LEE

Although sugar components in complex glycoproteins can be determined by a number of methods, such as colorimetry, partition chromatography, or gas chromatography, automated liquid chromatography perhaps is the most versatile and reliable. It provides speed, accuracy, sensitivity, and specificity and is especially useful in glycosidase digestions of complex carbohydrates because the digestion mixture can be analyzed without prior treatment. Two systems, one for neutral sugars and the other for amino sugars, will be described here.

Determination of Neutral Sugars

This is a slightly modified version of the previously published method.[1] An accelerated system has been recently reported.[2]

Reagents

Sodium borate buffer, 0.15 M, pH 7.40. Dissolve 9.28 g of boric acid in 1 liter of distilled water and titrate with 2 N NaOH to pH 7.40 ± 0.02.

Sodium borate buffer, 0.4 M, pH 10.0. Dissolve 24.74 g of boric acid in 1 liter of distilled water. Adjust the pH of the solution to 9.6–9.8 with NaOH pellets, and finally titrate to pH 10.00 ± 0.02 with saturated NaOH solution.

Potassium tetraborate solution (10%, w/v). Dissolve 100 g of $K_2B_4O_7 \cdot 4H_2O$ in about 800 ml of distilled water, and finally make up to 1 liter.

Orcinol–H_2SO_4 reagent

(a) 70% H_2SO_4. To make one batch, a 4-liter Pyrex beaker containing 750 ml of chilled water is immersed in an ice bath. Gently pour in 1750 ml of 98% H_2SO_4. As the heat of dilution provides enough convection, no additional stirring is necessary at this stage. After the temperature of the diluted mixture has come down to near room temperature, transfer and store at room temperature in a tightly capped bottle until it is ready for dissolving orcinol. Three to four batches of dilution can be carried out conveniently in a sink filled with crushed ice. For the storage of diluted acid, the original containers for 98% H_2SO_4 are most suitable.

(b) Orcinol. Commercial orcinol is decolorized with charcoal and recrystallized from benzene. Some commercial orcinol can be used without recrystallization, if a high background of the colorimetric reaction can be tolerated. The orcinol–H_2SO_4 reagent is prepared as needed by dissolving 2.5 g of recrystallized orcinol in one batch of the 70% H_2SO_4 (see above). This reagent should be protected from light.

Anion exchange resin. Several choices are available among the specially prepared chromatographic resins. However the resins from different sources do not behave identically and require different chromatographic conditions. The method described here is for Technicon Chromo-bead S (low-pressure type). A micro-

[1] Y. C. Lee, J. F. McKelvy, and D. Lang, *Anal. Biochem.* **27**, 567 (1969).
[2] Y. C. Lee, G. S. Johnson, B. White, and J. Scocca, *Anal. Biochem.* **43**, 640 (1971).

fine anionic resin, MFA-6 (J. T. Baker), after careful sizing
(20 ± 10μ) gives results very similar to Chromo-bead S.

Apparatus for Chromatography

Columns. Water-jacketed thick-wall columns, 75 cm × 3 mm
(i.d.), can be purchased from B. R. Glass, Inc. (P.O. Box 1040,
Pasadena, Maryland) or other suitable sources.

High pressure pump. Many models of positive displacement pump
are commercially available. Main requirements are accuracy and
capability to pump against high pressure (up to 600 psi). Technicon Micropump and Milton Roy Mini-pump have been used
in the author's laboratory.

Gradient mixer. Any two-chamber gradient mixer suitable for
generating a linear gradient with 100 ml of each component
solution can be used.

Constant-temperature circulating water bath. The size of the water
bath depends on the number of columns it serves simultaneously.
For two columns, a 1–2 gal bath is adequate. Temperature is
kept at 55° ± 0.5°.

Pressure gauge. Preferably a stainless steel diaphragm type capable
of measuring up to 1000 psi.

Flowmeter. A 5-ml pipette affixed to a three-way microstopcock
is suitable. The flowmeter is placed between the pump and the
buffer reservoir.

In measuring the flow rate, first turn the stopcock to fill the pipette
upward. Then turn the stopcock to pump buffer to the column in such a
way that the pump withdraws the buffer only from the pipette. Measure
displaced liquid over 3–4 minutes for flow rate determination.

Apparatus for Colorimetry. The following description is for a system
of the Technicon autoanalyzer type. Other comparable systems also can
be used.

Proportionating pump. This is a peristaltic pump to mix column
effluent and reagents prior to heating for color reaction. A wide
variety of tubings are available for different pumping rates. The
manifold tubings of the proportionating pump were set up according to Table I.

Oil bath and reaction coil. The temperature of the oil bath is
kept at 95° ± 0.5°. A single length of reaction coil, 1.6 mm
(i.d.) ×40 ft, is used.

Colorimeter and recorder. A single-channel colorimeter with 420-nm
filters and a single pen recorder are used. For the neutral sugar

TABLE I

MANIFOLD TUBING ARRANGEMENT FOR NEUTRAL SUGAR ANALYSIS

Fluid	Tubing type	I.d. (inch)	Nominal flow rate (ml/min)
Effluent	Standard	0.030	0.32
Air	Standard	0.025	0.23
Orcinol	Acidflex	0.056	0.92
Waste	Acidflex	0.056	0.92

procedure described here, chart speed is usually set for 30 minutes per inch.

Sensitizing lamp. A fluorescent lamp bulb F4T5/CW inserted into the interior of a coil (22 turns, 2 mm i.d.). The sensitizer is not mandatory, but it increases the color yield by 25–30%.

Figure 1 illustrates the arrangement of analytical components. The components are placed as close as possible, and are linked with minimal lengths of glass tubing of the same outside and inside diameters as the

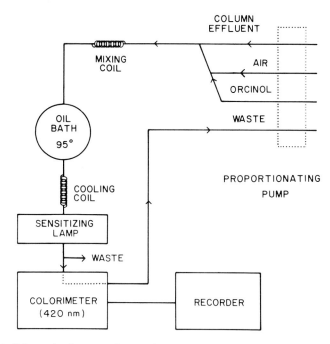

FIG. 1. Schematic diagram of analytical system for neutral sugar analysis. Tubings used in the proportionating manifold are described in Table I. The mixing coil is 2 mm (i.d.), 28 turns.

coils. Acidflex tubing is used to join the ends of glass tubings which are placed as close as possible.

An inexpensive, yet versatile system can also be built with components not specifically made for autoanalysis. For example, a Spectronic 20 colorimeter (Bausch and Lomb) with a flow-through cell (A. H. Thomas) linked to a VOM-5 recorder (Bausch and Lomb) or a Heath recorder (EU-20B) have been used in the author's laboratory. Both light wavelength and chart speed can be changed by turning knobs rather than changing filters or gears. These features are desirable if the same basic components are to be used for both neutral and amino sugar analysis (see below), or if other modifications are to be made.

Procedure

Preparation and Application of Sample. The sample for analysis is hydrolyzed with 1 N sulfuric acid for 4–6 hours at 100°, and deacidified with Dowex 1-X8 (200–400 mesh, carbonate form). The neutral hydrolyzate is concentrated *in vacuo* at a temperature below 40°, if necessary. An appropriate aliquot containing 0.01–0.08 μmole of each sugar is less than 200 μl is applied to the column under pressure,[3] and the column top is washed twice with 50 μl of the starting buffer.

Acid hydrolyzates can also be directly applied to the column if the total amount of acid does not exceed 0.2 meq. In this case sugar standards must be applied with the same amount of acid as the samples, because peak positions as well as peak areas vary as a function of the quantity of acid present in the sample.

Certain amounts of salt in the sample can be tolerated. For example, samples containing up to 200 μl of glycosidase digest in 0.1 M acetate buffer (pH 4–5) have been applied directly with satisfactory results.

Elution. A linear gradient is generated with 100 ml each of the 0.15 M (pH 7.40) buffer and the 0.40 M (pH 10.00) buffer. The gradient is started after the sample is applied, and the flow rates are checked. The column is eluted at the rate of 0.50 ml per minute. Thirty minutes after the start of the gradient, when the baseline is established, the chart drive on the recorder is turned on. After each analysis the column is regenerated and equilibrated by pumping first with 10% potassium tetraborate for 1 hour then with the 0.15 M borate buffer for 2 hours, at the rate of 0.50 ml per minute. Two columns of identical dimensions can be used to facilitate semicontinuous operation. Under these conditions, one column can be regenerated and equilibrated while the other is used for analysis.

---

[3] Either a nitrogen tank or a small air pump (such as Cole-Palmer Dyna-Vac pump, Model 3) can be used.

The back pressure is normally between 200 and 250 psi, and the columns are repacked when the back pressure exceeds 300 psi. The manifold tubings are changed after every 18–20 analyses.

The system should be thoroughly flushed with water before shutting down. The flushing can be accelerated by temporarily replacing the "air" tubing of the manifold with a piece of larger capacity tubing, connected to a water reservoir.

Calculation. Quantitation of sugar components is made with reference to standards analyzed under the same conditions. Peaks are measured as follows: The net peak height (H) in absorbance units is read from the chart. A line is drawn at the net half-peak height, and the peak width (W) in millimeters is measured at this crossline with a comparator (for example a 6× Edscorp comparator, Edmund Scientific), or a vernier caliper. Peak areas ($H \times W$) are calculated from these parameters. An arbitrary constant, C, is defined as $(H \times W)/\mu$mole, and is determined from the peak area given by a known quantity of standard. The quantity of unknown is in turn determined from the C value of the standard. Although C values change little even when different batches of reagents and buffers are used, they are sensitive to the state of the manifold tubings and the pumping rate. It is suggested to repeat standard runs after every 10 runs.

Comments

Resolution, Precision, and Linearity. A typical chromatogram is shown in Fig. 2. In the described system, mannose–ribose and fucose–arabinose are not completely separated. Improved separation of these pairs can be obtained, however, by using different gradients. For example, sufficient separation of fucose and arabinose (but not mannose and ribose) can be obtained with a 3-component gradient (50 ml each), $0.2\,M$ sodium borate (pH 7.0)/$0.2\,M$ sodium borate (pH 7.0)/$0.5\,M$ sodium borate (pH 10.0).

Precision of the method is quite high. Standard deviations are less than 2.5%. Peak areas under the described conditions are linear up to at least 0.08 μmole for most sugars. Each sugar has a different color yield, therefore a C value must be determined for each sugar. Generally speaking, the color yields are found to decrease in the order of pentose, hexose, and 6-deoxyhexose.

Other Applications. In addition to the sugar components of complex carbohydrates, this system can be used to separate oligosaccharides (e.g., α-1,2-D-mannobiose from α-1,6-D-mannobiose) and partially methyl-

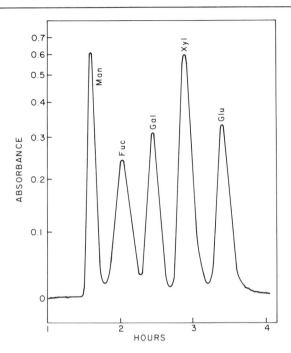

Fig. 2. A typical chromatogram of sugar standards: mannose (Man), 0.122 μmole; fucose (Fuc), 0.140 μmole; galactose (Gal), 0.111 μmole; xylose (Xyl), 0.139 μmole; and glucose (Glu), 0.111 μmole.

ated sugars (for example dimethyl and monomethyl ethers of D-arabinose[4]).

Final purification of glycosidases is often carried out by electro-focusing in sucrose. Even after dialysis, small amounts of glucose and fructose (as well as sucrose) may still remain with the enzymes to interfere with the monitoring of digestive course, such as the measurement of increase in reducing power. This system can overcome such difficulties if the sugars being released are not related to sucrose.

A faster system than is described here has been published.[2] In this system, which differs only in the chromatographic conditions from the system described here, DA-4 resin (Durrum Chemical Corp., Palo Alto, California 94303) in a 6 mm (i.d.) × 30 cm column is used. A linear gradient formed with 70 ml each of 0.4 M sodium borate (pH 8.0) and

[4] P. Mied and Y. C. Lee (1971), Anal. Biochem., in press. Presented in part at the Sixth Middle Atlantic Regional Meeting of American Chemical Society, February 3–5, 1971.

0.4 M sodium borate (pH 10.0) is pumped at the rate of 0.75 ml per minute. Mannose, fucose, galactose, xylose, and glucose are eluted within 2.5 hours. This system gains speed at the expense of some resolution.

Determination of Amino Sugars

Amino sugars can be determined with an amino acid analyzer or with the autoanalytical system as described above for the neutral sugars.

With Amino Acid Analyzer

If glucosamine and galactosamine are the only amino sugars under consideration, the existing amino acid analyzer based on the principles of Spackman et al.[5] can be used directly for the analysis. Either a short column for basic amino acids or a long column for acidic and neutral amino acids can be used with the same buffers and same ninhydrin reagent as with analysis of amino acids. The long-column system is more satisfactory when a large quantity of amino acids are also present. In this case, the column can be eluted with the second buffer (0.20 N Na^+, pH 4.25) from the beginning.

Glucosamine and mannosamine can be separated by this system, but mannosamine and galactosamine are not completely separated. For complete separation of all amino sugars, a more elaborate system, such as that proposed by Brendel et al.,[6] is necessary.

With Technicon-Type Autoanalyzer

The basic components described for neutral sugars can be used for the analysis of amino sugars by changing the manifold tubings and color filters.[7] The principle of this method is chromatographic separation of amino sugars on cation exchange resin and coloration with neocuproine reagent.

Apparatus and Reagents

Columns. Either a 9-mm or a 3-mm (i.d.) × 30 cm, water-jacketed column is used. The former is better for resolution, the latter, for speed.

Resin. The cation exchange resins specially prepared for amino acid analyzers give the best results. Alternatively, fine-mesh bead-polymerized cation exchange resin, such as MFC-6 (J. T. Baker),

[5] D. H. Spackman, W. H. Stein, and S. Moore, Anal. Chem. 30, 1190 (1958).
[6] K. Brendel, R. S. Steele, R. W. Wheat, and E. A. Davidson, Anal. Biochem. 18, 161 (1967); K. Brendel, N. O. Roszel, R. W. Wheat, and E. A. Davidson, Anal. Biochem. 18, 147 (1967).
[7] Y. C. Lee, J. R. Scocca, and L. Muir, Anal. Biochem. 27, 559 (1969).

can be carefully sized $(20 \pm 10 \mu)$ by repeated sedimentation in a cylinder. These inexpensive resins give somewhat less satisfactory resolution than the specially prepared commercial resins. However, for most purposes, they may be adequate.

Sodium citrate buffer $(0.35 N$ in Na^+, pH $5.28 \pm 0.02)$. Dissolve 49.1 g of citric acid·2H₂O, 28.8 g of NaOH, 13.6 ml of concentrated HCl, and 0.2 ml of octanoic acid in distilled water to a final volume of 2 liters. The pH of the buffer is adjusted with either concentrated HCl or 50% NaOH. The buffer is passed through a Millipore filter $(0.8 \mu$ pore size) before use.

Reagent A—Copper solution. Dissolve 40 g of anhydrous Na_2CO_3, 16 g of glycine, and 0.45 g of $CuSO_4 \cdot 5H_2O$ in distilled water to make 1 liter. This reagent is stable at room temperature for at least a year.

Reagent B—Neocuproine reagent. Dissolve 1.20 g of neocuproine (2,9-dimethyl-1,10-phenanthroline) hydrocloride in 1 liter of distilled water. This reagent is also stable at room temperature for at least a year.

Arrangement of Components. The arrangement of the manifold tubing is shown in Table II. A schematic diagram of the analytical system is shown in Fig. 3.

Preparation and Application of Sample. The sample is hydrolyzed in a sealed ampoule with 4–6 N HCl for 4–8 hours at 100°, and dried at about 40° in a flash evaporator. The dried hydrolyzate is dissolved in distilled water or 0.01 N HCl, and an aliquot of the solution containing 1–25 μg of hexosamine in 25–100 μl is applied to the column. The column top is washed twice with 50–100 μl of the eluting buffer.

TABLE II

ARRANGEMENT OF THE MANIFOLD TUBINGS FOR AMINO SUGAR ANALYSIS

	Procedure I		Procedure II	
Fluid	I.d.[a] (inch)	Nominal flow rate (ml/min)	I.d.[a] (inch)	Nominal flow rate (ml/min)
Air	0.051	1.00	0.025	0.23
Reagent A	0.035	0.42	0.035	0.42
Reagent B	0.035	0.42	0.035	0.42
Column effluent	0.040	0.60	0.045	0.80
Waste	0.051	1.00	0.060	1.40

[a] All tubings are of transparent standard type.

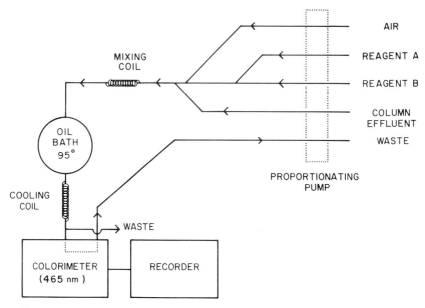

Fig. 3. Schematic diagram of analytical system for hexosamine analysis. Tubings used in the proportionating manifold are described in Table II.

Procedure I. The procedure described here is based on the published method using Aminex Q-15S cation exchange resin (Bio-Rad). In this procedure, the 3-mm column maintained at 55° is used. Elution rate is 0.72 ml per minute, and chart speed is set at 3 inches per 10 minutes. Glucosamine peak appears 20–25 minutes after the elution is started. Usually a second sample can be applied even before the first sample is completely recorded on the chart. By such staggered application of samples, efficiency can easily be doubled. Procedure I is suited for samples containing only one kind of amino sugar, and its aim is to eliminate interference due to neutral sugars and amino acids. After a series of analyses, both the column and the reaction coil is thoroughly flushed with water to avoid clogging (due to microbial growth in the former and due to salt deposit in the latter). Measurements of peak areas and calculation of results are similar to those used for the neutral sugar system.

Procedure II. This procedure is superior to Procedure I with respect to resolution. The 9-mm (i.d.) column is used here with a pumping rate of 1.00 ml per minute. Other details are similar to Procedure I.

Linearity and Standard Deviation. Linearity of peak area covers the range of 0.005–0.120 μmole. Standard deviations for standards and hydrolyzed biological samples are $\pm 2\%$ and $\pm 3\%$, respectively.

Other Applications. Methyl ethers of D-glucosamine can be analyzed by this system using a 6 × 700 mm column of Rexyn 101.[8]

[8] Y. C. Lee and J. R. Scocca, *Anal. Biochem.* **39**, 24 (1971).

[7] Isolation and Characterization of Connective Tissue Polysaccharides

By Lennart Rodén, John R. Baker, J. Anthony Cifonelli, and Martin B. Mathews

A great number of methods are now available for the isolation and characterization of connective tissue proteoglycans and their polysaccharide components (mucopolysaccharides, glycosaminoglycans). In this chapter is described a selected group of methods that have been used extensively by the authors and found to be satisfactory for these purposes. Although primarily developed for the study of mammalian tissues, the methods are usually applicable to homologous tissues of other vertebrates but not, generally, to invertebrate tissues.

Many of the techniques employed in the study of glycoproteins in general may also be applied to investigations of proteoglycans and polysaccharides. Therefore, only methods with special relevance to mucopolysaccharide analysis will be described here; information concerning general carbohydrate analysis should be sought elsewhere.[1] Furthermore, certain methods for the analysis of connective tissue polysaccharides are described in this volume[2] and in Volume 8[3] of this series and will therefore be treated only briefly.

I. General Characteristics of Connective Tissue Polysaccharides

Connective tissue proteoglycans are distinguished from other mammalian carbohydrate–protein compounds by the presence of relatively large polysaccharide chains containing repeating disaccharide units as their most characteristic feature. These are usually composed of a uronic acid and a hexosamine, and Table I gives the composition of the repeating disaccharide units of the polysaccharides which have been recognized so far. It is seen that two uronic acids, D-glucuronic acid and

[1] See Vol. 8 [1], [2], [7]; this volume [1], [4], [6], [10].
[2] This volume [2], [3], [123], [124], [125].
[3] Vol. 8 [3], [112], [113], [114].

TABLE I

COMPOSITION OF CONNECTIVE TISSUE POLYSACCHARIDES

Polysaccharide	Components of repeating disaccharide units	Other monosaccharide components	N-Acetyl groups	O-Sulfate groups	N-Sulfate groups
Hyaluronic acid	D-Glucosamine and D-glucuronic acid	L-Arabinose? D-Galactose? D-Glucose?	+	–	–
Chondroitin	D-Galactosamine and D-glucuronic acid	D-Xylose D-Galactose	+	–	–
Chondroitin 4-sulfate	D-Galactosamine and D-glucuronic acid	D-Xylose D-Galactose	+	+	–
Chondroitin 6-sulfate	D-Galactosamine and D-glucuronic acid	D-Xylose D-Galactose	+	+	–
Dermatan sulfate	D-Galactosamine and L-iduronic acid or D-glucuronic acid	D-Xylose D-Galactose	+	+	–
Heparan sulfate	D-Glucosamine and D-glucuronic acid or L-iduronic acid	D-Xylose D-Galactose	+	+	+
Heparin	D-Glucosamine and D-glucuronic acid or L-iduronic acid	D-Xylose D-Galactose	+[a]	+	+
Corneal keratan sulfate	D-Glucosamine and D-galactose	L-Fucose Sialic acid D-Mannose	+	+	–
Skeletal keratan sulfate	D-Glucosamine and D-Galactose	D-Galactosamine L-Fucose Sialic acid D-Mannose	+	+	–

[a] A minor proportion of the amino groups are acetylated, while most are sulfated.

L-iduronic acid, and two hexosamines, D-glucosamine and D-galactosamine, are the constituent sugars of these polysaccharides. In addition,
several other monosaccharides may be present as integral components,
including sialic acid, mannose, fucose, galactose, and xylose. With the
exception of galactose (in keratan sulfate), these sugars are not part of
the characteristic repeating disaccharide and occur either as side branches
or as constituents of the specific carbohydrate-protein linkage regions
(see below).

The hexosamine residues are commonly N-acetylated, although
most of the amino groups in heparin are N-sulfated, as are a considerable
proportion of those in heparan sulfate. Sulfate is also found as O-sulfate
in several other polysaccharides and is then generally located on carbon
4 or 6 of the hexosamine moiety. However, sulfate ester groups may also
be present in other positions, i.e., on iduronic acid residues of heparin
and heparan sulfate, on the glucuronic acid component of chondroitin
sulfate D, and on galactose in keratan sulfate.

In the tissues, the polysaccharides are, almost without exception,
covalently bound to protein. So far, three types of carbohydrate-protein
linkages have been recognized in the connective tissue proteoglycans:
(1) an O-glycosidic linkage between xylose and serine hydroxyl groups;
(2) an O-glycosidic linkage between N-acetylgalactosamine and the
hydroxyl groups of serine or threonine; and (3) an N-glycosylamine
linkage between N-acetylglucosamine and the amide group of asparagine.
The first linkage type is found in the chondroitin sulfates, dermatan
sulfate, heparin, and heparan sulfate; the second in skeletal keratan
sulfate (keratan sulfate II); and the third in corneal keratan sulfate
(keratan sulfate I). The existence of a covalent linkage between hyaluronic acid and protein has not yet been established with certainty.

The xylose-linked polysaccharides also contain two galactose residues and one glucuronic acid residue as components of the specific linkage
region (Fig. 1), the latter unit being bound to the first regular repeating
disaccharide of the polysaccharide chain.

It is not the purpose of this brief introduction to describe the complete
structural details of the connective tissue polysaccharides, and reference

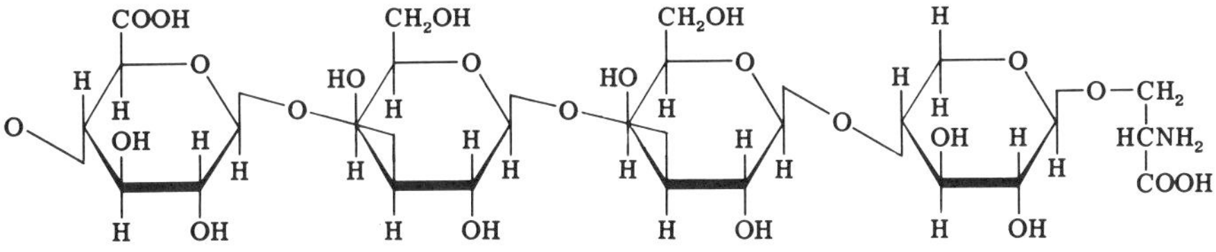

FIG. 1. Carbohydrate–protein linkage region of chondroitin 4-sulfate.

is made to other reviews,[4-7] where the subject is covered comprehensively. However, the preceding summary provides a basis for discussing some of the principles employed in fractionation procedures and the criteria of purity which should be applied to the isolated polysaccharides.

II. General Preparative Procedures

The macromolecular proteoglycans are often firmly associated with other tissue components. In the past, it has been difficult to extract the native proteoglycans efficiently, but recent advances in methodology now permit the isolation of several of these substances in nearly quantitative yields. In particular, the method developed by Sajdera and Hascall[9] for the isolation of cartilage proteoglycan by extraction with guanidine represents a valuable technique which has given new impetus to studies of the native proteoglycans. Methods for proteoglycan preparation will not be described here, and reference is made to other recent discussions of this subject.[7-10]

The quantitative isolation of a given polysaccharide from a tissue is a simpler task, since more drastic methods may be employed which degrade the protein core of the proteoglycan or cleave the carbohydrate–protein bond. Most often, the polysaccharides are liberated by proteolytic digestion, but extraction with water, salt solutions, or alkali is sometimes preferred in certain situations. A few tissues contain only one major polysaccharide (e.g., hyaluronic acid in vitreous body and chondroitin sulfate in cartilage), and the isolation procedure may then be relatively simple, but in general, several polysaccharides are present, and further fractionation of the polysaccharide mixture is necessary. The procedures that have been of greatest value are based on differences between the various polysaccharides with regard to (1) charge density, and (2) solubility in ethanol. The first group includes methods such as electrophoresis, ion exchange chromatography, and fractionation of quaternary ammonium complexes. The latter procedure, developed by Scott, is the

[4] J. S. Brimacombe and J. M. Webber, "Mucopolysaccharides." Elsevier, Amsterdam, 1964.
[5] R. W. Jeanloz, in "The Carbohydrates" (W. Pigman and D. Horton, eds.), 2nd ed., Vol. 2B, pp. 589–625. Academic Press, New York, 1970.
[6] L. Rodén, in "Metabolic Conjugation and Metabolic Hydrolysis" (W. H. Fishman, ed.), Vol. 2, pp. 345–442. Academic Press, New York, 1970.
[7] E. A. Balazs (ed.), "The Chemistry and Molecular Biology of the Intercellular Matrix." Academic Press, New York, 1970.
[8] C. P. Tsiganos and H. Muir, Biochem. J. 113, 885 (1969).
[9] S. W. Sajdera and V. C. Hascall, J. Biol. Chem. 244, 77 (1969).
[10] M. Schubert, in "Methods in Carbohydrate Chemistry" (R. L. Whistler, ed.), Vol. 5, pp. 106–109. Academic Press, New York, 1965.

single most useful procedure for the isolation and fractionation of anionic polysaccharides.

It is sometimes advisable to reduce the fat content of certain tissues, e.g., brain and skin, prior to extraction or digestion. A common procedure is to extract the ground tissue with several changes of acetone at room temperature over a period of several days. Alternatively, the tissue may be refluxed with chloroform–methanol (2:1). After drying, the defatted powder is subjected to proteolytic digestion or other methods for the extraction of polysaccharides.

A. Extraction from Tissues

1. Nondegradative Methods

Whereas mild extraction methods are necessary for the isolation of native proteoglycans, they are of limited use, if one wishes to isolate the entire polysaccharide pool from any given tissue, since difficulties are often encountered in extracting the proteoglycans quantitatively. However, one of the connective tissues polysaccharides, i.e., hyaluronic acid, may be obtained in reasonable yield from several tissues by extraction with water and salt solutions, and procedures for its isolation from vitreous body,[11] umbilical cord,[12] and synovial fluid[13] are well established. (Details of these methods will not be given here.)

It should be emphasized that it is essential to isolate hyaluronic acid by relatively mild methods if the integrity of the native structure is to be maintained. During digestion with papain at elevated temperatures, hyaluronic acid is degraded to some extent by cleavage of a small number of glycosidic linkages, and the isolated product is of considerably lower molecular weight than material obtained after salt extraction.

2. Degradative Methods

a. Extraction with Alkali

Treatment with alkali has in the past often been used for the complete extraction of polysaccharides from tissues. For example, this procedure has been applied successfully to the isolation of chondroitin sulfate from cartilage. After extraction of the tissue at 4° with $0.5 M$ alkali overnight, the extract is neutralized with acetic acid and dialyzed. Contaminating protein is removed by treatment with kaolin, Lloyd's

---

[11] T. C. Laurent, M. Ryan, and A. Pietruszkiewicz, *Biochim. Biophys. Acta* **42**, 476 (1960).

[12] A. Dorfman and J. A. Cifonelli, Vol. 1 [4].

[13] B. N. Preston, M. Davies, and A. G. Ogston, *Biochem. J.* **96**, 449 (1965).

reagent, or other adsorbents, and the chondroitin sulfate is recovered by precipitation with ethanol. The product is virtually free of contaminating protein and is generally of high purity, but it should be noted that some cleavage of glycosidic linkages presumably occurs during alkali treatment.

The basis for the release of chondroitin sulfate by alkali treatment is a β-elimination reaction by which the carbohydrate–protein linkage between xylose and the serine hydroxyl groups is cleaved.[14] Since this leaves the molecule open to further degradation by alkali from the reducing end, alkali treatment should be avoided if a preparation is desired in which the specific linkage region is intact. Alternatively, the alkali treatment may be carried out in the presence of borohydride, which reduces the xylose residue to xylitol, thereby effectively preventing further degradation.

Although alkali extraction is no longer used for preparative purposes to any great extent, the alkali sensitivity of the xylose–serine linkage and other linkages to the hydroxyamino acids provides a valuable diagnostic tool in the differentiation of carbohydrate–protein linkages. This subject has been discussed at length in this volume ([1] and [3]).

b. PROTEOLYTIC DIGESTION

Extraction with alkali has been largely superseded by digestion with proteolytic enzymes as the most common procedure for releasing polysaccharides from tissues. Usually, extensive proteolysis with a protease of broad specificity is desirable and treatment with papain or Pronase yields single polysaccharide chains with only a small residual peptide. (The authors have some preference for the use of papain in initial tissue proteolysis, since this enzyme is generally more effective than Pronase in achieving complete solubilization of various tissues.) Details of digestion procedures will be given in a subsequent section dealing with the preparation of individual polysaccharides. For certain purposes, particularly in studies of the structure of the protein moieties of various proteoglycans, it may rather be advantageous to use a protease which gives a more limited degradation. For instance, treatment of cartilage proteoglycan with trypsin and chymotrypsin yields "doublets" in which two polysaccharide chains are attached to the same peptide (the size of the peptide being approximately 30–40 amino acids).[15,16] Further degradation of such doublet preparations by specific chemical or enzymatic methods permits the study of a larger segment of the protein

[14] B. Anderson, P. Hoffman, and K. Meyer, *J. Biol. Chem.* **240**, 156 (1965).
[15] M. Luscombe and C. F. Phelps, *Biochem. J.* **102**, 110 (1967).
[16] M. B. Mathews, *Biochem. J.* **125**, 37 (1971).

core than is feasible with the smaller peptides obtained after papain digestion.

Whereas it would often be desirable to achieve complete proteolysis to the point where only one amino acid remains attached to the polysaccharide, this is rarely feasible. Exhaustive digestion of cartilage with papain yields chondroitin sulfate preparations in which the residual peptides, on average, consist of 5 amino acids. Similarly, the product obtained after digestion with Pronase contains peptides with several amino acids. When applied to the digestion of smaller glycopeptides from which the bulk of the polysaccharide chain has been removed, Pronase is more effective than papain in cleaving the peptide bonds around the carbohydrate–protein linkage. For instance, in chondroitin sulfate glycopeptides, obtained after digestion of the proteoglycan with papain and testicular hyaluronidase, serine constituted 20% of the total amino acids; prolonged digestion with papain increased the serine content only to 25%, whereas subsequent treatment with Pronase yielded a product in which the serine content was close to 50% of the total amino acids.[17] Almost complete removal of amino acids other than serine may be obtained by continued digestion with other enzymes, e.g., leucine aminopeptidase and carboxypeptidase.

Despite the difficulties in achieving complete proteolysis in the presence of a large, inhibitory polysaccharide chain, this goal may nevertheless be reached under favorable circumstances. Analysis of several lots of crude heparin, stage XIV, from the Wilson Laboratories, Chicago, has shown that virtually only serine remains after the commercial digestion procedure, which reportedly involves treatment with crude proteases of animal origin. This heparin preparation is extremely useful for the isolation of hydrolytic fragments from the carbohydrate-protein linkage region, such as O-β-D-xylosyl-L-serine.

Proteolytic treatment of glycoproteins containing the N-glycosylamine linkage between asparagine and N-acetylglucosamine has in several instances yielded a product with only asparagine attached to the oligosaccharide component. This same type of linkage occurs in corneal keratan sulfate, and exhaustive proteolysis with papain and Pronase has resulted in considerable enrichment of asparagine, up to about 60% of the total amino acids.[18] Again, the difficulties in completely removing the last few amino acids around the carbohydrate-protein linkage appear to be greater when a large polysaccharide chain is present than when the carbohydrate unit is relatively small.

In a subsequent section describing the isolation of individual polysac-

[17] J. D. Gregory, T. C. Laurent, and L. Rodén, *J. Biol. Chem.* **239**, 3312 (1964).
[18] J. R. Baker, J. A. Cifonelli, and L. Rodén, *Biochem. J.* **115**, 11P (1969).

charides, the exact details of proteolysis in specific instances will be given.

B. Fractionation Methods

After proteolysis, it is generally advantageous to remove low-molecular weight digestion products and small amounts of residual proteins prior to fractionation of the polysaccharides. This is accomplished by precipitation with trichloroacetic acid, followed by dialysis of the supernatant liquid, or alternatively, by dialysis alone. After addition of trichloroacetic acid to a final concentration of 5%, the mixture is kept in the cold for a few hours or overnight, the precipitate is removed by centrifugation, and the solution is neutralized and dialyzed against several changes of distilled water.

Losses of low-molecular weight polysaccharides, particularly heparin, heparan sulfate, and keratan sulfate, may occur on dialysis. This may be prevented in part by heating the dialysis tubing in a dry oven at 85° for 3 days before use. As an alternative to dialysis, ultrafiltration through membranes of graded retentivity based upon molecular size (Diaflo membranes, Amicon Corporation) may be preferred.

Although dialysis of the digest facilitates the subsequent purification of the polysaccharides, it is by no means necessary and becomes impractical in large-scale preparations where the volumes are too large for convenient handling. Instead, the polysaccharides may then be precipitated directly, usually with ethanol or a quaternary ammonium salt. Further purification is carried out using either or both of these reagents, as discussed in detail below.

1. Precipitation with Ethanol

a. GENERAL COMMENTS

Precipitation with ethanol is used as a simple means of recovering polysaccharides quantitatively from solution as well as for the fractionation of different polysaccharide species. Although simple in theory, the method sometimes presents certain practical problems that deserve further comment.

The polysaccharide concentration should preferably be 1–2%, but almost complete precipitation may be obtained with solutions as dilute as 0.1%, if a sufficient excess of ethanol is used. Higher concentrations may sometimes be necessary in large-scale preparations to avoid excessive volumes, but precipitates then tend to become syrupy and more difficult to handle. Also, at high polysaccharide concentration, fractionation tends to be less complete.

A sufficient concentration of salt is necessary for complete precipitation. This is seldom a problem with tissue digests, which generally contain buffer salts, but if a salt-free water solution of the polysaccharide is mixed with ethanol, no precipitation occurs, and the solution may remain completely clear. The polysaccharide is then brought out of solution by addition of sodium acetate or potassium acetate. A final salt concentration of less than 5% is sufficient. It is convenient to gradually add a 4:1 mixture of ethanol and saturated sodium acetate with stirring until precipitation is complete. The acetates have an advantage over many other salts in that they are highly soluble in ethanol, and there is consequently no risk of precipitating the salt even with a large excess of ethanol.

Any of the connective tissue polysaccharides will be precipitated completely by 4–5 volumes of ethanol, provided that the polysaccharide concentration is high enough and that sufficient salt is present. In the course of purification of crude proteolytic digests, however, it is preferable not to exceed about 2 volumes so as to avoid simultaneous precipitation of unwanted digestion products. Similarly, if a purified preparation is precipitated from a solution containing a high concentration of an ethanol-insoluble salt, repeated precipitation with 2 volumes of ethanol will result in complete desalting of the polysaccharide, whereas a larger excess of ethanol may bring down considerable amounts of the salt. These guidelines should be followed with some caution, however, since one of the polysaccharides, i.e., keratan sulfate, is quite soluble in ethanol and may remain, in large part, in the supernatant fluid after precipitation with only 2 volumes of alcohol.

If the presence of keratan sulfate in a tissue digest is known or suspected, a larger excess of ethanol should be used and special care should be taken to analyze the alcohol supernatant, e.g., by the anthrone method, to evaluate possible losses of this polysaccharide. In order to ensure complete precipitation of a polysaccharide after addition of ethanol, the mixture is stirred (conveniently with a magnetic stirrer) until the supernatant liquid is water-clear upon centrifugation. (This may require stirring for several hours.) Completeness of precipitation is then most easily checked by the addition of more ethanol and sodium acetate to an aliquot of the clear supernatant, or by a more specific procedure, such as uronic acid analysis by the carbazole method. (It is sometimes difficult to obtain a clear supernatant on precipitation of a crude digest, but the recovery of polysaccharide may nevertheless be satisfactory.)

The initial ethanol precipitate from a crude tissue digest may often be syrupy and difficult to redissolve owing to the presence of interfering digestion products; this is particularly characteristic of hyaluronic acid precipitates. On repeated dissolution and precipitation from water or salt solution, these impurities are removed in part, and the preparation becomes easier to dissolve. (Many hours may still be required, particularly if the hyaluronic acid content of the preparation is high.) In order to ensure complete removal of salt, it may be preferable to continue the process of dissolution and precipitation, until the polysaccharide remains in solution on addition of ethanol. A finely dispersed, flocculent precipitate is then obtained by addition of sodium acetate in ethanol (see above).

After the last reprecipitation, the precipitate is suspended twice in absolute ethanol and is recovered by centrifugation. (The second ethanol wash may be opalescent and contain a little polysaccharide owing to the absence of salt, but losses are usually negligible. If complete recovery is essential, some sodium acetate is added to this supernatant.)

b. Ethanol Fractionation

Fractional precipitation with ethanol is one of the classical methods for the separation of polysaccharide mixtures and is still among the most suitable procedures for the large-scale fractionation of several polysaccharides. In favorable situations, the method can be applied directly to a tissue digest. As described in detail in Section III, precipitation of a cartilage digest with 1.25 volumes of ethanol yields an almost pure preparation of chondroitin sulfate, and keratan sulfate and some chondroitin sulfate remain in the ethanol supernatant. If several polysaccharides are present, it may be more advantageous to fractionate the mixture with quaternary ammonium salts prior to ethanol fractionation. A group of similar polysaccharides (e.g., the galactosaminoglycans—chondroitin 4- and 6-sulfate and dermatan sulfate) which are difficult to resolve by the former procedure may then be fractionated with ethanol.

Fractionation with ethanol is most successfully carried out in the presence of divalent metal ions, such as calcium,[19] barium,[20] or zinc.[21] The procedure recommended by Meyer et al.[19] has been applied with excellent results in many cases and is carried out as follows: ethanol is added slowly, with stirring, to a 1–2% solution of polysaccharide in 5% calcium acetate–0.5 M acetic acid. The mixture is kept at 4° overnight, and any precipitate formed is collected by centrifugation. Further precipitation with higher concentrations of ethanol is carried out in the

[19] K. Meyer, E. Davidson, A. Linker, and P. Hoffman, *Biochim. Biophys. Acta* **21**, 506 (1956).

[20] S. Gardell, *in* "Methods in Carbohydrate Chemistry," (R. L. Whistler, ed.), Vol. 5, pp. 9–14. Academic Press, New York, 1965.

[21] R. Marbet and A. Winterstein, *Helv. Chim. Acta* **34**, 2311 (1961).

same fashion. The precipitates are washed with 80% aqueous ethanol and dried with absolute ethanol and ether. If desired, the polysaccharide is converted to the sodium salt by passage over a cation exchange resin in the sodium form or by dissolution in sodium chloride and reprecipitation with ethanol.

The concentration increments for each ethanol addition are chosen according to the nature of the mixture to be fractionated, but it may be noted that increments smaller than 5% are not likely to yield much improvement in resolution, and normally, larger steps are used. As an example of a typical fractionation, a mixture of dermatan sulfate and chondroitin sulfate from pig skin was resolved by precipitation at 18, 25, 40, and 50% ethanol concentration[22]; dermatan sulfate precipitates in the first two fractions, and the chondroitin sulfates at the higher concentrations. Other mixtures which have been similarly fractionated are chondroitin 4-sulfate/chondroitin 6-sulfate from cartilage[19] (precipitating, with considerable overlap, between 30–40% and 40–50%, respectively) and chondroitin sulfate/keratan sulfate from nucleus pulposus.[20]

c. Fractionation on Cellulose Columns

The ethanol fractionation procedure has been refined by the use of supporting media, such as cellulose or cellulose–Celite mixtures, on which the polysaccharide is precipitated with ethanol, followed by elution with stepwise decreasing concentrations of ethanol. In the procedure of Gardell,[20] a solution of the polysaccharide mixture in 0.3% barium acetate is applied to a column of cellulose which has been equilibrated with 80% ethanol containing 0.3% barium acetate. The column is then eluted with decreasing concentrations of ethanol–0.3% barium acetate. This procedure has been used, e.g., to fractionate polysaccharide mixtures from nucleus pulposus and cornea, and the description by Gardell[20] should be consulted for further details.

In a recent application of this useful method to the study of dermatan sulfate preparations,[23] it was concluded that the fractionation occurred almost exclusively on the basis of uronic acid composition, i.e., the relative proportions of iduronic acid and glucuronic acid, whereas molecular weight and sulfate content were of little or no importance over the range investigated. (In other situations, where molecular weight differences are greater, this factor does indeed influence the fractionation pattern.)

In summary, fractionation with ethanol is an extremely useful pro-

[22] L.-Å. Fransson and L. Rodén, *J. Biol. Chem.* **242**, 4161 (1967).

[23] L.-Å. Fransson, A. Anseth, C. A. Antonopoulos, and S. Gardell, *Carbohyd. Res.* **15**, 73 (1970).

cedure, but it suffers from the same drawbacks that are inherent in many other types of separation processes. The close similarity between the various members of the group as well as their individual heterogeneity—with regard to molecular weight, degree of sulfation, and uronic acid composition—often makes it impossible to achieve complete fractionation. Although a portion of a certain polysaccharide may be obtained free of contamination with other species, mixed fractions are not necessarily resolved even by careful refractionation over a narrower concentation range.

2. Precipitation of Polysaccharides with Quaternary Ammonium Compounds

a. GENERAL COMMENTS

Polyanions form water-insoluble salts with certain detergent cations, such as cetylpyridinium (CP) and cetyltrimethylammonium (CTA). The complexes are dissociated and dissolved by inorganic salts at certain concentrations (critical electrolyte concentrations) which depend largely on the charge density of the polymer.[24,25] These observations have been developed by Scott[24] into one of the most useful methods for the fractionation of complex polysaccharide mixtures. In many cases, fractionation with quaternary ammonium compounds is the only method needed to achieve complete purification of the individual components of a particular mixture. A comprehensive and penetrating review of the subject has appeared,[24] which covers its theoretical as well as practical aspects.

Besides their value in fractionation procedures, the quaternary ammonium compounds are also used to advantage for the recovery of polysaccharides in bulk from tissue digests or other solutions. The low solubility of the complexes makes it possible to precipitate polysaccharides even from solutions as dilute as 0.01% or less. Other uses include the quantitative determination of acidic groups in a polysaccharide by titration. At the equivalence point, or rather at a slight excess of detergent, a flocculent precipitate is formed, and the end point is therefore easily detectable. If the titration is carried out in $0.02\,M$ sulfuric acid, the ionization of the carboxyl groups is repressed, and separate quantitation of carboxyl and sulfate groups can thus be obtained. Under appropriate conditions, small amounts of polysaccharide may be quan-

[24] J. E. Scott, in "Methods of Biochemical Analysis" (D. Glick, ed.), Vol. VIII, pp. 145–197. Wiley (Interscience) New York, 1960.
[25] J. E. Scott, in "Methods in Carbohydrate Chemistry" (R. L. Whistler, ed.), Vol. 5, pp. 38–44. Academic Press, New York, 1965.

titated by measurement of the turbidity formed on addition of a quaternary ammonium compound to a polysaccharide solution. Details of
these procedures are given by Scott.[24]

Certain practical details are of critical importance for the successful
use of the quaternary ammonium compounds in fractionation procedures
as well as for precipitation of polysaccharides in bulk. The following
comments pertain to the use of cetylpyridinium chloride (CPC) but are
equally applicable to other compounds of the same category.

The quality of CPC varies somewhat from one manufacturer to another, particularly in regard to the degree of admixture with the neighboring homologs of the cetyl group. If desired, the detergent may be
recrystallized from water or acetone. The authors have found the materials supplied by K & K, Plainview, New York, and AB Recip, Stockholm, Sweden, quite satisfactory. CPC is used as a 1–10% solution in
water, or, occasionally, in salt of the same concentration as in the polysaccharide solution to be precipitated. In order to ensure a sufficient excess
of detergent, the polysaccharide is precipitated with 3 parts of CPC (on
a weight basis). The final mixture should contain at least 0.05% of free
CPC; this is particularly important to remember when a polysaccharide
mixture is fractionated by the stepwise dilution method (see below). In
large-scale preparations, it may be convenient to estimate the amount
of CPC needed by titrating an aliquot of the polysaccharide solution to a
flocculation end point.

The polysaccharide concentration should preferably be between 0.1
and 1% for optimal precipitation; at higher concentration, some occlusion
of unwanted impurities may occur, and at lower concentrations, some
material may be lost, mainly because of difficulties in recovering the
minute precipitates quantitatively, but also due to the slight solubility
of the cetylpyridinium complexes.

For complete precipitation of all polysaccharides, including hyaluronic acid, the ionic strength must be kept at or below 0.1. However,
the presence of some salt greatly aids the aggregation of the cetylpyridinium complexes, and a sodium chloride concentration of $0.03\,M$
is usually satisfactory for this purpose. Sulfate ions are considerably
more effective than chloride, and if difficulties are encountered in obtaining a flocculent precipitate, sodium sulfate should be added to a final
concentration of 0.02–$0.04\,M$ (in the absence of other salts).

Flocculation or aggregation also occurs more readily if the precipitation is carried out at elevated temperature, and it is advantageous to
keep the mixture at about 40° for 0.5–1 hour, or occasionally overnight,
before collecting the precipitate.

Normally, the precipitation with CPC from dilute salt solutions yields flocculent precipitates which are easily sedimented by low speed centrifugation. At higher salt concentrations, it is sometimes necessary to centrifuge at high speed (10,000–20,000 rpm or more) to recover a precipitate quantitatively. Centrifugation should be carried out at room temperature or higher to avoid crystallization of CPC. The density of the cetyl-pyridinium complex of heparin is lower than that of 1.4 M NaCl, from which the purest heparin fractions are precipitated, and the syrupy heparin complex floats to the top of the centrifuge tube.

As an alternative method of collecting the CPC precipitates, it is often useful to filter the suspension through a pad of Celite which has been washed with the appropriate salt solution on a Büchner funnel or a sintered-glass filter. The polysaccharide is recovered by extracting the Celite at 40° with a mixture of 2 M NaCl and ethanol (100:15, v/v). It should be noted that considerable losses may be incurred in the extraction step, unless care is taken to extract the Celite repeatedly, with stirring in a beaker, until the filtrate is free of polysaccharide as determined, e.g., by the carbazole method. The use of Celite is of particular value in collecting small precipitates from very dilute suspensions. A small amount of Celite is added to the suspension, the mixture is stirred for a short time, and the Celite with the adsorbed precipitate is sedimented by centrifugation.

Cetylpyridinium complexes are converted to sodium salts by dissolving the precipitate in 2 M NaCl–ethanol (100:15, v/v) as above, and the polysaccharide is precipitated by addition of 2 volumes of ethanol (or more in the case of the more ethanol-soluble polysaccharides). This cycle is repeated once to remove residual CPC and the polysaccharide is subsequently precipitated from water solution.

Removal of the quaternary ammonium ion can be accomplished by several alternative procedures, such as precipitation with thiocyanate and extraction with organic solvents. These methods have been reviewed by Scott.[24,25]

b. Fractionation of Polysaccharide Mixtures

Table II gives the critical electrolyte concentrations for the cetyl-pyridinium complexes of a number of polyanions, including some of the connective tissue polysaccharides (quoted from Scott[25]). It is evident from these data that precipitation with CPC at selected salt concentrations will separate the connective tissue polysaccharides into several groups. Three main groups may easily be distinguished, reflecting the fact that the charge densities of these polysaccharides fall within three main ranges: Group 1: hyaluronic acid and chondroitin; Group 2: the chondroitin sulfates, dermatan sulfate, and heparan sulfate; Group 3;

TABLE II
CRITICAL ELECTROLYTE CONCENTRATIONS (NORMALITY)
OF CETYLPYRIDINIUM COMPLEXES

	KCl	MgCl$_2$	Na$_2$SO$_4$	MgSO$_4$	NaCl
Pectin	0.14	0.04	0.03	—	0.13
Hyaluronate	0.22	0.19	0.15	0.22	0.21
Alginate	0.33	0.30	0.60	—	0.38
Polyacrylate	—	—	—	—	0.46
Polyglutamate	—	—	—	—	0.475
DNA	0.36	0.35	0.75	0.35	0.45
Chondroitin sulfate	0.9	1.00	—	0.7	—
Heparin, 28 units/mg	0.95	—	—	—	—
Heparin, 75 units/mg	1.10	—	—	—	—
Heparin, 100 units/mg	1.20	—	—	—	1.55
Dextran sulfate I (low MW)	1.00	4.1	—	—	1.75
Dextran sulfate II (high MW)	2.0	4	—	—	2.9
Carrageenan	2.5	4	—	0.7	—

heparin. Keratan sulfate behaves in an extraordinary fashion and is not readily classified in either of these groups. This polysaccharide will be discussed separately below.

From a practical standpoint, large amounts of polysaccharides are best fractionated in the following way. The polysaccharide is dissolved in a salt solution of a concentration above the critical electrolyte concentration; an excess of CPC is added (dissolved in a salt solution of the same concentration, or as a concentrated solution, e.g., 10%), and the solution is diluted to a concentration below the critical electrolyte concentration of the first group to be precipitated. The precipitate is collected by centrifugation or filtration through Celite, and the clear solution is diluted further. With sodium chloride as the electrolyte, it is suitable to use an initial concentration of $2 M$, and the solution is successively diluted to 1.4, 1.2, 0.5, and 0.1 or 0.05 M. It is recommended to use 0.05% CPC as diluent for each step, although water may be used if a sufficient excess of CPC is present (at least 3 times the weight of the polysaccharide). The completeness of precipitation may also be checked by addition of more CPC after each dilution step.

The above fractionation scheme yields the following results with a mixture of all the common connective tissue polysaccharides. Heparin of high purity is precipitated at 1.4 M; material of lower anticoagulant activity and lower sulfate content, including some heparan sulfate, precipitates on further dilution to 1.2 M; remaining heparan sulfate, chondroitin 4- and 6-sulfate, and dermatan sulfate precipitate on dilution to $0.5\,M$ (critical electrolyte concentration about $0.9\,M$), and, finally, hyaluronic acid is obtained on dilution to $0.1\,M$ (critical electrolyte concentration about $0.2\,M$).

In simpler mixtures of known composition, it may be preferable to choose an initial salt concentration at which polysaccharides of one group are precipitated on addition of CPC, leaving others in solution. These are then most conveniently recovered by ethanol precipitation of the supernatant fraction.

Preparations of individual polysaccharides are described in Section III, where exact experimental details of typical fractionation procedures will be found.

The behavior of some of the individual polysaccharides and the further subfractionation of the major groups require additional comments.

Group 1. Chondroitin is not normally found in most tissue digests. It occurs, in low concentration, together with low-sulfated chondroitin sulfate in cornea. In most cases, hyaluronic acid is therefore the only polysaccharide to be considered in this group.

Because of the relatively low critical electrolyte concentration of hyaluronic acid, preparations of this polysaccharide are sometimes contaminated with glycoproteins carrying acidic groups, such as sialic acid. However, the critical electrolyte concentrations of these glycoproteins are generally lower than that of hyaluronic acid, and reprecipitation from $0.1\,M$ NaCl or fractionation over a narrower concentration range provides material of sufficient purity. It should be noted, however, that trace amounts of neutral sugars, indicative of glycoprotein contamination, are always found in preparations of hyaluronic acid despite extensive purification by CPC fractionation or by other means.

In certain tissues with a high content of nucleic acids, these are usually present in the hyaluronic acid fraction. Since the critical electrolyte concentration of nucleic acids is higher than that of hyaluronic acid, they may be removed by refractionation at suitable salt concentration. Almost complete removal of nucleic acids from the hyaluronic acid fraction from pig skin may be achieved by precipitation with CPC from $0.3\,M$ NaCl which precipitates the nucleic acids and leaves the polysaccharide in the supernatant fraction.[22] If nucleic acids are present in large excess over hyaluronic acid, it is preferable to subject the proteolytic tissue digest to further treatment with nucleases prior to fractionation with CPC.[26]

Group 2. The similarity between the members of this group makes

---

[26] J. Knecht, J. A. Cifonelli, and A. Dorfman, *J. Biol. Chem.* **242**, 4652 (1967).

the subfractionation by the CPC procedure a difficult task. Chondroitin 4- and 6-sulfate differ only in the position of the sulfate groups, and the hybrid nature of dermatan sulfate results in partial identity between this polysaccharide and the chondroitin sulfates.

Chondroitin 4- and 6-sulfate may be separated by ethanol fractionation as described above, but a method has also been developed which is based on differences in solubility of their cetylpyridinium salts in organic solvents.[27] In the CPC-cellulose column procedure described below, the chondroitin 4-sulfate complex is dissolved and eluted by 40% propanol–20% methanol–1.5% acetic acid, whereas chondroitin 6-sulfate remains on the column and is subsequently eluted by other solvents.[27,28]

Dermatan sulfate may be separated, at least partially, from the chondroitin sulfates, by taking advantage of the difference in pK values of the iduronic acid and glucuronic acid residues (3.6 and 3.1, respectively). Whereas the solubility curves of the cetylpyridinium salts of dermatan sulfate and the chondroitin sulfates are quite similar at neutral pH, the dermatan sulfate complex becomes considerably less soluble than the chondroitin sulfate salts, if the ionization of the iduronic acid carboxyl groups is preferentially repressed by acidification to an appropriate pH. Thus, dermatan sulfate is precipitated by CPC from 0.7 M MgCl$_2$ containing 0.1 M acetic acid, while the chondroitin sulfates remain in solution. This procedure has been applied successfully to the separation of dermatan sulfate and chondroitin sulfate from human aorta[29]; however, under the same conditions, a large proportion of the dermatan sulfate from pig skin remains in the supernatant fraction.[22]

Group 3. The purification of heparin by precipitation with CPC offers a particularly favorable situation, since the critical electrolyte concentration of this substance is higher than that of any other mucopolysaccharide. Nevertheless, certain practical difficulties arise in the process which are discussed in Section III.

It could be assumed that the high salt concentration used in the precipitation of heparin would ensure the complete removal of contaminants, but certain acidic peptides may be present even after repeated precipitations,[30] and caution should therefore be exercised in the interpretation of amino acid analyses of the purified product.

Keratan Sulfate. This polysaccharide behaves differently from all other connective tissue polysaccharide on precipitation with CPC. If

[27] C. A. Antonopoulos and S. Gardell, *Acta Chem. Scand.* **17**, 1474 (1963).

[28] J. Svejcar and W. Van B. Robertson, *Anal. Biochem.* **18**, 333 (1967).

[29] C. A. Antonopoulos, S. Gardell, and B. Hamnström, *J. Atheroscler. Res.* **5**, 9 (1965).

[30] U. Lindahl, *Biochem. J.* **113**, 569 (1969).

CPC is added to a salt-free solution of keratan sulfate, an increasing turbidity is observed; however, a flocculation end point is never reached, and in excess detergent the polysaccharide again goes completely in solution. In the presence of a low concentration of salt, e.g., 0.03 M NaCl, careful titration will yield a flocculent precipitate at the equivalence point, but again, the precipitate is largely dissolved in an excess of CPC. Complete precipitation may be obtained more easily if the polysaccharide is dissolved in 0.05 M borate buffer, pH 9.5, but this procedure is not specific for keratan sulfate, since some neutral polysaccharides may also be precipitated under the same conditions.

The isolation of keratan sulfate on a large scale from tissue digests is best carried out by a combination of ethanol fractionation and ion exchange chromatography, as described in Section III.

c. FRACTIONATION ON CELLULOSE COLUMNS

The CPC procedure has been developed into a column chromatographic method which may be used for microanalytical as well as preparative purposes.[27,28,31,32] The polysaccharide mixture is applied to a cellulose column which has been washed with CPC, and the precipitated compounds are then fractionally eluted by stepwise increasing concentrations of salt. The details of a typical separation of hyaluronic acid, chondroitin sulfate, and heparin will be given as an illustration, essentially quoting the description by Gardell.[20]

A 40-g portion of cellulose powder (e.g., Whatman CF11) is suspended in 500 ml of water and deaerated *in vacuo* in a suction flask. The slurry is poured into a 2 × 50 cm column and, after settling, the cellulose is washed with about 1 liter of water followed by 500 ml of a 1% solution of CPC. The polysaccharides, 20 mg of hyaluronic acid, 35 mg of chondroitin sulfate, and 35 mg of heparin, are dissolved in 5 ml of water and applied to the column. After rinsing the walls with a small amount of water, the column is eluted with 200 ml of 1% CPC, followed by 200 ml each of 0.15 M, 0.50 M, and 1.00 M $MgCl_2$ containing 0.05% CPC. The polysaccharides are recovered by concentration and precipitation with ethanol. Analysis of the isolated fractions shows that complete separation has occurred.

With a suitable series of eluants, the seven common connective tissue polysaccharides may be largely separated from each other by chromatography on a CPC-cellulose column. Svejcar and Robertson[28] have

[31] C. A. Antonopoulos, E. Borelius, S. Gardell, B. Hamnström, and J. E. Scott, *Biochim. Biophys. Acta* **54**, 213 (1961).
[32] C. A. Antonopoulos, S. Gardell, J. A. Szirmai, and E. R. De Tyssonsk, *Biochim. Biophys. Acta* **83**, 1 (1964).

described such a procedure, which is slightly modified from the methods developed by Antonopoulos, Gardell, Scott, and their co-workers.[27,29,31,32] Fractionation of 50–150 μg of a polysaccharide mixture is carried out in a 3 × 60 mm column, which is successively eluted with the seven solvents indicated in Table III (references cited in footnotes **27, 28, 31,** and **32** should be consulted for further details). For several reasons, the separations are unfortunately not always as clear-cut as would appear from the table. The method has not yet been perfected to the point where there is absolutely no overlap between any of the fractions. Incomplete resolution is most likely to occur between fractions **3** and **4,** **4** and **5,** and **5** and **6,** whereas fractions **1, 2,** and **7** are expected to be of higher purity. Furthermore, as has been pointed out previously, a near-perfect separation of "pure compounds" may result from the chromatography of certain selected fractions, which are not necessarily representative of the entire pools of the respective polysaccharides. The biological heterogeneity with regard to molecular weight, degree of sulfation, and uronic acid composition makes it unrealistic, even on theoretical grounds, to strive for a fractionation of the polysaccharide components of tissue digests into seven distinct molecular species. The practical consequence of this situation is that the separation method will have to be validated in each single application, since it cannot be assumed that the polysaccharide mixture present in a particular tissue digest will behave in the same way as the model compounds.

The properties of keratan sulfate again deserve special comment. Being soluble in an excess of CPC, keratan sulfate appears in the first fraction from a CPC-cellulose column, eluted with 1% CPC. This technique therefore cannot be used for the purification of keratan sulfate from crude tissue digests, since the impurities which are not precipitable

TABLE III
ELUENTS USED IN FRACTIONATION OF MUCOPOLYSACCHARIDES
ON CPC-CELLULOSE COLUMNS

Eluent	Polysaccharide eluted
1. 1% CPC	Keratan sulfate
2. 0.3 M NaCl	Hyaluronic acid
3. 0.3 M MgCl$_2$	Heparan sulfate
4. 40% propanol–20% methanol– 1.5% acetic acid	Chondroitin 4-sulfate
5. 0.75 M MgCl$_2$–0.1 M acetic acid	Chondroitin 6-sulfate
6. 0.75 M MgCl$_2$	Dermatan sulfate
7. 1.25 M MgCl$_2$	Heparin

will also be present in this fraction. It should be noted that a certain proportion of the keratan sulfate is precipitated on the column, however, and this fraction is only eluted at high salt concentration.[31] It remains to be established in what way this material differs from the CPC-soluble fraction.

3. Ion Exchange Chromatography

Several methods have been described for the separation of connective tissue polysaccharides by ion exchange chromatography (see references cited in footnotes 33–38). The application to the isolation of individual polysaccharides is described in detail in Section III.

A number of different ion exchange materials have been used, including Dowex 1-X2, ECTEOLA-cellulose, DEAE-cellulose, DEAE-Sephadex, and Deacidite FF. It is not possible to express a distinct preference for one exchanger over the others, although Pearce[37] has found Dowex 1 somewhat better than the cellulose exchangers with regard to sharpness of resolution and other properties.

The general techniques of ion exchange chromatography are well known and need no further comment. Polysaccharides are usually applied to a column in water solution, but sometimes a considerable proportion of the material is not adsorbed, and it is then advisable to include a low concentration of salt, e.g., 0.03–0.05 M NaCl, either initially or when reapplying the unadsorbed fraction. The capacity of Dowex 1-X2 for small ions is in the order of 3.5 meq per gram of dry resin which would correspond to about 2 g of polysaccharide. However, the capacity for polysaccharides is much lower and ranges from 10 to 240 μmoles of repeating disaccharide units per gram of resin. With an average disaccharide weight of approximately 500, this corresponds to 5–120 mg of polysaccharide, the exact amount depending in part on the nature of the polysaccharide.[37]

Elution is carried out either with a salt gradient or in a stepwise fashion with increasing concentrations of salt. An example of a stepwise elution is given in Fig. 2, which shows the separation of hyaluronic acid, heparan sulfate, chondroitin sulfate, and heparin by elution with 0.5 M, 1.25 M, 1.5 M, and 2.0 M NaCl, respectively.[35] In a thorough survey of elution conditions, Pearce et al.[37] determined the concentration ranges for the elution of a series of polysaccharides, as shown in Table IV. As

[33] N. R. Ringertz and P. Reichard, *Acta Chem. Scand.* **13**, 1467 (1959).

[34] N. R. Ringertz and P. Reichard, *Acta Chem. Scand.* **14**, 303 (1960).

[35] S. Schiller, G. A. Slover, and A. Dorfman, *J. Biol. Chem.* **236**, 983 (1961).

[36] E. R. Berman, *Biochim. Biophys. Acta* **58**, 120 (1962).

[37] R. H. Pearce, J. M. Mathieson, and B. J. Grimmer, *Anal. Biochem.* **24**, 141 (1968).

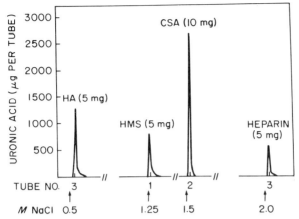

Fig. 2. Elution diagrams of hyaluronic acid (HA), heparin monosulfuric acid (HMS), chondroitinsulfuric acid A (CSA), and heparin chromatographed on Dowex 1-chloride columns, 0.9 × 44 cm. From S. Schiller, G. A. Slover, and A. Dorfman, *J. Biol. Chem.* **236**, 983 (1961).

TABLE IV

Salt Concentrations Required for Elution of Polysaccharides from Dowex 1[a,b]

	Sodium chloride concentration (M) corresponding to the elution of	
Glycosaminoglycan	5%	95%
Hyaluronate	0.27	0.78
	0.30	0.73
Chondroitin	0.39	0.94
Chondroitin 4-sulfate	0.75	1.14
	0.79	1.07
Chondroitin 6-sulfate	0.89	1.39
Dermatan sulfate A	0.89	1.25
Dermatan sulfate B	0.86	1.34
Keratan sulfate	0.76	1.35
Heparitin sulfate	0.38	1.44
Heparin	1.15	1.47
	1.10	1.45

[a] From R. H. Pearce, J. M. Mathieson, and B. J. Grimmer, *Anal. Biochem.* **24**, 141 (1968).

[b] Three to 5 μmoles of each polysaccharide was applied to a 400 mg Dowex 1 column and eluted with a linear NaCl gradient in 8 M urea. The proportion of each polysaccharide eluted was plotted against the NaCl concentration for each fraction collected. The molarities of NaCl corresponding to the elution of 5% and 95% of the polysaccharide were recorded.

seen from these data, it is not possible to obtain clear-cut separations in all cases, and the same considerations apply to the use of ion exchange chromatography that have been previously expressed regarding CPC fractionations; mainly, it is necessary to establish the validity of the method in every single application, and the effect of biological heterogeneity of the polysaccharides must be taken into account in the evaluation of elution patterns.

In certain cases, ion exchange chromatography has definite advantages over the CPC technique. Corneal keratan sulfate is obtained in high purity by chromatography on Dowex 1-X2, since a large proportion of this polysaccharide is strongly adsorbed and requires 3–$4 M$ NaCl for elution, whereas most other polysaccharides are desorbed at a lower concentration. Skeletal keratan sulfate, on the other hand, is eluted at lower salt concentration with considerable overlap with chondroitin sulfate. Additional fractionation with ethanol or CPC should therefore also be used for the purification of this keratan sulfate.

Another problem concerns the isolation of dermatan sulfate. This polysaccharide is not normally separable from the chondroitin sulfates by the usual ion exchange procedures. However, Barker et al.[38] have reported that chromatography on Deacidite FF (100–200 mesh; 7–9% cross-linking) by stepwise elution with sodium chloride separates dermatan sulfate, chondroitin 4-sulfate, and chondroitin 6-sulfate from each other.

The recovery of polysaccharides from ion exchange resins is usually adequate, but sometimes a considerable proportion of the applied material may remain adsorbed even on elution with a high concentration of salt. This is especially observed with preparations of hyaluronic acid, and although the reason for the incomplete recovery is not immediately obvious, it may be suggested that the shrinking of the resin, which occurs during elution, results in trapping of the larger molecules which are bound to interior binding sites.

III. Preparation of Individual Polysaccharides

In the following, details will be given for some of the procedures used by the authors for the isolation of various connective tissue polysaccharides. This will illustrate how the methods outlined in the preceding section may be applied in a specific preparative procedure, and the preparations described are to be regarded only as examples of how a certain polysaccharide may be isolated. Many alternative routes may easily be

---

[38] S. A. Barker, C. N. D. Cruickshank, and T. Webb, *Carbohyd. Res.* **1**, 52 (1965).

designed within the general framework of the established methods, and besides the tissue sources chosen here, others may be equally good or better in a particular case.

It is clear from the short outline of the general properties of the connective tissue polysaccharides in Section I that these substances are families of related compounds rather than single homogeneous compounds. As a result, the properties of preparations from a single tissue source may vary considerably depending on the isolation procedures, e.g., with regard to molecular weight, sulfate content, and, in the case of hybrid polysaccharides, in the relative contents of glucuronic acid and iduronic acid. Another parameter is the content of residual covalently bound peptides, which may vary not only as a result of the extent of proteolysis but which also depends on such factors as the use of alkaline conditions or, in the case of dermatan sulfate, digestion of the partially purified material with hyaluronidase. The influence of these factors on the properties of the product will be discussed in connection with the description of the individual preparations.

The composition and some physicochemical properties of representative polysaccharide samples are given in Table V. These preparations have been isolated as part of a program (supported by U.S. Public Health Service Grant 5 RO1 HE 11083) intended to provide investigators in the connective tissue field with a collection of suitable compounds for reference purposes. Since absolute analytical standards cannot be formulated for any of the mucopolysaccharides, the data in Table V may only be taken as typical of the results that may be expected under certain specified conditions of isolation.

A small number of samples of these preparations are still available from the authors.

1. Sodium Hyaluronate

Human umbilical cord is a convenient source for the preparation of both hyaluronate and chondroitin 6-sulfate.

Reagents

Phosphate–cysteine buffer, pH 6.5: 190 ml of 0.3 M Na_2HPO_4, 6.4 ml of 4.0 M NaH_2PO_4, 1.80 g of cysteine·HCl, 7.4 g of disodium ethylenediaminetetraacetate, and H_2O to make 1 liter

Crystalline papain (Nutritional Biochemical Corporation, Cleveland, Ohio 44128)

Cetylpyridinium chloride, 2% (K and K Laboratories, Inc. Plainview, New York)—abbreviation, CPC

TABLE V

ANALYTICAL AND PHYSICAL DATA FOR SODIUM SALTS OF ACID MUCOPOLYSACCHARIDES

	Hyaluronic acid	Chondroitin 6-sulfate	Chondroitin 4-sulfate	Dermatan sulfate	Heparin	Heparan sulfate (Ca salt)	Keratan sulfate I	Keratan sulfate II
Nitrogen[a,b]	3.0	2.7	2.6	2.3	2.6	2.5	2.7	3.2
Hexuronic acid[a,c]	47.2	34.6	34.1	12.8	38.7	44.1	1.9	2.0[k]
Hexosamine[a,d]	38.3	26.2	27.2	23.0	23.8	24.6	29.3	19.8
Galactose[a,j]	—	—	—	—	—	—	36.9	22.5
Sulfate[e,f]	0.00	0.98	0.97	1.29	2.33	0.99	1.17	1.30
Galactosamine[g]	0.001	1.000	1.000	1.000	0.001	0.002	0.014	0.119
Glucosamine[g]	1.000	0.007	0.001	0.029	1.000	1.000	1.000	1.000
Asp[g]	0.001	0.006	0.030	0.010	0.028	Trace	0.071	0.035
Ser[g]	0.001	0.017	0.067	0.015	Trace	Trace	0.013	0.067
Thr[g]	Trace	0.008	0.012	Trace	Trace	Trace	0.021	0.093
Glu[g]	0.001	0.015	0.023	Trace	Trace	Trace	0.014	0.129
Gly[g]	0.001	0.015	0.052	Trace	Trace	Trace	0.012	0.069
$[\alpha]_D^{24}$	−69°	−11°	−25°	−70°	+52°	+73°	+3°	−10.5°
η[h]	5.4	1.20	0.32	0.59	0.16	0.39	0.41	0.23
MW[i]	230,000	40,000	12,000	27,000	11,000	—	16,000	—

[a] Percent by weight.

[b] By Kjeldahl-nesslerization method.

[c] Carbazole method [Z. Dische, J. Biol. Chem. 167, 189 (1947)].

[d] By modified Elson–Morgan method [L. A. Elson and W. T. J. Morgan, Biochem. J. 27, 1824 (1933)] with correction for loss on hydrolysis for hyaluronic acid, chondroitin 6-sulfate, and chondroitin 4-sulfate only.

[e] Molar ratio of ester sulfate to hexosamine.

[f] By method of H. Muir, Biochem. J. 69, 195 (1958).

[g] By amino acid analyzer; molar ratio to major hexosamine.

[h] Intrinsic viscosity measured in 0.38 M NaCl with 0.01 M phosphate buffer, pH 7.0. [M. B. Mathews and A. Dorfman, Arch. Biochem. Biophys. 42, 41 (1953)]; based upon anhydrous weight of sample.

[i] Estimated from relationship between η and molecular weight [see Table 2 of M. B. Mathews, Biol. Rev. Cambridge Phil. Soc. 42, 499 (1967)].

[j] By method of Z. Dische, Methods Biochem. Anal. 2, 313 (1955).

[k] Anomalous yellow color, not characteristic of hexuronic acid. [M. B. Mathews and J. A. Cifonelli, J. Biol. Chem. 240, 4140 (1965).]

Procedure

Fresh human umbilical cords are washed to remove blood and are freed from extraneous tissues. They are cut into 1-cm segments, dried in several changes of acetone, and air-dried. Two hundred grams are mixed with 2 liters of phosphate–cysteine buffer and 100 mg of papain and set in an oven at 60° for 16–20 hours. The resultant solution is filtered by gravity through paper overnight in the cold, dialyzed for 2 days in cold, running tap water and filtered by suction through a Celite pad; 70 g NaCl are added, and the volume is made to 3 liters with distilled water.

The solution is warmed to about 35°, 700 ml of 2% CPC are added slowly with stirring, and the mixture is allowed to stand at room temperature overnight. After centrifugation, the supernatant is saved for the subsequent isolation of hyaluronate, while the precipitate containing chondroitin 6-sulfate and dermatan sulfate is washed with 500 ml of 0.3 M NaCl and recovered by centrifugation. Solution of the precipitate is effected with a mixture of 400 ml of 2.0 M NaCl and 80 ml of methanol; a small, insoluble residue is removed by centrifugation. Then 1.2 liters of 75% (v/v) ethanol are added with stirring, and the mixture is left overnight. The supernatant is withdrawn by suction, and the precipitate is recovered by centrifugation, washed successively with 95% ethanol, absolute ethanol, and ether, and dried *in vacuo* over P_2O_5. Yield is about 3.0 g of crude chondroitin 6-sulfate.

The supernatant from the above precipitation with CPC is dialyzed against cold running tap water for 2–3 days. A gummy aggregate, which may form, is saved. The fluid is made 0.03 M in NaCl, and 300 ml of 2% CPC are added with stirring. After several hours, the gummy aggregate is recovered by decantation and centrifugation, and the fluid is discarded. The combined gummy aggregates are dissolved by prolonged shaking or stirring with a mixture of 500 ml of 2 M NaCl and 500 ml of methanol, the solution is clarified by centrifugation, and added with stirring to 6 liters of 75% ethanol. The precipitate is recovered, washed, and dried as described above. The yield of hyaluronate is 6–7 g, containing a maximum of 3% chondroitin 6-sulfate based upon sulfur content. Further purification can be achieved by alcohol fractionation.

Ethanol Fractionation

Dissolve 10 g of crude hyaluronate in 2.5 liters of 0.5 M NaCl. At room temperature, add 2.1 liters of absolute ethanol with vigorous stirring. (Note that the first 1.5 liters may be added rapidly but that the remainder must be added slowly in order to avoid local precipitation.) Transfer the container to a 4° environment and continue stirring overnight. Centrifuge at 4°, and carefully decant supernatant. The precipitate (1–2 g) may be discarded. The supernatant is warmed to room temperature and 300 ml of absolute ethanol added slowly with vigorous stirring. Stirring is continued overnight at 4°. The precipitate is recovered by centrifugation at 4°, and the supernatant is removed by decantation and discarded. The precipitate is dissolved in 1.5 liters of 0.1 M NaCl and reprecipitated by adding the solution slowly to 4.5 liters of absolute ethanol at room temperature with stirring. The precipitate is allowed to settle overnight, whereupon most of the supernatant can be removed by siphon. The precipitate is recovered by centrifugation, washed successively with 95% ethanol, absolute ethanol, and ether and finally dried *in vacuo* over P_2O_5. The recovery is 5–7 g of hyaluronate with a composition shown in Table V.

2. Chondroitin 4-Sulfate

Chondroitin 4-sulfate is the major polysaccharide component of many cartilaginous structures. Most frequently, it occurs together with chondroitin 6-sulfate, which is sometimes the major polysaccharide, e.g., in embryonic chick cartilage. Only rarely is it considered necessary to remove the 6-sulfated isomer completely, and most preparations of "chondroitin 4-sulfate" contain in the order of 10–20% chondroitin 6-sulfate. The procedure below utilizes bovine nasal cartilage which is a readily accessible source and gives material of adequate purity. If highest purity is essential, further fractionation is necessary, or another tissue source may be chosen, e.g., sturgeon notochord.

Reagents

Crystalline papain
0.01 M EDTA–0.01 M cysteine buffer, pH 6.5
Cetylpyridinium chloride, 5%
NaCl, 2 M

Procedure

Bovine nasal septa are obtained from Wilson Laboratories, Chicago, Illinois, or the St. Louis Serum Co., East St. Louis, Illinois, and are stored at −20° until used. The septa are cleaned of soft tissues, ground, and dried in acetone; 500 g of the dry powder are suspended in 3 liters of buffer and digested with 250 mg of papain at 65° for 24 hours. A small insoluble residue is removed by centrifugation or filtration through Celite, and the polysaccharide is precipitated from the combined digest and washings (4 liters) by addition of 5 liters of ethanol (i.e., 1.25 volumes). After standing overnight, the chondroitin 4-sulfate is again dissolved in 4 liters of water and reprecipitated with 2 volumes of ethanol in the presence of 0.5% sodium acetate. The precipitate is washed twice with absolute ethanol and once with ether. The yield is 155 g. Analytical data for a typical preparation were: uronic acid, 29.9%; hexosamine, 24.3%; and sulfur, 4.5%. Quantitative separation of galactosamine and glucosamine by amino acid analyzer showed that glucosamine constituted 0.7% or less of the total hexosamine.

Further purification is achieved by precipitation with CPC in the following way. A solution of 50 g of chondroitin 4-sulfate in 5 liters of 0.5 M NaCl is mixed with 5 liters of 5% CPC in 0.5 M NaCl. The precipitate is allowed to settle overnight and is collected by centrifugation. Following dissolution of the ethylpyridinium complex in a mixture of 2 M NaCl and ethanol (100:15, v/v), 3 volumes of ethanol are added. The precipitate is again dissolved in the same mixture and reprecipitated with ethanol. Finally, the polysaccharide is precipitated from a water solution (1 liter) with 3 volumes of ethanol. The product is dried with absolute ethanol twice and with ether once. The yield is 39 g. Analyses of a typical preparation were: uronic acid, 33.3%; hexosamine, 26.2%. Glucosamine constituted 0.5% of the total hexosamine.

Comments

Preparation of an acetone-dried powder is not necessary, and the septa may equally well be digested directly. The digestion is so efficient that it is not even necessary to cut the cartilage into smaller pieces.

In view of the considerable expense in large-scale preparations, the authors usually prepare crystalline papain from crude material (e.g., Type II; Sigma) by the simple procedure of Kimmel and Smith.[38a] From 800 g of crude enzyme, over 40 g of crystalline papain is obtained.

As indicated by the analytical data, chondroitin 4-sulfate of high purity is obtained by the simple precipitation with 1.25 volumes of ethanol. The yield is not quantitative, but the recommended procedure has the advantage that nearly all the keratan sulfate remains in the supernatant fluid. Since the proportion of the chondroitin sulfate which precipitates under these conditions is somewhat variable and depends not only on the ethanol concentration but also on polysaccharide concentration, it is advisable to estimate roughly the amount of residual chondroitin sulfate in the supernatant fraction, e.g., by the carbazole method for uronic acid analysis. If necessary, more ethanol is then added to increase the yield.

[38a] J. R. Kimmel and E. L. Smith, *J. Biol. Chem.* **207**, 515 (1954).

3. Chondroitin 6-Sulfate

The crude chondroitin 6-sulfate fraction obtained from human umbilical cords (see Section III, 1) is purified by alcohol precipitation.

Reagents

Calcium acetate buffer, pH 4.6 ± 0.05: dissolve 125 g of calcium acetate·H$_2$O in 2.4 liters of water, add 71 ml of glacial acetic acid, and adjust pH.

Procedure

Prepare a solution of 8.0 g crude chondroitin 6-sulfate in 1.6 liters of calcium acetate buffer. At room temperature and with vigorous stirring, add very slowly 465 ml of absolute ethanol. A precipitate will appear. Set the flask in 4° environment and continue stirring slowly overnight. Centrifuge at 4° and decant supernatant, which may be slightly hazy. The precipitate, fraction A, contains the bulk of the impurities including nucleic acids and dermatan sulfate. Warm supernatant to room temperature, and with vigorous stirring add slowly 320 ml of absolute ethanol. Set the flask in 4° environment and continue stirring slowly. Centrifuge the precipitate (fraction B) at 4° and decant supernatant. Warm supernatant to room temperature and with vigorous stirring add slowly 400 ml absolute ethanol. Set in 4° environment and stir slowly overnight. Centrifuge at 4° and decant supernatant. Dissolve this last preecipitate (fraction C) in 500 ml of water and pass over Dowex 50 (Na$^+$ form). Add solid NaCl to make the solution 0.1 M in the salt and add solution slowly to 3 volumes of absolute ethanol with vigorous stirring. Allow precipitate to settle overnight and recover by centrifugation. Wash precipitate successively with 95% ethanol, absolute ethanol, ether, and dry *in vacuo* over P$_2$O$_5$.

Recovery is about 4.0 g of chondroitin 6-sulfate with a composition shown in Table V. Analysis following digestion with *Proteus vulgaris* chondroitinases showed the following disaccharide composition: 80% chondroitin 6-sulfate, 10% chondroitin 4-sulfate, and 10% unsulfated disaccharide.

4. Dermatan Sulfate

Principle

Dermatan sulfate (chondroitin sulfate B) is isolated from pig skin by proteolytic digestion with papain and precipitation with cetylpyridinium chloride. The crude polysaccharide is further purified (a) by precipitation as the copper complex or (b) by treatment with hyaluronidase and fractionation of the calcium salt with ethanol. A flow diagram illustrating the fractionation procedure is presented in Fig. 3.

Materials

Pig skin is obtained as an acetone-dried and ground preparation from the Wilson Laboratories, Chicago, Illinois. Cetylpyridinium chloride, crystalline papain, qualitative Benedict's solution, and other reagents are obtained from commercial sources (papain is more economically prepared from crude material; see Section III, 2). Testicular hyaluronidase with an activity of approximately 20,000 international units per milligram is obtained from AB Leo, Hälsingborg, Sweden.

Procedure

Dried pig skin (1 kg) is suspended in 5 liters of buffer, pH 6.5, containing 0.5 M NaCl, 0.01 M EDTA, and 0.01 M cysteine hydrochloride; 500 mg of crystalline papain are added. Digestion is carried out for 50 hours at 65°–70° with gentle stirring. After addition of 100 g of Celite analytical filter aid, the cooled digest is filtered through a pad of Celite (300 g) in a 27-cm o.d. Büchner funnel. The pad is rinsed with 2 liters of 0.5 M NaCl and the washings are combined with the filtrate. Mucopoly-

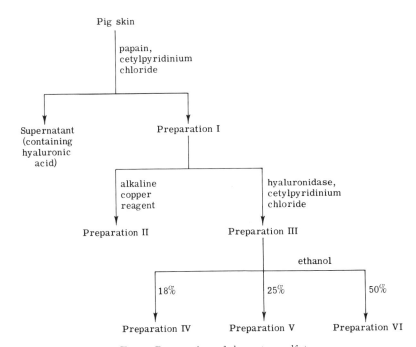

Fig. 3. Preparation of dermatan sulfate.

saccharides are precipitated by addition of 75 ml of a 10% solution of cetylpyridinium chloride. (Hyaluronic acid is not precipitated under these conditions and may be recovered from the filtrate as the cetylpyridinium complex by dilution with water.) The mixture is kept overnight at room temperature, Celite (25 g) is added, and the precipitate is collected by filtration through a 14-cm Büchner funnel containing a pad of Celite (10 g). The cetylpyridinium complex is dissolved by suspending the Celite pad in 250 ml of a solution containing 100 parts of $2 M$ NaCl and 15 parts of absolute ethanol. The mixture is stirred at 40° for 30 minutes and filtered through a sintered-glass filter. This procedure is repeated twice with 100-ml volumes of solution, and the combined filtrates are diluted with 3 volumes of water. The flocculent precipitate is collected by centrifugation, dissolved in 100 ml of the sodium chloride–ethanol mixture described above, and 3 volumes of absolute ethanol are added. After centrifugation the sediment is dissolved in a mixture of 200 ml of $1 M$ NaCl and 40 ml of ethanol, and the mucopolysaccharides are reprecipitated with 3 volumes of ethanol. Finally, the material is dissolved in 100 ml of water and the precipitation with ethanol is repeated. The preparation is washed twice with ethanol, once with ether, and dried under reduced pressure over phosphorus pentoxide. Yield 2.4 g (preparation I).

The crude preparation (I) is purified further either by precipitation as the copper complex or by digestion with hyaluronidase and ethanol fractionation as follows.

Method A. Copper Precipitation. A solution of preparation I (1.0 g) in 100 ml of water is mixed with 10 ml of saturated sodium hydroxide and 80 ml of qualitative Benedict's solution. After 10 minutes the gelatinous precipitate is centrifuged at $1000 g$ for 10 minutes. The precipitate is washed twice with 150 ml portions of a solution containing water, saturated sodium hydroxide, and Benedict's solution in the proportions 10:1:2. The washed precipitate is dissolved at once in a minimum amount of $4 M$ acetic acid. After neutralization with sodium hydroxide the solution is diluted to 100 ml and dialyzed overnight against running tap water. The remaining copper is removed by passage of the dialyzed material over a column (2.2×20 cm) of Dowex 50-X4 (H⁺) (50–100 mesh). The effluent is collected at a flow rate of approximately 5 ml per minute in a flask containing 10 ml of 5% $NaHCO_3$. The solution is adjusted to pH 4–5 with acetic acid, concentrated to approximately 20 ml, and 2 volumes of ethanol are added. (Occasionally the addition of small quantities of $2 M$ NaCl aids in the precipitation of dermatan sulfate at this stage.) The centrifuged precipitate is washed twice with absolute ethanol, twice with ether, and dried under vacuum at room temperature over phosphorus pentoxide. Yield 0.85 g (preparation II).

Method B. Digestion with Hyaluronidase and Fractionation with Ethanol. Preparation I (1.0 g) dissolved in 25 ml of 0.15 M NaCl–0.10 M sodium acetate, pH 5.0, is digested with 10 mg of hyaluronidase at 37°. After 75 hours the digest is diluted with water to 400 ml and 50 ml of 10% cetylpyridinium chloride are added. The mixture is stirred occasionally over a period of 1 hour and then centrifuged at 2000 g for 20 minutes. (Oligosaccharides produced by hyaluronidase digestion remain in the supernatant solution in the presence of excess cetylpyridinium chloride.) The sedimented material is washed with 1% cetylpyridinium chloride and then dissolved in 50 ml of 2 M NaCl. This solution is diluted to 80 ml with water to give a final NaCl concentration of 1.25 M. A small amount of precipitate which forms is removed by filtration through a Celite pad (1 × 3 cm) in a medium-porosity sintered-glass filter. (This precipitate is predominantly heparin and represents approximately 5% of preparation I.) The pad is washed with 10 ml of 1.25 M NaCl and the washings are pooled with the filtrate. Dermatan sulfate is recovered by precipitation with 3 volumes of ethanol followed by reprecipitation from NaCl-ethanol mixture and finally from water as described above. Yield 0.61 g (preparation III).

This fraction (III) is purified further in the following way. To a solution of 0.7 g in 70 ml of 5% calcium acetate–0.5 M acetic acid, ethanol (15 ml) is added to give a final concentration of 18% and the mixture kept overnight at 4°. The precipitate is collected by centrifugation at 4°, washed twice with ethanol, once with ether and dried under reduced pressure over phosphorus pentoxide. Yield 0.42 g (preparation IV).

A second fraction is obtained by addition of ethanol (8 ml) to the supernatant liquid to give a concentration of 25% ethanol. After standing overnight in the cold, the precipitate is collected as described above. Yield 0.15 g (preparation V). Finally, a fraction precipitating with 50% ethanol is obtained in a similar way. Yield 0.24 g (preparation VI).

Comments

Several different criteria are generally applied to the evaluation of the purity of dermatan sulfate preparations, since no single specific method is available.

As indicated in Section IV, A, 1, the determination of the uronic acid content of dermatan sulfate by the carbazole method gives values significantly lower than those obtained for all other uronic acid-containing mucopolysaccharides. Determination of the ratio of uronic acid to hexosamine therefore provides a means of estimating, to some degree, the purity of a dermatan sulfate preparation.

A more sensitive index of purity is afforded by the determination of the uronic acid content by both the carbazole and the orcinol methods (see Section IV, A, 1). Carbazole:orcinol ratios of about 0.25 are obtained for highly purified materials, whereas polysaccharides containing exclusively glucuronic acid give ratios of 2 or higher.

Other methods often employed for establishing purity of dermatan sulfate preparations include determination of optical rotation, sensitivity to hyaluronidase, and glucosamine content. A special comment is in order regarding the behavior of dermatan sulfate on treatment with testicular hyaluronidase. Although resistance to hyaluronidase digestion has been regarded as one of the characteristic properties of this polysaccharide, the finding that dermatan sulfate is a hybrid containing a certain amount of glucuronic acid has led to a revision of this notion. The presence of the glucuronic acid residues makes the molecule susceptible to attack by hyaluronidase at the hexosaminidic bonds involving these uronic acid units, and a partial degradation may therefore occur. As a consequence, dermatan sulfate preparations which have been treated with testicular hyaluronidase differ from more native materials in certain respects: (1) the glucuronic acid content is generally lower due to removal of glucuronic acid-containing segments of the molecule; (2) the molecular weight is lower; and (3) the content of covalently bound peptides is usually lower, with losses of up to 70% of the peptides initially present. It should be mentioned in regard to this last point that the glucuronic acid-containing segments of the molecule are often located close to the linkage to peptide and that the small glycopeptide fragments released by the enzymatic cleavage tend to be lost during the reisolation of the polysaccharide.

Analytical data for the dermatan sulfate preparations described here are listed in Table VI. In preparation I, obtained in the first stage of purification, glucosamine constitutes 8–10% of the total hexosamine, although hyaluronic acid has been largely removed.

Preparation II, obtained by precipitation of dermatan sulfate as the copper complex, contains less than 1% glucosamine. The carbazole:orcinol ratio of 0.34 suggests the presence of small amounts of glucuronic acid-containing mucopolysaccharides. The isolation of dermatan sulfate by this procedure (method A) has the advantage of simplicity and rapidly leads to a product of reasonable purity. Since the strongly alkaline conditions may cleave some of the protein-polysaccharide bonds, alternative methods are preferred if an intact peptide moiety is desired.

Preparation III, obtained after treatment with hyaluronidase to remove contaminating chondroitin sulfate, appears to be comparable in purity to preparation II, as judged by the carbazole:orcinol ratio.

Preparation IV has the lowest carbazole:orcinol ratio of the preparations described here, indicating that it contains almost exclusively iduronic

TABLE VI

ANALYTICAL DATA OF DERMATAN SULFATE PREPARATIONS[a]

Preparation	Uronic acid[b] (%)	Hexos- amine[c] (%)	Sulfate[d] (%)	$[\alpha]_D^{20}$	Carbazole: orcinol[e]	Gluco- samine[f] (%)	Yields[g]
I	19.8	25.0	—	—	—	8.0	2.4
II	18.0	26.7	15.3	−66°	0.33	0.8	2.0
III	15.1	25.7	17.3	−60°	0.32	2.2	1.4
IV	11.0	22.0	14.0	−55°	0.24	0.8	0.9
V	11.9	22.8	14.0	−49°	0.34	2.3	0.3
VI	16.1	20.8	—	—	0.41	4.2	0.5

[a] Data not corrected for moisture and salt contents.
[b] Determined by the carbazole method.
[c] Estimated by the Elson–Morgan method.
[d] Determined by the method of Lloyd and Dodgson as modified by H. Muir, *Biochem. J.* **69**, 195 (1958).
[e] Ratios calculated according to P. Hoffman, A. Linker, and K. Meyer, *Science* **124**, 1252 (1956).
[f] Calculated as percent of total hexosamine. Determined by Technicon amino acid analyzer after hydrolysis in 6 M HCl at 100° for 20 hours.
[g] Expressed as grams per kilogram of acetone-dried pig skin. The total yield of preparations IV–VI is 116% of preparation III. This is due to the lower contents of uronic acid and hexosamine in the purified fractions. When calculated on the basis of hexosamine content, the total yield is 98%.

acid and that mucopolysaccharides other than dermatan sulfate are absent.

Preparation V is of reasonable purity, as indicated by the carbazole: orcinol ratio of 0.34, and is adequate for most purposes, although the presence of more than 2% glucosamine makes repurification necessary for obtaining a product comparable to preparation IV.

Preparation VI is the least pure of the fractions and contains 4% glucosamine.

It should be strongly emphasized that preparation III and its subfractions IV–VI must all be regarded as partially degraded, since they have been treated with hyaluronidase. However, this procedure is to be preferred if a preparation of high iduronic acid content is desired. On the other hand, if the integrity of the native polysaccharide structure and the peptide component is essential, treatment with hyaluronidase should be avoided. The fractionation with ethanol or by other procedures is then carried out directly on preparation I or equivalent materials.

5. Heparin

Heparin preparations of a purity adequate for most laboratory uses are now available from a number of commercial sources. However, except for the most highly purified materials, the commercial preparations generally contain small amounts of other polysaccharides which may be removed by fractionation with cetylpyridinium chloride. The purification of two commercial preparations, from pig intestinal mucosa and beef lung, respectively, will be described. The preparations chosen were not of the highest purity available in order to illustrate the purification obtained by the procedures used.

Materials

Heparin from pig intestinal mucosa with an anticoagulant activity of 116 USP units/mg (preparation I), obtained from the Wilson Laboratories, Chicago, Illinois

Beef lung heparin with an activity of 120 USP units/mg (preparation 2), obtained from the Upjohn Company, Kalamazoo, Michigan. Both preparations had been bleached.

Procedure

To a solution of 1 g of heparin in 70 ml of $4 M$ NaCl are added 30 ml of 10% CPC and 100 ml of water to give a final concentration of $1.4 M$ NaCl. After 10 minutes in a 45°–50° water bath the mixture is centrifuged. An oily precipitate is obtained which usually floats to the top of the solution at this salt concentration. The clear solution is siphoned off or decanted over a cotton filter. The precipitate is rinsed gently with 25 ml of $1.4 M$ NaCl and dissolved at 40°–45° in 25 ml of a mixture containing 10 parts of $2 M$ NaCl and 2 parts of ethanol. Heparin is precipitated from this solution by the addition of 3 volumes of ethanol. After centrifugation the precipitate is redissolved in the salt–ethanol solution and again precipitated with ethanol. Finally, the material is dissolved in water, and the precipitation with ethanol is repeated. The heparin is washed twice with 50-ml portions of ethanol, twice with ether, and dried *in vacuo* at room temperature over P_2O_5 (fraction A).

The supernatant solution remaining after isolation of fraction A is diluted with water to a concentration of $1.2 M$ NaCl. The precipitate is isolated and converted into the sodium salt (fraction B) as described above. Dilution of the supernatant liquid from fraction B with 1 volume of water yields fraction C. The precipitate is collected by centrifugation and worked up as for fraction A. The supernatant is finally dialyzed, concentrated under vacuum at 40° to a small volume, and the residual material is precipitated with 3 volumes of ethanol (fraction D).

Analytical results and yields for the fractions obtained from the mucosal heparin (preparation 1) and the beef lung material (preparation 2) are given in Table VII.

TABLE VII

Composition of Fractions Obtained from Beef Lung and Hog Mucosal Heparin Preparations[a]

	Uronic acid (%)	Hexosamine (%)	Carbohydrate (%)	Nitrogen (%)	Sulfate: hexosamine[b]	$[\alpha]_D^{20}$	Anticoagulant activity,[c] (USP units/mg)	Yield[d]
Preparation 1	26.7	16.4	24.0	4.6	2.20	$+43°$	116	—
Fraction A	38.4	21.8	16.9	2.7	2.23	$+43°$	180	43
B	37.5	21.3	17.1	2.8	2.34	$+41°$	157	24
C	34.0	20.1	17.9	3.0	—	$+37°$	82	10
D	7.9	9.8	29.7	10.8	1.63	$+32°$	15	11
Preparation 2	36.9	20.0	17.9	1.7	—	—	120	—
Fraction A	39.0	23.2	16.8	—	2.46	$+43°$	180	69
B	38.4	20.9	14.8	—	2.10	$+38°$	109	8
C	34.4	20.8	16.0	—	2.24	$+37°$	67	11
D	50.9	26.0	24.0	—	1.37	—	13	2

[a] Values are not corrected for moisture.
[b] Expressed as molar ratio of sulfate to hexosamine.
[c] Performed by Dr. H. H. R. Weber, Wilson Laboratories, Chicago, Illinois.
[d] Yield based on hexosamine recoveries.

6. Heparan Sulfate

The most practical source for this polysaccharide is by-products obtained in the preparation of heparin from beef lung. Several fractions varying mainly in sulfate content are obtained by fractionation of the by-product mixture on Dowex 1. Heparan sulfate is also present in umbilical cord. Since this tissue may be used for the isolation of hyaluronic acid, chondroitin 6-sulfate, and dermatan sulfate, procedures for obtaining heparan sulfate from this source will also be given.

Materials

By-products of heparin preparation from beef lung are obtained from the Upjohn Co., Kalamazoo, Michigan (courtesy of Dr. L. L. Coleman).

Procedure

1. The heparin by-products (4 g) are dissolved in 400 ml of water, and the yellowish solution is applied, over a period of several hours, to a column (4.5 × 40 cm) of Dowex 1-X2 (Cl⁻, 200–400 mesh). (Prior to use, the resin is washed with 3 M HCl until the absorbancy at 260 nm drops below 0.1.)

After the addition of the crude mucopolysaccharide solution, the column is washed with approximately 2 bed volumes of water, and heparan sulfate is eluted in a stepwise fashion with 1.25 M, 1.5 M, and 3.0 M NaCl. Each eluate except the 3.0 M fraction is collected in three separate portions, the volumes of which are given in Table VIII. Elution with each solvent is continued, until the effluent gives a weak or negative Molisch

TABLE VIII
COMPOSITION OF HEPARAN SULFATE FRACTIONS FROM HEPARIN BY-PRODUCTS[a]

Sodium chloride elution			Uronic acid (%)	Hexosamine (%)	Sulfate: hexosamine molar ratios
NaCl Molarity	Volume (ml)	Yield (g)			
1.25: 1[b]	400	0.23	37.4	23.5	0.94
2	600	0.60	37.6	21.3	1.02
3	400	0.28	31.9	18.6	1.12
1.5: 1	400	0.17	36.8	22.7	1.21
2	500	0.37	39.4	26.6	1.43
3	400	0.25	29.1	17.2	1.62
3.0	900	0.52	33.6	24.4	2.08

[a] Eluates are monitored by the Molisch reaction: 1 drop of Molisch reagent (0.3% α-naphthol in ethanol) is added to 1 or 2 drops of eluate and mixed with 0.8 ml of sulfuric acid, followed by 2 drops of water and mixing.

[b] Three separate portions were collected from the 1.25 and 1.5 M fractions.

reaction. The eluates are dialyzed, concentrated to 20–30 ml, and 2 ml of
4 M NaCl and 2 volumes of ethanol are added. The precipitates are
washed three times with ethanol, twice with ether, and dried *in vacuo*.

Analytical data and yields of the various fractions are shown in
Table VIII. These results are representative of several different fractionations, although some by-product preparations gave yields of individual
fractions which varied significantly from those shown.

The composition of the several fractions indicates that these vary
mainly in sulfate content, and electrophoretic mobilities show corresponding differences. The predominant fraction eluted with 1.25 M NaCl, i.e.,
fraction 1.25:2, contains approximately equimolar quantities of sulfate and hexosamine, with equal amounts of N-sulfate and ester sulfate. The appearance of a single spot after electrophoresis and staining with acridine orange and of a single peak on gel chromatography
on Sephadex G-200 suggests reasonable homogeneity for this product.
Similarly, fraction 1.5:2 appeared reasonably homogeneous on electrophoresis. However, this material often shows a minor degree of polydispersity by gel chromatography on Sephadex G-200, and in such cases
further fractionation is required.

It is noted that the sulfate content of the fractions increases with increasing salt concentration of the eluents, the substances with highest
sulfate content being eluted with 3 M NaCl. The 3.0 M eluate and fraction 1.5:3 contain some dermatan sulfate which may be removed by precipitation as the copper complex, as described for the purification of
dermatan sulfate.

2. Umbilical cord extracts provide a convenient source for the isolation of a low-sulfated heparan sulfate. Fraction B, obtained during the
isolation of chondroitin 6-sulfate (Section III, 3) is fractionated as follows. One gram is dissolved in 100 ml of water and mixed with 10 ml of
saturated NaOH and 75 ml of Benedict solution. The gelatinous precipitate, containing hyaluronic acid and dermatan sulfate, is centrifuged at
1000 g for 10 minutes, and the sediment is washed with 150 ml of a solution composed of water, saturated NaOH, and Benedict solution (10:1:2,
by volume). The supernatant and wash solutions are pooled, neutralized
with acetic acid, and dialyzed overnight against tap water. The dialyzed
fluid is passed over a column of Amberlite IR-120 (H+ form). After neutralization, the solution is concentrated to 20 ml and 3 volumes of ethanol
are added. The yield in a typical preparation is 350 mg.

This sample is dissolved in 50 ml of water and applied to a 2.2 ×
35 cm column of Dowex 1-X2 (Cl−, 200–400 mesh). About two-thirds of
the material is not bound to the resin, and the water effluent is therefore
reapplied to the column several times. Approximately 90% of the poly-

saccharide is then eluted between 1.0 and 1.25 M NaCl and recovered by precipitation with ethanol as described.

The purified preparation has the following composition and properties: uronic acid, 44.9%; hexosamine, 29.0%; N-sulfated glucosamine, 11%; molar ratio of total sulfate to hexosamine, 0.54; $[\alpha]_D^{25}$, +65°. The galactosamine content is less than 1% of the total hexosamine.

7. Keratan Sulfate

Keratan sulfate I is most often prepared from bovine cornea, and keratan sulfate II from skeletal tissues is conveniently isolated from bovine nasal septum cartilage.

Materials

Bovine corneas may be obtained from several commercial sources, e.g., Pel-Freez Biologicals, Rogers, Arkansas.

The 1.25-volume ethanol supernatant, obtained after precipitation of chondroitin 4-sulfate from the papain digest of bovine nasal septum cartilage, is used for isolation of keratan sulfate II (see Section III, 2).

Procedure

Keratan Sulfate I. Bovine corneas (600 g wet weight) are digested with 1 g of papain in 1 liter of 0.01 M cysteine–0.01 M EDTA buffer, pH 6.5, at 65° for 24 hours. After filtration through Celite the digest is concentrated to 500 ml and 4 volumes of ethanol are added. The sirupy precipitate is dissolved in water and, after removal of a small insoluble residue, the solution is dialyzed for 24 hours against several changes of distilled water. A small precipitate is removed by filtration through Celite, and the polysaccharide mixture is fractionated on Dowex 1-X2 (Cl⁻).

The solution is applied, in a volume of about 2 liters, to a column (11 × 30 cm) of Dowex 1-X2 (Cl⁻). (The capacity of this column is at least 10 g of polysaccharide.) Elution is carried out successively with water (3–4 liters), 1.0, 1.5, 2.0, 3.0, and 4.0 M NaCl. The eluates are monitored by the Molisch reaction (see p. 108), and elution with each solvent is continued until the Molisch reaction becomes weak or negative. If better resolution is desired, the eluate at each salt concentration is collected in several portions, which are then worked up separately. The total volume used for each step is in the order of 3 to 6 liters.

The fractions eluted by 2, 3, and 4 M NaCl contain keratan sulfate and are dialyzed in large dialysis bags, concentrated to small volume, and precipitated with 4 volumes of ethanol. If necessary, salt is added for complete precipitation. The precipitates are washed with ethanol and ether and dried as described.

Yields and composition of the fractions are shown in Table IX.

The material eluted at 3 and 4 M salt concentration is essentially pure keratan sulfate. The 2 M fraction, however, may contain an appreciable proportion of chondroitin sulfate; this may be removed by ethanol fractionation as described below for a typical preparation.

The 2 M fraction (1.19 g) is dissolved in 112 ml of 5% calcium acetate containing 0.5 M acetic acid, ethanol is added slowly with stirring to a final concentration of 40%, and the mixture is kept at 4° for at least 3 hours. After centrifugation, the supernatant liquid is brought to a concentration of 66% ethanol, and the precipitate is collected and washed with 80% ethanol, absolute ethanol, and ether. Analysis of this material shows a negligible contamination with chondroitin sulfate.

Keratan Sulfate II. The 1.25-volume ethanol supernatant liquid (see Materials) obtained after removal of the bulk of chondroitin sulfate from digests of bovine nasal septa contains essentially all the keratan sulfate,

TABLE IX

COMPOSITION OF KERATAN SULFATE PREPARATIONS FROM BOVINE CORNEAL (KS I)
AND NASAL SEPTUM (KS II) TISSUES

Sodium chloride (*M*)	Uronic acid (%)	Hexosamine (%)	Carbohydrate (%)	Sulfate: hexosamine ratio	Yield[f]
KS I[a]					
2.0	4.0[b]	27.8	36.3	1.04	21
3.0[d]	1.8[c]	29.1	38.9	1.23	74
4.0	2.2[c]	24.2	27.0	1.29	5
KS II					
1.0	6.5[e]	18.7	20.5	0.51	10
1.5	7.8[e]	19.7	22.1	0.75	40
2.0	4.1[b]	22.1	25.0	0.93	30
3.0	2.5[c]	18.8	21.0	1.07	10
4.0	2.5[c]	18.5	21.4	1.08	10
Ethanol supernatant (starting material)	5.1	25.3	32.6	—	—

[a] The 1.0 and 1.5 M eluates contained chondroitin and chondroitin sulfate, respectively.

[b] Yellow-pink color given in the carbazole–uronic acid reaction. Based on spectral analysis, true uronic acid was estimated to be approximately half the values given in the table.

[c] Only yellow color in the carbazole reaction. See footnote k of Table V.

[d] The 3 M fraction from corneal tissue extracts was refractionated with ethanol to produce the KS I used as a reference standard (shown in Table V).

[e] True uronic acid contents estimated as approximately 70% of values given in the table.

[f] Yields are based on glucosamine contents of total recovered keratan sulfate.

amounting to approximately 4% of the total mucopolysaccharide content of this tissue. (The chondroitin sulfate content is generally considerably lower than that of keratan sulfate in the ethanol supernatant.)

The supernatant fluid is concentrated to remove most of the ethanol and then dialyzed against distilled water. An amount of polysaccharide equivalent to 1.0–1.2 g of hexosamine in 100 ml of water is then applied to a column (4.5 × 40 cm) of Dowex 1-X2 (chloride form). Elution with sodium chloride gradients is carried out as described for keratan sulfate I, except for the use of 0.5 M sodium chloride after elution with water in order to remove appreciable amounts of glycoprotein material. The skeletal keratan sulfates are eluted at a lower concentration than keratan sulfate I, and the bulk of the polysaccharide is found in the 1.5 and 2.0 M fractions; the 3 and 4 M fractions contain in the order of 10% each of the total keratan sulfate. Since considerable overlap may exist with chondroitin sulfate as well as glycoprotein in the 1.0 and 1.5 M eluates, additional fractionation by ethanol precipitation, as described for the corneal polysaccharide, may be necessary, if it is desired to obtain more highly purified keratan sulfate. Since the ratio of keratan sulfate II to chondroitin sulfate will vary from preparation to preparation, depending on a number of factors, the example given is to be regarded as an illustration of a typical fractionation.

IV. Characterization of Connective Tissue Polysaccharides

Various aspects of the characterization of connective tissue polysaccharides have been covered elsewhere in this series, and this section will therefore largely be limited to a discussion of some recent analytical developments.

A partial characterization of a polysaccharide preparation is provided already by the method of isolation which may occasionally be so selective as to yield only one single polysaccharide species. Particularly the methods based on differences in charge density, i.e., fractionation on ion exchange materials or by the CPC technique, are in themselves a means of characterizing the isolated materials. The CPC-cellulose column method has been used more extensively for analytical than for preparative purposes and has been refined by Gardell and his collaborators to the point where rather small differences may be detected within a population of similar polysaccharide molecules.[23,27,39]

The extent to which it is necessary to characterize a certain polysaccharide preparation naturally varies according to the needs of the problem under study. Qualitative and quantitative analyses of the component sugars and other constituents are often enough to establish the

[39] J. A. Szirmai, E. Van Boven-De Tyssonsk, and S. Gardell, *Biochem. Biophys. Acta* **136**, 331 (1967).

identity of a polysaccharide, at least tentatively, and further analyses including enzymatic degradation and physicochemical characterization may only be carried out if a more complete picture of the properties of the product is required.

A. Qualitative and Quantitative Analysis of Components

1. Uronic Acids

In the analysis of glycoproteins, it is generally possible to liberate the monosaccharide components in reasonable yield by acid hydrolysis under appropriate conditions. The individual sugars may then be quantitated either by direct enzymatic analysis of the hydrolyzate or after chromatographic separation (see this volume [1]). Unfortunately, this approach is not successful in the quantitative analysis of glucuronic acid-containing polymers, since glucuronidic linkages are notoriously much more stable to acid hydrolysis than most other glycosidic linkages and, once liberated, the free uronic acid is partially destroyed before hydrolysis of remaining residues is complete. Although the iduronidic linkages of dermatan sulfate are more acid-labile than glucuronidic bonds, it is likewise impossible to release iduronic acid quantitatively without simultaneous destruction of a considerable proportion of this sugar. Some progress has recently been made in the development of hydrolysis conditions which are more suitable for the analysis of uronic acid-containing polymers, but an entirely satisfactory solution to this problem is yet to be found.

Despite the difficulties inherent in procedures involving hydrolysis and subsequent separation and quantitation of the liberated uronic acids, such methods are essential for the unambiguous differentiation between glucuronic acid and iduronic acid and occupy an important position in current methodology. This group of methods includes analysis of hydrolyzates by (a) paper chromatography and electrophoresis, (b) ion exchange chromatography, and (c) gas–liquid chromatography. However, by far the most convenient procedures are the colorimetric methods which can be applied directly to the polysaccharide without prior hydrolysis, particularly the carbazole method[40] and the orcinol method.[41]

a. ANALYSIS OF HYDROLYZATES

Hydrolysis Conditions

A number of different hydrolysis conditions have been used, and a few examples are given below.

Polysaccharide samples (1–3 mg) are dissolved in 2 ml of $1 M$

---

[40] Z. Dische, J. Biol. Chem. 167, 189 (1947).
[41] A. H. Brown, Arch. Biochem. 11, 269 (1946).

sulfuric acid and heated at 100° for 2 hours. The hydrolyzate is neutralized with barium carbonate, and the suspension is filtered through a Celite pad on a medium-pore sintered-glass filter. The filtrate and washings are concentrated to a small volume (1–2 ml) and applied to a column for automated ion exchange chromatography.[12] Under these conditions, close to 50% of the uronic acid in dermatan sulfate is liberated, but only about 10% from chondroitin 4-sulfate.

Considerably higher yield of free glucuronic acid from chondroitin 4-sulfate is obtained by the procedure of Jeffrey and Rienits.[43] Chondroitin 4-sulfate (10 mg) is mixed with 100 mg of Dowex or AG 50-X8 (50–100 mesh, H^+ form) and 1 ml of 0.05 M HCl. The mixture is heated in a sealed tube at 100° for 24 hours and shaken every few hours. Hexosamine is removed by passing the hydrolyzate over a 2-ml column of Dowex 50 (H^+ form). The effluent is again passed through the column, which is then washed with 20 ml of water. Removal of neutral sugars may be accomplished by passing the effluent and washings over a column (2 ml) of Dowex 1-X4, followed by elution of the uronic acid with 30 ml of 2 M acetic acid. The yield of free glucuronic acid in this procedure was 74%.[43] However, only about 50% of the iduronic acid from dermatan sulfate was present in a 24-hour hydrolyzate.

In preparing samples for uronic acid analysis by gas–liquid chromatography, Radhakrishnamurthy et al.[44] hydrolyze the polysaccharide (0.2–0.5 mg) in 1 ml of 90% formic acid under nitrogen at 100–105° for varying periods of time. After hydrolysis the samples are taken to dryness under reduced pressure, the residue is dissolved in 1.0 ml of water, and aliquots are analyzed. Maximum amounts of uronic acid are released from hyaluronic acid and chondroitin 4-sulfate after hydrolysis for 24 hours, while iduronic acid release from dermatan sulfate reaches a maximum at 16 hours. The yield of uronic acid from chondroitin 4-sulfate and dermatan sulfate is about 20% of the dry weight of polysaccharide. Even after 40 hours, heparin yields only 30% of the theoretical uronic acid value.

A procedure has recently been described by Lindahl et al.[45] by which about 70% of the uronic acid content of heparin and N-acetylheparin may be liberated as the free acid or lactone. Samples (0.5 mg) are dissolved in 60 μl of 2 M trifluoroacetic acid and are heated at 100° for 3

[42] L.-Å. Fransson, L. Rodén, and M. L. Spach, Anal. Biochem. 21, 317 (1968).
[43] P. L. Jeffrey and K. G. Rienits, Biochim. Biophys. Acta 141, 179 (1967).
[44] R. Radhakrishnamurthy, E. R. Dalferes, Jr., and G. S. Berenson, Anal. Biochem. 24, 397 (1968).
[45] U. Lindahl, personal communication.

hours, resulting in complete removal of N-sulfate and N-acetyl groups. The hydrolyzate is evaporated to dryness and treated with 30 μl of 3.9 M sodium nitrite in 0.29 M acetic acid. Deamination is essentially complete after 10 minutes at room temperature, and the reaction mixture is diluted with 0.6 ml of 1 M acetic acid, passed through a column (1 × 3 cm) of Dowex 50-X8, equilibrated with 1 M acetic acid and finally evaporated to dryness with several additions of methanol. The residue which consists largely of uronosylanhydromannose disaccharides, is dissolved in 1.0 ml of 2 M trifluoroacetic acid, and the solution is heated at 100° for 4 hours. The hydrolyzate is evaporated to dryness and subjected to further analyses.

Paper Chromatography and Electrophoresis

Electrophoresis is not used to any great extent for the separation of glucuronic acid and iduronic acid. However, it is occasionally advantageous to separate the uronic acids in bulk from other hydrolysis products, and this is conveniently done by electrophoresis, e.g., on Whatman No. 3MM paper in 0.08 M pyridine–0.05 M acetic acid, pH 5.3, at 70 V per cm for 20–30 min. The uronic acid area is then eluted and subjected to paper chromatography as described in Vol. VIII [3]. In addition to the solvent suggested in Vol. VIII [3] (butanol–acetic acid–water, 50:15:35), the following solvents may be used: *tert*-amyl alcohol–isopropanol–water (4:1:2), *tert*-amyl alcohol–water–propanol–ethanol (4:2:1.3:0.5), *tert*-amyl alcohol–formic acid–water (4:1:1), and ethyl acetate–acetic acid–water (3:1:1).

Since iduronic acid is not readily available, a hydrolyzate of dermatan sulfate may serve as a source of the standard compounds (free iduronic acid and lactone).

Automated Chromatography on Anion Exchange Resin

The Technicon sugar chromatography system may conveniently be used, with some modifications, for the separation of glucuronic acid and iduronic acid [42] (see also this volume [1]).

Reagents

Orcinol purchased from Eastman Organic Chemicals and recrystallized from benzene.
Formic acid and sulfuric acid, reagent grade (Baker "analyzed")

Preparation of Resin

The resin (AG 1-X4 or AG 1-X8, minus 400 mesh, Cl⁻) is converted to the formate form by washing successively with 9 volumes of $1 M$ NaOH, 4 volumes of water, 2 volumes of $1 M$ formic acid, and 4 volumes of water. The resin is then suspended in 10 volumes of water and allowed to settle by gravity for 70 minutes; the settling distance is 27–29 cm. The finer particles remaining in suspension are used for packing the column; they are decanted and refractionated twice by the same procedure. Finally, the resin is sedimented overnight to remove colloidal material. This is repeated several times until the supernatant liquid is clear.

Procedure

A column of the dimensions 0.63×140 cm is filled to 120 cm with resin that has been well deaerated under reduced pressure in a suction flask. The column temperature is maintained at $37°$.

Samples containing 50–200 μg of uronic acid are applied to the column in water solution (1–2 ml) under slight air pressure. For the separation of uronic acids and nonsulfated aldobiuronic acids the column is eluted with 0.3 or $0.5 M$ formic acid at a rate of 30 ml per hour, which requires a pressure of 200–300 psi. The effluent from the column is analyzed by the automated orcinol-sulfuric acid method of the Technicon sugar chromatography system. In this procedure, the effluent is mixed with 1% aqueous orcinol and 70% v/v sulfuric acid. After passage through a mixing coil the effluent and reagents are heated at $95°$ for 24 minutes, passed through a cooling coil, and subsequently exposed to fluorescent light in order to increase the color yield. The optical density is measured at 420 nm and recorded with a single-pen recorder.

In the present procedure, a considerable upward shift of the baseline occurs as the effluent becomes acidic. This is readily adjusted by changing the aperture in the colorimeter from 0.2 to 0.6.

The separation obtained is illustrated in Fig. 4. If satisfactory resolution is not achieved with $0.5 M$ formic acid, the strength of the acid is decreased to $0.3 M$.

Essentially the same system may be used for the fractionation of oligosaccharides from enzymatic digests of mucopolysaccharides, but the eluting system is then changed to a LiCl gradient.[42] After the completion of a run with either eluent system, distilled water is pumped through the column for approximately 2 hours.

The sensitivity of the method is such that as little as 3 μg of uronic acid may be detected and estimated, although larger amounts are desirable. For reliable quantitation of glucuronic acid in dermatan sulfate at least 1 mg of polysaccharide is used, since this component may be present to the extent of only 5% of the total uronic acid.

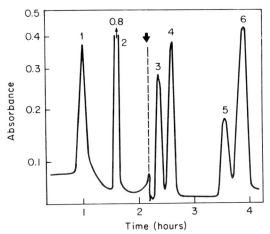

Fig. 4. Chromatography of uronic acids and uronic acid-containing disaccharides. Column, AG 1-X8 (formate form); eluent, 0.5 M formic acid. 1, Chondrosine; 2, glucuronolactone; 3, N-acetylchondrosine; 4, glucuronosylgalactose; 5, L-iduronic acid; 6, D-glucuronic acid. The bold arrow indicates the breakthrough of formic acid. From L.-Å. Fransson, L. Rodén, and M. L. Spach, *Anal. Biochem.* **21,** 317 (1968).

Calibration of the system is carried out with known amounts of glucuronic acid (as sodium salt). The color yield of iduronic acid is the same as that of glucuronic acid.

This system has been used extensively for the analysis of the uronic acid composition of dermatan sulfate preparations (for hydrolysis conditions, see above), and the correlation with results obtained by determination of carbazole:orcinol ratios is good.

Uronic Acid Analysis by Gas–Liquid Chromatography

A method for the separation of glucuronic acid and iduronic acid by gas–liquid chromatography has been described by Radhakrishnamurthy et al.[44]

Reagents

Hexamethyldisilazane. Distillation Products Industries, Rochester, New York

Trimethylchlorosilane, Distillation Products Industries

Pyridine, purchased from Baker; these 3 reagents are used without purification.

Apiezon M and Gas-Chrom-CLZ, mesh 80–100, purchased from Applied Science Laboratories, State College, Pennsylvania

Procedure

Trimethylsilylation of hexuronic acids and neutral sugars is carried out by the method of Sweeley *et al.*,[46] using hexamethyldisilazane reagent (HMDS, 20% in pyridine v/v) and trimethylchlorosilane (TMCS). Samples of glucuronic acid, iduronic acid, galactose, and xylose (10–100 μg) are prepared by evaporating 0.01% sugar solutions to dryness in small tubes with Teflon-lined screw caps. To each sample is added 0.1 ml of α-methyl-D-glucoside (100 μg) as an internal standard. When the samples are dry, they are trimethyl-silylated with 0.2 ml of HMDS reagent and 0.02 ml of TMCS. The reaction mixture is shaken for 1 minute, and 5 minutes later it is injected into the chromatographic column. In the case of polysaccharide hydrolyzates, the samples are warmed for 2 minutes at 80° to facilitate the reaction.

The conditions employed for chromatography of the TMS derivatives are as follows:

Instrument, F & M Model 402

Column, 6 inch glass, 15% Apiezon M on Gas-Chrom CLZ, mesh 80–100

Detector, control setting, 8; temperature 210°

Oven, isothermal 190° (slight variations in temperature occur, ±10°)

Flash heater, control setting, 4; temperature 200°

Range, setting 10

Attenuation, 4 and 8

Carrier gas, He, psig 30, rotameter reading 3.0

H_2, psig 15, rotameter reading 2.5

Air, psig 20, rotameter reading 3.0

Recorder chart speed, low × 1, 4 minutes/inch

Areas under peaks are calculated by triangulation and by planimetry.

A chromatogram of iduronolactone (an equilibrium mixture of lactone and acid) shows two well-resolved peaks with retention times relative to α-methyl-D-glucoside of 0.54 and 0.64. Glucuronolactone gives an asymmetric peak with a relative retention time of 0.73. Since the presence of small amounts of galactose and xylose in mucopolysaccharides could conceivably interfere with the uronic acid determinations, these sugars were also investigated.[44] It was found that their positions in the chro-

[46] C. C. Sweeley, R. Bentley, M. Makita, and W. W. Wells, *J. Amer. Chem. Soc.* **85**, 2497 (1963).

matogram are sufficiently removed from those of the hexuronolactones as to preclude any interference with the uronic acid analysis.

b. ANALYSIS WITHOUT PRIOR HYDROLYSIS

Since the quantitative release of uronic acids from a polymer-bound form cannot yet be accomplished, other methods have been developed that are not dependent on the isolation of the free sugars after hydrolysis. The most reliable results are probably obtained by the method of Tracey[47] which is based on the manometric determination of the carbon dioxide liberated by decarboxylation of uronic acids on strong acid hydrolysis (12% HCl at 111° for 5 hours). Glucuronic acid and iduronic acid both give stoichiometric yields of carbon dioxide, and the reaction is not dependent on the identity of the polysaccharide. The method is obviously not entirely specific and should therefore be used only for the analysis of purified polysaccharides in order to rule out interference from unrelated substances. The time required for the analysis of one hydrolyzate is in the order of 15–20 minutes, and relatively large amounts of material are needed (5–10 mg). As a consequence, the method is no longer popular and has been superseded by the more rapid—but sometimes less reliable—colorimetric methods, which require less material.

The most common colorimetric procedures are the carbazole method of Dische[40] and the orcinol reaction.[41] These have been described in detail in Volume 8 [3] together with an often used modification of the carbazole method.[48]

The carbazole method gives excellent results when applied to the analysis of polysaccharides which contain only glucuronic acid, i.e., hyaluronic acid, chondroitin 4-sulfate, and chondroitin 6-sulfate. However, the hybrid polysaccharides which are composed of both iduronic acid and glucuronic acid (dermatan sulfate, heparin, and heparan sulfate) present problems which have only partly been overcome. The difficulties encountered are due largely to the low color yield of iduronic acid in the carbazole reaction which is in the order of 25% of that of glucuronic acid. If it could be assumed that the polymer-bound iduronic residues behave similarly to the free monosaccharide, the total uronic acid content of a polysaccharide containing exclusively iduronic acid residues could be accurately determined with iduronic acid as standard or by simply applying a known correction factor to compensate for the difference between the two uronic acids. Under ideal circumstances, the difference in color yield could also be used to advantage for the determination of the relative proportions of the two uronic acids in a mixed

[47] M. V. Tracey, *Biochem. J.* **43**, 185 (1948).
[48] T. Bitter and H. M. Muir, *Anal. Biochem.* **4**, 330 (1962).

polymer, if the results are correlated with hexosamine analyses. At the extremes of the scale, the ratio of uronic acid to hexosamine would theoretically be 1:1 in a glucuronic acid-containing polysaccharide and 0.25:1 in a molecule with iduronic acid as the only uronic acid.

In actuality, it appears that the iduronic acid residues of dermatan sulfate behave roughly in the same fashion as the free monosaccharide in the carbazole method. However, exact quantitative relationships of the kind indicated above have not been established, partly because dermatan sulfate preparations always contain a certain amount of glucuronic acid and partly because of difficulties in preparing pure iduronic acid. An additional uncertainty factor is introduced by the inevitable minor variations in the hexosamine analysis, which make the determination of the minor uronic acid component quite unreliable. In summary, the uronic acid to hexosamine ratio gives an approximate estimation of the relative proportions of the two uronic acids but cannot presently be relied upon for exact determinations. Instead, the quantitation is carried out either by the chromatographic methods described previously or by determination of the "carbazole:orcinol ratio" which is discussed below.

It should be noted in this context that in the modified carbazole method of Bitter and Muir,[48] the color yield of dermatan sulfate is over 80% of that observed for chondroitin 4-sulfate. This method also has the added advantage of greater sensitivity and shorter reaction time and is therefore to be preferred over the original version for the estimation of total uronic acid.

In contrast to dermatan sulfate, heparin gives an anomalously high color yield in the carbazole reaction despite the fact that more than half of the uronic acid of this polysaccharide is iduronic acid. The reason for this difference is not yet known, although it may be speculated that the position of the iduronidic linkages (1,3 in dermatan sulfate and 1,4 in heparin) may influence the reaction. In any case, the anomalous behavior of heparin points up the necessity to exercise caution in the interpretation of the data derived from the carbazole reaction.

Carbazole to Orcinol Ratios. In the orcinol reaction, iduronic acid (free or bound) gives about 15% higher color yield than glucuronic acid. Used in conjunction, the carbazole and orcinol methods can therefore provide information concerning the relative proportions of glucuronic acid and iduronic acid in a mixture of the free sugars or in dermatan sulfate. The uronic acid content is determined by both methods, with glucuronic acid as standard, and the ratio of the amounts observed is then calculated. (Since the practice varies from one laboratory to another, it should be mentioned that the authors use a 20-minute heating time in the orcinol reaction, whereas a 40-minute period is recommended in Volume 8 [3].) For the free iduronic acid, the carbazole:orcinol ratio is 0.23, whereas, by definition, the value for free glucuronic acid is 1.0.

When the carbazole:orcinol ratio is determined for a preparation of dermatan sulfate which contains almost exclusively iduronic acid, a value of close to 0.23 is observed. However, the mucopolysaccharides containing D-glucuronic acid as their only uronic acid component have ratios greater than 2. This is due to incomplete liberation of D-glucuronic acid during the 20-minute heating time used by the authors for the analysis by the orcinol method. If the heating period is extended beyond 20 minutes, the color yield increases and the carbazole:orcinol ratio is reduced. Nevertheless, the shorter time is preferable, since it gives a greater difference in ratios between dermatan sulfate (0.2 or higher) and the glucuronic acid-containing polysaccharides (>2) and consequently a more accurate quantitation of the relative proportions of the two uronic acids.

2. Hexosamines

The procedures for hexosamine analysis have been described extensively in Volume 8 [3] and [1] and elsewhere in this volume [1]. It may merely be pointed out here that the quantitative determination of the hexosamine components of connective tissue polysaccharides meets with the same problem as the uronic acid analysis; i.e., owing to the stability of glucuronidic linkages, the quantitative release of the hexosamines requires more drastic hydrolytic conditions than are commonly needed in glycoprotein analysis. Although the hexosamines are much more resistant than uronic acids to destruction by strong acids, quantitative yields are nevertheless difficult to achieve.

An additional, specific problem is encountered in the analysis of heparin. The ready liberation of the N-sulfate groups upon acid hydrolysis renders the adjacent glucosaminidic linkage extremely resistant to hydrolysis owing to protonation of the free amino group. As a consequence, the uronidic linkages of N-desulfated heparin are cleaved more easily than the glucosaminidic bonds. A more complete liberation of hexosamine may be achieved if the amino groups are N-acetylated prior to continued hydrolysis.

3. Sulfate

In this section, a method is described which has been found satisfactory for estimating the total sulfate content of glycosaminoglycans. Also, procedures for identifying and determining N-sulfate and the 4- and 6-O-sulfates are given.

a. ESTIMATION OF TOTAL SULFATE[49,50]

Principle

Sulfate is completely liberated from the glycosaminoglycan by acid hydrolysis. The inorganic sulfate is then precipitated as barium sulfate, which gives a stable turbidity in the presence of gelatin and can be determined spectrophotometrically.

Reagents

HCl, 1 M

Barium chloride–gelatin reagent. Dissolve Difco Bacto Gelatin (1.5 g) in 300 ml of water at 60–70°. Leave at 4° overnight. After warming to room temperature, add barium chloride (1 g) to 200 ml of this solution, stir to dissolve, and leave to stand for 2–3 hours. Some barium sulfate which precipitates during this time may be removed by centrifugation. Stored at 4°, this reagent can be used for at least 1 week.

Gelatin reagent. The remaining 100 ml of 0.5% aqueous gelatin is stored likewise. Both gelatin reagents are warmed to room temperature before use.

Trichloroacetic acid, 4% aqueous solution

Sulfate standards. Prepare solutions containing up to 400 μg of sulfate per milliliter, using K_2SO_4 which has been dried at 105–110°.

All glassware used in the assay should be cleaned in hot, concentrated HNO_3 and rinsed in deionized water.

Procedure

Samples (1–3 mg) of sulfated polysaccharides are hydrolyzed in 1 ml of 1 M HCl in sealed tubes at 100° for at least 6 hours. After cooling, the tubes are opened and the contents are taken to dryness by rotary evaporation at approximately 40°. Dissolve the residues in 1 ml of water and mix.

To 0.1-ml aliquots of the hydrolyzates in 75 × 10 mm tubes, add 4% trichloroacetic acid (1.4 ml) and barium chloride–gelatin reagent (0.5 ml). Sulfate standards (10–40 μg of sulfate) and a water blank are similarly treated. After swirling, the mixtures are allowed to stand for 15–20 minutes; the turbidities are measured at 360 nm in silica micro cells of 2-cm light path.

[49] K. S. Dodgson, *Biochem. J.* **78**, 312 (1961).
[50] K. S. Dodgson and R. G. Price, *Biochem. J.* **84**, 106 (1962).

Another 0.1-ml aliquot of each solution is mixed with 4% trichloro-acetic acid (1.4 ml) and 0.5 ml of gelatin reagent (without barium chloride). These mixtures are also read at 360 nm and serve to estimate any UV-absorbing substances present in the hydrolyzed samples. Thus, any such absorption should be subtracted from the value obtained in the presence of barium chloride.

Other methods employing benzidine,[51] chloranilate,[52] and rhodizonate[53] may be used for the determination of ester sulfate.

b. O-Sulfate

A number of different procedures, which involve determination of rate of acid hydrolysis, infrared spectra, and susceptibility to enzymatic digestion, are available for identifying the site of attachment of O-sulfate groups.

Rate of Acid Hydrolysis

Ester sulfate groups linked to equatorial and axial secondary hydroxyls and to primary hydroxyls are hydrolyzed at different rates. In 0.25 M HCl at 100°, Rees[54] determined half-lives of 0.1–0.4 hour, 1–1.5 hours, and >1.5 hours, respectively. Considerable uncertainty attends the calculation of half-lives when two or more types of sulfate ester groups are present in a preparation. Furthermore, difficulty in estimating free sulfate in the presence of partially sulfated glycosaminoglycans may be experienced.

Infrared Spectra

Samples (1–2 mg) of sulfated glycosaminoglycans are examined as pellets with KBr.[55] An intense band at ~1240 cm^{-1}, attributable to the S–O bond-stretching vibration, is common to all. Peaks in the 800–860 cm^{-1} region may serve to distinguish isomeric sulfate esters. A peak in the region 810–820 cm^{-1} is characteristic of a 6-sulfate ester [with the sugar in the C1 (D) conformation], at ~830 cm^{-1} for a secondary equatorial conformation (e.g., glucose 4-sulfate), and at 850–860 cm^{-1} for a secondary axial sulfate group (e.g., galactose 4-sulfate).[56]

[51] Vol. 3 [146].
[52] Vol. 8 [114].
[53] T. T. Terho and K. Hartiala, Anal. Biochem. 41, 471 (1971).
[54] D. A. Rees, Biochem. J. 88, 343 (1963).
[55] Vol. IV [3].
[56] J. R. Turvey, Advan. Carbohyd. Chem. 20, 183 (1965).

Enzymatic Degradation[57]

For the purposes of distinguishing and determining the various isomeric sulfate esters of the galactosaminoglycans, the methods of Suzuki
and collaborators[57] offer considerable advantages and have superseded
acid hydrolysis methods.

The polysaccharide (dermatan sulfate, chondroitin 4-sulfate or chondroitin 6-sulfate) is digested with chondroitinase-ABC or chondroitinase-
AC, and the disaccharide products are chromatographed in order to
separate the 4- and 6-sulfated isomers. Alternatively, the digest may be
assayed by a Morgan-Elson method.[58] Only the 6-sulfated disaccharide
gives a chromogen in this procedure.[59]

If the glycosaminoglycan is susceptible to digestion by chondroitinase-
ABC, the resulting digest may be further incubated with chondro-4-
sulfatase or chondro-6-sulfatase. Liberation of inorganic sulfate by the
latter enzyme, for example, provides confirmation of the presence of
chondroitin 6-sulfate.

c. N-SULFATE

N-Sulfate groups may be determined following mild acid hydrolysis
or by treatment of the glycosaminoglycan with nitrous acid.

Acid Hydrolysis

The N-sulfate bond is readily hydrolyzed in dilute acid. For example,
the N-sulfate group of heparin has a hydrolysis constant of 0.062 min^{-1}
in 0.1 M HCl at 99.5°.[60] The rapid release of inorganic sulfate during
acid hydrolysis assists in establishing the presence and amount of N-sulfate groups. Commonly, hydrolysis in 0.04 M HCl at 100° for 90 minutes[61] is employed to effect complete N-desulfation. Some O-desulfation
also can be expected under these conditions.

As estimation of inorganic sulfate in the presence of partially hydrolyzed glycosaminoglycans is subject to error, it is preferable to determine N-sulfate following nitrous acid treatment.

---

[57] See this volume [124], [125]. See also, T. Yamagata, H. Saito, O. Habuchi, and
S. Suzuki, *J. Biol. Chem.* **243**, 1523 (1968) ; H. Saito, T. Yamagata, and S. Suzuki,
J. Biol. Chem. **243**, 1536 (1968) ; S. Suzuki, H. Saito, T. Yamagata, K. Anno, N.
Seno, Y. Kawai, and T. Furuhashi, *J. Biol. Chem.* **243**, 1543 (1968).
[58] Vol. 8 [3].
[59] M. B. Mathews and M. Inouye, *Biochim. Biophys. Acta* **53**, 509 (1961).
[60] R. A. Gibbons and M. L. Wolfrom, *Arch. Biochem. Biophys.* **98**, 374 (1962).
[61] G. J. Durant, H. R. Hendrickson, and R. Montgomery, *Arch. Biochem. Biophys.*
99, 418 (1962).

Estimation of Total N-Sulfated Hexosamine Residues by Treatment with Nitrous Acid

Nitrous acid reacts with the free amino groups and the N-sulfated hexosamine units of glycosaminoglycans. Susceptible glucosamine residues are converted to 2,5-anhydro-D-mannose during the reaction, and concurrently, cleavage of adjacent glycosidic bonds occurs, so that oligosaccharide products bear anhydromannose residues at their reducing ends. Therefore, this reaction has found application in analytical and structural studies of heparan sulfate and heparin.

Nitrous acid treatment of the glycosaminoglycan is carried out as described by Dische and Borenfreund[62] with the exception that treatment is prolonged to 80 minutes.[63] Then, estimation of anhydromannose reducing end groups is carried out by the indole method of Dische and Borenfreund.[62] Hexosamine residues bearing free amino groups are also estimated in this procedure.

The reaction may be made specific for attack at N-sulfate groups by carrying out the nitrous acid treatment at $-20°$. This method[64] will be described.

Reagents

1,2-Dimethoxyethane is freed of peroxide by shaking with Dowex 1 (carbonate form, 20–50 mesh) and then passing through a small column of the same resin. Store in the dark at $-20°$.

Nitrous acid reagent. To 10 ml of 3.5% aqueous sodium nitrite are added 20 ml of 1,2-dimethoxyethane. Cool to $-20°$ and pass through a column (2×8 cm) of Dowex 50-X8 (H+), which has been equilibrated with 60% aqueous 1,2-dimethoxyethane at $-20°$. Wash the column with 20 ml of 60% 1,2-dimethoxyethane, and pool the effluents. This nitrous acid reagent is normally used within a few hours of preparation, although it is known to be stable for 1–2 weeks if stored in Dry-Ice.

Procedure

First, the glycosaminoglycan (heparan sulfate or heparin) must be converted to the acid form by treatment with Dowex 50 (H+). An aqueous solution (250 mg in 15 ml) is passed through a column (2.2×5 cm) of Dowex 50-X12 (H+) at $2°$. Wash the column with 3×5 ml of water and pool the effluents. Add 60 ml of 1,2-dimethoxyethane and cool to $-20°$ before adding 40 ml of the nitrous acid reagent.

[62] Vol. 3 [12].
[63] D. Lagunoff and G. Warren, *Arch. Biochem. Biophys.* **99**, 396 (1962).
[64] J. A. Cifonelli, *Carbohyd. Res.* **8**, 233 (1968).

The rate of reaction may be followed conveniently by removing 0.1 ml aliquots at known time intervals, transferring to tubes containing 0.5 ml of 12% ammonium sulfamate and 1.4 ml of water at 2°, and then proceeding to estimate anhydromannose formation by the indole method[62] (without the initial nitrous acid treatment). For complete reaction under these conditions, approximately 8 hours are required for heparan sulfate and 2 hours for heparin. Glycosidically linked anhydromannose residues derived from nitrous acid treatment of heparan sulfate and heparin have been found to give approximately 60% more color in the indole method than an equimolar amount of free anhydromannose. (Normally for standard estimations of anhydromannose, known amounts of glucosamine are treated with nitrous acid.)

Recently, Cifonelli and King[65] have introduced 3-methyl-1-butyl nitrite and butyl nitrite for the selective N-desulfation of heparin and heparan sulfate.

4. N-Acetyl Groups

The determination of N-acetyl groups has been described in Vol. 8 [1].

5. Components of Carbohydrate-Protein Linkage Regions

Since proteolytic digestion is the method of choice for the isolation of connective tissue polysaccharides—rather than methods which cleave carbohydrate-protein bonds, small amounts of residual peptides are normally associated with the purified substances. The composition of these peptides may be determined by the usual procedures for amino acid analysis.

As indicated in Section I, three different monosaccharides are involved in the linkages of the mucopolysaccharides to protein, i.e. xylose, glucosamine, and galactosamine. These may all be determined by methods described elsewhere in this series (Vol. 8 [1], [3], and [4]; this volume [1]). However, a simple colorimetric procedure has recently been developed specifically for the determination of xylose in mucopolysaccharides.[66] This method, which can also be used for the determination of methylpentoses, is described below.

a. Determination of Xylose by an Anthrone Method

Reagents

Anthrone, 1%, in sulfuric acid of specific gravity 1.84 is prepared just before use. (Anthrone obtained from Hopkin & Williams

---

[65] J. A. Cifonelli and J. King, *Carbohyd. Res.* **21**, 173 (1972).
[66] C. P. Tsiganos and H. Muir, *Anal. Biochem.* **17**, 495 (1966).

Ltd., Chadwell Heath, Essex, England, is suitable without recrystallization.)

D-Xylose standard, 5–50 μg/ml, is dissolved in water saturated with benzoic acid.

Glassware: Thin-walled test tubes, 0.75 × 6 inches, are washed in hot dilute sulfuric acid prior to use because traces of detergents used in routine washing interfere greatly.

Procedure

Samples (3 ml) of the anthrone reagent are delivered from a burette into each clean test tube, frozen in acetone–solid CO_2 mixture, and allowed to stand in an ice–water bath. Triplicate 1-ml samples of unknown or standard are layered on top of the frozen reagent. Immediately afterward and while immersed in the ice water, each tube is first swirled gently while thawing and then shaken vigorously to mix the viscous liquid thoroughly. The temperature of the solutions during mixing and subsequently should not exceed that of the ice bath. Uncontrolled heating will increase interference from other sugars present.

The tubes are then transferred from the ice bath to a constant temperature bath at 40° and kept immersed for exactly 10 minutes for xylose or 30 minutes for other pentoses or methylpentoses. At the end of this time they are cooled by immersion in ice water for a few minutes and then left to stand for 20 minutes in the dark. Their optical density is read at 615 nm in the case of pentoses and 640 nm in the case of methylpentoses in a 1-cm cell.

Blanks are set up by replacing the sample or standards by 1 ml of water saturated with benzoic acid.

The absorption spectra of xylose and fucose overlap considerably, but differential analysis after heating for both 10 and 30 minutes permits quantitation of each sugar even in mixtures. The interference from hexoses has been reduced by comparison with previous methods and is in the order of 5%. Negligible interference was observed with glucuronolactone, glucosamine, iduronic acid, and sialic acid (in the order of 1–2%).

The highest color yield given by any sugar was observed with fructose. However, since this sugar is not present in mucopolysaccharide hydrolyzates, this lack of specificity does not present a problem.

In the experience of the authors, the accuracy of the analysis can be somewhat improved if a small correction is introduced for the color contribution of the bulk of the polysaccharide under examination. This can be done, e.g., in the analysis of chondroitin sulfate, by analyzing separately a sample of xylose-free oligosaccharides which may be prepared by digestion of cartilage proteoglycan with testicular hyaluronidase and separation of the oligosaccharides on Sephadex G-75.[17]

b. Determination of Galactose

In the xylose-containing mucopolysaccharides, two galactose residues are also present as specific components of the linkage region (see Fig. 1). One of these is linked to glucuronic acid and is liberated relatively slowly on acid hydrolysis owing to the stability of the glucuronidic linkage. Whereas the other galactose residue is released completely after hydrolysis in $1 M$ HCl for 1 hour at 100°, prolonged hydrolysis for periods of 10 hours or more is necessary to obtain maximum liberation of the second residue.[67]

The quantitation of galactose in hydrolyzates may be carried out by any of the general methods for monosaccharide analysis. It is convenient, however, to use a more specific enzymatic assay with galactose dehydrogenase from *Pseudomonas saccharophila*, if this enzyme is available. The procedure for the assay of galactose in heparin preparations will be described as an example.

Reagents

HCl, $1 M$

Tris·HCl buffer, $0.08 M$, pH 8.6, containing 8 mM glutathione and 0.8 mM NAD$^+$

D-Galactose dehydrogenase. The enzyme is prepared as described previously (Vol. V [40]) and is purified through the first ammonium sulfate precipitation

D-Galactose standards, 0.1–0.4 mM in aqueous solution

Procedure

Samples of heparin (10 mg) are dissolved in 2.0 ml of $1.0 M$ HCl and hydrolyzed in sealed tubes at 100° for varying periods of time up to 20 hours. After neutralization to pH 6 or 7 with NaOH, the hydrolyzates are evaporated to dryness, dissolved in 2.9 ml of water, and filtered through a sintered-glass filter. A 1.45-ml sample of each filtrate is transferred to a 1-cm quartz cuvette and mixed with 1.25 ml of Tris·HCl buffer. After warming the mixture to 30°, the reaction is started by the addition of 0.3 ml of D-galactose dehydrogenase solution.

The increase in absorbancy at 340 nm is followed, and the amount of galactose is calculated from the difference between the point of maximal absorbance and a zero time value obtained by extrapolation or from the appropriate blanks.

---

[67] U. Lindahl and L. Rodén, *J. Biol. Chem.* **240**, 2821 (1965).

Galactose standards containing from 0.1 to 0.4 μmoles are treated in a similar fashion.

Comments

Under the conditions of assay, 0.1 μmole of D-galactose should give an absorbance of 0.207, based on a molar absorbance of 6.22×10^3 for NADH. This value is approached in 30–45 minutes with concentrated enzyme solutions. With more dilute solutions the absorbance reaches a maximum value of 0.150–0.180 in 50–70 minutes, after which a slow decrease occurs. A more rapid reaction may be obtained by using a larger amount of more highly purified enzyme and a higher concentration of substrate.

The galactose dehydrogenase method has an advantage over the enzymatic assay with galactose oxidase inasmuch as the reaction is not influenced by the presence of rather large concentrations of salt. No interference was observed in the presence of 1 mmole of NaCl in the reaction mixture, and the hydrolyzates may thus be assayed without desalting. Nor is the reaction influenced by heparin itself or its hydrolysis products.

Because of color formation during hydrolysis, it is necessary, however, to decolorize samples which have been hydrolyzed more than 3 hours. Decolorization is carried out by passing the hydrolyzate over a column (1 × 10 cm) of Dowex 1-X2 (chloride form, 200–400 mesh). The effluent and washings are evaporated to dryness, dissolved in water, and analyzed as described.

In the presence of D-galactosamine, a control sample may be desired, since this sugar is oxidized at about 1% the rate of D-galactose.

Owing to the difficulties in hydrolyzing glucuronidic bonds completely, it would be preferable to determine polysaccharide-bound galactose by a colorimetric method such as the anthrone–sulfuric acid reaction. This can be done with small glycopeptides, but polysaccharide preparations cannot be reliably analyzed in this way.

B. Enzymatic Characterization of Polysaccharides

Enzymatic methods are an important part of the analytical process used for the characterization of connective tissue polysaccharides. A variety of glycosidases are now available, some of which are endoglycosidases with specificity for a certain polysaccharide, whereas others are exoglycosidases which may be used for stepwise removal of glycosyl groups from the nonreducing terminus of polysaccharides and oligosaccharides.

Hyaluronidases from various sources have long been used for analytical as well as preparative purposes. In recent years, the purification of bacterial chondroitinases and chondrosulfatases has provided a new set of valuable enzyme tools. The use of chondroitinase-ABC and chondroitinase-AC permits differentiation between dermatan sulfate and the chondroitin sulfates, and analysis of the disaccharide products by chromatographic and other methods easily establishes the position of the sulfate groups in these compounds. Also, a hybrid polysaccharide or a mixture may be similarly analyzed to determine the relative distribution of sulfate groups between positions 4 and 6 of the galactosamine residues.

The enzymatic methods have been described elsewhere in this series and will therefore not be treated in detail here (see Volume 8 [112], [113], and [114]; this volume [124] and [125]). It should merely be emphasized that the application of these methods is to be regarded as an almost indispensable part of the analytical process without which a full spectrum of information cannot be obtained.

C. Molecular Weight Determination

The weight-average molecular weight (M_w) and the number-average molecular weight (M_N) have been determined for glycosaminoglycans by ultracentrifugal methods,[68-71] light scattering,[72] and osmometry.[68,73] Since these methods require highly specialized equipment and methodology, several secondary procedures employing relatively simple and commonly available equipment have been devised.

Particular precautions should be taken in determination of the molecular weights of glycosaminoglycans for the following reasons: (1) cations are bound because of a high linear charge density; (2) because of chain flexibility and high anionic charge, the average chain configuration is dependent upon counterion (cation) concentration; (3) large contour lengths of chains contribute to nonideal effects at low polymer concentrations. The polymer chain has its most extended configuration in pure aqueous solutions. With increasing salt concentrations the average configuration approaches that of a theoretical random coil while viscous

[68] Å. Wasteson, *Biochem. J.* **122**, 477 (1971).
[69] R. L. Cleland and J. L. Wang, *Biopolymers* **9**, 799 (1970).
[70] S. E. Lasker and S. S. Stivala, *Arch. Biochem. Biophys.* **115**, 360 (1966).
[71] T. C. Laurent and A. Anseth, *Exp. Eye Res.* **1**, 99 (1961).
[72] M. B. Mathews, *Arch. Biochem. Biophys.* **61**, 367 (1956).
[73] M. B. Mathews and A. Dorfman, *Arch. Biochem. Biophys.* **42**, 41 (1953).

interactions with solvent decline. The greatest extent of change in configuration occurs up to $0.05\,M$ NaCl.[74] Relatively little change occurs between $0.2\,M$ NaCl and $0.4\,M$ NaCl.

The methods to be described discriminate between homologous polymers that differ in chain length by measurement of a physical parameter that depends upon frictional characteristics of the macromolecule. Consequently, it is necessary to maintain constant electrolyte composition. Generally, comparable results are obtained in monovalent–monovalent electrolyte solutions between $0.2\,M$ and $0.4\,M$. It should be noted that the significant factor influencing chain configuration is cation concentration, not ionic strength. In addition, it is necessary, in the viscosity method, to utilize an empirical relationship between the physical parameter measured and the molecular weight for each type of polymer.

Major uncertainty in the estimation of molecular weight is due to difficulties in accurate determination of the dry weights of samples. The materials are very hygroscopic and as ordinarily prepared contain 10–20% water. About 10% water is held extremely firmly and cannot be removed, short of producing polysaccharide decomposition. The resultant uncertainties in dry weight or solution concentrations are reflected in current literature values of molecular weight. It is recommended that a suitable reference preparation be employed when concentrations are determined by the common colorimetric procedures to ensure relative accuracy.

Finally, it is essential to utilize only well purified preparations of known structural type since polymer chain characteristics differ among the types of glycosaminoglycans. The methods to be described are based upon differences in polymer chain conformations and dimensions for homologous polymers. Since most preparations of purified glycosaminoglycans show some heterogeneity of composition, i.e., variation of molar proportions of sulfate to glucosamine or galactosamine, iduronate to glucuronate, N-acetyl to N-sulfate, etc., it is essential to ensure that these preparations do not differ greatly in composition from the reference compounds.

1. Viscosity Method

The intrinsic viscosity of the polymer is determined by means of a capillary or other type of viscometer, and the molecular weight is calculated from the empirically established Mark–Houwink relationship: $[\eta] = KM^{a}$, where $[\eta]$ is the intrinsic viscosity, M is the molecular

[74] M. B. Mathews, Arch. Biochem. Biophys. 43, 181 (1953).

weight, and K and a are the empirical constants.[75,76] The molecular weight that is thus calculated is termed the viscosity average molecular weight (M_v) and is usually close to M_w for a well fractionated polymer when M_w's were utilized in establishing the Mark–Houwink relationship.

These averages are defined:

$$M_w = \Big(\sum_{i=1}^{\infty} N_i M_i^2 \Big) \Big/ \Big(\sum_{i=1}^{\infty} N_i M_i \Big)$$

$$M_v = \Big[\Big(\sum_{i=1}^{\infty} N_i M_i^{1+a} \Big) \Big/ \Big(\sum_{i=1}^{\infty} N_i M_i \Big) \Big]^{1/a}$$

where M_i = molecular weight of species i, N_i = number of moles of species i, and a = exponent in the Mark–Houwink relationship. It is apparent that only if $a = 1$ is $M_v = M_w$. However, M_v is always considerably closer to M_w than to M_n of a polydisperse sample.

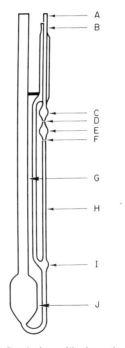

FIG. 5. Semimicro dilution viscometer.

[75] H. B. Bull, "An Introduction to Physical Biochemistry." Davis, Philadelphia, Pennsylvania, 1964.
[76] C. Tanford, "Physical Chemistry of Macromolecules." Wiley, New York, 1961.

Apparatus

Cannon–Ubbelohde semimicro dilution viscometer, size 100, un-calibrated, from Cannon Instrument Co. (P. O. Box 812, State College, Pennsylvania 16801). This instrument has a flow time for water of about 70 seconds at 25°, does not require a kinetic energy correction, and requires a minimum sample of 1 ml (Fig. 5).

Constant-temperature water bath, glass, minimum working depth of 28 cm, maintained at 25.0 ± 0.02°. The unit may be pur-chased: constant temperature bath, infrared, catalog No. 9926-D, Arthur H. Thomas Co., Philadelphia, Pennsylvania, or assem-bled from component parts.

Stopwatch, 0.1 second subdivision marks

Procedure

Step 1. Clean viscometer by soaking in 1 M KOH in 50% ethanol, previously filtered through a coarse-frit glass funnel, and rinse thoroughly with filtered distilled water. Rinse several clean 1-ml pipettes similarly. Protect openings with Al foil and dry in oven.

Step 2. Prepare a suitable buffer (Table X) and dissolve sample at a concentration to yield a flow time from 30–50 seconds above that of buffer. At least 2–3 ml of solution are required. Compute the an-hyrous concentration in grams per 100 ml.

Step 3. Clamp viscometer in water bath and with openings protected by Al foil except when adding liquids, pipette 1.00 ml of filtered buffer directly into bulb J avoiding loss on sides of viscometer tube G. Adjust position of viscometer so that capillary column is within ±2° of vertical. This can be done conveniently by aligning the capillary column (H) with a weighted thread hung outside the bath. Allow at least 10 minutes for temperature equilibration.

Step 4. Attach length of clean rubber tubing to tube A in order to permit application of suction by mouth. With moistened finger placed tightly over tube B, apply suction to tube A until liquid reaches the center of bulb C. Remove suction from tube A. Sample should quickly drop away from lower end of capillary into bulb I. Measure efflux time, allowing the liquid sample to flow freely past etch mark D, measuring time for the meniscus to pass etch mark D to etch mark F. Make at least four successive readings of the flow time until the average deviation from the mean is less than ±0.1 second.

Step 5. Remove, clean, and dry viscometer.

Step 6. Filter solution through a small fritted funnel and transfer 1.00

TABLE X
CONSTANTS OF THE RELATIONSHIP[a]: $[\eta] = KM^a$

Polysaccharide	Buffer	$K \times 10^4$	Exp[c]	Reference
Hyaluronate	0.2 M NaCl	2.28	0.816	d
Chondroitin 4-sulfate	0.2 M NaCl in 0.15 M PO₄ buffer, pH 7.0	3.1	0.74	e, f
Chondroitin 4-sulfate[b]	0.2 M NaCl	0.050	1.14	g
Chondroitin 6-sulfate	0.2 M NaCl in 0.15 M PO₄ buffer, pH 7.0	3.1	0.74	f, h
Dermatan sulfate	0.2 M NaCl in 0.15 M PO₄ buffer, pH 7.0	3.1	0.74	e, f
Keratan sulfate (cornea)	0.2 M NaCl	1.5	0.83	i
Heparin	0.5 M NaCl	0.355	0.90	j

[a] Approximate molecular weight ranges of validity are $10^5 - 10^6$ for hyaluronate and $10^4 - 5 \times 10^4$ for others.
[b] Obtained with paucidisperse preparations from bovine nasal cartilage consisting of mixed polymer of chondroitin 4-sulfate and chondroitin 6-sulfate.
[c] Exponent (a) in Mark-Houwink relationship.
[d] R. L. Cleland and J. L. Wang, *Biopolymers* **9**, 799 (1970).
[e] M. B. Mathews, *Arch. Biochem. Biophys.* **61**, 367 (1956).
[f] M. B. Mathews, *Biochim. Biophys. Acta* **35**, 9 (1959).
[g] Å. Wasteson, *Biochem. J.* **122**, 477 (1971).
[h] M. B. Mathews, unpublished observations.
[i] Approximate values calculated from data of T. C. Laurent and A. Anseth, *Exp. Eye Res.* **1**, 99 (1961).
[j] P. A. Liberti and S. S. Stivala, *Arch. Biochem. Biophys.* **119**, 510 (1967).

ml of filtrate into dry viscometer positioned as in step **3**. After equilibration, determine mean flow time.

Step 7. Add 1.00 ml of filtered buffer to viscometer and mix contents either by removing viscometer from bath and shaking or, more conveniently, by blowing gently through attached tubing while keeping a finger over tube B. After an equilibration period, measure efflux time as in step (**4**).

Step 8. Repeat step **7** two more times.

Step 9. Calculate results as follows: Under conditions where kinetic energy and other corrections are negligible, the relative viscosity of the solution is given by the relation

$$\eta_r = \eta/\eta_0 = dt/d_0 t_0$$

where η, d, and t are the coefficient of viscosity, the density and the efflux time of the solution, respectively; η_0, d_0, and t_0 are the corresponding quantities for the buffer. Additional terms are defined: specific viscosity

$$\eta_{sp} = dt/d_0 t_0 - 1 = \eta_r - 1$$

intrinsic viscosity,

$$[\eta] = \lim_{c \to 0} (dt/d_0 t_0 - 1)/c = \lim_{c \to 0} \eta_{\text{sp}}/c$$

If one is interested only in the intrinsic viscosity, the last expression may be simplified by neglecting the density ratio, since $\lim_{c \to 0} d/d_0 = 0$ then

$$\eta'_{\text{sp}} = t/t_0 - 1$$
$$[\eta] = \lim_{c \to 0} \eta'_{\text{sp}}/c$$

To obtain $[\eta]$, plot η'_{sp}/c versus c, and extrapolate linearly to $c = 0$. At low concentrations of glycosaminoglycans, the line obtained is given by the relation

$$\eta'_{\text{sp}}/c = [\eta] + k'[\eta]^2 c$$

where $k' \cong 0.5$. Calculate M_v from the appropriate relationship of Table X. For multiple determinations, obtain M_v graphically from a linear plot of the appropriate relationship on a log-log scale.

2. Electrophoresis Method

Electrophoresis in polyacrylamide gels can be utilized to estimate the molecular size of ultramicro samples of glycosaminoglycans. The migration distance is a linear function of log M for a homologous series.[77]

Apparatus

Polyanalyst electrophoresis apparatus with 5×75 mm electrophoresis tubes (Buchler Instrument Division Nuclear-Chicago Corp. 1327 Sixteenth Street, Fort Lee, New Jersey 07024)

Power supply, regulated

Destainer, Model 170 (Biorad Laboratories, 32nd and Griffin Ave., Richmond, California 94804)

Constant-temperature circulating water bath at 25°

Materials

Reagents, electrophoresis grade, may be purchased from Eastman Kodak Co. and elsewhere

Acrylamide

N,N'-Methylene bisacrylamide

Ammonium persulfate

N,N,N',N'-Tetramethylethylenediamine (TEMED)

Alcian blue 8GX (Fisher Chemical Co.)

---

[77] M. B. Mathews and L. Decker, *Biochim. Biophys. Acta* **244**, 30 (1971).

TABLE XI
REAGENTS (ML) USED IN PREPARATION OF GELS[a]

Gel concn. (%)[b]:	3	4	5	8	10
Acrylamide	1.50	2.00	2.50	4.00	5.00
Bisacrylamide	2.70	2.30	1.60	1.00	1.00
Buffer	3.30	3.20	3.40	2.50	1.50
Persulfate	2.50	2.50	2.50	2.50	2.50
TEMED	0.014	0.014	0.014	0.014	0.014

[a] Acrylamide stock solution contains 20 g in 100 ml of buffer; bisacrylamide stock solution contains 2.0 g in 100 ml of buffer; buffer is 0.05 M phosphate (pH 7.5); persulfate stock solution contains 0.48 g of ammonium persulfate in 100 ml of buffer; TEMED, N,N,N',N'-tetramethylethylenediamine.

[b] The 3% and 4% gels are supported by a 5-mm plug of 8% gel.

Procedure

Apply samples of 10–20 μg in 0.01 ml of 40% sucrose-buffer solutions to gels prepared as described in Table XI. Maintain a voltage of 27.5 ± 0.5 V for 120 minutes with water circulation at 25°, corresponding to a voltage drop within gel of approximately 3.5 V/cm. Remove gels and place in small porcelain boats containing 0.5% Alcian blue in 3% acetic acid for 60 minutes. Destain overnight in 7% acetic acid and store in 7% acetic acid. Measure the distance of migration of the peak of stain density visually or after spectrophotometric scan. Include at least two reference samples of known M. Plot distance of migration versus log M for reference samples and estimate M of unknown by linear interpolation. Linearity of the relationship between migration distance and log M and reasonably small standard errors of estimate in log M (about 15% in M) are found for molecular weight ranges and gel concentrations shown in Table XII. The distance of migration is a func-

TABLE XII
MOLECULAR WEIGHT LIMITS OF VALIDITY OF ELECTROPHORESIS METHOD

Glycosaminoglycan	Gel concentration (%)	Range, $M \times 10^{-3}$
Hyaluronate	3	21–285
Hyaluronate	5	3.7–285
Hyaluronate	10	3.7–136
Chondroitin 4-sulfate	5	12–58
Chondroitin 4-sulfate	10	12–58
Chondroitin 6-sulfate	5	4.3–72
Chondroitin 6-sulfate	10	4.3–72

tion of electrophoretic charge as well as frictional characteristics and dimensions of the polyanions. However, small deviations ($\pm 10\%$ maximum) in molar ratio of sulfate to hexosamine from the reference standards contribute little to error in estimations of M.

3. Gel Permeation Method

Analytical gel chromatography was adapted[78] for estimation of the molecular weight of microgram quantities of chondroitin sulfate employing reference standards prepared[79] from bovine nasal cartilage. A similar method on a larger scale was employed for molecular weight estimations of chondroitin sulfate, dermatan sulfate, and heparan sulfate.[80] The microgram scale method[78] will be described.

Apparatus

Polyethylene tubes (internal diameter 1.7–3.0 mm) or glass capillaries (internal diameter 0.8–1.2 mm) and 60–100 cm long (total volume, 0.5–7 ml).
Fraction collector

Materials

Sephadex G-200 (Pharmacia Fine Chemicals, Inc., 800 Centennial Ave., Piscataway, New Jersey 08854)
Blue Dextran 2000 (same source)

Procedure

Step 1. Equilibrate gel with 1.0 M NaCl (3 days at room temperature or 5 hours in boiling water bath, per Pharmacia Technical Data Sheet No. 11). Stabilize column by passing through 1.0 M NaCl at a hydrostatic head of about 100 cm of H_2O for 24–28 hours. Check quality of column packing and obtain void volume by chromatography of sample of Blue Dextran 2000 in 100 μl of 1.0 M NaCl. Apply sample of 100 μg in 50–100 μl of saline and elute with 1.0 M NaCl at a hydrostatic pressure of about 100 cm of water. Collect fractions and assay for uronic acid. Include at least two reference samples. Draw elution curves by plotting concentration versus elution volume.

Step 2. Estimate molecular weight of unknown sample by linear interpolation from a plot of the elution volumes of the reference samples versus log M. The elution volume (V_e) of each preparation is defined

[78] Å. Wasteson, *Biochim. Biophys. Acta* **177**, 152 (1969).
[79] Å. Wasteson, *Biochem. J.* **122**, 477 (1971).
[80] G. Constantopoulos, A. S. Dekaban, and W. R. Carrol, *Anal. Biochem.* **31**, 59 (1969).

as the position of the maximum of the eluted peak. The relationship of V_e and log M is linear for chondroitin 4-sulfate in the range of 12×10^3 to 30×10^3 daltons and deviates only slightly from linearity to 40×10^3 dalton.[78] Relationships for other glycosaminoglycans are not known as yet.

D. Differentiation of Carbohydrate–Protein Linkage Types

Carbohydrate–protein linkage analysis is an important part of the characterization of a proteoglycan. The same methods are used that have been applied successfully to the study of glycoprotein structure, and either the entire proteoglycan or the polysaccharide component, isolated after proteolytic digestion, may be analyzed. An extensive discussion of linkage differentiation is found elsewhere in this volume [1], and a detailed review of this aspect of proteoglycan structure has also recently appeared.[81]

Summarizing the methods used in the study of polysaccharide-protein linkages, two major approaches have proved fruitful.

1. The polysaccharide is subjected to exhaustive proteolysis, aimed at removing all but the carbohydrate-bound amino acid; partial acid hydrolysis, alone or in combination with enzymatic degradation, then permits the isolation of the monosaccharide–amino acid fragment representing the carbohydrate–protein linkage. This approach has been successful with many glycoproteins and has been used to establish the linkages of heparin, chondroitin 4- and 6-sulfate, dermatan sulfate, and corneal keratan sulfate.

2. Another extremely useful method for linkage analysis is based on the alkali lability of the linkages involving the hydroxyamino acids, serine and threonine. These amino acids are lost by conversion to unsaturated derivatives in the β-elimination reaction which results in cleavage of the carbohydrate–protein bonds. If the reaction is carried out in the presence of borohydride, the newly released reducing end group is simultaneously reduced to a sugar alcohol moiety and can be identified after acid hydrolysis of the polysaccharide. Since all but one of the known polysaccharide-protein linkages (i.e., that of keratan sulfate 1) involve serine or threonine, this approach has been of particular value in the study of the connective tissue proteoglycans.

Neither of the two general methods outlined above is ideal, and a number of problems are encountered in their execution. The first method suffers from the drawback that satisfactory removal of amino acids other than the one involved in the polysaccharide linkage is often difficult to achieve. Furthermore, a method involving partial acid hydrolysis will

[81] A. Gottschalk (ed.), "Glycoproteins," pp. 491–517. Elsevier, Amsterdam, 1972.

by its very nature result in a rather low yield of the linkage fragment to be investigated.

The β-elimination reaction is fraught with certain pitfalls in the interpretation of the data, and one must be aware of the fact that ambiguous results may sometimes be obtained as a consequence of the nature of the substance under study. In particular, the alkaline cleavage does not occur to any appreciable degree if the amino or carboxyl groups of the carbohydrate-linked serine or threonine residues are free. This and other factors that influence the reaction have been discussed more fully elsewhere.[81]

Ideally, the nature of a carbohydrate–protein linkage should be established by proper isolation of the monosaccharide–amino acid linkage fragment. However, this is often a difficult task, and, since most of the connective tissue polysaccharides are bound to protein by alkali-labile linkages, the second approach offers great advantages in terms of speed and simplicity. It is thus preferable to use this method in the initial studies; if a rigorous identification of a linkage type is needed, the investigation should be complemented by isolation of the linkage fragment.

E. Characterization by Electrophoresis

Electrophoresis on cellulose acetate membranes provides a simple, rapid, and sensitive method of identifying most of the acid glycosaminoglycans of connective tissues. It is important to ensure that the sample has been thoroughly digested with proteolytic enzymes (e.g., papain) and freed of contaminating protein by fractionation with CPC or ethanol.

The general procedures of cellulose acetate electrophoresis are sufficiently well known[82] not to require description here, but the particular conditions which have been found to give satisfactory separations of glycosaminoglycans will be described.

A small spot (less than 1 μl) of the polysaccharide solution (0.5–2.0 mg/ml) is applied to the origin line of the cellulose acetate strip (Oxoid, 15 × 5 cm, or Sepraphore III, 2.5 × 15 cm), which has been soaked in the buffer (0.025 M pyridine formate, pH 3.0). Similar quantities of standard glycosaminoglycans are spotted on either side of the test sample. A potential of 17 V/cm is applied for 1 hour. To locate the separated glycosaminoglycans, the strip is immersed in a 1% solution of Alcian blue in 1% aqueous acetic acid for 5–10 minutes. (Alternatively, a 1% aqueous solution of acridine orange often is used.) Excess dye is readily cleared from the strip by subsequently washing in a large volume of 1% acetic acid. Under these conditions, the glycosaminoglycan of highest

[82] J. Kohn, *in* "Chromatographic and Electrophoretic Techniques" (I. Smith, ed.), Vol. 2, p. 84. Heinemann, London, 1968.

mobility, i.e., heparin, is seen as a discrete spot at 8–10 cm from the origin. Other glycosaminoglycans have mobilities in the order: chondroitin sulfate < heparan sulfate < dermatan sulfate < keratan sulfate < hyaluronic acid. Heparan sulfate, which contains fractions of different charge densities, may be seen as a broad band overlapping the chondroitin sulfate and dermatan sulfate spots.

Certain separations are improved by employing alternative buffer systems.[83,84] Electrophoresis in 0.1 M HCl[85] separates glycosaminoglycans strictly according to charge density of ester sulfate groups.

[83] V. Näntö, Acta Chem. Scand. **17**, 857 (1963).
[84] A. Gardair, J. Picard, and C. Tarasse, J. Chromatogr. **42**, 396 (1969).
[85] E. Wessler, Anal. Biochem. **41**, 67 (1971).

[8] Isolation and Characterization of Glycosphingolipids[1]

By WALTER J. ESSELMAN,[2] ROGER A. LAINE, and CHARLES C. SWEELEY

Isolation of Glycosphingolipids[3]

The basis for extraction of glycosphingolipids from biological sources is their solubility in chloroform–methanol mixtures. Gangliosides (glycosphingolipids containing neuraminic acids) and glycosphingolipids with five or more carbohydrate residues are not only soluble in chloroform–methanol mixtures but also easily form molecular aggregates that are "soluble" in water. Glycosphingolipids with one to four residues, on the other hand, form emulsions in water. This fact is the basis for the Folch partition[4] (chloroform–methanol (2:1) with one-fifth volume of water or dilute KCl solution) in which gangliosides are partitioned into the upper water–methanol layer and neutral glycosphingolipids remain in the lower chloroform–methanol layer. Most glycosphingolipids are easily extracted from tissue or other material with chloroform–methanol (2:1) but quantitative extraction of gangliosides requires more polar extraction mixtures such as chloroform–methanol (1:1) or chloroform–methanol

[1] From the Department of Biochemistry, Michigan State University, East Lansing, Michigan 48823. Supported in part by the following grants from the U.S. Public Health Service: AM-12434 and RR-00480. (See R. A. Laine, W. J. Esselman and C. C. Sweeley, this volume [10].)
[2] National Institutes of Health Postdoctoral Fellow (5 FO2 HE36835-02).
[3] Abbreviations: TMS, trimethylsilyl group; NAc, N-acetyl group; NANA, N-acetylneuraminic acid; gal, galactose; glc, glucose; GLC, gas–liquid chromatography; TLC, thin-layer chromatography.
[4] J. Folch, M. Lees, and G. H. Sloane Stanely, J. Biol. Chem. **226**, 497 (1957).

(1:2).[5] Metal ions also affect the distribution of gangliosides in biphasic systems, and these problems have been recently reviewed by Wiegandt.[6] Glycosphingolipids are separated from a total lipid extract by silicic acid column chromatography[7] followed by thin-layer chromatography.[8] Ion exchange cellulose (DEAE) column chromatography is used to separate acidic compounds, such as sulfatide and gangliosides, from less acidic or nonpolar compounds.[9]

Extraction and Partition

All solvents used in the following procedures are redistilled from glass to remove nonvolatile compounds. Chloroform is stabilized by the addition of methanol (after distillation) to a final concentration 0.25% (by volume). Chloroform–methanol and other mixed solvents are given as volume/volume ratios unless otherwise noted.

A weighed portion of the tissue to be extracted is vigorously homogenized with seven volumes of methanol (w/v) in a blender or homogenizer. Fourteen volumes of chloroform are added and the mixture homogenized again. The final solvent ratio is chloroform–methanol (2:1). The material is filtered with a Büchner funnel using an aspirator and a coarse-grade solvent-washed filter paper. The residue is reextracted with 10 volumes (based on weight of the original material) of chloroform–methanol (2:1). After filtration, the residue is extracted a third time with 5 volumes (v/v) of chloroform–methanol (1:1) or chloroform–methanol (1:2). The third extraction is only necessary for quantitative removal of gangliosides.[5]

The final combined extract is adjusted by addition of chloroform so that the final proportion is chloroform–methanol (2:1). A volume of 0.1 M KCl equivalent to one-fifth that of the final solvent extract is added, mixed vigorously, and allowed to stand at 4° overnight or until the layers are completely separated. If the volumes are small, the layers may be separated by centrifugation. The upper and lower layers are washed three times with theoretical lower and upper phases, respectively, prepared by shaking a mixture of 1 volume of chloroform–methanol (2:1) and 0.2 volume of 0.1 M KCl in water and letting the phases separate. When dealing with large volumes of combined extracts,

---

[5] L. Svennerholm, *J. Neurochem.* **10**, 613 (1963).
[6] H. Wiegandt, *Advan. Lipid Res.* **9**, 249 (1971).
[7] C. C. Sweeley, Vol. 14 [254].
[8] E. Stahl (ed.), "Thin-Layer Chromatography," 2nd ed., English translation. Springer-Verlag, Berlin and New York, 1969.
[9] G. Rouser, G. Kritchevsky, A. Yamamoto, G. Simon, C. Galli, and A. J. Bauman, Vol. 14 [272].

the solvents can be evaporated *in vacuo* and the residue redissolved in a convenient volume of chloroform–methanol (2:1) for the partition and washing steps described above.

The combined lower phases (original and washes) are collected and reduced *in vacuo* to a small volume with gentle warming (<50°) on a rotary evaporator (fraction I). The combined upper aqueous phases are dialyzed against cold tap water for 24 hours and then lyophilyzed (fraction II). The lyophilyzed material, usually containing some insoluble protein, is extracted with chloroform–methanol–water (10:5:1), filtered and reduced to a small volume on a rotary evaporator. Fraction II generally contains only gangliosides and may be analyzed by thin-layer chromatography without further purification (see below). The extraction and isolation procedures are summarized in Table I.

Silicic Acid Column Chromatography

The lipids from the chloroform–methanol layer (fraction I) are fractionated into neutral lipids, glycolipids, and phospholipids on a column of Unisil silicic acid (Clarkson Chemical Co., Williamsport, Pennsylvania).[10] This procedure is useful for glycolipids containing from one to four glycosyl residues and for sulfatides. Unisil (20–40 g per gram of lipid) is activated at 80° for several hours and is slurried with chloroform as quickly as possible after removal from the oven and poured into a column. The adsorbent is washed with about three bed volumes of chloroform or until it is translucent. The column must not be allowed to run dry. A 20 mg/ml solution of the sample is applied in chloroform, but this volume is not critical. Neutral lipids are eluted with about five bed volumes of chloroform and then glycosphingolipids are eluted with 8 to 10 bed volumes of acetone–methanol (9:1). Phospholipids are eluted with 5 bed volumes of methanol.

A different procedure has been used to isolate a particular glycolipid on a preparative scale, as illustrated in the purification of trihexosyl ceramide, gal(1 → 4)gal(1 → 4)glc-ceramide, from a kidney of a patient with Fabry's disease (trihexosylceramidosis). A crude glycolipid and phospholipid mixture was obtained from fraction I by addition of 200 ml of ether and filtration of the resultant glycolipid-phospholipid precipitate at room temperature. The glycolipid mixture (3 g in one experiment) was then subjected to mild alkali-catalyzed methanolysis (see below). A silicic acid column (400 g) was prepared in chloroform–methanol (19:1), and the sample was applied in chloroform–methanol (19:1). The column was eluted successively with 1500 ml each of 12%, 14%, 16%, 20%, 30%, and 50% methanol in chloroform. Fabry tri-

[10] D. Vance and C. C. Sweeley, *J. Lipid Res.* **8**, 621 (1967).

TABLE I
SUMMARY OF GLYCOSPHINGOLIPID ISOLATION PROCEDURE

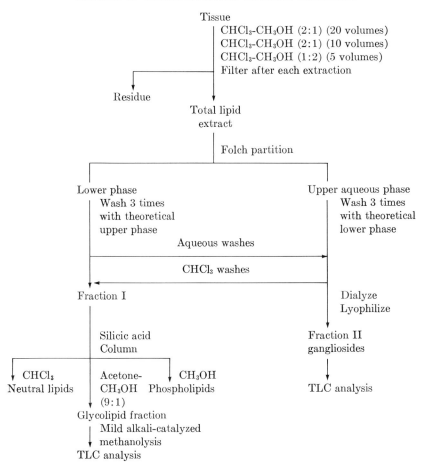

hexosylceramide (1.5 g) was eluted as a pure compound in the 20% fraction. The 16% and 30% fractions also contained some (1 g) of the Fabry lipid mixed with other glycolipids. A similar mixture of glyco-lipids was fractionated on a silicic acid column using a continuous gradient from 5% to 50% methanol in chloroform. A detailed description of the gradient maker and the results are given in an earlier volume of "Methods in Enzymology."[7]

Mild Alkali-Catalyzed Methanolysis

The glycolipid fraction from the silicic acid column is treated with mild base to remove contaminating phospholipids. This treatment does not affect glycolipids or gangliosides unless they contain an O-acyl group. The following quantities are used for 1–10 mg of glycolipid fraction. Add 1 ml of chloroform and 1 ml of 0.6 N NaOH in methanol to the dry fraction and allow the mixture to react at room temperature for 1 hour. Then add 1.2 ml of 0.5 N HCl in methanol, 1.7 ml of water, and 3.4 ml of chloroform, mix well, centrifuge, and remove the lower layer containing the glycolipids. Wash the lower layer three times with methanol:water (1:1) and then evaporate it to dryness *in vacuo*. If a ganglioside fraction is to be methanolyzed, the sample is treated in the same way except that after neutralization with methanolic HCl the sample is dried *in vacuo*, emulsified in water and dialyzed against tap water at 4° for 24 hours. The nondialyzable material is lyophilyzed and applied to TLC plates as described below.

Thin-Layer Chromatography

Glycosphingolipids are separated on thin-layer plates of silica gel G, H, or HR. The plates are prepared according to Stahl[8] and activated at 100°C for 2–4 hours. Plates of 0.25 mm thickness are used for general work and plates of 0.75 mm thickness are used for preparation of large quantities of material. Thin-layer tanks are lined with paper and equilibrated with solvent for 4 or more hours before use. Various commercial pre-prepared TLC plates (Quantum Industries, Fairfield, New Jersey, Brinkman Instruments Inc., Westbury, New York, and Analtek Inc., Wilmington, Delaware) have been used successfully for qualitative analysis of glycosphingolipids. Separation on these plates, however, is not usually as great as on plates made in the laboratory and contaminants are often obtained when silica gel is removed from pre-prepared plates and eluted with solvents.

A glycolipid mixture obtained from a column can be separated into various components on a silica gel H plate (0.25 mm) using a chloroform–methanol–water (100:42:6) solvent system (Fig. 1). Some hematoside, NANA$(2 \rightarrow 3)$gal$(1 \rightarrow 4)$glc-ceramide, (Fig. 1, lane A) is usually partitioned into the lower phase of a Folch wash and is separated from human or porcine globoside, galNAc$(1 \rightarrow 3)$gal$(1 \rightarrow 4)$gal$(1 \rightarrow 4)$glc-ceramide (Fig. 1, lane F) in this system. Monohexosyl ceramide, glc- and gal-ceramide, and dihexosyl ceramide, gal$(1 \rightarrow 4)$glc-ceramide and gal-$(1 \rightarrow 4)$gal-ceramide, often appear as two spots because of the presence of α-hydroxy fatty acids in the ceramide (Fig. 1, lane B). Otherwise the two forms of monohexosyl and dihexosylceramide are not separated on silica gel alone. Glucosylceramide and galactosylceramide, however, have been resolved on borate-impregnated thin-layer plates.[11] Sulfatide

[11] E. L. Kean, *J. Lipid Res.* **7**, 499 (1966).

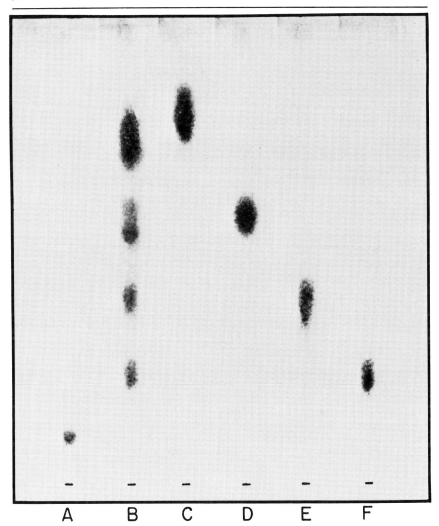

Fig. 1. Thin-layer chromatogram of hematoside, $NANA(2 \rightarrow 3)gal(1 \rightarrow 4)$ glc-ceramide (A); mixture of compounds in lane C through F (B); monohexosyl ceramide, gal- and glc-ceramide (C); dihexosyl ceramide, $gal(1 \rightarrow 4)$ glc-ceramide and $gal(1 \rightarrow 4)gal$-ceramide (D); trihexosyl ceramide, $gal(1 \rightarrow 4)gal(1 \rightarrow 4)glc$-ceramide (E); and tetrahexosyl ceramide $galNAc(1 \rightarrow 3)gal(1 \rightarrow 4)gal(1 \rightarrow 4)glc$-ceramide (F) on a silica gel H (0.25 mm) plate developed with chloroform–methanol–water (100:42:6) and visualized with α-naphthol spray.

(galactosylceramide sulfate, not shown) is not usually completely separated from dihexosylceramide, but these compounds can be completely separated by DEAE chromatography (see below). Gangliosides larger than hematoside remain very near the origin in this system.

Gangliosides and neutral glycosphingolipids with more than four glycosyl residues are separated by more polar solvent systems such as chloroform–methanol–water (60:45:10) or chloroform–methanol–2.5 N NH$_4$OH (65:45:9)[12] (Fig. 2). In the latter case, when gangliosides are

FIG. 2. Thin-layer chromatogram of trisialoganglioside, NANA (2 → 3)gal(1 → 3)galNAc(1 → 4)[NANA(2 → 8)NANA(2 → 3)]gal(1 → 4)glc-ceramide (A); disialo-ganglioside, NANA(2 → 3)gal(1 → 3)galNAc(1 → 4)[NANA(2 →3)]gal(1 → 4)glc-ceramide (B); and monosialoganglioside, galNAc(1 → 4)[NANA(2 → 3)]gal(1 → 4)glc-ceramide (C) on a silica gel G (0.25 mm) plate developed two times with chloroform–methanol–2.5 N NH$_4$OH (65:45:9) and visualized with α-naphthol spray.

[12] J. R. Wherrett and J. N. Cumings, *Biochem. J.* **86**, 378 (1963).

involved, the plate is developed two times with thorough drying (4 hours at room temperature) between developments. Hematoside is well separated from globoside on silica gel G plates with this system.[13]

Glycosphingolipids can be visualized with iodine vapor or by spraying with a 2% α-naphthol solution (in ethanol) followed by concentrated H_2SO_4 spray and heating for 10 minutes at 100°. The α-naphthol spray gives deep red-purple spots with carbohydrate-containing compounds and brown spots with phospholipids or neutral lipids. As little as 1–10 μg of material may be visualized in this way. Gangliosides are specifically visualized by spraying with the following solution[14]: mix 10 ml of 3% resorcinol (stored in refrigerator) with 80 ml of concentrated HCl, 0.25 ml of 0.1 M $CuSO_4$ and enough water to make 100 ml of solution. The sprayed plate is placed horizontally in a closed jar and heated in an oven at 125° for 20 minutes. Gangliosides appear as black or purple areas and other compounds appear as light brown areas.

Preparative thin-layer chromatography is carried out by streaking the sample on a 0.75 mm thick plate and developing as outlined above. Only the edges of the streak are visualized with I_2 or α-naphthol and areas containing neutral glycolipids are removed from the plate and the silica gel is eluted with chloroform–methanol–water (100:50:10). Gangliosides are eluted from silica gel with more polar solvents such as chloroform–methanol–water (50:50:15).

DEAE Column Chromatography

Water-soluble oligoglycosylceramides are separated from gangliosides from fraction II by the following procedure. Diethylaminoethyl cellulose (DEAE) in the acetate form is washed and columns are prepared exactly as described by Rouser et al.[9] The sample is applied in chloroform–methanol (7:3) and neutral glycolipids are eluted with 8 bed volumes each of chloroform–methanol (7:3) and (1:1). Gangliosides are retained on the column and may be eluted with 10 bed volumes of chloroform–methanol (2:1) saturated with aqueous 58% NH_4OH.

Dihexosylceramide and sulfatide isolated from a preparative TLC plate as described earlier may be separated on a DEAE column as described by Rouser et al.[9] The sample is applied in chloroform–methanol (9:1) and neutral dihexosylceramide is eluted with 10 bed volumes each of chloroform–methanol (9:1) and chloroform–methanol (7:3). The sulfatide is eluted with chloroform–methanol (4:1) made 10 mM with respect to ammonium acetate, to which is added 20 ml of 28% aqueous

[13] R. V. P. Tao and C. C. Sweeley, *Biochim. Biophys. Acta* **218**, 372 (1970).
[14] L. Svennerholm, *Biochim. Biophys. Acta* **24**, 604 (1957).

ammonia per liter.[9] Sulfate analysis of lipid fractions has been described previously.[15]

Florisil Column Chromatography

Hakamori has recently introduced an alternative method[16] of isolation of glycosphingolipids, which has been used successfully in our laboratory. By Florisil column chromatography of the peracetylated glycolipids, all the glycophingolipids (except polysialylgangliosides) may be isolated in one fraction. Briefly, the procedure consists of peracetylation of the total lipid extract with pyridine–acetic anhydride (3:2) (1 ml per 50 mg of dry total lipid). The pyridine and acetic anhydride are removed *in vacuo* with additions of toluene, and the products are applied to a Florisil column (40 g per gram of lipid), and neutral lipids and cholesterol are eluted with dichloroethane (8 bed volumes). Peracetylated glycosphingolipids are eluted with 8 bed volumes of dichloroethane–acetone (1:1), and phospholipids are eluted with 5 bed volumes of dichloroethane–methanol–water (2:8:1). Acetyl groups are removed from the glycolipids with 0.25% sodium methoxide in chloroform–methanol (1:1) (1 ml per 25 mg of lipid) at 25° for 30 minutes. The mixture is neutralized with acetic acid, emulsified in water and dialyzed overnight at 4°. The glycolipid fraction, analyzed by TLC, is free of contaminating phospholipids.

Characterization of Glycosphingolipids

The first step in the characterization of glycosphingolipids is the complete cleavage of the lipid into its component parts. This is carried out in our laboratory by methanolysis with 0.75 N methanolic HCl,[10] and this procedure has recently been studied in detail by Chambers and Clamp.[17] The products of methanolysis of a glycosphingolipid are sphingolipid bases and their O-methyl derivatives, fatty acid methyl esters, and methyl glycosides. These components can be separated by solvent extraction and analyzed by gas–liquid chromatography.

Methanolysis

A solution of a glycosphingolipid (up to 1 mg), isolated from columns or thin-layer chromatography plates, is evaporated to dryness in an 8-ml screw-capped culture tube fitted with a Teflon-lined cap. Three milliliters of 0.75 N methanolic HCl (prepared by bubbling gaseous HCl into methanol) is added to the sample, and the capped tube is heated at 80°

[15] J. C. Dittmer and M. A. Wells, Vol. 14 [482].
[16] T. Siato and S.-I. Hakomori, *J. Lipid Res.* 12, 257 (1971).
[17] R. E. Chambers and J. R. Clamp, *Biochem. J.* 125, 1009 (1971).

for 12–20 hours. At the end of this period, 0.05–0.3 μmole of mannitol (in methanol) is added as an internal standard. The sample is extracted three times with 1 ml hexane to remove fatty acid methyl esters. The hexane solution of methyl esters is retained for GLC analysis.

Approximately 100 mg of solid Ag_2CO_3 is added to each tube and carefully mixed until neutral. Methyl glycosides of amino sugars and neuraminyl methyl ester, and sphingosines are N-acetylated by addition of 1 ml of acetic anhydride. The remaining Ag_2CO_3 and AgCl act as catalyst for this reaction.[18] The mixture is allowed to react for 6–16 hours at room temperature, after which the sample is centrifuged and the precipitate is washed with methanol several times. About 0.25 ml of H_2O is added to decompose excess acetic anhydride, and the sample is evaporated under a stream of nitrogen. If N-acetylation is not performed, the neutralized sample is centrifuged, washed, and evaporated to dryness under nitrogen. The methanolysis procedure is summarized in Table II.

Trimethylsilylation and Gas–Liquid Chromatography of Methyl Glycosides

Dry samples of methyl glycosides are converted to trimethylsilyl (TMS) derivatives by addition of pyridine–hexamethyldisilazane–tri-

TABLE II
SUMMARY OF METHANOLYSIS PROCEDURE

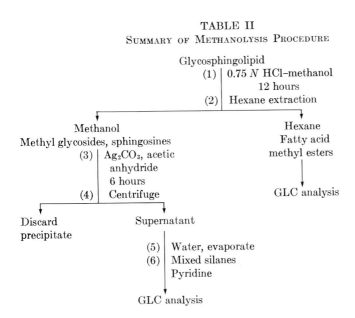

[18] J. R. Clamp, G. Dawson, and L. Hough, *Biochim. Biophys. Acta* **148**, 342 (1967).

methylchlorosilane (8:2:1) (about 50 μl for 500 μg of lipid). The mixture
is allowed to stand for 15 minutes at room temperature and an aliquot
is injected into the gas–liquid chromatograph. Other available reagents
for trimethylsilylation such as bis(trimethylsilyl)trifluoroacetamide are
not used because of partial silylation of N-acetyl groups. The mixed
silane solution is cloudy and may be used without centrifugation, but
exposure to water vapor must be avoided. If very small amounts of
sugars are present, the sample is evaporated under nitrogen and redissolved in a convenient solvent such as pyridine or CS_2.

An aliquot of the solution of TMS derivatives is injected into a gas–
liquid chromatographic column (2 m by 3 mm) of 3% SE-30 or OV-1 on
Supelcoport (80/100 mesh, Supelco Inc., Bellefonte, Pennsylvania) at
160° with nitrogen carrier gas (25 ml/minute). Programming from 150°
to 250° at 3° per minute is useful when sialic acids are present. A
chromatogram of a methanolyzed sample of globoside, galNAc($1 \rightarrow 3$)gal
($1 \rightarrow 4$)gal($1 \rightarrow 4$)glc-ceramide, is shown in Fig. 3. There are three peaks
for TMS methyl-D-galactoside (γ, α, and β forms); two peaks for methyl-
D-glucoside (α and β forms); and two major peaks for methyl-2-acet-
amido-2-deoxy-D-galactoside. Gas–liquid chromatography of methyl glycosides is described in detail elsewhere in this volume.[19] This method is

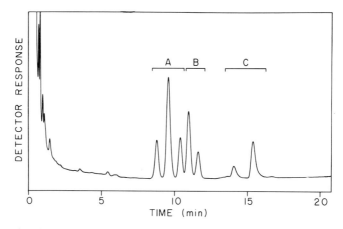

FIG. 3. Gas–liquid chromatography of trimethylsilyl methyl glycosides of D-galactose (A), D-glucose (B), and 2-acetamido-2-deoxy-D-galactose (C) derived from
globoside, galNAc($1 \rightarrow 3$)gal($1 \rightarrow 4$)gal($1 \rightarrow 4$)glc-ceramide and run on a 2 m ×
3 mm column of 3% SE-30 on 80/100 Supelcoport (Supelco Inc., Bellefonte, Penn-
sylvania) temperature programmed from 150° to 250° at 3° per minute. A. Hewlett-
Packard F & M 402 Gas Chromatograph, flash heater 250°, flame ionization detector
250° and a N_2 carrier gas flow rate of 25 ml/min was used.

[19] R. A. Laine, W. J. Esselman, and C. C. Sweeley, this volume [10].

suitable for identification of glycolipids containing fucose, glucose, galactose, galactosamine, glucosamine, and sialic acid. Mannose exhibits peaks overlapping with galactose and if these two sugars are present, the method employing alditol acetates is preferred.[20]

The ratios of glucose and galactose are determined without conversion factors by simply comparing the ratio of the total peak areas of each methyl glycoside. Since many glycosphingolipids contain only one glucose, ratios are usually expressed in relation to glucose and for globoside the ratio of galactose to glucose is 2. The ratio of galactosamine to glucose calculated in this way is usually about 0.65 for globoside. Methanolysis, N-acetylation, and trimethylsilylation must be carried out very carefully to obtain reproducible ratios for hexosamines. The mass ratio obtained for N-acetylneuraminic acid to glucose is usually 1.0 to 1.2, but these values should be compared with those obtained from known gangliosides treated in the same way. The absolute quantity of galactose and glucose are determined by comparison to the internal standard mannitol with the use of the following equation.[10]

$$\mu\text{moles glucose} = \frac{\text{area of glucose peaks}}{\text{area of mannitol peak}} \times$$

$$1.25 \times \mu\text{moles of mannitol added}$$

The mannitol peak (not shown in Fig. 3) falls between the second glucose peak and the first galactosamine peak and does not interfere with either compound. The area of peaks is calculated by triangulation.

Fatty Acids and Sphingosines

Normal fatty acids and α-hydroxy fatty acids are determined qualitatively and quantitatively by gas–liquid chromatography of the fatty acid methyl esters obtained from the hexane extract of the methanolyzate. Separation techniques, gas chromatographic columns and retention times are described elsewhere.[21-23] Sphingosines are determined by hydrolysis of the glycolipid with aqueous HCl followed by N-acetylation and GLC of the TMS derivatives.[24,25] A colorimetric assay[15] and a method involving GLC of aldehydes produced by $NaIO_4$ cleavage of sphingosine[26] are also available.

[20] J. S. Sawardeker, J. H. Sloneker, and A. Jeanes, *Anal. Chem.* **37**, 1602 (1965).
[21] Y. Kishimoto and N. S. Radin, *J. Lipid Res.* **4**, 130 (1963).
[22] K. Puro and A. Keranen, *Biochem. Biophys. Acta* **187**, 393 (1969).
[23] L. Svennerholm and S. Ställberg-Stenhagen, *J. Lipid Res.* **9**, 215 (1968).
[24] R. C. Gaver and C. C. Sweeley, *J. Amer. Chem. Soc.* **88**, 3643 (1966).
[25] H. E. Carter and R. C. Gaver, *J. Lipid Res.* **8**, 391 (1967).
[26] C. C. Sweeley and E. A. Moscatelli, *J. Lipid Res.* **1**, 40 (1959).

Enzymatic Degradation of Glycosphingolipids

The use of specific glycosidases is an important technique and has recently been used in this laboratory for sequence determination and anomeric analysis of glycolipids.[27] Glycosyl residues are released sequentially from globoside (cytolipin R reacts in the same way) by stepwise treatment with the following glycosidases; β-hexosaminidase from jack bean,[28] α-galactosidase from fig ficin,[29] and β-galactosidase from jack bean.[30] Reactions are carried out with 100 μg of lipid in 0.1 ml of 0.1 M sodium citrate buffer at pH 5, containing 100 μg of crude ox bile sodium taurocholate. After 18 hours at 37°, reaction mixtures are frozen and lyophilized. One milliliter of chloroform–methanol (2:1) is added and the mixture is sonicated for 5 minutes. After centrifugation, the supernatant fraction is dried, taken up in a small amount of chloroform–methanol (2:1) and spotted on a silica gel HR plate. The plate is developed in chloroform–methanol–water (100:42:6) and visualized with I_2 vapors or α-naphthol spray (see above). Products are identified by cochromatography with standards (Fig. 1) and by elution, methanolysis and GLC analysis.

Mass Spectrometry of TMS Glycosphingolipids

Mass spectrometry of intact TMS derivatives of glycolipids gives information about the sugar groups, the fatty acid and the sphingosine portion of glycosphingolipids.[31-33] Bis(trimethylsilyl)trifluoroacetamide (100 μl) and pyridine (50 μl) are added to 20–200 μg of the purified glycosphingolipid in a small capped vial and heated at 60° for about 30 minutes. An aliquot containing 10–20 μg of the TMS glycolipid is evaporated to dryness under nitrogen in a mass spectrometer direct probe tube. The samples are volatilized in the mass spectrometer ion source at temperatures ranging from 100° to 180° depending on the size of the oligosaccharide unit.

The following information can be obtained by comparison of the resulting mass spectra with those of reference samples: (1) whether the terminal residue is a hexose or hexosamine; (2) the number of and

[27] R. Laine, C. C. Sweeley, Y.-T. Li, M. Kisic, and M. M. Rapport, *J. Lipid Res.* in press.
[28] S.-C. Li and Y.-T. Li, *J. Biol. Chem.* **245,** 5153 (1970).
[29] S.-I. Hakomori, B. Siddiqui, Y.-T. Li, S.-C. Li, and C. G. Hellerqvist, *J. Biol. Chem.* **246,** 2271 (1971).
[30] Y.-T. Li and S.-C. Li, *J. Biol. Chem.* **243,** 3994 (1968).
[31] C. C. Sweeley and G. Dawson, *Biochem. Biophys. Res. Commun.* **37,** 6 (1969).
[32] G. Dawson and C. C. Sweeley, *J. Lipid Res.* **12,** 56 (1971).
[33] K. Samuelsson and B. Samuelsson, *Biochem. Biophys. Res. Commun.* **37,** 15 (1969).

nature of N-acetylneuraminic acid groups (i.e., terminal or branched); (3) whether N-acetyl and/or N-glycolylneuraminate is present; (4) information regarding the number of glycosyl residues present and the fatty acid and sphingosine composition. It is essential, because of the limitations of this technique (e.g., the inability to distinguish between hexoses), that it be used in conjunction with other techniques, such as permethylation analyses, and studies with specific glycosidases.

Ozonolysis of Glycosphingolipids

The carbohydrate portion of glycosphingolipids is cleaved from the lipid portion by the method of Wiegandt.[34] The glucose-sphingosine linkage is broken but there is no hydrolysis of other glycosidic linkages, including those of sialic acid residues. The glycolipid (100 mg) is ozonized in 50 ml of methanol at room temperature. Ozone consumption is monitored by bubbling the effluent gas through a KI-starch solution which blackens when excess ozone is present. The solution is dried *in vacuo* and the compound is hydrolyzed with 10 ml of $0.2 M$ Na_2CO_3 for 12 hours at 20°. Sodium ion is removed by stirring with Dowex 50 (H^+) and the resin is filtered. After a Folch partition, the upper aqueous phase is lyophilized and the resultant oligosaccharides (about 80% yield) are stored in a desiccator. The procedure can be changed for microscale operation (1 mg of lipid).

Permethylation of Glycosphingolipids

Permethylation, hydrolysis, and gas–liquid chromatography of glycosphingolipids is used to determine linkage of glycosyl residues. Permethylation is carried out in our laboratory[27] according to the method of Hakomori using methyl sulfinyl carbanion.[35] Sodium hydride (0.88 g of 57% in oil) is washed six times under nitrogen with dry hexane, drained thoroughly, and stirred with dimethyl sulfoxide (10 ml) under a stream of nitrogen at 70° for 3 hours or until bubbling ceases and the solution turns a dark clear green. Any dark precipitate is removed at this point by centrifugation.

The carbanion solution (about 0.5 ml) is added under a stream of nitrogen to the glycolipid sample in 0.5 ml of dimethyl sulfoxide in a screw-capped vial, and the mixture is sonicated briefly. After standing at room temperature for 2–6 hours, 1.5 ml of CH_3I is carefully added dropwise under nitrogen, and the mixture is allowed to react for 1 hour. After this step, it is not necessary to keep the reaction dry. The permethylated glycolipids are extracted into chloroform, the chloroform

[34] H. Wiegandt and G. Bashang, *Z. Naturforsch.* **B20**, 164 (1965).

[35] S.-I. Hakomori, *J. Biochem.* **55**, 205 (1964).

layer is washed once with 1% $Na_2S_2O_3$ to remove I_2, and four times with water. The chloroform fraction is mixed with absolute ethanol and is evaporated under nitrogen. Two milliliters of 1 N H_2SO_4 is added, and after heating at 105° for 12 hours the hydrolyzate is neutralized with $BaCO_3$, diluted, and filtered on Celite and filter paper, washing the Celite twice with water (5 ml). The sample is concentrated to 5 ml and percolated onto a small Dowex 50 H^+ column. Neutral sugars are eluted with water and methanol–water (1:3) (10 ml each) and amino sugars are eluted with 0.3 N NH_4OH. The resulting partially methylated sugars are reduced with $NaBH_4$, the products are acetylated and gas–liquid chromatography is carried out as described elsewhere in this volume.[36]

Partial Hydrolysis of Glycosphingolipids

The presence of N-acetylneuraminic acid or N-glycolylneuraminic acid in a ganglioside sample may be determined by mild acid hydrolysis of the neuraminic acid followed by TLC or GLC analysis.[37] The N-acylneuraminic acids are released from gangliosides (0.1–0.2 mg) with 1 ml of 0.05 M HCl (aqueous HCl) at 80° for 1 hour. The solution is extracted with chloroform and the aqueous phase is percolated through about 2 g of Dowex 1 (acetate form) in a small column. The column is washed with 8 ml of water, and the neuraminic acids are eluted with 10 ml of 1 M formic acid. The sample is analyzed on silica gel G plates using n-propanol–water–concentrated ammonia (6:2:1).[38] The R_f of N-acetylneuraminic acid is 0.41 and that of N-glycolylneuraminic acid is 0.28.[13] This sample may also be derivatized with bis(trimethylsilyl) trifloroacetamide and analyzed by GLC.[19]

Neuraminic acids are usually analyzed by GLC as their TMS-methyl ketoside methyl ester derivatives. A ganglioside sample (0.1 mg) or mixture of glycolipids is methanolyzed with 2 ml of 0.05 M methanolic HCl (1 part 12 N HCl and 240 parts methanol) at 80° for 1 hour.[39] The cooled solution is extracted three times with 3 ml hexane and the methanolic layer is evaporated to dryness under nitrogen. Pyridine–hexamethyldisilazane–trimethylchlorosilane (8:4:2) (50 μl) is added and the TMS derivatives are analyzed on a 3% OV-1 column (2 m × 2 mm) at 205°.[39]

The sulfate moiety of sulfatide (up to 2 mg) is released by reaction with 2 ml of 0.05 N methanolic HCl for 4 hours at room tempera-

[36] B. Lindberg, this volume [12].
[37] K. Puro, *Biochim. Biophys. Acta* 189, 401 (1969).
[38] E. Cranzer, *Hoppe-Seyler's Z. Physiol. Chem.* 328, 277 (1962).
[39] R. K. Yu and R. W. Ledeen, *J. Lipid Res.* 11, 506 (1970).

ture.[40,41] The reaction mixture is neutralized with aqueous NaOH and the glycolipid product is extracted by Folch partition with 4 ml of chloroform. The chloroform layer is washed with 2 ml of methanol–water (1:1) and is analyzed by TLC as described above.

The carbohydrate sequence of glycosphingolipids can be determined by partial degradation under mild acidic conditions. The glycolipid (1 mg) is hydrolyzed with 1 ml of 0.3 N HCl in chloroform–methanol (2:1) at 60° for various times up to 2 hours.[42] After Folch partition (with addition of 0.2 volume of water) the lower phase is dried $in\ vacuo$ and the upper phase is deionized with Amberlite CG-4B resin (OH form). The glycolipids (from lower phase) are purified by TLC and analyzed by GLC after complete hydrolysis. The upper phase is analyzed for sugars (and polysaccharides) by GLC.[19] Trihexosyl ceramide, gal(1 → 4)gal (1 → 4)glc ceramide, hydrolyzed for 2 hours in this way[42] did not yield ceramide but yielded cerebroside containing only glucose, and a dihexosyl ceramide containing glucose and galactose in a 1:1 molar ratio. The water-soluble fraction contained galactose and a disaccharide but no glucose. These data provide evidence for a linear arrangement of the hexose units as given above.

Characterization of the oligosaccharide moieties of glycolipids by partial degradation has also been carried out by treatment of glycolipids with 0.1 N aqueous HCl for 30 minutes to 3 hours,[43-45] and by treatment of gangliosides with 0.01 N H$_2$SO$_4$ at 85° for 1 hour and 0.1 N H$_2$SO$_4$ at 100° for 1 hour.[46]

Mixed Molecular Species of Glycosphingolipids

The purification and characterization of glycosphingolipids is complicated by the fact that naturally occurring glycosphingolipids which are homogeneous in the carbohydrate portion are heterogeneous in the sphingosine and fatty acid portions. Complete hydrolysis and subsequent analyses of the fatty acids and long-chain bases provide information about the extent of this heterogeneity. Additional heterogeneity may result when there are glycosphingolipids with identical composition but which differ in glycosidic linkage and/or anomeric configuration of one

[40] E. Mårtensson, $Biochim.\ Biophys.\ Acta$ **116**, 521 (1966).
[41] P. Stoffyn and A. Stoffyn, $Biochim.\ Biophys.\ Acta$ **70**, 218 (1963).
[42] C. C. Sweeley and B. Klionsky, $J.\ Biol.\ Chem.$ **238**, 3148 (1963).
[43] J. Kawanami, $J.\ Biochem.\ (Tokyo)$ **64**, 625 (1968).
[44] J. Kawanami, $J.\ Biochem.\ (Tokyo)$ **62**, 105 (1967).
[45] T. Yamakawa, S. Yokoyama, and N. Handa, $J.\ Biochem.\ (Tokyo)$ **53**, 28 (1963).
[46] R. Kuhn and H. Wiegandt, $Chem.\ Ber.$ **96**, 866 (1963).

or more glycosidic bonds. Such mixtures probably cannot be separated by TLC alone. Whether or not products which are homogeneous by TLC actually contain such mixtures can only be determined with techniques such as permethylation and sequential enzymatic degradation.

[9] Release of Oligosaccharides from Glycolipids

By SEN-ITIROH HAKOMORI and BADER SIDDIQUI

The original method[1] has been modified in the removal of osmic acid and in the detection of oligosaccharides released.

Reagents

Pyridine dried over barium oxide and distilled
Acetic anhydride, reagent grade
Dioxane, analytical grade
Sodium metaperiodate, $0.2\,M$, dissolved in 70% methanol
Osmium tetroxide (O_sO_4), 0.5%, solution in diethylether
Florisil (magnesia–silica gel, Floridin Co., Tallahassee, Florida, distributed by Fischer Scientific Co.)
Organic solvents: chloroform, methanol, 1,2-dichloroethane (all reagent grade)
Solution of sodium borohydride (^{3}H-labeled) and sodium borohydride (cold): 1.9 mg of sodium borotritium (NaB[3]H$_4$, 5 mCi) and 17.1 mg of sodium borohydride (NaBH$_4$) dissolved in 2.5 ml of 0.01 N sodium hydroxide.[2]

Procedure of Degradation

Glycosphingolipid (0.2–1 mg) was dissolved to 0.3 ml of pyridine and 0.2 ml of acetic anhydride. The solution was allowed to stand overnight and then evaporated under nitrogen after addition of 2 ml of toluene. The residue was dissolved in 0.4 ml of dioxane, and 0.1 ml of $0.2\,M$ sodium metaperiodate in 70% methanol and 10 µl of 0.5% osmium tetroxide in diethylether were added to the solution. The mixture was kept in the cold (4°) for 18 hours (overnight). One drop of glycerol, 6 ml of chloroform–methanol 2:1, and 1.5 ml of water were added, then the mixture was shaken vigorously. The upper layer was discarded, and the lower phase was shaken with 2.5 ml of the mixture of chloroform–methanol–water (1:10:10 v/v). Again the upper phase was

[1] S. Hakomori, *J. Lipid Res.* **7**, 789 (1966).
[2] M. Murakami and R. J. Winzler, *J. Chromatog.* **28**, 344 (1967).

discarded and the lower phase was dried under nitrogen. The residue was dissolved in 0.2 ml of 1,2-dichloroethane and chromatographed on a Florisil column (0.6 × 6 cm) prepared in 1,2-dichloroethane. The column was washed with 5 ml of 1,2-dichloroethane, then the oxidized glycolipid acetate was eluted with 6 ml of 1,2-dichlorethane–methanol (9:1), collected and evaporated under nitrogen to a small bulk and transferred to a conical test tube and further evaporated to dryness. The dried residue was dissolved in 0.2 ml of methanol, and 0.05 ml of 0.5% sodium methoxide was added. Degradation and simultaneous deacetylation were completed within 30 minutes at room temperature (reaction mixture A).

Separation of Oligosaccharides and Detection on Paper or Thin-Layer Chromatography

Direct Chromatography of Oligosaccharides and Test for Indirect Ehrlich Reaction (Morgan–Elson Test). The "reaction mixture A" was neutralized with 10% aqueous acetic acid, the bulky precipitate was centrifuged, and the supernatant was applied on paper (Whatman 3 MM) and developed with either of the following solvents with reference oligosaccharides: ethyl acetate–pyridine–acetic acid–water (5:5:1:3), ethyl acetate–pyridine–water (12:5:4), or methyl ethyl ketone–acetic acid–water (7:1:3). The spots were visualized with benzidine–trichloroacetic acid[3] or with silver nitrate–ethanolic sodium hydroxide.[4] The Morgan–Elson test can be also applied directly on paper[5] or after elution from paper.[6] The result of this test has great value for the differentiation of linkages of amino sugars[7] in oligosaccharides. Without the release of oligosaccharide from ceramide, this reaction cannot be performed. The reaction based on reducing properties of carbohydrates were much lower than that of monosaccharides and required at least 50 μg of oligosaccharide per spot.

Detection of Oligosaccharides by Radiochromatography after Reduction with Sodium Borohydride (³H). If a limited amount of glycolipid

[3] J. S. D. Bacon and J. Edelmann, *Biochem. J.* **48,** 114 (1951).

[4] E. F. L. J. Anet and T. M. Reynolds, *Nature (London)* **174,** 930 (1954).

[5] The paper strip was sprayed with 0.05 M sodium tetraborate in 50% ethanol (fully moistened but not flooded), and placed in a "steam rice cooker" (Toshiba or Hitachi) for 5 minutes followed by spraying with the Ehrlich reagent [a modified method of M. R. J. Salton, *Biochim. Biophys. Acta* **34,** 308 (1959)].

[6] Test solution in water, 100 μl, was mixed with 20 μl of 0.7 N sodium carbonate heated for 4 minutes at 100°, followed by addition of 700 μl of glacial acetic acid and 100 μl of the Ehrlich reagent; the color yield was read after 2 hours (a modified microadaptation of Morgan-Elson reaction).

[7] H. H. Baer, *in* "The Amino Sugars" (R. W. Jeanloz and E. A. Balazs, eds.), Vol. IA, p. 296. Academic Press, New York, 1969.

is available, the following procedure is recommended. The "reaction mixture A" is mixed with 0.1 ml of 0.01 M sodium hydroxide, 50 μl of sodium borohydride (^{3}H), and sodium borohydride (cold) solution prepared as indicated under Reagents, which contain 10 μmoles of sodium borohydride and 100 μCi radioactivity. The mixture was allowed to stand overnight at room temperature; 1 ml of chloroform was added and then shaken. The chloroform layer was discarded. A drop of glacial acetic acid and 5 ml of methanol were added to the aqueous layer, which was then evaporated under nitrogen to dryness.[7a] The dried residue was dissolved in 50 μl of water and 20-μl aliquots were applied on paper or on a thin-layer plate. After development with a suitable solvent (see preceding section), the chromatogram was scanned with a radiochromatography apparatus (Packard Instrument Co.). The radioactivity of the separated oligosaccharides and the amount of the glycolipids subjected to the degradation showed a stoichiometric relationship.

Comments

The principle of this degradation reaction is the same as that described by Wiegandt and Baschang with ozonolysis followed by alkaline degradation.[8,9] A possible reaction mechanism via the Lobry de Bruyn–van Eckenstein transformation and β-elimination of glycosides has been discussed.[1,9] In the modified method the acetylation of the C-3 hydroxyl group of sphingosine and any hydroxyl group of carbohydrates was performed prior to the formation of an aldehyde group at C-4 of the sphingosine; therefore, the aldehyde group of the intermediate product (I) is stable unless the O-acetyl group adjacent to the carbonyl is eliminated.

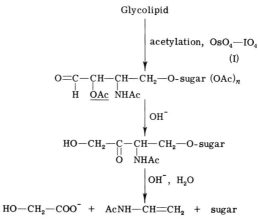

[7a] *Note added in proof:* Better separation of oligosaccharides was obtained when

The degradation is easily controlled and can be applied to a few milligrams of glycolipid, or to as little as 100 μg of glycolipid when combined with tritium labeling. The method is not applicable to the release of oligosaccharides from glycolipid containing an O-acetyl or O-acyl group.

---

the residue was desalted before being subjected to chromatography. The aqueous solution was passed through a column of 200 mg of mixed bed residue AG501x8, washed with 1 ml water, and the eluate and washings were lyophilized.

[8] H. Wiegandt and G. Baschang, Z. Naturforsch. **B20**, 164 (1965).
[9] H. Wiegandt, Angew Chem. Int. Ed. **7**, 87 (1968).

[10] Gas–Liquid Chromatography of Carbohydrates[1]

By ROGER A. LAINE, WALTER J. ESSELMAN, and CHARLES C. SWEELEY

Principle

Gas–liquid chromatography (GLC) can be routinely applied to the analysis of carbohydrates by the use of volatile derivatives, giving both information on identity (by relative retention time as compared to known standards) and quantitation (by integration of the peak areas for each sugar). The speed at which analyses can be performed and the ease of preparation of derivatives offer definite advantages over other methods of saccharide analysis.

Trimethylsilyl (TMS) derivatives[2] and alditol acetates[3] are the most widely used volatile forms of sugars, and will be the only ones considered here. Trimethylsilyl derivatives of monosaccharides are sometimes more conveniently analyzed as the methyl glycosides,[4,5] while oligosaccharides can be analyzed as the TMS alditols.[6] Selection of the derivative depends on the mixture to be resolved, and on whether the sugars are derived from a hydrolysis or a methanolysis procedure. In the latter case, or when the product needs to be recovered, such as in preparative work, the TMS derivative is used because of ease of preparation and subsequent removal

[1] This article is based on work supported in part by the Grants AM-12434 and RR-00480 from the U.S. Public Health Service.
[2] C. C. Sweeley, R. Bentley, M. Makita, and W. W. Wells, J. Amer. Chem. Soc. **85**, 2497 (1963).
[3] J. S. Sawardeker, J. H. Sloneker, and A. Jeanes, Anal. Chem. **37**, 1602 (1965).
[4] C. C. Sweeley and B. Walker, Anal. Chem. **36**, 1461 (1964).
[5] D. C. DeJongh, T. Radford, J. D. Hribar, S. Hanessian, M. Bieber, G. Dawson, and C. C. Sweeley, J. Amer. Chem. Soc. **91**, 1728 (1969).
[6] J. Kärkkäinen, Carbohyd. Res. **11**, 247 (1969).

of TMS functions.[7] Several peaks for each saccharide are generally obtained because of solvent equilibrium mixtures of pyranosidic, furanosidic, and anomeric forms of the methyl glycosides.[4-6] If a mixture of free sugars needs to be analyzed, such as that derived from a hydrolysis procedure, the alditol acetates may be more convenient because single peaks are observed for each saccharide.

Trimethylsilylating Reagents

Reagent 1

Pyridine (redistilled, stored over KOH), 10 ml
Hexamethyldisilazane (commercial reagent), 4 ml
Trimethylchlorosilane (commercial reagent), 2 ml

Solutions are added to a 20-ml screw-capped test tube with a Teflon-lined cap, mixed, and centrifuged. The supernatant solution is usable for at least a week, provided moisture is excluded.

Reagent 2

Bis-trimethylsilyltrifluoroacetamide (BSTFA). Available commercially neat, or as a mixture with 1% trimethylchlorosilane.
Dimethyl formamide, spectral grade: Add 50 μl of dimethyl formamide to a sample of 2 mg of sugars or less, then add 50 μl of BSTFA.

BSTFA is best handled in 5-ml vials with septa, using a 100 μl syringe for transfer. The reactions are easily manipulated in 1-2-dram vials with Teflon-lined caps.

Preparation of TMS Derivatives

For a sample containing 2-2000 μg of sugar, 100 μl of reagent 1 or 2 is added and the mixture is allowed to react at room temperature for 15-30 minutes. If crystals of sugars are used, or for oligosaccharides, heating at 80° may be required. Solution 1 will not silylate amino nitrogens at room temperature, whereas solution 2 will replace amino hydrogens, especially when warmed.

When a minimum solvent front is required, samples may be dried under a stream of N_2 after derivatization and redissolved in a suitable amount of CS_2 for injection.

---

[7] C. C. Sweeley, W. W. Wells, and R. Bentley, Vol. 8 [7].

Preparation of Acetate Derivatives

Fifty microliters each of acetic anhydride and dry pyridine are added to 1 mg of sample or less. The reaction is carried out at 80° for 1 hour in a sealed tube or vial with Teflon-lined cap. Redistilled toluene (2 ml) is added and the mixture is dried under a stream of nitrogen at 50°. A suitable amount of chloroform is added to redissolve the sample for injection.

Reduction of Mono- and Oligosaccharides

Sodium borohydride (10 mg) is added to a sample of the sugar in 1 ml of water. If amino sugars are present, the reaction should be carried out at 0°, otherwise room temperature is satisfactory. After 2 hours, Dowex 50-X8, H⁺ form, is added batchwise until an acid pH is reached (using pH paper). The solution is filtered from the ion-exchange resin, which is washed with water. Amino sugars are eluted from the resin with 0.3 N HCl. Solutions are taken to dryness *in vacuo* at 80°, and the boric acid is removed by 3 successive additions and evaporations of 20 ml of methanol.

Methanolysis Procedures and Preparation of Methyl
  Glycoside Standards

Complete release of sugars from many biological materials is effected by methanolysis with 1–2 N anhydrous methanolic HCl at 85–100° in 3 hours.[8] Negligible losses are shown after 24 hours at tnese conditions.[8] To a sample containing 1–3 mg of saccharides, 3 ml of 1 N anhydrous methanolic HCl is added, and the sample is heated at 85° for 3 hours. Powdered silver carbonate is added to neutralize, and if amino sugars are present (or sialic acid), 0.3 ml of acetic anhydride is added to the tubes containing the methanol and silver chloride. Re-N-acetylation is allowed to take place at room temperature for 6 hours with occasional mixing. The sample is centrifuged, and the supernatant is removed to another tube and taken to dryness under a stream of nitrogen. Authentic sugars are treated by the same procedure. The N-acetyl forms of amino sugars should be used for standards in methanolysis procedure to obtain the same ratios of glycosidic peaks as obtained for N-acetyl sugars from biological samples. Different relative peak areas are obtained for anomers when the amino sugar hydrochlorides are used as standards, as

---

[8] R. E. Chambers and J. R. Clamp, *Biochem. J.* **125**, 1009 (1971).

compared with equilibrium mixtures of anomers using the N-acetyl derivatives.[9]

If an internal standard is required for quantitation, mannitol may be added prior to the methanolysis procedure or to the sugar mixture. Mannitol elutes as the hexatrimethylsilyl derivative between glucose and the hexosamines. Care should be taken in the preparation of suitable standards in molar ratios for each sugar to determine the particular column losses and detector response with a given instrument, especially with amino-containing sugars. In our experience, the detector mass response for N-acetyl hexosamines is 0.6 to 0.7 when compared with the response for glucose. Charred residues in the first part of the column may reduce this further.[10] Sialic acid usually gives a mass response of 1.2 vs. an expected 1.6 when compared with that of glucose.

Column Packings for Gas–Liquid Chromatography

Commercially prepared packings for GLC columns give excellent results. Although the cost is somewhat higher than laboratory preparation of coated supports, the uniform quality of the available packings provides reproducible results at a great saving of time. The liquid phases SE-30, OV-1, and OV-101 give comparable results for trimethylsilyl derivatives of sugars as a 3% coating on 80/100- or 100/120-mesh silanized diatomaceous supports, such as Gas-Chrom Q, Supelcoport, Chromosorb G (AW-DCMS) or comparable products. Column dimensions of 2 m × 3 mm are generally used. For oligosaccharides, 0.05% OV-101 on 100/140-mesh DC 110 textured glass beads is used on columns of dimensions 0.6 m × 3 mm, but 3% packings of OV-1, SE-30, and Dexil-300 GC also give good results. Alditol acetates are chromatographed on 3% ECNSS-M on 100/120-mesh silanized diatomaceous supports. Fluorosilicones such as OV-210 and SP-2401 also give good separations for alditol acetates. Nitrogen is generally used as the carrier gas, although helium gives better resolution for the alditol acetates on ECNSS-M and for the oligosaccharides on OV-101. Coiled columns are packed with vibration with an aspirator connected to the outlet end. U-shaped columns are packed by vertical tapping, adding a few inches at a time to each arm. Silanized glass wool is used to contain the packing at each end of the column. Glass columns are washed with chromic acid, then water, followed by methanol and chloroform. Dimethyldichlorosilane (5% in hexane, 50 ml) is then aspirated through the column, and dry nitrogen is used to remove the last traces of solvent before packing materials are introduced.

[9] R. Laine and C. C. Sweeley, unpublished results.
[10] G. Dawson, private communication.

*Gas–Liquid Chromatography of Methyl Glycosides as
Trimethylsilyl Derivatives*

The lower tracing in Fig. 1 shows the separation (as trimethylsilyl methyl glycosides) of L-fucose (A), D-galactose (B), D-glucose (C), 2-acetamido-2-deoxy-D-glucose (D), and *N*-acetylneuraminic acid (E) on 3% SE-30. The upper record in Fig. 1 demonstrates the comparable retention behavior and peak patterns for D-mannose (F), and 2-acetamido-2-deoxy-D-galactose (G). Thus all saccharides in this group can be separated except for D-mannose and D-galactose, which show some peak overlap. The equilibrium mixture of anomeric and ring forms of each sugar, represented by its pattern of peaks, is useful for recognizing individual sugars in addition to their retention times. This is especially helpful in programmed runs where retention behavior may not be

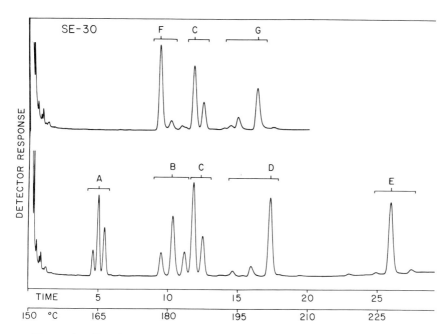

Fig. 1. Gas–liquid chromatography of the trimethylsilyl methyl glycosides of L-fucose (A), D-galactose (B), D-glucose (C), 2-acetamido-2-deoxy-D-glucose (D), *N*-acetylneuraminic acid (E), D-mannose (F), and 2-acetamido-2-deoxy-D-galactose (G) on a 2 m × 3 mm column of 3% SE-30 on 100/120 Supelcoport (Supelco, Inc.) temperature programmed from 150° to 250° at 3° per minute. An F & M Hewlett-Packard 402 gas chromatograph with flame ionization detectors was used. Flash heater and detector temperatures were maintained at 280°, and the carrier gas was nitrogen at a flow rate of 35 ml per minute.

strictly duplicable. Isothermal chromatography is necessary for exact retention time data. Table I gives the relative retention times ($R_{xylitol}$) for the TMS derivatives of free sugars (from aqueous equilibrium mixtures of anomers) and of the TMS derivatives of methyl glycosides.

Separation of N-Acetyl- and N-Glycolylneuraminic Acid as Trimethylsilyl Ester Glycosides

Figure 2 demonstrates isothermal separation of N-acetylneuraminic acid (A, B) and N-glycolylneuraminic acid (C) on 3% SE-30. N-Acetyl-neuraminic acid exhibits 2 peaks in this procedure which are completely separated from the N-glycolyl peak.

Separation of Oligosaccharides as Trimethylsilyl Derivatives of Their Reduced Forms

Using a fast programming rate (10° per minute), resolution of mono-to pentasaccharides can be easily accomplished as trimethylsilyl deriva-tives after reduction, as shown in Fig. 3.[9] TMS derivatives of reduced D-glucose (A), maltose (B), maltotriose (C), maltotetraose (D), and cellopentaose (E) are shown separated on OV-101 in 25 minutes. Re-duction eliminates separate peaks for the anomers of oligosaccharides

TABLE I

RELATIVE RETENTION TIMES ($R_{xylitol}$) FOR TRIMETHYLSILYL (TMS) DERIVATIVES OF A SERIES OF FREE SUGARS (IN AQUEOUS EQUILIBRIUM MIXTURE) AND TMS DERIVATIVES OF METHYL GLYCOSIDES ON 3% SE-30 AT 165°

Component	Relative retention times					
	TMS free sugar			TMS methyl glycosides		
D-Xylitol	1.00			1.00		
D-Erythrose	0.22[a]	0.25[b]	0.30			
2-Deoxy-D-ribose	0.29[b]	0.34				
D-Ribose	0.72[a]	0.78				
D-Arabinose	0.62	0.70	0.80[b]			
L-Fucose	0.68[b]	0.80	0.95[a]	0.50	0.56[a]	0.63
D-Xylose	0.93	1.20				
D-Fructose	1.51	1.62[a]	2.22			
D-Mannose	1.55[a]	2.42		1.31[a]	1.51	
D-Galactose	1.68[b]	1.97	2.42[a]	1.34	1.57[a]	1.81
D-Glucose	2.22	3.60		2.00[a]	2.24	
2-Deoxy-2-acetamido-D-galactose	3.89[b]	4.13[b]	4.84[a]	2.98[b]	3.43	4.20[a]
2-Deoxy-2-acetamido-D-glucose	5.38			3.15[b]	3.91	4.80

[a] Largest peak in group.
[b] Minor component.

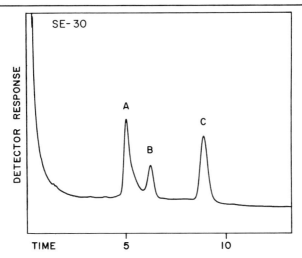

FIG. 2. Gas–liquid chromatography of the trimethylsilyl derivatives of *N*-acetyl-neuraminic acid (A, B) and *N*-glycolylneuraminic acid (C) on a 2 m × 3 mm column of 3% SE-30 on 100/120 mesh Supelcoport. Column temperature was maintained at 224° isothermally. Other instrument conditions were the same as described in the legend for Fig. 1.

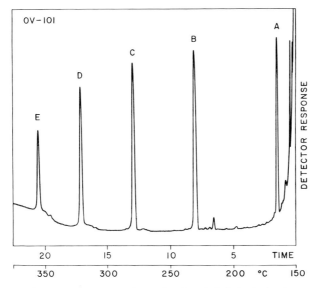

FIG. 3. Gas–liquid chromatography of trimethylsilyl derivatives of reduced D-glucose (A), maltose (B), maltotriose (C), maltotetraose (D), and cellopentaose (E) on a 0.6 m × 3 mm column of 0.05% OV-101 on 120/140-mesh DC 110 (textured glass beads) programmed from 150° to 350° at 10° per minute. Chromatography was performed on a Perkin-Elmer Model 900 instrument with flame ionization detectors. Flash heater and detector temperatures were 350°, and the carrier gas was helium at a flow rate of 40 ml per minute.

when a reducing end is present. For preparative work, where recovery of the intact oligosaccharides is necessary, the nonreduced forms can be used.

Gas–Liquid Chromatography of Alditol Acetates

Excellent separations can be obtained for the reduced forms of aldoses as the peracetyl derivatives on liquid phases such as ECNSS-M, OV-210, and SP-2401. Each alditol component is resolved as a single peak, and the common pentoses and hexoses can be separated easily in a single run.[3] Figure 4 illustrates the separation of mannitol (A), galactitol (B), and glucitol (C) on 3% ECNSS-M in 35 minutes. OV-210 separates glucitol and galactitol, but not galactitol and mannitol. Table II lists relative retention times for a series of alditols on ECNSS-M. Weight-to-area ratios approaching unity are achieved with this derivative. A slight disadvantage is that the isomeric pairs D-glucose and D-gulose, D-altrose and D-talose, and D-lyxose and D-arabinose give identical DL-isomers when reduced, and thus cannot be resolved. Mixtures of these pairs of sugars, however, are extremely rare in nature.

A very useful adaptation of the above method was developed by

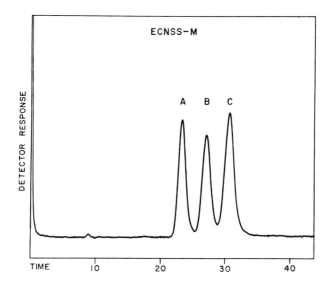

Fig. 4. Gas–liquid chromatography of D-mannitol (A), D-galactitol (B), and D-glucitol as hexaacetate derivatives on a 2 m × 3 mm column of 3% ECNSS-M on 100/120-mesh Supelcoport maintained isothermally at 200°. Instrumental conditions on a Hewlett-Packard F & M Model 402 gas chromatograph were as follows: flash heater, 250°; flame ionization detector, 250°; carrier gas helium at 35 ml per minute.

TABLE II

RELATIVE RETENTION TIMES OF ALDITOL ACETATES[a,b]

Alditol	Relative retention time $(R_{mannitol})$ 190°
Glycerol	0.021
Erythritol	0.102
L-Rhamnitol	0.218
L-Fucitol	0.242
Ribitol	0.340
L-Arabinitol	0.382
Lyxitol	0.382
Xylitol	0.530
D-Mannitol	1.000
Galactitol	1.160
Glucitol	1.340

[a] Adapted from J. S. Sawardeker, J. H. Sloneker, and A. Jeanes, *Anal. Chem.* **37,** 1602 (1965).

[b] Copper tubing, 10.0 ft × 0.25 inch o.d., packed with 3% ECNSS-M on Gas-Chrom Q, 100/120 mesh.

Björndal *et al.*[11] to separate the partially methylated forms of alditol acetates derived from permethylation and hydrolysis of oligosaccharides for identification of positions of linkages between sugars. Combined GLC–mass spectrometry was used to confirm these results.[11,12,13] This method has been very useful and rapid in our hands and should be seriously considered by anyone seeking to characterize carbohydrate structure (see this volume [12]).

Isolation of sugars from plasma, urine and tissues has been discussed in Volume 8[7] and in "Methods in Medical Research."[14]

[11] H. Björndal, B. Lindberg, and S. Svensson, *Acta Chem. Scand.* **21,** 1801 (1967).

[12] H. Björndal, B. Lindberg, and S. Svensson, *Carbohyd. Res.* **5,** 433 (1967).

[13] H. Björndal, C. G. Hellerqvist, B. Lindberg, and S. Svensson, *Angew. Chem. Int. Ed.* **9,** 610 (1970).

[14] "Methods in Medical Research," Vol. 12, pp. 115, 396, 400. Year Book Medical Publ., Chicago, Illinois, 1970.

[11] Gas Chromatography of Carbohydrates as Butaneboronic Acid Esters

By FRANK EISENBERG, JR.

Principle

Certain requirements have been set out by Sweeley et al.[1] in the selection of a volatile derivative for the gas–liquid chromatography (GLC) of carbohydrates which by and large have been met by the trimethylsilyl (TMS) ethers. Recently, n-butaneboronic acid (BBA) has been introduced,[2] a compound which reacts with diols to give cyclic diesters that also satisfy the criteria for routine GLC of carbohydrates.[3] Thus the carbohydrate BBA esters are easily prepared at room temperature, volatilize at 150–250°, separate on common liquid phases, and are readily dissociated for recovery of the carbohydrate. Arene and alkane boronates have been used for many years in carbohydrate chemistry (reviewed by Brooks and Maclean[2]); from their composition they are clearly formed according to the general equation

| Diol | Alkane-boronic acid | | Diol alkaneboronate |

Characterization of the carbohydrate BBA esters $(R = C_4H_9)$ shows that they are likewise synthesized by this reaction.

Although all classes of carbohydrates form BBA esters, there are some limitations in their applicability to GLC. Since BBA is a bifunctional reagent, compounds with even numbers of reactive groups (OH, NH_2, and COOH) give the better chromatographic response, as indicated by symmetrical peaks; those with odd numbers migrate as diffuse or unsymmetrical bands, presumably because of a remaining free polar group. Pentoses (4 OH), for example, show symmetrical, compact peaks while hexose (5 OH) peaks are poorly defined; as with other volatile derivatives, anomeric forms are chromatographically separable. Conversely, pentitols (5 OH) behave poorly while hexitols (6 OH) migrate

[1] C. C. Sweeley, W. W. Wells, and R. Bentley, see Vol. 8 [7] and this volume [10].
[2] C. J. W. Brooks and I. Maclean, J. Chromatogr. Sci. **9**, 18 (1971).
[3] F. Eisenberg, Jr., Carbohyd. Res. **19**, 135 (1971).

as single, sharply defined peaks. BBA esters of the particular hexitols representing the biologically important sugars mannose, glucose, and galactose are completely resolved within a few minutes, a separation impossible to achieve with the TMS ethers. Consistent with this behavior is the excellent response of the hexosaminitols (5 OH, 1 NH_2), hexonolactones (4 OH), and inositols (6 OH). Generally, then, the BBA esters are suitable for the GLC of carbohydrates with even numbers of reactive substituents and are especially useful for the separation of reduced six-carbon forms: alditols, aldosaminitols, and aldonolactones. Since the hexoses, hexosamines, and uronic acids are readily reduced to these alcohols, respectively, the method is highly suited to the analysis of hydrolytic products of biologically important polysaccharides.

Boronation of phosphorylated[4] sugar derivatives is possible only after suppression of the acidity of the phosphoric acid group by methylation. Possessing an even number (4) of OH groups, the hexose phosphate dimethyl esters are readily esterified and separable by GLC. Hexonic acid phosphates (4 OH), after methylation of both acidic terminals, also conform to this pattern.

Although all odd-number compounds tested respond poorly, an even number of reactive groups is not a sufficient requirement for good chromatographic performance. Additionally there must be no impediment to the formation of a cyclic diester. Scyllitol (6 OH), among the five cyclitols tested, is inert and methyl glucoside (4 OH) and sucrose (8 OH) respond poorly, all containing only trans OH groups. These same compounds resist the formation of boric acid esters[5,6] as well, indicating clearly that trans configuration between neighboring OH groups is prohibitive to cyclic diester formation. Cis configuration is favorable, since the other cyclitols react rapidly with both boric acid[6] and BBA. N-Acetylglucosamine, N-acetylgalactosamine, N-acetylmannosamine, and 2-deoxyglucose likewise give only a weak chromatographic response; although all contain four OH groups they can be considered odd-number compounds since only the OH groups at C-3, C-4, and C-6 are favorably situated for cyclic diester formation. Without exception, straight chain, even-number compounds behave well, so that a compound such as 3 O-methylxylose (3 OH), recently discovered in bacteria,[7] should be readily detectable as the BBA ester of 3 O-methylxylitol (4 OH).

[4] J. Wiecko and W. R. Sherman, 1972, unpublished; personal communication from Dr. Sherman.

[5] J. Böeseken, *Advan. Carbohyd. Chem.* **4**, 189 (1949).

[6] S. J. Angyal and L. Anderson, *Advan. Carbohyd. Chem.* **14**, 135 (1959).

[7] J. Weckesser, G. Rosenfelder, H. Mayer, and O. Lüderitz, *Eur. J. Biochem.* **24**, 112 (1971).

TABLE I

RETENTION TIMES[a] FOR BUTANEBORONATES OF VARIOUS SUGARS AND
SUGAR DERIVATIVES

Compound	Column temperature (°C)	Retention time (min)
Aldoses		
D- and L-Arabinose	150	5.4,[b] 8.0
D- and L-Xylose	150	6.5,[b] 7.6,[b] 8.8
D- and L-Fucose	150	6.8
D-Rhamnose	165	3.5,[b] 4.1, 5.2[b]
		3.5,[b] 4.1, 5.2[c]
D-Mannose	200	2.5,[b] 4.5[d]
D-Glucose	200	4.0, 4.9,[b] 5.8[b,e]
		4.0, 4.9,[b] 5.8[c,e]
D-Galactose	200	5.0[d]
Alditols		
Xylitol	178	7–11[f]
L-Fucitol[g]	187	4–6[f]
D-Mannitol	200	4.0
D-Glucitol	200	4.5
	210	14.0[h]
	240	7.1[h]
D-Allitol	200	4.3
L-Iditol	200	5.0
	210	15.5[h]
D-Galactitol	200	5.5
Hexosaminitols		
D-Mannosaminitol	200	5.5
	240	7.9[h]
D-Glucosaminitol	200	7.8
	215	4.5
	240	9.7[h]
D-Galactosaminitol	200	8.2
	215	4.7
	240	10.3[h]
Hexonolactones		
D- and L-Gulonolactone	190	7.2
L-Idonolactone	190	8.2
D-Gluconolactone	200	5.1, 5.5,[b] 6.0, 6.8[b]
L-Mannonolactone	200	5.7
L-Ascorbic acid	200	5.7

TABLE I *(Continued)*

Compound	Column temperature (°C)	Retention time (min)
Cyclitols		
Myoinositol	200	5.7
	210	17.0[h]
Mucoinositol	200	6.0
Neoinositol	200	6.5
	210	18.5[h]
D-Chiroinositol	200	9.8
Reagent		
n-Butaneboronic acid	92	10.0

[a] Barber–Colman Model 10 gas chromatograph equipped with an argon ionization detector at 225–250°, flash heater at 225–250°.
[b] Minor peak.
[c] Heated at 100° for 10 minutes during preparation of the derivative.
[d] Unsymmetrical.
[e] Weak response.
[f] Flat, diffuse peak.
[g] Retention time previously published [F. Eisenberg, Jr., *Carbohyd. Res.* **19**, 135 (1971)] is incorrect and applies instead to BBA.
[h] 6 m column.

In an effort to overcome some of the limitations just described, Wood and Siddiqui[8] have sequentially butaneboronated and trimethylsilylated various sugars and sugar derivatives. Hexoses gave single, sharp peaks as the double derivative; pentoses were unaffected by the silylation step since no free OH group remained after boronation. Unaccountably, D-glucitol, among other compounds tested, failed to give a single, major peak by their procedure.

Preparation and Properties of BBA Esters

Pyridine is the solvent of choice for this esterification. A solution of 5 mg BBA in 1 ml of commercial pyridine is prepared and to this is added about 1 mg of sugar. Concentrations are not critical provided reagent is present in excess; unreacted reagent is innocuous since it elutes with the solvent at column temperatures above 130°. The reaction rate is limited by the solubility of the sugar. Heating at 100° until solution is complete may be necessary, as is required for the inositols. The

[8] P. J. Wood and I. R. Siddiqui, *Carbohyd. Res.* **19**, 283 (1971).

hexitols, on the other hand, are esterified instantaneously at room temperature.

Reagent[9] can vary in purity from one lot to another and should be examined by GLC on the column described below for the presence of isobutaneboronic acid which reacts with diols in the same way to give a forepeak. At a column temperature of 92° BBA migrates as a sharp peak with a retention time of 10 minutes; isobutaneboronic acid emerges from the column at 7 minutes. The normal isomer can be freed of the contaminant by two recrystallizations from the minimal volume of hot water.

Although the esters are easily hydrolyzed, under the conditions of esterification the effect of water is negligible; since water is always present, both as a product of the reaction and accompanying the reagent as a preservative, no special attention is necessary to the drying either of pyridine or sugar. The derivatives are stable protected from excessive atmospheric moisture.

Columns

Among three coatings tested, EGSS-X, SE-30, and OV-17, all at 3% on 100–120-mesh GasChrom Q,[9] OV-17 gave the best response and was used exclusively for all compounds listed in Table I. Peaks produced on EGSS-X were not well formed, and SE-30 failed to separate glucosaminitol and galactosaminitol. Variation in column temperature, column length, and gas flow were sufficient in latitude to separate all classes of carbohydrates on OV-17 alone.

A glass U-shaped column 2 m × 4 mm i.d. was generally used, but for difficult separations a 6-m column consisting of two 3-m U-tubes connected in tandem was used where indicated in Table I.

Separations

Hexitols. Figure 1 shows the rapid and complete separation of mannitol, glucitol, and galactitol BBA esters. The identical pattern is seen if the alditols are obtained by reduction of the parent sugars, so that the method serves as a means of separating and identifying aldoses (or ketoses). To prepare a mixture of mannose, glucose, and galactose for GLC 2 mg of each sugar is dissolved in 0.1 ml H_2O. A solution of 12 mg of $NaBH_4$ in 0.1 ml of H_2O is added, and after 15 minutes washed Rexyn 101(H) cation-exchange resin is added until H_2 evolution ceases. The suspension is placed on a column of an equal amount of the same resin.

---

[9] Reagent and column packings were obtained from Applied Science Laboratories, Inc., State College, Pennsylvania.

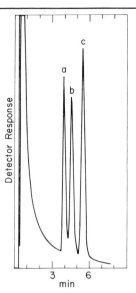

FIG. 1. Gas–liquid chromatogram of a mixture of BBA esters of: (a) D-mannitol, (b) D-glucitol, and (c) D-galactitol; 1 μg of each hexitol. 2 m column, 200°.

Combined effluents from several washes are evaporated to dryness and methanol is added and evaporated twice to remove boric acid. Pyridine is added and evaporated to remove methanol followed by the addition of 1 ml of pyridine containing 15 mg of BBA. A small aliquot of the solution is injected into the gas chromatograph.

Hexosaminitols. Figure 2 shows the separation of glucosaminitol and galactosaminitol BBA esters, representing two important amino sugars found in natural polysaccharides. Although they differ in retention time by 0.4 minute (Table I), they are not separable on a 2-m column but require a 6-m column for separation. Increasing the column temperature to 240° compensates for the longer path. Reduction of the parent hexosamines to the alcohols gives the same pattern. The aldosamine hydrochlorides are first neutralized with aqueous Na_2CO_3 and then reduced with $NaBH_4$, but instead of cation-exchange resin which would retain the amino sugar alcohol, glacial acetic acid is added to destroy excess borohydride. The procedure is identical to the previous beyond this stage.

Hexonolactones. Figure 3 shows the separation of 1,4-gulonolactone and 1,4-idonolactone BBA esters prepared by reduction of D-glucuronic and L-iduronic acids. Both uronic acids are important polysaccharide constituents. Presumably, because of their odd number of reactive substituents, hexuronic acids (4 OH, 1 COOH) show poor chromatographic response as BBA esters. The aldonic acids, however, give multiple BBA

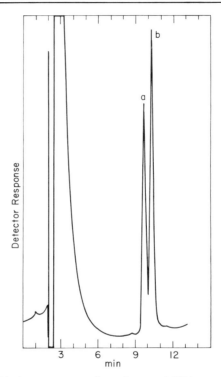

FIG. 2. Gas–liquid chromatogram of a mixture of BBA esters of: (a) D-glucosaminitol and (b) D-galactosaminitol; 2 μg of each hexosaminitol. 6 m column, 240°.

ester peaks owing to mixed lactone formation as seen with gluconolactone (Table I). By introducing the specific 1,4 lactonization step used by Perry and Hulyalkar[10] to eliminate multiple TMS aldonolactone peaks, single 1,4-lactone BBA ester peaks are produced. To prepare uronic acids for GLC, 6 mg of glucuronic acid, for example, is dissolved in 2 ml of H_2O and warmed with excess $BaCO_3$ to give the soluble barium salt, free of lactone. The solution is then chilled in ice and treated with 5 mg $NaBH_4$ for 1 hour. Low temperature reduction retards alditol formation from uronate; in spite of this precaution a slight reduction of the carboxyl group occurs as indicated by the minor glucitol peak seen in Fig. 3. After removal of boric acid as described under *Hexitols* the dry residue is dissolved in 0.5 ml of conc. HCl, evaporated to dryness, and freed of residual HCl by heating for 5 minutes at 100° under oil pump vacuum. Formation of the BBA derivative follows the previous procedures.

[10] M. B. Perry and R. K. Hulyalkar, *Can. J. Biochem.* **43,** 573 (1965).

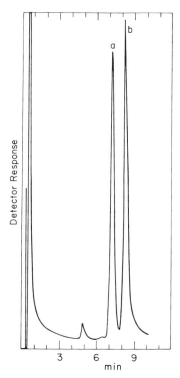

FIG. 3. Gas–liquid chromatogram of a mixture of BBA esters of: (a) L-1,4-gulonolactone and (b) L-1,4-idonolactone; 2 μg of each lactone. 2 m column, 190°.

Hexose and Hexonic Acid Phosphates. Using the column described above, Wiecko and Sherman[4] have extended this method to the phosphorylated 6-carbon sugars and sugar acids. In addition they have introduced methaneboronic acid which yields a similar series of cyclic diesters of higher volatility. Sugar phosphates are converted into the free acid form and lyophilized. They are then taken up in ether–methanol, 2:1, and treated with diazomethane. After evaporation to dryness, the residue is treated for about 1 hour at room temperature with either reagent in pyridine. Table II summarizes their results.

Collection of the Derivative

BBA esters can be condensed intact from the effluent gas and are stable indefinitely if protected from atmospheric moisture. Myoinositol BBA ester was condensed in 50% yield as a colorless liquid in a glass capillary attached to the exit port of the gas chromatograph. When dis-

TABLE II

RETENTION TIMES[a] FOR METHANE- AND BUTANEBORONATES OF CARBOHYDRATE
DIMETHYL PHOSPHATES

	Methaneboronates		Butaneboronates	
Dimethyl ester	Column Temp. (°C)	Retention time (min)	Column Temp. (°C)	Retention time (min)
Glucose 6-phosphate	200	4.5	220	12
Mannose 6-phosphate	200	7.7	220	18.2
Galactose 6-phosphate	190	7.7	225	9.6
Gluconic acid 6-phosphate[d]	185	15.2	220	12.4
Fructose 1-phosphate	200	3.8	220	9.8
Fructose 6-phosphate	187	8.2, 12.5[b]	225	15[c]

[a] Hewlett-Packard Model 402 gas chromatograph; flame ionization detector.
[b] Minor peak.
[c] Major peak following several small peaks.
[d] Carboxymethyl dimethylphosphate.

solved in chloroform and rechromatographed, the compound migrated unchanged from the original.

Mass Spectrometry

The carbohydrate BBA esters display simple mass spectra, clusters of peaks around a particular m/e reflecting the normal isotopic composition of the elements; boron contributes ^{11}B and ^{10}B in the ratio 4:1. Figure 4 shows the mass spectrum of myoinositol BBA ester isolated from the gas chromatographic effluent; the highest m/e detectable is 378, theoretical for the molecular weight (M) of $C_6H_6O_6B_3(C_4H_9)_3$, myoinositol trisbutaneboronate. Loss of a butyl radical, M-57, leads to m/e 321. Scission of the cyclohexane ring results in m/e 237, theoretical for $C_3HO_4B_2(C_4H_9)_2$, and the principal fragment m/e 139, $C_3H_3O_2BC_4H_9$, the two fragments together accounting for 2 H less than the intact molecule.

Recovery of the Sugar

Pure ester dissolved in acidified methanol is rapidly hydrolyzed, permitting recovery of the sugar unchanged. Myoinositol trisbutaneboronate, collected from the gas chromatograph, was dissolved in methanol and treated with a drop of 3 M HCl. Within a short time a crystalline precipitate appeared, identified by melting point as myoinositol.

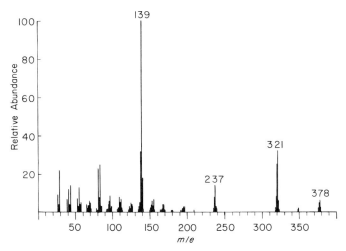

Fig. 4. Mass spectrum of myoinositol trisbutaneboronate determined with an LKB GC-MS mass spectrometer at 70 eV ionization voltage.

Structure of the Inositol BBA Esters

Although only one pair of OH groups in myoinositol is in the cis configuration, the results of these studies indicate clearly that the remaining trans OH groups are likewise esterified to form a single species of trisdiester. If, as in scyllitol, trans OH groups are inert, why does myoinositol combine with three molecules of BBA instead of one? A proposed mechanism of reaction that explains this behavior invokes an initial attack by the reagent at the 1,2 (cis) position to form a planar 5-member diester ring, skewing the cyclohexane ring and aligning the other (trans) OH groups favorably for esterification. A molecular model shows that the OH pairs at 3,5 and 4,6 become so aligned, giving rise to two 6-member diester rings. Myoinositol trisbutaneboronate is thus probably a 1,2;3,5;4,6 trisdiester. Since neo- (2,5-diaxial), D-chiro- (2,3-diaxial), and muco- (2,3,4-triaxial) inositol BBA esters behave chromatographically like myo- it is assumed that they are also trisdiesters. Again after an initial cis attack, the model suggests that neo- and D-chiro- are 1,2;3,4;5,6 trisbutaneboronates and muco- is 1,2;3,6;4,5, involving a 7-member diester ring. Scyllitol remains inert since no initial cis attack is possible. Further study is needed to establish these structures.

Quantitative Analysis of Carbohydrates in Biological Materials

The reproducible chromatographic behavior of the BBA esters has made the adaptation of this method to quantitative analysis possible. For the study of inositol biosynthesis from glucose 6-phosphate, currently under investigation in this laboratory, the method has proved eminently suitable. Inositol, the product, and glucitol, representing the substrate, are measured in terms of an internal standard, mannitol; as shown in Fig. 1 and Table I the three polyols are readily separated as BBA esters. The enzymatic reaction is stopped by heating the incubation mixture, sugar phosphates are hydrolyzed by phosphatase, and aldoses are reduced to alditols with $NaBH_4$. A known volume of standard aqueous mannitol is added roughly equivalent to the sugars present. Further processing to convert all polyols into BBA esters follows the procedure described under *Hexitols*. The solution is chromatographed, and from the ratio of peak height[11] of glucitol and inositol to mannitol BBA esters, compared with standard curves relating peak height to mass of the same compounds, the absolute concentrations of glucitol and inositol are estimated.

[11] Peak height instead of area can be used for narrow peaks.

[12] Methylation Analysis of Polysaccharides

By BENGT LINDBERG

Methylation analysis is one of the most important methods in structural polysaccharide chemistry. It involves methylation of all hydroxyl groups in the polysaccharide, hydrolysis of the fully methylated polysaccharide to a mixture of partially methylated sugars, and qualitative and quantitative analysis of this mixture. The free hydroxyls in the partially methylated sugars mark the positions in which the sugar residues are substituted. The method therefore gives information about all the building blocks of the polysaccharide, but not on their mutual arrangement or on the anomeric nature of the sugar residues. Information about the mutual arrangement of sugar residues may be obtained by comparing the results of methylation analyses of original and chemically modified polysaccharides.

Several different procedures for the various steps of the methylation analysis have been described in the literature.[1,2] In the following, only the procedures currently used in the author's laboratory will be described in detail. Other procedures from the literature may also give satisfactory results.

A serious limitation is that a complete procedure for the methylation

[1] H. O. Bouveng and B. Lindberg, *Advan. Carbohyd. Chem.* **15**, 53 (1960).
[2] H. Björndal, C. G. Hellerqvist, B. Lindberg, and S. Svensson, *Angew. Chem. Int. Ed.* **9**, 610 (1970).

analysis of polysaccharides containing aminodeoxy- or acetamidodeoxy-sugar residues has not been worked out. This part of the subject will therefore not be discussed in detail. It is probable, however, that the special problems associated with the methylation analysis of these polysaccharides will be solved in the near future. Sulfate and phosphate ester groups in polysaccharides also cause complications. These groups should preferably be removed before the methylation analyses, and these aspects will not be discussed.

Methylation of Polysaccharides

Many polysaccharides are soluble or swell in methyl sulfoxide (DMSO) and are preferably methylated by the Hakomori procedure.[3] The polysaccharide, in DMSO, is first treated with the strong base methylsulfinyl sodium and then with methyl iodide. Complete methylation is generally obtained in one step. When there is no shortage of material it is convenient to methylate 5–10 mg, but a complete methylation analysis has been performed with 0.2 mg.

Because of the practice closing bottles with plastic covers even solvents of "analytical purity" contain considerable amounts of plastisizers (e.g., phthalates) which will show up later in the analysis. In order to avoid contamination all solvents used in this and later steps should therefore be freshly distilled.

Polysaccharides not soluble in DMSO may be methylated according to Haworth,[4] that is, with strong aqueous base and methyl sulfate. Complete etherification is not obtained in a single methylation, but the product is soluble in DMSO and can be fully methylated by the Hakomori procedure.

Because of the small amount of methylated polysaccharide obtained, it is impractical to do a methoxyl analysis. Incomplete methylation, which is almost always due to the fact that part of the polysaccharide has not gone into solution, is revealed in the analysis of the hydrolyzate. Nonequivalence between terminal and branched residues, and especially the presence of unmethylated sugars or sugars with more than two free alcoholic hydroxyl groups are criteria of undermethylation.

O-Acyl groups (e.g., O-acetyl) present in the original polysaccharide are split off by the action of the strong base, but N-acetyl groups are stable but become N-methylated. Acetal and ketal groups, which may be natural (e.g., ketals of pyruvic acid) or introduced (e.g., acetals prepared by reaction with methyl vinyl ether) are also stable. Complications, expected to occur because methyl esters of uronic acid residues

[3] S. Hakomori, *J. Biochem. (Tokyo)* **55**, 205 (1964).
[4] W. N. Haworth, *J. Chem. Soc.* **107**, 8 (1915).

react with the base (β-elimination[5] and formation of methylsulfinyl-methyl ketone[6]) have proved to be insignificant. Several methylation analyses of acidic polysaccharides have, within the experimental errors, given almost quantitative recoveries of the uronic acid derivatives (compare, e.g., references cited in footnotes 7 and 8).

Preparation of Methylsulfinyl Sodium.[9] A 50% dispersion of sodium hydride in mineral oil (2 g) and dry (3 Å molecular sieves) DMSO (20 ml) is stirred and kept in an ultrasonic bath (40 kc/sec) at 50°. In about 1 hour a clear, green solution is obtained. Stirring and ultrasonic treatment is interrupted and a 1-cm layer of mineral oil is added in order to protect the solution from air. The reagent, which is approximately 2 M, solidifies at about 10° and may be stored frozen for at least 1 month. The reaction is exothermic, and precautions should be observed, especially when larger amounts of base are prepared.

Hakomori Methylation.[3] The polysaccharide (5 mg) in a 5-ml serum bottle sealed with a rubber cap, is dissolved in dry DMSO (1 ml). Agitation in the ultrasonic bath may facilitate the dissolution. Nitrogen is then flushed through the bottle (the rubber cap is punctured by two syringe needles), and 2 M methylsulfinyl sodium in DMSO (1 ml) is added with a syringe. The solution formed is agitated in the ultrasonic bath for 1 hour, kept for a further 6 hours and methyl iodide (1 ml) is then added dropwise with external cooling in ice water. The resulting solution is agitated in the ultrasonic bath for 30 minutes, poured into water (10 ml), dialyzed against tap water overnight, and concentrated to dryness in a vacuum rotator at 40°. Alternatively the reaction mixture is poured into water (10 ml) and extracted with chloroform (3 × 10 ml), the combined chloroform solutions are washed with water (3 × 10 ml), dried (MgSO$_4$), and concentrated to dryness.

Haworth Methylation.[4] The polysaccharide (100 mg) is dissolved, under nitrogen, in 45% (w/v) sodium hydroxide (15 ml) containing sodium borohydride (10 mg). The solution is cooled with ice, and methyl sulfate (6 ml) is added with stirring during 8 hours. Stirring is continued for an additional 12 hours; the solution is then neutralized with 5 M sulfuric acid, dialyzed, and concentrated.

Hydrolysis of Fully Methylated Polysaccharides

Methylated polysaccharides are often not very soluble in hot water, and hydrolysis is therefore generally preceded by formolysis. The condi-

[5] D. M. W. Anderson and G. M. Cree, *Carbohyd. Res.* **2**, 162 (1966).
[6] E. J. Corey and M. Chaykovsky, *J. Amer. Chem. Soc.* **87**, 1345 (1965).
[7] P. A. Sandford and H. E. Conrad, *Biochemistry* **5**, 1508 (1966).
[8] H. Björndal and B. Lindberg, *Carbohyd. Res.* **12**, 29 (1970).
[9] K. Sjöberg, *Tetrahedron Lett. 1966*, 6383 (1966).

tions which will ensure complete hydrolysis vary with the types of linkages involved. The conditions recommended below are thus far too severe for methylated polysaccharides containing furanosidic linkages only. On the other hand, uronidic and glycosaminidic linkages are hydrolyzed only to a limited extent under the same conditions.

Uronic acid residues are reduced to neutral sugar residues. This can be done on the original polysaccharide, on the fully methylated polysaccharide, or on a methanolyzate of the latter, containing a mixture of methyl glycosides and methyl ester-methyl glycosides of aldobiouronic acids. It is a risk, in the last procedure, that volatile methyl glycosides are lost. The two latter procedures are described below. When the reduction is made with lithium aluminum deuteride (e.g., I → II), two deuterium atoms are introduced at C-6. This facilitates the identification, by mass spectrometry, of the methylated sugars deriving from uronic acid residues.

(I)   (II)

Hydrolysis of Methylated Polysaccharides. Methylated polysaccharide, from 5 mg of polysaccharide, is dissolved in 90% formic acid (3 ml) and kept at 100° for 2 hours. The solution is concentrated to dryness in a vacuum rotator at 40°, dissolved in 0.25 M sulfuric acid (1 ml) and kept at 100° for 12 hours. The cooled solution is neutralized with barium carbonate, filtered, and concentrated in a vacuum rotator at 40°.

Reduction of Uronic Acid Residues in Methylated Polysaccharide. (A) The methylated polysaccharide from 5 mg of polysaccharide is dried in a vacuum over P_2O_5 overnight and dissolved in dry tetrahydrofuran (5 ml). Lithium aluminum hydride or deuteride (30 mg) is added, and the mixture is refluxed for 4 hours. Excess reducing agent is destroyed by adding in turn a few drops of ethyl acetate, ethanol, and water, and the solution is neutralized with 2 M aqueous phosphoric acid. Salts are removed by filtration, and the residue is concentrated and hydrolyzed as described above.

(B) Methylated polysaccharide from 5 mg of polysaccharide is refluxed in 2% methanolic hydrogen chloride (3 ml) for 4 hours, the solution is neutralized with silver carbonate, filtered and concentrated. The residue is refluxed with lithium aluminum hydride, or deuteride (50 mg), in tetrahydrofuran (5 ml) for 4 hours. The product is worked up as above and hydrolyzed with 0.25 M sulfuric acid for 12 hours at 100°. *Preparation of Partially Methylated Alditol Acetates.* The mixture of methylated sugars obtained from hydrolysis of a methylated polysaccharide (5–10 mg) is dissolved in water (5 ml). Sodium borohydride, or deuteride (10 mg) is added, and the mixture is kept for 2 hours at room temperature. Dowex 50 (H⁺, 1 ml) is added, and the mixture is kept for 5 minutes with occasional shaking, filtered, and concentrated to dryness in a vacuum rotator at 40°. Boric acid is removed by codistillations, in the vacuum rotator, with methanol (3 × 5 ml). The residue is treated with acetic anhydride–pyridine, 1:1 (2 ml) for 10 minutes at 100°. Toluene (5 ml), which gives an azeotrope with acetic anhydride, is added, and the mixture is distilled as above, until the rate of distillation decreases, when a new portion of toluene (5 ml) is added, and the solution is concentrated to dryness. The residue is dissolved in acetone (0.2 ml).

Gas–Liquid Chromatography of Partially Methylated Sugars as Their Alditol Acetates

The best method for separating and quantifying the partially methylated sugars obtained on hydrolysis of a fully methylated polysaccharide is gas–liquid chormatography (GLC). The partially methylated sugars must be converted to more stable derivatives, such as the methyl glycosides or the alditol acetates. Each methylated sugar will give at least two, possibly four, methyl glycosides (α- and β-pyranoside and/or α- and β-furanoside). This may facilitate identification of components in simple mixtures but is a definite disadvantage when complicated mixtures are analyzed. Each methylated aldose gives a single alditol acetate, and it is immediately evident from the retention time if, for example, a hexitol derivative contains one, two, three, or four methoxyl groups. All these, and also the nonmethylated alditol acetate, may be separated on the same chromatographic column. The mass spectra (MS) of these derivatives are readily interpreted. In order to avoid eventual ambiguities when interpreting the mass spectra, it is convenient to reduce the sugars with sodium borodeuteride as will be discussed below.

Several stationary phases may be used for the GLC. One of these is ECNSS-M, a nitrile silicone–polyester copolymer (manufactured by Applied Science Laboratory Inc., P.O. Box 140, State College, Pennsylvania). The retention times (T values) on this phase, for a number of alditol acetates derived from methylated hexoses, 6-deoxyhexoses, and pentoses, relative to that of 1,5-di-O-acetyl-2,3,4,6-tetra-O-methyl-D-glucitol, have been determined (Table I). Provided that the T values are determined by interpolation, e.g., between the 2,3,4,6-tetra- and the

TABLE I

RETENTION TIMES (T VALUES) ON ECNSS-M OF PARTIALLY METHYLATED SUGARS, IN THE FORM OF THEIR ALDITOL ACETATES, RELATIVE TO 1,5-DI-O-ACETYL-2,3,4,6-TETRA-O-METHYL-D-GLUCITOL

Position of OCH$_3$	Parent sugar							
	Ara	Rib	Xyl	Gal	Glc	Man	Fuc	Rha
2			2.92	8.1	7.9	7.9	1.67	1.52
3			2.92	11.1	9.6	8.8	2.05	1.94
4			2.92	11.1	11.5	8.8		1.72
6	—	—	—	5.10	5.62		—	—
2,3			1.54	5.68	5.39	4.82	1.18	0.98
2,4	1.40		1.34	6.35	5.10	5.44	1.12	0.99
2,5	1.10			3.70				
2,6	—	—	—	3.65	3.83	3.35	—	—
3,4	1.38	—	1.54	6.93	5.27	5.37		0.92
3,5		0.77	1.08	6.35		5.44		
3,6	—	—	—	4.35	4.40	4.15	—	—
4,6	—	—	—	3.64	4.02	3.29	—	—
2,3,4	0.73		0.68	3.41	2.49	2.48	0.65	0.46
2,3,5	0.48	0.40		3.28			0.62	
2,3,6	—	—	—	2.42	2.50	2.20	—	—
2,4,6	—	—	—	2.28	1.95	2.09	—	—
2,5,6	—	—	—	2.25			—	—
3,4,6	—	—	—	2.50	1.98	1.95	—	—
2,3,4,6	—	—	—	1.25	1.00	1.00	—	—
2,3,5,6	—	—	—	1.15			—	—

2,3-di-O-methyl-D-glucitol derivatives ($T = 1.00$ and $T = 5.39$, respectively), the reproducibility is good, generally better than ±3%. This is also true for different columns and for different temperatures. Partially methylated alditol acetates are generally separated at 160–200°, fully acetylated alditols at 200–210°.

One disadvantage with ECNSS-M is that the phase is not very stable but leaks out on use. The columns, therefore, do not last very long, which is inconvenient, especially for the highly efficient but expensive S.C.O.T. columns (capillary columns manufactured by Perkin-Elmer). On combined GLC-MS, this leakage results in increased background and contamination of the ionization chamber. The silicone polymer OV-225, containing methyl, phenyl, and cyanopropyl groups (manufactured by Applied Science Laboratory, Inc.), is stable up to 250°. It gives as good separations of partially methylated alditol acetates as ECNSS-M[10] and is replacing this phase in the author's laboratory.

[10] J. Lönngren and Å. Pilotti, Acta Chem. Scand. 25, 1144 (1971).

Some retention times on this column are listed in Table II. Most of these values were determined at 170°.

Some components, which are not separated on these phases may be separated on other phases, e.g., OS-128 or OS-138 (polyphenol ethers manufactured by Applied Science Laboratory, Inc.). Some retention times on OS-138 are listed in Table III. Retention times of alditol derivatives that are not available in the pure state may be determined by partial methylation of a glycoside or a derivative thereof and analysis of the reaction mixture, as the alditol acetates, by GLC-MS.

Separations, by GLC, of a number of methylated sugars derived from 2-acetamido-2-deoxy-D-glucose and 2-acetamido-2-deoxy-D-galactose have recently been reported.[11]

When the flame ionization detector is used, there is a linear relation between response and concentration. Moreover, there seems to be no need to use response factors for the different partially methylated alditol

TABLE II

RETENTION TIMES (T VALUES) ON OV-225 OF PARTIALLY METHYLATED SUGARS, IN THE FORM OF THEIR ALDITOL ACETATES, RELATIVE TO
1,5-DI-O-ACETYL-2,3,4,6-TETRA-O-METHYL-D-GLUCITOL

Position of OCH₃	Parent sugar							
	Ara	Rib	Xyl	Gal	Glc	Man	Fuc	Rha
2			2.15		6.6	5.4	1.43	1.37
3			2.15		7.6	6.8		1.67
4			2.15			6.8		1.57
6	—	—	—		5.0		—	—
2,3	1.07		1.19		4.50	3.69		0.92
2,4	1.10		1.06	5.10	4.21	4.51	1.02	0.94
2,5								
2,6	—	—	—	3.14	3.38		—	—
3,4			1.19	5.50	4.26	4.36		0.87
3,5				5.10		4.51		
3,6	—	—	—		3.73	3.67	—	—
4,6	—	—	—		3.49	2.92	—	—
2,3,4			0.54	2.89	2.22	2.19	0.58	0.35
2,3,5	0.41			2.76				
2,3,6	—	—	—	2.22	2.32	2.03	—	—
2,4,6	—	—	—	2.03	1.82	1.90	—	—
2,5,6	—	—	—	1.95			—	—
3,4,6	—	—	—		1.83	1.82	—	—
2,3,4,6	—	—	—	1.19	1.00	0.99	—	—
2,3,5,6	—	—	—	1.10				

[11] P. A. J. Gorin and R. J. Magus, *Can. J. Chem.* **49**, 2583 (1971).

TABLE III

RETENTION TIMES (T VALUES) ON OS-138 OF PARTIALLY METHYLATED SUGARS, IN
THE FORM OF THEIR ALDITOL ACETATES, RELATIVE TO
1,5-DI-O-ACETYL-2,3,4,6-TETRA-O-METHYL-D-GLUCITOL

Sugar[a]	T	Sugar	T
2,5-Rib	0.77	2,3,4,6-Glc	1.00
2,3,6-Gal	1.74	2,3,4-Man	1.93
2,4,6-Gal	1.88	2,3,4,6-Man	1.06[b]
2,3,4,6-Gal	1.16	2,3-Rha	0.86
2,3-Glc	3.0	2,4-Rha	0.94
2,3,4-Glc	1.88	3,4-Rha	0.80
2,3,6-Glc	1.75	2,3,4-Rha	0.49
2,3,6-Glc	1.63		

[a] 2,5-Rib = 2,5-di-O-methylribose, etc.

[b] We have recently failed to separate the 2,3,4,6-tetra-O-methyl-D-glucose and
-D-mannose derivatives on OS-138, possibly because the composition of the phase
has been changed. The derivatives are not separated on OS-128 either, but may be
separated on an OV-225 S.C.O.T. column at 190°.

acetates. The assumption that the molar response is the same for all
these derivatives is probably true within ±5% as demonstrated by several
methylation analyses of oligosaccharides and of polysaccharides with
regular structures. The errors introduced by this assumption are probably
less serious than those due to degradation during hydrolysis and losses
of the more volatile methylated sugars and their derivatives during
concentrations.

Alditol acetates deriving from acetamindodeoxy sugars have higher
retention times and probably different response factors from those of the
corresponding neutral derivatives.

For GLC, 2-m glass columns of 2 mm internal diameter or 16-m
S.C.O.T. columns are used. The glass columns contain 3% stationary
phase on Gas-Chrom Q, 100–120 mesh. The flow rate of the carrier gas
is about 20 ml per minute for the packed columns and 3–4 ml per minute
for the S.C.O.T. columns. The solution of partially methylated alditol
acetates (0.5 μl) is injected on the packed column, 0.1 μl on the S.C.O.T.
column.

Mass Spectra of Partially Methylated Alditol Acetates

As is evident from Tables I, II, and III, it is not always possible,
even in the methylation analysis of a homoglycan, to identify all methylated sugars from the T values of their alditol acetates. Moreover, nonsugar components from impurities in the original sample or in solvents
and reagents sometimes give strong peaks which may cause confusion.

With combined GLC-MS nonsugar components are easily detected and the substitution patterns of the partially methylated alditol acetates are determined. Systematic studies of MS of different partially methylated alditol acetates and their deuterated analogs have led to the following generalizations.[12-15]

1. Derivatives with the same substitution pattern (e.g., alditol acetates derived from 2,3,4-tri-O-methylhexoses) give very similar mass spectra typical of that substitution pattern. The small differences observed in relative intensities of peaks for stereoisomers are insufficient for unambiguous identification.

2. The base peak of the spectrum is generally m/e 43 (CH_3—$C\overset{\oplus}{\text{≡}}O$). The molecular ion is generally too weak to be observed.

3. Primary fragments are formed by fission between carbon atoms in the chain. Fission between a methoxylated and an acetoxylated carbon atom is preferred over fission between two acetoxylated carbon atoms. The fragment with the methoxyl group gives the positive ion. The alditol acetate (III) derived from 3-O-methyl-D-glucose therefore gives only two main primary fragments, m/e 189 and m/e 261.

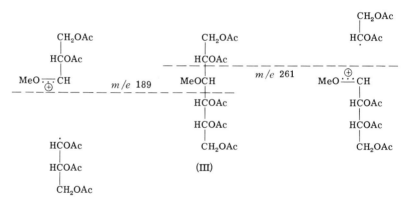

4. When a molecule contains two adjacent methoxylated carbon atoms, e.g., the alditol acetate derived from 2,3-di-O-methyl-D-glucose (IV), fission between these is preferred over fission between one of these and an acetoxylated carbon atom. Both fragments from this fission

[12] H. Björndal, B. Lindberg, and S. Svensson, *Carbohyd. Res.* **5**, 433 (1967).
[13] H. Björndal, B. Lindberg, Å. Pilotti, and S. Svensson, *Carbohyd. Res.* **15**, 339 (1970).
[14] H. B. Borén, P. J. Garegg, B. Lindberg, and S. Svensson, *Acta Chem. Scand.* **25**, 3299 (1971).
[15] K. Axberg, H. Björndal, Å. Pilotti, and S. Svensson, *Acta Chem. Scand.* **26**, 1319 (1972).

are found as positive ions. One important exception to this rule is the

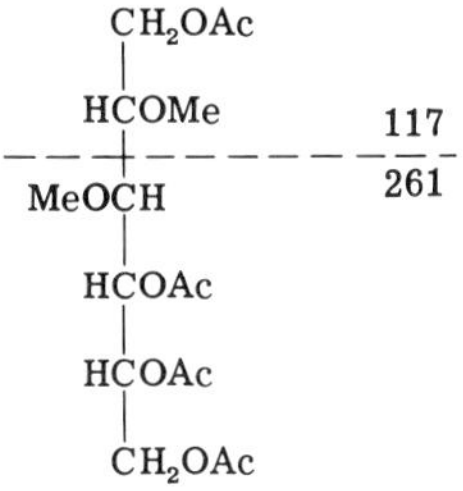

(IV) 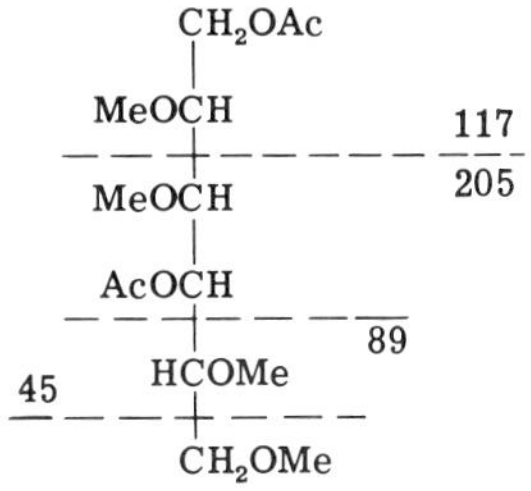   (V)

1,2-di-O-methylalditol acetates, e.g., that derived from 2,3,5,6-tetra-O-methyl-D-galactose (V), which give the primary fragment m/e 89 as a strong peak.

5. Secondary fragments are formed from the primary by single or consecutive loss of acetic acid (m/e 60), methanol (m/e 32), ketene (m/e 42), and formaldehyde (m/e 30). The only important elimination of formaldehyde is from the fragment m/e 89.

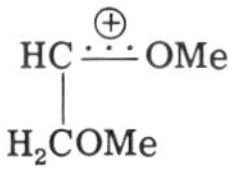

m/e 89 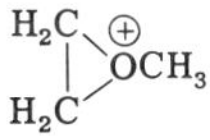 m/e 59

Methanol and acetic acid are generally lost by β-elimination, as exemplified for the two m/e 161 fragments, (VI) and (VII). Ketene is then eliminated when an acetoxyl is attached to an unsaturated carbon atom.

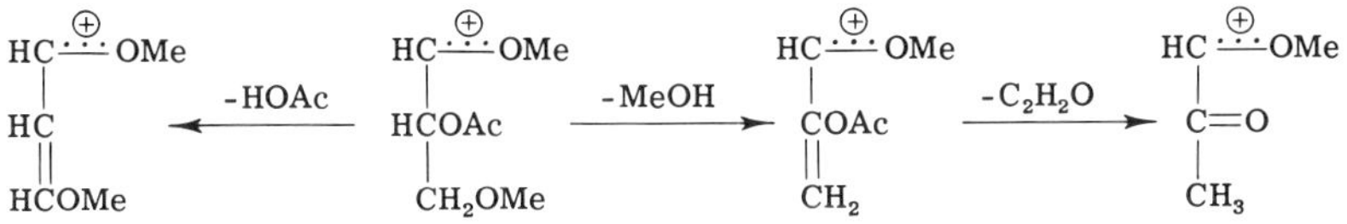

m/e 101 m/e 161 m/e 129 m/e 87

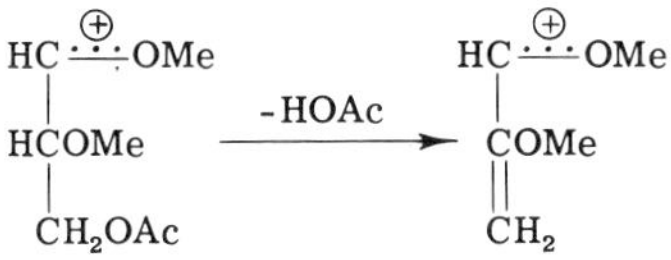

                 (VI)

m/e 161 m/e 101

(VII)

α-Elimination of acetic acid is observed for (VI) but no corresponding elimination of methanol from (VII). Thus (VI) gives the secondary fragments m/e 129, 101, and 87, but (VII) gives only one main secondary fragment, m/e 101.

6. Deoxy groups in partially acetylated alditol acetates are readily recognized from the primary fragments formed, as indicated for (VIII), (IX), and (X). The primary fragment m/e 69, formed from (IX)

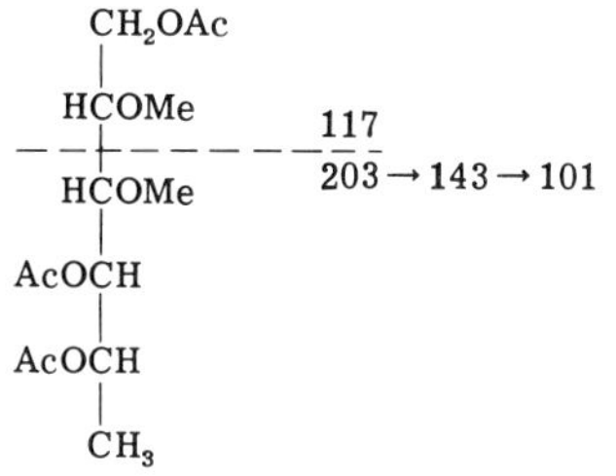

(VIII) 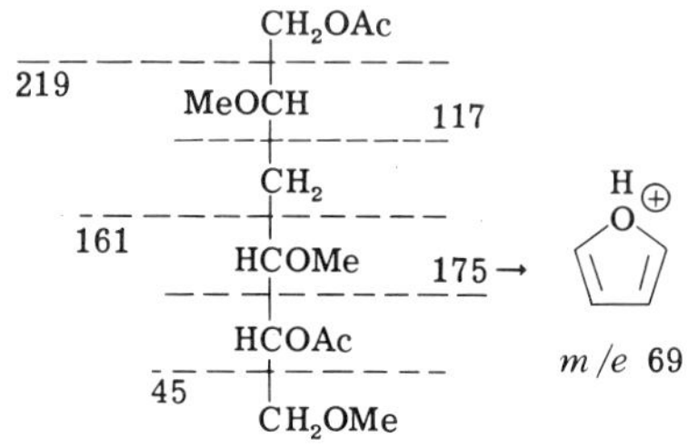   (IX)

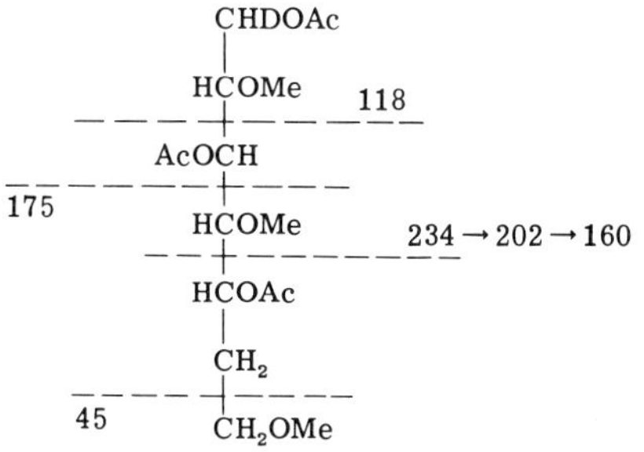

(X) 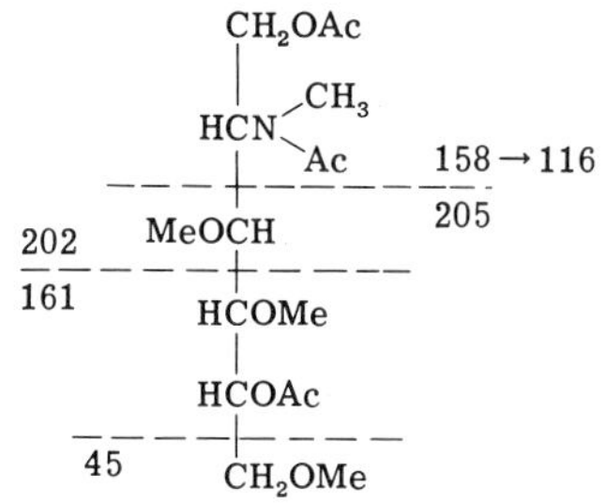   (XI)

via the primary fragment m/e 175 is one of the strongest ions of the spectrum.

7. Preliminary studies on alditol acetates, derived from acetamido-deoxysugars, indicate that their fragmentation follows the same principles. In the MS of the alditol acetate derived from 2-N-methylacetamido-2-deoxy-3,4,6-tri-O-methyl-D-glucose (XI), the strongest primary fragment is m/e 158, formed by fission between C-2 and C-3. The secondary fragment m/e 116, formed from m/e 158 by loss of ketene, is the base peak, even stronger than m/e 43. Mass spectra of other aminodeoxy derivatives of carbohydrates are discussed by Kochetkov and Chizhov.[16]

Using these principles, the substitution pattern of any partially methylated alditol acetate should be readily determined. The intensities of the primary fragments decrease with increasing molecular weight.

[16] N. K. Kochetkov and O. S. Chizhov, *Advan. Carbohyd. Chem.* **21**, 39 (1966).

Some of the highest primary fragments, m/e 261, 305, and 333, may therefore be rather weak, but are always considerably stronger than the ions at adjacent mass numbers. The primary fragment of m/e 45 ($CH_2\overset{\oplus}{=}OMe$) is typical for derivatives containing a primary methoxyl. A fragment of m/e 45, however, is often obtained from derivatives lacking a primary methoxyl group. This is seldom stronger than 10% of the base peak, whereas the primary fragment is about 20% of the base peak or stronger. In doubtful cases, an analysis of the spectrum generally solves the problem. Thus alditol acetates from 2,3,4-tri-O-methylhexoses, e.g., (XII), give a fairly strong m/e 45 peak. From the retention time, the substance should contain three methoxyls, and the substitution pattern is fully defined by the primary fragments m/e 233, 189, 161, and 117.

Alditol acetates derived from 3-O-methylpentoses (XIII) and 3,4-di-O-methylhexoses (XIV) give essentially the same mass spectra, which is expected, as they give one and the same primary fragment, m/e 189. As the retention times on GLC differ considerably for these compounds, this should cause no confusion.

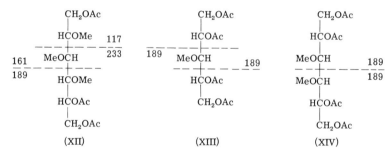

Alditol acetates, e.g., (XV), derived from 6-O-methylhexoses give only one primary fragment (m/e 45) derived from fission at a methoxylated carbon atom. Other primary fragments are formed by fission between acetoxylated carbon atoms. The fragments are consequently rather weak, and the MS is similar to that of a fully acetylated alditol.

On reduction, some pairs of methylated sugars, e.g., a 2,3- and 3,4-di-O-methylpentose, or a 3- and a 4-O-methylhexose give alditols with the same substitution pattern. These may even be identical, as in the mannose series, or enantiomorphs, as in the galactose series, and thus inseparable by GLC. The ambiguity in identifying the methylated sugar is, however, avoided if the reduction is performed with borodeuteride, as exemplified by the alditols derived from 2,3- and 3,4-di-O-methyl-D-xylose (XVI and XVII).

It is thus always possible to determine the substitution pattern of a partially methylated sugar from the MS of its alditol acetate. In

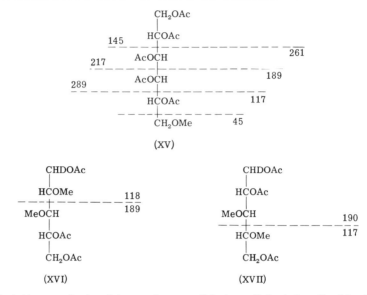

(XV)

(XVI)   (XVII)

methylation analysis of homoglycans, this is sufficient for the identification of all the methylated sugars. In methylation analysis of heteroglycans
containing different sugars of the same class (e.g., D-glucose, D-mannose,
and D-galactose), the MS evidence has to be combined with knowledge
of the T values of pertinent alditol derivatives, which moreover should
differ. It is usually possible, with combined GLC-MS, to identify all
components in the mixture of methylated sugars formed in methylation
analysis of polysaccharides.

It is not intended to describe the combined gas chromatograph–mass
spectrometer and its operation. In the author's laboratory two instruments, LKB 9000 and Perkin-Elmer 270, have been used, but other
commercial low-resolution instruments should also be satisfactory.

Although mass spectra should be essentially the same when determined
with different instruments, differences in relative intensities of peaks
may be observed even for different runs with the same instrument. These
minor differences, however, do not affect the identifications. Some
typical mass spectra are given in Figs. 1–3.

Suitable conditions for GLC of the mixture of methylated sugars
should be worked out, and GLC-MS should preferably be performed
using the same column. Mass spectra are taken at the maxima of the
small peaks and at the beginning, maximum, and end of the large
peaks. Identification of compounds that are not resolved or only partially
resolved is generally possible. For the quantitative analysis, complete
separation of compounds should be attempted. Separation of two unre-

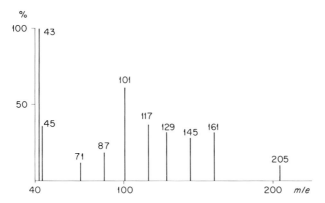

Fig. 1. Mass spectrum of 1,5-di-*O*-acetyl-2,3,4,6-tetra-*O*-methyl-D-glucitol.

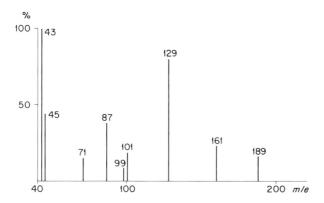

Fig. 2. Mass spectrum of 1,2,4-tri-*O*-acetyl-3,5-di-*O*-methyl-D-ribitol.

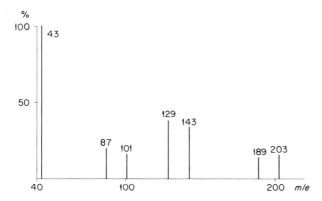

Fig. 3. Mass spectrum of 1,2,4,5-tetra-*O*-acetyl-3-*O*-methyl-L-rhamnitol.

solved peaks may be obtained by working at a lower temperature, using a S.C.O.T. column, or using another phase. Only as a last resource should the relative proportions of two unresolved components be estimated from the intensities of pertinent peaks in the MS.

Evaluation of Results

Methylation analysis of a polysaccharide of reasonably high molecular weight should give equimolecular amounts of methylated sugars derived from branched and nonreducing terminal residues. Double-branched residues are rare, but have been observed. Often the percentage of sugars apparently deriving from branched residues is higher than that deriving from terminal residues, and this discrepancy may be accounted for in several ways.

1. Incomplete methylation gives several components of low degrees of methylation, generally also unmethylated sugars, and is thus easily revealed.

2. Some demethylation may occur during hydrolysis of the fully methylated polysaccharide.[17] This source of error is not very important.

3. Some methylated sugars may be partially or totally degraded during the hydrolysis. This is true for such less common sugars as 2-deoxy sugars, 3-deoxyulosonic acids, and 3,6-dideoxyhexoses.

4. Some methylated sugars, especially those deriving from terminal 3,6-dideoxyhexose, 6-deoxyhexose, or pentose residues and their alditol derivatives, are volatile and may be lost during concentration of solutions. Penta-O-methylhexitol derivatives, obtained on methylation analysis of oligosaccharide alditols, are also volatile. This source of error may be eliminated[18] if the methylated polysaccharide is subjected to methanolysis in a closed tube and the reaction mixture, without concentration, is injected into the chromatographic column.

5. The polysaccharide may contain substituents which are stable during the methylation but are hydrolyzed off during the treatment with acid. Acetals of formaldehyde and acetaldehyde, which have been observed only once,[19] and ketals of pyruvic acid, which are common, show this behavior. An apparently inconsistent methylation analysis may indicate the presence of these or other substituents that have been overlooked in the preceding studies.

Aldose derivatives that are not methylated at C-4 and C-5 could

[17] I. Croon, G. Herrström, G. Kull, and B. Lindberg, *Acta Chem. Scand.* **14**, 1338 (1960).

[18] M. Zinbo and T. E. Timell, *Sv. Papperstidn.* **68**, 647 (1965).

[19] P. J. Garegg, B. Lindberg, T. Onn, and I. W. Sutherland, *Acta Chem. Scand.* **25**, 2103 (1971).

derive from pyranosidic or furanosidic residues. D-Ribose and D-fructose have been found only as furanosides, L-arabinose and D-galactose are found both as furanosides and as pyranosides, and most other sugars as pyranosides. Pyranosides are about 100 times more resistant to acid hydrolysis than furanosides, and it is possible, by a mild hydrolytic treatment, to decide whether a sugar component is pyranosidic or furanosidic. The situation may be complicated when, as sometimes happens, a sugar occurs both as pyranosidic and furanosidic residues in the same polysaccharide.

Special Applications

Methylation analysis of polysaccharides that have been modified by mild acid hydrolysis, enzymatic hydrolysis, Smith degradation, or other treatments may give structural information complementary to that obtained on methylation analysis of the original polysaccharide. These possibilities have not been utilized often in the past, but have become more attractive as the methylation analysis has become less laborious.

O-Methylated sugars are components of some polysaccharides, and are often accompanied by their parent sugars. When the polysaccharide is methylated with trideuteriomethyl iodide, it is possible to distinguish by MS between methylated sugars derived from a natively methylated sugar and its parent nonmethylated sugar. Thus the MS of (XVIII) shows that it derives from nonreducing terminal 3-O-methyl-L-rhamnose residues.

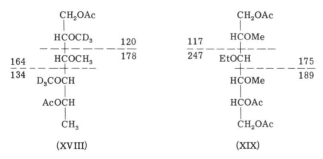

(XVIII)   (XIX)

Combined methylation–etylation analysis has been used in connection with structural studies on a dextran.[20] A methylated dextran derivative was subjected to a specific degradation, and the new hydroxyl groups formed thereby were ethylated. The technique has the advantage over methylation–trideuteriomethylation that derivatives with the same pattern of etherification may be separated by GLC. The fragmentation

[20] O. Larm, B. Lindberg, and S. Svensson, *Carbohyd. Res.* **20**, 39 (1971).

of the derived alditol acetates, e.g., (XIX), follows the same principles as for the methylated analogs.

Some polysaccharides contain O-acyl groups, most aften O-acetyl groups. As was already mentioned, these are split off during a Hakomori methylation. Their locations may be determined by a procedure devised by de Belder and Norrman[21] in which the free hydroxyl groups in the polysaccharide are protected by reaction with methyl vinyl ether and an acidic catalyst (scheme 1). Methylation of this protected material

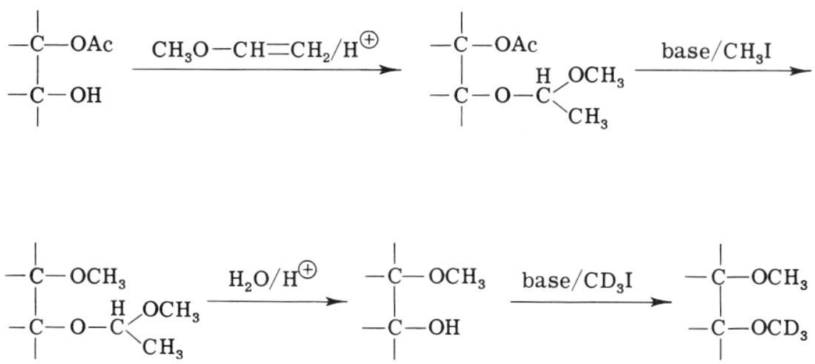

SCHEME 1. Location of O-acetyl groups.

and acid hydrolysis gives a mixture of sugars and methylated sugars in which the methoxyl groups mark the positions of the original O-acyl groups. It is also possible to hydrolyze selectively the protecting acetal groups and methylate the product with trideuteriomethyl iodide.[22] When the O-acyl group is linked to a sugar which occurs in different modes of substitution, more detailed information is obtained by the latter procedure.

Location of O-Acetyl Groups. A solution of the polysaccharide (20 mg) and p-toluenesulfonic acid (20 mg) in dry DMSO (2 ml) is placed in a serum flask, which is closed with a rubber cap and cooled to 15°. Methyl vinyl ether (2 ml), which has been condensed at −30°, is added in portions by means of a syringe. The mixture is kept for 3 hours with occasional shaking. The red mixture is then added to the top of a Sephadex LH-20 column (27 × 2 cm) which is irrigated with dry acetone. The optical rotation of the eluate is determined, and the fractions containing the polysaccharide, which is eluted before the low-molecular weight reagents, are combined and concentrated. This material is subjected to methylation analysis as described above. When the optical rotation of the polysaccharide is insignificant, the column may be cali-

[21] A. N. de Belder and B. Norrman, *Carbohyd. Res.* **8**, 1 (1968).
[22] H. Björndal and B. Lindberg, *Acta Chem. Scand.* **25**, 1281 (1971).

brated with, for example, fully methylated dextran. A somewhat different procedure for preparing the acetalated polysaccharide has been devised by Bhattacharjee *et al.*[23]

In the alternate procedure the àcetalated and methylated poly-saccharide is treated with 50% acetic acid for 2 hours at 100°, concentrated, dried in a vacuum over P_2O_5 overnight, and methylated with trideuteriomethyl iodide. The subsequent analysis is performed as described above.

[23] S. S. Bhattacharjee, R. H. Huskins, and P. A. J. Gorin, *Carbohyd. Res.* **13,** 235 (1970).

[13] Sugar Analysis of Cells in Culture by Isotope Dilution

By LAURA C. SHEN and VICTOR GINSBURG

The sugars of mammalian cells occur in small amounts, and their estimation by conventional methods is difficult. The present paper describes an isotope dilution method for the measurement of these sugars in cells grown in culture.[1]

Principle of Determination

Cells are grown in media containing uniformly labeled [^{14}C]glucose to constant specific activity. After extraction with 10% trichloroacetic acid, known amounts of sugars to be analyzed are added as carriers to the cell residues, which are then hydrolyzed with acid. The carrier sugars are then reisolated from the hydrolyzate, and from their specific activities, the total ^{14}C- activity in each sugar is determined. Assuming that the specific activities of the cell sugars before dilution with carrier are approximately equal to the specific activity of ribose derived from RNA (which is determined directly without added carrier), the absolute amount of each sugar liberated from the residue by hydrolysis can be calculated.

Growth and Labeling of Cells

The following procedure describes the method as applied to HeLa cells: The cells grown in suspension culture in Eagle's spinner No. 2 media supplemented with 5% fetal calf serum and 4 mM glutamine. Uni-

[1] L. Shen and V. Ginsburg, *Arch. Biochem. Biophys.* **122,** 474 (1967).

formly labeled [¹⁴C]glucose is added to make the specific activity of the glucose in the media 150 cpm/μg.[2]

An exponentially growing culture of cells is harvested by centrifugation, and the cell pellet is suspended at a concentration of 2 to 3×10^5 cells/ml in 200 ml of labeled media in a 250-ml rubber-stoppered centrifuge bottle equipped with a hanging magnetic stirring bar. The cells are grown at 37° with constant stirring. When the cells double (approximately 24 hours), 100 ml of culture is removed and the remaining cells are harvested by centrifugation and resuspended in 200 ml of fresh, labeled medium. This procedure is repeated until the cells attain a constant specific activity. With HeLa cells 3–4 generations are required (see Figs. 2–4). The cells removed at each generation are collected by centrifugation, and the resulting cell pellets are washed with two 30-ml portions of unlabeled media. The washed cells are then extracted three times with 15-ml aliquots of 10% trichloroacetic acid at 0°. The extracts are discarded, and the cell residues are washed with ether to remove residual trichloroacetic acid and then dried at room temperature. From 6×10^7 cells (the yield for 100 ml of culture), about 22 mg of dry residue is obtained.

Chromatography of Sugars

Sugars are isolated by descending paper chromatography. The following solvent systems are employed:

    I. Pyridine–ethylacetate–H_2O (1.0:3.6:1.15).
    II. Butylacetate–acetic acid–H_2O (3.0:2.0:1.0).
    Sugars on paper are located with $AgNO_3$ reagent.[3]

Colorimetry

After separation by paper chromatography the sugars are eluted and estimated by colorimetric analysis. Glucose, galactose, and mannose are determined with anthrone reagent[4]; fucose with cysteine-sulfuric acid[5]; lyxose, arabinose, and ribose with phloroglucinol reagent[6]; and N-acetylneuraminic acid with thiobarbituric acid reagent.[7] All analytical values are routinely corrected for the contribution of filter paper.

[2] With this specific activity, approximately 3×10^7 cells can be conveniently analyzed for the seven different sugars. If fewer cells are available or subcellular fractions are to be assayed, the specific activity of the glucose in the media should be increased.

[3] W. E. Trevelyan, D. P. Procter, and J. S. Harrison, *Nature* (*London*) **166**, 444 (1950).

[4] J. II. Roe, *J. Biol. Chem.* **212**, 335 (1955).

[5] Z. Dische and L. B. Shettles, *J. Biol. Chem.* **175**, 595 (1948).

[6] Z. Dische and E. Borenfreund, *Biochim. Biophys. Acta* **23**, 639 (1957).

[7] L. Warren, *J. Biol. Chem.* **234**, 1971 (1959).

Determination of Radioactivity

Radioactivity on paper is located with a strip scanner. Radioactivity of eluted sugars is determined on aliquots dried on planchets and counted in a low background gas flow counter or with a scintillation counter.

Determination of Neutral Sugars. Part of the dry residue, 10 mg is suspended in 1 ml of 1 N H_2SO_4; to the suspension is added 300 μg each of galactose, glucose, mannose, and fucose. The suspension is hydrolyzed at 100° for 4 hours. The hydrolyzate is neutralized with saturated Ba(OH)$_2$, with phenolphthalein as an indicator, and the resultant BaSO$_4$ precipitate is removed by centrifugation. The clear supernatant solution is then deionized by passage through a column of Amberlite MB-3 (2.0-ml bed volume). The eluate is evaporated to a small volume and applied to Whatman 3 MM paper as a 5-inch band and chromatographed in solvent I for 20 hours. When the chromatograph is scanned

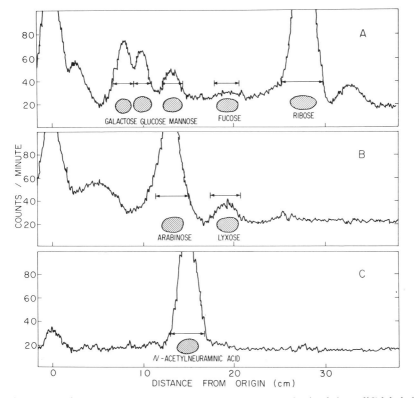

FIG. 1. Radioactive scans of chromatograms of sugars obtained from ^{14}C-labeled HeLa cells by hydrolysis: (A) neutral sugars; (B) pentoses derived from hexosamine; (C) sialic acid.

for radioactivity, there is a good correspondence between radioactive
peaks and marker sugars, as shown in Fig. 1A. The labeled sugars in
the areas indicated by arrows in Fig. 1A are eluted and rechromatographed in with the same solvent system: galactose and glucose for
64 hours, mannose for 48 hours, and ribose and fucose for 24 hours. The
sugars obtained after the second chromatography are used for specific
activity determination. The recovery of added carrier sugars is 20–40%.

Enough ribose is obtained from the hydrolyzate of 10 mg of dry
residue so that its specific activity can be determined directly without
added carrier. As shown in Fig. 2, by the third generation the specific
activity of ribose reaches about 65% of that of the labeled glucose in
the media (average of generations 3–6) and thereafter remains constant.
The difference in specific activity between ribose and glucose represents
dilution of the labeled carbons of glucose with unlabeled sources of carbon
present in the complex growth media. This dilution appears fairly
constant, as in separate experiments the figures were 68 and 70%,
respectively.

Determination of N-Acetylgalactosamine and N-Acetylglucosamine.
Part of the dry residue, 4 mg, is suspended in 1.5 ml of 4 N HCl, and
to the suspension is added 300 μg each of N-acetylglucosamine and
N-acetylgalactosamine. The suspension is hydrolyzed in a sealed tube at
100° for 6 hours. The hydrolyzate is lyophilized, and residual HCl is
removed in a vacuum desiccator over NaOH. The residue is dissolved
in 1 ml of H_2O and passed through a column of Dowex 50-X8 (H⁺ form)

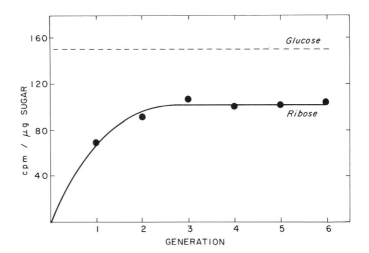

FIG. 2. Specific activity of ribose isolated from the hydrolyzate of HeLa cells
grown on uniformly labeled [¹⁴C]glucose. - - -, Specific activity of glucose in medium;
●, specific activity of ribose isolated from cells.

(2-ml bed volume). The column is washed with H_2O, then hexosamine is eluted with 5 ml of 2 N HCl and dried as described above. The hexosamines are then degraded to pentoses with ninhydrin[8] as follows: The sample is dissolved in 0.5 ml of 0.8 M sodium acetate buffer, pH 5.5, and 3.5 ml of 2% aqueous ninhydrin containing 4% pyridine, and heated at 100° for 30 minutes in a sealed tube. The reaction mixture is then deionized by passage through a column of Dowex 1-X8 (formate form) (3-ml bed volume) and then a column of Dowex 50-X8 (H^+ form) (3-ml bed volume). The eluate is collected, evaporated to a small volume, and applied to Whatman 3 MM paper as a 2-inch band. The sample is then chromatographed with solvent I for 20 hours with lyxose and arabinose as markers. Figure 1B shows the resulting chromatogram. The lyxose and arabinose areas of the paper indicated by arrows are eluted, and the specific activity of the pentoses is determined. The recovery of pentose based on added N-acetylhexosamine is about 20%.

Determination of N-Acetylneuraminic Acid. Part of the dry residue, 4 mg, is suspended in 1 ml of 0.05 N H_2SO_4, and to the suspension is added 100 μg of N-acetylneuraminic acid. The suspension is then heated

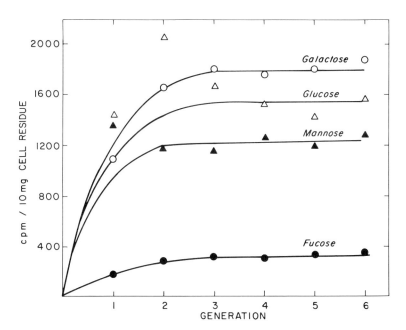

Fig. 3. Total ^{14}C activity in neutral sugars isolated from 3×10^7 cells grown on uniformly labeled [^{14}C]glucose.

[8] R. G. Spiro, Vol. 8 [1].

at 80° for 1 hour. The hydrolyzate is neutralized with saturated $Ba(OH)_2$, and $BaSO_4$ is removed by centrifugation. The clear supernatant fluid is passed through a column of Dowex 50-X8 (H^+ form) (1.5-ml bed volume) and then through a column of Dowex 1-X8 (formate form) (1.5-ml bed volume) to absorb the N-acetylneuraminic acid. The latter column is washed with H_2O, and N-acetylneuraminic acid is eluted with 5 ml of 0.3 N formic acid. The eluate is lyophilized, and the sample is dissolved in H_2O, applied to Whatman 3 MM paper as a 1.5-inch band, and chromatographed for 17 hours in solvent II. The resulting chromatogram is shown in Fig. 1C. N-Acetylneuraminic acid is eluted from the area indicated by the arrow, and its specific activity is determined. About 40% of the added carrier N-acetylneuraminic acid is recovered.

Calculation of the Absolute Amounts of Each Sugar

From the specific activities of the isolated sugars and the amounts of carrier added originally, the total ^{14}C activity liberated from the residues by hydrolysis can be calculated. The results for different generation times are given in Figs. 3 and 4. After four generations the ^{14}C activity in each sugar remains constant. Assuming the specific activity

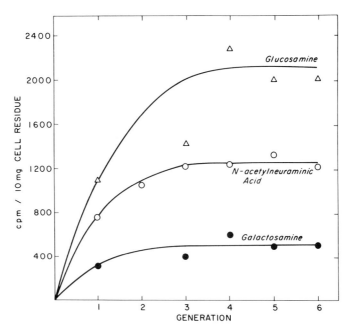

FIG. 4. Total ^{14}C activity in aminosugars isolated from 3×10^7 cells grown on uniformly labeled glucose [^{14}C]glucose.

SUGAR CONTENT[a] OF HeLa S₃ CELLS

| Sugar | Generations in [¹⁴C]glucose | | | |
	4	5	6	Average
Fucose	3.0	3.4	3.6	3.3
Mannose	12.7	12.0	12.9	12.5
Galactose	17.5	18.0	18.8	18.0
Glucose	15.1	13.8	15.8	14.9
Glucosamine	23.1	20.0	20.7	21.3
Galactosamine	6.0	5.5	5.5	5.7
N-Acetylneuraminic acid	12.2	13.2	12.0	12.5

[a] Values are expressed as micrograms per 10.0 mg of dry residue (3×10^7 cells).

of each sugar before added carrier to be equal to that of ribose,[9] their absolute amounts can be calculated. The table shows the sugar content of the cell residues after 4, 5, and 6 generations (a value of 103 cpm/μg derived from Fig. 2, was used for the specific activity of ribose). The analyses were repeated twice with identical results on different batches of HeLa cells.

Scope of Use

The method has been used to study the release of sugars from cells by trypsin[10]; to compare the level of sugars in cells grown in suspension culture versus monolayer[10]; to determine the distribution of fucose in the glycolipids and glycoproteins of HeLa cells[1]; and to investigate changes in the level of sugars after viral transformation.[11]

[9] Since ribose derives from the hexose phosphate pool, its specific activity is probably the same as that of the other sugars that arise from the same intermediates. With fucose, mannose, and sialic acid at least, this assumption is correct as evidenced by the good agreement between their levels in cells, as estimated by isotope dilution, and as estimated by direct colorimetric determination.[1]

[10] L. Shen and V. Ginsburg, *Wistar Inst. Symp. Monogr.* **8**, 67 (1968).

[11] H. C. Wu, E. Meezan, P. H. Black, and P. W. Robbins, *Biochemistry* **8**, 2509 (1969).

Section II

Preparations

[14] Tritium-Labeling of Glycoproteins That Contain Terminal Galactose Residues

By ANATOL G. MORELL and GILBERT ASHWELL

Principle. Glycoproteins containing galactose as the terminal non-reducing sugar of the carbohydrate chain(s) can be made radioactive by sequential exposure to galactose oxidase and tritiated borohydride. The enzymatically produced 6-aldehyde intermediate is thereby reconverted, in good yield, to galactose bearing a tritium atom on carbon 6. Since many of the naturally occurring glycoproteins of biological interest are terminated by the disaccharide, sialylgalactose, it is usually essential to remove the sialic acid moiety prior to treatment with galactose oxidase. This is readily accomplished by incubation with neuraminidase. In principle, glycoproteins bearing N-acetylgalactosamine or sialyl-N-acetyl-galactosamine on the nonreducing termini should be equally susceptible to these procedures.

Specific details are outlined below for the preparation of radioactive asialoceruloplasmin, the first protein to be so labeled.[1] The same technique, with minor modifications, has been utilized for the preparation of a variety of glycoproteins and glycopeptides including orosomucoid, fetuin, haptoglobin, α_2-macroglobulin, transferrin, and thyroglobulin.[2] In addition, several hormones, such as bovine thyrotropin and exophthal-mogenic factor as well as mouse thyroid-stimulating hormone, all of which contain a terminal galactose residue, have been successfully tritiated by this method with full retention of their biological activity.[3]

Reagents

Ceruloplasmin, prepared from Cohn fraction IV-1 of human plasma[4] was freed from small amounts of contaminating carbohydrate-rich protein by crystallization to yield material in which the ratio of absorbance at 610 nm to that at 280 nm was between 0.044 and 0.047. For the labeling of alternate glycoproteins, correspondingly appropriate purification procedures should be employed.

[1] A. G. Morell, C. J. A. Van Den Hamer, I. H. Scheinberg, and G. Ashwell, *J. Biol. Chem.* **241**, 3745 (1966).

[2] A. G. Morell, G. Gregoriadis, I. H. Scheinberg, J. Hickman, and G. Ashwell, *J. Biol. Chem.* **246**, 1461 (1971).

[3] R. J. Winand and L. D. Kohn, *J. Biol. Chem.* **245**, 967 (1970).

[4] A. G. Morell, P. Aisen, and I. H. Scheinberg, *J. Biol. Chem.* **237**, 3455 (1962).

Neuraminidase: Two of the commercially available preparations, *Vibrio cholerae* (California Biochemicals) and *Clostridium perfringens* (Worthington), have been used successfully in these studies without apparent interference by potentially contaminating enzymatic activities. Where highly purified neuraminidase preparations may be required, a simplified purification procedure has been described recently which utilizes the principle of affinity column chromatography.[5]

Galactose oxidase (20 units/mg): The commercial preparation obtained from General Biochemicals can be used directly as provided by the manufacturer. However, it is usually desirable to remove the pigment and the endogenous carbohydrate contaminants found in some of the crude enzyme preparations. This can be readily accomplished as follows[2]: The contents of one vial, 125 units, are dissolved in 5 ml of 0.025 M sodium phosphate buffer, pH 7.0, and applied to a column of DEAE-cellulose (1 × 5 cm) which has been equilibrated with the same buffer. The filtrate (10 ml) was made 1% in NaCl and precipitated by the addition of 3 volumes of ethanol:chloroform (9:1, v/v). After brief centrifugation, the pellet was dissolved in a small amount of the above buffer, and any insoluble material was removed by further centrifugation. The specific activity of this preparation increased 9-fold, and the original hexose content of 38% decreased 20-fold. The partially purified enzyme was used shortly after preparation; its stability has not been determined.

Horseradish peroxidase, 400 units/mg, was supplied by Worthington, and sodium borotritide was obtained from New England Nuclear with a nominal specific activity of 200 mCi/mmole. Owing to the greater stability of the potassium salt, potassium borotritide has been utilized in later studies and is preferable to the sodium salt.

Procedure

Preparation of the Asialoprotein. Ceruloplasmin (64 mg) was dissolved in 2 ml of 0.1 M sodium citrate-phosphate buffer, pH 5.6, and 1 unit of *C. perfringens* neuraminidase was added. After incubation for 30–35 minutes at 37°, approximately 95% of the sialic acid had been hydrolyzed as determined by the thiobarbituric acid assay of Warren.[6]

[5] P. Cuatrecasas and G. Illiano, *Biochem. Biophys. Res. Commun.* **44,** 178 (1971); see this volume [122].
[6] L. Warren, *J. Biol. Chem.* **234,** 1971 (1959).

The protein solution was diluted with an equal volume of water, the protein was precipitated by the addition of 12 ml of ethanol:chloroform (9:1), and the suspension was stored at room temperature for 1–2 hours. After centrifugation, the supernatant fluid was decanted and the blue pellet was suspended in 4 ml of 0.05 M sodium acetate buffer, pH 5.6, containing 0.15 M NaCl. The insoluble residue was removed by centrifugation, and the clear blue supernatant solution was dialyzed overnight at 0° against 100 volumes of the same buffer. The yield of asialoceruloplasmin ($A_{610}:A_{280}$, 0.0424) was 89%. If desired, this material can be crystallized.[7]

Enzymatic Oxidation of the Terminal Galactose Residues. Asialoceruloplasmin, galactose oxidase, and horseradish peroxidase, in final concentrations of 0.54%, 12.7 units/ml and 18 units/ml, respectively, were incubated at 25° for 66 hours at pH 7.0 in a buffer containing 0.02 M sodium phosphate, 0.045 M sodium acetate, 0.15 M NaCl, and 1% toluene. After dilution with 2 volumes of water, the pH was adjusted to 5.6 with 1.0 M acetic acid and the solution was passed through a column of one-third of its volume of DEAE-cellulose equilibrated with 0.05 M sodium acetate, pH 5.6, containing 0.05 M NaCl. The column was washed with the same buffer, and the blue protein was eluted by increasing the NaCl concentration of the above acetate buffer to 0.15 M. Treatment of the eluate with 3 volumes of ethanol-chloroform and dialysis, as in the preparation of asialoceruloplasmin, effected further purification. Examination of the final preparation ($A_{610}:A_{280}$, 0.041) revealed that 94% of the galactose residues originally present in the asialoceruloplasmin had been oxidized.[8]

Reduction with Tritiated Borohydride. The above oxidized asialoceruloplasmin was diluted to 0.2% protein with a buffer containing 0.05 M sodium phosphate, pH 7.8, and 0.05 M NaCl. To this solution was added 1.6 μmoles of tritiated sodium borohydride for each milligram of protein. After 10 minutes at room temperature, excess borohydride was discharged by lowering the pH to 5.6. The pH was then readjusted to 7.0, and the mixture was applied to a column of DEAE-cellulose (6 ml/10 mg of protein), equilibrated with a buffer containing 0.05 M sodium phosphate, pH 7.0, and 0.05 M NaCl. The adsorbed protein was washed on the column with this buffer and eluted by increasing the NaCl concentration to 0.22 M. The eluate was again treated with ethanol-chloroform and dialysis as above. The labeled material ($A_{610}:A_{280}$, 0.045) ex-

[7] A. G. Morell, I. Sternlieb, and I. H. Scheinberg, *Science* **166**, 1293 (1969).
[8] A. G. Morell, R. A. Irvine, I. Sternlieb, I. H. Scheinberg, and G. Ashwell, *J. Biol. Chem.* **243**, 155 (1968).

hibited a specific activity of 3.45×10^6 dpm/mg. After acid hydrolysis, more than 94% of the tritium was recovered in the neutral sugar fraction; galactose was identified as the sole radioactive product. In terms of its spectral properties, copper and galactose content, enzymatic activity, and immunological properties, the final preparation was indistinguishable from the starting material.[8]

Comments

In applying this procedure to glycopeptides or glycoproteins other than ceruloplasmin, it is clear that the specific purification steps employed must be adapted to the properties of the protein being labeled. However, the procedure can be simplified in that it is usually not necessary to isolate the product after each step providing that a means is available for purification of the final preparation.

For those proteins tolerating precipitation with ethanol–chloroform, this treatment is helpful in removing small quantities of added neuraminidase. It is not, however, a foolproof procedure. An alternate and preferable method for removing neuraminidase would appear to be by affinity gel chromatography.[5]

The amount of galactose oxidase to be used will vary with the terminal galactose content of the glycoprotein treated. In general, it has been found that 20 units of enzyme will usually oxidize about 5 μmoles of such protein bound galactose in 24 hours. In many cases, it is possible to monitor the oxidation step by utilizing the colorimetric assay embodied in the "galactostat" kit marketed by Worthington Corporation, Freehold, New Jersey. This assay is not applicable, however, to ceruloplasmin.

Owing to the lability and the losses frequently encountered in working with small quantities of radioactive borohydride, this material is usually added in 4- to 5-fold molar excess of the amount of aldehyde to be reduced. However, for larger-scale preparations, the cost of this reagent may become appreciable and the borotritide can be used stoichiometrically. In order to ensure completeness of reduction under these conditions, it is advisable to add an excess of cold borohydride at some point prior to isolation of the labeled protein.

[15] Tritium-Labeling of Glycoproteins That Contain Sialic Acid

By LEE VAN LENTEN and GILBERT ASHWELL

Principle. The recent demonstration that sialic acid plays a critical role in determining the catabolism of circulating glycoproteins revealed that labeling procedures requiring prior treatment with neuraminidase were generally unsuitable for most sialic acid-containing glycoproteins.[1] This objection has been overcome in the present method, which describes the preparation of a radioactive derivative of sialic acid remaining covalently bound to the intact glycoprotein.[2] Exposure to periodic acid, under carefully controlled conditions, permits selective cleavage of the two distal exocyclic carbons of sialic acid and results in the formation of an aldehyde product. Reduction with tritiated borohydride yields the derivative, 5-acetamido-3,5-dideoxy-L-arabino-2-heptulosonic acid, bearing a tritium atom on carbon 7 and with retention of the glycosidic linkage to the penultimate galactosyl residue (Fig. 1).

A detailed description of the preparation of tritiated orosomucoid is provided here. This technique has been applied successfully to ceruloplasmin[2] and to the glycoprotein hormones, human chorionic gonadotropin[3] and follicle-stimulating hormone.[4,5] In the latter two cases, the biological activity of the radioactive hormones remained essentially intact in marked contrast to the loss of activity incurred by the removal of only a fraction of the sialic acid residues.[6]

Reagents

Orosomucoid prepared by the method of Whitehead and Sammler, scaled up for larger quantities[2]

Potassium borohydride-T obtained from New England Nuclear with a nominal activity of 324 mCi per mmole.

Buffer A: 0.1 M sodium acetate, pH 5.6, containing 0.15 M NaCl

[1] A. G. Morell, G. Gregoriadis, I. H. Scheinberg, J. Hickman, and G. Ashwell, *J. Biol. Chem.* **246**, 1461 (1971).

[2] L. Van Lenten and G. Ashwell, *J. Biol. Chem.* **246**, 1889 (1971).

[3] J. Vaitukaitis, J. Hammond, G. Ross, J. Hickman, and G. Ashwell, *J. Clin. Endocrinol. Metab.* **32**, 290 (1971).

[4] J. Vaitukaitis, R. Sherins, G. T. Ross, J. Hickman, and G. Ashwell, *Endocrinology* **89**, 1356 (1971).

[5] M. Suttajit, L. E. Reichert, Jr., and R. J. Winzler, *J. Biol. Chem.* **246**, 3405 (1971).

[6] E. V. Van Hall, J. Vaitukaitis, G. T. Ross, J. W. Hickman, and G. Ashwell, *Endocrinology* **88**, 456 (1971).

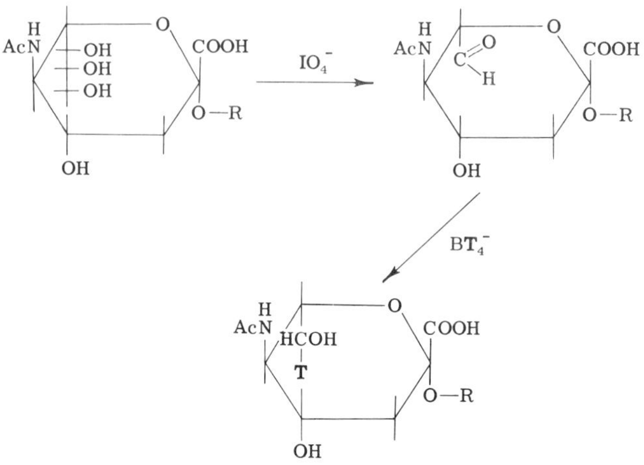

FIG. 1. Chemical conversion of glycosidically bound sialic acid to the tritiated (T) 5-acetamido-3,5-dideoxy-L-arabino-2-heptulosonic acid derivative.

Procedure. To 18.6 ml of a solution containing 41 mg of orosomucoid (15.8 μmoles of sialic acid) dissolved in cold buffer A was added 12.3 ml of $0.012 M$ sodium periodate. After incubation for 10 minutes at $0°$, the oxidation was stopped by the addition of an excess of ethylene glycol and the solution was dialyzed overnight at $4°$ against $0.05 M$ sodium phosphate, pH 7.4, containing $0.15 M$ NaCl.

To the cold, oxidized protein solution was added 1 mg of tritiated potassium borohydride dissolved in 0.2 ml of $0.01 M$ NaOH. After mixing, the solution was warmed to room temperature and the reduction was allowed to continue for 30 minutes with stirring. To ensure completeness of reduction, 10 mg of nonradioactive potassium borohydride was then added, and the incubation was continued for another 30 minutes. Excess borohydride was removed by dialysis against buffer A for several hours prior to exhaustive dialysis against water. After lyophilization, 38.5 mg of the modified protein was recovered.

At the end of the oxidation step, 14.1 moles of formaldehyde had been released and 28.9 moles of periodate had been consumed per mole of orosomucoid. These ratios were calculated on a dry weight basis and an assumed molecular weight of 40,000.[7] The specific activity of the final product was 11.5×10^6 dpm per mg. The specific activity of orosomucoid, prepared in the absence of periodate, was 0.55×10^6 dpm per mg. This nonspecific incorporation of tritium, amounting to 4–5% of the total

[7] R. W. Jeanloz, *in* "Glycoproteins" (A. Gottschalk, ed.), p. 362. American Elsevier, New York, 1966.

radioactivity, had been observed previously with ceruloplasmin.[8] In neither case was the tritium associated with the carbohydrate residues.

Comments. In the design of this method, care was exercised to define conditions such that maximal oxidation of the sialyl residues was achieved with minimal oxidation, or none, elsewhere in the molecule.

Similar conditions have been used in preparing the corresponding derivative of ceruloplasmin.[2] In both cases, the reaction was carried out at 0° with a periodate concentration of 4.75 mM and a 9- to 10-fold molar excess of periodate to sialic acid. The initially rapid consumption of periodate was essentially completed within 10–15 minutes. Excessive oxidation with periodate, because of the possible destruction of the underlying carbohydrate chains of the glycoprotein, or incomplete reduction with borohydride, is potentially harmful.

Thus, nonreduced preparations of both human chorionic gonadotropin and follicle-stimulating hormone proved to be largely inert as opposed to the almost full retention of biological activity of the completely reduced and tritiated material. It is clear, however, that conversion of the normal sialic acid residues to the 7-carbon analog is not without effect upon specific properties of the glycoprotein; *Clostridium perfringens* neuraminidase cleaves the modified sialic acid residues at a significantly diminished rate.[9] Furthermore, the capacity of such altered glycoproteins to inhibit agglutination of human erythrocytes by influenza virus has been shown to be markedly reduced.[9]

It is worthy of note, however, that the present method of labeling when combined with previously described incorporation of tritium into the terminal galactose of asialoproteins[10] permits a direct examination of the metabolic sequelae specifically related to the removal of sialic acid and the consequent exposure of the underlying carbohydrate moiety.

---

[8] A. G. Morell, C. J. A. Van Den Hamer, I. H. Scheinberg, and G. Ashwell, *J. Biol. Chem.* **241**, 3745 (1966).

[9] M. Suttajit and R. J. Winzler, *J. Biol. Chem.* **246**, 3398 (1971).

[10] This volume [14].

[16] Carbohydrate Antigens: Coupling of Carbohydrates to Proteins by Diazonium and Phenylisothiocyanate Reactions

By CHERYL R. McBROOM, COLLEEN H. SAMANEN, and I. J. GOLDSTEIN

Haptens are defined as small molecules that are incapable of stimulating antibody formation unless they are chemically linked to a macromolecule. Karl Landsteiner was instrumental in originating procedures such as diazotization whereby low molecular weight haptens may be rendered immunogenic by their covalent conjugation to protein.[1]

Goebels and Avery and their colleagues extended the pioneering work of Landsteiner by coupling the diazonium salts of *p*-aminobenzyl and *p*-aminophenyl glycosides of a variety of sugars to horse serum globulin.[2] Antibodies to scores of carbohydrates have been prepared by this procedure (see the table). A modification of this method by Westphal and Feier has become a standard technique in immunochemistry.[3]

The diazotization coupling reaction, however, rarely proceeds with more than 60% efficiency of coupling and is not specific for the side chain of any single specific amino acyl residue. Diazonium salts attack primarily tyrosyl, histidyl, and lysyl residues of proteins, diazo linkages to tryptophanyl and arginyl residues also being formed when the diazonium reactant is used in large excess.[4-6]

A means whereby the specificity and efficiency of coupling reactions could be increased was suggested by a consideration of various commonly used and well characterized protein reagents, such as phenylisothiocyanate (Edman reagent).[7] Under properly controlled conditions, phenylisothiocyanate reacts almost exclusively and nearly quantitatively with primary amines; under more vigorous conditions, it will also alkylate thiol and hydroxy groups before denaturing the protein.[8] The principle utilized

---

[1] K. Landsteiner, "The Specificity of Serological Reactions." Harvard University Press, Cambridge, Massachusetts, 1945.

[2] W. F. Goebel and O. T. Avery, *J. Exp. Med.* **50**, 521 (1929).

[3] O. Westphal and H. Feier, *Chem. Ber.* **89**, 582 (1956).

[4] A. N. Howard and F. Wild, *Biochem. J.* **65**, 651 (1957).

[5] E. W. Gelewitz, W. L. Riedeman, and I. M. Klotz, *Arch. Biochem. Biophys.* **53**, 411 (1954).

[6] M. Tabachnick and H. Sobotka, *J. Biol. Chem.* **235**, 1051 (1960).

[7] P. Edman, *Acta Chem. Scand.* **4**, 277 (1950).

[8] H. Fraenkel-Conrat, *in* "The Enzymes" Vol. 1 (P. D. Boyer, H. Lardy, and K. Myrbäck, eds.), p. 597. Academic Press, New York, 1959.

in this procedure involves the synthesis of carbohydrate derivatives containing the phenylisothiocyanato functional group[9] which serves to alkylate the amino terminal and lysyl ϵ-amino groups of a protein.

This article describes the coupling reactions of carbohydrates to proteins using either the diazotization reaction or the phenylisothiocyanate procedure recently developed in this laboratory.

The Diazonium Coupling of Aminophenyl Glycosides to Protein

Principle

The principle involved in this method is illustrated by the series of reactions below:

$$\text{Sugar-O}-\underset{}{\bigcirc}-\text{NH}_2 \xrightarrow[\text{HCl, O}^\circ]{\text{NaNO}_2} \text{Sugar-O}-\underset{}{\bigcirc}-\text{N}\equiv\text{N}^+\ \text{Cl}^-$$

$$\text{Sugar-O}-\underset{}{\bigcirc}-\text{N}\equiv\text{N}^+\ \text{Cl}^- + \text{Protein} \xrightarrow[\text{pH}]{\text{alkaline}} \text{Sugar-O}-\underset{}{\bigcirc}-\text{N}=\text{N-Protein}$$

Procedure

Preparation of p-Aminophenyl α-D-Galactopyranoside. To a methanol solution[10] (50 ml) of p-nitrophenyl α-D-galactopyranoside (0.473 g, 1.57 mmole; Cyclo Chemical Co., Los Angeles) in a round-bottom flask equipped with a magnetic stirrer is added platinum oxide (Adams catalyst; ca. 50 mg). The flask is connected to a hydrogenation apparatus which operates at atmospheric pressure. After three alternate evacuations and flushings with H_2, the hydrogenation is allowed to proceed with stirring at room temperature until the calculated volume of H_2 is consumed (approximately 1 hour). The resulting clear solution is filtered to remove catalyst (caution: the catalyst is pyrophoric, and the filter must not be allowed to dry) and, after evaporation of the solvent, p-aminophenyl α-D-galactopyranoside is obtained as needles, mp 177–178°, yield, 0.29 g, 68%.

Diazotization of p-Aminophenyl α-D-Galactopyranoside and Coupling to Bovine Serum Albumin. The p-aminophenyl glycoside (0.2 g, 0.74 mmoles) is dissolved in ice cold 0.1 M HCl (10 ml), and ice cold aqueous

---

[9] D. H. Buss and I. J. Goldstein, *J. Chem. Soc. C* **1968**, 1457 (1968).

[10] It is generally preferable to dissolve the sugar glycoside in water (1 ml) before the addition of methanol. Some glycosides may require the addition of more water for solution.

PHYSICAL CONSTANTS OF PHENYLGLYCOSIDES USED IN PROTEIN CONJUGATION

Sugar derivative	Melting point (°C)	$[\alpha]_D$ (°C)	$[\alpha]_D$ solvent	Protein used for conjugation
p-Nitrophenyl β-D-cellobioside	245–246[a]	−85.1	40% CH3OH	BSA,[a,f] human γ-globulin[a]
p-Aminophenyl β-D-cellobioside	255–256d[b]; 238–239[a]; 245d[b]	+51.3; −52.9	50% CH3OH; 50% CH3OH	
p-Nitrobenzyl β-D-cellobioside	199–200[c]	−32.3	H2O	Horse serum globulin[c]
p-Aminobenzyl β-D-cellobioside	188–190d[c]	−35.2	H2O	Horse serum globulin[c]
p-Nitrobenzyl β-D-cellobiuronic acid methyl ester	188–189[c]	−48.1	CH3OH	
p-Aminobenzyl β-D-cellobiuronide	—	—	—	Horse serum globulin[c]
p-Aminophenyl β-L-colitoside	75[d]	+99	CH3OH	BSA, ovalbumin[d]
p-Aminophenyl α-L-colitoside	180[d]	−166	CH3OH	BSA, ovalbumin[d]
p-Nitrophenyl α-L-fucoside	197[e]			
p-Aminophenyl α-L-fucoside	175[e]			
p-Nitrophenyl β-D-galactoside	180–182[g,h]; 173–175[i]	−204	CH3OH	Horse serum albumin,[e] BSA[f]
p-Aminophenyl β-D-galactoside	158–159[g,h]; 153[i]	−40.5; −40.3	CH3OH; CH3OH	Horse serum globulin, ovalbumin[a]; BSA,[i] porcine and bovine γ-globulin[h]
p-Nitrophenyl α-D-galactoside	169[e]	+225	H2O	
p-Aminophenyl α-D-galactoside	178[e]	+224	CH3OH	Horse serum albumin[e]
p-Nitrobenzyl β-D-galactoside	161–162[j]	−32.9	CH3OH	
p-Aminobenzyl β-D-galactoside	89–90[j]	−50.5	CH3OH	Horse serum globulin,[j] BSA[f]
p-Aminophenyl β-D-galacturonic acid methyl ester	108–110[j]	−75.8	H2O	Horse serum globulin[j]

Compound	$[\alpha]$	M.p.	Solvent	Protein
p-Nitrophenyl β-D-glucoside	−79.6	165[a,g]	CH₃OH	BSA,[a,f] human γ-globulin,[a] ovalbumin[a]
p-Aminophenyl β-D-glucoside	+58.6	156–157[a]	CH₃OH	
		160d[h]		Horse serum globulin,[g] porcine and bovine γ-globulin[h]
p-Nitrophenyl α-D-glucoside	+227.9	216–217[k]	CH₃OH	
p-Aminophenyl α-D-glucoside	+194.1	185–186[k]	CH₃OH	Horse serum globulin,[k] BSA[f]
p-Nitrobenzyl β-D-glucoside	−47.7	156–157[l]	CH₃OH	Horse serum globulin[l]
p-Aminobenzyl β-D-glucoside	−61.8	142–143[l]	H₂O	Horse serum globulin[l]
p-Aminobenzyl β-D-glucuronide				
p-Nitrophenyl N-acetyl-β-D-glucosaminide	−25.5	204[m]	Pyridine	
p-Aminophenyl N-acetyl-β-D-glucosaminide	+12.6	228[m]	H₂O	Various albumins[m]
o-Aminophenyl β-D-glucuronide				Hemocyanin[n]
p-Nitrophenyl β-gentiobioside	−79.8	221–223[b]	H₂O	
p-Aminophenyl β-gentiobioside	−46.8	237–238[b]	H₂O	BSA[f]
p-Nitrobenzyl β-gentiobioside	−49.7	120[c]	H₂O	
p-Aminobenzyl β-gentiobioside				Horse serum globulin[o]
p-Nitrophenyl β-lactoside	−36.4	260d[h]		
p-Aminophenyl β-lactoside	+6	242d[h]	H₂O	Horse serum globulin[p]
		233[b]	H₂O	Porcine and bovine γ-globulin,[h] BSA[f]
p-Nitrophenyl β-maltoside	+35.3	221[b]	H₂O	
p-Aminophenyl β-maltoside		91–92[b]	50% CH₃OH	Horse serum globulin,[p] BSA[f]
p-Nitrophenyl α-D-mannoside	+161	182[e]	CH₃OH	
p-Aminophenyl α-D-mannoside	+128	164[e]	CH₃OH	Horse serum globulin,[e] BSA[f]
p-Nitrophenyl β-panoside	+40[t]		H₂O	
p-Aminophenyl β-panoside				BSA[q]

(Continued)

TABLE (Continued)

Sugar derivative	Melting point (°C)	[α]D (°C)	[α]D solvent	Protein used for conjugation
p-Nitrophenyl α-L-rhamnoside	179[e]	−144	CH$_3$OH	
p-Aminophenyl α-L-rhamnoside	166–167[e]			Horse serum albumine[e]
p-Nitrophenyl β-sophoroside	261–262[r]	−67.9	H$_2$O	
p-Aminophenyl β-sophoroside	211–212[r]			BSA[s]

[a] G. J. Gleich and P. Z. Allen, *Immunochemistry* **2**, 417 (1965).
[b] F. H. Babers and W. F. Goebel, *J. Biol. Chem.* **105**, 473 (1934).
[c] W. F. Goebel, *J. Exp. Med.* **68**, 469 (1938).
[d] O. Lüderitz, O. Westphal, A. M. Staub, and L. LeMinor, *Nature (London)* **188**, 556 (1960).
[e] O. Westphal and H. Feier, *Chem. Ber.* **89**, 582 (1956).
[f] I. J. Goldstein and R. N. Iyer, *Biochim. Biophys. Acta* **121**, 197 (1966).
[g] W. F. Goebel and O. Avery, *J. Exp. Med.* **50**, 521 (1929).
[h] J. Yariv, M. M. Rapport and L. Graf, *Biochem. J.* **85**, 383 (1962).
[i] S. M. Beiser, *J. Mol. Biol.* **2**, 125 (1960).
[j] W. F. Goebel, *J. Exp. Med.* **66**, 191 (1937).
[k] W. F. Goebel, F. H. Babers, and O. T. Avery, *J. Exp. Med.* **55**, 761 (1932).
[l] W. F. Goebel, *J. Exp. Med.* **64**, 29 (1936).
[m] O. Westphal and H. Schmidt, *Justus Liebigs Ann. Chem.* **575**, 84 (1951).
[n] I. Cornell and L. Wofsy, *Immunochemistry* **4**, 183 (1967).
[o] W. F. Goebel, *J. Exp. Med.* **72**, 33 (1940).
[p] W. F. Goebel, O. T. Avery, and F. H. Babers, *J. Exp. Med.* **60**, 599 (1934).
[q] R. S. Martineau, P. Z. Allen, I. J. Goldstein, and R. N. Iyer, *Immunochemistry* **8**, 705 (1971).
[r] R. N. Iyer and I. J. Goldstein, *Carbohyd. Res.* **11**, 241 (1969).
[s] P. Z. Allen, I. J. Goldstein, and R. N. Iyer, *Biochemistry* **6**, 3029 (1967).
[t] R. N. Iyer, Ph.D. dissertation, State University of New York (1968).

0.05 M NaNO$_2$ (15 ml) is added dropwise with stirring. Diazotization of the aromatic amine requires one molar equivalent of HNO$_2$ per mole of amine. NaNO$_2$ is therefore added until a slight excess of HNO$_2$ is formed as determined by the formation of a blue color with starch-iodide indicator paper. Isolated diazonium salts are unstable and hence are best handled in aqueous solution. The solution of p-diazophenyl α-D-galactopyranoside is added with stirring to an ice cold solution (50 ml) of crystalline bovine serum albumin (BSA, 1.0 g) in 0.15 M NaCl, which has been adjusted to pH 9.0 with 0.5 M NaOH. The reaction, which becomes deep orange in color, is allowed to proceed at 0° for 2 hours, during which time the pH is maintained at 9.0 by the addition of 0.5 M NaOH. Following neutralization with 0.05 M HCl, the orange-red sugar-BSA conjugate is dialyzed exhaustively against distilled water, lyophilized, and the freeze dried powder stored in the refrigerator.

The Conjugation of Phenylisothiocyanato Glycosides to Proteins

Principle

Procedure

Preparation of p-Isothiocyanatophenyl β-D-Glucopyranoside. Thiophosgene (0.13 ml, 1.69 mmoles) is added to a magnetically stirred solution of p-aminophenyl β-D-glucopyranoside (202 mg) in 80% aqueous ethanol (25 ml), and the reaction mixture is allowed to stand at room temperature for 1.5 hours. (This procedure is performed in a well ventilated hood.) Within 45 minutes the pH generally drops to less than 2.0, and the orange solution becomes almost colorless. At this point, thin-layer chromatography (chloroform–methanol, 1:1, v/v) shows that all the starting material has reacted and that a single product has formed. Concentration almost to dryness leaves a solid to which water is added. Filtration, and washing the product with water, yields the phenylisothiocyanate derivative (0.200 g, 86%), mp 195–197°, which has a broad infrared band centered at 2130 cm^{-1}. Recrystallization from ethanol gives needles, mp 195–197°, $[\alpha]_D^{22}$ − 34.7° (ca. 1.09 in N,N-dimethylformamide).

Coupling of p-Isothiocyanatophenyl β-D-Glucopyranoside to Bovine Serum Albumin. A solution of bovine serum albumin (200 mg) in 0.15 M NaCl (20 ml) is brought to pH 9.0 by the addition of 0.1 N NaOH. *p*-Isothiocyanatophenyl β-D-glucoside (150 mg) is added in small portions to the magnetically stirred protein solution. The pH is readjusted to 9.0 with 0.1 N NaOH, and the reaction is allowed to continue at room temperature for 6 hours with further additions of 0.1 N NaOH solution as required to maintain the pH at 9.0. Although many of the aromatic isothiocyanate derivatives are sparingly soluble in aqueous solution, the reaction appears to proceed smoothly, perhaps because of continued solubilization of the phenylisothiocyanate derivative. The solution is refrigerated overnight, and on the following day the pH is adjusted to 7.0. Unreacted carbohydrate derivative is removed from the protein by dialysis against 0.15 M NaCl followed by passage through a Sephadex G-25 column (30 × 2.3 cm), eluting with 0.15 M NaCl while monitoring the optical density at 280 nm. Because the aromatic carbohydrate moiety contributes significantly to the molar extinction coefficient of the conjugate, it is necessary either to dilute the more highly concentrated fractions or to monitor their optical density at another wavelength. The subsequent elution of the unreacted *p*-isothiocyanatophenyl β-D-glucopyranoside may also be followed at 280 nm. Thin-layer chromatography (chloroform–methanol, 1:1 v/v) on the pooled fractions of the low molecular weight material shows the free hapten to be altered, and hence not recoverable.

Exhaustive dialysis does not result in the complete removal of the phenylisothiocyanato derivative. Contaminating free hapten may have no deleterious effect on the antigen to be used for immunization, but it may seriously interfere with the antigen used to determine the presence of precipitating antibody in the serum of an immunized animal since the soluble hapten molecules may compete with the conjugated protein for the antibody combining sites. In addition, the presence of free hapten may interfere with the determination of the extent of protein modification.

Determination of the Number of β-D-Glucopyranosido Residues Incorporated per Mole of Bovine Serum Albumin. The carbohydrate content of the conjugate is quantitated by the phenol–sulfuric acid method.[11] The dry weight of the protein–carbohydrate conjugate is estimated from its total nitrogen content, using the ninhydrin method of So and Goldstein.[12] From the nitrogen and carbohydrate analyses, it is estimated

---

[11] J. Hodge and B. Hofreiter, *in* "Methods in Carbohydrate Chemistry," Vol. 1 (R. L. Whistler and M. L. Wolfrom, eds.), p. 388. Academic Press, New York, 1962.

[12] L. L. So and I. J. Goldstein, *J. Biol. Chem.* **242**, 1617 (1967).

that 54 molecules of β-D-glucopyranosido residues are incorporated per molecule of bovine serum albumin, which has 57 lysyl groups.[13] Assuming only N-terminal and lysyl modification, the coupling efficiency is about 90%.

Comment

We have used the phenylisothiocyanate procedure to couple p-isothiocyanatophenyl β-D-glucopyranoside, p-isothiocyanatophenyl β-L-glucopyranoside, and p-isothiocyanatophenyl α-D-mannopyranoside to bovine serum albumin. However, the method may readily be extended to a variety of noncarbohydrate haptens as well as to other natural or synthetic macromolecules. For example, p-isothiocyanatophenyl derivatized haptens may be reacted with solid matrices (i.e., aminoethyl or p-aminobenzyl cellulose) to afford solid column adsorbents for affinity chromatography.

[13] T. Peters, Jr., *Clin. Chem.* **14**, 1147 (1968).

[17] Carbohydrate Antigens: Coupling of Carbohydrates to Proteins by a Mixed Anhydride Reaction

By GILBERT ASHWELL

Principle

The preparation of steriod-protein conjugates was described originally by Erlanger and his colleagues,[1] who, employing the mixed anhydride technique, established the covalent nature of the conjugate wherein the carboxyl group of the steroid was linked by an amide bond to the lysine residues of the protein. Subsequently, Arakatsu *et al.*[2] utilized the same basic procedure to couple isomaltose and isomaltotriose to bovine serum albumin. In this case, however, it was necessary to convert the neutral oligosaccharides to the corresponding aldonic acid derivatives. The mixed anhydride, formed by the condensation of the sugar acids with isobutylchloroformate, decomposed in the presence of protein at an alkaline pH with the formation of an oligosaccharide–protein conjugate and the release of CO_2. This material was antigenic in rabbits and gave rise to specific antibodies with cross reactivity toward dextran.[2]

[1] B. F. Erlanger, F. Borek, S. M. Beiser, and S. Lieberman, *J. Biol. Chem.* **228**, 713 (1957).
[2] Y. Arakatsu, G. Ashwell, and E. A. Kabat, *J. Immunol.* **97**, 858 (1966).

Materials and Methods

Oligosaccharides of the isomaltose series can be prepared by either enzymatic or acid hydrolysis of the dextran obtained from *Leuconostoc mesenteroides*, NRRL-B-512. This dextran, containing approximately 96% of α-1 $\rightarrow$ 6 linkages, was hydrolyzed at 100° in 0.33 N H_2SO_4 until the apparent conversion to glucose was 52%—a value appropriate for a maximum yield of the tri- to pentasaccharides. The neutralized hydrolyzate was fractionated on a charcoal–Celite column and further purified by large-scale paper chromatography.[3]

Isobutylchloroformate (stabilized with calcium carbonate) was obtained from Eastman Organic Chemicals. All other chemicals were of reagent grade quality and were purchased from readily available commercial sources.

Analytical procedures employed for quantitation of the carbohydrate content included the following colorimetric assays: orcinol, phenol–sulfuric, anthrone, and cysteine–sulfuric acid. The assays were performed as described previously.[4]

Descending paper chromatography was carried out at room temperature using Schleicher and Schuell paper, grade 589 Green Ribbon-C. Sugars and sugar acids were detected by means of the alkaline silver nitrate reagent.[5]

The following solvent systems were employed: solvent A, butanol–pyridine–water (6:4:3); solvent B, butanol–acetic acid–water (4:1:5).

Procedure

Preparation of Isomaltonic Acid. Bromine oxidation of the various oligosaccharides of the isomaltose series was carried out essentially as reported by Glattfeld and Hanke.[6] The procedure used for the preparation of isomaltonic acid is given in detail. To 0.5 g of isomaltose, dissolved in 5.0 ml of water, were added 0.42 g of lead carbonate and 0.17 ml of bromine. The flask was stoppered, shaken for 40 minutes and stored in the dark at room temperature until oxidation was complete. After 6 days, paper chromatographic examination of the reaction mixture in solvent systems A and B revealed no trace of the starting material. Excess bromine was removed by aeration, and the suspension was filtered. The colorless solution was stirred with a small amount of silver carbonate, refiltered, and passed over a column containing 2 g of Amberlite IR-

[3] J. R. Turvey and W. J. Whelan, *Biochem. J.* **67**, 49 (1957).
[4] E. A. Kabat, *in* Kabat and Mayer's "Experimental Immunochemistry," 2nd ed. Thomas, Springfield, Illinois, 1961.
[5] E. F. L. T. Anet and T. M. Reynolds, *Nature (London)* **174**, 930 (1954).
[6] J. W. E. Glattfeld and M. T. Hanke, *J. Amer. Chem. Soc.* **40**, 973 (1918).

120(H⁺). Upon lyophilization, 0.426 g of product was recovered. The lyophilized preparation was shown to consist of a mixture of intra- and intermolecular lactones as judged by the appearance of numerous silver nitrate reacting components upon paper chromatography in solvent B. Elution of each of these areas, followed by neutralization with dilute sodium hydroxide, resulted in the formation of a single product. Consequently, the entire preparation was converted to the sodium salt for storage. Analysis of the glucose content of this material gave the following values: orcinol, 38%; phenol–sulfuric acid, 42%; anthrone, 44%; and cysteine–sulfuric acid, 40%; theoretical, 47%. Values obtained for the free acid were: orcinol, 44%; phenol–sulfuric acid, 53%; and anthrone, 54%; theoretical, 50%.

Conjugation of Isomaltonic Acid to Protein. Bovine serum albumin, 3 μmoles, was dissolved in a mixture of 3.5 ml of water and 3.0 ml of dimethyl formamide (DMF). The pH was adjusted to 11.5 with 0.1 N NaOH, and the mixture was chilled in an ice bath. Approximately 600 μmoles of the free aldonic acid, dissolved in 2 ml of cold DMF containing 600 μmoles of tri-N-butylamine, was mixed with 600 μmoles of isobutylchloroformate. The mixture was kept at 0° for 15 minutes, at which time a second addition of 1.2 mmoles of tri-N-butylamine was made and the contents of the flask promptly added to the above solution of bovine serum albumin. The reaction mixture, adjusted to pH 10.5 with 0.1 N NaOH, was maintained at 0° for 1 hour and then allowed to stand at room temperature for 15–20 hours. The resulting suspension, dialyzed against distilled water with daily changes for 4 days, was centrifuged and both the soluble and the insoluble fractions were lyophilized. The aldonic acid content of the soluble product was 8 moles of isomaltonic acid per mole of BSA, as determined by anthrone assay (see Comments). A sample of bovine serum albumin was subjected to the conditions of the mixed anhydride reaction with omission of the oxidized isomaltose oligosaccharide.

Comments

The above technique provides a rapid and facile adjunct to the more widely used method of coupling diazo-sugar compounds to proteins and offers the advantage of conjugation of any available oligosaccharide without prior chemical synthesis of the appropriate aromatic glycoside. The extent of substitution of the 59 theoretically available lysine residues in bovine serum albumin, however, has been generally low, and further studies of the reaction conditions are needed to improve the yield of bound carbohydrate. This disadvantage may be more apparent than real since the product prepared by the condensation of isomaltotriose, and higher homologs, to BSA has been shown to give rise to specific antibodies with cross reactivity toward dextran.

It should also be pointed out that recent experience with this method[7] has revealed that a significant amount of noncovalently bound sugars may remain associated with the product even after prolonged dialysis. This loosely bound material can be readily removed by chromatography on Biogel-P-6, and this step is essential for accurate determination of the covalently attached sugar residues. In this context, it is clear that the carbohydrate content of the isomaltose– and isomaltotriose–BSA conjugates may be somewhat lower than the reported values.[2]

[7] G. Ashwell and E. A. Kabat, unpublished observations.

[18] Carbohydrate Antigens: Coupling of Carbohydrates to Proteins by Diazotizing Aminophenylflavazole Derivatives

By KARL HIMMELSPACH and GERD KLEINHAMMER

Flavazoles were first described by Ohle and co-workers,[1] who found that reducing monosaccharides, when allowed to react with o-phenylenediamine and phenylhydrazine under moderately acid conditions, readily form 1-phenylflavazoles (1-phenyl-$1H$-pyrazolo[3,4-b]quinoxalines). The reaction has been shown to work with di- and trisaccharides[2] and with higher oligosaccharides,[3-6] provided that the oligosaccharide carries no substituent at positions 2 and 3 adjacent to the reducing end.

Recently it was found that the flavazole reaction opens a route for coupling oligosaccharides to proteins,[7] thus converting them to haptenic groups on antigens, if m-nitrophenylhydrazine is used instead of the un-

[1] H. Ohle and G. A. Melkonian, *Ber. Deut. Chem. Ges.* **74**, 279 (1941); **74**, 398 (1941).

[2] G. Neumüller, *Ark. Kemi* **21A**, 1 (1945).

[3] P. Nordin and D. French, *J. Amer. Chem. Soc.* **80**, 1445 (1958).

[4] P. W. Robbins and T. Uchida, *Fed. Proc., Fed. Amer. Soc. Exp. Biol.* **21**, 702 (1962).

[5] T. Kobayashi, T. Haneishi, and M. Saito, *Nippon Nogei Kagaku Kaishi* **36**, 189 (1962); *Chem. Abstr.* **61**, 10759f (1964).

[6] P. W. Robbins, J. M. Keller, A. Wright, and R. L. Bernstein, *J. Biol. Chem.* **240**, 384 (1965).

[7] K. Himmelspach, O. Westphal, and B. Teichmann, *Eur. J. Immunol.* **1**, 106 (1971).

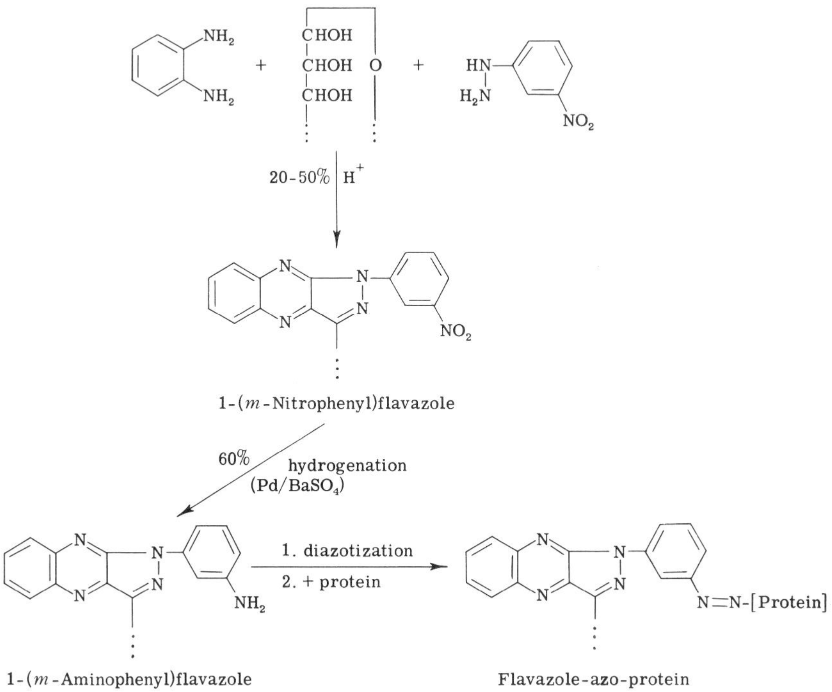

Fig. 1. Synthesis of oligosaccharide-flavazole-azo-protein conjugates.

substituted phenylhydrazine. This reaction, as shown in Fig. 1, yields the 1-(m-nitrophenyl)flavazole of the respective oligosaccharide, which is subsequently reduced to the corresponding 1-(m-aminophenyl)flavazole by catalytic hydrogenation. Finally, the latter is diazotized and allowed to react with protein to yield a conjugate, in which sugar moiety and protein are covalently linked.

The procedure does not necessitate protection of the sugar hydroxyl groups at any step. This especially makes it applicable to oligosaccharides having a degree of polymerization ≥3 and thus is superior to the classical method of coupling aminophenylglycosides as introduced by Goebel and Avery.[8] A method more comparable to the flavazole technique has been described by Arakatsu et al.[9] and consists of coupling oligosaccharide aldonic acids to the free amino groups of proteins by a mixed anhydride reaction with the aid of isobutylchlorocarbonate. However, azo coupling, as applied in the flavazole technique appears to be

[8] W. F. Goebel and O. T. Avery, J. Exp. Med. 50, 521 (1929).

[9] Y. Arakatsu, G. Ashwell, and E. A. Kabat, J. Immunol. 97, 858 (1966).

less hazardous, and, moreover, it can be carried out in aqueous solution without addition of organic solvents.

Previous studies[7] with flavazole–protein conjugates have been mainly concerned with their applicability as immunogens, i.e., for eliciting antibodies to oligosaccharides upon injection into animals. Conjugates of oligosaccharide flavazoles with the hemp seed protein edestin (MW 310,000) were found to be efficient immunogens in rabbits, giving rise to specific antioligosaccharide antibodies in all instances. These results and results obtained by Hämmerling[10] suggest edestin to be a useful carrier protein for carbohydrate haptenic groups.

Properties of the Flavazole Intermediates

All flavazoles concerned are yellow compounds, which can be safely stored at room temperature if moisture is excluded and long exposures to daylight are avoided. Those derived from oligosaccharides with a degree of polymerization ≥ 3 are readily soluble in water and some other polar solvents such as dimethyl formamide or pyridine. They are almost insoluble in ethyl alcohol and insoluble in dimethyl ether.

Chromatographic Behavior. Nitro- and aminophenylflavazoles derived from different oligosaccharides consisting of about 3–8 monosaccharide units are efficiently separable by paper chromatography on Whatman 3 MM paper with n-butanol–pyridine–water (6:4:3, by volume) as eluent. By virtue of their yellow color and of their fluorescence under UV light (nitrophenylflavazoles: bright yellow; aminophenylflavazoles: dark bluish) the zones containing flavazoles can be spotted on the chromatograms most conveniently without staining. This is exemplified in Fig. 2, showing the fluorescent bands of the series of isomaltose type oligosaccharide-nitrophenylflavazoles on a circular paper chromatogram.

After elution from paper, further purification of single flavazoles is achieved by passing them through Sephadex G-25 columns. During this process flavazoles are considerably retarded by adsorption to the gel matrix and thus freed from nonflavazole impurities emerging from the column earlier. However, this type of gel chromatography is not very effective in separating different flavazoles from each other. Absorption chromatography on charcoal, which is excellent for separating free oligosaccharides, cannot be applied to flavazoles, as these are too strongly (almost irreversibly) absorbed.

[10] U. Hämmerling, Dissertation, University of Freiburg (1965). The carrier properties of several proteins conjugated with *Salmonella typhimurium* polysaccharide were compared with respect to titers of antibodies formed in rabbits upon injection. Edestin proved to be far better than any other protein under investigation.

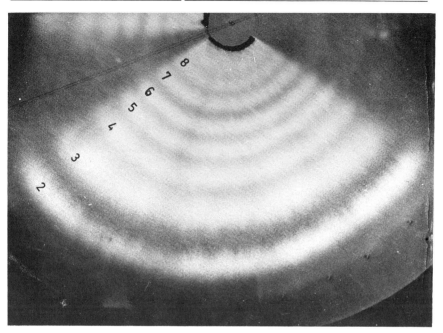

FIG. 2. Fluorescent zones of isomaltose type oligosaccharide 1-(m-nitrophenyl)-flavazoles on a circular paper chromatogram after development with n-butanol–pyridine–water (6:4:3, by volume) photographed under UV light. The numbers indicate the degree of polymerization of the respective parent oligosaccharide (e.g., 2 = isomaltose nitrophenylflavazole, 3 = isomaltotriose nitrophenylflavazole, etc.).

UV-Absorption Spectra. All flavazoles thus far investigated (including phenylflavazoles unsubstituted in the phenyl ring[11]) exhibit a typical light absorption pattern with marked peaks near 400 nm, at 335 nm, and in the region of 247–267 nm. In studies with isomaltose type oligosaccharide nitro- and aminophenylflavazoles,[7] whose absorption spectra are shown in Fig. 3, the molar absorptivities were found to obey Lambert–Beer's law and to be independent of the sugar moiety of the respective flavazole.

The latter finding appears to have general validity. Therefore colorimetric estimations of the flavazole intermediates may conveniently be carried out by measuring the light absorption at 335 nm or at 267 nm using the extinction coefficients given in the legend of Fig. 3. In the case of aminophenylflavazoles, the pH dependence of their spectra has to be considered; i.e., for readings at 267 nm the pH of the solution should be brought to a value below 3, for readings at 247 nm to a value above 7.

[11] P. Nordin and M. Doty, *Science* **134**, 112 (1961).

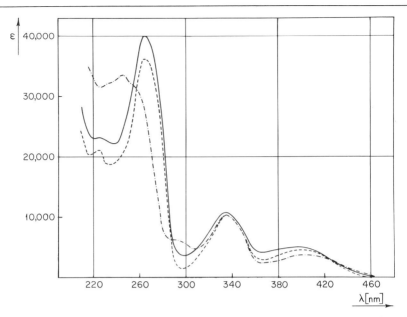

Fig. 3. Absorption spectra of nitro- and aminophenylflavazoles, measured with Cary 15 spectrophotometer. ——, 1-(m-nitrophenyl)flavazoles (IM4–IM7), solvent: water; $\epsilon_{267} = 40,000$; $\epsilon_{335} = 10,800$. - - - -, 1-(m-aminophenyl)flavazoles (IM4–IM7), solvent: 0.1 N HCl; $\epsilon_{267} = 36,300$, $\epsilon_{335} = 10,400$. · — · —, 1-(m-aminophenyl)flavazoles (IM4–IM7), solvent: 0.05 N NaOH; $\epsilon_{247} = 33,700$, $\epsilon_{335} = 10,400$. Abbreviations: IM4 = isomaltotetraose; IM7 = isomaltoheptaose. (From K. Himmelspach et al. (7); reproduced with permission of Verlag Chemie GmbH, Weinheim/Bergstr., Germany.)

By virtue of their characteristic bands the spectra are also suited for identifying flavazoles as such and for controlling their purity.

Procedures

Chemical Preparations

1-(m-Nitrophenyl)flavazoles. In an ampoule, 0.5 mmole of the oligo-saccharide, 72 mg of o-phenylenediamine (0.66 mmole), 672 mg of m-nitrophenylhydrazine sulfate (3.3 mmoles), and 6 ml of water are mixed and covered with nitrogen. The ampoule is sealed and heated in a steam bath (100°) for 4.5 hours. After cooling to room temperature, the mixture is diluted with water to about 250 ml, neutralized (NaOH), and extracted with ether in a Kutscher-Steudel extractor. The aqueous phase is then evaporated to dryness under diminished pressure. Subsequently, the residue is chromatographed on Whatman 3 MM paper [de-

scending, using n-butanol–pyridine–water (6:4:3, by volume) as eluent];
6–8 sheets (57 × 46 cm) are needed for the above quantity. In general
one strongly yellow zone and a number of faintly yellow zones with
higher R_f values are obtained on each chromatogram. The strongly
yellow zone containing the flavazole of the original sugar, is cut out and
eluted (the others contain lower molecular weight flavazoles due to hydrolytic cleavage of glycosidic linkages under the conditions of flavazole
formation). It is important that during the whole process of chromatography including drying of the paper sheets and elution, daylight, at
least for longer periods, should be fairly well excluded. The eluted material is rechromatographed on paper and then fed to the top of a
Sephadex G-25 column (bed volume about 600 ml) and eluted with
water containing 1% n-butanol. The flavazole fraction, emerging from
the column later than the bulk of impurities, is collected and lyophilized,
yielding the product as a fluffy yellow material. If the preparation is
to be followed by conversion of the nitro- to the aminophenylflavazole,
Sephadex chromatography and lyophilization may eventually be omitted,
the eluates from paper being subjected to hydrogenation immediately.

NOTES: The m-nitrophenylhydrazine sulfate used should have analytical grade purity. Commercial technical grade preparations may be
purified by recrystallization from ethanol with addition of charcoal before filtration.

The purification steps for nitrophenylflavazoles described above pre-
sume water solubility of the respective products, which is lacking in
the case of disaccharide nitrophenylflavazoles. These are better purified
by chromatography on silica gel (eluent: ethyl acetate–pyridine gradient)
followed by crystallization.

1-(m-Aminophenyl)flavazoles. Nitrophenylflavazole, 0.125 mmole, is
dissolved in 30 ml of water. To the solution, 30 ml of glacial acetic acid
and 20 mg of palladium on barium sulfate (10% Pd) are added. The
hydrogenation is carried out at room temperature and atmospheric pressure[12] and is normally completed within 2 hours. The catalyst is then
filtered off, the filtrate evaporated to dryness under diminished pressure
and the residue chromatographed on Whatman 3 MM paper (3–4 sheets)
using the procedure described for nitrophenylflavazoles. The zones containing aminophenylflavazole show a strong yellow color in daylight
and a dark bluish fluorescence under UV light. Elution and rechromatography on paper are followed by chromatography on a Sephadex G-25
column, using water containing 1% n-butanol as eluent. After some dark

---

[12] For a description of a suitable apparatus see A. I. Vogel, "Practical Organic
Chemistry," 2nd ed., p. 458 f. Longmans, Green, New York, 1951.

material has emerged from the column, the yellow aminophenylflavazole fraction is collected. From this the product is obtained by lyophilization. The yields are about 60%.

Diazotization. Thirty micromoles of 1-(*m*-aminophenyl)flavazole are dissolved in 4 ml of water. The solution is placed in an ice bath and stirred magnetically; 1.5 ml of 0.1 *N* HCl is added and then 1.2 ml of a solution of sodium nitrite in water containing 2 mg/ml. Cooling and stirring are continued for 30 minutes.

Azo Coupling to Protein. Generally, the coupling is effected by adding the above diazonium salt solution with cooling (ice bath) and stirring to a solution of an appropriate quantity of the protein in enough dilute NaOH or buffer (e.g., glycine–NaOH) to maintain a pH >9 during the reaction (exceptionally, for proteins sensitive to alkalic treatment or for cell bound proteins, the pH may be as low as 8.0). The following procedure for coupling to edestin is provided as an example. This protein requires rather strongly alkaline conditions for dissolution. The proposed procedure yields conjugates containing about 90–100 hapten groups per molecule, which is a favorable range for using them for immunizations. *Coupling of flavazole to edestin:* A solution of 50 mg of edestin in 12 ml of 0.05 *N* NaOH is placed in an ice bath. To this the diazonium salt solution prepared as described above is added with stirring. The mixture is left to stir for 3 hours, dialyzed against 4 liters of glycine–NaOH buffer (0.05 *M*; pH 10), then concentrated by vacuum filtration and filtered through a Sephadex G-75 column (bed volume 300–500 ml). The elution is followed spectrophotometrically, and the first (slightly yellow) fraction emerging from the column is collected and adjusted to a concentration of about 1 mg of protein per milliliter by vacuum filtration. For *determination of the coupling ratio* in the products, the sugar content of the solution may be estimated by the anthrone method[13] and the protein content by the Folin procedure[14] using calibration curves established with the respective sugar or protein.

Immunological and Serological Procedures

Immunization of Animals. There is great variability with respect to species of animal and immunization schedule used. With rabbits as experimental animals, experiments have shown that injections of sugar–protein conjugates frequently repeated in the early stages of immunization may lead to low titers of anti-sugar antibodies. The following pro-

---

[13] E. A. Kabat and M. M. Mayer, "Experimental Immunochemistry," 2nd ed., p. 528. Thomas, Springfield, Illinois, 1961.

[14] E. A. Kabat and M. M. Mayer, *ibid.*, p. 556.

cedure,[15] leaving a period of 5 weeks between the first and the second injection, proved to be distinctly superior to a variety of other schedules tried.[7] It may lead to titers of several milligrams of anti-sugar antibodies per milliliter of serum and is recommended.

In 1 ml of an emulsion containing 0.66 ml of complete Freund's adjuvant (FCA, cf. note below this description), 250 μg of antigen is injected into the thighs of the hind legs intramuscularly (doses are given per animal). After a 37-day interval, the animals receive 200 μg of antigen in 1 ml, containing 0.5 ml of FCA, intraperitoneally (or, alternatively, intramuscularly into the thighs of the hind legs). After 20 more days, 20 μg of antigen in 0.2 ml, containing 0.1 ml of FCA, is injected intraperitoneally. The animals are bled 11 days after the last injection.

NOTE: In order to achieve an even distribution of antigen solution in FCA, mixing has to be effected with the aid of an efficient mechanical blendor under cooling in an ice bath. It should be carried out until a stiff pasty emulsion is obtained. The degree of mixing may be considered satisfactory, if a sample brought onto the surface of water does not decompose rapidly into oil droplets and an aqueous phase.

Serological Tests. In the majority of cases a polysaccharide containing the oligosaccharide determinant of the immunizing antigen will be available and may be used for testing the sera. *Precipitin tests* are carried out by mixing serial dilutions of the polysaccharide in saline with equal volumes of the sera to be tested and incubating the mixture for several days at about 4°. The maximum amount of precipitated antibody is found by measuring the protein contents of all the precipitates and plotting these against polysaccharide dilution. A standard procedure for this and procedures for determining protein by the Kjeldahl method or by the Folin method are described by Kabat and Mayer.[16] Use of hemagglutination for the detection of anti-sugar antibodies precludes a suitable method for erythrocyte sensitization. A technique described by Hämmerling and Westphal[17] is based on the observation that polysaccharides, after partial esterification with fatty acids, are readily absorbed by red cell surfaces. Esterification with stearic acid has proved to be highly effective in this sense. Sensitization of erythrocytes is achieved by incubation of a 1% cell suspension with minute amounts of O-stearoyl polysaccharides ($\sim$10 μg/ml) for 30 minutes at 37°. For testing the sera, these are serially diluted in saline. To each serum dilution an equal volume of a 1 or 0.5% suspension of the sensitized cells is

---

[15] This procedure was originally suggested by N. A. Mitchison (London, 1968; personal communication).

[16] See footnote 13, *ibid.*, pp. 72–78, 476, 556.

[17] U. Hämmerling and O. Westphal, *Eur. J. Biochem.* **1**, 46 (1967).

added. After mixing, the cells are allowed to sediment (30 minutes at 37°, overnight at 2°). After this period the end point of titration, i.e., the last serum dilution showing agglutination, is read.

Scope and Limitations

All restrictions on applications of the flavazole technique for preparing oligosaccharide–protein conjugates apply to the first step of the procedure. The major limitation is that only *reducing* oligosaccharides, which are *not substituted in the positions 2 and 3* adjacent to the reducing end can form flavazoles.

If these requirements are fulfilled, the yields obtainable depend largely on the stability of the glycosidic linkages of the oligosaccharide toward hydrolysis under the conditions required for flavazole formation (pH 2 at 100° for about 4 hours). Usually the yields decrease with increasing molecular weight of the sugar. Thus within the series of isomaltose-type oligosaccharides, the yield was 50% for isomaltotetraose and

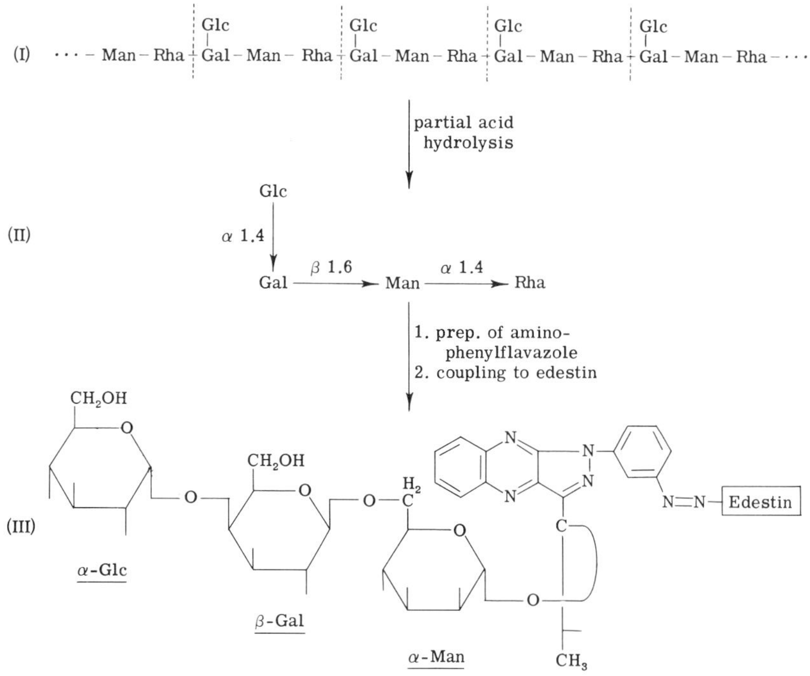

Fig. 4. Preparation of a flavazole-azo-edestin conjugate (III) from the tetra-saccharide II representing the repeating unit of the *Salmonella illinois* lipopolysaccharide O-specific side chain (I).

20% for isomaltooctaose. Experiments with maltose- and cellobiose-type oligosaccharides led to similar results.

An example for application of the method to an oligosaccharide of bacterial origin is shown in Fig. 4.[18] The oligosaccharide, a tetrasaccharide (II) composed of glucose, galactose, mannose, and rhamnose, represents the repeating unit of the *Salmonella illinois* lipopolysaccharide O-specific side chain (I)[19] and was prepared from the polysaccharide by partial acid hydrolysis. The latter was effected using a device described by Galanos *et al.*[20] in which oligosaccharides already formed are removed from the hydrolysis mixture by dialysis and thus protected from further fragmentation. Subjected to the flavazole technique, 50 mg of purified tetrasaccharide yielded 18 mg of nitrophenylflavazole (27.5% of theory) from which 5.4 mg of amine (31%) was obtained and allowed to react with edestin to give 10 mg of conjugate (III) containing 2.5 mg of the hapten (105 hapten groups per molecule of edestin).

The sera obtained upon injection of this antigen into rabbits were tested by bacterial agglutination and by passive hemagglutination (erythrocytes sensitized with *S. illinois* lipopolysaccharide), the titers averaged 1:1280 and 1:512, respectively. Bacterial cross agglutinations showed that the antibodies obtained were predominantly directed against the grouping Glc-Gal rather than against Gal-Man-Rha. Dextran-specific antibodies, the amounts of which averaged 1–5 mg per milliliter of serum, were obtained in rabbits immunized with conjugates of edestin with flavazoles derived from isomaltose-type oligosaccharides.[7]

Flavazole-azo-protein conjugates may also be used as test antigens for serological reactions, especially if water-soluble carrier proteins are chosen. By coupling of flavazoles to appropriate insoluble carrier materials, immunadsorbents may be prepared. Thus diazotized aminophenylflavazoles were shown to couple readily to the N-(p-hydroxyphenethyl) derivative of polyacrylamide, which is a commonly used matrix for specific biochemical adsorbents.[21]

[18] G. Kleinhammer, Dissertation. University of Freiburg (1972).
[19] O. Lüderitz, A. M. Staub, and O. Westphal, *Bacterial Rev.* **30**, 192 (1966).
[20] C. Galanos, O. Lüderitz, and K. Himmelspach, *Eur. J. Biochem.* **8**, 332 (1969).
[21] J. K. Inman and H. M. Dintzis, *Biochemistry* **8**, 4074 (1969).

[19] Preparation of Antisera against Glycolipids

By SEN-ITIROH HAKOMORI

Purified glycolipids with rare exceptions are not immunogenic. Therefore antisera directed against glycolipid have been prepared by injection of glycolipids complexed with heterologous macromolecules. Three methods can be distinguished according to the nature of the complex used as immunogen: (1) Cell membranes or subcellular particulate fraction in which glycolipids are naturally complexed with membrane proteins; (2) a chemically modified glycolipid which is covalently linked to a carrier polymer; and (3) a mixture of micellar solution of purified glycolipid and a heterologous carrier polymer.

The antibodies prepared by the first method are directed not only to multiple number of glycolipids present in membrane, but also to other membrane components. The method is useful, however, to detect the natural antigenicity of heteroglycans that are organized in membrane.[1] The reported examples are listed in the table.

The second method, based on Landsteiner's classic principle, has been used for sphingolipid as listed in the table. The specificity of antibodies directed against such a simple glycolipid–protein compound may not be identical to that directed against glycolipid complexed on natural membrane. The method requires large quantities of purified glycolipid, therefore it is not useful for minor glycolipid components.

The third method has been widely used because of its simplicity, wide applicability, and requirement for only a few milligrams of glycolipids; the method is useful not only for production of antiglycolipid, but also for production of antiphospholipid antibodies.[2,3] As the carrier polymers, crystalline bovine serum albumin,[4-8] methylated bovine serum albumin,[9] erythrocyte membrane protein,[5] and mycoplasma membrane protein,[7] have been used.

Examples only of the third method will be described in this article,

[1] P. Häyry and V. Defendi, *Virology* **41**, 22 (1970).
[2] K. Inoue and S. Nojima, *Biochim. Biophys. Acta* **144**, 409 (1967).
[3] T. Kataoka and S. Nojima, *J. Immunol.* **105**, 502 (1970).
[4] J. Koscielak, S. Hakomori, and R. W. Jeanloz, *Immunochemistry* **5**, 441 (1968).
[5] D. M. Marcus and R. Janis, *J. Immunol.* **104**, 1530 (1970).
[6] B. Siddiqui and S. Hakomori, *J. Biol. Chem.* **246**, 5766 (1971).
[7] S. Hakomori, E. Wasserman, and L. Levine, *Immunochemistry* in preparation.
[8] S. Hakomori, C. Teather, and H. D. Andrews, *Biochem. Biophys. Res. Commun.* **33**, 563 (1968).
[9] T. Yokoyama, E. G. Trams, and R. O. Brady, *J. Immunol.* **90**, 372 (1963).

Methods for Preparation of Antisera against Glycolipids

Method 1: Injection of cells and subcellular particulate fraction
  Anticytolipin K, R, and H[a−d]
  Antigalactosylcerebroside[e]
  Antigloboside[f]
  Antimycoplasmic glycolipid[g]
  Antigloboside, CTH, and hematoside[h]

Method 2: Injection of glycolipid covalently linked to protein or artificial polymers
  N-p-Nitrophenyl (glycosyl sphingosine diazo-coupled to protein[i, j]
  Glycosylsphingosine coupled to synthetic copolymer through carbodiimide[k]

Method 3: Injection of a mixture of glycolipid and heterologous proteins
  Antigloboside[l, m]
  Anti-Forssman glycolipid[n]
  Antisulfatide[o]
  Antihematoside[p]
  Antimycoplasmic glycolipid[q]
  Anti-"lacto-N-fucopentaose III" glycolipid[r]
  Antigangliosides[s−v]

---

[a] M. M. Rapport, L. Graf, V. P. Skipski, and N. F. Alonzo, *Nature (London)* **181,** 1803 (1958).

[b] M. M. Rapport and L. Graf, *Cancer Res.* **21,** 1225, (1961).

[c] M. M. Rapport, L. Graf, and H. Schneider, *Arch. Biochem. Biophys.* **105,** 431 (1964).

[d] M. M. Rapport and L. Graf, *Biochim. Biophys. Acta* **137,** 409 (1967).

[e] S. Joffe, M. M. Rapport, and L. Graf, *Nature (London)* **197,** 60 (1963).

[f] K. Inoue, T. Kataoka, and S. C. Kinsky, *Biochemistry* **10,** 2574 (1971).

[g] B. L. Beckman and G. E. Kenny, *J. Bacteriol.* **96,** 1171 (1968).

[h] M. Naiki and T. Taketomi, *Jap. J. Exp. Med.* **39,** 549 (1969).

[i] T. Taketomi and T. Yamakawa, *Lipids* **1,** 31 (1966).

[j] T. Taketomi and T. Yamakawa, *J. Biochem. (Tokyo)* **54,** 444 (1963).

[k] R. Arnon, M. Sela, E. S. Rachaman, and D. Shapiro, *Eur. J. Biochem.* **2,** 79 (1967).

[l] J. Koscielak, S. Hakomori, and R. W. Jeanloz, *Immunochemistry* **5,** 441 (1968).

[m] D. M. Marcus and R. Janis, *J. Immunol.* **104,** 1530 (1970).

[n] B. Siddiqui and S. Hakomori, *J. Biol. Chem.* **246,** 5766 (1971).

[o] S. Hakomori, E. Wasserman, and L. Levine, *Immunochemistry* in preparation.

[p] S. Hakomori, C. Teather, and H. D. Andrews, *Biochem. Biophys. Res. Commun.* **33,** 563 (1968).

[q] S. Razin, B. Prescott, and R. M. Chanock, *Proc. Nat. Acad. Sci. U.S.* **67,** 590 (1970).

[r] H. Yang and S. Hakomori, *J. Biol. Chem.* **246,** 1192 (1971).

[s] T. Yokoyama, E. G. Trams, and R. O. Brady, *J. Immunol.* **90,** 372 (1963).

[t] A. L. Sherwin, J. A. Londen, and L. S. Wolfe, *Can. J. Biochem.* **42,** 1964 (1964).

[u] T. A. Pascal, A. Saifer, and J. Gitlin, "Inborn Disorders of Sphingolipid Metabolism" (S. M. Aronson and B. W. Volk, eds.), p. 289. Pergamon, Oxford, 1967.

[v] M. M. Rapport, L. Graf, and R. Ledeen, *Fed. Proc., Fed. Amer. Soc. Exp. Biol.* **29,** 573 (1970).

since it is the most practical method for most cases. The reader should refer to references listed in the table for the first and second methods.

Preparation of Antigloboside Antiserum

Material

Crystalline bovine serum albumin (BSA)

Globoside, prepared from human erythrocytes and purified by crystallization

Freund's adjuvant (Difco)

Tubercle bacilli (killed and dried, Difco)

p-Nitrobenzoyl cellulose (Cellex, Calbiochem)

New Zealand white rabbit

Method of Immunization. Globoside, 1–2 mg, is suspended in 1 ml of distilled water and is dissolved with heating until a completely transparent solution is formed [10] (solution 1). BSA, 5 mg, is dissolved in 1 ml of 1.8% sodium chloride solution,[11] and 5 mg of tubercle bacilli are added (solution 2).

Solutions 1 and 2 and 2 ml of incomplete Freund adjuvant are aspirated into a 5-ml injection syringe and make a heavy emulsion by reciprocal movement between two syringes through a double connecting tube.

Aliquots of about 0.3 ml of the emulsion are injected into both flanks in the region of gluteal and shoulder muscles and also into two footpads. After 3–4 weeks, the animal is bled from the central artery of ear. If high titer serum is necessary, a second injection at week 3 is recommended, with bleeding 2 weeks after the second injection. The composition and preparation of globoside–BSA mixture for the second injection is the same as that used for the first injection except that no tubercle bacilli were added.

Purification of Antiserum. The antiserum can be used directly for precipitin reaction with pure globoside but cannot be used for complement fixation and for the reaction with cells and cell membranes because the antisera contain anti-BSA, which interferes with complement fixation by cross-reacting with some components of cells and cell membranes.

The anti-BSA antibodies can be removed by adsorption with BSA. A preliminary titration is necessary to obtain optimal ratio of BSA and

[10] Many glycolipids give a transparent aqueous solution in distilled water but are insoluble in the presence of salt.

[11] The addition of salt is necessary for making a stable heavy emulsion with Freund's adjuvant.

anti-BSA, and the actual amount of BSA added to the antiserum should be slightly less than is indicated by the optimal ratio. The mixture is incubated at 37° for 1 hour and at 2° for 2 days; this is followed by centrifugation at about 20,000 rpm. The trace amount of anti-BSA present in the supernatant is removed by passing the serum through a column of BSA coupled to p-aminobenzoyl cellulose.[12] A column of 0.5 g of cellulose-BSA (wet weight) containing 20 mg of BSA is prepared. Four milliliters of the antiserum is passed through the column, and the column is washed with three column volumes of 0.9% saline. The eluate and washings are dialyzed against 0.01% saline and pervaporated at 4° in a dialysis tube to the original volume (4 ml). The sample is ready for use.

Preparation of Anti-Forssman Glycolipids

The same method as that used for the preparation of antigloboside serum, but only a single injection of 1 mg of Forssman glycolipid, 5 mg of BSA, and incomplete Freund's adjuvant, is sufficient to get a high titer antiserum (Siddiqui and Hakomori, unpublished observation, 1970).

Preparation of Antihematoside and Antisulfatide Antiserum

The same method as that used for the preparation and antigloboside serum is also capable of producing antihematoside and antisulfatide antiserum in rabbit except that two booster injections are used. The antibodies formed in both cases are not precipitating but are complement fixing.[7,8] Anti-N-glycolylhematoside and anti-N-acetylhematosides are distinguishable by a qualitative complement-fixation test. The capability of producing antisulfatide and antihematoside is not greatly enhanced by complexing with methylated BSA as compared to a mixture of glycolipid and nonmethylated BSA. The quantitative complement fixation curve between antisulfatide and sulfatide differ from that between antisulfatide and cerebroside 6-sulfate, and the reaction is completely abolished by desulfation.[7]

Antiganglioside

Gangliosides absorbed on heterologous erythrocytes were reported to be immunogenic to heterologous animals although immunochemical specificities were not known.[9]

The complement-fixing antibody can be formed by injection of a mixture of monosialoganglioside (GM_1) and BSA according to the same method described above for preparation of the antisera. However antibody was not formed by injection of disialoganglioside (GD_{1a} or GD_{1b})

[12] R. V. Davis, R. M. Blanken, and R. J. Beagle, *Biochemistry* **8**, 2706 (1969).

or trisialoganglioside (GT) by the same method. The anti-GM$_1$ antibody is not directed against sialic acid part, but is directed against terminal β-galactosyl $(1 \rightarrow 3)$ N-acetylgalactosaminosyl residue.[13] The complement fixation reaction between monosialoganglioside and its rabbit antiserum was effectively inhibited by phenyl-β-galactoside and lactose (Levine and Hakomori, unpublished observation, 1967).

[13] M. M. Rapport, L. Graf, and R. Ledeen, *Fed. Proc., Fed. Amer. Soc. Exp. Biol.* **29**, 573 (1970).

[20] Isolation of Glycoproteins from Red Cell Membranes Using Phenol[1]

By CALDERON HOWE, KENNETH O. LLOYD,[2] and LUCILLE T. LEE

Background and Principle

The phenol method, based on that originally described by Westphal *et al.*[3] for the extraction of bacterial lipopolysaccharide antigens, has been applied by a number of investigators[4-6] to the preparation of glycoproteins from erythrocytes of various species. The subject has been reviewed elsewhere.[7] While these applications vary somewhat in detail, the underlying principle in all instances is the partition of membrane material between phenol and water. Lipid, and most of the membrane and contaminating nonmembrane proteins, go into phenolic solution, leaving in the aqueous phase a major glycoprotein component. The latter may be further purified by various procedures as indicated hereinafter. The glycoprotein fraction accounts for all the sialic acid of the cell and bears antigenic determinants of the ABO, MN, and perhaps other blood group systems, as well as the sialic acid (N-acetylneuraminic acid, NANA)-containing receptors for myxoviral hemagglutinins. These several biological markers characteristic of the cell surface are useful adjuncts to the assessment of purity of membrane fractions with respect

[1] Supported by NIH Grant No. AI-03168.
[2] NIH Research Career Development Awardee (K4-AI-38823).
[3] O. Westphal, O. Lüderitz, and F. Z. Bister, *Z. Naturforsch. B* **7**, 148 (1952).
[4] R. H. Kathan and R. J. Winzler, *J. Biol. Chem.* **238**, 21 (1963).
[5] C. Howe, S. Avrameas, C. deV. St. Cyr, P. Grabar, and L. T. Lee, *J. Immunol.* **91**, 683 (1963).
[6] G. F. Springer, Y. Nagai, and H. Tegtmeyer, *Biochemistry* **5**, 3254 (1966).
[7] C. Howe and L. T. Lee, *J. Immunol.* **102**, 573 (1969).

to the glycoprotein component, which alone among the several recognized membrane proteins exhibits these properties.

Isolation Procedures

The most advantageous starting material for the isolation of human erythrocyte glycoprotein is hemoglobin-free membranes prepared by hypotonic lysis under controlled conditions of pH and osmolarity, as described in this volume [21]. With this procedure, hemoglobin and most of the other cytoplasmic constituents and some of the nonglycoproteins are lost, yielding ghosts still retaining the full complement of lipids and all the aforementioned biological markers found in the intact cells. Alternatively, relatively crude stroma may be used. For preparation of the latter, packed washed erythrocytes are lysed in 10 volumes of distilled water and the membranes are precipitated at pH 5.5 with 0.01 M acetate buffer, pH 4.0. The resulting flocculum, although heavily contaminated with hemoglobin and other nonmembrane proteins, may serve as starting material for the phenol extraction.[5,7] Application of the same procedures to erythrocytes of other mammalian species is limited by the quantities of blood obtainable, either from individual animals or in the form of pooled samples. Small animals are exsanguinated either by carotid section (guinea pigs) or decapitation (mice). Whole blood is collected in acid citrate dextrose (ACD) solution or in EDTA (0.75 ml of 30% EDTA per 10 ml of whole blood). Five hundred mice (25–30 gm) yield approximately 100 ml of packed erythrocytes.

Extraction with Phenol

Step a. Colorless liquefied phenol[8] is diluted with an equal volume of distilled water, and NaCl to a final concentration of 0.5% is added. Twenty volumes of this phenol–saline mixture are then mixed with one volume of packed sediment (hemoglobin-free ghosts or crude acid-precipitated stroma). The mixture is held at 65–68° (ascertained by direct measurement) for 10 minutes with constant stirring.

Step b. The phenol-water mixture is centrifuged for 2 hours at 1000 g at room temperature. The phenol layer may be extracted successively twice more with 0.45% aqueous NaCl, and these extracts are pooled with the first aqueous layer. The total extract is dialyzed against distilled water until free of phenol, and lyophilized. At this stage, most or all of the detectable sialic acid is found in the aqueous layer. Phenol-soluble material concentrated by precipitation with excess acetone contains at most only traces of acid-hydrolyzable NANA.

---

[8] Phenol, Liquid, 88% Fisher Certified A-931.

Step c. Lyophilized aqueous fraction is taken up in chloroform–methanol (2:1, v:v) at a concentration of 0.3%, and an equal volume of water is added. The whole mixture is dialyzed 16 hours against distilled water, clarified by centrifugation at 100,000 *g* at 4°, and lyophilized. The yield of glycoprotein (virus receptor substance, VRS) is approximately 2% of the dry weight of lyophilized ghosts.

Step d. After the first lyophilization (step c), VRS may give a slightly cloudy or opalescent solution in water indicative of residual lipid. After one further centrifugation at 100,000 *g* for 2 hours at 4°, lyophilization of the supernatant yields a completely water-soluble and colorless product which retains all the biological markers. Analyses of VRS quoted herein were performed on completely soluble samples, certain of them dried to constant weight over phosphorus pentoxide, and redissolved in water to precisely known concentration.

Step e. For the preparation of glycoprotein said to retain maximum M and/or N blood group activity, material is obtained by phenol extraction essentially as described in steps a and b above, except that the temperature of extraction is held at 23–25° rather than 65–68°, and the first low speed centrifugation is done at 4°. The product at this stage may be further purified by differential high speed centrifugation, ethanol fractionation, agar gel and Sephadex G-200 column chromatography as described by Springer *et al.*[6]

Assays for Biological Activity (Blood Group Antigens and Virus Receptors)

Quantitation of blood group activity in erythrocyte glycoprotein is based on titered inhibition of agglutination of erythrocytes of corresponding blood groups by specific isoagglutinin. Such assays are limited to material from donors of groups A, B, AB, MM, MN, and NN because the four isoantigens represented are the only ones present in sufficient quantity to be detectable in soluble form by hemagglutination-inhibition, a relatively insensitive method. Antigenic determinants for other blood groups, which are readily recognized by agglutination of whole erythrocytes with corresponding isoagglutinins, may also be present in soluble erythrocyte subfractions but for quantitative or other reasons remain undetectable by inhibition.

Dilutions of the test antigen in known concentration in $0.15\,M$ NaCl are prepared in a standard volume and are mixed with antiserum at a dilution sufficient to give a strong agglutination reaction in the absence of soluble isoantigen. Tubes are held at room temperature for 1 hour, then centrifuged lightly. The sediment is gently resuspended and examined for macroscopically visible agglutination. Standard antigens of known potency should be included for reference. For this purpose, blood group A or B substances derived from human saliva, gastric mucosa or ovarian cyst fluid, and M or N substances from erythrocytes may be used.[9]

Assays for reactivity with hemagglutinating viruses are performed in an analogous manner. Dilutions of glycoprotein in known concentration in 0.5 ml of phosphate-buffered saline[10] are mixed with 0.1 ml virus at a final dilution representing 4–5 times the hemagglutination titer of the virus alone (4–5 hemagglutinin units). After 15–30 minutes at room temperature, 0.1 ml of washed chicken erythrocytes (2%) is added and the tubes are left undisturbed until the characteristic agglutination pattern of cells becomes visible to the unaided eye. The titer is taken as the highest dilution (lowest concentration) of glycoprotein showing complete inhibition of viral hemagglutination (button pattern). It is desirable to use a strain of influenza virus of known sensitivity to inhibition by glycoproteins. Other agglutinating viruses and appropriately sensitive erythrocytes may be used in inhibition assays, the conditions varying with the specific systems to be examined.[11]

Composition of Erythrocyte Glycoprotein

The composition of erythrocyte glycoprotein as reported by various investigators is summarized in Table I. Although the designations differ (e.g., VRS, MN, or NN substances, stromal inhibitor), all refer to the same macromolecule derived by phenol extraction of erythrocyte membranes. Subsequently applied methods of purification may account for some of the minor differences among the analytical values. The VRS preparations obtained from the blood of group A_1 donors were shown to inhibit specifically hemagglutination of A_1 cells by anti-A_1[7]; MN and NN substances likewise specifically inhibited the corresponding blood group isoagglutinin.[4,7,12] All the human and the simian materials were potent inhibitors of myxovirus hemagglutination, 0.1–0.2 μg VRS inhibiting 4–5 hemagglutinating units of influenza virus (strain FM-1).[7] The murine material showed 1% or less of the activity of the human VRS.

The glycoproteins of three species differed from one another in their content of neutral sugars and hexosamines (Table II). While glucose was absent from the human and murine glycoproteins, it was present in substantial quantity in the rhesus material. The ratio of glucosamine

[9] E. A. Kabat, "Blood Group Substances." Academic Press, New York, 1956.
[10] Phosphate (0.25 M) buffered saline, pH 7.4–7.6: Na$_2$HPO$_4$ 2.22 g; KH$_2$PO$_4$ 0.41 g; NaCl 15 g; water in 2 liters.
[11] C. Howe and L. T. Lee, Advan. Virus Res. **17**, 1 (1972).
[12] R. H. Kathan and A. Adamany, J. Biol. Chem. **242**, 1716 (1967).

TABLE I

Composition of Glycoprotein Isolated from Erythrocyte Membrane

Glycoprotein	% Protein[a]	% NANA[b]	% Neutral sugars[c]	% Hexo amine[d]	Reference
Human					
VRS (4328)	46	15	16[e]	10	h
VRS (9038)	53	17	14[e]	12	h
VRS (4494)	41	21	12.8	11	i
MN Substance	44	24	15.6[f]	14.6	j
NN Substance	44	15	16.8	9.8	k
Stromal inhibitor of myxo-	36	28	17.0[f]	16	l
viral hemagglutinin	ND[g]	23	14.6[f]	13	m
Simian (rhesus) VRS	25	8	36	16	Unpublished
Murine (Swiss) VRS	44	11	21.2	15	Unpublished

[a] Lowry method [O. H. Lowry, N. J. Rosebrough, A. L. Farr, and R. J. Randall, J. Biol. Chem. **193**, 265 (1951)] except value for NN substance derived from amino acid analysis.

[b] Thiobarbituric acid procedure [L. Warren, J. Biol. Chem. **234**, 1971 (1959)] on samples hydrolyzed 20–30 minutes at 80° in 0.1 N H_2SO_4. Value for NN substance averaged from results of 3 methods (cf. Table II).

[c] Cf. Table II.

[d] Elson and Morgan method [L. A. Elson and W. T. J. Morgan, Biochem. J. **27**, 1842 (1933)] on hydrolyzed samples.

[e] From reducing sugar value.

[f] By orcinol method [R. J. Winzler, Method. Biochem. Anal. **2**, 279 (1955)].

[g] ND = not done.

[h] C. Howe and L. T. Lee, J. Immunol. **102**, 573 (1969).

[i] C. Howe, O. O. Blumenfeld, L. T. Lee, and P. C. Copeland, J. Immunol. **106**, 1035 (1971).

[j] R. H. Kathan and A. Adamany, J. Biol. Chem. **242**, 1716 (1967).

[k] G. F. Springer, Y. Nagai, and H. Tegtmeyer, Biochemistry **5**, 3254 (1966).

[l] R. J. Winzler, in "Red Cell Membrane. Structure and Function" (G. A. Jamieson and T. J. Greenwalt, eds.), p. 157. Lippincott, Philadelphia, Pennsylvania, 1969.

[m] R. H. Kathan and R. J. Winzler, J. Biol. Chem. **242**, 1716 (1967).

to galactosamine in the human materials varied; but the hexosamine: galactose ratio was almost always 1. The strikingly lower gluco-samine:galactosamine ratio found in the murine substances may be related to their higher aspartic acid and lower threonine values as compared with either the human or the simian glycoproteins (Table III). The findings suggest that the murine glycoprotein contains a high proportion of carbohydrate-protein linkages involving aspartic acid and glucosamine as compared with the glycoproteins of the other two species, In the latter, the predominant linkages are probably galactosamine–

TABLE II

Neutral Sugars[a] and Hexosamines in Erythrocyte Glycoproteins

Compound	Human												Simian VRS		Murine (Swiss) VRS	
	VRS (4494)[b]		MN substance[c]		NN substance[d]		Stromal inhibitor[e]									
	%	Mole ratio	%	Mole ratio	%	Mole ratio	%	Mole ratio	%	Mole ratio	%	Mole ratio
Fucose	1.0	0.11	1.4[f]	0.11	0.8[f]	0.08	1.2	0.10	1.8	0.08	2.7	0.27
Mannose	2.1	0.20	0	0	5.4	0.49	2.6	0.20	6.2	0.26	7.5	0.72
Galactose	10.7	1.00	14.2	1.00	11.1	1.00	13.2	1.00	23.7	1.00	10.4	1.00
Glucose	0	0	0	0	0.3	0.03	ND[k]	ND	4.3	0.18	0.6	0.06
Hexosamine[g]	11.0	1.05	14.6	1.03	9.8	0.89	15.5	1.18	16.0	0.68	15.0	1.44
Glucosamine[h]	4.3	0.40	4.7[i]	0.33	4.2[i]	0.38	5.3	0.41	12.9	0.55	13.5	1.30
Galactosamine[h]	6.7	0.62	9.9[i]	0.67	4.8[i]	0.44	10.2	0.77	3.1	0.13	1.5	0.13

[a] Determined by gas–liquid chromatography (GLC) of alcohol acetates [J. S. Sawardeker, J. H. Sloneker, and A. Jeanes, *Anal. Chem.* **37**, 1602 (1965); W. F. Lehnhardt and R. J. Winzler, *J. Chromatogr.* **34**, 471 (1968); K. O. Lloyd, *Biochemistry* **9**, 3446 (1970)] except for MN and NN substances analyzed by paper chromatography. VRS = virus receptor substances.

[b] C. Howe, O. O. Blumenfeld, L. T. Lee, and P. C. Copeland, *J. Immunol.* **106**, 1035 (1971).

[c] R. H. Kathan and A. Adamany, *J. Biol. Chem.* **242**, 1716 (1967).

[d] G. F. Springer, Y. Nagai, and H. Tegtmeyer, *Biochemistry* **5**, 3254 (1966).

[e] R. J. Winzler, *in* "Red Cell Membrane. Structure and Function" (G. A. Jamieson and T. J. Greenwalt, eds.), p. 157. J. B. Lippincott, Philadelphia, Pennsylvania, 1969.

[f] Determined by cysteine–sulfuric acid method [Z. Dische and L. B. Shettles, *J. Biol. Chem.* **175**, 595 (1948)].

[g] Determined by Elson and Morgan method [L. A. Elson and W. T. J. Morgan, *Biochem. J.* **27**, 1842 (1933)].

[h] Determined by GLC except where indicated.

[i] Determined by amino acid analyses.

[j] Average of values by 3 methods: amino acid analyses on two different ion exchange resins; paper chromatography.

[k] ND = not done.

TABLE III

AMINO ACID ANALYSES OF ERYTHROCYTE GLYCOPROTEIN
(RESIDUES PER THOUSAND RESIDUES)

	Human				Murine VRS[f]
	VRS (4494)[a,b]	MN substance[c]	NN substance[d]	Stromal inhibitor[e]	
Lysine	45	24	45	36	
Histidine	44	30	39	38	
Arginine	46	41	42	40	
Aspartic	58	76	63	59	103
Threonine	125	127	103	138	63
Serine	144	135	98	135	94
Glutamic	107	121	82	100	52
Proline	69	69	73	66	94
Glycine	54	67	56	67	46
Alanine	55	78	72	67	71
Valine	69	81	75	77	29
Isoleucine	71	67	55	44	21
Leucine	58	69	89	45	
Tyrosine	17	13	28	36	
Phenylalanine	41	28	33	35	

[a] Amino acid analyses were kindly performed by Dr. O. O. Blumenfeld. VRS = virus receptor substance.

[b] C. Howe, O. O. Blumenfeld, L. T. Lee, and P. C. Copeland, *J. Immunol.* **106,** 1095 (1971).

[c] R. H. Kathan and A. Adamany, *J. Biol. Chem.* **242,** 1716 (1967).

[d] G. F. Springer, Y. Nagai, and H. Tegtmeyer, *Biochemistry* **5,** 3254 (1966).

[e] R. H. Kathan and R. J. Winzler, *J. Biol. Chem.* **238,** 21 (1963).

[f] Amino acids for which no values are given were not present in quantity sufficient for accurate analyses.

serine and/or galactosamine–threonine,[13] although carbohydrate linkages to aspartic acid may also occur.

Criteria of Purity

The virus receptor substance (VRS) obtained by the hot phenol method has been found by several criteria to be homogeneous. In poly-

[13] R. J. Winzler, *in* "Red Cell Membrane. Structure and Function" (G. A. Jamieson and T. J. Greenwalt, eds.), p. 157. J. B. Lippincott, Philadelphia, Pennsylvania, 1969.

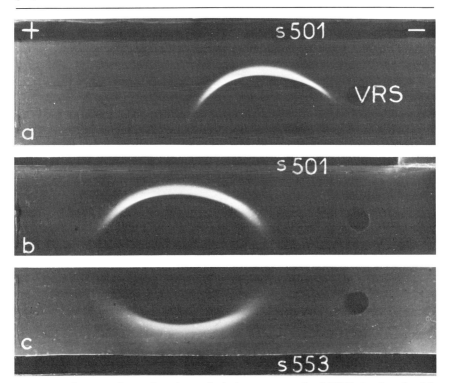

Fɪɢ. 1. Immunoelectrophoretic analysis in agarose gel of VRS developed with antibody to hemoglobin-free ghosts (serum 501, a,b) and to purified VRS (serum 553, c), showing a single component which stained strongly with Schiff reagent. (a) 1.5 hours; (b) and (c) 2 hours. From C. Howe and L. T. Lee, *J. Immunol.* **102**, 573 (1969).

acrylamide gel electrophoresis (PAGE), VRS migrated as a single band.[7] In certain instances, however, as reported elsewhere in this volume (Blumenfeld), a lipid component was visualized as a Schiff-positive band close to the gel front. The question is still unsettled as to what proportions of the various biological activities are carried by this lipopolysaccharide. Unpublished observations have suggested that delipidated VRS retains most of the viral inhibitor, and appreciable blood group activity. With the methods used, it was not possible to quantitate blood group activity in the lipid fraction.

In immunoelectrophoretic analyses of VRS, a single polysaccharide-containing antigen of relatively high mobility was developed with antiserum either to whole membranes or to purified VRS (Fig. 1).[7] The same antigenic determinants were held responsible for the cross-reactivity of

VRS and the component in the sialoprotein fraction derived from erythrocyte membranes solubilized with aqueous pyridine.[14] In quantitative precipitin analysis with the same antisera, VRS behaved as a single antigen, giving zones of antibody and antigen excess.[7,15] Supernatant tests further revealed that, with increasing precipitation, agglutinins for group O erythrocytes were removed from the antiserum. These agglutinins were considered to be of a specificity broader than that of the individual blood group antigens contained in the polysaccharide moiety of VRS, and to be directed mainly to determinants in the protein portion. Accordingly, there was complete cross-reactivity, in double diffusion analyses, between preparations of VRS from individual blood donors regardless of major blood groups. This constitutes another important criterion for the purity of the macromolecule.[7]

Critical factors in the preparation of VRS and MN substances appear to be the temperature at which the initial phenol extraction is carried out, as well as the rigor of subsequent manipulations.[6] The preparations designated herein as VRS were extracted at 65–68°. If extraction is carried out at 23–25° (step e, above) a glycoprotein of greater blood group N potency is obtainable, which is also of larger molecular weight. Some evidence has been adduced to suggest a close relationship between molecular size, NANA content, and biological activity, particularly inhibition of myxovirus hemagglutination and blood group MN antigenic activity.[6,16,17] However, VRS and MN substances were of comparable potency both in viral hemagglutination inhibition and in quantitative precipitin tests with antisera to membrane or to VRS.

The yield of glycoprotein was reported to be enhanced if membranes were extracted first with lithium diiodosalicylate (LIS) and the soluble fraction was extracted with 25% aqueous phenol.[18] Glycoprotein purified in this manner is antigenically identical to VRS and MN substances previously described. Recent electrophoretic studies of various glycoproteins, including the LIS–phenol-derived erythrocyte material, have shown that apparent molecular weights vary with the degree of acrylamide cross-linking (see this volume [5]). In the light of these findings, some of the molecular weights previously determined for phenol-

[14] L. Warren, *J. Biol. Chem.* **234**, 1971 (1959).

[15] C. Howe, O. O. Blumenfeld, L. T. Lee, and P. C. Copeland, *J. Immunol.* **106**, 1035 (1971).

[16] G. F. Springer, *Biochem. Biophys. Res. Commun.* **28**, 510 (1967).

[17] G. F. Springer, H. G. Schwick, and M. A. Fletcher, *Proc. Nat. Acad. Sci. U.S.* **64**, 634 (1969).

[18] V. T. Marchesi and E. O. Andrews, *Science* **174**, 1248 (1971); see also this volume [22].

derived glycoproteins (VRS and MN substances) by PAGE will have to be reevaluated.

[21] Isolation of Glycoproteins from Red Cell Membranes Using Pyridine

By OLGA O. BLUMENFELD and BINA ZVILICHOVSKY

The method depends on the unique solubility properties of glycoproteins which differentiate them from other membrane components. A procedure described below was developed primarily for the isolation of glycoproteins from human erythrocyte membranes.

Procedures for Isolation of Glycoprotein Fraction

Reagents

Isotonic sodium phosphate buffer pH 7.6: Mix 10 volumes of 0.103 M Na_2HPO_4 with 1.13 volumes of 0.155 M NaH_2PO_4. Sterilize by autoclaving.

Hypotonic sodium phosphate buffer pH 7.6: Dilute the isotonic buffer 66 ml to 1 liter. Sterilize by autoclaving.

Sodium acetate buffer, 0.5 M pH 5.0

Acetic acid, 0.5 M

Pyridine:mercaptoethanol (100:0.3): Pyridine is distilled in the presence of 1 g of ninhydrin per liter of pyridine; mercaptoethanol is added to pyridine immediately before use, 0.3 ml/100 ml of pyridine.

Preparation of Human Erythrocyte Membranes. The procedure of Dodge *et al.*[1] was used as follows: Erythrocytes were prepared from 1 unit of either freshly drawn human blood or from blood, not more than 1 week outdated, obtained from the blood bank. The blood was centrifuged at 2000 rpm (595 g) for 20 minutes at 4°, plasma and buffy coat were removed by aspiration, and erythrocytes were washed three times by centrifugation with the isotonic buffer as above; a RC2-B Sorvall

---

[1] J. T. Dodge, C. Mitchell, and D. J. Hanahan, *Arch. Biochem. Biophys.* **100**, 119 (1963).

TABLE I
TYPICAL COMPOSITION OF HUMAN ERYTHROCYTE MEMBRANES AND ITS
VARIOUS FRACTIONS

| | | Water-soluble proteins | | | |
| | | | | | |
Component	Membranes (μg/mg^f protein)	% Total found in membranes	μg/mg^f protein	Glycoprotein fraction (μg/mg^f protein)	Lipid-free glycoprotein (μg/mg^f protein)
Protein^a		35–40			
Lipid				Trace^b	
Cholesterol	200	0.3	2	—	—
Organic phosphate	26	5 or less	3	—	—
Sialic acid^c	27	82	60	940^d	700^d
Hexoses^c (galactose equiv.)	80	25	53	920^d	500^d
N-Acetylhexosamines^c	60	42	70	460	ND^e
Fucose^c	ND	ND	ND	90	ND

^{a} A typical membrane suspension contains about 2.5 mg of protein per milliliter (determined by Kjeldahl nitrogen or amino acid analysis).
^{b} Can be detected by polyacrylamide gel electrophoresis or thin-layer chromatography (see text).
^{c} Procedures used for determinations: see R. G. Spiro, Vol. 8 [1].
^{d} Glycoprotein fraction, average two preparations; in the lipid-free glycoprotein observed considerable variation in five preparations from individual donors, range 560–800 μg per milligram of protein for sialic acid; 380–680 μg per milligram of protein for hexoses.
^{e} ND = not done.
^{f} One milligram protein is equivalent to the sum of residue weights of constituent amino acids ($\sim$ 10 μmoles); values obtained by amino acid analysis.

Centrifuge Rotor GS-3 was usually used. The cells were lysed with the hypotonic buffer in 500-ml transparent centrifuge bottles using 30 ml of cells to 300 ml of buffer. After centrifugation at 7000 rpm (7250 g) for 60 minutes at 4°, the supernatant was removed, and membranes were washed similarly 5–6 times with hypotonic buffer, until free of hemoglobin and creamy white in appearance. About 200 ml of membrane suspension corresponding to approximately 500 mg of protein were obtained from 1 unit of blood. A typical composition of membranes is shown in Table I.

Solubilization of Membranes in Aqueous Pyridine; Isolation of Water-Soluble Proteins.[2,3] Membrane suspensions were dialyzed overnight against

[2] O. O. Blumenfeld, *Biochem. Biophys. Res. Commun.* **30**, 200 (1968).
[3] O. O. Blumenfeld, P. M. Gallop, C. Howe, and L. T. Lee, *Biochim. Biophys. Acta* **211**, 109 (1970).

distilled water prior to pyridine solubilization. Freezing of solutions or the presence of salt during subsequent steps of the procedure should be avoided. To one volume of membrane suspension (the volume may vary from 200 ml, as obtained from the erythrocytes to 2 ml or less when other cells are used) at 4°, is added 0.5 volume of ice-cold redistilled pyridine–mercaptoethanol (100:0.3). The membrane suspension clarifies instantaneously. Occasionally, with membranes which have not been freshly prepared, a slight turbidity remains, but this can be disregarded. The solution is then dialyzed for 16 hours at 4°, against 10 volumes of distilled water. Too prolonged a dialysis against larger volumes of water is avoided, as a complete removal of pyridine and a further decrease of the pH value of the solution can result in the precipitation of most of the proteins. After dialysis, the turbid solution is centrifuged at 90,000 g for 90 minutes, at 4°. The precipitate consists mainly of water-insoluble lipoproteins. In the case of the erythrocyte membranes, the supernatant consists of about 40% of the protein which includes the sialoglycoproteins. The remaining protein and bulk of the lipid are present in the precipitate; some carbohydrate remains with these insoluble components (see Table I). A similar distribution of components is observed after treatment of hepatic cell plasma membranes with aqueous pyridine. In this case, however, the sialic acid-containing components are divided between the soluble and insoluble protein fractions.[4]

Isolation of Glycoprotein Fraction.[5] The supernatant is concentrated about 3-fold by pressure dialysis (membrane ultrafilter UM 10, Diaflo) to contain 2–3 mg protein per milliliter. In the case of other membrane systems where such concentrations might be difficult to attain, solutions of lower concentrations may be used.

STEP I. To one volume of supernatant (volume can vary from 100 ml as obtained in erythrocyte membranes to 5–10 ml when using other membrane systems), at room temperature, are added in the following order: 0.05 volume of redistilled pyridine (without mercaptoethanol), 0.46 volume of absolute ethanol, and 0.12 volume of 0.5 M sodium acetate buffer, pH 5.0. The resulting pH value should be 6.3–6.4 (if necessary, adjustment to this value is made with 0.5 M acetic acid or pyridine). A precipitate forms immediately, and after 10 minutes at room temperature it is centrifuged at 90,000 g for 30 minutes. Initially the rotor and the chamber of the centrifuge should be at room temperature, but the refrigeration is turned on at the start of the centrifugation. The supernatant (called supernatant I) is saved, and the precipitate is discarded.

[4] F. R. Simon, O. O. Blumenfeld, and I. M. Arias, *Biochim. Biophys. Acta* **219**, 349 (1970).
[5] B. Zvilichovsky, P. M. Gallop, and O. O. Blumenfeld, *Biochem. Biophys. Res. Commun.* **44**, 1234 (1971).

STEP II. To one volume of supernatant I is added 0.9 volume of absolute ethanol, and the pH value is adjusted to pH 5.8–5.9 with 0.5 M acetic acid (about 0.07 volume). The solution is left overnight at 4° and centrifuged in the cold for 1 hour at 90,000 g; the precipitate is discarded. The supernatant, called supernatant II, contains the glycoprotein, is free of other proteins, but contains trace amounts of lipid components (see below). It is dialyzed exhaustively against distilled water and then lyophilized to yield the glycoprotein. Usually 15–20 mg of material is obtained from 1 unit of blood (450 ml).

Identification

The presence and purity of this glycoprotein component is determined by sodium dodecyl sulfate polyacrylamide gel electrophoresis.[3,6] Conditions of electrophoresis are usually 5% polyacrylamide gels performed in the presence of 0.1% sodium dodecyl sulfate with 0.1 M sodium phosphate buffer of pH 7.1, at 8 V per centimeter. Gels are prerun for 1 hour. Samples are preincubated in a solution of 1% sodium dodecyl sulfate, 1% mercaptoethanol. 0.02 M sodium phosphate buffer, pH 7.1, for 1 hour at 37° and applied to gels directly, without dialysis. Sucrose is added to samples prior to application to gels at a final concentration of 10%. Pyronine Y is usually used as a front marker in the Coomassie-stained gels. Parallel gels are prepared on each sample, one for detection of protein bands with the Coomassie Blue stain, and the other for detection of carbohydrate components with the periodic acid–Schiff stain. In the former case the gels are fixed in 25% trichloroacetic acid overnight, and stained for 1–1.5 hours in 0.1% Coomassie Blue in 25% methanol and 7% acetic acid; the gels are destained by frequent washings with 7% acetic acid. The parallel gels are fixed overnight in 15% acetic acid, reacted 2 hours with 0.4% periodic acid, washed with 15% acetic acid for 24 hours (with several changes), and stained with the Schiff reagent. The sensitivity of these staining procedures is often not adequate to detect proteins or glycoproteins present at low concentrations, and in other membrane systems the use of the radioactive labels may be required.

A typical gel pattern of the glycoprotein fraction is shown in Fig. 1. Here about 0.25 mg of material was used per gel; this excessive quantity was intentionally employed to ascertain the purity of the preparation by allowing detection of small amounts of nonglycoprotein components. The pattern usually seen consists of a major Coomassie and Schiff stain-

[6] A. L. Shapiro, E. Viñuela, and J. V. Maizel, *Biochem. Biophys. Res. Commun.* **28**, 815 (1967).

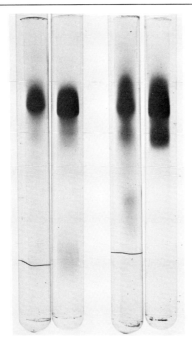

Fɪɢ. 1. Electrophoresis on 5% polyacrylamide gels in 0.1% sodium dodecyl sulfate (pH 7.1). From left to right: (a) glycoprotein preparation; (b) delipidated glycoproteins. Parallel gels stained with Coomassie Blue (left on each pair) and Schiff–periodic acid stain (right of each pair).

ing band, frequently accompanied by one or more Coomassie and Schiff staining bands of lesser intensity which are present close to the major band. Neither aggregated material is present on top of gels, nor are bands staining uniquely with the Coomassie stain present. A faint Schiff staining band may be found moving close to the front of the gel and represents trace amounts of lipid components (see below).

Properties of the Glycoprotein Fraction

Most glycoprotein preparations are slightly turbid in water, and the insoluble material usually consists partly of lipid. Lipid-free glycoprotein (see below) is water soluble. In contrast to other membrane fractions, this glycoprotein preparation shows no aggregation, and its molecular weight determined by sedimentation equilibrium method of Yphantis[7] is 22,000, assuming a partial specific volume of 0.68.

[7] D. A. Yphantis, *Biochemistry* **3**, 297 (1964).

TABLE II
AMINO ACID COMPOSITION

Residue	Glycoprotein preparation[a]	Delipidated glycoprotein[a]
Lys	62	38
His	40	35
Arg	49	43
Cysteic acid	2	ND
Asp	70	69
Thr	101	113
Ser	122	133
Glu	123	113
Pro[b]	66	63
Gly	50	51
Ala	52	58
Met	13	ND
Val	80	81
Ile	73	75
Leu	63	65
Tyr	26	17
Phe	9	14

[a] Values are expressed as residues per 1000 residues. ND = not done.

[b] In certain preparations a high level of proline or a component eluting with proline is found (close to 100 residues/1000 residues).

The glycoprotein component has a unique composition (Tables I and II). The content of carbohydrate is about 60–70%, distributed by weight about equally between sialic acid and other carbohydrate components. Serine, threonine, and glutamic acid are predominant amino acids, and the content of isoleucine is higher than that of leucine. Most membrane protein fractions contain higher levels of leucine. Trace levels of lipid components are also present (see below).

This glycoprotein preparation exhibits the M, N, and A blood group activities (the presence of B blood group activity was not assessed); it is closely related to the preparation isolated by extraction with phenol[8,9,10] in amino acid and carbohydrate compositions, as well as the presence of trace levels of lipids, appearance on polyacrylamide sodium dodecyl sulfate gels (however, the phenol-extracted glycoprotein contains aggregated material on top of gels), and in the presence of antigenic

[8] R. H. Kathan, R. J. Winzler, and C. H. Johnson, J. Exp. Med. 113, 37 (1961).

[9] R. J. Winzler, in "Red Cell Membrane Structure and Function" (G. A. Jamieson and T. J. Greenwalt, eds.), p. 157. Lippincott, Philadelphia, Pennsylvania, 1969.

[10] C. Howe, K. O. Lloyd, and L. T. Lee, this volume [20].

determinants. The preparation is immunologically reactive with phenol extracted glycoprotein.

Comments

The glycoprotein at this stage contains trace amounts of lipid components visualized on gels as a periodate–Schiff positive band close to the front of the gel. Such lipid components can be extracted into ethanol–ether (see below) and detected by thin-layer chromatography of the extract. Some lipid components are more tenaciously retained than others, and several extractions are needed to completely delipidate the glycoprotein. Although common phospholipids and glycolipids are most likely present, the nature, role, and the specificity of association of all the lipid with the glycoprotein is not yet clear. The question as to whether all the blood group activity resides in the delipidated glycoprotein is still not resolved.

Separation of Glycoprotein from Lipid Components

A partial separation of protein and lipid components can be achieved by centrifugation of the glycoprotein preparation in water at 90,000 g for 30 minutes. The clear supernatant contains the glycoproteins. To achieve complete delipidation, the glycoprotein preparation in water (1–2 mg/ml) is extracted at 0° with 5 volumes of ether–ethanol (4:1). The aqueous phase is then lyophilized, redissolved in water, and extracted as above until free of lipid components as assessed by the chromatography of an aliquot of the concentrated ethanol–ether phase, on thin layers of silica gel H developed in chloroform–methanol–water (75:25:4).

To obtain the lipid components, the ethanol–ether phases are evaporated to dryness *in vacuo*, 1 ml of water is added and extraction with ethanol–ether is repeated several times (usually twice).

Properties of Lipid-Free Glycoprotein

The lipid-free glycoprotein is soluble in water; on sodium dodecyl sulfate gel electrophoresis it shows a band pattern identical to that of the glycoprotein fraction except for the absence of the periodate–Schiff staining band close to the front; however, on completely delipidated protein new, faint Coomassie staining bands moving ahead of the major glycoprotein band (see Fig. 1) are now seen; their origin is still unclear.

The delipidated glycoprotein appears to contain less carbohydrate (see Table I), but variability among several preparations from individual donors has been observed, and this observation remains to be investigated.

However, its amino acid composition is very similar to that of the glycoprotein before delipidation. In view of the fact that variability of carbohydrate analysis is found in the delipidated samples, the possible variability of carbohydrate components in the glycoprotein before delipidation should be kept in mind.

[22] Isolation of Membrane-Bound Glycoproteins with Lithium Diiodosalicylate

By V. T. MARCHESI

Many investigators have found that the neutral salt solutions which are commonly used as protein solvents are not suitable for the isolation of glycoproteins from membranes.[1] As a result a wide variety of unusual salts, detergents, and organic solvents have been tested for their capacity to solubilize this class of molecules. Since solutions of lithium diiodosalicylate (LIS) were found to be effective for dissociating model peptides,[2] we tested their capacity to disrupt red cell membranes and found that low concentrations of this salt solubilize glycoproteins from red cell and other membranes.[3] The procedure involving the use of this salt for the extraction of membranes is simple and reproducible and results in almost quantitative recovery of the major glycoprotein of the red cell membrane in a water-soluble and biologically active form.

Isolation Procedures

Preparation of Lithium Diiodosalicylate. LIS is prepared by neutralizing 3,5-diiodobenzoic acid (Eastman No. 2166) with lithium hydroxide (Fisher Scientific) and then crystallizing the product from hot (50°C) water 1–2 times.

3,5-Diiodobenzoic acid should be crystallized twice with methanol before the neutralization step. This will ensure that the lithium salt forms a clear, colorless solution (at 40 g per 100 ml of water) and has an ultraviolet absorbance maximum at 323 nm. LIS can also be obtained from Eastman (No. 11187). LIS can be dissolved in water, Tris·HCl buffer, sucrose, or urea for the extraction of membranes, but most of the other commonly used buffers are not suitable as extracting media since LIS precipitates in the presence of many neutral salts.

[1] S. A. Rosenberg and G. Guidotti, *J. Biol. Chem.* **243**, 1985 (1968).
[2] D. R. Robinson and W. P. Jencks, *J. Amer. Chem. Soc.* **87**, 2470 (1965).
[3] V. T. Marchesi and E. P. Andrews, *Science* **174**, 1247 (1971).

Preparation of Membranes. LIS can be used to extract glycoproteins from red cell membranes prepared by osmotic lysis using either PO_4 [4] or Tris[5] buffers. We routinely extract freeze-dried membranes, but those suspended in water, Tris·HCl buffer, or dilute phosphate buffer are also suitable. The only precaution necessary is to make sure that the concentration of neutral salts is not high enough to precipitate the LIS.

Extraction Procedure. Membranes are suspended in 0.3 M LIS (12 g/100 ml) and 0.05 M tris(hydroxymethyl)aminomethane (Tris) hydrochloride, pH 7.5, at a concentration of approximately 25 mg of membrane protein per milliliter, and stirred at room temperature for 5–10 minutes. Two volumes of distilled water are added, and the suspension is stirred for 5 more minutes at 4°. The suspension is then centrifuged at 45,000 g for 90 minutes at 4°, and the supernatant, which contains most of the membrane proteins, is decanted and mixed with an equal volume of freshly prepared 50% phenol in water. This mixture is stirred vigorously for 15 minutes at 4°, and centrifuged at 4000 g for 1 hour at 4° in a swinging-bucket rotor. The centrifuged material separates into two phases; the upper (aqueous) phase contains most of the soluble glycoprotein. This is removed without disturbing the lipid layer at the interface and is dialyzed against several changes of distilled water at 4° over a period of 24 hours, and then freeze-dried. The dry material is suspended in cold ethanol and mixed for 1–2 hours in the cold, then centrifuged to collect the precipitate. Three ethanol washes are carried out. The sediment is then dissolved in distilled water and dialyzed against water overnight. The material is centrifuged at 10,000 g for 30 minutes at 4° to obtain a clear supernatant containing the soluble glycoprotein.

Purification and Properties of the Isolated Glycoprotein

The glycoprotein isolated from human red blood cell membranes by the above procedure may contain a small amount of contaminating protein and some bound lipid. The nonglycoprotein contaminants can be removed by passing the material through a phosphocellulose column equilibrated with 0.02 M sodium citrate buffer at pH 3.6. Since this glycoprotein is 25% sialic acid by weight and has an isoelectric point below pH 3.0, it emerges as a single peak in the breakthrough. Other glycoproteins which do not have a high content of sialic acid will have to be purified by other methods.

The eluted material is pooled and dialyzed against distilled water overnight and then freeze-dried. The dry protein is then treated with

[4] J. T. Dodge, C. Mitchell, and D. J. Hanahan, *Arch. Biochem. Biophys.* **100**, 119 (1963); see also this volume [21].

[5] V. T. Marchesi and G. E. Palade, *J. Cell Biol.* **35**, 385 (1967).

chloroform–methanol (2:1) three times to remove any residual lipid, and then redissolved in water or buffer.

Approximately 35 to 50 mg of glycoprotein can be extracted from 450 ml of human blood. This amounts to 3–4% of the original dry membranes and represents at least 70–80% of the total glycoprotein of the red cell membrane.

The glycoprotein isolated from human red cells is a single polypeptide chain which is 60% carbohydrate and 40% protein by weight, and has a monomeric molecular weight of approximately 50,000.[6] This molecule carries blood group antigens (MN, A, B, I) and receptors for influenza virus, phytohemagglutinin (kidney bean), and wheat germ agglutinin.

This method has also been used to extract glycoproteins from red blood cells of a variety of different species, and from membranes of lymphoid cells, platelets, and various human tumors.

[6] J. P. Segrest, R. L. Jackson, E. P. Andrews, and V. T. Marchesi, *Biochem. Biophys. Res. Commun.* **44**, 390 (1971).

[23] Isolation of Lipopolysaccharides from Bacteria

By LORETTA LEIVE and DAVID C. MORRISON

Several well known methods are available for isolation of lipopolysaccharides (LPS)[1] from bacteria.[2-6] Two new methods, applicable mainly to *Escherichia coli*, are described. The EDTA method results in a somewhat lower yield than other methods but is very rapid. The aqueous butanol method results in yields equal to or better than other methods. Both procedures would be expected to have a less drastic effect on LPS structure than the commonly used other methods.

[1] Abbreviations used: LPS, lipopolysaccharide; EDTA, ethylenediaminetetraacetic acid; Tris, tris(hydroxymethyl)aminomethane.
[2] O. Westphal, O. Lüderitz, and F. Bister, *Z. Naturforsch. B* **7**, 148 (1952). The method is described by O. Westphal and K. Jann, *in* "Methods in Carbohydrate Chemistry," Vol. V (R. L. Whistler, ed.), p. 83. Academic Press, New York, 1965.
[3] A. Boivin and L. Mesrobeanu, *C. R. Soc. Biol.* **112**, 76 (1933). The method is described by A. M. Staub, *in* "Methods in Carbohydrate Chemistry," Vol. V (R. L. Whistler, ed.), p. 92. Academic Press, New York, 1965.
[4] E. Ribi, K. C. Milner, and T. D. Perrine, *J. Immunol.* **82**, 75 (1959).
[5] M. J. Osborn, S. M. Rosen, L. Rothfield, and B. L. Horecker, *Proc. Nat. Acad. Sci. U.S.A.* **48**, 1831 (1962). The method is described by M. J. Osborn, Vol. 8 [21].
[6] C. Galanos, O. Lüderitz, and O. Westphal, *Eur. J. Biochem.* **9**, 245 (1969).

EDTA Method

Principle. Brief exposure of whole cells of *E. coli* to EDTA releases about 50% of the cell's LPS.[7,8] A very high proportion of the high molecular weight material released from the cells is LPS, so little further purification is necessary. Material that is 70–95% pure by weight can be obtained in two steps: treatment of cells with EDTA and separation of the supernatant fluid by centrifugation, followed by one other step, such as dialysis. Further purification results in material of 98% purity. Various species of *Salmonella* also release LPS under these conditions, but the yield is usually lower (15–45%).

Reagents

Tris·HCl, 0.12 M, pH 8.5 (measured at room temperature)
Sodium-EDTA, 0.5 M, adjusted to pH 8.5 with NaOH
MgCl$_2$, 2 M

Growth of Bacteria. Although bacteria may be grown in any minimal or rich medium, the use of a Tris-based medium, such as the alkaline phosphatase induction medium of Levinthal *et al.*,[9] is recommended since Tris is sometimes toxic to cells not adapted to its presence. This toxicity can cause loss of soluble materials or lysis during EDTA treatment, thus increasing contamination of the isolated LPS. This medium can be made into a minimal medium by omitting peptone and adding 2 mM K$_2$HPO$_4$.

EDTA Treatment. The following method is applicable to between 1 mg and 100 g wet weight of cells. Cells are harvested at a density of between 3 and 8 × 10^8 cells per milliliter and are washed at room temperature with 0.12 M Tris·HCl at pH 8.5. This buffer is used for the following steps unless this pH is observed to cause cell lysis, in which case buffer at pH 8.0 is used instead. When small volumes of cell suspensions (1–500 ml) are used, the undisturbed cell pellet is rinsed with a small volume of buffer; in the case of larger volumes of cells, the cells are resuspended in 0.01–0.1 volume of buffer and washed once by centrifugation. The cells are then resuspended in buffer at a concentra-

[7] L. Leive, *Biochem. Biophys. Res. Commun.* **21**, 290 (1965).
[8] L. Leive, V. K. Shovlin, and S. E. Mergenhagen, *J. Biol. Chem.* **243**, 6384 (1968).
[9] The composition is as follows: 0.12 M Tris·HCl, pH 7.5; 0.08 M NaCl; 0.02 M KCl; 0.02 M NH$_4$Cl; 3 mM Na$_2$SO$_4$; 1 mM MgCl$_2$; 0.2 mM CaCl$_2$; 2 μM ZnCl$_2$; 0.5% glucose; and 0.5% Difco Bacto peptone, pH 7.5. C. Levinthal, E. R. Signer, and K. Fetherolf, *Proc. Nat. Acad. Sci. U.S.A.* **48**, 1230 (1962).

tion of 10–200 mg wet weight/ml[10] and brought to 37°; EDTA is then added to yield a concentration of 0.05 mmoles per gram wet weight of cells. Note that the critical parameter is the amount of EDTA per cell; the concentration per milliliter of either cells or EDTA is immaterial. After 5 minutes of gentle agitation at 37°, $MgCl_2$ is added to yield a concentration 1.2 times the EDTA concentration, and the cells are centrifuged at room temperature.

When working with dense cell suspensions, i.e., >50 mg wet weight per milliliter, a time and speed of centrifugation should be chosen to just bring down the cells since at high concentration LPS solutions may aggregate and, if centrifuged more vigorously, may sediment with the cells. It is, therefore, advisable to centrifuge the supernatant a second time to remove residual cells, rather than centrifuging more vigorously initially. Since "rough" (incomplete) LPS can precipitate in the presence of Mg^{2+},[5] EDTA treatment of cells containing such LPS should be terminated by adding Mg^{2+} exactly equivalent to, not in excess of, the EDTA.

The cell-free supernatant can be freed from small molecules by several methods: (a) Dialysis for 10–24 hours against several changes of any desired buffer. The first several hours of dialysis should be against buffer of high ionic strength to ensure that nucleotides and other charged molecules pass the dialysis barrier. (b) Alcohol precipitation. Ethanol is added slowly at 4° to yield a final concentration of 90% ethanol; after 1–4 hours at 4° the precipitated LPS is collected by centrifugation. (c) Gel filtration. The material is chromatographed on a column of Sephadex G-25 or Bio-Gel P-20, equilibrated with any desired buffer. LPS, which may be assayed by any suitable method, for instance, the general phenol–sulfuric carbohydrate assay,[11] is excluded and the smaller molecules retarded. If necessary the extract may first be concentrated by any method, for instance, in vacuo or by ultrafiltration through collodion tubing under vacuum. (d) Ultrafiltration in an Amicon filter cell. This method is detailed below in the section on Aqueous Butanol Method.

The above procedure yields LPS which is 70–95% pure, depending upon the strain. The main contaminant is protein, with small amounts of phospholipid; if lysis of cells has been avoided (see below), no nucleic

[10] For small numbers of cells, the weight is calculated assuming that 10^9 cells = 1 mg wet weight; for larger preparations the pelleted cells are drained, the top of the pellet and sides of the tube are wiped free of moisture, and weighed.

[11] M. Dubois, K. Gilles, J. K. Hamilton, P. A. Rebers, and F. Smith, Nature (London) 168, 167 (1951).

acid should be present. The LPS may be stored frozen or be lyophilized after removal of salt.[12]

Further Purification. To obtain even purer LPS, the following steps can be added before the removal of small molecules: (1) Phenol extraction: the cell-free supernatant is extracted with 45% phenol at 68° exactly as described by Westphal *et al.*[2] The extraction usually needs to be done only once. The aqueous phase is dialyzed as described above. No further purification is necessary since no RNA is present. (2) Pronase digestion. This procedure is described below in the section on Aqueous Butanol Method; the digest is thereafter subjected to concentration and ultrafiltration as described. Material with less than 2% contamination can be obtained by either of these methods.

Comments. One source of contamination in the product is cell lysis. It is advisable to monitor for lysis during the experiment by checking whether the optical density of the suspension decreases after addition of EDTA. Possible causes of lysis include (a) growth of organisms in medium that does not include Tris (see above); (b) too high a pH during EDTA treatment. Strains that lyse at pH 8.5 usually show good LPS release without lysis at pH 8.0. However, pH 8.5 is preferred for strains which do not lyse, because better yields are obtained; and (c) too long an exposure to EDTA.

Another source of contamination is leakage of intracellular material. Since Tris at 4° has been shown to leach out soluble pools and cause RNA breakdown and excretion,[13] the cells are never chilled during the preparation.

In some cases prolonged dialysis after the EDTA treatment may result in degradation of the LPS as detected by production of smaller carbohydrate fragments; this phenomenon may be due to an as yet uncharacterized enzyme or enzymes that degrade LPS.[14] Such degradation does not occur after phenol extraction which presumably destroys this activity. If such degradation occurs, other more rapid methods of removing small molecules, such as gel filtration or diafiltration, should be used.

The conditions described above usually result in comparable, or

---

[12] Lyophilization markedly and irreversibly changes the physical properties of LPS preparations, as measured by the dependence of sedimentation velocity on concentration (D. C. Morrison and L. Leive, manuscript in preparation). It is therefore recommended that lyophilization be avoided if such changes would be detrimental.

[13] L. Leive and V. Kollin, *Biochem. Biophys. Res. Commun.* **28**, 229 (1967).

[14] S. B. Levy and L. Leive, unpublished observations; also G. Weinbaum and S. Okuda, unpublished observations.

slightly lower, yields of LPS than other methods. For instance, EDTA extraction of *E. coli* 0111:B4 yields 40–50% of the LPS whereas phenol extraction followed by RNase digestion yields 65–90%.[8] However, if centrifugation to sediment LPS, rather than RNase digestion, is used to purify phenol-extracted LPS,[2] yields are lowered close to those of the EDTA method. EDTA-released material from this strain was comparable to phenol-extracted material in chemical composition and both biological and immunological properties (as measured by toxicity in mice and antibody production in rabbits) and therefore can be considered a representative sample of the total LPS.[8]

In general, the above method is not as useful for *Salmonella* species as for *E. coli*, since a smaller portion of LPS from *Salmonella* species is solubilized by EDTA. However, reasonably good yields of LPS may be obtained from *Salmonella anatum* by increasing the calcium concentration during growth to 2 mM (the medium mentioned above contains 0.2 mM $CaCl_2$). This increases the subsequent percent release of LPS by EDTA from 30% of the total to about 45%.[15] However, adding calcium to the growth medium does not improve the yields for *S. typhimurium*, which releases only about 20% of its LPS in the EDTA method. The effect of growth with added calcium on subsequent extraction of LPS by EDTA from other strains of *Salmonella* has not been tested.

Aqueous Butanol Method

Principle. A biphasic system of 0.15 M NaCl (saline) and butanol-1 is used to extract LPS from concentrated suspensions of whole bacteria. Nearly all the LPS is extracted into the aqueous phase, as is some of the cellular protein and nucleic acid. Most of the cells are not lysed, as indicated by observation in the light microscope. The protein is digested with pronase and both protein and RNA removed from the LPS by gel filtration or diafiltration.[16] Better yields are obtained with *E. coli* than with *Salmonella* species. This organic extraction method is analogous to the aqueous ether method of Ribi *et al.*[4] However, use of butanol appears to make subsequent purification easier.

Reagents

H_2O-saturated butanol-1 (reagent grade)
NaCl, 0.15 M (saline)
Tris·HCl buffer, 0.12 M, pH 8.0 (adjusted at 25°)
Pronase (B grade, Calbiochem). A solution of approximately 0.5

[15] D. A. Lawrence and L. Leive, manuscript in preparation.
[16] D. C. Morrison and L. Leive, manuscript in preparation.

mg/ml is prepared in sodium phosphate buffer, 0.2 M, pH 7.0, just prior to use.

Growth of Cells. Cells may be grown in either minimal or enriched media. The bacteria are harvested in mid-log phase by centrifugation and may be stored in a pellet at $-20°$.

Preparation of Crude Cell Extract. All procedures are carried out at 0–4° unless otherwise indicated. Cells are suspended in saline in a beaker at a concentration of about 2–3 g wet weight of cells per 10 ml of saline. The concentration of cells is relatively critical since too dilute suspensions result in poor yields; if too concentrated, poor phase separation will result after centrifugation. After thorough dispersion by mixing on a magnetic stirrer, an equal volume of butanol-1 is added slowly with continuous stirring. After 15 minutes of mixing, the phases are separated by centrifugation. Centrifugation in the Spinco Model L preparative ultracentrifuge at 35,000 g average for 20 minutes gives excellent phase separation, but much lower speeds may be used provided satisfactory phase separation is obtained.

The bottom aqueous phase is removed by aspiration and saved. One-half the initial volume of saline is added to the residual butanol phase and insoluble precipitate, and the above extraction is repeated twice. The aqueous extracts are then combined and centrifuged at approximately 13,000 g for 5 minutes to remove any residual insoluble material.

Purification of LPS. The procedure involves digestion with pronase and separation of the LPS from small molecules by diafiltration, or by gel filtration chromatography on agarose columns.[17]

The aqueous extracts are warmed to 37° and pronase freshly prepared in phosphate buffer (see above) is added to a final concentration of approximately 20 μg/ml. Several hours at 37° suffice for subsequent removal of most of the protein; for instance, after 3 hours of pronase digestion, purification as described below removes 85% of the protein present in the initial extract. Overnight incubation has been used routinely since we have been unable to observe any major change in physical properties (as measured by sedimentation velocity, and isopycnic centrifugation) or chemical composition of LPS after even prolonged treatment with pronase. Control experiments have also indicated that the presence of butanol dissolved in the aqueous extracts does not inhibit the action of pronase.

During pronase digestion, a flocculent white precipitate appears; this

---

[17] A brief digestion with pancreatic ribonuclease may be included prior to the pronase step if desired. However, even without such a digestion the subsequent purfication steps remove nucleic acid completely, probably because of their digestion by endogeneous ribonucleases.

precipitate has been characterized as almost exclusively protein. The extract is cooled to 4° and centrifuged for 10 minutes at 17,000 g. In addition to the white precipitate, which is sedimented, a thin layer of insoluble material floats at the top of the centrifuge tube. This material is gently removed by aspiration and the clear yellow supernatant decanted.

The pronase-digested extract can be further purified by either of two methods. In the first method the extract is intially concentrated at 4° to about one-fourth its volume by ultrafiltration in an Amicon pressure cell (Amicon Corporation, Lexington, Massachusetts) at a pressure of 4–5 psi. Any convenient size of cell can be used: e.g., 12 ml for small volumes (less than 10 ml) and 400 ml for larger volumes. The choice of filter depends to some extent on the size of the bacterial LPS. Almost complete retention of the complete ("smooth") LPS of *E. coli* 0111:B4 is obtained with PM-30 filters (exclusion molecular weight > 30,000); however, some of the remaining protein is also retained with this size filter. If much larger pore size filters are used, essentially none of the protein is retained, but LPS retention is decreased. Optimal LPS retention with minimal protein retention has been obtained for this strain using the XM-100A filter size. These data are summarized in the table.

An undetectable amount of nucleic acid (measured by absorption at 260 nm) is retained by any of these filters. Although these data were obtained for *E. coli* 0111:B4, LPS preparations from other coliforms should be retained by XM-50 or XM-100A filters since such preparations, no matter what their true molecular weight, form aggregates in solution. For instance, we have obtained equally good retention of *E. coli* K12 (semirough) LPS by the XM-100A filter.

RETENTION OF LIPOPOLYSACCHARIDE AND PROTEIN FROM *Escherichia coli*
0111:B4 BY AMICON FILTERS[a]

Filter size	Exclusion molecular weight	Percent of initial material retained by filter	
		Lipopoly-saccharide[b]	Protein[c]
PM-30	30,000	95	38
XM-50	50,000	94	19
XM-100A	100,000	92	5
XM-300	300,000	47	2

[a] The crude extract was digested with pronase and filtered as described in the text.

[b] Assayed by the thiobarbituric acid assay for colitose as described by M. A. Cynkin and G. Ashwell, *Nature (London)* **186,** 155 (1960).

[c] Assayed with the Folin–Ciocalteau reagent; see Vol. 3 [73].

After concentration, the material is further purified by a wash at constant volume in the same filtration apparatus (termed "diafiltration" by the Amicon Company). It can be calculated[18] that diafiltration with 5 volumes of buffer is sufficient to remove more than 99% of any material not retained by the filter. After concentration of the extract, the pressure cell is connected in series to a reservoir containing any desired buffer. Pressure is applied to the system,[19] and diafiltration is allowed to proceed until no UV absorbing material can be detected in the effluent. (This step, in addition to removing all small molecular weight material, essentially "dialyzes" the LPS into the desired buffer.) A similar result can be achieved, but with less efficiency, by reducing the initial volume to 10% or less of the filter cell volume, adding buffer to fill the cell, and then reducing the volume to the original level. This procedure is repeated 2 or 3 times. The purified LPS solution is centrifuged briefly to remove any insoluble material, and the LPS is stored at $-20°$.

The second method of purification of the pronase-digested extract is by gel filtration. If necessary to reduce the volume relative to the column volume the extract may be concentrated by any method (for instance, *in vacuo*, or as described above, or by ultrafiltration through collodion tubing under vacuum). A column of Sepharose 4B (Pharmacia, Uppsala, Sweden) is packed under gravity (42×1.5 cm and 85×7.5 cm columns have both been used). The extract, containing 0.1–10 mg LPS/ml, is applied and chromatographed with any desired buffer at a pressure of 50–75 cm H_2O. The LPS elutes in advance of degraded protein, RNA, and small molecules, all of which fall within the included volume; its exact position (or positions if there are different types of LPS molecules within the same organism)[20] depends on the strain used. The position of the LPS peak(s) may be monitored by the general phenol–sulfuric carbohydrate assay,[11] RNA by $OD_{260 \text{ nm}}$, and protein with the Folin–Ciocalteau reagent.[21] The fractions containing LPS are combined, then stored in solution at $-20°$.

Yield, Purity, and Properties. Approximately 65–90% of the LPS of *E. coli* is recovered after extraction and purification by the above

---

[18] Amicon Publication No. 403, p. 36, Amicon Corporation, Lexington, Massachusetts, 1970.

[19] An increase in sample volume will occur until the pressure in the reservoir and the cell are equalized. This can be avoided if the pressure cell and the reservoir are independently brought to operating pressure before being connected.

[20] Further description of the chromatographic properties of different LPS fractions from *E. coli* 0111:B4 are given by D. C. Morrison and L. Leive, manuscript in preparation.

[21] See Vol. 3 [73].

procedure in several *E. coli* strains tested; the yields were thus essentially as good as by phenol extraction. Yields of LPS from *Salmonella typhimurium* and *Salmonella anatum* are lower, amounting to between 30 and 50% of the total LPS. Greater purity can be achieved than is easily obtained with phenol-extracted material: preparations are usually 96–98% pure; they contain no detectable nucleic acid (<0.1%) or free lipid (assayed by chloroform–methanol extraction[22]) and 2–4% of protein. When the properties of butanol- and phenol-extracted LPS from *E. coli* 0111:B4 were compared,[16] it was found that they had similar toxicity when injected into mice, and similar chemical composition.

[22] J. Kanfer and E. P. Kennedy, *J. Biol. Chem.* **238**, 2919 (1963).

[24] Isolation of Oligosaccharides from Human Milk

By AKIRA KOBATA

Oligosaccharides of human milk (see the table) are used in studies on the acceptor specificity of glycosyltransferases, the substrate specificity of glycosidases, and the structure of antigenic determinants.[1-10]

Principle. After removal of lipids, proteins, and most of the lactose, the oligosaccharides are isolated by Sephadex gel filtration, followed by paper chromatography and in some cases by paper electrophoresis. As the oligosaccharides present in individual samples of milk vary with the ABO and Lewis blood type of the donor,[6,7,11,12] samples from individuals

[1] A. Kobata, E. F. Grollman, and V. Ginsburg, *Arch. Biochem. Biophys.* **124**, 609 (1968).
[2] V. M. Hearn, Z. G. Smith, and W. M. Watkins, *Biochem. J.* **109**, 315 (1968).
[3] A. Kobata and V. Ginsburg, *J. Biol. Chem.* **245**, 1484 (1970).
[4] A. Kobata, E. F. Grollman, and V. Ginsburg, *Biochem. Biophys. Res. Commun.* **32**, 272 (1968).
[5] C. Race, D. Ziderman, and W. M. Watkins, *Biochem. J.* **107**, 733 (1968).
[6] E. F. Grollman, A. Kobata, and V. Ginsburg, *J. Clin. Invest.* **48**, 1489 (1969).
[7] A. Kobata and V. Ginsburg, *J. Biol. Chem.* **247**, 1525 (1972).
[8] D. Aminoff and K. Furukawa, *J. Biol. Chem.* **245**, 1659 (1970).
[9] G. Y. Wiederschain and E. L. Rosenfeld, *Biochem. Biophys. Res. Commun.* **44**, 1008 (1971).
[10] W. M. Watkins, Blood-group specific substances, *in* "Glycoproteins" (A. Gottschalk, ed.), p. 462. Elsevier, Amsterdam, 1966.
[11] E. F. Grollman and V. Ginsburg, *Biochem. Biophys. Res. Commun.* **28**, 50 (1967).
[12] A. Kobata, M. Tsuda, and V. Ginsburg, *Arch. Biochem. Biophys.* **130**, 509 (1969).

with certain blood types are better than others, for the isolation of specific oligosaccharides.

Preliminary Treatment of Milk.[7,13] A liter of milk is centrifuged for 15 minutes at 100 *g* at 2°, and solidified lipid is removed by filtration through loosely packed glass wool in the cold. To the clear filtrate, ethanol is added with stirring to a final concentration of 68% and left overnight in a refrigerator. Lactose and protein thus precipitated are removed by centrifugation at 0°, and the precipitate is washed twice with 200-ml portions of 68% ethanol at 0°. The supernatant solution and washings are combined and evaporated to a syrup under reduced pressure. The syrup is dissolved in 100 ml of H_2O, and insoluble material is removed by centrifugation for 15 minutes at 100 *g* at 2°.

For enzyme studies an alternative procedure can be adopted as follows: Skimmed milk prepared as above is saturated with ammonium sulfate at 0°. Precipitated protein and excess ammonium sulfate are removed by centrifugation for 30 minutes at 10,000 *g* at 0°, and the precipitate is washed with 200 ml of saturated ammonium sulfate solution. The protein fraction is suitable for enzyme studies. The washing and supernatant solution are combined and concentrated to 100 ml under reduced pressure.

Pyridine, 400 ml, is added drop by drop with stirring to the ice-cooled concentrate. Precipitated ammonium sulfate is removed by filtration and washed with 100 ml of pyridine. The filtrate and washing are combined and evaporated to dryness under reduced pressure. The residue is shaken with 100 ml of H_2O for 30 minutes at room temperature, and the suspension is filtered after cooling at 0° for several hours. The precipitate on the filter is washed with 100 ml of 68% ethanol. The filtrate and washing are combined and evaporated under reduced pressure to a syrup. The syrup is dissolved in 100 ml of H_2O, and insoluble material is removed by centrifugation at 100 *g* for 15 minutes.

Sephadex Gel Filtration.[7] The crude oligosaccharides solution obtained by the preliminary treatment of milk described above are separated into three fractions by gel filtration as follows: The oligosaccharides fraction (100 ml) from 1 liter of human milk is mounted on a column of fine grade Sephadex G-25 (8 × 160 cm) that has been washed with H_2O overnight and eluted with H_2O. After the void volume (2.6 liters) of H_2O has passed through, the effluent is collected in 20-ml fractions.

Aliquots (50 μl) are taken from every other tube and assayed for total hexose content with phenol sulfuric acid reagent[5]; the sample is diluted with water to 0.2 ml and colored with 0.2 ml of 5% aqueous phenol and 1 ml of concentrated sulfuric acid. The tubes are pooled into

[13] A. Kobata and V. Ginsburg, *J. Biol. Chem.* **244**, 5496 (1969).

OLIGOSACCHARIDES OF HUMAN MILK

Trivial name	Structure	Reference
2'-Fucosyllactose	Fuc-α-(1 → 2)-Gal-β-(1 → 4)-Glc	a
3-Fucosyllactose	Gal-β-(1 → 4) \ Fuc-α-(1 → 3) ⟩ Glc	b
Lactodifucotetraose	Fuc-α-(1 → 2)-Gal-β-(1 → 4) \ Fuc-α-(1 → 3) ⟩ Glc	c
Lacto-N-tetraose	Gal-β-(1 → 3)-GlcNAc-β-(1 → 3)-Gal-β-(1 → 4)-Glc	d
Lacto-N-neotetraose	Gal-β-(1 → 4)-GlcNAc-β-(1 → 3)-Gal-β-(1 → 4)-Glc	e
Lacto-N-fucopentaose I	Fuc-α-(1 → 2)-Gal-β-(1 → 3)-GlcNAc-β-(1 → 3)-Gal-β-(1 → 4)-Glc	f
Lacto-N-fucopentaose II	Gal-β-(1 → 3) \ Fuc-α-(1 → 4) ⟩ GlcNAc-β-(1 → 3)-Gal-β-(1 → 4)-Glc	g
Lacto-N-fucopentaose III	Gal-β-(1 → 4) \ Fuc-α-(1 → 3) ⟩ GlcNAc-β-(1 → 3)-Gal-β-(1 → 4)-Glc	h
Lacto-N-difucohexaose I	Fuc-α-(1 → 2)-Gal-β-(1 → 3) \ Fuc-α-(1 → 4) ⟩ GlcNAc-β-(1 → 3)-Gal-β-(1 → 4)-Glc	i
Lacto-N-difucohexaose II	Gal-β-(1 → 3) \ Fuc-α-(1 → 4) ⟩ GlcNAc-β-(1 → 3)-Gal-β-(1 → 4) \ Fuc-α-(1 → 3) ⟩ Glc	j
3'-Sialyllactose	NANA-α-(2 → 3)-Gal-β-(1 → 4)-Glc	k
6'-Sialyllactose	NANA-α-(2 → 6)-Gal-β-(1 → 4)-Glc	l
LS-tetrasaccharide a	NANA-α-(2 → 3)-Gal-β-(1 → 3)-GlcNAc-β-(1 → 3)-Gal-β-(1 → 4)-Glc	m
LS-tetrasaccharide b	NANA-α-(2 → 6) \ Gal-β-(1 → 3) ⟩ GlcNAc-β-(1 → 3)-Gal-β-(1 → 4)-Glc	m
LS-tetrasaccharide c	NANA-α-(2 → 6)-Gal-β-(1 → 4)-GlcNAc-β-(1 → 3)-Gal-β-(1 → 4)-Glc	n
Disialyllacto-N-tetraose	NANA-α-(2 → 6) \ NANA-α-(2 → 3)-Gal-β-(1 → 3) ⟩ GlcNAc-β-(1 → 3)-Gal-β-(1 → 4)-Glc	o

Lacto-N-hexaose	$\begin{array}{l}\text{Gal-}\beta\text{-}(1 \to 4)\text{-GlcNAc-}\beta\text{-}(1 \to 6)\\ \text{Gal-}\beta\text{-}(1 \to 3)\text{-GlcNAc-}\beta\text{-}(1 \to 3)\end{array}\bigg\rangle\text{Gal-}\beta\text{-}(1 \to 4)\text{-Glc}$	p
Lacto-N-neohexaose	$\begin{array}{l}\text{Gal-}\beta\text{-}(1 \to 4)\text{-GlcNAc-}\beta\text{-}(1 \to 6)\\ \text{Gal-}\beta\text{-}(1 \to 4)\text{-GlcNAc-}\beta\text{-}(1 \to 3)\end{array}\bigg\rangle\text{Gal-}\beta\text{-}(1 \to 4)\text{-Glc}$	q
S-5	$\begin{array}{l}\text{NANA-}\alpha\text{-}(2 \to 6)\text{-Gal-}\beta\text{-}(1 \to 4)\text{-GlcNAc-}\beta\text{-}(1 \to 6)\\ \text{Gal-}\beta\text{-}(1 \to 3 \text{ and } 4)\text{-GlcNAc-}\beta\text{-}(1 \to 3)\end{array}\bigg\rangle\text{Gal-}\beta\text{-}(1 \to 4)\text{-Glc}$	q
S-6	Fucosyl S-5	q
N-2	Fucosyllacto-N-hexaose and fucosyllacto-N-neohexaose	q
N-3	Difucosyllacto-N-hexaose and difucosyllacto-N-neohexaose	q

[a] R. Kuhn, H. H. Baer, and A. Gauhe, Chem. Ber. 89, 2513 (1956).

[b] J. Montreuil, C. R. Acad. Sci. 242, 192 (1956).

[c] R. Kuhn and A. Gauhe, Justus Liebigs Ann. Chem. 611, 249 (1958).

[d] R. Kuhn and H. H. Baer, Chem. Ber. 89, 504 (1956).

[e] R. Kuhn and A. Gauhe, Chem. Ber. 95, 518 (1962).

[f] R. Kuhn, H. H. Baer, and A. Gauhe, Chem. Ber. 89, 2514 (1956).

[g] R. Kuhn, H. H. Baer, and A. Gauhe, Chem. Ber. 91, 364 (1958).

[h] A. Kobata and V. Ginsburg, J. Biol. Chem. 244, 5496 (1969).

[i] R. Kuhn, H. H. Baer, and A. Gauhe, Justus Liebigs Ann. Chem. 611, 242 (1958).

[j] R. Kuhn and A. Gauhe, Chem. Ber. 93, 647 (1960).

[k] R. Kuhn and R. Brossmer, Chem. Ber. 92, 1667 (1959).

[l] R. Kuhn, Naturwissenschaften 46, 463 (1959).

[m] R. Kuhn and A. Gauhe, Chem. Ber. 98, 395 (1965).

[n] R. Kuhn and A. Gauhe, Chem. Ber. 95, 513 (1962).

[o] L. Grimmonprez and J. Montreuil, Bull. Soc. Chim. Biol. 50, 843 (1968).

[p] A. Kobata and V. Ginsburg, J. Biol. Chem. 247, 1525 (1972).

[q] A. Kobata and V. Ginsburg, Arch. Biochem. Biophys. 150, 273 (1972).

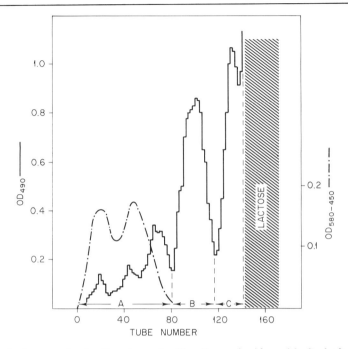

Fig. 1. Large-scale fractionation of milk oligosaccharides with Sephadex G-25. The procedure is described in the text. Aliquots 30 μl, of every other fraction were assayed for total sugar with phenol–sulfuric acid reagent (——); total assay volume 1.4 ml. For sialic acid (− · − ·), 0.2-ml aliquots were assayed with resorcinol reagent [L. Svennerholm, *Biochim. Biophys. Acta* **24**, 604 (1957)]; total assay volume was 2 ml.

three fractions (Fig. 1); A (tubes 1–80), B (tubes 81–116) and C (tubes 117–140). As is shown in Fig. 1, almost all the sialic acid derivatives are included in fraction A.

Isolation of 2′- and 3-Fucosyllactose and Lactodifucotetraose from Fraction C. Fraction C is lyophilized and dissolved again in water to make 100 mg per milliliter of solution. The solution is applied on Whatman No. 3 MM papers as a streak (20 μl per centimeter of width).

The papers are subjected to descending chromatography with ethyl acetate–pyridine–H$_2$O 12:5:4 (solvent I) as a solvent for 2 days. Fucosyllactoses ($R_{lactose}$ = 0.64 for 3-fucosyllactose and 0.66 for 2′-fucosyllactose) and lactodifucotetraose are located by staining guide strips with alkaline-AgNO$_3$ reagent[14] (Fig. 2A), and eluted with water.

The two isomers of fucosyllactose can be separated paper chromato-

---

[14] E. F. L. J. Anet and T. M. Reynolds, *Nature* (*London*) **174**, 930 (1956).

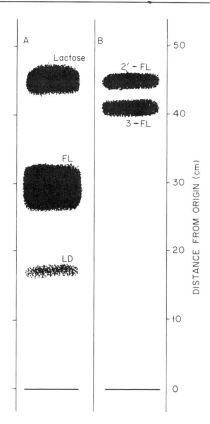

FIG. 2. (A) Paper chromatogram of oligosaccharides from fraction C. (B) Separation of 2′-fucosyllactose and 3-fucosyllactose by rechromatography. FL, fucosyllactose; LD, lactodifucotetraose.

graphically by using phenol–isopropanol–formic acid–water, 85:5:10:100 lower layer, reported by J. Montreuil.[15]

Pure 3-fucosyllactose is more easily obtained from milk of "nonsecretor" donors which does not contain 2′-fucosyllactose.[11]

Isolation of Tetra- to Hexasaccharides from Fraction B. Fraction B is occasionally contaminated with 3′- and 6′-sialyllactose which tail from fraction A. Since 3′-sialyllactose and 6′-sialyllactose have almost similar mobility with lacto-N-tetraose and lacto-N-fucopentaose I, respectively, in the solvent used to separate neutral sugars, they are removed before paper chromatographic separation as follows: fraction B is passed through a mixed-bed resin column (1 × 10 cm) containing Amberlite AG 50 (H⁺) and AG 3 (OH⁻) (referred in this paper as

[15] J. Montreuil, *Bull. Soc. Chim. Biol.* **42**, 1399 (1960).

"mixed-bed column"). The column is washed with 30 ml of H_2O. The eluate and washing are combined and lyophilized. The white residue is dissolved in water to make 100 mg/ml solution, and applied on Whatman No. 3 MM paper as in the case of fraction C. The papers are subjected to descending chromatography with solvent I as a solvent for 6 days.

Milk from the donor with Le^b blood type (about 80% of the population) gives four major sugar bands together with two or three minor bands when the guide strip is stained with alkaline $AgNO_3$ reagent (Fig. 3A). The sugar band I of Fig. 3A is eluted with water and passed through a small mixed-bed column (0.5 × 3 cm) to remove yellowish contaminant, and lyophilized. Five hundred milligrams of the white powder is dissolved in 10 ml of H_2O. Ethanol, 25 ml, is added slowly with stirring. After it has stood overnight at room temperature, the crystalline *lacto-N-tetraose* is collected by filtration (yield 410 mg).

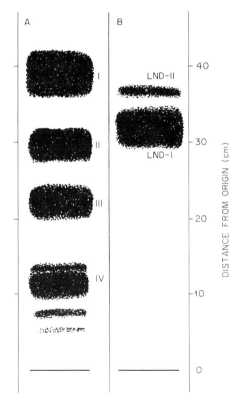

Fig. 3. (A) Paper chromatogram of oligosaccharides from fraction B. (B) Separation of lacto-*N*-difucohexaose I (LND-I) and lacto-*N*-difucohexaose II (LND-II).

The mother liquid is concentrated to 2 ml and applied as a band on Whatman No. 1 paper (10 μl per centimeter of width) and developed with solvent I for 6 days. Staining of guide strips with alkaline $AgNO_3$ reagent reveals two sugar bands: one corresponding to lacto-N-tetraose and a second, slower component corresponding to lacto-N-neotetraose ($R_{\text{lacto-}N\text{-tetraose}} = 0.93$).

Lacto-N-neotetraose is eluted with H_2O; the resulting solution is passed through a mixed-bed resin column used as above and evaporated to a syrup. *Lacto-N-neotetraose* is crystallized by adding ethanol (yield 60 mg).

Band II of Fig. 3A is almost pure *lacto-N-fucopentaose I.*

About 90% of band III is lacto-N-fucopentaose II, and the remaining part is lacto-N-fucopentaose III. *Lacto-N-fucopentaose II* is purified by reprecipitation of the sugar with acetone four times after elution of band III.

Band IV of Fig. 3 A is eluted from papers with water and rechromatographed with ethyl acetate–pyridine–acetic acid–H_2O, 5:5:1:3 (solvent II) as a solvent for 9 days. The guide strip shows one major sugar band together with a weak band with higher mobility (Fig. 3B). The major band is almost pure *lacto-N-difucohexaose I,* and is obtained as a white powder from a water eluate by addition of acetone. The main component of the minor band is *lacto-N-difucohexaose II,* but it is contaminated with about 10 ~ 15% of lacto-N-hexaose, lacto-N-neohexaose and a fucose containing hexasaccharide.[16] Therefore, repeated acetone precipitation from aqueous solution (generally five times will be adequate) is required to get a satisfactory pure lacto-N-difucohexaose II, from this fraction.

Milk from donors of Lewis negative blood types lacks lacto-N-fucopentaose II, lacto-N-difucohexaose I, and lacto-N-difucohexaose II.[6] As a result, *lacto-N-fucopentaose III, lacto-N-hexaose,* and *lacto-N-neohexaose* are best isolated from milk obtained from these individuals, who comprise about 5% of the population.[13] The band III portion of Fig. 3A obtained from milk of the donor of Lewis-negative blood types is eluted with H_2O; the resulting solution is evaporated to dryness after passing through a mixed-bed column (2 × 0.5 cm), giving about 50 mg of lacto-N-fucopentaose III from 1 liter of milk.

The band IV portion of the paper is eluted with water, and the

---

[16] This sugar is an isomer of lacto-N-difucohexaose I and lacto-N-difucohexaose II with the probable structure:

$$\text{Fuc-}\alpha\text{-}(1 \rightarrow 2)\text{-Gal-}\beta\text{-}(1 \rightarrow 3)\text{-GlcNAc-}\beta\text{-}(1 \rightarrow 3)\text{-Gal-}\beta\text{-}(1 \rightarrow 4) \diagdown \text{Glc}$$
$$\text{Fuc-}\alpha\text{-}(1 \rightarrow 3) \diagup$$

resulting solution is evaporated to dryness. The residue is dissolved in
0.01 N HCl to make 10 mg/ml concentration and heated at 100° for
1 hour. A hexasaccharide-containing fucose[16] is completely hydrolyzed to
lacto-N-tetraose and fucose by this treatment. The hydrolyzate is then
neutralized by passage through a mixed-bed column (0.5 × 4 cm). After
concentration, the effluent is applied as a band on Whatman No. 3
MM paper (0.5 mg of sugar per centimeter) and chromatographed
for 10 days using solvent II. Two sugar bands are detected by staining
a guide strip with alkaline silver nitrate. Lacto-N-hexaose (slower moving
band) and lacto-N-neohexaose (faster moving band) are eluted from
the chromatogram. Each eluate is passed through a mixed-bed column
(0.5 × 2 cm) and evaporated to dryness. About 15 mg of lacto-N-
hexaose and 5 mg of lacto-N-neohexaose can be obtained from 1 liter
of milk.

*Isolation of Sialic Acid Containing Oligosaccharides and Higher
Neutral Oligosaccharides from Fraction A.* After lyophilization, fraction
A is dissolved in water to make a solution of 100 mg/ml. The solution

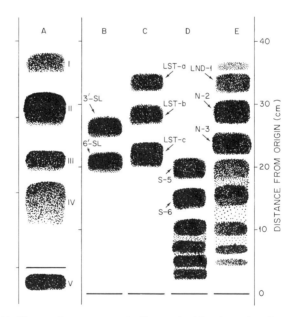

Fig. 4. (A) Electrophoretogram of oligosaccharides from fraction A. (B) Paper
chromatographic separation of band II; (C) of band III; (D) of band IV; and (E)
of band V, respectively. 3'-SL, 3'-sialyllactose; 6'-SL, 6'-sialyllactose; LST-a, b, c,
LS-tetrasaccharide a, b, and c, respectively; LND-I, lacto-N-difucohexaose I.

is applied on Whatman No. 3 MM paper as a streak (20 μl per centimeter of width). The papers are subjected to electrophoresis using H_2O–pyridine–glacial acetic acid, 3870:30:11.5 (pH 5.4).

As shown in Fig. 4A, five bands are obtained by staining a guide strip with alkaline $AgNO_3$ reagent. Pure disialyl lacto-N-tetraose can be recovered from band 1. Band 2 contains *3'-sialyllactose* and *6'-sialyllactose*, and band 3 contains *LST-a*, *LST-b*, and *LST-c*.

The mixtures can be separated as follows. Band 2 and band 3 are subjected to paper chromatography with solvent II for 3 days and for 6 days, respectively (Fig. 4B and C). Pure sialyl containing oligosaccharides are recovered by elution from the paper. Developing for 14 days with the same solvent on Whatman 3 MM, sugars in the band 4 are fractionated as is shown in Fig. 4D. Complete structure of *S-5* and partial structure of *S-6* are now elucidated, as shown in the table.

The neutral sugars (band 5) are separated into their components, as shown in Fig. 4E, by paper chromatography with solvent II for 14 days. The structures of *N-2* and *N-3* are partially elucidated as is shown in the table.

[25] UDP-Galactosamine, UDP-N-Acetylgalactosamine, UDP-Glucosamine, and UDP-N-Acetylglucosamine

By FRANK MALEY

Assay Method

UDP-GalN

The commerical availability of Gal-1-P uridyl transferase facilitated the preparation of this nucleotide by the following sequence of reactions:

$$GalN\text{-}1\text{-}P + UDPG \rightleftharpoons UDP\text{-}GalN + Glc\text{-}1\text{-}P$$
$$Glc\text{-}1\text{-}P \rightleftharpoons Glc\text{-}6\text{-}P$$

Principle. The assay procedure for the Gal-1-P uridyl transferase reaction is essentially that already presented,[1] except for the substitution of GalN-1-P. This procedure measures the rate of Glc-1-P formation by coupling the reaction with phosphoglucomutase and Glc-6-P dehydrogenase. The rate of NAD reduction as a consequence of the latter enzyme is measured at 340 nm. The GalN-1-P used was prepared by the

---

[1] See Vol. 9 [127].

procedure of Carlson *et al.*[2] with galactose-adapted yeast obtained from the Sigma Chemical Co., St. Louis, Missouri.

Reagents

    UDP-Glc, 10 mM

    GalN-1-P, 7.85 mM

    Glc-1,6-P$_2$, 14 μM

    Tris·HCl, 70 mM, pH 8.0

    Dithiothreitol, 7 mM

    Phosphoglucomutase, 110 units (crystalline, from the Sigma Chemical Co.)

    Gal-1-P uridyl transferase, 3.3 units, prepared from calf liver as a partially purified enzyme (Boehringer-Mannheim Corp., New York)

Procedure

The reagents are present at the concentrations indicated in a volume of 14 ml. Two drops of toluene are added to the reaction mixture, which is incubated at 37° in a screw-cap vial.

Determination of Extent and Rate of Reaction. At about 2-hour intervals, 0.1-ml aliquots are removed and added to 0.9 ml of H$_2$O, which is then heated at 100° for 2 minutes. Aliquots from this solution are added to a cuvette containing 50 mM Tris·HCl (pH 8.5), 0.2 mM NADP, and water to a volume of 1.0 ml followed by 5 μl of Gl-6-P dehydrogenase (0.35 unit). With this procedure, the rate of Glc-6-P formation can be followed. It is necessary to delete Mg^{2+} from the incubation mixture since it promotes the nonspecific hydrolysis of UDPG and as a consequence greatly reduces the yield of the desired product. The reaction is linear for about 15 hours, and no additional Glc-6-P formation was detectable after 30 hours.

Isolation of UDP-GalN. From the amount of Glc-6-P formed, it is estimated that 64% of the GalN-1-P had been converted to UDP-GalN. The reaction mixture is then passed through a column (2 × 5 cm) of Dowex 1-formate and the column is eluted by the following schedule: 50 ml of 0.1 N formic acid, 10 ml of 1 N formic acid, five 20-ml fractions of 1 N formic acid. The latter fractions contained the UDP-GalN and were lyophilized to a white powder.

UDP-GalNAc-1-^{14}C. Forty microliters of 1 mM triethylamine in methanol is added to a solution of 19.5 μmoles of UDP-GalN in water followed by the addition of 36.8 μmoles of acetic anhydride-1-^{14}C (13.6 μCi/

[2] D. M. Carlson, A. L. Swanson, and S. Roseman, *Biochemistry* **3**, 402 (1964).

μmole) in 1.5 ml of methanol. After 1 hour at room temperature, the solution is stored at 4° overnight. The addition of 5 ml of water to this solution precedes its passage through a column (1 × 5 cm) of Dowex-50 in the pyridinium form. The column is eluted with 20 ml of water, and the combined eluent containing 8.6×10^8 dpm is lyophilized; this provides a residue with 3.1×10^8 dpm. Relyophilization of the residue does not remove additional radioactivity. Accurate specific activity determinations require the removal of the 260 nm absorbing pyridine; therefore, a solution of the residue should be passed through a Dowex-50 column in the cold and neutralized to the desired salt. Improved yields can be obtained by raising the acetic anhydride: UDPGal ratio to 3:1. Alternatively, the compound can be placed on a Dowex 1 × 8 (200–400 mesh) bicarbonate column (1 × 10 cm) and eluted with a convex gradient consisting of a 500-ml H_2O mixing chamber and 1 M triethylamine bicarbonate in the reservoir. On elution of the major UV absorbing region, this area is concentrated *in vacuo* to remove the volatile salt.

UDP-Glucosamine and UDP-N-Acetylglucosamine

UMP-morpholidate + GlcN-1-P → UDP-GlcN

$$\text{UDP-GlcN} \xrightarrow{\text{Ac}_2\text{O}} \text{UDP-GlcNAc}$$

Principle. Other than the enzymatic synthesis of UDP-GlcN by susbtituting GlcN-1-P for Glc-1-P in the UDPG pyrophosphorylase reaction,[3] a method for the direct chemical synthesis of UDP-GlcN has not been described. Undoubtedly, one of the reasons has been the poor yields that are incurred when the morpholidate procedure[4] is employed. Since the solubility of GlcN-1-P in pyridine, even as the tri-*n*-octylamine salt, appears to be one of the limiting features of this synthesis, other solvents were sought. Dimethyl sulfoxide resolves this problem in part.

Reagents

GlcN-1-P, tri-*n*-octylamine salt
UMP-morpholidate (Sigma Chemical Co., St. Louis, Missouri)
Dimethyl sulfoxide, anhydrous
Pyridine, anhydrous

Procedure

UDP-GlcN. A solution of 100 μmoles of the monopotassium salt of GlcN-1-P[3] is converted to the pyridinium salt by passage through a

---

[3] F. Maley, G. F. Maley, and H. A. Lardy, *J. Amer. Chem. Soc.* **78**, 5303 (1956).
[4] J. G. Moffat and H. G. Khorana, *J. Amer. Chem. Soc.* **83**, 649 (1961).

column (1 × 5 cm) of Dowex-50 in the pyridinium form. The eluate is concentrated to dryness *in vacuo* in a 50-ml round-bottom flask with about 100 μmoles of tri-*n*-octylamine (50 μl) and dried by repeating the concentration 3 to 4 times with 10 ml of anhydrous pyridine (dried over calcium hydride). Then 41 mg of UMP-morpholidate in 2 ml of anhydrous pyridine is added, and the pyridine is evaporated. The residue is suspended in 5 ml of anhydrous dimethyl sulfoxide (dried over a Linde 4X molecular sieve) and stirred magnetically in the tightly stoppered round-bottom flask. After 5 days the solution is taken up in 50 ml of water and passed through a column (2 × 6 cm) of Dowex 1-formate (8X, 200–400 mesh). The column is washed thoroughly with water to remove most of the dimethyl sulfoxide and eluted first with seven 20-ml fractions of 0.1 N formic acid, then with seven similar fractions of 1 N formic acid. The latter contains 17 μmoles of UDP-GlcN, which was obtained as a white powder on lyophilizing. The yields are consistently between 30 and 35%.

UDP-GlcNAc. Acetylation of UDP-GlcN with labeled or unlabeled acetic anhydride by the procedure described for the synthesis of UDP-GalNAc, but with 4 moles of acetic anhydride per mole of UDP-GlcN, provides almost quantitative yields of UDP-GlcNAc.

[26] UDP-[14C]N-Acetylgalactosamine[1]

By Don M. Carlson and Saul Roseman

$$\text{Galactosamine} + \text{ATP} \xrightarrow{\text{Galactokinase}} \text{galactosamine-1-P} + \text{ADP}$$

$$\text{Galactosamine-1-P} \xrightarrow[\text{anhydride}]{\text{[14C]acetic}} \text{[14C]}N\text{-acetylgalactosamine-1-P} + \text{[14C]acetate}$$

$$\text{UMP-morpholidate} + \text{[14C]}N\text{-acetylgalactosamine-1-P} \rightarrow$$
$$\text{UDP-[14C]}N\text{-acetylgalactosamine} + \text{morpholine}$$

Principle. Incubation of galactosamine, ATP, and Mg^{2+} with crude yeast galactokinase preparations yields galactosamine 1-phosphate. This product is isolated by ion-exchange chromatography and crystallized as the free acid. After N-acetylation with [14C]acetic anhydride, [14C]N-acetylgalactosamine-1-P is crystallized as the dipotassium salt. Condensation of [14C]N-acetylgalactosamine with UMP-morpholidate yields UDP-[14C]N-acetylgalactosamine.

[1] D. M. Carlson, A. L. Swanson, and S. Roseman, *Biochemistry* **3**, 402 (1964).

Reagents[2]

    Saccharomyces fragilis, galactose-adapted, Sigma Chemical Co.
    Sodium bicarbonate, 0.1 M
    Galactosamine·HCl
    Phosphoenolpyruvate, Na
    ATP, Na$_2$
    Tri-n-octylamine (redistilled)
    UMP-morpholidate
    Anhydrous pyridine
    3-Phosphoglycerate, Na$_3$ (3-PGA)
    Potassium phosphate buffer, pH 7.8
    Magnesium chloride
    Sodium hydroxide, 2 N
    Toluene
    Dowex 50 H$^+$, 100–200 mesh, X12.
    Acetic anhydride, or [14C]acetic anhydride
    Sodium borohydride, 1.0 M
    Acetone 2.0 M
    Acetic anhydride 0.5 M, in water
    Hydrochloric acid, 2 N
    Phenolphthalein
    Potassium borate, 0.2 M, pH 9.0
    Ehrlich reagent: 1.0 g of p-dimethylaminobenzaldehyde is added
        to 1.25 ml of 10 N HCl and diluted to 100 ml with glacial acetic
        acid.

Procedures

    Enzyme Preparation. The crude extract containing galactokinase is
prepared by autolyzing 2 g of *S. fragilis* in 6 ml of 0.1 M NaHCO$_3$ for
12 hours at 25°. The supernatant fluid obtained after centrifuging for
30 minutes at 32,000 g is decanted and the residue is suspended in an
additional 6 ml of NaHCO$_3$. The mixture is again centrifuged, and the
supernatant fluids are combined. This crude extract contains the galactokinase; the protein concentration is approximately 50 mg/ml.

    Enzyme Assay. The principle of the enzyme assay is based on the
conversion of a reducing sugar into a glycoside. Excess substrate is reduced to the sugar alcohol with sodium borohydride. Acid hydrolysis
removes the aglycon moiety, and the resulting sugar is assayed by conventional procedures. The procedure, devised for the assay of galacto-

---

[2] Unless otherwise indicated, reagents are commercially available.

kinase, is generally applicable to enzyme-catalyzed reactions where reducing sugars are converted into glycosides.

The assay mixture contains the following components (micromoles) in a final volume of 0.14 ml: potassium phosphate buffer, pH 7.8, 200; $MgCl_2$, 2; galactosamine·HCl, 40; ATP, 40; PEP, 10; and 3-PGA, 10. The reaction is initiated by adding 0.01 ml of the galactokinase preparation. The incubation is terminated after 15 minutes at 30° by heating at 100° for 2 minutes. Denatured protein is removed by centrifugation, and galactosamine-1-P is measured in 0.05 ml of the supernatant fluid by the following procedures.

Reduction of Remaining Galactosamine. To prevent excessive foaming, one drop of capryl alcohol is added, then 0.025 ml of 1.0 M $NaBH_4$ is added. After mixing, the solution is held at room temperature for 5 minutes with occasional shaking, and the $NaBH_4$ addition is repeated. After 5 minutes the excess $NaBH_4$ is destroyed by the addition of 0.025 ml of 2.0 M acetone; it is mixed and kept at room temperature for 5 minutes. Alternatively, the borohydride can be destroyed by heating the mixture at 100° for 3 minutes.

Acetylation. Maintain the pH between 7 and 9 with saturated $NaHCO_3$ while adding 0.1 ml of 0.5 M acetic anhydride.[3] The N-acetylation is complete in a few minutes.

Hydrolysis. N-Acetylgalactosamine-1-P is hydrolyzed by adding 0.2 ml of 2 N HCl, then heating at 100° for 10 minutes. After cooling, the mixture is neutralized to a phenolphthalein end point with 2 N NaOH. The final volume is adjusted to 1.0 ml, and 0.5 ml is assayed directly for N-acetylgalactosamine by a modified Morgan–Elson method.[4]

Preparation of Galactosamine-1-P. After the activity of the galactokinase has been ascertained, an incubation mixture containing the following components (millimoles) is prepared in a final volume of 140 ml and incubated at 30°: galactosamine·HCl, 6.0; PEP, 10.0; ATP, 6.0; 3-PGA, 10.0; potassium phosphate buffer, pH 7.8, 30.0; $MgCl_2$, 3.0; and 30 ml of the crude galactokinase preparation.[5] The ATP generating system is necessary to negate in part the phosphatases present in the crude extract. The course of phosphorylation is most conveniently measured by using a pH stat; 2 N NaOH is added as required to maintain the pH at 7.8. Although the reaction proceeds very rapidly, 0.2 ml of toluene

[3] J. J. Distler, J. M. Merrick, and S. Roseman, *J. Biol. Chem.* **230**, 497 (1958).
[4] C. Spivak and S. Roseman, *J. Amer. Chem. Soc.* **81**, 2403 (1959).
[5] Changes are suggested for increasing the enzyme concentration and amount of Dowex 50 H⁺ resin and for the assay of galactosamine-1-P eluted from the column. See Checker's comments in D. M. Carlson, A. L. Swanson, and S. Roseman, *Biochem. Prep.* **13**, 3 (1971).

is added after 6 hours and incubation is continued for an additional 12 hours for maximum yield of galactosamine-1-P. The reaction is terminated by heating 100° for 5 minutes, and the precipitate is removed by centrifugation. As assayed by the above procedure, an 80% yield of galactosamine-1-P is achieved (4.8 mmoles). The solution is placed on a column of Dowex 50 H$^+$, 420 ml,[5] and the column is washed with water to eluate galactosamine-1-P. The product is eluted from the column following the nucleotides and other phosphorylated compounds and is located by assaying for total phosphorus and galactosamine-1-P.[5] Fractions containing the product are combined (3.9 mmoles) and lyophilized. The resulting white powder (1.1 g) is dissolved in 20 ml of water, and 20 ml 95% ethanol is added. Crystals are harvested after 4–6 days. The first crop of material contains 0.95 g; about 0.1 g is obtained in the second crop of crystals after more ethanol has been added.

Preparation of N-Acetylgalactosamine-1-P. Crystalline galactosamine-1-P is N-acetylated as described by Distler *et al.*[3] This procedure is generally applicable for preparing relatively small quantities of [^{14}C]N-acetylgalactosamine-1-P or larger amounts of the unlabeled material. The relative concentrations of acetic anhydride and galactosamine-1-P are varied as dictated by circumstances; excess galactosamine-1-P for the complete utilization of [^{14}C]acetic anhydride if preparing the ^{14}C-labeled compound and excess acetic anhydride for complete N-acetylation of galactosamine-1-P if preparing the ^{12}C-labeled compound.

For preparing unlabeled N-acetylgalactosamine-1-P, a solution containing 1 mmole (250 mg) of galactosamine-1-P, equivalent (84 mg) NaHCO$_3$, and 3 ml of methanol in 20 ml of water is cooled in ice. Five millimoles of acetic anhydride (0.47 ml) is added over a 30-minute period with continuous stirring, while the pH is maintained between 7 and 9 by adding NaHCO$_3$. The solution is stirred for 1 hour, although the reaction is essentially complete within a few minutes. Dowex 50 H$^+$, 200–400 mesh, X8, is added to remove the Na$^+$, the mixture is filtered, and the resin is washed with water. The acetic acid is removed by ether extraction. Control experiments with [^{14}C]acetate added prior to the Dowex 50 showed that 6 or 7 extractions (ether:water, 2:1 v/v) completely removed acetic acid. The pH is adjusted to 9.0 with KOH, then the product is concentrated on a rotary evaporator to a colorless syrup that is insoluble in 95% ethanol, but very soluble in methanol. The methanolic solution is treated with a mixture containing equal volumes of ethanol and acetone until a faint persistent turbidity is observed. The addition of the ethanol–acetone mixture is repeated over a period of 3 weeks. The resulting crystals are harvested by centrifugation, washed twice with 95% ethanol, twice with absolute ethanol, and finally with ether and dried to a constant weight. The first crop of crystals represents a 35–40% yield.

For economic purposes, galactosamine-1-P can be in 2- to 3-fold excess when acetylating with [^{14}C]acetic anhydride, especially when very high specific activity is used. Separation of the remaining galactosamine-1-P from the desired product can be performed with a column of Dowex 50 H⁺, 100–200, X12 as described above. The ^{14}C-labeled product, not adsorbed by the column, is detected by liquid scintillation counting. Degree of separation of [^{14}C]N-acetylgalactosamine-1-P and galactosamine-1-P is ascertained by the ninhydrin reaction which is negative for the acetylated compound, positive with the latter compound.[4] The radioactive fractions are pooled and evaporated to a small volume on a rotary evaporator. Acetic acid (4 N) is added, and the evaporation is repeated to remove possible contamination with [^{14}C]acetic acid. Conversion into UDP-[^{14}C]N-acetylgalactosamine can be accomplished as described here or as outlined earlier for small quantities of sugar phosphates.[6]

Preparation of UDP-N-Acetylgalactosamine. N-Acetylgalactosamine-1-P (1.0 mmole) is converted to the bis(tri-n-octylammonium) salt, and this product is allowed to react with uridine 5'-phosphoromorpholidate (0.5 mmole) in anhydrous pyridine for 5 days.[7] UDP-N-acetylgalactosamine is isolated as described previously.[7]

Comments. Galactokinase catalyzes the synthesis of the α-anomer of D-galactosamine-1-P, and thus the desired α-anomer of UDP-N-acetylgalactosamine is obtained. D-Galactosamine-1-P, prepared as described above, serves as a substrate for the enzymatic synthesis of UDP-galactosamine, which can be N-acetylated with [^{14}C]acetic anhydride.[8] The oxidation of UDP-acetylgalactosamine (prepared essentially as described above) with galactose oxidase followed by reduction with NaB^{3}H$_4$ introduces tritium label into the C-6 position[9] UDP-N-acetylgalactosamine prepared as described here is active in systems which synthesize blood group A substance.[10]

[6] Vol. 8 [16].

[7] Vol. 8 [15].

[8] F. Maley, *Biochem. Biophys. Res. Commun.* **39**, 371 (1970).

[9] G. L. Nelsestuen and S. Kirkwood, *Anal. Biochem.* **40**, 359 (1971); see also this volume [31].

[10] D. M. Carlson, 4th Int. Conf. on Cystic Fibrosis of the Pancreas (Mucoviscidosis), Berne/Grindelwald 1966, Part II, pp. 304–307. Karger, Basel, 1968; *in* "Blood and Tissue Antigens" (D. Aminoff, ed.), p. 304. Academic Press, New York, 1970.

[27] ADP-[^{14}C]Glucose

By Jack Preiss and Elaine Greenberg

Principle. ADP-[^{14}C]glucose of high specific activity is prepared from glucose-^{14}C by coupling the hexokinase (reaction 1) and phosphoglucomutase (reaction 2) reactions to the ADP-glucose pyrophosphorylase reaction (reaction 3).

$$\text{Glucose} + \text{ATP} \rightarrow \underset{(1)}{\text{glucose 6-P}} \rightleftharpoons \underset{(2)}{\text{glucose 1-P}} + \text{ATP} \rightarrow \underset{(3)}{\text{ADP-glucose}}$$

Reagents

Bicine [N,N'-bis(2 hydroxyethyl)glycine]–NaOH buffer, 1 M pH 8.0

MgCl$_2$, 0.1 M

Bovine plasma albumin (BPA), 10 mg/ml

ATP, sodium salt, 0.05 M

[^{14}C]Glucose, uniformly labeled (specific activity, 360 mCi/mmole) 1.4 μmoles/ml

Yeast hexokinase, 1000 units/ml (Sigma Type F-300)

Phosphoglucomutase from rabbit muscle, 10 mg/ml; commercial preparation (specific activity 250 units per mg)

Glucose 1,6-diP, sodium salt, 1.3 μmoles/ml

Inorganic pyrophosphatase from yeast 0.18 mg/ml; diluted from commercial preparations with H$_2$O (specific activity, 800 units/mg)

Disodium ethylenediaminetetraacetate (EDTA), 0.1 M, pH 7.2

Pyruvate, sodium salt, 0.18 M

ADP glucose pyrophosphorylase from *Rhodospirillum rubrum*[1,2] 18 units/ml (specific activity 343 units/mg)

Alkaline phosphatase from *Escherichia coli*, 10 mg/ml: obtained from commercial sources (specific activity 10 units/mg)

All enzyme units indicated above are defined as micromoles of products formed in 1 minute at 37°.

Procedure. The bicine buffer (0.14 ml), MgCl$_2$ (0.11 ml), BPA (0.035 ml), ATP (0.11 ml), [^{14}C]glucose (0.5 ml, 250 μCi), phosphoglucomutase (0.01 ml), and glucose 1,6-diP (0.14 ml), are mixed together and the reaction is initiated with hexokinase (0.005 ml) at room temperature

[1] C. E. Furlong and J. Preiss, *J. Biol. Chem.* **244**, 2539 (1969).

[2] M. R. Paule and J. Preiss, *J. Biol. Chem.* **246**, 4602 (1971).

(25°). After 15 minutes, when more than 95% of the glucose is converted to a mixture of glucose 6- and 1-phosphates, ADP-glucose pyrophosphorylase (0.14 ml), inorganic pyrophosphatase (0.07 ml), and sodium pyruvate (0.18 ml) are added. The reaction mixture is then incubated further for 30–40 minutes at room temperature. At that time 80–90% of the radioactivity is converted to ADP-glucose. The reaction mixture is heated for 2 minutes in a boiling H_2O bath; after cooling, alkaline phosphatase (0.03 ml) is added and the reaction mixture is incubated for 20 minutes at 37°. This last procedure converts all the glucose phosphate to free glucose plus orthophosphate while leaving the ADP-glucose intact. EDTA (0.1 ml) is added to the reaction mixture, and it is streaked on Whatman No. 1 paper and chromatographed (descending) for 16 hours in the ethanolic–ammonium acetate–pH 7.5 solvent system [95% ethanol/ 1 M ammonium acetate, pH 7.5 (5:2)]. The ADP-glucose spot is located by UV and radioactivity and is cut out and eluted with H_2O. The eluted ADP-glucose is restreaked on Whatman No. 1 paper and chromatographed (descending) for 16 hours in the ethanolic–ammonium acetate– pH 3.8 solvent system [95% ethanol/1 M ammonium acetate, pH 3.8 (5:2)]. The ADP-glucose is located as indicated above and cut out and placed in 800 ml of absolute ethanol for 3–4 hours to wash off the residual ammonium acetate. After drying, the ADP-[^{14}C]glucose is eluted off the paper with H_2O. The yield of ADP-[^{14}C]glucose from the initial amount of [^{14}C]glucose ranges from 70 to 83%.

Characterization of the Product. Descending paper chromatography on Whatman No. 1 paper in three solvent systems [95% ethanol/1 M ammonium acetate, pH 7.5 (5:2); 95% ethanol/1 M ammonium acetate, pH 3.8 (5:2); isobutyric acid/1 M NH_3/0.1 M EDTA, pH 7.2 (10.0: 6.0:0.16)] revealed that only one UV absorbing spot and one radioactive peak (detected by scanning with a Nuclear-Chicago Actigraph II strip counter) were present and that both migrated the same as authentic ADP-glucose. Both radioactive and UV-absorbing spots migrated the same as did ADP-glucose in paper electrophoresis in 0.05 M citrate buffer, pH 3.9. Hydrolysis of the radioactive material in 0.01 N HCl gave rise to [^{14}C]glucose as determined by chromatography in three solvent systems [88% aqueous phenol solution; ethyl acetate/pyridine/ H_2O (3.6:1.0:1.15); 1-butanol/pyridine/H_2O (6:4:3)]. Spectral analysis of the product at pH 7 showed that it was typical of an adenosine derivative with an absorption maximum at 259 nm ($A_{280}:A_{260} = 0.15$ and $A_{250}:A_{260} = 0.83$). Calculation of the glucose content from the original specific activity of the starting [^{14}C]glucose and of the adenosine content by using the extinction coefficient of ADP at pH 7.0 (15.4 × 10^3 at 259 nm) indicated that there was 1 mole of adenosine per mole of glucose.

The ADP-[¹⁴C]glucose product could be quantitatively incorporated into glycogen primer using the ADP-glucose: α,4-glucan-4-glucosyl transferase from *Escherichia coli*.[3]

Comments on Procedure. The reaction mixtures absolutely required hexokinase and phosphoglucomutase. Omission of glucose 1,6-diphosphate reduced the yield of ADP-glucose to 53%, and omission of yeast inorganic pyrophosphatase reduced the yield to 30–40%. The yield of ADP-glucose did not depend on the presence of the activator of *R. rubrum* ADP-glucose pyrophosphorylase, sodium pyruvate.[1] It was added to ensure full activity of the pyrophosphorylase. Although ADP-glucose pyrophosphorylases from other sources may be used in the procedure described above, that obtained from *E. coli*[4] cannot be used since it is inhibited by the sulfate present in the commercial preparations of phosphoglucomutase and by ADP, a product of the hexokinase reaction.

ADP-glucose is the glucosyl donor for starch synthesis in plants[5–11] and for glycogen synthesis in bacteria.[12–14]

[3] J. Preiss and E. Greenberg, *Biochemistry* **4**, 2328 (1965).

[4] J. Preiss, L. Shen, E. Greenberg, and N. Gentner, *Biochemistry* **5**, 1833 (1965).

[5] E. Recondo and L. F. Leloir, *Biochem. Biophys. Res. Commun.* **6**, 85 (1961).

[6] R. B. Frydman, *Arch. Biochem. Biophys.* **102**, 242 (1963).

[7] R. B. Frydman and C. E. Cardini, *J. Biol. Chem.* **242**, 313 (1967).

[8] R. B. Frydman and C. E. Cardini, *Arch. Biochem. Biophys.* **116**, 9 (1966).

[9] A. K. Doi, K. Doi, and Z. Nikuni, *Biochim. Biophys. Acta* **113**, 312 (1966).

[10] T. Murata, T. Sugiyama, and T. Akazawa, *Biochem. Biophys. Res. Commun.* **18**, 371 (1965).

[11] J. Preiss and E. Greenberg, *Arch. Biochem. Biophys.* **118**, 702 (1967).

[12] E. Greenberg and J. Preiss, *J. Biol. Chem.* **239**, 4314 (1964).

[13] L. C. Gahan and H. E. Conrad, *Biochemistry* **7**, 3979 (1968).

[14] G. J. Walker and J. E. Builder, *Eur. J. Biochem.* **20**, 14 (1971).

[28] GDP-[¹⁴C]Mannose

By JACK PREISS and ELAINE GREENBERG

GDP-[¹⁴C]mannose may be prepared from mannose by using protamine-treated, cell-free extracts of *Arthrobacter viscosus* NRRL B1973 fortified with yeast hexokinase.[1] This method enables one to prepare GDP-mannose of high specific activity and in yields better than those reported in previously published chemical procedures.

[1] J. Preiss and E. Greenberg, *Anal. Biochem.* **18**, 464 (1967).

Reagents

ATP, sodium salt, 0.1 M, pH 6.0

[^{14}C]Mannose, 2 μmoles/ml (obtained from commercial preparations, specific activity 1.6 to 5×10^7 cpm/μmole)

Tris·HCl buffer, 1 M pH 8.0

MgCl$_2$, 0.1 M

Yeast hexokinase (Sigma Type F-300) 1000 units/ml

GTP, ammonium salt, 100 μmoles/ml

α-Glucose-1,6-diP, sodium salt 10 μmoles/ml

Yeast inorganic pyrophosphatase 0.9 mg/ml; diluted from commercial preparations with H$_2$O (specific activity 800 units/mg)

Arthrobacter extract (see below for preparation)

Escherichia coli alkaline phosphatase, 10 mg/ml; obtained from commercial sources (salt fractionated grade, Worthington Biochemical Corporation, Freehold, New Jersey, specific activity 10 units/mg)

All enzyme units indicated above are defined as micromoles of product formed in 1 minute at 37°.

Growth of Organism. The organism used, *Arthrobacter viscosus* NRRL B1973, was obtained as a lyophilized culture from Dr. A. Jeannes and R. F. Anderson of the Northern Regional Research Laboratories in Peoria, Illinois. It was maintained on 2% agar containing 1% glucose, 0.5% Difco bactopeptone, 0.3% Difco malt extract, and 0.3% Difco yeast extract. Cells were grown in 60 ml of media containing 1% galactose, 0.5% bactopeptone, 0.3% malt extract, and 0.3% yeast extract for 2 days at 25° on a New Brunswick shaker. This culture was then transferred to 1.5 liters of culture broth that contained 3% galactose, 0.3% enzyme-hydrolyzed casein (Nutritional Biochemicals Corporation), 0.4% K$_2$HPO$_4$, 0.08% MgSO$_4$, and 0.005% MnSO$_4$·4H$_2$O. The broth was adjusted to pH 7.0 with 5 N H$_2$SO$_4$. The bacteria was grown 4–5 days with shaking at 25°. The cells were harvested by centrifugation at 20,000 g for 30 minutes at 3°. The bacterial paste, 9–10 g was washed with about 100 ml of cold 0.9% NaCl and centrifuged as above. Cells were then stored for at least overnight at -12° and were still active in forming GDP-[^{14}C]mannose for at least 1 month.

Preparations of Crude Extract. Frozen *Arthrobacter* cells, 4 g, were thawed and suspended in 12 ml of 0.05 M Tris·HCl, pH 7.5, containing 0.01 M MgCl$_2$ and 0.005 M dithiothreitol, and then ruptured in a French pressure cell at 20,000 psi. The broken cell preparation was centrifuged for 10 minutes at 30,000 g. The 30,000 g supernatant fraction was then centrifuged at 105,000 g for 1 hour. To 10 ml of the 105,000 g supernatant fluid was added, with stirring, 6.0 ml of a 1% protamine sulfate solution. After 10 minutes, the resultant suspension was centrifuged at 30,000 g for 10 minutes. The supernatant fluid was used as the enzyme source for the preparation of GDP-[^{14}C]mannose. The protein content of the fraction was measured by the procedure of Lowry *et al.*[2]

Procedure. The Tris·HCl buffer (1 ml), MgCl$_2$ (0.4 ml), ATP (0.4 ml), [^{14}C]mannose (6 ml), and yeast hexokinase (0.03 ml) are incubated in a volume of 11 ml for 15 minutes at 37°. Then glucose-1,6-diP (0.1 ml), yeast inorganic pyrophosphatase (0.09 ml), and GTP (0.20 ml) are added to the reaction mixture. The protamine-treated *Arthrobacter* extract (15 ml, 104 mg of protein) is added, and the reaction mixture is incubated for 60–90 minutes. At this time about 50–70% of the [^{14}C]mannose is converted to GDP-[^{14}C]mannose and the reaction is terminated by heating in a boiling water bath for 2 minutes. To the heat-denatured mixture is added 0.5 ml of the *E. coli* alkaline phosphatase and the mixture is incubated for 1 hour at 37°. The reaction mixture is centrifuged at 10,000 g for 10 minutes and the heat-denatured protein precipitate is discarded. The resultant supernatant fluid is adsorbed onto a Dowex 1-X8 formate column (1 × 12 cm) and the column is washed with 20 ml of H$_2$O to remove almost all the [^{14}C]mannose. The column is eluted with 2 N formic acid until no more 260 nm absorbing material appears in the eluate. This usually occurs after 100–150 ml of the acid has passed through the column. The column is then eluted with 4 N formic acid. A radioactive peak and an ultraviolet peak, which are coincident with each other, are eluted in 135–160 ml. The fractions containing the radioactivity are combined and extracted four times with 2 volumes of ether. The remaining formic acid is extracted with ether overnight in a liquid–liquid extractor as described by Smiley and Ashwell.[3] The radioactive fraction is adjusted to pH 6 with 1 N NH$_4$OH and concentrated by flash evaporation. The overall yield of GDP-[^{14}C]mannose from [^{14}C]mannose in 5 separate preparations ranged from 48 to 62%.

Characterization of the Product. Descending paper chromatography on Whatman No. 1 filter paper in four solvent systems [95% ethanol/1 M ammonium acetate, pH 7.5 (5:2); 95% ethanol/1 M ammonium acetate, pH 3.8 (5:2); isobutyric acid/1 M NH$_3$/0.1 M EDTA, pH 7.2 (10.0:6.0:0.16); 600 g of ammonium sulfate in 1 liter of 0.1 M sodium phosphate, pH 6.8, and 20 ml of 1-propanol] revealed that only one ultraviolet-absorbing spot and radioactive peak (detected by scanning with a Nuclear-Chicago Actigraph II strip counter) was present and

[2] O. H. Lowry, N. J. Rosebrough, A. L. Farr, and R. J. Randall, *J. Biol. Chem.* **193,** 265 (1951).
[3] J. D. Smiley and G. Ashwell, *J. Biol. Chem.* **263,** 357 (1961).

that they both migrated the same as did GDP-mannose. Both the radio-active spot and the ultraviolet-absorbing spot migrated the same as did GDP-mannose in paper electrophoresis in 0.05 M citrate buffer, pH 3.9. Hydrolysis of the radioactive material in 0.01 N HCl gave rise to [^{14}C]mannose as determined by chromatography in three solvent systems [88% aqueous phenol solution; ethyl acetate/pyridine/H_2O (3.6: 1.0:1.15); 1-butanol/pyridine/H_2O (6:4:3)]. Spectral analysis of the product at pH 7 showed that it was typical of a guanosine derivative with an absorption maximum at 252 nm ($A_{280}:A_{260} = 0.65$ and $A_{250}: A_{260} = 1.15$). Calculation of the mannose content from the original specific activity of the starting [^{14}C]mannose and of the guanosine content by using the extinction coefficient of GDP at pH 7.0 (13.7 × 10^3 at 252 nm) indicated that there was 1 mole of guanosine per mole of mannose. Furthermore, the compound could be quantitatively converted to GDP-[^{14}C]mannuronate with GDP-mannose dehydrogenase and DPN.[4]

Comments on Procedure. Presumably biosynthesis of GDP-[^{14}C]mannose from [^{14}C]mannose catalyzed by yeast hexokinase and *Arthrobacter* extracts occurs by the following reactions:

$$\text{Mannose} + \text{ATP} \xrightarrow{} \underset{(1)}{\text{mannose 6-P}} \rightleftarrows \underset{(2)}{\text{mannose-1-P}} \xrightarrow{\text{GTP}} \underset{(3)}{\text{GDP-mannose}}$$

Reaction (1) is catalyzed by the yeast hexokinase, and reactions (2) and (3) are catalyzed by enzymes present in the *Arthrobacter* extracts.[5]

Omission of the inorganic pyrophosphatase or the α-glucose 1,6-di-phosphate decreased the yield of sugar nucleotide by 50%. The alkaline phosphatase treatment was used in preparing the GDP-[^{14}C]mannose since it degraded all nucleoside mono-, di-, and triphosphates that would be present in the reaction mixture to the nucleoside level. These compounds would therefore not be contaminants of the GDP-mannose fraction that was isolated by Dowex 1 chromatography. The phosphate formed from the phosphatase hydrolysis is eluted from the column by 2 N formic acid. The preparation of the crude extracts, the formation of GDP-mannose, and the purification and isolation of the sugar nucleotide take about 2 days.

[4] J. Preiss, *J. Biol. Chem.* **239**, 3127 (1964).
[5] J. Preiss and E. Wood, *J. Biol. Chem.* **239**, 3119 (1964).

[29] GDP-L-[14C]Fucose[1,2]

By HARRY SCHACHTER, HANAKO ISHIHARA, and EDWARD C. HEATH

GDP-L-[14C]fucose is prepared by the utilization of L-fucose kinase[3] to synthesize 14C-β-L-fucose 1-phosphate; the isolated radioactive fucose phosphate derivative is then used as substrate with GTP in the GDP-L-fucose pyrophosphorylase reaction[4] to synthesize the sugar nucleotide.

$$\text{L-fucose} + \text{ATP} \xrightarrow{\text{Mg}^{2+}} \beta\text{-L-fucose 1-phosphate} + \text{ADP}$$

$$\beta\text{-L-fucose 1-phosphate} + \text{GTP} \rightleftharpoons \text{GDP-L-fucose} + \text{P-P}_i$$

Synthesis and Isolation of β-L-[14C]Fucose 1-Phosphate

The extensively purified preparation of L-fucose kinase from porcine liver may be utilized for this procedure although the ammonium sulfate fraction is satisfactory. The method outlined here utilizes the latter preparation, thus permitting both the kinase and the pyrophosphorylase (see below) to be prepared from the same extract.

Procedure. The ammonium sulfate (0–30%) pellet obtained from 250 g of porcine liver is dissolved in approximately 15 ml of 0.13 M sodium phosphate buffer, pH 7.4. An incubation mixture is prepared that contains the following (in micromoles) in a final volume of 100 ml: L-[14C]fucose, 67 (8.7×10^6 cpm/μmole); ATP, 2000; MgCl$_2$, 1000; KF, 1000; Tris, pH 8, 6700; and 10 ml of the ammonium sulfate fraction described above. After 3 hours of incubation, 2 volumes of ethanol are added, and the mixture is placed in a water bath at 55° for 5 minutes and then in an ice bath. After cooling for 30 minutes, the mixture is centrifuged for 10 minutes at 2000 g, the precipitate is washed three times with 100-ml portions of 70% ethanol, and the supernatant and wash solutions are combined. The solution is concentrated under reduced pressure and fractionated on a Dowex 1-HCO$_3^-$ column (2.5 × 27 cm) in a manner identical with that described for the small scale preparation.[5] The fractions eluted from the column between 0.3 and 0.4 M ammonium bicarbonate are combined and adjusted to pH 4 with Dowex 50, H$^+$ form; colorimetric assay of the solution for 6-deoxyhexose [6] indicates the presence of 120 μmoles of fucose. After removal of resin by filtration, the filtrate is con-

[1] I. Jabbal and H. Schachter, *J. Biol. Chem.* **246**, 5154 (1971).
[2] H. Ishihara and E. C. Heath, *J. Biol. Chem.* **243**, 1110 (1968).
[3] H. Ishihara, H. Schachter, and E. C. Heath, this volume [48].
[4] H. Ishihara and E. C. Heath, this volume [49].
[5] H. Ishihara, D. J. Massaro, and E. C. Heath, *J. Biol. Chem.* **243**, 1103 (1968).
[6] Z. Dische and L. B. Shettles, *J. Biol. Chem.* **175**, 595 (1948).

centrated under reduced pressure to approximately 3 ml. To the filtrate is added 0.5 ml of a 2 M solution of barium acetate, and the small precipitate of barium carbonate is removed by centrifugation. The precipitate is washed twice with 1-ml portions of cold water, and to the combined supernatant solutions are added 10 volumes of ethanol. The suspension is allowed to stand overnight at 4°, and the white, flocculent precipitate is collected by centrifugation, washed with cold ethanol and ether, and dried under vacuum. Analysis of the crude barium fucose phosphate indicates that it is 54% pure on a dry weight basis. Reprecipitation of the barium salt with ethanol increases the purity to 66%, with a yield of approximately 40% based on the original amount of fucose used in the incubation mixture.

The fucose phosphate preparation is further purified by conversion to the lithium salt as follows. A solution (5 ml) containing 30 μmoles of barium fucose phosphate is shaken with 2 ml (bed volume) of Dowex 50, Li$^+$ form (200–400 mesh) resin for 5 hours at room temperature. The resin is removed by filtration and washed with a small amount of water; the filtrate is concentrated under reduced pressure to dryness and dissolved in 1 ml of methanol, and lithium fucose phosphate is precipitated by the addition of 10 ml of cold acetone. The precipitate is collected by centrifugation, washed successively with cold acetone and ether, and dried under vacuum. Analysis indicates that the lithium salt is approximately 75% pure on a dry weight basis.

Synthesis of GDP-L-[^{14}C]Fucose

The pork liver supernatant remaining after precipitation of L-fucose kinase activity at 30% ammounium sulfate is adjusted to an ammonium sulfate concentration of 50% and centrifuged at 20,000 g for 20 minutes. The resultant pellet is dissolved in about 10 ml of 0.03 M sodium phosphate buffer at pH 7.4 containing 0.1 mM dithiothreitol. This crude preparation of GDP-L-fucose pyrophosphorylase is further purified by gel filtration on a Sephadex G-100 column (5 × 85 cm) equilibrated with 0.03 M sodium phosphate at pH 7.4, containing 0.1 mM dithiothreitol. An incubation mixture is prepared which contains the following (in micromoles) in a final volume of 300 ml: β-L-[^{14}C]fucose 1-phosphate, 27; GTP, 210; MgCl$_2$, 1200; KF, 2400; Tris, pH 8.0, 12,000; and 150 ml of GDP-L-fucose pyrophosphorylase from the Sephadex G-100 column. After incubation at 37° for 2 hours, the reaction is stopped by addition of 2 volumes of ethanol. The suspension is centrifuged at 4000 g for 15 minutes, and the pellet is reextracted with ethanol. The combined ethanol extracts are evaporated under reduced pressure and the residue is dissolved in 300 ml of water. The solution is applied to a column (5 × 9 cm) of Dowex 1-X2, chloride form, 50–100 mesh, and the column is washed with 300 ml of 0.005 M Tris·HCl buffer, pH 7.5. The adsorbed materials are fractionated by elution with a linear gradient (3000 ml) of 0–1.0 M KCl in 0.005 M Tris·HCl, pH 7.5. Fractions are assayed for radioactivity and for ultraviolet light-absorbing materials; GDP-L-[¹⁴C]fucose is eluted after L-[¹⁴C]fucose 1-phosphate and is the main ultraviolet light-absorbing peak. The GDP-fucose peak fractions are pooled, evaporated under reduced pressure, and desalted by passage through a column (5 × 80 cm) of Sephadex G-10 equilibrated with 0.05 M triethylamine-bicarbonate, pH 7.5. Yields of GDP-[¹⁴C]fucose are usually better than 60%. The final GDP-L-[¹⁴C]fucose preparation has a specific activity of 8.3 × 10⁶ cpm/μmole, exhibits an ultraviolet absorption spectrum characteristic of a guanosine nucleotide, and moves as a single radioactive and ultra-violet absorbing peak on high voltage electrophoresis at pH 6.5 in pyridinium acetate and on paper chromatography in the following solvent systems: I, ethanol–1 M ammonium acetate at pH 7.5 (7:3); and II, ammonium hydroxide–water–ethanol (1:10:80). Hydrolysis of the nucleotide sugar with 0.01 M HCl for 10 minutes at 100° releases a radioactive compound which migrates with standard L-fucose on high voltage electrophoresis in 1% sodium tetraborate and on paper chromatography in several standard solvent systems.

[30] TDP-[3-³H]Glucose and TDP-[4-³H]Glucose[1,2]

By OTHMAR GABRIEL

Methods for the synthesis of TDP-glucose specifically tritium-labeled at carbon 3 or carbon 4 of the hexose moiety are described. It is nec-essary first to prepare labeled glucose-*3T* and glucose-*4T*, respectively; this is followed by enzymatic or chemical phosphorylation[3] to sugar 1-phosphate and condensation to the corresponding sugar nucleotide. It should be noted that once the specifically labeled parent sugar is prepared, the condensation of glucose 1-phosphate to any purine or pyrimidine

---

[1] The terms TDP-glucose-*3T*, TDP-glucose-*4T*, glucose-*3T*, or glucose-*4T* as used in this paper refer to tritiated compounds labeled specifically at carbon 3 or 4, respectively, of the hexose.

[2] This work was supported by Grant AI-07241 from the National Institute of Al-lergy and Infectious Diseases, National Institutes of Health, United States Public Health Service.

[3] D. L. MacDonald, *J. Org. Chem.* **27**, 1107 (1962).

base can be accomplished by organic synthesis.[4,5] The use of enzymes for the latter purpose is restricted to the specificity of the pyrophosphorylase.

Synthesis of TDP-Glucose-*3T*[6]

Principle. Agrobacterium tumefaciens converts sucrose in good yield to a disaccharide in which carbon 3 of the glucose moiety has been oxidized to a keto group (α-D-ribohexopyranosyl-3-ulose-β-D-fructopyranoside).[7] Reduction of the keto group with tritiated borohydride yields the expected two epimeric disaccharides α-D-glucosyl-*3T*-β-D-fructoside and α-D-allosyl-*3T*-β-D-fructoside. The specificity of sucrose phosphorylase cleaves only the glucose containing disaccharide resulting in formation of α-D-glucose 1-phosphate-*3T*. The allosyl disaccharide remains intact. The specifically labeled 1-phospho-α-D-glucopyranosyl-*3T* can easily be separated from the unchanged α-D-allosyl-*3T*-β-D-fructoside by ion-exchange chromatography. Conversion to the sugar nucleotide is accomplished by enzymatic catalysis.

Procedures

Preparation of α-D-Glucose 1-Phosphate-3T. Agrobacterium tumefaciens (I.A.M.-1525, Japanese Federation of Culture Collection of Microorganisms) is grown aerobically at 27° into the late logarithmic growth phase in the following medium: sucrose 20 g; urea, 0.5 g; K_2HPO_4, 3 g; ammonium sulfate, 1 g; sodium chloride, 0.2 g; calcium chloride, 0.1 g, $MgSO_4 \cdot 7H_2O$ 0.2 g, ferric ammonium sulfate, 0.005 g. The last three items are sterilized separately and added after the medium has been autoclaved for 15 minutes; 750 ml of medium is used in a 4-liter flask. After about 3 days' shaking at 27° all the sucrose is utilized and 3-ketosucrose appears as the only major component. A cell-free 0.1-ml aliquot can be tested by the addition of 3 ml of $0.1\,N$ NaOH. This solution is allowed to stand 3 minutes at room temperature and the increase of optical density at 340 nm is measured (about 0.2 reading per micromole of 3-ketosucrose in a 1-cm light path). The cells are removed by centrifugation at 20,000 g for 30 minutes, and the clear supernatant is concentrated under reduced pressure to 80 ml. Addition of 2 volumes of ethanol to this solution and storage overnight at $-10°C$ results in the formation of a precipitate. The precipitate is removed by centrifugation and discarded. The alcohol is removed by

[4] J. G. Moffatt, Vol. 8 [15].
[5] A. D. Elbein, Vol. 8 [16].
[6] O. Gabriel and G. Ashwell, *J. Biol. Chem.* **240**, 4123 (1965).
[7] Referred to herein as 3-ketosucrose.

evaporation under reduced pressure. The aqueous solution is deionized on a mixed-resin bed (Rexyn I-300). The solution (about 2000 μmoles of 3-ketosucrose) is adjusted to pH 8.0 with 1 N ammonium hydroxide (volume 50 ml), and 2 ml of a solution containing 25 mg of sodium borohydride (10 mCi in 0.1 M NaHCO₃) is added and kept at room temperature for several hours. The reduction with sodium borohydride is known to produce preferentially the allosyl derivative, and about 100 μmoles (5%) of α-D-glucosyl-$3T$-β-D-fructofuranoside is expected to be formed under these conditions. The reaction mixture is deionized on a mixed-resin bed, and the clear percolate is concentrated to a syrup *in vacuo*.

Sucrose phosphorylase has been prepared from *Pseudomonas sac-charophila* (ATCC) according to Doudoroff[8] with the exception that the 0–50% ammonium sulfate fraction was found to contain the enzymatic activity rather than the 50–70% fraction reported.

The above syrup is dissolved in 0.033 M phosphate buffer pH 6.7 containing about 250 units of sucrose phosphorylase in about 10 ml. Incubation at 30°C is carried out for 30 minutes. Several aliquots are taken to assay for α-D-glucose 1-phosphate by enzymatic analysis.[9] A total formation of 82 μmoles of glucose 1-phosphate has been found. The reaction is stopped by addition of 60 ml of boiling ethanol. The denatured protein is removed by centrifugation and washed with 75% ethanol, and the wash is added to the original supernatant solution. Enzymatic assay[9] has indicated the presence of 82 μmoles of α-D-glucose-1-phosphate-$3T$.

The alcoholic solution is adjusted to pH 7.0 with ammonium hydroxide and the ethanol is removed at room temperature by distillation under reduced pressure. Removal of inorganic phosphate is accomplished as the magnesium ammonium phosphate complex. To 25 ml of the solution, 2.5 ml of concentrated ammonia and 4 ml of 10% magnesium acetate are added. The preparation is allowed to stand for 1 hour at 0°, the precipitate is removed by centrifugation and washed with 1.5 N ammonia containing 14 mg of ammonium acetate per milliliter. The supernatant solutions are combined, concentrated to dryness in vacuum, and dissolved in a small volume of water. The solution is passed through a small column of Dowex 50 H⁺, and the column is washed with cold water to a final volume of 20 ml. The solution is adjusted to pH 8.0 and applied to a Dowex 1-formate column (25-ml bed volume) set up

[8] M. Doudoroff, Vol. 1 [28].
[9] H. U. Bergmeyer, Ed. "Methods of Enzymatic Analysis." Academic Press, New York, 1965.

in a cold room at 0–3°. The resin is washed with cold distilled water until the cysteine-carbazole reaction[10] is negative. In addition, aliquots are assayed for radioactivity, and washing is continued until no more radioactivity appears in the eluate. About 1000 ml of water is necessary for washing the column, containing free fructose and about 1600 μmoles of α-D-allosyl-$3T$-β-D-fructofuranoside.

The column is eluted with 1 N formic acid, and each fraction is examined for radioactivity. A single radioactive peak is recovered in a 500-ml fraction. The formic acid is removed by extraction with ether overnight in a liquid–liquid extractor at 0°C. The aqueous phase is adjusted to pH 8.0 with ammonium hydroxide and concentrated to about 10 ml at room temperature *in vacuo*. The brown concentrate is treated with a small amount of Norite A and filtered to yield a clear, colorless solution. This method of preparation has revealed the presence of a single, radioactive compound indistinguishable from authentic glucose 1-phosphate. Enzymatic assay[9] indicated a total recovery of 68 μmoles α-D-glucose-1-phosphate-$3T$. The specific activity was 6.9×10^6 cpm/μmole (at about 30% counting efficiency).

Preparation of TDP-Glucose-3T. The TDP-glucose pyrophosphorylase preparation used is the 50–70% ammonium sulfate fraction of *P. aeruginosa* 7700 described by Kornfeld and Glaser.[11] The incubation mixture contains 5 μmoles of Tris·chloride buffer pH 8.0, 10 μmoles $MgCl_2$, 1 μmole of EDTA, 30 μmoles of TTP, 10 units of TDP-glucose pyrophosphorylase, and 8 μmoles of α-D-glucose-$3T$-1-phosphate in a total volume of 4.4 ml. The reaction is allowed to proceed for 45 minutes at 37°. The reaction mixture is cooled to 0°, and 1 ml of 1 M $HClO_4$ is added. The resulting precipitate is removed by centrifugation and reextracted with 10 ml of ice cold 0.15 M $HClO_4$. Norite A, 600 mg, is added to the supernatant fraction, and the mixture is kept at 0° for 30 minutes. The charcoal is filtered on a sintered glass funnel, washed with water, and eluted three times with 5 ml of 50% ethanol containing 0.5% ammonium hydroxide. The eluate is concentrated *in vacuo* to a small volume, applied to Whatman No. 1 filter paper, and subjected to descending chromatography in ethanol–1 M ammonium acetate (7:3). A single radioactive and ultraviolet absorbing band migrates with the same mobility as authentic TDP-glucose. The product is eluted with water and behaves in every respect like TDP-glucose. No evidence for the presence of TDP-4-keto-6-deoxy glucose has been found. A total of 7.8 μmoles of TDP-glucose-$3T$ were recovered by this

[10] Z. Dische and E. Borenfreund, *J. Biol. Chem.* **192**, 583 (1951).
[11] S. Kornfeld and L. Glaser, *J. Biol. Chem.* **236**, 1791 (1961).

method. The specific activity of the product was 6.9×10^6 cpm/μmole, in good agreement with the specific activity of the starting material α-D-glucose-1-phosphate-*3T*. The TDP-glucose-*3T* prepared was analytically identical to authentic TDP-glucose and was also active when assayed with TDPG-pyrophosphorylase.

Properties and Radiopurity of Products

It should be noted that TDP-glucose-*3T*, like other sugar-nucleotides, is sensitive to treatment with alkali. TMP and cyclic glucose-1,2-phosphate-*3T* are the degradation products. Proper care has to be taken to avoid exposure of the nucleotide to alkaline conditions for any extended period of time.

In order to test the radiopurity, TDP-glucose-*3T* was converted to methyl-D-glucopyranoside-*3T*. The methyl glucoside was then subjected to periodate degradation followed by isolation of the degradation products.[6] The results of the degradation procedure established that glucose-*3T* prepared by this method has as a minimum value of 98% tritium located at carbon 3.

Storage of compounds with high specific activity is limited by radioactive autodestruction. During the course of one year, in which α-D-glucose-1-phosphate-*3T* was stored frozen, no evidence for redistribution of label was detected.

Synthesis of TDP-Glucose-*4T*[11a]

Principle. Methyl-α-D-galactopyranoside (I) is converted into methyl 2,3,6-tri-*O*-benzoyl-α-D-galactopyranoside (II). Oxidation of (II) with methyl sulfoxide leads to methyl 2,3,6-tri-*O*-benzoyl-α-D-xylohexopyranoside-4-ulose (III). Reduction of (III) with sodium borotritiide results in the formation of the expected 4-tritiated methyl 2,3,6-tri-*O*-benzoyl-α-D-glucopyranoside (IV) and α-D-galactopyranoside (V), respectively. On debenzoylation with sodium methoxide, methyl-α-D-glucopyranoside-*4T* (VI) and methyl-α-D-galactopyranoside-*4T* (VII) are formed. Finally, acid hydrolysis yields the free sugars D-glucose-*4T* (VIII) and D-galactose-*4T* (IX). Conversion to the sugar phosphate and to TDP-glucose is accomplished by enzymic reactions.[11a]

Chromatographic Methods. Descending paper chromatography is carried out on Whatman No. 1 paper in the following solvent system: Solvent (1)—pyridine–ethyl acetate–water (1:3.6:1.15).

Thin-layer chromotography (TLC) is carried out on silica gel G (Merck) in the following solvent systems: (a) ethyl acetate–pentane

[11a] O. Gabriel and L. C. Lindquist, *J. Biol. Chem.* **243**, 1479 (1968).

(3:1 v/v); (b) benzene–methanol (99:1 v/v); (c) benzene–methanol (95:5 v/v); (d) benzene–methanol (90:10 v/v).

Solvent systems (a) to (d) are used with plates spread with a suspension of 25 g silica gel in 62 ml of water and activated for 30 minutes at 110°. The time for development is about 1 hour. Microscopic slides (75 × 25 mm) coated with silica gel G are used for analytical purposes.

The following detection methods on thin-layer plates are used: (1) examination with UV light; (2) spray with 10% sulfuric acid in methanol followed by heating on a hot plate: brown to black spots; (3) spray with 2.4-dinitrophenylhydrazine (0.1 g 2.4-dinitrophenylhydrazine dissolved in 5 ml of HCl conc. and 95 ml of ethanol); (4) detection of benzoylated sugar derivatives by spraying with water. Melting points are determined on a Kofler hot stage.

Procedure for the Synthesis of D-Glucose-4T[12]

Methyl-α-D-galactopyranoside (I).[13] Reagent grade anhydrous D-galactose (50 g) was heated under reflux with methanolic hydrochloric acid (2% v/v, 400 ml) for 12 hours. The solution was neutralized by addition of lead carbonate (50 g) and kept for 3 hours under stirring. The lead salts are removed by filtration with Celite and washed with 70 ml methanol. The combined filtrate is evaporated *in vacuo* to a syrup. The warm syrup is mixed with 15 ml water and crystallization occurs for 20 hours at room temperature and for 24 hours at 5°C. The mother liquor is removed by vacuum filtration (cooled 7 cm sintered glass funnel). The solid crystals are washed with 80% ethanol (5°, about 25 ml) and two times with 10 ml of ethanol. The air-dried product 24.6 g showed $[\alpha]_D^{23}$ of +174°.

Conversion of (I) to Methyl-2,3,6-tri-O-benzoyl-α-D-galactopyranoside (II).[14] A solution of 6.81 g (35.1 mmoles) methyl-α-D-galactopyranoside in 70 ml of dry pyridine was cooled to 0° in an ice bath. Benzoylchloride (14 ml, 121 mmoles) was added dropwise under stirring. The reaction mixture was kept at room temperature for 3 days. The unreacted benzoylchloride was decomposed by dropwise addition of the reaction mixture of 200 ml of ice-cold saturated aqueous NaHSO₃ solution under stirring. Extraction of the aqueous suspension with 250 ml of chloroform is carried out; the chloroform layer is washed with water, dried, and evaporated to dryness *in vacuo*, yielding 10.4 g of

[12] O. Gabriel, *Carbohyd. Res.* **6**, 319 (1968).

[13] J. L. Frahn and J. A. Mills, *Aust. J. Chem.* **18**, 1303 (1965).

[14] E. J. Reist, R. R. Spencer, D. F. Calkins, B. R. Baker, and L. Goodman, *J. Org. Chem.* **30**, 2312 (1965).

a white solid product. Recrystallation from methanol result in 10 g of the tribenzoate (57%); m.p. 132–136°; TLC in solvents (a) to (d) show one single component.

Oxidation of (II) to Methyl-2,3,6-tri-O-benzoyl-α-D-xylohexopyrano-side-4-ulose (III). Compound (II) (250 mg, 500 μmoles) is dissolved in 1.5 ml of dimethyl sulfoxide and 1.0 acetic anhydride in a glass-stoppered test tube and kept for 19 hours at room temperature. TLC in solvent (b) indicates that all the starting material (R_f 0.47) is converted into two new compounds (R_f 0.62 and 0.70). Only the component with R_f 0.62 reacts with the 2.4 dinitrophenylhydrazine reagent. The reaction mixture is fractionated by column chromatography on a silica acid column (California Biochemical Corporation) (dimensions 3 × 19 cm), and 1-ml fractions are collected at 10-minute time intervals. The solvent used is benzene–methanol (99:1 v/v). Aliquots (5 μl) of each fraction are spotted on a silica gel plate, and the compounds present are detected with reagent 2. Positive reaction is noticed in fractions 35–38 and 40–43. After combination of fractions 35–38 and 40–43, TLC in solvent (b) shows the presence of the expected phenylhydrazine-reactive component in fractions 40–43.

Reduction of (III) to Methyl-2,3,6-tri-O-benzoyl-α-D-glucopyrano-side-4T (IV) and to Methyl-2,3,6-tri-O-benzoyl-α-D-galactopyranoside-4T (V). The combined fractions 40–43 are concentrated *in vacuo* at room temperature to dryness and the residue is dissolved in 5 ml of ethanol. The solution is cooled in an ice bath, and under stirring 200 μl of sodium borotritiide (1000 μmoles/ml, about 10 mCi, New England Nuclear Corporation) are added. After 10 minutes, the heterogeneous reaction mixture containing some crystalline material is adjusted to pH 4.0 with glacial acetic acid. The solution is evaporated to dryness under reduced pressure, 1 ml of acetone is added, and the suspension is again evaporated to dryness. The residue is suspended in 2 ml of methylene chloride and the organic layer is washed 3 times with 1 ml of water. The aqueous phases are discarded. The methylene chloride solution is evaporated to dryness and the residue dissolved in 1 ml acetone. Examination by TLC in solvent (c), shows three components: The first spot (R_f 0.68) is radioactive and has the same mobility as (II). The second component (R_f 0.79) also contains tritium whereas the third component (0.86) is free of radioactivity.

The reaction mixture is streaked onto five 20 × 20 cm TLC plates and is subjected to chromatography in solvent (c). A reference standard of methyl-2,3,6-tri-O-benzoyl-α-D-galactopyranoside is included on each plate. The compounds are located by examination in UV light, and the area corresponding to the mobility of (II), (IV), and (V) is eluted with acetone. The eluate is concentrated to dryness *in vacuo* and stored over silica gel in a vacuum desiccator overnight.

Debenzoylation of (IV) to Methyl-α-D-glucopyranoside (VI) and (V) to Methyl-α-D-galactopyranoside-4T (VII). The dry residue containing the mixture of benzoylated derivatives (IV) and (V) is dissolved in 1 ml of dry methanol, and a freshly prepared solution of 0.5% sodium methoxide (60 μl) is added. After 30 minutes at room temperature, an aliquot is examined by TLC in solvent (c). All the product remains at the origin, indicating quantitative debenzoylation. The mixture of tritiated α-methylglycosides (VI) and (VII) is subjected to descending paper chromatography on Whatman No. 1 paper in pyridine-ethyl acetate–water solvent. After 48 hours the well separated compounds (VI) and (VII) are located by their radioactivity and eluted with water. About two-thirds of the total material corresponds to methylglucoside, one-third to methylgalactoside.

Acid Hydrolysis of Methyl-α-D-glucopyranoside-4T (VI) to D-Glucose-4T (VIII) and Methyl-α-D-galactopyranoside-4T (VII) to D-Galactose-4T (IX). Aliquots of the eluted pure methylglycosides (VI) and (VII) (250 μl) containing about 3–5 μmoles each are evaporated to dryness. The residues are dissolved in 0.5 M sulfuric acid (0.5 ml), and the solution is heated for 1 hour in a boiling water bath. The samples are chilled, neutralized with 0.15 M Ba(OH)$_2$. The BaSO$_4$ is removed by centrifugation and washed with a small volume of water. The supernatant and the water wash are passed through a 0.5 cm × 1.0 cm column of Dowex 50 (H$^+$). The hydrolyzate of (VI) results in the formation of D-glucose-4T (VIII), that of (VII) in the release of D-galactose-4T (IX). Paper chromatography shows in each instance one single radioactive peak with identical mobility of the expected monosaccharides D-glucose and D-galactose, respectively. Both radioactive sugars prepared by the above method were active when used in the appropriate enzymatic assay. (D-glucose-4T is a substrate for hexokinase, D-galactose-4T is oxidized by galactose oxidase.) The specific activity for D-glucose-4T was determined to be 2 × 10^6 cmp/μmole (about 30% counting efficiency).

Proof of Radiopurity of Methyl-α-D-glucopyranoside-4T (VI)[12]

Periodate oxidation of carrier diluted methyl-α-D-glucopyranoside-4T (VI) was used to provide experimental evidence for the selective tritiation of the sugar at carbon 4. Isolation, identification and measurement of radioactivity in the degration products indicated that a minimum of 99% tritium was located on carbon 4.

Procedure for Synthesis of TDP-Glucose-4T[11a]

Principle. The synthesis is performed in two steps: First, enzymatic phosphorylation of D-glucose-4T by hexokinase is carried out. In a second incubation, conversion to α-D-glucose-1-phopshate-4T catalyzed by phosphoglucomutase followed by formation of TDP-glucose-4T in the presence of TDP-glucose pyrophosphorylase is achieved. Inorganic pyrophosphatase is included in the second incubation mixture to assure quantitative conversion to the desired end product TDP-glucose-4T.

Preparation of D-Glucose-6-phosphate-4T. D-Glucose-4T is converted quantitatively to D-glucose-6-phosphate-4T with hexokinase. The incubation mixture has the following composition in a total volume of 1 ml: 1 μmole of D-glucose-4T (2×10^6 cpm), 2.5 μmoles of ATP, 50 μmoles of Tris·chloride buffer pH 7.5, 5 μmoles of $MgSO_4$, and 1 μl of hexokinase (yeast, Boehringer). Incubation of 30 individual samples is carried out at room temperature for 15 minutes. The samples are boiled for 1 minute and pooled. An aliquot of the reaction mixture (50 μl) is tested with glucose-6-phosphate-dehydrogenase,[9] and quantitative conversion to glucose-6-phosphatate-4T is found. The rest of the sample is applied to Dowex 1-X2 column (formate, 200–400 wash, 1 cm × 14 cm), and the column is washed with water. A linear gradient is used consisting of 500 ml of water in the mixing flask. The flow rate is adjusted to 0.6 ml per minute, and 6-ml fractions are collected. Fractions 100–165 are combined and lyophilized. Quantitative recovery of D-glucose-6-phosphate-4T (30 μmoles) is obtained. The specific activity is determined to be 2×10^6 cmp/μmole. The compound obtained is identical with authentic D-glucose 6-phosphate.

Preparation of TDP-glucose-4T. The same TDP-glucose pyrophosphorylase preparation[11] described for the synthesis of TDP-glucose-*3T* is used. The incubation mixture has the following composition in a total volume of 1.0 ml: 1.0 μmole of TTP, 25 μmoles of Tris·chloride buffer pH 7.5, 2.5 μmoles of $MgCl_2$, 0.5 μmole of EDTA, 0.2 μmole of D-glucose-6-phosphate-4T, 10 μl of phosphoglucomutase (Boehringer), 50 μl of TDP-glucose pyrophosphorylase, 5 μl of glucose 1,6-diphosphate ($2 \times 10^{-4} M$) and 2 μl of inorganic pyrophosphatase (Sigma).

Incubation of 10 individual samples of the above composition is carried out for 2 hours at 37°. The samples are combined, a 100-μl aliquot is removed, heated for 1 minute at 100°, and assayed for residual glucose 6-phosphate.[9] The conversion is quantitative. The remaining sample is cooled to 0°, and 2.5 ml of 1 N $HClO_4$ is added. The precipitate is removed by centrifugation and reextracted with two 0.5-ml portions of cold $0.1 N$ HClO$_4$. To the pooled supernatant fractions, 100 mg of Norite A is added and the suspension is kept for 30 minutes at 0°. The charcoal is filtered on a sintered glass funnel, washed extensively with cold water, and eluted with four 4-ml portions of 50% ethanol containing 0.5% ammonium hydroxide. The eluate is concentrated to a small volume *in vacuo*, applied to Whatman No. 1 paper, and subjected to descending chromatography in ethanol–1 M ammonium acetate (7:3). A single radioactive and UV absorbing band with a mobility corresponding to authentic TDP-glucose can be observed. The radioactive band is eluted with water, and the product obtained is identical to TDP-glucose. The specific activity of the isolated TDP-glucose-*4T* is 2×10^6 cpm/μmole.

[31] UDP-[4-³H]Galactose and UDP-*N*-[6-³H] Acetylgalactosamine

By S. KIRKWOOD

UDP-[4-³H]Galactose

The procedure outlined below describes the synthesis of UDP-[4-³H] galactose.[1] However, the procedure obviously lends itself to the preparation of samples of both UDP-galactose and UDP-glucose labeled with tritium in any desired position. The synthesis commences with an appropriately labeled sample of hexose, and [4-³H]galactose, [1-³H]galactose, [1-³H]glucose, [2-³H]glucose, [3-³H]glucose, [4-³H]glucose, [5-³H]glucose, and [6-³H]glucose are all available commercially (Amersham/ Searle). Further, the nucleotide diphosphate sugar resulting from any one of these labeled hexoses can be converted to the corresponding 4-epimer by the action of UDP-glucose 4-epimerase.

It is also worthy of comment that the enzyme D-glucoside-3-dehydrogenase[2] has been observed to act directly on UDP-glucose to produce the 3-ketoglucose derivative.[3] Reduction of this by NaB³H$_4$ should lead to UDP-[3-³H]glucose which can be epimerized to UDP-[3-³H]galactose.

[1] R. D. Bevill, E. A. Hill, F. Smith, and S. Kirkwood, *Can. J. Chem.* **43**, 1577 (1965).
[2] K. Hayano and S. Fukoi, *J. Biol. Chem.* **242**, 3665 (1967).
[3] Luis Glaser, Department of Biochemistry, Washington University Medical School, St. Louis, Missouri 63110, personal communication. Dr. Glaser will be pleased to send the experimental details of this procedure to interested parties.

Reagents

D-[4-³H]Galactose (available from Amersham/Searle)
Crystalline phosphoric acid (available from Matheson, Coleman and Bell)
Uridine 5′-phosphoromorpholidate (available from Sigma Chemical Co.)
Anhydrous pyridine maintained over CaH₂
Tri-*N*-octylamine
Lithium hydroxide, 1 *N*
Tetrahydrofuran
Acetic anhydride

Procedure. The [4-³H]galactose is first converted to the 1-phosphate by the procedure of MacDonald.⁴ The sugar (11.3 mg) is dissolved in 2.7 ml of acetic anhydride containing 200 mg of anhydrous sodium acetate, and the solution is heated at 95–100° for 2 hours and then washed into 50 ml of ice water. The aqueous solution is extracted twice with 25-ml portions of chloroform; the chloroform extract is dried over CaCl₂ and then concentrated *in vacuo*. To the dried syrup (which is the β-pentaacetate of glucose) is added 36 mg of crystalline H₃PO₄, and the flask is evacuated by an oil pump (0.2 mm). The temperature is raised to 50°, the contents of the flask are thoroughly mixed and then maintained at 50°, under vacuum, for 2 hours. The dark brown reaction mixture is then dissolved in 10 ml of tetrahydrofuran and added to 50 ml of ice-cold 1 *N* LiOH. A white precipitate forms immediately, and after 16 hours the solution is centrifuged to remove lithium phosphate and then passed over an IR-120 column (H⁺ form); the eluate is neutralized with tri-*n*-octylamine and evaporated to dryness *in vacuo*. The overall yield at this point, based on radioactivity, is 38%.

The tri-*n*-octylammonium salt of the sugar 1-phosphate can be converted to UDP-galactose on a scale of anywhere between 2 and 100 μmoles by reaction in pyridine solution with uridine 5′-phosphoromorpholidate. If synthesis on a scale above 10 μmoles is desired, the procedure of Roseman *et al.* can be conveniently used.⁵,⁶ A typical synthesis on this scale was described by Bevill *et al.*,¹ the overall yield was 70%. If it is desired to carry out the synthesis on a scale between 2 and 10

⁴ D. L. MacDonald, Vol. 8 [11].
⁵ S. Roseman, J. J. Distler, J. G. Moffatt, and H. G. Khorana, *J. Amer. Chem. Soc.* **83**, 659 (1961).
⁶ J. G. Moffatt, Vol. 8 [15].

μmoles, then either the procedure of Nordin *et al.*[7] or that of Kochetkov *et al.*[8] as described by Elbein,[9] can be used.

UDP-N-[6-³H]Acetylgalactosamine

The procedure outlined below[10] is a simple and economical method for the preparation of UDP-N-[6-³H]acetylgalactosamine. It involves the oxidation of UDP-N-acetylgalactosamine to the corresponding 6-aldehydro derivative by galactose oxidase, followed by reduction of this material with NaB^3H_4. The method has the further advantage that it is carried out on a microscale and yields a product of almost any desired specific activity because of the very high specific activity of NaB^3H_4 which is commercially available. The procedure will be found particularly useful by those workers who desire a sample of UDP-N-acetylgalactosamine labeled specifically in the 6-position. It will also be useful to those whose purposes are served by a sample bearing a ³H-label at some point in the galactosamine moiety. Further, since the method is applicable to UDP-galactosamine and UDP-deoxygalactose, 6-³H-labeled samples of these materials can also be prepared.

In principle any method for the synthesis of UDP-N-acetylgalactosamine can be used to prepare labeled material. It is worth drawing attention to the method of Maley,[11] which also works well on a microscale and lends itself to the preparation of samples of UDP-N-acetylgalactosamine labeled in various positions with a variety of isotopes. Further, the observation of Maley and Maley [12] that UDP-glucose 4-epimerase will act on UDP-galactosamine permits the extension of the present procedure to the synthesis of the corresponding 6-³H-labeled derivatives of glucosamine through the use of a combination of the procedure of Maley[11] and the one described here.

Reagents. The following materials are available from commercial sources.

Galactose oxidase, 10 units/mg protein (Sigma Chemical Co.)
UMP-phosphoromorpholidate (Sigma Chemical Co.)
Catalase, 3000 units/mg protein (Worthington Biochemical Corp.)

---

[7] J. H. Nordin, W. L. Salo, R. D. Bevill, and S. Kirkwood, *Anal. Biochem.* **13**, 405 (1965).
[8] N. K. Kochetkov, E. I. Budowsky, V. N. Shibaev, and M. A. Grachev, *Biochim. Biophys. Acta* **59**, 747 (1962).
[9] A. D. Elbein, Vol. 8, p. 142.
[10] G. L. Nelsestuen and S. Kirkwood, *Anal. Biochem.* **40**, 359 (1971).
[11] F. Maley, *Biochem. Biophys. Res. Commun.* **39**, 371 (1970); see also this volume [25].
[12] F. Maley and G. F. Maley, *Biochim. Biophys. Acta* **31**, 577 (1959).

Streptomycin sulfate, B grade (Calbiochem)

NaB³H₄, 5.3 Ci/mmole (Amersham/Searle Corp.)

Other materials can be prepared in the laboratory.

UDP-N-Acetylgalactosamine: by synthesis from N-acetylgalactosamine 1-phosphate by the procedure of Nordin et al.[7] or of Kochetkov et al.[8,9]

N-Acetylgalactosamine: by a modification of the procedure of Carlson et al.[13] developed by Edstrom.[14]

Procedure. UDP-N-Acetylgalactosamine (1.54 μmoles), 32 units of galactose oxidase, 80 μg of streptomycin sulfate, and 0.85 mg of catalase are incubated at room temperature in 0.23 ml of 15 mM potassium phosphate buffer (pH 7.0) for 70 hours. Concentrated ammonium hydroxide (20 μl) is then added, followed by 10 μg of NaB³H₄ (5.3 Ci/mmole), and the solution is allowed to stand at room temperature for 3 hours. NaBH₄ (0.25 mg) is then added followed, after the solution has been allowed to stand 1 hour, by 50 μl of acetone. The entire solution is applied to a 0.8 × 8 cm column of DEAE-cellulose; the column is then washed with water and eluted with a gradient of 0 to 0.12 M ammonium bicarbonate (pH 7.0). The fractions containing the nucleotide sugar (as judged by 260 nm absorption) are combined, concentrated to dryness *in vacuo* and streaked on Whatman 3 MM paper. The paper is irrigated for a period of 40 hours, using the descending technique, with the solvent system 1 M ammonium acetate (pH 3.8):95% ethanol (3:7 by volume). The paper is then dried; it is scanned for radioactivity, and the large symmetrical peak of radioactivity is located. The chromatogram is washed with absolute ethanol, and the area containing the radioactive peak is cut out and eluted with water.

This procedure gives a yield of radioactivity in the region of 30%. The precise yield is difficult to determine because of uncertainty in the specific activity of the NaB³H₄ sample used. The preparation contains a single component as judged by chromatography in two further solvent systems[10]; N-acetylgalactosamine is the only radioactive substance released by mild acid hydrolysis,[10] and within experimental error 100% of the radioactivity is located at carbon-6 of the N-acetylgalactosamine moiety.[10]

[13] D. M. Carlson, A. Swanson, and S. Roseman, *Biochem. Prep.* **13**, 3 (1971); see also this volume [26].

[14] R. D. Edstrom, Department of Biochemistry, College of Medical Sciences, University of Minnesota, Minneapolis, Minnesota, personal communication. Dr. Edstrom indicates he would be glad to provide the details of this procedure to interested persons.

[32] Labeled Galactosyl Ceramide and Lactosyl Ceramide

By Norman S. Radin

Galactosyl Ceramide

Principle

Galactosyl ceramide (galactocerebroside or cerebroside) is oxidized enzymatically in the 6-position of the galactose moiety, then reduced with labeled sodium borohydride.[1]

Preparation

A mixture of brain cerebrosides can be obtained commercially from a number of suppliers specializing in lipids. It is still quite expensive, so a simple preparation method is offered here. Prepare a lipid extract of whole brain or brain white matter in C–M (chloroform–methanol) 2:1 and wash the extract with 0.2 volume of water.[2] Using a rotary vacuum evaporator, remove most of the solvent, adding generous amounts of benzene occasionally. Freeze and lyophilize the mixture. This yields the dry total lipids which can then be dissolved in 40 volumes of C–M 4:1 and applied to a column of Florisil and eluted with the same solvent, as described previously.[3] Alternatively, the lipids can be dissolved in 50 volumes of C–M 98:2 and applied to a column of silica gel (35 g per gram of lipids, preferably Unisil 200/325 mesh, Clarkson Chemical Co., Williamsport, Pennsylvania). In this case, elution is carried out with C–M 98:2 (25 ml per gram of silica gel) and then with an equal volume of C–M 90:10. With either column, one should collect fractions and identify the fractions that are rich in cerebroside.

Small amounts of contaminating ester-type lipids are then methanolyzed by dissolving the pooled, dried fractions in C–M 2:1 (40 ml per gram of lipid) containing $0.07\,M$ NaOH. After 1 hour add 0.2 volume of acetic acid in water 1:50 (v/v) and remove the upper layer that forms. Wash the lower layer with an equal volume of methanol–water 1:1 and lyophilize the lower layer as above, using benzene.

The crude cerebroside is then purified with a silica gel column as above. Mature rats (retired breeders) yield about 24 mg per brain, and white matter yields about 30 mg per gram.

---

[1] N. S. Radin, L. Hof, R. M. Bradley, and R. O. Brady, *Brain Res.* **14**, 497 (1969).
[2] N. S. Radin, Vol. 14 [44].
[3] N. S. Radin, Vol. 14 [46].

Brain cerebrosides consist of a mixture containing different fatty acids, especially 2-hydroxy acids. In bovine brain, an appreciable portion contains h18:0, which is revealed on thin-layer chromatograms as a distinct spot just below the two spots for the hydroxy and nonhydroxy acid cerebrosides. While all these forms appear to act as substrates for cerebroside galactosidase, it looks as though the stearoyl form is a somewhat more active substrate than the mixture. This can be made by acylation of galactosyl sphingosine.[4]

Galactosyl sphingosine (psychosine) is readily made by a modification of the procedure of Taketomi and Yamakawa.[5] Place 1.5 g of galactosyl ceramide, 45 ml of n-butanol, 5 ml of water, and 3.5 g of KOH into a 250-ml flask and reflux the system for 4 hours. The ground joint should be wiped clean to prevent freezing from the KOH; a long neck is preferable. Dialyze the mixture against water to remove the KOH and butanol, using a system which agitates the sample (such as a large Erlenmeyer flask on a gyratory shaker). This should be done expeditiously, with frequent changes of water, as prolonged dialysis seems to lower the yield. To reduce the danger of bursting, fill the dialysis sac only partially and squeeze out most of the air before tying off the sac. If some butanol is still present after 8 hours of dialysis, complete its removal by vacuum evaporation and then lyophilize the aqueous part. Lyophilize the sac contents, warm the powder with 25 ml of C–M 2:1, and filter off the insoluble matter. Replace the solvent by benzene by repeated evaporation to a small volume, lyophilize the mixture, and purify the resultant crude psychosine with a column of Unisil, as above. The sample solvent and starting eluent should be C–M 90:10 and the second solvent should be C–M 85:15 or 80:20. The fatty acid comes off first, then a bit of unhydrolyzed cerebroside, then the psychosine (about 350 mg). For optimal separation use a pump and fraction collector. Psychosine is readily detected on thin-layer plates by spraying with alkaline bromothymol blue in water; this yields a blue spot (with a faint spot just below for galactosyldihydrosphingosine). A suitable thin-layer solvent is C–M–water–ammonium hydroxide 70:30:4:1, used with silica gel.

The above procedure works equally well with the cerebroside found in Gaucher spleens, glucosyl ceramide.

For the acylation, dissolve 1 mmole of psychosine (462 mg) in a small test tube containing 3 ml of tetrahydrofuran (distilled from KOH) and 5 ml of sodium acetate solution, containing 2.5 g of the trihydrate.

[4] K. C. Kopaczyk and N. S. Radin, J. Lipid Res. 6, 140 (1965).
[5] T. Taketomi and T. Yamakawa, J. Biochem. 54, 444 (1963).

While stirring the mixture vigorously with a magnetic stirring bar, add 1.2 mmoles of stearoyl chloride in 4 ml of tetrahydrofuran. Let the reaction continue at least 30 minutes and recover the cerebroside by partitioning with 30 ml of C–M 2:1 and 3 ml of 6 N HCl. Wash the lower layer with two 10-ml portions of methanol–water 1:1 and lyophilize the lipid layer. The cerebroside is purified as with the natural mixture of cerebrosides.

Reagents

Tetrahydrofuran, freshly distilled from KOH pellets
Phosphate buffer, 10 mM, pH 7
Galactose oxidase, from AB Kabi, Stockholm, or from Worthington Biochemical Corp., Freehold, New Jersey. Store in freezer.
Thin-layer chromatography plates, silica gel with or without binder. These are used with C–M–water 24:7:1 in paper-lined jars, and the spots are visualized with any general indicator, such as iodine, alkaline bromothymol blue in water, or a charring reagent. Iodine is preferred for samples to be counted but some adsorbents retain the iodine strongly and the scintillation solution must be decolorized with a bit of ascorbic acid or sodium borohydride.

Oxidation of the Cerebroside. Dissolve 33 mg of galactosyl ceramide in 33 ml of tetrahydrofuran and add 33 ml phosphate buffer. Add 0.5 ml galactose oxidase in water (500,000 Kabi units or 250 Worthington units) and shake the mixture gently at room temperature for 4 hours. Add an additional portion of oxidase (0.2 ml of the same suspension) and shake the mixture overnight.

The cerebroside aldehyde is recovered by adding 66 ml of C–M 4:1 and washing the resultant upper layer with two 44-ml portions of the same C–M mixture. Pool the two washes with the initial lower layer and wash this with 44 ml of methanol–water 1:1. Evaporate the lower layer to dryness and check the product by thin-layer chromatography, as described above. The aldehyde moves distinctly faster than the original cerebroside, some of which is still present as a result of incomplete oxidation. The aldehyde seems to be quite stable to storage in the freezer, probably because it exists in the cyclic form.

Reduction of the Aldehyde. Dissolve the mixture of aldehyde and cerebroside in 5 ml of tetrahydrofuran, with some warming. Add 0.35 ml of sodium borohydride (1 mg, 6 mCi) in 1 mM NaOH and leave the mixture (capped) overnight. To complete the reduction, add 10 mg of nonradioactive borohydride and leave the mixture 1 hour. Regenerate the labeled cerebroside with 7 ml of 1 N acetic acid and isolate the lipid by partitioning with 25 ml C–M 2:1. Wash the lower layer with four or more portions of methanol–water 1:1 containing 1 mg of NaCl per milliliter. The tritium in the washes can be monitored by counting a small portion; it is not necessary to continue the washing until zero activity enters the upper layer, as the labeled impurities are removed in the next step.

The cerebroside can be purified on a small, narrow column of silica gel as described above, but with a higher proportion of adsorbent (3 g). A pump is desirable to improve the flow rate when a narrow column is used. Count the fractions and check the identity of the active materials by cochromatography with some of the original cerebroside on a thin-layer plate. Locate the carrier spots with iodine and count them by liquid scintillation in an eluting system as described in this volume [111]. It is wise to count also the regions just above and just below the carrier spots in case a related impurity is present.

Pool the cerebroside fractions and weigh the material after evaporation with benzene and lyophilization as above. Dissolve it in a known volume of C–M 2:1 and determine the specific activity. The activity on a molar basis is a bit indefinite when a natural mixture of cerebrosides is used. The 24:1 cerebroside has a molecular weight of 810; h24:0, 828; 18:0, 728. These are the most common acids in brain cerebroside, so 800 could be taken as a reasonable estimate of molecular weight without actually determining the fatty acid distribution.

Hydrolysis of the labeled cerebroside shows virtually all of the tritium to be in the galactose moiety. The specific activity is very high, and dilution with the same batch of cerebroside is ordinarily required

Notes on the Method

The labeled cerebroside can also be purified by thin-layer chromatography, especially if a smaller batch is used.[1] Streak about 3 mg of lipid on a 20×20 cm plate 0.5 mm thick and develop the plate as above. Possibly a solvent a little less polar would be preferable, depending on the adsorbent. Silica gel of high purity should be used, and even this should be washed by a preliminary run with C–M 4:1. Reactivate the plate by heating for 1 hour at 100–110° before use with the lipid.

The cerebroside band is most sensitively located with a spray of alkaline bromothymol blue, 0.4 mg/ml in water. Let the plate dry, scrape off the powder with a razor blade, and disperse the powder in C–M–water 7:7:1 with the aid of an ultrasonic bath of the type used for cleaning. (A mortar and pestle, used with care, will do as well.) Pack the slurry into a column and elute with the same solvent, 25 ml per gram of silica gel. The bromothymol blue and impurities from the adsorbent are removed by partitioning with chloroform (0.47 ml per milliliter of eluent) and dilute aqueous NaCl (0.28 ml per milliliter of eluent). Enough ammonium hydroxide should be added to keep the indicator in the blue form. Wash the lower layer with methanol–water–salt as above until all the blue color is gone, and lyophilize the lipid in the lower layer as above. Probably the scale of working could be doubled with the use of currently available precoated plates with 1 mm-thick silica gel. A yellow color in the lower layer is a sign that more ammonia is needed to convert the indicator to the blue form.

Tetrahydrofuran is toxic and should be handled in a hood. Excess solvent should be returned to the stock bottle, which is stored (well closed) in a freezer to retard peroxide formation. Sodium borotritide vials sometimes contain gaseous tritium and should be opened cautiously. A purple color in the solid tritide is normal, and the dry material seems to keep fairly well in the freezer. We order relatively small lots at a time, paying extra for this service, and dissolve the borohydride in an appropriate amount of nonradioactive borohydride in NaOH. It is probably more economical to label several aldehydes or ketones simultaneously, then purify the individual labeled substances from the reaction mixture.

At present the galactose oxidase from AB Kabi is somewhat cheaper than the American product. However, since no detailed study was made of the relation between amount of enzyme and yield of radioactivity, it is possible that one could use somewhat less enzyme. It would probably be feasible to oxidize a smaller proportion of the galactoside and reduce it with borotritide of higher specific activity.

Lactosyl Ceramide

This galactoside ("lactoside," "cytolipin H," D-galactopyranosyl-$(1 \rightarrow 4)$-D-glucopyranosyl-$(1 \rightarrow 1')$ ceramide) is labeled by a similar procedure. The stearoyl form is available commercially (Miles Laboratories, Inc., Research Division, Kankakee, Illinois) but the long-chain base moiety is DL-dihydrosphingosine. Perhaps this is suitable as a substrate; all work with labeled lactoside thus far has been done only with material derived from natural sources, which contains D-sphingosine as the major long-chain base. Various workers have isolated lactoside on a preparative basis from spleen, red cells, tumors, and organs of patients dying from certain lipidoses. The technique for such an isolation is very much like the one described above for brain cerebrosides, except for the use of a more polar set of C–M mixtures. The procedure described below, involving partial hydrolysis of gangliosides, is a modification of methods used in several laboratories.[1]

Gangliosides may be purchased (Type III, Sigma Chemical Co., St.

Louis, Missouri) or prepared in crude form from brain gray matter. In the latter case, the procedure described by Kanfer[6] can be used. It is probably sufficient to purify the gangliosides through alkaline methanolysis, partitioning, and dialysis. Heat the lyophilized gangliosides in 1 N sulfuric acid for 2 hours at 90° and make the mixture slightly alkaline with gaseous ammonia while maintaining the temperature at 0°. Add 5 volumes of C–M 2:1 and wash the lower layer twice with methanol–water–salt as described for cerebrosides. Lyophilize the lower layer and purify the lactoside with a silica gel column as for the cerebrosides. Products of overhydrolysis, ceramide and glucosyl ceramide, are eluted with C–M 96:4, and lactoside is eluted with C–M 90:10. If a relatively tall, narrow column is used with a pump and fraction collector, high purity lactoside can be isolated.

Our yield from 1 kg of whole pig brain was about 185 mg. There is some disagreement in the literature as to the precise optimal hydrolytic conditions. It is possible to elute the silica gel column with a more polar solvent (C–M 3:1) and recover the incompletely hydrolyzed lipids, then reprocess them through the hydrolysis and isolation.

Gaseous ammonia is used rather than aqueous ammonia simply to keep down the volumes handled in the partitioning.

The lactoside is labeled in much the same way as galactosyl ceramide, as described above. Because of the higher solubility of the dihexoside, a higher concentration is used during the enzymatic oxidation. This is helpful because the enzyme does not seem to act as rapidly on the lactoside. Dissolve 50 mg of lactoside in 12 ml of tetrahydrofuran and add 12 ml of buffer and 1 ml of oxidase (10^6 Kabi units or 500 Worthington units). After 4 hours, add an additional 0.5 ml of enzyme and shake the mixture overnight. The isolation of the aldehyde is carried out as with the cerebroside aldehyde, with 25 ml of C–M 4:1 and 15-ml washes. The same thin-layer solvent can be used.

The reduction step is carried out just as described above, with the same amount of borohydride and solvents. The specific activity from our preparation was 10,000 cpm/nmole, calculated for a molecular weight of 890. This is primarily the stearoyl lactoside, since it is made from ganglioside. The brains of large farm animals (the customary source) are from relatively young animals, so the content of 20:1 sphingosine is rather low. If any glucosyl ceramide is seen as a contaminant prior to enzymatic oxidation, it will probably be removed in the final purification. In any event, it is harmless, as this lipid does not become labeled.

Lactoside can be labeled very effectively by reduction with tritium gas,

[6] J. N. Kanfer, Vol. 14 [62].

in which case the tritium goes primarily into the double bond of the sphingosine.[7] If the labeled product is to be used in assaying for lactoside galactosidase (see this volume [113]), the product of enzymatic action, glucosyl ceramide, will have to be separated chromatographically from the unhydrolyzed substrate.[7] However, exposure of organic compounds to tritium gas ordinarily results in some indiscriminate labeling, and it would be necessary to determine the specific activity in the glucosyl ceramide rather than in the lactoside in order to define the enzyme activity in moles hydrolyzed. Another possible complication is that the galactosidase might be contaminated with glycosyl ceramide glucosidase and thereby yield a certain amount of labeled ceramide. However, the glucosidase could be blocked with gluconolactone.

[7] S. Gatt and M. M. Rapport, *Biochem. J.* **101**, 680 (1966).

[33] C_{55}-Isoprenyl Pyrophosphate[1]

By K. J. STONE and J. L. STROMINGER

C_{55}-Isoprenyl pyrophosphate is an essential intermediate in bacterial peptidoglycan synthesis. In the bacterium, it functions as part of a cyclic mechanism in which N-acetyl sugar peptide units are transferred from uridine nucleotides and utilized to form the linear peptidoglycan.[2] Hydrolysis by bacterial C_{55}-isoprenyl pyrophosphate pyrophosphatase liberates C_{55}-isoprenyl phosphate, which is required to initiate the cycle. This hydrolysis is inhibited by bacitracin.[3]

Preparation of High Specific Activity [^{32}P]C_{55}-Isoprenyl Pyrophosphate

Principle. Cells of *Micrococcus lysodeikticus* (ATCC 4689) accumulate [^{32}P]C_{55}-isoprenyl pyrophosphate when grown in the presence of [^{32}P]P_i and subsequently inhibited with bacitracin.[4] Organic solvent extraction of such cells followed by mild alkaline hydrolysis of the lipid results in a fraction in which approximately 80–95% of the radioactivity

[1] Supported by research grants from the U.S. Public Health Service (AM-13230 and AI-09152) and the National Science Foundation (GB-29747X).
[2] J. S. Anderson, P. M. Meadow, M. A. Haskin, and J. L. Strominger, *Arch. Biochem. Biophys.* **116**, 487 (1966).
[3] G. Siewert and J. L. Strominger, *Proc. Nat. Acad. Sci. U.S.* **57**, 767 (1967).
[4] K. J. Stone and J. L. Strominger, *J. Biol. Chem.* **247**, 5107 (1972).

is C_{55}-isoprenyl pyrophosphate. The remaining radioactive contaminants, C_{55}-isoprenyl phosphate and a mixture of C_{55}-isoprenyldiphosphate N-acetyl sugar peptide derivatives, can be removed by chromatography if necessary.

Procedure. One liter of cells are grown in a medium containing 1% Difco Bactopeptone, 0.1% yeast extract, and 0.5% NaCl adjusted to pH 7.5 with NaOH. At one-third of maximum growth, they are transferred to a medium containing 25 mCi of inorganic phosphate and one-tenth the original quantity of yeast extract. Bacitracin (160 μg/ml) is added after 4 hours, and the cells are harvested after a further 3 hours and then washed 2–3 times with 200 ml of 0.1 N Tris buffer, pH 7. The cell paste is suspended in 20 volumes of cold ($-15°$) acetone for 10 minutes and collected on a sintered glass filter. Silanized glassware is used throughout. This residue is extracted by mixing with 4 volumes of chloroform–methanol (2:1) at room temperature for 20 minutes. Filtration and reextraction with 4 volumes of n-butanol–pyridinium acetate (pH 4.2, made by adjusting the pH of 6 M acetic acid with pyridine) (2:1) is repeated three times and is facilitated by sonicating the suspension in short 10-second bursts (6 or 7 times) in the presence of 1 ml of glass beads (5 μ diameter). The butanol–pyridinium acetate extracts are combined, then washed 4 times each with 0.5 volume of water; the resulting aqueous phases are backwashed once with n-butanol. It is important that the pH 4.2 extract be kept cold during sonication and extraction since the compound is relatively unstable at this pH at room temperature. All butanol layers are combined, made alkaline with pyridine, and evaporated to dryness under vacuum.

The resulting lipid is dissolved in 1 ml of CCl_4, to which is added 9.4 ml of ethanol, 0.8 ml of water, and 0.3 ml of 1 N NaOH. This mixture is incubated at 37° for 30 minutes when the pH is still alkaline. Ethyl formate (0.4 ml) is added, and the mixture is incubated for a further 5 minutes at 37°. The sample at pH 7 is evaporated to dryness under vacuum, dissolved in 6 ml of chloroform–methanol, 2:1, and 4 ml water is added. Brief sonication of this mixture eliminates the formation of an interphase. The lower organic phase is washed four times with 5 ml of methanol–water, 1:1, to remove the hydrolyzed products of phosphatides (approximately 8 × 10⁸ cpm), and the aqueous washings are back-washed twice with chloroform. The combined organic phase is evaporated to dryness under vacuum, suspended in acetone by sonication and passed through a 1 ml column of basic silica gel G (Woelm) which has been washed with acetone. Elution with 3 ml of acetone is followed by 3 ml of chloroform–methanol (1:1) and 3 ml of methanol. The last two fractions are combined.

Properties of the Product

The major radioactive component (80–95%) of the lipid product is C_{55}-isoprenyl pyrophosphate, as estimated by TLC and radioautography, and contains 4×10^7 to 10×10^7 cpm. Enzymatic hydrolysis (using prenyl pyrophosphate pyrophosphatase present in bacterial membranes) shows that 55% of the label is in the terminal phosphate group and 45% is in the β-phosphate group. The radioactive material shows a strong tendency to bind to glass unless the apparatus is previously silanized with a 5% solution of dimethylchlorsilane in toluene.

Preparation of Milligram Quantities of C_{55}-Isoprenyl Pyrophosphate

Principle. The procedure involved in isolating milligram quantities of C_{55}-isoprenyl pyrophosphate represents a scaled-up version of the procedure described above. The larger amount of material available, however, enables a much more rigorous purification to be made.

Preparation. Bacitracin (25 g) is added to 500 liters of one-third maximum grown cells 12 hours after seeding with 5 liters of fully grown bacteria. The absorbance at 700 nm levels off, and the cells are harvested at 14 hours when the absorbance begins to fall. An acetone powder made from the 660 g of yellow cells using 7.5 liters of acetone at $-20°$ is extracted once with 2.3 liters of $ChCl_3$-methanol (2:1) and three times with butanol–pyridinium acetate. Sonication in the presence of 100 ml of glass beads is used in the last three extractions, and the organic phase from these is combined, washed, and taken to dryness as above. This material is dissolved in 20 ml of carbon tetrachloride and saponified as above to give approximately 1.25 g of a yellow-brown oil. The corresponding fraction from a preparation of $[^{32}P]C_{55}$-isoprenyl pyrophosphate is added to facilitate the following purification procedure.

The total material is suspended in 20 ml of acetone and applied to a column of silica gel H (250 g) which is packed using acetone. Elution with 1 liter of acetone removes nonpolar lipid which is discarded. The column is then eluted with 1 liter of each of the following solvents: (1) chloroform–methanol, 2:1; (2) absolute methanol; (3) 98% methanol; (4) butanol–pyridinium acetate, pH 4.2, 2:1; 14-ml fractions are collected. Radioactive material is eluted in two main peaks which are combined and applied in a small volume of chloroform–methanol (18:1) to a column of silica gel H (30 g) packed in chloroform. This second column is eluted with a linear gradient of chloroform–methanol (565 ml, 18:1) to chloroform–methanol–water (725 ml, 65:25:4). A single major radioactive component is eluted and peak tubes are combined and taken to dryness (29.9 mg of a yellow oil). This chromatography may be repeated using a more shallow gradient if appreciable quantities of other lipids which absorb at 425 nm are present.

Final purification is effected by dissolving the oil in 0.4 ml of chloroform–methanol (2:1), 5 ml of absolute ethanol is added, and the mixture is cooled to 0° for 5 minutes. Centrifugation is used to isolate the white precipitate of lipid pyrophosphate which contains over 96% of the radioactivity originally in solution and weighs 12.0 mg. The overall yield of product during chromatography and precipitation is 60%.

Properties of the Product

Enzymatic hydrolysis using C$_{55}$-isoprenyl pyrophosphate pyrophosphatase present in membranes from *Streptococcus fecalis* liberates quantative amounts of inorganic phosphate and C$_{55}$-isoprenyl phosphate. Mild acid hydrolysis liberates inorganic pyrophosphate and total phosphate analysis typically indicates a value of 1.87 moles of phosphate per mole of product calculated as C$_{55}$-isoprenyl pyrophosphate. The lipid products of acid hydrolysis are typically those of a prenyl pyrophosphate and consist of hydrocarbon and primary and tertiary alcohols. Thin-layer chromatography, gas–liquid chromatography, and mass spectrometry confirm the isoprenoid nature of these hydrolysis products and show that although the major prenolog is a C$_{55}$ derivative, C$_{60}$ and C$_{65}$ derivatives are also present.

Section III

Purification and Properties of
Carbohydrate-Binding Proteins

[34] Concanavalin A, the Jack Bean (*Canavalia ensiformis*) Phytohemagglutinin

By B. B. L. AGRAWAL and I. J. GOLDSTEIN

Concanavalin A[1] (con A), a globulin from the jack bean (*Canavalia ensiformis*) was first described by Sumner[1a] and isolated in crystalline form by dialysis of the sodium chloride extract of jack bean meal. Later, Sumner *et al.*[2] reported the ability of con A to precipitate certain carbohydrates including glycogen and starch from corn and rice, and to agglutinate erythrocytes of the horse, dog, and cat.

Con A has been shown to interact to form a precipitate with biopolymers containing multiple α-D-glucopyranosyl (or its 2-acetamido-2-deoxy derivative), α-D-mannopyranosyl or β-D-fructofuranosyl residues as nonreducing termini.[3,4] These include polysaccharides such as glycogens,[2,5] dextrans,[3,6] α-mannans,[2,7] fructans,[8] and pneumococcal polysaccharides[9]; glycoproteins, e.g., immunoglobulins,[10] blood group substances,[11] and other phytohemagglutinins[12]; and lipopolysaccharides[13] and teichoic acids.[14]

More recently con A has been shown to interact with bacterial[15] and animal cells, to distinguish between normal and tumor cells on the basis of membranous carbohydrate-containing structures,[16-19] and to initiate differentiation and cell division.[20,21]

[1] No attempt has been made here to provide full documentation of the voluminous literature on concanavalin A that has been published during the last several years.

[1a] J. B. Sumner, *J. Biol. Chem.* **37**, 137 (1919).

[2] J. B. Sumner and S. F. Howell, *J. Bacteriol.* **32**, 227 (1936).

[3] I. J. Goldstein, C. E. Hollerman, and J. M. Merrick, *Biochim. Biophys. Acta* **97**, 68 (1965).

[4] I. J. Goldstein and L. L. So, *Arch. Biochem. Biophys.* **111**, 407 (1965).

[5] J. A. Cifonelli, R. Montgomery, and F. Smith, *J. Amer. Chem. Soc.* **78**, 2485 (1956).

[6] I. J. Goldstein, R. D. Poretz, L. L. So, and Y. Yang, *Arch. Biochem. Biophys.* **127**, 787 (1968).

[7] L. L. So and I. J. Goldstein, *J. Biol. Chem.* **243**, 2003 (1968).

[8] L. L. So and I. J. Goldstein, *Carbohyd. Res.* **10**, 231 (1969).

[9] J. A. Cifonelli, P. Rebers, M. B. Perry, and J. K. N. Jones, *Biochemistry* **5**, 3066 (1966).

[10] M. A. Leon, *Science* **158**, 1325 (1967).

[11] K. O. Lloyd, E. A. Kabat, and S. Beychok, *J. Immunol.* **102**, 1354 (1969).

[12] I. J. Goldstein, L. L. So, Y. Yang, and Q. C. Callies, *J. Immunol.* **103**, 695 (1969).

[13] I. J. Goldstein and A. M. Staub, *Immunochemistry* **7**, 135 (1970).

[14] W. J. Reeder and R. D. Ekstedt, *J. Immunol.* **106**, 334 (1971).

[15] J. S. Tkacz, E. Barbara Cybulska, and J. O. Lampen, *J. Immunol.* **105**, 1 (1971).

Thus, con A may serve as a versatile model for studies of protein–carbohydrate and antibody–antigen interaction, as a tool for investigating the glycosyl composition and structure of certain biopolymers, and as a probe for studying the carbohydrate-containing constituents of cell surfaces.

Studies on the primary[22] and X-ray crystallographic structure[23,24] of con A are in progress.

Method of Isolation

Principle

The isolation procedure is based on the principle of affinity chromatography and depends on the ability of con A to form a precipitate with dextrans.[3] The method involves extraction of jack bean meal with $0.15 M$ NaCl, clarification of the extract and concentration of the proteins by $(NH_4)_2SO_4$ fractionation, followed by adsorption of con A on a suitable cross-linked dextran gel[25,26] (Sephadex G-50). After elution of inert proteins, adsorbed con A is displaced from the dextran gel by D-glucose, a competitive inhibitor of polysaccharide–con A interaction.[27,28] The con A may also be displaced from the Sephadex bed by $0.02 M$ glycine·HCl buffer as reported by Olson and Liener[29] or prepared by the original isolation and crystallization procedure of Sumner.[2]

Procedure

Salt Extraction and Preliminary Fractionation. Defatted, finely ground jack bean meal (General Biochemicals, Chagrin Falls, Ohio;

---

[16] M. Inbar and L. Sachs, *Nature* (*London*) **223**, 710 (1969).

[17] M. J. Cline and D. C. Livingston, *Nature* (*London*) *New Biol.* **232**, 155 (1971).

[18] B. Ozanne and J. Sambrook, *Nature* (*London*) *New Biol.* **232**, 156 (1971).

[19] M. M. Burger and K. D. Noonan, *Nature* (*London*) **228**, 512 (1970).

[20] M. Wecksler, A. Levy, and W. G. Jaffé, *Acta Cient. Venezo.* **19**, 154 (1968).

[21] A. E. Powell and M. A. Leon, *Exp. Cell. Res.* **62**, 315 (1970).

[22] F. A. Quiocho, G. N. Reeke, Jr., J. W. Becker, W. N. Libscomb, and G. M. Edelman, *Proc. Nat. Acad. Sci. U.S.* **68**, 1853 (1971).

[23] K. D. Hardman, M. K. Wood, M. Schiffer, A. B. Edmundson, and C. F. Ainsworth, *Proc. Nat. Acad. Sci. U.S.* **68**, 1393 (1971).

[24] J. Waxdal, J. L. Wang, M. N. Pflumm, and G. E. Edelman, *Biochemistry* **10**, 3345 (1971).

[25] B. B. L. Agrawal and I. J. Goldstein, *Biochem. J.* **96**, 23C (1965).

[26] B. B. L. Agrawal and I. J. Goldstein, *Biochim. Biophys. Acta* **147**, 262 (1967).

[27] L. L. So and I. J. Goldstein, *J. Immunol.* **99**, 158 (1967).

[28] I. J. Goldstein, C. E. Hollerman, and E. E. Smith, *Biochemistry* **4**, 876 (1965).

[29] M. O. J. Olson and I. E. Liener, *Biochemistry* **6**, 105 (1967).

Sigma Chemical Co., St. Louis, Missouri), 300 g, is suspended in 0.15 M NaCl (1.5 liters), and stirred overnight in the cold (ca. 4°). The resulting suspension is filtered through cheesecloth, the filtrate being saved, and the residue is reextracted for 8 hours with an additional 1.5-liter portion of 0.15 M NaCl. The suspension is filtered as before, and the residue is discarded. The combined filtrates are centrifuged for 1 hour at 9500 rpm (14,600 g, using the GSA rotor of a Sorvall RC-2 refrigerated centrifuge), the residue being discarded. The clear yellow supernatant solution is made 30% saturated with $(NH_4)_2SO_4$ by gradual addition of the solid salt (176 g/liter). The pH is adjusted to 7.0 with concentrated NH_4OH, and the mixture is stirred gently at room temperature (ca. 25°) for 4 hours. Precipitated proteins are removed by centrifugation (GSA rotor, 9500 rpm, 1 hour) and discarded. The bulk of the proteins including con A is now precipitated by adding $(NH_4)_2SO_4$ (356 g/liter) to 80% saturation. The precipitate is collected by centrifugation as above (the supernatant solution being discarded), dissolved in H_2O (500 ml), and dialyzed in the cold against several changes of water, and finally against 1.0 M NaCl. The dialyzed protein solution is centrifuged, and the clear solution is used for dextran gel adsorption.

Adsorption of Concanavalin A to Sephadex.[25,26] A column (4 × 60 cm) of Sephadex G-50 (Pharmacia, fine or medium mesh) is prepared and equilibrated with 1.0 M NaCl. A flow rate of 30–40 ml/hour is suitable.

The protein solution from above, in 1 M NaCl, is applied to the column at a rate of 30–40 ml/hour. After addition of the sample the column is connected to a reservoir of 1 M NaCl. Fractions (ca. 10 ml) are collected every 15 minutes and monitored for protein by absorbance at 280 nm. Washing the column with 1 M NaCl to elute inert protein normally requires 24–48 hours. After the absorbance at 280 nm $\leqq$ 0.1, a 0.10 M solution of glucose in 1 M NaCl is added to elute con A from the dextran gel. Fractions having an absorbance at 280 nm of 0.20 or greater are combined.

Removal of Glucose. Glucose can be removed from con A by dialysis or gel filtration.

Dialysis. A solution can be freed of glucose by dialysis against large volumes of 1.0 M NaCl. About 18–20 changes over the period of a week are usually sufficient to remove all bound glucose as detected by the phenol-H_2SO_4 test.[30]

Gel Filtration. Excess glucose can also be removed from con A by

[30] M. Dubois, K. A. Gilles, J. K. Hamilton, P. A. Rebers, and F. Smith, *Anal. Chem.* **28**, 350 (1956).

gel filtration on a Sephadex G-25 column equilibrated with $1.0 M$ NaCl. Except for small quantities of con A, this method is not recommended as it requires repetitive cycling to remove all glucose bound to con A.

For freeze drying, solutions of con A containing glucose can be dialyzed free of salt and sugar by extensive dialysis against distilled water. A few drops of toluene are included in the dialysis sac to prevent microbial growth. Freeze drying can also be accomplished from $0.15 M$ NaCl.

Assay for Concanavalin A

Qualitative. For rapid screening of con A activity we have routinely used a gel diffusion method[4] utilizing the ability of con A to precipitate specific polysaccharides (e.g., glycogens, dextrans). The first fraction emerging from the Sephadex G-50 column was shown to be devoid of con A activity. Only the fraction eluted with glucose will precipitate glycogen (or dextran, etc.) as evidenced by the precipitation arc formed in agar gel.

Quantitative. The quantitative precipitation method of So and Goldstein[31] may be used for the quantitative determination of con A.

The assay is conducted in 3-ml calibrated glass centrifuge tubes (Bellco Glass, Inc., Vineland, New Jersey). The contents are mixed, and the tubes are capped and allowed to stand at room temperature for 24 hours. The precipitates are collected by centrifuging the tubes at top speed, 2700 rpm, in a desk model International clinical centrifuge for 15 minutes. The precipitates are washed twice with 0.5-ml portions of $1.0 M$ NaCl containing $0.10 M$ phosphate, pH 7.2, the wash solution being carefully decanted after centrifugation. The washed precipitates

TABLE I

COMPOSITION OF CONCANAVALIN A ASSAY MEDIUM

Component	Volume (ml)
Protein in $1.0 M$ NaCl	0.10
$4.1 M$ NaCl	0.22
$0.10 M$ phosphate, pH 7.4	0.18
H_2O	0.5–0
Dextran NRRL 1355-S[a] 2 mg/ml H_2O	0–0.5
Total volume	1.0 ml

[a] Obtained from Dr. Allene Jeanes, Northern Utilization Research and Development Division, USDA, Peoria, Illinois.

[31] L. L. So and I. J. Goldstein, *J. Biol. Chem.* **242**, 1617 (1967).

TABLE II
STAGES IN THE PURIFICATION OF CONCANAVALIN A

Isolation stage	Volume (ml)	Nitrogen (μg/ml)	Total nitrogen (mg)	Recovery of nitrogen (% of the total)	% of Total nitrogen precipitated by dextran
0.15 M NaCl extract	2760	1941	5357	100	23.8
$(NH_4)_2SO_4$	530	8125	4316	80.6	26.8
Fraction 0.30–0.80					
Sephadex G-50	1500	1973	2961	55.2	—
Fraction I					
Fraction II	450	2436	1096	20.4	97.6

are digested with $H_2SO_4:H_2O$ (1:5, v/v), followed by treatment with H_2O_2 to complete the oxidation. Color development is with the ninhydrin reagent. The purity of the con A is read from the peak of the curve of protein precipitated upon examining the graph relating μg protein precipitated vs. mg dextran added. Glycogen[32] (2 mg/ml) may also be used as the precipitating polysaccharide.

Yield and Purification. The total yield of con A from the jack bean meal (100 g) is ca. 2.4 g. However, the yield varies depending on the source of jack bean, harvest time, fineness of the powder, etc. The chromatographic recovery (Sephadex G-50) is approximately 94%. The recovery and purification data on con A from a typical preparation are presented in Table II.

General Comments

Method. In principle, the clarified sodium chloride extract of jack bean meal may be used directly for the selective adsorption of con A on Sephadex G-50. However, this is not always efficient because the extract often contains sugars which serve as inhibitors of concanavalin A-polysaccharide interaction and result in lower recoveries due to decreased adsorption on Sephadex G-50 gels.

Ammonium sulfate fractionation is employed primarily to concentrate the proteins and to remove debris and other substances which often interfere with the adsorption step. However, for speed of preparation, this step may be omitted.

The selection of Sephadex G-50 for adsorption of con A is based on two factors. First, it has the lowest exclusion limit of all the common Sephadex gels (G-50 through 200) which bind con A (the more highly

---

[32] L. L. So and I. J. Goldstein, *J. Immunol.* **102**, 53 (1969).

cross-linked Sephadex G-25 does not bind con A[25,26]); and second, it also has the highest flow rate of these gels. These factors contribute to a rapid removal of extraneous proteins and a selective adsorption of con A to the dextran gel. Subsequent elution with 0.1 M glucose elutes con A in a very concentrated solution (the peak tube may afford a 4% solution of con A).

Stability and Solubility of Concanavalin A. Concanavalin A is best stored in 1 M NaCl at 4° and as such is stable over many months or even years. A small portion (0–1%) will tend to precipitate from solution on long standing (over months). Certain low molecular weight carbohydrates (e.g., D-glucose or methyl α-D-glucopyranoside)[31] stabilize the protein.

The freeze-dried protein is almost equally active in terms of precipitating specific polysaccharides. It may, however, contain aggregates (<5%) which after solution may be removed by centrifugation.

Commercial Preparations. Most biological supply houses prepare and market con A (e.g., Calbiochem, La Jolla, California).

[35] Lima Bean (*Phaseolus lunatus*) Lectin

By WILLIAM GALBRAITH and I. J. GOLDSTEIN

The hemagglutinin (lectin) of the lima bean (*Phaseolus lunatus*) exhibits type A blood group activity.[1] The lectin reacts strongly with type A, slightly with type A_2, very slightly with type B, and not at all with type O red blood cells.[2] The same specificity is displayed toward blood group substance and human ovarian cyst or hog gastric mucin (A and O).[2]

Lima bean lectin occurs as two apparently noninterconvertible molecular species,[2,3] one with twice the molecular weight of the other. Component II, the higher molecular weight lima bean lectin species, has about 4 times the agglutinating activity of component III, the lower molecular weight species. Component II will precipitate type B blood group substance in agar gel whereas component III will not.[2]

[1] W. C. Boyd and R. M. J. Reguera, *J. Immunol.* **62**, 333 (1949).
[2] W. Galbraith and I. J. Goldstein, *FEBS Lett.* (*Fed. Eur. Biochem. Soc.*) **9**, 197 (1970).
[3] N. R. Gould and S. L. Scheinberg, *Arch. Biochem. Biophys.* **137**, 1 (1970).

Assay Method

Principle. The lima bean hemagglutinin has been assayed routinely by two procedures: Hemagglutination[4] and quantitative precipitation.[5] Hemagglutination offers a rapid, convenient and reproducible method to test for lectin activity. This method depends on determining the highest dilution of a stock lectin solution which still retains the capacity to bring about the agglutination or clumping of type A erythrocytes.

The quantitative precipitin method provides a more precise measure of activity and is more suitable for hapten inhibition studies and for evaluating the many parameters that affect the lectin's activity. This procedure depends on the establishment of a precipitin curve between a constant amount of the lectin and varying quantities of type A blood group substance. It is analogous to the classical precipitin curve afforded in antibody–antigen systems.

Hemagglutination. Hemagglutination assays are conducted by serial dilution of lectin in PBS (0.01 M sodium phosphate in 0.15 M sodium chloride) pH 7.0. A 2% suspension (v/v) of freshly washed type A red blood cells in PBS serves as a source of cells. Equal volumes (approximately 0.05 ml) of diluted lectin and red blood cell suspension (in 10 × 75 mm test tubes) are mixed on a vortex mixer. After standing for 24 hours at 22°, the tubes are examined for agglutination. The activity is expressed as titer, the reciprocal of the highest dilution of lectin at which agglutination still occurs. A rapid estimate of hemagglutinating activity may be obtained after 30 minutes' incubation by centrifuging the tubes for 15 seconds at top speed in an International Desk Top Centrifuge. Specific activity is defined as titer per milligram of protein per milliliter. Protein content of partially purified lectin may be determined by the biuret method[6] using bovine serum albumin as standard. The pure lectin may be quantitated by its absorbance[3] at 280 nm using $E_{1\ cm}^{1\%} = 12.3$.

Precipitin Analyses. Quantitative precipitin analyses are performed in tapered 3-ml glass centrifuge tubes (Bellco Glass, Inc., Vineland, New Jersey) in a total volume of 200 μl. Incubation mixtures contain 2 μmoles of sodium phosphate buffer pH 7.0, 28.5 μmoles of NaCl, and varying amounts of blood group substance. For a typical precipitin curve, 5–100 μg of blood group substance are added to a series of tubes, each

[4] E. A. Kabat, *in* Kabat and Mayer's "Experimental Immunochemistry," 2nd ed., p. 97. Thomas, Springfield, Illinois, 1961.

[5] *In* "Methods in Immunology and Immunochemistry" (C. A. Williams and M. W. Chase, eds.), Vol. 3, p. 1. Academic Press, New York, 1971.

[6] See Volume 3 [73].

containing 100 μg of lectin. Incubation is conducted at 37° for 2 days in tubes capped with rubber stoppers. At the end of this time the tubes are centrifuged for 15 minutes at top speed in an International Desk Top Centrifuge, and the supernatant solution is carefully poured off. The tubes are inverted and allowed to drain for 15 minutes. The precipitates are washed twice by adding 0.1 ml of PBS to the drained tubes followed by vigorous agitation with a vorter mixer. After centrifugation the washings are decanted and the precipitates are drained as before. The washed precipitates are dissolved in 0.05 N NaOH (0.3–0.9 ml), and a 0.3-ml aliquot is used for a protein determination by a semimicro Lowry procedure.[7] Because of its low content of aromatic amino acids, blood group substance does not contribute significantly to the total protein determined (100 μg type A blood group substance gave a Lowry value equivalent to 4.2 μg of protein); the Lowry value represents, therefore, essentially the precipitated lectin.

Purification of the Lima Bean Lectin

Principle. A partially purified lima bean extract is passed over a column of type A blood group substance, insolubilized by treatment with N-carboxyanhydroleucine. After elution of inert protein, adsorbed lima bean lectin is displaced by N-acetyl-D-galactosamine.

Carolina or Sieva Pole lima beans, *Phaseolus lunatus* (W. Atlee Burpee Co., Clinton, Iowa) are ground once on the coarse setting and twice on the fine setting of an electric coffee mill. The finely ground meal is extracted with four times its weight of 0.15 M NaCl at 4° for 2 hours. The extracts are filtered through cheesecloth, and the residue is reextracted as before. The combined extracts are centrifuged in a Sorvall Model RC-2 refrigerated centrifuge at 14,600 g for 20 minutes to remove an insoluble residue. The supernatant solution is titrated with 6 N HCl to pH 4.0 and allowed to stand overnight at 4°. The precipitate obtained by centrifugation is discarded and the supernatant solution returned to pH 7.0 with 7 N NaOH. The acid-soluble protein fraction precipitating between 40% and 60% saturated ammonium sulfate at 4° contains most of the activity and has the highest specific activity by hemagglutination. The 40–60% ammonium sulfate precipitate is dissolved in 0.15 M NaCl and dialyzed against 0.15 M NaCl to remove $(NH_4)_2SO_4$. A small amount of insoluble material is removed by centrifugation.

Subsequently the lectin is specifically absorbed to type A blood group substance by passing the 40–60% ammonium sulfate fraction over a column of type A human ovarian cyst blood group substance

[7] R. Mage and S. Dray, *J. Immunol.* **95**, 525 (1965).

insolubilized by treatment with N-carboxyanhydroleucine.[8] The insolubilized blood group substance is mixed with 5 parts of Celite to obtain reasonable flow properties and poured into a column [1.8 × 8 cm, polyleucyl type A column (PLA column)]. The PLA column is washed with a buffer consisting of 0.01 M sodium phosphate and 0.15 M NaCl at pH 7.0 (PBS buffer). The partially purified extract in PBS buffer is added to the column, and most of the protein (96%) is washed through with PBS. All the hemagglutinating activity remains on the column and is displaced with 0.4 M N-acetyl-D-galactosamine in 0.15 M NaCl (4 ml).

The lectin eluted from the specific absorption column is dialyzed against 0.15 M NaCl to remove N-acetyl-D-galactosamine. Resolution of the components is achieved by recycling chromatography at 20° on a column (2.5 × 94 cm) of Sephadex G-200. Fractionation of the lectin gives a small quantity of an inactive high molecular weight material (component I) and two active fractions numbered in order of elution from the Sephadex column (components II and III). Two cycles are sufficient for separation, but component III is bled off during the first cycle to obtain pure component II. Components II and III are collected individually and concentrated on a Model 52 Amicon ultrafiltration cell (UM-10 or PM-10 membrane). The yield of the two lima bean lectins is 0.07% by weight of the dry meal.

Another purification of the lima bean lectin has been described by Gould and Scheinberg.[3] In our hands the Gould and Scheinberg purification, which does not utilize affinity chromatography, does not yield lectin as pure as the protein from the method described here. At the stage of purification prior to separation of the lectin into components II and III, the Gould and Scheinberg preparation gave minor bands on disc gel electrophoresis that do not correspond to components I, II, or III.

Several different gel filtration media were investigated for separating the lectin components in a single pass without resorting to recycling gel filtration. None of the following gels—Biogel A 0.5 m, Biogel P-200, and Biogel P-300—gave the quality of separation obtained by recycling chromatography on Sephadex G-200. However, recently we have found that a column of Biogel P-200 (4.5 × 45 cm) provides a reasonable resolution of a 2-ml sample containing 10 mg of protein per milliliter of 0.15 M NaCl into the two lima bean lectin components; pooling of the peak tubes from the two protein fractions give components II and III, each pure by disc gel electrophoresis at pH 4.3.

---

[8] M. E. Kaplan and E. A. Kabat, *J. Exp. Med.* **123**, 1061 (1966). See article on preparation of the phytohemagglutinin from *Dolichos biflorus*, this volume [39].

The purification of choice depends upon the intended use of the lectin. For highest purity and ease of separation of the components, the purification scheme outlined here is recommended. For purposes where a less pure preparation is acceptable, the Gould and Scheinberg purification is more rapid. However, for diagnostic use of the lectin as an anti-A or N-acetyl-D-galactosamine-binding reagent, there is no need to separate the components of the lectin. Under these conditions, lima bean lectin of reasonable purity and quantity can be obtained in two working days.

Properties of the Lima Bean Lectin [2,3,9]

Lima bean lectin binds N-acetyl-D-galactosamine and its glycoside.[2] These sugars will therefore inhibit the interaction between lima bean lectin and type A red blood cells and type A blood group substance.

Lima bean lectin occurs as two protein species, designated components II and III. These species have similar amino acid compositions and subunit structures and differ only in molecular weight: component II, $s_{20,w}$ 9.6, mw 247,000; component III, $s_{20,w}$ 6.0, mw 124,000. The subunits of both components appear to be identical, with a molecular weight of 31,000. The amino acid composition of the lima bean lectin is similar to other lectins with a high content of acidic and hydroxy amino acids and a low content of sulfur-containing amino acids. The lima bean lectin has no methionine and only 2 moles of half-cystine per mole of subunits— one bound in a disulfide linkage and the other free. At 280 nm, $E_{1cm}^{1\%}$ = 12.3 (Ref. 3).

The lima bean lectin is a glycoprotein[2] containing 6–7 moles of mannose, 2 moles of glucosamine, and 1 mole of fucose per 31,000 g of protein.[2] It is strongly precipitated by concanavalin A, indicating terminal α-linked D-mannopyranosyl residues.[2]

As a metal-requiring hemagglutinin,[2] the lima bean lectin was shown to contain both Mn^{2+} and Ca^{2+} with activity dependent upon Mn^{2+} which can be removed by dialysis against EDTA. The current data suggest 4 moles Ca^{2+} and 1 mole Mn^{2+} per mole of component III.

Factors That Affect the Activity of Lima Bean Lectin

Several factors affect the precipitation reaction between lima bean lectin and blood group A substance. The lectin contains bound Ca^{2+} and Mn^{2+}, the latter being required for the activity of the lectin. As isolated, the lectin generally displays full activity. A solution of 0.1 mM EDTA inhibits completely the precipitation reaction; addition of Mn^{2+} or

[9] N. R. Gould and S. L. Scheinberg, *Arch. Biochem. Biophys.* **141**, 607 (1970).

Ca²⁺ will reverse this inhibition. Metal chelating buffers, e.g., carbonate, give·less activity in the alkaline range than nonmetal chelating buffers.

Sulfhydryl reagents, e.g., 5,5'-dithiobis-(2-nitrobenzoic acid)[2,3,9] destroy the lectin's activity although β-mercaptoethanol up to 10% (v/v) does not inhibit the precipitation reaction.

Lima bean lectin has a broad pH optimum[2] extending from pH 4.5 to 8.5. More precipitation occurs at 37° than at 20° or 4° and the precipitation reaction is therefore routinely conducted at 37°. The precipitation reaction is independent of salt concentration from 0.05 M to 0.25 M NaCl, KCl, or KBr, but KI or KSCN partially inhibit the reactivity of the lima bean lectin.

The precipitation reaction between lima bean lectin and type A blood group substance is also sensitive to reaction volume; for this reason, the volume of the incubation mixture should be kept as small as is convenient.

[36] Gorse (*Ulex europeus*) Phytohemagglutinins

By Toshiaki Osawa and Isamu Matsumoto

The crude extract of *Ulex europeus* seeds has been routinely used for the diagnosis of secretors[1] and the determination of subgroups of A and AB blood groups.[2] It has been demonstrated that there exist two kinds of anti-H(O) phytohemagglutinins in the seeds of *Ulex europeus*.[3] One is inhibited by L-fucose (eel serum type; hemagglutinin I), and the other is inhibited most by di-*N*-acetylchitobiose (*Cytisus* type; hemagglutinin II). These two phytohemagglutinins have been separated and purified by the procedures as described below.[3,4]

Assay Method

Titration. The diluent and the erythrocyte suspending solution in all tests is 0.15 M aqueous NaCl containing 0.025 M sodium phosphate buffer (pH 7.0). Human erythrocytes freshly obtained from a donor are washed 3 times and used for routine assays as a 3% suspension. To each 0.05 ml of 2-fold serial dilutions of hemagglutinin solution, in tubes, an equal volume of the human erythrocyte suspension is added.

[1] W. C. Boyd and E. Shapleigh, *Blood* **9**, 1195 (1954).
[2] W. C. Boyd and E. Shapleigh, *J. Lab. Clin. Med.* **44**, 235 (1954).
[3] I. Matsumoto and T. Osawa, *Biochim. Biophys. Acta* **194**, 180 (1969).
[4] I. Matsumoto and T. Osawa, *Arch. Biochem. Biophys.* **140**, 484 (1970).

The mixture is kept for 1 hour at room temperature and then examined for agglutination.

Sugar Inhibition. The inhibition assays are carried out in tubes as follows. To each 0.05 ml of 2-fold serial dilution of human saliva or to each sugar solution, prepared with the same accuracy as in the quantitative chemical analysis, is added an equal volume of hemagglutinin solution carefully diluted to have four to five minimum hemagglutinating doses. After incubation for 2 hours at room temperature, 0.05 ml of human O erythrocyte suspension is added. The mixture is kept for 1 hour at room temperature and then examined for agglutination.

Purification Procedures

Extraction and Ammonium Sulfate Fractionation. One hundred grams of finely powdered seeds (F. W. Schumacher, Sandwich, Massachusetts) are suspended in 1.1 liter of 0.9% NaCl and allowed to stand overnight at 4° with occasional stirring. To the clear yellow supernatant obtained by centrifugation, solid ammonium sulfate is added, and after standing overnight in a cold room, the precipitate is dialyzed against distilled water until free of NH_4 and then lyophilized. Protein precipitating at 0–40% (crude hemagglutinin I) and 40–70% (crude hemagglutinin II) saturation with ammonium sulfate has hemagglutinating activity.

Purification of Hemagglutinin I

Step 1. CM-Cellulose. Crude hemagglutinin I, 1.4 g, is dissolved in 20 ml of 5 mM citrate buffer (pH 3.8, starting buffer), dialyzed overnight against the same buffer, and then applied to a column (2 cm × 40 cm) equilibrated with the same buffer. Gradient elution is performed with 1 liter of starting buffer in the mixing vessel and 1 liter of 0.2 M phosphate buffer (pH 6.8) in the reservoir. Fractions of 10 ml were collected at a flow rate of 20 ml/hour at 4°. The absorbancy of the fractions is measured at 280 nm, and the hemagglutinating activity is determined. Two protein peaks are detected (fractions A and B), and most of the hemagglutinating activity is observed in a retarded fraction (fraction B).

Step 2. Sephadex G-200. Fraction B, 150 mg, is dissolved in 4 ml of 0.2 M phosphate buffer (pH 6.8) containing NaCl (0.2 M) and applied to a column (2 cm × 110 cm) equilibrated with the same buffer. Elution is carried out with the same buffer and fractions of 4 ml are collected at a flow rate of 15 ml/hour at 4°. Four protein peaks are detected, and fraction B-2, which has most hemagglutinating activity, is pooled and lyophilized.

Step 3. Biogel P-200. A Biogel column (1.2 cm × 110 cm) is equilibrated and operated in the same fashion as in step 2 loading 20 mg of fraction B-2. The main protein peak which has a strong hemagglutinating activity is pooled and lyophilized (purified hemagglutinin I).

Purification of Hemagglutinin II

Step 1. DEAE-Cellulose. Crude hemagglutinin II, 500 mg, is dissolved in 10 ml of 5 mM phosphate buffer (pH 7.7, starting buffer) and dialyzed overnight against the same buffer, and then applied to a column (1.5 cm × 30 cm) of DEAE-cellulose equilibrated with the same buffer. Parabolic chloride gradient elution is performed as follows. Three open flasks are connected in series by means of tubes at their bottom. Flasks 1 and 2, which are magnetically stirred, each contains 700 ml of the limit buffer (2.0 M sodium chloride, 5 mM phosphate buffer, pH 7.7). Fractions of 10 ml are collected at a flow rate of 20 ml/hour at 4°. The hemagglutinating activity is recovered in two protein peaks (fractions A and B). Fraction A, apparently not retained by the column, is inhibited by di-N-acetylchitobiose, whereas fraction B is inhibited by L-fucose. The tubes corresponding to fraction A are collected, and the solution is dialyzed and lyophilized.

Step 2. Pevikon Block Electrophoresis. A solution of 30 mg of fraction A from DEAE-cellulose column in 0.5 ml of Veronal buffer (pH 8.6, $\mu = 0.05$) is mixed with sufficient Pevikon (polyvinyl chloride) particles to give a slurry which is then deposited into a cut (0.5 cm wide) made in a block (2.5 cm × 3.9 cm × 40 cm) of the supporting material in the same buffer. A voltage of 300 V is applied and electrophoresis is run for 3 hours at 4°. The block is sliced into 1-cm segments, and each one is eluted with 2 ml of 5 mM phosphate buffer (pH 7.5) containing 0.2 M NaCl. The eluates, which show strong hemagglutinating activity, are pooled, dialyzed, concentrated by ultrafiltration.

Step 3. Biogel P-200. The concentrated hemagglutinin solution obtained above is dialyzed against 5 mM phosphate buffer (pH 7.7) containing 0.2 M NaCl and applied to a column (1.2 cm × 110 cm) equilibrated with the same buffer. Elution is carried out with the same buffer, and fractions of 2 ml are collected at a flow rate of 15 ml/hour at 4°. The hemagglutinating activity is found only in the first peak in absorption at 280 nm. The fractions having hemagglutinating activity are pooled, dialyzed, and lyophilized (purified hemagglutinin II). The details of purification of hemagglutinin I and II are shown in Table I.

Properties of Purified Hemagglutinins

Physical and Chemical Properties. The purified hemagglutinins are homogeneous in a disc electrophoresis on polyacrylamide gel. A single

TABLE I
DETAILS ON PURIFICATION OF *Ulex europeus* HEMAGGLUTININ I AND II

Fractions	Yield from 100 g seeds (mg)	Activity[a] (μg/ml)			Sugars (mg/ml)[b]	
		O	A	B	L-Fucose	Di-N-acetyl-chitobiose
Crude extract	11700	1000	—	—	0.63	>10
(NH$_4$)$_2$SO$_4$ fractions						
0–0.4 satn.	670	250	1000	500	0.63	>10
0.4–0.7 satn.	1850	625	2500	1250	>10	2.5
Hemagglutinin I						
Step 1. (CM-cellulose)						
Fraction A	220	1000	—	—	—	—
Fraction B	130	125	500	250	0.32	>10
Step 2. (Sephadex G-200)						
Fraction B-1	2	500	—	—	—	—
Fraction B-2	22	15	30	30	0.32	>10
Fraction B-3	43	500	—	—	—	—
Fraction B-4	6	500	—	—	—	—
Step 3. (Biogel P-200, purified hemagglutinin I)	20	15	30	30	0.32	>10
Hemagglutinin II						
Step 1. (DEAE-cellulose)						
Fraction A	310	625	2500	1250	>10	2.5
Fraction B	230	625	2500	1250	0.63	>10
Step 3. (Biogel P-200, purified hemagglutinin II)	2	2	6	4	>10	2.5

[a] Minimum hemagglutinating dose against human O, A, or B cells.

[b] Minimum amounts completely inhibiting four hemagglutinating doses.

TABLE II
CARBOHYDRATE COMPOSITION OF PURIFIED HEMAGGLUTININ I AND II

Sugar	g/100 g Hemagglutinin	
	I	II
Mannose	1.6	8.7
Fucose	0.6	0.4
Glucose	0.6	0.8
Xylose	0.6	0.2
Arabinose	0.3	3.5
Galactose	0.1	6.3
Glucosamine	1.4	1.8
Total	5.2	21.7

TABLE III
INHIBITION ASSAY OF PURIFIED HEMAGGLUTININ I AND II WITH SUGARS

Sugar	Minimum amounts (mg/ml) completely inhibiting 4 hemagglutinating doses	
	Purified hemagglutinin I	Purified hemagglutinin II
D-Glucose	>10	>10
D-Galactose	>10	>10
L-Fucose	0.63	>10
N-Acetyl-D-glucosamine	>10	>10
N-Acetyl-D-galactosamine	>10	>10
Salicine	>10	10
Phenyl β-D-glucopyranoside	>10	10
Benzyl β-D-glucopyranoside	>10	>10
Methyl N-acetyl-β-D-glucosaminide	>10	>10
Phenyl N-acetyl-β-D-glucosaminide	>10	10
Maltose	>10	>10
Cellobiose	>10	10
Gentiobiose	>10	>10
Lactose	>10	>10
N-Acetyllactosamine	>10	>10
Di-N-acetylchitobiose	>10	2.5
Tri-N-acetylchitotriose	>10	1.3
Tetra-N-acetylchitotetraose	>10	2.5
Penta-N-acetylchitopentaose	>10	2.5

band is obtained at pH 8.9 in each case. When examined in an ultracentrifuge, the purified hemagglutinins are sedimented as a single symmetrical peak. The sedimentation velocity method gives a sedimentation coefficient $s_{20,w}^0$ of 6.5 for both of the purified hemagglutinins. These purified hemagglutinins are found to be glycoproteins. The carbohydrate compositions of the purified hemagglutinins are listed in Table II.

Specificity. The results of the inhibition assays on the purified hemagglutinins with simple sugars are shown in Table III.

Remarks. The preparations after step 1 from both crude hemagglutinin I and II have almost the same specificity as purified hemagglutinin I and II, respectively. These preparations, therefore, can be conveniently used for most of serological purposes.

[37] *Wistaria floribunda* Phytomitogen

By Toshiaki Osawa and Satoshi Toyoshima

Since Nowell's discovery of the mitogenic activity against human peripheral lymphocytes in the extracts of *Phaseolus vulgaris* seeds,[1] the extracts of many plant seeds have been extensively surveyed for this biological activity.[2-4] The mitogenic activity of the extracts of *Wistaria floribunda* seeds are first reported by Barker and Farnes.[5] This phytomitogen has been purified and characterized as described below.[6]

Assay Methods

Hemagglutination Assays. The titration and the inhibition assays using human erythrocytes freshly obtained from a donor are carried out as described elsewhere in this volume (*Ulex europeus* phytohemagglutinins).

Mitogenic Assays. Normal human venous blood is withdrawn into syringes previously treated with heparin. The heparinized blood is transferred to glass cylinders, and the erythrocytes are allowed to sediment by gravity at 37°. The leukocyte-rich plasma is withdrawn, and a 1-ml aliquot is transferred to a sterile culture bottle to which 2 ml of NCTC-109 (Difco), 0.3 ml of calf serum and 0.06 ml of mitogen solution (20 mg/ml) are previously added. The cultures are incubated at 37°. A negative control and a positive control with PHA-M (20 mg/ml, Difco) are included in each experiment. The culture is performed 3 times on the same sample.

Morphological examination of lymphocyte transformation is carried out by determining the percentage of transformed cells from Giemsa-stained preparations, counting approximately 1000 cells on each mitogen sample, after incubation for 72 hours.

Radioactivity assay of [3H]thymidine incorporation is performed

[1] P. C. Nowell, *Cancer Res.* **20**, 462 (1960).

[2] H. J. Downing, G. C. M. Kemp, and M. A. Denborough, *Nature (London)* **217**, 654 (1968).

[3] M. Krüpe, W. Wirth, D. Nies, and A. Ensgraber, *Z. Immunitaetsforsch. Allerg. Klin. Immunol.* **137**, 442 (1969).

[4] J. W. Parker, J. Steiner, A. Coffin, R. J. Lukes, K. Burr, and L. Brilliantine, *Experientia* **25**, 187 (1969).

[5] B. E. Barker and P. Farnes, *Nature (London)* **215**, 659 (1967).

[6] S. Toyoshima, Y. Akiyama, K. Nakano, A. Tonomura, and T. Osawa, *Biochemistry* **10**, 4457 (1971).

by adding [³H]thymidine (0.5 μCi, The Radiochemical Centre, Amersham, England) to each tube (3 to 4×10^5 lymphocytes) after the appropriate incubation time. Fifteen hours later, the cells are collected and washed 3 times with cold 0.02 M phosphate-buffered saline (pH 7.0). To this residue is added 2 ml of cold 5% trichloroacetic acid, the precipitates are collected on glass fiber paper (Whatman GF/C), washed with 10 ml of methanol and dried. The dried residue is then mixed with 5 ml of scintillation fluid [2,5-diphenyloxazole (3 g), 1,4-bis[2-(5-phenyloxazolyl)]benzene (0.1 g) in 1000 ml of toluene], and the counts per minutes of each sample are determined with a Packard automatic scintillation counter.

Inhibition assay with sugars is carried out as follows. To 0.05 ml of sugar solution in Eagle's Minimal Essential Medium (MEM) is added an equal volume of mitogen solution in MEM. After incubation for 10 minutes at room temperature, 0.9 ml of cell suspension (4×10^5 cells/ml) is added. The mixture is incubated for 72 hours and [³H]-thymidine incorporation is determined as described above.

Purification Procedures

Step 1. Extraction and Ammonium Sulfate Fractionation. Finely powdered *W. floribunda* seeds, 100 g, are suspended in 1 liter of 0.9% NaCl and allowed to stand overnight at 4° with occasional stirring. To the clear yellow supernatant obtained by centrifugation, solid $(NH_4)_2SO_4$ is added to give 40%, 70%, and 100% saturation. The fractions thus obtained are dialyzed against distilled water until free of NH_4^+, centrifuged and lyophilized. Strong hemagglutinating activity is observed in the fractions precipitating at 100% saturation (crude nitrogen).

Step 2. SE-Sephadex C-50. Crude mitogen, 150 mg, is dissolved in 5 ml of 0.05 M phosphate buffer (pH 5.0, starting buffer) and dialyzed overnight against the same buffer, and then applied to a column (2 cm × (30 cm) equilibrated with the same buffer. Elution is carried out with the same buffer and, after the first large peak is eluted out, gradient elution is performed with 100 ml of starting buffer in the mixing vessel and 100 ml of the same buffer containing NaCl (0.5 M) in the reservoir. Fractions of 5 ml are collected at a flow rate of 8 ml/hour at 4°. Protein concentration is determined at 280 nm. Two large protein peaks (fraction A and B) are detected. Although the bulk of the hemagglutinating and leukoagglutinating activities are recovered in fraction A (the first large peak), the mitogenic activity is observed only in fraction B which has relatively weak agglutinating activity against both erythrocytes and leukocytes.

Step 3. Sepharose 6B. Fraction B, 15 mg, is dissolved in 1 ml of

TABLE I

DETAILS OF PURIFICATION OF *Wistaria floribunda* MITOGEN

Fraction	Yield from 100 g of seeds (mg)	Hemagglutinating activity (μg/ml)[a]			Leuko-agglutinating activity (μg/ml)[b]	Mitogenic activity	Sugars (mg/ml)[c]	
		O	A	B			D-Galactose	N-Acetyl-D-galactosamine
Step 1. [(NH₄)₂SO₄ fractionation]								
0–0.4 satn.	300	>10000	>10000	>10000		(−)		
0.4–0.7 satn.	1600	630	630	630		(−)	>10	>10
0.7–1.0 satn. (crude mitogen)	450	160	160	160		(+)	2.5	0.08
Step 2. (SE-Sephadex)								
Fraction A	160	80	80	80	320	(−)	2.5	0.08
Fraction B	150	5000	5000	5000	2500	(+)	2.5	0.63
Step 3. (Sepharose 6B, purified mitogen)	100	5000	5000	5000	2500	(+)	2.5	0.63

[a] Minimum hemagglutinating dose against human O, A, or B cells.
[b] Minimum agglutinating dose against human leukocytes.
[c] Minimum amounts completely inhibiting four hemagglutinating doses.

0.02 M phosphate buffer (pH 5.0) containing NaCl (0.15 M) and applied to a column (2.5 cm × 30 cm) equilibrated with the same buffer. Elution is carried out with the same buffer and fractions of 2.5 ml are collected at a flow rate of 8 ml/hour at 4°. The strong mitogenic activity concomitant with very weak hemagglutinating activity is eluted as a single peak (purified mitogen).

The details of the purification of *W. floribunda* mitogen are summarized in Table I.

Properties of Purified Mitogen

Physical and Chemical Properties. The purified mitogen is homogeneous in a disc electrophoresis on polyacrylamide gel. A single band is obtained at pH 8.9. When examined in an ultracentrifuge, the purified mitogen is sedimented as a single symmetrical peak. The sedimentation velocity method gives a sedimentation coefficient $s_{20,w}^{0}$ of 4.5. The purified mitogen is found to be a glycoprotein, and its chemical composition is listed in Table II.

Specificity. [³H]Thymidine incorporation of human peripheral lymphocytes exposed to the purified mitogen is appreciably inhibited by N-acetyl-D-galactosamine and di-N-acetylchitobiose.

TABLE II

CHEMICAL COMPOSITION OF PURIFIED *W. floribunda* MITOGEN

Amino acid	g/100 g	Carbohydrate	g/100 g
Asp	10.66	Mannose	5.80
Thr	6.05	Fucose	0.91
Ser	7.10	Arabinose	0.84
Glu	9.08	Xylose	0.33
Pro	3.48	Glucosamine	3.50
Gly	5.01	Total	11.38
Ala	3.76		
Cys	0.62		
Val	4.76		
Met	0.28		
Ile	4.92		
Leu	5.67		
Tyr	2.35		
Phe	5.82		
Trp	3.05		
Lys	4.67		
His	3.68		
NH₃	1.91		
Arg	4.65		
Total	87.52		

TABLE III

INHIBITION OF HEMAGGLUTINATION OF FRACTION A IN STEP 2 AND PURIFIED
MITOGEN WITH SIMPLE SUGARS

| | Minimum amounts (mg/ml) completely inhibiting 4 hemagglutinating doses | |
Sugars[a]	Fraction A	Purified mitogen
D-Galactose	2.5	2.5
L-Arabinose	10	5
N-Acetyl-D-galactosamine	0.08	0.63
Lactose	0.63	1.25
Melibiose	1.25	1.25
Raffinose	2.5	2.5

[a] The following sugars are not inhibitory at a concentration of 20 mg/ml: D-Glucose, L-Fucose, D-Mannose, N-Acetyl-D-glucosamine and Maltose.

The results of the hemagglutination-inhibition assays on the purified mitogen as well as fraction A in step 2 with simple sugars are listed in Table III.

[38] Common Lentil (*Lens culinaris*) Phytohemagglutinin

By HARVEY J. SAGE[1] and ROBERT W. GREEN

A common lentil, *Lens culinaris* or *Lens esculenta*, has been shown to contain hemagglutinating activity for erythrocytes from a number of animal species.[2] By extended starch-gel electrophoresis[3] or disc-gel electrophoresis,[4] this activity has been found to be in two very closely related proteins. Both proteins have been shown to bind to simple saccharides, such as glucosides and mannosides,[2,5] polysaccharides, such as glycogen[3] and phosphomannans,[6] and human serum glycoproteins.[6]

[1] Career Development Awardee of the U.S. Public Health Service. Supported by NIH Grant AI-06710.

[2] I. K. Howard and H. J. Sage, *Biochemistry* 8, 2436 (1969).

[3] G. Entlicher, M. Tichá, J. Kŏstíř, and J. Kocourek, *Experientia* 25, 17 (1969).

[4] I. K. Howard, H. J. Sage, M. D. Stein, N. M. Young, M. A. Leon, and D. F. Dyckes, *J. Biol. Chem.* 246, 1590 (1971).

[5] M. D. Stein, I. K. Howard, and H. J. Sage, *Arch. Biochem. Biophys.* 146, 353 (1971).

[6] N. M. Young, M. A. Leon, T. Takahashi, I. K. Howard, and H. J. Sage, *J. Biol. Chem.* 246, 1596 (1971).

Purification of the two hemagglutinins (called LcH-A and LcH-B) can be readily achieved by a two-step process involving (a) isolation of a mixture of LcH-A and LcH-B from a crude saline extract of lentils and (b) separation of the two proteins by CM-cellulose chromatography using a shallow salt gradient.

Assay Procedure

Hemagglutinin activity is measured by a standard dilution titer assay. Typically a 1:2 serial dilution of hemagglutinin is prepared in pH 7.2 buffered isotonic saline. A 2% suspension of washed red cells (human, or rabbit) is prepared in the same buffer. Equal volumes (0.1 ml) of red cell suspension and hemagglutinin dilution are mixed in small tubes and the cells allowed to settle for 30 minutes at room temperature. The titer is then read from the settling pattern.[7] The assay is about four times more sensitive with rabbit red cells than with human erythrocytes. A qualitative assay for hemagglutinin is reported[4] which utilizes a modified Ouchterlony gel-double diffusion procedure with serum glycoproteins.

Preparation of LcH Mixture

A mixture containing the two LcH proteins but devoid of other proteins is readily obtained from a crude saline extract by either of two methods: (A) dialysis against distilled water followed by DEAE-celluose chromatography,[2] or (B) specific absorption to Sephadex G-100[4] or G-150[3] and elution with D-glucose.[3,4,8] Method B is a modification of the method of Agrawal and Goldstein.[9] One variation of each method will be presented below. For modifications of these procedures the reader is referred to published methods.[3,4]

Crude Saline Extract. Dried lentils (1 kg), obtained from food shops, are soaked overnight in cold water. The excess water is drained off and the swollen lentils blended with 1 liter of chilled 0.15 N NaCl for 5 minutes in a high speed blender. The slurry is centrifuged at high speed (10,000–20,000 g), for 15 minutes and the sediment is discarded. The clear yellow supernatant (1–1.2 liters) should have a titer of 1–16 versus human red cells. The crude extract can be used directly for the purification methods described below or a partial purification can be obtained by

---

[7] D. H. Campbell, J. S. Garvey, N. E. Cremer, and D. H. Sussdorf, "Methods in Immunology," pp. 155–165. W. A. Benjamin, New York, 1963.

[8] M. Tichá, G. Entlicher, J. Kŏstíř, and J. Kocourek, *Biochim. Biophys. Acta* **211**, 282 (1970).

[9] B. B. L. Agrawal and I. J. Goldstein, *Biochim. Biophys. Acta* **147**, 262 (1967).

ammonium sulfate fractionation at neutral pH[4,8] and the partially purified product used as below.

Purification Method A. Crude extract is dialyzed against several
changes of cold distilled water adjusted to pH 5.8. Most of the yellow
color diffuses out and a heavy white precipitate forms in the dialysis
bag. The precipitate contains most of the hemagglutinin activity. The
precipitate is centrifuged down at 4° and washed twice with cold distilled water adjusted to pH 5.8. It is then dissolved in a minimal volume
of 0.015 M phosphate, pH 7.5 buffer and applied to a 500-ml column of
DEAE-cellulose (DE-32, Whatman) equilibrated with the same buffer
at room temperature. The hemagglutinins are washed directly through
the column and all other contaminating proteins remain adhered to the
column. The resultant purified LcH-mixture is ultracentrifugally homogeneous and is 100% reactive with red cells.[2] Yields are 200–400 mg
of LcH-mixture per kilogram of dried lentils with an activity recovery
of considerably better than 50% from the crude saline extract.

Purification Method B. Crude extract is directly applied to a 3-liter
column of Sephadex G-100 (Pharmacia) equilibrated with 0.15 N NaCl
at room temperature. The sample is washed through with 3–4 liters of
0.15 N NaCl. The washed through material will be devoid of agglutinin
activity and contain most of the protein and other materials present in
the crude extract. The column is then washed through with 3–4 liters
of 0.05 N D-glucose in 0.15 N NaCl, samples are collected and tested for
hemagglutinin activity (the 0.05 N D-glucose present in the eluted LcH-
mixture is not inhibitory enough to mask the activity assay with the
concentrations of hemagglutinins ordinarily used). The resultant LcH
mixture is indistinguishable from that obtained by purification Method
A, and the yields are at least as good. Of the two methods described,
Method B is slightly more convenient. If it is desired to prepare LcH
mixture in the absence of binding sugars, Method A is the one of
choice. LcH mixture can be eluted from Sephadex columns with acid[3]
instead of glucose. However, there is evidence that the acid-eluted product
may be partially denatured.[10]

Separation of LcH-A and LcH-B

Purified LcH mixture is dialyzed against 0.02 M acetate–0.075 N
NaCl pH 5.0 buffer and applied to a 50-ml column of CM-cellulose (CM-
52, Whatman) equilibrated with the same buffer at room temperature.
After washing through with several column volumes of equilibration
buffer, a shallow 0.075 N–0.150 N NaCl gradient is applied to the

---

[10] M. Tichý, M. Tichá, and J. Kocourek, *Biochim. Biophys. Acta* **229**, 63 (1971).

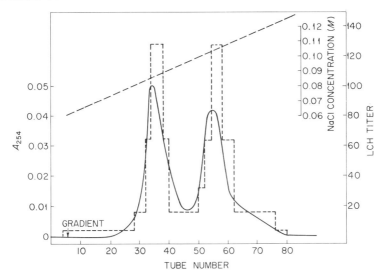

FIG. 1. Separation of lentil hemagglutinins LcH-A and LcH-B by CM-cellulose chromatography. ———, A_{254}; ----, hemagglutinin titer; and ———, NaCl concentration plotted against tube number. Taken from I. K. Howard, H. J. Sage, M. D. Stein, N. M. Young, M. A. Leon, and D. F. Dyckes, *J. Biol. Chem.* **246**, 1590 (1971), with the editor's permission.

column in the acetate buffer. Usually a linear gradient is applied with 900 ml of equilibration buffer in the mixing chamber and 900 ml of 0.02 M acetate–0.15 N NaCl pH 5.0 buffer in the elution chamber. The protein content of the column effluent is followed with a UV monitor and hemagglutinin activity measured as described above. Figure 1 shows a typical column run. The first hemagglutinin eluted from the column is LcH-A and the second protein is LcH-B. Yields are 100% of the activity of the LcH-mixture applied to the column.

Purity of the LcH-A and LcH-B solutions is determined by disc gel electrophoresis using standard procedures.[11,12] Figure 2 shows a disc gel electrophoresis pattern of LcH mixture. Purified LcH-A shows a single band corresponding to the anodic component of the LcH mixture and purified LcH-B shows a single band corresponding to the cathodic component of the mixture.

The relative amounts of LcH-A and LcH-B in different brands of lentil is very different. In 6 brands tested, 3 had equal amounts of LcH-B and LcH-A and 3 had about 90–95% LcH-A. Figures 1 and 2 show patterns typical of the brands with equal amounts of the two proteins.

[11] L. Ornstein, *Proc. N.Y. Acad. Sci.* **121**, 321 (1964).
[12] B. J. Davis, *Proc. N.Y. Acad. Sci.* **121**, 404 (1964).

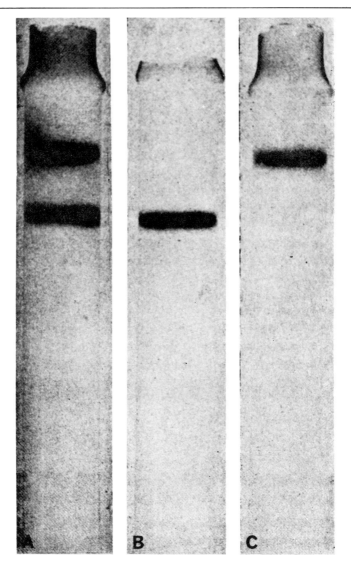

FIG. 2. Disc gel electrophoresis of (A) LcH-mixture, (B) LcH-A, and (C) LcH-B. Cathode is at the top. Taken from I. K. Howard, H. J. Sage, M. D. Stein, N. M. Young, M. A. Leon, and D. F. Dyckes, *J. Biol. Chem.* **246,** 1590 (1971), with the editor's permission.

Concentration of LcH-A and LcH-B (as well as LcH mixture prepared by any of the above procedures) is best done by ultrafiltration and storage of the concentrates frozen at $-20°$. Little activity, if any, is lost in months of storage and the materials are completely in solution upon thawing. In contrast dialysis against water and lyophilization causes some activity losses and insoluble material.

Properties of LcH-A and LcH-B

Table I shows some of the physical-chemical properties of LcH-A and LcH-B. In most respects the two proteins are indistinguishable. They have the same size, sedimentation constant, UV spectra, and both dissociate into half molecules in the presence of certain solvents.[4] LcH mixture is relatively heat stable,[2] particularly in the presence of binding sugars. The reactions of LcH-A and LcH-B with simple sugars, polysaccharides, glycoproteins, and cell surfaces are indistinguishable.[4] Tichá et al.[8] have detected a difference in their strength of binding to Sephadex G-150 and have used this to isolate one of the hemagglutinins in pure form. This component (called hemagglutinin I) is the stronger binding of the two and appears to be identical to LcH-A. These authors have also demonstrated that hemagglutinin I dissociates in acid media. The compositions of LcH-A (hemagglutinin I) and LcH-B have been reported.[4,8] Their amino acid compositions are practically identical except for lysine where LcH-B contains four more residues per molecule. This would account for their different electrophoretic properties. In addition to amino acids and a small amount of bound carbohydrate, these proteins contain bound Ca^{2+} and Mn^{2+}.[8,10] It has been suggested that LcH-A and LcH-B each are molecules consisting of two non-covalently bonded identical polypeptide chains.[4] LcH-A and LcH-B are immunochemically identical[4] suggesting that they have identical, or nearly identical, conformations. This is also shown in Fig. 3, which shows the optical rotatory dispersion spectra of LcH-A and LcH-B. There is no detectable difference in their ORD spectra. LcH-A has been shown to have two identical saccharide binding sites.[5]

LcH-A and LcH-B each bind to materials containing Makela's[13] group III sugars.[2-6] Although they bind to Sephadex gels, they do not react detectably with a large number of dextrans[2,6] for reasons which are not apparent. Both proteins bind more strongly to mannosides than to glucosides, and both proteins bind more strongly to α-glycosides than to β-glycosides. Table II shows some binding constants of LcH-A to simple saccharides and to rabbit red cells. The binding constant for red cells is 4–5 orders of magnitude greater than for simple sugars, indicating that the red cell receptor(s) is not a simple sugar (certainly not a mannoside or glucoside). Although a saccharide-containing red cell receptor is implicated, the nature of the receptor is unknown. LcH-A

[13] O. Mäkelä, Ann. Med. Exp. Biol. Fenn. 35, 3 (1957).

TABLE I

SOME PROPERTIES OF LcH MIXTURE: LcH-A (HEMAGGLUTININ I) AND LcH-B

Property	Measurement conditions	Value			Reference
		LcH mixture	LcH-A (hemagglutinin I)	LcH-B	
Sedimentation constant	pH 7.4 Tris buffer	$s^0_{20,w}$	$s^0_{20,w} = 3.78$	—	8
Molecular weight (sedimentation equilibrium)	pH 7.2 Buffered saline		$s^0_{20,w} = 4.0$	$s^0_{20,w} = 4.0$	2, 4
	pH 7.2 Buffered saline	48–50,000	48,000	48,000	4
	6 M Guanidine hydrochloride–0.1 M mercapto-ethanol	25,000	25,000	25,000	4
Molecular weight (gel filtration)	pH 7.2 Buffered saline	44,000	39,000	39,000	2, 4
	pH 7.4 Tris buffer	—	42,000	—	8
Molecular weight (disc gel electrophoresis)	Sodium dodecyl sulfate	—	24,000	24,000	4
UV absorption extinction coefficient	pH 7.2 Buffered saline	$E^{1\%}_{280\ nm} = 12.6$	$E^{1\%}_{280\ nm} = 12.6$	$E^{1\%}_{280\ nm} = 12.6$	2, 4

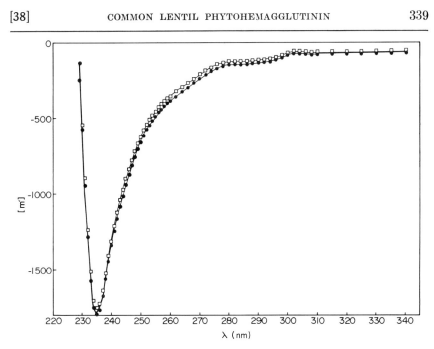

Fig. 3. Optical rotatory dispersion spectrum of lentil hemagglutinins LcH-A and LcH-B in pH 7.2 phosphate-buffered saline at 25°. Measured on a Cary 60 recording spectropolarimeter. Mean residue rotation, [m′]; versus wavelength in nanometers, λ; □ LcH-A, ● LcH-B.

and LcH-B also bind to lymphocytes in culture and stimulate these cells to mitose,[6] as evidenced by the uptake of a DNA precursor from the culture medium. A saccharide-containing receptor has been implicated in the binding and subsequent stimulatory process.

TABLE II

BINDING OF LcH-A TO SIMPLE SACCHARIDES AND RABBIT ERYTHROCYTES

Binding substance	K_a	Maximal number of LcH-A molecules bound per molecule of binding substance
Methyl D-glucopyranoside[a]	1.0×10^2	1
D-Mannose[a]	2.5×10^2	1
Rabbit erythrocytes[b]	1.1×10^7	5.0×10^5

[a] M. D. Stein, I. K. Howard, and H. J. Sage, *Arch. Biochem. Biophys.* **146**, 353 (1971).

[b] M. D. Stein, to be published and M. D. Stein, H. J. Sage, and M. A. Leon, *Arch. Biochem. Biophys.* in press.

[39] Horse Gram (*Dolichos biflorus*) Lectin

By MARILYNN ETZLER

The seeds of *Dolichos biflorus* contain a lectin (plant hemagglutinin) that specifically agglutinates type A erythrocytes[1-4] and precipitates blood group A substance.[2-4] The ability of the lectin to combine with blood group A substance is due to the specificity of its combining sites for terminal nonreducing α-linked N-acetyl-D-galactosamine (GalNAc) residues.[4] The lectin can thereby serve as a valuable tool for detecting terminal nonreducing α-linked GalNAc residues in polysaccharides and glycoproteins.

Assay Method

The *Dolichos biflorus* lectin can be assayed by hemagglutination or precipitin methods. The quantitative precipitin assay is the more sensitive and precise method; however, the hemagglutination assay is rapid and is the method of choice when working with crude extracts of seeds; these extracts give a high background in the precipitin assay due to spontaneous precipitation of other substances in the extract. Hemagglutination and precipitin procedures have been extensively reviewed.[5,6]

Hemagglutination Procedure. A convenient hemagglutination procedure is to prepare serial dilutions of the lectin solution with 0.9% NaCl using a hemagglutination plate and Takatsy microtitrator with 0.025-ml loops. A drop of a 2% suspension of type A_1 erythrocytes is added to each dilution, and the hemagglutination plate is covered and kept at room temperature for 1 hour. The titer is the reciprocal of the highest dilution showing detectable agglutination.

Inhibition of hemagglutination[7] can be used for testing the reactivity of the lectin with soluble substances.

Quantitative Precipitin Procedure. A microprecipitin technique[8] using

---

[1] G. W. G. Bird, *Current Sci.* **20**, 298 (1951).
[2] W. C. Boyd and E. Shapleigh, *J. Immunol.* **73**, 226 (1954).
[3] G. W. G. Bird, *Brit. Med. Bull.* **15**, 165 (1959).
[4] M. E. Etzler and E. A. Kabat, *Biochemistry* **9**, 899 (1970).
[5] E. A. Kabat, "Kabat and Mayer's Experimental Immunochemistry," 2nd ed. Thomas, Springfield, Illinois, 1961.
[6] E. A. Kabat, "Structural Concepts in Immunology and Immunochemistry." Holt, Rinehart, and Winston, New York, 1968.
[7] W. M. Watkins, Vol. 8 [119].
[8] G. Schiffman, E. A. Kabat, and W. Thompson, *Biochemistry* **3**, 113 (1964).

5–6 μg of nitrogen of purified *Dolichos biflorus* lectin gives very precise and reproducible results.[4] The precipitin curve is set up in conical 3-ml tubes by combining a constant amount of lectin with various concentrations of antigen (5–200 μg). The final salt concentration is adjusted to 0.9% NaCl, and the final volume is 250 μl. It is advisable to keep the final volume as low as possible since larger volumes increase the possibility of soluble antigen–lectin complexes.[4] The tubes are incubated for 1 hour at 37° and kept at 4° for 2 days.

The tubes are centrifuged at 7000 rpm in the cold for 15 minutes, and the precipitates are washed twice with 0.25 ml of 0.9% NaCl. Nitrogen in the washed precipitates can be determined by a modified ninhydrin method[8] as described below.

The precipitates (in the conical 3-ml tubes) are digested with 25 μl of $0.9 M$ H_2SO_4 for 45 minutes in a sand bath at 160°. The tubes are cooled to room temperature, 15 μl of 30% H_2O_2 is added, and the tubes are again heated for 45 minutes. If charring is still visible in the tubes, a second heating with H_2O_2 is required. The tubes are cooled to room temperature, and 0.2 ml H_2O is added to each tube.

A ninhydrin solution is freshly prepared by mixing 240 mg of ninhydrin, 4.5 ml of ethyleneglycol monomethyl ether and 1.5 ml of $4 M$ acetate buffer, pH 6.5. The ninhydrin solution is activated by adding 100 μl of $0.01 M$ KCN per 4 ml of ninhydrin solution. Of the activated ninhydrin mixture, 100 μl is added to each tube, and after thorough mixing, the tubes are heated at 95° for 20 minutes. The contents of the tubes are quantitatively transferred to 10-ml volumetric flasks and made up to volume with 50% ethanol. Readings are made at 570 nm, and the amount of nitrogen is calculated on the basis of an ammonium sulfate standard.

Activity of the lectin with monovalent substances can be tested by determining the ability of the substances to inhibit precipitation.

Purification Procedure

Several workers have utilized alcohol precipitation[3] and methods based on charge[9] to purify the *Dolichos biflorus* lectin; however, purification can be achieved rapidly and specifically by the method of affinity chromatography utilizing the insoluble polyleucyl hog blood group A + H substance[10] as the immunoadsorbent. This procedure yields a very homogeneous preparation of the lectin and is described in detail below.

[9] J. Kocourek and G. A. Jamieson, *Abstr. 7th Int. Congr. Biochem.* Vol. 4, 716D-61 (1967).

[10] M. E. Kaplan and E. A. Kabat, *J. Exp. Med.* **123**, 1061 (1966).

Materials. Dolichos biflorus seeds are obtained from S. B. Penick, Co., 100 Church Street, New York, New York.

Hog A + H blood group substance is purified by ethanol precipitation[11] from a pepsin-digested hog gastric mucin preparation (obtained from Wilson Laboratories, Chicago, Illinois).

Polyleucyl hog A + H substance (PL hog A + H) is prepared by the copolymerization of blood group substance with the N-carboxyanhydride of L-leucine.[12] A 4 mg/ml solution of hog A + H blood group substance in 0.07 M $NaHCO_3$ is stirred in the cold. To this solution an equal amount of the N-carboxyanhydride of L-leucine is added very slowly with vigorous stirring. Stirring in the cold is continued for 3 days, and the insoluble PL hog A + H is centrifuged, washed with H_2O, and lyophilized.

Procedure. Dolichos biflorus seeds are ground in a mortar and suspended to a concentration of 10% (weight/volume) in 0.01 M phosphate-buffered saline, pH 7.2 (PBS). (An alternative procedure for making the 10% extract is to soak the seeds in the buffer and grind them for 2 minutes in a Waring Blendor.) The 10% suspension is either stored at 4° overnight or incubated 1 hour at 37° and centrifuged at 15,000 rpm. The supernatant (crude extract) has a hemagglutination titer of about 32–64. Quantitative precipitin analysis of the extract shows 8–9% of the total nitrogen to be lectin.[4]

A column is prepared by mixing PL hog A + H and coarse Celite in a ratio of 1:5. Approximately 200 mg of PL hog A + H are required to purify 60 mg of lectin. The PL hog A + H is quite hydrophobic and tends to stop up sintered disks of columns; it is therefore advisable to layer the bottom of the column with Celite. The flow rate can be increased by increasing the amount of Celite. The seed extract is applied to the column, and saturation of the column can be detected by testing the hemagglutination titer of the eluates with type A_1 erythrocytes.

If large amounts of the lectin are to be prepared, the PL hog A + H can be added to the seed extract and stirred in the cold overnight. This suspension is then poured into a large-diameter column layered with Celite. (A Büchner funnel with coarse sintered disk is appropriate for a column. A flow rate of 1 ml per minute can be achieved although care must be taken to prevent channeling.)

The column is washed extensively in the cold with PBS until the optical density at 280 nm of the eluate is less than 0.050.

[11] E. A. Kabat, "Blood Group Substances: Their Chemistry and Immunochemistry." Academic Press, New York, 1956.

[12] H. Tsuyuki, H. von Kley, and M. A. Stahmann, *J. Amer. Chem. Soc.* **78**, 764 (1956).

A solution of 0.01 M N-acetyl-D-galactosamine in PBS is then used to specifically elute the lectin from the column.

The lectin is concentrated and applied to a Bio-gel P-10 column to separate the sugar from the lectin.

Properties

Homogeneity. The purified *Dolichos biflorus* lectin forms a single peak in the ultracentrifuge when tested at concentrations as high as 10 mg/ml and forms a single precipitin band in immunodiffusion and immunoelectrophoresis against rabbit antisera to the crude extract. The lectin forms a single, diffuse band in acrylamide gel electrophoresis under acid and alkaline conditions. The lectin forms a single peak in isoelectrofocusing, and it is totally precipitated by human blood group A substance.[4]

Stability. The lectin is stable when stored at concentrations of 1–3 mg/ml in 0.01 M PBS, pH 7.2, at 4°C. When stored at higher concentrations, an appreciable precipitate forms with time, which appears to represent an aggregated form of the lectin. The lectin has been kept at room temperature for several days with no detectable loss of activity.

Composition. Amino acid analyses show a high concentration of aspartic acid and serine. No cysteine has been detected.[4,9] The lectin contains about 2% hexose and 1.5% N-acetylhexosamine.

Molecular Weight. The molecular weight of the lectin as determined by sedimentation velocity and viscosity measurements is 141,000.[4] The lectin dissociates in urea and sodium dodecyl sulfate to about 4 components.[13]

Isoelectric Point. The lectin has an isoelectric point of pH 4.5 as determined by isoelectrofocusing.

Specificity. The *Dolichos biflorus* lectin precipitates with blood group A_1 and A_2 substances as well as with streptococcal group C polysaccharide and desialylated ovine submaxillary mucin.[4,13,14] Each of these polysaccharides has terminal nonreducing α-linked N-acetyl-D-galactosamine residues.

Of a variety of monosaccharides tested, only N-acetyl-D-galactosamine was able to inhibit precipitation of the lectin with blood group A substance. Galactosamine hydrochloride and N-acetyl-D-glucosamine have no inhibitory activity thus showing the importance of the N-acetyl group on C-2 and the stereochemistry of the hydroxyl on C-4. The methyl-α-glycoside of N-acetyl-D-galactosamine is a better inhibitor

---

[13] W. G. Carter and M. E. Etzler, unpublished observations, 1971.
[14] G. Uhlenbruck, I. Sprenger, A. M. Leseney, J. Font, and R. Bourrillon, *Vox Sang.* 19, 488 (1970).

than the monosaccharide, and the β-ethyl glycoside gives only weak inhibition thereby showing a strong specificity of the lectin for the α-linkage.[4]

There is still some uncertainty as to the size of the combining site of the lectin. Inhibition studies with blood group A active oligosaccharides show the disaccharide (α-D-GalNAc-(1 → 3)-D-Gal) and the trisaccharide (α-D-GalNAc-(1 → 3)-β-D-Gal-(1 → 3)-D-GNAc) to be equal in inhibitory power to methyl-α-D-GalNAc. These data indicate the combining site to be no larger than an α-linked D-GalNAc; however, the blood group A active pentasaccharide

$$(\alpha\text{-D-GalNAc-}(1 → 3)\text{-}\beta\text{-D-Gal-}(1 → 4)\text{-}\beta\text{-D-GNAc-}(1 → 6)\text{-R})$$
$$\uparrow 2$$
$$| 1$$
$$\alpha\text{-L-Fuc}$$

is a better inhibitor than methyl-α-D-GalNAc.[4] Further study is necessary before conclusions can be made as to the size of the combining site.

[40] Red Kidney Bean (*Phaseolus vulgaris*) Phytohemagglutinin

By Rosalind Kornfeld, Walter T. Gregory, and Stuart A. Kornfeld

Phytohemagglutinin (PHA) preparations derived from the red kidney bean (*Phaseolus vulgaris*) have three properties: (1) They agglutinate erythrocytes without regard to blood group type; (2) they agglutinate leukocytes; and (3) they stimulate small lymphocytes to transform into blast cells and undergo mitosis. The ability of red kidney bean extracts to agglutinate red blood cells has been known since the early 1900's.

Rigas and Osgood[1] in 1955 described the purification of PHA from red kidney beans, and a preparation similar to theirs is commercially available from Difco Laboratories, Detroit, Michigan, as PHA-P. Subsequently, in 1960, Nowell[2] described the mitogenic effect of PHA on lymphocytes. Weber, Nordman, and Gräsbeck[3] then showed that PHA-P could be separated into two distinct entities—an erythroagglutinating PHA (E-PHA) and a leukoagglutinating PHA (L-PHA), both of which possess mitogenic activity.

The procedure described here for the purification of E-PHA is based

[1] D. A. Rigas and E. E. Osgood, *J. Biol. Chem.* **212**, 607 (1955).
[2] P. C. Nowell, *Cancer Res.* **20**, 462 (1960).
[3] T. Weber, C. T. Nordman, and R. Gräsbeck, *Scand. J. Haematol.* **4**, 77 (1967).

on the method of Weber *et al.*[3] with additions and modifications developed in our laboratory. In the course of purification one also obtains in other fractions the L-PHA.

Assay Method

Principle. Activity of E-PHA is determined by measuring its ability to agglutinate human erythrocytes.

Reagents

Human erythrocytes 4% suspension, thrice washed in saline–bicarbonate buffer

Saline–bicarbonate buffer–0.9% NaCl in 0.01 M NaHCO$_3$

Procedure. On either a glass microscope slide or, preferably, a glass plate, which can accommodate more spots, sample and reagents are deposited in the following order to form a droplet of 40 μl: (1) an aliquot of the E-PHA solution, diluted in saline–bicarbonate to approximately 0.1 mg of protein per milliliter; (2) saline–bicarbonate buffer—enough to bring the volume of the drop to 25 μl; (3) 15 μl of 4% suspension of erythrocytes. The droplet is mixed with the tip of a small glass stirring rod to form a spot approximately 1 cm in diameter. Generally, a series of such spots are laid out containing increasing amounts of the E-PHA (0, 2 μl, 5 μl, 10 μl, etc.). The rod is wiped between each mixing, and then the glass plate is tipped and tilted to make the red cells in each spot rotate slowly. The degree of agglutination is scored from 1+ (barely discernible agglutination) to 4+ (all red cells clumped strongly) over a period of 3 minutes.

Definition of Unit and Specific Activity. One hemagglutinating unit (HU) is that amount of E-PHA which causes 4+ agglutination in 3 minutes in the assay. Specific activity is expressed as HU per milligram of protein as measured by the method of Lowry *et al.*[4]

Purification Procedure

All procedures were carried out at 0–4°.

Step 1. Dialysis of Difco PHA-P. Three vials of lyophilized PHA-P from Difco Laboratories, Detroit, Michigan, are dissolved in 5 ml of 0.066 M phosphate buffer pH 4.5 and dialyzed against the same buffer overnight.

Step 2. SE-Sephadex Chromatography. The dialyzed PHA-P is loaded on a column (1.5 × 16 cm) of SE-Sephadex-C50 equilibrated with

[4] O. H. Lowry, N. J. Rosebrough, A. L. Farr, and R. J. Randall, *J. Biol. Chem.* **193**, 265 (1951).

0.066 M phosphate buffer, pH 4.5. The column is then eluted batchwise with (1) 60 ml of equilibrating buffer, which elutes an inactive protein peak; (2) 60 ml of 0.066 M phosphate buffer pH 6.0, which elutes leuko-agglutinating PHA; and then (3) 60 ml of 0.066 M phosphate buffer pH 8.0, which elutes the erythroagglutinating PHA.

Fractions of about 3 ml are collected in an automatic fraction collector and the absorption at 280 nm measured to locate the protein peaks. Fractions comprising the third peak are pooled and concentrated to about 5 ml with an Amicon pressure concentrator using a UM-10 membrane filter. A small amount of 280 nm absorbing material passes the membrane filter, but all the hemagglutinating activity is recovered in the concentrate.

Step 3. Sephadex G-150 Chromatography. The concentrate from the previous step is loaded on a Sephadex G-150 column (2.5 × 77 cm) in 0.9% NaCl–0.01 M NaHCO$_3$, the column is eluted with 0.9% NaCl–0.01 M NaHCO$_3$, and fractions of 4.9 ml are collected. The fractions containing E-PHA activity are pooled. Often this step results in material that gives a single broad band on disc gel electrophoresis. In some preparations minor bands are still seen, and one may need to carry out the phosphocellulose chromatography described below.

Step 3 Alternate or Step 4. Phosphocellulose Chromatography. Usually either Sephadex G-150 chromatography or phosphocellulose chromatography will suffice to remove contaminants still present in the SE-Sephadex

TABLE I

PURIFICATION OF E-PHA

Step	Volume (ml)	Protein (mg)	Activity (HU)	Recovery (%)	Specific activity (HU/mg)
1. Dialyzed PHA-P	5.69	260	284,000	100	1090
2. SE-Sephadex					
Peak I	6.6	25.1	0	—	—
Peak II	15.8	114	31,600	11.1	—
Peak III	13.0	58.5	130,000	45.8	2220
3.[a] Sephadex G-150	30.5	18.3	61,000	21.5	3330
3.[a] Alternate phosphocel- lulose	15	23.5	77,800	27.4	3310

[a] Material from peak III step 2 was divided; a portion was run over Sephadex G-150 and a portion over phosphocellulose. The numbers have been adjusted as if the entire peak III had been subjected to either one or the other procedure so that they may be compared directly.

material from step 2. In Table I the results of a preparation in which the step 2 material was divided and a portion put over Sephadex G-150 and another portion over phosphocellulose are shown for comparison. The sample is dialyzed against $0.02\,M$ phosphate buffer pH 5.0 and loaded on a Whatman P11 phosphocellulose column (1×24 cm) equilibrated in the same buffer. The column is washed with 10 ml of that buffer and elution is achieved with a linear gradient composed of 50 ml of $0.02\,M$ phosphate, pH 5 in the mixing chamber and 50 ml of $0.2\,M$ phosphate pH 8 in the reservoir. The peak of E-PHA emerges when the phosphate concentration reaches approximately $0.1\,M$ and the pH is about 6.5.

The procedure for purification is shown in Table I. In the preparation shown, fractions were always pooled only from the center of the peaks with the object of achieving the highest purity, as judged by the absence of contaminating bands on disc gel electrophoresis, rather than the best yield.

Allen, Svenson, and Yachnin[5] have also described a procedure for separating PHA-P into a low hemagglutinating mitogen (L-PHAP) and a high hemagglutinating mitogen (H-PHAP). Their H-PHAP is apparently the same material here called erythroagglutinating PHA.

Preparation of [131]*I-Labeled E-PHA*

By the chloramine-T method,[6] with very brief exposure, E-PHA can be iodinated with ^{131}I and retain full activity. To 1.0 ml of $0.1\,M$ phosphate buffer, pH 7.5 containing E-PHA (0.1–2 mg) and biochemical grade $Na^{131}I$ (50–100 μCi) is added 0.04 ml of chloramine-T (4 mg/ml); the preparation is mixed for 5–10 seconds and then quickly 0.16 ml of sodium metabisulfite (2.4 mg/ml) and 0.2 ml of KI (10 mg/ml) are added. The reaction mixture is passed over a column of Sephadex G-25 ($V_0 = 5.0$ ml) and the ^{131}I-E-PHA is recovered in the excluded volume.

Properties

The E-PHA of red kidney beans is a glycoprotein of about 128,000 molecular weight that has its isoelectric point near pH 6.5.[7]

Binding Affinity. E-PHA binds to both erythrocytes and lymphocytes and a variety of other cell types as well. Using ^{131}I-E-PHA the associa-

[5] L. W. Allen, R. H. Svenson, and S. Yachnin, *Proc. Nat. Acad. Sci. U.S.* **63**, 334 (1969).

[6] W. M. Hunter, *in* "Handbook of Experimental Immunology" (D. M. Weir, ed.), p. 608. Blackwell, Oxford, 1967.

[7] T. H. Weber, *Scand. J. Clin. Lab. Invest.*, Suppl. 111, 8 and 26 (1969).

TABLE II
BINDING PROPERTIES OF E-PHA

Cell type	Receptor sites/cell	Association constant (K)
Normal human lymphocytes	2.7×10^6	$1.1 \times 10^7\ M^{-1}$
Chronic lymphocytic leukemic lymphocytes	1.15×10^6	$1.3 \times 10^7\ M^{-1}$
Human erythrocytes	5.2×10^5	$5.7 \times 10^6\ M^{-1}$
Isolated erythrocyte glycopeptide	—	$4.7 \times 10^3\ M^{-1}$

tion constant for binding and the number of binding sites per cell has been determined for the cell types shown in Table II.[8] Also shown is the association constant of E-PHA for a glycopeptide isolated from the surface of erythrocytes.

E-PHA Receptor Site on Erythrocytes. Trypsin treatment of erythrocytes decreases the number of E-PHA binding sites in the cells by about 40% and releases a glycopeptide of about 10,000 molecular weight into the medium. This glycopeptide has the property of inhibiting agglutination of erythrocytes by E-PHA in the usual hemagglutination assay and presumably contains E-PHA receptor sites. Following alkaline borohydride treatment and Pronase digestion, the small glycopeptide referred to in Table II was obtained.[9] It binds specifically to E-PHA with the affinity shown and also is a potent inhibitor of E-PHA-induced red cell agglutination. It consists of a branched-chain oligosaccharide linked *N*-glycosidically to asparagine in the peptide portion and was shown to

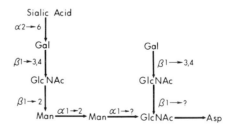

FIG. 1. Structure of the erythrocyte glycopeptide with E-PHA receptor activity. Gal = galactose; GlcNAc = *N*-acetylglucosamine; Man = mannose; Asp = asparagine.

[8] S. Kornfeld and R. Kornfeld, *in* "Glycoproteins of Blood Cells and Plasma" (G. A. Jamieson and T. J. Greenwalt, eds.), p. 50. Lippincott, Philadelphia, Pennsylvania, 1971.

[9] R. Kornfeld and S. Kornfeld, *J. Biol. Chem.* **245**, 2536 (1970).

have the structure depicted in Fig. 1. Removal of the sialic acid residue does not affect the activity of the glycopeptide, but removal of the galactose residues with β-galactosidase abolishes its activity. The galactose residues are essential but not sufficient to confer activity since oligosaccharides which posses the galactose $\xrightarrow{\beta}$ N-acetylglucosamine linkage are not good inhibitors of E-PHA-induced red cell agglutination. The mannose residues in the core region of the glycopeptide appear to be involved in the binding of E-PHA. For example, the glycopeptide of fetuin and the glycopeptide of γG-immunoglobulin both of which contain oligosaccharide chains with galactose $\rightarrow$ N-acetylglucosamine sequences in the periphery and mannose residues in the core region are relatively good inhibitors.[8]

Mitogenicity. As mentioned before erythroagglutinating PHA also stimulates lymphocytes to undergo mitosis. The mitogenic potency of E-PHA can be determined by measuring incorporation of [³H]thymidine into DNA by lymphocytes after 3 days of culture in the presence of E-PHA.[10] Such measurements indicate that E-PHA can stimulate the rate of DNA synthesis 25- to 50-fold over the rate in resting lymphocytes. The concentration of E-PHA required to achieve half-maximal stimulation of DNA synthesis varies from 4 to 8 μg/ml (3 to $6 \times 10^{-8} M$). The red cell glycopeptide is also a potent inhibitor of E-PHA-induced DNA synthesis in lymphocytes.

[10] S. Kornfeld and R. Kornfeld, *Proc. Nat. Acad. Sci. U.S.* **63**, 1439 (1969).

[41] *Sophora japonica* Hemagglutinin

By R. D. PORETZ

The hemagglutinin (lectin) from *Sophora japonica* is capable of agglutinating type A and B human red blood cells[1] as well as precipitating with A- and B-active blood group glycoproteins. These interactions are inhibited by a variety of saccharides related to D-galactose.

Assay Method

The lectin from *Sophora japonica* may be quantitated on a relative scale by hemagglutination titration or in absolute terms by the quantitative precipitin reaction.

[1] W. T. J. Morgan and W. M. Watkins, *Brit. J. Exp. Pathol.* **34**, 94 (1953).

Hemagglutination

Principle. The lectin is capable of specifically causing the agglutination of human red blood cells from type B and, to a lesser degree, from type A individuals.

Procedure. Twofold serial dilutions of 0.025 ml of lectin solution are made in a microtitration plate employing microtiter equipment (Cooke Engineering Co.) and PBS[2] as diluent. The last well should contain only PBS. To each well is added 0.025 ml of a 2% suspension of type B human red blood cells in PBS. After incubation of 30 minutes at room temperature, slightly agitated aliquots are removed (with a Pasteur pipette) from the reaction mixture and examined microscopically for agglutination. The degree of agglutination is scored from trace to 3+. The red blood cells in the well containing no lectin should display no agglutination. The potency of the lectin preparation is taken as the concentration of protein (μg/ml) required for trace agglutination of the cells. Accurate comparison of the potency of various lectin preparations may be made only when the titrations of these preparations are conducted simultaneously with the same red blood cell preparation.

Quantitative Precipitation

Principle. The lectin is capable of being specifically and quantitatively precipitated by B-active human blood group substance; this reaction displays equivalance curve characteristics.

The amount of protein maximally precipitated represents the quantity of lectin in the preparation.

Reagents

PBS[2]

B-active human blood group substance dissolved in PBS (400 μg/ml)

Lowry reagent C: 1 ml of 0.5% $CuSO_4 \cdot 5H_2O$ in 1% sodium tartrate, added to 50 ml of 2% sodium carbonate in 0.1 N sodium hydroxide

Phenol reagent diluted: 2 N Folin–Ciocalteau phenol reagent diluted with an equal volume of water

Procedure. Step 1. To a series of 3-ml conical centrifuge tubes are added increasing amounts of B-active human blood group substance (0–100 μg) dissolved in PBS; the volume in each tube is brought to 0.3 ml with PBS. Exactly 0.2 ml of test solution containing 100–400 μg of

---

[2] PBS, sodium phosphate buffer (pH 7.8, 0.02 M) containing 0.14 M sodium chloride.

lectin per milliliter is added to each tube and the contents are mixed with a Vortex mixer. The tubes are covered with Parafilm and incubated for 1 hour at room temperature and 24 hours at 4°. The reaction mixtures are centrifuged at 4° and 800 g for 20 minutes (International Centrifuge Model PR-2 head No. 259, cup No. 384, adaptor No. 2740), the supernatants are removed, and the precipitates are resuspended (agitation by a Vortex mixer) in 0.5 ml of cold PBS. The tubes are centrifuged, and the supernatants are discarded. The wash procedure is repeated for a total of three times. After removal of the last supernatant the tubes are inverted on absorbent paper to drain.

Step 2. The protein content of the precipitates is determined by the method of Mage and Dray.[3] The drained precipitates are dissolved by the addition to each tube of 0.1 ml of water and 0.3 ml of 0.05 M NaOH. This is followed by the addition of 1.5 ml of Lowry reagent C. The reaction mixtures are mixed with a Vortex mixer and allowed to stand. After 20 minutes, 0.15 ml of diluted phenol reagent is added, the tubes are agitated and allowed to stand for 30 minutes. The blue color is quantitated at 700 nm. A standard curve of at least 5 concentrations of bovine serum albumin (0–100 μg) in a volume of 0.1 ml is prepared in a parallel manner. The protein content is calculated for each precipitate, and the protein content of the tube containing no blood group substance is subtracted from these values. An equivalance curve is constructed by plotting the quantity of protein precipitated in each tube with respect to the amount of blood group substance added and a smooth curve is drawn through the points. The maximum protein precipitated represents the total lectin present in 0.2 ml of test solution.

Purification Procedure

Step 1. Crude Extract. Sophora japonica seeds (100 g, F. W. Schumacher, Sandwich, Massachusetts) are freshly ground to a coarse consistency in a small bottle of an Oster blender and suspended by gentle stirring in 1 liter of PBS at 4°. After 15 hours the cloudy brown suspension is filtered through a double layer of cheesecloth, the retentate is washed with 100 ml of PBS, and the filtrate is centrifuged at 15,000 g and 4° for 45 minutes. The supernatant, fraction F-1 is retained and the sediment is discarded.

Step 2. Ethanol Fractionation. Precipitation of the lectin by ethanol is conducted at −20°C by the slow addition of 95% ethanol to gently stirred fraction F-1 so that the final concentration of ethanol is 50%; that is, the volume of ethanol added is 1.1 times the volume of fraction

---

[3] R. Mage and S. Dray, *J. Immunol.* **95**, 525 (1965).

F-1. After complete addition of ethanol the suspension is stirred for 1 hour and centrifuged at 15,000 g and 4° for 45 minutes. The supernatant, fraction F-2 is decanted and the precipitate is suspended in 200 ml of PBS. This is dialyzed against 3 × 4 liters of PBS. Brown dialyzable material is noted. The dialyzed suspension is centrifuged at 25,000 g and 4° for 1 hour, and the supernatant is carefully filtered through a glass wool plug, yielding fraction F-3.

Step 3. Specific Adsorption of the Lectin. To 1.0 g of insoluble poly-leucyl hog gastric mucin (PLHGM) is added fraction F-3, and this suspension is stirred at 4° for 3–15 hours. The suspension is centrifuged at 4° and 25,000 g for 1 hour, and the supernatant is carefully decanted (fraction F-4). The precipitate is suspended in 100 ml of cold PBS for 15 minutes and centrifuged, and the supernatant is removed. This is repeated (2 or 3 times) until the OD_{280} of the supernatant is below 0.05 unit. The sediment is suspended in the cold with 100 ml of cold PBS containing 2% D-galactose. After 1 hour the suspension is centrifuged at 4° and 25,000 g for 1 hour and the supernatant is carefully decanted through a glass wool plug. Exhaustive dialysis of the supernatant yields the purified lectin as fraction F-5. The results of the purification are shown in the table.

Preparation of Purified Hog Gastric Mucin. Impure A-active hog gastric mucin is obtained from Wilson laboratories as granular mucin type 1701-W. This is partially purified by the procedure of Morgan and King.[4] Crude hog gastric mucin (50 g) is ground in a mortar with 1 liter of 90% aqueous phenol, first producing a paste and then a thick sus-

CHARACTERISTICS OF FRACTIONS FROM THE PURIFICATION OF THE
Sophora japonica HEMAGGLUTININ

Fraction	Total protein[a]	HA[b]	Functional activity[c]	Percent functional purity[d]
F-1	2400	12	200	8.4
F-2	735	>1200	0	—
F-3	1145	7.5	195	17
F-4	795	670	<6	<1
F-5	149	1.1	144	97

[a] As milligrams of bovine serum albumin.

[b] Agglutinin activity as minimum concentration (μg/ml) of protein required to agglutinate type B red blood cells.

[c] Maximum protein precipitated with blood group substance (BGS).

[d] $\dfrac{\text{Maximum protein precipitated with B-active BGS}}{\text{total protein}} \times 100.$

[4] W. T. J. Morgan and H. K. King, *Biochem. J.* **37**, 640 (1943).

pension. This suspension is stirred for 36 hours at room temperature and centrifuged at 1000 g for 3 hours. The supernatant (super 1) is decanted, and the residue is resuspended in 250 ml of fresh 90% phenol. After 18 hours the suspension is centrifuged at 1000 g for 3 hours and the supernatant removed and combined with super 1. To the well stirred combined supernatants is slowly added enough ethanol–phenol solution (1:1) so that the final concentration of ethanol is 10%. The suspension is stirred for 18 hours, centrifuged at 1000 g for 2 hours, and the supernatant is decanted. Additional ethanol–phenol mixture (1:1) is added to the stirred supernatant so as to bring the ethanol concentration to 15%. This suspension is stirred for 18 hours and centrifuged at 1000 g for 2 hours; the supernatant is removed. The residue is triturated with 200 ml of 95% ethanol and centrifuged at 1000 g for 30 minutes. The supernatant is removed, and the trituration of the precipitate is twice repeated. The precipitate is then dissolved in 1 liter of water, exhaustively dialyzed against water, and lyophilized, yielding 19.5 g of white powder.

Preparation of Polyleucyl Hog Gastric Mucin. The method of Kaplan and Kabat[5] is described. To a cold (ice bath) well stirred water solution (500 ml) containing 2.0 g of purified hog gastric mucin and 2.94 g sodium bicarbonate is slowly added 2.0 g of N-carboxy L-leucine anhydride (New England Nuclear). The suspension is stirred for an additional 18 hours at 4° than centrifuged at 25,000 g for 1 hour. The supernatant is decanted, and the residue is resuspended in 500 ml of 0.07 M sodium bicarbonate and centrifuged. This is repeated twice, followed by 4 washes with water. A suspension of the washed precipitate is lyophilized yielding 2.1 g of insoluble PLHGM. Phenol–sulfuric analysis of the acid-hydrolyzed PLHGM indicated it contained 52% hog gastric mucin and presumably 48% leucyl residues.

Properties of the Purified Lectin

Purity and Physiochemical Characterization. Specific precipitation of various preparations of the purified protein with B-active blood group substance indicated a functional purity greater than 97%. The lectin is homogeneous in polyacrylamide gel disc electrophoresis at pH 4.4. However, multiple forms in equilibrium with each other may be detected during electrophoresis at pH 8.4.[6] The lectin displays a pI of 5.47, a molecular weight of 135,000 g/mole, and a single SDS subunit of 33,000 g/mole.[7]

[5] M. E. Kaplan and E. A. Kabat, *J. Exp. Med.* **123**, 1961 (1966).
[6] R. D. Poretz and P. Niu, unpublished results.
[7] J. Timberlake and R. D. Poretz, in preparation.

Complete activity is retained by the protein after lyophilization or when stored at $-10°$.

Specificity.[8] The lectin is capable of readily precipitating from solution with all B-active and A-active, but no H-active human blood group substances examined. The protein does not precipitate with various galactose-containing plant gums, including gum arabic, karaya gum, gum tragacanth, locust bean gum, and guar gum, or with BSA-diazophenyl β-galactoside and lactoside conjugates. However, it can bind to insolubilized A-active human blood group substance and hog gastric mucin as well as Sepharose-β-galactoside adsorbents. Competitive (haptenlike) inhibitors of the lectin in the order of decreasing potency are: N-acetyl-D-galactosamine, lactose, methyl β-D-galactopyranoside, methyl α-D-galactopyranoside, melibiose, D-galactose, D-fucose, and L-arabinose.

[8] S. Chien and R. D. Poretz, unpublished results.

[42] Wheat Germ (*Triticum vulgaris*) Agglutinin

By V. T. MARCHESI

Aub and co-workers first reported that crude wheat germ lipase preparations selectively agglutinated certain types of tumor cells.[1] The active principle of this preparation was subsequently isolated and partially characterized by Burger and co-workers,[2] and it has been used extensively to study the surface properties of isolated cells. The molecular properties of the receptors on cell membranes that react with this agglutinin are not yet known, but N-acetylglucosamine is probably one of the determining sugars since this hexosamine or polymers of it (e.g., chitobiose) inhibit the binding of the agglutinin to cell membranes or to isolated membrane glycoproteins.[2,3] A simple method for the purification of this agglutinin is described here using ovomucoid-Sepharose as an affinity chromatography system with dilute acetic acid as the eluting solvent.

Preparation of Ovomucoid-Sepharose Conjugate

Ovomucoid (Miles Laboratories) was conjugated to Sepharose 4-B (Pharmacia) according to the procedure described by Cuatrecasas.[4]

[1] J. C. Aub, C. Tieslau, and A. Lankester, *Proc. Nat. Acad. Sci. U.S.* **50**, 613 (1963).
[2] M. M. Burger and A. R. Goldberg, *Proc. Nat. Acad. Sci. U.S.* **57**, 359 (1967).
[3] V. T. Marchesi and E. P. Andrews, *Science* **174**, 1247 (1971).
[4] P. Cuatrecasas, *J. Biol. Chem.* **245**, 3059 (1970).

Eight hundred milligrams of ovomucoid was dissolved in 40.0 ml of 0.2 M sodium bicarbonate, pH 8.6, and added to 80 ml of Sepharose 4-B beads immediately after the beads were activated by 24 g of cyanogen bromide (Eastman) at pH 11–11.5. After stirring at 4° for 24 hours, the Sepharose beads were washed successively with 0.2 M sodium chloride, 6 M guanidine·HCl, and 0.1 N acetic acid. Eighty percent of the ovomucoid remained bound to the Sepharose after these washes. The beads were stored at 4° in 0.1 M sodium chloride containing 0.025% sodium azide without apparent loss of binding capacity.

Preparation of Crude Wheat Germ Agglutinin

Crude "lipase" preparations were obtained from Miles Laboratories in the form of a dry powder. This was dissolved either directly in 0.25 M sodium chloride in 0.05 M phosphate, pH 7.0, and applied to the affinity column or dissolved first in distilled water and heated to 58° for 10 minutes as suggested previously.[2] After heating, the suspension was chilled to 4° and centrifuged, and the precipitate was discarded. The supernatant was filtered through glass wool, and then an equal volume of

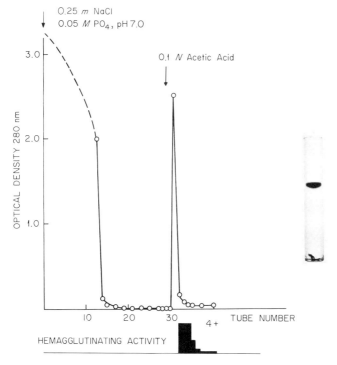

FIG. 1. Elution profile of agglutinating activity of wheat germ lipase extract.

0.5 M sodium chloride in 0.1 M phosphate pH 7.0 was added to raise the salt concentration to 0.25 M. This was run onto the ovomucoid–Sepharose column equilibrated with 0.25 M NaCl–0.05 M PO$_4$. Although it is simpler to bypass the heat-precipitation step, we have found that heating precipitates lipids and other contaminants which otherwise bind tightly to the ovomucoid-Sepharose columns and limit their use. Columns which are exposed only to heat-treated wheat germ extracts can be regenerated by washing with 8 M urea and 0.25 M sodium chloride and used repeatedly without apparent loss in binding capacity.

Procedure

An extract obtained from 10 g of crude wheat germ lipase can be applied to a 1.0 × 10.0 cm ovomucoid–Sepharose column in 0.25 M NaCl–0.05 M PO$_4$ pH 7.0 at room temperature. The same buffer solution is used to wash the column until the optical density (280 nm) of the effluent is 0.020 or lower. Elution of agglutinating activity is then achieved by adding 0.1 N acetic acid. A typical elution profile is illustrated in Fig. 1. Tubes containing agglutinating activity are pooled, dialyzed against distilled water or 0.01 N acetic acid at 4° for 18 hours and then freeze-dried. The yield from 10 g of crude lipase is approximately 40–45 mg of freeze-dried agglutinin. This material migrates as a single polypeptide chain when the reduced and alkylated form is electrophoresed on acrylamide gels containing sodium dodecyl sulfate (0.1%) and 0.1 M phosphate, pH 7.0 (Fig. 1-B). The native molecule (not reduced or alkylated) forms a clear, colorless solution in water and is bound quantitatively by human red blood cells or soluble glycoprotein isolated from such cells.

[43] *Lotus tetragonolobus* L-Fucose-Binding Proteins

By JOSEPH YARIV, A. JOSEPH KALB, and SHMARYAHU BLUMBERG

An extract of *Lotus tetragonolobus* seeds has been used in the study of blood group specificity.[1] High specificity toward L-fucose of the hemaglutinating factor in this extract was established.[1-3] Extracts of other seeds which agglutinate human O red blood cells and which are

[1] W. T. J. Morgan and W. M. Watkins, *Brit. J. Exp. Pathol.* **34**, 94 (1953).

[2] O. Mäkelä, Academic Dissertation, University of Helsinki, 1957.

[3] G. F. Springer and P. Williamson, *Biochem. J.* **85**, 282 (1962).

inhibited preferentially by L-fucose are those of *Ulex europaeus*, *Ulex nanus*, and *Lotus siliquosus*.[2]

Isolation Procedure

Principle. L-Fucose-binding protein is adsorbed by specific interaction with a resin prepared by coupling an L-fucosyl derivative to Sepharose.[4] The three L-fucose-binding proteins of *L. tetragonolobus* are eluted with a buffer containing L-fucose and are resolved by chromatography on DEAE-cellulose.[5]

Preparation of Specific Adsorbent. β-L-Fucopyranosylamine[6] is prepared by keeping a solution of L-fucose (150 mmoles) in 200 ml of methanolic ammonia (30% ammonia by volume) until a heavy precipitate is formed (8 days, room temperature). Yield of filtered and washed product is 55%; m.p. 146–147° (decomp); $[\alpha]_D^{29}$ −50° (ca. 1 in H_2O).

N-(*N*-Benzyloxycarbonyl-ε-aminocaproyl)-β-L-fucopyranosylamine is prepared as follows. Isobutylchloroformate (30 mmoles) and triethylamine (30 mmoles) are added to a solution of *N*-benzyloxycarbonyl-ε-aminocaproic acid (30 mmoles) in dimethyl formamide (50 ml), stirred for 20 minutes at −5°, and filtered. The filtrate is added immediately to a suspension of β-L-fucopyranosylamine (25 mmoles) in dimethyl formamide (50 ml), and the mixture is stirred at room temperature until the solution is clear (1 hour). The solution is allowed to stand at room temperature for 16 hours and the solvent is removed *in vacuo*. The residue is crystallized twice from methanol–ether. Yield 70%; m.p. 169–170°; $[\alpha]_D^{29}$ +1.5° (ca. 2 in H_2O).

N-(*N*-Benzyloxycarbonyl-ε-aminocaproyl)-β-L-fucopyranosylamine (10 mmoles) is hydrogenated over palladium on charcoal (10%, 100 mg) in aqueous methanol (100 ml of 80% methanol, by volume) at atmospheric pressure for 8 hours to give *N*-(ε-aminocaproyl)-β-L-fucopyranosylamine. The product is crystallized from methanol–ether. Yield 85%; m.p. 170–171°; $[\alpha]_D^{29}$ +0.5° (ca. 2 in H_2O).

Sepharose 4B (Pharmacia, Uppsala) is activated with cyanogen bromide and then allowed to react with *N*-(ε-aminocaproyl)-β-L-fucopyranosylamine by the general method of Axén *et al.*[7] Cyanogen bromide (7 g) is stirred with water (70 ml) for 10 minutes, until most of it goes into solution. A Sepharose slurry (70 ml), washed with water, is then

[4] S. Blumberg, J. Hildesheim, J. Yariv, and K. J. Wilson, *Biochim. Biophys. Acta* **264**, 171 (1972).

[5] A. J. Kalb, *Biochim. Biophys. Acta* **168**, 532 (1968).

[6] E. W. Thomas, *J. Med. Chem.* **13**, 755 (1970).

[7] R. Axén, J. Porath, and S. Ernback, *Nature* (*London*) **214**, 1302 (1967).

added to the cyanogen bromide–water mixture. The pH of the reaction mixture is adjusted to 11 and maintained at this pH for 6 minutes by adding $4 N$ NaOH. The reaction mixture is kept below 26° by addition of crushed ice. The activated Sepharose is washed rapidly with 15 volumes of cold water on a Büchner funnel. The wet Sepharose is then added quickly to a solution of the N-(ϵ-aminocaproyl)-β-L-fucopyranosylamine (840 μmoles in 30 ml 0.5 M NaHCO$_3$) and the mixture is stirred gently for 16 hours at 4°. The resin is filtered and washed thoroughly with 0.1 M NaHCO$_3$ and water.

The amount of L-fucose derivative covalently bound to the Sepharose is determined by hydrolysis of the resin in $6 N$ hydrochloric acid (22 hours, 110°) and estimation of the liberated ϵ-aminocaproic acid on a Beckman-Spinco amino acid analyzer. In five successive preparations the fucosyl resin was found to contain about 2 μmoles of covalently bound N-(ϵ-aminocaproyl)-β-L-fucopyranosylamine groups per milliliter. The resin is stored at 4° and can be used repeatedly.

Preparation of Seed Extract. One hundred grams of freshly milled seed (marketed as 'Asparagus pea' by Thompson and Morgan, Ltd., Ipswich) is extracted with a cold salt solution (2–3 l of 0.85% sodium chloride, 0.02 M sodium phosphate, pH 6.8) by stirring for 3 minutes in a homogenizer immersed in an ice bath. The clarified extract is fractionated by addition of solid ammonium sulfate, to 30% saturation (4°, overnight). The supernatant solution is brought to 60% saturation (4°, overnight) and the precipitate is collected and dispersed in 40 ml of 0.05 M sodium phosphate buffer pH 6.8, and dialyzed extensively against many changes of same buffer.

Affinity Chromatography. A volume of extract which corresponds to 50 g of seed is applied to a column of the L-fucosyl resin (1.3 × 25 cm) equilibrated with 0.05 M sodium phosphate, pH 6.8, at 4° (flow rate 1 ml/minute). The column is washed with buffer until no 280 nm absorbing material is detected in the effluent (approximately 200 ml). Elution of protein is carried out with 40 mM L-fucose in the same buffer. Protein (50 mg) emerges in a single peak after the appearance of L-fucose in the effluent. It is dialyzed in the cold against many changes of buffer until free of L-fucose and is stored frozen at −20°. The protein can be concentrated by ammonium sulfate precipitation, pressure dialysis, or lyophilization.

DEAE-Cellulose Chromatography. Chromatographic separation of the L-fucose-binding proteins in 0.01 M sodium phosphate (pH 7.6) is accomplished on a DEAE-cellulose column (capacity, 0.69 meq/g) equilibrated with the same buffer at 3°. Protein A is eluted with the starting buffer and proteins B and C are then eluted with a linear gradient of NaCl in the same buffer. Electrophoretically pure fractions are pooled and concentrated.

Characterization of the Separated Proteins

Electrophoresis. The separated proteins, A, B, and C, are analyzed for purity by electrophoresis on cellulose acetate in a Beckman Microzone model R-100 electrophoresis system. The order of mobility from cathode to anode is A, B, C. The supporting electrolyte is $0.02\,M$ sodium phosphate, pH 6.8. A constant voltage of 400 V is applied for 20 minutes, and the strip is developed with 0.002% nigrosin in 2% acetic acid.

Extinction Coefficients. Extinction coefficients, $E_{1\,cm}^{1\%}$ (280 nm) in $0.05\,M$ sodium phosphate, pH 6.8, are for A, 17.8; for B, 20.9; and for C, 17.4.

Molecular Weights. Weight-average molecular weights of the proteins in $0.05\,M$ sodium phosphate (pH 6.8) as determined by sedimentation-equilibrium ($\bar{V} = 0.75$, assumed) are for A, 120,000; for B, 58,000; and for C, 117,000.

Binding of L-*Fucose.* Binding of L-fucose is determined by the equilibrium dialysis method, using L-fucose labeled with carbon-14 (0.41 mC/mmole, Calbiochem, Los Angeles). In each dialysis cell (Technilab Instruments, Los Angeles) one compartment is filled with 1 ml of protein solution (1–3 mg/ml in $0.05\,M$ sodium phosphate buffer, pH 6.8), and the other with 1 ml of buffer. The membrane is made of Visking dialysis tubing. Labeled L-fucose solution (0.5–10 μl) is added to the protein or to the buffer. After 20 hours of intermittent stirring on a rotating device at 4°, duplicate samples of 100 μl from both compartments of the dialysis cell are transferred to 15 ml of Bray's solution for determination of counting rates (Packard TriCarb scintillation counter). Counting rates of a blank and of a standard solution are determined as well. The concentration of a stock solution of labeled L-fucose is determined by the method of Park and Johnson[8] or by any alternative colorimetric procedure and the specific counting rate determined. The concentration of protein is determined by absorption at 280 nm.

The results of binding experiments are plotted according to Scatchard,[9]

$$r/f = -Kr + NK$$

where r is the number of moles of L-fucose bound per gram of protein, f is the molar concentration of free L-fucose, K is the intrinsic binding constant, and N is the number of moles of binding sites per gram of

[8] J. T. Park and M. J. Johnson, *J. Biol. Chem.* **181**, 149 (1949).
[9] G. Scatchard, *Ann. N.Y. Acad. Sci.* **51**, 660 (1949).

protein at saturation. The equivalent weight of protein which binds 1 mole of L-fucose at saturation is therefore $1/N$.

The association constants, K, at 4° for binding of L-fucose by protein A is 0.9 to $1.2 \times 10^4 M^{-1}$, by protein B is 0.5 to $0.6 \times 10^4 M^{-1}$, and by protein C is 2.1 to $3.7 \times 10^4 M^{-1}$. The equivalent binding weights are: A, 2.8 to 3.1×10^4 g; B, 2.9 to 3.1×10^4 g; and C, 3.0 to 3.2×10^4 g.

[44] Soy Bean (*Glycine max*) Agglutinin

By HALINA LIS and NATHAN SHARON

The ability of soybean (*Glycine max*) seeds to strongly agglutinate erythrocytes has been known since the beginning of the century.[1] The protein responsible for this hemagglutinating activity was first isolated in purified form and characterized by Liener and his co-workers,[2-4] who named it soyin and subsequently renamed it soybean hemagglutinin.[4] More recently, with the finding that this protein also agglutinates cells other than erythrocytes, its name was changed to soybean agglutinin (SBA).[5] SBA is one of many cell agglutinating proteins, or lectins, which are widely distributed in nature and which are finding increasing use in investigations of the structure of the cell surface and of carbohydrate-containing biopolymers.[6]

Assay Method

Principle. The activity of soybean agglutinin is assayed by measuring its ability to agglutinate trypsinized rabbit erythrocytes. This can be done by the conventional procedure of serial 2-fold dilution, with visual estimation of the degree of agglutination, either in test tubes[7] or with the Takásy microtitrator,[8] which provides a method for making rapid, accurate dilutions using minute quantities of material. However, the

[1] G. C. Toms and A. Western, *in* "Chemotaxonomy of the Leguminosae" (J. Harborne, D. Boulter and B. L. Turner, eds.), p. 369. Academic Press, New York, 1971.

[2] I. E. Liener and M. J. Pallansch, *J. Biol. Chem.* **197**, 29 (1952).

[3] I. E. Liener, *J. Nutr.* **49**, 527 (1953).

[4] S. Wada, M. J. Pallansch, and I. E. Liener, *J. Biol. Chem.* **233**, 395 (1958).

[5] H. Lis, B. A. Sela, L. Sachs, and N. Sharon, *Biochim. Biophys. Acta* **211**, 582 (1970).

[6] N. Sharon and H. Lis, *Science* **177**, 949 (1972).

[7] E. A. Kabat and M. M. Mayer, "Experimental Immunochemistry," 2nd ed., pp. 114–115. Thomas, Springfield, Illinois (1961).

[8] J. L. Sever, *J. Immunol.* **88**, 320 (1962).

visual method is not very precise and does not permit the detection of small differences in hemagglutinating activity. A quantitative procedure has been developed by Liener,[9] in which the degree of agglutination is evaluated photometrically, by measuring the absorbance of the layer of unsedimented erythrocytes. This method is described below.

Reagents

Saline: 0.9% solution of NaCl

Phosphate-buffered saline (PBS): 0.006 M K Na phosphate buffer, pH 7.4, in saline

Alsever's solution: 2.05 g of glucose, 0.8 g of sodium citrate, and 0.42 g of NaCl, dissolved in 100 ml of H_2O and brought to pH 6.1 by the addition of solid citric acid

Anticoagulant: 8 g of sodium citrate, 54 ml of 37% formaldehyde, 100 ml saline

Trypsin: 1% solution of Bacto-trypsin 1:250 in PBS

Stock blood suspension: venous whole rabbit blood, added to an equal volume of Alsever's solution containing 1/30 volume of the anticoagulant. This suspension can be stored as long as 2 weeks at 4°.

Preparation of Standard Trypsinized Erythrocyte Suspension. Trypsinized erythrocytes should be prepared on the day of the assay. Erythrocytes are collected from the stock blood suspension by centrifugation at room temperature in a clinical table centrifuge (2000 rpm, 5 minutes) and washed 3–4 times with saline (5 ml of saline for each milliliter of packed erthrocytes). The washed erythrocytes are added to PBS (about 4 ml of cells per 100 ml of PBS) to give a suspension with an absorbance of 2 at 620 nm. To 10 parts of this suspension is added 1 part of 1% trypsin solution, and the mixture is incubated at 37° for 1 hour. The trypsinized erythrocytes are then washed 4–5 times with saline as above to remove the last traces of trypsin and are finally suspended in sufficient saline to give a standard erythrocyte suspension with an adsorbance of 1 at 620 nm (1.2–1.5 ml packed cells/100 ml). About 80 ml of standard erythrocyte suspension is obtained from 5 ml of stock blood suspension.

Hemagglutination Assay. The material to be tested is dissolved in saline. Serial 2-fold dilutions of the starting solution are made in a final volume of 1 ml of saline in 10 × 75 mm test tubes. To each tube is added 1 ml of the standard erythrocyte suspension, the contents of each tube are mixed by inversion and the tubes are placed in

[9] I. E. Liener, *Arch. Biochem. Biophys.* **54**, 223 (1955).

a rack that holds them in an exactly vertical position. After 2.5 hours at room temperature, the tubes are read in the photometer, due care being taken not to agitate the contents. Each experiment should include a set of 2–4 control tubes, containing 1 ml of saline and 1 ml of standard blood suspension.

Measurements of Absorbance. All measurements of absorbance are performed at 620 nm in a Coleman Junior spectrophotometer, Model 6A, equipped with a special adapter to hold the 10×75 mm test tubes. The opening of the adapter should be masked with black plastic tape to leave an aperture of 1 cm^2 only, the center of the aperture being located 5 cm from the top of the adapter. The size and position of the aperture are so calculated that in the control tubes the absorbance of the erythrocyte suspension at the level of the aperture remains unchanged after 2.5 hours, thus eliminating the effect of the spontaneous sedimentation of the erythrocytes.

Calculation of Hemagglutinating Activity. One hemagglutinating unit (HU) is arbitrarily defined as that amount of material which is required to cause a decrease of 50% in the absorbance of the erthrocyte suspension in 2.5 hours under the conditions described above. The reciprocal of dilution (x) corresponding to one HU is calculated from the readings of the two tubes that have optical densities nearest to half the absorbance of the control $(E_{50} \sim 0.25)$, one of the readings (E_A) being lower and the other (E_B) being higher than E_{50}. The following equation (where A is the reciprocal of dilution of the tube with E_A) is then used:

$$\log x = \log A + \frac{E_{50} - E_A}{E_B - E_A} \cdot \log 2$$

The specific hemagglutinating activity of the material tested, HU/mg protein, is calculated from the value of x and from the concentration in the starting solution.

The method is highly reproducible (within $\pm 5\%$) when the assay is done with the same preparation of standard erythrocyte suspension. With different preparations of erythrocytes, the variations can be higher. It is therefore advisable to include in each assay of activity a sample of SBA with known specific activity, for comparison.

Purification of Soybean Agglutinin–Method I

Soybean meal contains four closely related agglutinins (isolectins), one major and three minor ones.[10,11] The procedure described below is

[10] H. Lis, C. Fridman, N. Sharon, and E. Katchalski, *Arch. Biochem. Biophys.* **117**, 301 (1966).
[11] N. Catsimpoolas and E. W. Meyer, *Arch. Biochem. Biophys.* **132**, 279 (1969).

for the purification of the major component and is based on the method described by Liener[3] and modified by us.[12]

Isolation of Crude SBA. Five hundred grams of untoasted defatted soybean meal[13] are suspended in 6 liters of distilled water at room temperature and extracted for 1 hour with constant stirring. The suspension is acidified to pH 4.6 with concentrated HCl (18 ml) and allowed to settle overnight at 4°. Most of the clear yellow supernatant fluid is siphoned off, and the remainder is collected by centrifugation (10 minutes at 6000 rpm in a Sorval RC 2 centrifuge). To each liter of supernatant, 300 g of $(NH_4)_2SO_4$ are gradually added while stirring; the precipitate is removed by filtration[14] and discarded. To each liter of the supernatant is added 270 g of $(NH_4)_2SO_4$ while stirring. After allowing the precipitate to settle overnight at 4°, most of the supernatant is siphoned off and the precipitate is collected by centrifugation (10 minutes, 5000 rpm). It is suspended in 150–200 ml of water, and the suspension is dialyzed against distilled water for 24 hours in the cold room with two changes of water. Any insoluble material which is present after dialysis is removed by centrifugation and discarded. The pH of the solution is adjusted to 4.6 with 1 N HCl. After addition of $(NH_4)_2SO_4$ (56 g/100 ml of solution), the precipitate is collected by centrifugation (10 minutes, 5000 rpm) and dissolved in 50 ml of 0.05 M phosphate buffer, pH 6.1. The above solution is dialyzed against 60% ethanol at −15° for 48 hours. The precipitate which forms inside the dialysis bag is collected by centrifugation in the cold[15]; it is suspended in approximately 25 ml of distilled water and dialyzed overnight against a large volume of distilled water in the cold. After removal by centrifugation of any insoluble material, the solution is lyophilized to give 700–800 mg of crude SBA.

Chromatography on Calcium Phosphate. A column (2.8 × 15 cm)

---

[12] H. Lis, N. Sharon, and E. Katchalski, *J. Biol. Chem.* **241**, 684 (1966).

[13] We now use meal kindly supplied by Etz Hazait factory, Petah Tikva, Israel. The meal is taken out from the processing line after extraction of the oil with hexane. Removal of solvent is carried out in the laboratory by drying in a hood at room temperature.

[14] Sometimes difficulty is encountered at this stage: upon addition of $(NH_4)_2SO_4$ a very fine suspension may be formed, which sediments only after prolonged high speed centrifugation. This occurred, for example, in our laboratory when certain batches of Soyafluff 20 were used as starting material for the isolation of SBA The reason for this is not clear, but may in part be the result of the industrial procedures used for the production of the meal, the exact details of which are not always available. In such cases, it is advisable to try other sources of untoasted meal.

[15] The supernatant is enriched with respect to the minor soybean agglutinins. They can be obtained from this supernatant in purified form as described by Lis *et al.*[10]

of calcium phosphate (hydroxylapatite), prepared according to Tiselius *et al.*,[16,17] is equilibrated with 0.001 M phosphate buffer, pH 6.8, at room temperature. Crude SBA (ca. 750 mg) is dissolved in 35 ml of the same buffer and applied to the column. The column is washed with 200 ml of the 0.001 M buffer, followed by 0.072 M buffer. Elution with this buffer continues until no material absorbing at 280 nm is detected in the effluent (about 500 ml). The purified agglutinin is then eluted from the column with 0.2 M phosphate buffer, pH 6.8. Fractions of 10 ml are now collected at a flow rate of 40 ml/hour. The fractions are analyzed for protein by measuring the absorbance at 280 nm and are assayed for hemagglutinating activity. The fractions containing the activity (tubes 9–15) are pooled, dialyzed against distilled water at 4° and lyophilized, to give 100–150 mg of purified SBA with a specific activity of 5000–6000 units per milligram of protein. At this stage the product is sufficiently pure for most purposes. However, it still contains trace amounts of the three other agglutinins, which can be removed by chromatography on DEAE-cellulose.

Chromatography on DEAE-Cellulose. SBA (50 mg) is dissolved in 10 ml of 0.01 M phosphate buffer, pH 6.8 (starting buffer), and applied to a column (1.2 × 40 cm) of DEAE-cellulose in equilibrium with the same buffer. The column is washed with 200 ml of starting buffer, whereupon the three minor agglutinins are eluted. The column is then connected to a closed mixing chamber containing 200 ml of starting buffer, and elution of the main agglutinin is performed by passing into the mixing chamber 0.4 M NaCl in starting buffer. Five milliliter fractions are collected at a rate of 25–30 ml per hour, monitored at 280 nm and assayed for hemagglutinating activity. The fractions containing activity (tubes 10–25 after application of the gradient) are pooled, dialyzed and lyophilized. Yield 35 mg; specific activity 5000–6000 HU/mg.

Properties

Chemical Properties.[12] SBA is a glycoprotein with a molecular weight of about 110,000. It contains 4.5% of neutral sugars and about 1% of amino sugars. The neutral sugar has been identified as D-mannose and the amino sugar as N-acetyl-D-glucosamine. The amino acid composition of SBA is characterized by its relatively high content of aspartic acid, serine, and threonine, and by the absence of cysteine and the very low content of methionine. SBA is stable in dry form at room temperature and in solution in frozen form.

---

[16] A. Tiselius, S. Hjertén, and Ö. Levin, *Arch. Biochem. Biophys.* **65**, 132 (1956).
[17] Ö. Levin, see Vol. 5 [2].

Biological Properties. SBA agglutinates different types of cell; however, the concentrations of SBA required for the agglutination vary markedly with the type of cell.[5] Thus, rabbit erythrocytes which have been treated with trypsin are highly sensitive, requiring only 0.1–0.2 µg of SBA per milliliter for detectable agglutination. These cells are used therefore for the routine assay of SBA. Untrypsinized rabbit erythrocytes are much less sensitive, the concentration of SBA required for their agglutination being 100–200-fold higher than that for the trypsinized cells. SBA agglutinates human erythrocytes of all types (A > O > B), although the concentrations required for their agglutination are rather high (0.2–1 mg/ml). In this case, too, the susceptibility to agglutination is increased about 100-fold by trypsinization. In addition to erythrocytes, SBA also agglutinates somatic cells, grown in culture, which have been transformed by viral or chemical carcinogens or by irradiation; the untransformed parent cells are not agglutinated under the same conditions, unless they have been treated with trypsin.[18] In all cases, the agglutination is specifically inhibited by N-acetyl-D-galactosamine and to a lesser extent by D-galactose[5,18] indicating that N-acetyl-D-galactosamine-like residues are present on cell surfaces. SBA can therefore be used for the detection and quantitation of such residues[19] and for the study of the changes that cell surfaces undergo upon malignant transformation.

The toxicity of SBA has been studied under a variety of conditions.[2,3,20] The LD_{50} of SBA administered intraperitoneally to young rats is about 50 mg/kg. When administered to rats by stomach tube no lethal effect was observed up to a level of 500 mg/kg.

ADDENDUM

Method II: Purification of SBA by Affinity Chromatography[21]

By JULIUS A. GORDON, SHMARYAHU BLUMBERG, HALINA LIS, and NATHAN SHARON

Principle. The four soybean agglutinins are adsorbed from solution to a column made of D-galactose covalently attached to Sepharose. The

[18] B. A. Sela, H. Lis, N. Sharon, and L. Sachs, *J. Membrane Biol.* **3**, 267 (1970).
[19] B. A. Sela, H. Lis, N. Sharon, and L. Sachs, *Biochim. Biophys. Acta* **249**, 564 (1971).
[20] I. E. Liener and J. E. Rose, *Proc. Soc. Exp. Biol. Med.* **83**, 539 (1953).
[21] J. A. Gordon, S. Blumberg, H. Lis, and N. Sharon, *FEBS Letters,* in press (1972).

active material is specifically eluted from the column by a solution of D-galactose.

Preparation of Column. The column is prepared as described by Blumberg *et al.*[22] except that D-galactose is used instead of L-fucose.

Preparation of β-D-Galactopyranosylamine (1-β-Amino-1-deoxy-D-galactopyranoside). This compound is prepared by the procedure employed for the synthesis of other glycosylamines.[23,24] D-Galactose (27 g, 150 mmoles) is dissolved in 200 ml of liquid ammonia in methanol (30% v/v). After the preparation has stood in a closed vessel at room temperature 1 week, the precipitate, which consists of α-D-galactopyranosylamine, is discarded. The solution is kept at room temperature for another 3–4 days. During this period the vessel is kept in a hood and opened every day for 1–2 hours to permit the evaporation of excess ammonia. Under these conditions, the desired β isomer of D-galactopyranosylamine crystallizes. It is collected by filtration, washed with absolute methanol, and dried in a vacuum desiccator over NaOH. Yield, ca. 25%; m.p. 136–137°.

Preparation of N-(N-Benzyloxycarbonyl-ε-aminocaproyl)-β-D-galactopyranosylamine. Isobutylchloroformate (30 mmoles) and triethylamine (30 mmoles) are added to a solution of N-benzyloxycarbonyl-ε-aminocaproic acid (30 mmoles) in dimethyl formamide (50 ml), kept at −5°. The mixture is stirred for 20 minutes (at −5°) and then filtered. The filtrate is immediately added to a suspension of β-D-galactopyranosylamine (25 mmoles) in dimethyl formamide (50 ml), and the mixture is stirred at room temperature until the solution is clear (1 hour). The solution is allowed to stand at room temperature for 16 hours and the solvent is removed *in vacuo.* The residue is crystallized twice from ethanol. Yield 60%; m.p. 159–160°.

Preparation of N-(ε-Aminocaproyl)-β-D-galactopyranosylamine. N-(N-Benzyloxycarbonyl-ε-aminocaproyl)-β-D-galactopyranosylamine (10 mmoles) in 80% aqueous methanol (100 ml) is hydrogenated over palladium on charcoal (10%, 100 mg) at atmospheric pressure, for 8 hours. The mixture is filtered and the filtrate concentrated *in vacuo.* The residue is crystallized from methanol–ether. Yield 75%; m.p. 206–208°.

Preparation of the Galactosyl Resin. Sepharose 4B is activated with cyanogen bromide and then allowed to react with N-(ε-aminocaproyl)-β-D-galactopyranosylamine by the general method of Axén *et al.*,[25] follow-

---

[22] S. Blumberg, J. Hildesheim, J. Yariv, and K. J. Wilson, *Biochim. Biophys. Acta* **264**, 171 (1972).

[23] C. A. Lobry de Bruyn and F. H. van Leent, *Rec. Trav. Chim. Pays-Bas* **14**, 134 (1895).

[24] H. S. Isbell and H. L. Frush, *J. Org. Chem.* **23**, 1309 (1958).

[25] R. Axén, J. Porath, and S. Ernback, *Nature* (*London*) **214**, 1302 (1967).

ing the detailed procedure described by Blumberg *et al.*[26] Cyanogen bromide (10 g) is stirred with water (100 ml) for 10 minutes, when most of it goes into solution. A slurry of Sephadex (100 ml) washed with water, is then added to the cyanogen bromide–water mixture. The pH of the reaction mixture is adjusted to 11 and maintained at this pH for 6 minutes by adding 4 N NaOH. The reaction mixture is kept below 26° by the addition of crushed ice. The activated Sepharose is washed rapidly with 15 volumes of cold water on a Büchner funnel. The wet Sepharose is then added quickly to a solution of N-(ϵ-aminocaproyl)-β-D-galacto-pyranosylamine (1.6 mmoles in 40 ml of 0.5 M NaHCO$_3$) and the mixture is stirred gently for 16 hours at 4°. The resin is filtered and washed thoroughly with 0.1 M NaHCO$_3$ and water. The amount of D-galactose derivative covalently bound to the Sepharose is determined by hydrolysis of the resin in 6 N hydrochloric acid (22 hours, 110°) and estimation of the liberated ϵ-aminocaproic acid on a Beckman-Spinco amino acid analyzer. The galactosyl resin was found to contain 2 μmoles of covalently bound N-(ϵ-aminocaproyl)-β-D-galactopyranosylamine groups per milliliter.

Purification of SBA. Untoasted defatted soybean meal (50 g) is extracted with saline (250 ml) for 1 hour at room temperature with constant stirring. The insoluble residue is removed by centrifugation (10 minutes at 6000 rpm in a Sorvall RC 2 centrifuge), and the supernatant is cooled to 4°. The following steps are all carried out in the cold room. To the supernatant is added (NH$_4$)$_2$SO$_4$ (30 g/100 ml), the precipitate is removed by centrifugation (10 minutes, 6000 rpm) and discarded. An additional amount of (NH$_4$)$_2$SO$_4$ is added to the supernatant (25 g/100 ml) and the precipitate is collected as above. It is dissolved in a minimal volume of water and dialyzed extensively, first against distilled water and finally against saline. The dialyzed solution is centrifuged to remove any precipitate formed and is applied to the galactosyl column (2.0 × 30 cm), previously washed with 1 liter of saline. The column is washed with saline (approximately 750 ml) until no significant amount of material absorbing at 280 nm is detected in the effluent (absorbance ~0.1). Elution of the agglutinins is carried out with 200 ml of a solution of D-galactose in saline (5 mg/ml). Fractions of 10 ml are collected at a rate of 60 ml/hour and monitored at 280 nm. The fractions containing UV absorbing material (tubes 7–13) are pooled, dialyzed against distilled water, and lyophilized; yield about 75 mg (specific activity 7000–8000 units/mg).

Comments. Method II gives a higher yield of SBA than does method I, and the product has a somewhat higher specific activity. In

[26] S. Blumberg, I. Schechter, and A. Berger, *Eur. J. Biochem.* **15**, 97 (1970).

all properties tested, including amino acid composition, carbohydrate content, migration on acrylamide gel, and biological specificity, the agglutinins prepared by both methods appear to be identical. However, SBA prepared by method II is a mixture of all four agglutinins. The major SBA can be obtained free of the three minor components, by chromatography on DEAE-cellulose as described under method I.

[45] Snail (*Helix pomatia*) Hemagglutinin

By STEN HAMMARSTRÖM

The albumin gland of the snail *Helix pomatia* (class Gastropoda, phylum Mollusca) contains a protein (*Helix pomatia* A hemagglutinin) which agglutinates human A erythrocytes but not human B or O erythrocytes.[1,2] The hemagglutinin is also found in the eggs but not in the hemolymph.[3] The albumin gland, which is part of the sexual apparatus, contains relatively large amounts of the protein (approximately 8% of total protein in the combined supernatants after PBS extraction and ultracentrifugation, see below). The hemagglutinin seems to be evenly distributed over the entire gland as visualized by indirect immunofluorescence staining with rabbit antisera against the purified hemagglutinin.[3] Similar hemagglutinins have been detected in total extracts of snails from the species *Helix hortensis*[4] and *Otala lactea*.[5] These agglutinin-containing extracts and *Helix pomatia* A hemagglutinin seem to have approximately the same specificity as judged from their ability to agglutinate normal and enzyme treated erythrocytes of different vertebrates.[2] Only *Helix pomatia* A hemagglutinin has been purified and studied in greater detail.

Purification

The hemagglutinin from *Helix pomatia* is easily purified by immunospecific adsorption to insoluble human or hog blood group A substance followed by elution with D-GalNAc.[6] The procedure is as follows: The albumin glands from about 200 snails (approximately 90 g wet weight)

[1] O. Prokop, D. Schlesinger, and A. Rackwitz, *Z. Immun. Forsch.* **129**, 402 (1965).
[2] O. Prokop, G. Uhlenbruck, and W. Köhler, *Vox Sang.* **14**, 321 (1968).
[3] S. Hammarström, unpublished observations.
[4] O. Prokop, A. Rackwitz, and D. Schlesinger, *J. Forensic Med.* **12**, 108 (1965).
[5] W. C. Boyd and R. Brown, *Nature (London)* **208**, 593 (1965).
[6] S. Hammarström and E. A. Kabat, *Biochemistry* **8**, 2696 (1969).

are dissected out, homogenized in a blender, and extracted in the cold with 200 ml of 0.15 M phosphate-buffered saline, pH 7.2 (PBS). The extract is centrifuged at 12,000 rpm for 20 minutes at 4°, and the precipitate is reextracted with the same volume of buffered saline. The combined supernatants are then spun in the ultracentrifuge at 40,000 rpm for 2.5 hours. Three fractions are obtained; a clear, light green supernatant, a light precipitate, and a heavy gelatinous precipitate. The supernatant contains by far the highest concentration of hemagglutinin and is used for further purification. The heavy precipitate consists essentially of galactan.

The clear supernatant is then passed repeatedly through a column of carefully washed insoluble hog blood group A + H substance, mixed with Celite (1:1). Saturation is monitored by measuring the absorbance at 280 nm and the hemagglutinating activity of the solution before and after passage through the column. At saturation the column is washed extensively with PBS until the absorbance at 280 nm is below 0.050. Specific elution is effected by 0.005–0.015 M D-GalNAc in PBS. The eluted material is concentrated by ultrafiltration and passed twice through a column of Biogel P-10 to remove free and bound D-GalNAc.

Insoluble hog blood group A + H substance is prepared by copolymerization with the N-carboxyanhydride of L-leucine[7] as described by Kaplan and Kabat.[8] The hemagglutinin binding capacity of polyleucyl hog blood group A + H substance is approximately 30% on a weight basis.[6]

The crude extract contains in addition to *Helix pomatia* A hemagglutinin a second component[6] (Fig. 1B), which also combines with hog blood group A + H substance. This protein, which does not agglutinate human A erythrocytes, can be removed by preeluting the column with 0.05 M D-Glc or by extensive washing with PBS.

In a typical experiment[6] 85.5% of the adsorbed hemagglutinin was recovered after elution with 0.005 M D-GalNAc. Very little additional nitrogen, however, was recovered by raising the D-GalNAc concentration to 0.05 M (0.01%) or by elution with 2 M KSCN (0.7% of total eluted nitrogen). The latter fraction contained only nonhemagglutinin components, indicating that the hemagglutinin is completely eluted.

The hemagglutinin can also be purified by adsorption to Sephadex G-200 and elution with D-GalNAc, D-Gal, or even D-Glc in the same manner as described for insoluble hog blood group A + H substance.[9,10]

[7] H. Tsuyuki, H. von Kley, and M. A. Stahmann, *J. Amer. Chem. Soc.* **78**, 764 (1956).
[8] M. E. Kaplan and E. A. Kabat, *J. Exp. Med.* **123**, 1061 (1966).
[9] O. Kühnemund and W. Köhler, *Experientia* **25**, 1137 (1969).
[10] I. Ishiyama and G. Uhlenbruck, *Z. Klin. Chem. Klin. Biochem.* **9**, Heft 5 (1971).

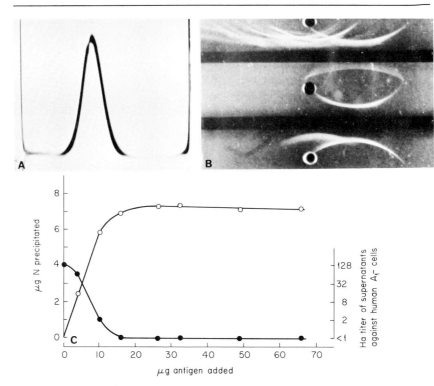

Fig. 1. Ultracentrifugal, immunoelectrophoretic, and precipitin pattern of immunosorbent purified *Helix pomatia* A hemagglutinin. (A) Schlieren pattern of purified snail hemagglutinin [fraction II of S. Hammarström and E. A. Kabat, *Biochemistry* **8**, 2696 (1969)] (1.30 mg of N per milliliter in 0.9% NaCl). The photomicrograph was taken after 144 minutes at 50740 rpm. (B) Immunoelectrophoretic analysis of purified snail hemagglutinin, fraction II, 1.30 mg of N per milliliter (center well), and of crude extract of albumin gland (upper and lower wells); rabbit antiserum against crude extract of albumin gland (upper trough) and hog blood group A + H substance (lower trough) were added to develop reactions. (C) Precipitation of purified snail hemagglutinin fraction II, 6.47 μg of N per tube, by human blood group A substance (cyst MSM) (○—○). The total volume was 200 μl. Hemagglutination titer of supernatants against human A_1-erythrocytes (●—●).

It is likely that nonreducing α-linked D-Glc end groups in dextran are responsible for the binding of the hemagglutinin. This is a very weak interaction since D-Glc can displace the hemagglutinin from the Sephadex column. Precipitin inhibition studies furthermore show that D-Glc or Me-α-D-Glc are very poor inhibitors as compared with D-GalNAc (see below). The abundance of D-Glc end groups in dextran and the high valency of the hemagglutinin seem, however, to compensate for the weakness of the interaction. A similar situation has been described for certain group C antistreptococcal antibodies.[11]

Purity

Immunosorbent purified *Helix pomatia* A hemagglutinin is homogeneous on gel filtration on Biogel P-150 and P-300 and gives only one symmetrical peak in the analytical ultracentrifuge[6] (Fig. 1A). It gives only one line in immunoelectrophoresis even when tested at very high protein concentration with antiserum to the crude extract[6] (Fig. 1B). With this antiserum at least 15 components were detected in the crude extract (Fig. 1B). It is furthermore completely precipitated by human blood group A substance[6] (Fig. 1C).

Polyacrylamide gel electrophoresis at alkaline pH revealed, however, a certain degree of heterogeneity. Although approximately 90% of the material eluted with 0.005–0.015 M D-GalNAc appeared as one band, three additional bands with closely similar electrophoretic mobilities were seen. The relative amount of these bands was different in fractions eluted earlier from the immunosorbent column than in fractions eluted later. In the last fraction (1.3% of total eluted hemagglutinin) three bands (including the major band) of approximate equal density were obtained. Since this fraction, as well as the earlier fractions, were completely precipitated by human blood group A substance[6] it follows that all bands must represent the *Helix pomatia* A hemagglutinin. Polyacrylamide gel electrophoresis at different gel concentrations showed furthermore that these bands differed in charge but not in size.[3] This may either reflect a heterogeneity of the starting material—a pool of albumin glands from several hundred snails was used—or may be due to secondary chemical changes.

Chemical and Physicochemical Properties

The $s_{20,w}^0$ value for the hemagglutinin was determined to 5.3 s (0.9% NaCl). From this, the intrinsic viscosity, and the partial specific volume, a molecular weight of 1.0×10^5 was calculated.[6] Further ultracentrifugation studies,[12] using the meniscus depletion sedimentation equilibrium method, showed that this value was too high. A mean molecular weight of 79,000 was obtained.[12]

Amino acid analysis shows that the hemagglutinin contains approximately 18 moles of half-cystine and 10 moles of methionine per molecu-

[11] T. J. Kindt, C. W. Todd, K. Eichmann and R. M. Krause, *J. Exp. Med.* **131**, 343 (1970).

[12] S. Hammarström, A. Westöö, and I. Björk, *Scand. J. Immunol.* (1972) in press.

lar weight of 79,000.[6] Carbohydrate analysis revealed the presence of D-galactose (4.0%) and D-mannose (3.3%).[6] Trace amounts of hexosamine were also detected. It is, however, likely that the latter represents residual D-GalNAc not completely removed by dialysis and gel filtration.

Subunit Structure

Direct alkylation of the hemagglutinin at pH 8.6 either in Tris buffer alone or in 6 M guanidine·HCl showed that free sulfhydryl groups were absent[12] (Table I). Reduction with excess dithiothreitol (DTT) in 6 M guanidine·HCl lead, on subsequent alkylation, to the incorporation of approximately 18 moles of [14C]-acetate per molecular weight of 79,000[12] (Table I). Thus under these conditions the hemagglutinin was completely reduced. Gel filtration of the alkylated material on Sephadex G-100 in 6 M guanidine·HCl gave the elution pattern shown in Fig. 2A. One symmetrical included peak, in addition to aggregated material in the exclusion volume, was obtained.[12] Under these conditions the unreduced hemagglutinin is included.[12] The mean molecular weight of the included component (*Helix pomatia* A hemagglutinin subunit) was 13,000 as determined from a calibration curve for the same column, using completely reduced and alkylated proteins of known molecular weights[12] (Fig. 2B).

The amino acid composition showed that the hemagglutinin per molecular weight 13,000 contains 12 moles of lysine and arginine and 3 moles of half cystine. Thus, if the protein consists of identical subunits, 13 peptides with maximally 3 containing cysteine would be expected on tryptic digestion. In fact, 13–14 major peptides (2 acid, 6 neutral, and 5 or 6 basic peptides) were found after combined electrophoretic and chromatographic analysis[13] of tryptic digest, prepared from reduced and 14C-carboxymethylated or performic acid oxidized hemagglutinin.[12] Three of these peptides were radioactive. One of them (acid peptide) gave however a more intense radioactive spot.[12] These data suggest that the hemagglutinin consists of only one type of subunit with an approximate molecular weight of 13,000.

Reduction in the absence of unfolding agents followed by alkylation revealed that disulfide bonds were not cleaved[12] (Table I).

Treatment of the hemagglutinin with 6 M guanidine·HCl at pH 4.0 or with 6 M guanidine·HCl alone for 15 minutes up to 2 days gave one component with a molecular weight of 26,000–31,000 both on gel filtration in 6 M guanidine·HCl + 0.1 M D-GNAc and on ultracentrifugation (meniscus depletion sedimentation equilibrium method).[12] Moreover the

---

[13] H. Jörnvall, *Eur. J. Biochem.* **16**, 41 (1970).

TABLE I

Degree of Alkylation of *Helix pomatia* A Hemagglutinin after Reduction under Various Conditions

Reducing agent	Molar excess of DTT-SH over protein-SH	Conditions during reduction	Moles [^{14}C]acetate per protein mw of 79,000		No. of experiments
			Mean	Range	
—	—	PBS; 0.1 M Tris pH 8.7; 7 M guanidine·HCl	<0.65	<0.15–0.65	4
Dithiothreitol (DTT)	3–85	7 M guanidine·HCl–0.5 M Tris, pH 8.7	18.3	16.7–20.1	5
DTT	2.5	0.1 M Tris, pH 8.7	0.7	—	1
DTT	1.5–1000	8 M urea–0.5 M Tris, pH 8.7	7.9	6.3–8.6	4

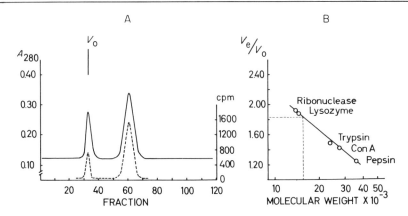

FIG. 2. Gel filtration profile of completely reduced and [^{14}C]alkylated *Helix pomatia* A hemagglutinin on Sephadex G-100 column in 6 M guanidine·HCl (A) and calibration curve for the same column (B). (A) The solid line denotes absorbance at 280 nm, and the dashed line denotes radioactivity. The column was 2.5 × 40 cm. The exclusion volume (V_0) is indicated in the figure. (B) Completely reduced and alkylated proteins were used for calibrations. The dashed lines denote the V_e/V_0 value and the molecular weight for the major component of Fig. 2A. Con A, concanavalin A.

hemagglutinin did not dissociate further when treated with 7 M guanidine·HCl containing 1 M propionic acid as determined by gel filtration experiments in this medium.[12] Control experiments demonstrated that the hemagglutinin was not reduced under these conditions (0.2 mole [^{14}C]acetate per molecular weight 79,000).

Conditions were also found that gave rise to partial reduction of the hemagglutinin (e.g., reduction with excess DTT in 8 M urea ± 0.1 M D-GNAc or in 6 M guanidine·HCl containing 0.1–0.2 M D-GalNAc or D-GNAc). 3–4 disulfide bonds were cleaved under these conditions[12] (Table I). Gel filtration in dissociating agents in the presence of D-GNAc gave only one symmetrical component with a molecular weight of 12,500–16,700 (determined from standard curves for reduced and unfolded proteins or non-reduced but unfolded proteins respectively). This component furthermore contained an active carbohydrate binding site.[12]

Taken together these data suggest that the hemagglutinin is made up of six identical or closely similar polypeptide chains, each containing one intrachain disulfide bond and a carbohydrate binding site. The subunits are arranged in pairs in which they are linked together by a single disulfide bond. Native hemagglutinin is formed by the interaction of non-covalent forces between three dimers.

Homogeneity of Carbohydrate Binding Site

A blood group A active pentasaccharide, AR_L 0.52, α-D-GalNAc $(1 \rightarrow 3)$-[α-L-Fuc-$(1 \rightarrow 2)$]-β-D-Gal-$(1 \rightarrow 4)$-β-D-GNAc-$(1 \rightarrow 6)$-3-hexenetetrol(s) labeled with tritium in the hexenetetrol residue[14] was used in equilibrium dialysis experiments to investigate the binding properties of *Helix pomatia* A hemagglutinin. This structure constitutes the determinant that is responsible for precipitation of blood group A substance by the hemagglutinin (see below).

Figure 3A demonstrates a linear relationship when the binding data[15] are plotted according to Scatchard.[16] A heterogeneity index, a, of 1.04 was obtained when the binding data were plotted according to Sips distribution[17,18] (Fig. 3B). Thus, within the experimental error the binding sites are homogeneous.

On the basis of a molecular weight of 1.0×10^5 for the hemagglutinin, the equilibrium dialysis data indicate that the hemagglutinin contains one combining site/molecular weight of 16,000–17,000.[15] This value is in reasonable agreement with the molecular weight of the subunit.

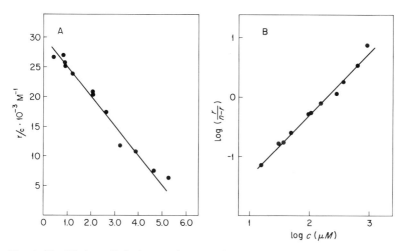

Fig. 3. Equilibrium dialysis experiments with purified *Helix pomatia* A hemagglutinin and tritium-labeled A active pentasaccharide [³H]AR_L 0.52. The data are plotted both according to Scatchard (A) and according to Sips (B). Approximately 12 mg of snail protein was used for each determination.

[14] C. Moreno and E. A. Kabat, *J. Exp. Med.* **129**, 871 (1969).
[15] S. Hammarström and E. A. Kabat, *Biochemistry* **10**, 1684 (1971).
[16] G. Scatchard, *Ann. N.Y. Acad. Sci.* **51**, 660 (1949).
[17] R. Sips, *J. Chem. Phys.* **16**, 490 (1948).
[18] F. Karush, *Advan. Immunol.* **2**, 1 (1962).

Further evidence for homogeneity of binding sites has been obtained by equilibrium dialysis displacement experiments[15] (Fig. 4). In this experiment the capacities of the cross reactive haptens Me-α-D-GalNAc, Me-α-D-GNAc, and Et-β-D-GNAc to displace the labeled A active pentasaccharide, [^{3}H]AR$_L$0.52, from the binding site were investigated. Linear displacement curves essentially parallel with the self-displacement curve, AR$_L$0.52, were found in the plot. Thus over the range tested (from 10 to 70–80% displacement) there is no substantial heterogeneity of association constants for the binding of the displacing haptens.

Specificity of Carbohydrate Binding Site

The specificity of the binding site of *Helix pomatia* A hemagglutinin has been investigated by the following procedures.

Equilibrium Dialysis and Displacement. The intrinsic association constant (K_0) for the interaction of the hemagglutinin site with Me-α-D-GalNAc or the blood group A active pentasaccharide is 5.0×10^3 l/mole at 25° and pH 7.2[15] (Table II; Figs. 3 and 4). These compounds are the best inhibitors of precipitation studied so far. The K_0 values for Me-α-D-

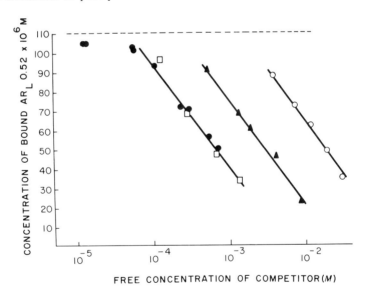

FIG. 4. Equilibrium dialysis displacement experiments with unlabeled ligands. Approximately 12 mg of hemagglutinin and a concentration of [^{3}H]AR$_L$ 0.52 which gives $110 \times 10^{-6} M$ bound radioactive hapten were employed for each determination. Displacing haptens were AR$_L$0.52 (●—●), self displacement curve; Me-α-D-GalNAc (□—□), Me-α-D-GNAc (▲—▲), and Et-β-D-GNAc (○—○). The dashed horizontal line signifies the concentration of bound radioactive hapten in the absence of competitor.

GNAc and Et-β-D-GNAc were approximately 4.5 and 27 times lower than for Me-α-D-GalNAc, respectively[15] (Table II; Fig. 4). Although the intrinsic association constants for the best inhibitors are relatively low, they are of the same order of magnitude as for other carbohydrate-anticarbohydrate systems.[14,19,20]

Inhibition of Precipitation. The most complete data on the specificity of the hemagglutinin site have been obtained by inhibition of precipitation using either human blood group A substance or *Salmonella typhimurium* SH 180 lipopolysaccharide (LPS) as test antigens.[6,15,21] These two systems were used since they differ markedly in sensitivity. With the same amount of hemagglutinin, approximately 1000 nmoles of added D-GalNAc is needed for 50% inhibition in the former system whereas 25 nmoles of added D-GalNAc is required for the same degree of inhibition in the latter. The reason for this difference is that the hemagglutinin reacts with α-linked D-GalNAc in blood group A substance,[15] whereas the structure, α-D-Gal $(1 \rightarrow 6)$-D-Glc..., is "immunodominant" in *S. typhimurium* SH 180 LPS.[21]

Table III shows the concentrations of various monosaccharides, methylglycosides, and oligosaccharides needed for 50% inhibition. The data are based on complete inhibition curves.

In summary the results show: (1) Me-α-D-GalNAc is the best inhibitor of all compounds investigated so far. (2) In the series of α-D-GalNAc containing oligosaccharides and derivatives, only the terminal nonreducing D-GalNAc residue seems to bind significantly to the site. This may indicate that the combining site is relatively small. (3) Me-α-D-GalNAc is only about 4 times more active than Me-α-D-GNAc, indicat-

TABLE II

INTRINSIC ASSOCIATION CONSTANTS (K_0) AND STANDARD FREE-ENERGY CHANGE ($-\Delta F°$) FOR THE INTERACTION OF *Helix pomatia* A HEMAGGLUTININ WITH VARIOUS HAPTENS AT pH 7.3 AND 25.0 $\pm$ 0.1°

Hapten	$K_0 \times 10^{-3}$ (M^{-1})	$-\Delta F°$ (kcal/mole)
AR$_L$ 0.52[a]	5.0	5.04
Me-α-D-GalNAc	5.0	5.04
Me-α-D-GNAc	1.1	4.14
Et-β-D-GNAc	0.18	3.08

[a] See text.

[19] L. L. So and I. J. Goldstein, *J. Biol. Chem.* **243**, 2003 (1968).
[20] M. Katz and A. M. Pappenheimer, Jr., *J. Immunol.* **103**, 401 (1969).
[21] S. Hammarström, A. A. Lindberg, and E. S. Robertson, *Eur. J. Biochem.* **25**, 274 (1972).

TABLE III

CONCENTRATION OF MONOSACCHARIDES, METHYLGLYCOSIDES, AND OLIGOSACCHARIDES NEEDED FOR 50% INHIBITION OF PRECIPITATION OF PURIFIED *Helix pomatia* A HEMAGGLUTININ WITH DIFFERENT CARBOHYDRATE ANTIGENS[a]

Inhibitor	Micromoles for 50% inhibition of precipitation of hemagglutinin		
	With blood group A substance	With germfree rat colon antigen[b]	With *S. typhimurium* SH 180 LPS
Me-α-D-GalNAc	0.76	—	—
α-D-GalNAc-(1 → 3)-D-Gal	>0.14	—	—
α-D-GalNAc-(1 → 3)-β-D-Gal-(1 → 3)-D-GNAc	0.96	—	—
AR$_L$0.52[c]	2.00	—	—
Ph-α-D-GalNAc	1.65[g]	—	—
o+p-NO$_2$Ph-α-D-GalNAc	5.37[g]	—	—
Et-β-D-GalNAc[d]	>3.60	—	—
p-NO$_2$Ph-β-D-GalNAc[d]	>0.99[g]	—	—
D-GalNAc[d]	1.65	0.130	0.024
D-GalNH$_2$[e]	>9.73	—	—
D-GalNAc-ol[f]	>11.6	—	—
2-O-Ac-Me-α-D-Gal	—	0.050	—
2-O-Ac-Me-β-D-Gal	—	0.098	—
Me-α-D-GNAc	4.00	—	0.070
Et-β-D-GNAc	10.0	—	—
D-GNAc	9.00	0.52	0.150
D-GNH$_2$[e]	>18.8	—	—
D-GNAc-ol[f]	>11.6[g]	—	—
2-O-Ac-Me-α-D-GNAc	—	1.60	—
2-O-Ac-Me-β-D-GNAc	—	3.50	—
Me-α-D-Gal	>10.9	—	7.00
α-D-Gal-(1 → 6)-D-Glc	—	>5.40	3.20 .
Me-β-D-Gal	>10.7	—	—
β-D-Gal-(1 → 3)-D-GalNAc	>0.59	—	—
β-D-Gal-(1 → 3)-D-GNAc	>1.95	—	—
β-D-Gal-(1 → 4)-D-GNAc	>2.09	—	—
β-D-Gal-(1 → 6)-D-GNAc	>1.31	—	—
D-Gal	>15.8	>2.40	10.5
Me-α-D-Glc	>8.50	—	24.1
Me-β-D-Glc	>17.9	—	—
D-Glc	>18.4	—	40
Me-α-D-Man	>39.0	—	55
D-Man	>22.3	—	80
Me-α-D-ManNAc	>5.32	—	—
Me-β-D-ManNAc	>4.08	—	—
D-ManNAc	>11.3	—	—
L-Fuc	>15.8	—	>150

TABLE III (Continued)

| | Micromoles for 50% inhibition of precipitation of hemagglutinin | | |
Inhibitor	With blood group A substance	With germfree rat colon antigen[b]	With S. typhimurium SH 180 LPS
β-L-Fuc-(1 $\rightarrow$ 3)-D-GNAc	>1.79	—	—
D-Xyl	—	—	95
D-Ara	—	—	>150
S. typhimurium TV 160 core fragment	—	—	0.70
S. typhimurium SH 180 core fragment	—	—	4.50

[a] Precipitation was performed in a total volume of 200 μl except for S. typhimurium SH 180 LPS where 400 μl was used. The following amounts of hemagglutinin and carbohydrate antigens (= equivalence point) were used: 12.8 μg of human blood group A substance (cyst MSM) + 5.2 μg N of hemagglutinin; 11.8 μg of germfree rat colon antigen + 3.5 μg N of hemagglutinin; 88.7 μg of S. typhimurium SH 180 LPS + 3.4 μg N of hemagglutinin.

[b] Rat colon antigen was obtained by phenol–water extraction of feces from germfree rats. The preparation was purified by ethanol precipitation (55–65% ethanol fraction) and treated with neuraminidase.

[c] See text.

[d] The capacities of these three compounds to give 50% inhibition of precipitation of human blood group H substance, first stage periodate oxidation and Smith degradation (cyst JS; 14.9 μg) with the hemagglutinin (3.9 μg N) were also studied [S. Hammarström and E. A. Kabat, Biochemistry 10, 1684 (1971)]. The concentrations needed were 0.031, 0.072, and 0.010 μmole/200 μl for Et-β-D-GalNAc, p-NO$_2$Ph-β-D-GalNAc, and D-GalNAc, respectively.

[e] D-GalNH$_2$, 2-amino-2-deoxy-D-galactose; D-GNH$_2$, 2-amino-2-deoxy-D-glucose.

[f] D-GalNAc-ol, 2-acetamido-2-deoxy-D-galactitol, D-GNAc-ol, 2-acetamido-2-deoxy-D-glucositol.

[g] These values are calculated from inhibition data obtained with 9.4 μg of human blood group A substance (cyst MSM) and 3.9 μg N of hemagglutinin. With this amount of hemagglutinin 0.92 μmoles of D-GalNAc was needed for 50% inhibition as compared to 1.65 μmoles in the system shown in the table.

ing that the steric orientation of the hydroxyl group on C-4 is of lesser importance. (4) The presence of an equatorially oriented N-acetyl or O-acetyl group on C-2 is essential for strong binding to the site. (5) All α-glycosidically linked compounds bind more strongly to the site than the corresponding β-linked derivatives. (6) Sugars that are in any way structurally related to D-GalNAc can be shown to interact with the hemagglutinin site provided the assay system is made sufficiently sensitive.

Direct Precipitation. Human blood group A substance, desialized ovine submaxillary mucin, group C streptococcal polysaccharide, hog blood group A + H substances, and purified intestinal mucin from germfree rats are precipitated by the hemagglutinin.[3,6,15] In these macromolecules, multiple nonreducing α-linked D-GalNAc end groups are exclusively or predominantly responsible for the interaction with the hemagglutinin site. For human blood group A substance this has been rigorously shown by means of an N-deacetylase from *Clostridium tertium*.[6] This enzyme specifically removes the N-acetyl group on nonreducing D-GalNAc of the blood group A determinant.[22] Enzyme treatment reduced the precipitating ability of blood group A substance to approximately one-tenth that of the original preparation[6] (Fig. 5A). Re-N-acetylation restored its precipitating ability completely[6] (Fig. 5A).

The hemagglutinin also precipitates teichoic acids from *Staphylococcus aureus* containing α-linked D-GNAc nonreducing end groups, but not those with β-linked nonreducing D-GNAc[6] (Fig. 5B). The content of α-D-GNAc containing teichoic acids could be determined quantitatively in strains that have a mixture of both by using the 3528 strain as reference (this strain contains teichoic acid with α-linked D-GNAc nonreducing end groups exclusively).

The hemagglutinin did not precipitate group A streptococcal poly-

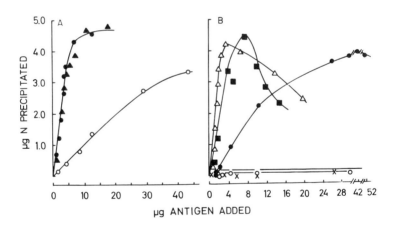

FIG. 5. Precipitation studies on hemagglutinin of fraction II of S. Hammarström and E. A. Kabat [*Biochemistry* **8**, 2696 (1969)]. (A) Precipitation of purified *Helix pomatia* A hemagglutinin, 3.83 μg of N of fraction II, by human blood group A substance prior to (●) and after (○) treatment with blood group A N-deacetylating enzyme and of enzyme-treated and re-N-acetylated (▲) A substance; total volume 200 μl. (B) Precipitation of teichoic acids from different strains of *Staphylococcus aureus* (△, strain 3528; ■, NYH-6; ●, Copenhagen; ×, A1) and of S. albus (Prengel) (○) with purified snail hemagglutinin; 3.83 μg of N of fraction II, total volume of 200 μl.

[22] D. M. Marcus, E. A. Kabat, and G. Schiffman, *Biochemistry* **3**, 437 (1964).

saccharide.[15] In this polysaccharide D-GNAc is β-linked to the rhamnosyl-$(1 \rightarrow 3)$-rhamnose backbone. From these studies it would appear that macromolecules with sterically accessible nonreducing α-linked D-GNAc end groups, but not with β-linked D-GNAc end groups, are precipitated by the hemagglutinin.

The hemagglutinin also precipitates blood group substances of B, H, Le^a, or precursor specificity.[6] The amount needed to precipitate a given amount of hemagglutinin varied within and between serotypes but was in all cases higher than that needed for blood group A substance.[6] Human blood group H substance also precipitates with the hemagglutinin after periodate oxidation and Smith degradation. The material was actually more active than untreated H-substance.[23] A second cycle of oxidation and degradation (2nd periodate step) decreased the activity markedly. However, some activity was still left.[23] The group(s) responsible for precipitation in blood group substances other than A is (are) not known. A possible candidate is α-D-GalNAc linked directly on serine or threonine in the peptide backbone.[24] In light of the data obtained with LPS from *Salmonella typhimurium* rough mutants (see below) other interpretations are however possible.

Recent studies[21] have shown that lipopolysaccharides from certain rough mutants of *S. typhimurium* are precipitated by the hemagglutinin. While the LPS from the parent smooth strain was inactive, LPS from mutants of chemotype Ra and Rb and to some extent also chemotype SR were precipitated by the hemagglutinin. LPS from mutants of chemotypes Rc, Rd, Re as well as isolated lipid A were inactive. The reaction was immunologically specific and could also be demonstrated by independent techniques. Lipopolysaccharide core fragments prepared by weak acid hydrolysis and purified by gel filtration from the Rb mutant TV160 and the Ra mutant SH 180 both inhibited precipitation between the hemagglutinin and SH 180 LPS (Table III). The inhibition powers of these fragments were comparable to melibiose (α-D-Gal-$(1 \rightarrow 6)$-D-Glc). The precipitating capacity of TV 160 lipopolysaccharide was abolished after treatment with the enzyme D-galactose oxidase. It was suggested that the structural element, α-D-Gal-$(1 \rightarrow 6)$-D-Glc . . . , present in all LPS from active mutants is the "immunodominant" structure responsible for precipitation.

This finding is rather unexpected considering the poor inhibition obtained with Me-α-D-Gal or melibiose. However, in a situation where both reagents are polyvalent even very weak interactions may be sufficient for precipitation.

---

[23] S. Hammarström and E. A. Kabat, unpublished observations.
[24] K. O. Lloyd and E. A. Kabat, *Proc. Nat. Acad. Sci. U.S.* **61**, 1470 (1968).

Utility

Helix pomatia A hemagglutinin may be a useful tool for the detection of carbohydrate end groups on macromolecules and on cell surfaces. It is already used routinely in blood group serology as a specific agglutinin for human A erythrocytes.[25] Its absolute specificity for human A erythrocytes as compared to B- and O-erythrocytes can be shown by trace-labeling it with [125]iodine[3] (Fig. 6). It has furthermore been shown to be useful in group determination of β-hemolytic streptococci. Thus group C and H streptococci are specifically agglutinated by the hemagglutinin.[26] Likewise *Escherichia coli* 026, 086, 0111, and 0126 are agglutinated by the crude snail extract.[27]

However, when used as a tool for the detection of particular carbohydrate end groups, mere demonstration of an interaction with an unknown macromolecule or cell is obviously not sufficient. Complementary analysis is in most cases essential for identification of the end group. In such analysis inhibition with haptens may be performed in order to compare the relative strengths of the reaction with an already established system involving the same anticarbohydrate reagent. The unknown macromolecule may furthermore be chemically or enzymatically treated to remove or modify the end group in question. The relatively broad

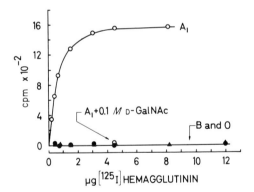

Fig. 6. Binding of radioactive *Helix pomatia* A hemagglutinin to human A-, B-, and O-erythrocytes. Approximately 40×10^6 cells were incubated with different amounts of ^{125}I-labeled hemagglutinin (specific activity = 2700 cpm/μg) in a volume of 2.0 ml. The cells were washed two times with PBS, and cell-bound radioactivity was determined by γ-counting.

[25] C. Högman, personal communications.
[26] W. Köhler and O. Prokop, Z. *Immunol. Forsch.* **133**, 50 (1967).
[27] G. Uhlenbruck, O. Prokop, and W. Haferland, *Zentralbl. Bakteriol. I Orig.* **199**, 271 (1966).

specificity of the hemagglutinin tends to limit its usefulness in this respect.

More useful may be the possibility to separate polysaccharides, glycoproteins, oligosaccharides or even cells with the aid of the hemagglutinin, made insoluble through polycondensation or otherwise.

[46] Eel Serum Anti-Human Blood-Group H(O) Protein[1]

By Parimal R. Desai and Georg F. Springer

Eel anti-human blood-group H(O) protein, a potent hemagglutinin of human blood-group O erythrocytes,[2] is a 7 S globulin.[3,4] Originally it was thought that the combining sites of the eel antibody for bloodgroup H(O) specific structures are complementary to α-L-fucopyranose (6-deoxy-α-L-galactopyranose),[5,6] but we found that monosaccharides with both L- and D-galactose configuration are complementary to these sites and function as inhibitory haptens provided they possess at least one methyl group on either C-3 or C-5.[7-10] Numerous other hexoses, pentoses and their derivatives are inactive.[7] The eel anti H(O) protein specifically precipitates not only with blood-group H(O) active macromolecules of human, animal, and plant origin, but surprisingly also with 3-O-methyl-D-fucose (D-digitalose) and 3-O-methyl-D-galactose.[8,9,11] The minimum combining structure which shows inhibitory activity with the eel anti-H(O) antibody is smaller than a monosaccharide. It consists of a methyl substituent attached equatorially to a pyranose ring, an ether oxygen adjoining this methyl group, and an axial, oxygencarrying substituent cis to the methyl group on a contiguous C atom.[8,10]

[1] This investigation was supported by National Science Foundation Grant GB-8378. The Department is maintained by the Susan Rebecca Stone Fund for Immunochemistry Research.
[2] S. Miyazaki, *Nagasaki Idai Hoigaku Gyoho* **2**, 542 (1930).
[3] A. Bezkorovainy, G. F. Springer, and P. R. Desai, *Biochemistry* **10**, 3761 (1971).
[4] G. F. Springer and P. R. Desai, *Vox Sang.* **18**, 551 (1970).
[5] W. M. Watkins and W. T. J. Morgan, *Nature (London)* **169**, 825 (1952).
[6] R. Kuhn and H. G. Osman, *Hoppe-Seyler's Z. Physiol. Chem.* **303**, 1 (1956).
[7] G. F. Springer and P. Williamson, *Biochem. J.* **85**, 282 (1962).
[8] G. F. Springer, P. R. Desai, and B. Kolecki, *Biochemistry* **3**, 1076 (1964).
[9] G. F. Springer, T. Takahashi, P. R. Desai, and B. J. Kolecki, *Biochemistry* **4**, 2099 (1965).
[10] P. R. Desai and G. F. Springer, *Proc. 10th Congr. Int. Soc. Blood Transfus.*, Stockholm, p. 500 (1965).
[11] G. F. Springer and P. R. Desai, *Biochemistry* **10**, 3749 (1971).

While the overall molecular contours of the presumed combining sites of the inhibitors and the precipitinogens are closely similar, there are some additional specific requirements for monosaccharides to function as precipitinogens. They apparently consist, within this contour, of three vicinal oxygens protruding from a C_1 pyranose ring. The oxygen at C-3 must carry an apolar group, and the two oxygens flanking this group must be capable of hydrogen bonding. One of these latter oxygens must be equatorial and trans to the oxygen at C-3 and the other axial and cis. Additional apolar groups are not compatible with the precipitating activity of the pyranose structure; they transform it into either an inhibitory hapten or an inactive compound.[11]

Isolation Procedure

Principle. The isolation of the eel anti-H(O) serum protein is based on its ready and specific precipitability with the 2 monosaccharides mentioned above and the ease of dissociation of the precipitated monosaccharide–protein complex upon dialysis with simultaneous removal of the precipitinogen.

Procedure. Sera, 3–8 ml per fish, are obtained from live eels (*Anguilla rostrata*) weighing >1.5 lb each, after anesthesis with urethane.[7] These eels are caught in fresh or brackish water. We found that on the eastern seaboard of North America their serum has maximum anti-human blood-group H(O) protein content between September and December. The precipitinogen D-digitalose may be obtained from the cardiac glycosides strospesid or panstrosid as well as from the antibiotic chartreusin by repeated hydrolysis of 10% solutions with Kiliani's mixture[12] followed where necessary by preparative paper chromatography.[7,8]

The anti-human blood-group H(O) antibody is precipitated from the eel serum with an equivalent amount of D-digitalose (2–6 μmoles per 0.5 ml of serum) as described under Assay Methods. The protein content of the precipitates of two tubes from each experiment is determined colorimetrically with Folin–Ciocalteau reagent[13] and used in calculation of percent recovery of the isolated anti-H(O) protein. The washed precipitates are either worked up immediately or stored at −20° in the freeze-dried state. The precipitates are suspended at approximately double their original protein concentration in buffered saline (0.1 M NaCl, 0.05 M PO_4^{-3}, pH 7.3) and dialyzed on a mechanical shaker at 4° against 20 volumes of buffered saline which is gradually replaced by deionized, distilled water until no more Cl^- is demonstrable with $AgNO_3$. This stepwise electrolyte decrease minimizes antibody precipitation. In-

---

[12] H. Kiliani, *Ber. Deut. Chem. Ges.* **63**, 2866 (1930).
[13] M. Heidelberger and C. F. C. MacPherson, *Science* **97**, 405 (1943); **98**, 63 (1943).

soluble matter amounting on the average to 3.8% of the total recovered is removed by centrifugation at 1500 g and 4° for 1 hour. Both the soluble and the insoluble fractions are freeze-dried and then dried to constant weight at 22–25° over P_2O_5 at 10^{-1} mm Hg. The average recovery of the soluble eel anti-blood-group H(O) protein, by weight, is about 94% of that expected from the colorimetric analyses of the precipitates. Prior to chemical analyses the antibody is electrodialyzed at 4° through a dialysis membrane at 200 V[9] and dried as described above.

Assay Methods

Hemagglutination and Hemagglutination Inhibition Assays.[14-16] These are carried out at 22–25° in 10 × 75 mm tubes with buffered saline as diluent and solvent. The voulme of all reagents in a test is kept constant; either 0.05 or 0.02 ml is used throughout. Twofold serial dilutions of the test material are made using a different 0.1 ml serological pipette for each dilution. In hemagglutination assays, 0.5% suspension of human O erythrocytes is added to the titrated solutions of the eel anti-H(O) serum; the mixtures are then shaken and allowed to stand for 90 minutes before reading. In hemagglutination inhibition assays, prior to addition of red cells 4 minimum hemagglutinating doses of eel anti-H(O) serum are added to the titrated solutions of the inhibitor; the mixtures are shaken, then incubated for 2 hours. In both assays agglutination is read microscopically by at least two individuals, and the materials are tested at least twice. In hemagglutination assays an appropriately diluted suspension of human O erythrocytes serves as negative control; in hemagglutination inhibition assays, an additional positive control consisting of an eel serum standard diluted to 4 hemagglutinating doses and titrated in 2-fold serial dilutions is used. L-Fucose is included as a standard in all inhibition assays.

Precipitin and Precipitation Inhibition Tests. These quantitative procedures, based on those of Heidelberger and Kendall,[17] are used with some modifications[8,9]; 0.5-ml portions of all reagents are pipetted into graduated, glass-stoppered, 10-ml centrifuge tubes using Ostwald-Folin pipettes. The samples are incubated for 30 minutes in an ice bath and then for 7–10 days at 4° with daily agitation.

In inhibition studies haptens are added in 0.5-ml volume to 0.5 ml

[14] G. F. Springer, R. E. Horton, and M. Forbes, *J. Exp. Med.* **110**, 221 (1959).

[15] G. F. Springer and R. E. Horton, *J. Clin. Invest.* **48**, 1280 (1969).

[16] G. F. Springer, *in* "Methods in Immunology and Immunochemistry" (C. A. Williams and M. W. Chase, eds.), Vol. IV. Academic Press, New York, in press, 1972.

[17] E. A. Kabat, *in* "Kabat and Mayer's Experimental Immunochemistry," 2nd ed., pp. 22–96. Thomas, Springfield, Illinois, 1961.

of undiluted serum and incubated in an ice bath for 30 minutes; thereafter 0.5 ml of precipitinogen solution of concentration corresponding to that at the beginning of equivalence is added, and the procedure is continued as above. Inhibition is calculated as the percent decrease of protein precipitated in presence of the inhibitory hapten as compared to that precipitated in its absence. Negative controls in all assays consist of eel serum or isolated antibody alone and antigen or hapten alone, each adjusted to appropriate volume.

Precipitates obtained with antigens are recovered by centrifugation and washed twice with 1.5 ml of ice-cold buffered saline. Because of their greater solubility, precipitates obtained with monosaccharides are washed only once. Washed precipitates are dissolved in 2.5 ml of water containing 0.01 meq of NaOH, and their protein content is determined colorimetrically[13] with four different concentrations of human γ-globulin as standard. The extinction given by the eel antibody is the same as that given by the human γ-globulin between 15 and 200 μg of protein with a deviation of $\pm 1\%$ below and $\pm 3\%$ above 100 μg protein.[11]

Properties

Appearance, Solubility, and Purity.[11] The isolated eel anti-bloodgroup H(O) protein is a white, fluffy powder that gives a clear or slightly opalescent solution up to at least 0.2% in buffered saline (pH 7.3) and up to at least 0.1% in buffers of pH 5.5–9.0. The average wet ash content of this antibody is 3.7% before and $<1\%$ after electrodialysis; maximum weight loss on drying to constant weight at 80° is 8%.

Stability.[11] Eel antibody solutions in buffered saline or deionized water are stable for at least 1 month at 4°. Repeated freezing and thawing severely damages the antibody. A 12-day incubation of 0.1% solutions of eel antibody in buffers of pH 5.5–9.0 at 4°, followed by adjustment of the solution pH to 7.3 by dialysis, does not alter the extent of their human O erythrocyte agglutinating activity.

Immunochemical Reactivity and Specificity.[4,11] The hemagglutinating activity of the isolated antibody before and after electrodialysis is the same as that of the original eel serum. The isolated eel antibody also gives typical precipitin curves, in the same proportions as those when it is still in serum, with all the blood-group H(O) active macromolecules and the precipitating monosaccharides. Blood-group H(O) active ovarian cyst glycoprotein, *Sassafras* polysaccharide, D-digitalose and 3-*O*-methyl-D-galactose each precipitate in the equivalence zone from 68.7 to 95.9% of the total antibody protein; the average precipitation from different pools is about 75%. An average of 2.3% protein is found in the washes while about 23% remains in the supernatants, whose agglutinin titers parallel their protein content. After adjustment of the protein content of dialyzed supernatants of monosaccharide precipitin tests to its preprecipitation concentration, the same proportion of protein is precipitable by the precipitinogens as from the original antibody solution. The results are the same when whole eel serum is used. At equivalence the precipitation of eel antibody with D-digitalose is optimal between pH 6.70 and 7.15. It should be noted that >98% of the D-digitalose remains in the supernatant in all areas of the precipitin curve.

Immunochemical homogeneity of the isolated eel antibody is indicated by the shape of the precipitin curves and by the single symmetric arc obtained in immunoelectrophoresis using rabbit anti-isolated eel anti-H(O) protein. Also, agar gel diffusion studies on the eel anti-H(O) protein with the H(O)-specific human ovarian cyst glycoprotein, *Taxus*, and *Sassafras* antigens as well as with the blood-group H(O)-specific antigen from *E. coli* O_{128}[18] show only one sharp band; all these bands fuse completely with one another.

Removal, in the equivalence area, of >85% of the eel anti-H(O) antibody by monosaccharide precipitation and inhibition of this monosaccharide precipitation to >80% by H(O) specific haptens such as L-fucose, methyl α-L-fucopyranoside and 2,3-di-O-methyl-D-galactose but not by the H(O) inactive 3-O-methyl-D-glucose attest to the specificity of the reaction.

Chemical Characteristics.[11] The isolated electrodialyzed eel anti-H(O) protein contains ca. 15.7% N (corrected). Asp, Gly, Glu, Ala, Ser, and Thr are the predominant amino acids, and there is a scarcity of Met, Trp, and Phe. No significant amounts of carbohydrate are demonstrable except GlcN, of which 0.39% (3 moles per mole of antibody) is found. End-group analyses show approximately equal quantities of Ser and Ala at the NH$_2$-termini, and Ser and Gly in the same ratio at the COOH-termini and suggest the presence of two types of polypeptide chains in the eel anti-H(O) protein.

Physicochemical Characteristics.[3,19] The isolated eel anti-H(O) protein is usually homogeneous by ultracentrifugal and electrophoretic criteria. Ultracentrifugal analyses of some preparations show a more rapidly migrating component, apparently a multimer of the eel antibody molecules, which may account for up to 10% of the total material; it disappears in presence of $8 M$ urea. The eel antibody migrates like a human α_2-globulin at pH 8.6 and its isoelectric point is at pH 5.25 ±0.05. Its $s_{20,w}^0$ is 7.2 S, $D_{20,w}^0$ is 5.0×10^{-7} cm^2/sec. Its partial specific

[18] G. F. Springer, P. Williamson, and W. C. Brandes, *J. Exp. Med.* **113**, 1077 (1961).
[19] B. Jirgensons, G. F. Springer, and P. R. Desai, *Comp. Biochem. Physiol.* **34**, 721 (1970).

volume is 0.705 ml/g and its intrinsic viscosity is 3.4 ml/g. Its molecular weight is 123,000 and its β-value is 2.0×10^{-6} suggesting a nearly spherical shape. Its $A^{1\%}_{278 \text{ nm}}$ is 12.696 in water and its $[\alpha]^{29}_D$ is $-77°$ (C 0.5, water, 1 dm). The anti-H(O) antibody preparations from individual eels give closely similar circular dichroism spectra between 185 and 310 nm which differ significantly from those of individual human 7 S immunoglobulins.

Succinylation and/or reduction-alkylation of the eel anti-H(O) antibody followed by sedimentation and diffusion studies as well as disc electrophoresis in polyacrylamide–sodium dodecyl sulfate indicate that this protein consists of three physically bonded subunits of apparently identical molecular weight of 40,000 each of which in turn consists of four polypeptide chains of identical size; these chains, whose molecular weight is about 10,000, are joined by disulfide linkages.

[47] Homogeneous Mouse Immunoglobulins (Myeloma Proteins) That Bind Carbohydrates

By MICHAEL POTTER and C. P. J. GLAUDEMANS

Plasma cell tumors are derived from highly specialized cells that are restricted to making single molecular species of immunoglobulin (homogeneous immunoglobulin, myeloma proteins). These tumors can be induced in strain BALB/c mice by the intraperitoneal injection of mineral oil (0.5 ml given 3 times when the mouse is 2, 4, 6 months of age). Although these homogeneous immunoglobulins are synthesized and secreted by tumors, a modest percentage (ca. 5%) of them can be shown to bind antigens. This is usually demonstrated by a random screening process in which a myeloma protein is reacted with a battery of test antigens. The antigens are macromolecules that contain multiple haptenic repeating units, e.g., derivatized proteins, or polysaccharides. The multivalent IgA and IgM myeloma proteins can be tested for precipitating activity to polyvalent antigens in agar-gel diffusion tests. This permits the rapid screening of large numbers of samples. When a myeloma protein–antigen system is identified, it can be further studied to establish the chemistry of the hapten, specificity of related haptens, and the binding affinity of the myeloma protein for the hapten (K_A). When large numbers of mouse myeloma proteins are screened (i.e., several hundred) often more than one protein can be found that binds the same antigen. Immunochemical characterization of myeloma proteins that bind the same antigen usually reveals that they differ from each other in primary structure. Most of the differences are attributed to variations in the so-called variable parts of the immunoglobulin molecules. In a few rare cases evidence has been obtained that myeloma proteins derived from tumors of independent origin may be chemically and antigenically similar. A summary of the available myeloma proteins that bind polysaccharides is given in the table.

Isolation of Myeloma Proteins

Myeloma proteins are isolated from the serum or ascites of mice bearing syngeneically transplanted plasma cell tumors. Two methods of transplantation are employed. The subcutaneous method of transplantation by trochars is used to maintain all stock tumors and may also be used to obtain serum. Pieces of healthy tumor tissue (avoiding areas of necrosis) are removed with sterile instruments and placed in a watch glass containing sterile physiological salt solution or a suitable basic tissue culture medium. Two or three fragments are minced with scissors and then loaded into a 2.5–3 inch 13-gauge trochar (trochars can be made from paracentesis needles that have been rebeveled at the appropriate length and fitted with a smoothed-end obturator that protrudes about 2 mm). The trochar is introduced into the inguinal region and inserted laterally toward the axilla, and the tumor tissue is deposited as the trochar is withdrawn. The tumor is allowed to grow until 3–5 cm in diameter when the blood (1.5–2.0 ml) is harvested by cardiac puncture (a 2-ml tuberculin syringe fitted with a 1-inch, 23-gauge needle has been optimal in our experience). Transplantation to syngeneic F_1 hybrids which attain a larger size than inbred mice, enables better serum yields.

The ascites method provides a much better source of myeloma protein. The ascites forms of plasma cell tumors are often unstable in regard to protein production during continuous intraperitoneal passage and should be checked at each transfer generation for protein. Nonproducing ascites lines are discarded and new lines reinitiated from the continuous subcutaneous lines as required. Ascites lines are initiated by transferring intraperitoneally suspensions of 10^6 to 10^7 cells. Cell suspensions can be obtained by finely mincing the tumor fragments and aspirating the suspensions during the mincing procedure with a syringe and needle (21–25 gauge). Plasma cell tumor cells are fragile, and for this reason tissue homogenizers are not used. The ascites that develops can be used for the next transfer generations. Ascites is harvested from mice by inserting a sterile 19-gauge needle intraperitoneally and allowing the ascites to drain into a test tube. It is often possible to obtain 2 or more paracenteses from the same mouse several days apart. A single mouse may produce

MOUSE MYELOMA PROTEINS THAT BIND POLYSACCHARIDE ANTIGENS

Specificity	Myeloma proteins immunoglobulin chain class[a]	K_A M^{-1}	Immunoabsorbent			Polysaccharide antigens precipitated	References
			Matrix	Specificity group	Eluent		
Phosphorylcholine	S63 (α κ)[b] McPC603 (α κ) MOPC 167 (α κ)	1 × 10⁶ 1 × 10⁵ 1 × 10⁵	Sepharose 4B	Glycyltyrosine + p-diazonium phenyl phosphorylcholine	Phosphorylcholine	Pneumococcus C polysaccharide Extracts of Aspergillus, Trichoderma, Lacto-bacillus-4 Ascaris suum	d–i
	MOPC 511 (α κ)	1 × 10⁴	Sepharose 2B	Pneumococcus C polysaccharide			
α-1 → 3 Linked D-glucose	J558 (α λ)		8 or 10% Acrylamide gel (25% bis)	Dextran B 1355 S1,3		Leuconostoc mesenteroides	j–l
	MOPC 104E (μ λ)	3.1 × 10⁴				Dextran B 1355S	
β-1 → 6 Linked D-galactose	SAPC 10 (α κ)		Sepharose 4B	p-Aminophenyl β-D-galactoside	p-NO₂-phenyl β-D-galactoside	Gum ghatti, arabino-galactan	m–o
	TEPC 191 B (α κ) J 539 (α κ)	2.2 × 10⁸	Affinose			Wheat extract antigen	
β-Linked N-acetyl-D-mannosamine	MOPC 406 (α κ)					Salmonella weslaco lps	p–q
β-Linked N-acetyl-D-glucosamine	S117 (α κ)					Blood group H substance,[c] Staphylococcus aureus teichoic acid, group A streptococcal polysaccharide	r

| α-Methyl-D-galactoside | MOPC 384 (α κ) | 8 or 10% Acrylamide gel | Proteus mirabilis sp-2 lps | α-Methyl-D-galacto-side | Salmonella tel aviv lps, Salmonella tranoroa lps, Proteus mirabilis sp 2 lps | p |
| β-2 → 6 linked fructofuranose | J606 (γ3 κ) | | | | β-2 → 6 linked levan, Leuconostoc mesenteroides inulin | s |

[a] The class of heavy and light chain are given.

[b] Six other proteins probably of similar structure but of independent mouse origin have been found.

[c] After first and third periodate oxidating and Smith degradation.

[d] B. Chesebro and H. Metzger, Biochemistry 11, 766 (1972).

[e] M. Cohn, G. Notani, and S. A. Rice, Immunochemistry 6, 111 (1969).

[f] M. A. Leon and N. M. Young, Biochemistry 10, 1424 (1971).

[g] H. Metzger, B. Chesebro, N. Hatler, J. Lee, and N. Otchin, in "Progress in Immunology" (B. Amos, ed.), pp. 253–267. Academic Press, New York (1971).

[h] M. Potter and R. Lieberman, J. Exp. Med. 132, 737 (1970).

[i] A. Sher, E. Lord, and M. Cohn, J. Immunol. 107, 1226 (1971).

[j] M. A. Leon, N. M. Young, and K. R. McIntire, Biochemistry 9, 1023 (1970).

[k] M. Weigert, M. Cesari, S. J. Yonkovich, and M. Cohn, Nature (London) 228, 1045 (1970).

[l] N. M. Young, I. B. Jocius, and M. A. Leon, Biochemistry 10, 3457 (1971).

[m] M. Potter, Ann. N.Y. Acad. Sci. 190, 306 (1971).

[n] M. Potter, E. B. Mushinski, and C. P. J. Glaudemans, J. Immunol. 108, 295 (1972).

[o] A. Sher and H. Tarikas, J. Immunol. 106, 1227 (1971).

[p] M. Potter, Fed. Proc., Fed. Amer. Soc. Exp. Biol. 29, 85 (1970).

[q] L. Rovis, A. E. Kabat, and M. Potter, Fed. Proc., Fed. Amer. Soc. Exp. Biol. 31, 749 (1972).

[r] G. Vicari, A. Sher, M. Cohn, and A. E. Kabat, Immunochemistry 7, 829 (1970).

[s] H. M. Grey, J. W. Hirst, and M. Cohn, J. Exp. Med. 133, 289 (1971).

as much as 10 ml of ascitic fluid, and if the tumor is a good producer the ascites can contain circa 20 mg/ml of myeloma protein. Not all lines are such good producers.

The myeloma proteins are optimally purified from ascites or serum by affinity chromatography, i.e., immunoabsorption and elution with a hapten. The preparation of the specific immunoabsorbents depends upon the hapten involved and whether it can be linked to an insoluble substance. A model system in this respect has been developed by Chesebro and Metzger[1] in which they coupled phosphorylcholine by a diazo linkage to glycyltyrosine, which in turn had been connected to Sepharose. These workers found that the glycyltyrosine spacer was necessary for facile dissociation with haptens of the myeloma protein bound to the phosphorylcholine. Cuatrecasas and co-workers[2,3] have shown that a spacer between ligand (hapten) and basic column material is imperative for adequate binding of enzymes. Sher and Tarikas[4] and Wofsy and Burr[5] have reported the preparation of phenyl-β-D-galactoside attached to Sepharose by a diazo linkage. Recently a novel absorbent has been reported. It was prepared by polymerizing acrylamide with ultraviolet radiation in the presence of riboflavin and a dextran, thus trapping the dextran antigen within the network of the polyacrylamide gel. Acrylamide absorbents in general can be made by incorporating proteins,[6] polysaccharides,[7] and lipopolysaccharides into gel containing acrylamide (75%) and bisacrylamide (25%) with a total solid content of 8–10%. With lipopolysaccharides, the concentration of the antigens to be incorporated is critical; 8 mg/ml has proved to be optimal. Gels can also be polymerized with ammonium persulfate and N,N,N',N'-tetramethyl ethylenediamine.

The valence of the immunoglobulin may be an important factor in successful absorption. Univalent (Fab fragments), bivalent, 4-chain monomer, and poly multivalent polymer (for IgA, IgM immunoglobulins) forms of the same protein apparently each have different binding energies. Each immunoabsorbent system must be individually analyzed. In our hands, derivatized agarose gels are not as efficient in the purification of anti galactoside myeloma proteins as are derivatized Sepharose

[1] B. Chesebro and H. Metzger, *Biochemistry* 11, 766 (1972).
[2] P. Cuatrecasas and C. B. Anfinsen, *Annu. Rev. Biochem.* 40, 259 (1971).
[3] P. Cuatrecasas, M. Wilchek, and C. B. Anfinsen, *Proc. Nat. Acad. Sci. U.S.* 61, 636 (1968).
[4] A. Sher and H. Tarikas, *J. Immunol.* 106, 1227 (1971).
[5] L. Wofsy and B. Burr, *J. Immunol.* 103, 380 (1969).
[6] S. Carrel and S. Bavantum, *Immunochemistry* 8, 39 (1971).
[7] N. M. Young, I. B. Jocius, and M. A. Leon, *Biochemistry* 10, 3457 (1971).

gels. The former frequently bound albumin as well as myeloma proteins. Nevertheless, their preparation, which follows the general procedure of Cuatrecasas and Anfinsen,[2] will be given below also.

Preparation of Immunoabsorbents

Sepharose-Based Immunoabsorbents. p-Aminophenyl-β-D-thiogalactoside (168 mg) was dissolved in water (5.2 ml). Aqueous hydrochloric acid (0.98 N, 1.13 ml) previously cooled in ice was added, followed by an ice-cold solution of sodium nitrite (40 mg) in water (1.8 ml). This yellow solution was added slowly and with stirring to a solution of bovine serum albumin (BSA) (1.0 g) in ice cold 0.04 N NaOH (56 ml). The resulting red solution was left for 2 hours in ice after which the pH was adjusted to 7.3 with 1 N HCl, and the whole was dialyzed vs cold distilled water. In this concentration (ca. 1.6% solids) the material was extremely potent in precipitating the β 1 → 6 galactose linked specific TEPC 191 mouse myeloma protein. The BSA-phenylgalactoside can be conveniently kept after lyophilization.

Coupling of the above soluble albumin–phenylgalactoside to Sepharose was done as follows: Sepharose 2B (120 ml, settled volume) was admixed with 200 ml of distilled water and cooled to about 10°. Cyanogen bromide (10 g) was added all at once, and with the help of a pH meter the suspension was kept at pH 10.5 by the uninterrupted, dropwise addition of 4 N aqueous sodium hydroxide solution. When the pH showed almost no tendency to drop, and was thus fairly constant, the whole was filtered through a sintered glass funnel and washed with 2 liters of ice-cold Na_2HPO_4–citrate buffer pH 6.4. The Sepharose gel thus reacted, was immediately treated with the bovine serum albumin–diazophenyl-β-D-thiogalactoside dissolved in ice cold phosphate–citrate buffer (150 ml), the pH remaining at 6.3. The suspension was left stirring, *very gently* to prevent rupturing the Sepharose beads, at +5° overnight. The material was next washed on a sintered glass filter with much ice-cold distilled water over a period of several days. If the resulting gel is red, and the filtrate virtually colorless, it is a good indication that binding took place. The material is preserved by the presence of sodium azide in 0.025% overall concentration. In our case, if all the p-aminophenyl-β-D-galactoside was linked to the BSA, and if all the BSA-galactoside was in turn successfully linked to the Sepharose, it can be calculated that there is 0.0048 mmole of ligand per milliliter (settled volume) of Sepharose-ligand gel (or 4.8 μmoles/ml). If only *one* site per IgA monomer (after reduction and alkylation) needs to be bound for retention by the gel, and if every ligand can take part in the binding, it follows that theoretically one could bind 750 mg of myeloma protein per milliliter of this Sepharose-ligand gel (assuming a molecular weight of 156,000 per 2 binding sites for the myeloma protein). It should, of course, be borne in mind that many ligand groupings may be buried within the gel network and may be inaccessible. Each column material prepared should, therefore, be checked individually for its capacity with the serum in question. This column material retained monomeric IgA (reduced and alkylated) from myeloma serum that reacted with a β $1 \rightarrow 6$ galactan. The protein could be eluted with pH 7.4 buffer solutions containing p-nitrophenyl-β-D-galactoside $(0.05\,M)$ to yield a single substance.

Agarose-Based Immunoabsorbents. Affinose AF 202 (agarose-$NH(CH_2)_3NH(CH_2)_3NHCO(CH_2)_2COOH$, 8 μmoles of ligand per milliliter, commercially available from the Affitron Corporation, 4737 Muscatel Avenue, Rosemead, California 91770), 50 ml, was stirred with 80 ml of distilled water. p-Aminophenyl-β-D-thiogalactoside (50 mg, Calbiochem) was added, and the pH was adjusted to 4.7. Next, 1-ethyl-3(3-dimethylaminopropyl) carbodiimide hydrochloride (1.5 g) in water (5 ml) was added, and the whole was stirred gently for 2 hours while maintaining the pH at 4.7. It was then left for 3 hours (pH reached 5.5) and washed on a sintered filter with 0.1 M aqueous NaCl (10 liters).

This absorbent could bind IgA monomer from a myeloma serum originally containing IgA which was precipitable by a β $1 \rightarrow 6$ galactan. Elution with sodium tetraborate buffer pH 10 would yield the IgA monomer, sometimes contaminated by the presence of albumin only.

Lipopolysaccharide-Acrylamide Gel Absorbent (Modified from Carrel et al.[6]). The lipopolysaccharide of *Proteus mirabilis* sp 2 (of mouse origin) was prepared from acetone-killed organisms. The bacteria were heated in 0.85% NaCl, and for each 50 ml of suspension, 50 ml of phenol was added; the preparation was brought to a single phase by heating at 70° for 15 minutes. The mixture was cooled, centrifuged, and the aqueous phase was removed, and then dialyzed against distilled water. The lipopolysaccharide was recovered from the dialyzed aqueous phase by lyophilization. To each ml of 10% acrylamide monomer solution (containing 75 mg of acrylamide and 25 mg bisacrylamide per ml), 8 mg of lipopolysaccharide was added. The mixture was stirred over gentle heat until clear or opalescent, and then polymerized by the addition of 0.5 μl N,N,N',N'-tetramethyl ethylenediamine and 10.0 μl of fresh 10% ammonium persulfate. Polymerization was rapid at room temperature, but the whole was allowed to stand overnight at 4°. The gel was then extensively washed with 0.85% NaCl, and homogenized in a Potter-Elvehjem tissue homogenizer. "Fines" were removed from the wash and columns with good flow rates were obtained. The MOPC 384 IgA protein was added to the column and was eluted with 0.2 M methyl-α-D-galactoside. The columns could be washed and reutilized.

For unknown reasons, this method has occasionally not worked for lipopolysaccharides. Nevertheless, it provides a promising approach for this type of absorption.

Section IV

Enzymes of Sugar Activation

[48] L-Fucose Kinase from Pig Liver[1,2]

By Hanako Ishihara, Harry Schachter, and Edward C. Heath

$$\text{L-Fucose} + \text{ATP} \xrightarrow{\text{Mg}^{2+}} \beta\text{-L-fucose 1-phosphate} + \text{ADP}$$

Assay Method

Principle. For assay of the kinase in relatively impure preparations (Method A), L-[^{14}C]fucose is employed as substrate and the reaction is followed by the increase in barium-insoluble radioactivity.[3] For purer enzyme preparation (Method B), the kinase reaction may be followed by spectrophotometric determination of ADP formation according to the procedure of Kornberg and Pricer.[4]

Method A

Reagents

L-[1-^{14}C]Fucose, 10 mM (5.4 × 10^5 cpm per μmole)
Potassium ATP, 100 mM
MgCl$_2$, 100 mM
KF, 100 mM
Tris buffer, pH 8, 500 mM
Zinc sulfate, 0.33 N
Barium hydroxide, 0.33 N

Procedure. The reaction mixture consists of 10 μl of L-[1-^{14}C]fucose, 10 μl of potassium ATP, 10 μl of MgCl$_2$, 25 μl of KF, 25 μl of Tris buffer, and appropriate dilutions of enzyme in a final volume of 300 μl. The mixture is incubated at 37°, and, at zero time and at 10-minute intervals, 100-μl aliquots are withdrawn and placed in clean tubes; 100 μl each of zinc sulfate and barium hydroxide solutions are added, and the mixture is thoroughly mixed. After standing for 5 minutes, the suspension is centrifuged at 2000 g for 5 minutes, and the radioactive content of a 100-μl aliquot of the supernatant fluid is determined. Differences in the radioactive content of the supernatant fluids between the zero time

[1] H. Ishihara, D. J. Massaro, and E. C. Heath, *J. Biol. Chem.* **243**, 1103 (1968).
[2] I. Jabbal and H. Schachter, *J. Biol. Chem.* **246**, 5154 (1971).
[3] R. E. Trucco, R. Caputto, L. F. Leloir, and N. Mittelman, *Arch. Biochem.* **18**, 137 (1948).
[4] A. Kornberg and W. C. Pricer, Jr., *J. Biol. Chem.* **193**, 481 (1951).

controls and the incubated samples are regarded as the amount of
L-fucose that is phosphorylated.

Method B

Reagents

L-Fucose, 100 mM
Potassium ATP, 100 mM
MgCl$_2$, 100 mM
KF, 100 mM
Tris buffer, pH 8, 500 mM
Sodium phosphopyruvate, 10 mM
NADH, 10 mM
Lactic dehydrogenase (Worthington), 2× crystallized, 1 mg/ml

Procedure. Incubation mixtures contained 10 μl of L-fucose, 25 μl
of potassium ATP, 10 μl of MgCl$_2$, 25 μl of KF, 25 μl of Tris buffer, and
appropriate dilutions of the enzyme in a final volume of 500 μl. The
mixtures are incubated at 37° for 20 minutes, and the reaction is terminated by placing the tubes in a boiling water bath for 1 minute. The
mixture is cooled and centrifuged at 2000 g for 5 minutes. Aliquots
(50–100 μl) of the supernatant fluid are added to cuvettes that contain
2.5 μl of sodium phosphopyruvate, 10 μl of NADH, 1 μl of MgCl$_2$, 50 μl
of Tris buffer, and 50 μl of lactic acid dehydrogenase in a final volume
of 1 ml. The mixtures are incubated until no further oxidation of NADH
is detected spectrophotometrically (340 nm), and the total change in
absorbance is regarded as proportional to the amount of ADP formed.
Control incubation mixtures which contain no L-fucose are handled in
an identical manner to compensate for residual ATPase activity in the
preparations.

Definition of Unit and Specific Activity. One unit of L-fucose kinase
activity is defined as that amount which is required to catalyze the
formation of 1 nmole of L-fucose 1-phosphate per 20 minutes under either
of the assay conditions described above. Unless otherwise indicated, all
procedures of the purification are carried out at 4°.

Purification Procedure

Preparation of Crude Extract. Porcine liver (110 g), obtained immediately after slaughtering at a local packing house, is cut into small
pieces and suspended in 4 volumes of 0.25 M sucrose solution. The suspension is mixed gently in a Waring Blendor for about 1 minute, ad-

justed to pH 7.6 with 1 N NaOH, and blended vigorously for 3 minutes. The homogenate is filtered through cheesecloth, and the filtrate is centrifuged at 11,000 g for 40 minutes.

Protamine Sulfate. To the crude extract (380 ml) is added 95 ml of a 2% protamine sulfate solution (pH 7.6); the mixture is allowed to stand for 15 minutes, then centrifuged at 18,000 g for 20 minutes. The supernatant fluid is retained for further fractionation.

Ammonium Sulfate. To 450 ml of the protamine sulfate supernatant fraction is added 74 g of solid ammonium sulfate; the mixture is stirred for 15 minutes and centrifuged at 18,000 g for 15 minutes. The precipitate is dissolved in 50 ml of 0.03 M phosphate buffer, pH 7.4, and retained for further fractionation.

Sephadex G-100. A column (3 × 80 cm) containing Sephadex G-100 is equilibrated for several days with 0.03 M phosphate buffer, pH 7.4. An aliquot (55 ml) of diluted ammonium sulfate fraction (10 mg of protein per milliliter) is applied to the column, and the gel is washed with 0.03 M phosphate buffer; 5-ml fractions are collected in an automatic refrigerated fraction collector. L-Fucose kinase activity is recovered usually in fractions 9 through 32 in a relatively symmetrical peak of activity. These fractions are pooled (120 ml) and stirred mechanically while 58 g of solid ammonium sulfate is added; stirring is continued for an additional 15 minutes, and the suspension is centrifuged for 15 minutes at 18,000 g. The precipitate is dissolved in approximately 10 ml of 0.03 M phosphate buffer, pH 7.4, and dialyzed against 500 ml of the same buffer for 3 hours.

Alumina C_γ Gel. The dialyzed fraction from the preceding step is diluted with 0.03 M phosphate buffer to give a final protein concentration of 10 mg/ml. One-half volume of alumina C_γ gel (Calbiochem; 13 mg of solids per milliliter, stored in 0.03 M phosphate buffer, pH 7.4) is added to the diluted protein solution; the suspension is chilled in an ice bath, stirred for 20 minutes, and centrifuged at 18,000 g for 20 min. The supernatant fluid contains insignificant L-fucose kinase activity and is discarded. The precipitate is washed successively (by centrifugation as above) with 1.5-ml portions of 0.01 M and 0.05 M phosphate buffer, pH 7.4, and finally with 0.1 M phosphate buffer, pH 8. Although there is some variability in the distribution of the enzyme activity in these fractions, optimal recovery usually is obtained by combining all three of the gel eluates.

With this fractionation procedure, L-fucose kinase is purified approximately 70-fold, with recovery of 14% of the activity. These results are summarized in the table.

PURIFICATION OF L-FUCOSE KINASE

Fraction	Volume (ml)	Protein (mg/ml)	Specific activity (units/mg protein)	Overall recovery (%)
Crude extract	380	32.0	0.95	100
Protamine sulfate	450	10.6	2.11	87.9
Ammonium sulfate (0–30% saturation)	50	29.5	4.97	65.2
Sephadex G-100	11.6	10.4	21.6	22.6
Alumina C$_\gamma$	6.2	3.82	65.5	13.5

Properties

Stability. L-Fucose kinase is a relatively unstable enzyme under the conditions used in these studies. Thus, the purified enzyme preparation loses 32% and 60% of its activity after storage at −16° for 24 and 72 hours, respectively. This instability is partially overcome by addition of ovalbumin to the preparation to a final concentration of 10 mg/ml; under these conditions, 82% and 70% of the activity remains after storage at −16° for 24 and 48 hours, respectively. Neither L-fucose nor ATP, when added to the kinase preparation, affects its stability. The ammonium sulfate fraction is relatively stable and loses only approximately 10% of its activity when stored at −16° for 1 week.

Specificity. The relative rates of formation of ^{32}P-labeled sugar phosphates (using unlabeled sugar and [^{32}P]ATP as substrates) are as follows: L-fucose, 100; D-glucose, 107; D-ribose, 87; L-rhamnose, 41; D-arabinose, 28; L-arabinose, 31; and D-mannose, D-galactose, D- and L-xylose, D-fucose, and L-fuculose, less than 5.

Several nucleoside triphosphates are capable of partially substituting for ATP as a phosphate donor in the kinase reaction. The relative rates of formation of L-fucose 1-phosphate with various nucleoside triphosphates as substrates are as follows: ATP, 100; CTP, 60; UTP, 59; GTP, 56; and TTP, 3.

[49] GDP-L-Fucose Pyrophosphorylase from Pig Liver[1]

By HANAKO ISHIHARA and EDWARD C. HEATH

β-L-Fucose 1-phosphate + GTP ⇌ GDP-L-fucose + P-P$_i$

Assay Method

Principle. The assay procedure used in these studies to measure GDP-L-fucose pyrophosphorylase activity is based on the exchange of radioactive inorganic pyrophosphate with GTP.[2] Thus, incubation of inorganic [32P]labeled pyrophosphate and unlabeled GDP-L-fucose with the pyrophosphorylase results in the formation of [32P]GTP, which may be determined as charcoal-adsorbable radioactivity. The enzyme may alternatively be assayed in a similar manner except that L-fucose 1-phosphate, GTP, and inorganic [32P]pyrophosphate are substituted for GDP-L-fucose and inorganic [32P]pyrophosphate. Similar results are obtained with either type of incubation mixture.

Reagents

GDP-L-fucose, 1 mM
Sodium [32P]pyrophosphate, 50 mM (approximately 2 × 10⁴ cpm/μmole)
MgCl₂, 100 mM
KF, 100 mM
Tris buffer, pH 8, 500 mM

Procedure. Incubation mixtures contain 20 μl of GDP-L-fucose, 20 μl of sodium [32P]pyrophosphate, 10 μl of MgCl₂, 25 μl of KF, 20 μl of Tris buffer, and appropriate dilutions of enzyme in a final volume of 220 μl. After incubation at 37° for 5 minutes, 1.5 ml of an ice-cold solution of 5% trichloroacetic acid containing 0.03 M unlabeled sodium pyrophosphate are added to the mixture. The precipitate is removed by centrifugation, and 0.1 ml of charcoal (Norit A) suspension (150 mg/ml) is added to the supernatant fluid. The charcoal is collected by centrifugation and washed with 3 ml of cold 5% trichloroacetic acid, followed by three 3-ml portions of cold water. The charcoal is then suspended in 0.3 ml of 50% ethanol containing 3% (by volume) of concentrated ammonium hydroxide; 0.2 ml of the suspension is removed for the determination of radioactivity.

---

[1] H. Ishihara and E. C. Heath, *J. Biol. Chem.* **243**, 1110 (1968).
[2] A. Munch-Petersen, *Acta Chem. Scand.* **10**, 928 (1956).

Definition of Unit and Specific Activity. One unit of GDP-L-fucose pyrophosphorylase is defined as the amount of enzyme that catalyzes the incorporation of 1 nmole of inorganic [^{32}P]pyrophosphate into GTP per 5 minutes under the conditions described above. Specific activity is defined as units per milligram of protein. Unless otherwise noted, all procedures are conducted at 4°.

Purification Procedure

Preparation of Crude Extract. Porcine liver (120 g) is obtained immediately after slaughtering at a local packing house, cut into small pieces, and suspended in 480 ml of 0.25 M sucrose solution. The suspension is stirred gently in a Waring Blendor for 2 minutes, adjusted to pH 7.6 by the addition of 1 N ammonium hydroxide, and then blended at high speed for 5 minutes. The homogenate is filtered through cheesecloth, and the filtrate is centrifuged at 11,000 g for 40 minutes. To 410 ml of the supernatant fluid (crude extract) is added 100 ml of a 2% solution of protamine sulfate (pH 7.6). The suspension is stirred for 5 minutes, and the precipitate is removed by centrifugation at 18,000 g for 15 minutes.

Ammonium Sulfate Fractionation. To 465 ml of the protamine sulfate supernatant fraction is added 76.3 g of solid ammonium sulfate, and the solution is stirred mechanically for 15 minutes. The suspension is centrifuged at 18,000 g for 15 minutes, and to the supernatant fraction is added 55 g of solid ammonium sulfate; the suspension is stirred for 15 minutes, then centrifuged at 18,000 g for 10 minutes; the precipitate is dissolved in 0.03 M phosphate buffer, pH 7.6, in a final volume of 105 ml.

Calcium Phosphate Gel. One-fourth volume of calcium phosphate gel (14 mg of solids per milliliter) is added to 105 ml of the ammonium sulfate fraction. The suspension is stirred for 10 minutes and centrifuged at 18,000 g for 10 minutes. The supernatant fluid is retained for further fractionation.

Sephadex G-100 and DEAE-Cellulose Fractionation. Thirty milliliters of the calcium phosphate gel fraction, containing approximately 550 mg of protein, are applied to a column (3 × 80 cm) of Sephadex G-100 which is equilibrated for several days against 0.03 M phosphate buffer, pH 7.6, containing 0.01 M 2-mercaptoethanol. The column is eluted with the same buffer at the rate of 25 ml per hour, and 5-ml fractions are collected. The enzymatic activity is recovered essentially in the void volume of the column (105 ml, fractions 8 through 28); these fractions are pooled and immediately applied to a column (2.5 × 20 cm) of DEAE-cellulose which is equilibrated with the phosphate-mercaptoethanol buffer. The column is washed with 50 ml of buffer, and then the enzyme is eluted with the same buffer containing $0.1\,M$ sodium chloride; the flow rate is 20 ml per hour, and 5-ml fractions are collected. The enzyme activity is recovered primarily in fractions 12 through 17, and these fractions are pooled and retained for further fractionation.

Alumina $C\gamma$ Fractionation. The DEAE-cellulose fraction is diluted with phosphate-mercaptoethanol buffer to a final protein concentration of 10 mg/ml. To 35 ml of the diluted solution are added 35 ml of alumina C_γ gel suspension (13 mg of solids per milliliter); the mixture is stirred for 15 minutes and centrifuged at 18,000 g for 10 minutes. The supernatant fluid is discarded, and the gel is washed successively with 4 ml each of $0.01\,M$, $0.05\,M$, and $0.1\,M$ phosphate buffer containing $0.01\,M$ mercaptoethanol. These supernatant solutions are pooled and dialyzed against $0.03\,M$ phosphate–mercaptoethanol buffer for 3 hours.

With this purification procedure, GDP-L-fucose pyrophosphorylase is purified approximately 46-fold relative to the protamine sulfate supernatant fraction, with an overall recovery of about 45% of the activity. These results are summarized in the table. The purified enzyme preparation is unstable to storage at either $-16°$ or $0°$; thus, 80–85% of the activity is lost when the enzyme is stored for 24 hours in the absence of mercaptoethanol. The presence of $0.01\,M$ mercaptoethanol in the buffer reduces losses on storage to about 50% in a 24-hour period, while inclusion of $0.01\,M$ dithioerythritol in the buffer further decreases the loss in activity to approximately 30–35% in a 24-hour period. These conditions are the most satisfactory of those tested in preserving the pyrophosphorylase activity. Thus, further addition of albumin, GDP-L-fucose, GTP, or L-fucose 1-phosphate to the storage buffer, or variation of the pH of the buffer, does not enhance the stability of the enzyme.

PURIFICATION OF GDP-L-FUCOSE PYROPHOSPHORYLASE

Fraction	Volume (ml)	Protein (mg/ml)	Specific activity (units/mg protein)	Recovery (%)
Crude extract	410	40.0	—[a]	—[a]
Protamine sulfate	465	14.1	0.39	100
Ammonium sulfate	105	24.2	0.93	92.2
Calcium phosphate gel	126	18.2	1.01	91.5
Sephadex G-100 and DEAE-cellulose	18.5	18.0	6.45	84.5
Alumina C_γ	23	2.8	18.1	45.8

[a] Interfering reactions in the crude extract prevented accurate estimation.

Specificity. GDP-L-fucose pyrophosphorylase exhibits a relatively high degree of specificity for GTP and β-L-fucose 1-phosphate. When GTP is the nucleoside triphosphate substrate, only α-D-mannose 1-phosphate exhibits a partial capacity (17.5%) to substitute for L-fucose 1-phosphate; synthetic α-L-fucose 1-phosphate and the 1-phosphate derivatives of D-glucose and D-galactose are inactive. Similarly, in the presence of β-L-fucose 1-phosphate, only UTP partially substitutes (18.7%) for GTP; CTP; TTP, and ATP are inactive.

[50] ADP-Glucose Pyrophosphorylase from *Escherichia coli* B

By HACHIRO OZAKI and JACK PREISS

$$\text{ATP} + \text{glucose-1-P} \rightleftarrows \text{ADP-glucose} + \text{PP}_i$$

Assay Method

Principle. Enzymatic activity is determined by measuring the synthesis of [^{32}P]ATP from ADP-glucose and P-[^{32}P]P$_i$ in the presence of the activator fructose-1,6-diphosphate (FDP). The ATP is isolated by adsorption onto Norit A and estimated by measuring the radioactivity contained in the Norit. The procedure given below corresponds to the maximum of activity in presence of activator.

Reagents

ADP-glucose, 10 mM

Sodium [^{32}P]pyrophosphate, 10 mM (specific activity 5–40 × 10^5 cpm/μmole)

Tris·HCl buffer, 1 M, pH 8.0

MgCl$_2$, 0.1 M

Bovine plasma albumin (BPA), 10 mg/ml

Fructose diphosphate (FDP), sodium salt, 20 mM

Sodium fluoride, 0.5 M

Trichloroacetic acid, 5%

Sodium pyrophosphate, 0.1 M, pH 8.0

Norit suspension, 150 mg per milliliter of water

Ethanol 50%, containing 1 ml of concentrated NH$_3$ per liter of solution

Procedure. The ADP-glucose (0.02 ml), pyrophosphate-^{32}P (0.05 ml), Tris·HCl buffer (0.02 ml) MgCl$_2$ (0.02 ml), BPA (0.01 ml), FDP (0.02

ml), NaF (0.005 ml) solutions are mixed together, and reaction is initiated (in a volume of 0.25 ml) with ADP-glucose pyrophosphorylase (0.00015–0.0015 unit). After incubation at 37° for 10 minutes, the reaction is terminated by addition of 3 ml of cold 5% trichloroacetic acid, followed by 0.1 ml of the sodium pyrophosphate solution. The nucleotides are then adsorbed by the addition of 0.1 ml of the Norit A suspension. The suspension is centrifuged, and the supernatant fluid discarded. The charcoal is washed twice with 3-ml portion of 5% cold trichloroacetic acid and once with 3 ml of cold distilled water. The washed Norit is suspended in 2 ml of an aqueous solution of 0.1% NH_3 and 50% ethanol. One-milliliter portions are pipetted into planchets and counted in a gas flow counter.

One unit of enzyme activity is defined as that amount of enzyme catalyzing the formation of 1 μmole of ATP per minute at 37° under the above conditions. Specific activity is designed as units per milligram of protein. Protein concentration is determined by the method of Lowry *et al.*[1]

The above assay procedure has proved to be useful for assay of crude extracts both from bacteria and plants.   .

Activity-Stain Method

This method is qualitative and is used for localization of the enzyme activity on the disc acrylamide gel and among the fractions obtained from column chromatography.

Principle. ADP glucose pyrophosphorylase activity is visualized by coupling it with hexokinase and glucose-6-phosphate dehydrogenase, and measuring reduction of nitroblue tetrazolium (NBT).

Reagents

ADP-glucose, 10 mM
Sodium pyrophosphate, 0.1 M
Tris·HCl buffer, 1 M, pH 8.0
$MgCl_2$, 0.1 M
Fructose diphosphate (FDP), sodium salt, 20 mM
Glucose, 10 mM
TPN, 10 mg/ml
Phenazine methosulfate (PMS), 1 mg/ml
Nitroblue tetrazolium (NBT) 5 mg/ml
Hexokinase 1000 IU/ml (Sigma Type F-300)

[1] O. H. Lowry, N. J. Rosebrough, A. L. Farr, and R. J. Randall, *J. Biol. Chem.* **193**, 265 (1951).

Glucose-6-phosphate dehydrogenase (ammonium sulfate free) 1000 IU/ml

Procedure. The ADP-glucose (0.16 ml), pyrophosphate (0.04 ml), Tris·HCl buffer (0.2 ml), MgCl₂ (0.16 ml), FDP (0.12 ml), glucose (0.2 ml), TPN (0.04 ml), PMS (0.1 ml), NBT (0.1 ml) hexokinase (0.01 ml), and glucose-6-phosphate dehydrogenase (0.01 ml) solutions are mixed together in a volume of 2.0 ml. For localization of the enzyme activity on the acrylamide gel (0.5 × 5 cm) after electrophoresis the gel is incubated in 2 ml of the reaction mixture in a small test tube (0.8 × 7.5 cm). After 20 minutes' incubation activity, bands will have formed where there is about 0.05 unit of enzyme, and the gel is rinsed with water several times to remove excess dyes.

For staining the column effluent, 0.05–0.1 ml of the reaction mixture is placed on a spot-plate and 0.005 ml of the enzyme fraction is added, mixed well, and left at room temperature. Activity of more than 0.1 unit/ml can be detected in a few minutes of incubation.

Growth of Cells

The *E. coli* B strain used was a mutant, SG3, derepressed in the levels of ADP-glucose pyrophosphorylase and glycogen synthetase. Cells were grown on minimal media containing per liter: 11 mg of CaCl₂, 246 mg of MgSO₄·7H₂O, 2.4 grams of (NH₄)₂SO₄, 6.8 grams of KH₂PO₄ 14.2 grams of Na₂HPO₄ and 12 grams of D-glucose. The media was adjusted to pH 7.2 with NaOH. The cells were grown at 37°, harvested 8 hours after the stationary phase was attained, and stored at −12°.

Purification Procedure

Step 1. Extraction of the Enzyme. Thawed cells, 449 g, are combined with 1200 g of chilled Super-brite 100 (Minnesota Mining and Manufacturing Company) glass beads and 320 ml of 0.05 M glycylglycine buffer, pH 7.0, containing 5 mM GSH. This mixture is homogenized at medium speed in a Waring Blendor for 25 minutes. The temperature is kept below 10° by strapping slabs of Dry Ice to the sides of the stainless steel blender with rubber tubing. The mixture is stirred again for 10 minutes after addition of 1200 ml of the glycylglycine–GSH buffer. Then the glass beads are allowed to settle for 5 minutes and the supernatant solution is decanted. Another 800 ml of the buffer is used to rinse the glass beads with slow stirring for 10 minutes. The glass beads are allowed to settle; the wash is then decanted and added to the first supernatant solution. The resulting crude extract is used for the purification of the enzyme.

Step 2. Digestion by Deoxyribonuclease (DNase) and Heat Treatment. To the crude extract (2330 ml) are added 30 mg DNase (bovine pancreatic DNase-1, Sigma Chemical Company) and 12 ml of 1 M MgCl₂ to give a final concentration of 5 mM. The resulting solution is mixed well by stirring gently with a glass rod and left at room temperature for 30 minutes. After digestion, 70 ml of 1 M phosphate, pH 7.0, is added to the solution to give a final concentration of 0.03 M. The solution is then heated with stirring (in 600-ml portions in a 3-liter Erlenmeyer flask) in a water bath maintained at 70° until the temperature of the solution rises to 66° (for 12–15 minutes). The solutions are then quickly chilled in an ice bath, and centrifuged at 16,000 g. The supernatant solution is used for ammonium sulfate fractionation.

Step 3. Ammonium Sulfate Fractionation. Solid ammonium sulfate is added to the supernatant fraction to give 60% saturation, stirred for 20 minutes and centrifuged at 16,000 g for 15 minutes. The precipitate is dissolved in 0.05 M Tris·HCl, pH 7.2, containing 5 mM dithiothreitol (DTT) and dialyzed overnight against 40 times the volume of the same buffer. The dialyzed solution is centrifuged at 75,000 g for 90 minutes in order to remove insoluble material (mostly glycogen).

Step 4. DEAE-Cellulose Chromatography. The supernatant solution in Step 3 is adsorbed onto a DEAE-cellulose column (3.5 × 45 cm, Whatman DE-52) which is equilibrated with 0.015 M potassium phosphate buffer, pH 7.5. The column is washed with 1 liter of 0.015 M potassium phosphate buffer, pH 7.5, containing 2.5 mM DTT. Then the enzyme is eluted by a linear gradient that contains 5 liters of 0.015 M potassium phosphate–2.5 mM DTT buffer, pH 7.5, in the mixing chamber and 5 liters of 0.2 M potassium phosphate buffer, pH 7.0, containing 2.5 mM DTT and 0.3 M KCl in the reservoir chamber. Fractions containing the enzyme activity are pooled, and the enzyme is precipitated with ammonium sulfate crystals which are added to a concentration of 75% saturation. The precipitate is dissolved in 0.05 M Tris·HCl buffer, pH 7.2, containing 2.5 mM DTT, and dialyzed against this buffer overnight. The resulting insoluble material is removed by centrifugation.

Step 5. DEAE-Sephadex Chromatography. The supernatant solution is adsorbed onto a DEAE-Sephadex column (2.5 × 36 cm, Type A-50, Pharmacia) which is equilibrated with 0.015 M potassium phosphate buffer, pH 6.1, containing 0.1 M KCl and 2.5 mM DTT. The column is washed with 100 ml of this buffer. Then the enzyme is eluted by a linear gradient that contains 2 liters of this buffer in the mixing chamber and 2 liters of 0.015 M potassium phosphate buffer, pH 6.1, containing 0.45 M KCl and 2.5 mM DTT in the reservoir chamber. Fractions containing the enzyme activity are pooled, adjusted to pH 7.1 adding 1 M Tris·HCl pH 8.5 and concentrated in an Amicon ultrafiltration apparatus to 50 ml. Enzyme is precipitated with ammonium sulfate crystals at 75% saturation. The precipitate is dissolved in $0.05\,M$ Tris·HCl, pH 7.2, containing 2.5 mM DTT, and dialyzed against this buffer overnight. The resulting insoluble material is removed by centrifugation.

Step 6. DEAE-Cellulose Chromatography at pH 8.5. The supernatant solution is adsorbed onto a DEAE-cellulose column (1.5×22 cm) which is equilibrated with $0.05\,M$ Tris·HCl, pH 8.5, containing 2.5 mM DTT. The column is washed with 50 ml of $0.05\,M$ Tris·HCl, pH 8.5, containing 2.5 mM DTT and $0.1\,M$ KCl. Then the enzyme is eluted by a linear gradient that contains 700 ml of the same buffer used for washing in the mixing chamber and 700 ml of the same buffer which contains $0.35\,M$ KCl instead of $0.1\,M$ KCl in the reservoir chamber. Fractions containing the enzyme activity are pooled, concentrated and fractionated with ammonium sulfate as in step 5. The resulting precipitate is dissolved in $0.01\,M$ potassium phosphate buffer, pH 6.8, containing 2.5 mM DTT, and dialyzed against 1 liter of this buffer overnight.

Step 7. Hydroxyapatite Chromatography. The dialyzed enzyme is adsorbed onto a hydroxyapatite column (1.5×9.5 cm, Bio-Gel HT, a product of Bio-Rad Laboratories) which is equilibrated with $0.01\,M$ potassium phosphate buffer, pH 6.8, containing 2.5 mM DTT. The column is washed with 30 ml of this buffer. The enzyme is eluted by a linear gradient that contains 200 ml of the same buffer used for washing in the mixing chamber and 200 ml of $0.2\,M$ potassium phosphate buffer, pH 6.8, containing 2.5 mM DTT in the reservoir. Flow rate of the buffer is approximately 2 ml per hour. Fractions containing the enzyme activity are pooled, concentrated by the precipitating enzyme with ammonium sulfate at 75% saturation, dissolved in $0.05\,M$ Tris·HCl buffer, pH 7.2, containing 2.5 mM DTT, and dialyzed against 500 ml of this buffer overnight. This step resulted in an overall purification of 650-fold over the crude fraction with a 27% yield (Table I). This enzyme preparation is stored in this buffer at $-20°$. It is stable also at $0°C$ (in ice bath) for at least several months.

Properties of the Enzyme

Homogeneity and Molecular Weight. Analysis of the ADP-glucose pyrophosphorylase on analytical disc gel electrophoresis in several gel concentration (6–9%) using the Tris–glycine buffer system[2] and

[2] L. Ornstein and B. Davis, *Ann. N.Y. Acad. Sci.* **121**, 321 (1964).

TABLE I
PURIFICATION OF ADP-GLUCOSE PYROPHOSPHORYLASE

Step	Volume (ml)	Protein (mg)	Specific activity (units/mg)	Total activity (units)
1. Crude extract	2,330	68,300	0.159	10,825
2. DNase and heat treatments	1,990	9,440	1.156	10,890
3. Ammonium sulfate	292	4,540	1.90	8,615
4. DEAE-cellulose	79	720	9.25	6.676
5. DEAE-Sephadex (pH 6.1)	14	154	22.5	3,466
6. DEAE-cellulose (pH 8.5)	11.5	43.8	72.9	3,195
7. Hydroxyapatite (pH 6.8)	8.0	25.9	103.0	2,670

Hedrick and Smith's system[3] and Coomassie Blue as the protein stain[4] indicated only one protein band as being present. The protein band was shown to contain the enzyme activity by the activity-stain method described above. The molecular weight of the enzyme was determined by disc gel electrophoresis by the method of Hedrick and Smith.[3] The molecular weight was estimated to be 211,000 ± 6000.

Sedimentation velocity was measured with analytical ultracentrifuge equipped with schlieren interference. The enzyme (0.75–3.0 mg protein per milliliter) in 0.1 M Tris·HCl, pH 7.2, containing 2.5 mM DTT exhibited a single symmetrical peak. By extrapolation of the $s_{20,w}$ values to zero protein concentration, the $s_{20,w}$ at infinite dilution was determined to be 10.9 S.

The molecular weight by sedimentation equilibrium was determined by carrying out a meniscus depletion run at three different concentrations of enzyme (0.75, 1.5, and 3.0 mg/ml) as described by Yphantis.[5] The molecular weight of the enzyme was determined to be 207,000 ± 7000, assuming a partial specific volume for the enzyme as 0.74.

The molecular weight was also determined by sucrose-density gradient centrifugation according to Martin and Ames.[6] Beef liver catalase (molecular weight is 240,000) and pig heart lactate dehydrogenase (molecular weight is 135,000) were used as standard. The molecular weight of the enzyme was determined to be 204,000 ± 6000.

Subunit and Its Molecular Weight. Electrophoresis in polyacrylamide gel (10%) containing 0.1% sodium dodecyl sulfate was carried

[3] J. L. Hedrick and A. J. Smith, *Arch. Biochem. Biophys.* **126**, 155 (1968).
[4] R. A. Chrambach, M. Reisfeld, M. Wyckoff, and J. Zacceri, *Anal. Biochem.* **20**, 150 (1967).
[5] D. A. Yphantis, *Biochemistry* **3**, 297 (1964).
[6] R. G. Martin and B. N. Ames, *J. Biol. Chem.* **236**, 1372 (1961).

out as described by Weber and Osborn,[7] except that samples were heated at 65° for 15 minutes and were not dialyzed prior to electrophoresis. After electrophoresis the gels were stained with Coomassie Blue by the method of Weber and Osborn,[7] except that the electrophoretically destaining procedure was omitted. Electrophoresis of ADP glucose pyrophosphorylase by this method revealed the presence of one protein component with a molecular weight of 53,000 ± 2000 using bovine serum albumin, catalase, ovalbumin, yeast alcohol dehydrogenase, and pig heart lactate dehydrogenase as standards.

The enzyme preparation was dialyzed for 48 hours at room temperature against 6 M guanidine hydrochloride solution containing 0.102 M/β-mercaptoethanol and 0.02 M NaCl as described by Kawahara and Tanford.[8] The dialyzed enzyme in guanidine–HCL was then subjected to a meniscus depletion sedimentation equilibrium according to the procedure of Yphantis.[5] Results obtained after 96-hour run with enzyme at 0.73 mg protein per milliliter at 29,440 rpm showed a linear plot of the log of the fringe displacement vs. (radial displacement),[2] indicating the homogeneity of the dissociated protein. The molecular weight of the dissociated protein was determined to be 47,600 assuming the value for partial specific volume as 0.74.

Based on these observations, it is concluded that the ADP glucose pyrophosphorylase from *E. coli* B consists of four subunits of identical molecular weight.

Kinetic Properties. The enzyme from *E. coli* B is activated by fructose diP, TPNH and pyridoxal 5'-P (PLP) and to lesser extents by other glycolytic intermediates.[9-11] The activation curves are sigmoidal in shape and at pH 8.0 the concentrations of FDP, TPNH, and PLP required for 50% of maximal activation are 0.12, 0.14, and 0.011 mM, respectively. The effect of FDP (1.5 mM) on the apparent affinities of the substrates is shown in Table II. FDP decreased the concentration of substrates required for 50% of maximal velocity about 5- to 8-fold and increased the V_{max} of ADP-glucose pyrophosphorolysis and synthesis about 2- and 6-fold, respectively. The pyrophosphate saturation curves are hyperbolic. However, the ATP, ADP-glucose, and MgCl$_2$ curves are sigmoidal in shape. In the presence of activator, the glucose-1-P curve is hyperbolic, but in the absence of FDP the glucose-1-P saturation

[7] K. Weber and M. Osborn, *J. Biol. Chem.* **244**, 4406 (1969).
[8] K. Kawahara and C. Tanford, *J. Biol. Chem.* **241**, 3228 (1966).
[9] J. Preiss, L. Shen, E. Greenberg, and N. Gentner, *Biochemistry* **5**, 1833 (1965).
[10] N. Gentner and J. Preiss, *J. Biol. Chem.* **243**, 5882 (1968).
[11] N. Gentner, E. Greenberg, and J. Preiss, *Biochem. Biophys. Res. Commun.* **36**, 373 (1969).

TABLE II
KINETIC PARAMETERS OF *Escherichia coli* B ADP-GLUCOSE PYROPHOSPHORYLASE
IN TRIS · HCl BUFFER, pH 8.0

	$S_{0.5}$ (mM)[a]		V_{max} (μmoles/min/mg)	
Substrate	−FDP	+FDP	−FDP	+FDP
ADP-glucose	0.90	0.11		
Pyrophosphate	0.61	0.11	59	103
ATP	2.5	0.39		
Glucose-1-P	0.24	0.032	17	104
MgCl$_2$	11.5	2.0		

[a] $S_{0.5}$ is defined as the concentration required to give 50% of maximal velocity.

curve exhibits negative cooperativity. Inhibitors of the enzyme are 5′-adenylate, ADP, P$_i$, and sulfate.[9-12] The concentration of activator modulates the sensitivity of the enzyme to inhibition. In the presence of 1.5 mM FDP, 60 μM 5′-AMP is required for 50% inhibition while at 0.5 mM FDP, 50% inhibition occurs at 25 μM 5′-AMP. These phenomena have formed the basis for a hypothesis on the regulation of the biosynthesis of ADP-glucose and glycogen in *E. coli*.[13]

[12] N. Gentner and J. Preiss, *Biochem. Biophys. Res. Commun.* **27**, 417 (1967).
[13] J. Preiss, *in* "Current Topics in Cellular Regulation" (B. L. Horecker and E. R. Stadtman, eds.), Vol. 1, p. 125. Academic Press, New York, 1969.

[51] CMP-Sialic Acid Synthetase of Nuclei

By EDWARD L. KEAN

$$\text{Sialic acid} + \text{CTP} \xrightarrow[\text{}]{\text{Mg}^{2+}} \text{CMP-sialic acid} + \text{pyrophosphate}$$

Several reports have described the properties of this enzyme as it is obtained in soluble extracts of tissue.[1-5] However, present evidence demonstrates that CMP-sialic acid synthetase is located in the nucleus of the cell.[5,6] The source of the nuclei used in the following description

[1] S. Roseman, *Proc. Nat. Acad. Sci. U.S.* **48**, 437 (1962).
[2] L. Warren and R. S. Blacklow, *J. Biol. Chem.* **237**, 3527 (1962).
[3] E. L. Kean and S. Roseman, *J. Biol. Chem.* **241**, 5643 (1966).
[4] E. L. Kean and S. Roseman, Vol. 8 [31].
[5] E. L. Kean, *Exp. Eye Res.* **8**, 44 (1969).
[6] E. L. Kean, *J. Biol. Chem.* **245**, 2301 (1970).

of the properties and preparation of this enzyme is rat liver, although nuclei from a variety of tissues have been used.

Assay Method

Principle. Unused sialic acid in the incubation medium is reduced with sodium borohydride, while CMP-sialic acid, a glycoside, is resistant to this treatment.[2-4] After the excess borohydride is oxidized, CMP-sialic acid is hydrolyzed to sialic acid by the acidity of the reagents used in the thiobarbituric acid (TBA) assay[7] which is used to measure the free sialic acid. Controls compensate for extraneous color derived from reduced sialic and other components of the incubation mixture.

Reagents

Tris·MgCl₂ buffer mixture composed of 9 parts of $1.0\,M$ Tris·HCl buffer, pH 8.5, and 1 part of $1.0\,M$ MgCl₂

NaBH₄, 100 mg/ml H₂O

Acetone

2-Thiobarbituric acid (TBA), 0.6% in water. Heat may be used to aid in dissolving the compound, which will remain in solution for about 1 month when stored at room temperature.

Sodium metaperiodate, $0.2\,M$ in $9.0\,M$ H₃PO₄

Sodium arsenite, 10% in $0.5\,M$ Na₂SO₄ containing $0.1\,N$ H₂SO₄

Saturated Na₂SO₄

Cyclohexanone (reagent grade, distill before use if yellow)

Sialic acid: N-acetylneuraminic (NAN), $0.05\,M$, or N-glycolyl-neuraminic (NGN), $0.05\,M$, the pH is adjusted to 5.5 with KHCO₃ and the solutions stored frozen.

CTP, $0.05\,M$, adjusted to pH 7 with KHCO₃

Procedure. Each of the following components is pipetted into 15 × 125 mm test tubes: 0.025 ml of NAN or NGN; 0.025 ml of CTP; 0.05 ml of Tris·Mg buffer; about 0.6 unit of enzyme; and water to a total volume of 0.25 ml. Incubation is carried out at 37° for 10 minutes. A control which contains all the above components and boiled enzyme is included with each incubation. After incubation, 0.05 ml of cold NaBH₄ solution is added and the solutions are mixed. After 15 minutes at room temperature, 0.05 ml of acetone is added, the solution is mixed vigorously with a Vortex, and allowed to stand at room temperature for an additional 15 minutes. Periodate, 0.1 ml, is added to each tube with mixing. After 20 minutes at room temperature, 1.0 ml of sodium arsenite is added and

[7] L. Warren, *J. Biol. Chem.* **234**, 1971 (1959).

the solution is mixed until the yellow-brown color disappears.[8] TBA solution (1.5 ml) is added, and, after mixing, the tubes are capped with a marble and placed in boiling water for 15 minutes. The solutions are then cooled for 5 minutes in an ice bath, 0.25 ml of saturated sodium sulfate is added, and the tubes are allowed to stand at room temperature for 10 minutes. The chromophore (whose color disappears in the cold) is extracted with 4.0 ml of cyclohexanone. The extraction may be accomplished conveniently by mixing with a Vortex for about 20 seconds using sufficient vigor to ensure adequate mixing. The tubes are centrifuged at about 1000 g at room temperature for 5 minutes, and the optical density of the upper phase is measured at 549 nm. Too vigorous mixing may result in cloudy solutions which may require recentrifugation of the upper phases before reading. Standard curves are prepared with either NAN or NGN by adding 0.005–0.05 μmole of sialic acid to tubes containing water, buffer, sodium borohydride, and acetone. The assay is then continued from the point of addition of periodate. Due to the differences in extinction coefficients, standard curves are prepared using the sialic acid that was employed as substrate.

The difference in optical density between the complete incubation tube and its boiled control is used in the calculations. Essentially identical values are obtained with standard curves prepared as above or when prepared in the presence of excess reduced sialic acid.[4]

Definition of Unit and Specific Activity. One unit of enzyme activity is defined as that quantity producing 1 μmole of CMP-sialic acid per hour when incubated at 37°. Specific activity is defined as units per milligram of protein.

Preparation of Enzyme

Isolation of Nuclei. The procedures of Chaveau, Moulé, and Rouiller[9] or that of Blobel and Potter[10] can be used to obtain purified nuclei.[11]

---

[8] If necessary, the assay may be interrupted at this point. The tubes are covered and stored overnight at 4°.
[9] J. Chaveau, Y. Moulé, and C. H. Rouiller, *Exp. Cell Res.* **11**, 317 (1956). See also H. Busch, Vol. 12A [51].
[10] G. Blobel and V. R. Potter, *Science* **154**, 1662 (1966). It is felt that this method is easier to perform than that of Chaveau *et al.*[9] due to the absence of the step requiring homogenization in dense sucrose.
[11] The purity of the nuclei used in these studies from rat liver was established on the basis of enzymatic, chemical, and morphological criteria. Little or no activity of the following marker enzymes can be detected in the nuclei purified by the procedures described: glucose-6-phosphate and 6-phosphogluconate dehydrogenases, glucose 6-phosphatase, NADH-cytochrome c reductase, succinic dehydrogenase, galactosyl transferase (*N*-acetylglucosamine as acceptor), and sialyl transferase

Male, Sprague-Dawley rats, 200–300 g, are fasted overnight and then killed by decapitation. The liver are removed, rinsed in cold 0.15 M NaCl, blotted on filter paper, and weighed quickly.

Nuclei Prepared by the Procedure of Chaveau et al.[9]

Reagents

Sucrose, 2.4 M, containing 3.3 mM CaCl$_2$
Sucrose, 1.0 M, containing 1.0 mM CaCl$_2$
Sucrose, 0.25 M, containing 3.3 mM CaCl$_2$
Tris·HCl, 0.01 M, pH 7.5, containing 1% (v/v) mercaptoethanol

Procedure. All of the following operations are performed at about 4°. The livers are minced and then homogenized with a solution of 2.4 M sucrose–3.3 mM CaCl$_2$ (10 ml of medium per gram wet weight of liver) with six passages of a glass–Teflon homogenizer having a 0.01-inch clearance. The homogenate is filtered through glass wool and centrifuged at 40,000 g for 60 minutes. The plug of tissue floating on the surface is skimmed off, the solution of dense sucrose over the white pellet of nuclei is removed, and the sides of the tube are wiped with gauze. The pellet is resuspended with a solution of 1.0 M sucrose–1.0 mM CaCl$_2$ (1 ml per gram of fresh tissue) and rehomogenized by hand with a Dounce homogenizer (large clearance pestle). The suspension is centrifuged at 3,000 g for 5 minutes, the supernatant fluid is decanted, and the nuclear pellet is resuspended with a Dounce homogenizer in a solution of 0.25 M sucrose–3.3 mM CaCl$_2$ (1 ml per gram fresh tissue).

Most of the studies reported here were performed with nuclei prepared by the procedure described above. Another excellent method for obtaining highly purified nuclei is that of Blobel and Potter.[10] Similar results were obtained for CMP-sialic acid synthetase with nuclei prepared by either

---

(desialyzed porcine submaxillary mucin which was used as acceptor was kindly donated by Dr. Don M. Carlson and Mr. J. Plantner). It has been suggested that the glycosyl transferases are located on the Golgi apparatus. [H. Schachter, I. Jabbal, R. L. Hudgin, L. Pinteric, E. J. McGuire, and S. Roseman, *J. Biol. Chem.* **245**, 1090 (1970)]. The ratios of RNA:DNA and of protein:DNA are in accord with values in the literature for purified nuclei [D. B. Roodyn, *Biochem. Soc. Symp.* **23**, 20 (1963)], for example, 0.22 and 2.9, respectively. In addition, little or no activity of CMP-sialic acid hydrolase is present in purified nuclei [E. L. Kean and S. R. Coghill, *Fed. Proc., Fed. Amer. Soc. Exp. Biol.* **30**, 1117 (1971), and in this volume [138]. When examined by phase contrast and electron microscopy, the appearance and morphological integrity of the nuclei are similar to previous reports describing nuclei isolated by these procedures.

method. Somewhat higher yields of nuclei are obtained with the procedure of Blobel and Potter.

Nuclei Prepared by the Method of Blobel and Potter[10]

Reagents

Tris, potassium, magnesium buffer (TKM): 0.05 M Tris·HCl, pH 7.5; 0.025 M KCl; 0.005 M MgCl$_2$
Sucrose, 0.25 M, in TKM buffer
Sucrose, 2.3 M, in TKM buffer
Tris·HCl, 0.01 M, pH 7.5, containing 1% (v/v) mercaptoethanol

Procedure. Minced livers are homogenized (using a glass-Teflon homogenizer having a 0.01-inch clearance) with 0.25 M sucrose–TKM buffer in a ratio of 2 ml of buffer to 1 g of liver. The homogenate is filtered through four layers of gauze. Into each of three 1 inch × 3 inch cellulose nitrate centrifuge tubes are added 6.5 ml of the filtered homogenate and 13 ml of a solution of 2.3 M sucrose–TKM. The tube is covered with parafilm and the contents mixed gently by inversion several times. (Addition of the dense sucrose is made with a syringe rather than a pipette.) The solution is underlaid with 10 ml of the 2.3 M sucrose–TKM buffer using a syringe to which is attached a 13-gauge needle, 120 mm in length (the tip of the needle should be ground down and the edge blunted). The tubes are centrifuged for 60 minutes at 25,000 rpm (63,581 g_{av}) using the Spinco SW 25.1 rotor. After centrifugation, a layer of material floating at the top of the tube is skimmed off and the dense sucrose supernatant solution is removed with a Pasteur pipette together with any floating particulate material. The last few milliliters of sucrose are decanted, and the sides of the tube wiped with gauze. The white pellet of nuclei at the bottom of the tube is suspended in 0.25 M sucrose–TKM and rehomogenized with a Dounce homogenizer. The suspension is centrifuged at 1000 g for 10 minutes. The supernatant is decanted, and the pellet of purified nuclei is resuspended with 0.25 M sucrose–TKM (1 ml per gram of fresh tissue).

Dialysis. Sucrose interferes in the TBA-NaBH$_4$ assay used for the determination of enzymatic activity and is removed by dialysis. The suspension of nuclei is dialyzed against 300-fold volume excess of 0.01 M Tris·HCl buffer, pH 7.5, containing 1% mercaptoethanol for 16 hours with one change of dialysis medium. The retentate is removed from the bag; particulate and viscous material that may be present is retrieved. The bag is rinsed with a small volume of fresh dialysis medium, and the wash is added to the retentate. The suspension of disrupted, dialyzed nuclei is suspended by homogenization by hand before sampling. With nuclei prepared by the procedure of Blobel and Potter,[10] the dialyzed product is usually in the form of a uniform suspension that does not require homogenization. With a ratio of about 0.6 ml of the dialyzed preparation of nuclei per gram of fresh tissue, 0.025 ml incubated for 10 minutes will provide amounts of enzymatic activity convenient for analysis.

Properties

Activity and Yield

The yields of CMP-NAN synthetase in nuclei purified from several tissues by the procedure of Chaveau et al.[9] varied from 18% to 53% of the activities present in the crude homogenates (Table I). The yield is somewhat higher when the nuclei are prepared by the method of Blobel and Potter.[10] When corrected for the yield of nuclei (i.e., the yield of DNA) from 54% to 90% of the total CMP-NAN synthetase activity is nuclear.

Stability. The enzyme is more stable when prepared and stored in the presence of mercaptoethanol or glutathione. After dialysis the enzyme is stable for about 2 weeks when stored at 0°. Although it is inactivated by freezing and thawing or lyophilizing, it is stable to these operations when carried out in the presence of 0.15 M glutathione, pH 7.0.

TABLE I

ACTIVITIES AND YIELDS OF CMP-NAN SYNTHETASE IN
NUCLEI FROM SEVERAL TISSUES

Tissue	Homogenate (units/g)[a]	Enzyme activity Nuclei (units/g)[a]	Homogenate (SA)[b]	Nuclei (SA)[b]	Yield of enzyme in recovered nuclei[c] (%)	Corrected enzyme yield (%)
Hog retina	18	3.2	0.29	1.8	18	90
Rat tissues						
Liver	95	26	0.73	9.7	28	54
Kidney cortex	26	9.4	0.49	4.5	36	59
Brain	19	3.5	0.38	3.0	18	60
Spleen	30	16	0.42	1.3	53	70

[a] Units per gram, wet weight, of tissue.
[b] SA, specific activity.
[c] The yield of enzyme in "recovered nuclei" refers to the percentage of the activity that is observed in the whole homogenate, while the corrected enzyme yield is this value corrected for the recovery of DNA.

Kinetic Properties. The following apparent K_m values are observed for the nuclear enzyme. Values obtained previously for the soluble enzyme from the hog submaxillary gland are given in parentheses[3,4]: NAN, 0.72 mM (0.8); NGN, 1.40 mM (2.3); CTP, 0.48 mM (0.6); Mg^{2+}, 6.75 mM.

pH Optimum. In Tris·HCl buffers, the pH optimum is between 8.3 and 8.6.

Nucleotide Specificity and the Effect of Metals. ATP, ITP, and TTP cannot substitute for CTP, while GTP, dCTP, and UTP are about 4, 5, and 7% as active as CTP, respectively. The enzyme shows an absolute requirement for divalent cation. At 20 mM, a concentration where Mg^{2+} shows maximal stimulation, the following metal chlorides had the indicated activities compared to Mg^{2+}: Mn^{2+}, 30%; Ca^{2+}, 15%; Cd^{2+}, 3%; Cu^{2+}, Co^{2+}, and Zn^{2+}, inactive.

Solubilization. The enzyme is solubilized by extracting intact purified nuclei with 0.1 M potassium phosphate buffer, pH 7.5, containing 1% mercaptoethanol. Measurement of this activity at the optimal pH can be accomplished after dialyzing the phosphate extract against 0.01 M Tris, pH 7.5, containing 1% mercaptoethanol. The suspension of disrupted nuclei resulting from the dialysis step, described previously, to remove sucrose from the purified nuclei, contains enzyme partially solubilized and enzyme still associated with particulate material.

Activity of Other Nuclei and Nucleated Cells. As described above, CMP-sialic acid synthetase is recovered in high yield from purified nuclei of a number of mammalian tissues. Activity has also been detected in nuclei isolated from baby hamster kidney cells in culture.[12] In addition to these studies dealing with the enzyme in purified nuclei, a considerable amount of presumptive evidence is available that is consistent with the assignment of the nucleus of the cell as a site for this enzyme, for example: most of the CMP-sialic acid synthetase activity of the unfertilized sea urchin egg is located in the nucleate fragments after the egg is disrupted by relatively mild procedures (centrifugation).[13,14] Activity is found only in the regions of the ocular lens of the hog that contain nucleated cells (capsule-epithelial region).[5] Although the red cell is inactive,[3] lymphocytes from tonsillar tissue[15] and bovine leukocytes[16] contain CMP-sialic acid synthetase.

[12] E. L. Kean and M. Adashi, unpublished experiments (1970).
[13] E. L. Kean and W. E. Bruner, *Biol. Bull.* **139**, 426 (1970).
[14] E. L. Kean and W. E. Bruner, *Exp. Cell. Res.* **69**, 384 (1971).
[15] E. L. Kean and A. E. Powell, unpublished experiments (1970).
[16] W. Gielen, R. Schaper, and H. Pink, *Hoppe-Seyler's Z. Physiol. Chem.* **351**, 768 (1970).

Subcellular Distribution. The distribution of CMP-sialic acid synthetase among the subcellular fractions of rat liver obtained by the procedures of Schneider and Hogeboom[17] and of de Duve *et al.*[18] is presented in Table II. Most of the enzymatic activity is associated with the nuclear fraction. A similar distribution is observed in the retina of the hog.[5] The nuclear fractions that are obtained by these procedures are not homogeneous, however. The distribution of enzyme shown by these techniques is given greater validity by the investigations performed with purified nuclei, described above.

Paper Chromatography and Electrophoresis

Rigorous control of experimental conditions is essential for reproducible paper chromatographic data. These conditions have often been less than adequately described for the chromatography of CMP–sialic acid. Reproducible migrations of CMP-NAN are obtained when paper

TABLE II
SUBCELLULAR DISTRIBUTION OF CMP-SIALIC ACID SYNTHETASE

| | Fractionation procedure | | | |
| | de Duve *et al.*[a] | | Schneider and Hogeboom[b] | |
Fraction	Enzyme units per gram liver	Relative[c] yield (%)	Enzyme units per gram liver	Relative yield (%)
Crude homogenate	87	—	117	100
Cytoplasmic extract	2.8	—	—	—
Nuclear	97	99	97	91
Mitochondria—heavy	0.013	0.01	0.27	0.25
Mitochondria—light	0.044	0.04		
Microsomes	0.97	1.0	9.0	8.4
Supernatant	0.0	0.0	0.44	0.41
Recovery[d] (%)	98	—	91	—

[a] C. de Duve, B. C. Pressman, R. Gianetto, R. Wattiaux, and F. Appelmans, *Biochem. J.* **60**, 604 (1955).
[b] W. C. Schneider and G. H. Hogeboom, *J. Biol. Chem.* **183**, 123 (1950).
[c] Relative yield is the percentage of total activity recovered.
[d] Recovery (%) in the procedure of de Duve *et al.*[a] is based upon the summation of activities in the cytoplasmic extract and the nuclear fraction taken as 100%.

[17] W. C. Schneider and G. H. Hogeboom, *J. Biol. Chem.* **183**, 123 (1950).
[18] C. de Duve, B. C. Pressman, R. Gianetto, R. Wattiaux, and F. Appelmans, *Biochem. J.* **60**, 604 (1955).

TABLE III
PAPER CHROMATOGRAPHY OF CMP-NAN

	Solvent systems[a]			
	1	2	3	4
Enzyme source				
Nuclei (rat liver) (R_f)	0.45	0.30	0.45	0.22
Hog submaxillary gland (R_f)	0.45	0.30	0.44	0.21
Standard NAN (R_f)	0.63	0.56	0.58	0.33
Irrigation time (hours)	16	12	16	22

[a] Solvent systems:

1. 95% ethanol:1.0 M Tris · HCl, pH 7.5 (65:35).
2. 95% ethanol:1.0 M NH₄Ac, pH 7.3 (7:3).
3. n-propanol:0.1 M Tris · Ac, pH 7.6 (1:1).
4. t-butanol:2 M NH₄ formate, pH 6.0 (65:35).

   Paper chromatography was performed at room temperature with Whatman No. 1 paper by the descending technique. The containers were lined with Whatman 3 MM paper soaked with the respective solvents. The spotted papers were equilibrated in the chambers for at least 2 hours prior to irrigation. R_f values were calculated from the position of the peaks of these radioactive compounds as revealed by scanning the paper strips.

chromatography is performed as described in Table III. [1-¹⁴C]NAN was used as the substrate is preparing CMP-[1-¹⁴C]NAN for these studies.[19] CMP-NAN synthesized by the enzyme from purified nuclei, after isolation and purification,[20] migrates in several solvent systems identical to the compound synthesized by the enzyme from hog submaxillary gland (Table III). After paper electrophoresis on Whatman 3 MM paper in 1% sodium tetraborate, pH 9.4, at 4000 V for 25 minutes, CMP-NAN migrates 16.5 cm toward the anode. Anodic migration in 0.05 M phosphate buffer, pH 7.1, of 14.4 cm is observed after electrophoresis at 2400 V for 45 minutes.

[19] [1-¹⁴C]NAN, is prepared by the NAN-aldolase procedure [P. Brunetti, G. W. Jourdian, and S. Roseman, *J. Biol. Chem.* 237, 2447 (1962)], using [1-¹⁴C], pyruvate *N*-acetyl D-mannosamine and bacterial NAN-aldolase (Type III from Sigma Chemical Corp.). See also Vol. 8 [31].

[20] CMP-[1-¹⁴C]NAN is isolated from the incubation mixture and purified by ion exchange and paper chromatographic techniques.³,⁵,⁶,¹⁴ See also Vol. 8 [31].

[52] UDP-D-Xylose 4-Epimerase from Wheat Germ[1]

By DER-FONG FAN and DAVID SIDNEY FEINGOLD

Assay Method

Principle. A two-step assay procedure is used to estimate UDPXyl 4-epimerase, with UDP-U-^{14}C-Xyl as substrate. After incubation, the sugar nucleotides present in the reaction mixture are hydrolyzed and the free sugars are separated by paper chromatography. The relative amount of labeled D-xylose and L-arabinose released from the UDP-pentose formed in the reaction is determined with a four II strip scanner.

In an alternative assay, unlabeled UDPXyl is used as substrate. After hydrolysis of UDPpentose present in the reaction mixture, the D-xylose and L-arabinose released are converted to their trimethylsilyl (TMS) derivatives, which are estimated by gas–liquid chromatography (GLC).

Reagents

UDPXyl, 0.01 M
UDP-U[14C]Xyl and UDP-U[14C]Ara[2]
Trimethylsilylating reagent (TMS reagent):pyridine (reagent grade, dried over KOH):hexamethyldisilazane[3]:trimethyl chlorosilane[3] (10:4:2 v/v/v).

Procedure. Reaction mixtures, containing 0.05 μCi UDP-U[14C]Xyl and 40 μl of enzyme solution are incubated in a capillary tube at 30° for 20 minutes. The contents of the tube are made 0.1 N in HCl, the sugar nucleotides are hydrolyzed by heating for 15 minutes at 100°, and the hydrolyzate is taken to dryness in a stream of N_2 at 25°. In order to avoid interference by salt during the subsequent paper chromatography, the sugars are extracted from the dry residue with pyridine and applied to Whatman 3 MM filter paper, which is then developed in *n*-butanol:pyridine:water (10:3:3 v/v/v). The radioactivities of the separated D-xylose and L-arabinose are determined with the strip scanner.

In an alternative assay, the reaction mixture contains 0.2 M sodium phosphate buffer, pH 8.0 (0.20 ml) and 0.01 M UDPXyl (0.1 ml) in a total volume of 0.4 ml. The reaction is started by addition of 0.1 ml of enzyme preparation containing from 2 to 20 mg of protein. A boiled

---

[1] D. F. Fan and D. S. Feingold, *Plant Physiol.* **46**, 592 (1970).
[2] Obtained from New England Nuclear Corp., Boston, Massachusetts.
[3] Obtained from Aldrich Chemical Co., Inc., Milwaukee, Wisconsin.

control without substrate is run at the same time. After an appropriate period of incubation at 30°, the reaction mixture is held at 100° for 2 minutes to stop the reaction. The precipitated protein is removed, and the supernatant fluid is made 0.1 N in HCl, heated at 100° for 15 minutes, and then taken to dryness in a vacuum desiccator. L-Arabinose and D-xylose in the reaction mixture are converted to their TMS derivatives by adding 1.5 ml of TMS reagent and holding at 60° for 15 minutes. The mixture is evaporated to dryness as described above; TMS derivatives are extracted from the residue with 1.5 ml of dry ethyl acetate, and the extract is evaporated to dryness and taken up in 20 μl of TMS reagent. After 30 minutes with occasional shaking, 0.5 μl of the mixture is chromatographed isothermally at 150° on a column of 3% OV-1 on Gas-Chrom Q (60–80 mesh).[4] The area of peaks with the same retention time as standard TMS-D-xylose and TMS-L-arabinose is determined, and the quantity of L-arabinose in the reaction mixture is calculated from the ratio of these areas. Under these conditions, D-xylose and L-arabinose can be clearly differentiated; the area obtained from L-arabinose is 1.05 that obtained from an equimolar quantity of D-xylose. A unit of enzyme is defined as the amount of enzyme required to convert 1 μmole of UDP-D-xylose to UDP-L-arabinose per minute under the conditions of assay.

Purification of UDPXyl 4-Epimerase

All purification procedures are carried out at 0 to 4°. Centrifugations are performed at 10,000 g for 20 minutes.

Preparation of Crude Extract. Raw wheat germ, 2 kg, is stirred for 1.5 hours with 13 liters of 0.1 M sodium phosphate, pH 7.0, containing 0.5 g of EDTA and 0.5 ml of 2-mercaptoethanol per liter (buffer A). The thick slurry is squeezed through four layers of cheesecloth to yield approximately 10 liters of turbid supernatant fluid (crude extract).

$MnCl_2$ Treatment. Under vigorous stirring, 0.5 M $MnCl_2$ is added to the crude extract to a final concentration of 0.015 M. After 5 minutes, the mixture is centrifuged and the supernatant fluid is retained ($MnCl_2$ supernatant). UDPXyl 4-epimerase cannot be assayed accurately by GLC in either crude extract or $MnCl_2$ supernatant, because of the high endogenous L-arabinose content and low specific activity of these fractions. However, enzymatic activity can be demonstrated with UDP-U-^{14}C-Xyl as described above.

$(NH_4)_2SO_4$ Fractionation. To the $MnCl_2$ supernatant (10 liters),

---

[4] Obtained from Applied Science Laboratories, State College, Pennsylvania.

solid ammonium sulfate is added to 50% saturation, and the precipitated protein is discarded. The supernatant fluid is then brought to 65% saturation; the resultant precipitate is collected by centrifugation and dissolved in buffer A to a final volume of 250 ml ($(NH_4)_2SO_4$ fraction).

Dialysis. The $(NH_4)_2SO_4$ fraction is dialyzed for 18 hours against 19 liters of buffer A; the retentate is diluted with buffer A to 800 ml and refractionated with ammonium sulfate. Protein which precipitates between 50 and 70% $(NH_4)_2SO_4$ saturation is collected and dissolved in 120 ml of buffer A. This solution is dialyzed against 18 liters of buffer A as above, and the retentate is diluted with buffer A to 500 ml and again fractionated with solid ammonium sulfate; the fraction which precipitates between 50 and 70% saturation is dissolved in 90 ml of buffer A (dialyzed fraction).

Sephadex G-75 Column Chromatography. The dialyzed fraction is placed on a 5.5-cm diameter column containing 1900 ml of Sephadex G-75 equilibrated with buffer A. The protein is eluted with buffer A; 15 ml fractions are collected, and active fractions are pooled and fractionated with solid ammonium sulfate. The protein which precipitates between 50 and 70% $(NH_4)_2SO_4$ saturation is collected and dissolved in 50 ml of buffer A (Sephadex G-75 fraction).

DEAE-Cellulose Column Chromatography. The Sephadex G-75 fraction is dialyzed against 10 liters of 0.01 M sodium phosphate, pH 7.0, containing 0.5 g of EDTA and 0.5 ml of 2-mercaptoethanol per liter (buffer B) for 5 hours and then placed on a 4.5 cm diameter column containing 1500 ml of DEAE-cellulose equilibrated with buffer B. The column is eluted with buffer B, and 20-ml fractions are collected. Active fractions (600 ml) are pooled and brought to 75% $(NH_4)_2SO_4$ saturation. After centrifugation, the protein is present as a felt floating at the top of the solution. This material is collected with a glass rod and dissolved in 24 ml of buffer A (DEAE-cellulose fraction).

Sephadex G-100 Column Chromatography. DEAE-Cellulose fraction is chromatographed on a Sephadex G-100 column (4 cm in diameter); buffer A is used as eluent and 10-ml fractions are collected. Active fractions are pooled and brought to 75% saturation with solid ammonium sulfate. The protein felt is collected as above and dissolved in 7 ml of buffer A.

The enzyme purification is summarized in the table.

Properties

Stability. The enzyme is stable for at least 3 weeks at 0 to 4°, but it loses 50% of its initial activity after 2 months at this temperature or when stored frozen for 12 hours. However, the $(NH_4)_2SO_4$ fraction

PURIFICATION OF UDPXYL 4-EPIMERASE FROM WHEAT GERM

Purification step	Volume (ml)	Total activity (units)	Total protein (g)	Specific activity (units/mg protein $\times 10^{-3}$)	Recovery (%)
Crude extract	10,000	—	350	—	—
MnCl$_2$ supernatant	10,000	—	275	—	—
(NH$_4$)$_2$SO$_4$ fraction	250	13.7	25.7	0.5	100
Dialyzed fraction	90	11.3	13.5	0.8	82
Sephadex G-75	50	7.7	7.3	1.1	56
DEAE-cellulose	24	6.0	2.5	2.4	44
Sephadex G-100	7	2.3	0.2	11.0	17

can be stored frozen with no loss of activity. Purified enzyme loses 80% of its initial activity when held at 50° for 1 minute.

Absence of Other Enzyme Activity. The Sephadex G-100 fraction is free from UDPGal-, dTDPGlc-, and UDPGalacturonate 4-epimerase activity. When tested shortly after elution from Sephadex G-100, the enzyme preparation contains a small amount of UDP-glucuronate carboxy-lyase activity, which, however, disappears upon storage at 0° to 4° for 10 days.

pH Optimum. UDPXyl 4-epimerase has a pH optimum in the vicinity of 8.0.

Kinetic Constants. K_m is $1.5 \times 10^{-3} M$ for UDP-D-xylose and $5 \times 10^{-4} M$ for UDP-L-arabinose. The equilibrium constant, K, for the reaction UDP-L-arabinose $\leftrightharpoons$ UDP-D-xylose is 1.25.

Cofactor Requirement. NAD, NADP, NADH, NADPH, UMP, UDP, UTP, UDPGlc, UPDGal, UDPGlcUA, and dTDPGlc at concentrations of 0.5 mM or 2.0 mM do not affect reaction rate when tested in the standard assay. Prolonged dialysis of Sephadex G-100 fraction against a large excess of buffer A or treatment of the enzyme with nicotinamide adenine dinucleotidase (EC 3.2.2.5) or charcoal does not decrease enzyme activity as compared with untreated controls.

Effect of Sulfhydryl Reagents. The enzyme is inactivated by p-mercuriphenylsulfonate; full activity can be restored by subsequent treatment with cysteine. p-Mercuriphenyl sulfonate irreversibly inactivates any contaminating UDPGlcUA carboxy-lyase which may be present in the preparation.

[53] UDP-Glucuronic Acid 4-Epimerase from *Anabaena flos-aquae*

By MARY A. GAUNT, HELMUT ANKEL, and JOHN S. SCHUTZBACH

UDP-glucuronic acid $\rightleftharpoons$ UDP-galacturonic acid

UDP-glucuronic acid 4-epimerase has been identified in plants,[1] bacteria,[2] and blue-green algae.[3] The enzyme from *Anabaena flos-aquae* has been partially purified, and some of its properties have been determined.[4] The algae enzyme, similarly to the plant enzyme, does not require exogenous NAD for activity, whereas the enzyme from *Diplococcus pneumoniae* has a requirement for exogenous NAD.

Assay Method

Principle. Enzyme is incubated with UDP-[14C]glucuronic acid. The product, UDP-galacturonic acid-14C, is separated from remaining substrate by paper electrophoresis. Radioactive compounds are detected by autoradiography, excised, and counted.

Reagents

UDP-[14C]glucuronic acid, 10 μCi/ml (62 mCi/mmole)
Enzyme in 0.1 M potassium phosphate buffer, pH 7.5
Buffer for paper electrophoresis, 0.15 M ammonium acetate, pH 5.8

Procedure. Reaction mixtures containing 5 μl of UDP-[14C]glucuronic acid and 10 μl of enzyme are incubated at 25° in sealed glass capillary tubes. The reaction is terminated after 5 minutes by spotting the contents of the capillary on Whatman No. 1 paper and drying in a stream of air. Electrophoresis in ammonium acetate buffer, pH 5.8, at 15–20 V/cm for 2.5 hours separates substrate from product.[5] Radioactive compounds are located by radioautography, excised, and counted in a liquid scintillation spectrometer. One unit of activity is defined as that amount of enzyme that catalyzes the formation of 1 nmole of product in 10 minutes.

[1] D. S. Feingold, E. F. Neufeld, and W. Z. Hassid, *J. Biol. Chem.* **235**, 910 (1960). See also E. F. Neufeld, Vol. 8 [46].
[2] E. E. B. Smith, G. T. Mills, H. P. Bernheimer, and R. Austrian, *Biochim. Biophys. Acta* **29**, 640 (1958).
[3] H. Ankel and R. G. Tischer, *Biochim. Biophys. Acta* **178**, 415 (1969).
[4] M. A. Gaunt, Ph.D. Thesis, Marquette University, Milwaukee, Wisconsin, 1971.
[5] The relative electrophoretic mobilities are UDP-glucuronic acid, 1.0; UDP-galacturonic acid, 0.91.

Purification

The following procedure results in an 11-fold purification of the enzyme with a 37% yield. Most major contaminating enzyme activities are removed by this procedure.

Growth of Cells. Anabaena *flos-aquae*[6] is grown in the medium of Kratz and Meyer.[7] Starter cultures are prepared by washing the entire contents of a slant of cells into 200 ml of medium contained in a 500-ml Erlenmeyer flask. The cultures are closed with cotton plugs and maintained at 25° under static conditions for 5 days. Cultures are continuously illuminated from above by three 20 W fluorescent lamps at a distance of 40 cm.

For large-scale preparations, the algae are grown in a 30-liter glass jar fermentor (Fermentation Design, Inc., Allentown, Pennsylvania). Ten liters of starter containing approximately 1 g of cells/liter (wet weight) are added to 15 liters of medium in the fermentor. The cells are grown at 40°, 15 psi aeration, and 150 rpm agitation. During growth the cells are illuminated on 3 sides by a light bank consisting of twelve 20-W fluorescent lamps at a distance of 12 cm. The cells are harvested by centrifugation at 12,000 g for 15 minutes when they reach an optical density of 1.0 (measured at 400 nm in a 1-cm light path; approximately 24 hours). All subsequent operations are carried out at 0–4°.

Preparation of Cell-Free Extract. The packed cells are washed twice with 10 volumes of distilled water and twice with 10 volumes of 0.1 M potassium phosphate buffer, pH 7.5, containing EDTA at a concentration of 1 mM (buffer). The packed washed cells can be stored frozen at $-20°$ until needed.

After thawing, the cells are suspended in buffer to approximately 200 mg/ml (wet weight). One hundred milliliters of the cell suspension is mixed with 10 g of glass powder (5 μ) and the cells are disrupted by ultrasonic treatment with a Branson Sonifier (Model 125) for 20 minutes at an output of 11 amp. Cell debris is removed by centrifugation at 28,000 g for 15 minutes. The resulting supernatant (fraction I) is centrifuged at 100,000 g for 30 minutes, and the precipitate is discarded. The high speed supernatant (fraction II) is then fractioned with ammonium sulfate.

---

[6] Anabaena *flos-aquae* was obtained from the Culture Collection of Algae, Indiana State University, Department of Botany, Bloomington, Indiana.

[7] The culture medium contains the following: (in g/liter): MgSO$_4$·7H$_2$O, 0.15; K$_2$HPO$_4$, 1.0; Ca(NO$_3$)$_2$ 4H$_2$O, 0.01; NaNO$_3$, 1.0; EDTA, 0.05; FeSO$_4$·6H$_2$O, 0.004; microelements, 1 ml/liter. The microelement stock solution contains the following (in g/liter): ZnSO$_4$·7H$_2$O, 8.82; MnCl$_2$·4H$_2$O, 1.44; MoO$_3$, 0.71; CuSO$_4$·5H$_2$O, 1.57; Co(NO$_3$)$_2$·6H$_2$O, 0.49. W. A. Kratz and J. Myers, *Amer. J. Bot.* **42**, 282 (1955).

Ammonium Sulfate Fractionation. To this solution (70 ml) is added 12 g of $(NH_4)_2SO_4$, and the precipitate is discarded. To the supernatant (64 ml) is added 12 g of $(NH_4)_2SO_4$; the precipitate is removed by centrifugation and taken up in 5 ml of buffer (fraction III).

Sephadex G-100. Fraction III (4.5 ml) is applied to a column of Sephadex G-100 (1 × 140 cm) equilibrated with 0.02 M potassium phosphate buffer, pH 7.5, containing EDTA at a concentration of 1 mM. The column is eluted with the same buffer and fractions of 1.7 ml are collected. Fractions containing enzyme activity are pooled (fractions 70–82; 22 ml) and enzyme is precipitated by the addition of 8.2 g of $(NH_4)_2SO_4$. The precipitate is removed by centrifugation and is taken up in 1.1 ml of buffer (fraction IV). The purification is summarized in Table I.

Properties

Stability. The enzyme retains activity for at least 6 months when stored frozen at −20°.

Contaminating Activities. The purified enzyme is free of UDP-glucose dehydrogenase and UDP-glucuronic acid carboxy-lyase activities but still contains UDP-galactose 4-epimerase activity. That the latter activity is due to a separate enzyme is demonstrated by a change in the ratios of specific activities of the two 4-epimerases during purification and by different rates of heat inactivation at 58°.

The only detectable radioactive compounds in standard reaction mixtures incubated with the purified enzyme are UDP-[¹⁴C]glucuronic acid and UDP-[¹⁴C]galacturonic acid; judged by autoradiography after electrophoresis in ammonium acetate buffer, pH 5.8, or after chromatography in 1 M ammonium acetate–95% ethanol (3:7).

TABLE I
PURIFICATION OF UDP-GLUCURONIC ACID 4-EPIMERASE

Fraction	Total units	Specific activity[a] (units/mg)	Purification	Recovery (%)
I. Extract	828	0.88	—	100
II. 100,000 g supernatant	903	2.8	3.2	100
III. $(NH_4)_2SO_4$	518	3.1	3.5	63
IV. Sephadex G-100	140	9.6	11	17

[a] As measured under conditions of the standard assay.

Molecular Weight. The molecular weight of the enzyme is approximately 54,000 as judged by its elution volume from a calibrated Sephadex G-100 column.

pH Optimum. Maximum activity is obtained at pH 8.5. The enzyme is routinely assayed at pH 7.5 because of its greater stability at the lower pH.

Equilibrium Constant. The equilibrium constant for the reaction UDP-glucuronic acid $\rightleftharpoons$ UDP-galacturonic acid is 2.6 at 25° and at 37°.

Kinetic Properties. The substrate concentrations required for half-maximal activity are $3.7 \times 10^{-5} M$ for UDP-glucuronic acid and $1.2 \times 10^{-5} M$ for UDP-galacturonic acid.

Activators and Inhibitors. The enzyme is not activated by the addition of divalent cations or by the addition of NAD. Treatment with charcoal, 1 mM p-hydroxymercuribenzoate, or NADase has no effect on activity.

NAD and NADP, in fact, inhibit the enzyme. The inhibition is competitive with respect to UDP-glucuronic acid and the values for K_i are 6.8 mM for NAD and 2.4 mM for NADP. A number of nucleotides and sugar nucleotides also inhibit the enzyme (Table II). UDP-glucose is a competitive inhibitor with respect to substrate with a K_i of 0.15 mM. UMP, UDP, and also UDP-xylose inhibit the enzyme but the kinetics do not follow those of simple competitive inhibitors. In the absence of inhibitor, the apparent K_m for UDP-glucuronic acid is $3.7 \times 10^{-5} M$

TABLE II

EFFECT OF NUCLEOTIDES AND SUGAR NUCLEOTIDES ON THE ACTIVITY OF
UDP-GLUCURONATE 4-EPIMERASE[a]

Addition ($3 \times 10^{-3} M$)	% Inhibition
UMP	100
UDP	100
UTP	100
UDP-glucose	100
TDP	95
TDP-glucose	83
CMP	53
CDP-glucose	53
AMP	50
ADP-glucose	60
GMP	41
GDP	39

[a] The final UDP-[^{14}C]glucuronic acid was $3.3 \times 10^{-5} M$.

and the value for the Hill constant is 1.0. In the presence of UMP (0.4 mM or 1.0 mM) the value for the apparent K_m increases and the value of the Hill constant is 1.6. At constant UDP-glucuronic acid concentration with varying UMP, UDP, or UDP-xylose, the Hill constants for inhibitor vary from 1.7 to 2.1.

[54] UDP-Glucose Dehydrogenase from Beef Liver

By Juris Zalitis, Marc Uram, Anna Marie Bowser, and David Sidney Feingold

$$UDPGlc + 2 NAD^+ + H_2O \rightarrow UDPGlcUA + 2NADH + 2 H^+$$

In 1954 Strominger et al. demonstrated the NAD-linked conversion of UDPGlc to UDPGlcUA catalyzed by an enzyme from beef liver.[1] The enzyme has subsequently been demonstrated in prokaryotic[2] as well as in other eukaryotic organisms.[3] UDPGlc dehydrogenase from bovine liver has been purified to homogeneity, and a number of its physical and chemical properties have been determined.[4]

Assay Method

Principle. The enzymatic oxidation of UDPGlc to UDPGlcUA with the concomitant reduction of 2 moles of NADH is assayed by measuring the amount of NADH formed spectrophotometrically.

Reagents

Glycylglycine buffer, 0.5 M, pH 8.7 (adjusted with 0.5 N NaOH)
NAD, 0.01 M
UDPGlc, 0.005 M

Procedure. Into a 1-ml cuvette are pipetted 0.2 ml glycylglycine buffer, 0.1 ml NAD, 0.1 ml UDPGlc, and 0.6 ml distilled water. The cuvette and its contents are tempered at 30° in the constant temperature cell holder of the spectrophotometer, 10 μl of the sample to be assayed, containing from 0.001 to 0.01 unit of enzyme is added and the contents of the cuvette are mixed. The absorbance at 340 nm is recorded

[1] J. L. Strominger, H. M. Kalckar, J. Axelrod, and E. C. Maxwell, *J. Amer. Chem. Soc.* **76**, 6411 (1954).
[2] A. Bdolah and D. S. Feingold, *J. Bacteriol.* **96**, 1144 (1968).
[3] G. J. Dutton, *in* "Gluconic Acid: Free and Combined" (G. J. Dutton, ed.), p. 210. Academic Press, New York, 1966.
[4] J. Zalitis and D. S. Feingold, *Arch. Biochem. Biophys.* **132**, 457 (1969).

at 15-second intervals, starting no later than 30 seconds after addition of enzyme. The change in absorbance is constant with time for at least 2 minutes and is proportional to enzyme concentration over the range indicated.

Definition of Unit and Specific Activity. One unit of UDPGlc dehydrogenase is defined as the amount of enzyme necessary to catalyze the oxidation of 1 μmole of UDPGlc to UDPGlcUA per minute at 30° under the specified conditions. Specific activity is defined as units per milligram of protein. Protein concentrations are determined by the method of Lowry *et al.*[5]

Purification Procedure

All operations are carried out between 0 and 4° unless otherwise specified. The term "buffer," when not specified means 0.01 M sodium acetate, pH 5.5, 0.002 M in EDTA- and 0.01 M in 2-mercaptoethanol. (In all ammonium sulfate precipitations, 30 minutes are allowed to elapse between termination of ammonium sulfate addition and centrifugation.)

Step. 1. Preparation of Crude Extract. Livers are obtained from the slaughterhouse within minutes of killing the steer and are cooled immediately in crushed ice. They are sliced into large pieces, and adhering fat and connective tissue is dissected out. The tissue then is cut into 4-cm cubes; 300-g lots are homogenized with 300 ml of buffer for 60 seconds at full speed in a Waring Blendor. The pooled homogenates are adjusted to pH 4.9 with 0.1 N acetic acid. The thick suspension is centrifuged for 30 minutes at 10,000 g, and the clear red supernatant fluid is decanted through a pad of cotton to remove floating fat particles.

Step 2. Ammonium Sulfate Fractionation—I. Solid ammonium sulfate is slowly added to the well stirred crude extract to 30% saturation (167 g of ammonium sulfate per liter), and the suspension is centrifuged at 10,000 g for 30 minutes. The precipitate is discarded, and the clear supernatant solution is brought to 50% ammonium sulfate saturation by addition of 121 g of solid ammonium sulfate per liter of solution and centrifuged as before. The precipitate is dissolved in a minimum volume (approximately 1 liter) of buffer to yield a solution containing approximately 100 mg of protein per milliliter.

Step 3. Heat Treatment. (For efficient heat exchange, this step should be conducted with continuous vigorous mechanical stirring.) The solution is adjusted to pH 4.9 with 0.1 N acetic acid, and portions of 200 ml in a stainless steel beaker are quickly heated to 60° by immersion in an 80° bath (for maximum reproducibility; the solution should reach 60° in

[5] O. H. Lowry, N. J. Rosebrough, A. L. Farr, and R. J. Randall, *J. Biol. Chem.* **193**, 265 (1951). See also Vol. 3 [73].

2–3 minutes). The beaker is held at 60° for one additional minute, then rapidly cooled in crushed ice. The resulting thick suspension is centrifuged for 30 minutes at 10,000 g; the supernatant fluid is retained. The precipitate is suspended in a volume of buffer equal to that of the supernatant fluid, and the suspension is centrifuged. The two supernatant solutions are combined.

Step 4. Ammonium Sulfate Fractionation—II. The ammonium sulfate concentration in the heat-treated supernatant fluid is determined conductrimetrically with a Barnstead conductivity meter, using 50% saturated ammonium sulfate as a standard. The solution is then fractionated with solid ammonium sulfate between 35 and 45% saturation as previously described. The protein which precipitates at 45% saturation is dissolved in a minimal volume of buffer (approximately 200 ml).

Step 5. Calcium Phosphate Gel Titration. It is best to establish the optimal conditions by a small-scale pilot run, titrating the enzyme onto calcium phosphate gel as described by Oliver.[6] This is necessary because of variation in the properties of the calcium phosphate gel and also because of variations in the enzyme preparation, although usually a protein to calcium phosphate dry weight ratio of 1:1.25 is sufficient to adsorb all the activity. The appropriate quantity of calcium phosphate gel[7] (aged for at least 2 weeks) is added to the solution. After standing for 5 minutes, the suspension is centrifuged at 1000 g for 10 minutes and the inactive supernatant fluid is discarded. The gel is extracted with several batches of 0.10 M potassium phosphate buffer, pH 7.4, and then with 0.20 M potassium phosphate buffer. The eluates with the highest specific activity are pooled. Concentrated solutions of EDTA and 2-mercaptoethanol are added to final concentrations of 0.002 M and 0.01 M, respectively, and solid ammonium sulfate is added to a final concentration of 60% of saturation (371 g/liter). The suspension is centrifuged at 10,000 g for 20 minutes and the precipitate is dissolved in buffer and dialyzed overnight against 50 volumes of the same buffer. The retentate is centrifuged at 10,000 g for 20 minutes, and the supernatant fluid (calcium phosphate gel eluate, concentrated and dialyzed) is retained.

Step 6. Carboxymethyl Cellulose Chromatography. Carboxymethyl cellulose columns are prepared as follows. Twenty-gram batches of coarse fibrous Whatman carboxymethyl cellulose (Reeve Angel and Co., New Jersey) are stirred with 1 liter of 0.2 M NaOH, 0.1 M in EDTA. The slurry is filtered under vacuum and washed on the filter with distilled water until the pH of the filtrate reaches 8–10. The procedure is repeated

[6] I. T. Oliver, *Nature (London)* **190**, 222 (1964).
[7] D. Keilin and E. F. Hartree, *Proc. Roy. Soc. Ser. B*, **124**, 397 (1938). See also Vol. 1 [12].

twice, the material is then suspended in 0.2 N acetic acid and the pH of the suspension is brought to 5.5 with 0.2 M NaOH, 0.05 M in EDTA. This suspension is used to pour a 5.5 × 80 cm column. The column bed is washed with 0.1 M sodium acetate, pH 5.5, 0.02 M in EDTA, until the pH of the effluent reaches 5.5. The column then is washed with 0.01 M sodium acetate, pH 5.5, 0.02 M in EDTA and 0.01 M in 2-mercapto-ethanol, until the pH of the effluent is constant at 5.5. Calcium phosphate gel eluate, concentrated and dialyzed, is applied to the 5.5 × 80 cm column of carboxymethyl cellulose and the column is eluted with a linear gradient of NaCl, obtained by using a mixing chamber containing 2 liters of buffer and a reservoir containing 0.1 M sodium acetate, pH 6.0, 0.75 M in NaCl, 0.002 M in EDTA, and 0.01 M in 2-mercaptoethanol. The flow rate is adjusted to 300–400 ml per hour, and 18-ml fractions are collected. The enzyme should be in the last peak to emerge from the column after a number of peaks of highly colored protein.

The fractions of highest specific activity are pooled, brought to 60% saturation with solid ammonium sulfate and centrifuged at 10,000 g for 30 minutes. The precipitate is dissolved in 5–10 ml of buffer and dialyzed for 12 hours against 5 liters of buffer; precipitated protein is removed by centrifugation for 20 minutes at 10,000 g (carboxymethyl cellulose fractions, concentrated and dialyzed).

Step 7. Sephadex G-200 Chromatography. The enzyme sample is applied to a column of Sephadex G-200 (3 × 130 cm), prepared in buffer. The column is then eluted with buffer at a flow rate of approximately 14 ml per hour, and 10-ml fractions are collected. UDPGlc dehydrogenase elutes in the void volume (previously established using Blue Dextran 2000, Pharmacia, Inc., Piscataway, New Jersey, as a standard). Fractions of highest specific activity are pooled and concentrated by addition of solid ammonium sulfate to 60% saturation. The resultant precipitate is dissolved in a minimal volume of buffer (usually 1 ml per 30 units of enzyme) and then dialyzed for 12 hours against 2 liters of buffer (Sephadex G-200 fraction, concentrated and dialyzed).

A large-scale purification is summarized in the table.

Properties

Physical Characteristics. When examined in the ultracentrifuge, the purified enzyme sediments as a single symmetrical peak with a sedimentation coefficient, $s_{20,w}$ of 12.8.[4] The molecular weight determined by the method of sedimentation equilibrium is 3×10^5.[8] A solution containing 1 mg of enzyme per milliliter gives an absorbance of 0.98 at 277 nm.

[8] M. Uram, F. Lamy, A. M. Bowser, and D. S. Feingold, *An. Asoc. Quim. Argent.* in press.

PURIFICATION OF UDPGlc DEHYDROGENASE

Fraction	Volume (ml)	Activity (units/ml)	Protein (mg/ml)	Specific activity	Purification (—fold)	Total activity
pH 4.9 extract	4800	0.45	72.8	0.0062	—	2160
Ammonium sulfate—I	1540	1.5	100	0.015	2.5	2350
Heat-treated supernatant	1080	1.5	44	0.034	5.5	1650
Ammonium sulfate—II	320	4.3	81	0.053	8.5	1360
Calcium phosphate gel eluate, concentrated and dialyzed	113	6.4	10	0.64	103	724
Carboxymethyl-cellulose fractions, concentrated and dialyzed	10.5	29.5	15	2.00	323	310
Sephadex G-200 fraction concentrated and dialyzed	6.2	28.3	8.0	3.52	563	175

The purified enzyme is homogeneous as judged by polyacrylamide gel electrophoresis, immunoelectrophoresis, and agar immunodiffusion.[4] It gives a single symmetrical peak when examined by isoelectric focusing in a gradient of sucrose; the isoelectric point is 6.74.[9] Physical and chemical evidence suggests that the enzyme is made up of six identical subunits of molecular weight 50,000.[9] On the basis of the specific activity of the purified preparation and molecular weight 3×10^5, a turnover number of 1050 moles of UDPGlc oxidized per minute per mole of enzyme at 30° and pH 8.7 is obtained.

pH Optimum. The optimum pH is 9.25; activity falls off symmetrically on either side of this point, being approximately 80% of maximum at pH 8.7 and 9.8.[10] (Since the enzyme loses activity very rapidly at pH 9.25, the assay is routinely run at pH 8.7.)

Stability. UDPGlc dehydrogenase is most stable at pH 5.0. At this pH it exists as an inactive dimer of molecular weight 6×10^5 which dissociates at pH 7.0 to yield active enzyme.[4] At 0–4° a solution of enzyme in $0.01\,M$ sodium acetate, pH 5.5, $0.002\,M$ in EDTA, $0.01\,M$ in 2-mercaptoethanol, and $0.02\,M$ in sucrose retains full activity for at least 6 months.[8]

Specificity. UDPGlc dehydrogenase is relatively unspecific for the nucleoside portion of the substrate but is highly specific for the D-glucosyl moiety. Thus, in addition to UDPGlc, dTDPGlc, CDPGlc, GDPGlc, and ADPGlc, all are substrates for the enzyme, although K_m values for the

[9] M. Uram, *Fed. Proc., Fed. Amer. Soc. Exp. Biol.* **30**, 1553 (1971).
[10] J. Zalitis and D. S. Feingold, unpublished data (1968).

last 4 compounds are at least 2 orders of magnitude greater than that of UDPGlc. GDPMan, UDPGal, α-D-glucopyranosyl phosphate, or free D-glucose are not oxidized.[11] In addition to NAD, deamino- and acetyl-pyridine analogs of NAD are active as H acceptors but NADP is not.[12]

Kinetic Constants. K_m for UDPGlc is 0.013 mM; for NAD, 0.1 mM; NADH is an inhibitor competitive with NAD, $K_i = 0.006$ mM.[11] UDPGlcUA is an inhibitor competitive with UDPGlc ($K_i = 0.05$ mM) and noncompetitive with NAD. At low concentrations, UDP-D-xylose acts as an inhibitor competitive with UDPGlc ($K_i = 0.004$ mM). Examination of inhibition patterns at higher UDPXyl concentrations shows that a cooperative interaction exists between several inhibitor binding sites on the enzyme, the apparent number of which is 2.3.[13] UDPGal is an inhibitor competitive with UDPGlc, at pH 8.3, $K_i = 1.3$ mM; K_m for UDPGlc = 0.1 mM.[14]

Sulfhydryl Groups. UDPGlc dehydrogenase contains 72 sulfhydryl groups and no disulfide bonds. Enzymatic activity is abolished by reaction of one sulfhydryl group per subunit with 5,5′ dithiobis(2-nitrobenzoic acid), *p*-mercuribenzoic acid, or iodoacetamide. UDPGlcUA, NADH, or UDP-D-xylose, but not NAD or UDPGlc, partially protect the active sulfhydryl group.[9]

[11] J. Zalitis and D. S. Feingold, *Biochem. Biophys. Res. Commun.* **31**, 693 (1968).
[12] N. D. Goldberg, Ph.D. Thesis, University of Wisconsin, 1963.
[13] E. F. Neufeld and C. W. Hall, *Biochem. Biophys. Res. Commun.* **19**, 456 (1965).
[14] G. Salitis and I. T. Oliver, *Biochim. Biophys. Acta* **81**, 55 (1964).

[55] UDP-*N*-Acetylglucosamine Dehydrogenase from *Achromobacter georgiopolitanum*[1]

By DER-FONG FAN and DAVID SIDNEY FEINGOLD

$$\text{UDPGlcNAc} + 2\text{NAD}^+ + \text{H}_2\text{O} \rightarrow \text{UDPGlcUANAc} + 2\text{NADH} + 2\text{H}^+$$

Assay Method

Principle. In the presence of NAD and UDPGlcNAc dehydrogenase (see below) UDPGlcNAc is oxidized to UDPGlcUANAc. Two moles of NAD are reduced per mole of UDPGlcNAc oxidized. The amount of NADH formed is measured spectrophotometrically.

[1] D. F. Fan, C. E. John, J. Zalitis, and D. S. Feingold, *Arch. Biochem. Biophys.* **135**, 45 (1969).

Reagents

Glycine buffer, 0.5 M, pH 9.0.

Sodium phosphate buffer, 0.1 M, pH 7.5, containing 0.5 g EDTA and 0.5 ml of 2-mercaptoethanol per liter (buffer).

UPDGlcNAc, 0.01 M

NAD, 0.1 M

Procedure. UDPGlcNAc dehydrogenase is assayed by following the reduction of NAD at 340 nm at 30° in a 1-cm light path cuvette in a Gilford multiple sample absorbance recorder. The assay mixture contains: glycine buffer (pH 9.0), 0.45 ml; NAD, 0.1 ml; buffer, 0.3 ml; and enzyme 0.05 ml (containing 1–4 mg of protein) in a total volume of 0.9 ml. The reaction is started by addition of 0.10 ml of 0.01 M UDPGlcNAc. A unit of enzyme activity is the quantity that converts 1 μmole of UDPGlcNAc to UDPGlcUANAc per minute under these conditions.

Preparation of UDPGlcNAc Dehydrogenase

Achromobacter georgiopolitanum mutant strain A (ATCC 25020) is cultured at 25° for 16 hours with aeration in 10 liters of glucose–salts–yeast extract medium containing the following components: 0.3% yeast extract, 0.15% $(NH_4)_2SO_4$, 0.5% NaCl, 0.005% $MgSO_4 \cdot 7H_2O$, 0.35% KH_2PO_4, 0.44% K_2HPO_4 and 1.0% glucose.[2] One liter of log phase cells in the same medium is used as inoculum. (All subsequent operations are performed at 0–4°.) The cells are harvested, washed once with distilled water, suspended in 2.5 volumes of buffer, and disrupted with the Branson sonifier.[3] The sonicate is centrifuged at 100,000 g for 3 hours and the colorless supernatant fluid, in which no UDPGlc dehydrogenase activity can be detected, is fractionated with solid ammonium sulfate. The fraction which precipitates between 50 and 80% saturation is collected and dissolved in 1 ml of buffer. This UDPGlcNAc dehydrogenase preparation is free from NADH-oxidase and had a specific activity in a range of 5–10 × 10^{-3} unit.

Properties

The enzyme has a pH optimum of 9.0 in glycine buffer. In Tris buffer at pH 9.0 the enzyme is only half as active in glycine buffer, and it has no activity in sodium borate at pH 9 or 10.

The K_m value for UDPGlcNAc is 5 × 10^{-4} M. K_m for NAD is 1.5 × 10^{-3} M.

[2] E. J. Smith, *Biochem. Biophys. Res. Commun.* **15**, 593 (1964).

[3] Model S-110, Branson Instruments, Danbury, Connecticut.

NADH is an inhibitor strictly competitive with NAD; K_i is $5 \times 10^{-5} M$. Under conditions of assay, inorganic phosphate is an activator; K_a is $1 \times 10^{-3} M$. The enzyme preparation is moderately specific for UDPGlcNAc and NAD. Under assay conditions, UDPGlc is oxidized at 0.1 the rate of UDPGlcNAc; with UDPGlcNAc as substrate, NADP is reduced at 0.1 the rate of NAD.

UDPGlcNAc dehydrogenase activity can also be demonstrated in the wild-type strain of *A. georgiopolitanum* (ATCC 23203), but the latter is less convenient as an enzyme source because of difficulty in harvesting cells in the presence of polysaccharide.

Use of Enzyme as a Preparative Tool

UDPGlcNAc dehydrogenase can be used in the preparation of UDPGlcUANAc. Since NADH formed during the dehydrogenation inhibits the enzyme, it is most convenient to eliminate the 100,000 g centrifugation step in order to obtain a preparation which contains NADH oxidase. The clear brownish supernatant fluid obtained upon centrifugation of the sonicate at 25,000 g is directly fractionated between 50 and 80% $(NH_4)_2SO_4$ saturation and the precipitate is dissolved in 3 ml of buffer. A typical reaction mixture contains (μmoles) UDPGlcNAc, 20; NAD, 40; glycine buffer (pH 9.0), 750; buffer, 250; and enzyme in a final volume of 3.0 ml. After 1 hour at 30°, the reaction mixture is held at 100° for 2 minutes; the precipitated protein is removed by centrifugation. The UDPGlcUANAc is purified by paper chromatography in 95% ethanol: 1 M ammonium acetate (pH 7.0), 7:3 (v/v).[4] The compound is washed on the paper with absolute ethanol, eluted with water, desalted on a column of Sephadex G-10, and finally separated from contaminating NAD by paper electrophoresis in 0.2 M ammonium formate (pH 3.6).[5]

[4] A. C. Paladini and L. F. Leloir, *Biochem. J.* **51**, 426 (1952).
[5] D. S. Feingold, E. F. Neufeld, and W. Z. Hassid, Vol. 6 [108].

[56] UDP-Galacturonic Acid Decarboxylase from *Ampullariella digitata*[1]

By DER-FONG FAN and DAVID SIDNEY FEINGOLD

UDP-D-galacturonic acid → UDP-L-arabinose + CO_2

Assay Method

Since all radioactive L-arabinose in the acid-hydrolyzed reaction mixture originates from the substrate, UDP-U-[14C]GalUA, the total radioactivity of L-arabinose found in the acid hydrolyzates is used as a measure of the formation of UDPAra from UDP-U-[14C]GalUA.

Reagents

Sodium phosphate, 0.2 M, pH 7.0, containing 0.5 g of EDTA and 0.5 ml of 2-mercaptoethanol per liter (buffer)
UDP-[14C]GalUA and UDP-[14C]GlcUA.
UDPXyl, 0.01 M
UDPGalUA, 0.01 M
UDPGlcUA, 0.01 M

Procedure. UDP-U-[14C]GalUA[2] (1 μl, 0.05 μCi), NAD (4 μl), enzyme (10 μl), and buffer (10 μl) are mixed and taken up in a capillary tube. After incubation at 30° for 60 minutes, the reaction mixture is held at 100° for 1 minute; then it is made 0.1 N in HCl and hydrolyzed by heating at 100° for 15 minutes. The radioactive xylose and arabinose in the hydrolyzate are separated by paper chromatography in *n*-butanol–pyridine–water (10:3:3, v/v/v), and the quantity of radioactive arabinose is determined with a strip counter.

Preparation of Enzyme. *Ampullariella digitata* strain 399 (ATCC 15349) is maintained on peptone-Czapek agar. The organism is cultured at 25° for 2 days with aeration in 7 liters of peptone-Czapek broth containing the following components: 3% sucrose, 0.3% NaNO₃, 0.1% K₂HPO₄, 0.05% MgSO₄, 0.05% KCl, 0.001% FeSO₄, and 0.5% peptone. One liter of a 2-day culture of the organism in the same medium is used as starter. (All subsequent operations are performed at 0–4°.) The cells are harvested by centrifugation, washed twice with distilled water, and suspended in 200 ml of ice-cold buffer. The cells are disrupted with the Branson sonifier[3] at maximal power output, and the sonicate is cen-

---

[1] D. F. Fan and D. S. Feingold, *Arch. Biochem. Biophys.* **148**, 546 (1972).
[2] See Vol. 6 [46].
[3] Model S-110, Branson Instrument, Danbury, Connecticut.

trifuged at 25,000 g for 20 minutes. The clear orange supernatant fluid is diluted to 350 ml with buffer, brought to 0.01 M MnCl$_2$ by addition of 0.5 M MnCl$_2$ and the precipitate is removed by centrifugation. The fraction which precipitates from the supernatant fluid between 40 and 70% (NH$_4$)$_2$SO$_4$ saturation is collected and dissolved in buffer to yield 10 ml of enzyme preparation.

Properties

The enzyme retains full activity for at least 1 month when stored frozen at $-20°$. The pH range for activity is from 7 to 9. Contaminating activities include UDPGlcUA carboxy-lyase and UDPGlcUA 4-epimerase.

UDPXyl 4-epimerase is not present. UDPGlcUA 4-epimerase present in the enzyme preparation is completely inhibited by 1.5 mM UDPXyl, which, however, has no effect on either UDPGlcUA or UDPGalUA carboxy-lyase.

[57] Formation of UDP-Apiose from UDP-Glucuronic Acid

By HANS GRISEBACH, DIETHARD BARON, HEINRICH SANDERMANN, and ECKARD WELLMANN

The branched chain pentose D-apiose (3-C-hydroxymethyl-D-*erythro*-furanose) is of widespread occurrence in higher plants.[1,2] D-Apiose originates from the metabolism of D-glucuronic acid.[3] Partially purified enzyme preparations which catalyze the formation of uridine diphospho-apiose (UDP-Api) and uridine diphospho-D-xylose (UDP-Xyl) from uridine-diphospho-D-glucuronic acid (UDP-GlcUA) have been prepared from duckweed (*Lemna minor* L.)[4,5] and from cell suspension cultures of parsley.[6,7]

Assay Method

Principle. The enzymatic synthesis of apiose is determined by assay for the incorporation of radioactivity from UDP-[U-^{14}C]GlcUA. The mixture of UDP-sugars formed in the incubation is hydrolyzed to the free sugars and the radioactivity in the sugars is determined after separation by paper chromatography.

Reagents

 Buffer, Tris hydrochloride, 0.5 M, pH 7.8
 Dithioerythritol, 0.01 M
 NAD, 0.021 M
 UDP-[U-^{14}C]GlcUA (302 μCi/μmole, Radiochemical Centre
  Amersham), 10 μl containing 0.025 μCi
 Aniline phthalate, 1.66 g of phthalic acid and 0.93 g of aniline
  in 100 ml of water-saturated butanol.

Procedure. Enzyme is added to a solution composed of 10 μl UDP-[U-^{14}C]GlcUA,[8] 10 μl of NAD, 10 μl of dithioerythritol, and 20 μl of Tris·HCl. In the case of the enzyme from *Lemna* and in crude extracts of parsley 1 $\times$ 10^{-3} M KCN is added.

The protein concentration should be such that not more than 5% of the substrate reacts in 10 minutes. The mixture is incubated at 30° for 10 minutes. At the end of this period 20 μl of acetic acid is added. The solution is then heated for 15 minutes at 100° to hydrolyze the nucleotide sugars. The reaction mixture is then applied as a 7 cm-wide band to one sheet of paper and chromatographed together with apiose,[9,10]

---

[1] H. Grisebach, *Helv. Chim. Acta* **51**, 928 (1968).

[2] H. Grisebach and R. Schmid, *Angew. Chem.* **84**, 192 (1972). Int. Ed. **11**, 159 (1972).

[3] H. Grisebach and H. Sandermann, *Biochem. Z.* **346**, 322 (1966).

[4] H. Sandermann and H. Grisebach, *Biochim. Biophys. Acta* **208**, 173 (1970).

[5] E. Wellmann and H. Grisebach, *Biochim. Biophys. Acta* **235**, 389 (1971).

[6] D. Baron, E. Wellmann, and H. Grisebach, *Biochim. Biophys. Acta* **258**, 310 (1972).

[7] E. Wellmann, D. Baron, and H. Grisebach, *Biochim. Biophys. Acta* **244**, 1 (1971).

[8] Though the substrate concentration (2.6 $\times$ 10^{-6} M) is lower than the optimal concentration (ca. 2 $\times$ 10^{-5} M), the reaction is linear with protein concentration and time under the conditions specified.

[9] The standard procedure[10] for the preparation of D-apiose from the Australian marine plant *Posidonia australis* has been modified (H. Sandermann, unpublished procedure): *Lemna minor* L. (duckweed) is thoroughly extracted with ethanol. The powdered residue is hydrolyzed with 1 M trifluoroacetic acid (15 minutes, 100°). By lyophilization of the hydrolyzate D-apiose is obtained with arabinose as a minor impurity.

[10] D. J. Bell, *in* "Methods in Carbohydrate Chemistry," Vol. I (R. L. Whistler and M. L. Wolfrom, eds.), p. 260. Academic Press, New York, 1962.

xylose and arabinose as reference sugars for approximately 12 hours with the solvent system ethyl acetate–pyridine–water (8:2:1, by v/v/v). The reference sugars are detected with aniline phthalate. The apiose and the xylose/arabinose zones are cut into strips 2 cm wide and counted in toluene-2,5-diphenyloxazole (5 g/liter). The background is determined by counting a zone between the zones of apiose and xylose which is equal in width to the pentose zone.

Separation of Nucleotide Sugars. The incubation mixture of the enzyme assay before hydrolysis with acetic acid is separated by descending paper chromatography at 4° with ethanol–1 M ammonium acetate (pH 7.5) (5:2, v/v). The paper is prewashed with 0.01 M EDTA at pH 7.0 and water.

After 24 hours, 4 radioactive zones are detected with parsley enzyme and 5 with *Lemna* enzyme. These zones correspond to (R_{UDPG} values in parentheses): (a) UDP-GlcUA (0.45); (b) UDP-apiose + UDP-xylose + UDP-arabinose (1.0; the latter compound only with parsley); (c) minor unidentified product (2.1; with *Lemna* only); (d) cyclic apiose 1,2-phosphate (3.0) [11]; and (e) minor unidentified product (3.5).

For separation of UDP-sugars the chromatogram is allowed to run for 5 days. After this time 5 radioactive zones are detected with parsley enzyme and 3 with *Lemna* enzyme: UDP-GlcUA (0.45); minor unidentified compound (0.64, only with parsley); UDP-arabinose (0.88, only with parsley); UDP-xylose (1.0), and UDP-apiose (1.2). [12,13]

The cyclic apiose phosphate runs off the paper during the longer run. Elution of the UDP-apiose/UDP-xylose zone and rechromatography with the same solvent system leads again to conversion of part of the UDP-apiose to cyclic apiose phosphate.

Preparation and Purification of Enzyme

General Remarks. The enzyme system has been purified about 25-fold from *Lemna minor* [5] and from cell cultures of parsley. [6] Young parsley leaves can also be used as source of the enzyme, but it must be noted

---

[11] Of the apiose formed in the incubation, 20–40% is found after chromatography as cyclic apiose phosphate and only 60–80% as UDP-apiose. Both compounds have been characterized by treatment with phosphoesterases. [4] The amount of cyclic apiose phosphate formed is diminished neither by preequilibration of the tank with ethanol–water [12] nor by using ammonium acetate of pH 3.8 instead of pH 7.5. However, nearly 100% transfer of apiose is found when 7-O-glucosylapigenin and the specific UDP-apiose: 7-O-glucosylapigenin transferase (this volume [61]) are added to the reaction mixture before chromatography. This proves that all the apiose must have been originally present as UDP-apiose.

[12] K. C. Tovey and R. M. Roberts, *J. Chromatogr.* **47,** 287 (1970).

[13] The zones of UDP-xylose and UDP-apiose are not completely separated.

that maximal activity of the enzyme is reached at about day 24 after sowing of the seedlings and then declines.[14]

Cultures of Lemna minor[15]

Stock solution	g/liter
NH_4NO_3	6.6
$Ca(NO_3)_2 \cdot 4H_2O$	8.3
$MgSO_4 \cdot 7H_2O$	16.6
$ZnSO_4$	2.1
$MnSO_4$	0.5
$CoSO_4 \cdot 7H_2O$	0.03
$CuSO_4 \cdot 5H_2O$	0.13
$Na_2MoO_4 \cdot 2H_2O$	0.83
H_3BO_3	0.5
Na-EDTA	16.6
KH_2PO_4	8.6
$Na_2HPO_4 \cdot H_2O$	0.2
$FeSO_4$	0.8
KCl	1.0

The stock solution is adjusted to pH 5.5 with $1.5 N$ KOH; 5 ml of the solution is pipetted into a 1-liter Fernbach flask, and 655 ml of distilled water is added. The sterilized flasks are inoculated with about 20 mother fronds and kept at 27° under continuous light (~2000 lux). After about 4 weeks the surface of the medium is completely covered with plants.

Cell Suspension Cultures of Parsley

MEDIUM (slightly modified medium B5 of Gamborg *et al.*[16])

$NaH_2PO_4 \cdot H_2O$	150 mg
KNO_3	2500 mg
$(NH_4)_2SO_4$	134 mg
$MgSO_4 \cdot 7H_2O$	250 mg
Sucrose	20 g
Stock solutions: A	1 ml
B	1 ml
C	1 ml
D	5 ml
E	10 ml
F	10 ml

[14] K. Hahlbrock, A. Sutter, E. Wellmann, R. Ortmann, and H. Grisebach, *Phytochemistry* **10**, 109 (1971).
[15] M. A. M. Lacor, *Acta Bot. Neerl.* **17**, 357 (1968).
[16] O. L. Gamborg, R. A. Miller, and K. Ojima, *Exp. Cell Res.* **50**, 151 (1968).

STOCK SOLUTIONS

A:	$CaCl_2/100$ ml H_2O	15 g
B:	$KJ/100$ ml H_2O	75 mg
C:	$MnSO_4 \cdot 1H_2O$	1120 mg
	H_3BO_3	300 mg
	$ZnSO_4 \cdot 7H_2O$	300 mg
	$Na_2MoO_4 \cdot 2H_2O$	25 mg
	$CuSO_4 \cdot 5H_2O$	25 mg
	$CoCl_2 \cdot 6H_2O$	25 mg
	H_2O to 100 ml	
D:	$FeSO_4 \cdot 7H_2O$	278 mg
	Na_2-EDTA	372 mg
	H_2O to 100 ml	
E:	Nicotinic acid	10 mg
	Pyridoxine·HCl	10 mg
	Thiamine·HCl	100 mg
	Myoinositol	1000 mg
	H_2O to 100 ml	
F:	2,4-Dichlorophenoxy	10 mg/
	acetic acid	100 ml
		H_2O

The medium is adjusted to pH 5.5 with 0.5 N KOH.

The tissue for the batch-propagated suspension cultures is originally derived from leaf petioles of *Petroselinum hortense*.[17] Suspension cultures from parsley cells are obtained by transferring a piece of callus tissue from solid "D" medium to the liquid medium. Cells, 2 g wet weight, are transferred to 40 ml of fresh medium every 10 days by means of a sterile sieve spoon. Cells are grown in 200-ml Erlenmeyer flasks[18] at 26° on a rotary shaker at 120 rpm and with an amplitude of 2.5 cm for 9–10 days in the dark and illuminated for 24 hours with white light (ca. 27,000 lux) from fluorescent lamps (Philips TL 40 W/18) before enzyme extraction. Cells are harvested by vacuum filtration through a porous glass filter.

Partial Purification of Enzyme from Lemna minor. All operations are carried out at 4° unless otherwise stated.

About 250 g (wet weight) of plants is suspended in 500 ml of 0.1 M Tris·HCl buffer (pH 7.5) containing $1 \times 10^{-2} M$ 2-mercaptoethanol and $1 \times 10^{-3} M$ KCN (buffer A) and homogenized with 250 g of quartz sand in a precooled mortar. The resulting slurry is filtered through four layers of cheesecloth and the residue is again extracted with 250 ml of the same buffer. The filtrate is centrifuged at 35,000 g for 20 minutes. The supernatant is subjected to $(NH_4)_2SO_4$ fractionation at pH 7.5 by

[17] I. K. Vasil and A. D. Hildebrandt, *Planta* **68**, 69 (1966).
[18] The cultures can easily be scaled up to 1-liter flasks containing 400 ml of medium.

addition of the solid salt. Protein which precipitates between 35 and 70% saturation is collected by centrifugation at 35,000 g for 10 min and dissolved in 5 ml of buffer A.

The solution is chromatographed with $1 \times 10^{-2} M$ Tris·HCl buffer (pH 7.5) containing $1 \times 10^{-3} M$ dithioerythritol and $1 \times 10^{-3} M$ KCN on a column (20 cm × 3 cm) of Sephadex G-25 equilibrated before use with the same buffer.

The protein fraction is then chromatographed with $1 \times 10^{-2} M$ Tris·HCl buffer (pH 7.5) containing $1 \times 10^{-3} M$ dithioerythritol (buffer B) on a column (50 cm × 2.5 cm) of Sephadex G-200 at a flow rate of 25 ml per hour. The fractions with enzymatic activity (ca. 15 ml) are absorbed on a DEAE-cellulose column (2 g) equilibrated before use with buffer B. The column is then washed with 50 ml of $0.09 M$ KCl in the same buffer, and the protein is eluted with 3 ml of $0.15 M$ KCl in the same buffer. The fractions with enzymatic activity are then desalted by chromatography on a column of Sephadex G-25 equilibrated before use with $5 \times 10^{-3} M$ buffer B. The solution is stored at $-20°$.

Approximately 26-fold purification of enzyme activity is obtained by this procedure (Table I).

In the crude extract the reaction is not linear with protein concentration. Also, the enzymatic activity cannot be determined in the presence of $(NH_4)_2SO_4$ (see below). After desalting with Sephadex G-25 the reaction is still inhibited at protein concentration >1 mg/ml. After gel filtration on Sephadex G-200, no inhibition is observed at high protein concentration.

Purification of Enzyme from Parsley Cell Cultures. The enzyme puri-

TABLE I

PARTIAL PURIFICATION OF APIOSE/XYLOSE-SYNTHESIZING ENZYME SYSTEM FROM *Lemna minor*

Purification step	Protein (mg)	Specific activity[a] (units $\times 10^6$/mg protein)		Purification (-fold)	
		Xylose	Apiose	Xylose	Apiose
Crude extract	1300	—	—	—	—
$(NH_4)_2SO_4$ fractionation (0.35–0.7)	530	—	—	—	—
Sephadex G-25 column	450	0.52	0.79	1	1
Sephadex G-200 column	39	4.5	6.5	8.6	8.3
DEAE-cellulose column[b]	7.8	14.8	20.6	29	26

[a] One enzyme unit is defined as the enzyme quantity which catalyzes the conversion of 1 μmole of UDP-GlcUA per minute at 30° in the enzyme assay.

[b] After removal of salts with Sephadex G-25.

fication is carried out as described for the enzyme from *Lemna minor* with the modification that the column chromatography on Sephadex G-200 and DEAE-cellulose are carried out in reverse order.

Approximately 24-fold purification of enzyme activity is obtained by this procedure (Table II).

In the crude extract the enzymatic reaction is partially inhibited. Reaction with the enzyme from the Sephadex G-200 column is linear with protein concentration up to 0.1 mg/ml and with time.

Properties of Enzyme

The properties of enzyme from *Lemna* and from parsley cell cultures are identical within the limits of experimental error.

Attempts to separate the enzymatic activities for apiose and xylose formation by chromatography on either DEAE-cellulose or Sephadex G-200, by isoelectric focusing on Sephadex G-75 thin-layer plates, and by analytical disc electrophoresis have so far been unsuccessful.[5,6]

Enzyme Contamination. With the enzyme from parsley, arabinose is also formed besides apiose and xylose because of the presence of the enzyme UDP-L-arabinose 4-epimerase (EC 5.1.3.5), which is only partly separated from the apiose/xylose enzyme by the procedure of Table II. Parsley and cell cultures of this plant contain a second enzyme which catalyzes the synthesis of UDP-D-xylose alone from UDP-GlcUA.[7]

TABLE II

PARTIAL PURIFICATION OF APIOSE/XYLOSE-SYNTHESIZING ENZYME
SYSTEM FROM PARSLEY CELL CULTURES

Purification step	Protein (mg)	Specific activity[c] (units $\times 10^6$/mg protein)		Apiose/ Xylose[d]	Purification (-fold)	
		Xylose	Apiose		Xylose	Apiose
Crude extract	1000	—	—	—	—	—
Treatment with Dowex 1 X-2	760	7.0	10.3	1.47	1	1
(NH$_4$)$_2$SO$_4$ fractionation (0.4–0.5)[b]	135	22.2	35.0	1.58	3.2	3.4
DEAE-cellulose column	63	91.0	145.0	1.60	13	14
Sephadex G-200 column	40[a]	167	247.0	1.48	24	24

[a] This protein value determined by the method of Warburg and Christian is too high.

[b] After removal of salts with Sephadex G-25.

[c] For definition of enzyme unit, see Table I.

[d] This represents the sum of xylose plus arabinose.

This enzyme is separated from the apiose/xylose synthetase on DEAE-cellulose.[7]

Cofactor Requirement. The enzyme has a requirement for NAD⁺. The optimal NAD⁺ concentration is about $3 \times 10^{-3} M$.

Activators and Inhibitors. Dithioerythritol has a stabilizing effect on the enzyme, whereas 2-mercaptoethanol inhibits the reaction. In the presence of 10^{-3} to $10^{-2} M$ NH₄⁺ apiose synthesis is inhibited and xylose synthesis is stimulated when the pH of the medium is between pH 8.2 and 7.5, whereas at pH 7.0 NH₄⁺ also stimulates apiose formation.[6]

Both apiose and xylose formation are inhibited to about the same extent by UDP, UTP, UDP-xylose, and UDP-glucose. The strongest inhibitors are UDP (about 50% inhibition at $10^{-5} M$) and UDP-D-xylose (about 50% inhibition at $10^{-4} M$).

Stability. In the presence of $10^{-3} M$ dithioerythritol the enzyme is stable for about a month at $-35°$, whereas at $2°$ it loses about 50% of its activity after 96 hours. Repeated thawing and freezing leads to strong losses of activity.

Optimum pH. The enzyme has a pH optimum of about 8.2.

Substrate Affinity. The K_m value for UDP-GlcUA is $2 \times 10^{-6} M$ at pH 8.2 for apiose and xylose formation.

[58] Formation of TDP-L-Rhamnose from TDP-D-Glucose

By Luis Glaser, Harold Zarkowsky, and L. Ward

$$\text{TDP-D-glucose} \xrightarrow{\text{(DPN)}} \text{TDP-4-keto-6-deoxy-D-glucose} \quad (1)$$

$$\text{TDP-4-keto-6-deoxy-D-glucose} + \text{TPNH} + \text{H}^+ \rightarrow \text{TDP-L-rhamnose} + \text{TPN} \quad (2)$$

These reactions were described in Volume 8 of this series.[1] Significant advances have been made since that time in the purification of these enzymes, and in understanding their mechanism. The preparation of pure TDP-D-glucose oxidoreductase, catalyzing reaction (1), and the preparation of the two protein fractions that together catalyze reaction (2) are described herein.

TDP-D-Glucose Oxidoreductase

Two independent methods for the purification of this enzyme from *Escherichia coli* B have been described,[2,3] yielding enzyme of essentially

[1] L. Glaser and S. Kornfeld, Vol. 8 [54].
[2] H. Zarkowsky and L. Glaser, *J. Biol. Chem.* **244**, 4750 (1969).
[3] S. F. Wang and O. Gabriel, *J. Biol. Chem.* **244**, 3430 (1969).

the same purity. The crystallization of the enzyme has been reported by Wang and Gabriel.[3] The purification method of Zarkowsky and Glaser[2] will be described in detail.

Assay Method

The method is based on the formation of a chromogen adsorbing at 320 nm,[4] after incubation of TDP-4-keto-6-deoxy-D-glucose in alkali at 37°. The extinction coefficient[2] for this chromogen is 6.5×10^3 liter M^{-1} cm^{-1}. One unit is defined as the quantity of enzyme catalyzing the formation of 1 μmole of TDP-4-keto-6-deoxy-D-glucose per hour.

Reagents

Tris·HCl, 0.5 M, pH 8.0
TDP-D-glucose, 3 mM
NaOH, 0.5 N

Procedure. The reaction mixture (0.7 ml) contains 0.1 ml of 0.5 M Tris·HCl, pH 8.0, 0.1 ml of 3 mM TDP-D-glucose, and enzyme. After incubation at 37° the reaction is stopped by the addition of 0.3 ml of NaOH, and after an additional incubation at 37° for 10 minutes the absorbancy at 320 nm is measured. An identical reaction mixture containing enzyme but no TDP-D-glucose is used as a control. The assay should be used with great caution in crude extracts, to ensure linearity with time and enzyme concentration.

Purification Procedure

Commercial frozen *E. coli* B obtained from Grain Processing Co., Muscatine, Iowa, as washed ¾ log cells grown in rich medium are used as starting material. Unless otherwise stated, all steps in the purification are carried out at 3°. Percentage saturation of $(NH_4)SO_4$ is calculated from a standard monogram.[5] All precipitates are collected by centrifugation at 16,000 g for the times indicated. Disc gel electrophoresis is carried out by the method of Davis[6] using 7% gel with the addition of $5 \times 10^{-3} M$ 2-mercaptoethanol to the upper buffer.

Enzyme Extraction. Five pounds of frozen *E. coli* B are allowed to thaw overnight in 5 liters of 0.05 M Tris·HCl, pH 8.0, $10^{-4} M$ EDTA. The cells are disrupted in a Manton-Gaulin press, and cell debris is removed by centrifugation for 30 minutes.

Salt Fractionation. To the supernatant fluid is gradually added solid

[4] R. Okazaki, T. Okazaki, J. L. Stromioger, and A. M. Michelson, *J. Biol. Chem.* **237**, 3014 (1962).

[5] F. DiJeso, *J. Biol. Chem.* **243**, 2022 (1968).

[6] B. J. Davis, *Ann. N.Y. Acad. Sci.* **121**, 404 (1964).

$(NH_4)_2SO_4$ to 60% saturation. Thirty minutes after the last addition of $(NH_4)_2SO_4$, the precipitate is collected by centrifugation for 30 minutes, and the supernatant fluid is discarded. The precipitate is suspended in 0.05 M Tris·HCl, pH 8.0, and dialyzed against two 10-liter changes of the same buffer for 16 hours.

The dialyzed solution is brought to 25°, and 25 mg each of RNase and DNase are added. After 45 minutes of constant stirring at 25°, the solution is cooled to 3°, and again fractionated with $(NH_4)_2SO_4$. Since the concentration of $(NH_4)_2SO_4$ required to precipitate the enzyme at this stage is somewhat variable, the correct ammonium sulfate concentration should be determined by a small-scale fractionation of each preparation. In the preparation shown in Table I, the solution was adjusted to 40% saturation by the addition of solid $(NH_4)_2SO_4$, and after 30 minutes the precipitate was collected by centrifugation for 20 minutes suspended in 0.05 M Tris·HCl, pH 8.0, and dialyzed for 16 hours against three 6-liter volumes of 0.025 M Tris·HCl, pH 8.0.

First DEAE-Cellulose Chromatography. The dialyzed solution is put on a column (80 × 5.5 cm) of DEAE-cellulose, equilibrated with 0.025 M Tris·HCl, pH 8.0, until the absorbance of the eluate at 280 nm is 0.30. The column is washed with 0.1 M KCl in 0.025 M Tris·HCl, pH 8.0 (6 liters were required in the preparation shown in Table I). The eluting buffer is then changed to 0.3 M KCl in 0.025 M Tris·HCl, pH 8.0, and fractions are collected. The enzyme is eluted after 1.35 liters of this buffer have passed through the column.

Hydroxyapatite Chromatography. The fractions containing the en-

TABLE I
PURIFICATION OF dTDP-D-GLUCOSE OXIDOREDUCTASE

Procedure	Volume (ml)	Activity (units/ml)	Total activity (units)	Protein (mg/ml)	Specific activity (units/mg)
Ammonium sulfate, 60%	2220	3.8	8400	—	—
Ammonium sulfate, 0–40%	475	13.6	6400	91	0.15
DEAE-cellulose I	540	6.6	3560	7.8	0.85
Hydroxyapatite	175	12.3	2150	1.4	8.8
DEAE-cellulose II	270	4.8	1300	0.09	53
Pressure filtration	6.3	240	1510	2.8	86
Sephadex G-100					
Tube 25	5.0	41.2⎫		0.12	333
Tube 26	5.0	43.0⎬	615	0.15	287
Tube 27	5.0	34.8⎭		0.14	249

a Reprinted with permission from H. Zarkowsky and L. Glaser, *J. Biol. Chem.* **244**, 4750 (1969).

zyme are pooled, and put on a column (36 × 2.5 cm) of hydroxyapatite, equilibrated with 0.025 M Tris·HCl, pH 8.0. The enzyme, which does not stick to the column, is eluted by washing the column with 0.025 M Tris·HCl, pH 8.0. The fractions containing the enzyme are pooled and dialyzed against two changes of 2 liters each of 0.025 M Tris·HCl, pH 8.0.

Second DEAE-Cellulose Chromatography. The dialyzed enzyme is put on a DEAE-cellulose column (36 × 2.5 cm) equilibrated with 0.025 M Tris·HCl, pH 8.0. The enzyme is eluted with a linear gradient. The mixing flask contains 1 liter of 0.15 M KCl in 0.025 M Tris·HCl, pH 8.0, and the reservoir 1 liter of 0.35 M KCl in 0.025 M Tris·HCl, pH 8.0. The enzyme is eluted after 900 ml of buffer have passed through the column. The fractions containing the enzyme are pooled and concentrated by pressure filtration, with the use of an XM-50 Diaflo membrane.

Chromatography on Sephadex G-100. The enzyme solution from the previous step (6 ml) is put on a column (80 × 2.5 cm) of Sephadex G-100, equilibrated with 0.025 M Tris·HCl, pH 8.0, and eluted with the same buffer at a rate of 20 ml per hour. Fractions of 5 ml are collected. About 50% of the enzyme with a specific activity of 300 units/mg is recovered in tubes 25–27. The remaining enzyme is recovered as fractions of lower specific activity. Fractions 25–27 (Table I) showed one major band and several minor bands on gel electrophoresis of the preparation shown in Table I. A summary of the purification is shown in Table I. This enzyme can be further purified by preparative disc gel electrophoresis.

Preparative Gel Electrophoresis. Since the enzyme could be recovered in better than 80% yield from analytical polyacrylamide gels by elution, a simple procedure is used for preparative disc gel electrophoresis, with the use of a jacketed glass tube, 12 × 3 cm. A 5-cm analytical gel and a 1-cm stacking gel are used. They can be loaded with up to 7.5 ml of enzyme (5 mg of protein) in 10% glycerol. Electrophoresis is carried out at 5 amp for 15 hours at 0°. Higher currents led to a distortion of the protein bands due to thermal convection. At the end of the run, 2- to 3-mm slices of the gel are cut and the enzyme is eluted by homogenizing the slices in 7 ml of 0.025 M Tris·HCl, pH 8.0, $5 \times 10^{-3} M$ 2-mercaptoethanol in a TenBroeck homogenizer. Fractions containing the enzyme are dialyzed overnight against three changes of 500 ml each of 0.025 M Tris·HCl, pH 8.0. Enzyme eluted from the preparative disc gel shows a single band on analytical acrylamide gel and has a specific activity of 420 units/mg of protein. The purified enzyme can be stored for several months at 0° under toluene vapor, but decays if frozen and thawed repeatedly.

The same purification scheme has been successfully used starting with 50 lb of *E. coli* B.

Properties of the Enzyme

The K_m for TDP-D-glucose is $7 \times 10^{-5} M$. Of a variety of nucleotides tested, only dUDP-D-glucose is a substrate; the K_m is $2.2 \times 10^{-3} M$, but the maximal velocity is the same as that obtained with TDP-D-glucose as a substrate. The pH optimum is 8.0.

The molecular weight of the enzyme is 78,000 by sedimentation equilibrium and gel exclusion chromatography[2,7] and is composed of two subunits of molecular weight 40,000.[7] The enzyme contains 1 mole of tightly bound DPN per mole of enzyme.[2,3] Reaction of the enzyme with *p*-hydroxymercuribenzoate results in release of DPN, and dissociation of the enzyme into subunits. A variety of experimental observations[7-9] substantiate a mechanism for this reaction in which the initial step is conversion of TDP-glucose to TDP-4-ketoglucose an enzyme-bound DPNH, followed by loss of H_2O to yield a 5,6-glycoseen, which is then reduced to yield TDP-4-keto-6-deoxy-D-glucose and enzyme bound DPN. Evidence has been presented[7,10] that reduction of enzyme-bound DPN to enzyme-bound DPNH results in a conformational change in the protein such that substrate cannot be released at a significant rate from this form of the enzyme, thus the intermediates in the normal catalytic process remain tightly bound to the enzyme. The mechanism of this enzyme has been discussed in detail in reference 9. No reversibility of the overall reaction can be demonstrated. Reversibility of the first step, the conversion of TDP-glucose to TDP-4-ketoglucose has been demonstrated.

Preparation of Apoenzyme

The following procedure will yield apoenzyme in high yield.[7] Eighty-five units of enzyme are incubated with $5 \times 10^{-3} M$ *p*-hydroxymercuribenzoate in 2 ml of 20% glycerol–0.025 M Tris·HCl, pH 8.0, for 20 minutes at 30°. After addition of 50 ml of 1 M 2-mercaptoethanol, the solution is placed on a 1.5×92 cm column of Sephadex G-100 equilibrated with 0.1 M KCl–0.025 M Tris·HCl, pH 8.0–$5 \times 10^{-4} M$ dithio-

[7] H. Zarkowsky, E. Lipkin, and L. Glaser, *J. Biol. Chem.* **245**, 6599 (1970).

[8] S. F. Wang and O. Gabriel, *J. Biol. Chem.* **245**, 8 (1970).

[9] L. Glaser and H. Zarkowsky, *in* "The Enzymes" (P. D. Boyer, ed.), 3rd ed. Vol. 5, p. 465. Academic Press, New York, 1972.

[10] H. Zarkowsky, E. Lipkin, and L. Glaser, *Biochem. Biophys. Res. Commun.* **38**, 797 (1970).

threitol; 1.9 ml fractions are collected at 3°. Holoenzyme is assayed by the standard assay procedure. Apoenzyme, molecular weight 40,000, is assayed after preincubation with $2 \times 10^{-4} M$ DPN for 15 minutes at 25°. The recovery of apoenzyme is 50–75%. The apoenzyme is stable for several weeks at 4° when stored under nitrogen in the presence of dithiothreitol.

Reactivation of the apoenzyme by DPN occurs in two steps; the apoenzyme binds DPN weakly (K_m $5 \times 10^{-6} M$ at 16° and $5 \times 10^{-4} M$ at 37°),[7] and this complex undergoes a slow conformational change to yield active enzyme containing tightly bound DPN. All DPN analogs containing the ADP-ribose portion of DPN react with the apoenzyme to yield inactive protein of molecular weight 78,000. Hypoxanthine analogs of DPN do not bind tightly to the apoenzyme.

Epimerization and Reduction of TDP-4-keto-6-deoxyglucose

Assay Method

Two proteins designated E-I and E-II[11] are required to catalyze reaction (2) above. The assay is based on the oxidation of TPNH in the presence of E-I and E-II dependent on the addition of TDP-4-keto-6-deoxy-D-glucose.

Reagents

Tris·HCl, 0.5 M, pH 8.0

TPNH, 1 mM in 0.01 M Tris·HCl, pH 8.0. This solution is prepared fresh daily and stored at 4°.

TDP-4-keto-6-deoxy-D-glucose, 1 mM prepared daily from TDP-D-glucose with purified TDP-D-glucose oxidoreductase. Since pure oxidoreductase is free of E-I and E-II and TPNH oxidase activity, it need not be inactivated before this substrate can be used in the assay.

Procedure. The reaction mixture (1 ml) contained 0.2 ml of Tris·HCl buffer, 0.1 ml of TPNH, 0.1 ml of TDP-4-keto-6-deoxy-D-glucose and E-I and E-II the oxidation of TPNH is followed at 340 nm in a spectrophotometer thermostatted at 25°. A control is included without TDP-4-keto-6-deoxy-D-glucose to measure TPNH oxidase activity. To measure either E-I or E-II activity, the quantity of one enzyme is held constant and the other is varied over a range where the rate is propor-

---

[11] A. Melo and L. Glaser, *J. Biol. Chem.* **243**, 1475 (1968).

tional to enzyme concentration. A unit is defined as the quantity of each enzyme that will produce 1 μmole of product per minute.

Purification Procedure

The procedure described here is a modification of that described originally.[11] In the original method E-I and E-II are separated by chromatography on Sephadex G-100. In the present procedure this step has been replaced by ion exchange chromatography. All steps in the purification are carried out at 3°.

Cell Extraction. Pseudomonas aeruginosa ATCC 7700 is grown as previously described.[1] Washed cells can be stored as a frozen paste. The purification procedure has also been carried out using cells grown in the same medium in a fermentor to stationary phase with identical results. Cells, 250 g wet weight, are suspended in 600 ml of 0.05 M Tris·HCl–0.01 M MgCl$_2$–0.001 M EDTA, pH 8.0 with the aid of a Waring Blendor, and the cells are disrupted in a Manton-Gaulin press. Cell debris is removed by centrifugation at 16,000 g for 30 minutes.

Salt Fractionation. To the supernatant fluid (800 ml) is added 250 ml of freshly prepared 2% protamine sulfate, pH 7.5. After 10 minutes, the precipitate is collected by centrifugation at 16,000 g for 20 minutes and discarded.

To the supernatant fluid is added an equal volume of neutral saturated ammonium sulfate, after 30 minutes the precipitate is collected by centrifugation, dissolved in 80 ml of 0.01 M potassium phosphate, pH 7.0, and dialyzed for 16 hours against several changes of the same buffer. After dialysis the solution is cloudy and the precipitate is removed by centrifugation at 100,000 g for 2 hours; the precipitate is discarded, and the supernatant fluid is used for further purification of the enzyme(s) (volume, 220 ml).

DEAE-Cellulose Chromatography. The enzyme solution is put on a 3.5 × 36 cm column of DEAE cellulose 52 (Whatman) equilibrated with 0.01 M potassium phosphate, pH 7.0. The column is eluted with 2 liters linear gradient of potassium phosphate from 0.01 M potassium phosphate to 0.1 M potassium phosphate; 7.5 ml fractions are collected every 15 minutes.

Fractions 70–110 contain E-II, and fractions 189–279 contain E-I. These fractions are pooled, lyophilized, taken up in a minimum volume of H$_2$O, and dialyzed against 0.005 M potassium phosphate, pH 7.0. They can be stored frozen for several weeks without loss of activity. E-II is highly unstable during purification and the recovery of activity is very low. A summary of the purification is shown in Table II. Both fractions are free of TPNH oxidase activity.

TABLE II
PURIFICATION OF TDP-L-RHAMNOSE SYNTHETASE FRACTIONS

Fraction	Total enzyme units	Specific activity (units/mg)
Ammonium sulfate precipitate after dialysis and centrifugation	21.6[a]	0.0028
E-II DEAE fractions 70–110	1.1	0.12
E-I DEAE fractions 189–279	55.0	0.146

[a] This represents the total activity of E-I and E-II and is limited by the quantity of E-II in the extract.

Properties of the Enzymes

E-II catalyzes the sequential exchange of the hydrogens at C-3 and C-5 of TDP-4-keto-6-deoxy-D-glucose with the medium,[11,12] but free TDP-4-keto-6-deoxy-L-rhamnose is not formed by the enzyme. E-I is specifically protected from heat denaturation by TPNH and probably contains the TPNH binding site. The proposed mechanism for this reaction is shown in Fig. 1. E-II is postulated to be an epimerase which forms enzyme-bound TDP-4-keto-6-deoxy-L-rhamnose, which is stereo-

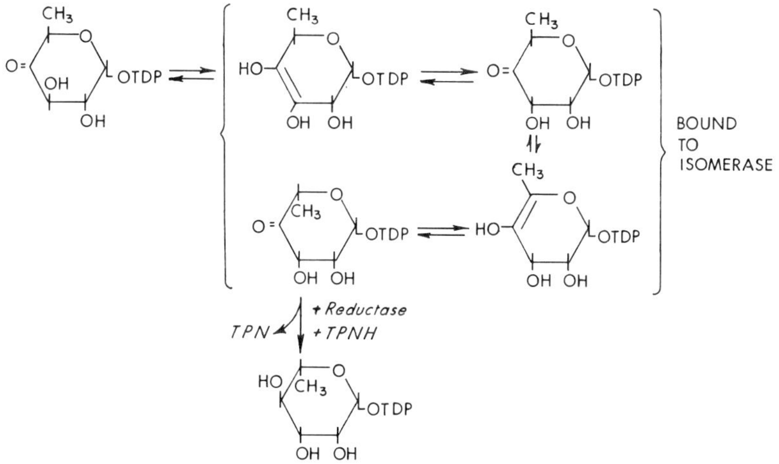

FIG. 1. Proposed mechanism for conversion of TDP-4-keto-6-deoxy-D-glucose to TDP-L-rhamnose. Isomerase is E-II and reductase is E-I.

[12] L. Glaser, in "The Enzymes" (P. D. Boyer, ed.), 3rd ed., Vol. 6, p. 355. Academic Press, New York, 1972.

specifically reduced by E-I and TPNH to TDP-L-rhamnose, which is then released from the enzyme.

No precise kinetic data have been obtained for the enzyme. The following data are approximate and have been determined under standard assay conditions. The K_m for TDP-4-keto-6-deoxy-D-glucose is $1 \times 10^{-4} M$ and independent of the ratio of E-I and E-II in the assay. TDP-L-rhamnose is a competitive inhibitor with respect to this substrate $K_i = 2 \times 10^{-4} M$. The K_m for TPNH is $5 \times 10^{-5} M$. TPN is a competitive inhibitor with respect to TPNH with K_i $7 \times 10^{-4} M$. Substitution of the hydrogen at C-3 or C-5 of TDP-4-keto-6-deoxy-D-glucose with deuterium decreases the rate of the reaction, 3.4- and 2.0-fold, respectively.[12] Analogous enzymes for the synthesis of other deoxy sugars have been described.[13]

[13] O. Gabriel, this volume [59].

[59] Formation of TDP-6-Deoxy-L-Talose from TDP-D-Glucose[1,2]

By OTHMAR GABRIEL

Preparative Procedures

Principle. Enzyme preparations obtained from *Escherichia coli 045*[3] catalyze the following sequence of reactions:

$$\text{TDP-D-glucose} \rightarrow \text{TDP-4-keto-6-deoxy-D-glucose (TDP-D-xylo-4-hexulose)} \quad (1)$$

$$\text{TDP-4-keto-6-deoxy-D-glucose} \rightarrow \text{TDP-4-keto-L-rhamnose}$$
$$\text{(TDP-4-keto-6-deoxy-L-mannose, TDP-L-arabino-4-hexulose)} \quad (2)$$

$$\text{TDP-4-keto-L-rhamnose} \xrightarrow{\text{NADPH}} \text{TDP-6-deoxy-L-talose} \quad (3)$$

For the preparation of TDP-6-deoxy-L-talose it is sufficient to use crude extracts of *E. coli 045* obtained by sonic disruption of cells. Incubation of TDP-glucose with this crude enzyme extract is presence of NADPH leads to formation of TDP-6-deoxy-L-talose. No isolation of intermediates is necessary and the progress of the reaction can be best followed by colorimetric assay for the newly formed 6-deoxy hexose. Isolation of TDP-6-deoxy-L-talose is accomplished by chromatography.

[1] R. W. Gaugler and O. Gabriel, *Fed. Proc., Fed. Amer. Soc. Exp. Biol.* **29**, 337 (1970).
[2] R. W. Gaugler and O. Gabriel, *J. Biol. Chem.* in press.
[3] W. H. O. International Escherichia Centre Statens Seruminstitut, Copenhagen.

Separation of the crude enzyme mixture into individual fractions catalyzing reactions (2) and (3), respectively, can be achieved by conventional methods of enzyme purification such as ammonium sulfate precipitation and DEAE chromatography.

For the assay of TDP-4-keto-L-rhamnose 3,5-epimerase (reaction 2) incubation with TDP-4-keto-6-deoxy-D-glucose-*3T* is employed: Tritium label is exchanged into the medium as a result of 3,5-epimerase activity.

TDP-6-deoxy-L-talose dehydrogenase activity (reaction 3) can be measured by direct spectrophotometric determination of NADPH oxidation at 340 nm.

Preparation of TDP-6-Deoxy-L-talose

Growth and Harvesting of Cells. A culture of *E. coli 045* kept on nutrient agar slants (Difco) is used for incubation of growth medium containing 17.5 g/liter Antibiotic Medium No. 3 (Difco). Four 1-liter Erlenmeyer flasks, each containing 250 ml medium, are placed in a gyratory shaker (New Brunswick) at 37°. The optical density is followed at 750 nm and the cells are grown up to early logarithmic growth phase (OD about 0.800). The cells are harvested by centrifugation, washed with ice cold water. The total yield of cells is about 2 g wet weight. The cells are suspended in ice cold 0.025 M Tris·HCl buffer, pH 8.0, containing 0.001 M MgCl$_2$, 0.01 M β-mercaptoethanol, and 0.001 M sodium EDTA. About 5 ml of buffer are used per gram wet weight of cells. The cells were disrupted by sonication (Heat Systems-Ultrasonics, Inc.) for a total of 60 seconds in 10-second pulses. The temperature of the solution remained below 10°C at all times. The call debris was removed by centrifugation at 15,000 g for 20 minutes and the supernatant solution (about 15–20 mg total protein per milliliter solution) was used directly for the experiments below.

Conditions for Incubation. Incubation mixtures with a total volume of 200 µl contained TDP-glucose (Sigma) 0.1 µmole, Tris·HCl buffer, pH 8.0, 2.5 µmole, β-mercaptoethanol 0.1 µmole and 25–50 µl of the above sonic enzyme mixture. A total of 0.3 µmole of NADPH (Boehringer) is added in 30-minute intervals to ensure a sufficient quantity of coenzyme throughout the 2-hour incubation period at 37°. Before a colorimetric sugar test for deoxyhexose[4] synthesis is carried out, the major part of proteins has to be removed. For this purpose, the reaction mixture is heated for 1 minute in a boiling water bath. The heat-de-

---

[4] Z. Dische, *in* "Methods in Carbohydrate Chemistry," R. L. Whistler and M. L. Wolfrom, eds., Vol. I, pp. 468–469, 501–503. Academic Press, New York, 1962.

natured proteins are removed by centrifugation. A control experiment without substrate addition is carried out parallel.

Preparation of TDP-4-Keto-6-deoxyglucose

For a study of reactions (2) and (3) it is useful to prepare TDP-4-keto-6-deoxyglucose. For this purpose, the same incubation mixture is used as described above but NADPH is omitted. Termination of the incubation period is carried out by addition of 600 μl (3 volumes) of ethanol and heating the sample in a boiling water bath for 1 minute. For preparative purposes, it is suggested to prepare up to 10 individual incubation mixtures simultaneously, each containing 0.1 μmole of sugar nucleotide. The samples are combined and evaporated to a small volume (200 μl) under reduced pressure. The solution is diluted to 5 ml with water, the pH is adjusted to 4.0 with 1 N acetic acid, and 10 mg of Norit is added. The charcoal is washed with water, and the nucleotide is eluted with 0.2% ammonia in 50% ethanol. The ethanol is removed by evaporation under reduced pressure.

Preparation of TDP-4-Keto-6-deoxyglucose-3T

The identical procedure is followed as outlined above except the starting material is TDP-glucose-3T.[5] The product obtained should have a specific activity of 1.25 × 10[7] cpm/μmole (about 30% counting efficiency).

Assay Procedures

6-Deoxyhexose Formation. The progress of the overall reaction is best followed in the crude system by direct colorimetric determination of 6-deoxyhexose according to Dische.[4] When fucose is used as a standard in the 3-minute cysteine-sulfuric acid determination, 6-deoxytalose gives 1.5 times the value of fucose for the same molar concentration.

TDP-Glucose-4,6-dehydratase (Reaction 1). The first step of the overall reaction results in formation of TDP-4-keto-6-deoxyglucose formation. This intermediate has a characteristic absorption peak at 318 nm (E_{318} = 4.8 × 10[3]) in 0.1 N NaOH. The incubation mixture is identical with the one described above with the exception that NADPH is omitted: the sample and a control are diluted with 0.1 N NaOH to 1.0 ml, kept for 15 minutes at 37° and the optical density is measured at 318 nm.

TDP-4-Keto-L-rhamnose-3,5-epimerase (Reaction 2). TDP-4-keto-6-deoxyglucose-3T (1.25 × 10[7] cpm/μmole) is obtained by incubation of TDP-glucose-3T with TDP-glucose-4,6-dehydratase (see above).

[5] O. Gabriel, this volume [30].

TDP-4-keto-6-deoxy glucose-*3T*, 2 nmoles are incubated with enzyme (50 μl) in 60 μl of 0.025 *M* Tris·HCl buffer pH 8.0. After 30 minutes at 37° the reaction is terminated by the addition of 10 μmoles of sodium borohydride. The mixture is allowed to stand for 30 minutes at room temperature. The solution is diluted to a total volume of 500 μl. An aliquot is taken and the tritium activity is determined in a liquid scintillation counter. Another aliquot is taken and evaporated to dryness. The dry samples are again dissolved in water and evaporated to dryness. This process is repeated once more. Finally the samples are counted. The difference in counts before and after repeated evaporation represent the amount of tritium exchanged by the enzymatic reaction. A control experiment containing the identical components but heat inactivated enzyme must be included in each experiment.

TDP-6-Deoxy-L-talose dehydrogenase (Reaction 3). The direct spectrophotometric assay for disappearance of NADPH by measuring the optical density at 340 nm is probably the easiest method. For this purpose microcuvettes are used and the incubation mixture is in a total volume of 300 μl in 0.025 *M* Tris·HCl buffer, pH 8.0, containing 0.03 μmole of TDP-4-keto-6-deoxyglucose and 0.04 μmole of NADPH. Especially in the crude enzyme mixture, it is important to establish first the endogenous oxidation of NADPH by a control experiment omitting the substrate TDP-4-keto-6-deoxyglucose.

Isolation of TDP-6-Deoxy-L-talose

For the conversion of 1 μmole of TDP-glucose, 10 individual incubation mixtures are prepared, each containing 0.1 μmole of TDP-glucose and the components outlined above. The incubation at 37° is terminated by addition of 600 μl ethanol and 1 minute heating in a boiling water bath. The denatured protein is removed by centrifugation and discarded. In the supernatant liquid, the main portion of ethanol is removed at room temperature by evaporation under reduced pressure (about 3 ml of total volume). The ice cold solution is diluted with water to about 10 ml adjusted to pH 4.0 with 1 *N* acetic acid. Adsorption of nucleotides on Norit A is accomplished in the cold by addition of 10 mg charcoal. The suspension is filtered on a sintered glass funnel, and the charcoal is washed extensively with distilled water. Elution of the nucleotide is accomplished by 3 × 1 ml portions of 0.1% ammonia in 50% ethanol. The eluate is neutralized in an ice bath with 1 *N* acetic acid, and the sample is concentrated at 30° under reduced pressure.

Separation of nucleotides is carried out by thin-layer chromatography. The sample is streaked onto one polyethyleneimine impregnated cellulose plate (Polygram Cel 300 PEI, Brinkman Instruments, Inc.). The solvent used is 10% aqueous ethylene glycol containing 1% $Na_2B_4O_7 \cdot 10H_2O$, 2% H_3BO_3 and 0.75% LiCl. The plates were developed at room temperature for 2.5 hours.

A single band was located by UV light. Isolation of the pure nucleotide is achieved by elution with 1.5 M LiCl in 0.025 M Tris·HCl buffer pH 8.0. Removal of salts is achieved after adjustment of pH to 4.0 with 1 N acetic acid and repetition of the Norit adsorption described above.

Properties of TDP-6-Deoxy-L-talose

All the analytical data obtained on this compound are in agreement with the expected values for this sugar nucleotide. The only sugar that can be released by mild acid hydrolysis is 6-deoxy-L-talose. A significant difference between TDP-6-deoxy-L-talose with other sugar nucleotides is its great sensitivity to alkali: For example, 1 N ammonia at 37° for 2 hours leads to total destruction. Consequently, steps along the isolation procedures, usually employed for isolation of sugar nucleotides, exposing the sugar nucleotide to alkaline conditions for any extended periods of time, have to be avoided. Degradation of TDP-6-deoxy-L-talose in alkali leads to formation of TMP and the 1,2-cyclic monophosphate of 6-deoxy-L-talopyranose.

Procedure for Separation of
TDP-4-Keto-L-rhamnose-3,5-epimerase (Reaction 2) and
TDP-6-Deoxy-L-talose dehydrogenase (Reaction 3)

For the separation of these two enzymes a large-scale preparation is necessary. The same medium and the same growing conditions are employed as described above. The cells grown in a 1-liter flask are transferred to a 10-liter fermentor (New Brunswick) and the cells are grown to early logarithmic growth phase under vigorous aeration. This in turn is used as an inoculum for a large-scale batch in a 300-liter fermentor, yielding 950 g of wet weight cells. The frozen cells are suspended in 0.05 M Tris·HCl buffer pH 8.0 containing 0.01 M β-mercaptoethanol and 0.001 M sodium EDTA. The suspension is passed twice through a Manton-Gaulin continuous-flow homogenizer at 10,000 psi. Cell debris is removed by centrifugation at 30,000 g for 30 minutes.

Enzyme Purification

All operations are carried out at 4° and 0.05 M Tris·HCl buffer pH 8.0 containing 0.01 M β-mercaptoethanol and 0.001 M sodium EDTA, used throughout the purification unless otherwise specified.

Step 1. Purification of Cell-Free Extract. The crude cell free extract is obtained by collecting the supernatant solution as described above (total volume about 8.7 liters).

Step 2. Protamine Sulfate Precipitate. To 8700 ml of supernatant crude extract, a 2% solution of protamine sulfate (Eli Lilly) (400 ml) is slowly added under stirring. The mixture is allowed to sit for several hours. After centrifugation for 20 minutes at 20,000 g the precipitate is discarded.

Step 3. Ammonium Sulfate Precipitation. To 8000 ml of the protamine supernatant 2070 g of solid ammonium sulfate (45% saturation) is added slowly under stirring. The solution is kept overnight before the precipitate is removed by centrifugation. To the 0–45% ammonium sulfate supernatant fluid, 1250 g solid ammonium sulfate is added (45–70% saturation) and the precipitate is removed by centrifugation. The main activity of all three enzymes (reactions 1, 2, and 3) is located in this fraction. It should be noted that in some preparations nucleic acids are not sufficiently removed in step 2 and considerable interference with the ammonium sulfate precipitation occurs. In this case a repetition of the ammonium sulfate precipitation is necessary. A total of about 60 g of protein is obtained in the 45–70% fraction.

Step 4. DEAE Column Chromatography. The 45–70% ammonium sulfate fraction is dissolved in the minimum amount of Tris buffer and is dialyzed against 10 liters of the same buffer for 12 hours. During that time period the outside buffer was replaced once. A column of DEAE-cellulose, DE-52, microgranular (5 cm × 40 cm, 785 ml bed volume) is equilibrated with Tris·HCl buffer. A sample (45–70%) containing 13 g protein in 785 ml is loaded on the column (flow rate 2 ml/minute). Fractions are collected at 12 minute time intervals. Elution is carried out with a linear gradient of 1 liter of 0.5 M KCl in the same Tris·HCl buffer and 1 liter of Tris buffer in the mixing flask. The individual fractions were examined for protein (UV at 280 nm) and for enzymatic activity.

TDP-4-keto-L-rhamnose-3,5-epimerase is located by the third assay procedure in fractions 25–50. Similarly, TDP-6-deoxy-L-talose reductase is located by the fourth assay procedure in fractions 51–80. The fractions containing 3,5-epimerase are pooled (total volume 610 ml). Fractions 51–80 containing reductase are also pooled (total volume 690 ml). To each of the two fractions obtained, ammonium sulfate is added up to 70% saturation, and the proteins are stored temporarily at −20°C as ammonium sulfate precipitates for further purification. Four DEAE columns like the one described are necessary to process all the 45–70% ammonium sulfate precipitate obtained in step 3.

Purification of TDP-4-Keto-L-rhamnose-3,5-epimerase

Material obtained in fractions 25–50 from four separate DEAE columns is combined and dialyzed against Tris·HCl buffer (total volume 490 ml, 13 g of protein). This solution is applied to a DEAE column (2.5 cm × 75 cm, 367 ml bed volume) at 2 ml/minute flow rate. A stepwise elution is carried out. First, 500 ml of Tris·HCl buffer is percolated through the column. No enzymatic activity is in this eluate. A second fraction (820 ml volume) containing 0.1 M KCl in Tris·HCl buffer is collected. All the 3,5-epimerase activity is located in this fraction. Finally, 0.2 M KCl in Tris·HCl buffer does not release any more enzyme from the column. Fraction two of this procedure, containing all TDP-4-keto-L-rhamnose-3,5-epimerase activity, was dialyzed, concentrated by ultrafiltration and stored at $-20°C$ in 10% glycerol (total volume 200 ml, 6.8 g protein).

Purification of TDP-6-Deoxy-L-talose dehydrogenase

The ammonium sulfate precipitate of the combined fractions 5–80 (step 4 DEAE chromatography) is dialyzed against Tris·HCl buffer (final volume 271 ml, 7.6 g protein). The dialyzed sample is applied to a DEAE column (2.5 cm × 80 cm, 390 ml bed volume). The flow rate is adjusted to 2 ml/minute, and fractions are collected at 10-minute intervals. A linear gradient of 1 liter of 0.3 M KCl in Tris·HCl buffer, and, 1 liter of Tris·HCl buffer in the mixing flask, is used as eluent. After the use of 500 ml of liquid in each of the two flasks, the slope of the KCl gradient is reduced by adding to the reservoir flask 500 ml of 0.3 M KCl in Tris buffer and 500 ml of 0.15 M KCl in Tris buffer to the mixing flask. Fractions 60–90 are found to contain TDP-6-deoxy-L-talose reductase and are pooled. After dialysis against Tris buffer and concentration by ultrafiltration, the sample is stored in 10% glycerol at $-20°C$. When all the protein of step 4 (fractions 51–80) is processed in this way a total of 1.4 g of protein in 160 ml is obtained in this fraction.

Properties

Specificity. The enzymes have strict specificity for their substrates, and only thymidine derivatives will show activity. TDP-6-deoxy-L-talose reductase (reaction 3) can use NADH instead of NADPH as the coenzyme.

Stability. The enzyme preparations are stable when stored at $-20°C$. Repeated freezing and thawing of solutions will inactivate the enzymes. It is suggested to store larger amounts of enzyme in 10% glycerol.

Purity. Both enzymes TDP-4-keto-L-rhamnose-3,5-epimerase and

TDP-6-deoxy-L-talose dehydrogenase are free of cross contamination. Thus only mixing of the two enzyme fractions will lead to 6-deoxyhexose formation. Examination of both fractions by disc gel electrophoresis reveals the presence of many other proteins, and further purification is necessary to obtain homogeneous enzyme proteins.

Acknowledgments

This work was supported by Grant AI-07241 from the National Institute of Allergy and Infectious Diseases, National Institutes of Health, United States Public Health Service.

[60] The Biosynthesis of 3,6-Dideoxy Sugars: Formation of CDP-4-Keto-3,6-dideoxyglucose from CDP-Glucose[1]

By Pedro Gonzalez-Porqué and Jack L. Strominger

The following reactions have been shown to lead to the formation of CDP-3,6-dideoxy sugars.[2]

$$
\alpha\text{-D-Glucose-1-P} \underset{\substack{\text{PP}_i \\ \text{CDP-D-glucose} \\ \text{pyrophosphorylase}}}{\overset{\text{CTP}}{\rightleftarrows}} \text{CDP-D-glucose} \tag{1}
$$

$$
\text{CDP-D-glucose} \xrightarrow[\substack{\text{CDP-D-glucose} \\ \text{oxidoreductase}}]{\text{NAD}} \text{CDP-4-keto-6-deoxy-D-glucose} \tag{2}
$$

$$
\text{CDP-4-keto-6-deoxy-D-glucose} \xrightarrow[\substack{E_1, E_3 \\ \text{pyridoxamine} \\ \text{phosphate}}]{\text{NADH}} \text{CDP-4-keto-3,6-dideoxy-D-glucose} \tag{3}
$$

$$
\text{CDP-4-keto-3,6-dideoxy-D-glucose} \xrightarrow[E_2]{\text{NADPH}} \text{CDP-3,6-dideoxy-D-glucose} \tag{4}
$$

All enzymes have been partially purified.[2-4] In this paper, the purification of the enzymes catalyzing reactions (2) and (3) are described. All three enzymes have been purified to homogeneity as judged by polyacrylamide gel electrophoresis in the presence of SDS.[5] The requirement of a new cofactor for reaction (3) is described.

[1] Supported by a research grant from the U.S. Public Health Service (AM-13230).
[2] H. Pape and J. L. Strominger, *J. Biol. Chem.* **244**, 3598 (1969).
[3] S. Matsuhashi, M. Matsuhashi, and J. L. Strominger, *J. Biol. Chem.* **241**, 4267 (1966).
[4] V. Ginsburg, P. J. O'Brien, and C. W. Hall, *Biochem. Biophys. Res. Commun.* **7**, 1 (1962).
[5] K. Weber and M. J. Osborn, *J. Biol. Chem.* **244**, 4406 (1969).

General Remarks. All operations were carried out at 0–4°. All buffers were 0.001 M in EDTA. Centrifugation was carried out at 15,000 g for 30 minutes. Whatman DE-52 was used for the DEAE-cellulose columns. All DEAE-cellulose columns were packed with a difference in height of 1 meter from the reservoir to the column. Protein concentrations are given in arbitrary units (optical density). An optical density unit (OD unit) is defined as the amount of protein that gives in solution at pH 7.5 a difference of absorbancy of 1.00 between the readings at 280 nm and 310 nm in a 1-cm light path cuvette.

CDP-D-Glucose Oxidoreductase (Reaction 2)

Assay Method

Principle. CDP-D-glucose oxidoreductase catalyzes the intramolecular hydrogen transfer from carbon atom 4 to carbon atom 6 in CDP-D-glucose.[6] The amount of product formed can be measured spectrophotometrically by its maximum absorption at 320 nm ($\epsilon_m = 6.5 \times 10^3$) in alkaline solution.[7]

Reagents

Tris·HCl, 0.5 M, pH 7.5
NAD, 0.001 M
CDP-D-glucose, 0.01 M
NaOH, 0.1 M

Procedure. The assay system contains 0.01 ml of NAD, 0.01 ml of CDP-D-glucose, 0.01 ml of Tris·HCl buffer, and enzyme in a total volume of 0.1 ml. The test tubes were incubated at 37° and the reaction stopped by addition of 0.4 ml of NaOH. After 15 minutes of incubation at 37° in alkaline medium, the absorbancy is measured in a 1-cm light path cuvette at 320 nm.

Definition of Unit and Specific Activity. A unit is defined as the amount of enzyme which transforms 1 μmole of substrate per hour under the above-described conditions. Specific activities are given in units per protein optical density unit.

Purification Procedure

Growth of Organism. Cells of *Pasteurella pseudotuberculosis* type V, 25 VO[2], are grown with aeration in a medium containing 3% Difco Tryptic Soy Broth at 30°. The medium is inoculated with a 10% in-

[6] A. Melo, W. H. Elliott, and L. Glaser, *J. Biol. Chem.* **243**, 1467 (1968).

[7] S. Matsuhashi, M. Matsuhashi, J. G. Brown, and J. L. Strominger, *J. Biol. Chem.* **241**, 4283 (1966).

oculum of an overnight preculture. After approximately 4 hours (end of the logarithmic phase), cells are harvested in a Sharples centrifuge. The yield is about 4 g of wet cells per liter of medium.

Step 1. Crude Extract. Cells from 110 liters of culture (468 g, wet weight) are suspended in four times their volume (1,872 ml) of 0.01 M potassium phosphate buffer, pH 7.5. The cell suspension is then sonicated for 2 minutes in a Branson sonicator at 70 output in batches of about 150 ml, taking care that the temperature does not rise over 10–12°. The extract is then centrifuged at 15,000 g. The supernatant solution (crude extract) contains about 100 OD units of protein per milliliter.

Step 2. Streptomycin Sulfate Precipitation. The crude extract is diluted to 65 OD per milliliter with 0.01 M potassium phosphate buffer, pH 7.5, and a solution of 5% streptomycin sulfate in distilled water is added drop by drop with stirring until a final concentration of 0.8% in streptomycin sulfate is reached. After standing for 1 hour with stirring, the solution is centrifuged for 20 minutes and the precipitate is discarded.

Step 3. Ammonium Sulfate Precipitation. The supernatant solution from step 2 is brought to 0.1 M potassium phosphate, pH 7.5, and solid ammonium sulfate is added to a final concentration of 65% saturation. The solution is then centrifuged and the precipitate collected and dissolved in 250 ml of 0.01 M potassium phosphate buffer, pH 7.5. The solution is then dialyzed for 48 hours against 4 liters of the same buffer with three changes of buffer.

Step 4. DEAE-Cellulose Gradients. The solution from step 3 (345 ml, 31 OD protein units per milliliter) is applied to a column of Whatman DE-52 (12 × 5.5 cm) which has been washed with 10 volumes of 0.5 M potassium phosphate, 0.01 M EDTA, pH 7.5, and then equilibrated with 30 volumes of 0.01 M potassium phosphate buffer, pH 7.5. The column is then eluted with the following linear gradients.

FIRST GRADIENT. Between 2.5 liters of 0.01 M and 2.5 liters of 0.05 M potassium phosphate buffer, pH 7.5. The fractions having cofactor activity (peak at about 0.02 M salt, see below) were pooled and the rest discarded.

SECOND GRADIENT. Between 2.5 liters of 0.05 M and 2.5 liters of 0.2 M potassium phosphate buffer, pH 7.5. In this gradient, E_3 activity and oxidoreductase activity are separated (E_3 peak at about 0.11–0.12 M salt and oxidoreductase peak at about 0.16 M salt).

THIRD GRADIENT. Between 2.5 liters of 0.2 M and 2.5 liters of 0.35 M potassium phosphate buffer, pH 7.5. E_1 activity (peak at about 0.28 M salt) is pooled.

The four activities separated in this column, cofactor, E_3, E_1, and oxidoreductase, are then purified separately.

Step 5. Ammonium Sulfate Fractionation. The fractions containing oxidoreductase activity from step 4 (second gradient) are pooled (1340 ml, 1160 OD total) and precipitated by addition of ammonium sulfate to 50% saturation. The precipitate is then collected by centrifugation and extracted successively with 100 ml of 50, 40, 35, 30, 25, and 20% saturation ammonium sulfate in 0.1 M potassium phosphate, pH 7.5. Most of the activity is extracted at 30 and 25% saturation. Occasionally some activity is found in 35 and 20% saturation fractions. By repeating the procedure with these two latter fractions, almost quantitative recovery can be achieved.

Step 6. Gel Filtration on Sephadex G-100. The active fractions from step 5 (200 OD) are pooled and precipitated by addition of ammonium sulfate to 60% saturation. The precipitate is collected by centrifugation, dissolved in 41 ml of 0.01 M potassium phosphate and 5% glycerol and layered on the top of a Sephadex G-100 column (9 × 150 cm) which is developed in the same buffer. Fractions of 50 ml are collected. The activity is close to the void volume but separated from it.

Step 7. Calcium Phosphate Gel. The active fractions from step 6 (67 OD in 715 ml) are concentrated by ultrafiltration in an Amicon cell using a PM-10 membrane to 37 ml (1.8 OD per milliliter) and 75 ml of calcium phosphate gel adjusted to 50 mg of solid per milliliter in 0.01 M potassium phosphate buffer, pH 7.5, is added with stirring. After 15 minutes, the solution is centrifuged and the supernatant solution (23.4 OD, 97 ml) contains over 90% of the activity.

Step 8. DEAE-Cellulose Gradient. From step 7, 15 OD are applied to a column of DEAE-cellulose (3.5 × 1.4 cm) which has been washed with 10 volumes of 0.5 M potassium phosphate buffer and equilibrated with 30 volumes of 0.01 M potassium phosphate buffer. Elution is achieved in a linear gradient between 250 ml of 0.075 M and 250 ml of 0.2 M potassium phosphate buffer, pH 7.5. The active fractions eluted at about 0.12–0.13 M are pooled.

A summary of the purification procedure is shown in Table I.

Properties of the Enzyme

Some of the properties of the enzyme have already been published, pH optimum of 7.5–8.0, $K_m = 1.7 \times 10^{-5} M$, and NAD requirement.[7] The enzyme shows a single band of about molecular weight 40,000 in 7.5% SDS gels.[5] According to its filtration behavior on Sephadex G-100, a molecular weight of near 80,000 can be estimated. This result suggests that the enzyme is composed of two identical subunits.

Stability and Storage. The enzyme is stable if frozen at −20° in 0.01 M potassium phosphate, pH 7.5, for months without appreciable loss of activity. Its inactivation when exposed to very low salt concentra-

TABLE I

SUMMARY OF THE PURIFICATION OF CDP-D-GLUCOSE OXIDOREDUCTASE[a]

Step	Volume (ml)	Total protein (OD units)	Total activity (μmoles formed/hour)	Specific activity (μmoles formed/hr/OD unit)
1. Crude extract	2770	180,000	12,300	0.068
2. Streptomycin sulfate	3500	25,300	—[c]	—[c]
3. Ammonium sulfate, 0–65%	345	10,700	11,700	1.1
4. 1st DEAE-cellulose column	1340	1,160	13,500	11.6
5. Ammonium sulfate, 20–35%	41	200	11,500	58
6. Sephadex G-100 column	37	67	7,950	118
7. Calcium phosphate gel	97	23.4	8,300	355
8. 2nd DEAE-cellulose column[b]	5.1	3.96	4,500[b]	1140

[a] Starting material: 468 g of wet cells.

[b] Only 6800 units from step 7 were used in step 8.

[c] Streptomycin sulfate interferes with the assay.

tions may be due to dissociation into subunits or denaturation. However, very dilute solutions of the enzyme are stable.

E_1 and E_3 Enzymes (Reaction 3)

Assay Method

E_1 and E_3 enzymes catalyze the conversion of CDP-4-keto-6-deoxy-glucose to CDP-4-keto-3,6-dideoxyglucose in the presence of a cofactor which has been identified as pyridoxamine 5′-phosphate.[8] The radioactive product formed in the reaction can be separated from substrate by thin-layer chromatography.

Reagents

NADH or NADPH, $0.075\,M$ in $0.25\,M$ potassium phosphate, pH 7.5, and $0.01\,M$ EDTA

Radioactive CDP-4-keto-6-deoxy-D-glucose, 0.8 mM in $0.01\,M$ potassium phosphate buffer at a specific activity of 36,000 cpm/nmole (prepared from CDP-[^{14}C]D-glucose with the use of CDP-glucose oxidoreductase).

The assay of each component is done in the presence of an excess of the other components.

[8] P. Gonzales-Porqué and J. L. Strominger, *Proc. Nat. Acad. Sci. U.S.* **69** (1972).

Procedure. In a final volume of 10 μl, add 1 μl of CDP-4-keto-6-deoxy-D-glucose and 1 μl each of NADH, E_1, E_3, and cofactor (pyridoxamine 5'-phosphate) at appropriate concentrations (see below). The tubes are incubated at 25°, and the reaction is stopped by immersion in an ice bath. The content of the tubes is then applied as a 2.8 cm band on a 20 × 20 cm plate of Cellulose MN-300 (0.4 mm thickness) and subjected to ascending chromatography for 3–5 hours in isobutyric acid:1 M ammonia, 5:3. The plates are then dried, worked free of the solvent by immersion in acetone for 2 minutes, and radioautographed overnight. The radioactive band corresponding to product is then scraped off with a razor blade and counted in 10 ml of a toluene-based scintillation fluid.

Purification Procedure

The first four steps described for the purification of the oxidoreductase are common for all the components of reaction (2). Because of the presence of another enzyme, E_2,[2] which catalyzes reaction (4), it is difficult to measure E_3, E_1, or cofactor activities separately, Table II shows the recoveries for the complete system in the four steps. In further puri-

TABLE II

SUMMARY OF ACTIVITIES AND RECOVERIES FROM THE FIRST THREE STEPS
OF THE PURIFICATION OF E_1 AND E_3

Step	Volume (ml)	Total protein (OD units)	Cofactor[b]	Total activity[c] (nmoles formed/ hour)	Specific activity (nmoles formed/ hour/ OD unit)
1. Crude extract[a]	3540	230,000	−	49,000	0.21
2. Streptomycin sulfate	4290	40,000	−	39,000	
			+	39,000	0.98
3. (a) Ammonium sulfate	210	12,100	−	4,500	
			+	10,300	0.85
(b) Dialysis	360	10,900	−	13,800	
			+	23,000	2.1

[a] From 524 g of bacteria.

[b] − or + refers to activity measured in the absence or in the presence of excess cofactor.

[c] The cofactor has no stimulatory effect until after ammonium sulfate precipitation, and the activity does not decay after prolonged dialysis. These facts speak for a protein-bound cofactor. An inhibitor which can be removed by dialysis appears to be present in the ammonium sulfate.

fication, the amount of activity recovered at step 4 (DEAE-cellulose column) was used as 100%.

Purification of Cofactor

The active fractions from step 4 (first gradient, see above) are pooled, concentrated by rotary evaporation, desalted on a column of Biogel P-2, and purified through a column of Dowex 1 (acetate form). At pH 7.5, it is eluted in a gradient of ammonium acetate at about 0.1 M. The spectrum of the active fractions in neutral, acid, and alkaline conditions is similar to pyridoxamine 5'-phosphate. It was also identified by thin-layer chromatography.[8] Moreover, authentic pyridoxamine 5'-phosphate has the same activity as the isolated cofactor.

Purification of E_1

Step 5. Gel Filtration on Sephadex G-100. The active fractions from step 4 (third gradient) previously described are concentrated by ultra-filtration in an Amicon cell on a PM-10 membrane to a volume of 58 ml and brought to a final concentration of 5% glycerol. The solution is then applied to a column of Sephadex G-100 (9 × 150 cm) which is developed in 0.01 M potassium phosphate buffer, pH 7.5. Fractions of 50 ml were collected. The void volume was at fraction 72 and the activity peak at fraction 98. The active fractions (fractions 93–105, 35.7 OD) are pooled.

Step 6. DEAE-Cellulose Chromatography. The active fractions from step 5 are applied to a 6 × 1.5 cm column of DEAE-cellulose (3.5 OD of protein per milliliter of anion exchanger) which has been washed and equilibrated as previously described. Elution is carried out with a linear gradient between 400 ml of 0.15 M and 400 ml of 0.3 M potassium phosphate buffer, pH 7.5. The activity is eluted at about 0.21 M buffer. The active fractions are pooled and concentrated by ultrafiltration in an Amicon cell using a PM-10 membrane. The sample is then diluted to approximately 0.01 M buffer concentration (8 ml).

Step 7. Preparative Polyacrylamide Gel Electrophoresis. The material from step 6 (11 OD) is subjected to electrophoresis in a Buchler Poly Prep 100 apparatus under the following conditions:

GELS. Lower, 40 ml of 10% acrylamide and 0.13% N,N'-methylene-bisacrylamide in 0.38 M Tris·HCl, pH 8.3; upper, 15 ml of 2.5% acrylamide and 0.25% N,N'-methylenebisacrylamide in 0.015 M Tris·potassium phosphate buffer, pH 6.8.

BUFFERS. Upper and lower, 0.025 M Tris·glycine, pH 8.3; chamber buffer, 0.1 M Tris·glycine, pH 8.3; eluting buffer, 0.05 M Tris·HCl, pH 7.5, 0.001 M EDTA.

Sample. 11 OD in 0.015 M Tris-potassium phosphate, pH 6.9 and 10% glycerol and 0.001 M dithiothreitol in a total volume of 10 ml.

The run is made at 400 V, which gives approximately a current of 50 mA. The elution rate is 1 ml/minute. Fractions of 5 ml are collected every 5 minutes. The activity and OD coincide in one of the protein peaks, and the fractions having the same specific activity (fractions 48–52) are pooled.

A summary of the purification procedure is given in Table III.

Properties of E₁

The molecular weight estimation in 7.5% SDS gels[5] shows a single band which corresponds to a molecular weight of about 60,000, which is in accord with the behavior of the enzyme on gel filtration.

Spectrum. The spectrum of the enzyme shows a maximum at 280 nm and a small shoulder at about 325–330 nm which may indicate bound residual cofactor. The absorbancy at 325–330 nm is about one-twentieth of the absorbancy at 280 nm.

Stability and Storage. The enzyme is stable to lyophilization and can be stored as a lyophilized powder for months without loss of activity.

Other Properties. The enzyme is not inhibited by N-ethylmaleimide at a concentration as high as $5 \times 10^{-3}\ M$, which suggests that it does not

TABLE III
PURIFICATION OF E₁[a]

Step[c]	Volume (ml)	Total protein (OD units)	Cofactor[c]	Total activity (nmoles formed/ hour)	Specific activity (nmoles formed/ hour/ OD unit)
4. DEAE-cellulose column	58	123	+	68,000[b]	550
			−	10,000	
5. Sephadex G-100 column	92	35.7	+	49,500	1390
			−	4,600	
6. Second DEAE-cellulose column	7.8	11	+	28,300	2570
			−	900	
7. Polyacrylamide gel electrophoresis	25	2.1	+	13,800	6600

[a] From 524 g of wet cells, see Table II.

[b] E₁ from steps 4, 5, and 6 are too dilute to determine the protein concentration with accuracy. All of them are concentrated in an Amicon cell using a PM-10 membrane to the volumes described. Activities were measured in the presence of an excess of E₃ (1 μg of enzyme from step 8, Table IV).

[c] + or − refers to activity measured in the presence of excess cofactor ($10^{-5}\ M$ pyridoxamine 5′-phosphate) or in its absence.

contain an essential thiol group. Binding experiments by the rapid flow dialysis technique[9,10] show that E_1 binds the substrate, but only in the presence of pyridoxamine 5'-phosphate.

Purification of E_3

Step 5. Ammonium Sulfate Fractionation and Gel Filtration on Sephadex G-100. The active fractions after step 4 (second gradient) previously described are pooled (510 ml, 520 OD) and precipitated with 65% ammonium sulfate and extracted twice with 50 ml of 55% saturated ammonium sulfate in 0.1 M potassium phosphate buffer, pH 7.5, and then three times with 50 ml of 40% saturated ammonium sulfate in the same buffer. The three extracts with 40% saturated ammonium sulfate are pooled and again precipitated with 65% saturated ammonium sulfate. The precipitate is collected, dissolved in 41 ml of 0.01 M phosphate buffer, pH 7.5, and 5% glycerol and layered on the top of a Sephadex G-100 column (9 × 150 cm) operated as above. The activity peak appears in fractions 122–134 at about two-thirds of the total column volume.

Step 6. DEAE-Cellulose Column: Phosphate Buffer Gradient. The active fractions from Sephadex G-100 (540 ml, 27 OD) are applied to a DEAE-cellulose column (5 × 2 cm) which has been washed and equilibrated with 0.01 M phosphate buffer as previously described. The activity is then eluted in a linear gradient between 250 ml of 0.05 M and 250 ml of 0.12 M potassium phosphate, pH 7.5. The activity peak appears at about 0.09 M buffer.

Step. 7. DEAE-Cellulose Column: Ammonium Acetate Gradient. The active fractions from step 6 are pooled (96 ml, 11.6 OD) and concentrated by ultrafiltration in an Amicon cell using a PM-10 membrane. The sample is diluted in 0.01 M ammonium acetate, pH 7.5, and concentration is repeated. The sample is then applied to a column of DEAE-cellulose (1.5 × 21 cm) which has been washed with 0.5 M ammonium acetate, pH 5.8, and finally equilibrated with 0.01 M of the same buffer. The activity is then eluted with a linear gradient between 0.075 M and 0.15 M ammonium acetate, pH 5.8. The activity peak is eluted at about 0.12 M salt. The active fractions (86 ml) are pooled and concentrated to 1.3 ml and desalted on a Sephadex G-25 column (1 × 20 cm).

A summary of the purification is given in Table IV.

Properties of E_3

Molecular Weight. The enzyme after step 7 appears as a single band on a 7.5% SDS gel electrophoresis[5] with molecular weight 40,000 in accord with the behavior of the enzyme on Sephadex G-100 filtration.

[9] S. P. Colowick and F. C. Womack, *J. Biol. Chem.* **244,** 774 (1969).
[10] P. Gonzalez-Porqué and J. L. Strominger, *J. Biol. Chem.* **247** (1972).

TABLE IV
PURIFICATION OF E_3

Step	Volume (ml)	Total protein (OD units)	Total activity (nmoles formed/ hour)	Specific activity (nmoles formed/ hour/OD unit)
4. DEAE-cellulose column	510	520	161,500[a]	310
5. Sephadex G-100 column	540	27	86,000	3,185
6. Second DEAE-cellulose column	96	11.6	34,111	2,940
7. (a) Third DEAE-cellulose column	86	1.0	2,400	2,400
(b) Concentration and de-salting[b]	1.9	0.6	11,694	19,480

[a] From 324 g (wet weight) of cells. Activities were measured in the presence of an excess of E_1 (1 μg from step 6, Table III, and a final concentration of 10^{-5} M pyridoxamine 5'-phosphate).

[b] The same effect described in Table II is observed here. The solution appears to contain an inhibitor which is removed by concentration on an Amicon cell using a PM-10 membrane followed by desalting on a column of Sephadex G-25. Increasing the amount of EDTA in the assay mix or doing it in the presence of dithiothreitol does not reverse the inhibition.

SH Groups. The amino acid analysis of the enzyme after performic acid oxidation shows one cysteine residue per 40,000 of enzyme, which indicates the presence of only one cysteine residue per mole of enzyme. The same result is obtained by titration of thiol groups with 5,5'-dithiobis-2-nitrobenzoic acid (DTNB) in 1% SDS. The enzyme is inhibited by sulfhydryl reagents.

Other Characteristics. The purified enzyme possesses NADH oxidase activity which is probably an aspect of its catalytic activity in reaction (3).

Synthesis of Complex Carbohydrates

[61] UDP-Apiose:7-O-(β-D-Glucosyl)Apigenin Apiosyltransferase from Parsley

By H. GRISEBACH and R. ORTMANN

A partially purified enzyme catalyzing the transfer of apiose from UDP-apiose[1] to 7-O-(β-D-glucosyl)apigenin to form apiin has been prepared from parsley[2] and from cell cultures of this plant.[3]

Assay Method

Principle. The enzymatic synthesis of apiin from 7-O-(β-D-glucosyl)-apigenin is determined by assay for the incorporation of radioactivity from UDP-[U14C]apiose.

---

[1] See this Volume [57].
[2] R. Ortmann, H. Sandermann, and H. Grisebach, *FEBS (Fed. Eur. Biochem. Soc.) Lett.* **7**, 164 (1970).
[3] R. Ortmann, Doctoral thesis. Freiburg, Germany (1972).

Reagents

Buffer, Tris·HCl 0.2 M, pH 7.5 and pH 7.0

NAD, 15 μmole/ml (10 mg/ml)

UDP-[U-^{14}C]D-glucuronic acid (302 μCi/μmole, Radiochemical Centre, Amersham) 10 μl containing 0.025 μCi

Apiin,[4] 8 mg in 1 ml ethyleneglycol monomethylether

Apigenin-7-O-β-D-glucoside,[6] 1.2 mg in 1 ml of ethyleneglycol monomethylether

Acetic acid, 15%

Toluene cocktail, 5 g of PPO in 1 liter of toluene

Schleicher and Schuell 2043 b paper

Procedure. PREPARATION OF UDP-APIOSE.[1] Ten microliters UDP-[U-^{14}C]GlcUA, 10 μl of NAD, and 10 μl of Tris·HCl (pH 7.5) are incubated with 50 μl of enzyme (about 3×10^{-5} unit) from *Lemna* or parsley cell cultures[1] for 30 minutes at 30°.

TEST FOR TRANSFERASE. Fifty microliters of the above incubation mixture (containing 11,000 to 16,000 dpm UDP-apiose, besides UDP-xylose, and unreacted substrate[1]), 5 μl of the solution of apigenin 7-glucoside, Tris·HCl (pH 7.0) and enzyme in a total volume of 105 μl are incubated for 15 minutes at 30°. The protein concentration should be such that not more than about 25% of the UDP-apiose reacts in 15 minutes. At the end of this period, 10 μl of the solution of apiin are added and the mixture is applied immediately as a 10 cm-wide band to the paper and chromatographed for 15 hours with 15% acetic acid. The apiin zone is detected under UV-light (350 nm) and counted in the toluene cocktail. The background is determined in an incubation without enzyme.

In a quicker but not so accurate modification of the assay the reaction products are separated by thin-layer chromatography. After addition of apiin, the solution is applied as a 8 cm-wide band to a polyamide-cellulose thin-layer plate (0.3 nm)[7] and dried for 15 minutes with hot air from a hair dryer. The plate is then developed with the solvent system chloroform–methanol–methylethylketone–acetylacetone

[4] Apiin can be isolated from parsley seeds.[5]

[5] H. Grisebach and W. Bilhuber, *Z. Naturforsch. B* **22**, 746 (1967).

[6] Apigenin 7-glucoside is obtained by partial hydrolysis of apiin with 2% trifluoroacetic acid. 2 mg of apiin are heated to 100° for 15 minutes with 0.25 ml of the acid. The acid is removed *in vacuo* and apigenin 7-glucoside is purified by paper chromatography (paper prewashed with methanol, 10% acetic acid and 0.01 M EDTA) with butanol–acetic acid-water (20:1:4, v/v/v); R_f = 0.49.

[7] Fifteen grams of polyamide (Woelm, Eschwege) and 6 g of cellulose powder (Macherey and Nagel) are homogenized for 1 minute in a Waring Blendor with 20 ml of chloroform and 80 ml of methanol.

(66:20:10:2, v/v/v/v). The apiin zone ($R_f = 0.34$) is counted in the toluene cocktail.

Preparation and Purification of Apiosyltransferase

General Remarks. The apiosyltransferase has been detected in young parsley leaves and in cell suspension cultures of this plant. It is not present in *Lemna minor.* In parsley, enzyme activity is highest in the primary leaves at about day 24 after sowing of the seedlings and then declines.[8] In cell suspension cultures of parsley,[9] the extractable enzyme activity is drastically increased by prior illumination of the cells. Maximum enzyme stimulation is observed at approximately day 10 after inoculation, and maximum enzyme activity is reached 24 hours after onset of illumination.[10]

A complete separation of the apiosyltransferase from the UDP-glucose:7-O-apigenin glucosyltransferase,[11] which is also present in parsley, can be achieved on hydroxyapatite.

Partial Purification of Enzyme from Cell Suspension Cultures of Parsley

The cell cultures are illuminated for 24 hours before enzyme extraction. All operations are carried out at 4°. The steps up to and including DEAE-cellulose column chromatography should be carried out without delay.

About 530 g (wet weight) of cells are suspended in 265 ml of 0.2 M Tris·HCl (pH 7.5) plus 265 μl of mercaptoethanol in a 1-liter beaker and homogenized in an ice bath for 2 minutes with an Ultra Turrax (Janke and Kunkel, type T 45 N, 10,000 rpm). After a stop of 1 minute homogenization is repeated and the procedure is continued for a total homogenization time of 15 minutes. The homogenate is centrifuged for 10 minutes at 27,000 g. The supernatant (610 ml) is stirred for 15 minutes with 33 g of Dowex 1 X2 (Cl⁻ form equilibrated with 0.2 M Tris·HCl, pH 7.5), and the Dowex is filtered off through glass wool. To the filtrate is added over a period of 10 minutes 60 ml of a cold (4°) solution of protamine sulfate (filtrate of a 2% solution adjusted to pH 7.5 with sodium hydroxide). The solution is stirred for 10 minutes and then centrifuged at 27,000 g for 10 minutes.

[8] K. Hahlbrock, A. Sutter, E. Wellmann, R. Ortmann, and H. Grisebach, *Phytochemistry* **10**, 109 (1971).

[9] See this volume [57].

[10] K. Hahlbrock, J. Ebel, R. Ortmann, A. Sutter, E. Wellmann, and H. Grisebach, *Biochim. Biophys. Acta* **244**, 7 (1971).

[11] A. Sutter, R. Ortmann, and H. Grisebach, *Biochim. Biophys. Acta* **258**, 71 (1972).

Protein in the supernatant is precipitated by addition of solid $(NH_4)_2SO_4$ to 100% saturation. The precipitate is collected by centrifugation at 27,000 g for 10 minutes and dissolved in 27 ml $5 \times 10^{-2} M$ Tris·HCl of pH 7.5 containing $1 \times 10^{-3} M$ dithioerythritol. This solution is chromatographed on a column (3.5 × 30 cm) of Sephadex G-25 with the same buffer.

The protein fraction (38 ml) is applied to a DEAE-cellulose column (2 × 16 cm, prewashed with 0.5 N NaOH, 0.5 N HCl and water) equilibrated with the above buffer, and the column is washed with the same buffer. Protein is eluted with a linear gradient of 500 ml $5 \times 10^{-2} M$ Tris·HCl, pH 7.5, containing $2 \times 10^{-3} M$ dithioerythritol and 500 ml 4×10^{-1} Tris·HCl, pH 7.5, containing $2 \times 10^{-3} M$ dithioerythritol at a flow rate of 80 ml per hour. The enzyme is eluted between 0.22 M and 0.27 M Tris buffer. Fractions with the highest enzymatic activity ($\sim$100 ml) are concentrated to 4 ml by filtration through a "Diaflo" concentrator (Amicon, Model 50, ultrafiltration cell).

The concentrated solution is then chromatographed with $2.5 \times 10^{-2} M$ sodium phosphate buffer, pH 7.0, containing $2 \times 10^{-3} M$ dithioerythritol on a column (2.5 × 40 cm) of Sephadex G-100, equilibrated with the same buffer, at a flow rate of 10 ml per hour.

Fractions with the highest enzymatic activity ($\sim$24 ml) are absorbed on a column (2.5 × 2 cm) of hydroxyapatite (Bio Gel HTP, Bio Rad Laboratories, Richmond, California) equilibrated with 2.5×10^{-3} sodium phosphate buffer, pH 7.0. The column is washed with the same buffer

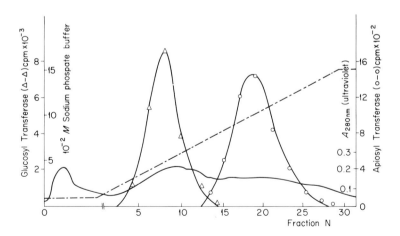

FIG. 1. Separation of glucosyl transferase (△——△) and apiosyl transferase (○——○) on hydroxyapatite with a linear gradient of sodium phosphate buffer (pH 7.0). ——, protein; – – – –, gradient.

containing $4 \times 10^{-3} M$ dithioerythritol until no more protein is eluted. Enzyme is then eluted with a linear buffer gradient prepared from 50 ml 2.5×10^{-2} sodium phosphate buffer, pH 7.0, containing $4 \times 10^{-3} M$ dithioerythritol and 50 ml of $1.5 \times 10^{-1} M$ of the same buffer at a flow rate of 50 ml per hour. The peak of enzyme activity appears at about $9.5 \times 10^{-2} M$ sodium phosphate buffer (Fig. 1).

Approximately 123-fold purification of enzyme activity is obtained by this procedure (see table).

Properties

Stability. After DEAE-cellulose chromatography the enzyme is stable for about 2 months at $-30°$ in $0.2 M$ Tris·HCl (pH 7.5) containing $2 \times 10^{-3} M$ dithioerythritol. At $4°$ it loses 60% of its activity after 5 days.

The enzyme from the hydroxyapatite column loses about 20% of its activity after 10 days at $4°$. Freezing of the solution results in 100% loss of activity.

Optimum pH. The enzyme has a pH optimum of about 7.0 in Tris· HCl buffer or sodium phosphate buffer.

Cofactors and Activators. Cofactors are not required. Bovine serum albumin (100 μg to 105 μl of the incubation) increases the activity by about 50%.

Donor Specificity. The enzyme is specific for UDP-apiose as glycosyl donor. No reaction takes place with UDP-D-glucose, UDP-D-glucuronic acid, UDP-D-xylose, UDP-D-arabinose, UDP-D-galactose, or TDP-D-glucose.

Acceptor Specificity. Apigenin-7-O-β-D-glucoside is the best acceptor.

PURIFICATION OF APIOSYLTRANSFERASE

Purification step	Protein (mg)	Specific activity[a] (units × 10^2/mg protein)	Purification (-fold)	Recovery (%)
1. Crude extract	1200	—	—	—
2. Dowex supernatant	1200	1.65	1	100
3. Protamine sulfate	610	3.3	2	100
4. Ammonium sulfate and Sephadex G-25	520	3.3	2	85
5. DEAE-cellulose	114	10	6.1	58
6. Sephadex G-100	26	29	17.4	38
7. Hydroxyapatite	4	202	123	35

[a] One enzyme unit is defined as the enzyme quantity which catalyzes the formation of 1 μmole of apiin at a UDP-apiose concentration of 0.03 nM during 1 minute at $30°$ in the enzyme assay.

7-O-glucosides of flavones and isoflavones can also serve as acceptors. No reaction takes place with 3-glucosides or apigenin-8-C-glucoside or with aglycons of flavonoids.

Substrate Affinity. The K_m value for apigenin-7-O-glucoside is 6 × 10^{-5} M at a UDP-apiose concentration of 2.7 × 10^{-7} M.

Other Properties. The isoelectric point determined by electrofocusing is 4.8, and the molecular weight estimated on a column of Sephadex G-100 is about 50,000 daltons.

[62] UDP-Glucose:β-Sitosterol Glucosyltransferase of Developing Cotton Seeds

By W. T. FORSEE, R. A. LAINE, J. P. CHAMBERS, and A. D. ELBEIN

UDP-D-glucose + β-sitosterol → β-sitosterol glucoside + UDP

A particulate enzyme fraction isolated from developing cotton fibers and seeds catalyzes the transfer of glucose from UDP-glucose to endogenous sterol acceptors. This enzyme preparation in turn catalyzes the esterification of the steryl glucosides with fatty acids from an unidentified acyl donor present in the enzyme fraction. However, the solubilized and partially purified glucosyltransferase is free of acylating activity and has an absolute requirement for an exogenous sterol acceptor. This enzyme catalyzed the reaction shown above.

Assay Method

Principle. Activity is measured by determining the amount of incorporation of UDP-glucose-^{14}C into steryl glucosides which can be extracted with chloroform–methanol.

Reagents

Tris·HCl buffer, 0.05 M, pH 8.5
UDP-glucose, 1.25 mM (25,000 cpm/20 μl)
β-Sitosterol, 6.25 mM [in a micellar solution containing 10 mg β-sitosterol, 80 mg L-α-lecithin (Sigma, commercial grade), and 250 mg of Triton X-100 all in a volume of 4 ml]. Lecithin and β-sitosterol are dissolved in a small volume of chloroform–methanol, and Triton X-100 is added.[1] Chloroform and methanol are evaporated to dryness under a stream of filtered air at 60°.

[1] A. Kaplan and Mei-Hui Teng, *J. Lipid Res.* **12**, 324–330 (1971).

Then H_2O is added dropwise while alternately vortexing and warming the test tube in 60° water bath. The resulting water-clear dispersion of β-sitosterol does not precipitate when stored at 0–4°C for 2 weeks.
Chloroform:methanol (2:1)
Chloroform:methanol:H_2O (3:48:47)[2]

Procedure. The reaction mixture contains 100 μl of Tris buffer, 20 μl of UDP-glucose, 30 μl of β-sitosterol, and enzyme in a final volume of 0.25 ml. The reaction is started by the addition of enzyme and incubated at 37°. The reaction is terminated after 20 minutes by the addition of 4 ml of chloroform:methanol (2:1), and 0.4 ml H_2O is added before vortexing and briefly centrifuging in a clinical centrifuge. The upper aqueous phase is removed with an aspirator and discarded. The lower phase is washed (vortexed) one time with 0.75 ml of chloroform:methanol: H_2O (3:48:47),[2] transferred to scintillation vials, and evaporated to dryness with a hair dryer. Ten milliliters of a toluene–Permafluor solution is added to each vial and the radioactivity is counted in a liquid scintillation spectrometer.

Definition of Unit and Specific Activity. One unit of enzyme is the amount of enzyme catalyzing the formation of 1 μmole of steryl glucoside per hour. Specific activity is expressed as units per milligram of protein. Protein is determined by the method of Lowry *et al.*[3]

Purification Procedure

Preparation of Particulate Enzyme Fraction. Cotton seeds are separated from the fibers of cotton bolls which had developed 25–35 days beyond flowering. Enzyme activity is found in both seeds and fibers, but larger quantities can be obtained from the seeds. Seeds from 6 cotton bolls are homogenized for 15 seconds at low speed in a chilled Waring Blendor containing 220 ml of 50 mM Tris, pH 7.3, 2 mM mercaptoethanol, and 3 g of polyvinylpyrrolidone (PVP). The homogenate is filtered through cheesecloth and centrifuged at 20,000 g for 10 minutes. The supernatant is discarded, and the upper portion of the pellet is washed twice in 75 ml of 50 mM Tris (pH 7.3), 2 mM mercaptoethanol, and 0.2% PVP.

Triton Solubilization. The washed pellet is resuspended in 80 ml of the above solution and 20% Triton X-100 is added to a concentration of 0.25%. After centrifugation at 30,000 g for 15 minutes, the super-

[2] J. Folch, M. Lees, and G. H. Sloane-Stanley, *J. Biol. Chem.* **226**, 497–508 (1957).
[3] O. H. Lowry, N. J. Rosebrough, A. L. Farr, and R. J. Randall, *J. Biol. Chem.* **193**, 265–275 (1951).

natant is retained and more Triton is added to give a final concentration of 2.4%.

DEAE-Cellulose Chromatography. The above solution is applied to a DEAE-cellulose (Cl⁻) column (3 × 10 cm) which had previously been equilibrated with 25 mM Tris, (pH 7.3) 2 mM mercaptoethanol, and 0.2% Triton. The column is then washed with 200 ml of 50 mM Tris (pH 7.3) containing 2 mM mercaptoethanol and then eluted batchwise with 0.5 M KCl in the same buffer solution. It is important to maintain the pH of the buffer at less than pH 7.5 in order to obtain a good yield of enzyme from DEAE-cellulose. After dialyzing against 25 mM Tris (pH 7.5) containing 8 mM mercaptoethanol, any precipitate is removed by centrifugation and discarded.

Ammonium Sulfate Precipitation. The supernatant from the above step is adjusted to pH 8.0 with Tris, then saturated $(NH_4)_2SO_4$ which has been adjusted to pH 8.0 with ammonium hydroxide is added dropwise, while stirring, to a final concentration of 55%. The precipitate is collected by centrifugation, resuspended in a small volume of 50 mM Tris (pH 8.5), and dialyzed overnight against 25 mM Tris (pH 8.5) containing 2 mM mercaptoethanol.

Comments. The removal of endogenous sterol acceptors from the particulate enzyme fraction during the solubilization process is shown in the purification table. As shown, the addition of Triton followed by DEAE chromatography yields an enzyme preparation with an absolute requirement for added sterol acceptors.

Properties

Stability. The enzyme retains almost 100% activity when stored at 0° for 2 weeks.

Specificity. The enzyme demonstrates a strong specificity for glucose over galactose and mannose, when added as the UDP-sugar nucleotides.

PURIFICATION OF UDP-GLUCOSE:β-SITOSTEROL GLUCOSYLTRANSFERASE

Fraction	Total protein (mg)	Activity in absence of added sterol (endogenous)	Total units	Specific activity
20,000 g pellet	680	3.1	15	0.022
0.25% Triton supernatant	89	2.1	11	0.12
2.4% Triton	88	0.11	4.8	0.054
DEAE-cellulose	6.5	0.00	2.6	0.40
Ammonium sulfate	3.9	0.00	2.0	0.51

The latter two sugars are incorporated less than 5% as efficiently as glucose. The enzyme also shows considerable preference for UDP-glucose over other sugar nucleotide donors. If activity with UDP-glucose is taken as 100, relative activities with other glucose nucleotides are TDP-glucose, 18; CDP-glucose, 10; ADP-glucose, 8; and GDP-glucose, 1. Cholesterol or stigmasterol substitute for β-sitosterol with approximately 60% efficiency.

Substrate Affinities. The apparent K_m for β-sitosterol varies depending on the amount of lecithin in the reaction mixture. An apparent K_m of about 4×10^{-5} is observed when the Triton:lecithin:β-sitosterol ratio is 24:1:1, whereas at a ratio of 24:8:1 the apparent K_m is 1×10^{-4}. The latter ratio gives maximum reaction rates. The apparent K_m values determined for UDP-glucose under similar conditions are 2.2×10^{-5} and 3.5×10^{-5}, respectively.

pH Optimum. Maximum activity occurs between pH 8.5 and 9.0 in both Tris buffer and glycine buffer, the activity being slightly higher in the latter.

Activators and Inhibitors. When lecithin is included in the Triton β-sitosterol micelles, the rate of the reaction is stimulated more than 3-fold. The enzyme is slightly stimulated by the divalent cations Ca^{2+} and Mg^{2+}, while Zn^{2+} and Mn^{2+} are somewhat inhibitory.

Characterization of Product. Direct information about the structures of steryl glycosides can be obtained from microgram quantities when the trimethylsilyl derivatives are analyzed by gas–liquid chromatography and mass spectrometry. Mass spectra can be recorded on each of the peaks to afford structural data for both the sugar and steroid moieties.[4] Further structural information can be derived after methanolysis by analyzing the separated moieties in the same manner. As little as 100 μg of a mixture of steryl glycosides is sufficient for this analysis. An analysis of the product formed from endogenous sterols in the particulate enzyme fraction revealed the presence of both steryl glucosides and acylated steryl glucosides. These two products are separated and purified by thin-layer chromatography, using 3 solvent systems composed of chloroform, methanol, and H_2O in ratios of (65:25:4), (95:5:0.2), and (85:15:0.5), respectively, on silica gel plates. Both products contain glucose as the only sugar, while the predominant sterol in both products is β-sitosterol. The acylated form which has been analyzed more completely is also composed of lesser amounts of 3 other sterols, one being a precholesterol species and the other two not identified. An analysis

[4] R. A. Laine and A. D. Elbein, *Biochemistry* **10**, 2547–2553 (1971).

of the product catalyzed by the solubilized enzyme has not been completed.

Acknowledgment

This work was supported by grants from the Robert A. Welch Foundation and Cotton Incorporated.

[63] UDP-Glucose:β-Xyloside Transglucosylase from Embryonic Chicken Brain

By Jack Distler and George W. Jourdian

$$\text{UDP-Glucose} + \text{xylosylserine} \rightarrow \text{glucosylxylosylserine} + \text{(UDP)}$$

Assay Method

Principle. UDP-glucose:β-xyloside transglucosylase activity is isolated from 13-day embryonic chicken brain and detected in a number of tissues, including other embryonic chicken tissues, guinea pig tissue, mouse ascites cells, cultural human fibroblasts, and L cells.[1] The enzyme system catalyzes the transfer of glucose from UDP-glucose to a limited number of acceptors including xylosyl- and arabinosylserine yielding β-linked glucosides. While the function of the enzyme is as yet unknown, it is distinct from transglucosylases involved in glycogen and collagen biosynthesis.

Enzymatic activity is determined by measuring the rate of formation of [14C]glucosylxylosylserine in assay mixtures containing a suitable acceptor and UDP-[14C]glucose. The 14C-labeled product is separated from other 14C-labeled compounds by high-voltage paper electrophoresis. Xylosylserine is routinely used as an acceptor during purification and tissue distribution studies.

Reagents

Sodium cacodylate buffer, 1.0 M, pH 5.7
UDP-[14C]glucose, 0.01 M (5×10^6 cpm/μmole)
Xylosylserine,[2] 1.0 M (adjusted to pH 5.7 with HCl)
MnCl$_2$, 0.3 M

---

[1] J. Distler and G. W. Jourdian, *Fed. Proc., Fed. Amer. Soc. Exp. Biol.* **26**, 345 (1967).
[2] Xylosylserine is prepared by the procedure of K. Kum and S. Roseman, *Biochemistry* **5**, 3061 (1966); and arabinosylserine by the method of G. W. Jourdian and J. Distler, *Carbohyd. Res.* **22**, 369 (1972).

Procedure. The incubation mixtures contain the following components in a final volume of 0.05 ml: 10 μmoles of sodium cacodylate, pH 5.7; 5 μmoles of xylosylserine; 0.05 μmoles of UDP-[^{14}C]glucose; 1.5 μmoles of MnCl$_2$, and 25 μl of enzyme. After incubation for 60 minutes at 37°, the reaction is stopped by heating at 100°C for 3 minutes. The denatured protein is removed by centrifugation, and an aliquot of the supernatant is subjected to electrophoresis on Whatman 3 MM paper saturated with 0.15 M formic acid–1.60 M acetic acid buffer, pH 2.0 (60 minutes at 80 V/cm). Glucosylxylosylserine migrates toward the cathode (approximately 9.5 cm), whereas free glucose remains near the origin; glucose 1-phosphate and UDP-glucose migrate toward the anode. The area of paper corresponding to the ^{14}C-labeled product is cut out and the radioactivity determined in a liquid scintillation spectrometer. The rate of the reaction is linear with time and protein concentration for 4 hours under the conditions described above.

Definition of Unit and Specific Activity. A unit of enzyme is that quantity of enzyme which will catalyze the synthesis of 1 nmole of glucosylxylosylserine in 60 minutes under the conditions described above. Specific activity is defined as millimicromoles of product formed per milligram of protein per 60 minutes. The method of Lowry *et al.*[3] is used to determine the protein concentration.

Preparation of Enzyme

Step 1. Preparation of Particulate Fraction. Fresh or frozen brains from 30 dozen 13-day chicken embryos are homogenized at 2–4° with an equal volume of Tyrode's medium (with MnCl$_2$ in place of MgSO$_4$) with a glass–Teflon homogenizer (clearance 0.006–0.009 inch). The homogenate is centrifuged for 20 minutes at 121 g. The supernatant is carefully separated from the loosely packed pellet and set aside. The pellet is homogenized in 75 ml of Tyrode's medium and recentrifuged. The supernatant fractions are combined and centrifuged at 105,000 g for 1 hour. The particulate fraction, containing the transferase activity, is suspended in 0.05 M imidazole·HCl buffer, pH 7.0, recentrifuged, and finally suspended in 30 ml of the same buffer. An acetone powder is prepared from this suspension by the procedure of Morton.[4] Generally, 1–1.5 g of tan acetone powder is obtained with a 20–50% enhancement of transglucosylase activity over that contained in the particulate suspension.

Step 2. "Solubilization." Acetone powder (1 g) is suspended in 20 ml

[3] See Vol. 8 [73].
[4] See Vol. 1 [6].

of cold 0.33 M imidazole·HCl buffer, pH 7.0, containing 0.15 M KCl and 1% Triton X-100. To obtain a uniform suspension, the mixture is gently homogenized several times during a period of 3 hours. After centrifugation at 105,000 g for 2 hours, the supernatant is collected and dialyzed against 20 volumes cold distilled water for 12 hours.

Step 3. DEAE-Cellulose Fractionation. The dialyzed fraction is applied to a 2 × 15 cm column of DEAE-cellulose (previously equilibrated with 0.05 M imidazole, pH 7.0, and then washed with water). After washing the column with 100 ml of water, the enzyme is eluted with 150 ml of 0.05 M imidazole buffer, pH 7.0, containing 0.075 M KCl (fraction A).[5] The eluent containing transferase activity may be concentrated by pressure dialysis, or precipitated with $(NH_4)_2SO_4$ at 50% saturation. Additional enzyme (fraction B) of lower specific activity may be eluted from the column by increasing the concentration of KCl to 0.3 M and then eluting with 1% Triton X-100 in the same buffer–salt mixture. Fraction A contains no detectable high molecular weight endogenous glucosyl acceptor(s) which is present in crude extracts.[1] Gelfiltration chromatography of fraction A on Sephadex G-200[6] suggests that the molecular weight of the active unit is of the order of 47,000.

Properties

All the studies below were performed with fraction A.

Effect of pH and Metal Ions. The enzyme exhibits transferase activity in cacodylate buffer from pH 5–7 with a sharp optimum at pH 5.7 with either xylosyl- or arabinosylserine as the glucosyl acceptor. The purified enzyme requires added Mn^{2+} for optimal activity. It is inactive in the presence of $1 \times 10^{-2} M$ EDTA, but complete activity is restored on addition of $4 \times 10^{-2} M$ $MnCl_2$; partial activation (less than 10%) is obtained with Co^{2+}, Zn^{2+}, and Mg^{2+}.

Enzyme Kinetics. The Michaelis constant for UDP-glucose is $7.0 \times 10^{-5} M$; for xylosylserine, $4.2 \times 10^{-2} M$; and for arabinosylserine, $8.0 \times 10^{-2} M$.

Specificity. UDP-Galactose, UDP-N-acetylglucosamine, UDP-N-acetylgalactosamine, UDP-glucuronic acid, GDP-mannose, glucose, or glucose 1-phosphate will not replace UDP-glucose as a glycosyl donor.[7]

---

[5] Elution of the DEAE-cellulose with a linear gradient of 0–0.075 M KCl serves only to elute the enzyme in a larger volume of buffer and does not achieve additional purification.

[6] P. Andrews, *Biochem. J.* **91,** 222 (1964).

[7] Crude extracts, however, will transfer galactose from UDP-galactose to xylosylserine, although not to arabinosylserine. The latter enzyme activity is removed during chromatography on DEAE-cellulose (step 3). A glucuronosyltransferase described by A. E. Brandt, J. Distler, and G. W. Jourdian, *Proc. Nat. Acad. Sci. U.S.* **64,** 374 (1969) is also removed in this step.

PURIFICATION PROCEDURE FOR UDP-GLUCOSE:β-XYLOSIDE TRANSGLUCOSYLASE
FROM EMBRYONIC CHICKEN BRAIN

Step	Volume (ml)	Protein (mg)	Total units	Specific activity (units/mg protein)	Recovery (%)
1. Particulate					
Crude extract	150	3450	1500	0.43	100
Particulate fraction	30	516	762	1.48	51
Acetone powder	20	576	922	1.60	61
2. Solubilization					
Supernatant of 105,000 g	20	301	608	2.02	41
3. DEAE-cellulose					
Fraction A	30	14	114	8.14	8
Fraction B	30	56	177	3.16	12

Transfer of glucose to arabinosylserine occurs at 55% of the rate to be found with xylosylserine. Each reaction is probably catalyzed by the same enzyme since the ratio of activity (arabinosylserine/xylosylserine) remains constant at all stages of purification and in all tissue extracts examined for transglucosylase activity. Other acceptors of glucosyl residues in this system are (in % activity relative to xylosylserine): serine-α-L-arabinopyranoside, 55; methyl-β-D-xylopyranoside, 32; serine-β-D-glucopyranoside, 30; and methyl-β-D-glucopyranoside, 9. Serine-β-D-galactopyranoside, D-xylose, D-glucose, maltose, methyl-α-D-xylopyranoside, N-acetyl-D-glucosamine and -D-galactosamine, L-serine, sphingosine, glycogen, and degraded collagen with terminal non-reducing galactosides[8] are not acceptors in this system.

Stability. The acetone powder (step 1) may be stored indefinitely at −20°. The 19-fold purified transglucosylase is stable for up to 2 weeks when stored in an ice bath. While the crude "particulate" fractions can be frozen or lyophilized without loss of activity, the "soluble" fractions (steps 2 and 3) are inactivated by this treatment.

---

[8] H. B. Bosman and E. H. Eylar, *Biochem. Biophys. Res. Commun.* 30, 89 (1968).

[64] Glucosyl Ceramide Synthetase from Brain

By Norman S. Radin

Uridine diphosphoglucose + ceramide → glucosyl ceramide + UDP

Assay Method[1]

Principle. The glucose donor contains radioactive glucose and is readily separated by solvent partitioning from the labeled glucocerebroside that is formed. The product is counted by liquid scintillation. The lipoidal acceptor, N-acylsphingosine, can contain 2-hydroxy fatty acid or nonhydroxy acid, although the normally observed product of the reaction contains only nonhydroxy acid. In order to bring the lipid substrate into close contact with the particulate enzyme, we use lyophilized tissue that is mixed with a benzene solution of the substrate.

Reagents

UDPGlc. The labeled material is available from New England Nuclear Corp., Boston, Massachusetts. It is made up with non-radioactive UDPGlc to 3.2 mM and a specific activity of roughly 2600 cpm/nmole. The molecular weight of the dipotassium salt (3.5 hydrate) is 705.6.

ATP, 13.3 mM, neutralized to about pH 7.4

The remaining reagents are prepared as described in this volume [65].

Procedure. The procedure in [65] is followed exactly, except for the use of ATP solution instead of water and the use of nonhydroxy ceramide instead of hydroxy ceramide.

Notes on the Assay Method. See the Notes in [65]. We have found our blanks to be higher with labeled UDPGlc, about 0.13% of the amount incubated.

A blank incubation run without added ceramide yields some glucocerebroside because of the presence of endogenous ceramide in whole brain or microsomes. This is higher than the activity observed with labeled UDPGal as the galactosyltransferase works poorly with nonhydroxy ceramide.

Properties

Stability. Although the enzyme appears to be somewhat stable in frozen brain, it is fairly labile in lyophilized rat brain. We observed a loss

---

[1] A. Brenkert and N. S. Radin, *Brain Res.* **36**, 183, 1972.

of 16% after 1 day, 41% after 7 days, and 46% after 27 days. This biphasic loss rate suggests two compartments for the enzyme. A test with mouse brain showed a 19% loss after storage of whole brain 20 days at $-70°$ and a 25% loss after 20 days of storage of lyophilized brain *in vacuo* over P_2O_5 at $-20°$.

Activators and Inhibitors. The lyophilized brain preparation apparently contains an inhibitory metal, as addition of EDTA affords some stimulation. Both Mg^{2+} and Mn^{2+} activate the preparation, Mn^{2+} being superior. Nicotinamide is not stimulating in this system but is included in the medium for convenience if galactosyltransferase is also being assayed. Some stimulation is afforded by the addition of ATP.

A similar chick brain glucosyltransferase was reported to have no need for divalent metal ions, an unusual feature for a glycosyl transferase.[2]

Kinetic Properties. The rat brain enzyme shows a pH optimum of 7.4 and the chick brain enzyme shows an optimum at 7.8, using Bicine buffer. The K_m of the chick brain transferase is about 0.12 mM with respect to UDPGlc and 0.08 mM for ceramide.

Specificity. The rat brain enzyme utilizes either hydroxy or nonhydroxy ceramide,[1,3] so it is not really necessary to separate the two types of ceramide in preparing the substrate. There is some evidence for utilization of sphingosine as a glucose acceptor, somewhat less efficiently.

Age Changes. The specific activity of the glucosyltransferase parallels to a certain extent the rate of ganglioside deposition in brain. (Glucosyl ceramide is the first glycolipid in the pathway of ganglioside biosynthesis.) The specific activity is fairly high in newborn rats, rises for a few days, then remains level until about 16 days of age. It then falls off somewhat for about 45 days and remains constant for a long time. The glucosyltransferase appears to occur only in neurons, not in glia.

[2] S. Basu, B. Kaufman, and S. Roseman, *J. Biol. Chem.* **243**, 5802, 1968.
[3] S. N. Shah, *J. Neurochem.* **18**, 395, 1971.

[65] Galactosyl Ceramide Synthetase from Brain

By Norman S. Radin

Uridine diphosphogalactose + ceramide → galactosyl ceramide + UDP

Assay Method[1]

Principle. The galactose donor contains radioactive galactose and is readily separated by solvent partitioning from the radioactive cerebroside that is formed. The product is counted by liquid scintillation. The lipoidal acceptor, N-acylsphingosine, must contain 2-hydroxy acids rather than nonhydroxy acids, as the galactosyltransferase shows strong preference for this type of ceramide. In order to bring the lipid into close contact with the particulate enzyme, we use lyophilized tissue that is mixed with a benzene solution of substrate.

Reagents

UDPGal. The radioactive material, labeled with tritium or ^{14}C in the galactose moiety, is available from New England Nuclear Corp., Boston, Massachusetts. It is made up with nonradioactive UDPGal to 3.2 mM and a specific activity of about 2600 cpm/nmole. The molecular weight of the anhydrous dipotassium salt is 642.5.

Tris·HCl buffer, 0.2 M, pH 7.4 (measured at room temperature). This also contains 10 mM nicotinamide, 4 mM EDTA, and 2 mM dithiothreitol.

MnCl$_2$, 50 mM

Ceramide. This is obtainable in the hydroxy form or as a mixture of the two forms from Supelco, Inc., Bellefonte, Pennsylvania and from Applied Science Laboratories, Inc., State College, Pennsylvania. However, the commercial material is very expensive and it may be necessary to prepare your own substrate. This can be done by degradation of galactosyl ceramide (see this volume [32] for the preparation of this lipid) and column separation.

The following is a modification of the method of Carter, Rothfus, and Gigg.[2] Dissolve 4.18 g of brain cerebroside (fairly pure) in 60 ml of chloroform and 218 ml of 95% alcohol, warming a little. Cool the

[1] A. Brenkert and N. S. Radin, *Brain Res.* **36**, 183, 1972.
[2] H. E. Carter, J. A. Rothfus, and R. Gigg, *J. Lipid Res.* **2**, 228, 1961.

solution and add 5.56 g of periodic acid in 120 ml of water. Leave the solution in the dark for 2 hours and decompose the excess periodic acid with 4 ml of ethylene glycol. Transfer the mixture to a 1000-ml separatory funnel with 400 ml of chloroform + 80 ml of methanol + 40 ml of water. Wash the resultant lower layer with 100 ml, 200 ml, and 200 ml of methanol–water 1:1 containing 2 mg/ml of salt. It is important that all the iodate ion be washed away, as it would otherwise block the next step.

The next step is a reduction of the dialdehyde formed as the result of periodate action on the galactose moiety. Evaporate the lower layer to a small volume with a rotary vacuum evaporator and add 120 ml of chloroform + 900 ml of 95% alcohol + 320 ml of water + enough 5 N NaOH to yield a pH of about 9 (pH paper). Rapidly weigh out roughly 1 g of sodium borohydride (avoiding moisture uptake in the bottle), dissolve this in 120 ml of 0.1 N NaOH, and add the solution cautiously to the lipid solution, with mixing. Do this step in a hood, since bubbles of hydrogen form. Stir the mixture, loosely capped, for 6 hours and then, adjust the pH to about 7 with 6 N HCl (more bubbles). Add 26.8 ml more HCl, which should yield a final concentration of about 0.1 N HCl. Leaving the mixture 24 hours cleaves the degraded sugar moiety, liberating ceramide.

The ceramide is isolated from the mixture by partitioning in a 2000-ml separatory funnel with the addition of 600 ml of chloroform. The total volume is over 2000 ml, but part of the upper layer can be discarded before transferring the mixture to the funnel. Wash the lower layer with two 400-ml portions of methanol–water–salt, as above, and then methanol–water without salt. Evaporate the lower layer to a small volume, with generous additions of benzene, and lyophilize the mixture. Examination by thin-layer chromatography (silica gel with chloroform–methanol–acetic acid 90:2:8) shows two major spots for hydroxy ceramide and nonhydroxy ceramide, as well as faster and slower running impurities.

Dissolve the ceramides in 100 volumes of chloroform, filter the mixture through a Celite-coated funnel, and concentrate the filtrate to about 200 ml. Prepare a column of 242 g of silica gel (preferably Unisil, see this volume [32]) in a tall cylinder, such as 2.2 × 122 cm, fitted with a pressure inlet. Pump chloroform through the column until the air is displaced and then pump in the sample solution. Elute with additional 2420 ml of chloroform at about 425 ml/hour, collecting fractions. Elute additionally with an equal volume of chloroform–methanol 99:1.

Pool the fractions containing pure ceramide, as evaluated by thin-layer chromatography. The shapes of the spots will depend in part on the distribution of fatty acids in the original cerebroside. Dissolve the hydroxyceramide in benzene, 1 mg/ml, with the aid of heat, and store it at $-20°$. The ceramides can be characterized by thin-layer chromatographic comparison with commercial samples, or by quantitative determination of the fatty acid or sphingosine content.

Brain preparation: While most of the galactosyltransferase is in the microsomes, there is some activity in the cytosol, so we assay the activity in whole brain. Homogenize the tissue sample in 2.5 volumes of water and lyophilize the mixture. The powder can be stored under vacuum over desiccant at $-20°$, but noticeable loss of activity is detectable after 1 day.

Procedure. Pipette 0.2-ml portions of warm ceramide solution into screw-cap culture tubes, 13×100 mm. To the tubes add 0.1-ml aliquots of dried brain, homogenized in 100 volumes of benzene with a Teflon–glass motor-driven homogenizer. Since the powdered brain settles rapidly, one should pipette only a single aliquot at a time. Mix the tube contents and remove the solvent with a stream of nitrogen, but do not use a water bath to speed the evaporation.

Immerse the tubes in ice and add 0.2 ml of incubation cocktail, which is prepared while the evaporations are taking place. The cocktail is made by mixing 1 ml of the buffer, 0.1 ml of the labeled UDPGal, 0.6 ml of the $MnCl_2$, and 0.3 ml of water. Cover the tubes with Teflon-lined caps and shake them by hand in an ice-filled ultrasonic bath of the type used in cleaning until a creamy suspension is produced. Then incubate the tubes at $37°$ in a fast-moving shaker for 90 minutes.

The radioactive cerebroside thus produced is isolated by first adding 4 ml of chloroform–methanol 2:1 containing 0.1 mg of a brain sphingolipid mixture, as carrier. Add 0.8 ml of 0.88% KCl, which forms two liquid phases. Centrifuge the mixture and wash the lower layer with 1 ml of methanol–water 1:1 containing 0.44% KCl. Repeat the wash but with omission of the KCl. Insoluble material at the interface is removed by filtering the lower layer through a sintered glass funnel with a 2 cm medium-porosity disk coated with Celite. The filtrate is caught in a scintillation vial, evaporated to dryness with a stream of warm air, and counted in almost any scintillation solvent. We use a rather polar mixture; see this volume [113].

Notes on the Assay Method. Since the specific activity of the UDPGal is known, the number of nanomoles synthesized can be calculated. It is not necessary to correct for counting efficiency if the precursor and product are counted in the same scintillation mixture.

A blank without enzyme should be run occasionally. Apparently some UDPGal or an impurity in the radioactive material remains in the lower layer with the cerebroside. It can be eliminated by purifying the cerebroside on a thin-layer plate before counting. A few percent of the lipoidal activity seems to be in an ester-linked compound that can be removed by mild alkaline methanolysis.

The carrier sphingolipids can be prepared from brain lipids by acetone and ether extraction, which take out most of the polyunsaturated lipids. Any type of preparation is suitable as long as some cerebroside is present.

Removal of insoluble material during the work-up is most conveniently done with a Büchner funnel which has been sealed at the top with a 28/15 ball joint. This allows one to apply a small pressure of air to the top of the liquid by means of a 28/15 socket joint and a spring clamp. This technique is useful because handling the filtrate is simplified: it goes directly into the counting vial. A single funnel can be used for the day's run, as long as a chloroform–methanol–water 60:35:8 rinse is used between samples.

The reaction rate is linear with respect to time up to 90 minutes and with respect to dry brain weight up to 1 mg. However, these questions were examined with brains from 16-day-old rats, which exhibit approximately maximal specific activity, and brains or other tissues from other species or other ages should be checked for linearity.

A blank incubation run without added hydroxy ceramide will yield some radioactive cerebroside, but this is of the nonhydroxy acid type. Brain normally contains some nonhydroxy ceramide (hydroxy ceramide may occur in very low concentrations) which acts as a galactose acceptor. It is likely that a single enzyme utilizes both types of ceramides, as indicated by the observation that exogenous hydroxy ceramide represses formation of nonhydroxy cerebroside.

A weakness of the above assay method is the degree of variability in the results, possibly due to the difficulty of preparing a good suspension of the incubation system. Triplicate samples are recommended.

Properties

Stability. The enzyme appears to be quite stable in frozen brain, but the exact extent has not been determined. Lyophilizing brain seems to activate a decay system, and rat brain loses 11% of its initial activity in 7 days and 25% in 27 days.

Activators and Inhibitors. Mg^{2+} activates the enzyme about as well as Mn^{2+}, and nicotinamide, dithiothreitol, and EDTA produce small stimulatory effects. In experiments with wet microsomes[3] we found ATP to be stimulatory, but it is inhibitory in the system described above. There is some evidence for the presence of a readily solubilized inhibitor

[3] P. Morell and N. S. Radin, *Biochemistry* 8, 506, 1969.

in brain. A synthetic analog of hydroxy ceramide, N-bromoacetyl DL-erythro-3-phenyl-2-amino-1,3-propanediol, produces 71% inhibition at 0.3 mM concentration. Preincubation of the assay system prior to addition of the labeled UDPGal shows that the transferase is rather unstable in the absence of UDPGal.

In our hands, the enzyme (in wet microsomes or dry brain) was severely inhibited by a variety of detergents. However Shah[4] has reported that Tween 20 had little effect, under different incubation conditions. A study of embryo chick brain transferase[5] was carried out with a high level of detergent, but this enzyme appears to be far more active than the rat brain enzyme. Whether the difference between species is due to different sensitivities to detergent or to other factors remains to be determined.

When wet tissue is used in the assay, it is helpful to evaporate the ceramide solution onto Celite, a diatomaceous earth which yields a thinly dispersed lipid deposit.[3]

The chick enzyme utilizes Ca^{2+}, as well as Mg^+ or Mn^{2+}, and is strongly inhibited by 0.5 mM sphingosine. It was partially purified by centrifugal fractionation of the crude mitochondrial fraction, or by fracturing the mitochondrial fraction with a French press and dialyzing against Bicine buffer. The dialyzed preparation was stable if 0.1% mercaptoethanol was included in the medium.

Kinetic Properties. The rat brain enzyme shows a pH optimum of 7.4 with Tris·HCl buffer but almost as much activity at 7.8, while the chick enzyme shows an optimum at 7.7, measured with Bicine buffer and Mg^{2+}.

The K_m with respect to UDPGal is 0.21 mM for the rat enzyme and 0.04 mM for the chick enzyme. The K_m with respect to ceramide is 2.1 mM for the rat enzyme and 0.11 mM for the chick enzyme.

Specificity. Experiments with the chick enzyme with glucosyl ceramide and Tay-Sachs ganglioside (sialyl trihexosyl ceramide) as galactose acceptors indicated that three different transferases are involved. Lactosyl ceramide did not act as an acceptor at all. ADPGal could not substitute as a galactose donor. Ceramides containing nonhydroxy fatty acid act as acceptors, and enhanced activity has been obtained by including lecithin with the substrate.[6] Sphingosine is galactosylated by UDPGal in brain, but it is not yet known whether the same enzyme is

[4] S. N. Shah, *J. Neurochem.* **18**, 395, 1971.
[5] S. Basu, A. M. Schultz, M. Basu, and S. Roseman, *J. Biol. Chem.* **246**, 4272, 1971.
[6] P. Morell, E. Costantino-Ceccarini, and N. S. Radin, *Arch. Biochem. Biophys.* **141**, 738, 1970.

involved; the rate of reaction is somewhat less than with hydroxy ceramide.

Age Changes. The specific activity of the cerebroside-forming enzyme is very dependent on the age of the animal.[1,4,5] It is low prior to myelination, increases greatly during the brief period of maximal myelin deposition, then drops somewhat. In the case of chickens, the activity is quite low in the "adult"; in the case of rats, the level at 1 year of age is still one-fourth of the maximal activity (at 16 days). The enzyme occurs in neurons as well as in glial cells, but we do not know what portion of the total brain activity is attributable to the different cell types.

The specific activity of the galactosyltransferase in kidney that makes galactosyl ceramide is much higher in male rats than in female rats; the difference is overcome by administration to females of testosterone.[7]

[7] G. M. Gray, *Biochim. Biophys. Acta* **239**, 494, 1971.

[66] Biosynthesis of Mannophosphoinositides in *Mycobacterium phlei*

By Clinton E. Ballou

1-Phosphatidyl-D-myoinositol + GDPMan →
$\qquad$ 1-phosphatidyl-D-myoinositol 2-*O*-α-D-mannopyranoside   (1)

1-Phosphatidyl-D-myoinositol 2-*O*-α-D-mannopyranoside + GDPMan →
$\qquad$ 1-phosphatidyl-D-myoinositol 2,6-di-*O*-α-D-mannopyranoside   (2)

1-Phosphatidyl-D-myoinositol 2,6-di-*O*-α-D-mannopyranoside + palmityl-CoA →
$\qquad$ 1-phosphatidyl-D-myoinositol 2,6-di-*O*-α-D-mannopyranoside monopalmitate   (3)

1-Phosphatidyl-D-myoinositol 2,6-di-*O*-α-D-mannopyranoside monopalmitate +
$\qquad$ palmityl-CoA →
$\qquad$ 1-phosphatidyl-D-myoinositol 2,6-di-*O*-α-D-mannopyranoside dipalmitate   (4)

Reactions (1)–(4) are catalyzed by a particulate fraction from *Mycobacterium phlei*.[1-3] The palmityl group in reaction (3) is probably esterified to position 6 of the mannose, which is attached to position 2 of the myoinositol, whereas in reaction (4) the palmityl group is probably esterified to position 3 of the myoinositol ring.[4] In the crude system, additional mannose units are transferred to that mannose at position 6 of

[1] D. L. Hill and C. E. Ballou, *J. Biol. Chem.* **241**, 895 (1966).
[2] P. Brennan and C. E. Ballou, *J. Biol. Chem.* **242**, 3046 (1967).
[3] P. Brennan and C. E. Ballou, *J. Biol. Chem.* **243**, 2975 (1968).
[4] C. Prottey and C. E. Ballou, *J. Biol. Chem.* **243**, 6196 (1968).

the myoinositol ring to yield a homologous series of compounds with the chain lengthened up to the pentasaccharide at this position. Glycolipids of this type are found in *Nocardia* and *Corynebacterium* which presumably possess the same enzymatic activities, while *Propionibacterium* species catalyze only reaction 1.[5] The major inositol-containing lipid in the latter bacteria is the closely related neutral compound 1-acyl-2-(6′-acyl-α-D-mannopyranosyl)-L-myoinositol.[4]

Principle

These glycolipids are found in the membrane fraction of mycobacteria, although there is some evidence that the more highly glycosylated forms are concentrated in the lipid material associated with the cell wall, while the less highly glycosylated forms are in the microsomal fraction.[6] When [14C]GDPMan is supplied to the crude particulate or soluble enzyme preparation, several different [14C]mannophosphoinositides are formed from endogenous acceptors, and the acylation steps may also occur through utilization of endogenous acyl donors. Acetone extraction of the particulate enzyme eliminates the endogenous reactions and the incorporation of [14C]mannose becomes dependent on added phosphatidylmyoinositol. The product of the enzymatic reaction may be extracted into chloroform and chromatographed on silicic acid thin-layer plates for identification and estimation of the individual [14C]-glycolipids. Phosphatidylmyoinositol dimannoside and its monoacyl derivative are acceptors of long-chain fatty acids from acyl-CoA, a reaction that is catalyzed by the same enzyme preparation described above. Again, the complexity of the reaction requires separation of the components by thin-layer chromatography before their estimation. In *Propionibacterium* species, the single initial mannosyl transfer may be studied without the complications of the further glycosylation and acylation steps.

Growth of Bacteria

Mycobacterium phlei (ATCC 346) is grown at 37° by aeration or under shaking conditions on the following medium: 20 g of glycerol; 5 g of casamino acid; 1 g of fumaric acid; 1 g of K_2HPO_4; 300 mg of $MgSO_4 \cdot 7H_2O$; 20 mg of $FeSO_4 \cdot 7H_2O$; KOH to give pH 7; and water to 1 liter. The cells are harvested by centrifugation and stored at $-10°$.

Preparation and Characterization of Phosphoinositides

Extraction of Lipids. Mycobacterium phospholipids are obtained by extracting cells with chloroform–methanol–water (16:6:1, v/v/v) at

[5] P. Brennan and C. E. Ballou, *Biochem. Biophys. Res. Commun.* **30**, 69 (1968).
[6] Y. Akamatsu, Y. Ono, and S. Nojima, *J. Biochem. (Tokyo)* **59**, 176 (1966).

30° for 2 days,[2] then refluxing the residue with the same solvent for 3 hours, and finally by extracting the remaining residue with absolute methanol at 60° for 2 days. The clear filtrates are combined and concentrated to dryness under reduced pressure. The dried phospholipid is triturated repeatedly with acetone and the residue is extracted overnight with an excess of absolute ethanol. The ethanol-insoluble lipid is suspended in chloroform–methanol–water (3:48:47, v/v/v), converted to the sodium form with EDTA-Na$_2$ and washed.[7] Finally the lipid is emulsified with water and the mixture is dialyzed against water for several days and lyophilized. The colorless solid is stored at $-10°$ under nitrogen. The product from *M. phlei* and *M. tuberculosis* contains about 30 mg of organic phosphorus per gram of lipid.

Column Chromatography. Mycobacterium phospholipid (880 mg) is applied to a 2×43 cm column of DEAE-cellulose (acetate form) packed in chloroform–methanol–water (20:9:1) and developed with a linear gradient of ammonium acetate $(0–0.1\ M)$ in the same solvent mixture[8]

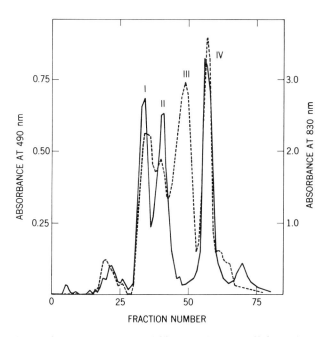

FIG. 1. Separation of phosphoinositides on DEAE-cellulose (acetate form). Fractions of 10 ml were collected. The solid line is carbohydrate, and the dashed line is phosphate. A description of the peaks is given in the text.

[7] M. C. Pangborn and J. A. McKinney, *J. Lipid Res.* **7**, 627 (1966).
[8] H. S. Hendrickson and C. E. Ballou, *J. Biol. Chem.* **239**, 1369 (1964).

to give the pattern shown in Fig. 1. Peak I contains diacyl phosphatidyl-myoinositol dimannoside, peak II is monoacyl phosphatidylmyoinositol dimannoside, peak III is phosphatidylmyoinositol, and peak IV is mainly phosphatidylmyoinositol dimannoside plus phosphatiylmyoinositol pentamannoside. To each of the pooled fractions is added 100 ml of water and the organic solvents are removed by rotary evaporation under vacuum. The aqueous emulsion is dialyzed at 4° against several changes of water and then lyophilized. Each fraction can be purified further by rechromatography in the same manner or by thin-layer chromatography.

Thin-Layer Chromatography. Preparative thin-layer chromatography is done on 20 × 40 cm glass plates coated with silica gel H, 0.5 mm thick. Up to 35 mg of phospholipid, when prepared as described, can be applied to each chromatogram. Plates are developed for 8 hr with solvent A and the lipids are located by spraying the surface with water or exposing the plate to iodine vapor.[9] The R_f values of the various phosphoinositides are given in the table. For the location of radioactive lipids, the plate is dried and a strip is divided into segments which are then analyzed for radioactivity by scintillation counting. Lipids are recovered from silica gel by elution first with chloroform–methanol (1:1, by volume), then methanol, and finally chloroform–methanol (2:1, v/v), in that order. The combined eluates are evaporated to dryness, the residue is dissolved in a small volume of chloroform–methanol–water (190:10:1, v/v/v), and the solution is passed through a 1.2 × 9 cm column of Sephadex G-25, previously equilibrated with methanol,[10] to remove salts extracted from the silicic acid. The lipid is washed through with 50 ml of the same sol-

CHROMATOGRAPHIC PROPERTIES OF PHOSPHOINOSITIDES AND THEIR
DEACYLATION PRODUCTS

Material	Intact lipid[a] (Solvent A)	Deacylated lipid[b] (Solvent B)
Diacyl dimannophosphoinositide	0.61	0.62
Monoacyl dimannophosphoinositide	0.52	0.62
Dimannophosphoinositide	0.28	0.61
Phosphatidylmyoinositol	0.57	1.00
Glycerylphosphorylmyoinositol		1.00
Glycerlphosphorylmyoinositol dimannoside		0.62

[a] These R_f values vary with the ages of the chromatogram plates.

[b] These values are expressed as $R_{glycerylphosphorylmyoinositol}$.

[9] M. E. Tate and C. T. Bishop, *Can. J. Chem.* **40**, 1043 (1962).

[10] A. N. Siakotos and G. Rouser, *J. Amer. Oil Chem. Soc.* **42**, 913 (1965).

vent. The solvent is removed by evaporation, the residue is emulsified with water, and the mixture is dialyzed and lyophilized. Generally, the lipids purified on silicic acid are less active as acceptors than those prepared by ion exchange chromatography, probably owing to the extraction of inhibitors from the silica gel.

Solvent A was used for thin-layer chromatography [chloroform–methanol–acetic acid–water (30:15:4:2, v/v/v/v)] and solvent B for paper chromatography [2-propyl alcohol–concentrated NH_4OH (2:1, v/v)].

Preparation of Extracts for Enzymatic Studies

Log phase cells of *M. phlei*, grown under shaking conditions, are harvested, washed with distilled water, lyophilized, and stored at $-10°$ for no longer than 2 weeks. About 3 g of cells, dry weight, is suspended in 50 ml of 0.02 M Tris·HCl buffer, pH 7.45, chilled to 1°, and disrupted for 9 minutes in an MSE-Mullard 60 W ultrasonic disintegrator fitted with a ⅜-inch diameter probe, the temperature being maintained at about 1°. The disintegrated material is centrifuged at 3000 g for 10 minutes. The supernatant is removed and centrifuged at 7000 g for 10 minutes. The resulting supernatant (total cell-free extract) is centrifuged at 100,000 g for 60 minutes. The supernatant "soluble" enzyme retains full activity after storage at $-10°$ for 2 months, but is inactivated by repeated freezing and thawing.

The pellet obtained from the high speed centrifugation is washed once with 0.02 M Tris·HCl, pH 7.45, and suspended in 20 ml of the same buffer (particulate enzyme). This preparation is stable at 0° or $-10°$ for at least 1 week.

Acetone powders yield an active "solubilized" particulate enzyme preparation significantly free of endogenous mannose acceptor. The suspended particulate enzyme preparation (20 ml) is added dropwise to 120 ml of stirred acetone at $-10°$. This suspension is centrifuged at 1000 g, and the supernatant is discarded. The pellet is dried in a desiccator over P_2O_5 with continuous aspiration in the cold for several hours. From 6 g of dried cells, 250 mg of dried acetone powder is obtained which shows little loss of activity after 2 weeks at $-10°$. The dry acetone powder, 50 mg, is ground in a chilled mortar and pestle for 2 minutes with twice its weight of glass beads (Superbrite, Minnesota Mining and Manufacturing Company, St. Paul, Minnesota), and 2 ml of 0.02 M Tris·HCl buffer, pH 7.45. The resulting suspension is centrifuged at 30,000 g for 30 minutes, and the supernatant is removed and centrifuged at 35,000 g for 1 hour. The clear yellow supernatant (solubilized particulate enzyme) is used within 2 hours after preparation because of its extreme lability. All activity is lost after 12 hours at 0° and also on freezing and thawing.

Incorporation of Label from GDP-[^{14}C]Mannose into Mannophosphoinositides of *M. phlei*

The Assay. The reaction mixture contains, in a final volume of 1.0 ml, 80 μmoles of Tris·HCl buffer (pH 7.35), 5 μmoles of MgCl$_2$, 7.8 nmoles of GDP-[^{14}C]mannose (46,000 dpm), 1.5 mg of Cutscum, 1.4 mg of phosphatidylmyoinositol, and enzyme protein. Incubation is at 37° for 1 hour. The reaction is stopped by the addition of 20 ml of chloroform–methanol (2:1; v/v) and the water soluble material is removed by washing the organic layer first with 3 ml of 0.29% NaCl and then three times with 4-ml portions of chloroform–methanol–water (3:48:47 by volume) containing 0.29% NaCl. Incorporation of mannose is determined by counting the lower phase, and the products of the reaction are determined by thin-layer chromatography.

Specificity. Phosphatidylmyoinositol from *M. phlei* and from yeast are active acceptors of mannose, but the lipid from wheat germ is inactive. That from *M. phlei* is inhibitory above 1 mg/ml while the activity with yeast phosphatidylmyoinositol is still increasing linearly at 3 mg/ml. This difference probably reflects differences in the micellar properties of the two lipids. Phosphatidylmyoinositol, purified by thin-layer chromatography on silica gel, is inactive.

Nature of the Product. Phosphatidylmyoinositol dimannoside is rarely the major product, since endogenous acyl donors lead to the formation of significant amounts of the mono- and diacyl derivatives. *Propionibacterium* species yield only the monomannoside.

Optimum pH. While somewhat variable, maximum activity is obtained around pH 7.

Activators and Inhibitors. The enzyme requires Mg^{2+} ion (5 mM) and is activated by Cutscum at a level of 0.1–0.2%. It is inhibited by high levels of Mg^{2+} and phosphatidylmyoinositol, as well as by GTP, GDP, and P$_i$.

Dimannophosphoinositide-Acylating System

The Assay. The incubation mixture contains 80 μmoles of Tris·HCl buffer (pH 7.35), 1.5 mg of *M. phlei* phospholipid, 0.6 nmole of [^{14}C]palmityl-CoA (26,000 dpm) and solubilized particulate enzyme in a final volume of 1.0 ml. Incubation is at 37° for 1 hour.

Reaction is terminated by the addition of 20 ml of chloroform–methanol (2:1, by volume), followed by 2 ml of 0.025 N HCl containing 2% acetic acid. The tube is swirled vigorously on a Vortex mixer, and the aqueous layer is discarded. The infranatant solution is washed with 2 ml of water followed by 2 ml of 0.05 N NaOH in 50% ethanol to remove most of the free fatty acid, and finally with 2 ml of water to remove the last trace of labeled fatty acyl-CoA. The organic phase is evaporated to dryness, and the residue is dissolved in 2 ml of chloroform–methanol (2:1, by volume). An aliquot (0.2 ml) from the reaction mixture is applied to a Silica Gel G plate, which is developed twice with diethyl ether and twice with diethyl ether–benzene–ethanol–acetic acid (200:250:10:1, v/v/v/v). This treatment removes free fatty acid and endogenous glycerides from the origin of the chromatogram, leaving only the phospholipid products, which are analyzed by scintillation counting.

The pattern of labeling of dimannophosphoinositides in these experiments is determined by thin-layer chromatography of a portion of the lipid solution on silica gel G. The origin is cut out and the labeled phospholipid is eluted. The eluate is applied to a plate of silica gel H which is developed in solvent A. The gel is divided into 0.5-cm bands, and each is analyzed for radioactivity. A typical chromatogram is shown in Fig. 2.

Specificity. The acylation occurs equally well with myristyl-, palmityl-, stearyl-, oleyl-, and tuberculostearyl-CoA. The most active acceptor is the crude *M. phlei* phospholipid preparation. Phosphatidylmyoinositol

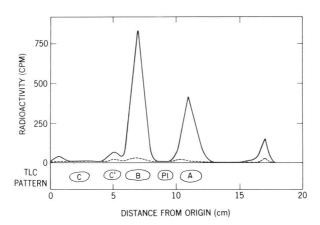

Fig. 2. Thin-layer chromatography (TLC) of labeled dimannophosphoinositides formed by incubation of [^{14}C]palmityl-CoA and *M. phlei* phospholipid. Solid line is the pattern of labeling with *M. phlei* phospholipid as acyl acceptor. Dashed line is the pattern of labeling with the endogenous acceptor in the solubilized particulate enzyme. A, Phosphatidylmyoinositol dimannoside dipalmitate; B, the monopalmitate; C and C′, isomeric forms of phosphatidylmyoinositol dimannoside; PI is phosphatidylmyoinositol.

dimannoside and monoacyl phosphatidylmyoinositol dimannoside were better acceptors than diacyl phosphatidylmyoinositol dimannoside, and there is some suggestion that the activity of the latter may reflect an exchange reaction rather than a net synthesis.

Nature of the Product. A thin-layer chromatogram of the ^{14}C-labeled products of a typical reaction is shown in Fig. 2. The mono- and dipalmityl derivatives account for nearly all the incorporated fatty acid.

Optimum pH. Optimal activity was obtained at about pH 7.3.

Inhibitors. The enzyme was inhibited 50% by 5 mM Mg^{2+} and lost activity on dialysis.

[67] Lactose Synthetase (UDP-D-Galactose:Acceptor β-4-Galactosyltransferase) from Bovine Milk[1]

By K. E. EBNER, R. MAWAL, D. K. FITZGERALD, and B. COLVIN

$$\text{UDP-D-Galactose} + \text{D-glucose} \xrightarrow[\alpha\text{-lactalbumin}]{\text{Mn}^{2+}} \text{lactose} + \text{UDP} \qquad (1)$$

$$\text{UDP-D-Galactose} + \text{GlcNAc} \xrightarrow{\text{Mn}^{2+}} N\text{-acetyllactosamine} + \text{UDP} \qquad (2)$$

$$\text{UDP-D-Galactose} + \text{GlcNAc} \xrightarrow{\beta,\,1-4} \text{CHO-Protein} \xrightarrow{\text{Mn}^{2+}} \text{Gal-GlcNAc} \xrightarrow{\beta,\,1-4} \text{CHO-Protein} \qquad (3)$$

The galactosyl transferase from milk and other tissues transfers galactose to a variety of carbohydrate acceptors.[2] Reaction (1) requires α-lactalbumin for significant rates (this reaction is catalyzed in the absence of α-lactalbumin, but the K_m for glucose is 1.4 M whereas the K_m for glucose is 5 mM in the presence of α-lactalbumin).[3] Reaction (2) is inhibited markedly by α-lactalbumin[4] whereas reaction (3) is only slightly inhibited by α-lactalbumin. The transfer of galactose to an N-acetylglucosaminyl residue on the carbohydrate side chain of a glycoprotein represents the physiological reaction in most tissues except the lactating mammary gland, when Reactions (1) and (3) occur concomi-

[1] Department of Biochemistry, Oklahoma State University. Journal Article 2393 of the Agricultural Experiment Station, Oklahoma State University, Stillwater, Oklahoma 74074. This research was supported by Grant AM 10764 from the National Institute of Health, National Science Foundation and a Career Development Award, 1 KO4 GM 42396 to K. E. Ebner from the National Institutes of Health.
[2] F. L. Schanbacher and K. E. Ebner, *J. Biol. Chem.* **245**, 5057 (1970).
[3] D. K. Fitzgerald, U. Brodbeck, I. Kiyosawa, R. Mawal, B. Colvin, and K. E. Ebner, *J. Biol. Chem.* **245**, 2103 (1970).
[4] K. Brew, T. C. Vanaman, and R. L. Hill, *Proc. Nat. Acad. Sci. U.S.A.* **59**, 491 (1968).

tantly. Reaction (2) is of little apparent physiological significance since the concentration of GlcNAc is very low in tissue.

Assay Method

Principle. Two major types of assays are available.[5] The first, or incorporation assay, measures the incorporation of UDP-[^{14}C]galactose into an appropriate acceptor (e.g., glucose or GlcNAc) and the product is separated from the reactants by chromatography on a Dowex 1-formate column. It is necessary to use this assay in tissue extracts. The second assay, spectrophotometric, measures the formation of UDP by coupling with pyruvate kinase and lactate dehydrogenase and measuring the oxidation of NADH. This assay generally cannot be used in tissue extracts or the early stages of enzyme purification because of the endogenous oxidation of NADH. However, it is ideally suited for kinetic and mechanistic studies.

Galactosyl Transferase

INCORPORATION ASSAY

Reagents

> [^{14}C]Lactose, 10 mM, 0.25 μCi/ml
> Lactose, 10 mM
> Glycine, 1 M, pH 8.5 and 9.0
> D-Glucose, 0.5 M
> N-Acetyl-D-glucosamine, 0.5 M
> α-Lactalbumin, 10 mg/ml
> MnCl$_2$, 50 mM
> UDP-[^{14}C]galactose, 6 mM, 0.8 μCi/ml
> Tween 80, 10% v/v
> EDTA, 1 M, pH 7.0
> Columns, Dowex 1-X8, 50–100 mesh, formate form, 6 × 70 mm in Pasteur pipettes, washed with 10 mM lactose

Procedure for Tissue Extracts. Tissue extracts or homogenates are centrifuged at 1000 g for 10 minutes, filtered through glass wool and the supernatant solution (used for assay) is made 0.2% in Tween 80. In the standard assay the reaction is run in 12-ml conical centrifuge tubes containing in the order of addition 5 μl of 1 M glycine, pH 9.0, 5 μl of glucose, 10 μl of α-lactalbumin, 10–50 μl of tissue extract, 10 μl of MnCl$_2$, 10 μl of UDP-[^{14}C]galactose, and water in a final volume of

[5] D. K. Fitzgerald, B. Colvin, B. Mawal, and K. E. Ebner, *Anal. Biochem.* **36**, 43 (1970).

0.1 ml. The incubation blank contains all reagents except glucose, and this corrects for the endogenous loss of UDP-[^{14}C]galactose due to hydrolysis and is treated in the same manner as the assay tubes. The reagents (less extract) are kept at 37°. The usual time of incubation is 4 minutes, and the use of short incubation times reduces the relative amount of endogenous UDP-[^{14}C]galactose loss. The reaction is stopped by placing the reactant tubes in boiling water for 1 minute (or by the addition of 5 μl of 1 M EDTA, pH 7.0). The reaction mixture is quantitatively transferred with 10 mM lactose to the Dowex columns, the columns are eluted with 10 mM lactose, and the first 4 ml eluted from the column are collected into a marked tube containing 5 μmoles of EDTA, pH 7.0, to reduce the oxidation of Mn^{2+}. A 2-ml sample is placed in a scintillation vial, 10 ml of Bray's solution are added and the vials are counted in a Packard TriCarb liquid scintillation counter. The following controls are run: (a) a zero time blank used to check the efficiency of the Dowex columns where the UDP-[^{14}C]galactose is added after the complete reaction is stopped: (b) background—a complete assay without UDP-[^{14}C]galactose; (c) lactose recovery run in duplicate—a complete assay with no UDP-[^{14}C]galactose; after the reaction is stopped, 0.01 ml of 10 mM [^{14}C]lactose is added and the mixture is passed through the Dowex columns and compared to the standard [^{14}C]lactose which is added to a mixture after passage through the Dowex column; the efficiency of the columns is usually 90–95%; (d) UDP-[^{14}C]galactose standard—the UDP-[^{14}C]galactose (as in assay) is added to a reaction mixture after passage through the Dowex column.

Under the conditions of the assay, the rate of lactose formation is proportional to the amount of extract up to 3 nmoles of lactose per minute of incubation.

The same assay is used when N-acetylglucosamine is the substrate (galactosyl acceptor). In this case 5 μl of N-acetylglucosamine but no α-lactalbumin is in the reaction mixture and the concentration of the other reactants and assay conditions are described above.

Procedure for Purified Preparations. The procedure is the same as described in (a) except that the buffer is 5 μl of 1 M glycine, pH 8.5, and the α-lactalbumin concentration is reduced (4 μl of 2.5 mg/ml solution) so that the final concentration is 100 μg/ml.

SPECTROPHOTOMETRIC ASSAY

Reagents

Glycine, 0.25 M, pH 8.5
MnCl$_2$, 0.1 M in 2.0 M KCl

NADH-PEP mix (NADH, sodium salt, 1.66 mM; phosphoenolpyruvate, potassium salt, 15.9 mM) pH 8.0

Pyruvate kinase (Sigma, Type 1 from rabbit muscle) 1:10 dilution in 20 mM glycylglycine, pH 8.5

α-Lactalbumin, 10 mg/ml

D-Glucose, 0.4 M

UDP-galactose, potassium salt, 4.67 mM

N-Acetylglucosamine, 0.4 M

Procedure. The reaction mixture in order of addition contains 0.2 ml
of glycine, 0.05 ml of MnCl$_2$, 0.1 ml of the NADH–PEP mix, 10 μl of
α-lactalbumin, 50 μl of pyruvate kinase, 50 μl of glucose, 50 μl of UDP-
galactose, and water and enzyme to initiate the reaction in a final
volume of 1 ml at room temperature. The pyruvate kinase should be
checked to ensure that it does not oxidize NADH. This preparation
contains sufficient lactate dehydrogenase to reduce the pyruvate formed
in the reaction, but this should be checked with each preparation by the
addition of crystalline lactate dehydrogenase to the reaction mixture. The
blank does not contain substrates. The above assay measures the formation of UDP and indirectly the formation of lactose. The galactosyltransferase activity also may be measured by using 50 μl of N-acetylglucosamine as the galactosyl acceptor instead of glucose, and no α-lactalbumin
since α-lactalbumin inhibits this reaction.[3,4] The reaction rate is measured
on a recording spectrophotometer at 340 nm. The A_{340} of the NADH–PEP
mixture used in the assay should be between 0.9 and 1.1 when read against
water. This solution decomposes upon storage. The pyruvate kinase solution is susceptible to bacterial contamination; it is best stored at $-20°$
and may be frozen and thawed a few times without affecting the assay.
The suitability of the indicator enzymes may be checked by a standard
solution of UDP (0.03–0.08 μmole of UDP per milliliter) to ensure they
are not limiting. The assay is linear with respect to protein to about
10 nmoles UDP per minute at 22°. For precise kinetic studies[6] a fixed-
time assay is recommended where the reaction is stopped with acid-
EDTA, neutralized, and the resulting UDP formed is measured enzymatically. The units are nanomoles of UDP formed per minute at 22°.

Purification Procedure

All procedures are at 0–4° unless otherwise specified.

Step 1. Portions, 2–3 liters, of raw skim milk are made 2.34 mM in
mercaptoethanol and warmed to 12–14°, care being taken not to exceed
15°. The pH is adjusted to between 5.2 and 4.6 by slowly adding 2 M

[6] J. F. Morrison and K. E. Ebner, *J. Biol. Chem.* **246**, 3977, 3985, and 3992 (1971).

HCl while stirring slowly with a glass rod until large curds are formed. The curded milk is immediately strained through a filter of glass wool overlaid by four layers of cheesecloth. The filtrate was collected in 3-liter beakers sitting in ice-water and when about half had filtered (10 minutes), 20 ml of 1 M Tris is added to the filtrate to raise the pH. After filtration, the pH is adjusted immediately to 7.4 with 1 M Tris. The volume is about 2500 ml per 3-liter portion of raw skim milk.

Step 2. Ammonium sulfate (390 g/liter) was added[7] over a 30-minute period while stirring, and after completion the solution was stirred for 1 hour. The mixture was centrifuged for 30 minutes at 11,000 g (Sorvall RC2-B with GS-3 head, 500 ml bottles) and the precipitate was dissolved in a minimum volume (100 ml) of 20 mM Tris, 5 mM MgCl$_2$, 2 mM mercaptoethanol, pH 7.4. The preparation was dialyzed against (2 liters) 10 mM Tris, 2.5 mM MgCl$_2$, 1 mM mercaptoethanol, pH 7.4, for 5–6 hours with buffer changes at every 1–1.5 hours. This preparation may be stored frozen.

Step 3. The sample (from 12 liters of milk) was thawed and centrifuged at 38,000 g for 20 minutes, and the supernatant solution was filtered through a small plug of glass wool. The solution was diluted with 1.5–1.8 liters of cold deionized water, and the pH was adjusted to 6.0 with 1 M acetic acid. The diluted solution was poured slowly, with stirring, into 150 g of moist phosphocellulose (equilibrated with 20 mM sodium acetate, 5 mM MgCl$_2$, 2 mM mercaptoethanol, pH 6.0, and stored moist until used), and the mixture was stirred for 10 minutes. The mixture was poured into a large column (15 × 30 cm) and allowed to settle overnight. The excess liquid was allowed to nearly drain off (500 ml/hr), and the column was washed with 2 liters of the same buffer. The enzyme was eluted off the column with 3 liters of 200 mM sodium acetate, 5 mM MgCl$_2$, 2 mM mercaptoethanol, pH 6.0, during the night.

Step 4. The enzyme was precipitated with (NH$_4$)$_2$SO$_4$ (561 g/liter), stirred for an hour, and centrifuged at 11,000 g for 30 minutes; the precipitate was dissolved in a minimum volume (100–150 ml) of 20 mM Tris, 5 mM MgCl$_2$, 2 mM mercaptoethanol, pH 7.4. The enzyme may be stored frozen at this stage.

Step 5. The enzymatic solution was desalted on a Sephadex G-25 column (4.5 × 50 cm) using 20 mM Tris, 5 mM MgCl$_2$, 2 mM mer-

---

[7] A convenient method for adding the ammonium sulfate is as follows. About 20–25 holes (1 mm in diameter) are drilled in a symmetrical pattern in the bottom of a 600-ml plastic beaker. The rate of flow of ammonium sulfate is controlled by a regulated mechanical stirrer whose metal blade sweeps the bottom of the beaker.

captoethanol, pH 8.5, as the buffer and 20-ml fractions were collected. The fractions containing the protein were pooled and were divided into portions containing 600–700 mg of protein assuming that an absorbance of 1.0 A_{280} was equal to 1 mg of protein. These pooled samples may be stored frozen.

Step 6. A sample containing 600–700 mg of protein was placed on a 2.3 × 21 cm column containing 12–15 cm of hydroxyapatite,[8] and 4–5 ml fractions were collected; the column was washed with 100 ml of 20 mM Tris, 5 mM MgCl$_2$, 2 mM mercaptoethanol, pH 8.5, followed by 100 ml of the previous buffer plus 10 mM ammonium sulfate and finally 300–400 ml of the same buffer plus 40 mM ammonium sulfate, which elutes the enzyme. The A_{220} is used to monitor protein, and the enzyme is located by assay. Tubes containing more than 0.025 unit/ml were pooled, and the enzyme was precipitated by the slow addition of ammonium sulfate (561 g/liter). The suspension was stirred for 45 minutes, then centrifuged at 38,000 g for 20 minutes, and the precipitate was dissolved in a minimum volume (5–10 ml) of 20 mM Tris, 5 mM MgCl$_2$, 2 mM mercaptoethanol, pH 7.4, and stored frozen.

Step 7. About 8–10 samples from step 6 were desalted on a Sephadex G-25 column and rechromatographed on hydroxyapatite, precipitated with ammonium sulfate and dissolved in a minimum amount of buffer as described in step 6; 8–10 mg of purified protein with a specific activity of about 3 is obtained from 80 liters of skim milk.

It is convenient to store the enzyme as a lyophilized powder. Enzyme from step 7 was precipitated with ammonium sulfate (561 g/liter) and dissolved in a minimum volume of 20 mM MgCl$_2$, 2 mM mercaptoethanol, pH 8.5 (10 mg of enzyme, 5–8 ml of buffer), and dialyzed for 6–8 hours against 500 ml of water containing 2 mM mercaptoethanol. The latter solution is changed every hour. The dialysis procedure removes most of the salt and is readily lyophilized. Complete removal of salt results in a product with very low activity. The lyophilized enzyme is stored dry at −15° and is stable for over a year.

Recent work has shown that the galactosyl transferase from bovine[9] and human[10] milk may be purified by affinity chromatography whereby α-lactalbumin is chemically coupled to Sepharose. This procedure is useful for purification at the latter stages of purification (step 6), since

[8] Prepared by the procedure of H. W. Siegelman, G. A. Wieczorek, and B. C. Turner, *Anal. Biochem.* **13**, 402 (1965). The preparation is aged for at least a month. It is important to test each batch, since some do not perform properly.

[9] J. P. Trayer, P. Mattock, and R. L. Hill, *Fed. Proc., Fed. Amer. Soc. Exp. Biol.* **29**, 597 (1970).

[10] P. Andrews, *FEBS (Fed. Eur. Biochem. Soc.) Lett.* **1**, 297 (1970).

other proteins appear to interfere at earlier stages of purification and lowers the capacity of the column. The procedure is functional since α-lactalbumin in the presence of substrates forms inactive dead-end complexes.[6,11] Frozen solutions of the galactosyl transferase loose activity, and this loss is accompanied by the formation of an inactive aggregate which may be removed by passage through an α-lactalbumin-Sepharose affinity column.

Properties

Stability. The purified enzyme is best stored dry as a lyophilized powder at $-15°$. The enzyme is more stable in a concentrated form and in the presence of mercaptoethanol and high ionic strength.

Specificity. The enzyme is specific for UDP-D-galactose as the galactosyl donor and dUDP-D-galactose is also effective, but ADP-D-galactose, TDP-D-galactose, CDP-D-galactose, GDP-D-galactose, TDP-D-glucose, GDP-D-glucose, UDP-D-glucose, and dUDP-D-glucose are ineffective.[12] The specificity of the galactosyl acceptor is broader in that a variety of β-1,4 glycosides of Glc-NAc, glucose and the carbohydrate side chains of glycoproteins containing a free N-acetylglucosaminyl β-1-4 linked residue are galactosyl acceptors.[2,4,12]

Activators and Inhibitors. The purified enzyme has an absolute requirement for Mn^{2+} and other divalent cations are neither active nor inhibit.[6] A previous report on activation by Mg^{2+} in a less pure form of the enzyme probably reflects the ability of Mg^{2+} to displace Mn^{2+} from other proteins. The enzyme is inhibited by reagents which complex Mn^{2+} (e.g., EDTA) and those which inactivate -SH groups.

Effect of pH and Temperature. The pH optimum[3,5] is between 7.5 and 9.0, and this is a function of the type of buffer, purity, and degree of solubilization of the galactosyl transferase. The temperature optimum[12] is reported to be 42°.

Kinetics. The steady-state kinetic analysis indicates that the reaction is of the equilibrium-ordered type,[6] where Mn^{2+} adds first, followed by UDP-galactose, and the galactosyl acceptor. The products are released in the order of galactose-acceptor and UDP. The true kinetic constants for the substrates were determined, and it was proposed that α-lactalbumin is a protein modifier of the galactosyl transferase.

[11] R. Mawal, J. F. Morrison, and K. E. Ebner, *J. Biol. Chem.* **246**, 7106 (1971).
[12] H. Babad and W. Z. Hassid, *J. Biol. Chem.* **241**, 2672 (1966).

[68] α-Lactalbumin and the Synthesis of Lactose[1]

By K. E. EBNER, R. MAWAL, D. K. FITZGERALD, and B. COLVIN

$$\text{UDP-D-Galactose} + \text{D-glucose} \xrightarrow[\text{α-lactalbumin}]{\text{Mn}^{2+}} \text{lactose} + \text{UDP}^2$$

Assay Methods

Principle. The assays for α-lactalbumin are similar to those described for galactosyltransferase except that α-lactalbumin is limiting and the galactosyltransferase is saturating. The galactosyltransferase by itself can utilize glucose[3] as a poor substrate, and thus excessive amounts of glucose or galactosyltransferase in the reaction will result in a high endogenous rate; as a consequence, conditions are chosen to minimize the endogenous rate. It is also necessary to construct a standard curve for each set of assays to ensure accurate results. To date, α-lactalbumins from various sources[4] will react with the bovine galactosyltransferase and results of such assays must be expressed in terms of a standard curve prepared from bovine α-lactalbumin since there is no assurance that the various α-lactalbumins react in the same quantitative manner with bovine-galactosyltransferase.

Incorporation Assay

Reagents

Glycylglycine, 1 M, pH 7.5
Glucose, 0.5 M
Galactosyltransferase[3] (HA$_I$ preparation so that 5–10 μl gives a rate of 0.1–0.2 nmole of product formed per minute)
α-Lactalbumin, 2.5 μg/ml prepared fresh from a 10 mg/ml stock solution

[1] Department of Biochemistry, Oklahoma State University. Journal Article 2394 of the Agricultural Experiment Station, Oklahoma State University, Stillwater, Oklahoma 74074. This research was supported by Grant AM 10764 from the National Institutes of Health, National Science Foundation and a Career Development Award 1 KO4 GM42396 to K. E. Ebner from the National Institutes of Health.
[2] D. K. Fitzgerald, B. Colvin, R. Mawal, and K. E. Ebner, *Anal. Biochem.* **36**, 43 (1970).
[3] D. K. Fitzgerald, U. Brodbeck, I. Kiyosawa, R. Mawal, B. Colvin, and K. E. Ebner, *J. Biol. Chem.* **245**, 2103 (1970).
[4] N. Tanahashi, U. Brodbeck, and K. E. Ebner, *Biochim. Biophys. Acta* **154**, 247 (1968).

MnCl$_2$, 30 mM

UDP-[^{14}C]galactose (6.0 mM, 0.8 μCi/ml)

Procedure. This assay is used in tissue extracts or other extracts which rapidly oxidize NADH and hence do not permit use of the spectrophotometric assay. The assay is done at 37° in the same manner as the incorporation assay for the galactosyltransferase (this volume [67]). The reaction mixture, in the order of addition of reagents, contains 5 μl of glycylglycine, pH 7.5, 5 μl of glucose, 5–10 μl of galactosyltransferase, extract containing 0.02–0.1 μg of α-lactalbumin, 10 μl of MnCl$_2$, 10 μl of UDP-[^{14}C]galactose, and water in a final volume of 0.1 ml. A standard curve is prepared from 0, 5, 10, 20, 30, and 40 μl of α-lactalbumin. A tube containing all reactants except α-lactalbumin (extract) is used to measure the endogenous rate, which is substracted from the standards in preparing the standard curve.[2] In homogenates of extracts, the incorporation of UDP-[^{14}C]galactose into lactose per minute is relative to the concentration of bovine α-lactalbumin if it is used to prepare the standard curve, and it should be expressed in these terms.

Spectrophotometric Assay

This assay is used in more purified systems where the endogenous oxidation of NADH is minimal. The assay is similar to the assay described for galactosyltransferase (this volume [67]).

Reagents

Glycylglycine, 1 M, pH 8.0

MnCl$_2$, 0.1 M in 2.0 M KCl

NADH, 1.66 mM; phosphoenolpyruvate 15.9 mM, pH 8.0

Pyruvate kinase[5] (Sigma Type 1, 1:10 dilution in 20 mM glycylglycine pH 8.0)

Galactosyltransferase (20 units)

Glucose, 0.4 M

UDP-galactose, 4.67 mM

Procedure. The procedure is the same as the one described for the galactosyltransferase (this volume [67]). The reaction mixture in order of addition contains 50 μl of glycylglycine, pH 8.0, 50 μl of MnCl$_2$, 0.10 ml of NADH-phosphoenolpyruvate, 50 μl pyruvate kinase, sample, 50 μl of glucose, 50 μl of UDP-galactose, and water in a total volume of 1.0 ml. The sample should contain 0.2–1.0 μg of α-lactalbumin, and a

---

[5] Some preparations of pyruvate kinase catalyze the endogenous oxidation of NADH, but this is minimized by Mn^{2+}, and hence it is added first. Pyruvate kinase is best stored at -20° and may be frozen and thawed a few times without affecting the assay.

standard curve is prepared which contains 0–1.0 μg of purified bovine α-lactalbumin/ml.

Purification of Bovine α-Lactalbumin

Most commercial preparations of α-lactalbumin contain contaminating amounts of other proteins and require further purification. The following procedure is a combination of previous methods and involves acid precipitation[6] and chromatography on Bio-Gel P-30 and DEAE cellulose.[7] The acid precipitation removes most proteins except β-lactoglobulin, which is removed by further purification.

Procedure. Six liters of raw skim milk are warmed to room temperature and divided into two 3-liter portions. Anhydrous sodium sulfate (20 g/100 ml) is added slowly while stirring over a 1-hour period; after the addition, the mixture is stirred for 15 minutes. The mixture is filtered through 4 layers of cheesecloth and filter paper on top of three glass rods (0.5 cm in diameter). The filtration is slow and is allowed to proceed overnight. The filtrate should be clear, but during filtration the sodium sulfate tends to crystallize out, and this should be redissolved by stirring and gentle heating prior to the next step. The filtrates are combined (volume, 4300 ml), and concentrated HCl (about 1 ml/100 ml filtrate) is added slowly (30 minutes) while stirring to lower the pH to 2.0. The mixture is centrifuged at 10,000 g for 20 minutes at 25°, and the supernatant solution is discarded. The precipitate is dissolved in (1/10 of the volume of the previous filtrate) 0.01 M NH$_4$OH and the pH of this solution (about 3.0) is adjusted to 3.5 with 1 M NH$_4$OH to remove residual traces of β-lactoglobulin. The mixture is centrifuged at 10,000 g for 20 minutes at 4°, and the supernatant solution is discarded. The precipitate is dissolved in 0.01 M NH$_4$OH (one fourth of the volume of the previous solubilizing step). The pH is adjusted to 4.0 with 1 N HCl and covered; the α-lactalbumin is allowed to precipitate overnight at 4°.

The mixture is centrifuged at 10,000 g for 20 minutes at 4° and *both* the precipitate of crude α-lactalbumin and the supernatant solution are saved. Ammonium sulfate (11.5 g/100 ml) is added to the supernatant solution to precipitate additional α-lactalbumin, which is removed by centrifugation as described above. The combined precipitates of the impure α-lactalbumin are dissolved in a minimum volume of 50 mM potassium phosphate, 0.1 M KCl, pH 7.4 (40–50 ml). The solution is stored at −15° or placed directly on the P-30 column.

Between 15 and 25 ml of the crude α-lactalbumin solution are placed

[6] R. Aschaffenburg and J. Drewry, *Biochem. J.* **65**, 18 (1957).
[7] U. Brodbeck, W. L. Denton, and K. E. Ebner, *J. Biol. Chem.* **242**, 1391 (1967).

on a 4.8 × 110 cm Bio-Gel P-30 column equilibrated and eluted with 50 mM potassium phosphate, 0.1 M KCl, pH 7.4. Fractions of 8–10 ml are collected at a flow rate of 60 ml per hour, and A_{280} is read for each tube. Two major protein peaks are eluted, and the lower molecular weight peak is pooled (150 ml), assayed and stored frozen or placed directly onto a DE-32 column (2.5 × 10 cm), which is equilibrated with 5 mM potassium phosphate, pH 8.0. The sample of α-lactalbumin off the P-30 column is diluted 10-fold with water prior before being put on the column. About 0.5 g of α-lactalbumin (1 mg/ml = A_{280} of 2.0) is put on the column and eluted with a linear gradient (500 ml each) from 5 to 200 mM potassium phosphate, pH 8.0. Ten-milliliter fractions are collected, and the A_{280} is read. The large single peak containing α-lactalbumin (assay) is pooled (500 ml), placed into cellulose casing[8] and dialyzed exhaustively (4 days with repeated changes) against distilled water at 4° until the conductance of the solution is near that of the distilled water. The solution is lyophilized, and the α-lactalbumin is stored dry at −15°. The final product exhibits a single band (100 μg) on disc gel electrophoresis[9] and does not appear to contain minor protein contaminants nor stain for carbohydrate. The yield is about 125 mg per liter of skim milk.

Commercial preparations should be examined by disc gel electrophoresis and, if impure be purified by chromatography on Bio-Gel P-30 and DEAE-cellulose. Similar procedures are used to prepare α-lactalbumin from other sources.[10]

[8] Cellulose casing, size 18, from Union Carbide, Chicago, is recommended. α-Lactalbumin can pass through other types of dialysis tubing.

[9] According to the instructions on the Canalco Model 6 instrument. Gels were pre-electrophoresed for 30 minutes at 5 mA per gel, and protein separations (up to 100 μg) were at 5 mA per gel, using 7% separating gel, pH 9.5.

[10] D. V. Schmidt and K. E. Ebner, *Biochim. Biophys. Acta* **243**, 273 (1971).

[69] An N-Acetylgalactosaminyltransferase from Human Milk; A Product of the Gene That Determines Blood Type A

By Akira Kobata

$$\text{UDP-GalNAc} + \underset{\underset{\text{Fuc-}\alpha1}{|}}{\underset{2}{\text{Gal-}\beta\text{-}(1-R)}} \xrightarrow{\text{Mn}^{2+}} \underset{\underset{\text{Fuc-}\alpha1}{|}}{\underset{2}{\text{GalNAc-}\alpha\text{-}(1-3)\text{Gal-}\beta\text{-}(1-R)}} + [\text{UDP}]$$

An N-acetylgalactosaminyltransferase[1] which occurs in human milk from donors of blood type A or AB but not in milk from donors of blood type B or O has been purified 55-fold.[2] The enzyme catalyzes the reaction shown above and is responsible for the formation of the structural determinants of blood type A. Enzymes of similar activity were found in human serum,[3] human submaxillary gland,[4] and gastric mucosa[5] as well as in the organs of pigs.[6]

Assay Method

Principle. The "A" enzyme has a strict acceptor requirement in that it will transfer N-acetylgalactosamine to terminal galactose only if the galactose is substituted at the C-2 position with L-fucose. Therefore, the "A" enzyme activity can be assayed specifically by using several oligosaccharides with Fuc-α-$(1 \rightarrow 2)$-Gal-β-structure as acceptors. Because of the availability and proper molecular size, 2'-fucosyllactose is used for the study.[7]

[1] The transferase will be referred in this paper as "A" enzyme.

[2] A. Kobata and V. Ginsburg, *J. Biol. Chem.* **245**, 1484 (1970).

[3] Y. S. Kim, J. Perdomo, A. Bella, and J. Nordberg, *Proc. Nat. Acad. Sci. U.S.* **68**, 1753 (1971); T. Sawicka, *FEBS Lett. (Fed. Eur. Biochem. Soc.)* **16**, 346 (1971).

[4] V. M. Hearn, Z. G. Smith, and W. M. Watkins, *Biochem. J.* **109**, 315 (1968).

[5] H. Tuppy and H. Schenkel-Brunner, *Eur. J. Biochem.* **10**, 152 (1969).

[6] D. M. Carlson, *in* "Biochemistry of Glycoproteins and Related Substances" (E. Rossi and E. Stoll, eds.), Karger, Basel, 1968.

[7] A more convenient assay method using 2'-fucosyllactobiuronic acid as an acceptor is now under study. The radioactive product of "A" enzyme from this acceptor has a negative charge at pH 5.4 and can be separated from UDP-N-acetylgalactosamine, N-acetylgalactosamine-1-phosphate, N-acetylgalactosamine and the unrelated product described in the text. 2'-Fucosyllactobiuronic acid is prepared as follows: Dissolve 100 μmoles of 2'-fucosyllactose in 3 ml of ice cold 0.2 M Na$_2$CO$_3$ solution. Add 250-μl aliquots of 0.1 M I$_2$-0.1 M KI solution at 20-minute intervals until the brown color remains for 1 hour (generally five times). The reaction mixture is concentrated under vacuum to 0.5 ml and spotted on Whatman No. 3 MM

Reagents

[^{14}C]UDP-GalNAc, 2 mM (acetyl label), with specific activity of 90 μCi/μmole.

2'-Fucosyllactose, 70 mM, is isolated from human milk by the method described in this volume [24].

Tris·HCl buffer, 0.05 M, pH 7.5

MnCl$_2$, 0.025 M and 0.1 M

Enzyme solution, dialyzed against 0.05 M Tris·HCl buffer, pH 7.5

Pyridine-acetic acid buffer, pH 5.4, is prepared by adding glacial acetic acid to the mixture of pyridine 3 ml and water 387 ml to adjust the pH as above.[8]

Ethyl acetate–pyridine–acetic acid–water (5:5:1:3, v/v/v/v)

Procedure. The incubation mixture contains the following: 2 mM [^{14}C]UDP-GalNAc, 5 μl; 70 mM 2'-fucosyllactose, 3 μl; 0.25 M MnCl$_2$, 2 μl; and the enzyme solution and 0.05 M Tris·HCl buffer, pH 7.5, to make total volume of 50 μl. The mixture is mixed well with a Vortex mixer; 10 μl of toluene is added, and the mixture is incubated at 37° for 14 hours. The reaction is stopped by heating in a boiling water bath for 2 minutes. After centrifugation, the solution is transferred to Whatman No. 3 MM paper as a 2-cm band and subjected to electrophoresis in pyridine–acetic acid buffer, pH 5.4, at 50 V/cm for 60 minutes. The paper is cut 10 cm from the anodal end to remove the residual UDP-N-acetylgalactosamine and N-acetylgalactosamine 1-phosphate which migrates there, and the remaining paper is chromatographed for 16 hours using ethyl acetate–pyridine–acetic acid–water (5:5:1:3) as a solvent to separate the reaction product

$$GalNAc-\alpha-(1 \to 3)\text{-}Gal\text{-}\beta\text{-}(1 \to 4)\text{-}Glc$$
$$Fuc\text{-}\alpha\text{-}(1 \to 2)\nearrow$$

from the free N-acetylgalactosamine and the unrelated product which remains at the origin. All the radioactivity remains near the origin during electrophoresis. The section of the chromatogram containing the product ($R_{2'\text{-fucosyllactose}} = 0.66$) is assayed for ^{14}C activity in 5 ml of Bray's solution[9] in a scintillation counter.

---

paper as a 40-cm band. After electrophoresis in pyridine–acetic acid buffer pH 5.4, at 50 V/cm for 60 minutes, guide strip is stained with periodate-benzidine reagent [H. T. Gordon, W. Thornburg, and L. N. Werum, *Anal. Chem.* **28**, 849 (1956)]. 2'-Fucosyllactobiuronic acid migrates about 20 cm to the anodal end and is recovered by elution with water. The syrup (yield 70% of theory) can be used as an acceptor without further purification.

[8] L. Grimmonprez and J. Montreuil, *Bull. Soc. Chim. Biol.* **50**, 843 (1968).

[9] G. Bray, *Anal. Biochem.* **1**, 279 (1960).

Definition of Units and Specific Activity. A unit of activity is defined as the amount of enzyme required for the synthesis of 1 pmole of product per hour under the standard assay conditions. Specific activity is expressed as units per milligram of protein as measured by the method of Lowry *et al.*[10]

Purification Procedure

All procedures unless otherwise specified are carried out at 0–4°.

Fraction 1. Human Milk. One liter of milk (either fresh or frozen) from donors of blood type A is centrifuged for 10 minutes at 100 g at 2°. The solidified lipid layer is removed by filtration through a packed glass wool in the cold. The clear solution is dialyzed against 20 volumes of 0.05 M Tris·HCl buffer, pH 7.5, overnight.

Fraction 2. Ammonium Sulfate Fractionation. Solid ammonium sulfate is added to make 35% of saturation. The precipitate is discarded after centrifugation, and the supernatant solution is made 50% saturation. The 35–50% precipitate is collected by centrifugation for 20 minutes at 10,000 g, suspended in 50 ml 0.05 M Tris·HCl buffer, pH 7.5, and placed in a dialysis bag. The preparation is then dialyzed with stirring overnight against 2 liters of the same buffer. The dialyzed solution is centrifuged for 15 minutes at 10,000 g, and the small precipitate is discarded.

Fraction 3. Sephadex G-200 Fractionation. The dialyzate (5 ml, containing 30–40 mg of protein per milliliter) is passed through a column of Sephadex G-200 (2 × 90 cm) that has been washed overnight with 0.05 M Tris·HCl buffer, pH 7.5. The column is eluted with the same buffer and after 125 ml had passed through, effluent is collected in 5-ml fractions. A major protein peak comes out at the void volume followed by two (occasionally three) minor peaks. The enzyme activity appears at the front edge of the first minor peak. Under above conditions, the enzyme appears around tube 15 and is usually completely eluted by tube 25 (between 200 and 250 ml). The enzyme fractions are pooled and concentrated to 4 ml by filtration through a collodion bag (Carl Schleicher and Schuell, Keene, New Hampshire).

Fraction 4. MnCl₂ Fractionation. To 4 ml of the concentrated enzyme at 0°, 1 ml of 0.1 M MnCl₂ is added and, after mixing, the solution is warmed up to 37° for 2 minutes. A precipitate of inactive protein is removed by centrifugation at 1000 g for 5 minutes at room temperature. The supernatant solution is dialyzed overnight against 1 liter of 0.05 M Tris·HCl buffer, pH 7.5 and stored at −20°. The enzyme at this stage

---

[10] See Vol. III [73].

HUMAN MILK "A" ENZYME PURIFICATION

Fraction	Protein (mg)	Total units ($\times 10^{-3}$)	Specific activity (units/mg protein)	Recovery of activity (%)
Defatted, dialyzed milk	10,500	210	20	100
Ammonium sulfate, 35–50% saturation	1,639	164	100	78
Sephadex G-200 fractionation	194	103	531	49
MnCl$_2$ fractionation	90	98	1095	47

is mostly free from nucleotide pyrophosphatase activity and the overall yield of enzyme is about 40–50%.

A typical purification is summarized in the table.

Properties

Optimum pH and Effect of Metal Ions. The enzyme shows a pH optimum at 7.5, and requires Mn^{2+} at an optimum concentration of 0.01 M. Cd^{2+}, Fe^{2+} and Co^{2+} at the same concentration, slightly stimulate the reaction, while Ni^{2+}, Zn^{2+}, Cu^{2+}, or Ca^{2+} has no effect.

Stability. The purified enzyme can be stored frozen at $-20°$ for at least two years without loss of activity.

Acceptor Specificity. The enzyme transfers N-acetyl-D-galactosamine to β-galactosyl residues that are substituted with L-fucosyl residue on the *C-2*. Galactosyl residues without this substitution are not acceptors. Lactodifucotetraose and lacto-N-difucohexaose I, even though their penultimate galactosyl residues are substituted on the *C-2* position with L-fucose (for structures, see this volume [24]), cannot be acceptors. This is presumably because of steric hindrance by the second fucose residue on the adjacent sugar.

Substrate Affinities. K_m: UDP-GalNAc = $2.5 \times 10^{-4} M$; 2′-fucosyllactose = $4.0 \times 10^{-4} M$; lacto-N-fucopentaose I = $3.5 \times 10^{-4} M$; 2-O-α-L-fucopyranosyl-D-galactose = $7.4 \times 10^{-4} M$; 2-O-α-L-fucopyranosyl-methyl-β-D-galactopyranoside = $5.0 \times 10^{-4} M$.

[70] Trehalose Phosphate Synthetase from *Mycobacterium smegmatis*[1]

By DAVID F. LAPP, BETTY W. PATTERSON, and ALAN D. ELBEIN

$$\text{Purine sugar nucleotide} + \text{G-6-P} \xrightarrow{\text{Mg}^{2+}} \text{trehalose-P} + \text{purine nucleoside diphosphate}$$

$$\text{Pyrimidine sugar nucleotide} + \text{G-6-P} \xrightarrow[\text{polyanion}]{\text{Mg}^{2+}} \text{trehalose-P}$$
$$+ \text{pyrimidine nucleoside diphosphate}$$

The synthesis of trehalose-P from a number of different sugar nucleotide glucosyl donors appears to be catalyzed by a single enzyme. However this enzyme is able to distinguish between those sugar nucleotides with a pyrimidine base and those with a purine base. Synthesis of trehalose-P from the pyrimidine sugar nucleotides requires activation of the enzyme by a macromolecular polyanion, such as RNA isolated from the organism itself, or chondroitin sulfate or heparin. The purine sugar nucleotides can serve as effective glucosyl donors in the absence of polyanion, and in these cases activity is only increased 2- to 4-fold by polyanion.

Assay Methods

Trehalose-P formation can be determined either colorimetrically or by a radioactive assay.[2,3]

Principle: The colorimetric method is based on the relative stability of trehalose to mild acid and alkaline hydrolysis, whereas sugar nucleotides and glucose-6-P are destroyed by such treatment. Trehalose content of the assay mixture is then measured using the anthrone reagent.[4]

The radioactive procedure is based on the isolation of ^{14}C-labeled trehalose-P by paper electrophoresis, or the isolation of ^{14}C-labeled trehalose by paper chromatography, after treatment of the incubation mixture with alkaline phosphatase and mixed-bed resin.

The colorimetric and the radioactive chromatographic assays are actually measures of trehalose content. These methods are more convenient to use with crude enzyme preparations since such extracts contain phosphatase activity, and some of the trehalose-P formed during the incubation is converted to trehalose.

[1] This work was supported by grants from the Robert A. Welch Foundation and the National Institute of Allergy and Infectious Diseases.
[2] C. Liu, B. W. Patterson, D. Lapp, and A. D. Elbein, *J. Biol. Chem.* **244**, 3728 (1969).
[3] D. Lapp, B. W. Patterson, and A. D. Elbein, *J. Biol. Chem.* **246**, 4567 (1971).
[4] F. A. Loewus, *Anal. Chem.* **24**, 219 (1962).

Reagents

ADP-D-glucose, GDP-D-glucose, UDP-D-glucose, TDP-D-glucose, CDP-D-glucose, 0.1 M (40,000 cpm per micromole [^{14}C]glucose nucleotide when the radioactive assays are used)

Glucose-6-P, 0.1 M

MgCl$_2$, 0.05 M

Tris·HCl, 0.1 M, pH 7.0

1 N HCl

2% NaOH

Anthrone reagent

Alkaline phosphatase (Worthington Biochemical Corp.)

Mixed-bed resin: Dowex 50-H$^+$ and Dowex 1-CO$_3^{2-}$

Procedure

Colorimetric Assay. The assay mixture contains the following components in a final volume of 0.2 ml: 0.5 μmole of sugar nucleotide; 1.0 μmole of G-6-P; 0.75 μmole of MgCl$_2$; 10 μmoles of Tris, pH 7.0; 20–40 μl of enzyme, and an appropriate amount of polyanion when the pyrimidine sugar nucleotides are used as substrate. For each preparation of enzyme, the optimal concentration of polyanion has to be determined by assaying the enzymatic activity with increasing amounts of polyanion. The optimal concentration of the polyanions, heparin, chondroitin sulfate, and the RNA fraction isolated from *M. smegmatis* falls in the range of 0.5–0.07 μg per 80 μg of the DEAE-cellulose enzyme preparation.

The assay mixtures are incubated at 37° for 20 minutes. The reaction is stopped by the addition of 50 μl of 1 N HCl, and the tubes are heated at 100°C for 15 minutes to hydrolyze any remaining sugar nucleotide. Then 0.3 ml of 2% NaOH is added and the mixtures are heated at 100° for an additional 15 minutes to destroy the sugars having free carbonyl groups. An aliquot (0.3 ml) of the reaction mixture is removed and assayed for trehalose content by the anthrone procedure.

Radioactive Assays. (i) CHROMATOGRAPHIC. Assay mixtures are as described in the colorimetric procedure except that ^{14}C-labeled sugar nucleotides are used. The enzymatic reaction is terminated by heating the assay tubes at 100° for 5 minutes. One milligram of alkaline phosphatase in 0.1 ml of 0.1 M Tris, pH 8.0, is added to convert any trehalose-P to trehalose. The tubes are incubated at 37° for 1 hour. The reaction is stopped by heating at 100° for 5 minutes. After treatment with mixed-bed ion-exchange resin, the neutral sugars are separated by descending paper chromatography in 1-propanol–ethyl acetate–water (7:1:2). The trehalose area is located, and the radioactive content is determined by counting the paper in a liquid scintillation spectrometer.

(ii) ELECTROPHORETIC. The reaction mixture is terminated by the addition of 50 μliters of 1 N HCl and the tubes are heated at 100°C for 15 minutes to destroy the remaining sugar nucleotide. The mixture is streaked on 2-inch strips of Whatman No. 3 MM paper and electrophoresed in 0.2 M ammonium formate, pH 3.6. The trehalose-P area is located, and its radioactive content is determined. Trehalose-P migrates just behind the yellow picric acid marker.

Preparation of Enzyme

Growth of Cells. M. smegmatis is grown at 37° in Trypticase Soy broth for 48–72 hours on a rotary shaker. Cells are harvested by centrifugation and washed once with distilled water. These cells may be stored at −20° until needed.

Preparation of Cell-Free Extract. Forty grams of cell paste is suspended in 200 ml of 0.01 M Tris, pH 7.5, containing 0.005 M β-mercaptoethanol. The cells are disrupted by sonic oscillation, and the cellular debris is removed by centrifugation at 20,000 g for 20 minutes.

Ammonium Sulfate Fractionation. Solid ammonium sulfate (313 g/liter) is slowly added to the crude extract to give a 50% saturated solution. The mixture is allowed to stand at 4° for 20 minutes, and the precipitate is collected by centrifugation, dissolved in a minimal amount of 0.01 M Tris, pH 7.5, containing β-mercaptoethanol, and dialyzed overnight against 4–6 liters of the same buffer.

DEAE-Cellulose Column Chromatography. DEAE-cellulose is washed with 0.5 N NaOH and then 0.5 N HCl; it is stored in 1 M KCl. A column (1.5 × 12 cm) is prepared and washed with 0.01 M Tris buffer, pH 7.5, containing 0.005 M β-mercaptoethanol, until the eluate is free of Cl⁻ ions. Ten milliliters of the ammonium sulfate fraction is applied to the column, which is then washed with 500 ml of 0.01 M Tris buffer. The column is eluted with a linear gradient of KCl from 0 to 1 M. Five milliliter fractions are collected and assayed for enzymatic activity. The enzymatic activity elutes between tubes 65 and 80 (0.15–0.3 M KCl). Active fractions are pooled, and the protein is precipitated by the addition of ammonium sulfate (to 70% saturation). The precipitate is collected by centrifugation, dissolved in a minimum of 0.1 M Tris buffer containing 0.005 M β-mercaptoethanol, and dialyzed overnight against the same buffer. Alternatively, a single-step elution procedure can be utilized to obtain larger quantities of enzyme. About 15 ml of ammonium sulfate fraction is added to the column, which is then washed with 0.01 M Tris buffer. The column is then eluted with 0.15 M KCl and fractions of 12 ml are collected. Enzymatic activity is usually located in tubes 20–30.

The RNA fraction is eluted from DEAE-cellulose after the enzyme.

It usually emerges from the column between 0.5 and 0.8 M KCl. Active fractions of RNA are pooled and heated at 100° for 30 minutes and then centrifuged at 27,000 g. The supernatant liquid is concentrated to a small volume in an Amicon ultrafiltration cell, using a 50,000 molecular weight cutoff filter. The concentrated fraction is extracted with phenol, and the aqueous phase is separated and extracted with ether to remove any residual phenol. The aqueous phase is made 0.2 M with respect to sodium acetate, and the RNA is precipitated from solution by the addition of two volumes of ethanol. The polynucleotide is collected by centrifugation, dissolved in water, lyophilized, and stored as a powder at −20° until used.

A typical purification is summarized in the table.

Properties

Stability. The DEAE-cellulose enzyme is stable for about 24 hours at 0°. However, the enzyme is stable to freezing, and less than 10% of its original activity is lost after storage for 1 week at −20°. Repeated freezing and thawing causes inactivation.

pH Optimum. The optimum pH of the enzyme with both purine and pyrimidine sugar nucleotides in Tris·HCl is 7.0.

Specificity. The DEAE-cellulose enzyme is nonspecific with respect to the sugar nucleotide used as the glucosyl donor. Ranked in their order of effectiveness, ADP-Glc, GDP-Glc, UDP-Glc, TDP-Glc, and CDP-Glc are all glucosyl donors. However there is a significant difference between those nucleotides with pyrimidine bases and those with purine bases.

PURIFICATION OF TREHALOSE PHOSPHATE SYNTHESIS FROM
Mycobacterium smegmatis[a]

| | | Specific activity (nmoles trehalose/mg protein/min) | | | Yield (%) | |
| | | | UDP-Glc | | | |
Step	Protein (mg)	GDP-Glc +CS[b]	−CS	+CS	GDPG	UDPG
1. Crude sonicate	4630	0.8	1.1	1.5	100	100
2. Ammonium sulfate fractionation (0–50%)	1092	7.6	8.3	8.1	224	127
3. DEAE-cellulose	13.99	27.4	4.8	31.6	10	6

[a] Data in this table was obtained using the colorimetric assay. Activity measurements in crude extracts are difficult to make because of rapid degradation of substrates by other enzymes and also because of high concentrations of endogenous trehalose in these extracts. Therefore these numbers represent only approximations.
[b] CS = chondroitin sulfate.

When UDP-Glc, TDP-Glc, and CDP-Glc are used, there is an almost complete dependence on the presence of a polyanion for the synthesis of trehalose-P. The polyanion increases the activity 20-fold or more. There is significant activity in the absence of polyanion with the purine sugar nucleotides ADP-Glc and GDP-Glc. The activity in the presence of polyanion is stimulated only about 3-fold.

Kinetic Constants. The apparent K_m at $37°$ for UDP-Glc in the presence of polyanion is $1 \times 10^{-4} M$. In the absence of polyanion very little trehalose-P is synthesized, therefore a K_m cannot be obtained. The apparent K_m for GDP-Glc is $1 \times 10^{-3} M$ both in the presence and absence of polyanion. The apparent K_m for G-6-P with GDP-Glc is $2 \times 10^{-3} M$ both in the presence and absence of polyanion; whereas with UDP-Glc as substrate, the K_m for G-6-P was $1 \times 10^{-3} M$ in the presence of polyanion.

Activators and Inhibitors. The enzyme exhibits no absolute requirement for a divalent cation; however, enzymatic activity is increased about 2-fold in the presence of $5 \times 10^{-3} M$ MgCl$_2$. Mn^{2+} and Fe^{2+} can replace Mg^{2+}.

A wide variety of high molecular weight polyanions can activate the DEAE-cellulose enzyme to utilize the pyrimidine sugar nucleotides for the synthesis of trehalose-P. Their effectiveness as activators appears to depend on their molecular weight and on the distribution of their charges. The three most effective activators are heparin, chondroitin sulfate and the polynucleotide fraction isolated from the organism. Chondroitin sulfate and heparin are routinely used in this laboratory as the activators because they are available commercially and are not subject to RNase degradation. Only slight activation of enzymatic activity can be observed in the crude cell-free extract and ammonium sulfate fraction, presumably because the RNA activator is already present in these fractions.

Compounds such as salts and nucleoside mono-, di-, and triphosphates at fairly high concentrations ($2.5 \times 10^{-2} M$) inhibit the synthesis of trehalose-P from the pyrimidine sugar nucleotides when tested in the presence of polyanion. These compounds appear to inhibit the binding of the polyanion activator to the enzyme.

[71] Trehalose Phosphate Phosphatase from Mycobacterium smegmatis [1]

By Mike Mitchell, Mike Matula, and Alan D. Elbein

Trehalose 6-phosphate $+ H_2O \rightarrow$ trehalose $+ P_i$

A soluble enzyme from *Mycobacterium smegmatis* catalyzes the dephosphorylation of trehalose phosphate to trehalose. Since the enzyme is reasonably easy to isolate and purify and is reasonably specific for trehalose phosphate, it is a useful reagent for determining trehalose phosphate.[2]

Assay Method

Principle. The activity of the enzyme can be measured by following the release of inorganic phosphate from trehalose 6-phosphate. Inorganic phosphate is determined by the method of Chen *et al.*[3]

Reagents

> Trehalose 6-phosphate, Na or K salt, 0.05 M. Ba trehalose 6-phosphate is prepared by the method of McDonald and Wong[4] and is converted to the Na or K salt.
> MgCl₂, 0.05 M
> Tris·HCl buffer, 0.1 M pH 7.0

Procedure. Incubation mixtures contain the following components (in μmoles) in a final volume of 0.25 ml: trehalose 6-phosphate, 1; MgCl₂, 1; Tris buffer, 10 and an appropriate amount of enzyme. Control tubes are also prepared in which boiled enzyme (100°, 5 minutes) is used or in which trehalose phosphate is added at the end of the incubation period. Incubations are usually for 15 minutes at 37°. Reactions are stopped by placing tubes in a boiling water bath for 5 minutes. The complete reaction mixture or an aliquot thereof is then tested for its content of inorganic phosphate by the method of Chen *et al.*[3]

When acid labile phosphorylated compounds are tested phosphate is assayed by the method of Bartlett,[5] since this method does not involve the use of strong acid.

---

[1] This work was supported by grants from the Robert A. Welch Foundation and the National Institute of Allergy and Infectious Diseases.
[2] M. Matula, M. Mitchell, and A. D. Elbein, *J. Bacteriol.* **107**, 217 (1971).
[3] P. S. Chen, T. Y. Toribara, and H. Warner, *Anal. Chem.* **28**, 1756 (1956).
[4] D. L. MacDonald and R. Y. K. Wong, *Biochim. Biophys. Acta* **86**, 380 (1964).
[5] G. R. Bartlett, *J. Biol. Chem.* **234**, 466 (1958).

Definition of Unit. One unit is defined as the amount of enzyme that will catalyze the release of 1 μmole of inorganic phosphate from trehalose phosphate in 1 minute at 37°.

Purification Procdure

Step 1. Growth of Cells and Preparation of Crude Extract. Mycobacterium smegmatis is grown in Trypticase Soy Broth at 37° for 2–4 days on a rotary shaker. Cells are harvested by centrifugation, washed with water, and stored at −20° until used. Twenty grams of cells paste is suspended in 100 ml of 0.01 M Tris buffer, pH 7.0, and subjected to sonic oscillation with a Sonifier cell disruptor for a total of 10 minutes at 0° (2.5 minutes of sonic oscillation followed by a 5-minute period of cooling). Cell debris is removed by centrifugation at 30,000 g for 20 minutes. The supernatant fraction contains the enzymatic activity.

Step 2. Manganese Precipitation and Calcium Phosphate Gel Treatment. A solution of 1 M MnCl$_2$ is added to the crude extract to a final concentration of 0.05 M. The mixture is allowed to stand for 5 minutes in an ice bucket, and the precipitate is removed by centrifugation and discarded. To 100 ml of supernatant liquid, 40 ml of calcium phosphate gel (15 mg/ml) is added slowly with stirring. The mixture is allowed to stand in an ice bucket for several minutes, and the precipitate is removed by centrifugation and discarded.

Step 3. Ammonium Sulfate Fractionation. To 100 ml of ice cold calcium phosphate gel supernatant liquid, 27 g of solid ammonium sulfate (0–45% saturation) is added slowly with stirring. The precipitate is removed by centrifugation and discarded, and to the supernatant fluid an additional 26 g of solid ammonium sulfate (45–80% saturation) is added with stirring. The precipitate is collected by centrifugation, dissolved in 10 ml of water, and dialyzed overnight against several liters of 0.01 M Tris buffer, pH 7.0.

Step 4. DEAE-Cellulose Column Chromatography. Ten milliliters of ammonium sulfate fraction is applied to a 1.5 × 10 cm column of DEAE-cellulose (Cl$^-$), which is then washed with Tris buffer. Columns are then eluted with a linear gradient of 0 to 1 M KCl in 0.01 M Tris buffer. The enzyme, which is eluted at about 0.2 M KCl, is concentrated by precipitation with ammonium sulfate and dialyzed against 0.01 M Tris buffer.

The results of a typical purification procedure are shown in the table.

Properties

pH Optimum. The pH optimum of the phosphatase in Tris buffer is 7.0.

Specificity. The enzyme is fairly specific for trehalose 6-phosphate although it does show slight activity on mannose-6-P and fructose-6-P.

PURIFICATION OF TREHALOSE PHOSPHATE PHOSPHATASE FROM
Mycobacterium smegmatis

Step	Specific activity (U/mg protein)	Total units
Crude extract	0.0185	92.5
Ammonium sulfate	0.266	53.2
DEAE-cellulose	0.87	37.7

Glucose-6-P, trehalose 6,6'-diphosphate, and a variety of other phosphorylated compounds are inactive.

Activators and Inhibitors. The enzyme is inactive in the absence of a divalent metal. The requirement is fairly specific for Mg^{2+}; the only other cation that shows some activity is Co^{2+} (60% of the activity of Mg^{2+}). Citrate is a competitive inhibitor of the enzyme, probably owing to the chelation of Mg^{2+}. EDTA also inhibits the enzyme, but in this case the kinetics are not competitive and appear complex. NaF is a noncompetitive inhibitor of the enzyme.

Stability. The enzyme slowly loses activity in the freezer over a period of several weeks. However, the enzyme can be kept in ice as a suspension of the enzyme in an 80% ammonium sulfate solution without any apparent loss activity over a period of several weeks.

Kinetic Parameters. The K_m value for trehalose phosphate is $1.5 \times 10^{-3} M$. The optimum concentration of Mg^{2+} is about $1.5 \times 10^{-3} M$.

[72] Synthesis of Raffinose-Type Sugars

By L. LEHLE and W. TANNER

Sufficient *in vivo*[1-3] and *in vitro*[3-7] evidence has accumulated to show that the biosynthesis of the most important members of the raffinose sugars has to be formulated in the following way:

$$\text{Galactinol} + \text{sucrose} \rightleftarrows \text{raffinose} + \text{myoinositol} \qquad (1)$$

$$\text{Galactinol} + \text{raffinose} \rightleftarrows \text{stachyose} + \text{myoinositol} \qquad (2)$$

$$\text{Galactinol} + \text{stachyose} \rightleftarrows \text{verbascose} + \text{myoinositol} \qquad (3)$$

[1] M. Senser and O. Kandler, *Z. Pflanzenphysiol.* **57**, 376 (1967).
[2] M. Senser and O. Kandler, *Phytochemistry* **6**, 1533 (1967).
[3] W. Tanner, *Ann. N. Y. Acad. Sci.* **165**, 726 (1969).
[4] W. Tanner and O. Kandler, *Plant Physiol.* **41**, 1540 (1966).
[5] W. Tanner, L. Lehle, and O. Kandler, *Biochem. Biophys. Res. Commun.* **29**, 166 (1967).

Galactinol[8,9] itself, an α-galactoside of myoinositol (L-1-(O-α-D-galactopyranosyl)myoinositol), is synthesized by the usual nucleotide pathway via UDP-galactose.[10] Although the enzymes catalyzing reactions (1)–(3) possess hydrolytic activity, they differ greatly from α-galactosidases[11,12] as far as efficiency of the transfer and substrate specificity is concerned. The enzymes described herein, therefore, have to be grouped among the glycosyl transferases (EC 2.4.1).

Synthesis of Raffinose by Galactinol:Sucrose 6-Galactosyltransferase

Galactinol + sucrose ⇌ raffinose + myoinositol

Assay Method

Principle. The transfer of the galactosyl moiety from galactinol to sucrose is assayed by determining the amount of [14C]raffinose formed from [14C]sucrose. The reaction components are separated by paper chromatography.

Reagents

Tris·HCl buffer, 0.5 M, pH 7.2
Galactinol, 0.1 M [13]
[14C]Sucrose, 0.001 M (35 μCi/μmole)
Enzyme

Procedure. The reaction mixture contains 10 μl of Tris, 10 μl of galactinol, 10 μl of [14C]sucrose, and 20 μl of enzyme. After incubation for 1–4 hours at 32°, the reaction is stopped with 0.2 ml of ethanol and the preparation is centrifuged; the supernatant fluid is separated on paper in the solvent system n-butanol/pyridine/water/acetic acid (60:40:30:3). Radioactive spots are located with a strip scanner, cut out, and meas-

---

[6] L. Lehle, W. Tanner, and O. Kandler, *Hoppe-Seylers Z. Physiol. Chem.* **351**, 1494 (1970).

[7] J. A. Webb and S. Pathak, *Plant Physiol.* Suppl. 46, p. 27 (1970).

[8] R. J. Brown and R. F. Serro, *J. Amer. Chem. Soc.* **75**, 1040 (1953).

[9] G. A. Kabat, D. L. MacDonald, C. E. Ballou, and H. O. L. Fischer, *J. Amer. Chem. Soc.* **75**, 4507 (1953).

[10] R. B. Frydman and E. F. Neufeld, *Biochem. Biophys. Res. Commun.* **12**, 121 (1963).

[11] J. E. Courtois, *in* "Carbohydrate Chemistry of Substances of Biological Interest" (M. L. Wolfrom, ed.), p. 140. Pergamon, Oxford, 1959.

[12] J. E. Courtois and F. Petek, *Bull. Soc. Chim. Biol.* **39**, 715 (1957).

[13] Galactinol can be isolated according to R. M. McCready, J. B. Stark, and A. E. Goodban, *J. Amer. Soc. Sugar Beet Technol.* **14**, 127 (1966); J. V. Dutton, *Int. Sugar J.*, Aug., p. 235; Sept., p. 261 (1966).

ured directly on paper in a scintillation counter in toluene/2.5-diphenyl-oxazole (efficiency 70%). This test proceeds linearly up to 5 hours and a protein content of 3 mg of crude extract.

Enzyme Purification

All procedures are carried out at about 4°.

Step 1. Preparation of Crude Extract. Ripe seeds, 200 g, from *Vicia faba* are powdered at room temperature in a Waring Blendor and then extracted in a chilled mortar in two portions each with 200 ml of 0.1 M Tris·HCl buffer, pH 7.3, containing 0.005 M Cleland reagent (DTE). The homogenate is centrifuged for 30 minutes at 27,000 g giving a clear supernatant of about 250 ml.

Step 2. Treatment with Protamine Sulfate. The supernatant is brought to a protein concentration of about 50 mg/ml; 2% protamine sulfate is slowly added to a final ratio of 9 mg per 100 mg of protein. After 30 minutes of stirring, the resulting precipitate is removed by centrifugation at 27,000 g for 15 minutes and discarded.

Step 3. Ammonium Sulfate Fractionation. To the protamine-treated supernatant cold, saturated $(NH_4)_2SO_4$ solution, pH 7.3, is slowly added with continuous stirring to give 33% saturation. After 30 minutes, the precipitate is separated by centrifugation as above and the supernatant is brought to 55% saturation. After further stirring for 30 minutes, the mixture is centrifuged for 10 minutes at 13,000 g. The resulting pellet is dissolved approximately in 70 ml 0.1 M Tris·HCl, pH 7.3, containing 0.005 M DTE and dialyzed overnight against 3 liters of 0.05 M Tris·HCl, pH 7.5, with 0.001 M DTE.

Step 4. DEAE-Cellulose Chromatography. The enzyme solution is adsorbed on a DEAE-cellulose column (2.5 × 30 cm) which had been equilibrated with 0.01 M Tris, pH 7.5, containing 0.05 M KCl and 0.001 M DTE. The column is washed with equilibration buffer until all protein not bound is removed. Desorption is performed with a 1-liter linear gradient from 0.05 M KCl to 0.2 M KCl in 0.01 M Tris·HCl with 0.001 M DTE. Fractions of 6 ml are collected and those with the highest specific activity are pooled and concentrated to a small volume in an Amicon ultrafiltration cell with filter No. XM-50. The enzyme appears from the column after about 270 ml of elution volume.

Step 5. Sephadex G-200 Gel Chromatography. The pooled and concentrated fractions are loaded onto a column (2.5 × 80 cm) of Sephadex G-200 equilibrated with 0.01 Tris buffer, pH 7.5, containing 0.1 M KCl and 0.002 M DTE. The column is eluted at a flow rate of 4 ml/hour; 2-ml fractions are collected and the active fractions (100–120) are pooled and concentrated as described before.

Step 6. Hydroxyapatite Chromatography. After dialysis against 0.01 M

Tris·HCl buffer with $0.002\,M$ DTE, pH 7.5, the enzyme solution is placed on a column (2.5×13 cm) of hydroxyapatite, which had been equilibrated with $0.01\,M$ potassium phosphate buffer, pH 7.5, containing $0.002\,M$ DTE. Elution is carried out stepwise. The following potassium phosphate buffers, pH 7.5, are used: (a) 100 ml of $0.01\,M$; (b) 100 ml of $0.05\,M$; (c) 100 ml of $0.1\,M$; (d) 100 ml of $0.2\,M$. The enzyme purified 400-fold is eluted by $0.2\,M$ buffer. The active fractions are again concentrated as described above. A peak with much lower specific activity appears also at $0.1\,M$ buffer.

With the total procedure outlined above it is possible to separate the enzyme responsible for reaction (1) i.e., the synthesis of raffinose, from the *Vicia faba* enzyme catalyzing reactions (2) and (3) (see Section III).

A typical purification is summarized in Table I.

Properties

Stability. The crude extract loses 50% of its original activity within 3 days at $4°$. The activity of the purified enzyme when frozen is preserved for at least a month.

pH Optimum. The enzyme shows a broad pH dependance (pH 5.5–8.0 in phosphate buffer) with an optimum at 6.8.

Kinetics. The Michaelis constant for galactinol has been determined as $7 \times 10^{-3}\,M$ at pH 7.2 in the presence of $4 \times 10^{-4}\,M$ sucrose. K_m for sucrose is $9 \times 10^{-4}\,M$ under the assay conditions given above.

Specificity. The acceptor specificity has been tested with [^{14}C]galactinol (labeled galactinol has been isolated by paper chromatography from the water-soluble extract of *Lamium* leaves after photosynthesis in $^{14}CO_2$ according to Kandler[14]). Out of 10 tested substrates (sucrose, raffinose, stachyose, fructose, glucose, galactose, cellobiose, melibiose, lactose,

TABLE I

PURIFICATION PROCEDURE FOR GALACTINOL:SUCROSE 6-GALACTOSYLTRANSFERASE

Fraction	Total protein (mg)	Total activity (cpm $\times$ 10^{-6}/ hour)	Specific activity (cpm $\times$ 10^{-3}/ hour/mg)
1. Supernatant	22,000	85.2	3.8
2. Protamine sulfate	9,350	94.6	10.2
3. Ammonium sulfate	2,688	75.6	28.1
4. DEAE-cellulose	182	30.2	165.8
5. Sephadex G-200	50	14.2	285.0
6. Hydroxyapatite	3	4.8	1600.0

[14] O. Kandler, *Ber. Deut. Bot. Ges.* **77**, 62 (1964).

glycerol) only a transfer to sucrose could be observed. As donors only galactinol, p-nitrophenyl-α-D-galactopyranoside, and raffinose work to a significant extent. The latter reaction corresponds to the exchange reaction between raffinose and sucrose described by Moreno and Cardini.[15] UDP-galactose functions as galactosyl donor neither with the crude extract nor the purified enzyme.

Effect of Metals and SH Compounds. The reaction does not require metals. However, heavy metal ions like Ag^{2+}, Hg^{2+}, Zn^{2+}, Al^{+3}, inhibit the reaction. The enzyme is highly dependent on SH-protecting agents. Thus, no activity can be found when Cleland reagent (DTE) is omitted during extraction and dialysis procedure; $0.001\,M$ N-ethylmaleimide or iodoacetamide cause 90% and 50% inhibition, respectively.

Galactinol:Raffinose 6-Galactosyltransferase

Galactinol + raffinose $\rightleftarrows$ stachyose + myoinositol

Assay Method

Principle. The transferase activity can be measured in two different ways (Test I and Test II).

Test I is based on the transfer of the ^{14}C-labeled galactosyl moiety from [^{14}C]galactinol to raffinose. The radioactivity in stachyose is determined after paper chromatographic separation.

In Test II, the amount of myoinositol set free due to the galactosyl transfer is determined with myoinositol dehydrogenase according to Weissbach.[16] Since, however, myoinositol is also released to a small extent by the hydrolytic activity of the enzyme, the above result has to be corrected. This is done by measuring the amount of galactose set free, following essentially the procedure of Wallenfels and Kurz.[17] The difference between the amount of inositol and galactose is a minimal measure for the activity of the transfer reaction. The correction, however, amounts to less than 10% of the total myoinositol set free.[18]

Test I

Reagents

Phosphate buffer, $0.5\,M$, pH 7.0
Raffinose, $0.08\,M$

---

[15] A. Moreno and C. E. Cardini, *Plant Physiol.* **41**, 909 (1966).
[16] A. Weissbach, *Biochim. Biophys. Acta* **27**, 608 (1958).
[17] K. Wallenfels and G. Kurz, *Biochem. Z.* **335**, 559 (1962).
[18] W. Tanner and O. Kandler, *Eur. J. Biochem.* **4**, 233 (1968).

[^{14}C]Galactinol 0.015 M (12.6 μCi/μmole)
Enzyme

Procedure. The following solutions are mixed together in a final volume of 0.05 ml: 10 μl of buffer, 10 μl of raffinose, and 10 μl of [^{14}C]galactinol. Enzyme, 20 μl, is added to initiate the reaction, which is left at 32° for 1–4 hours. The reaction is stopped with 0.1 ml of ethanol and the supernatant fraction is chromatographed in ethyl acetate/butanol/acetic acid/water (30:40:35:40).

The radioactive areas are located with a strip counter and the radioactivity measured directly on paper with a methane flow counter or in toluene/2.5-diphenyloxazole with a scintillation counter.

Test II

Reagents

Phosphate buffer, 0.5 M
Galactinol, 0.05 M
Raffinose, 0.1 M
Enzyme

Procedure. The incubation mixture contains 0.05 ml of phosphate buffer, 0.05 ml of galactinol, 0.025 ml of raffinose, and 0.1 ml of enzyme. The incubation is carried out at 32° for 30–60 minutes. The reaction is stopped by boiling for 30 seconds; 2.4 ml of water is added, the precipitate is removed by centrifugation, and myoinositol is determined in an aliquot of 1 ml. Another milliliter is used to determine the amount of D-galactose liberated for correction (see above). Myoinositol is measured with myoinositol dehydrogenase (EC 1.1.1.18) prepared from *Aerobacter aerogenes* according to Weissbach[16] and D-galactose with D-galactose dehydrogenase (EC 1.1.1.48) according to Wallenfels and Kurz.[17]

Enzyme Purification

All procedures are carried out at about 4°.

Step 1. Crude Extract. Eighty grams of ripe seeds from the dwarf bean (*Phaseolus vulgaris*) are homogenized in 240 ml of 0.1 M sodium phosphate buffer, pH 7.0, with a blender (Ultra Turrax) for 2 minutes. The brei is squeezed through cheese cloth and centrifuged for 10 minutes at 17,000 g.

Step 2. Ammonium Sulfate Precipitation. The supernatant is subjected to ammonium sulfate fractionation. A cold, saturated, neutral solution is added successively until a concentration of 35 and 50% saturation, respectively, is reached. Each fraction is stirred for 30 minutes and then cleared by centrifugation. The pellet arising between 35 and 50% is dissolved in a minimum of $0.1\,M$ phosphate buffer, pH 7.0, and dialyzed for 18 hours in $0.05\,M$ phosphate buffer, pH 7.3.

Step 3. Calcium Phosphate Gel Fractionation. To the redissolved fraction obtained between 35 and 50% $(NH_4)_2SO_4$ saturation calcium phosphate gel, prepared according to Kunitz[19] is slowly added until a ratio of 80 mg of gel per 100 mg of protein is reached. After 15 minutes the gel is centrifuged off; the activity stays in the supernatant. The gel is rewashed with 30 ml of $0.05\,M$ phosphate buffer, pH 7.0.

Step 4. Ammonium Sulfate Fractionation. From a second ammonium sulfate precipitation the fraction between 42 and 50% saturation is used. Generally a 20- to 30-fold purification is achieved.

A typical purification is summarized in Table II.

Properties

Stability. The enzyme is stable for several months when kept unfrozen at $0°$.

pH Optimum. The pH optimum for the formation of stachyose determined in phosphate buffer is between pH 6.0 and 7.0.

Specificity. The enzyme exhibits a high acceptor specificity. No transfer has been observed to glycerol, fructose, sucrose, maltose, cellobiose, trehalose, gentiobiose, melezitose, stachyose. Slight acceptor activity is obtained with glucose (the product is identical to melibiose), galactose, and lactose. As galactosyl donor, only galactinol works of the physiological substances tested. No transfer from UDP-galactose on ADP-galactose can be shown.

Since stachyose does not serve as an acceptor, the enzyme from *Phaseolus vulgaris* does not catalyze the synthesis of verbascose. This agrees with the observation that *P. vulgaris* seeds contain only traces of verbascose, whereas stachyose is the main oligosaccharide.

Kinetics. The K_m value for galactinol is $7.3 \times 10^{-3}\,M$ in the presence

TABLE II

PURIFICATION PROCEDURE FOR GALACTINOL:RAFFINOSE 6-GALACTOSYLTRANSFERASE

Fraction	Total protein (mg)	Total activity (μmoles/hour)	Specific activity (μmoles/hour/mg)
Supernatant, 17000 g	8490	131.5	0.015
$(NH_4)_2SO_4$, first	1235	83.8	0.067
Calcium phosphate gel	360	80.0	0.222
$(NH_4)_2SO_4$, second	101	37.4	0.370

[19] M. Kunitz, *J. Gen. Physiol.* **35**, 423 (1952).

of 25 mM raffinose. The K_m value for raffinose is $8.4 \times 10^{-4} M$ when 0.3 mM [^{14}C]galactinol is present. An equilibrium constant, K (stachyose) (myoinositol)/(raffinose) (galactinol) of 4 has been obtained.

Effect of SH Compounds. The transfer is very sensitive to sulfhydryl poisons: 10 μM p-chloromercuribenzoate inhibits to 37%, and 50 μM almost completely.

Purification of an Enzyme from *Vicia faba* Catalyzing the Synthesis of Stachyose and Verbascose

Assay Method

The enzyme activity is measured in the same way as described for stachyose synthesis, with the only exception that stachyose is used as the acceptor instead of raffinose. Both tests can be used; the conditions are identical to those given above.

Preparation and Purification

Step 1. Crude Extract. One hundred grams of ripe seeds from *Vicia faba* are homogenized with an Ultra Turrax blender for 3 minutes with 240 ml of 0.1 M sodium phosphate buffer, pH 7.0. The extract is filtered through cheesecloth and the filtrate is centrifuged at 17,000 g for 10 minutes.

Step 2. Ammonium Sulfate Precipitation. To the supernatant, cold, saturated neutral ammonium sulfate solution is slowly added to yield a 33% saturation. After 30 minutes the precipitate is removed by centrifugation and the supernatant is brought to 50% saturation. The precipitate obtained, is taken up in 0.1 M sodium phosphate buffer pH 7.0 and dialyzed overnight against 0.05 M sodium phosphate buffer, pH 7.0.

Step 3. Calcium Phosphate Gel Fractionation. Calcium phosphate gel[19] is added (80 mg/100 mg of protein). The mixture is stirred for 15 minutes and centrifuged. The pellet is rewashed with 0.05 M phosphate buffer, and the supernatants containing the activity are pooled.

Step 4. Ammonium Sulfate Fractionation. A second ammonium sulfate precipitation yields further purification. The fraction between 40% and 48% saturation contains most of the activity and is used without further dialysis. A purification of 40- to 50-fold is achieved.

A typical purification is summarized in Table III.

Properties

pH Optimum. Maximum rate of verbascose synthesis occurs at pH 6.3.

Specificity. Besides stachyose, the best galactosyl acceptors are

TABLE III

PURIFICATION OF AN ENZYME FROM *Vicia faba* CATALYZING THE SYNTHESIS
OF STACHYOSE AND VERBASCOSE

Fraction	Total protein (mg)	Total activity (μmoles/hour)	Specific activity (μmoles/hour/mg)
Supernatant, 17,000 g	12,444	57	0.0046
$(NH_4)_2SO_4$, first	1,188	55	0.0460
Calcium phosphate gel	313	32	0.1020
$(NH_4)_2SO_4$, second	39	9.8	0.2510

raffinose and melibiose. For these sugars the transfer from galactinol is better than the hydrolysis of galactinol. As compared to raffinose as the best acceptor, less than 5% transfer to lactose, D-galactose, and D-glucose takes place. There is no transfer to glycerol, fructose, glucose 1-phosphate, glucose 6-phosphate, sucrose, maltose, cellobiose, trehalose, gentiobiose, melezitose. Verbascose itself does not serve as a galactosyl acceptor, which means that the enzyme cannot catalyze the biosynthesis of ajugose the following member of the raffinose family. *Vicia faba* seeds contain verbascose as the main oligosaccharide.

Kinetics. The K_m value for galactinol with raffinose as acceptor (2.5 × $10^{-2} M$) is 1.1 × $10^{-2} M$. The K_m for raffinose and stachyose is 8.5 × 10^{-4} and 3.3 × $10^{-3} M$, respectively; the determinations were carried out in the presence of 0.3 mM ^{14}C-galactinol.

Inhibition. A strong mutual inhibition of raffinose and stachyose is observed when both substrates are present, indicating that one and the same enzyme catalyzes both the synthesis of stachyose and of verbascose.

[73] Glycogen Synthetase (Synthase) from Skeletal Muscle: *D* and *I* Forms[1]

By CARL H. SMITH, CARLOS VILLAR-PALASI, NORMAN E. BROWN, KEITH K. SCHLENDER, ADDISON M. ROSENKRANS, and JOSEPH LARNER

$$\text{UDP-glucose}^{2-} + (\text{glucose})_n \rightarrow \text{UDP}^{3-} + (\text{glucose})_{n+1} + \text{H}^+$$

UDPG:α-1,4-glucan, α-4-glucosyltransferase (glycogen synthase), is present in two interconvertable forms in most organisms and tissues.[2]

[1] Supported by grants from the National Institutes of Health, U.S. Public Health Service AM 14334 and NIAMD AM 14436.
[2] J. Larner and C. Villar-Palasi, *Curr. Top. Cell. Regul.* **3**, 195 (1971).

These two forms differ in their state of phosphorylation and in their dependence on glucose 6-phosphate for activity. In skeletal muscle, in the presence of an anion activator, the completely phosphorylated (D) form is totally dependent on glucose 6-phosphate, and the dephosphorylated (I) form is totally independent.[3,4]

Assay Method

Principle. Activity measurements are based on the incorporation of uniformly labeled [^{14}C]glucose from UDPG into glycogen primer and determination of the radioactivity after selective precipitation of the labeled glycogen in individual filter paper squares.[5]

Reagents

    (A) UDP-[^{14}C]glucose, 10 mM, 0.03–0.09 μCi/mmole

    (B) Tris·HCl, 0.5 M, pH 7.8: Na$_2$SO$_4$ 0.15 M, EDTA, 0.2 M; KF 0.25 M

    (C) Rabbit liver glycogen, 8%; glycogen solutions are deionized by passage through columns of MB-3 Amberlite, followed by precipitation of the glycogen by ethanol in the presence of a trace of LiBr

    (D) Glucose 6-phosphate, 0.1 M, pH 7.8

    (E) Tris·HCl, 50 mM, pH 7.8: EDTA 5 mM, glycogen 1 mg/ml, and mercaptoethanol 50 mM, prepared daily.

Substrate Mixtures. Probably because of contamination with microorganisms, the background radioactivity of the mixtures of glycogen and [^{14}C]glucose UDPG increases with time. Therefore, it is necessary to store the reagents separately until needed. The assay mixtures are prepared by mixing 8 ml of solution A, 1.2 ml of solution B, 1.5 ml of solution C, and 1.3 ml of either distilled water (for assay of synthase I) or 1.3 ml of solution D (for total synthase $I + D$ assay).

Procedure. Enzyme samples are usually diluted in solution E. If frozen solutions are to be assayed, they should be preincubated 30 minutes at 30° after dilution. To small test tubes containing 60 μl of assay mixture, placed in a water bath at 30°C, 30 μl of the appropriately diluted enzyme is added. A tube containing 60 μl of assay mixture and 30 μl of distilled water is used as blank. After 10 minutes of incubation, a 75-μl aliquot of each tube is transferred to a folded numbered 2 × 2 cm square of Whatman 31 ET filter paper, and immediately dropped into a beaker containing 66% ethanol which has been cooled to −10°. A

[3] N. E. Brown and J. Larner, *Biochim. Biophys. Acta* **242**, 69 (1971).
[4] C. H. Smith, N. E. Brown, and J. Larner, *Biochim. Biophys. Acta* **242**, 81 (1971).
[5] J. A. Thomas, K. K. Schlender, and J. Larner, *Anal. Biochem.* **25**, 486 (1968).

magnetic bar, shielded by a stainless steel screen, is used to agitate the ethanol solution. The papers are washed with three changes of 66% ethanol for 5, 30, and 20 minutes, respectively, using no less than 10 ml of ethanol per paper sample. After this treatment, the papers are rinsed in acetone for 5 minutes, dried under an infrared lamp, and placed in vials containing 0.5% PPO in toluene. Radioactivity is determined in a liquid scintillation counter and enzyme activity calculated from the specific radioactivity of the UDP-[^{14}C]glucose. Radioactive glycogen remains within the paper. Therefore, the filter papers may be removed, the backgrounds briefly recounted, and the vials of scintillant are reused.

One unit of activity is the amount of enzyme that catalyzes the incorporation of 1 μmole of glucose into glycogen in 1 minute under the specified conditions. Determinations in the presence of glucose 6-phosphate represent total synthase activity $(D + I)$. Fractional activity of synthase I is equal to the ratio of activity in the absence of glucose 6-phosphate to total activity.

Purification

General Outline. Glycogen synthase D and glycogen synthase I are prepared by similar procedures. In the initial stage, the enzyme is purified 100- to 600-fold to yield a glycogen protein complex. Conversion to the desired form is accomplished by incubation utilizing the endogenous kinase or phosphatase. This glycogen–enzyme complex, while not homogeneous, is stable and useful for many studies. Fractionation of synthase D to this stage will first be described in detail. The procedure for preparation of synthase I to the glycogen complex will then be outlined with the points of difference from the preparation of the D form given in detail. In contrast to other procedures which have been described,[6] the techniques given here are designed to stabilize the enzyme by keeping it in the presence of glycogen as much as possible.

A virtually homogeneous preparation of either form may be obtained from the glycogen complex by amylase digestion and gel filtration. These procedures are the same for the two forms and are described at the conclusion of this section.

Buffer Solutions

TE = Tris·HCl 50 mM, EDTA 5 mM, pH 7.8

TEM = Tris·HCl 50 mM, EDTA 5 mM, mercaptoethanol 50 mM (pH 7.8 unless otherwise specified)

---

[6] T. R. Soderling, J. P. Hickenbottom, E. M. Reiman, F. L. Hunkler, D. H. Walsh, and E. G. Krebs, *J. Biol. Chem.* **245**, 6317 (1970).

TEM-sulfate = TEM buffer with 2 mM Na$_2$SO$_4$

TEF = Tris·HCl 50 mM, EDTA 5 mM, KF 100 mM, pH 8.2

Solutions are conveniently prepared as a 10-fold concentrated stock mixture. The pH of all Tris solutions is measured in such a stock at room temperature unless otherwise specified. Mercaptoethanol should be added immediately before use.

Glycogen Synthase D, Glycogen Complex[3]

Initial Fractionation. A rabbit is anesthetized with intravenous pentobarbital, killed, and bled. The hind leg and back muscles are removed and cut into 2-cm pieces. Unless otherwise described, all subsequent steps are performed at 4° in the cold room.

The muscle is homogenized 90 seconds with 2.5 volumes of TEF buffer in a Waring Blendor at medium speed. The thick homogenate is then centrifuged 60 minutes at 10,000 g in a large capacity head of a refrigerated centrifuge; the supernatant is filtered through glass wool.

The filtered supernatant solution is then cooled to 0° to 3° in an ice-salt bath. With a thermometer in the solution and with constant stirring, chilled alcohol ($-70°$) is added rapidly to give a final concentration of 30% by volume. The temperature should not rise to more than 9° during addition. The solution is allowed to stand approximately 10 minutes in the ice-salt bath, while the temperature falls to $-5°$ to 0°. The preparation is then centrifuged 15–20 minutes at 10,000 g at $-10°$.

The supernatant, which should be pink and nearly clear, is discarded. The precipitate is redissolved in TEF solution. For this purpose, it is convenient to use a motor-driven Teflon–glass homogenizer. The redissolved precipitate is diluted to a volume of 300 ml per 600 g of rabbit muscle. The mixture is centrifuged 150 minutes at 75,000 g in a vacuum ultracentrifuge, and the supernatant is discarded. The precipitated glycogen pellet may be stored at $-70°$ for as long as 4 months. Material from several rabbits may be combined during the ethanol precipitation, or at this step.

Column Chromatography. A DEAE-cellulose column is prepared and washed with TEM buffer solution. The bed volume should be 120 ml per 600 g of rabbit muscle. The glycogen pellets are dissolved with homogenization in TEM buffer. The resulting thick tan suspension is diluted to a volume of 250 ml per 600 g of rabbit muscle, and incubated 30 minutes at 30° to solubilize the glycogen–protein complex. The suspension is then cooled and centrifuged 30 minutes at 50,000 g to remove coarse material that will otherwise clog the DEAE column. The slightly opalescent supernatant is applied to the column. The column is then washed with two or more volumes of TEM buffer and then with 10 bed volumes of 0.1 M NaCl in TEM solution to remove phosphorylase. Without this exhaustive washing, phosphorylase will persist into the final preparation.

Synthase and synthase I kinase are eluted from the column with 0.3 M NaCl in TEM, and an ultraviolet monitor is used to locate the protein peaks. Since the kinase binds the DEAE-cellulose more tightly than synthase, it is important to continue the elution until the absorbance no longer decreases. The dilute solution is usually iridescent due to the endogenous glycogen in the enzyme complex. Glycogen is added to give a concentration of 0.25–0.5 mg/ml beyond that present originally. Synthase and synthase kinase are precipitated together with ethanol, as described earlier, with a final concentration of 30 volumes percent.

Conversion to the Dependent Form. The sedimented precipitate is redissolved in TEM solution to give approximately 10 ml per 600 g of muscle tissue. MgCl₂, ATP, and cyclic AMP are added to final concentrations of 12 mM, 5 mM, and 0.01 mM, respectively. (The effective concentration of Mg^{2+} will be only 7 mM owing to chelation with EDTA.) As eluted from the column, the enzyme is approximately 70% dependent. For conversion to total dependence, the preparation is incubated 48–72 hours at 4°, and the extent of conversion is monitored by assay. Although significant conversion will occur within 12–16 hours, incubation is terminated only when the preparation reaches 95–99% dependence. The incubation temperature should not be appreciably lower than 4°, or conversion will be too slow. Incubation at higher temperatures for shorter periods results in incomplete conversion.

Final Ethanol Precipitations. After conversion, the mixture may be diluted 2-fold with TEM solution. Using a final concentration of 15 volumes percent, the enzyme is precipitated with ethanol as described previously. The precipitate is collected by centrifugation at −10° and redissolved in TEM solution. This procedure is repeated with the same ethanol concentration. The resulting final precipitate is redissolved in TEM buffer, pH 7.8, in a volume of about 5 ml per 600 g of original rabbit muscle and is dialyzed several hours against 200 volumes of TE buffer. Relative purification at various stages is shown in Table I.

Glycogen Synthase I Glycogen Complex

Initial Fractionation. This procedure is performed exactly as described for synthase D up to the initial ethanol precipitation, except that TE buffer is used in place of TEF buffer to avoid inhibiting the synthase phosphatase.

Conversion to the Independent Form. The fresh ethanol precipitate

TABLE I
PURIFICATION OF GLYCOGEN SYNTHASE D FROM SKELETAL MUSCLE[a]

Step	Total activity	Specific activity[b]	Recovery (%)	Purification (-fold)
Homogenate	2156	0.018	100	—
Initial supernatant	2010	0.037	93	2
First ethanol (30%)	1285	0.092	60	5
Pellet (78,000 g)	732	0.110	34	6
0.25 M NaCl eluate	535	2.70	25	149
Second ethanol (30%)	674	5.19	31	285
Final ethanol (15%)	357	12.01	17	660

[a] Muscle, 1500 g, was treated as described in the text.
[b] Units of enzyme activity per milligram of protein.

is resuspended with homogenization in TEM sulfate buffer, pH 6.8, to give a volume of 250 ml per 600 g of rabbit muscle. The thick suspension is transferred to wide dialysis sacs and dialyzed 7 hours at 4° against 15 volumes of the same buffer solution. The bath is changed once during the procedure. During this time, conversion to the glucose 6-phosphate independent form (95–100%) will occur. This should be checked by assay before proceeding. Incubation for appreciably longer times will lead to losses of activity. Incubation is terminated by adding TEM-sulfate buffer, pH 8.2, to bring the pH to 7.8.

Column Chromatography. A DEAE column is washed with TEM-sulfate buffer. The preparation is incubated 30 minutes at 30° to solubilize the enzyme and centrifuged 30 minutes at 50,000 g at 4° for clarification.

The supernatant is applied to the column which is washed with TEM sulfate and 10 bed volumes of 0.1 M NaCl in TEM sulfate. The enzyme is eluted from the column with 0.25 M NaCl in TEM sulfate buffer. To minimize contamination with synthase kinase, only the major absorbance peak is saved.

Final Fractionation. Immediately after elution from the column, glycogen (0.25 mg/ml) is added to stabilize the enzyme. It is then fractionated twice with 15% ethanol. The resulting final precipitate is treated as described for synthase D. Relative purification at various stages is shown in Table II.

Amylase Digestion and Gel Filtration[3,4]

The glycogen protein complex may be purified to homogeneity by amylase digestion and gel filtration. After removal of the glycogen, the enzyme protein tends to form high molecular weight aggregates with

TABLE II

PURIFICATION OF GLYCOGEN SYNTASE I FROM SKELETAL MUSCLE[a]

Step	Total activity	Specific activity[b]	Recovery (%)	Purification (-fold)
Homogenate	639	0.015	—	—
Initial supernatant	588	0.030	92	2.0
First ethanol (30%)	565	0.11	88	7.3
After conversion	506	0.091	79	6.1
Before DEAE	347	0.083	54	5.5
0.25 M NaCl eluate	183	1.1	29	73
Final ethanol (15%)	228	5.2	36	347

[a] Rabbit muscle, 405 g, was treated as described in the text.
[b] Units of enzyme activity per milligram of protein.

low specific activity. Such aggregation decreases the yield of these steps.

The amylase digestion and gel filtration may each be performed either at room temperature or in the cold room. The comparative advantages of each alternative procedure is discussed below.

The glycogen enzyme complex (200 units of activity, 20–40 mg of protein) is diluted to 10 ml with TEM pH 6.8. Salivary amylase[7] (200–500 μg of protein) is added and the mixture is dialyzed against 1 liter TEM, pH 6.8. The progress of the digestion may be followed by testing the contents of the bath qualitatively for carbohydrate.[8] Incubation is continued 12–20 hours in the cold room or 3–5 hours at room temperature. The bath is changed 2–4 times during incubation. Carbohydrate release should diminish but may not disappear entirely. The contents of the sac will become distinctly lighter, but a gray color persists.

After incubation, the contents of the sac are transferred to a tube and the pH is adjusted to 7.8 with 2 M Tris base. The preparation is incubated 30 minutes at 30° to solubilize as much protein as possible and may be centrifuged 5 minutes in the clinical centrifuge to remove remaining insoluble material.

A Sepharose 4 B column 2.5 × 85 cm is equilibrated with TEM buffer for preparations of synthase D or with TEM sulfate for the preparation of synthase I. The material is then applied to the Sepharose column, and fractions of 5 ml are collected Aggregated enzyme and other material are eluted at the void volume (0.25 bed volume). This material is cloudy, of low specific activity, and quite heterogeneous. Active enzyme follows at 0.5–0.7 bed volume. This may produce a slight deflection of

[7] P. Bernfeld, Vol. 1 [17].
[8] M. Dubois, K. A. Gilles, J. K. Hamilton, P. A. Rebers, and F. Smith, *Anal. Chem.* **28**, 350 (1956).

an ultraviolet monitor but must be detected by assay. After the active enzyme, nucleotide previously bound to the enzyme is eluted. Enzyme eluted from the Sepharose column has been satisfactory for a variety of studies. However, should additional fractionation be required, Sephadex G-200 may be used.[3]

For maximum specific activity, the amylase digestion and gel filtration (especially the latter step) should be performed in the cold room. The degree of aggregation, however, is greater at lower temperatures, and this factor will decrease the yield.

We have observed that glycogen synthase I tends to aggregate more readily than synthase D.[9] Synthase I may be difficult to fractionate at cold temperatures, and the yield will be very low (see Table III). Therefore, if maximum specific activity is not required, it is desirable to perform the digestion and gel filtration steps at room temperature in the preparation of synthase I.

To concentrate the enzyme, the fractions comprising the eluted activity peak are transferred to a dialysis sac which is placed in a cylinder of powdered Ficol in the cold room. After 18–24 hours the volume has decreased to approximately 20% of the original. If it will not interfere with further studies, it is desirable to add glycogen to the preparation in the dialysis sac during the concentration step. After concentration, the sac is then retied, transferred to a bath, and dialyzed 4–6 hours against TE, pH 7.8, to remove mercaptoethanol. The preparation is then stored at $-75°$.

Properties

Composition and Stability. The two forms of the enzyme as obtained before treatment with amylase are fairly stable for 4–6 months when stored at $-75°$. They are heterogeneous in protein composition and con-

TABLE III
AMYLASE DIGESTION AND GEL FILTRATION

Sample	% Yield[a]	Specific activity[b]
Synthase D, 4°	45	15–20
Synthase I, 4°	5	15–18
20°	35	10

[a] Based on total of glycogen–enzyme complex subjected to digestion.
[b] Units of enzyme activity per milligram of protein.

[9] C. H. Smith and J. Larner, in preparation.

tain in addition a relatively large amount of glycogen. However, they appear to be essentially free of phosphorylase and synthase converting enzymes.

Essentially homogeneous preparations of both forms of synthase are obtained by amylase treatment and gel filtration. Large losses of activity occur, however, during this treatment, and the final products are rather unstable unless glycogen is added to the enzyme solutions.

Chemistry and Physical Structure. The highly purified forms of the synthase show only one band in electrophoresis in sodium dodecyl sulfate. The molecular weight of this enzyme subunit is approximately 90,000[4,6] as calculated by the method of Weber and Osborn.[10] The molecular weight of the intact enzyme has been reported as 250,000[3] and 400,000.[6] The D form of synthase contains 6 moles of alkali labile (serine) phosphate per 90,000 g. Little or no labile phosphate was found in the I form.[4] Both forms of muscle synthase contain 6 thiol groups per subunit of 90,000 molecular weight.[4] Recent electron microscopy studies of the purified I form demonstrated flattened hexagonal and ring forms.[11] Stacks of four rings were frequently seen.

Chymotryptic peptides of ^{32}P-labeled synthase D purified to this stage show clear differences with the chymotryptic peptides of ^{32}P-labeled phosphorylase a.[12] Therefore, while the hexapeptide surrounding the phosphorylated serine group of both enzymes is identical,[13] this identity does not extend to larger fragments.

Kinetics. The D form is totally inactive in the absence of glucose 6-phosphate. Both D and I, however, are activated by glucose 6-phosphate. The I form is also activated by sulfate and a variety of anions (the basis for the inclusion of sulfate in the recommended assay mixtures).[14] Both forms are inhibited by adenine nucleotides; the D to a considerably greater extent than the I.[15] Studies of the two substrate kinetics have been performed for the D form and have indicated a "Ping-Pong" type mechanism.[3] In the presence of glucose 6-phosphate, the K_m of UDPG ranged from 0.08 to 0.2 mM and that of glycogen from 1.32×10^{-5} to 5.7×10^{-4} mg/ml, depending on the concentration of the co-substrate.

Synthase I has been studied only as the glycogen-bound enzyme.[16]

[10] K. Weber and M. Osborn, *J. Biol. Chem.* **244**, 4406 (1970).

[11] L. Rebhun, C. Smith, and J. Larner, in preparation (1972).

[12] A. Rosenkrans and J. Larner, in preparation (1972).

[13] J. Larner and F. Sanger, *J. Mol. Biol.* **11**, 491 (1965).

[14] J. A. Thomas, K. K. Schlender, and J. Larner, in preparation.

[15] R. Piras, L. B. Rotham, and E. Cabib, *Biochemistry* **7**, 56 (1968).

[16] K. K. Schlender and J. Larner, in preparation.

Two kinetics states distinguished by their K_m for UDPG, were observed. In the absence of sulfate, values of 0.11 mM and 0.72 mM were obtained. Sulfate lowered the K_m's of the two forms to 0.05 mM and 0.27 mM, and increased the $V_{\max}$ of each form approximately 2-fold.

[74] Glycogen Synthetase from *Escherichia coli* B

By JEFFREY FOX, SYDNEY GOVONS, and JACK PREISS

Assay Method

Principle. Enzymatic activity is determined by measuring the transfer of [^{14}C]glucose from ADP-[^{14}C]glucose to glycogen. The glycogen is isolated by methanol precipitation, redissolved in H_2O, and radioactivity estimated by liquid scintillation spectrometry. (The procedure given below corresponds to standard primed conditions.)

Reagents

ADP-[^{14}C]glucose, 7 mM (specific activity 500 cpm/nmole)
Rabbit liver glycogen (RLgly), 10 mg/ml
Tris·acetate buffer, 1 M, pH 8.5
Potassium acetate, 0.5 M
Magnesium acetate, 0.01 M
Glutathione (GSH), 0.2 M
Bovine serum albumin (BSA), 10 mg/ml
Methanol, 75%–KCl, 1%
Aquasol (obtained from New England Nuclear, Boston, Massachusetts)

Procedure. The ADP-[^{14}C]glucose (0.02 ml), RLgly (0.05 ml), Tris·acetate buffer (0.01 ml), KAc (0.01 ml), MgAc$_2$ (0.01 ml), GSH (0.01 ml), and BSA (0.01 ml) are combined. The reaction is initiated with glycogen synthetase (10^{-4} to 2×10^{-3} unit) in a final volume of 0.2 ml. After incubation at 37° for 15 minutes, the reaction is terminated with the addition of 2 ml of 75% MeOH–1% KCl. After 5 minutes, the mixture is centrifuged for 5 minutes in a clinical centrifuge, and the supernatant fluid is discarded. The glycogen precipitate is washed once with 2 ml of 75% MeOH–1% KCl, decanted, and allowed to drain. The glycogen is dissolved in 1 ml of H_2O and a 0.5 ml portion is mixed with 5 ml of Aquasol and counted in a liquid scintillation spectrometer.

One unit of enzyme activity is that amount of enzyme catalyzing the transfer of 1 μmole of glucose per minute under the above conditions. Specific activity is units per milligram protein (as determined by Lowry et al.[1]).

Activity Stain Method

This method is a qualitative means for localizing the enzyme on poly acrylamide gels.

Principle. Glycogen synthetase activity is visualized by coupling production of ADP to myokinase, hexokinase, and glucose-6-phosphate dehydrogenase to reduce nitroblue tetrazolium (NBT).

Reagents

ADP-glucose, 10 mM
Rabbit liver glycogen, 10 mg/ml
Tris·acetate buffer, 1 M, pH 8.0
MgCl$_2$, 0.1 M
Potassium acetate, 0.5 M
Glucose, 0.1 M
TPN, 10 mg/ml
Phenazine methosulfate (PMS), 1 mg/ml
Nitroblue tetrazolium (NBT), 5 mg/ml
Myokinase (dialyzed), 80 IU/ml (pig muscle, 360 IU/mg)
Hexokinase, 1000 IU/ml (Sigma Type F-300)
Glucose-6-phosphate dehydrogenase (ammonium sulfate free) 1000 IU/ml

Procedure. The ADP-glucose (0.20 ml), RLgly (0.50 ml), Tris·acetate (0.20 ml), MgCl$_2$ (0.01 ml), potassium acetate (0.2 ml), glucose (0.10 ml), TPN (0.040 ml), PMS (0.10 ml), NBT (0.10 ml), myokinase (0.010 ml), hexokinase (0.005 ml), and glucose-6-phosphate dehydrogenase (0.010 ml) are mixed in a final volume of 2.0 ml. The enzyme activity is localized by incubating the polyacrylamide gel (0.5 × 5 cm) with the reaction mixture in a small test tube at 37°. Dye bands form within a few minutes when approximately 0.5 unit of enzyme is present. The gel is rinsed several times with water to remove any excess dye.

Growth of Cells

The *E. coli* B strain used was a mutant, SG3, derepressed in the levels of ADP-glucose pyrophosphorylase and glycogen synthetase. Cells were grown on minimal media containing per liter: 11 mg of CaCl$_2$, 246

---

[1] O. H. Lowry, N. J. Rosebrough, A. C. Farr, and R. J. Randall, *J. Biol. Chem.* **193**, 265 (1951).

mg of $MgSo_4 \cdot 7H_2O_4$, 2.4 grams $(NH_4)_2SO_4$, 6.8 grams of KH_2PO_4, 14.2 grams of Na_2HPO_4 and 12 grams of D-glucose. The media was adjusted to pH 7.2 with NaOH. The cells were grown at 37°, harvested 3 hours after stationary phase was attained and stored at $-12°$.

Purification Procedure

Step 1. Extraction of the Enzyme. Thawed cells (450 g) are combined with 1200 g of chilled Superbrite 100 (Minnesota Mining and Manufacturing Company) glass beads and 360 ml of 0.05 M glycylglycine buffer, pH 7.0, that contains 5 mM DTT. This mixture is homogenized at medium speed in a Waring Blendor for a total of 30 minutes with intermittent cooling in a rock salt–ice bath to keep the temperature below 20°. Another 1350 ml of glycylglycine buffer is added, and the mixture is slowly stirred for 10 minutes. The glass beads are allowed to settle for 5 minutes, and the supernatant solution is decanted. The glass beads are rinsed again as above with 900 ml of the buffer. Both supernatant fluids are combined, and this constitutes the crude extract.

Step 2. Isolation of 70,000 g Particulate Fraction. The crude extract is centrifuged at 16,000 g for 10 minutes to remove unbroken cells and the remaining glass beads. The supernatant fluid that is obtained is centrifuged at 70,000 g for 90 minutes. The precipitate is resuspended in 0.05 M potassium phosphate buffer, pH 6.8, containing 5 mM DTT and 0.1 M NaCl with the aid of a Waring Blendor. The volume is about one-fourth that of the 16,000 g supernatant solution.

Step 3. α-Amylase Treatment. The 70,000 g precipitate is treated with α-amylase to solubilize the glycogen synthetase. The α-amylase used (porcine pancreas, phenylmethylsulfonylfluoride treated, Worthington, Freehold, New Jersey) must be treated with diisopropylfluorophosphate in order to inhibit a proteolytic contaminant that is phenylmethylsulfonylfluoride insensitive. Two international units of α-amylase are added per milliliter of resuspended 70,000 g precipitate. The resulting mixture is incubated for 30 minutes at 37° and then for 12–14 hours overnight in the cold room with gentle shaking. Next, the mixture is diluted 1:4 with 0.05 M Tris·HCl buffer, pH 7.5, that contains 5 mM DTT. The diluted mixture is centrifuged at 70,000 g for 90 minutes. The glycogen synthetase is now found in the supernatant fluid fraction. Attempts to solubilize this enzyme with salts, chaotropic agents,[2] and detergents have not been successful.

Step 4. Ammonium Sulfate Fractionation. The 70,000 g supernatant fluid from step 3 is made 5% in sucrose by adding solid (90 g) sucrose.

[2] Y. Hatefi and W. G. Hanstein, *Proc. Nat. Acad. Sci. U.S.* **62**, 1129 (1969).

Crystalline ammonium sulfate is added to give 60% saturation, and the mixture is stirred for at least 20 minutes and centrifuged at 23,000 g for 15 minutes. The precipitate is dissolved in 0.05 M Tris·HCl buffer, pH 7.5, containing 5 mM DTT and 5% sucrose and dialyzed against 50 volumes of the same buffer overnight. Insoluble material is removed from the dialyzed solution by centrifugation at 23,000 g for 10 minutes.

Step 5. DEAE-Cellulose Chromatography. The supernatant solution from step 4 is adsorbed onto a DEAE-cellulose column (2.5 × 40 cm, Whatman DE-52) which is equilibrated with 0.05 M Tris·HCl buffer, pH 7.5, containing 3 mM DTT and 5% sucrose. The column is washed with 400 ml of this buffer. The enzyme is eluted with a linear gradient composed of 1 liter of the same buffer in the mixing chamber and 1 liter of that buffer containing 0.6 M NaCl in the reservoir chamber. Fractions containing the enzyme activity are pooled, reduced to 50% of the pool volume by ultrafiltration (PM-30 membrane, Amicon), and further concentrated by precipitating the enzyme by addition of solid ammonium sulfate to 75% saturation. The precipitate is dissolved in 0.05 M Tris·HCl buffer, pH 7.5, containing 3 mM DTT and 5% sucrose and dialyzed against 50 volumes of this buffer overnight. The resulting insoluble material is removed by centrifugation. Most of the remaining α-amylase added in step 3 is separated from the glycogen synthetase in this step.

Step 6. Acid Precipitation at pH 5.7. The concentrated, dialyzed enzyme from step 5 is titrated to pH 6.1–6.2 with approximately an equal volume of 1 M imidazole·HCl buffer, pH 5.7. This mixture is dialyzed against 50–100 volumes of 0.06 M imidazole·HCl, pH 5.7, containing 3 mM DTT, and 5% sucrose for a period of 4–12 hours, during which time a heavy precipitate forms. The mixture is removed from the dialysis bag and centrifuged at 27,000 g for 15 minutes. The supernatant fluid is decanted, and the precipitate is rinsed with a small amount of the above 0.06 M imidazole·HCl buffer, pH 5.7, and centrifuged again. The second supernatant fluid is removed. The precipitate which contains the enzyme is redissolved in 0.05 M Tris·HCl buffer, pH 7.5, containing 3 mM DTT and 5% sucrose and dialyzed overnight against this buffer. Insoluble material is removed by centrifugation.

Step 7. Sephadex G-200 Chromatography. The enzyme is further purified by gel filtration. A Sephadex-G-200 (Pharmacia) column, 40 × 2.5 cm, is used. The column is equilibrated with 0.05 M Tris·HCl buffer, pH 7.5, that contains 5 mM DTT and 5% sucrose. The enzyme fraction from step 6 is added to the column and then eluted with this buffer. The glycogen synthetase is considerably diluted (from 5 to 15-fold) on Sephadex, suggesting that the enzyme has some affinity for the dextran.

Fractions containing enzyme are pooled and concentrated by ultrafiltration (Amicon, PM-30 membrane).

Step 8. Second Acid Precipitation. The Sephadex G-200 concentrate is dialyzed against 50–100 volumes of 0.06 M imidazole·HCl buffer, pH 5.7, containing 3 mM DTT and 5% sucrose for a period of 4–12 hours. The precipitate that forms is collected by centrifugation and treated as in step 6 above. This acid precipitation has been repeated a third time without an increase in specific activity.

A typical purification is summarized in the table.

Properties of the Enzyme

Purity. At this state, the glycogen synthetase is freed of contaminating traces of α-amylase (<0.1 unit, total). Also, most of the *E. coli* branching enzyme[3] has been removed. Preparations with higher specific activity (40–80) units/mg have been obtained with this procedure starting with smaller quantities of *E. coli* B cells. A single activity band was seen on acrylamide gels using the Ornstein–Davis system.[4] Several protein bands staining with Coomassie Blue were present in addition to that corresponding to the activity band, indicating that the enzyme was not pure.

A preliminary determination of the molecular weight has been made using the technique of sucrose density gradient centrifugation according to Martin and Ames.[5] Rabbit muscle pyruvate kinase (molecular weight

PURIFICATION OF GLYCOGEN SYNTHETASE FROM *Escherichia coli*

Step	Fraction	Volume (ml)	Protein (mg)	Specific activity (units/mg)	Total activity (units)
1.	Crude extract	2440	74,000	0.1	7400
2.	16,000 g supernatant	2310	62,500	0.11	6930
	70,000 g precipitate	2000	28,000	0.21	6150
3.	70,000 g supernatant	1830	4,060	1.23	4600
4.	Ammonium sulfate	57	1,615	2.02	3300
5.	DEAE-cellulose	45	666	3.15	2100
6.	Acid precipitate	11.2	239	7.5	1820
7.	Sephadex G-200	7.1	104	12.5	1310
8.	Second acid precipitate	8.4	48	18.4	880
	Third acid precipitate	5.2	43	18.1	780

[3] N. Sigal, J. Cattaneo, J. P. Chambost, and A. Favard, *Biochem. Biophys. Res. Commun.* **20**, 616 (1965).

[4] L. Ornstein and B. Davis, *Ann. N.Y. Acad. Sci.* **121**, 321 (1964).

[5] R. G. Martin and B. N. Ames, *J. Biol. Chem.* **236**, 1372 (1961).

is 237,000) and beef heart lactate dehydrogenase (molecular weight is 135,000) were used as standards. The molecular weight of the glycogen synthetase was approximately 115,000 by this method.

Reaction without Primer. The enzyme catalyzes the synthesis of a polyglucoside from ADP-glucose in the absence of primer. The reaction is similar to those described by Gahan and Conrad[6] in *Aerobacter aerogenes* and by Ozbun *et al.*[7,8] in spinach leaves. In the present system, the enzyme is stimulated by 0.25 M sodium citrate and BSA. Other proteins can substitute for BSA. The reaction is characterized by an initial lag phase which is followed by a linear polymerization phase.

Specificity. The glycogen synthetase is specific for ADP-glucose and deoxy-ADP-glucose.[9] The K_m's were 2.5 × 10^{-5} M and 3.8 × 10^{-5} M, respectively. The V_{max} for deoxy-ADP-glucose was 72% of that observed for ADP-glucose. The enzyme will use α-1,4-glucans as primers from various sources, including *E. coli*, shellfish, and rabbit liver glycogen, corn amylopectin, amylose, and potato starch. Maltose, maltotriose, and higher α-1,4-oligosaccharides will also serve as primers.

Inhibitors. ADP appeared to inhibit by competing with ADP-glucose. The K_i for ADP was 1.5 × 10^{-5} M.[9] Deoxy-ADP-glucose also was a competitive inhibitor for ADP-glucose with a K_i of 3.4 × 10^{-5} M.[9]

Activators. Various salts stimulate the enzyme as much as 2.5-fold. These include ammonium sulfate, sodium fluoride, sodium citrate, potassium phosphate (all at 0.25 M), and EDTA (0.125 M).

pH Optimum. The optimum is in the alkaline range extending from pH 7 to 9.5.

Stability. The particulate enzyme is stable for at least 6 months when kept on ice. The solubilized enzyme is much less stable and begins to lose activity after several weeks at 0°. Its lifetime may be extended at this temperature by periodic additions of DTT. The solubilized enzyme could be stored for at least 4 months (after rapid freezing in an ethanol–Dry Ice bath) at −80° and withstands repeated freezing and thawing.

[6] L. C. Gahan and H. E. Conrad, *Biochemistry* **7**, 9379 (1968).
[7] J. L. Ozbun, J. S. Hawker, and J. Preiss, *Biochem. Biophys. Res. Commun.* **43**, 631 (1971).
[8] J. L. Ozbun, J. S. Hawker, and J. Preiss, *Biochem. J.* **126**, 953 (1972).
[9] J. Preiss and E. Greenberg, *Biochemistry* **4**, 2328 (1965).

[75] Starch Synthetase from Spinach Leaves

By J. L. Ozbun, J. S. Hawker, and Jack Preiss

ADP-glucose + α-glucan → α-1,4-glucosyl glucan + ADP

Assay Method

Principle. Enzymatic activity is determined by following the rate of incorporation of the radioactive glucosyl moiety of ADP-[^{14}C]glucose into a methanol-insoluble product. The reaction rate may also be followed by measuring the amount of ADP formed.

Reagents

Bicine   [N,N-bis(2-hydroxyethyl)glycine]–NaOH   buffer $1 M$, pH 8.5

Potassium acetate, $1 M$

Glutathione, sodium salt, $0.2 M$

EDTA, sodium salt $0.1 M$, pH 7.2

ADP-[^{14}C]glucose, 7 μmoles/ml (specific activity 500 cpm/nmole) prepared as described in this volume [27]

Corn amylopectin, 10 mg/ml

Sodium citrate, $1 M$

Bovine serum albumin (BSA), 10 mg/ml

Methanol, 75%–KCl, 1%.

Procedure. PRIMED REACTION. The Bicine buffer (0.02 ml), potassium acetate (0.005 ml), glutathione (0.01 ml), EDTA (0.01 ml), amylopectin (0.1 ml), and ADP–glucose (0.02 ml) are mixed together in a volume of 0.165 to 0.19 ml. The reaction is then initiated with 0.01–0.045 ml (0.001–0.02 unit) of the transferase fraction (final volume of the reaction mixture is 0.2 ml) and incubated at 37° for 15 minutes. The reaction is terminated by addition of 2 ml of the 75% methanol–1% KCl solution. After waiting 5 minutes for precipitation of the amylopectin, the reaction mixture is centrifuged in a clinical centrifuge for 5 minutes. The supernatant fraction is decanted and the amylopectin precipitate is washed two times with 2 ml of the methanol–KCl solution and then dissolved in 1 ml of H$_2$O (some heating in a boiling water bath may be required to effect complete solution of the precipitate). An aliquot, 0.5 ml, is counted in 10 ml of Bray's solution[1] which has 0.1% BBOT (2,5-bis[2(5-*tert*-butyl-benzoxazolyl)]thiophene) as the scintillation fluor.

---

[1] G. Bray, *Anal. Biochem.* **1**, 279 (1960).

UNPRIMED REACTION. The reaction mixture for the unprimed reaction is the same as for the primed reaction except that potassium acetate and amylopectin are omitted and sodium citrate (0.1 ml) and BSA (0.01 ml) are added. The reaction is stopped by heating for 1 minute in a boiling water bath, and carrier amylopectin (0.1 ml) is added before addition of the methanol–KCl solution. The washing of the amylopectin precipitate and determination of the product formed is the same as described above.

Definition of Unit and Specific Activity. One unit of enzyme as that amount which catalyzes the incorporation of 1 μmole of glucose into polymer per 15 minutes at 37° under the conditions of the assay for the primed reaction. The specific activity is expressed as units of enzyme activity per milligram of protein. Protein concentration is determined by the method of Lowry *et al.*[2] with crystalline bovine serum albumin as a standard.

Application of Assay Method to Crude Tissue Preparations. The method described above for the primed reaction is satisfactory for assay of the crude spinach leaf extracts. However, there is amylase activity present in the crude extract which appears to inhibit the unprimed reaction. This reaction is most sensitive to α-amylase activity.

Purification Procedure

Step 1. Preparation of Crude Extract. Fresh deveined spinach leaves (150 g) from the local supermarket are washed and homogenized in a Waring Blendor with 200 ml of a solution containing 0.1 M phosphate buffer, pH 7.5, 0.01 M EDTA, and 0.005 M GSH for 2 minutes. All operations are carried out at 0–4°. The extract is centrifuged at 45,000 g for 20 minutes, and the supernatant fraction which contains almost all the activity, is used for further purification.

Step 2. Ammonium Sulfate Fractionation. The supernatant fraction from step 1 is made to 40% saturation with solid ammonium sulfate and centrifuged at 30,000 g for 15 minutes. The precipitate is dissolved in 25 ml of 0.05 M HEPES buffer (N-2-hydroxyethylpiperazine-N^1-2-ethanesulfonic acid), pH 7.0, containing 0.005 M dithiotheitol (DDT) and 0.01 M EDTA and dialyzed overnight against the same solution.

Step 3. DEAE-Cellulose Chromatography. The dialyzed solution is placed on a 2.5 × 36 cm DEAE-cellulose column which has been equilibrated with 10 resin bed volumes of a 0.05 M Tris·acetate buffer, pH 8.5, containing 0.005 M EDTA, 0.002 M DDT, and 10% (w/v) sucrose. After absorption of the enzyme solution, the column is washed with

[2] O. H. Lowry, N. J. Rosebrough, A. L. Farr, and R. J. Randall, *J. Biol. Chem.* **193**, 265 (1951).

1 volume of the equilibrating buffer. Then 4 liters of the equilibrating buffer with increasing KCl concentration (linear gradient 0 to 0.2 M KCl) is passed through the column and collected in 26.5-ml fractions. Appropriate fractions (Table I, Fig. 1) are combined and concentrated to 30 ml, using an Amicon micropore ultrafiltrator with a P-30 membrane, and then further reduced to about 2 ml by precipitation with solid $(NH_4)_2SO_4$ (40% saturation). The precipitates are dissolved in 0.05 M HEPES buffer, pH 7.0, containing 0.005 M DTT, 0.01 M EDTA, and 10% (w/v) sucrose and dialyzed against 1 liter of the same buffer overnight.

The yields and specific activities obtained in each fraction of a typical purification are summarized in Table I.

Properties of the Transferase Fractions

Kinetics. The 4 fractions containing transferase activity have been studied with respect to K_m for primers and for ADP-glucose.[3] These are summarized in Table II. The apparent affinities for fraction III for

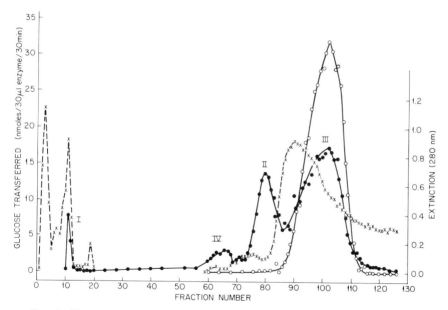

FIG. 1. Chromatography of starch synthetases from spinach leaves on DEAE-cellulose. Fractions were assayed with amylopectin (●——●) and without amylopectin under unprimed conditions (sodium citrate plus BSA) (○——○). Absorption at 280 nm is also indicated (x——x).

[3] J. L. Ozbun, J. S. Hawker, and J. Preiss, *Biochem. J.* **126**, 953 (1972).

TABLE I

PURIFICATION OF SPINACH LEAF ADP-GLUCOSE:α1,4-GLUCAN α4-GLUCOSYLTRANSFERASE

Purification step	Volume (ml)	Protein (mg/ml)	Activity				Specific activity	
			Units/ml		Total units			
			Primed	Unprimed	Primed	Unprimed	Primed	Unprimed
1. Crude extract	305	7.7	1.2	0.09	363	26.3	0.16	0.011
2. (NH$_4$)$_2$SO$_4$	25.5	46	13.3	9.0	339	230	0.29	0.20
3. DEAF-cellulose								
Fraction I (11–12)[a]	2.4	3.85	1.87	0.	4.5	0.0	0.5	—
Fraction II (75–83)	2.5	4.3	21.6	2.45	54	6.1	5.0	0.57
Fraction III (99–110)	2.6	14.0	2.5	58.5	65	152	1.8	4.2
Fraction IV (61–68)	2.5	1.65	2.6	0.12	6.6	0.3	1.6	0.07

[a] Fraction numbers are from Fig. 1.

TABLE II

K_m VALUES FOR THE STARCH SYNTHETASE FRACTIONS FROM SPINACH LEAVES

Substrate	Synthetase fraction			
	I	II	III	IV
ADP-Glucose (mM)	0.20	0.29	0.15	0.25
Soluble amylose, corn (mg/ml)	0.06	0.02	a	0.04
Amylopectin, corn (mg/ml)	0.05	0.03	a	0.05
Glycogen, rabbit liver (mg/ml)	1.0	0.80	a	0.88
Glycogen, oyster (mg/ml)	1.54	1.18	6.66	1.54
β-Limit dextrins				
Amylopectin, corn (mg/ml)	0.06	0.02	0.22	0.03
Glycogen, *Escherichia coli* (mg/ml)	1.0	2.0	1.0	1.4

a Nonlinear kinetics at low substrate concentrations.

amylose, amylopectin, and rabbit liver glycogen are high, but a K_m value could not be determined because of the nonlinear kinetics observed at low primer concentrations. The relative activities for various primers are shown in Table III. Fraction III, which catalyzes the unprimed reaction, has slightly greater activity V_{max} with rabbit liver glycogen as a primer than the other fractions. Fraction III also has much less activity with the primers maltose, maltotriose, and β-limit dextrins than the other 3 fractions. The other fractions generally have the same properties with

TABLE III

ACTIVITIES OF STARCH SYNTHETASES FROM SPINACH LEAVES TOWARD DIFFERENT PRIMERS

Primer	Concentration	Synthetase fraction			
		I	II	III	IV
Amylopectin, corn	5 mg/ml	100[a]	100[a]	100[a]	100[a]
Soluble amylose, corn	5 mg/ml	114	116	103	123
Glycogen, rabbit liver	5 mg/ml	153	136	188	159
Glycogen, oyster	5 mg/ml	122	143	39	150
β-Limit dextrins					
Amylopectin, corn	5 mg/ml	17	21	25	17
Amylopectin, corn	0.5 mg/ml	13	25	6	15
Glycogen, *Escherichia coli*	5 mg/ml	25	38	7	28
Maltose	0.40 M	81	85	24	102
Maltotriose	0.05 M	102	78	30	110

[a] The rates of reactions obtained with the primer corn amylopectin were arbitrarily assigned a value of 100. The values obtained with the other primers are expressed as a percentage of the rate observed with amylopectin.

respect to relative activity towards various primers and with respect to K_m values for the primers and for ADP-glucose.

Unprimed Reaction

Fraction III is able to catalyze the formation of an α-glucan in the absence of primer. This reaction is greatly stimulated by 0.5 M sodium citrate.[3,4] Bovine serum albumin in the presence of 0.5 M sodium citrate will further stimulate the reaction and allows the reaction to be linear with time and protein concentration.[4] In the presence of salts only, the reaction rate has an initial lag period and is autocatalytic. Other salts may substitute for sodium citrate[3,4]; these are sodium or potassium acetate, ammonium sulfate, and sodium fluoride. However KCl, KBr, $NaClO_4$, and KCNS are inactive in stimulating unprimed synthesis.[4] Carbowax 6000 at 250 mg/ml can substitute for the salt requirement. Fraction 1 does not catalyze the unprimed reaction. The small amount of unprimed reaction present in fractions II and IV (Table I) may possibly be due to contamination of these fractions with fraction III.

The product formed in the unprimed reaction appears to be branched (i.e., it contains α-1,6 glucosidic linkages in addition to α-1,4 linkages).[4] The formation of the α-1,6 linkages is most probably due to the branching enzyme activity present in fraction III (unpublished results).

[4] J. L. Ozbun, J. S. Hawker, and J. Preiss, *Biochem. Biophys. Res. Commun.* **43**, 631 (1971).

[76] Xylosyltransferases from *Cryptococcus laurentii*

By JOHN S. SCHUTZBACH and HELMUT ANKEL

Enzyme I

$$n\text{UDP-xylose} + \text{soluble acceptor} \rightarrow \text{soluble} - (\text{xylosyl})_n\text{-acceptor} + n\text{UDP} \quad (1)$$

Enzyme II[1]

$$n\text{UDP-xylose} + \text{insoluble acceptor} \rightarrow \text{insoluble} - (\text{xylosyl})_n\text{-acceptor} + n\text{UDP} \quad (2)$$

$$\text{UDP-xylose} + \text{Man-3-mannose} \rightarrow \text{Man-3-mannose} + \text{UDP} \quad (3)$$
$$\underset{\beta\text{-Xyl-2}}{|}$$

A cell-free particulate preparation from *Cryptococcus laurentii* contains at least two different xylosyltransferases.[2] One of these catalyzes

[1] The abbreviations used are: Man-3-mannose, O-α-D-mannopyranosyl-$(1 \rightarrow 3)$-D-mannose; Man-3-[β-Xyl-2]-mannose, O-α-D-mannopyranosyl-$(1 \rightarrow 3)$-O-[β-D-xylopyranosyl-$(1 \rightarrow 2)$]-D-mannose.

the transfer of D-xylose from UDP-xylose to an acidic extracellular poly-saccharide composed of D-mannose, D-xylose, and D-glucuronic acid.[3] The acidic polysaccharide acts as a xylosyl acceptor in cell-free systems only after it has been pretreated by mild acid hydrolysis to remove terminal xylosyl residues. The second enzyme transfers D-xylose from UDP-xylose to endogenous insoluble acceptor at a very slow rate (Reaction 2). The endogenous acceptor is composed of D-mannose, D-xylose, and D-galactose and is believed to be a glycoprotein associated with the cell wall complex of the organism. The second enzyme can also catalyze the transfer of D-xylose to Man-3-mannose (Reaction 3). The latter re-action can be used as a convenient assay for enzyme II.

Assay Method

Assay for Enzyme I. See Vol. 8 [70].

Principle. Assay methods are based on the determination of radio-activity transferred from UDP-[^{14}C]xylose to specific acceptors. Radio-active products are separated from remaining substrate by chromato-graphic or ion-exchange procedures.

Reagents

UDP-[^{14}C]xylose, 10 μCi/ml (172 mCi/mmole)
Man-3-mannose,[4] 0.02 M
Solvent for paper chromatography: 95% ethanol–1 M ammonium acetate, 7:3

Procedures. ASSAY 1. Assay 1 is used to measure D-xylosyl transfer from UDP-[^{14}C]xylose to endogenous acceptor. Reaction mixtures con-tain 5 μl of UDP-[^{14}C]xylose, 15 μl of water, and 20 μl of particulate enzyme and are incubated at 25° in small test tubes. The reaction is terminated after 4 hours by applying the reaction mixture to the origin of a chromatogram on Whatman No. 1 paper. After chromatography in the ethanol–ammonium acetate solvent, appropriate areas at the origin are excised and counted in a gas-flow counter.

ASSAY 2. Assay 2 measures the rate of transfer of D-xylose from UDP-[^{14}C]xylose to Man-3-mannose. The reaction mixture contains 5 μl of UDP-[^{14}C]xylose, 5 μl of water, 10 μl of Man-3-mannose, and 20 μl of enzyme. Incubations are carried out at 25° for 1 hour in small test tubes.

[2] J. S. Schutzbach and H. Ankel, *Fed. Proc., Fed. Amer. Soc. Exp. Biol.* **30**, 1186 (1971).

[3] A. Cohen and D. S. Feingold, Vol. 8 [70].

[4] Prepared from the acidic extracellular polysaccharide of *C. laurentii* (see refer-ence 3) by acetolysis according to the procedures of Y. C. Lee and C. E. Ballou, *Biochemistry* **4**, 257 (1965).

The reaction is terminated by applying the reaction mixture to the upper one-third of a 4×10 cm strip of Whatman No. DE81 paper. The paper is allowed to dry and is then eluted with 1 ml of water onto planchets. The uncharged product is eluted by this procedure while UDP-[^{14}C]-xylose remains on the paper. The eluates are dried under reduced pressure, and the eluted radioactivity is counted in a gas-flow counter. Controls are run without acceptor and with boiled enzyme.

Preparation of Enzyme

The procedures for cell growth and preparation of the particulate enzyme are carried out as described for mannosyl transferases from *Cryptococcus laurentii* (this volume [77]). The brown middle layer contains all the xylosyl transferase activity and is carefully isolated and resuspended in 2 volumes of $0.1 M$ Tris·HCl buffer, pH 7.3. The final pH of the enzyme suspension is approximately 6.3–6.5.

Properties

Activity. Under conditions of the assay, D-xylosyl transfer to endogenous acceptor is very weak (100–200 cpm in 4 hours). Incorporation into endogenous acceptor can be increased 3- to 6-fold by the addition of GDP-mannose (0.1 μmole) and MgCl$_2$ (2.0 μmoles) to the reaction mixture. The stimulation is presumably due to the formation of additional α-D-mannosyl-1,3-D-mannosyl acceptor sites on the endogenous acceptor.[5] Activity as measured with Assay 2 is in the order of 1000–2000 cpm of [^{14}C]xylosyl transferred per hour. At saturating concentrations of UDP-xylose, the rate of transfer of D-xylose to Man-3-mannose is 4 nmoles per hour per milligram of protein.

Stability. The xylosyl transferase activity of both enzymes is stable for at least 1 week at 4°. Enzyme II loses 90% activity at 38° for 30 minutes whereas enzyme I retains full activity under the same conditions.

Specificity. Enzyme I is active with the acidic polysaccharide from *C. laurentii* and has lower activity toward similar polysaccharides from related organisms.[3] Enzyme II is specific for an α-D-mannosyl-1,3-D-mannose linkage. The disaccharide, Man-3-mannose, as well as some tri- and tetrasaccharides containing this linkage at nonreducing ends, are acceptors. Xylose is transferred to the mannose pentultimate to the nonreducing end with formation of a branched product containing a β-1,2-xylosyl linkage.

Inhibitors. Xylosyl transfer to the acidic polysaccharide is inhibited

[5] H. Ankel, E. Ankel, J. S. Schutzbach, and J. C. Garancis, *J. Biol. Chem.* **245**, 3945 (1970).

by a number of different nucleotides.[6] The percentage of inhibition of enzyme I by 0.6 mM nucleotides under standard assay conditions is ATP, 40%; AMP, 29%; GTP, 40%; CTP, 88%. Inhibition of enzyme II by the same nucleotides at the same concentrations is ATP, 9%; GTP, 10%; AMP and CTP, no inhibition.

pH Optimum. The pH optimum for enzyme I is 7.5 in 0.1 M Tris·HCl buffer. The pH optimum of enzyme II is 6.3.

Kinetic Properties. The K_m values for UDP-xylose are 0.7 mM for enzyme I and 3.3 mM for enzyme II.

Identification of Product. The product formed in Reaction (3) (R_{Man}, 0.28) is separated from Man-3-mannose (R_{Man}, 0.53) by ascending chromatography in 1-propanol-ethyl acetate–H_2O, 7:1:2. Treatment with Sweet Almond Emulsion (Worthington) that contains both α-mannosidase and β-xylosidase activity at pH 5.0 results in quantitative cleavage of xylosyl moiety. When the xylosidase activity of Emulsin is inhibited with 0.5 M methyl-β-D-xylopyranoside, only the terminal D-mannose is removed, resulting in the formation of O-β-D-xylopyranosyl-$(1 \rightarrow 2)$-D-mannose. Free D-xylose is liberated upon hydrolysis in 1 N HCl at 100° for 2 hours.

[6] A. Cohen and D. S. Feingold, *Biochemistry* **6**, 2933 (1967).

[77] Mannosyltransferases from *Cryptococcus laurentii*[1,2]

By JOHN S. SCHUTZBACH and HELMUT ANKEL

$$\text{GDP-mannose} + \text{D-mannose} \xrightarrow{\text{Mn}^{2+}} \text{Man-2-mannose} + \text{GDP} \qquad (1)$$

$$\text{GDP-mannose} + \text{Man-6-mannose} \xrightarrow{\text{Mn}^{2+}} \text{Man-6-Man-6-mannose} + \text{GDP} \qquad (2)$$

$$\text{GDP-mannose} + \text{Man-CH}_3 \xrightarrow{\text{Mg}^{2+}} \text{Man-3-Man-CH}_3 + \text{GDP} \qquad (3)$$

$$\text{GDP-mannose} + \text{Xyl-CH}_3 \xrightarrow{\text{Mn}^{2+}} \text{Man-Xyl-CH}_3 + \text{GDP} \qquad (4)$$

A particulate enzyme preparation from *Cryptococcus laurentii* catalyzes the transfer of mannose from GDP-mannose to a number of

[1] J. S. Schutzbach and H. Ankel, *J. Biol. Chem.* **246**, 2187 (1971).
[2] The abbreviations used are: Man-2-mannose, O-α-D-mannopyranosyl-$(1 \rightarrow 2)$-D-mannose; Man-3-mannose, O-α-D-mannopyranosyl-$(1 \rightarrow 3)$-D-mannose; Man-4-mannose, O-α-D-mannopyranosyl-$(1 \rightarrow 4)$-D-mannose; Man-6-mannose, O-α-D-mannopyranosyl-$(1 \rightarrow 6)$-D-mannose; Man-2-Man-2-mannose, O-α-D-mannopyranosyl-$(1 \rightarrow 2)$-O-α-D-mannopyranosyl-$(1 \rightarrow 2)$-D-mannose; Man-6-Man-6-mannose, O-α-D-mannopyranosyl-$(1 \rightarrow 6)$-O-α-D-mannopyranosyl-$(1 \rightarrow 6)$-D-mannose; Man-CH$_3$, methyl-α-D-mannopyranoside; Xyl-CH$_3$, methyl-α-D-xylopyranoside; β-Xyl-CH$_3$, methyl-β-D-xylopyranoside.

exogenous saccharide acceptors.[3] At least four different mannosyl transferases are present in the preparation, each catalyzing the formation of a specific mannosyl-acceptor linkage. The mannosyl transferases, in conjunction with a xylosyl transferase present in the same preparation, can be utilized for the stepwise synthesis of a branched heteropentasaccharide.

Assay Method

Principle. Assay methods are based on the determination of radioactivity transferred from GDP-[^{14}C]mannose to specific acceptors. Radioactive products are separated from remaining substrate by paper chromatography or ion-exchange procedures.

Reagents

GDP-mannose-^{14}C, 10 μCi/ml (52 mCi/mmole)
$MnCl_2$, 0.1 M
$MgCl_2$, 0.2 M
Tris·HCl, 0.1 M, pH 7.3
D-Mannose, 0.5 M
Methyl-α-D-mannopyranoside, 0.2 M
Methyl-α-D-xylopyranoside, 0.5 M
Man-6-mannose,[4] 0.1 M
Sodium dodecyl sulfate, 10% (w/v)
Solvents for paper chromatography: 1-propanol–ethyl acetate–H_2O, 7:1:2 (by volume), solvent 1; 95% ethanol–1 M ammonium acetate, 7:3, solvent 2

Procedures. ASSAY 1. Reaction mixtures containing 5 μl of GDP-[^{14}C]mannose, 5 μl of $MnCl_2$ or $MgCl_2$, 10 μl of acceptor, 20 μl of Tris buffer, and 10 μl of enzyme are incubated at 25° in small test tubes. The reaction is terminated after 30 minutes by the addition of 10 μl of sodium dodecyl sulfate. The formation of product is determined by autoradiography after ascending chromatography in solvent 1. Appropriate areas

---

[3] The same enzyme preparation also catalyzes the transfer of mannose from GDP-[^{14}C]mannose to endogenous acceptor. This transferase activity requires divalent metal ions and results in the formation of α-1,2- and α-1,3-mannosylmannosyl linkages. The ^{14}C-labeled product has very similar properties to a neutral polysaccharide isolated from the cell wall of the organism. H. Ankel, E. Ankel, J. S. Schutzbach, and J. C. Garancis, *J. Biol. Chem.* **245**, 3945 (1970). Because of the low transferase activity to the endogenous acceptor and because of the formation of a mixed product, it has been difficult to identify this activity with the mannosyltransferases described in this article.

[4] Prepared from yeast mannan according to the procedure of G. H. Jones and C. E. Ballou, *J. Biol. Chem.* **244**, 1043 and 1052 (1969). The *Arthrobacter* used to remove side chains from the mannan was supplied by Dr. C. E. Ballou.

on the chromatogram are excised and counted in a liquid scintillation spectrometer. Controls are run without acceptor and with boiled enzyme. This chromatographic assay is applicable for all four transferases and separates all the reaction products, but is quite time consuming.

ASSAY 2. A faster, less specific assay involves separation of the uncharged products from GDP-mannose on diethylaminoethyl cellulose paper. Reaction mixtures are prepared as above. The reaction is terminated by applying the reaction mixture directly to the upper one-third of a 4 × 10 cm strip of Whatman No. DE81 paper. The paper is allowed to dry and is then eluted with approximately 1 ml of water onto planchets. The eluates are dried under reduced pressure, and the eluted radioactivity is counted in a gas-flow counter. Controls are as above. Assay 2 is not suitable when more than one uncharged soluble product is formed in the reaction.

Preparation of Enzyme

Growth of Cells. C. laurentii var. *flavescens* (NRRL Y-1401) is grown at 28° in a medium containing 2% (w/v) glucose, 0.1% urea, 0.1% KH_2PO_4, 0.05% $MgSO_4 \cdot 7H_2O$, and 0.2 mg/liter thiamine·HCl.[5,6] The medium, 125 ml, is contained in a 500-ml Erlenmeyer flask equipped with two stainless steel baffles and is inoculated with 3 ml of a 24-hour culture (OD at 400 nm, 1.0–1.4). The culture is incubated with vigorous rotary shaking for 16–18 hours. When the optical density of the culture reaches 0.7, the cells are harvested by centrifugation. For the separation of particulate enzyme it is practical to use at least 10 flasks as described.

Preparation of Particulate Enzyme. All operations are carried out at 0–4°. The packed cells are washed consecutively with 10 volumes of 1% NaCl, distilled water and 0.1 M Tris·HCl buffer, pH 7.5, the latter containing reduced glutathione and EDTA at a concentration of 1 mM. The washed residue is suspended in 1 volume of the same buffer and the cells are disrupted by ultrasonic treatment with a Branson Sonifier (Model 125) for 7 minutes at an output of 11 amp. Cell debris is removed by centrifugation at 12,000 g for 15 minutes and the particulate fraction is isolated by centrifugation at 105,000 g for 30 minutes. The high speed pellet is resuspended in approximately 10 volumes of the same buffer and again centrifuged. This step is repeated once. The particulate material separates into three distinct layers. The brown middle layer contains all the transferase activities and is carefully isolated and resuspended in approximately 3 volumes of the same buffer. The transferase activities are usually stable for at least 1 week when stored at 4°.

[5] J. Mager, *Biochem. J.* **41**, 603 (1947).
[6] H. Ankel and D. S. Feingold, *Biochemistry* **5**, 182 (1966).

Transferase Activities Present in the Enzyme Preparation

Under standard assay conditions (Assay 1), the ratios of activities of mannosyl transfer from GDP-mannose to the following acceptors at a concentration of 20 mM are: D-mannose, 1.0; Man-2-mannose, 1.6; Man-2-Man-2-mannose, 0.2; Man-6-mannose, 3.2; Man-6-man-6-mannose, 2.0; Man-CH$_3$, 16; Man-CH$_3$ (Mg^{2+}), 2.0; Xyl-CH$_3$, 1.1; β-Xyl-CH$_3$, 0.11; and mannoheptulose, 0.25. The metal ion used throughout was Mn^{2+} except where indicated. The following compounds are not acceptors for mannosyl transfer in the presence of either Mn^{2+} or Mg^{2+}: D-xylose, L-arabinose, D-galactose, D-glucose, D-mannitol, myoinositol and 2-deoxy-D-glucose.

Properties

Enzyme I

Specificity. Enzyme I catalyzes the formation of α-1,2-mannosyl-mannose linkages. This enzyme can utilize free D-mannose, Man-2-mannose, and Man-CH$_3$ as acceptors. Mannose is the acceptor of choice in the assay system because under assay conditions mannose does not react at significant rates with any of the other transferases. Man-2-mannose, prepared by acetolysis of *Saccharomyces cerevisiae* mannan,[7] is a substrate specific for enzyme I.

Activators. Divalent metal ions are required for activity, optimal activation occurring with MnCl$_2$ at a concentration of 6–12 mM. Only 9% of the maximum rate is obtained in the presence of 10 mM MgCl$_2$. The chlorides of the following ions at 10 mM activate less than 5%: Ca^{2+}, Co^{2+}, Cu^{2+}, Ni^{2+}, and Zn^{2+}.

pH Optimum. Maximum activity is obtained at pH 7.0 to 7.5.

Kinetic Properties. The K_m values are 0.15 mM for GDP-mannose, 160 mM for D-mannose, 31 mM for Man-2-mannose and 27 mM for Man-CH$_3$. Values of V_{max} for D-mannose, Man-2-mannose, and Man-CH$_3$ are 25, 25, and 50 nmoles of mannosyl transferred per hour per milligram of protein, respectively.

Enzyme II

Specificity. Enzyme II catalyzes the formation of α-1,6-mannosyl-mannosyl linkages. Of all the acceptors listed above, the enzyme utilizes only Man-6-mannose as a substrate. The major product formed is Man-6-Man-6-mannose, although a minor product with a faster chromatographic

---

[7] Y. C. Lee and C. E. Ballou, *Biochemistry* **4**, 257 (1965).

mobility in solvent 1 is also produced. The minor product has not been characterized.

Activators. Divalent metal ions are required for activity, $MnCl_2$ at 12 mM being optimal. $MgCl_2$ at 12 mM is only 8% as effective. The chlorides of the following divalent ions activate less than 2%: Ca^{2+}, Co^{2+}, Cu^{2+}, Ni^{2+}, and Zn^{2+}.

pH Optimum. The pH optimum of enzyme II is 7.5.

Kinetic Properties. The K_m values are 0.4 mM for GDP-mannose and 16 mM for Man-6-mannose.

Enzyme III

Specificity. Enzyme III catalyzes the formation of Man-3-Man-CH_3 from GDP-mannose and Man-CH_3. Under standard assay conditions, mannosyl transfer to Man-CH_3 is the only detectable activity of the enzyme. Free D-mannose is an acceptor for mannosyl transfer only at high concentration (0.5 M) and at a slow rate.

Activators. Divalent metal ions are required for activity. Enzyme III is optimally activated by $MgCl_2$ at a concentration of 60 mM. Taking the reaction rate at 10 mM $MgCl_2$ as 100, the chlorides of other divalent cations at the same concentrations give the following activities: Co^{2+}, 68; Mn^{2+}, 41; Ca^{2+}, 10; Ni^{2+}, 5; Cu^{2+} and Zn^{2+}, less than 1.

pH Optimum. Maximum activity occurs in the pH range 7.5–8.5.

Kinetic Properties. The K_m values are 0.4 mM for GDP-mannose and 200 mM for Man-CH_3.

Enzyme IV

Specificity. Enzyme IV catalyzes the transfer of mannose from GDP-mannose to Xyl-CH_3. Free D-xylose is not an acceptor for this enzyme. The linkage formed is resistant to cleavage by both α- and β-mannosidase and has not been further characterized.

Activators. The enzyme requires $MnCl_2$ at a concentration of 3 mM for optimal activity. $MgCl_2$ at the same concentration is only 8% as effective and the enzyme has no detectable activity with the following cations: Ca^{2+}, Co^{2+}, Cu^{2+}, Ni^{2+}, and Zn^{2+}.

pH Optimum. Enzyme IV has a pH optimum of 7.5–8.5.

Identification of Products

The mannosyl-mannose linkages formed in the transferase reactions are all hydrolyzed by α-mannosidase (Sweet Almond Emulsin, Worthington), but not by β-mannosidase (hemicellulase from *Rhizopus* mold, Sigma). Free [^{14}C]mannose can be identified by chromatography with solvent 1. The position of the linkage is identified by chromatographic and electrophoretic methods that separate the four possible α-linked disaccharides. Carbon-14 labeled disaccharides are isolated from each of the reaction products after partial acid hydrolysis in $0.1\,N$ HCl for 2 hours at $100°$ followed by chromatography in solvent 1. Man-6-mannose (R_{Man}, 0.46) can be differentiated from the three other α-linked mannobioses (R_{Man}, 0.53) with this solvent system. Electrophoresis on Whatman No. 1 paper at 30 V/cm for 8 hours in $0.04\,M$ sodium borate, pH 9.2, separates the remaining mannobioses. The electrophoretic mobilities relative to mannose in this system are: Man-2-mannose, 0.82; Man-3-mannose, 0.91; Man-4-mannose, 1.0; and Man-6-mannose, 0.91.

The table presents a summary of the properties of the four mannosyl transferases in *C. laurentii*.

Synthesis of a Branched Pentasaccharide

The particulate enzyme preparation described herein can be used for the stepwise synthesis of a branched pentasaccharide.

Step 1. A large scale reaction mixture is prepared using GDP-[^{14}C]-mannose (50 μl), Man-CH$_3$ (50 μl), MgCl$_2$ (40 μl), Tris buffer (50 μl), and enzyme (250 μl) at reagent concentrations as described for the assay. Reaction mixtures are incubated for 16 hours and the labeled Man-3-Man-CH$_3$ is isolated by paper chromatography in solvent 1. The particulate material in the reaction mixture is removed prior to chromatography by centrifugation at 6000 g for 15 minutes.

Step 2. The radioactive product isolated from step 1 is dissolved in 50 μl of water and is incubated with 50 μl of 20 mM UDP-xylose and 200 μl of the particulate enzyme for 24 hours.[8] The product is isolated

MANNOSYL TRANSFERASES IN *Cryptococcus laurentii*

	Enzyme I	Enzyme II	Enzyme III	Enzyme IV
Substrate or substrates	D-Mannose Man-2-mannose Man-CH$_3$	Man-6-mannose	Man-CH$_3$	Xyl-CH$_3$
Linkage formed	α-1,2	α-1,6	α-1,3	
K_m for GDP-mannose (mM)	0.15	0.4	0.4	0.5
pH optimum	7.0–7.5	7.5	7.5–8.5	7.5–8.5
Divalent cation	Mn^{2+}	Mn^{2+}	Mg^{2+}	Mn^{2+}
Optimal cation concentration (mM)	6–12	12	60	3

[8] This volume [76].

by chromatography in solvent 1 as before. Under these conditions there is almost complete conversion of the labeled acceptor to a branched trisaccharide. Steps 1 and 2 can be combined in a single reaction mixture allowing the direct isolation of the methyl-trisaccharide. The GDP-[^{14}C]mannose can also be replaced by an equivalent volume of 20 mM GDP-mannose and the UDP-xylose can be replaced by labeled UDP-xylose resulting in the formation of a methyl-trisaccharide labeled only in the xylosyl moiety. Yields are somewhat reduced, however, when both sugar-nucleotides are combined in the same reaction mixture.

Analysis of Products. Xylosyl-^{14}C residues can be removed from the [^{14}C]xylosyl product with β-xylosidase (Sweet Almond Emulsin) at pH 5.0. When the xylosidase activity of Emulsin is inhibited with 0.5 M β-Xyl-CH$_3$ only the terminal mannose is removed. Under these conditions a nonreducing methyl disaccharide (Xyl-Man-CH$_3$) with the chromatographic mobility of mannose in solvent 1 is formed. Treatment of the [^{14}C]xylosyl labeled methyl disaccharide with β-xylosidase results in the formation of free [^{14}C]xylose.

Step 3. The reaction catalyzed at this step results in the transfer of mannose from GDP-mannose to the terminal nonreducing mannosyl residue of the trisaccharide with the formation of an α-1,6-linkage. The methyl trisaccharide isolated after step 2 is used as the acceptor in a reaction similar to that in step 1 except that MgCl$_2$ is replaced by an equal volume of MnCl$_2$. Either labeled or unlabeled GDP-mannose can be used as the mannosyl donor. The reaction product is isolated after chromatography in solvent 2 and is further purified by chromatography in solvent 1. When GDP-mannose is present in limiting concentrations or when the reaction time is limited to approximately 4 hours, a methyl-tetrasaccharide is the major product. Under conditions where GDP-mannose is in excess during 16-hour incubations, methylpentasaccharide is the major or sole product formed (see step 4).

Analysis of Product. The terminal nonreducing mannose of the tetrasaccharide is removed by treatment with α-mannosidase although at a very slow rate. Partial acid hydrolysis as before of the methyltetrasaccharide labeled with [^{14}C]mannose only at the terminal nonreducing end yields ^{14}C-labeled Man-6-mannose.

Step 4. When a reaction mixture similar to that in step 3 is prepared, using either methyl tetrasaccharide as the acceptor or methyl trisaccharide in the presence of excess GDP-mannose, a pentasaccharide containing a terminal α-1,2-mannosyl-mannosyl linkage is formed. The reaction goes to completion during 16-hour incubations. The terminal linkage is hydrolyzed by α-mannosidase and is characterized after partial acid hydrolysis as before.

The final product has the following structure:

$$\alpha\text{-Man-1,2-}\alpha\text{-Man-1,6-}\alpha\text{-Man-1,3-}\alpha\text{-Man-CH}_3$$
$$|$$
$$\beta\text{-Xyl}$$

The pentasaccharide is very similar in structure and composition to oligosaccharides isolated after acetolysis of cell wall polysaccharides of *C. laurentii*.[3]

[78] Synthesis of a Cell Wall Glucomannan in Mung Bean Seedlings[1]

By ALAN D. ELBEIN

GDP-D-glucose[14C] + GDP-D-mannose + acceptor → [14C]glucomannan + GDP

GDP-[14C]mannose + acceptor → [14C]glucomannan + GDP

A particulate enzyme fraction from mung bean seedlings catalyzes the incorporation of mannose from GDP-[14C]mannose into an alkali-insoluble polymer with the chemical properties of a β-$(1 \rightarrow 4)$-glucomannan. In the presence of GDP-D-mannose the particulate preparation also catalyzes the incorporation of glucose from GDP-D-[14C]glucose into the alkali-insoluble glucomannan.[2,3] Although the polymer described here is insoluble in alkali, the particulate enzyme also incorporates mannose from GDP-D-[14C]mannose into the hemicellulose A and B fractions.[4]

Assay Method

Principle. The assay is based on the fact that the glucomannan is insoluble in hot 2% NaOH. Therefore, the incorporation of radioactivity from GDP-D-[14C]mannose into this product can be followed. Although the particulate enzyme also incorporates glucose from GDP-D-[14C]glucose into cellulose,[5] there is an increased incorporation of radioactivity from GDP-D-[14C]glucose in the presence of GDP-D-mannose. This increased incorporation is into the glucomannan.

[1] This work was supported by a grant from the Robert A. Welch Foundation.

[2] A. D. Elbein and W. Z. Hassid, *Biochem. Biophys. Res. Commun.* 23, 311 (1956).
[3] A. D. Elbein, *J. Biol. Chem.* 244, 1608 (1969).
[4] S. S. Brar and A. D. Elbein, *Photochemistry* 11, 651 (1972).
[5] G. A. Barber, A. D. Elbein, and W. Z. Hassid, *J. Biol. Chem.* 239, 4056 (1964).

Reagents

GDP-D-[^{14}C]glucose and GDP-D-[^{14}C]mannose, uniformly labeled in the sugar moiety. Both of these nucleotides are available commercially or can be synthesized either chemically[6] or enzymatically.[7,8]

GDP-D-mannose, 0.1 M

MgCl, 0.05 M

Tris·HCl buffer, 0.1 M, pH 7.5

NaOH, 2%

Procedure. A typical incubation mixture for the incorporation of mannose contains 0.1 μmole of GDP-D-[^{14}C]mannose (15,000 cpm); 2 μmoles of MgCl$_2$; 5 μmoles of Tris buffer, pH 7.5; and 0.1 ml of particulate enzyme (100–200 μg of protein) in a final volume of 0.2 ml.

For determining the incorporation of glucose from GDP-D-[^{14}C]glucose into the glucomannan, assay mixtures contain 0.02 μmole of GDP-D-[^{14}C]glucose (15,000 cpm); 0.1 μmole of GDP-D-mannose; 2 μmoles of MgCl$_2$; 5 μmoles of Tris·HCl buffer, pH 7.5; and 0.1 ml of particulate enzyme in a final volume of 0.2 ml. The incorporation of radioactivity from GDP-D-[^{14}C]glucose into cellulose is determined in the absence of GDP-D-mannose.

Control tubes are also prepared in which boiled enzyme (100°, 5 minutes) is used or in which enzyme is added just before the reaction is stopped. Incubations are at 37° for 15 minutes. Reactions are stopped by adding 1 ml of water and placing the tubes in a boiling water bath for 5 minutes. The precipitate is isolated by centrifugation and extracted several times with 1 ml of hot 2% NaOH. After each extraction, tubes are cooled and the precipitate is reisolated by centrifugation. The alkali-insoluble precipitate is then washed several times with water and its radioactive content is determined.

Preparation of Enzyme

Mung bean seeds (*Phaseolus aureus*) are germinated and grown for 3 or 4 days at room temperature in a moist chamber. At the end of this time, roots and hypocotyls are removed and ground in a chilled mortar with sea sand and an equal weight of 0.1 M Tris buffer. The homogenate is squeezed through two layers of cheesecloth and centrifuged at

[6] S. Roseman, J. J. Distler, J. G. Moffat, and H. G. Khorana, *J. Amer. Chem. Soc.* **83**, 659 (1961).

[7] G. A. Barber and W. Z. Hassid, *Biochem. Biophys. Acta* **86**, 397 (1964).

[8] S. M. Rosen, L. D. Zelecznick, D. Fraenkel, I. Werner, M. J. Osborn, and B. L. Horecker, *Biochem. Z.* **342**, 375 (1965).

1000 g for 5 minutes. The residue is discarded and the supernatant solution is centrifuged at 20,000 g for 20 minutes. The pellet is gently resuspended in a small volume (3–4 ml) of Tris buffer and gently homogenized. This enzyme preparation is stable for 1 or 2 days at 0°, but slowly loses activity on prolonged storage.

Properties of the Product

The product formed from GDP-D-[¹⁴C]mannose, or GDP-D-[¹⁴C]glucose in the presence of GDP-D-mannose has the same solubility properties as cellulose. Thus, it is insoluble in 2% NaOH but soluble in Schweitzer's reagent and syrupy phosphoric acid. A series of oligosaccharides can be obtained from the glucomannan either by enzymatic hydrolysis with a β-mannanase or by acetolysis. The oligosaccharides are then isolated by paper chromatography in *n*-propanol–ethyl acetate–water (7:1:2). A radioactive disaccharide isolated from the polymer synthesized from GDP-D-[¹⁴C]glucose (plus GDP-D-mannose) has the same chromatographic mobility in three solvent systems as the radioactive disaccharide from the polymer synthesized from GDP-[¹⁴C]mannose. The $1 \rightarrow 4$ linkage in the glucomannan is demonstrated by methylation and isolation of 2,3,6-trimethylmannose and also by Smith degradation and isolation of [¹⁴C]erythritol. The polymer as well as the oligosaccharides are susceptible to hydrolysis by enzymes that cleave β-glycosidic linkages.

Properties of the Enzyme System

Specificity. The glucomannan synthesizing system shows a complete specificity for GDP-D-mannose and GDP-D-glucose. Thus, the following showed less than 5% of the activity exhibited by GDP-D-[¹⁴C]mannose or GDP-D-[¹⁴C]glucose (in the presence of GDP-D-mannose): TDP-D-[¹⁴C]mannose, D-[¹⁴C]mannose-6-P, D-[¹⁴C]mannose-1-P, D-[¹⁴C]mannose, UDP-D-[¹⁴C]glucose, TDP-D-[¹⁴C]glucose, ADP-D-[¹⁴C]glucose or CDP-D-[¹⁴C]glucose.

Occurrence. This reaction has not been demonstrated as such in other plant systems. However, in many plant systems which synthesize cellulose from GDP-D-[¹⁴C]glucose, it has been found that GDP-D-mannose stimulates the incorporation of radioactivity into alkali-insoluble products. Thus, these enzyme systems may also be synthesizing glucomannans.

Activators and Inhibitors. It has not been possible to separate the transferase activity from endogenous acceptor. Therefore, nothing is known about the nature of the acceptor. Furthermore, it is not known whether there are two (or more) different transferases present in the particles.

The incorporation of mannose from GDP-D-[¹⁴C]mannose was

strongly dependent on the presence of Mg^{2+} with the optimum concentration of this cation being about $1 \times 10^{-2}\,M$. The incorporation of glucose from GDP-D-[^{14}C]glucose, either in the presence or the absence of GDP-D-mannose, was neither stimulated nor inhibited by the addition of Mg^{2+}. The pH optimum for both reactions was 7.5 in Tris·HCl buffer. The incorporation was slightly inhibited by phosphate buffer.

GDP-D-mannose stimulates the incorporation of radioactivity from GDP-D-[^{14}C]glucose. A 3- to 5-fold increase in incorporation is observed when the GDP-D-mannose to GDP-D-glucose ratio is 3:1. GDP-D-glucose, on the other hand, inhibits the incorporation of radioactivity from GDP-D-[^{14}C]mannose. GDP-D-galactose also inhibits the incorporation of both GDP-D-[^{14}C]mannose and GDP-D-[^{14}C]glucose.

[79] Synthesis of Mannan in *Micrococcus lysodeikticus;* Lipid-Bound Intermediates

By MALKA SCHER and W. J. LENNARZ

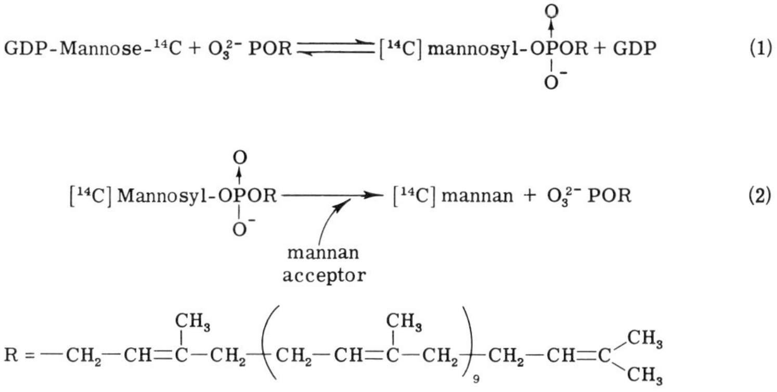

$$GDP\text{-Mannose-}^{14}C + O_3^{2-}\,POR \rightleftharpoons [^{14}C]\,mannosyl\text{-}\overset{\overset{O}{\uparrow}}{\underset{\underset{O^-}{|}}{O}}POR + GDP \tag{1}$$

$$[^{14}C]\,Mannosyl\text{-}\overset{\overset{O}{\uparrow}}{\underset{\underset{O^-}{|}}{O}}POR \xrightarrow[\text{mannan}\atop\text{acceptor}]{} [^{14}C]\,mannan + O_3^{2-}\,POR \tag{2}$$

$$R = -CH_2-CH=\overset{\overset{CH_3}{|}}{C}-CH_2\left(CH_2-CH=\overset{\overset{CH_3}{|}}{C}-CH_2\right)_9 CH_2-CH=C\overset{CH_3}{\underset{CH_3}{\diagdown}}$$

FIG. 1. Role of mannosyl-1-phosphorylundecaprenol (MPU) in mannan synthesis.

A particulate preparation that catalyzes the synthesis of mannan and mannosyl-1-phosphorylundecaprenol was obtained by fractionation of a cell-free extract of *Micrococcus lysodeikticus*. Kinetic studies and selective inhibition of the individual reactions outlined above indicate that MPU[1] is an obligatory intermediate in the synthesis of mannan from GDP-mannose. Initially MPU synthetase catalyzes the transfer of mannosyl units from GDP-[^{14}C]mannose to undecaprenyl phosphate, the

---

[1] The abbreviation used is: MPU, mannosyl-1-phosphorylundecaprenol.

acceptor lipid, to form [^{14}C]MPU (Fig. 1, Eq. 1). Mannosyl units are then transferred from the lipid intermediate to a mannan acceptor to form [^{14}C]mannan (Eq. 2). The method of assay for production of both [^{14}C]MPU and [^{14}C]mannan will be described.

Assay for [^{14}C]MPU

Principle. MPU synthetase activity is determined by measuring the incorporation of radioactive mannose, from GDP-[^{14}C]mannose, into lipid. The reaction mixture is extracted with $CHCl_3$–CH_3OH, 2:1, and, since the enzyme preparation is free of endogenous diglycerides, no mannosyldiglycerides are formed. Therefore, the sole radioactive product in the chloroform phase is [^{14}C]MPU.

Reaction Mixture.[2] The reaction mixture contains (in micromoles) Tris·HCl, pH 7.3, 25; $MgCl_2$, 20; undecaprenyl phosphate, 0.0045; GDP-[^{14}C]mannose, 0.0067 (10,000 dpm); enzyme, 21 μg of protein; phosphatidylglycerol, 0.040, in a final volume of 0.3 ml. GDP-[^{14}C]mannose can either be purchased from New England Nuclear or prepared according to the method of Preiss and Greenberg.[3] Immediately prior to use in enzymatic assays, transfer an aliquot of purified phosphatidylglycerol and an aliquot of purified undecaprenyl phosphate dissolved in $CHCl_3$ to each of two small tubes and remove the solvent from the lipid samples under a stream of nitrogen. After removal of residual $CHCl_3$ under reduced pressure, add water to each tube to yield final concentrations of 0.1–1.0 μmole of lipid per milliliter, and disperse each mixture separately by sonic oscillation with the small probe of a Biosonik (Bronwill Scientific Division, Rochester, New York) sonic oscillator at intensities up to 20% of maximum.

Procedure. After incubation at 37° for 30 minutes, 5 ml of $CHCl_3$–CH_3OH, 2:1, is added to stop the reaction.[4] Reaction tubes are mixed on a Vortex mixer and then placed in a bath at 55° for 3 minutes. After this time, the tubes are mixed again and allowed to remain at 55° for 7 minutes. Denatured protein is removed by filtration of the mixture through a small funnel containing a wad of glass wool. The tube and funnel are rinsed with 0.5 ml of $CHCl_3$–CH_3OH, and 2.5 ml of 0.9% NaCl is added to the filtrate. The mixture is mixed thoroughly for at least 1 minute, and is chilled on ice for 3–5 minutes. After centrifugation for 5–10 minutes the upper 0.9% NaCl layer is removed and discarded. For routine assays for MPU synthesis, the lower layer is transferred directly to a scintillation vial for counting. The organic solvent is evap-

[2] M. Lahav, T. H. Chiu, and W. J. Lennarz, *J. Biol. Chem.* **244**, 5890 (1969).

[3] J. Preiss and E. Greenberg, *Anal. Biochem.* **18**, 464 (1967).

[4] W. J. Lennarz and B. Talamo, *J. Biol. Chem.* **241**, 2707 (1966).

orated with a stream of nitrogen and the residue is counted in 15 ml of toluene scintillation fluid. Approximately 0.4 nmoles of MPU are formed under the conditions described above.

Preparation of Enzyme

M. lysodeikticus is the source of the acceptor lipid, undecaprenyl phosphate, as well as the enzyme.

Growth of Cells.[4] One-liter cultures of *M. lysodeikticus* ATCC 4698 are grown overnight at 30° in 2-liter flasks on a rotary shaker. The growth medium consists of 1% Bactopeptone (Difco), 0.5% NaCl, and 0.1% yeast extract. Cells are harvested by centrifugation when the optical density reached 400 Klett units (No. 66 filter), and they are washed once in 0.02 M Tris·HCl, pH 7.6. The yield of packed cells is generally 10–12 g, wet weight, per liter.

Preparation of Crude Extract.[4] Unless noted otherwise, all subsequent operations are carried out at 0–5°. The cells are suspended in an equal volume of 0.02 M Tris·HCl, pH 7.6, and are broken by passage through a French pressure cell at 28,000 psi. The breakage process is repeated once, then the resulting suspension is treated with a small amount of DNase at 0°. After a decline in viscosity becomes evident (5 minutes), the suspension is centrifuged at 7000 g for 10 minutes. The supernatant is removed and recentrifuged at 10,000 g for 10 minutes. The resulting supernatant (crude extract) is obtained in a yield of approximately 100 ml per 100 g of packed cells.

Preparation of Lipid-Depleted Enzyme.[4] The crude extract is centrifuged at 100,000 g for 2 hours, and the pellet so obtained is suspended in a volume of 0.02 M Tris·HCl (pH 7.6) equal to four-tenths the volume of the crude extract. The concentration of protein, determined by the Lowry method,[5] in this preparation should be about 18 mg/ml. This suspension is rapidly added dropwise to 6 volumes of stirred acetone at −10°. The resulting suspension is centrifuged for 1–2 minutes in a clinical centrifuge, and the supernatant is discarded. The pellet is quickly resuspended in one-third the original volume of acetone at −10° and recentrifuged for 1–2 minutes. Discard the supernatant and, after spreading the pellet on the walls of the bottom part of the centrifuge tube by gentle tapping, dry the pellet under reduced pressure at 0° for at least 3 hours. The dry acetone powder is removed from the tubes, quickly ground to a fine powder with a mortar and pestle, and stored in a vacuum at −20°. From 120 ml of crude extract, 950 mg of acetone powder (45–55% protein, by weight) can be obtained. The dry powder retains activity for

[5] O. H. Lowry, N. J. Rosebrough, A. L. Farr, and R. J. Randall, *J. Biol. Chem.* 193, 265 (1951).

at least 6 months. For routine use, dissolve the acetone powder in 0.02 M Tris·HCl (pH 7.6) at a final concentration of 10 mg of protein per milliliter.

Preparation of Lipids

Extraction and Preparation of Crude Lipid.[4] Total lipid is isolated from packed cells by extraction in 20 volumes of $CHCl_3$–CH_3OH, 2:1. The mixture is stirred at room temperature for 4–12 hours, after which time insoluble material is removed by filtration through a Büchner funnel. The filtrate is washed with 0.20 volume of 0.9% NaCl solution; after both phases have cleared, the $CHCl_3$ layer is concentrated under reduced pressure. From 100 g of cells, wet weight, approximately 1.0 g of crude lipid is obtained.

Purification of Undecaprenyl Phosphate.[2] DEAE-cellulose columns (20 × 100 mm) are prepared according to the method of Rouser *et al.*[6] Crude lipid (50 mg) from *M. lysodeikticus* is dissolved in a minimal volume of $CHCl_3$–CH_3OH, 2:1, and applied to the DEAE-cellulose column which has been equilibrated with $CHCl_3$–CH_3OH. The column is sequentially eluted at a flow rate of approximately 150 ml per hour with the following solvents: 200 ml of chloroform–methanol, 2:1; 200 ml of methanol; 750 ml of 0.0075 M ammonium acetate in 99% methanol containing 1% water; and 600 ml of 0.05 M ammonium acetate in 99% methanol. The ammonium acetate solutions are prepared from a stock solution of 5 M ammonium acetate in methanol containing 3% acetic acid. Preliminary experiments have been performed in order to establish the chromatographic elution pattern of the total phospholipids and of undecaprenyl phosphate. Undecaprenyl phosphate was found to be eluted with 0.0075 M ammonium acetate. Fractions of 25 ml of this solvent are collected, and the lipid in each fraction is isolated for analysis of acceptor activity in MPU synthesis. Each fraction is evaporated to dryness and dissolved in 5 ml of $CHCl_3$–CH_3OH, 2:1, and washed with 2 ml of water. The solution is mixed vigorously and centrifuged. The aqueous layer is discarded and the $CHCl_3$ layer is evaporated to dryness. The lipid extract from each fraction is redissolved in $CHCl_3$ and stored at −20°. The most active fractions are well separated from the majority of phospholipid. In fact less than 1% of the total phospholipids is found in the region that has high acceptor activity.

For the routine preparation of undecaprenyl phosphate, the fraction that is eluted between 400 and 600 ml of 0.0075 M ammonium acetate is collected. This fraction is evaporated to dryness under reduced pres-

[6] G. Rouser, G. Kritchevsky, D. Heller, and E. Lieber, *J. Amer. Oil Chem. Soc.* **40**, 425 (1963).

sure, dissolved in 5 ml of $CHCl_3$–CH_3OH and washed with 2 ml of water. The $CHCl_3$ layer is concentrated to dryness. The yield of undecaprenyl phosphate from 50 μmoles of crude phospholipid is 0.3 to 0.5 μmole. The undecaprenyl phosphate fraction from the DEAE-cellulose column is further purified by gel filtration on a Sephadex LH-20 column (2.5 × 80 cm) prepared in 0.001 M ammonium acetate in $CHCl_3$–CH_3OH, 1:1. The lipid dissolved in the same solvent is applied to the column which is then eluted at a flow rate of 0.5 ml per minute; fractions of 2.5 ml are collected. Fractions containing lipid phosphate are pooled, evaporated to dryness, and washed as indicated above. The final yield in this step should be about 0.1 μmole.

In order to destroy any traces of contaminating phosphoglycerides, the undecaprenyl phosphate fraction obtained by Sephadex LH-20 chromatography is treated with dilute NaOH in $CHCl_3$–CH_3OH.[7] To 0.1–1.0 μmole of lipid dissolved in 1 ml of chloroform–methanol, 1:4, add 0.1 ml of 1 N NaOH. The mixture is incubated at 37° for 10 minutes, and then neutralized with 0.1 ml of 1 N acetic acid. Then 2 ml of $CHCl_3$–CH_3OH, 9:1, 1 ml of isobutyl alcohol, and 2 ml of water are added, and the solution is mixed vigorously. The aqueous layer is discarded and the $CHCl_3$ layer is washed with 1 ml of H_2O–CH_3OH, 2:1, and evaporated to dryness. At this stage of purification, the undecaprenyl phosphate can be dissolved in $CHCl_3$ and stored at −20° prior to use in the assay for MPU synthesis. Although not necessary for enzymic studies, the lipid can be further purified by rechromatography on DEAE-cellulose after the alkaline hydrolysis to remove any free fatty acids released from contaminating phosphoglycerides.

Purification of Phosphatidylglycerol. Purified phosphatidylglycerol is prepared as described by Hopfer *et al.*[8] However, it should be emphasized that it is not essential to use phosphatidylglycerol in the assay. A variety of other surfactants also exhibit some degree of stimulation of MPU synthesis.

Characteristics of Reaction

Requirements of Enzyme.[2] In addition to the substrates, GDP-mannose and undecaprenyl phosphate, MPU synthetase has an absolute requirement for a divalent metal ion, with Mg^{2+} being more effective than Ca^{2+}, Co^{2+}, or Mn^{2+}. The optimum Mg^{2+} concentration with purified undecaprenyl phosphate is 67 mM. With the lipid-depleted acetone powder preparation in the presence of purified undecaprenyl phosphate,

[7] R. M. C. Dawson, *Biochem. J.* **75**, 45 (1960).
[8] U. Hopfer, A. L. Lehninger, and W. J. Lennarz, *J. Membrane Biol.* **2**, 41 (1970).

a surfactant such as phosphatidylglycerol must be added to obtain optimal enzymatic activity. There is an optimum ratio of undecaprenylphosphate to phosphatidylglycerol of 1:9. Optimum activity is obtained between pH 6.9 and 7.5.

Reaction Products. In the presence of enzyme, GDP-mannose labeled with ^{14}C in the mannose moiety, undecaprenyl phosphate, Mg^{2+}, and phosphatidylglycerol, [^{14}C]MPU is the labeled product. The reversibility of the reaction was demonstrated by inhibition of MPU synthesis with the addition of exogenous GDP. Moreover, the addition of GDP to an incubation mixture containing enzyme, cofactors, and [^{14}C]MPU resulted in production of GDP-[^{14}C]mannose.[9] Further evidence that GDP is a product of the reversible reaction was obtained using GDP-mannose labeled in both the guanosine and mannose moieties. The major radioactive nucleotide product of the enzymatic reaction was GDP, which was identified chromatographically. The amounts of GDP and MPU produced were stoichiometrically equivalent.[2]

Preparation and Purification of Large Quantities of [^{14}C]MPU[9]

To prepare large quantities of ^{14}C-MPU, the reaction conditions already described for assay purposes must be modified to obtain the maximum yield.

Enzyme Preparation. M. lysodeikticus crude cell-free extract (120 ml) prepared as already described, is heated at 40° for 20 minutes. The pH of the solution (at 0°) is adjusted to 5.5 with 1 N acetic acid, and 31.4 g of $(NH_4)_2SO_4$ is added with stirring. After 30 minutes, the precipitate is collected by centrifugation, dissolved in 0.02 M Tris·HCl, pH 7.6, and dialyzed overnight against 4 liters of the same buffer. The dialyzed preparation (128 ml) is added with rapid stirring to 1100 ml of acetone at −25°. The resultant precipitate is washed successively with 600 ml each of acetone and diethyl ether at −25° and then dried *in vacuo* for 2 hours. Preparation of enzyme under these conditions yields about 2.70 g, dry weight, of acetone powder. Prior to use, 2.25 g of acetone powder is suspended in 60.0 ml of 0.02 M Tris·HCl, pH 7.6, and stirred at 0° for 45 minutes. Insoluble material is removed by centrifugation at 6000 g for 5 minutes.

Preparation and Purification of MPU. The reaction mixture contains (in millimoles) Tris·maleate, pH 8.5, 12.5; $MgCl_2$, 10; Tris·HCl, pH 7.6, 0.5, in 60 ml of water. To complete the reaction mixture, 60 ml of a solution of acetone powder enzyme, 75 ml of a suspension of 1.45 g of *M. lysodeikticus* crude lipids dispersed in H_2O by sonic oscillation, and 2 ml of GDP-[^{14}C]mannose (14.72 μmoles, specific activity 3.24 × 10^5

[9] M. Scher, W. J. Lennarz, and C. C. Sweeley, *Proc. Nat. Acad. Sci. U.S.* **59**, 1313 (1968).

dpm per micromole) are added. After incubation at $37°$ for 2 hours, the reaction is terminated with the addition of 20 volumes of $CHCl_3$–CH_3OH, 2:1. The lipid extract is prepared as already described for preparation of crude lipid from whole cells. In one preparation 2.7×10^6 dpm (8:15 μmoles) of [^{14}C]mannose-containing lipid was obtained in the lipid extract.

The lipid from the reaction mixture is applied to a silicic acid column (60 g) and eluted successively with 1000 ml of $CHCl_3$, 600 ml of acetone, and 600 ml of $CHCl_3$–CH_3OH, 1:1. Crude MPU, quantitatively recovered in the last fraction, is dissolved in $CHCl_3$–CH_3OH, 2:1, and applied to a DEAE-cellulose column (4.5×30 cm) prepared in $CHCl_3$–CH_3OH, 2:1. The column is eluted with 2400 ml of $CHCl_3$–CH_3OH, 2:1, 700 ml of CH_3OH, and 700 ml of $CHCl_3$–CH_3OH, 2:1, containing 84 ml of concentrated NH_4OH. All the radioactive lipid is recovered in the acidic lipid fraction eluted with the last solvent. This fraction is evaporated to dryness and dissolved in 16 ml of $CHCl_3$–CH_3OH, 1:4. After addition of 1.5 ml of 1 N NaOH, the solution is incubated at $37°$ for 15 minutes. Then 1.5 ml of 1 N acetic acid, 30 ml of $CHCl_3$–CH_3OH, 9:1, 15 ml of isobutanol, and 30 ml of H_2O are added and the solution is mixed vigorously. The aqueous layer is discarded and the $CHCl_3$ layer is washed with 15 ml of H_2O–CH_3OH, 2:1, and then evaporated to dryness. The [^{14}C]MPU, 2.30×10^6 dpm (7.0 μmoles), is applied to a DEAE-cellulose column and eluted with $CHCl_3$–CH_3OH, 2:1, and CH_3OH, as described above. The column is then eluted with 0.005 M ammonium acetate in 99% CH_3OH and 10-ml fractions are collected. MPU is eluted in a single radioactive peak between fractions 144 and 160. The pooled fractions are evaporated to dryness, dissolved in 2 ml of 1% CH_3OH in $CHCl_3$, and applied to a Unisil silicic acid column (1×7 cm). The column is eluted with 10 ml each of 1% CH_3OH in $CHCl_3$, 5% CH_3OH in $CHCl_3$, and 50% CH_3OH in $CHCl_3$. The last effluent contains 1.71×10^6 dpm (5.2 μmoles) of MPU. Examination of the MPU fraction by thin-layer chromatography on silica gel (eluent $CHCl_3$, CH_3OH, H_2O, 12:6:1) revealed the presence of one compound that was positive in tests for phosphate and lipid (Rhodamine). This compound ($R_f = 0.22$) contained greater than 95% of the radioactivity. Two other minor components ($R_f = 0.48$, 0.58) were detected with Rhodamine; both were free of phosphate and radioactivity.

MPU is further purified by gel filtration on a Sephadex column. To a column (2.5×81 cm) of Sephadex LH-20 prepared in 0.001 M ammonium acetate in $CHCl_3$–CH_3OH, 1:1, is added 0.855×10^6 dpm (2.62 μmoles) of MPU. The column is eluted with the above solvent at a flow rate of 0.5 ml per minute, and 2-ml fractions are collected. The MPU is eluted in a single radioactive peak in fractions 100–108, the recovery being 0.765×10^6 dpm (2.43 μmoles). Mannose and phosphate determinations, as well as NMR and mass spectral analyses, were performed on MPU purified in this manner. MPU is stored at $-20°$ in chloroform. For use in an enzymatic reaction mixture as the mannosyl donor for mannan synthesis the solvent is evaporated and traces of residual chloroform are removed by drying under reduced pressure for 30–60 minutes. Water or buffer is added to the dry residue, and the mixture is subjected to sonic oscillation with the small probe of a Biosonik instrument at intensities up to 20% of maximum.

Properties of [^{14}C]*MPU*. MPU is alkali stable and acid labile. The ratio of phosphate to mannose in the aqueous phase of an acid hydrolyzate of [^{14}C]MPU is total P:mannose (^{14}C):mannose (anthrone) = 1.00:1.08:0.98. The mass spectrum of the lipid residue in the ether phase of the hydrolyzate consists of polyprenols, mainly the C_{55} alcohol, undecaprenol, and its dehydration products.

Assay for [^{14}C]Mannan

Principle. Either GDP-[^{14}C]mannose or [^{14}C]MPU which is prepared enzymatically can be used as mannosyl donors with a particulate enzyme preparation which catalyzes the synthesis of mannan.[10] The chromatographic method of assay takes advantage of the fact that the polysaccharide product is immobile, whereas the radioactive lipid intermediate and sugar nucleotide substrate move away from the origin of the paper chromatogram.

Reaction Mixture.[10] The reaction mixture contains (in micromoles): Tris·HCl (pH 7.6), 1.8; Tris·maleate (pH 8.5), 40; GDP-[^{14}C]mannose, 0.029 (1.23×10^3 dpm/nmole) or MPU-^{14}C, 0.0048 (1.0×10^3 dpm/nmole); and particulate enzyme, 0.82 mg of protein; in a volume of 0.20 ml.

Procedure.[10] The reaction is stopped after 60 minutes at 37° by boiling for 3 minutes. An entire reaction mixture is spotted as a 1.5-inch band on Schleicher and Schuell green ribbon paper, No. 589. After elution at 30° with ethanol–1.0 M ammonium acetate, pH 7.3 (5:2) for 18 hours, [^{14}C]mannan synthesis is estimated by cutting the origin zone from a chromatogram and counting it in toluene scintillation fluid.

Preparation of Enzyme[10]

M. lysodeikticus cultures are grown and harvested as already described. The washed cells are used directly or after having been frozen at $-20°$ for periods up to 2 months. In either case there is no evidence

[10] M. Scher and W. J. Lennarz, *J. Biol. Chem.* **244**, 2777 (1969).

of any difference in the yield of activity of the enzyme. Cells are suspended in a volume of 0.02 M Tris·HCl, pH 7.6, equal to the wet weight of cells and this suspension is passed twice through a French pressure cell at 28,000 psi. The broken cell suspension is centrifuged at 8000–14,000 g. The supernatant preparation is further fractionated by ultracentrifugation in a Spinco model L ultracentrifuge at 100,000 g for 1 hour. The pellet produced by ultracentrifugation is suspended in a quantity of 0.02 M Tris·HCl, pH 7.6, equal to approximately 0.016 of the volume of the crude extract before centrifugation. The protein concentration in the pellet suspension should be approximately 47 mg/ml, and about 360 mg of protein in such a suspension can be obtained from 100 g, wet weight, of cells.

Characteristics of Reaction

With GDP-[^{14}C]mannose as substrate, [^{14}C]MPU and [^{14}C]mannan are formed. In the presence of EDTA, both MPU and mannan synthesis are markedly inhibited, although mannan synthesis from MPU is not inhibited. On the other hand, Triton X-100 prevents the formation of mannan from either GDP-mannose or MPU, but MPU synthesis from GDP-mannose is not altered. These inhibition studies and the kinetics of incorporation of mannosyl units from GDP-mannose into mannan and MPU have furnished proof that MPU is an obligatory intermediate in the transfer of mannosyl units from GDP-mannose to mannan. Mannan labeled *in vitro* using GDP-[^{14}C]mannose or [^{14}C]MPU as the mannosyl donor was extracted from the reaction mixture and partially purified by dialysis and gel filtration. The distribution of radioactive mannosyl units in the product was found to be concentrated in the non-reducing termini of the polysaccharide. Thus under the conditions described, both GDP-mannose, indirectly, and MPU, directly, serve as mannosyl donors to endogenous mannan in the enzyme preparation, and little or no *de novo* synthesis of the mannan from MPU is observed.

[80] Chitin Synthetase System from Yeast[1]

By Enrico Cabib

Chitin synthetase zymogen

activating factor ----→ | X ---- activating factor inhibitor

active chitin synthetase

$$n\text{UDP-}N\text{-acetylglucosamine} + \text{particle} \xrightarrow{} n\text{UDP} + (N\text{-acetylglucosamine})_n\text{-particle}$$

Yeast chitin synthetase, a particulate preparation, can be isolated in an inactive, or zymogen, form. By incubation either with trypsin or with an enzyme from yeast (activating factor) the zymogen is transformed into active forms. It is not known whether the active forms obtained with trypsin or activating factor are different or identical.

The action of trypsin can be blocked with soybean trypsin inhibitor and that of activating factor with a heat-stable protein from yeast. In what follows the preparation of chitin synthetase both in the zymogen and in the active form, of the activating factor and of the activating factor inhibitor will be described.

I. Chitin Synthetase Zymogen

Assay Method

Principle. Zymogen is first converted into an active form by incubation with trypsin. The incorporation of radioactivity from UDP-N-acetyl-D-glucosamine into insoluble particles is then measured.

Reagents

UDP-N-acetylglucosamine, 10 mM, labeled with ^{14}C in the acetyl-glucosamine moiety (specific activity 200,000 cpm/μmole), can be purchased commercially or prepared in the laboratory[2]

Imidazole chloride, 0.5 M, pH 6.5

Crystalline trypsin, 2 mg/ml, in 0.05 M potassium phosphate, pH 7

Soybean trypsin inhibitor, 3 mg/ml, in 0.05 M potassium phosphate, pH 7

[1] F. A. Keller and E. Cabib, *J. Biol. Chem.* **246**, 160 (1971); E. Cabib and F. A. Keller, *J. Biol. Chem.* **246**, 167 (1971). Since the nature of the acceptor is not yet known, the noncommittal "chitin synthetase" is preferred to a more specific name.
[2] See Vol. 8 [18].

N-Acetyl-D-glucosamine, 0.8 M
Ethanol, absolute
Ethanol, 66%
Ethanol, 66%, containing 0.1 M ammonium acetate
Scintillation mixture: toluene, 1 liter; 2,5-diphenyloxazole, 5.35 g; 1,4-bis[2-(5-phenyloxazolyl)]benzene, 0.32 g, and CAB-O-SIL thixotropic gel (Packard Instrument Company), 43 g

Procedure. ACTIVATION STEP. The incubation mixture contains 20 μl of zymogen suspension, 3 μl of imidazole buffer, varying amounts of trypsin (see below), and water in a total volume of 43 μl. After incubation for 15 minutes at 30° the reaction is stopped by adding a volume of trypsin inhibitor solution identical to the volume of trypsin used in the corresponding reaction mixture, and the tubes are immediately cooled in ice.

The amount of trypsin necessary for maximal activation varies somewhat from preparation to preparation of zymogen. Also, with aged preparations, the optimal amount of trypsin is lower and an excess of the proteolytic enzyme leads to a substantial loss in chitin synthetase activity. Therefore, amounts of trypsin solution from 1 to 4 μl are tried. With a fresh preparation of zymogen, the optimum amount of trypsin is usually about 2 μl.[3]

ASSAY STEP. To each incubation mixture from the activation step, 2 μl of N-acetylglucosamine and 5 μl of UDP-[14C]N-acetylglucosamine are added. Incubation is for 30 or 60 minutes at 30°. The reaction is stopped by adding 1 ml of 66% ethanol, and the tubes are centrifuged for 5 minutes at 1500 g. The pellets are washed twice with 1 ml of 66% ethanol containing 0.1 M ammonium acetate, and resuspended in 0.4 ml of absolute ethanol by stirring on a Vortex mixer at maximum speed. The suspension is poured into a scintillation vial, followed by a washing of the tube with 0.4 ml of absolute ethanol. Scintillation mixture (12 ml) is added, and the vials are vigorously stirred on the Vortex mixer and counted in a scintillation spectrometer.

Note. It should be noticed that the Mg^{2+} necessary for chitin synthetase action is contained in the enzyme preparation, which is suspended in buffer with 2 mM $MgSO_4$. If less enzyme is used, or the particles are suspended in Mg^{2+}-free buffer, the divalent cation should be added separately.

---

[3] The activation step can also be carried out with an excess of activating factor instead of trypsin. For details of the procedure see Assay of Activating Factor in Section III. The maximal activity obtained is usually about two-thirds of that induced by trypsin.

Definition of Unit. One unit of zymogen is that amount which will yield 1 unit of active chitin synthetase (see Section II) when maximally activated.

Preparation of Enzyme

All operations are carried out at 0–2°, unless otherwise stated.

Yeast Growth. *Saccharomyces cerevisiae* S288C, a haploid of the α mating type (available from Dr. G. Fink, Department of Genetics, Cornell University, Ithaca, New York) is grown at 28–30° in a medium containing 1% yeast extract, 2% peptone, and 2% glucose. The cells are harvested by centrifugation during the logarithmic phase, at an absorbance of 0.3, as measured at 660 nm with the Coleman Junior spectrophotometer and 1-cm diameter cuvettes. The yeast is washed twice with distilled water and stored in the refrigerator until used. No decrease in enzyme activity was observed after 48 hours of storage.

Spheroplast Preparation. Spheroplasts are obtained and washed as described in Vol. 22 [14]. The final spheroplast dilution is such that, for each gram of yeast (wet weight) used, 0.67 ml of spheroplast suspension is obtained.

Lysis of Spheroplasts and Preparation of Particles. The spheroplast suspension is added dropwise, while stirring with a Vortex mixer, to 5 volumes of Buffer A[4] containing 0.385 M mannitol. At brief intervals further additions are made of 3 and 5 volumes, respectively, of Buffer A.[4] The suspension is then submitted to sonic oscillation in 15-ml portions, using the microtip of a Branson Sonifier Model S75 for three 10-second periods at setting 1 and two 5-second periods at setting 2, both at optimal tuning, while cooling in ice. If a different sonic oscillator is used, the criterion to be followed should be to use the minimal treatment, both in intensity and time, that will lead to almost complete disappearance of round forms of spheroplast size, as observed in the phase contrast microscope.

The sonically treated suspension is centrifuged for 10 minutes at 20,000 g, and the pellet is washed twice with an amount of Buffer A[4] equivalent to about 4 volumes of the original spheroplast suspension. If the zymogen is going to be used immediately the final pellet is resuspended, in Buffer A,[4] to give a total volume equal to that of the original spheroplast suspension. If the zymogen must be stored, the final suspension, in the same buffer, should contain 33% glycerol. This suspension can be stored at −10° without freezing. Before use, glycerol is eliminated by washing the particles twice with Buffer A[4] by centrifugation and

---

[4] Imidazole chloride, 50 mM, at pH 6.5, containing 2 mM MgSO$_4$.

resuspending them in the same buffer. The preparation contains about 10 mg of protein per milliliter. Storage with glycerol results in a loss of about 30% of the activity in a week.

Zymogen can also be prepared successfully from *Saccharomyces carlsbergensis* 74 S (National Collection of Yeast Cultures, England), using the above described technique.

II. Active Chitin Synthetase

Assay Method

The reagents and procedure are the same as used in the assay step of the measurement of zymogen (see Section I). The incubation mixture contains 20 μl of active chitin synthetase, 3 μl of 0.5 M imidazole chloride, pH 6.5, 5 μl of 10 mM UDP-[^{14}C]N-acetylglucosamine, 2 μl of 0.8 M N-acetylglucosamine, and water in a total volume of 52 μl.

Definition of Unit. One unit of chitin synthetase is that amount which catalyzes the incorporation of 1 μmole of acetylglucosamine per minute into chitin.

Enzyme Preparation

Active forms of chitin synthetase can be obtained by incubating zymogen (see Section I) either with trypsin or with activating factor. The conditions for the former are given in Section I and for the latter in Section III; the incubation mixtures can be scaled up as required. After incubating with trypsin or activating factor and stopping the reaction with soybean trypsin inhibitor or with activating factor inhibitor, respectively, the activated particles may be centrifuged, washed with Buffer A[4] and resuspended in the same buffer.

Properties

Most properties were examined on preparations of zymogen contaminated with activating factor,[1,5] in which activation took place during the assay. Therefore, some of the kinetic data are subject to revision.

Stability. Chitin synthetase has not been stored in the active form. Data are only available for the zymogen (see Section I).

Kinetics. The pH optimum is around 6.2. The K_m value for UDP-N-acetylglucosamine is between 0.6 and 0.9 mM.

Activators. The enzyme has an absolute requirement for divalent cations. Mg^{2+} shows a broad optimum between 1 and 20 mM, whereas Mn^{2+} is maximally effective at 1 mM. Independently from the effect of

[5] E. Cabib and V. Farkas, *Proc. Nat. Acad. Sci. U.S.* **68**, 2052 (1971).

cations, free N-acetylglucosamine increases 3- to 5-fold the enzymatic activity, with a K_m of 4.7 mM. Other polyhydric compounds, such as glucose and glycerol, are also effective, but at much higher concentrations.

Inhibitors. The enzyme is strongly and competitively inhibited by the antibiotic Polyoxin A ($K_i = 5 \times 10^{-7} M$). UDP, a reaction product, is also inhibitory, whereas other nucleotides are less effective.

III. Chitin Synthetase Activating Factor

Assay Method

Principle. Chitin synthetase zymogen is incubated with different amounts of activating factor. After stopping the reaction with excess activating factor inhibitor, the amount of active chitin synthetase formed is measured.

Reagents

> Chitin synthetase zymogen, see Section I
> Activating factor inhibitor, see Section IV
> Imidazole chloride, pH 6.5, 50 mM

For other reagents see assay method for chitin synthetase zymogen (Section I).

Procedure. ACTIVATION STEP. The incubation mixture contains 20 µl of zymogen, 1 µl of 40 mM MgSO$_4$, 1 µl of 0.5 M imidazole chloride, pH 6.5, different amounts of activating factor (20 µl maximum) and 0.05 M imidazole chloride, pH 6.5 to complete a total volume of 42 µl. Controls without activating factor are included.

Incubation is carried out for 30 minutes at 30°, and the reaction is stopped by adding an excess of activating factor inhibitor in 3 µl (see Section IV), and cooling the tubes immediately in ice.

ASSAY STEP. This step is carried out exactly in the same way as the assay step for zymogen (see Section I above).

Definition of Unit. One unit of activating factor is that amount which causes the formation of 1 unit of chitin synthetase activity (see Section II) in 30 minutes.

Preparation of Enzyme

Principle. Activating factor accompanies the particles which contain chitin synthetase, when these are obtained without making use of sonic oscillation. A mild sonic treatment then releases the activating factor in the soluble state. All operations are carried out at 0–2°, except where otherwise stated.

Yeast Growth. *S. cerevisiae* S288C is grown at 28–30° in a medium containing 2% glucose and 0.7% Bacto Yeast nitrogen base, up to the onset of the stationary phase (absorbancy at 660 nm between 0.4 and 0.45, as measured in a Coleman Junior spectrophotometer with 1-cm diameter cuvettes). The cells are harvested by centrifugation, washed twice with distilled water, and stored in the refrigerator until used.

Spheroplast Preparation. See Section I. Since the cell wall of stationary phase cells is more resistant to the action of snail enzyme, the amount of the latter is doubled. It is also convenient to increase the concentration of mannitol to $1\,M$, both during incubation and washing, because of the fragility of spheroplasts from stationary phase yeast.

Solubilization and Ammonium Sulfate Precipitation. Spheroplast lysis, centrifugation and washing of the particulate fraction, and final resuspension in Buffer A,[4] are carried out as for the preparation of zymogen, but omitting the sonic oscillation. The resuspended particles are submitted to two 5-second periods of sonic oscillation, using the microtip of the Branson sonifier at setting 1 and optimal tuning, and centrifuged for 10 minutes at 20,000 g. The supernatant fluid, including a layer of loose particles, is removed and incubated for 20 minutes at 37°. This incubation increases the yield of activating factor for unknown reasons. The suspension is then centrifuged for 30 minutes at 100,000 g. The clear supernatant is removed and frozen overnight. After thawing, an abundant precipitate of denatured protein is removed by centrifugation for 10 minutes at 20,000 g. Activating factor is precipitated from the supernatant fluid by adding 475 mg of solid ammounium sulfate per milliliter. The pellet is redissolved in 0.05 M imidazole, pH 6.5, containing 1 mM EDTA, in a total volume one-third that of the original, and dialyzed against two changes of the same buffer for 3 hours. Glycerol is added to the dialyzate to a final concentration of 33%. The yield in the ammonium sulfate step is around 70% and the specific activity of the product ranges between 15 and 35 munits per milligram of protein.[6]

Properties

Stability. The purified activating factor is stable for at least a month when stored at $-20°$ in the presence of glycerol.

Specificity. Activating factor from *S. cerevisiae* is equally effective on zymogens from the same yeast or from *S. carlsbergensis* 74 S.

Kinetics. The transformation of chitin synthetase zymogen into active enzyme produced by the activating factor shows a linear dependence on

---

[6] Attempts to prepare activating factor from *S. carlsbergensis* 74 S, using the method described for *S. cerevisiae*, have been unsuccessful so far.

amount of activating factor added or time, until about two-thirds of zymogen has been converted. Whereas active chitin synthetase depends on a divalent cation, no such requirement was found for the activation of zymogen.

Inhibitor. Activating factor is stoichiometrically inhibited by a heat-stable protein from yeast, the preparation of which is described in Section IV. In the same Section, directions for titrating activating factor against the inhibitor are given.

IV. Activating Factor Inhibitor[1,5]

Assay Method

Principle. The inhibitor binds tightly to the activating factor for chitin synthetase zymogen. Addition of increasing amounts of inhibitor to an activation mixture produces a linear decrease in the amount of active chitin synthetase formed.

Reagents

    Chitin synthetase zymogen, see Section I

    Chitin synthetase activating factor, see Section III

    Other reagents are the same used in the assay methods for chitin
      synthetase zymogen and for activating factor.

Procedure. ACTIVATION STEP. The incubation mixture contains 20 μl of chitin synthetase zymogen, 1 μl of 0.5 M imidazole chloride, pH 6.5, 1 μl of 40 mM MgSO$_4$, 3 μl of an appropriate dilution of activating factor, variable amounts of inhibitor in a volume not exceeding 15 μl, and 0.05 M imidazole chloride, pH 6.5, to complete a volume of 42 μl. When dilutions of purified inhibitor are used, the diluting medium is 0.05 M MES,[7] pH 6.1, containing 2 mg/ml of bovine serum albumin. Controls lacking inhibitor or both inhibitor and activating factor are included. Incubation is for 30 minutes at 30°. The reaction is stopped by adding more inhibitor, in excess over the activating factor present and the tubes are immediately cooled in ice.

ASSAY STEP. This step is carried out as for the assay step for zymogen (see Section I above). The amount of activating factor used in the assay must be in the linear range of chitin synthetase activation, i.e., an amount which causes not more than 60–70% of the maximal activation. A plot of chitin synthetase activity as a function of added inhibitor should show a linear decrease.[5] The amount of inhibitor present is calculated from the linear portion of the graph. Since the determination involves the difference between relatively large numbers, the error is also

---

[7] MES, 2-(N-morpholino)ethanesulfonic acid.

large. It is recommended to include in the determination several amounts of inhibitor, each one in duplicate.

Obviously, the method can also be used to assay activating factor, if the concentration of the inhibitor solution is known.

Definition of Unit. One unit of inhibitor is that amount which will neutralize 1 unit of activating factor (see Section III).

Purification of Inhibitor from *S. carlsbergensis* 74 S

All operations are carried out between 0 and 5°, except where stated otherwise. A typical preparation is described below.

Yeast Growth. S. carlsbergensis 74 S is grown in a medium containing, per liter, 3 g of Bacto malt extract, 3 g of Bacto yeast extract, 5 g of Bacto peptone, and 20 g of glucose, at 28–30°, to the early stationary phase. From 9 liters of culture medium, about 144 g of yeast are obtained after centrifuging and washing three times with distilled water. The yeast is frozen overnight.

Step 1. Extraction and Concentration. The thawed cells are suspended in 191 ml of boiling 50 mM imidazole chloride, pH 6.5, and boiled for 3 minutes. After rapid cooling in ice, the suspension is centrifuged for 10 minutes at 27,000 g. The pellet is reextracted once with 96 ml of the same buffer, and both supernatant liquids are combined. The extract, 317 ml, is concentrated to a final volume of 27 ml in a rotary evaporator at room temperature and dialyzed against two 3-liter changes of distilled water for 16 hours. The faintly turbid solution remaining in the dialysis bag is centrifuged for 10 minutes at 27,000 g and the pellet is discarded.

Step 2. Treatment with DEAE-Cellulose. To the centrifuged dialyzate (34 ml) an equal volume of 50 mM imidazole chloride, pH 7.2, is added and the solution is passed, at room temperature, through a microgranular DEAE-cellulose (DE-52 from H. Reeve Angel Company, New York, New York) column (2.5 × 16 cm), which has been previously equilibrated with 25 mM imidazole chloride at pH 7.2. The column is washed with 152 ml of the same buffer, and the washing is combined with the first filtrate (total volume 220 ml).

Step 3. Chromatography on Carboxymethylcellulose. The DEAE eluate is concentrated to 20 ml in a rotary evaporator at room temperature and applied to a Sephadex G-25 (fine) column (2.5 × 36 cm) previously equilibrated with 25 mM MES at pH 6.1. The same buffer is used for subsequent washing of the column. After the void volume, the next 40 ml are collected[8] and directly applied to a microgranular carboxymethyl cellulose (CM-52, H. Reeve Angel Company, New York,

---

[8] In later experiments the Sephadex filtration, which sometimes results in low recoveries, has been substituted by dialysis against 25 mM MES, pH 6.1.

New York) column (2.5 × 10 cm), previously equilibrated with 25 mM MES, pH 6.1. The column is then washed with 36 ml of the same buffer and a linear sodium chloride gradient is applied. The mixing chamber contains 250 ml of 25 mM MES, pH 6.1, and the reservoir 250 ml of the same buffer, to which NaCl has been added to a final concentration of 0.4 M. Fractions of 10 ml are collected and assayed for inhibitor content. The fractions containing the bulk of the inhibiting activity (No. 21 to 27 in the preparation described), are pooled and concentrated to 3.5 ml by ultrafiltration through an Amicon Diaflo filter, type UM-10.

In the case described, the overall purification achieved was 65-fold and the protein concentration in the most purified fraction, 0.34 mg/ml.

Purification of Inhibitor from S. cerevisiae S288C

The procedure is the same as for S. *carlsbergensis*, with the following modifications: (1) The yeast is grown in a medium containing 1% yeast extract, 2% peptone, and 2% glucose. (2) In the DEAE-cellulose step, the inhibitor remains adsorbed on the exchanger after passage and washing. It is then eluted with 150 ml of 25 mM imidazole chloride at pH 7.2, containing 0.1 M NaCl, and the purification is continued as described in the preceding section.

Properties

Stability. The inhibitor appears to be quite stable when stored at −20°. Even the purified preparations can be boiled without loss of activity. Nevertheless, activity is lost when the purified inhibitor is diluted in the absence of albumin. This inactivation appears to be due to adsorption to glass, because it is much less marked when polypropylene material is used.

Purity. The purified preparations, both from S. *carlsbergensis* 74 S and from S. *cerevisiae* S288C, when submitted to disc gel electrophoresis,[9] give a single major band which coincides with the inhibitory activity. On increasing the amount of protein applied to the gel, several minor bands appear.

Specificity. An absolute quantitative comparison of the inhibitory activity of the preparations from S. *carlsbergensis* and S. *cerevisiae* has not yet been made. The inhibitor from each species is effective against the activating factors both of the same and of the other species.

---

[9] See Vol. 22 [39].

[81] Cellulose Synthetase from *Acanthamoeba*

By JANET L. POTTER and ROBERT A. WEISMAN

A particulate enzyme prepared from encysting cells of *Acanthamoeba castellanii* (Neff) catalyzes the incorporation of glucose from UDP-[¹⁴C]glucose into both alkali-soluble and alkali-insoluble β-(1,4)-glucans. As the amoeboid forms are transformed into cyst forms there is a 30-fold increase in the *in vitro* synthesizing activity of extracts prepared during the first 26 hours of encystment. During this same period the quantity of alkali-insoluble β-glucan (cellulose) synthesized *in vivo* increases 25-fold. Alkali-soluble and alkali-insoluble β-glucans corresponding to the enzymatically synthesized product have been isolated from cyst walls of *Acanthamoeba*.

Assay Method

Principle. The assay is based on the measurement of radioactivity incorporated from UDP-glucose labeled in the glucosyl residue into both alkali-soluble and alkali-insoluble products.[1]

Reagents

UDP-D-glucose uniformly labeled with ¹⁴C in the D-glucosyl moiety (containing approximately 4×10^3 cpm/nmole)
$MgCl_2$, 0.025 M
Tris-acetate buffer, 0.05 M, pH 7.4
Acid-swollen cellulose[2]
NaOH, 2%

Procedure. The reaction mixture consists of UDP-[¹⁴C]glucose (1.7 mM with 2×10^5 cpm), $MgCl_2$ (3.7 mM), acid swollen cellulose (1 mg) and the particulate fraction (containing approximately 1 mg of protein) in a final volume of 0.32 ml of 0.05 M Tris·acetate buffer, pH 7.4. The mixture is incubated in a conical test tube for 30 minutes at 30°, after which the reaction is terminated by adding 1 ml of water and heating the tubes in a 100° water bath for 10 minutes. The water-insoluble material is collected by centrifugation and washed two times with 1 ml of water. This water-washed pellet is then twice extracted with 1-ml portions of 2% NaOH for 5 minutes at 100°. The alkali-insoluble precipitate is

---

[1] A. D. Elbein, G. A. Barber, and W. Z. Hassid, Vol. 8 [72].

[2] G. Jayme and F. Lang, *in* "Methods in Carbohydrate Chemistry," Vol. III (R. L. Whistler, ed.), p. 75. Academic Press, New York, 1963.

washed three times with 1 ml of water. A 0.1-ml aliquot of the combined alkali supernatants and the entire alkali-insoluble precipitate are applied to 1-inch diameter Whatman No. 1 filter papers and dried under an infrared lamp. The filter paper are placed in vials containing 0.49% PPO and 0.1% POPOP in toluene; radioactivity is measured in a liquid scintillation spectrometer.

Preparation of the Enzyme

Growth of Cells. Cultures of *A. castellanii* are grown in 800 ml of proteose peptone medium in low-form culture flasks aerated by rotary shaking.[3,4] After 4 or 5 days the amoebae are collected by centrifugation at 600 *g*, washed once with a nonnutrient encystment medium[3] and then resuspended in 250 ml of encystment medium in 1-liter Erlenmeyer flasks. The encystment cultures are aerated by rotary shaking until approximately 50% of the cell population have begun to form obvious cyst walls as judged by light microscopy (about 24 hours).

Preparation of the Enzyme. The encysting cells of *A. castellanii* are collected by centrifugation at 600 *g*. The cell pad is twice washed with cold 0.05 *M* Tris·acetate buffer, pH 7.4. The final pellet is resuspended in 5–10 ml of buffer and passed through a French pressure cell at 18,000 psi until greater than 99% cell breakage has been achieved (generally requiring at least two passes). The broken-cell suspension is centrifuged at 1000 *g* at 4° for 15 minutes, and the supernatant is discarded. The pellet is resuspended in a small volume of buffer (with a protein concentration of approximately 5 mg/ml), frozen at −50° and stored at −20°.

Nature of the Product[5]

The partial acid hydrolysis (5–10 minutes in fuming HCl at 25°) of both the alkali-soluble and alkali-insoluble products results in a series of degradation products that are indistinguishable from those of authentic cellulose following descending paper chromatography in butan-1-ol:pyridine:0.1 *N* HCl (5:3:2) or propan-1-ol:ethyl acetate:water (7:1:2). Similarly, when subjected to acetolysis[6] both of the enzymatic products give rise to a series of acetylated derivatives indistinguishable from the acetolysis products of cellulose when separated by thin-layer

[3] R. A. Weisman and M. O. Moore, *Exp. Cell Res.* **54**, 17 (1969).

[4] E. D. Korn, *J. Biol. Chem.* **238**, 3584 (1963).

[5] J. L. Potter and R. A. Weisman, *Biochim. Biophys. Acta* **237**, 65 (1971).

[6] M. L. Wolfrom and A. Thompson, *in* "Methods in Carbohydrate Chemistry," Vol. III (R. L. Whistler, ed.), p. 143. Academic Press, New York, 1963.

chromatography on silica gel G with double development in benzene: methanol (96:4).

Properties of the Enzyme

Stability. Frozen preparations of the enzyme are stable for about 72 hours before activity begins to decrease. Less than 50% of the original activity remains after 1 week.

Specificity. The enzyme appears to be specific for UDP-glucose since little if any incorporation is observed with the glucose derivatives of the other nucleotides as substrates.

Optimum pH. Incorporation of radioactivity is maximal in the pH range of 6.5–7.5.

Activators and Inhibitors. To date compounds stimulating or inhibiting enzymatic activity affect incorporation into alkali-soluble and alkali-insoluble material approximately equally. Mg^{2+} (1.5–4.0 mM) increases incorporation 2-fold whereas EDTA (50 mM) decreases it by half. The addition of glucose (1 mM), glucose 1-phosphate (20 mM), and glucose 6-phosphate (2 mM) results in an activity increase of 1.7, 1.8, and 6-fold, respectively.[7]

p-Hydroxymercuribenzoate (2.5 mM), UDP (20 mM), and UTP (10 mM) completely inhibit incorporation. When added to cultures of whole cells both actinomycin D (100 μg/ml of encystment medium) and cycloheximide (20 μg/ml) block the normal development of the cellulose synthetase activity during the course of encystment.

---

[7] Although the incorporation of radioactivity into the alkali-insoluble material is stimulated by glucose 6-phosphate only β-1,4-glucans are synthesized as judged by analysis of the product (partial acid hydrolysis).

[82] Synthesis of Bacterial O-Antigens

By M. J. Osborn, M. A. Cynkin, J. M. Gilbert, L. Müller, and M. Singh

Biosynthesis of the O-antigen chains of the lipopolysaccharides of *Salmonella* occurs by a complex series of reactions involving participation of a C_{55} polyisoprenol-phosphate coenzyme. This coenzyme, called variously undecaprenyl phosphate,[1] antigen carrier lipid,[2] or glycosyl

---

[1] M. Scher and W. J. Lennarz, *J. Biol. Chem.* **244**, 2777 (1969).

[2] A. Wright, M. Dankert, P. Fennessey, and P. W. Robbins, *Proc. Nat. Acad. Sci. U.S.* **57**, 1798 (1967).

carrier lipid[3] (P-GCL[4]) acts as intermediate carrier of glycosyl units during assembly and polymerization of the oligosaccharide repeating unit of the O-antigen. The pathway of biosynthesis in *S. typhimurium* is as follows[3,5-9]:

$$\text{UDP-Gal} + \text{P-GCL} \rightleftharpoons \text{Gal-PP-GCL} + \text{UMP} \tag{1}$$

$$\text{Gal-PP-GCL} + \text{TDP-Rha} \rightarrow \text{Rha-Gal-PP-GCL} \, [+\text{TDP}] \tag{2}$$

$$\text{Rha-Gal-PP-GCL} + \text{GDP-Man} \rightarrow \text{Man-Rha-Gal-PP-GCL} \, [+\text{GDP}] \tag{3}$$

$$\text{Man-Rha-Gal-PP-GCL} + \text{CDP-Abe} \rightarrow \text{Abe-Man-Rha-Gal-PP-GCL} \, [+\text{CDP}] \tag{4}$$

$$n\text{Abe-Man-Rha-Gal-PP-GCL} \rightarrow \left(\begin{array}{c} \text{Abe} \\ | \\ \text{Man-Rha-Gal} \end{array} \right)_{n-1}\!\!\!\begin{array}{c} \text{Abe} \\ | \\ \text{Man-Rha-Gal-PP-GCL} \end{array}$$
$$+ \, n\text{-}1 \; \text{PP-GCL} \tag{5}$$

$$\left(\begin{array}{c} \text{Abe} \\ | \\ \text{Man-Rha-Gal} \end{array} \right)_{n}\!\!\!\begin{array}{c} \text{Abe} \\ | \\ \text{Man-Rha-Gal-PP-GCL} \end{array} + \text{core lipopolysaccharide} \rightarrow$$
$$\text{O-antigen-lipopolysaccharide}[+\text{PP-GCL}] \tag{6}$$

$$\text{PP-GCL} \rightarrow \text{P-GCL} + \text{P}_i \tag{7}$$

A similar reaction sequence is found[2,10-13] in *S. anatum*, in which the O-antigen is composed of Man-Rha-Gal trisaccharide units, except that reaction (4) is lacking.

All the enzymes of O-antigen synthesis, as well as P-GCL and the final acceptor lipopolysaccharide, are firmly bound to the membranous cell envelope fraction.

Assay of the Overall Reaction and Intermediate Steps Dependent on Endogenous P-GCL

Principle. The complete reaction sequence is catalyzed by particulate enzyme preparations from *S. typhimurium* in the presence of a mixture

[3] M. J. Osborn, *Annu. Rev. Biochem.* **38**, 501 (1969).

[4] Abbreviations used: P-GCL, phosphoryl glycosyl carrier lipid; PP-GCL, pyrophosphoryl glycosyl carrier lipid; Gal, D-galactose; Rha, L-rhamnose; Man, D-mannose; Abe, abequose (3,6-dideoxy-D-galactose).

[5] I. M. Weiner, T. Higuchi, L. Rothfield, M. Saltmarsh-Andrew, M. J. Osborn, and B. L. Horecker, *Proc. Nat. Acad. Sci. U.S.* **54**, 228 (1965).

[6] M. J. Osborn and I. M. Weiner, *J. Biol. Chem.* **243**, 2631 (1968).

[7] M. J. Osborn and R. Yuan Tze-Yuen, *J. Biol. Chem.* **243**, 5145 (1968).

[8] M. A. Cynkin and M. J. Osborn, *Fed. Proc., Fed. Soc. Exp. Biol.* **27**, 293 (1968).

[9] J. L. Kent and M. J. Osborn, *Biochemistry* **7**, 4396, 4419 (1968).

[10] A. Wright, M. Dankert, and P. W. Robbins, *Proc. Nat. Acad. Sci. U.S.* **54**, 235 (1965).

[11] M. Dankert, A. Wright, W. S. Kelley, and P. W. Robbins, *Arch. Biochem. Biophys.* **116**, 425 (1966).

[12] D. Bray and P. W. Robbins, *Biochem. Biophys. Res. Commun.* **28**, 334 (1967).

[13] S. Kanegasaki and A. Wright, *Proc. Nat. Acad. Sci. U.S.* **67**, 951 (1970).

of the 4 nucleotide sugar substrates and Mg^{2+} ion.[5] Intermediate steps can also be assayed by varying substrate composition and reaction conditions. Depending on the assay procedure employed, it is possible to determine differentially the incorporation of radioactivity from labeled nucleotide sugar substrate into the final lipopolysaccharide product and the mono- and oligosaccharide intermediates (reactions 1–4) in addition to determination of overall total incorporation. Although wild-type bacteria are suitable as source of the particulate enzyme fraction for most purposes, mutants blocked either in biosynthesis of a required nucleotide sugar (UDP-galactose or GDP-mannose), or in synthesis of the core region of lipopolysaccharide (rfa type) are of advantage in detection and assay of certain of the intermediate reactions.

Reagents

Tris·acetate buffer, 1.0 M, pH 8.5

2-(N-morpholino)ethanesulfonate (MES) buffer, 0.2 M, pH 6.0

$MgCl_2$, 0.1 M

EDTA, 0.1 M, pH 7.5

UDP-galactose, 2.5 mM

UDP-[^{14}C]galactose, 2.5 mM (2000 cpm/nmole or higher)

TDP-rhamnose, 2.5 mM (see Vol 8 [54] for preparation)

TDP-[^{14}C]rhamnose, 2.5 mM (2000 cpm/nmole or higher) (see Vol 8 [54] for preparation)

CDP-abequose, 0.5 mM (see below for preparation)

CDP-[^{14}C]abequose, 0.5 mM (2000 cpm/nmole or higher; see below for preparation)

[^{14}C]UMP or [^{3}H]UMP, 2.5 mM (2000 cpm/nmole or higher)

Alfonic 1012-6 detergent, 0.5% (w/v); obtained from Conoco Petrochemicals

Lipopolysaccharide purified from SL TV119[14] or other rfb mutant of *S. typhimurium*, 2 mg/ml (see Vol. 8 [21] for isolation and purification)

Acetic acid, 0.1 N containing 5 mM EDTA

Chloroform:methanol, 2:1 (v/v)

Pure solvent upper phase (PSUP[15]: 15 ml of $CHCl_3$ and 240 ml of methanol are added to a solution containing 1.83 g of KCl in 235 ml of H_2O.

NH_4OH, 1.0 N containing 5 mM $MgCl_2$. The mixture is shaken well and the suspension used as such.

Acetic acid, 5 N

[14] T. V. Subbaiah and B. A. D. Stocker, *Nature (London)* **201**, 1298 (1964).

[15] J. Folch, M. Lees, and G. H. Sloane-Stanley, *J. Biol. Chem.* **226**, 497 (1957).

Preparation of Particulate Enzyme Fractions. (i) GROWTH OF BAC-
TERIA. Cells are grown in a medium containing 1% Proteose Peptone
No. 3 (Difco), 0.1% beef extract, and 0.5% NaCl at 37° with vigorous
aeration. Cultures are harvested in mid-log phase (approximately one-
half of maximum growth), and washed once with cold 0.9% NaCl. The
washed cell pellet may be stored at −18° for several months.

(ii) PREPARATION OF THE CELL ENVELOPE FRACTION.[5] All steps are
carried out at 2–4°. Two grams of frozen cells are suspended in 10 ml of
50 mM Tris·acetate, pH 8.5, containing 1 mM EDTA and sonicated with
a Branson 20-kc sonifier or equivalent instrument. The cell suspension
is placed in a 2.5 × 10 cm cellulose nitrate tube (a 1 × 3.5 inch centrifuge
tube cut down to 10 cm length is used) in an ice–H_2O bath and sonicated
for three 15-second periods with 1–2 minutes cooling between bursts.
The sonicate is centrifuged at 50,000 rpm (R_{av} = 159,000 g) for 30 minutes, and the supernatant fluid is discarded. The pellet is suspended in
10 ml of sonication buffer and centrifuged for 10 minutes at 1200 g to
remove unbroken cells. The turbid supernatant fraction is centrifuged
at 50,000 rpm for 30 minutes as above, and the supernatant fluid is
discarded. The pellet is suspended in 4 ml of sonication buffer and employed as cell envelope fraction. The protein content is determined by
the Lowry procedure (Vol. 3 [73]). Except where indicated below, the
cell envelope fraction can be stored at −70° for at least 2 weeks without loss of activity.

Preparation of EDTA-Treated Cells. The EDTA-treated cells preparation of Robbins *et al.*[16] may be employed as alternative to the cell
envelope fraction. Cells are grown to mid-log phases as above, and
rapidly chilled by addition of ice to the culture flask. Cells are collected
by centrifugation, washed with cold 0.9% NaCl and resuspended in
1/20 the original volume of cold 10 mM EDTA, pH 8. The suspension
(approximately 2 mg/ml of protein) is frozen at −20° overnight or
longer and used directly as enzyme fraction.

Preparation of Nucleotide Sugars. CDP-abequose and CDP-[^{14}C]-
abequose are prepared from CDP-glucose and CDP-[^{14}C]glucose by
the method of Elbein,[17] using a crude soluble enzyme preparation from
Salmonella typhimurium G-30[18] (UDP-galactose-4-epimerase negative).
Cells are grown and sonicated as described above. The sonicate is
centrifuged at 4° for 30 minutes at 30,000 g, and the supernatant

[16] P. W. Robbins, A. Wright, and J. L. Bellows, *Proc. Nat. Acad. Sci. U.S.* **52**,
1302 (1964).

[17] A. D. Elbein, *Proc. Nat. Acad. Sci. U.S.* **53**, 803 (1965).

[18] M. J. Osborn, S. M. Rosen, L. Rothfield, and B. L. Horecker, *Proc. Nat. Acad.
Sci. U.S.* **48**, 1831 (1962).

fraction is centrifuged for 1 hour at 159,000 g. The high speed supernatant fraction is employed as enzyme. For synthesis of CDP-abequose, the reaction mixture contains 100 μmoles of CDP-glucose, 125 μmoles of TPNH, 125 μmoles of DPN, 250 μmoles of phosphate buffer pH 7.0, and 10 ml of enzyme in a total volume of 15 ml. After incubation for 2 hours at 37°, the reaction mixture is transferred to a separatory funnel and extracted twice with 2 volumes of $CHCl_3$. The organic phases are discarded. Two volumes of ethanol are added to the aqueous phase, and the precipitate is removed by centrifugation (15 minutes, 30,000 g, 2–4°), and washed twice with 10 ml of 70% ethanol. The combined supernatant fractions are concentrated to approximately 3 ml *in vacuo* at 10–15°, and any precipitate is removed by centrifugation. CDP-abequose is isolated by paper chromatography. The sample is streaked on 3 sheets of Whatman No. 3MM paper and chromatographed 18–20 hours in ethanol:1 M ammonium acetate, pH 7.0 (7:3). CDP-glucose is spotted as standard. The paper is washed with 0.02 M EDTA (pH 7.5) and then with H_2O before use. The CDP-abequose band ($R_{CDP-Glc} = 1.4$) is visualized under UV light and cut out. The strips are soaked for 1 hour in absolute ethanol to remove ammonium acetate (it is essential that the ethanol be absolute to avoid elution of the nucleotide sugar), dried in air, and eluted with H_2O. CDP-abequose is further purified by rechromatography for 18–24 hours on Whatman No. 40 paper in isobutyric acid:conc. $NH_4OH:H_2O$ (57:4:39). To avoid degradation of the labile nucleotide sugar during drying of the paper, the sheet is dipped 6 times through diethyl ether immediately after removal from the tank. The CDP-abequose band ($R_{CDP-Glc} = 1.75$) is eluted with H_2O, and the pH adjusted to 6–6.5 for storage. The final yield is approximately 20 μmoles. The product is degraded at a significant rate during storage at $-18°$ in solution; for long-term storage it is advantageous to keep the uneluted chromatography strips in a desiccator at $-18°$ and elute portions as needed.

Synthesis of ^{14}C-labeled CDP-abequose carried out similarly except that the reaction mixture is scaled down appropriately and Whatman No. 40 paper is employed in the first chromatography.

General Assay Methods

Procedures used for determination of radioactivity incorporated into total O-antigen-related products, lipid-bound mono- and oligosaccharide intermediates, and polymeric products are as follows.

Procedure A. Incorporation of Radioactivity into Total O-Antigen Products, Including Oligosaccharide Intermediates (Sum of Reactions 1–6).[5] Incubations are carried out as described below in 5-ml round-bottom centrifuge tubes (14 × 60 mm, Sorvall No. 118). The reaction is stopped by addition of 3 ml of cold 0.1 N acetic acid–5 mM EDTA, and the tubes are chilled. The insoluble cell envelope (to which the radioactive products are bound) is generally collected by membrane filtration (Millipore Type HA, 0.45 μ, 24 mm diameter or equivalent), and washed on the filter 3 times with 5 ml of cold acetic acid–EDTA. The filter is dried under a heat lamp and transferred to a scintillation vial containing 5–10 ml of scintillation fluid for counting in a liquid scintillation counter. If amounts of protein greater than 0.5 mg are employed in the assay, filtration is slow, and the cell envelope is collected and washed by centrifugation at 2–4° (3–5 minutes at 8000 rpm in the SE-12 rotor of the Sorvall RC-2B centrifuge). The pellet is washed 3 times by dispersing it in 2.5 ml of cold acetic acid–EDTA with the aid of a glass ball-tipped homogenizer which fits the inside of the tube. The washed pellet is suspended in 0.5 ml of H_2O and transferred to a scintillation vial. The tube is washed with an additional 0.5 ml of H_2O, the wash fluid added to the vial, and the sample is counted in a liquid scintillation counter.

Procedure B. Incorporation of Radioactivity into $CHCl_3$:*Methanol-Soluble Mono- and Oligosaccharide Intermediates (Reactions 1–4).* Lipid-linked intermediates up to the dimeric octasaccharide derivative are measured after extraction into $CHCl_3$:methanol (C:M).[5-7] Incubations are carried out in 12-ml conical centrifuge tubes, and the reaction is stopped by addition of 4 ml of C:M (2:1). The mixture is agitated briefly in a Vortex mixture in order to obtain a single phase, and is permitted to stand at room temperature for 10–15 minutes. PSUP (0.8 ml) is then added, and the mixture is agitated for 20 seconds on a Vortex mixer. The phases are separated by brief (0.5–1 minutes) centrifugation at room temperature in a clinical centrifuge, and the upper (aqueous) phase and interface precipitate are aspirated off and discarded. The organic (lower) phase is washed twice more with PSUP as above. The organic phase is then transferred to a scintillation vial with a Pasteur pipette, taking care to avoid transfer of any residual interface precipitate, and dried under a heat lamp. H_2O (0.5 ml) is added, and the sample is counted in a liquid scintillation counter.

Procedure C. Incorporation into Polymeric O-Antigen (i.e., Lipopolysaccharide plus Polymeric Lipid-Linked Intermediates Containing 3 or More Oligosaccharide Repeat Units). This can be estimated as the difference between total incorporation (Procedure A) and incorporation into C:M-soluble products (Procedure B) in duplicate incubation mixtures.

Procedure D. Incorporation into Lipopolysaccharide. O-antigen chains attached to lipopolysaccharide can be determined independently of oligo- and polysaccharide intermediates linked to P-GCL by differential alkaline degradation of the latter.[8,19] The galactosyl-PP-GCl linkages are hydrolyzed to give glycosyl-1,2 cyclic phosphates with liberation of the lipid. The reaction is stopped by addition of 2 ml of $1 N$ NH_4OH–5 mM $MgCl_2$, and the mixture is permitted to stand at room temperature for 90 minutes. The reaction mixture is then filtered through a Millipore filter (Type HA, 0.45 μ, 24 mm diameter), and the filter is washed 3 times with 5 ml of 0.1 N acetic acid–5 mM EDTA, dried and counted as in Procedure A. Lipopolysaccharide is quantitatively retained by the filter, while the products of alkaline hydrolysis of the lipid-linked intermediates pass through.

Overall Reaction: Incorporation of Sugars into O-Antigen (Sum of Reactions 1–7).[5] Cell envelope fractions and EDTA-treated cell preparations from wild-type bacteria and mutant strains blocked in biosynthesis of UDP-galactose, GDP-mannose, or TDP-rhamnose are able to carry out the entire reaction sequence from the mixture of 4 nucleotide sugar substrates. Reaction mixtures contain 20 μl of Tris·acetate buffer pH 8.5, 25 μl of $MgCl_2$, 5 μl of EDTA, 10 μl each of UDP-galactose, TDP-rhamnose, GDP-mannose, and CDP-abequose (one of which is radioactive) and 0.1–1 mg of enzyme protein in a final volume of 0.25 ml. A control tube is included in which the incubation mixture is heated in a boiling H_2O bath for 2 minutes before addition of radioactive substrate. At temperatures above 20°, incorporation of galactose, rhamnose, and mannose occurs equally well in the presence and the absence of CDP-abequose,[6] and this substrate can therefore be omitted from the incubation mixture if desired. Under such conditions, incubations are for 10–30 minutes at 37°. Incorporation is approximately linear with time for 10–15 minutes and continues thereafter at a somewhat reduced rate. The abequosyl transferase (reaction 4) is labile under the conditions of incubation at 37°, and when CDP-abequose is present, incubations are carried out at 22°.[6] Total incorporation of radioactive substrate into products of O-antigen synthesis is determined by Procedure A; Procedures B, C, and D may also be employed to measure incorporation into mono- and oligosaccharide-PP-GCL intermediates, total polymer, and lipopolysaccharide, respectively. The specific activity of the cell envelope fraction for total incorporation of galactose, rhamnose or mannose (assay Procedure A) is about 2 nmoles per milligram of protein in 10 minutes. Incorporation of abequose (at 22°) is approximately 1 nmole/mg per 30 minutes. Specific activities of EDTA-treated

[19] L. Eidels and M. J. Osborn, *Proc. Nat. Acad. Sci. U.S.* **68**, 1673 (1971).

cell preparations are somewhat higher.[16] The enzyme activities required for the overall reaction are stable to storage at $-18°$ for several weeks, except for the abequosyl transferase. Approximately 50% of this activity is lost within 3 days.

Formation of Galactose-PP-GCL by Galactose-PP-GCL Synthetase (Reaction 1).[7] This reaction is readily reversible and can be assayed either by incorporation of radioactive galactose from the nucleotide sugar into gal-PP-GCL or by exchange of labeled UMP into UDP-galactose. In the forward direction, galactose-PP-GCL synthesis is measured by incorporation of radioactivity from UDP-[14C]galactose into the $CHCl_3$:methanol-soluble product according to Procedure B. Incubations are carried out in 12-ml conical centrifuge tubes, and the reaction mixture is identical to that described above except that UDP-[14C]-galactose is the only nucleotide sugar present. Control tubes contain heat-inactivated enzyme. The reaction is extremely rapid and reaches apparent equilibrium within 2 minutes at $37°$ and 5 to 7 minutes at $20°$. With longer incubation times the yield of galactose-PP-GCL may decrease progressively due to reversal of the reaction by UMP generated from UDP-galactose by UDP-sugar hydrolase[20] and other enzymes present in the cell envelope fraction. The maximal amount of galactose incorporated is limited by the availability of endogenous P-GCL in the enzyme preparation, and corresponds to approximately 0.75 nmole per milligram of cell envelope protein. Initial rates of reaction are difficult to obtain with this procedure, and for this purpose either the UMP exchange reaction or addition of exogenous P-GCL (see Section II) are recommended.

Exchange of labeled UMP into UDP-galactose is determined as follows. The reaction mixture contains in a final volume of 50 μl: 3 μl of Tris·acetate pH 8.5, 5 μl of $MgCl_2$, 1 μl of EDTA, 2 μl of nonradioactive UDP-galactose, 2 μl of [14C] or [3H]UMP and cell envelope (0.01–0.1 mg of protein). A control lacking UDP-galactose is included. After incubation for 5–10 minutes at $25°$ or $37°$, 1 μl of 5 N acetic acid is added and the reaction mixture is heated for 2 minutes in a boiling H_2O bath. Carrier nonradioactive UDP-galactose, UDP, and UMP (0.1 μmole each) are added, and UDP-galactose is separated from UMP and UDP by high voltage paper electrophoresis. The entire reaction mixture is applied to Whatman 3 MM paper as a 2-cm streak and electrophoresed in 0.05 M ammonium formate, pH 3.5, for 1 hour at 50 V/cm. Nucleotides are visualized under UV light, and the band corresponding to UDP-galactose is cut out and placed in a scintillation vial.

[20] L. Glaser, A. Melo, and R. Paul, *J. Biol. Chem.* **242**, 1944 (1967).

The strip is wet with 0.5 ml of H_2O prior to addition of 10 ml of scintillation fluid.

Formation of Rhamnosyl-Galactose-PP-GCL (Reactions 1 and 2).[5] The procedure is identical to that employed for synthesis of galactose-PP-GCL except that the incubation mixture contains nonradioactive UDP-galactose and TDP-[14C]rhamnose. The reaction reaches completion within 10 minutes at 37°. The extent of reaction is again limited by the amount of endogenous P-GCL. This limitation can be overcome by use of an alternative assay method utilizing added P-GCL (Section II, below).

Formation of Tri- and Tetrasaccharide-PP-GCL Intermediates (Reactions 1-3 or 1-4).[5,6] At incubation temperatures of 20° and above, polymerization is rapid, and little accumulation of these oligosaccharide intermediates is observed. However, polymerization of man-rha-gal-PP-GCL can be inhibited differentially by lowering the temperature of incubation.[5] The reaction mixture is that described above for the overall reaction with GDP-[14C]mannose as the radioactive substrate. Incubations are carried out at 10° for 5 minutes, and trisaccharide-PP-GCL formation is determined by Procedure B. Under these conditions 50–80% of the total [14C]mannose incorporated into the particulate fraction is recovered as C:M-soluble oligosaccharide product. Synthesis of man-rha-gal-PP-GCL can also be determined in an assay system employing exogenous P-GCL (Section II, below).

Polymerization of the complete, abequose-containing tetrasaccharide intermediate is rapid even at reduced temperatures (5–10°), and over 80% of the total abequose incorporated is recovered in polymeric products under all conditions examined.[6]

Transfer of O-Antigen Chains from P-GCL to Lipopolysaccharide by O-Antigen Ligase (Reaction 6).[8] This reaction can be assayed independently of earlier reaction of the sequence by a 2-step procedure in which [14C]polysaccharide-PP-GCL substrate is generated from nucleotide sugar precursors in the first incubation step, and subsequently transferred to added acceptor lipopolysaccharide in a second incubation period. The cell envelope fraction is obtained from a double mutant blocked both in synthesis of the core region of lipopolysaccharide and in synthesis of UDP-galactose (rfa and UDP-galactose-4-epimerase negative). The rfa mutation prevents transfer of O-antigen claims to endogenous lipopolysaccharide during the first incubation step; however, rfa mutants accumulate the polysaccharide-PP-GCL intermediate *in vivo* and the amount of residual free P-GCL available for O-antigen synthesis *in vitro* is very low. The UDP-galactose epimerase mutation is therefore superimposed in order to prevent *in vivo* accumulation of O-antigen intermediates. Synthesis of polysaccharide-PP-GCL in the first incubation step is carried out as follows. The incubation mixture contains in a final volume of 2.50 ml: 200 μl of Tris·acetate, pH 8.5, 250 μl of MgCl$_2$, 50 μl of EDTA, 100 μl each of UDP-galactose, TDP-rhamnose, and GDP-[^{14}C]mannose (5000 cpm/nmole or greater), 50 μl of 0.5% Alfonic 1012-6 and cell envelope fraction (5–10 mg of protein) from a rfa epi$^-$ mutant. This amount of incubation mixture is sufficient for 16 assay tubes in incubation II. After incubation for 60 minutes at 25°, the incubation mixture is chilled and diluted with 10 ml of H$_2$O, and the cell envelope is recovered by sedimentation at 2–4° for 15 minutes at 159,000 g. The tube is drained and carefully wiped to remove residual supernatant fluid, and the cell envelope pellet is resuspended in 0.4 ml of cold H$_2$O for use in incubation II. Aliquots (20 μl) are taken for measurement of total [^{14}C]mannose incorporation (Procedure A) as well as incorporation into the polymeric intermediate (Procedure C). Assay of ligase activity in incubation II is as follows. Incubation mixtures initially contain 20 μl of acceptor lipopolysaccharide (2 mg/ml), 20 μl of 0.2 M MES buffer pH 6.0, 3 μl of 0.5% Alfonic 1012-6 and 50 μl of H$_2$O. The lipopolysaccharide is obtained from a rfb mutant, such as TV119, which lacks the first enzyme of O-antigen synthesis. This mixture is first incubated for 30 minutes at 37° in order to disaggregate the lipopolysaccharide, and 20 μl of the above cell envelope fraction containing [^{14}C]polysaccharide-PP-GCL (1500–5000 cpm) are then added. After incubation for 15–30 minutes at 37°, transfer of ^{14}C to lipopolysaccharide is determined by Procedure D. Acceptor lipopolysaccharide is omitted from control incubation mixtures. In contrast to earlier steps in the pathway, the ligase reaction has a pH optimum of 5.5–6.5, and does not require added metal ions under the incubation conditions employed. The reaction is linear with time for approximately 30 minutes, and continues thereafter at a reduced rate until 70–80% of the initial polysaccharide-PP-GCL has been transferred to lipopolysaccharide. Although a small fraction (10–15%) of the radioactivity incorporated into the cell envelope during the first incubation is usually present as trisaccharide-PP-GCL, transfer of the oligosaccharide to lipopolysaccharide is negligible under the condition of the ligase assay.

Assay Procedures Employing Exogenous Glycosyl Carrier Lipid and Lipid-Linked Intermediates

Principle. Although the procedures described above permit assay of several intermediate steps in O-antigen synthesis, these methods suffer several disadvantages: the formation of each successive intermediate depends on the integrity of all preceding enzymes in the pathway, and the extent of reaction is limited by the small amount of endogenous P-GCL present in the cell envelope. These drawbacks can be overcome by procedures in which purified P-GCL and isolated intermediates are employed as exogenous substrates. Free P-GCL is purified from crude lipid extracts of *S. typhimurium* or *E. coli;* glycosyl-PP-GCL intermediates are synthesized enzymatically in substrate amounts from the purified lipid and isolated as described below. Utilization of the added substrates by the particulate enzyme fraction is promoted by addition of a nonionic detergent to the reaction mixture.

Purification and Enzymatic Assay of P-GCL Purification

Reagents

   CHCl$_3$ (see Vol. 14 [47] for solvent specifications)
   Methanol (see Vol. 14 [47] for solvent specifications)
   CHCl$_3$:methanol (C:M), 2:1 (v/v)
   C:M, 1:1 (v/v)
   C:M, 2:1 (v/v) containing 1% conc. NH$_4$OH
   C:M, 3:1 (v/v)
   Pure solvent upper phase (PSUP), see above
   LiOH, 0.136 M in anhydrous methanol
   Ethanol (absolute)
   50% aqueous methanol (v/v) containing 0.05 M KCl
   DEAE-acetate (see Vol. 14 [47] for preparation)
   99% methanol; 10 ml of H$_2$O is added to 990 ml of methanol
   Ammonium acetate, 10 mM–acetic acid, 5 mM, in 99% methanol

Procedure. The procedure described has been employed successfully for isolation of P-GCl from *S. typhimurium, E. coli, Staphyloccus aureus*, and *Micrococcus lysodeikticus.*[21] Frozen cells (200 g wet weight) are dispersed in 1 liter of methanol in a Waring Blendor, and the suspension is poured into 2 liters of CHCl$_3$. The mixture is stirred 30 minutes at room temperature under N$_2$ and filtered through Whatman No. 1 paper (previously washed with CHCl$_3$:methanol (C:M), 2:1 v/v) in a Büchner funnel. The residue is reextracted with 1 liter of C:M (2:1) containing 1% conc. NH$_4$OH for 30 minutes with stirring, and is filtered. The combined filtrates are washed 2 times in a separatory funnel with 500 ml of PSUP (C:M:H$_2$O (3:48:47) containing 0.05 M KCl), and the washed organic phase is evaporated to dryness under reduced pressure at 0–5°. The residue is dissolved in 60 ml of C:M (3:1), and any insoluble material is removed by filtration through a plug of glass wool,

---

[21] M. J. Osborn, unpublished results.

prewashed with C:M (3:1). The crude lipid extract from *S. typhimurium* contains approximately 50 μmoles of lipid P per milliliter, and is stored at −18° under N₂.

Purification of P-GCL from the crude lipid extract is based on differential degradation of the bulk phospholipids by alkali; the poly-isoprenol phosphate coenzyme is stable to treatment under conditions resulting in essentially quantitative deacylation of glycerophosphatides. Aliquots of the crude lipid extract are deacylated by the method of Brockerhoff.[22] Twenty milliliters of extract (approximately 1 mmole of lipid phosphate) is added to a 1-liter separatory funnel containing 55 ml of 0.136 M LiOH is anhydrous methanol, and the mixture is allowed to stand 30 minutes at room temperature. Absolute ethanol (150 ml), H₂O (225 ml), and CHCl₃ (300 ml) are then added, the mixture is shaken well, and the phases are allowed to separate. Brief centrifugation may be necessary to complete phase separation. The upper aqueous phase is backwashed with 150 ml CHCl₃, and the CHCl₃ layer is combined with the original organic phase. The pooled organic phases (which contain P-GCL) are washed once by shaking with 150 ml of 50% (v/v) aqueous methanol containing 0.05 M KCl. The washed organic phase is stored in the cold while the remainder of the crude lipid extract is carried through the deacylation procedure. The washed organic phases are combined, evaporated to dryness under reduced pressure at 0 to 5° and dissolved in 50 ml of C:M (1:1). The solution is stored at −18° under N₂. The efficiency of the deacylation procedure is monitored by determination of total lipid P remaining after deacylation. No more than 5% of the original lipid phosphate should remain; if deacylation is incomplete, the residual glycerophosphatide is incompletely removed in the subsequent chromatographic step. If necessary, the deacylation procedure may be repeated after taking the organic phase to dryness under reduced pressure at 0 to 5°, and dissolving it in 20 ml of C:M (3:1).

Final purification of P-GCL is accomplished by chromatography on DEAE-cellulose by the method of Dankert *et al.*[11] A 2.5 × 20 cm column of DEAE-acetate is prepared by the method of Rouser *et al.* (Vol. 14, p. 292) and washed with 500 ml each of glacial acetic acid, methanol and C:M (1:1) before use. The product from the deacylation step is applied in 100 ml of C:M (1:1), and the column is washed with 200 ml of C:M (1:1) and 600 ml of 99% methanol to remove fatty acids, and residual phosphatidylethanolamine and other nonacidic lipids. Elution of P-GCL is then begun with 1 liter of 10 mM ammonium acetate-5 mM acetic acid in 99% methanol; 15 ml fractions are collected and

[22] H. Brockerhoff, *J. Lipid Res.* **4**, 96 (1963).

stored in the cold. The peak of P-GCL is detected by enzymatic assay in the galactose-PP-GCL synthetase reaction (see below), and total phosphate contents are determined in the region of the peak and the fractions preceding and following it. The elution pattern is illustrated in Fig. 1. The peak of P-GCL activity is immediately preceded by, and often incompletely separated from, a peak of inactive phospholipid. Fractions containing P-GCL are pooled and concentrated to approximately 5 ml under reduced pressure at 0°. Three volumes of CHCl₃ are added and the solution is washed 3 times with 5 ml of 50% aqueous methanol to remove salt. The washed organic phase is evaporated to dryness under reduced pressure at 0°, dissolved in 2 ml of CHCl₃:methanol (3:1), and stored at −18° under N₂. Purity of the product at this stage is approximately 50% on the basis of total phosphate, as determined by enzymatic assay in the galactose-PP-GCL synthetase reaction and thin-layer chromatography on silica gel H in solvent I, butan-1-ol:glacial acetic acid:H₂O (6:2:2), and solvent II, CHCl₃:methanol:H₂O:conc. NH₄OH (65:25:3.6:0.5).

This preparation is suitable for use without additional purification. If desired however, residual contaminating phospholipid can be separated

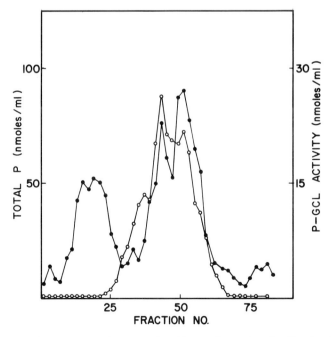

FIG. 1. Purification of P-GCL on DEAE-cellulose. ●—●, Total phosphate; ○—○, P-GLC activity. See text for details.

by rechromatography on a 0.9 × 10 cm column of DEAE-acetate. The sample is applied in 10 ml of C:M (1:1) and the column is washed with 50 ml of C:M (1:1) and 100 ml of 99% methanol. P-GCL is then eluted with a linear gradient of ammonium acetate. The reservoir contains 75 ml of 15 mM ammonium acetate–5 mM acetic acid in 99% methanol and the mixing chamber, 75 ml of 5 mM acetic acid in 99% methanol. One-milliliter fractions are collected and assayed for P-GCL and total phosphate as above. Fractions of highest specific activity are pooled, concentrated and desalted as above. The product is dissolved in 2 ml of C:M (3:1) and stored at −18° under N_2. The final product gives a single iodine-staining spot on thin layer chromatography in solvents I and II ($R_f = 0.66$ and 0.16, respectively) and is enzymatically converted to galactosyl-PP-GCL in a yield of 70–80%.

The overall yield of P-GCL is 1–2 μmoles per 100 g wet weight of cells. Estimation of recovery is somewhat uncertain, since P-GCL activity in the enzymatic assay is severely inhibited by bulk phospholipids in crude lipid extracts. However, the overall recovery appears to be over 50%.

Enzymatic Assay of P-GCL.[21] The procedure measures conversion of P-GCL to [^{14}C]galactosyl-PP-GCL, reaction (1), by a solubilized preparation of gal-PP-GCL synthetase which is dependent on added P-GCL for activity.

Reagents

Tris·acetate buffer, 1 M, pH 8.5
$MgCl_2$, 1 M
UDP-[^{14}C]galactose (5000–10,000 cpm/nmole)
Alfonic 1012-6 detergent, 0.5% (w/v). Obtained from Conoco Petrochemicals.
Soluble enzyme fraction (see below)

Procedure. Reaction mixtures contain, in a final volume of 0.10 ml: 5 μl of 1 M Tris·acetate buffer, pH 8.5, 1 μl of 1 M $MgCl_2$, 4 μl of 2.5 mM UDP-[^{14}C]galactose, 7 μl of 0.5% (w/v) Alfonic 1012-6 detergent, 10 μl of methanol, P-GCL (0.05–1 nmole) and 20–50 μl of soluble enzyme fraction (Poly-Tergent extract, see below). The solution of P-GCL in C:M (3:1) is first added to the bottom of a conical 12-ml centrifuge tube and taken to dryness under a stream of N_2. Methanol and Alfonic 1012-6 are then added and the tube is agitated on a Vortex mixture to disperse the lipid. The other components are then added, and the reaction mixtures are incubated for 15–30 minutes at 25°. A blank tube lacking added P-GCL is included. The reaction is stopped by addition of 2 ml of C:M (2:1), and incorporation of ^{14}C into the lipid product is measured by Procedure B. Conversion of purified P-GCL to the monosaccharide product is virtually quantitative. It should be noted, however, that prolonged incubation times may lead to reversal of the reaction due to hydrolysis of UDP-galactose to UMP by UDP-galactose hydrolase activity in the soluble enzyme fraction. Since the hydrolase activity varies from preparation to preparation, optimal enzyme concentrations and incubation times to achieve maximal yields of product should be checked with each preparation.

The assay is linear with concentration of added P-GCL within the designated limits when chromatographically purified P-GCL fractions are employed. However, the reaction is severely inhibited by bulk phospholipid in crude lipid extracts, and is approximately linear with concentration only at the lower limits of the range of sensitivity.

Enzymatic Synthesis of Lipid Intermediates from Purified P-GCL

Mono- and oligosaccharide intermediates are synthesized from purified P-GCL using a soluble enzyme fraction isolated by detergent extraction of the cell envelope, and are isolated by extraction into chloroform:methanol. Polysaccharide-PP-GCL((man-rha-gal)$_n$-PP-GCY) is synthesized from the trisaccharide-PP-GCL intermediate by the particulate cell envelope fraction, and is isolated as described below.

Reagents

    Tris·maleate buffer, 1 M, pH 6.0
    MgCl$_2$, 1 M
    Poly-Tergent S 305-LF, 10% (w/v); obtained from Olin Chemicals
    Alfonic 1012-6, 0.1% (w/v) (Conoco Petrochemicals)
    Alfonic 1012-6, 0.25% (w/v)
    Purified P-GCL (see above)
    Soluble glycosyl transferase fraction (see below)
    Cell envelope fraction obtained from a rfa, epi⁻ double mutant of
        S. typhimurium

Other reagents are as described in Section I.

Preparation of Soluble Glycosyl Transferase Fraction.[23] Twenty milliliters of cell envelope fraction (30–40 mg of protein per milliliter) in 50 mM Tris·acetate, pH 8.5, are stirred with 1.4 ml of 10% (w/v) Poly-Tergent S 305-LF for 5 minutes at 0° and the soluble fraction is recovered by centrifugation at 159,000 g for 1 hour at 2–5°. The supernatant

[23] L. Müller, J. M. Gilbert, M. Singh, and M. J. Osborn, manuscript in preparation.

solution contains 10–15% of the original protein and 60–70% of the total activity of the 3 glycosyl transferase activities leading to formation of man-rha-gal-PP-GCL. Abequosyl transferase activity is lost. O-antigen polymerase and ligase are not solubilized. The soluble fraction contains less than 10% of the original phospholipid and lipopolysaccharide of the envelope. The soluble transferase activities are labile to storage, but are effectively stabilized by addition of 1 volume of 40% (w/v) glycerol. The solution is divided into multiple aliquots for storage at $-70°$. Under these conditions the activities are stable for several weeks. Other nonionic detergents, including Triton X-100, are also effective in extracting the 3 glycosyl transferase activities, but a larger fraction of the total protein of the envelope is solubilized and the resulting extracts are considerably less stable.

Preparation of Mono-, Di- and Trisaccharide-PP-GCL Intermediates.[23] These are synthesized enzymatically from purified P-GCL and the appropriate nucleotide sugars. The solubilized glycosyl transferase fraction is employed as enzyme. The procedure is similar for the three intermediates (except for nucleotide sugar additions), and only the preparation of man-rha-gal-PP-GCL will be described in detail. Purified P-GCL (100 nmoles) is added to the bottom of a 40-ml glass-stoppered conical centrifuge tube and is evaporated to dryness under a stream of N_2. Methanol (0.12 ml) and 0.5% Alfonic 1012-6 (0.28 ml) are then added, and the lipid is dispersed by vigorous agitation on a Vortex mixer. The remaining components are then added as follows: 0.20 ml of $1 M$ Tris·acetate, pH 8.5, 0.40 ml of 0.1 M $MgCl_2$, 0.20 ml each of 2.5 mM UDP-galactose and 2.5 mM TDP-rhamnose, 0.16 ml of 2.5 mM GDP-[^{14}C]mannose (5000 cpm/nmole or greater), 1.0 ml of soluble glycosyl transferase enzyme (Poly-Tergent extract, approximately 1.1 mg/ml protein), and H_2O to a final volume of 4.0 ml. Incubation is at 25° for 2 hours. The product is extracted into 12 ml of C:M (2:1) and the organic phase is washed 3 times with 2.5 ml of PSUP. The washed organic phase is evaporated to dryness under a stream of N_2 in the cold, dissolved in 0.5 ml of 0.1% Alfonic 1012-6, and stored in small aliquots at $-70°$ under N_2. The yield of trisaccharide-PP-GCL is 60–75 nmoles. The product is stable to storage for at least 10 days. This preparation is suitable for use as substrate for O-antigen polymerase without further purification, but can be further purified by DEAE-cellulose chromatography if desired. The product is extracted from the reaction mixture into C:M, washed and taken to dryness as described above. The residue is dissolved in 5 ml of C:M (1:1) and applied to a 0.9 × 3 cm column of DEAE-acetate, prepared as described above. The column is washed successively with 15 ml each of C:M (1:1), 99% methanol, and 25 mM ammonium acetate–5 mM acetic acid in 99% methanol; these remove detergent, phospholipid, and free P-GCL. [^{14}C]Trisaccharide-PP-GCL is then eluted with 20 ml of 100 mM ammonium acetate–5 mM acetic acid in 99% methanol. One-milliliter fractions are collected and immediately chilled. Fractions containing [^{14}C]trisaccharide-PP-GCL are pooled and rapidly desalted by extracting into CHCl$_3$. This is necessary to minimize degradation of the labile galactosyl-PP linkage at the high salt concentration required for elution. Three volumes of CHCl$_3$ and 1 volume of H$_2$O are added to the pooled fractions, the mixture is shaken, and the phases are separated by brief centrifugation. The upper aqueous phase is backwashed once with 1 volume of CHCl$_3$, and the organic phases are pooled and washed twice with 50% methanol. The washed organic phase is taken to dryness in the cold under a stream of N$_2$, dissolved in C:M (3:1), and stored at −70° under N$_2$.

Preparation of Polysaccharide-PP-GCL.[24] Conversion of isolated man-rha-gal-PP-GCL to a GCL-linked polymer of trisaccharide repeating units is catalyzed by the particulate cell envelope fraction. The cell envelope is obtained from a rfa mutant (blocked in biosynthesis of the core region of lipopolysaccharide) in order to prevent transfer of the polymeric product to endogenous lipopolysaccharide during the reaction. The incubation mixture contains in a final volume of 3.25 ml: 0.05 ml of 1 M Tris·maleate buffer, pH 6.0, 0.40 ml of 0.1 M MgCl$_2$, 0.8 ml of [^{14}C]mannose-labeled trisaccharide-PP-GCL in 0.1% Alfonic 1012-6 (40–50 nmoles) and cell envelope (18 mg of protein). After incubation for 2 hours at 25° the reaction mixture is extracted with 12 ml of C:M (2:1), and the phases are separated by centrifugation. The polymeric product is recovered in the insoluble residue at the interface. The aqueous and organic phases are carefully removed with a Pasteur pipette, and the residue is suspended in 2 ml of 0.25% Alfonic 1012-6. The suspension is sonicated in an ice bath for three 30-second periods, maintaining the temperature of the sample below 15°, and insoluble material is removed by centrifugation at 105,000 g for 1 hour at 2–5°. The supernatant fraction contains the polysaccharide-PP-GCL and is stable to storage at −70° for several days. The yield of polysaccharide-PP-GCL is 60–80%. This preparation is free of P-GCL and the trisaccharide derivative, and is suitable for use as substrate in the O-antigen ligase reaction.

Utilization of Exogenous Lipid-Linked Intermediates: Assay of Individual Steps in the Pathway

Each reaction in the pathway (with the exception of abequosyl transferase, reaction (4)) can be assayed in the cell envelope fraction independently of preceding steps by use of isolated intermediates as substrate.

[24] M. Singh, M. A. Cynkin, and M. J. Osborn, manuscript in preparation.

Reagents

These are as described above.

Galactosyl-PP-GCL Synthetase–Reaction (1). Transfer of [¹⁴C]galactose-1-P from UDP-[¹⁴C]galactose to purified P-GCL is measured.[21] Incubations are carried out in 12-ml conical centrifuge tubes. Purified P-GCL (2.5 nmoles in C:M (3:1)) is added first and taken to dryness under a stream of N_2. Methanol (10 μl) and 15 μl of 0.5% Alfonic 1012-6 are added, and the lipid is dispersed by agitation on a Vortex mixer. The remaining components are then added as follows: 5 μl of 1 M Tris·acetate buffer, pH 8.5, 1 μl of 1 M $MgCl_2$, 8 μl of 2.5 mM UDP-[¹⁴C]galactose (6000 cpm/nmole) and cell envelope fraction (1–10 μg of protein). The final volume is 0.10 ml. After incubation at 25° for 10 minutes, the reaction is stopped by addition of 2 ml of C:M (2:1), and incorporation of ¹⁴C into galactosyl-PP-GCL is determined by Procedure B. Values are corrected for blank tubes containing heat-inactivated enzyme. The reaction is linear with time for 20–30 minutes and linear with enzyme concentration over the indicated range. The specific activity of the cell envelope fraction is approximately 5 to 8 nmoles/min per milligram of protein.

Rhamnosyl and Mannosyl Transferase–Reactions (2) and (3).[21] The procedures are identical to the above except as follows. For determination of rhamnosyl transferase, nonradioactive galactosyl-PP-GCL is substituted for P-GCL, and TDP-[¹⁴C]rhamnose replaces UDP-[¹⁴C]-galactose. Similarly, for assay of mannosyl transferase, nonradioactive rha-gal-PP-GCL and GDP-[¹⁴C]mannose are employed. Specific enzyme activities of 1–3 nmoles/min per milligram of protein are obtained.

O-Antigen Polymerase–Reaction (5).[23,25] Conversion of isolated [¹⁴C]man-rha-gal-PP-GCL to a chromatographically immobile polymeric product is measured. Incubation mixtures contain in a final volume of 45 μl: 15 μl of 2 M Tris·maleate buffer, pH 6.0, 1 μl of 0.5 M $MgCl_2$, 10 μl of ¹⁴C-labeled trisaccharide-PP-GCL in 0.1% Alfonic 1012-6 (1.0 nmole, 5000 cpm per nmole) and cell envelope (5–40 μg of protein). Control tubes contain heat-inactivated enzyme. After incubation for 60 minutes at 25°, the reaction is stopped by heating in a boiling H_2O bath

---

[25] An alternative procedure which allows polymerase activity to be measured in the absence of detergent has been described by Kanegasaki and Wright.[13] A mixture of an aqueous dispersion of trisaccharide-PP-GCL with particulate enzyme is subjected to several cycles of alternate freezing and thawing. This procedure appears either to "shock" the lipid intermediate into the membrane or to disrupt the membrane structure in such a way that the enzyme protein becomes accessible to the exogenous substrate. The method should be generally applicable when addition of detergent is undesirable.

for 3 minutes. The entire reaction mixture is applied to a sheet of Whatman No. 3 MM paper as a 2-cm streak and chromatographed 15–18 hours in ethanol: 1 M ammonium acetate, pH 7.3 (7:3). Polymeric products remain within 2 cm of the origin. This region is cut out and placed in a scintillation vial. The paper is wetted with 0.5 ml of H_2O, and 6 ml of scintillation fluid is added for counting. The reaction is linear with time for 2 hours and linear with protein concentration within the range indicated. The specific activity of the cell envelope fraction is 10–12 nmoles per milligram of protein per hour. With exogenous trisaccharide-PP-GCL as substrate, the polymerase reaction shows a sharp pH optimum of 6.0–6.5, a value considerably lower than that observed when the substrate is generated from endogenous P-GCL and nucleotide sugars. The reaction requires a divalent cation; Mg^{2+} and Zn^{2+} are somewhat more effective than Mn^{2+}, Ca^{2+}, Ba^{2+}, or Co^{2+}.

O-Antigen Ligase Reaction (6). Transfer of O-antigen polysaccharide chains to lipopolysaccharide is measured by conversion of alkali-labile [14C]polysaccharide-PP-GCL to an alkali-stable product in the presence of added acceptor lipopolysaccharide.[24] The acceptor lipopolysaccharide is obtained from a rfb mutant of *S. typhimurium* such as TV119. Acceptor is prepared by incubating 1 ml of lipopolysaccharide (11 μmoles of heptose) with 0.5 ml of 0.5 M Tris·maleate, pH 6.0, and 0.50 ml of 0.5% Alfonic 1012-6 for 30 minutes at 37°. Ligase incubation mixtures contain 40 μl of acceptor, 50 μl of [14C]mannose labeled polysaccharide-PP-GCL in 0.25% Alfonic 1012-6 (approximately 5000 cpm at a specific activity of 10,000 cpm/nmole [14C]mannose), and cell envelope fraction (20–150 μg of protein) in a total volume of 0.10 ml. After incubation for 60 minutes at 25°, the reaction is stopped by addition of 2 ml of 1 N NH_4OH–5 mM $MgCl_2$, and incorporation of 14C into lipopolysaccharide is determined by Procedure D. Values are corrected for blanks containing heat-inactivated enzyme. The reaction is linear with time for 75–90 min and linear with enzyme concentration within the limits designated. The concentration of polysaccharide-PP-GCL employed is approximately 0.5 nM based on an average chain length of 10 trisaccharide repeating units. Under these conditions, the observed rate of reaction is approximately 1.5 nmoles (based on [14C]mannose) per milligram per hour; however, the concentration of substrate employed in the routine assay procedure is well below saturation, and 2- to 3-fold higher rates can be attained by increasing the concentration of polysaccharide-PP-GCL. The reaction is similar to that observed when the polysaccharide intermediate is generated endogenously (see above) with respect to pH optimum (pH 5.5–6.5) and lack of stimulation by added metal ions. In both systems, the reaction is, however, partially inhibited by EDTA.

[83] Synthesis of Capsular Polysaccharides of Bacteria[1]

By FREDERIC A. TROY, FRANK E. FRERMAN, and EDWARD C. HEATH

The biosynthesis of complex heteropolysaccharides of bacterial cell walls has been shown to involve the preassembly of polysaccharide repeating units on a phospholipid carrier. This mode of biosynthesis has been well established for the *Salmonella* O-antigen[2,3] and the peptidoglycan of *Staphylococcus aureus* and *Micrococcus lysodeikticus*.[4] In each instance the phospholipid carrier was characterized as a (C_{55}) polyisoprenyl monophosphate and the repeating unit oligosaccharides are covalently bound to the lipid through a pyrophosphate linkage. This subject has been extensively reviewed by Osborn.[5]

More recently, this general mechanism of heteropolysaccharide biosynthesis was extended to yet another class of cell envelope polymers, the capsular polysaccharide of gram-negative bacteria. This report outlines the general procedures utilized for establishing the role of lipid-bound oligosaccharide intermediates in the biosynthesis of the capsular polysaccharide of *Aerobacter aerogenes*. These studies establish that the capsular polysaccharide of this organism is biosynthesized by the stepwise formation of the repeating structural unit of the polymer bound to undecaprenol by way of a pyrophosphate bridge.

$$\text{Gal} \to \text{Man} \to \text{Gal} \to \text{O}-\overset{\overset{\text{O}}{\|}}{\text{P}}-\text{O}-\overset{\overset{\text{O}}{\|}}{\text{P}}-\text{O}-\text{CH}_2-\overset{\overset{\text{H}}{|}}{\text{C}}=\overset{\overset{\text{CH}_3}{|}}{\text{C}}-\text{CH}_2-$$

Preparation of Particulate Cell Envelope Fraction

Organism. A heavily encapsulated strain of *Aerobacter aerogenes* was obtained from the culture collection of the Department of Microbiology,

[1] F. A. Troy, F. E. Frerman, and E. C. Heath, *J. Biol. Chem.* **246**, 118 (1971).

[2] A. Wright, M. Dankert, P. Fennessey, and P. W. Robbins, *Proc. Nat. Acad. Sci. U.S.* **59**, 1798 (1967).

[3] I. M. Weiner, T. Higuishi, L. Rothfield, M. Saltmarsh-Andrew, M. J. Osborn, and B. L. Horecker, *Proc. Nat. Acad. Sci. U.S.* **54**, 228 (1965).

[4] J. S. Anderson, M. Matsuhashi, M. A. Haskin, and J. L. Strominger, *J. Biol. Chem.* **242**, 3180 (1967).

[5] M. S. Osborn, *Annu. Rev. Biochem.* **38**, 501 (1965).

The Johns Hopkins School of Medicine. Detailed structural analysis of this polysaccharide indicated that it is a high molecular weight polymer composed of the following tetrasaccharide repeating unit[6]:

$$- \rangle \; 3\text{-Gal-}1 \; - \rangle \; 3\text{-Man-}1 \; -\rangle \; 3\text{-Gal-}1 \; - \rangle$$
$$\uparrow 1,2$$
$$\text{GlcUA}$$

Stock cultures of the organism were prepared by growing a single colony isolate to the midlogarithmic phase of growth in Trypticase soy broth. Sterile glycerol was then added to a final concentration of 15% (by volume), and 1.5-ml aliquots of the suspension were transferred aseptically to sterile screw-capped vials and maintained at $-15°$. The contents of one vial were thawed and used to inoculate 250 ml of synthetic medium contained in a 1-liter Erlenmeyer flask or 1 liter of Trypticase soy broth contained in a 2-liter Erlenmeyer flask. Synthetic medium contained the following components (grams per liter): $(NH_4)_2SO_4$, 10 (footnote 4); Na_2HPO_4, 10; KH_2PO_4, 3; K_2SO_4, 1; NaCl, 1; $MgSO_4 \cdot 7H_2O$, 0.2; $CaCl_2 \cdot 6H_2O$, 0.02; $FeSO_4 \cdot 7H_2O$, 0.001; Trypticase soy broth, 0.5; and glucose, 10. Glucose was prepared as a 25% solution, sterilized separately, and aseptically added to the sterile media just prior to use. Cultures were incubated on a rotary shaker at $37°$ until growth had reached the late logarithmic phase (12–14 hours); approximately 8 g (wet weight) of cells per liter were obtained from either medium. The cells were harvested by centrifugation and washed with 1 volume of $0.01 \, M$ Tris buffer, pH 7.1, containing $0.03 \, M$ NaCl. When large amounts of cells were required, organisms were grown in a New Brunswick fermentor at $37°$ with forced aeration.

Enzyme Preparation I. Washed cells (1 g wet weight) were subjected to cold osmotic shock according to the procedure of Nossal and Heppel.[7] After stage II of this procedure, the shocked cells were harvested by centrifugation for 15 minutes at 30,000 g. The upper half of the cell pellet (fluffy, lightly packed material) was selectively removed with a spatula and resuspended in 1–1.5 ml of distilled water; this suspension was then repeatedly (six to eight times) expelled from a syringe through a 21-gauge needle. This procedure apparently ruptured the osmotically fragile cells, as evidenced by the increased viscosity of the suspensions and a complete loss of cell viability after this treatment. On the other hand, prior to this treatment the shocked cell suspension exhibited a high degree of viability. Cell envelope fractions were prepared daily.

Enzyme Preparation II. Alternatively, a particulate cell envelope fraction possessing satisfactory glycosyltransferase activities could be

---

[6] E. C. Yurewicz, M. A. Ghalambor, and E. C. Heath, this volume [139].
[7] N. G. Nossal and L. A. Heppel, *J. Biol. Chem.* **241**, 3055 (1966).

prepared by sonic disruption of the bacteria. Washed cell pellets (8 g, wet weight) were suspended in an equal volume of distilled water and subjected to sonic oscillation with a Bronwill Biosonik probe at maximal power for 1 minute with constant cooling ($-5°$). The suspension was then centrifuged for 20 minutes at 4000 g to remove unbroken cells and large debris. The supernatant was removed and centrifuged at 30,000 g for 90 minutes and the resulting pellet was resuspended in distilled water. This enzyme preparation was stable for at least 4 weeks when stored at $-15°$.

Enzyme Preparation III. The 30,000 g pellet described above was used as the starting material for preparing a lipid-depleted cell envelope fraction. Extraction of cell envelopes was carried out at $-25°$. The pellet was suspended in distilled water and was added to 40 volumes of acetone and stirred for 5 minutes. After centrifugation at 5000 g for 5 minutes, the precipitate was suspended in butanol (20 volumes) and stirred for 2 minutes. The suspension was centrifuged and the pellet was washed successively with 40 volumes of acetone and 40 volumes of diethyl ether. Residual solvent was removed under reduced pressure and the dry residue was crushed to a fine powder and stored in a desiccator at $-15°$. Lipid-depleted enzyme preparations were quite stable; transferase activities decreased about 20% during 4 months.

Assay Procedures. The incorporation of individual sugars into polysaccharide and into the chloroform–methanol (C:M) soluble fraction was determined in the following standard incubation mixture: Tris buffer, pH 7.9, 25 μmoles; $MgCl_2$, 2.5 μmoles; UDP-galactose, UDP-glucuronic acid, and GDP-mannose, 0.05 μmole each; chloramphenicol, 12.5 μg; and cell particulate fraction, equivalent to 1.3–6 mg of protein. The volume of the incubation mixtures was 0.4 ml.

When the lipid-depleted enzyme was used, the protein was suspended in Tris buffer (0.25 M, pH 7.9) containing $MgCl_2$ (25 mM) to give the desired protein concentration. These incubation mixtures contained: Tris buffer, pH 7.9, 16 μmoles; $MgCl_2$, 1.6 μmoles; UDP-galactose, GDP-mannose, and UDP-glucuronic acid, 0.05 μmole each; and lipid-depleted particulate fraction, 1.0–1.6 mg of protein. The volume of the incubation mixtures was 0.26 ml. The final concentration of Triton X-100 was 0.12% (by volume). When highly purified preparations of carrier phospholipid were used, it was found that phosphatidylglycerol was a more effective surfactant. However, regardless of the purity of the carrier phospholipid, the addition of either surfactant was essential for recovery of the activity of the lipid-depleted enzyme preparation. Lipid fractions to be tested for acceptor activity were prepared immediately prior to assay. An aliquot of the lipid fraction dissolved in $CHCl_3$ was dried under a stream of nitrogen; water and a surfactant were then added to give the desired concentration of lipid and surfactant. The mixture was then dispersed by sonic oscillation with the small Biosonik probe at 20% maximum intensity. The dispersed lipid was incubated with the lipid-depleted enzyme for 10 minutes at 22° prior to the addition of substrates.

Incorporation of radioactivity into polysaccharide was determined by applying an aliquot of the incubation mixture directly onto Whatman No. 3 MM paper and developing the chromatogram in solvent I; the paper was dried and the origin was cut out, placed in a vial containing scintillation fluid, and counted. The incorporation of radioactive sugars into C:M-soluble components was determined by pipetting an aliquot of the incubation mixture into approximately 100 volumes of C:M, 2:1. After heating (10 minutes at 45°), the extract was thoroughly mixed on a Vortex mixer and nonlipid components were removed by partitioning the extract against 0.9% NaCl according to the procedure of Folch, Lees, and Sloane-Stanley.[8] The washing procedure was repeated, a portion of the organic phase was evaporated to dryness in a scintillation vial, and the radioactivity was determined.

Experimental Procedure

Extraction of Lipids. Total lipid was extracted from cells by suspending washed cells in 20 volumes of C:M, 2:1, the suspension was stirred for 3 hours at room temperature and the extract was filtered through glass wool. The organic phase was then partitioned against 0.2 volume of 0.9% NaCl as described by Folch *et al.*[8]

Chromatography. Paper chromatograms were developed in a descending manner using Whatman No. 3 MM paper unless otherwise specified. The following solvent systems were used: I, ethanol-1 M ammonium acetate, pH 7.5 (7:3); II, 1-butanol–pyridine–H_2O (9:5:4); III, 1-butanol–isopropyl alcohol–H_2O (14:2:4); and IV, ethyl acetate–acetic acid–H_2O (10:5:9). Unlabeled sugars and polyols were located on paper chromatograms with alkaline silver nitrate[9] after treatment of the paper with 5 mM periodic acid in acetone.

Analytical thin-layer chromatography was carried out on silica gel G plates (0.75 mm thick), which were activated at 100° for 3 hours and stored in a desiccator for not longer than 48 hours prior to use. Silica gel H was used for preparative thin-layer chromatography after washing the gel with chloroform–methanol–formic acid (2:1:1) followed by distilled water; the washed gel was dried for 2 days at 100°. Preparative

[8] J. Folch, M. Lees, and G. H. Sloane-Stanley, *J. Biol. Chem.* **226**, 497 (1957).

[9] W. E. Trevelyan, D. P. Procter, and J. S. Harrison, *Nature (London)* **166**, 444 (1950).

thin layers (1.7 mm thick) were activated for 1 hour at 100°, cooled, and used immediately. The developing solvents used were: I, chloroform–methanol–H_2O (65:25:4), and II, 2,6-dimethylheptanone–acetic acid–H_2O (65:45:6). Lipids were located on thin-layer chromatograms as follows: iodine vapor for total lipid; acid molybdate for phospholipid; and ninhydrin for lipids containing amino nitrogen.[10] Recovery of lipids from thin-layer chromatograms was carried out as described by White and Frerman.[11]

DEAE-cellulose columns (2 × 20 cm) were prepared by the method of Rouser et al.,[12] and lipids were eluted from the column as described by Dankert et al.[13] Sephadex LH-20 columns were prepared and lipids were eluted as described by Dankert et al.[13]

Silicic acid was suspended in chloroform prior to preparation of columns and the lipids were applied to columns in chloroform. Deacylated lipids were fractionated on a 60-g silicic acid column which was eluted successively with chloroform (1 liter), acetone (600 ml), and C:M, 1:1 (1 liter). Preliminary experiments indicated that essentially no deacylated phospholipid was eluted from the column with methanol.

For chromatography of uronic acid-containing oligosaccharides, columns (1 × 13 cm) of Dowex 1 (HCO_2^-) resin were prepared and eluted with a linear gradient of NH_4HCO_3 (0–0.06 M).

Paper electrophoresis was carried out on Whatman No. 3 MM paper at 60 V/cm in the following buffer systems: I, 0.05 M triethylamine carbonate, pH 6.5; II, 0.12 M pyridinium acetate, pH 5.3; and III, 0.09 M $Na_2MoO_4 \cdot 2H_2O$, pH 5.1.

Deacylation of Lipids. Lipid extracts were dried under reduced pressure or under a stream of nitrogen and deacylated by alkaline methanolysis according to the procedure described by White and Frerman.[11] Lipids were subjected to methanolysis for 1.5 hours at 0°. When large amounts (more than 60 μmoles) of phospholipid were deacylated the procedure was repeated at least once.

Hydrolysis and Isolation of Mono- and Oligosaccharides from Lipid. After the maximal accumulation of radioactive carbohydrate in the C:M-soluble fraction, phosphorylated mono- and oligosaccharide derivatives were isolated from the C:M extract (Method 1). Incubation

[10] J. C. Dittmer and R. L. Lester, *J. Lipid Res.* **5**, 126 (1964).

[11] D. C. White and F. E. Frerman, *J. Bacteriol.* **94**, 1854 (1967).

[12] G. Rouser, G. Kritchevsky, D. Heller, and E. Leiber, *J. Amer. Oil Chem. Soc.* **40**, 425 (1963).

[13] M. Dankert, A. Wright, W. S. Kelley, and P. W. Robbins, *Arch. Biochem. Biophys.* **116**, 425 (1966).

mixtures were inactivated by the addition of 3 volumes of 0.1 N acetic acid. The particulate fraction was then successively washed by centrifugation with 0.1 N acetic acid and water, and the lipid intermediates were isolated by extraction with C:M (3:1) according to the procedure of Osborn and Weiner.[14] The organic solvent was removed by evaporation and the residue was subjected to hydrolysis in 0.05 N KOH for 15 minutes at 100°. The yields of lipid-linked intermediates obtained by this method were sometimes poor; therefore, the intermediates were usually isolated by mild alkaline hydrolysis of the washed particulate fraction (Method 2, see below). The products isolated by either method were identical.

Alternatively, the phosphorylated derivatives could be obtained by first isolating the particulate fraction from the incubation mixtures by centrifugation (30,000 g, 15 minutes). The resulting particulate material was washed several times to eliminate contaminating sugar nucleotides. Treatment of the washed particulate fraction with 2 N NH$_4$OH at 37° for 20 minutes (Method 2) resulted in a quantitative loss of radioactivity from the lipid fraction (as judged by chromatography in solvent system I or organic solvent extraction) with the concomitant appearance of radioactivity in mono- or oligosaccharide cyclic 1,2-phosphate esters. The cyclic phosphate esters could be converted to the acid-stable 2-phosphate esters by treatment at pH 2 (10 minutes, 100°); treatment of these products with bacterial alkaline phosphomonoesterase yielded the corresponding free mono- or oligosaccharides.

The phosphorylated or free mono- and oligosaccharides yielded by these procedures were isolated and purified by either paper chromatography, electrophoresis, or Sephadex chromatography as described in the individual experiments.

Digestion of in Vitro Synthesized Capsular Polysaccharide with Depolymerase. Studies of the activity of the phage-induced depolymerase on the capsular polysaccharide established that it is an endogalactosidase[6] which catalyzes the hydrolysis of the capsular polysaccharide to two limit oligosaccharides, A and B. Each oligosaccharide is composed of galactose, mannose, and glucuronic acid in a molar ratio of 2:1:1; galactose is the terminal reducing sugar in both oligosaccharides. It has been established that oligosaccharide A possesses the structure; oligosaccharide B is an octasaccharide composed of 2 eq of oligosaccharide A,

$$\text{Gal}-1 \rightarrow 3-\text{Man}-1 \rightarrow 3-\text{Gal}$$
$$\overset{2}{\underset{\text{GlcUA}}{\uparrow 1}}$$

[14] M. J. Osborn and I. M. Weiner, *J. Biol. Chem.* **243**, 2631 (1968).

with two nonreducing glucuronate termini and one galactose nonreducing terminus.

Radioactive capsular polysaccharide was isolated from incubation mixtures by precipitation with 70% ethanol; the precipitate was dissolved in water and added to a polysaccharide depolymerase assay mixture as described above. Following exhaustive digestion (5 hours, 42°) of the polysaccharide (electrophoretically immobile) with 400 units of depolymerase, the incubation mixture was applied to Whatman No. 3 MM paper and subjected to electrophoresis in Buffer System I or II. After electrophoresis of the incubation mixture, the paper strips were scanned and the radioactivity at the origin and in the oligosaccharide region was quantitated by cutting out the appropriate areas and counting them in the scintillation counter. ^{14}C-Uniformly labeled capsular polysaccharide isolated from the wild-type strain of *A. aerogenes* was included as a control.

Specific Hydrolytic Enzymes. β-Glucuronidase from bovine liver was obtained from Worthington. With phenolphthalein-glucuronide (Sigma) as substrate, 0.2 mg of the enzyme preparation catalyzed the liberation of 0.038 μmole of glucuronic acid per 30 minutes at 37° when assayed as described by Fishman, Springer, and Brunetti.[15]

End-Group Analysis of Oligosaccharides by Sodium Borohydride Reduction. The reducing terminus of the dephosphorylated oligosaccharides isolated by the procedure described above was determined by reduction with sodium borohydride. Since these compounds all possessed galactose at the reducing terminus, the degree of polymerization of complex oligosaccharides which had been labeled *in vitro* with [^{14}C]-galactose could be determined by this procedure. Thus after borohydride reduction of the galactose reducing terminus and hydrolysis, the ratio of [^{14}C]galactitol to [^{14}C]galactose indicates the degree of polymerization of these oligosaccharides. The general procedure used for these analyses was as follows. The dephosphorylated oligosaccharides were dissolved in 0.5–1 ml of water and two 100-μmole portions of sodium borohydride were added at 60-minute intervals; an additional 200 μmoles of sodium borohydride was then added, and the reaction mixture was incubated at room temperature for 15–18 hours. Excess sodium borohydride was consumed by the addition of an excess of acetone. The sample was evaporated to dryness, redissolved in methanolic-HCl (pH 4), and again evaporated to dryness; this procedure was repeated four to six times to remove borate from the sample. The sample was then hydrolyzed in 3 N HCl for 3 hours at 100° and chromatographed in solvent system II or III. The region corresponding to galactose-galactitol was eluted and

[15] W. H. Fishman, B. Springer, and R. Brunetti, *J. Biol. Chem.* **173**, 449 (1948).

rechromatographed in solvent system III or analyzed by electrophoresis in buffer III; the areas of the paper corresponding to those of standard galactitol and galactose were removed from the chromatogram, and the radioactive content of each region was determined in a liquid scintillation counter.

Materials. Unless otherwise indicated, all chemicals were analytical reagent grade commercial preparations. UDP-[^{14}C]galactose (252 mCi/mmole), UDP-[^{14}C]glucuronic acid (233 mCi/mmole), GDP-[^{14}C]mannose (151 mCi/mmole), and [^{14}C]uridine 5'-monophosphate (370 mCi/mmole) were purchased from New England Nuclear. The 1,2-monophosphate cyclic ester of galactose (cyclic galactose 1,2-phosphate) was prepared from UDP-galactose by treatment with $2 N$ NH$_4$OH at 37° for 2 hours and characterized by its chromatographic behavior (R_f 0.53) with respect to cyclic glucose 1,2-phosphate in solvent system I as described by Paladini and Leloir[16]; the galactose to phosphate ratio was 1:1. Galactose 2-phosphate was prepared by treatment of cyclic galactose 1,2-phosphate at pH 2 for 10 minutes at 100° and characterized by its chromatographic behavior (R_f 0.21) in the same solvent system. Paladini and Leloir[16] reported R_f 0.55 for cyclic glucose 1,2-phosphate and R_f 0.22 for glucose 2-phosphate in solvent system I. Treatment of UDP-galactose under stronger conditions (0.1 N NaOH at 100° for 20 minutes or 0.05 N KOH at 100° for 15 minutes) resulted in the formation of a mixture of galactose 2-phosphate and galactose 1-phosphate. These results are similar to those reported by Paladini and Leloir[16] for the behavior of UDP-glucose in alkali.

Mass Spectrometry. Mass spectrometry was conducted with a Consolidated Electrodynamics Corporation 21–110 double focusing mass spectrometer. The ionization energy was 70 eV; the source temperature was 280°, and the probe temperature was 270°.

Analytical Methods. The following analytical procedures were used: organic phosphate, by the method of Chen, Toribara, and Warner[17] as modified by Ames and Dubin[18]; and protein, by the method of Lowry et al.[19] as modified by Miller.[20]

Results

Incorporation of Galactose into Capsular Polysaccharide. As shown in Table I, the incorporation of galactose from UDP-[^{14}C]galactose into

[16] A. C. Paladini and L. F. Leloir, *Biochem. J.* **51**, 426 (1952).
[17] P. S. Chen, T. Y. Toribara, and H. Warner, *Anal. Chem.* **28**, 1756 (1956).
[18] B. N. Ames and D. T. Dubin, *J. Biol. Chem.* **235**, 769 (1960).
[19] O. H. Lowry, N. J. Rosebrough, A. L. Farr, and R. J. Randall, *J. Biol. Chem.* **193**, 265 (1951).
[20] G. L. Miller, *Anal. Chem.* **31**, 964 (1959).

TABLE I

INCORPORATION OF GALACTOSE INTO CAPSULAR POLYSACCHARIDE

Conditions of incubation	Incorporation (nmole)
Complete system	0.22
Omit UDP-glucuronic acid	0.13
Omit GDP-mannose	0.04
Omit UDP-glucuronic acid and GDP-mannose	0.05

capsular polysaccharide was maximal in the presence of UDP-glucuronic acid and GDP-mannose. However, galactose incorporation was significantly more dependent on the presence of GDP-mannose than UDP-glucuronate. When Enzyme Preparation I was used, polysaccharide synthesis did not exhibit an absolute dependence on the presence of all three sugar nucleotide derivatives.

Incorporation of Galactose into Polysaccharide and into Chloroform:Methanol-Soluble Intermediate. When assays for polysaccharide biosynthesis were conducted at 37°, little radioactivity from the labeled sugar nucleotides was accumulated in the C:M-soluble fraction. However, when the incubations were conducted at 12°, significant accumulation of radioactivity in the C:M-soluble fraction resulted. The relationship between incorporation of [14C]galactose into the C:M-soluble fraction and into polysaccharide is shown in Fig. 1. In the incubation mixture containing only UDP-[14C]galactose, there was an accumulation of galactose in the C:M-soluble fraction, but only a small amount of polysaccharide was formed; as pointed out below, this polymer appears to be unrelated to capsular polysaccharide. If the incubation was supplemented with unlabeled GDP-mannose and UDP-glucuronic acid, galactose was incorporated into the C:M-soluble fraction at a linear rate for 45 minutes, whereas there was a lag of about 60 minutes in the incorporation of galactose into polysaccharide. In the latter incubation, the extent of galactose incorporation into the C:M fraction was diminished. Similarly, if only UDP-[14C]galactose was present initially in an incubation mixture and unlabeled GDP-mannose and UDP-glucuronic acid were added after 20 minutes, galactose was incorporated into the lipid fraction at a linear rate for 45 minutes, and polysaccharide was synthesized at the control rate only after a 60-minute lag. The inverse relationship between the accumulation of galactose in the C:M-soluble fraction and the formation of capsular polysaccharide implied a precursor-product relationship involving a lipid intermediate.

Characterization of Phosphogalactosyltransferase. The preceding ex-

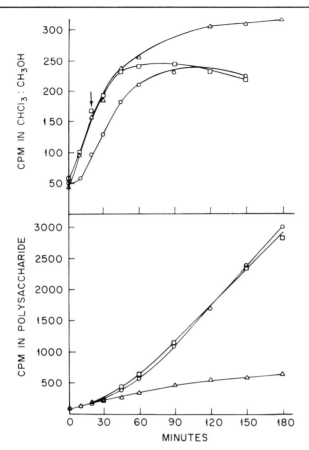

FIG. 1. Incorporation of galactose into capsular polysaccharide and lipid at 12°. Standard incubation mixtures were prepared containing enzyme preparation I (2.1 mg of protein). UDP-[¹⁴C]galactose (8330 cpm per nmole), 0.25 μmole, UDP-glucuronic acid and GDP-mannose, 0.25 μmole each, were added to incubation mixtures as indicated. At the intervals indicated, aliquots were removed from the incubation mixtures and assayed for the incorporation of [¹⁴C]galactose into capsular polysaccharide and into the C:M-soluble fraction. ○, Complete mixture; △, minus UDP-glucuronic acid, GDP-mannose; □, UDP-glucuronic acid, GDP-mannose added at 20 minutes (↓).

periments indicated that galactose was incorporated into a C:M-soluble component and kinetic studies suggested that this component was an intermediate in the biosynthesis of the capsular polysaccharide. Consequently, experiments were conducted to determine whether either of the other two monosaccharides of the capsular polysaccharide, mannose and glucuronic acid, could be incorporated into C:M-soluble components.

Incubation mixtures were prepared containing either UDP-[^{14}C]galactose, GDP-[^{14}C]mannose, or UDP-[^{14}C]glucuronic acid, and these mixtures were compared with respect to their ability to catalyze the incorporation of radioactivity from each of the three sugar nucleotides into the C:M-soluble fraction. Radioactivity was incorporated into a C:M-soluble component only in the incubation mixture containing UDP-[^{14}C]galactose.

To determine whether the enzymatic reaction responsible for the incorporation of [^{14}C]galactose into the C:M-soluble fraction was catalyzed by a phosphogalactosyltransferase, experiments were performed to determine the effect of unlabeled UMP and unlabeled UDP- galactose on the reaction.

Uridine-P-P-galactose + undecaprenyl-P $\rightleftharpoons$ galactose-P-P-undecaprenyl + UMP

Incubation mixtures were prepared containing UDP-[^{14}C]galactose as the only substrate; when [^{14}C]galactose had been incorporated maximally into the C:M-soluble fraction, a 10-fold excess of either unlabeled UDP-galactose or UMP was added to the incubation mixture. The results of these experiments are shown in Fig. 2. Addition of either unlabeled

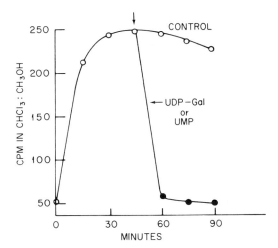

Fig. 2. Effect of unlabeled UDP-galactose and UMP upon the radioactivity in the [^{14}C]galactosylpyrophosphoryl lipid intermediate. Initially, each incubation mixture contained UDP-[^{14}C]galactose (7645 cpm/nmole) and particulate enzyme preparation I (1.67 mg of protein) in a final volume of 0.425 ml. At intervals, aliquots (25 μl) were removed and assayed for C:M-soluble radioactivity. After 45 minutes (↓), a 10-fold excess (1.0 μmole) of either unlabeled UDP-galactose or UMP was added separately to duplicate incubation mixtures; a third incubation mixture (control) received an equivalent volume (10 μl) of water. The incubations were conducted at 12°.

UDP-galactose or UMP to the equilibrium mixture resulted in a rapid loss of radioactivity from the C:M fraction. The rate of loss of radioactivity from the C:M-soluble fraction was essentially identical with the initial rate of incorporation of [^{14}C]galactose into the fraction, suggesting that the loss of radioactivity was the result of an imbalance imposed on the equilibrium of a single enzymatic reaction. Thus, these results support the contention that galactose is incorporated into the C:M-soluble fraction via a reversible phosphogalactosyltransferase reaction; under equilibrium conditions the addition of excess UMP displaces radioactive galactose from the lipid intermediate, whereas the addition of excess unlabeled UDP-galactose essentially dilutes the specific radioactivity of galactose. The net result of either addition would be a loss of radioactivity from the C:M fraction at a rate approximating that of incorporation and to an extent consistent with the quantity added. Substitution of UDP for UMP in this experiment resulted in no significant loss of radioactivity from the C:M-soluble fraction. When the experiment was conducted with unlabeled UDP-galactose and [^{14}C]UMP was added to the reaction mixture after 45 minutes of incubation, [^{14}C]UDP-galactose was isolated from the reaction mixture; mild acid hydrolysis of the isolated sugar nucleotide indicated that all the radioactivity resided in the uridine moiety.

Hydrolysis of the C:M-soluble, [^{14}C]galactose-containing component with 0.1 N NaOH for 20 minutes at 100° released 94% of the radioactivity as a mixture of galactose 1- and galactose 2-phosphate. Treatment of the [^{14}C]galactolipid bound in the particulate preparation (see "Experimental Procedure," Method 2) with 2 NH$_4$OH, at 37° for 20 minutes quantitatively released the radioactivity from the preparation; the resulting radioactive product was characterized as cyclic galactose 1,2-phosphate.

Mannosyltransferase Reaction. Evidence supporting the contention that the galactosylpyrophosphoryl-lipid derivative would serve as an acceptor of a mannosyl moiety of GDP-mannose was obtained by demonstration of the sequential transfer of first galactose and then mannose, in equal molar quantities, to the C:M-soluble derivative; the results of this experiment are illustrated in Fig. 3.

Following maximal formation of membrane-bound [^{14}C]galactose pyrophosphoryl lipid, the particulate fraction was washed repeatedly to free it of excess UDP-[^{14}C]galactose and placed in an incubation mixture which contained only GDP-[^{14}C]mannose. [^{14}C]Mannose was incorporated into the C:M-soluble fraction to essentially the same extent as galactose in the first stage of the incubation period; since the specific activities of the two sugar nucleotide derivatives were nearly identical,

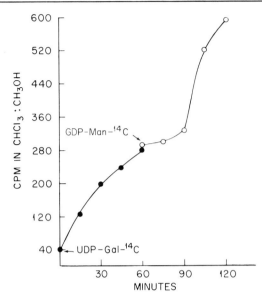

FIG. 3. Sequential transfer of galactose and mannose to carrier lipid. Following the incorporation of galactose into the lipid fraction (0–60 minutes) in a standard incubation mixture containing enzyme preparation I (4 mg of protein) and UDP-[^{14}C]galactose (28,600 cpm/nmole), the particulate fraction was isolated, washed, and suspended in a similar incubation mixture containing GDP-[^{14}C]mannose (25,900 cpm/nmole). Incubations were conducted at 15°, and 10-μl aliquots were removed as indicated and assayed for C:M-soluble radioactivity as described under "Experimental Procedure."

these results demonstrate the stoichiometric transfer of first galactose and then mannose to the lipid fraction. The product of this reaction was isolated as described under "Experimental Procedure" (Method 2) and treated successively with $2\,N$ NH_4OH for 20 minutes at 37°, and then pH 2 for 10 minutes at 100°, and alkaline phosphatase, yielding the results illustrated in Fig. 4. Frame 1 of this figure indicates that all the radioactivity in the lipid derivative initially moved with the solvent front on the chromatogram; frame 2 indicates the chromatographic mobility of the water-soluble product of mild alkaline hydrolysis of the lipid derivative, cyclic mannosylgalactosyl 1,2-phosphate, relative to a standard sample of cyclic galactose 1,2-phosphate; frame 3 indicates the chromatographic mobility of the product obtained after mild acid hydrolysis of the cyclic phosphate ester, mannosylgalactose 2-phosphate relative to standard galactose 2-phosphate; the electrophoretic mobility of the disaccharide 2-phosphate ester is shown in frame 4; and the dephosphorylated, neutral disaccharide mannosylgalactose is shown in

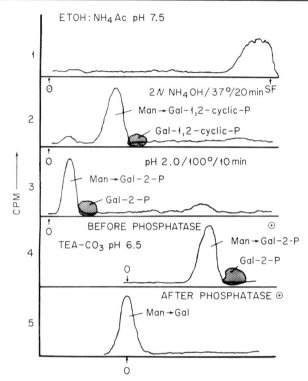

Fɪɢ. 4. Isolation of mannosylgalactose from lipid intermediate. Isolation of the disaccharide mannosylgalactose from the lipid fraction was carried out as described in the text following the sequential formation of mannosylgalactosyl-P-P-lipid (see Fig. 3). TEA, triethylamine.

frame 5. Characterization of the disaccharide as mannosylgalactose was determined by chromatographic analysis of the products of acid hydrolysis of this disaccharide before and after reduction with sodium borohydride. Complete acid hydrolysis of the disaccharide yielded two components, each containing equivalent radioactivity, which exhibited chromatographic mobilities identical with standard galactose and mannose. Reduction of the disaccharide with sodium borohydride prior to acid hydrolysis resulted in the formation of two components, each containing equivalent radioactivity, which corresponded in chromatographic mobility to galactitol and mannose.

Glucuronosyltransferase. It was found that polysaccharide biosynthesis catalyzed by Enzyme Preparation I (see "Experimental Procedure") did not exhibit an absolute requirement for UDP-glucuronic acid. Preliminary experiments indicated that this observation was the result of endogenous UDP-glucuronic acid which accumulates in the mutant organism when grown in the absence of galactose. Therefore, in the following experiments a particulate enzyme preparation obtained from sonically disrupted cells was used (Enzyme Preparation II, "Experimental Procedure"). As previously demonstrated, glucuronic acid is not incorporated into the C:M-soluble fraction when only UDP-glucuronic acid is present in the incubation mixture. The nucleotide sugar cosubstrate-dependence of the glucuronosyltransferase was investigated by preparing three incubation mixtures containing UDP-glucuronic acid-^{14}C as the only labeled substrate; one incubation mixture contained unlabeled UDP-galactose, a second, unlabeled GDP-mannose, and a third incubation mixture contained both unlabeled cosubstrates. Only the incubation mixture containing all three sugar nucleotides incorporated [^{14}C]glucuronic acid into the C:M-soluble fraction.

It was of interest to determine whether glucuronic acid could be transferred directly to the mannosylgalactosylpyrophosphoryl-lipid. The enzymatic transfer of glucuronic acid to mannosylgalactosylpyrophosphoryl-lipid was demonstrated in an experiment in which [^{14}C]glucuronic acid was enzymatically transferred to the preformed mannosylgalactosylpyrophosphoryl-lipid. In this experiment the particulate enzyme preparation initially was incubated with unlabeled UDP-galactose and GDP-mannose; a control incubation mixture received neither sugar nucleotide. After incubation for 75 minutes (the predetermined interval required for maximal accumulation of the disaccharide intermediate), the particulate fractions were washed three times (0.01 M Tris buffer, pH 7.1, containing 0.03 M NaCl) and resuspended in an incubation mixture containing only UDP-[^{14}C]glucuronic acid. Glucuronic acid was incorporated into the C:M fraction only in the incubation mixture which had been previously incubated with UDP-galactose and GDP-mannose. These results suggest that the cosubstrate requirements for glucuronic acid incorporation into the C:M-soluble fraction could be satisfied by preformed mannosylgalactosylpyrophosphoryl-lipid generated in the particulate fraction and that glucuronic acid was the third sugar transferred to the lipid intermediate.

The formation of glucuronosylmannosylgalactosylpyrophosphoryl-lipid was confirmed by isolation and characterization of the lipid-linked trisaccharide. The preceding experiment was scaled up 4-fold and all sugar nucleotide substrates labeled in the sugar moiety with ^{14}C. After accumulation of the radioactive mannosylgalactosylpyrophosphoryl-lipid, the particulate fraction was washed as described above and suspended in an incubation mixture containing only UDP-[^{14}C]glucuronic acid. After incubation for 50 minutes at 12°, the particulate fraction was thoroughly washed until the supernatant was essentially free of radio-activity and suspended in water. An aliquot was removed and sub-jected to the paper chromatographic assay (Solvent I) to determine the incorporation of radioactivity into polymeric material and into lipid. A similar aliquot was extracted with C:M to determine directly the in-corporation of radioactivity into the lipid fraction. The remaining par-ticulate material was then hydrolyzed directly with 2 N NH$_4$OH (Method 2) and particulate material was removed by centrifugation. This treat-ment released more than 98% of the radioactivity originally present in the lipid fraction into the aqueous phase, which was subjected to acid hydrolysis (pH 2, 10 minutes, 100°) and then treated with alkaline phos-phatase (pH 8, 2 hours, 37°). The alkaline phosphatase-treated material was isolated by gel filtration chromatography on a Sephadex G-25 column. Polymeric material eluted at the void volume and represents a 90% recovery of radioactive polymer determined by the paper chromatographic assay. The radioactivity which eluted in the trisac-charide region of the column represents a 92% recovery of radioactivity incorporated into the lipid fraction based on the paper chromatographic assay or the direct assay of the C:M-extracted material.

The oligosaccharide fraction was desalted by chromatography on Sephadex G-15, applied to a column of Dowex 1 (HCO$_3^-$), and eluted with a linear gradient of NH$_4$HCO$_3$. Approximately 20% of the radio-activity was not adsorbed by the resin and probably represents manno-sylgalactose. The remaining radioactivity eluted as a sharp peak at 8×10^{-3} M NH$_4$HCO$_3$; the radioactive fractions were pooled and treated with Dowex 50 (H$^+$) to remove NH$_4$HCO$_3$. This radioactive component ap-peared homogeneous since only a single radioactive component was ob-tained after chromatography in solvent IV and electrophoresis in buffer II.

After reduction with NaBH$_4$, the oligosaccharide preparation was desalted by chromatography on Sephadex G-15 and hydrolyzed. The products of oligosaccharide hydrolysis were chromatographed in solvent II. Radioactive areas corresponding to authentic mannose, galactose or galactitol, glucuronolactone, and uronic acid were cut out and counted in the liquid scintillation spectrometer. After assaying the galactose-galactitol region for radioactivity, the paper was washed several times with diethyl ether to remove scintillation fluid and then eluted with distilled water. Analysis of this radioactive component by electro-phoresis in Buffer III showed that all of the radioactivity migrated with the galactitol standard. Quantitative analysis of these products indicated that the oligosaccharide is composed of galactitol, mannose, and glu-curonic acid in a molar ratio of 1.00:1.03:0.88, respectively.

Linkage of the glucuronosyl moiety of the trisaccharide to mannose was confirmed by isolation of the aldobiouronic acid, glucuronosylmannose by controlled acid hydrolysis; the trisaccharide was hydrolyzed (2 N HCl, 1 hour, 100°) and subjected to gel filtration in distilled water. Two distinct peaks of radioactivity were observed. Fractions containing radioactivity were pooled and analyzed by chromatography in solvent II; peak I contained a single immobile component while peak II contained predominantly galactose. Strong acid hydrolysis (3 N HCl, 3 hours, 100°) of the compound in peak I yielded radioactive components which migrated with mannose, glucuronolactone, and uronic acid standards. The radioactive areas of the chromatogram were assayed for radioactivity and the results indicated a molar ratio of mannose and glucuronic acid of 1.00 to 1.02, respectively.

Because it was previously established that glucuronic acid incorporation was dependent on mannosylgalactosylpyrophosphoryl-lipid as acceptor, these results are consistent with the formation of a trisaccharide intermediate, glucuronosylmannosylgalactosylpyrophosphoryl-lipid.

Galactosyltransferase. Evidence for the transfer of galactose from UDP-galactose to the glucuronosylmannosylgalactosylpyrophosphoryl-lipid, representing completion of the tetrasaccharide repeating unit, was obtained in the following experiment. After maximal accumulation of the radioactive mannosylgalactosylpyrophosphoryl-lipid in an initial incubation mixture as described above, the particulate fraction was thoroughly washed and resuspended in an incubation mixture containing UDP-[^{14}C]-glucuronic acid and UDP-[^{14}C]galactose, the mixture was incubated for an additional period of 50 minutes at 12° and the particulate fraction was washed free of contaminating sugar nucleotides and suspended in distilled water. Aliquots were removed for chromatography in Solvent I and for extraction with C:M in order to determine the extent of incorporation of radioactivity into polymeric material and into the lipid fraction. Radioactive components in the lipid fraction were released from the particulate fraction by mild alkaline hydrolysis (Method 2) and following centrifugation (30,000 g, 15 minutes) the supernatant was hydrolyzed at pH 2 (10 minutes, 100°) and treated with alkaline phosphatase as described above. After these procedures, radioactivity originally present in the lipid fraction was quantitatively recovered from the tetrasaccharide region of a Sephadex G-25 column.

The oligosaccharide isolated on Sephadex G-25 was desalted on Sephadex G-15 and applied to a Dowex 1 (HCO$_3$⁻) column; the radioactive oligosaccharide was eluted with 12 mM NH$_4$HCO$_3$; the fractions containing radioactivity were pooled and NH$_4$HCO$_3$ was removed by treatment with Dowex 50 (H⁺). The radioactive material was then subjected to chromatography in Solvent IV and electrophoresis in Buffer II; in both systems radioactivity migrated as a single component with a mobility identical with the tetrasaccharide repeating unit of the polymer.

The oligosaccharide was hydrolyzed after prior reduction with $NaBH_4$ and desalting on Sephadex G-15. The hydrolyzate was subjected to paper chromatography in solvent II; radioactivity was associated with mannose, galactitol or galactose, glucuronolactone, and uronic acid standards. Further analysis of the galactose-galactitol region of the chromatogram by paper electrophoresis indicated an equimolar ratio of galactose and galactitol. Individual sugars were analyzed quantitatively by the methods previously described and the analyses showed that the oligosaccharide is a tetrasaccharide composed of galactose, mannose, and glucuronic acid with the molar ratios 2:1:1; galactose is the terminal reducing sugar. Further confirmation of the identity of the intermediate tetrasaccharide with oligosaccharide A was obtained by periodate oxidation studies; exhaustive treatment of the tetrasaccharide with periodic acid resulted in complete destruction of galactose and glucuronic acid with essentially complete recovery of mannose. These data are consistent with the location of the mannose moiety of the tetrasaccharide in an internal position substituted in a manner that would prevent periodate oxidation; the mannose moiety of oligosaccharide A is substituted by galactose at C-3 and glucuronic acid at C-2.

The lipid-derived tetrasaccharide was further characterized by isolation of the aldobiouronic acid, β-D-glucuronosylmannose, after acid hydrolysis (2 N HCl, 1 hour, 100°). The hydrolyzate was subjected to gel filtration on a Sephadex G-15 column which was eluted with distilled water. Acid hydrolysis (3 N HCl, 3 hours, 100°) followed by chromatography in solvent II yielded glucuronic acid and mannose in a molar ratio of 1.00 to 1.03; more than 95% of the radioactivity was identified as galactose. After the addition of 2 μmoles of unlabeled β-D-glucuronosylmannose, the aldobiouronic acid mixture was treated with β-glucuronidase (22 hours, 37°, 4 mg). After treatment with β-glucuronidase, the mixture was subjected to electrophoresis in buffer II to separate mannose, glucuronic acid, and any unhydrolyzed substrate; the strip was scanned to locate radioactive compounds. Under the conditions described, 71% of the substrate was enzymatically hydrolyzed, and mannose and glucuronic acid were recovered in a molar ratio of 1.00:1.05. Therefore, the disaccharide is of the glucuronosylmannose type and the anomeric configuration of the glycosidic linkage is β.

Based on its molar composition, terminal reducing sugar, and chromatographic and electrophoretic mobilities, the lipid-bound tetrasaccharide biosynthesized *in vitro* is identical with the limit tetrasaccharide which was isolated from depolymerase-treated capsular polysaccharide.[6]

Polymerization of Repeating Tetrasaccharide on Carrier Phospholipid. Although previous experiments had established that the lipid-bound oligosaccharides ultimately were polymerized to form high molecular weight polysaccharide, it was of interest to determine whether polymerization of the repeating unit occurred at the lipid intermediate stage of biosynthesis. It was observed that incubation of enzyme preparation I in the presence of all three sugar nucleotide substrates for more extended periods of time resulted in the accumulation of lipid-bound tetrasaccharide as well as a smaller amount of lipid-bound oligosaccharide, which was isolated and characterized as an octasaccharide composed of 2 eq of the tetrasaccharide repeating unit. A standard incubation mixture (final volume, 0.75 ml) was prepared containing UDP-[14C]galactose, UDP-[14C]glucuronic acid and GDP-[14C]mannose (approximately 10,000 cpm/nmole each), and enzyme preparation I (5.0 mg of protein). The mixture was incubated at 12° for 75 minutes, and at various intervals aliquots were removed and assayed for incorporation of radioactivity into the C:M fraction and into polysaccharide. After incorporation of radioactivity into the lipid fraction had reached equilibrium (75 minutes), the remainder of the incubation mixture (0.67 ml) was treated as described in Method 2 for the isolation of phosphorylated oligosaccharides from the carrier phospholipid. Dephosphorylation of the mixture as described under "Experimental Procedure," resulted in the isolation of most (>90%) of the radioactivity in two oligosaccharides (I and II) which exhibited the following properties. (a) Both were completely retained by Sephadex G-75; (b) electrophoretic and chromatographic mobilities of I and II were indistinguishable from limit depolymerase products A and B, respectively; (c) quantitative analysis indicated that each oligosaccharide contained galactose, mannose, and glucuronic acid in molar ratios of approximately 2:1:1; and (d) reduction with sodium borohydride followed by acid hydrolysis indicated a ratio of galactitol to galactose of 1:1 in I, and a ratio of 1:3 in II. These data are consistent with the conclusion that oligosaccharides I and II, isolated from the carrier phospholipid, correspond in structure to the tetrasaccharide repeating unit and an octasaccharide corresponding to a "dimer" of the repeating unit. Thus, while the complete and detailed structure of the octasaccharide derived from the phospholipid carrier has not been established, the evidence presented above strongly supports the contention that polymerization of the tetrasaccharide repeating unit of the polysaccharide occurs at the carrier phospholipid-bound stage of biosynthesis.

Characterization of Carrier Lipid. The particulate enzyme preparations (Enzyme Preparations I and II, see "Experimental Procedure") contain both the various glycosyltransferases which catalyze the biosynthesis of the tetrasaccharide as well as endogenous phospholipid carrier to which the oligosaccharide is covalently bound. In order to isolate and characterize the carrier phospholipid it was essential to obtain an enzyme preparation which retained catalytic activity but which required the addition of an exogenous source of the carrier lipid. As illustrated in Table II, organic solvent extraction of enzyme preparation II yielded an enzyme preparation which met these requirements. The particulate envelope preparation treated in this manner exhibited phosphogalactosyltransferase activity only when the incubation mixtures were supplemented with exogenous lipid. Further, it was observed (Experiment I) that the biologically active lipid was stable to alkaline methanolysis, as evidenced by no loss in acceptor activity of the lipid after deacylation. It was observed, however, that fractionation of the deacylated crude lipids on silicic acid yielded a phospholipid fraction that was a far more effective phosphogalactosyl acceptor (37,000 cpm of [^{14}C]galactose incorporated per micromole of lipid phosphorus) than

TABLE II

Lipid Dependence of Phosphogalactosyltransferase Reaction with Lipid-Depleted Particulate Enzyme Preparation[a]

Incubation mixture	Lipid phosphorus (μmole)	Galactose incorporation (cpm)
I. Supplementation with crude lipid		
Complete − crude lipid	None	26
Complete + crude lipid + heated enzyme	0.38	17
Complete + crude lipid	0.38	515
Complete + crude deacylated lipid	0.43	483
Complete + crude lipid − Triton X-100	0.38	93
II. Supplementation with lipid fractions from silicic acid column		
Complete + CHCl$_3$ fraction	0.002	748
Complete + acetone fraction	0.002	174
Complete + C:M, 1:1, fraction	0.48	17,800
III. Supplementation with lipid fractions from silica gel H		
Complete + lipid I	0.10	38,100
Complete + lipid II	0.11	232

[a] The standard incubation mixtures contained the lipid-depleted enzyme preparation III (2.0 mg of protein), exogenous lipid as indicated, and UDP-[^{14}C]galactose (15,200 cpm/nmole). After incubation at 12° for 60 minutes, the incubation mixtures were extracted with chloroform–methanol and assayed as described under Experimental Procedure.

the crude deacylated lipid (1300 cpm of [^{14}C]galactose incorporated per micromole of lipid phosphorus). The exact basis for this observation is not understood, although possibly the presence of large quantities of fatty acid methyl esters in the crude deacylated lipid mixture are potent inhibitors of the phosphogalactosyltransferase reaction.

The lipid-depleted enzyme preparation supplemented with alkali-stable phospholipid catalyzed the incorporation of galactose into the C:M soluble fraction via a phosphogalactosyl-transferase reaction identical with that observed in the crude envelope particulate fraction, as indicated by the following observations: (a) addition of an excess of unlabeled UMP or UDP-galactose to an incubation mixture after maximal incorporation of [^{14}C]galactose into the C:M-soluble fraction had been attained resulted in a loss of radioactivity from the lipid; (b) with unlabeled UDP-galactose as substrate in the incubation mixture, the addition of [^{14}C]UMP after maximal galactose incorporation resulted in the formation of [^{14}C]UDP-galactose; and (c) treatment of the C:M-soluble, [^{14}C]galactose-containing lipid with 0.1 N NaOH, as described above, quantitatively released galactose phosphate. Further, the participation of an alkali-stable phospholipid in a UDP-galactose-dependent transfer of mannose (from GDP-mannose) into the C:M-soluble fraction was established in the following manner. Four incubation mixtures were prepared: (a) contained alkali-stable phospholipid and UDP-[^{14}C]-galactose; (b) contained alkali-stable phospholipid and unlabeled UDP-galactose and, after 60 minutes of incubation, GDP-[^{14}C]mannose was added to this mixture; and (c) and (d) were prepared with the omission of either alkali-stable phospholipid or unlabeled UDP-galactose and, after 60 minutes, GDP-[^{14}C]mannose was added to both. The results of this experiment (Table III) confirmed that [^{14}C]mannose was incorporated into the lipid fraction in equimolar amounts to galactose only in the presence of UDP-galactose and alkali-stable phospholipid.

Analysis by thin-layer chromatography of the alkali-stable phospholipid fraction isolated by silicic acid column chromatography indicated the presence of two components (Table II, Experiment III). Lipid I exhibited an R_f of 0.05 in solvent I and 0.62 in solvent II; lipid II exhibited an R_f of 0.57 in solvent I and 0.02 in solvent II. To assess the activity of the two lipids in the phosphogalactosyltransferase reaction, each was isolated by preparative thin-layer chromatography in solvent I; a test strip was treated with iodine vapor to locate the lipids. The phospholipids were eluted from the silica gel H and assayed for their ability to stimulate [^{14}C]galactose incorporation into the lipid fraction catalyzed by the lipid-dependent enzyme preparation. The results presented in Table II, Experiment III, indicate that only lipid I stimulated

TABLE III

LIPID DEPENDENCE OF MANNOSYLTRANSFERASE REACTION WITH LIPID-DEPLETED
PARTICULATE ENZYME PREPARATION[a]

Additions		Incorporation (pmoles)
Sugar nucleotide	Phospholipid	
UDP-[14C]Gal	+	56.7
UDP-[12C]Gal GDP-[14C]Man	+	57.2
GDP-[14C]Man	−	0.01
UDP-[12C]Gal GDP-[14C]Man	−	0.01

[a] Standard incubation mixtures were prepared as described in the text and contained enzyme preparation III (1.0 mg of protein); crude deacylated phospholipid; 0.21 μmole of lipid phosphorus; and either UDP-[14C]galactose (29,600 cpm/nmole) or GDP-[14C]mannose (28,900 cpm/nmole) as indicated. Preliminary incubations with UDP-galactose were carried out as described in the text and incorporation of 14C-labeled sugars into the lipid fraction was determined after incubation (60 minutes, 12°) in the presence of the radioactive substrate as described under Experimental Procedure.

galactose incorporation. The chromatographic properties of lipid I on thin-layer chromatography in solvents I and II, as well as its behavior on DEAE-cellulose and silicic acid column chromatography, were similar to those reported for undecaprenyl phosphate.

Analysis of the carrier lipid by mass spectrometry after hydrolysis at pH 2 as described by Wright et al.[2] indicated that the dephosphorylated carrier lipid is a polyisoprenol compound. The mass spectrum is illustrated in Fig. 5 and is similar to those reported for the carrier lipid involved in peptidoglycan,[4] O-antigen,[2,3] and mannan biosynthesis.[21] More than 95% of the polyisoprenol is undecaprenol (C_{55}); the C_{60} homolog constitutes less than 5% of the total polyisoprenyl phosphate fraction. In a typical preparation, 0.3 μmole of polyisoprenyl phosphate was recovered from 100 μmoles of crude phospholipid.

In order to establish the activity of the purified polyisoprenyl phosphate in the capsular polysaccharide synthetase system, the purified lipid was used to synthesize the lipid-linked mannosylgalactose intermediate. The incubation mixture contained (in micromoles): Tris buffer (pH 7.9), 32; $MgCl_2$, 3.2; UDP-[14C]galactose (16,985 cpm/nmole), 0.1; GDP-[14C]mannose (14,505 cpm/nmole), 0.1; polyisoprenyl phosphate,

[21] M. Scher, W. J. Lennarz, and C. C. Sweeley, *Proc. Nat. Acad. Sci. U.S.* **59**, 1313 (1968).

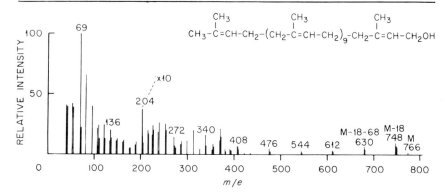

FIG. 5. Mass spectrum of the *Aerobacter* carrier lipid after treatment with mild acid. The purified carrier lipid (0.1 μmole of lipid phosphorus) was treated at pH 2 for 10 minutes at 100°, and the product was extracted into petroleum ether. The extract was then analyzed by mass spectrometry as described under Experimental Procedure.

0.07; phosphatidylglycerol, 0.23; and 2.8 mg of lipid-depleted enzyme protein. The mixture (0.52 ml) was incubated for 75 minutes at 30° and then extracted with 6 ml of C:M, 2:1. The extract was washed three times by the procedure of Folch et al.,[8] evaporated, and the lipid dispersed in water by sonic disruption. After hydrolysis with $2 N$ NH$_4$OH (37°, 20 minutes) and extraction with butanol, radioactivity was released quantitatively into the aqueous phase. Acid hydrolysis (pH 2, 10 minutes, 100°) yielded a radioactive component whose mobility in solvent I was consistent with a disaccharide phosphate; some free disaccharide was also observed. The radioactive compound was treated with alkaline phosphatase, reduced with NaBH$_4$, desalted with a mixed bed resin (equal parts of Dowex 1 (HCO$_3^-$) and Dowex 50 (H$^+$)), and hydrolyzed ($2 N$ HCl, 2 hours, 100°). The radioactive components of the hydrolyzate were isolated and characterized as described above for the characterization of mannosylgalactitol. Sugars and polyols were quantitatively analyzed as previously described and the results indicated a molar ratio of galactitol to mannose of 1.00 to 0.87.

Acid hydrolysis of the lipid-derived radioactive component after prior reduction yielded mannose and galactitol in equimolar amounts indicating that the purified polyisoprenyl phosphate serves as the lipid carrier of mannosylgalactose 1-phosphate.

[84] Synthesis of the Hydroxylysine-Linked Carbohydrate Units of Collagens and Basement Membranes

By ROBERT G. SPIRO and MARY JANE SPIRO

$$\text{UDP-Gal + Hyl-collagen (basement membrane)} \xrightarrow{\text{Mn}^{2+}}$$
$$O\text{-}\beta\text{-}\text{D-Gal-}(1 \to 5)\text{-Hyl-collagen (basement membrane)} + (\text{UDP}) \quad (1)$$

$$\text{UDP-Glc} + O\text{-}\beta\text{-}\text{D-Gal-}(1 \to 5)\text{-Hyl-collagen (basement membrane)} \xrightarrow{\text{Mn}^{2+}}$$
$$O\text{-}\alpha\text{-}\text{D-Glc-}(1 \to 2)\text{-}O\text{-}\beta\text{-}\text{D-Gal-}(1 \to 5)\text{-Hyl-collagen (basement membrane)}$$
$$+ (\text{UDP}) \quad (2)$$

Carbohydrate units linked to hydroxylysine are a characteristic structural feature of collagens and basement membranes which have been observed in proteins from both vertebrate and invertebrate sources.[1-3] These units occur in the form of 2-*O*-α-D-glucosyl-D-galactose disaccharides or single galactose residues linked by β-glycosidic bonds to the hydroxyl group of this amino acid.[4]

Their biosynthesis proceeds through the action of two specific glycosyltransferases[5,6] [Eqs. (1) and (2)], which have been found to occur in many tissues.[7]

UDP-Glucose : Galactosylhydroxylysine-Collagen (Basement Membrane) Glucosyltransferase

Assay

Principle. The assay[5] is based on the measurement of the radioactivity incorporated into glucosylgalactosylhydroxylysine (Glc-Gal-Hyl) after incubation of the enzyme with UDP-[¹⁴C]glucose and an acceptor containing galactosylhydroxylysine (Gal-Hyl). This specific measurement is made possible by the alkaline stability of the Gal-Hyl bond. After alkaline hydrolysis and desalting, the radioactive product is separated by paper chromatography and measured by radioscanning.

Preparation of Acceptor. Peptides containing Gal-Hyl are prepared by collagenase and Pronase digestion of basement membrane, followed

[1] R. G. Spiro, *J. Biol. Chem.* **244**, 602 (1969).

[2] R. G. Spiro, *Annu. Rev. Biochem.* **39**, 599 (1970).

[3] R. G. Spiro, *in* "Glycoproteins" (A. Gottschalk, ed.), 2nd ed. Part B, p. 964. Elsevier, Amsterdam (1972).

[4] R. G. Spiro, *J. Biol. Chem.* **242**, 4813 (1967).

[5] R. G. Spiro and M. J. Spiro, *J. Biol. Chem.* **246**, 4899 (1971).

[6] M. J. Spiro and R. G. Spiro, *J. Biol. Chem.* **246**, 4910 (1971).

[7] R. G. Spiro and M. J. Spiro, *J. Biol. Chem.* **246**, 4919 (1971).

by gel filtration, to obtain Glc-Gal-Hyl-peptides from which the glucose is then selectively removed by partial acid hydrolysis.

Basement membranes, prepared from bovine renal glomeruli by sonic treatment[8] are digested at 37° in 0.15 M Tris·acetate, pH 7.4, in the presence of 0.005 M calcium acetate with purified collagenase from *Clostridium histolyticum* (Worthington).[9] The membranes are suspended in the buffer (25 mg/ml), and the collagenase is added initially to equal 0.7% (w/w). At 24 and 48 hours, further additions of the enzyme equal to 0.35% and 0.1% of the weight of the substrate, respectively, are made and the pH is readjusted to 7.4 with 1 M Tris. The incubations are carried out with shaking in the presence of toluene for a total period of 72 hours. Upon completion of this incubation, the undigested material is removed by centrifugation (30,000 g for 30 minutes) and the clear supernatant, after adjustment to pH 7.8 with Tris, is further digested with Pronase (Calbiochem). This enzyme is added initially to equal 0.3% of the membrane weight with further enzyme additions of 0.1% being made at 24 and 48 hours. Incubation is carried out for a total of 72 hours at 37° in the presence of toluene and the digest is then lyophilized and fractionated by filtration on a column (2.1 × 80 cm) of Sephadex G-25 (fine).

This column is equilibrated with 0.1 M pyridine acetate buffer, pH 5.0, and the sample (representing 250 mg to 1 g of basement membrane) is applied in a small volume and eluted with the same buffer. Fractions of 5 ml are collected and aliquots are taken for analysis of hexose by the anthrone method (Vol. 8 [1]) and peptide by the ninhydrin procedure.[10] Two anthrone-positive glycopeptide peaks are obtained which emerge prior to the peak containing the large amount of peptide material. The second glycopeptide peak, which in the case of a column of this size reaches a maximum at tube 33 and has as its carbohydrate only the hydroxylysine-linked disaccharide unit, is pooled and the buffer removed by lyophilization.

For the selective removal of the glucose these Glc-Gal-Hyl-peptides are hydrolyzed in 0.1 N sulfuric acid at a concentration of 0.25 μmole/ml in sealed tubes at 100° for 28 hours. The resulting Gal-Hyl-peptides are then separated from the liberated glucose by adsorption on a column of Dowex 50-X2, 200–400 mesh, H⁺ form, extensively washed with water, and eluted from the column with 10 volumes of 1.5 N NH₄OH. The ammonia is removed by lyophilization and the amount of Gal-Hyl-

[8] R. G. Spiro, *J. Biol. Chem.* **242**, 1915 (1967).
[9] R. G. Spiro, *J. Biol. Chem.* **242**, 1923 (1967).
[10] S. Moore and W. H. Stein, *J. Biol. Chem.* **211**, 907 (1954).

peptides present in this fraction is determined by the anthrone reaction, with galactose as a standard. Direct measurement of the Gal-Hyl can be made after hydrolysis of an aliquot with $2 N$ NaOH at $105°$ for 24 hours in capped polypropylene tubes, followed by acidification to pH 3 with HCl and analysis on the Technicon amino acid analyzer (see this volume [1]). The yield of Gal-Hyl-peptides obtained by this procedure is 9–10 μmoles per 100 mg of basement membrane.

It has also been found possible to prepare the acceptor for the glucosyltransferase from a kidney fraction which contains both glomerular and tubular basement membranes. For this purpose, bovine kidney cortex is homogenized in $0.15 M$ NaCl in a Sorvall Omnimixer and then submitted to ultrasonic treatment in a manner similar to that employed for the disruption of the purified glomeruli.[8] Gal-Hyl-peptides can be obtained from this basement membrane fraction in the same manner as outlined for the glomerular basement membrane and in similar yield.

Native collagens containing unsubstituted Gal-Hyl units may be used directly as acceptors. Skin collagen from calf has been employed effectively in this capacity.[5] The Gal-Hyl content of these acceptors must be determined on the amino acid analyzer after alkaline hydrolysis. It is useful to solubilize the collagen acceptors by a brief treatment (30–60 minutes) at $37°$ in $0.1 N$ NaOH (12–15 mg/ml) followed by neutralization to pH 6.8.

For a standard assay procedure the Gal-Hyl-peptides which have a higher relative V_{max} are preferable to the collagens for use as acceptors. Moreover the collagens have to be added in large amounts to achieve suitable acceptor site concentration, leading to viscous incubation mixtures and a greater number of extraneous components on the radioscans.

Reagents

Tris·acetate buffer, $0.15 M$, pH 6.8, containing $2 \times 10^{-3} M$ 2-mercaptoethanol

Manganese acetate, $0.5 M$

UDP-glucose-^{14}C (5 μCi/μmole), $7.5 \times 10^{-4} M$

Galactosylhydroxylysine (Gal-Hyl) peptides, $2.5 \times 10^{-3}M$ (galactose equivalents)

Enzyme in $0.15 M$ Tris·acetate, pH 6.8, containing $2 \times 10^{-3} M$ 2-mercaptoethanol

NaOH, $2.5 N$

HCl, $1 N$

Dowex 50-X4, 200–400 mesh, H$^+$ form

NH$_4$OH, $1.5 N$

1-Butanol–acetic acid–water (4:1:5)

Procedure. The incubation is performed in 13 × 100 mm stoppered glass tubes for 2 hours at 37° with shaking. Prior to the addition of the enzyme, 0.2 ml of the UDP-glucose and 0.1 ml of the Gal-Hyl-peptides are pipetted into the tubes and taken to dryness under a stream of nitrogen at 30° (an appropriate aliquot of a collagen solution representing a similar number of acceptor sites can replace the Gal-Hyl-peptides and is similarly taken to dryness). Enzyme (0.2–2 units) and buffer are added to the dried substrates to give a volume of 100 μl, and 5 μl of the manganese acetate solution is then added. The final volume of 105 μl therefore contains 15 μmoles of Tris·acetate buffer, 2.5 μmoles of manganese acetate, 0.2 μmoles of 2-mercaptoethanol, 0.15 μmole of UDP-[^{14}C]glucose, and Gal-Hyl-peptides equivalent to 0.25 μmole of galactose. Control incubations without exogenous acceptor are also performed.

At the end of the incubation, 0.4 ml of the NaOH is added to each tube, and the sample is transferred to polypropylene tubes equipped with tightly fitting polypropylene caps which are fastened with metal clips to prevent evaporation. Hydrolysis is carried out for 20 hours at 105° to obtain Glc-Gal-Hyl. The samples are diluted with 8 ml of water and titrated to pH 3 with 1 N HCl. They are then placed on small columns (8.5 mm internal diameter) containing 2.5 g of Dowex 50 resin. The columns are washed with 45 ml of water, and the Glc-Gal-Hyl recovered by elution with 10 ml of the NH$_4$OH. The eluates are dried under reduced pressure on an Evapomix (Buchler) at 45° and then applied directly as 2-cm streaks on 3.7-cm strips of Whatman No. 1 paper. Descending chromatography is performed in the upper phase of the 1-butanol–acetic acid–water system for 6 days (the lower phase is placed in the bottom of the cabinet). Guide strips containing Glc-Gal-Hyl and Hyl are run in each chromatographic trough and are stained with ninhydrin. The R_{Hyl} of Glc-Gal-Hyl in this system is 0.52.[4] The radioactivity on each strip is localized with a Nuclear Chicago radioscanner (Actigraph III) (Fig. 1), and quantitation of the Glc-Gal-Hyl peak is accomplished with the use of an integrator attached to the scanner or by planimetry. The counting efficiency under these conditions is 13%.

It is essential that the activity of the product is determined by this specific assay involving chromatographic separation, as with all but the purest enzyme preparation significant amounts of other radioactive products are formed. Direct precipitation techniques employing collagen as substrate lead to erroneous results, the radioactivity determined being in excess of that incorporated into Glc-Gal-Hyl.[5]

Definition of Unit and Specific Activity. A unit is defined as the amount of enzyme that catalyzes the transfer of 1 nmole of glucose to

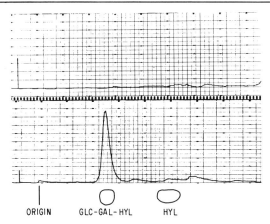

ORIGIN   GLC-GAL-HYL   HYL

FIG. 1. Radioscans from standard assay of UDP-glucose:galactosylhydroxylysine-collagen (basement membrane) glucosyltransferase showing the amount of ^{14}C-labeled Glc-Gal-Hyl formed during incubation of enzyme from the kidney cortex in the presence of Gal-Hyl-peptides (lower frame) and in the absence of these acceptors (upper frame). Chromatography was performed in 1-butanol–acetic acid–water (4:1:5) for 6 days. From R. G. Spiro and M. J. Spiro, *J. Biol. Chem.* **246**, 4899 (1971).

Gal-Hyl-peptides per hour. Specific activity is expressed as units per milligram of protein, as measured by the method of Lowry *et al.*[11]

Enzyme Preparation

Rat kidney cortex is homogenized in an all-glass Potter-Elvehjem homogenizer in 2 volumes of ice-cold $0.15\,M$ Tris·acetate buffer, pH 6.8, containing $0.002\,M$ 2-mercaptoethanol, and centrifuged at $100,000\,g$ for 60 minutes.[5] About 60% of the total activity of the homogenate is found in this high-speed supernatant under these conditions, and the highest activity per gram of tissue is found in 10-day-old rats. A $4.0\,M$ solution of ammonium sulfate, adjusted to pH 7.0, is added to the supernatant to give a final concentration of $1.0\,M$. The sediment obtained after centrifugation is discarded, the supernatant is brought to a concentration of $1.6\,M$, and the precipitate, dissolved in $0.15\,M$ Tris·acetate, pH 6.8, $0.002\,M$ mercaptoethanol, is dialyzed against this buffer for 20 hours with 3 changes.

Further purification can be achieved by filtration of this ammonium sulfate fraction (Table I) on a column of Bio-Gel A-1.5 m, 100–200

[11] O. H. Lowry, N. J. Rosebrough, A. L. Farr, and R. J. Randall, *J. Biol. Chem.* **193**, 265 (1951).

TABLE I
PURIFICATION OF GLUCOSYLTRANSFERASE FROM RAT KIDNEY CORTEX[a]

Fraction	Specific activity (units/mg protein)	Relative specific activity	Relative total activity
Supernatant, 100,000 g	0.45	1.0	1.0[b]
1.0–1.6 M Ammonium sulfate precipitate	1.80	4.0	0.79
Bio-Gel A-1.5 m, pool	5.59	12.4	0.37
Bio-Gel A-1.5 m, peak tube	7.80	17.3	0.06

[a] Tissue from 10-day-old male animals.

[b] This fraction contains 60% of the activity present in the homogenate. The remaining enzyme activity is present in particle fractions and can be released by ultrasonic treatment.

mesh. In a typical fractionation, 240 mg of protein are applied to a column (2.1 × 83 cm) equilibrated and eluted with the Tris-mercaptoethanol buffer, and 5-ml fractions are collected. The enzyme emerges as two components, the major peak reaching its maximum activity at an elution volume of 176 ml. When the fractions representing an elution volume of 155–187 ml are pooled, a purification of 12.4-fold compared to the ammonium sulfate precipitate is achieved.

All the above reactions are carried out at 2°.

The enzyme can also be purified in a similar manner from the high-speed supernatant of calf kidney or rat cartilage.

Properties of the Enzyme

Effect of pH. The enzyme has good activity over the pH range of 6.0 to 8.5, with maximal activity occurring at pH 7.4.

Metal Requirements. The enzyme has an absolute requirement for a divalent cation, manganese being the most effective. When 0.025 M manganese is replaced by the same concentration of magnesium or calcium, approximately 40% of the activity is obtained, whereas the activity with nickel, cobalt, or zinc is 20–30%. Iron, cadmium, or copper give no activity.

Specificity. UDP-glucose is the most effective glycosyl donor for this enzyme. TDP-glucose is 32% as effective, and CDP-glucose gives 25% of the activity obtained with the UDP-glucose. GDP- and ADP-glucose are essentially inactive.

The enzyme in general requires as its acceptor galactose attached to hydroxylysine, either by itself (Gal-Hyl) or in peptides or proteins (Table II). The ε-amino group of the hydroxylysine must be unsub-

TABLE II

ACCEPTOR SPECIFICITY OF GLUCOSYLTRANSFERASE[a]

Acceptor	Relative activity per micromole of acceptor site
Gal-Hyl-peptides	100
Gal-Hyl	112
N-Acetyl-Gal-Hyl-peptides	0
Glc-Gal-Hyl-peptides	0
Galactosylsphingosine (psychosine)	65
Thyroglobulin glycopeptides, unit B (NAN-free)	9
Earthworm cuticle collagen glycopeptides	0
Glomerular basement membrane	
Native	0
Glc-free	157
Glc-Gal-free	0
Calf skin collagen	
Alkali solubilized	223
Glc-free	152
Sclera collagen, rabbit	301
Squid fin collagen	93

[a] Data taken from R. G. Spiro and M. J. Spiro, *J. Biol. Chem.* **246**, 4899 (1971).

stituted. The proteins which may serve as acceptors include collagens from both vertebrate and invertebrate sources, those having a higher Gal-Hyl content being the better acceptors on a weight basis. Native basement membranes will not serve as an acceptor unless the glucose is selectively removed by partial acid hydrolysis. Galactosylsphingosine (psychosine) is the only exception to the galactosylhydroxylysine acceptor requirement.[5]

Kinetics. The K_m for UDP-glucose is $3.1 \times 10^{-5} M$. The K_m for the Gal-Hyl-peptides is $1.2 \times 10^{-3} M$, that for calf skin collagen, $1.3 \times 10^{-4} M$, and for basement membrane from which glucose has been selectively removed, $1.6 \times 10^{-4} M$. The relative V_{max} values for these three acceptors are respectively, 1.93, 0.88, and 0.54.

The reaction is linear with time over a period of 2 hours.

Inhibitors. The enzyme activity is completely inhibited by p-mercuribenzoate ($0.0125 M$) and by dialysis against $0.01 M$ EDTA at pH 6.8. Significant inhibition is also produced by UTP and ATP ($0.006 M$), Tween (0.025%), and sucrose ($0.3 M$).

Stability. The crude enzyme is quite stable to storage in the frozen state for a period of several months. After purification the enzyme loses activity during dialysis or storage in the frozen state. This activity can be partially or fully restored by inclusion of $0.15 M$ NaCl in the incubation mixture.

Identification of Product

The standard assay is specific for Glc-Gal-Hyl. It has been shown that the glucose is attached to the hydroxylysine-linked galactose by a 2-O-α-D-glycosidic bond, which is the naturally occurring linkage and configuration.[5]

Distribution

The enzyme has been found widely distributed in tissues of the rat (Table III). In most tissues the activity is considerably higher in young animals and in kidney it reaches its highest activity at about 10 days of age.

UDP-Galactose:Hydroxylysine-Collagen (Basement Membrane) Galactosyltransferase

Assay

Principle. The assay[6] is based on the measurement of radioactivity incorporated into Gal-Hyl when high molecular weight acceptors con-

TABLE III

DISTRIBUTION OF UDP-GLUCOSE:GAL-HYL-COLLAGEN (BASEMENT MEMBRANE) GLUCOSYLTRANSFERASE ACTIVITY IN RAT TISSUES[a]

Tissue	Enzyme activity	
	Units per gram tissue	Units per milligram protein
Cartilage	69.9	1.87
Kidney cortex	25.7	0.45
Thyroid	24.9	0.28
Lung	23.9	0.44
Spleen	20.9	0.26
Skin	15.9	0.42
Serum	13.0[b]	0.26
Liver	11.6	0.14
Brain	6.4	0.28
Heart	4.7	0.10
Adipose tissue (perirenal)[c]	5.2	1.29
Uterus[d]	50.3	0.83

[a] Activity in supernatants (10,000 g for 10 minutes) of the tissues from 10-day-old (25 g) male rats unless otherwise indicated.

[b] Expressed as units/ml.

[c] From 135-g rats.

[d] From 200-g rats.

taining unsubstituted hydroxylysine are incubated with ^{14}C-labeled UDP-galactose. The alkaline stability of the Gal-Hyl bond permits measurement of the radioactive product formed after hydrolysis of the incubation mixture in NaOH, desalting, paper chromatography, and radioscanning.

Preparation of Acceptor. Disaccharide-free glomerular basement membrane can be prepared by a single Smith periodate degradation.[4] Because a $1 \rightarrow 2$ linkage is present in the unit, only one such degradation is required; if a second oxidation is performed the newly uncovered hydroxylysine residues are destroyed, leaving no remaining acceptor sites.

Bovine glomerular basement membrane, which contains approximately 14.5 disaccharide units per 100 mg (or a basement membrane fraction prepared as described in the section on the glucosyltransferase) is suspended (5 mg/ml) in $0.06 M$ sodium acetate buffer, pH 4.5 containing $0.03 M$ sodium metaperiodate and shaken at 4° in the dark for 28 hours. The reaction is terminated by the addition of ethylene glycol in a 10-fold molar excess over the periodate. The suspension is then titrated to pH 8.0 with dilute NaOH and reduced by the addition of a 25-fold molar excess of sodium borohydride (in relation to the periodate) in the form of a $0.6 M$ solution of this reagent in $0.6 M$ sodium borate buffer, pH 8.0. The reaction is allowed to proceed for 16 hours at 2° and terminated by lowering the pH to 5 by the addition of acetic acid. The mixture is then dialyzed at 2° against $0.1 M$ sodium chloride followed by extensive dialysis against distilled water. The dialyzed material is carefully removed from the dialysis tubing, HCl is added to make a concentration of $0.1 N$, and hydrolysis in this acid is carried on at 80° for 2 hours. The hydrolyzate is titrated to pH 6.8 with NaOH and dialyzed in 18/100 Visking cellophane tubing at 2° against three changes of distilled water for 8 hours followed by lyophilization. Throughout this procedure some of the basement membrane may stay insoluble and must be transferred carefully from the dialysis tubing. The amount of unsubstituted hydroxylysine, which is equivalent to the acceptor sites, should be assessed by analysis on the amino acid analyzer of an alkaline hydrolyzate ($2 N$ NaOH, 105°, 24 hours).

Native fibrillar collagens, such as ichthyocol or skin and tendon collagens can also be used as acceptors. They should not be treated with periodate, as this destroys the unsubstituted hydroxylysine residues which are present in larger amounts than those involved in the linkage of the carbohydrate units.[1] These unsubstituted hydroxylysines in the collagens are as effective as acceptors as those which have been exposed by periodate treatment.[6]

The disaccharide-free basement membranes and native collagens are conveniently solubilized by brief incubation at 37° with 0.1 N NaOH (30–60 minutes) followed by neutralization with acetic acid to pH 6.8.

Reagents

> Tris·acetate, 0.15 M, pH 6.8, containing $2 \times 10^{-3} M$ 2-mercapto-ethanol
>
> Manganese acetate, 0.2 M
>
> UDP-[^{14}C]galactose (6 μCi/μmole), $7.5 \times 10^{-4} M$
>
> Disaccharide-free basement membrane or native ichthyocol, $1.5 \times 10^{-3} M$ acceptor sites (unsubstituted hydroxylysine)
>
> Enzyme in 0.15 M Tris·acetate, pH 6.8, containing $2 \times 10^{-3} M$ 2-mercaptoethanol
>
> NaOH, 2.5 N
>
> HCl, 1 N
>
> Dowex 50-X4, 200–400 mesh, H$^+$ form
>
> NH$_4$OH, 1.5 N
>
> 1-Butanol–acetic acid–water (4:1:5)

Procedure. Incubations are performed in 13×100 mm stoppered glass tubes for 2 hours at 37° with shaking. Prior to the addition of the enzyme, 0.2 ml of the nucleotide and 0.2 ml of acceptor are pipetted into the tubes and taken to dryness under a stream of nitrogen at 30°. The enzyme (0.2–2 units) and buffer are added to give a volume of 100 μl, and 5 μl of the manganese acetate are then added. The final volume of 105 μl therefore contains 15 μmoles of Tris·acetate, 0.2 μmole of 2-mercaptoethanol, 1 μmole of manganese acetate, 0.15 μmole of UDP-[^{14}C]-galactose, and 0.3 μmole of acceptor sites. Control incubations without exogenous acceptor are also performed.

At the end of the incubation, hydrolysis in NaOH, desalting on Dowex 50, paper chromatography, and radioscanning are performed in a manner identical to that described for the glucosyltransferase assay. The R_{Hyl} of galactosylhydroxylysine in the butanol–acetic acid–water system is 0.69.[4]

As in the case of the glucosyltransferase assay, it is essential that the incorporation of radioactivity into hydroxylysine-linked galactose be determined by this specific chromatographic assay, as precipitation techniques lead to erroneous results due to the incorporation of activity into components other than galactosylhydroxylysine.

Definition of Unit and Specific Activity. A unit is defined as the amount of enzyme that catalyzes the transfer of 1 nmole of galactose from UDP-galactose to the hydroxylysine of disaccharide-free glomerular basement membrane. The specific activity is expressed as units per milligram of protein as determined by the method of Lowry et al.[11]

Preparation

The enzyme can be prepared from 10-day-old rat kidney cortex or from calf kidney cortex by homogenization at 2° in an all-glass Potter-Elvehjem homogenizer in 2 volumes of 0.15 M Tris·acetate, pH 6.8, containing 0.002 M 2-mercaptoethanol.[6] The supernatant obtained after centrifugation at 100,000 g contains about 70% of the activity of the homogenate. Some purification (about 8-fold) can be achieved by filtration on Bio-Gel A-1.5 m of the material precipitated from this supernatant between 1.0 and 1.6 M ammonium sulfate. If fractionation of the enzyme is performed, the inclusion of 0.1 M galactose in the buffer used during homogenization and in all the solutions used in the subsequent steps of purification is necessary to preserve activity.

The enzyme can also be obtained from the cartilage of 10-day-old rats by homogenization in 2 volumes of the Tris acetate buffer in a Sorvall Omnimixer for five 15-second intervals. Bio-Gel A-1.5 m filtration of the 100,000 g supernatant from this tissue leads to a maximal purification of 17-fold.[7]

Properties

Effect of pH. The enzyme has good activity between pH 6 and 7.5 with the maximum activity occurring at pH 6.8.

Metal Requirements. The enzyme has an absolute requirement for a divalent cation, which is best met by manganese at a concentration of 0.01 M. Magnesium can replace the manganese to the extent of 45% at this concentration. Other cations, including nickel, cobalt, iron, copper, calcium, cadmium, and zinc are entirely ineffective.

Specificity. The enzyme is quite specific for UDP-galactose as the glycosyl donor. CDP- and TDP-galactose result in the transfer of less than 10% of the activity obtained with the UDP-galactose. ADP-galactose and GDP-galactose are completely inactive.

The enzyme will transfer galactose only to the hydroxyl group of hydroxylysine residues in high molecular weight acceptors and requires that the ε-amino group be unsubstituted (Table IV). In order for the glomerular basement membrane to serve as an acceptor, the disaccharide units must be removed. However, native collagens from various sources can serve as effective acceptors. Indeed, removal of the disaccharide units from the collagens by Smith degradation decreases

TABLE IV

ACCEPTOR SPECIFICITY OF GALACTOSYLTRANSFERASE[a]

Acceptor	Relative activity per micromole acceptor site	Relative activity per milligram acceptor protein
Glomerular basement membrane		
Native	0	0
Glc-free	0	0
Disaccharide-free	100	100
N-Acetyl-disaccharide-free	0	0
Trypsin peptides, disaccharide free	23	—
Collagenase–pronase peptides, disaccharide free	0	—
Calf skin collagen		
Native	132	59
Disaccharide free	51	9
Ichthyocol		
Native	136	97
Disaccharide free	69	7
Bovine tendon collagen	97	78
Rabbit sclera collagen	174	78
Squid fin collagen	147	74
Sawfly silk collagen	109	307
Hydroxylysine	0	—

[a] Data taken from M. J. Spiro and R. G. Spiro, *J. Biol. Chem.* **246,** 4910 (1971).

their acceptor capacity per protein weight due to the destruction of the unsubstituted hydroxylysine residues which are more numerous than those present in the glycosylated form.[6]

Kinetics. The K_m for the UDP-galactose is $3.1 \times 10^{-4} M$. The K_m values for disaccharide-free basement membrane and ichthyocol are, respectively, $4 \times 10^{-4} M$ and $6.4 \times 10^{-3} M$, expressed on the basis of moles of unsubstituted hydroxylysine residues.

Inhibitors. The enzyme is completely inhibited by dialysis against $0.01 M$ EDTA in $0.15 M$ Tris·acetate, pH 6.8. No activity is found in the presence of $0.02 M$ ATP, while 50% inhibition is effected by UDP $(0.02 M)$ and sucrose $(0.3 M)$.

Stability. The enzyme is very unstable when stored at 2° and loses activity rapidly stored in the frozen state. However, activity can be partially preserved if the enzyme is stored frozen in the presence of various neutral sugars or sugar alcohols, including $0.1–0.25 M$ glucose, galactose, N-acetylglucosamine or sorbitol, or 20% glycerol. For this reason, when the enzyme is fractionated by ammonium sulfate precipitation or Bio-Gel filtration, $0.1 M$ galactose is included in all solutions.

TABLE V
DISTRIBUTION OF UDP-GALACTOSE:HYL-COLLAGEN (BASEMENT MEMBRANE)
GALACTOSYLTRANSFERASE ACTIVITY IN RAT TISSUES[a]

| Tissue | Enzyme activity | |
	Units per gram tissue	Units per milligram protein
Cartilage	68.1	1.82
Lung	23.1	0.43
Spleen	18.9	0.23
Skin	11.5	0.30
Kidney cortex	8.7	0.15
Thyroid	6.1	0.07
Liver	4.3	0.05
Brain	0	0
Heart	0	0
Uterus[b]	11.9	0.20

[a] Activity in supernatants (10,000 g for 10 minutes) of tissues from 10-day-old (25 g) male rats unless otherwise indicated.

[b] From 200-g rats.

Identification of Product

The assay procedure is specific for galactosylhydroxylysine. It has been shown that the enzyme links the galactose to the hydroxylysine in the β-configuration, which is the naturally occurring form in basement membranes and collagens.[6]

Distribution

The enzyme has been found in several tissues of the rat (Table V) and, like the glucosyltransferase, is present in greater amounts in young animals, reaching its highest activity in kidney at approximately 10 days of age.

[85] Biosynthesis of Chondroitin Sulfate

By LENNART RODÉN, JOHN R. BAKER, TORSTEN HELTING,
NANCY B. SCHWARTZ, ALLEN C. STOOLMILLER,
SADAKO YAMAGATA, and TATSUYA YAMAGATA

I. Introduction

The biosynthesis of macromolecular carbohydrate-protein compounds of animal tissues usually follows a pattern which is remarkably similar from one group of substances to another. Whether we are dealing with a plasma glycoprotein synthesized by a liver cell or a connective tissue proteoglycan manufactured by a chondrocyte, the sequence of events is the same: the protein moiety is synthesized by the routes that are now well established for simple proteins, and, subsequently, the carbohydrate prosthetic group is added by stepwise transfer of the various monosaccharide units from the corresponding nucleotide sugars. Recently, a few exceptions to this general pattern have been found, and, for example, the biosynthesis of the iduronic acid residues of heparin has been shown to occur at the polymer level by epimerization of glucuronic acid units which have previously been incorporated from UDP-glucuronic acid.[1]

The means by which constancy in structure is achieved during the biosynthesis of a macromolecule are sometimes fundamentally different from one type of molecule to another. Whereas structural information for a protein is encoded in the corresponding messenger RNA, the regularity in the assembly of a carbohydrate molecule results from the specificity of the glycosyltransferases involved in this process. Certain generalizations can now be made concerning the mode of action of these enzymes which may be expressed most concisely in terms of the "one enzyme–one linkage" hypothesis which covers three different structural aspects: (1) a transferase is specific for the sugar transferred—in other words, for a particular nucleotide sugar; (2) the enzyme is also specific for the glycosyl acceptor; and (3) the transferase catalyzes formation of only one particular type of linkage, with respect to position and anomeric configuration. A notable exception from the second type of specificity is lactose synthetase, which catalyzes galactosyl transfer to either glucose or *N*-acetylglucosamine and is regulated in this respect by lactalbumin.

[1] U. Lindahl, G. Bäckström, A. Malmström, and L.-Å. Fransson, *Biochem. Biophys. Res. Commun.* **46**, 985 (1972).

The acceptor specificity is usually determined by the sugar at the nonreducing terminus. In some instances, the penultimate sugar and its linkage to the terminal sugar are also important. Some transferases show a preference for either large or small acceptors. The linkage synthesized is characteristic of the enzyme, and there is as yet no example of a single glycosyltransferase which can synthesize more than one type of linkage.

In view of these considerations it can be postulated from the structure of chondroitin sulfate (Fig. 1) that the biosynthesis of this polysaccharide is catalyzed by six distinct glycosyltransferases: (1) a xylosyltransferase which initiates polysaccharide chain formation by xylosyl transfer to the protein core of the proteoglycan; (2) a galactosyltransferase which catalyzes transfer to the xylosyl-protein formed by the first enzyme; (3) a second galactosyltransferase with the first galactose residue as acceptor; (4) a glucuronosyltransferase which completes the formation of the specific carbohydrate–protein linkage region by glucuronosyl transfer to the second galactose unit; (5) an N-acetyl-galactosaminyltransferase and (6) a second glucuronosyltransferase which are responsible for the synthesis of the characteristic repeating disaccharide units. The existence of these enzymes in cartilage and other polysaccharide-producing tissues is now well established, and in the following a brief survey of their properties will be given.

The Enzymes

The chondroitin sulfate glycosyltransferases are, with the exception of the chain-initiating xylosyltransferase, firmly associated with the membranes of the endoplasmic reticulum. This situation has been one of the major obstacles which has hindered progress in our knowledge of the individual enzymes. Some advances have recently been made, however, which hold promise for further rapid developments in this area. Helting[2,3] has succeeded in solubilizing several of the glycosyltransferases involved in heparin synthesis in mouse mastocytoma by treatment of the particulate enzyme system with ammonia in the presence of Tween 20. The solubilization is apparently caused to a large extent by the increase in ionic strength which occurs upon addition and subsequent neutralization of the ammonia, and a modified procedure has been developed, based on the use of Nonidet P-40 and potassium chloride, which is described in detail in Section VIII. The solubilized enzymes are no longer sedimentable at 100,000 g and appear to have

[2] T. Helting, *J. Biol. Chem.* **246**, 815 (1971).
[3] T. Helting, *J. Biol. Chem.* **247**, 4327 (1972).

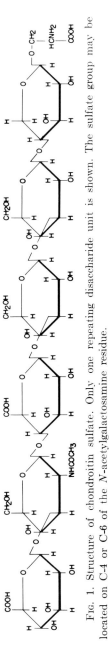

FIG. 1. Structure of chondroitin sulfate. Only one repeating disaccharide unit is shown. The sulfate group may be located on C-4 or C-6 of the N-acetylgalactosamine residue.

been released from the membranes in a monomolecular form, since they emerge in large part in retarded positions on gel chromatography.

None of the firmly membrane-bound glycosyltransferases has as yet been extensively purified, but some progress has been made in the purification of the more soluble xylosyltransferase which may be obtained in high yield in a 100,000 g supernatant fluid from a homogenate of embryonic chick cartilage. This enzyme has recently been purified over 1000-fold by affinity chromatography.[4]

The Substrates

The original detection of the glycosyltransferases participating in chondroitin sulfate synthesis was possible because of the presence of endogenous acceptors in homogenates and subcellular fractions of the tissues used as enzyme sources (e.g., embryonic chick cartilage and mouse mastocytoma). Initially, it was shown that all four monosaccharide components of chondroitin sulfate (glucuronic acid, N-acetylgalactosamine, galactose, and xylose) could be transferred from their respective nucleotide sugars to such endogenous acceptors, and partial characterization of the products demonstrated that the incorporated sugars were part of the expected chondroitin sulfate structures.[5-10] Ideally, the various glycosyl transfer reactions should be studied with the substrates that are present in $vivo$, but it is currently not profitable to pursue this approach beyond a certain stage. Although it is safe to assume that the endogenous acceptors consist of incomplete proteoglycan molecules in all stages of growth, it is not possible with present methodology to isolate and quantitate the multitude of molecular species which make up this acceptor pool. Much of the work on the properties and kinetic parameters of the glycosyltransferases is therefore better carried out with exogenous acceptors of well-defined structure. As an example, the assay of enzyme activity is totally unreliable, if it is not known whether enzyme concentration or acceptor concentration is rate-limiting in a given situation, and this uncertainty can be eliminated by using a known concentration of exogenous substrate.

[4] L. Rodén and N. B. Schwartz, $Biochem.$ $Soc.$ $Trans.$ **1**, 5 (1973).
[5] R. L. Perlman, A. Telser, and A. Dorfman, $J.$ $Biol.$ $Chem.$ **239**, 3623 (1964).
[6] J. E. Silbert, $J.$ $Biol.$ $Chem.$ **239**, 1310 (1964).
[7] E. E. Grebner, C. W. Hall, and E. F. Neufeld, $Biochem.$ $Biophys.$ $Res.$ $Commun.$ **22**, 672 (1966).
[8] E. E. Grebner, C. W. Hall, and E. F. Neufeld, $Arch.$ $Biochem.$ $Biophys.$ **116**, 391 (1966).
[9] H. C. Robinson, A. Telser, and A. Dorfman, $Proc.$ $Nat.$ $Acad.$ $Sci.$ $U.S.$ **56**, 1859 (1966).
[10] T. Helting and L. Rodén, $J.$ $Biol.$ $Chem.$ **244**, 2790 (1969).

A variety of exogenous substrates have been used in the study of chondroitin sulfate biosynthesis, and some of these are listed in Table I. Mostly, they consist of oligosaccharides with the appropriate acceptor monosaccharide at the nonreducing terminus, but it is seen that the free monosaccharide, D-xylose, may serve as an acceptor for the first galactosyltransferase (enzyme 2). It may also be noted that the acceptor in the first step of polysaccharide chain formation is a protein, and a cartilage proteoglycan from which the chondroitin sulfate chains have been removed by Smith degradation serves well in this capacity.

The acceptor specificities of the chondroitin sulfate glycosyl-transferases have recently been reviewed in detail,[11,12] and the pertinent findings are summarized below.

1. *Xylosyltransferase.* The best available substrate for this enzyme is the Smith-degraded proteoglycan mentioned above. Smaller peptides can also serve as acceptors, e.g., serylglycylglycine, but with much lower efficiency. The observation that a macromolecular substrate has

TABLE I

CHONDROITIN SULFATE-SYNTHESIZING GLYCOSYLTRANSFERASES
AND SOME OF THEIR EXOGENOUS SUBSTRATES

Enzyme	Exogenous acceptors
1. Xylosyltransferase	Smith-degraded cartilage proteoglycan, L-serylglycylglycine
2. Galactosyltransferase I	Xylose, O-β-D-xylosyl-L-serine
3. Galactosyltransferase II	4-O-β-D-Galactosyl-D-xylose, 4-O-β-D-galactosyl-O-β-D-xylosyl-L-serine
4. Glucuronosyltransferase I	3-O-β-D-Galactosyl-D-galactose, O-β-D-galactosyl-$(1 \rightarrow 3)$-O-β-D-galactosyl-$(1 \rightarrow 4)$-D-xylose
5. N-Acetylgalactosaminyltransferase	GlcUA-GalNAc-GlcUA-GalNAc-GlcUA-GalNAc (chondroitin hexasaccharide), GlcUA-(GalNAc-4S)-GlcUA-(GalNAc-4S)-GlcUA-(GalNAc-4S) (chondroitin 4-sulfate hexasaccharide)
6. Glucuronosyltransferase II	GalNAc-GlcUA-GalNAc-GlcUA-GalNAc (chondroitin pentasaccharide), (GalNAc-6S)-GlcUA-(GalNAc-6S)-GlcUA-(GalNAc-6S) (chondroitin 6-sulfate pentasaccharide)

[11] L. Rodén, in "Metabolic Conjugation and Metabolic Hydrolysis" (W. H. Fishman, ed.), Vol. 2, p. 345. Academic Press, New York, 1971.

[12] L. Rodén, J. R. Baker, N. B. Schwartz, A. C. Stoolmiller, S. Yamagata, and T. Yamagata, in "Biochemistry of the Glycosidic Linkage" (R. Piras and H. G. Pontis, eds.), p. 345. Academic Press, New York, 1972.

the highest acceptor activity is analogous to what has been found in other instances of glycosyl transfer to polypeptide acceptors. The protein·core of submaxillary mucin, obtained after enzymatic removal of sialic acid and N-acetylgalactosamine, was an excellent acceptor for N-acetylgalactosaminyl transfer, whereas small peptides from a Pronase digest were completely inactive.[13] A similar observation was made by Spiro and Spiro,[14] who showed that the galactosyl acceptor activity of tryptic collagen peptides was only 23% of that observed with a macromolecular collagen substrate.

2. Galactosyltransferase I. This enzyme only requires the monosaccharide acceptor, D-xylose, free or in glycosidic linkage. However, a substrate such as O-β-D-xylosyl-L-serine which bears greater resemblance to the native acceptor is a better substrate than the free monosaccharide.

3. Galactosyltransferase II. The second galactosyltransferase has an absolute requirement for a larger acceptor structure. Transfer does not occur to free galactose or to most galactose-containing disaccharides, while the disaccharide of appropriate structure, i.e., 4-O-β-D-galactosyl-D-xylose, is a good acceptor. Lactose also exhibits slight acceptor activity, but it should be noted that this disaccharide differs from 4-O-β-D-galactosyl-D-xylose only in having a primary alcohol group instead of a hydrogen at C-5 of the reducing terminal monosaccharide.

4. Glucuronosyltransferase I. The first glucuronosyltransferase similarly requires at least a disaccharide for maximal activity, e.g., 3-O-β-D-galactosyl-D-galactose, although minimal transfer to free D-galactose also occurs. The nature and linkage position of the penultimate group is qualitatively somewhat less important, and the enzyme also catalyzes transfer to 4-O-β-D-galactosyl-D-galactose, 6-O-β-D-galactosyl-D-galactose, 4-O-β-D-galactosyl-D-xylose, and lactose.

5. N-Acetylgalactosaminyltransferase. The substrates commonly used for the assay of this enzyme are oligosaccharides from chondroitin or chondroitin sulfate, containing a nonreducing terminal glucuronic acid unit and a penultimate N-acetylgalactosamine residue. The exact identity of the penultimate group is not crucial, since the reaction proceeds equally well with a hexasaccharide from hyaluronic acid which has an N-acetylglucosamine residue in this position. However, it is not known whether other monosaccharides can substitute for N-acetylhexosamine as the penultimate group.

6. Glucuronosyltransferase II. Like the preceding enzyme, the second glucuronosyltransferase is assayed with oligosaccharides from chondroitin and chondroitin sulfate. A nonsulfated or 6-sulfated N-acetyl-

[13] E. J. McGuire and S. Roseman, *J. Biol. Chem.* **242**, 3745 (1967).
[14] M. J. Spiro and R. G. Spiro, *J. Biol. Chem.* **246**, 4910 (1971).

galactosamine unit in the nonreducing terminal position is required for acceptor activity, and no transfer occurs to a 4-sulfated residue. Little or no activity is observed with free N-acetylgalactosamine. Suitable acceptor oligosaccharides have not yet been available to determine whether the nature of the penultimate group is of importance for transfer.

The process catalyzed by the six glycosyltransferases is summarized in Fig. 2. It should be pointed out that, eventually, it would be desirable to compare the *in vivo* process depicted in the figure with the artificial situation created in an assay with exogenous substrates. It is possible and even likely that the native, partially glycosylated protein molecules are far better acceptors than small oligosaccharides. This is suggested by the finding that the K_m values for the latter are generally higher than would be expected for most enzymatic reactions (about $10^{-2} M$). Even so, the use of exogenous acceptors has been invaluable for gathering information concerning substrate specificities which would otherwise have been impossible to obtain. More specifically, we have been able to investigate problems such as the minimum size of the acceptors, the related question of the influence of the penultimate glycosyl group on acceptor activity, and the effect of variations in the position of the gly-

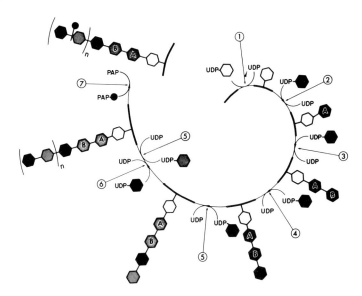

FIG. 2. Biosynthesis of chondroitin sulfate. (1) Xylosyltransferase; (2) galactosyltransferase I; (3) galactosyltransferase II; (4) glucuronosyltransferase I; (5) N-acetylgalactosaminyltransferase; (6) glucuronosyltransferase II; (7) sulfotransferase. Protein core, thick line; xylose, open hexagon; galactose, dotted hexagon; glucuronic acid, filled hexagon; N-acetylgalactosamine, hatched hexagon; sulfate, filled circle.

cosidic linkage between the nonreducing terminal monosaccharide and the penultimate group. A more complete discussion of these problems is found in recent reviews.[11,12]

II. Xylosyltransferase[15-17]

UDP-xylose + acceptor → xylosyl-acceptor + UDP

Assay Methods

Principle. Xylosyltransferase may be assayed with either (1) a low-molecular weight acceptor, such as Ser-Gly-Gly, or (2) a high-molecular weight acceptor, i.e., Smith-degraded cartilage proteoglycan (see Section IX). Conditions of the transferase assay with high- or low-molecular weight acceptors differ only in the isolation of the reaction products: (A) xylosyl peptides are purified by adsorption to Dowex 50 and subsequent elution with ammonia for measurement of incorporated radioactivity, and (B) the high-molecular weight product is isolated by precipitation with trichloroacetic acid in the presence of carrier protein or by paper chromatography.

Reagents

> Buffer A: 0.05 M KCl–0.05 M MES[18] buffer, pH 6.5, containing 0.006 M MnCl$_2$ and 0.012 M MgCl$_2$
> Buffer B: 0.05 M KCl–0.05 M MES buffer, pH 6.5
> Buffer C: 0.20 M KCl–0.05 M MES buffer, pH 6.5, containing 0.003 M MnCl$_2$ and 0.012 M MgCl$_2$
> Acceptors: (A) 0.075 M Ser-Gly-Gly (Fox Chemical Co.); (B) Smith-degraded cartilage proteoglycan, 10 mg/ml (see Section IX).
> Enzyme: Aliquots of particulate or soluble preparations (see below) containing from 0.02 to 0.2 mg of protein.
> UDP-[^{14}C]xylose: Ethanolic solutions of UDP-[^{14}C]xylose (New England Nuclear, specific activity 150 mCi/mmole) are evaporated to dryness under a stream of nitrogen, and the nucleotide sugar is dissolved in deionized water at a concentration of 20 μCi/ml. Solutions are stored at $-20°$ after freezing in Dry Ice–acetone (quick-freezing is important in order to prevent degradation). After prolonged storage (8 weeks or more) or repeated freezing and thawing, it may become necessary to repurify the UDP-[^{14}C]xylose by paper chromatography [water-washed

[15] A. C. Stoolmiller, A. L. Horwitz, and A. Dorfman, *J. Biol. Chem.* **247**, 3525 (1972).
[16] J. R. Baker, L. Rodén, and A. C. Stoolmiller, *J. Biol. Chem.* **247**, 3838 (1972).
[17] J. R. Baker, L. Rodén, and S. Yamagata, *Biochem. J.* **125**, 93P (1971).
[18] MES, (2-N-morpholino)ethanesulfonic acid.

Whatman No. 3MM paper; solvent, 95% ethanol–1 M ammonium acetate, pH 5.0 (7:3)].

UDP-xylose: UDP-xylose is purified by chromatography as described above and the concentration of aqueous solutions is determined by spectral analysis and adjusted to 12.0 ± 0.2 μmole/ml (A_m at 262 nm and neutral pH is 10 × 10³). Aliquots of the UDP-xylose stock solution are stored at −20° until used; these are not refrozen.

UDP-[14C]xylose substrate mixture: Buffer B, 0.066 ml, is mixed with 0.100 ml of UDP-[14C]xylose, 0.004 ml of UDP-xylose, and 0.030 ml of 1 M KF. The specific activity of the UDP-[14C]xylose in this substrate mixture is ~30 mCi/mmole.

Bovine serum albumin, 1% (Sigma, fraction V powder)

EDTA, 0.25 M, pH 8

Trichloroacetic acid, 10%–phosphotungstic acid, 4%

Trichloroacetic acid, 6%

Scintillation mixture: Aqueous samples of 0.5 ml are mixed with 10 ml of Permafluor solution (Packard) and 1 ml of Bio-Solv solubilizer (BBS-3, Beckman) and counted at 4°. Both acidic and basic samples are miscible with the scintillation fluid; however, use of acidic samples is preferable to avoid unwanted chemiluminescence.

Procedure. To a 6 × 50 mm tube are added 50 μl of enzyme in Buffer A, 20 μl of acceptor, and 5 μl of UDP-[14C]xylose substrate mixture. Control incubation mixtures may contain no added acceptor or are not incubated. After incubation for 15–60 minutes (commonly for 30 minutes), tubes are transferred to an ice–water bath and 5 μl of 0.25 M EDTA are added to stop the reaction. Depending on the nature of the acceptor, xylose incorporation is assayed by one of the following methods:

Method 1: Reaction mixtures in which Ser-Gly-Gly is the exogenous acceptor are centrifuged for 10 minutes in a bench-top centrifuge. A 50-μl aliquot of the supernatant fluid is diluted with 0.4 ml of 0.02 M HCl and applied to a column (0.5 × 5 cm) of AG 50W-X2 (H⁺, 200–400 mesh). After washing with 3–4 bed volumes of 0.01 M HCl, adsorbed reaction products are eluted with 3 bed volumes of 2 M ammonium hydroxide. The eluate is evaporated to dryness in a Buchler Evapo-Mix, taken up in 0.5 ml of 0.01 M HCl, and the incorporated radioactivity is measured.

Method 2: Xylosyl transfer to the Smith-degraded proteoglycan is assayed by any of the following procedures (a–c):

(a) After addition of 0.05 ml of 1% albumin, proteins are precipitated by addition of 0.15 ml of 10% trichloroacetic acid–4% phospho-tungstic acid. The precipitate is collected by centrifugation, washed twice with 0.3 ml of 5% trichloroacetic acid, and dissolved in 0.1 ml of 1 M NaOH. The sample is transferred to a scintillation vial, diluted to 0.5 ml with water, and the radioactivity is measured.

(b) 50-μl aliquots are spotted on 2 × 2 cm squares of Whatman No. 3MM paper which have been stapled by one edge only to a sheet of corrugated plastic, attached to a hard rubber mat. (In order to fa-cilitate subsequent removal of the staples, the stapling is carried out without using the base portion of the stapler.) After drying with a hair drier, the papers are moistened with a few drops of 10% trichloroacetic acid–4% phosphotungstic acid and left for 2–3 minutes; the plastic sheet is placed in a tray of suitable size, to which 500 ml of 6% trichloro-acetic acid are added. The tray is rocked gently for 5 minutes (either manually or in a rocking apparatus), the trichloroacetic acid is discarded, and the washing procedure is repeated twice. (To ensure complete wash-ing, the paper strips are bent away from the plastic support.) Finally, the strips are rinsed with three 500-ml portions of ethanol and air-dried. After transfer to scintillation vials, the strips are moistened with 0.5 ml of water prior to the addition of the scintillation fluid.

(c) The reaction mixture is applied to a 2 × 2 cm area of a sheet of Whatman No. 3MM paper and, after drying, the chromatogram is devel-oped with Leloir's solvent [95% ethanol–1 M ammonium acetate, pH 5.0 (7:3)] for 16 hours. The paper is dried, 3 × 3-cm areas including the origin are cut out, and the radioactivity of the immobile products is measured. Radioactivity of control reaction mixtures typically ranges from 60 to 180 cpm. When radioactivity in controls exceeds 200 cpm, the sensitivity of the assay becomes significantly impaired, and the UDP-[^{14}C]xylose should be repurified as described above.

Enzyme Preparation

Epiphyses from tibias and femurs of 13-day-old White Leghorn chick embryos are dissected free of soft tissues, chilled, and finely diced. (The epiphyses from five dozen eggs can be processed conveniently at one time.) Diced epiphyses are homogenized in 4 volumes of Buffer A in a glass Duall tissue grinder with a motor-driven (250 rpm) glass pestle. After 20 complete strokes (total time 5–7 minutes), the homogenate is centrifuged at 10,000 g for 10 minutes, and the supernatant fluid is then centrifuged at 105,000 g for 1 hour.

Between 50 and 75% of the total xylosyltransferase activity of the 10,000 g supernatant fluid is present in the 105,000 g supernatant frac-tion. Increasing the concentration of KCl in the 10,000 g supernatant fluid to 0.5–2.0 M before high speed centrifugation results in recovery of more than 90% of the transferase activity in the soluble fraction. Fractions containing high concentrations of KCl are dialyzed against Buffer A before assay. The enzyme may be stored frozen at $-20°$ for as long as one year without appreciable loss of enzymatic activity. The protein concentration in soluble fractions, obtained with or without salt, is approximately 2 mg/ml.

Further purification of the soluble xylosyltransferase is accomplished by the following procedures, which are performed at 4°.

Step 1. Ammonium Sulfate Fractionation. Solid ammonium sulfate is added to 200 ml of 105,000 g supernatant fraction (derived from approximately 1000 embryos) to 27% saturation. After 30 minutes, the suspension is centrifuged at 42,000 g for 1 hour, and the resulting small pellet is discarded. The ammonium sulfate concentration is brought to 68% saturation, and the precipitate is recovered by centrifugation as before and dissolved in 20 ml of Buffer C. After dialysis against the same buffer, the enzyme may be stored frozen.

Step 2. Gel Filtration on Sephadex G-200. A 3-ml sample of the material from the previous step is applied to a column (2.4 × 100 cm) of Sephadex G-200 which is eluted with 0.2 M KCl–0.05 M MES buffer, pH 6.5, at a flow rate of 8–10 ml per hour. Xylosyltransferase activity is assayed by Method 2 (c). Some activity is present in the void volume material, but most of the enzyme is retarded and emerges as a broad peak with an effluent volume of 200–300 ml. Fractions containing the transferase activity are concentrated by ultrafiltration (Amicon Diaflo filter UM-05).

Approximately 40-fold purification over the 105,000 g supernatant fraction is achieved by this two-step procedure.

Properties

The pH optimum is between 6.7 and 7.0. The enzyme requires divalent metal ion and exhibits full activity with 5 mM Mn^{2+}. Manganese can be partially replaced by Mg^{2+}, also with an optimum concentration of 5 mM.

K_m values for Ser-Gly-Gly and the Smith-degraded proteoglycan are 20 mM and 0.064 mM, respectively (the latter value based on total serine concentration, but should more appropriately be related to the proportion of serine residues which is available for glycosylation; however, reliable estimates of this value are not available). With the exception of Ser-Gly-Gly, acceptor peptides of defined structure are not known.

Characterization of the product of transfer to the Smith-degraded proteoglycan shows that xylose is bound by an alkali-labile linkage and can be released as xylitol upon treatment with alkaline borohydride. Exhaustive proteolytic digestion with papain and Pronase yields a compound with the electrophoretic and chromatographic properties of xylosylserine. The dansyl derivative of this compound has the same electrophoretic mobility as authentic dansyl-xylosylserine.

III. Galactosyltransferase I[2,10]

$$\text{UDP-galactose} + \text{D-xylose} \rightarrow 4\text{-}O\text{-}\beta\text{-}D\text{-galactosyl-}D\text{-xylose} + \text{UDP} \qquad (1)$$

$$\text{UDP-galactose} + O\text{-}\beta\text{-}D\text{-xylosyl-}L\text{-serine} \rightarrow$$
$$4\text{-}O\text{-}\beta\text{-}D\text{-galactosyl-}O\text{-}\beta\text{-}D\text{-xylosyl-}L\text{-serine} + \text{UDP} \qquad (2)$$

Assay Method

Principle. Xylose or xylosylserine may be used as exogenous substrates for the assay of galactosyltransferase I in the chick cartilage system or with enzyme from other tissues. The disaccharide product of Reaction (1) is isolated by electrophoretic separation from unreacted UDP-galactose and other charged compounds, followed by paper chromatography which separates the disaccharide from galactose and other neutral substances.

If available, xylosylserine is preferable as a substrate in view of its somewhat higher acceptor activity and the greater ease with which the reaction product can be isolated. The galactosylxylosylserine formed by Reaction (2) is adsorbed to Dowex 50 and subsequently eluted with ammonia, and the radioactivity is measured.

Reagents

Buffer A: Tris·HCl, 0.05 M, pH 7.5

Buffer B: MES, 0.05 M, pH 5.5, containing 0.2 M KCl

Buffer C: 0.080 M pyridine–0.046 M acetic acid, pH 5.3

Acceptors: D-xylose, 0.2 M; O-β-D-xylosyl-L-serine, 0.04 M

UDP-[^{14}C]galactose, 20 μCi/ml, specific activity 250–275 mCi/mmole

UDP-galactose, 0.05 M

UDP-[^{14}C]galactose substrate mixture: Buffer A, 200 μl, is mixed with UDP-[^{14}C]galactose, 200 μl, and UDP-galactose, 2 μl, thus yielding a solution containing 10 μCi/ml with a specific activity of approximately 35 μCi/μmole.

MnCl$_2$, 0.2 M

Ethyl acetate–acetic acid–water (3:1:1, v/v)

Aniline phthalate reagent: 0.75 ml of aniline and 1.66 g of o-phthalic acid are dissolved in 100 ml of water-saturated n-butanol.

Procedure 1. The reaction mixture contains in a final volume of 75 μl: 0.2 M xylose, 10 μl; 0.2 M MnCl$_2$, 5 μl; UDP-[^{14}C]galactose substrate mixture, 10 μl; and 50 μl of enzyme in Buffer B (containing 0.05–2.0 mg of enzyme protein). After incubation at 37° for 1 hour, the reaction is stopped by immersing the tubes in a boiling water bath for 2–3 minutes. The entire reaction mixture is applied to a 57-cm sheet of Whatman No. 3MM paper and subjected to electrophoresis in Buffer C at 80 V/cm for 30 minutes. The neutral reaction product remains close to the origin, and, after drying, this section of the paper is cut off (with a 2.5 cm margin toward the anode) and sewn onto another sheet of paper (about 42 cm long). Paper chromatography is then carried out in ethyl acetate–acetic acid–water (3:1:1) for 20–24 hours. Under these conditions, 4-O-β-D-galactosyl-D-xylose migrates 9–13 cm from the origin. The position of the disaccharide is determined by strip scanning and by aniline phthalate staining of a guide strip with the authentic compound (see Section IX). Quantitation of the radioactive product is performed by liquid scintillation counting of the galactosylxylose area.

Comments. The paper chromatography may be carried out on the same sheet that is used for electrophoresis, but it is then advisable to extend the electrophoresis time to 1 hour or more to ensure that charged radioactive compounds have moved off the paper. Only two major radioactive peaks are observed on the chromatogram when purified enzyme preparations are used, i.e., galactosylxylose and galactose. With particulate preparations, a considerable amount of radioactivity may be found at the origin; if a clear-cut separation from galactosylxylose is not obtained, the isolation of the product is modified slightly. After completed incubation, 3 volumes of ethanol are added, precipitated material is removed by centrifugation, and the disaccharide is purified from the supernatant liquid as described.

Procedure 2. Reaction conditions are the same as in Procedure 1 except that xylose is substituted by 0.4 μmole of xylosylserine. After completed incubation and heat inactivation, the samples are mixed with 0.6 ml of 0.01 M HCl and centrifuged; the supernatant solutions are applied to columns (0.6 × 2.5 cm) of Dowex 50-X2 (H$^+$ form, 200–400 mesh). The columns are washed with 6 ml of 0.01 M HCl, and the product is eluted with 2–3 bed volumes of 2 M NH$_4$OH. After evaporation to dryness, the samples are taken up in 0.5 ml of water, and the radioactivity is measured.

Enzyme Preparation

A 100,000 g pellet is prepared from a homogenate of embryonic chick cartilage as described in Section II and is suspended in Buffer B to give a protein concentration of 5–10 mg/ml. Soluble enzyme fractions may be prepared as described in Section VIII.

Properties

Galactosyltransferase I is tightly membrane bound, and 90–95% of the activity is found in the 100,000 g pellet of a homogenate of embryonic chick cartilage.

Stability. A 10,000 g supernatant fluid or a 100,000 g pellet may be stored at $-20°$ for several months without appreciable loss of activity. Solubilized fractions of the enzyme are also relatively stable on freezing.

pH Optimum. The particulate enzyme from chick cartilage has a pH optimum of 5.4, the activity falling rapidly below this pH and more slowly on the alkaline side. The solubilized enzyme from mouse mastocytoma has a pH optimum of 7.2.

Substrate Affinities. The K_m value for UDP-galactose is approximately $1 \times 10^{-4} M$. K_m for xylose is $2 \times 10^{-2} M$ with the enzyme from mouse mastocytoma and is of the same order of magnitude with the chick cartilage enzyme but has not been accurately determined. K_m for xylosylserine with the cartilage enzyme has been reported to be approximately $2 \times 10^{-3} M$.

Metal Requirement. Mn^{2+} is required for maximum activity and increases product formation up to 15 mM, where a plateau is reached. At higher concentrations, the activity gradually decreases. Manganese can be partially replaced by Mg^{2+}, Co^{2+}, and Ca^{2+}.

Specificity. Galactosyl transfer has been shown to D-xylose, O-β-D-xylosyl-L-serine, methyl β-D-xylopyranoside, and p-nitrophenyl β-D-xylopyranoside. Exact details of the enzyme specificity have not been established, since the enzyme has not yet been purified to homogeneity. It is likely that the natural substrate, which is presumably a xylosyl peptide or a xylosyl protein, is a better substrate than the monosaccharide or the small xylosides.

Product Identification. The product of galactosyl transfer to D-xylose has been identified as 4-O-β-D-galactosyl-D-xylose.

IV. Galactosyltransferase II[2,10]

UDP-galactose + 4-O-β-D-galactosyl-D-xylose →
$\qquad\qquad\qquad$ 3-O-β-D-galactosyl-4-O-β-D-galactosyl-D-xylose

Assay Method

Principle. Galactosyl transfer to 4-O-β-D-galactosyl-D-xylose is measured essentially as described for galactosyltransferase I with D-xylose as substrate. The trisaccharide product is isolated by paper electrophoresis and paper chromatography.

Reagents

Acceptor: 4-O-β-D-galactosyl-D-xylose, 0.2 M

Other reagents are identical with those used for the assay of galactosyltransferase I.

Procedure. The assay procedure is identical with Procedure 1 described for the assay of galactosyltransferase I with the exception that D-xylose is replaced by 10 μl of 4-O-β-D-galactosyl-D-xylose.

Properties

Many of the properties of galactosyltransferase II are similar to those of galactosyltransferase I.

pH Optimum. The enzyme from chick cartilage is active over a wide pH range, with an optimum at pH 5.4. The solubilized enzyme from mouse mastocytoma has an optimum at pH 7.5.

Substrate Affinities. K_m for UDP-galactose is $3 \times 10^{-4} M$. The solubilized enzyme from mouse mastocytoma gives a K_m value for 4-O-β-D-galactosyl-D-xylose of $2 \times 10^{-2} M$.

Metal Requirement. Maximum activity is observed at a Mn^{2+} concentration of approximately 15 mM. At this or higher concentrations, at least two endogenous products are formed which do not migrate on electrophoresis at pH 5.3 but separate from each other and migrate slightly ahead of 3-O-β-D-galactosyl-4-O-β-D-galactosyl-D-xylose on paper chromatography. The formation of these products is not dependent on the presence of added acceptor. If interference from these or other endogenous products is encountered, further purification of the trisaccharide product may be carried out by electrophoresis in 0.05 M borate buffer, pH 9.2, at 25 V/cm for 90 minutes.

Specificity. Besides 4-O-β-D-galactosyl-D-xylose, 4-O-β-D-galactosyl-O-β-D-xylosyl-L-serine serves as substrate for galactosyltransferase II with approximately equal efficiency. Galactosyl transfer also occurs to 3-O-β-D-galactosyl-D-xylose, but this substance is actually a substrate for galactosyltransferase I, since the galactose is transferred to carbon 4 of the xylose moiety. Lactose and raffinose show some acceptor activity, but it is not known whether these are truly substrates for galactosyltransferase II.

Product Identification. The product of galactosyl transfer to 4-O-β-D-galactosyl-D-xylose has been identified as 3-O-β-D-galactosyl-4-O-β-D-galactosyl-D-xylose.

V. Glucuronosyltransferase I[3,19,20]

3-O-β-D-galactosyl-D-galactose + UDP-glucuronic acid →
$$3\text{-}O\text{-}\beta\text{-}D\text{-glucuronosyl-}3\text{-}O\text{-}\beta\text{-}D\text{-galactosyl-}D\text{-galactose} + \text{UDP} \quad (1)$$

$$\text{Lactose} + \text{UDP-glucuronic acid} \rightarrow \text{glucuronosyllactose} + \text{UDP} \quad (2)$$

Assay Method

Principle. Glucuronosyl transfer to 3-O-β-D-galactosyl-D-galactose is measured after electrophoretic separation of the trisaccharide product.

Lactose may be used as an alternative substrate, but its acceptor activity is lower, and it is not established with certainty that lactose and 3-O-β-D-galactosyl-D-galactose are substrates for the same glucuronosyl-transferase.

Reagents

Buffer A: MES, 0.05 M, pH 5.5, containing 0.2 M KCl and 0.009 M MnCl$_2$

Buffer B: pyridine (0.080 M)–acetic acid (0.046 M), pH 5.3

Acceptors: 3-O-β-D-galactosyl-D-galactose, 0.15 M, or lactose, 0.5 M

KF, 0.3 M

EDTA, 0.1 M, pH 7

UDP-[14C]glucuronic acid, 20 μCi/ml (specific activity, 125 μCi/μmole)

UDP-glucuronic acid, 0.002 M

UDP-[14C]glucuronic acid substrate mixture; UDP-[14C]glucuronic acid, 125 μl, is mixed with 50 μl of UDP-glucuronic acid, 75 μl of KF, and 125 μl of lactose or 3-O-β-D-galactosyl-D-galactose.

Procedure. Reaction mixtures contain 50 μl of enzyme in Buffer A and 15 μl of the UDP-[14C]glucuronic acid substrate mixture. After incubation at 37° for 1 hour, the reaction is stopped by addition of 25 μl of 0.1 M EDTA, and the entire incubation mixture is spotted on a 57-cm strip of Whatman No. 3MM paper. Electrophoresis is carried out in Buffer B at 80 V/cm for 60–90 minutes, and the trisaccharide product is located by strip scanning and quantitated by liquid scintillation counting of the appropriate area. Glucuronosylgalactosylgalactose migrates at a rate of 0.57, relative to glucuronic acid. Glucuronosyllactose is slightly slower, with an $R_{\text{glucuronic acid}}$ value of 0.49. It should be noted that in control

---

[19] T. Helting and L. Rodén, *J. Biol. Chem.* **244**, 2799 (1969).

[20] A. E. Brandt, J. Distler, and G. W. Jourdian, *Proc. Nat. Acad. Sci. U.S.* **64**, 374 (1969).

experiments without added acceptor, there is usually no radioactivity in the trisaccharide area. If interference from endogenous radioactivity is observed in this position or in the vicinity of the origin, the reaction mixture may be precipitated with ethanol and the supernatant fluid analyzed, as described in Section III.

Enzyme Preparation

A 100,000 g pellet is prepared from a homogenate of embryonic chick cartilage as described in Section II and is suspended in Buffer A to give a protein concentration of 5–10 mg/ml. The enzyme has also been prepared from embryonic chick brain[20] and mouse mastocytoma.[3]

Properties

Glucuronosyltransferase I is found largely in the particulate fraction of a cartilage homogenate, which sediments between 10,000 g and 100,000 g. Approximately 15% of the activity remains in the 100,000 g supernatant fluid. A fraction of the particulate enzyme from mouse mastocytoma (18%) has been solubilized by detergent–alkali treatment as described in Section VIII and has been purified 330-fold by gel chromatography on Sepharose 4B.[3]

pH Optimum. The particulate enzyme from chick cartilage is active over a wide pH range (5–8) with an inconspicuous maximum at pH 5.4. Assays should routinely be carried out around this pH, since the breakdown of UDP-glucuronic acid to free glucuronic acid increases drastically at higher pH values.

The pH curve for the solubilized enzyme from mouse mastocytoma has a more usual shape with an optimum at pH 7.8.

Substrate Affinities. The K_m value of UDP-glucuronic acid is approximately $1 \times 10^{-4}\,M$ for the particulate chick cartilage enzyme and $2.5 \times 10^{-4}\,M$ for the solubilized enzyme from mouse mastocytoma. K_m for 3-O-β-D-galactosyl-D-galactose is $9 \times 10^{-3}\,M$ with the latter enzyme and is of the same order with the particulate chick cartilage enzyme.

Metal Requirement. The enzyme requires Mn^{2+} for full activity, a plateau being reached at 15 mM (20–25 mM with the solubilized mouse mastocytoma preparation).

Specificity. Glucuronosyltransferase I utilizes several galactose-containing oligosaccharides as substrates, including 3-O-β-galactosyl-D-galactose (100%), 4-O-β-D-galactosyl-D-galactose (126%), 6-O-β-D-galactose (27%), 3-O-β-D-galactosyl-4-O-β-D-galactosyl-D-xylose (97%), 3-O-β-D-galactosyl-4-O-β-D-galactosyl-O-β-D-xylosyl-L-serine (167%), and lactose (10–25%). Transfer to galactose occurs at 2.5% of the rate with 3-O-β-D-galactosyl-D-galactose.

The use of lactose as an alternative substrate requires further comment. In an embryonic chick brain preparation, the acceptor activity of lactose is of the same order as that of 3-O-β-D-galactosyl-D-galactose,[20] whereas the activity with the chick cartilage enzyme is much lower.[19] It is noteworthy that no product formation is observed with lactose as substrate for the solubilized mouse mastocytoma enzyme.[3] Differences between the two substrates are also evident in relation to the subcellular distribution of the transferase activity. An increase in total activity (to 135–180%) is observed in the 100,000 g pellet as compared to the 10,000 g supernatant fluid, when the enzyme is assayed with 3-O-β-D-galactosyl-D-galactose. In contrast, only 10–20% of the original activity is found in the particulate fraction, when the enzyme is assayed with lactose, and virtually no activity remains in the supernatant fluid.

The reasons for the observed discrepancies are not clear, and although competition experiments indicate that lactose and 3-O-β-D-galactosyl-D-galactose are substrates for the same enzyme (N. Schwartz, unpublished results), this remains to be established more firmly. Consequently, the more natural substrates from the chondroitin sulfate–protein linkage region are generally to be preferred over lactose and are sometimes indispensable, e.g., in the assay of the solubilized, purified mastocytoma enzyme and similar preparations which cannot utilize lactose.

Product Identification. Characterization of the products of transfer to 3-O-β-D-galactosyl-D-galactose and 3-O-β-D-galactosyl-4-O-β-D-galactosyl-D-xylose showed that glucuronic acid had been transferred to form a β-1, 3 linkage to galactose.

VI. N-Acetylgalactosaminyltransferase[21,22]

UDP-N-acetylgalactosamine + acceptor → N-acetylgalactosaminyl-acceptor + UDP

Assay Method

Principle. N-Acetylgalactosaminyl transfer from UDP-[14C]N-acetylgalactosamine to a suitable oligosaccharide acceptor is measured. Several oligosaccharides containing nonreducing terminal glucuronic acid residues may be used and are isolated from chondroitin 4- or 6-sulfate or chondroitin after digestion with testicular hyaluronidase; the present procedure utilizes the hexasaccharide from chondroitin 4-sulfate.

The labeled heptasaccharide product is isolated by precipitation with cetylpyridinium chloride on a cellulose column, followed by elution with 0.3 M NaCl.

[21] A. Telser, H. C. Robinson, and A. Dorfman, *Arch. Biochem. Biophys.* **116**, 458 (1966).
[22] A. Telser, Ph.D. thesis, University of Chicago, 1968.

Reagents and Materials

HEPES,[23] 0.05 M, pH 7.0, containing 0.05 M KCl

$MnCl_2$, 0.2 M

NaCl, 0.03 M and 0.3 M

Cetylpyridinium chloride, 1% and 0.25%

Hexasaccharide from chondroitin 4-sulfate, 15 mg/ml (see Section IX)

UDP-[^{14}C]N-acetylgalactosamine, 20 μCi/ml, specific activity 43 μCi/μmole

Cellulose powder (e.g., Whatman No. CF 11). The powder is repeatedly suspended in water and allowed to settle overnight to remove fines, and 6 × 50 mm columns are then prepared in disposable Pasteur pipettes. The columns are equilibrated with 0.25% cetylpyridinium chloride prior to use.

Procedure. The reaction is conveniently carried out in 12-ml conical centrifuge tubes which are kept in an ice bath prior to incubation. The reaction mixtures contain the following: enzyme in HEPES buffer, 0.3 ml; hexasaccharide, 10 μl; $MnCl_2$, 5 μl; and UDP-[^{14}C]N-acetylgalactosamine, 20 μl. After incubation at 37° for 1 hour, the tubes are heated in a boiling water bath for 3 minutes and centrifuged. The supernatant fluid from each tube is transferred to a 6 × 50 mm cellulose column, and the flow through the column is stopped momentarily with a finger, while 0.1 ml of 1% cetylpyridinium chloride is added and mixed with the incubation mixture. The column is rinsed with 1 ml of water and 12 ml of 0.03 M NaCl, and the product is eluted with 3 ml of 0.3 M NaCl. A 0.5-ml aliquot of the eluate is used for radioactivity measurement by liquid scintillation counting.

Enzyme Preparation

A 100,000 g pellet from a homogenate of embryonic chick cartilage is prepared as described in Section II and is suspended in HEPES buffer, pH 7.0, at a protein concentration of 5–10 mg/ml. Solubilized preparations may also be used, as described in Section VIII.

Properties

N-Acetylgalactosaminyltransferase is one of the more soluble glycosyltransferases catalyzing chondroitin sulfate synthesis and is approximately evenly distributed between pellet and supernatant fractions when a 10,000 g supernatant fluid from chick cartilage homogenate is centrifuged at 100,000 g.[15]

[23] HEPES, N-2-hydroxyethylpiperazine-N'-2-ethanesulfonic acid.

Stability. A 10,000 *g* supernatant fluid or a 100,000 *g* pellet from a homogenate of embryonic chick cartilage may be stored at $-20°$ for several months without appreciable loss of activity.

pH Optimum. The pH optimum of the particulate chick cartilage enzyme is 7.0, when assayed with chondroitin sulfate hexasaccharide as substrate. When polysaccharide formation is measured in the presence of both UDP-*N*-acetylgalactosamine and UDP-glucuronic acid, the pH optimum is 7.5.

Substrate Affinities. The K_m value for chondroitin 4-sulfate hexasaccharide is 0.18 mg/ml, i.e., approximately $1 \times 10^{-4} M$. K_m for UDP-*N*-acetylgalactosamine has not been determined.

Metal Requirement. Maximum activity is obtained in the presence of $0.003 M$ Mn^{2+}. Mg^{2+} can partially replace manganese, but at the optimal concentration, $0.03 M$, product formation is only 5% of that observed in the presence of manganese.

Product Identification. The radioactive product is precipitable with cetylpyridinium chloride and emerges in the position of a heptasaccharide on gel chromatography but has not been otherwise characterized.

Specificity. Several oligosaccharides, belonging to the various homologous series obtained from chondroitin 4-sulfate, chondroitin 6-sulfate, and chondroitin by digestion with testicular hyaluronidase, serve as substrates for *N*-acetylgalactosaminyl transfer (see Table IIA). These oligosaccharides all have glucuronic acid residues at their nonreducing termini and *N*-acetylgalactosamine or *N*-acetylgalactosamine sulfate in the penultimate position. The nature of the penultimate sugar in the acceptor is apparently not crucial for activity, since an analogous hyaluronic acid hexasaccharide, with *N*-acetylglucosamine in the penultimate position, is also a substrate for the enzyme.

VII. Glucuronosyltransferase II[21,22]

UDP-glucuronic acid + acceptor → glucuronosyl-acceptor + UDP

Assay Method

Principle. Glucuronosyl transfer from UDP-[^{14}C]glucuronic acid to suitable oligosaccharide acceptors is measured. Several oligosaccharides containing nonreducing terminal *N*-acetylgalactosamine or *N*-acetylgalactosamine 6-sulfate residues may be used; these are isolated from chondroitin or chondroitin 6-sulfate after digestion with testicular hyaluronidase and β-glucuronidase. The recommended procedure utilizes a heptasaccharide from chondroitin 6-sulfate.

The labeled octasaccharide product is isolated by precipitation with

TABLE IIA

ACCEPTOR AND DONOR SPECIFICITIES OF N-ACETYLGALACTOSAMINYLTRANSFERASE

Donor UDP-nucleotide sugar	Oligosaccharide acceptors with nonreducing terminal glucuronic acid[a]						
	Chondroitin		Chondroitin 4-sulfate		Chondroitin 6-sulfate hexasaccharide	Hyaluronic acid hexasaccharide	Desulfated dermatan sulfate
	Hexasaccharide	Tetrasaccharide	Hexasaccharide	Tetrasaccharide			
UDP-N-acetyl-galactosamine	+	+	+	+	+	+	—
UDP-glucuronic acid	—	—	—	—	—	—	
UDP-N-acetyl-glucosamine	—	—	—	—	—	—	

[a] Even-numbered oligosaccharides from chondroitin, the chondroitin sulfates, and hyaluronic acid were prepared by digestion with testicular hyaluronidase as described in Section IX, E. Odd-numbered oligosaccharides with N-acetylhexosamine at the nonreducing end were obtained from the even-numbered compounds by digestion with β-glucuronidase. The desulfated dermatan sulfate tetrasaccharide was isolated from an acid hydrolyzate of the polysaccharide.

cetylpyridinium chloride on a cellulose column, followed by elution with 0.3 M NaCl.

Reagents

HEPES, 0.05 M, pH 7.0, containing 0.05 M KCl and 0.003 M MnCl$_2$ (pH is adjusted to 7.0 before addition of MnCl$_2$)

Acceptors: chondroitin 6-sulfate penta- or heptasaccharide (see Section IX)

KF, 2 M

UDP-[^{14}C]glucuronic acid, 20 μCi/ml (specific activity, 125 μCi/μmole)

UDP-glucuronic acid, 0.01 M

UDP-[^{14}C]glucuronic acid substrate mixture: UDP-[^{14}C]glucuronic acid, 250 μl, mixed with 50 μl of UDP-glucuronic acid, 100 μl of KF, 100 μl of HEPES buffer, and 10 mg of penta- or heptasaccharide.

NaCl, 0.03 M and 0.3 M

Cetylpyridinium chloride, 1% and 0.25%

Absolute ethanol

Cellulose powder (Whatman No. CF 11)

Cellulose columns are prepared as described in Section VI.

Procedure. The assay procedure is similar to that described in Section VI with minor modifications. Reaction mixtures contain 190 μl of enzyme in HEPES buffer and 10 μl of the UDP-[^{14}C]glucuronic acid substrate mixture and are incubated at 37° for 30–60 minutes. The tubes are heated for 3 minutes in a boiling water bath, 20 μl of UDP-glucuronic acid are added, and, after centrifugation for 10 minutes, the supernatant fluid is applied to a cellulose column. Oligosaccharides are precipitated by addition of 0.1 ml of 1% cetylpyridinium chloride, and after the liquid has drained into the cellulose, the column is washed successively with 1 ml of water, 12 ml of 0.03 M NaCl, 6 ml of absolute ethanol, and the labeled octasaccharide product is then eluted with 3 ml of 0.3 M NaCl. An aliquot of this eluate is used for scintillation counting.

Enzyme Preparation

Same as in Section VI.

Properties

Many of the properties and kinetic parameters of glucuronosyltransferase II are similar to those of N-acetylgalactosaminyltransferase. Of the total activity in a 10,000 g supernatant fluid from a chick cartilage homogenate, 75% or more is sedimented by centrifugation at 100,000 g for 60 minutes.

Stability. A 10,000 g supernatant fluid or a 100,000 g pellet from a homogenate of embryonic chick cartilage may be stored at $-20°$ for several months without appreciable loss of activity.

pH Optimum. The pH optimum of the particulate chick cartilage enzyme is 7.0.

Substrate Affinities. K_m values have not been determined.

Metal Requirement. Maximum activity is obtained in the presence of 0.003 M Mn^{2+}.

Product Identification. The product of glucuronosyl transfer to the nonsulfated pentasaccharide from chondroitin has been characterized by several methods: (a) on paper electrophoresis, paper chromatography, and gel chromatography, the radioactive product migrates to the same position as the authentic hexasaccharide; (b) digestion with β-glucuronidase releases all the radioactivity as free glucuronic acid, indicating that transfer has occurred to the nonreducing terminal N-acetylgalactosamine residue of the pentasaccharide; (c) on digestion with chondroitinase from *Proteus vulgaris*, all the radioactivity is found in N-acetylchondrosine which is produced from the nonreducing terminal disaccharide unit, whereas no radioactivity is found in the unsaturated disaccharide fraction which represents the interior and reducing terminal portions of the molecule.

Specificity. Several oligosaccharides from chondroitin and chondroitin 6-sulfate may be used as acceptors for glucuronosyl transfer (see Table IIB). A condition for acceptor activity is the presence of a nonreducing terminal N-acetylgalactosamine or N-acetylgalactosamine 6-sulfate residue. It is noteworthy that the presence of a sulfate group on C-4 of

TABLE IIB

Acceptor and Donor Specificities of Glucuronosyltransferase II

Donor UDP-nucleotide sugar	Oligosaccharide acceptors with nonreducing terminal N-acetylhexosamine[a]			
	Chondroitin pentasaccharide	Chondroitin 4-sulfate pentasaccharide	Chondroitin 6-sulfate pentasaccharide	Hyaluronic acid pentasaccharide
UDP-N-acetyl-galactosamine	−	−	−	−
UDP-glucuronic acid	+	−	+	−

[a] See footnote to Table IIA.

the terminal N-acetylgalactosamine unit completely abolishes acceptor activity.

The relative acceptor activity is dependent on the size of the substrate. Although exact quantitative comparisons have not been carried out over a large range of molecular sizes, it has been established that the acceptor activity of the heptasaccharide from chondroitin 6-sulfate is approximately 7 times greater than that of the pentasaccharide, and the pentasaccharide is, in turn, a better substrate than the trisaccharide. It is not known whether the nature of the penultimate sugar influences acceptor activity, largely because appropriate potential acceptors with a penultimate residue other than glucuronic acid have not been available for study.

VIII. Solubilization of Glycosyltransferases[2,4,12]

Solubilization by Treatment with Detergent-Alkali. A particulate glycosyltransferase preparation (100,000 g pellet) is obtained from embryonic chick cartilage as described in Section II. The pellet is suspended at 4° by sonication in cold 0.05 M Tris·acetate buffer, pH 5.5, containing 0.001 M EDTA and 0.05 M KCl, to yield a protein concentration of approximately 20 mg/ml, and detergent (Tween 20 or Nonidet P-40) is added to a final concentration of 0.5% (by volume). This solution is brought to pH 10.4 by addition of concentrated ammonium hydroxide and quickly readjusted to pH 5.5 by addition of glacial acetic acid.[2] After centrifugation at 100,000 g for 1 hour, the glycosyltransferase activities in supernatant and pellet fractions are assayed as described in previous sections. The galactosyltransferases and glucuronosyltransferase I are assayed directly at pH 5.5, whereas xylosyltransferase and N-acetylgalactosaminyltransferase are assayed after dialysis against the appropriate buffers (see Sections II and VI).

Alternatively, a two-step procedure may be used. The 100,000 g pellet fraction is first incubated with 0.5% detergent at 4° for 30 minutes and centrifuged at 100,000 g for 1 hour. The pellet is resuspended in buffer and treated with detergent and alkali as described.

Solubilization by Treatment with Detergent-Salt. A 100,000 g pellet from a cartilage homogenate is suspended in buffer, treated with 0.5% detergent and 0.5 M KCl for 30 minutes at 4°, and centrifuged at 100,000 g for 1 hour.

As with the detergent-alkali solubilization, an alternative two-step procedure may be employed. The 100,000 g pellet is suspended in buffer, treated with 0.5% detergent in the cold for 30 minutes, and centrifuged at 100,000 g for 1 hour. The resulting pellet is resuspended in buffer containing 0.5% detergent, and KCl is added to a final concentration of

0.5 M. After incubation at 4° for 30 minutes, the mixture is centrifuged at 100,000 g for 1 hour. Typical results of such a solubilization procedure are shown in Table III.

IX. Preparation of Substrates

A. Smith-Degraded Cartilage Proteoglycan (Substrate for Xylosyltransferase)

Principle. Cartilage proteoglycan is extracted with guanidine chloride from bovine nasal septa[24,25] and is purified by precipitation with cetylpyridinium chloride and with ethanol. The chondroitin sulfate chains are subsequently removed by Smith degradation (periodate oxidation, borohydride reduction, and mild acid cleavage).[16]

A more highly purified preparation of the proteoglycan may be obtained by density gradient centrifugation in cesium chloride,[25] but if large quantities are desired, this step involves considerably greater expense and labor.

Procedure. Bovine nasal septa may be obtained from Wilson Laboratories, Chicago, Illinois, or from St. Louis Serum Company, East St. Louis, Illinois. They are stored at −20° until used. The cartilage is cleaned of soft tissues and sliced with a meat slicer into approximately 1 mm-thick slices. The slices (150 g) are suspended in 1 liter of 4 M guanidine chloride containing 0.05 M Tris·HCl, pH 7.5, and stirred at room temperature for 24 hours. After filtration through cheesecloth, the solution is dialyzed for 2 days against 0.05 M Tris buffer, pH 7.5, in the presence of thymol as a preservative. The dialyzed solution is mixed with an equal volume of 1 M NaCl and 100 ml of 10% cetylpyridinium chloride are added. The precipitate is sedimented by centrifugation at 5000 g for 30 minutes at room temperature. (If sedimentation is incomplete, centrifugation is carried out at higher centrifugal force and for a longer period of time.)

The cetylpyridinium complex is dissolved with stirring in 1 liter of 2 M NaCl containing 15% ethanol. (This step usually requires several hours.) The proteoglycan is then precipitated by addition of 2 volumes of ethanol, redissolved in 1 liter of water and again precipitated with 2 volumes of ethanol. This step is repeated once, and the final precipitate is washed twice with absolute ethanol and once with ether and dried in a desiccator over phosphorus pentoxide. (If addition of ethanol does not produce a precipitate, when the proteoglycan has been dissolved in

[24] S. W. Sajdera and V. C. Hascall, *J. Biol. Chem.* **244**, 77 (1969).
[25] V. C. Hascall and S. W. Sajdera, *J. Biol. Chem.* **244**, 2384 (1969).

TABLE III

SOLUBILIZATION OF GLYCOSYLTRANSFERASES[a]

Treatment	Activity (cpm)				
	Xylosyl-transferase	Galactosyl-transferase I	Galactosyl-transferase II	Glucuronosyl-transferase I	N-Acetylgalactos-aminyltransferase
None	554	41,013	13,940	3,753	2,280
Nonidet P-40					
Pellet	1282	35,687	11,596	3,339	2,616
Supernatant	454 (26%)	7,899 (18%)	4,751 (29%)	1,076 (24%)	228 (8%)
Nonidet P-40 + 0.5 M KCl					
Pellet	201	3,967	3,880	847	678
Supernatant	1131 (85%)	49,780 (93%)	9,823 (72%)	2,590 (76%)	1,542 (70%)

[a] A 100,000 g pellet from a homogenate of embryonic chick cartilage was treated for 30 minutes with Nonidet P-40 (final concentration 0.5%) in the presence of 0.5 M KCl. After centrifugation at 100,000 g for 1 hour, the glycosyltransferases in the supernatant and pellet fractions were assayed as described. Numbers in parentheses indicate the percentage of the enzyme activity that was not sedimentable at 100,000 g.

water for the second time, a solution of sodium acetate in ethanol is added, until flocculation occurs.)

Smith Degradation. One gram of proteoglycan is dissolved in 200 ml of 0.05 M sodium metaperiodate containing 0.25 M sodium perchlorate, and the solution is adjusted to pH 5.0 and kept in the dark at 25°. The pH is checked every day and readjusted to 5.0, if necessary. After 5 days, the mixture is dialyzed against several changes of distilled water for 24 hours. Potassium borohydride is added to a final concentration of 0.15%, and reduction is allowed to proceed for 12 hours at 25°. The solution is then acidified to pH 1.0 with HCl and kept at 25° for 3 days. (At this stage, part of the degraded proteoglycan may be insoluble at the acid pH but the precipitate is largely dissolved upon subsequent neutralization.) After neutralization and dialysis overnight against several changes of distilled water, the dialyzed material is concentrated to 10 ml and centrifuged to remove a small precipitate. The supernatant solution is applied to a column (2.1 × 36 cm) of Sephadex G-75 or G-200, which is eluted with 0.2 M NaCl at a rate of 4 ml per hour (fraction volume, 2 ml). The protein emerges with the void volume and is clearly separated from a large hexosamine-containing peak which consists of the degraded chondroitin sulfate chains. The fractions containing protein are pooled, dialyzed overnight against distilled water, and freeze-dried. Yield, 143 mg.

The degraded proteoglycan consists of approximately ⅔ protein and ⅓ keratan sulfate. This polysaccharide remains attached to the protein core, since it is largely resistant to periodate oxidation. The extent of removal of the chondroitin sulfate chains may be determined by analysis of the galactosamine content of the product. (This is most conveniently done in the course of amino acid analysis.) Even in the most extensively degraded preparations, galactosamine constitutes approximately 25% of the total hexosamine. However, this is due in large part to the galactosamine content of keratan sulfate which amounts to 10–20% of the total hexosamine of this polysaccharide. It may be noted that preparations in which the galactosamine content equals or exceeds the glucosamine content still exhibit adequate xylosyl acceptor activity.

B. *O*-β-D-Xylopyranosyl-L-serine (Substrate for Galactosyltransferase I)

Principle. The preparation of xylosylserine is carried out by the three-step procedure of Kum and Roseman[26] for the synthesis of *O*-

---

[26] K. Kum and S. Roseman, *Biochemistry* **5**, 3061 (1966).

serine glycosides. The synthesis involves Koenigs–Knorr condensation of 2,3,4-tri-O-acetyl-α-D-xylopyranosyl bromide with N-CBZ-L-serine benzyl ester,[27] hydrogenolysis to remove the benzyl and CBZ groups, and ammonolysis to remove the acetyl groups.

Reagents

N-CBZ-L-serine benzyl ester (Sigma)

2,3,4-Tri-O-acetyl-α-D-xylopyranosyl bromide (acetobromoxylose). This compound is prepared by the method of Bárczai-Martos and Körösy.[28]

Benzene. Commercial benzene (reagent grade) is purified and dried by standard procedures.[29]

Silver carbonate

Calcium sulfate, anhydrous

Silver perchlorate, anhydrous

Hydrogen, dry

Dioxane, reagent grade

Platinum oxide catalyst (Adams' catalyst)

Methanol. Commercial reagent grade methanol is treated with NaBH$_4$ (1 g/500 ml) at room temperature, distilled, and stored over anhydrous sodium sulfate.

Ammonia, anhydrous

Sephadex LH-20

Koenigs–Knorr Condensation. N-CBZ-L-serine benzyl ester (6.58 g) is dissolved in 100 ml of benzene and stirred for 10 minutes with 9.1 g of anhydrous silver carbonate and 20.4 g of calcium sulfate. Next, 6.78 g of acetobromoxylose and 0.2 g of silver perchlorate are added, and the mixture is stirred in the dark for 48 hours maintaining anhydrous conditions.

Progress of the condensation reaction may be monitored by thin-layer chromatography on Quanta/Gram Type Q1 precoated glass plates which are developed with 4% methanol in benzene (v/v). Esters are visualized with hydroxylamine-ferric chloride[30] and xylose derivatives

[27] CBZ, benzyloxycarbonyl.

[28] Acetobromoxylose may be prepared by any of a number of procedures; see, e.g., M. Bárczai-Martos and F. Körösy, *Nature* **165**, 369 (1950) or F. Weygand, *in* "Methods in Carbohydrate Chemistry" (R. L. Whistler and M. L. Wolfrom, eds.), Vol. 1, p. 182. Academic Press, New York, 1962.

[29] A. I. Vogel, "A Textbook of Practical Organic Chemistry," 3rd ed., p. 172. Wiley, New York, 1956.

[30] M. E. Tate and C. T. Bishop, *Can. J. Chem.* **40**, 1043 (1962).

with an orcinol reagent.[31] The following components are observed in the reaction mixture, with R_f values as indicated: acetobromoxylose, 0.85; xylosylated N-CBZ-L-serine benzyl ester, 0.75; N-CBZ-L-serine benzyl ester, 0.34. Other products including triacetylxylose and its dimer have R_f values of less than 0.35.

After completed reaction, the mixture is filtered, and the filtrate is diluted with 2 volumes of benzene and washed once with 100 ml of cold saturated $NaHCO_3$ solution and four times with 50-ml quantities of water. The organic phase is dried over anhydrous sodium sulfate and concentrated to a syrup under reduced pressure.

Purification of the Condensation Product. In the original procedure of Kum and Roseman[26] the glycosylated serine derivative was purified by repetitive chromatography on silicic acid columns. Modification of this step by substituting Sephadex LH-20 for silicic acid has proved advantageous and is described below.

The reaction mixture is diluted to 30 ml with benzene–methanol (1:1) and applied to a column (5 × 115 cm) of Sephadex LH-20 which has been equilibrated with the same solvent. Fractions of 8 ml are collected and the eluate is monitored by thin-layer chromatography of 20-μl aliquots from each fraction. The desired compound is the first to be eluted and emerges at an effluent volume of approximately 600 ml. Twelve fractions containing most of the product (approximately 6 g) together with smaller amounts of contaminating substances are pooled, concentrated to a small volume and diluted to 30 ml with benzene–methanol (1:1). Aliquots of 10 ml are rechromatographed in the same fashion, and fractions containing only product are pooled and concentrated to a syrup. Each 10-ml portion yields approximately 1.4 g of chromatographically homogeneous material which is subjected to hydrogenolysis without further purification.

2,3,4-Tri-O-acetyl-O-β-D-xylopyranosyl-L-serine. Approximately 1.4 g of xylosylated N-CBZ-L-serine benzyl ester are dissolved in a solution containing 80 ml of dioxane and 20 ml of water and reduced with hydrogen at 2 atmospheres in a Parr low pressure hydrogenation apparatus in the presence of 200 mg of platinum oxide catalyst for 4 hours. The catalyst is removed by filtration and upon evaporation of the solvent the product crystallizes spontaneously. Recrystallization of the product from methanol–ethanol–petroleum ether (30–60°) yields triacetylxylosylserine (m.p. 188–191°) in 85% yield.

O-β-D-Xylopyranosyl-L-serine. A 0.5-g quantity of the triacetyl derivative is suspended in 50 ml of anhydrous methanol, cooled to 0° and

[31] A. H. Brown, *Arch. Biochem.* **11**, 269 (1952).

mixed with 50 ml of methanol saturated with ammonia at 0°. The mixture is allowed to stand at room temperature for 5 hours with the exclusion of moisture. The solution is concentrated at reduced pressure at room temperature until the residue spontaneously crystallizes. The crude xylosylserine is recrystallized from minimal amounts of ethanol–water. Following two recrystallizations, 222 mg of the final product is obtained, m.p. 224–227°; yield, 70%.

C. 4-O-β-D-Galactopyranosyl-D-xylose (Substrate for Galactosyltransferase II)

This substrate is synthesized by the following series of reactions: D-Arabinose → benzyl β-D-arabinopyranoside → benzyl 3,4-O-isopropyl-idene-β-D-arabinopyranoside → benzyl 3,4-O-isopropylidene-2-O-tosyl-β-D-arabinopyranoside → benzyl 2-O-tosyl-β-D-arabinopyranoside → benzyl 2,3-anhydro-β-D-ribopyranoside; condensation of the latter compound with 2,3,4,6-tetra-O-acetyl-α-D-galactopyranosyl bromide in a Koenigs-Knorr reaction, followed by alkaline opening of the epoxide ring and simultaneous removal of the acetyl groups, yields the benzyl glycoside of the desired disaccharide; the protective group is finally removed by catalytic hydrogenolysis.

Benzyl β-D-Arabinopyranoside.[32,33] D-Arabinose (100 g) is mixed with benzyl alcohol (500 ml), the mixture is cooled in an ice–salt bath, and hydrogen chloride is introduced until the solution is saturated, which requires about 20 minutes. The mixture is then shaken overnight at room temperature, whereupon the product crystallizes out in part. One liter of ether is added slowly with constant stirring, and the mixture is left for 4 hours at 5° to ensure complete crystallization. The product is filtered, washed with ether, and air-dried; yield, 145 g (91%). Several recrystallizations from approximately 17 parts of ethanol give pure benzyl β-D-arabinopyranoside, m.p. 172–173° (corr.), $[\alpha]_D^{20}$ −209° (c 0.4, water).

Benzyl 3,4-O-Isopropylidene-β-D-arabinopyranoside.[33] A mixture of benzyl β-D-arabinopyranoside (20 g), dry acetone (1 liter), anhydrous cupric sulfate (60 g), and concentrated sulfuric acid (1 ml) is shaken at room temperature for 18 hours. The solution is then neutralized with gaseous ammonia, the solid is removed by filtration, and the filtrate is concentrated to a syrup. Ether (150 ml) is added to precipitate the un-reacted benzyl β-D-arabinopyranoside, which is removed by filtration, and the filtrate is again concentrated to a syrup. On high-vacuum distilla-

[32] C. E. Ballou, *J. Amer. Chem. Soc.* **79**, 165 (1957).
[33] H. G. Fletcher, Jr., *in* "Methods in Carbohydrate Chemistry" (R. L. Whistler and M. L. Wolfrom, eds.), Vol. 2, p. 386. Academic Press, New York, 1963.

tion (0.15 mm) the main fraction of benzyl 3,4-*O*-isopropylidene-*β*-D-arabinopyranoside distills over at 135–145° (bath temperature, 175°); yield, 16.8 g (72%). The product solidifies on standing and may then be recrystallized from ether; m.p. 55–58°, $[\alpha]_D^{20}$ −209° (c 2, ethanol). On storage at room temperature, it decomposes slowly and should, therefore, be used for the following step as promptly as possible.

Benzyl 3,4-O-Isopropylidene-2-O-tosyl-β-D-arabinopyranoside.[34] Benzyl 3,4-*O*-isopropylidene-*β*-D-arabinopyranoside (81 g) is dissolved in pyridine (500 ml) and treated with *p*-toluenesulfonyl chloride (112 g). The solution is allowed to stand at room temperature during 48 hours and is then poured on ice. The mixture is extracted with chloroform, and the chloroform solution is dried over anhydrous sodium sulfate, filtered, and concentrated to dryness. The syrup is used directly in the next step. (If crystallized from 96% ethanol, the product has m.p. 62–64°, $[\alpha]_D^{20}$ −194° (c 1, chloroform.)

Benzyl 2-O-Tosyl-β-D-arabinopyranoside.[34] The product from the previous step is dissolved in boiling acetone (900 ml) and 0.1 M formic acid (1800 ml) is added over a period of 20 hours under reflux. Concentration yields a syrup which crystallizes from aqueous ethanol. Recrystallization from ethanol/isopropyl ether yields benzyl 2-*O*-tosyl-*β*-D-arabinopyranoside (70 g) with m.p. 120–123°.

Benzyl 2,3-Anhydro-β-D-ribopyranoside.[34] Benzyl 2-*O*-tosyl-*β*-D-arabinopyranoside (70 g) is dissolved in methanolic sodium methoxide (1 liter, prepared from 13 g of sodium) and allowed to stand overnight. The solution is diluted with water (500 ml) and neutralized with sulfuric acid, the methanol is removed by vacuum distillation, and the aqueous solution is extracted with chloroform. The combined chloroform solutions are dried over anhydrous sodium sulfate and filtered; the filtrate is concentrated to dryness. Spontaneous crystallization occurs giving benzyl 2,3-anhydro-*β*-D-ribopyranoside (29 g) with m.p. 64–76°. Recrystallization from isopropyl ether gives 25 g with m.p. 74–76°. The pure substance has m.p. 76–77°, $[\alpha]_D^{20}$ −67° (c 0.8, chloroform).

2,3,4,6-Tetra-O-acetyl-α-D-galactopyranosyl Bromide. The procedure for the synthesis of this compound follows the directions of Bárczai-Martos and Kőrösy[28] as described by Lemieux.[35] In a 1-liter, 3-necked flask equipped with an efficient stirrer and a thermometer, 400 ml of acetic anhydride is cooled in an ice–water bath, and 2.4 ml of 60–70% perchloric acid are added dropwise. The solution is warmed to room temperature, and 100 g of anhydrous D-galactose are added to the stirred

---

[34] P. J. Garegg, *Acta Chem. Scand.* **14,** 957 (1960).

[35] R. U. Lemieux, *in* "Methods in Carbohydrate Chemistry" (R. L. Whistler and M. L. Wolfrom, eds.), Vol. 2, p. 221. Academic Press, New York, 1963.

mixture at such a rate that the reaction temperature remains between 30° and 40° (total time required for the addition is approximately 0.5 hour). Red phosphorus (30 g) is added after the reaction mixture is cooled to 20°, followed by 180 g (58 ml) of bromine at such a rate as to keep the reaction temperature below 20°. Water (36 ml) is added dropwise to the continuously stirred and cooled mixture over an about 0.5-hour period to prevent the temperature from rising above 20°. The mixture is kept for 2 hours at room temperature. Chloroform (300 ml) is added, and the mixture is filtered through a filter-bed of fine glass wool. The reaction flask and the filter funnel are washed with 50 ml of chloroform. The filtrate is poured into 800 ml of water (near 0°) in a 3-liter separatory funnel. After washing, the chloroform layer is drawn off into a 2-liter separatory funnel which contains 300 ml of 0° water. The operation is repeated by adding 50 ml of chloroform to the original aqueous mixture and combining the chloroform extracts. After vigorous shaking, the chloroform layer is poured into 500 ml of a stirred saturated aqueous solution of sodium hydrogen carbonate kept in a 2-liter beaker. The mixture is transferred to a 2-liter separatory funnel with the aid of a little chloroform and shaken vigorously. The chloroform layer is stirred 10 minutes with 10 g of dry silicic acid. The mixture is filtered, and the faintly yellow solution is evaporated under reduced pressure below 60° in a rotary evaporator to a hard crystalline mass. The solid is transferred to a mortar with the aid of 500 ml of a 2:1 (v/v) mixture of petroleum ether and ether and is ground in the solvent. The mixture is filtered, and the residue is washed with 50 ml of cold, dry ether. The crude 2,3,4,6-tetra-O-acetyl-α-D-galacto-pyranosyl bromide is dried at reduced pressure over sodium hydroxide. Recrystallization from ether or diisopopyl ether affords the pure product; m.p. 83–84°, $[\alpha]_D^{20}$ +215° (c 1, chloroform); yield 75%.

Koenigs–Knorr Coupling of 2,3,4,6-Tetra-O-acetyl-α-D-galactopyrano-syl Bromide and Benzyl 2,3-Anhydro-β-D-ribopyranoside.[36] Benzyl 2,3-anhydro-β-D-ribopyranoside (8.9 g; 0.04 mole), freshly prepared silver oxide (10 g; 0.044 mole), Drierite (40 g; preheated 2 hours at 240°), and 40 ml of anhydrous, ethanol-free chloroform are stirred in a three-necked round-bottom flask, wrapped in aluminum foil to protect the contents from light, and equipped with a sealed stirrer, a drying tube, and a dropping funnel. After 1 hour, a solution of iodine (2 g) and 2,3,4,6-tetra-O-acetyl-α-D-galactopyranosyl bromide (16.5 g; 0.04 mole) in 60 ml of dry, ethanol-free chloroform is added slowly over a 3-hour period, and stirring is continued for 20 hours. The reaction mixture is filtered

[36] B. Lindberg, L. Rodén, and B.-G. Silvander, *Carbohyd. Res.* **2**, 413 (1966).

through Celite, and the salts are washed with chloroform. The combined filtrate and washings are washed with aqueous sodium thiosulfate, dried over calcium chloride, and concentrated to a syrup (25 g).

The product is heated in 2 M aqueous NaOH (500 ml) at 100° for 16 hours, and the cooled solution is deionized by passage through columns of Dowex 50 (H⁺) and Dowex 3 (free base), and concentrated to a syrup (8.2 g).

The syrup is dissolved in 50% aqueous ethanol (250 ml) and stirred, in an atmosphere of hydrogen, with 10% palladium–charcoal (5 g). After 5 hours, when the uptake of hydrogen (about 360 ml) has ceased, the catalyst is removed by filtration, and the filtrate is concentrated to a syrup (4.5 g). As shown by paper chromatography in ethyl acetate–acetic acid–water (3:1:1), the syrup consists essentially of 4-O-β-D-galactopyranosyl-D-xylose, contaminated by xylose, galactose, and some unknown compounds. Further purification is obtained by cellulose column chromatography in ethyl acetate–acetic acid–water (3:1:1), yielding chromatographically pure disaccharide. The disaccharide, which has not been crystallized, has $[\alpha]_{5780}^{22}$ +15° (c 0.5, water).

D. 3-O-β-D-Galactopyranosyl-D-galactose   (Substrate for Glucuronosyltransferase I)

This disaccharide is most conveniently isolated from hydrolyzates of arabinogalactan from *Larix occidentalis*.[3,36,37]

Wood from *Larix occidentalis* is obtained from Crown Zellerbach Corp., Camas, Washington. Ground heartwood (580 g) is continuously extracted with ether for 30 hours and subsequently with methanol for 90 hours. The wood (540 g) is then extracted with cold water (4 + 2 + 2 liters) for 12–18 hours with continuous stirring. The suspension is filtered through cheesecloth, and the extract is clarified by filtration through a Celite pad. After concentration to a viscous solution, the polysaccharide is precipitated with 9 volumes of ethanol and dried with absolute ethanol and ether (yield, 46.5 g).

The polysaccharide (50 g) is dissolved in 2.5 liters of 0.02 M HCl and heated at 95° for 4 hours to remove arabinose. After addition of several volumes of ethanol, the precipitated polysaccharide is dissolved in 1 liter of 0.2 M HCl and heated at 100° for 1 hour. The hydrolyzate is cooled and neutralized with Amberlite IR 45 (OH⁻), concentrated to 200 ml, and 9 volumes of ethanol are added. Precipitated material is rehydrolyzed in 0.2 M HCl, and this hydrolysis–precipitation cycle is carried out a total of 4 times. The ethanol supernatant solutions from

---

[37] H. Bouveng and B. Lindberg, *Acta Chem. Scand.* **10**, 1515 (1956).

the 4 hydrolyzates are combined, concentrated to a syrup, dissolved in 1 liter of distilled water, and applied to a column (14 × 30 cm) of carbon–Celite.[38] After rinsing with 20 liters of distilled water, oligosaccharides are eluted with ethanol of stepwise increasing concentration (5–25% with 5% increments). In each step approximately 80 50-ml fractions are collected. The eluate is concentrated and analyzed by paper chromatography in ethyl acetate–acetic acid–water (3:1:1). 3-*O*-β-D-Galactosyl-D-galactose is eluted from the column by 10% ethanol with some slight overlap into the 5% fraction. After concentration of the eluate to a syrup, the disaccharide is crystallized from ethanol–water (m.p. 160–162°), $[\alpha]_D^{22}$ +64° (c 1.0, water).

E. Oligosaccharides Containing repeating Disaccharides from Chondroitin 4-Sulfate, Chondroitin 6-Sulfate, and Chondroitin

1. Even-Numbered Oligosaccharides with Nonreducing Terminal Glucuronic acid (Substrates for *N*-Acetylgalactosaminyltransferase)

Principle

The polysaccharide is digested with testicular hyaluronidase, and the oligosaccharide products are separated by gel chromatography on Sephadex G-25 or G-50. Additional purification is carried out by ion exchange chromatography.[39-42]

a. Hyaluronidase Digestion and Gel Chromotography

Reagents and Materials

Sodium acetate, 0.10 M, pH 5.0, containing 0.15 M NaCl
Testicular hyaluronidase (a highly purified enzyme is preferable, and a preparation with an activity of about 20,000 IU/mg may be obtained from AB Leo, Hälsingborg, Sweden)
NaCl, 0.2 M, containing 10% ethanol
Sephadex G-25 and G-50, superfine
Chondroitin 4-sulfate and chondroitin 6-sulfate (see this volume

[38] R. G. Spiro, see Vol. 8 [1].
[39] B. Weissmann, K. Meyer, P. Sampson, and A. Linker, *J. Biol. Chem.* **208**, 417 (1954).
[40] P. Hoffman, K. Meyer, and A. Linker, *J. Biol. Chem.* **219**, 653 (1956).
[41] P. Flodin, J. D. Gregory, and L. Rodén, *Anal. Biochem.* **8**, 424 (1964).
[42] L.-Å. Fransson, L. Rodén, and M. L. Spach, *Anal. Biochem.* **21**, 317 (1968).

[7]: III: 2, 3); these polysaccharides may also be purchased from Miles Laboratories, Elkhart, Indiana

Chondroitin; this polysaccharide is prepared by desulfation of chondroitin 4-sulfate as described by Kantor and Schubert.[43] Chondroitin 4-sulfate (5 g) is suspended in 800 ml of anhydrous methanol containing 0.5% acetyl chloride and is stirred in a closed vessel for 24 hours. The insoluble material is recovered by centrifugation; the same treatment is repeated twice. Subsequently, the polysaccharide is deesterified by treatment with 0.1 M KOH in the cold for 12 hours, and, following neutralization and decolorization with charcoal, the chondroitin is precipitated with ethanol and dried with absolute ethanol and ether.

Procedure. A 1-g sample of polysaccharide is dissolved in 25 ml of sodium acetate buffer and digested with 10 mg of hyaluronidase (200,000 units) at 37° for 24 hours. If necessary, the digest is clarified by centrifugation, and a 10-ml aliquot is then applied to a column (3 × 240 cm) of Sephadex G-25, superfine. The column is eluted with 0.2 M NaCl–10% ethanol at a rate of 15–20 ml per hour, and fractions are analyzed by the carbazole method.[44] The material from each peak is pooled, concentrated to a small volume and desalted on a column (2 × 120 cm) of Sephadex G-25, superfine, which is eluted with water. After evaporation, the oligosaccharide solution is lyophilized.

Comments. Sephadex G-25 is suitable for separation of the lower members of the homologous series, whereas larger oligosaccharides are fractionated on G-50.[45] For optimal resolution, the load should not exceed about 0.5 g of the oligosaccharide mixture, applied in a volume of 10–15 ml.

The identity of the separated oligosaccharides is readily inferred from their positions in the elution pattern. Also, the relative proportions of the lower members serve as an additional aid in identification. Under the conditions of digestion, the tetrasaccharide is the major product and should characteristically be preceded by a somewhat smaller peak containing hexasaccharide, whereas a minor peak containing disaccharide should be present in a more retarded position. Positive identification is obtained by hexosamine analysis before and after reduction of each oligosaccharide with borohydride.[41] Oligosaccharides from chondroitin and chondroitin 6-sulfate may be identified more conveniently by determination of the ratio of uronic acid or total hexosamine to reducing

[43] T. G. Kantor and M. Schubert, *J. Amer. Chem. Soc.* **79**, 152 (1957).

[44] Z. Dische, *J. Biol. Chem.* **167**, 189 (1947).

[45] L.-Å. Fransson and L. Rodén, *J. Biol. Chem.* **242**, 4170 (1967).

terminal N-acetylhexosamine (Morgan–Elson method).[46] This analysis cannot be applied to chondroitin 4-sulfate oligosaccharides, since they are Morgan–Elson negative.

The oligosaccharide fractions obtained by gel chromatography are usually not homogeneous. Overlapping with neighboring homologs may occur to some extent and can be overcome by refractionation or by ion exchange chromatography. Furthermore, owing to the nature of the starting material, each fraction contains a small proportion of incompletely sulfated components (e.g., the hexasaccharide fraction contains some mono- and disulfated components in addition to the major, trisulfated hexasaccharide). Yet another source of heterogeneity is the presence of covalently bound peptides in the polysaccharide starting material. Consequently, the hyaluronidase digest contains a small proportion of glycopeptides which are eluted in positions partially overlapping with the oligosaccharides. Separation of the oligosaccharides according to degree of sulfation as well as removal of glycopeptide contaminants can be accomplished by ion exchange chromatography as described below. Alternatively, the problem of removing glycopeptides can be circumvented by choosing a different preparative procedure. If cartilage proteoglycan (see Section IX:A) is used as a substrate for hyaluronidase digestion rather than the isolated polysaccharide, the resulting mixture of oligosaccharides and macromolecular core material can easily be separated on Sephadex G-75, and the peptide-free oligosaccharides can then be further fractionated as described.[47]

b. Ion Exchange Chromatography [39,40,42,48]

Ion exchange chromatography may be used either for further purification of oligosaccharide fractions obtained by gel chromatography or as the sole method for the fractionation of oligosaccharide mixtures. If available, a Technicon sugar chromatography system or similar equipment facilitates the procedure and may be used for analytical as well as preparative purposes (see [7]:IV, A, 1).

Nonsulfated oligosaccharides from chondroitin or hyaluronic acid are applied, in amounts not exceeding 100 mg, to a column (0.9×120 cm) of Dowex 1-X8 (200–400 mesh, Cl⁻) which is eluted with a nonlinear gradient of LiCl. The reservoir contains 300 ml of $0.2\,M$ LiCl, and the constant-volume mixing vessel initially contains 300 ml of water. The column is eluted at a rate of 30 ml per hour, and 3-ml fractions are collected and analyzed by the carbazole method. (Alternatively, a small

[46] E. A. Davidson, see Vol. 8 [3].
[47] J. D. Gregory, T. C. Laurent, and L. Rodén, *J. Biol. Chem.* **239**, 3312 (1964).
[48] L.-Å. Fransson, *Biochim. Biophys. Acta* **156**, 311 (1968).

portion of the effluent is shunted off by a split-stream device and subjected to automated analysis by the orcinol reaction as described elsewhere in this volume ([17]:IV, A, 1).) Oligosaccharide-containing fractions are pooled, concentrated to a small volume, and desalted on a
column (2 × 120 cm) of Sephadex G-25 or G-10.

Fractionation of sulfated oligosaccharides is carried out by the same
procedure with the modification that a gradient of 0–2.0 M LiCl is
used. The separation of the fully sulfated oligosaccharides from the
undersulfated compounds and the glycopeptides appearing in the same
position on gel chromatography should normally be complete, the two
latter groups being eluted well ahead of the fully sulfated substances on
resin chromatography. The position of the glycopeptides may be determined by the ninhydrin reaction.

For the preparation of larger amounts of oligosaccharides, the above
procedures are scaled up in an appropriate fashion, as exemplified by
the following description. A hyaluronidase digest of 2.5 g of chondroitin
6-sulfate in 250 ml of buffer is prepared as described above. After dilution with an equal volume of water, the solution is applied to a column
(4 × 30 cm) of Dowex 1-X2 (200–400 mesh, Cl⁻), and the oligosaccharides are eluted with a linear gradient of 0.2–2.0 M NaCl (2000 ml
in each vessel). The fractionation is monitored by uronic acid analysis.

2. Odd-Numbered Oligosaccharides with Nonreducing Terminal N-acetylgalactosamine and N-acetylgalactosamine Sulfate Residues (Substrates for Glucuronosyltransferase II)

Principle

Even-numbered oligosaccharides with nonreducing terminal glucuronic acid residues are digested with β-glucuronidase, and the products
are separated by gel chromatography or ion exchange chromatography.

Reagents

Sodium acetate, 0.05 M, pH 5.0
β-Glucuronidase (preparations of varying degrees of purity are
    available from several commercial sources)
Oligosaccharides from chondroitin, chondroitin 4-sulfate, and chon-
    droitin 6-sulfate, prepared as described in the preceding section

Procedure. A 1–2% solution of oligosaccharide in sodium acetate
buffer, containing 300 Fishman units per milligram of oligosaccharide,

is incubated at 37° for 24 hours. The digest is fractionated on a 200-cm column of Sephadex G-25, superfine, which is eluted with 0.2 M NaCl. (The diameter of the column is chosen according to the amount of oligosaccharide applied; e.g., 0.5–1.0 g of material is a safe load for a 3 × 200 cm column.) The separation is monitored by uronic acid analysis, and the oligosaccharide is recovered as described in the preceding section.

Comments. For complete removal of protein, particularly if a crude enzyme preparation has been used, it is advantageous to use precipitation with trichloroacetic acid (final concentration 5%) as an initial purification step, followed by neutralization of the supernatant fluid prior to gel chromatography. The purity of the β-glucuronidase is normally of little consequence, but it should be noted that the presence of N-acetylhexosaminidase in crude preparations results in continued degradation of the nonsulfated oligosaccharides from hyaluronic acid and chondroitin and necessitates more careful fractionation of the digest. (The sulfated N-acetylhexosamine residues in chondroitin 4- or 6-sulfate oligosaccharides are not attacked by N-acetylhexosaminidase.) If desired, the additional cleavage may be avoided by using a more highly purified β-glucuronidase or by carrying out the digestion in the presence of an N-acetylhexosaminidase inhibitor.

F. Other Oligosaccharide Substrates and Reference Compounds[10,19,49-54]

In addition to the substances discussed above, a number of others may be desired as substrates or as reference standards for comparison with the radioactive products formed in the various transferase reactions. This group includes, e.g., 4-O-β-D-galactosyl-O-β-D-xylosyl-L-serine, 3-O-β-D-galactosyl-4-O-β-D-galactosyl-D-xylose, and 3-O-β-D-glucuronosyl-3-O-β-D-galactosyl-D-galactose, which are formed from the substrates recommended for the two galactosyltransferases and glucuronosyltransferase I. These compounds have been isolated from partial acid hydrolyzates of chondroitin sulfate and heparin, and the original papers should be consulted for preparative details.[19,49,51-54] Similarly, three of the substrates described in the preceding sections have been isolated from hydrolyzates of these polysaccharides, i.e., 4-O-β-D-galactosyl-D-xylose, 3-O-β-D-galactosyl-D-galactose, and O-β-D-xylosyl-L-serine.[49,51-54]

[49] U. Lindahl and L. Rodén, *J. Biol. Chem.* **240**, 2821 (1965).
[50] L. Rodén and G. Armand, *J. Biol. Chem.* **241**, 65 (1966).
[51] U. Lindahl and L. Rodén, *J. Biol. Chem.* **241**, 2113 (1966).
[52] L. Rodén and R. Smith, *J. Biol. Chem.* **241**, 5949 (1966).
[53] U. Lindahl, *Biochim. Biophys. Acta* **130**, 361 (1966).
[54] U. Lindahl, *Ark. Kemi* **26**, 101 (1967).

It should be emphasized at this point that the choice of preparative method depends in part upon the amounts of material required. If substrate quantities are needed, a chemical synthesis is generally to be preferred. On the other hand, it is often less time-consuming to isolate smaller amounts by a procedure based on partial acid hydrolysis and subsequent fractionation of the hydrolyzate. If an oligosaccharide is needed only as a reference compound in chromatography, a partially purified hydrolyzate may even be adequate for this purpose; e.g., the nonionic fraction of a chondroitin sulfate hydrolyzate contains xylose, galactose, galactosylxylose, galactosylgalactose, and galactosylgalactosylxylose, which are all easily distinguishable on paper chromatography in ethyl acetate–acetic acid–water (3:1:1).[52]

A third category of preparative procedures is based on the use of specific glycosidases. When appropriate enzymes are available, enzymatic degradation permits the isolation of certain oligosaccharides in higher yield and with greater ease than is feasible with any other method. This approach has proved invaluable in the preparation of the oligosaccharides containing repeating disaccharide units which are derived from chondroitin sulfate by degradation with testicular hyaluronidase and β-glucuronidase.

[85a] Enzymes Involved in the Formation of the Carbohydrate Structure of Heparin

By ULF LINDAHL

Introduction

The polysaccharide backbone of the heparin molecule is composed of alternating units of D-glucosamine and a uronic acid, which may be either D-glucuronic acid or L-iduronic acid.[1] A microsomal fraction from a transplantable mast cell tumor has been shown to catalyze the incorporation of D-glucuronic acid and N-acetylglucosamine from the corresponding UDP-sugars into microsomal polysaccharide. A method for the quantitation of this process has been described by Silbert in a previous volume of this series.[2]

In the present account, methods will be described for the separate assay of each of the two glycosyltransferases involved in the polymerization reaction. In addition, some properties of an epimerase converting

[1] U. Lindahl and O. Axelsson, J. Biol. Chem. 246, 74 (1971).
[2] J. E. Silbert, Vol. 8 [84].

D-glucuronic acid residues, in the polysaccharide chain, to L-iduronic acid units will be presented.

I. N-Acetylglucosaminyltransferase

UDP-N-acetylglucosamine + acceptor → N-acetylglucosamine-acceptor + UDP

Assay Method

Principle. A microsomal preparation from mastocytoma tissue catalyzes the transfer of [^{14}C]N-acetylglucosamine from UDP-[^{14}C]N-acetylglucosamine to an acceptor with glucuronic acid in nonreducing terminal position. The resulting labeled product is isolated by paper electrophoresis and paper chromatography.[3]

Reagents

UDP-[^{14}C]N-acetylglucosamine ($\sim$40 μCi/μmole)
Tris·acetate, 0.05 M, pH 7.4, containing 0.07 M KCl and 1 mM EDTA
MnCl$_2$, 0.24 M, in Tris·acetate buffer of the above composition
Acceptor (fraction B$_1$)[4]

Procedure. Assay mixtures containing the following components are prepared, in a final volume of 0.06 ml: 0.05 M Tris·acetate (pH 7.4), 0.07 M KCl, 1 mM EDTA, 0.02 M MnCl$_2$, 0.05 μmole of fraction B$_1$, 0.05–0.1 μCi of UDP-[^{14}C]N-acetylglucosamine and enzyme preparation (0.05–0.15 mg of protein). After incubation at 37° for 60 minutes, the reaction mixtures are spotted on strips of Whatman No. 3MM paper, and components anionic at pH 2.0 are removed by electrophoresis in 0.5 M formic acid–1.4 M acetic acid (40 V/cm; 2 hours). The paper is dried and [^{14}C]N-acetylglucosamine-labeled carbohydrate–serine compounds are separated from radioactive breakdown products by paper chromatography for 10 hours in ethyl acetate–acetic acid–water (3:1:1,

---

[3] T. Helting and U. Lindahl, *Acta Chem. Scand.* **26** (1972), in press.

[4] The acceptor preparation is a mixture of carbohydrate–serine compounds, isolated after degradation of heparin with nitrous acid.[5] Fraction B$_1$ has the general structure (UA-GlcNAc)$_2$-GlcUA-Gal-Gal-Xyl-Serine,[6] where UA may be either glucuronic or iduronic acid. About half of the nonreducing terminal residues of fraction B$_1$ are glucuronic acid and may be removed by treatment with β-glucuronidase, yielding fraction B$_1$-β.[7]

[5] U. Lindahl, *Biochim. Biophys. Acta* **130**, 368 (1966).

[6] The following abbreviations are used: UDP, uridine diphosphate; GlcNAc, N-acetylglucosamine; GlcUA, glucuronic acid; IdUA, iduronic acid; UA, uronic acid (glucuronic acid or iduronic acid); Gal, galactose; Xyl, xylose; PAPS, 3′-phosphoadenylyl sulfate.

[7] T. Helting and U. Lindahl, *J. Biol. Chem.* **246**, 5442 (1971).

by volume). In this procedure, free N-acetylglucosamine migrates off the paper, whereas the reaction product does not move appreciably. The latter component is located with a strip scanner and quantitated, after elution with water, by use of a liquid scintillation spectrometer.

Preparation of Enzyme

Preparation of Microsomal Fraction. The heparin-producing FMS mast-cell tumor[8] is maintained in the solid state in (A/Sn × Leaden)F$_1$ mice, by subcutaneous and intramuscular transplantation in the hind legs every 10–14 days. The tumors collected (0.5–1 g from each mouse) are dissected free from adherent and necrotic tissue and immediately homogenized with a Potter-Elvehjem glass homogenizer in 2 volumes of ice-cold Tris·acetate buffer, of the composition given above. Differential centrifugation[9] of the homogenate is carried out at 0–4°. After centrifugation at 10,000 g for 10 minutes, the resulting supernatant is centrifuged at 100,000 g for 60 minutes. The supernatant is discarded, and the final pellet is suspended in Tris·acetate–KCl–EDTA buffer, to give a protein concentration of 1–3 mg/ml.

The microsomal preparation contains substantial amounts of hydrolytic enzyme(s) responsible for the conversion of UDP-N-acetylglucosamine to free N-acetylglucosamine. Thus, the nucleotide precursor is completely degraded after 30 minutes of incubation. Also the product obtained on incubating fraction B$_1$ with UDP-[14C]N-acetylglucosamine is unstable, as treatment of the labeled material with the particulate enzyme overnight results in the liberation of all radioactivity as N-acetylglucosamine.

Effect of Detergent and Alkali. To microsomal fraction from 15 g of tumor, suspended in 10 ml of cold Tris·acetate–KCl–EDTA buffer, Tween 20 is added to a final concentration of 2% (w/v). The pH of the solution is raised to 10.4–10.6 by the addition of ammonia and is then quickly readjusted to 7.4, by treatment with glacial acetic acid.[9] After centrifugation at 159,000 g for 2 hours, the top 10 ml of each tube (total volume, 13 ml) is withdrawn and used as solubilized enzyme.

A major portion of the hydrolytic enzyme(s) responsible for the formation of free N-acetylglucosamine from UDP-N-acetylglucosamine, but less than 10% of the UDP-N-acetylglucosamine:Fraction B$_1$ N-acetylglucosaminyltransferase, is brought into solution by the detergent-alkali treatment. Since 75% of the total microsomal protein is likewise solubilized, the particulate transferase may be purified by this procedure.

[8] J. Furth, P. Hagen, and E. I. Hirsch, *Proc. Soc. Exp. Biol. Med.* **95**, 824 (1957).
[9] T. Helting, *J. Biol. Chem.* **246**, 815 (1971).

Such purified preparations have been used to determine the kinetic parameters of the enzyme.

Properties

Kinetic Parameters. Product formation occurs essentially linearly with time for about 60 minutes, and is proportional to the concentration of protein within a range of at least 0–2.5 mg/ml. The enzyme is stimulated by Mn^{2+} ions, which give maximal product formation at 20 mM concentration. The Mn^{2+} may be completely replaced by Co^{2+} and partially by Mg^{2+} or Ca^{2+}. The N-acetylglucosamine transfer reaction occurs over a wide pH range, with optimum at pH 7.5.[3]

Specificity. Transfer of N-acetylglucosamine in the mastocytoma system requires an acceptor compound with a glucuronic acid residue in nonreducing terminal position.[3] Carbohydrate-serine compounds closely related to the acceptor component of fraction B_1, but with terminal N-acetylglucosamine or iduronic acid residues (fraction B_1-β^4), are essentially devoid of acceptor activity. Furthermore, the size of the acceptor molecule appears to be of critical importance, as smaller fragments derived from the heparin–protein linkage region, although with glucuronic acid in terminal position, are not acceptors under the assay conditions employed. It is notable that hyaluronic acid hexasaccharide (containing β-1 → 3 glucuronidic linkages) is a weak acceptor in the N-acetylglucosaminyltransferase reaction. This finding indicates that 1 → 4 glucuronidic linkages (as present in the heparin–protein linkage region[10]) are not mandatory for acceptor activity.

II. Glucuronosyltransferase

UDP-glucuronic acid + acceptor → glucuronosyl-acceptor + UDP

Assay Method

Principle. Particulate or solubilized enzyme from the FMS mast cell tumor[8] catalyzes the transfer of labeled glucuronic acid from UDP-[^{14}C]-glucuronic acid to an acceptor with an N-acetylglucosamine unit in nonreducing terminal position. The labeled product is isolated by paper electrophoresis.[7]

Reagents

UDP-[^{14}C]glucuronic acid (~30 μCi/μmole)
Tris·acetate, 0.05 M, pH 7.4, containing 0.07 M KCl and 1 mM EDTA

---

[10] U. Lindahl, *Biochim. Biophys. Acta* **156**, 203 (1968).

MnCl$_2$, 0.24 M, in Tris·acetate buffer of the above composition
Acceptor (fraction B$_1$-β)[4]

Procedure. An assay mixture identical to that described above for
the N-acetylglucosaminyltransferase is used, except that 0.03 μmole[11]
of fraction B$_1$-β and 0.1 μCi of UDP-[^{14}C]glucuronic acid, respectively,
are substituted for the acceptor and nucleotide sugar components indicated. After incubation at 37° for 60 minutes, the reaction mixtures are
spotted on Whatman No. 3MM paper and subjected to electrophoresis
in 0.08 M pyridine–0.046 M acetic acid, pH 5.3 (80 V/cm; 90 minutes).
Detection and quantitation of the labeled product, which migrates similarly to the undigested fraction B$_1$, is carried out as described above.

Preparation of Enzyme

The mastocytoma microsomal preparation described above is employed. UDP-Glucuronic acid is more stable in this enzyme preparation
than UDP-N-acetylglucosamine, and the liberation of free glucuronic
acid is comparatively small. Contrary to the N-acetylglucosaminyltrans-
ferase, most (about 70%) of the glucuronosyltransferase is solubilized
on treatment of the particulate enzyme with Tween and alkali.[3]

Properties

Kinetic Parameters. The rate of product formation remains constant
for more than 60 minutes and is proportional to the concentration of
protein, within the range tested (0–2.5 mg/ml). Although the metal ion
requirement for the transfer of glucuronic acid to fraction B$_1$-β has not
been investigated in detail, the presence of Mn^{2+} appears to be essential
for product formation. The pH dependence of the glucuronosyltransferase
differs slightly from that of the N-acetylglucosaminyltransferase, with
optimum at pH 7.0.[3]

Specificity. Fraction B$_1$[4] does not serve as an acceptor in the glucuronosyltransferase reaction, but acquires acceptor properties on treatment with β-glucuronidase (yielding fraction B$_1$-β[4]).[7] Glucuronic acid
is thus transferred to nonreducing, terminal N-acetylglucosamine residues. Furthermore, the anomeric configuration of these residues appears
to be of importance, as the pentasaccharide GlcNAc-GlcUA-GlcNAc-
GlcUA-GlcNAc, isolated from hyaluronic acid, does not function as a
substrate.[7,12] Finally, it may be noted that the acceptor size requirement

---

[11] Based on the amount of glucuronic acid liberated from fraction B$_1$ by treatment
with β-glucuronidase.

[12] Glucosamine is believed to occur in heparin linked via α-1 → 4 bonds, whereas the
glucosaminidic linkage in hyaluronic acid is β-1 → 4.

is apparently less exacting for the glucuronosyltransferase than for the
N-acetylglucosaminyltransferase. Whereas the latter enzyme is active
only with large substrates, such as fraction B_1 or hexasaccharide
from hyaluronic acid, transfer of glucuronic acid occurs with essentially equal efficiency to substrates ranging in size from trisaccharide to
heptasaccharide.[13]

III. Anhydroglucuronic Acid Epimerase

nonsulfated (polysaccharide-D-glucuronic acid-polysaccharide)

↓PAPS

sulfated (polysaccharide-L-iduronic acid-polysaccharide)

Assay Method

Principle. A microsomal system from the heparin-producing FMS
mast-cell tumor[8] is employed. [14C]Glucuronic acid is incorporated from
UDP-[14C]glucuronic acid into endogenous microsomal acceptor, in the
presence of an excess of unlabeled UDP-N-acetylglucosamine. The nonsulfated polysaccharide formed contains [14C]glucuronic acid as the
only labeled uronic acid component present in significant amounts. Continued incubation in the presence of PAPS affords a sulfated polymer,
in which [14C]iduronic acid constitutes a major proportion of the total
labeled uronic acid. During the sulfation process no further radioactivity
is incorporated, as an excess amount of unlabeled UDP-glucuronic acid
is added together with the PAPS. By this procedure incorporation and
C_5-epimerization of labeled glucuronic acid occur during separate incubation periods. The resulting ratio of [14C]iduronic acid: [14C]glucuronic
acid is taken as a measure of epimerase activity; it is determined after
degradation of the labeled polysaccharide and separation of the uronic
acid monosaccharides formed.[14]

Reagents

UDP-[14C]glucuronic acid (~200 μCi/μmole)
UDP-glucuronic acid, unlabeled
UDP-N-acetylglucosamine, unlabeled
PAPS,[15] unlabeled

[13] T. Helting, unpublished.
[14] U. Lindahl, G. Bäckström, A. Malmström, and L.-Å. Fransson, *Biochem. Biophys.
Res. Commun.* **46,** 985 (1972); U. Lindahl, M. Höök, A. Malmström, and L.-Å.
Fransson, unpublished observations.
[15] A. S. Balasubramanian, L. Spolter, L. I. Rice, J. B. Sharon, and W. Marx,
Anal. Biochem. **21,** 22 (1967).

Tris·HCl, 0.05 M, pH 7.3, containing 0.01 M $MnCl_2$, 0.01 M $MgCl_2$ and 5mM $CaCl_2$

Acetate buffer, 0.05 M, pH 5.5, containing 0.01 M EDTA and 0.01 M cysteine-HCl

Trifluoroacetic acid, 2 M

$NaNO_2$, 3.9 M, in 0.29 M acetic acid

Acetic acid, 1 M

Papain, purified from a crude preparation[16]

Heparin, purified from a crude preparation by precipitation with cetylpyridinium chloride[17]

D-Glucuronic acid

L-Iduronic acid, prepared from dermatan sulfate[5]

Procedure. The reaction mixtures contain, in a total volume of 0.3 ml: 0.05 M Tris·HCl, pH 7.3; 8 mM $MgCl_2$; 8 mM $MnCl_2$; 4 mM $CaCl_2$; UDP-[^{14}C]glucuronic acid, 4 μCi; UDP-N-acetylglucosamine, 0.75 μmole; and enzyme preparation (50 μl; 2–3 mg of protein). After 60 minutes of incubation at 37°, the mixtures are transferred to tubes containing 0.5 μmole of unlabeled UDP-glucuronic acid and 0.4 μmole of PAPS, and incubation is continued for a total period of 2 hours. Carrier heparin (0.5 mg) is added, and incubations are terminated by immersion in boiling water for 3 minutes.

The boiled mixtures are diluted to 1.0 ml with acetate buffer–EDTA–cysteine-HCl, pH 5.5, and are then adjusted to 1.0 M with respect to NaCl. After digestion with 2 mg of papain at 65° overnight, the samples are passed through columns (1 × 90 cm) of Sephadex G-50, eluted with pyridine–acetic acid–water (10:6:984) at a rate of 4 ml/hour. The labeled polysaccharides, emerging as distinct peaks at the void volume of the columns, are precipitated from concentrated solutions by the addition of 3 volumes of ethanol.

Degradation of the polysaccharide samples to yield free uronic acid is carried out as follows. N-Acetyl and N-sulfate groups are removed by hydrolysis in 0.5 ml of 2 M trifluoroacetic acid at 100° for 3 hours. The hydrolyzates are transferred to conical centrifuge tubes, evaporated to dryness, and deaminated by treatment at room temperature with 30 μl of 3.9 M $NaNO_2$ in 0.29 M acetic acid.[18] After 10 minutes, the deamination

[16] J. R. Kimmel and E. L. Smith, *J. Biol. Chem.* **207**, 515 (1954).

[17] U. Lindahl, J. A. Cifonelli, B. Lindahl, and L. Rodén, *J. Biol. Chem.* **240**, 2817 (1965).

[18] J. E. Shively and H. E. Conrad, *Biochemistry* **9**, 33 (1970).

mixtures are diluted with 0.6 ml of 1 M acetic acid, passed through columns (1 × 3 cm) of Dowex 50-X8 previously equilibrated with 1 M acetic acid, and finally evaporated to dryness with several additions of methanol. The deamination products, essentially uronosylanhydromannose disaccharides,[1] are dissolved in 1.0 ml of 2 M trifluoroacetic acid and heated at 100° for 4 hours. The resulting mixtures of mono- and disaccharides are evaporated to dryness, dissolved in small volumes of water, and applied to strips of Whatman No. 3MM paper along with standards of D-glucuronic acid and L-iduronic acid. After descending chromatography in ethyl acetate–acetic acid–water (3:1:1, by volume) for 8–10 hours, the strips are dried and either stained by a silver dip procedure[19] or analyzed for radioactivity with a strip scanner. The radioactive components corresponding to the two uronic acids and their lactones (generally about 70% of the total radioactivity) are eluted with water and quantitated with a liquid scintillation spectrometer.

Preparation of Enzyme

A microsomal fraction is prepared from the FMS mast-cell tumor,[8] according to the procedure of Silbert.[2]

Properties

The extent of epimerization of glucuronic acid residues depends on the concomitant sulfation process.[14] Whereas the nonsulfated polymer does not contain any detectable labeled iduronic acid, the sulfated incubation product shows a [14C]iduronic acid:[14C]glucuronic acid ratio of about 1:2.[20] Furthermore, polysaccharide synthesized at lower PAPS concentration contains less sulfate and less iduronic acid than that produced under optimal conditions. The nature of the relationship between sulfation and uronic acid epimerization is so far unknown.

Less than 10% of the total radioactivity of nonsulfated polysaccharide is removed as free [14C]glucuronic acid on treatment with β-glucuronidase, indicating that most of the labeled uronic acid units occupy internal positions in the polymer. The relatively high yield of labeled

---

[19] I. Smith, in "Chromatographic and Electrophoretic Techniques" (I. Smith, ed.), Vol. 1, p. 252. Wiley (Interscience), New York, 1960.

[20] This value is based on the radioactivities of both the free acids and the lactones. As measured on the paper chromatogram, glucuronic acid invariably shows a higher ratio of lactone:free acid than does iduronic acid. Owing to this difference, calculations based on free acids only yield higher proportions of labeled iduronic acid (see reference in footnote 14).

iduronic acid after sulfation of the polymer therefore suggests that such internal glucuronic acid residues may be attacked by the epimerase. No further information is available as to the specificity of the enzyme. Also, the metal ion requirement is unknown; the ions included in the assay mixture are primarily intended to provide optimal conditions for the glycosyltransferases.

[86] Glucanosyltransferase from *Bacillus subtilis*

By John H. Pazur

$$G_m + G_n \rightarrow G_{m-x} + G_{n+x}$$

where G = glucosyl residues linked by α-$(1 \rightarrow 4)$ linkages; m and n = 5, 6, 7 . . . ; x = 2, 3, 4, 5 or 6.

Assay Method

Principle. Glucanosyl transferase catalyzes the transfer of a segment of two or more contiguous glucose residues from one maltopentaose molecule to a second molecule to yield maltotriose and maltoheptaose.[1] The enzyme also acts on malto-oligosaccharides of high molecular weight, amylose, and amylopectin, by a similar mechanism. The amount of maltoheptaose synthesized from maltopentaose can be measured by radioactivity measurements when [14C]maltopentaose is employed as the substrate or by colorimetric procedure when nonlabeled maltopentaose is used. In both cases it is necessary to separate the maltoheptaose from the other oligosaccharides by suitable chromatographic procedures. The unit of transferase activity is defined as the amount of enzyme which effects the synthesis of 1 μmole of maltoheptaose per minute from maltopentaose.

Reagents

Maltopentaose, 0.01 M in 0.01 M sodium acetate buffer, pH 5
Maltoheptaose, 0.01 M
[1-14C]Malto-oligosaccharides, 0.01 M concentration of each oligosaccharide
Silver nitrate, saturated solution
Potassium hydroxide, 2 M
Sodium thiosulfate, 0.05 M

---

[1] J. H. Pazur and S. Okada, *J. Biol. Chem.* **243**, 4732 (1968).

Procedures. The assay based on radioactivity measurements was easily performed by use of an "oligosaccharide mapping" technique.[2] In this procedure, 10 μl of [1-^{14}C]malto-oligosaccharides solution was subjected to paper chromatography on 35 cm^2 Whatman No. 1 paper in one direction by three ascents of a solvent system of *n*-butyl alcohol–pyridine–water (9:5:7 by volume). One milliliter of the solution of the enzyme buffered to pH 5.0 was sprayed uniformly on the area of the dried chromatogram containing the oligosaccharides. After incubation at room temperature for 15 minutes, the chromatogram was developed in a second direction in the same solvent system by three ascents of the solvent. The radioactive products in the chromatogram were located by autoradiography. The area of the chromatogram containing the malto-heptaose was located and cut from the chromatogram. The paper containing the [1-^{14}C]maltoheptaose was placed in a scintillation vial with the appropriate amount of the phosphor solution and the counts per minute of the maltoheptaose were determined. From the counts per minute, the enzymatic activity of the sample was calculated.

In the second procedure, a sample of the enzyme solution was mixed with an equal volume of 0.01 M solution of maltopentaose[3] buffered at pH 5, and the resulting mixture was incubated at room temperature for varying periods of time. Samples of 5 μl of the digest and varying amounts of reference maltoheptaose[4] were placed on paper chromatograms. A separation of the maltoheptaose was effected on the paper by 5 ascents of the solvent of *n*-butyl alcohol–pyridine–water (9:5:7 by volume). The maltoheptaose was detected on the paper by dipping the chromatogram in silver nitrate solution (10 ml of saturated solution of silver nitrate and 90 ml of acetone) drying and developing the color by placing the chromatogram in alcoholic potassium hydroxide solution (10 ml of 2 N KOH and 200 ml of methanol) for approximately 2 minutes. The areas of the chromatogram containing oligosaccharide appeared as brown spots. The chromatogram was then washed in 0.05 N sodium thiosulfate for 15 minutes and allowed to air dry.[5] From a comparison of the color intensity of the maltoheptaose spot with the color standards of maltoheptaose, the amount of maltoheptaose synthesized can be calculated. From these values and the protein content[6] of the enzyme,

[2] J. H. Pazur and S. Okada, *J. Biol. Chem.* **241**, 4146 (1966).
[3] J. H. Pazur and T. Budovich, *J. Biol. Chem.* **220**, 25 (1956).
[4] D. French, M. L. Levine, and J. H. Pazur, *J. Amer. Chem. Soc.* **71**, 356 (1949).
[5] F. C. Mayer and J. Larner, *J. Amer. Chem. Soc.* **81**, 188 (1959).
[6] O. H. Lowry, N. J. Rosebrough, A. L. Farr, and R. J. Randall, *J. Biol. Chem.* **193**, 265 (1951).

solution the specific activities of the enzyme preparations can be calculated.

Purification Procedure

Step 1. Filtration through Starch and Celite. A sample of 5 g of a powdered enzyme preparation (HT concentrate derived from *Bacillus subtilis*[7] and obtainable from Miles Chemical Company, Elkhart, Indiana) was stirred with 100 ml of water for 0.5 hour at 4°, and insoluble material was removed by filtration. Solid ammonium sulfate was added to the filtrate to 30% saturation. The solution was passed through a column containing a mixture of 20 g of corn starch granules and 15 g of Celite. Assays of the initial solution and the fractions from the starch column showed that most of the α-1 → 4-glucan hydrolase (α-amylase) in the crude enzyme sample was adsorbed on the starch and that the transferase activity was not adsorbed but appeared in the filtrate.

Step 2. Precipitation with Ammonium Sulfate. The concentration of the ammonium sulfate in the filtrate from step 1 was increased to 70% saturation. A white precipitate which formed on refrigeration of the solution overnight was collected by centrifugation. The precipitate was dissolved in 20 ml of water, and the resulting solution was dialyzed against distilled water.

Step 3. Chromatography on DEAE-Cellulose. A 5-ml sample of the dialyzed solution from step 2 was introduced onto a DEAE-cellulose (0.86 meq/g capacity, medium mesh, Sigma Chemical Co., St. Louis, Missouri) column (20 × 1.8 cm) which had been thoroughly washed with 0.1 M sodium acetate buffer of pH 6. Gradient elution of the protein on the column was effected by use of sodium acetate gradient (0.1–0.5 M) at a flow rate of 0.1 ml per minute. Samples of 5 ml of the eluate were collected. Assays showed that the transferase was located in fractions 17–27.

Step 4. Precipitation with Ammonium Sulfate. Fractions 17–27 from step 3 were combined and dialyzed against distilled water. Sufficient ammonium sulfate was added to give 50% saturation. The precipitate which formed on refrigeration overnight was collected by centrifugation and was redissolved in 5 ml of water.

Step 5. Filtration through Sephadex. The solution of the glucanosyl transferase from step 4 was introduced onto a column (30 × 1.5 cm) of Sephadex G-200. The column was washed with water and fractions of

[7] N. E. Welker and L. L. Campbell, *J. Bacteriol.* **94**, 1124 (1967) have designated the organism from which the enzyme preparation (HT concentrate) was obtained as a strain of *Bacillus amyloliquefaciens*.

5 ml of the eluate were collected. The transferase was located in fractions 3–8. These samples were combined and analyzed for transferase activity. The protein content of this sample was low, approximately 0.05%, but the glucanosyl transferase was readily detectable in this solution by the two assay procedures.

Properties

The purified glucanosyl transferase yielded a single band on paper electrophoresis at several pH values. At pH 7.5, a voltage of 110 DC volts and a current of 12 mA, the enzyme migrated a distance of 5 cm toward the positive pole in a 15-hour period. On density gradient centrifugation the transferase yielded a pattern typical of a homogeneous preparation. By comparison of the sedimentation rate of the transferase to the rates of other proteins,[8] the molecular weight of the transferase was estimated to be 70,000–80,000. The pH optimum for the enzyme is 5. The transferase is stable at low temperatures but is inactivated at elevated temperatures.

The transferase is capable of transferring segments of 2–6 glucose residues linked by α-D$(1 \to 4)$ linkages from starch to low molecular weight oligosaccharides[1] in addition to the transfer reactions described above. Since the transferase acts on starch, it is possible that this enzyme is utilized for the metabolism of starch by the organism. Similar types of transferases have been detected in plant tissues,[9,10] and their role in the metabolism of starch and oligosaccharides needs to be assessed.

[8] R. G. Martin and B. N. Ames, *J. Biol. Chem.* **236**, 1372 (1961).
[9] M. Abdullah and W. J. Whelan, *Arch. Biochem. Biophys.* **112**, 592 (1965).
[10] D. French and M. Abdullah, *Cereal Chem.* **43**, 555 (1966).

[87] Peptidoglycan Transpeptidase in *Bacillus megaterium*

By GARY G. WICKUS and JACK L. STROMINGER

Assay Method

Principle. A particulate enzyme preparation from *Bacillus megaterium* catalyzes the utilization of the uridine nucleotides, UDP-N-acetylmuramyl-L-Ala-D-Glu-*meso*-Dap-D-Ala-D-Ala and UDP-N-acetylglucosamine, for peptidoglycan synthesis. A peptidoglycan transpeptidase(s) can be detected in this system by measuring the incorporation of free

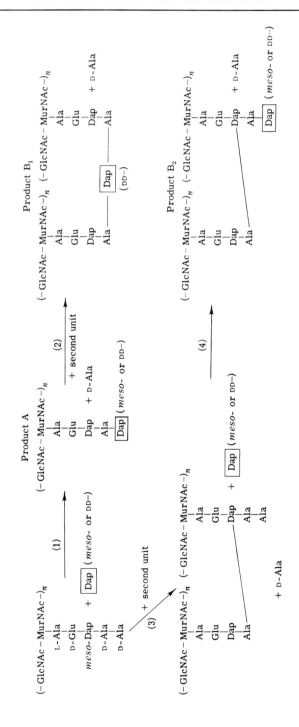

[^{14}C]*meso* and/or [^{14}C]DD-Dap into peptidoglycan polymer [Eqs. (1), (2), and (4)].[1]

Procedure. The preparation of the substrate UDP-MurNAc-L-Ala-D-Glu-*meso*-Dap-D-Ala-D-Ala has been described.[2] A typical incubation mixture contains in a total volume of 40 μl, 5.0 μmoles of Tris buffer, pH 8.5, 0.5 μmole of $MgCl_2$, 70 nmoles of UDP-GlcNAc, 30 nmoles of UDP-MurNAc-L-Ala-D-Glu-*meso*-Dap-D-Ala-D-Ala, 10 nmoles of race-mic [^{14}C]Dap (2.5×10^5 cpm, Calatomic) and 25 μl of a particulate enzyme preparation. The mixture is incubated at 25° for 1 hour and is then stopped by boiling for 1 minute. The amount of peptidoglycan labeled with [^{14}C]Dap is determined after descending paper chroma-tography. The entire reaction mixture is spotted on Whatman 3 MM chromatography paper and chromatographed for 20 hours in isobutyric acid:1 N NH_4OH (5:3). After drying, the peptidoglycan product at the origin labeled with [^{14}C]Dap is cut out and counted in a liquid scintil-lation spectrometer.

Preparation of Enzyme

Growth of Bacterial Cells. Bacillus megaterium QMB1551 was grown in modified Medium G.[3] The organism was cultured in a New Brunswick 25-liter Microferm at 30°C with an airflow of 15 liters/minute at 400 rpm. The cells were harvested during logarithmic growth, washed in 0.05 M Tris buffer, pH 7.8, and frozen until used.

Crude Membrane Preparation. Enzyme was prepared by grinding cells of *B. megaterium* with 120-μ glass beads (3M Company) in a Gifford-Wood Mini-mill. One part of cells (wet weight) was suspended in five parts by volume of 0.05 M Tris buffer, pH 7.8, containing 0.02 M $MgCl_2$. Two parts by volume of acid-washed glass beads were then added along with 0.05 ml of antifoaming agent, tri-*n*-butylcitrate, per 50 ml of solution. The apparatus was placed in an ice bath, and the cells were ground at top speed for 25 minutes at a gap setting of 15.

The ground mixture was fractionated in a Sorvall refrigerated cen-trifuge at 0°. The suspension was centrifuged at 6000 g for 10 minutes to remove whole cells and glass beads. The supernatant solution was then decanted and centrifuged at 40,000 g for 20 minutes. The dense brown membrane pellet was separated from the white flocculent cell wall layer by decanting. The membrane pellet was then resuspended in an equal volume of the above-mentioned Tris buffer containing $MgCl_2$. The protein content was approximately 25–35 mg/ml. Enzymes prepared by

[1] G. G. Wickus and J. L. Strominger, *J. Biol. Chem.* **247**, 5297 (1972).

[2] K. Izaki and J. L. Strominger, *J. Biol. Chem.* **243**, 3180 (1968).

[3] T. Hashimoto, S. H. Black, and P. Gerhardt, *Can. J. Microbiol.* **6**, 203 (1960).

sonication, grinding with alumina, or French press yielded little or no activity.

Properties

Dependence on Peptidoglycan Polymer Formation for Detection. The optimal conditions for the incorporation of free [^{14}C]Dap were the same as those for peptidoglycan formation. The reaction proceeds best at 25°, has a pH optimum of 8.5, and requires 0.015 M magnesium ion for maximum activity.[4] [^{14}C]Dap (0.3 mM) could not be incorporated into preformed unlabeled peptidoglycan. At this concentration of Dap, incorporation into peptidoglycan occurs only during active polymerization.[1]

Inhibition by Penicillins and Cephalosporins and the Reversal of This Inhibition. This reaction is inhibited by penicillins and cephalosporins at reasonably low concentrations (Table I). At a level where growth was inhibited approximately 50% by penicillins, the incorporation of free [^{14}C]Dap into peptidoglycan synthesized *in vitro* was also inhibited by about 50%. However, cephalosporins inhibited growth at concentrations which had only a slight effect on [^{14}C]Dap incorporation.[5]

The inhibition of [^{14}C]Dap incorporation by cloxacillin and penicil-

TABLE I
Effect of Penicillins and Cephalosporins on [^{14}C]Dap
Incorporation into Peptidoglycan[a]

Antibiotic added (μg/ml)	[^{14}C]Dap labeled peptidoglycan (cpm)
None	5570
Cloxacillin, 0.5	700
Dicloxacillin, 0.5	730
Methicillin, 0.5	2940
Penicillin G, 100.0	2540
Penicillin V, 10.0	1680
Ampicillin, 10.0	2070
Cephalothin, 0.1	3700
Cephalexin,[b] 1.0	1940
7-Phenoxyacetamido deacetoxycephalosporanic acid,[b] 1.0	2640

[a] The assay was carried out as described with the additions indicated.

[b] Generously given by Dr. J. Dolfini of the Squibb Institute for Medical Research.

[4] P. E. Reynolds, *Biochim. Biophys. Acta* **237**, 239 (1971).

[5] G. G. Wickus and J. L. Strominger, *J. Biol. Chem.* **247**, 5307 (1972).

lin G could be partially reversed by washing the particles and treating them with excess penicillinase (Calbiochem) capable of destroying all the antibiotic (Table II). In each case, only 55–65% of the original activity could be recovered from the totally inhibited particles.[5]

Specificity. Competition experiments revealed that many D-amino acids would compete with [¹⁴C]Dap for incorporation, but L-amino acids were inactive. The incorporation of [¹⁴C]D-alanine could be directly demonstrated. UDP-MurNAc-L-Ala-D-Glu-L-Lys-D-Ala-D-Ala would substitute for UDP-MurNAc-L-Ala-D-Glu-*meso*-Dap-D-Ala-D-Ala in the reaction.[1]

Distribution. A cell-free system from the gram-negative organism *Escherichia coli,* capable of synthesizing peptidoglycan in amounts similar to the system from *B. megaterium,* could not incorporate [¹⁴C]Dap into peptidoglycan. However, a cell-free system derived from *B. stearothermophilis* could incorporate large amounts of [¹⁴C]Dap into peptidoglycan.[1]

TABLE II

PARTIAL REVERSIBILITY OF CLOXACILLIN OR PENICILLIN G
INHIBITION OF [¹⁴C]DAP INCORPORATION[a]

Antibiotic (μg/ml)	Temperature and minutes of preincubation with antibiotic	Washing and treatment with penicillinase	[¹⁴C]Dap incorporated (cpm)	Percent recovery
None	—	—	7780	—
Cloxacillin 1.5	0°C, 10 min	—	790	—
1.5	0°C, 10 min	+	4900	63
1.5	0°C, 45 min	+	5300	68
1.5	0°C, 90 min	+	5200	67
Cloxacillin 2.0	25°C, 5 min	—	840	—
2.0	25°C, 5 min	+	4300	55
2.0	25°C, 10 min	+	4100	53
Penicillin G				
3000[b]	25°C, 15 min	—	490	—
3000	25°C, 0 min	+	4700	60
3000	25°C, 5 min	+	4050	52
3000	25°C, 15 min	+	4550	59
3000	(25°C, 5 min)[c]	(+)[c]	8050	100

[a] The assay for [¹⁴C]Dap incorporation was carried out as described with the additions and treatments noted.

[b] Massive amounts of penicillin G were used due to a particulate penicillinase present.

[c] Penicillin was preincubated with penicillinase before incubation with particulate enzyme.

Comments

Several additional transpeptidation reactions can be detected by more difficult means in cell-free preparations from *B. megaterium* [Eqs. (2), (3), and (4)]. At least one of these, Eq. (3) appears to be catalyzed by a different enzyme than the reaction described above.[5] The question of the existence of multiple transpeptidase enzymes can probably be answered only after the development of independent assays and subsequent solubilization and purification of the enzyme(s) responsible for these activities.

[88] D-Alanine Carboxypeptidase from *Bacillus subtilis*[1]

By JAY UMBREIT and JACK L. STROMINGER

UDP-MurNAc-L-Ala-D-Glu-*meso*-Dap-D-Ala-D-Ala →
UDP-MurNAc-L-Ala-D-Glu-*meso*-Dap-D-Ala + D-Ala

The D-alanine carboxypeptidase catalyzes the above reaction.[2] The substrate for the reaction is the precursor for cell wall biosynthesis. The enzyme is found exclusively in the membrane or particulate fraction of the cell. It is irreversibly inhibited by penicillin G and other β-lactam antibiotics.[2,3] Although the enzyme is the major penicillin-binding component of the cell, inhibition of its activity does not kill the cell.[4]

Assay Method

Reagents

Sodium cacodylate, 1 M, pH 6.0
UDP-MurNAc-pentapeptide 83 μM (labeled in the terminal two D-alanines with ^{14}C)[5]
ZnSO$_4$, 0.5 M

Procedure. A typical reaction mixture consists of 5 μl of 1 M sodium cacodylate, pH 6.0, 18 μl of water, 2 μl of [^{14}C]UDP-MurNAc-pentapeptide (83 μM), 2 μl of ZnSO$_4$ (0.5 M), and 1 μl of the enzyme prepara-

[1] Supported by research grants from the U.S. Public Health Service (AI-09152) and the National Science Foundation (GB-29747X).

[2] P. J. Lawrence and J. L. Strominger, *J. Biol. Chem.* **245**, 3660 (1970).

[3] J. N. Umbreit, Ph.D. Thesis, Harvard University (1972).

[4] P. M. Blumberg and J. L. Strominger, *Proc. Nat. Acad. Sci. U.S.* **68**, 2814 (1971).

[5] E. Ito, S. G. Nathenson, D. N. Dietzler, J. S. Anderson, and J. L. Strominger, Vol. 8 [58].

tion. After incubation at 37° for a given time (usually 30 minutes), the reaction is terminated by boiling for 2 minutes. The reaction mixture is spotted on 20×20 cm sheets of Whatman No. 3 MM paper and subjected to ascending chromatography for 3 hours in isobutyric acid:1 N NH$_4$OH, 5:3. If a rapid assay is desired, 5 μl of 0.2 M alanine is added to the reaction mixture just before spotting. After chromatography, the dried papers are sprayed with ninhydrin, and the alanine spots are cut out and counted. Radioautography without added carrier is used when a more accurate assay is desired.

Purification of the Enzyme

The enzyme is prepared by grinding the cells in a colloidal mill, collecting the membrane fractions, solubilizing the activity in Triton X-100, and purifying the solubilized extract by standard protein chemistry methods.

Step 1. Preparation of Membranes. B. subtilis strain Porton is employed.[2] Membranes are prepared after disintegration of cells in the "Micro-mill" (Gifford-Wood, Hudson, New York). One liter of 5 μ glass beads, previously washed extensively in *aqua regia* and dried, are added to 1 liter of 100 mM Tris·HCl, pH 7.5, containing about 250 g (wet weight) cells. Twelve drops of t-butyl citrate are added as antifoaming agent, and the cells are ground at 4° in ice water at top speed with a vernier setting of 5–7 for 15 minutes. The mixture was then decanted into centrifuge tubes and centrifuged for 30 minutes at 15,000 rpm (Sorvall centrifuge, SS-34 rotor). The supernatant solution is decanted and saved; the pellet is reground twice more in the "Micromill." The remaining pellet consists predominantly of cell walls. The pooled supernatant solutions are centrifuged for 1 hour at 45,000 rpm in an A-211 rotor in a Model B-60 ultracentrifuge (International Equipment Co.) at about 120,000 g. The supernatant solutions are carefully decanted, and the loose pellet is then homogenized in a Potter homogenizer with approximately 200 ml of 100 mM Tris·HCl, pH 7.5, 10^{-3} M β-mercaptoethanol, to a final protein concentration of about 15 mg/ml. The membrane fraction obtained can be stored in the freezer ($-20°$) indefinitely.

Step 2. Solubilization with Triton X-100. The material from step 1 is mixed with enough Triton X-100 to bring the final concentration to 1% in detergent at a protein concentration of 10 mg/ml. This is mixed at 4° for 30 minutes and then centrifuged at 120,000 g (42,000 rpm, A-211 rotor, International Equipment Co. B-60 ultracentrifuge) for 1 hour. The supernatant solution (about 300 ml) is decanted and used for the next step.

Step 3. Gel Filtration. Half of the supernatant solution is charged directly to a 1.5-liter 6% agarose column (Biogel 1.5 m) equilibrated in 1% Triton X-100, 100 mM Tris·HCl, pH 8.6. The column used is a Pharmacia K50/100 column (Pharmacia Fine Chemicals, Inc., Piscataway, New Jersey) fitted with flow adaptors and operated by descending flow from a Mariette flask at 11 cm containing the above buffer. About 120 fractions of 10 ml are collected. Usually only A_{280} is monitored after experience demonstrated the coincidence of this absorption with the protein as determined by alkaline ninhydrin.[6] Activity coincides with the main protein peak. The pooled fractions (about 600 ml) are then dialyzed against four changes of 4 liters of 10 mM Tris·HCl, pH 8.6, 1% Triton X-100, 10^{-3} M 2-mercaptoethanol for about 60 hours until the conductivity of the dialyzate is about 0.5×10^3 μMHOS. The other half of the supernatant solution from step 2 is processed similarly.

Step 4. ECTEOLA-Cellulose Column Chromatography. The dialyzate (1200 ml) is then charged to a 300-ml ECTEOLA column previously equilibrated in 10 mM Tris·HCl, pH 8.6, 1% Triton X-100, 10^{-3} M mercaptoethanol (0.2×10^3 μMHOS). The column is then washed with 1 liter of the same buffer and eluted with a linear gradient of 1 liter of 10 mM Tris·HCl, pH 8.6, 10^{-3} M mercaptoethanol to 1 liter of 10 mM Tris·HCl, pH 8.6, 10^{-3} M mercaptoethanol, 0.2 M NaCl, both containing 1% Triton X-100. Fractions of 10 ml are collected. Activity is eluted at about 0.11 M NaCl. The relevant fractions are pooled and dialyzed against 10 mM Tris·HCl, pH 8.6, 10^{-3} M mercaptoethanol, 1% Triton X-100 until the conductivity is again about 0.5×10^3 μMHOS.

Step 5. DEAE-Cellulose Column Chromatography. The pooled material from step 4 (100 ml) is charged to a 100-ml DEAE column (DE52, Whatman) equilibrated in the same buffer as that used for dialysis. After washing with 1 liter of the same buffer, the elution is carried out with a linear gradient from 1 liter of 10 mM Tris·HCl, pH 8.6, 1% Triton X-100, 10^{-3} M mercaptoethanol to 1 liter 100 mM Tris·HCl, pH 7.0, 10^{-3} mercaptoethanol, 1% Triton X-100. Fractions of 5 ml are collected. Activity is eluted at about pH 7.5.

Step 6. Calcium Phosphate Gel. The pooled fractions are then tested by use of a 1-ml aliquot for the minimum amount of calcium phosphate gel required to absorb 90% of the total activity. Typically 124 ml of pooled material from step 5 is mixed with 2.5 ml of calcium phosphate gel (0.8 g/ml solids). After incubation at 4° for 30 minutes, the calcium phosphate is recovered by centrifugation in a Sorvall centrifuge SS-34 rotor at 10,000 rpm for 30 minutes, and washed with 100 mM Tris·HCl,

[6] C. H. Hirs, Vol. 11 [35].

PURIFICATION OF D-ALANINE CARBOXYPEPTIDASE

Step	Volume (ml)	Total activity (cpm $\times 10^8$)	Yield (%)	Total protein (mg)	Specific activity (cpm/mg/min $\times 10^6$)	Purification (-fold)
1. Crude membranes	100	60	100	3000	2	—
2. Triton X-100 supernatant solution	325	26	43	1290	2	—
3. Agarose 1.5 m column	1200	21	35	348	6.3	3.2
4. ECTEOLA column	95	4.6	7.6	10.6	43	22
5. DEAE column	124	3.4	5.6	2.5	137	69
6. CaPO$_4$ gel	10	1.6	2.6	0.75	213	107

pH 8.6, $10^{-3}M$ mercaptoethanol, 1% Triton X-100, and eluted successively with 10-ml portions of 5, 7, 10, and 20% saturated ammonium sulfate (at 4°) in 100 mM Tris·HCl, pH 8.6, 1% Triton X-100. (Note that in order to form a clear solution at the higher ammonium sulfate concentrations, the solution must be cooled to about 4°.) The enzyme was usually eluted in the fraction containing 7% ammonium sulfate and could be stored in this buffer at 4% indefinitely.

A summary of the purification is given in the table.

Properties of the Enzyme

The molecular weight of the single polypeptide chain is 50,000 daltons as determined by SDS gel electrophoresis[7] and gel filtration in guanidine-HCl.[8] In solution in the presence of Triton, the enzyme is near the void volume of Sephadex G-200 and probably consists of an aggregate in the Triton detergent micelle.[3]

The enzyme requires Zn^{2+} with an optimum at 15 mM. The pH optimum of the purified enzyme in the presence of $ZnSO_4$ is pH 5.5, but the membrane-bound enzyme has a pH optimum slightly lower. The K_m for UDP-MurNAc-pentapeptide is $2.6 \times 10^{-5} M$, and the rate constant is 1.5×10^{-2} sec^{-1}.

The enzyme is irreversibly inhibited by penicillin G and other β-lactam antibiotics.[2] The enzyme reacts with penicillin G stoichiometrically, 1 mole of penicillin being required for each 50,000 dalton polypeptide chain for complete inactivation. Upon reacting with penicillin G, the enzyme loses one of its four titratable sulfhydryl groups. The binding constant for the reversible binding of penicillin G to the active site before acylation occurs is $9.6 \times 10^{-5} M$, and the rate constant for the irreversible acylation of the enzyme is 1.2×10^{-1} at 4°. Similar constants have been measured for other β-lactam antibiotics.[3]

The presence of Triton X-100, or a crude lipid extract, is required to retain the activity of the enzyme after purification. In the presence of Triton X-100, the enzyme can be stored at 4° indefinitely.

[7] K. Weber and M. J. Osborn, *J. Biol. Chem.* **244**, 4406 (1969).
[8] W. W. Fish, K. G. Mann, and C. Tanford, *J. Biol. Chem.* **244**, 4989 (1969).

Section VI

Degradation of Complex Carbohydrates

[89] α-Mannosidase, β-Glucosidase, and β-Galactosidase from Sweet Almond Emulsin

By Y. C. LEE

Commercially available sweet almond emulsin is a convenient source of glycosidases. Although it is mainly regarded as β-D-glucopyranosidase, activities of β-D-galactopyranosidase,[1] α-D-galactopyranosidase,[1] α-D-mannopyranosidase,[1] β-D-N-acetylglucosaminidase,[1] β-D-xylopyranosidase,[2] and α-L-arabinopyranosidase[3] have been detected. Some of these glycosidases have been substantially purified to be useful as a practical tool for structural studies of complex carbohydrates.[4,5]

α-D-Mannosidase

Assay

A sample of enzyme solution (5–25 μl) is added to 0.6 ml of 0.1 M sodium acetate buffer (pH 4.8) containing 10 mM p-nitrophenyl α-D-mannopyranoside. The reaction mixture is incubated for 10 minutes at 37°. Upon addition of 2 ml of 0.2 M Na_2CO_3, absorbance at 400 nm is measured in a 13 × 100 mm tube with a Spectronic 20 (Bausch and Lomb) colorimeter. Molar absorbance of p-nitrophenol under these conditions is 18,000. One unit of enzyme is defined as that quantity of enzyme capable of hydrolyzing 1 μmole of p-nitrophenyl α-D-mannopyranoside under the described conditions.

Purification

The following one-step chromatography was developed for isolating α-D-mannosidase free of β-N-acetylglucosaminidase. Commercial preparations of sweet almond emulsin (marketed under the name "β-glucosidase") are somewhat variable in the levels of α-D-mannosidase activity as well as other glycosidases. Slight changes in elution conditions from those described below (for example, pH of the gradient buffers) may be necessary for satisfactory results.

[1] H. Blaumann and W. Pigman, *in* "The Carbohydrates" (W. Pigman, ed.), p. 562. Academic Press, New York, 1957.
[2] D. J. Manners and J. P. Mitchell, *Biochem. J.* **103**, 43p (1967).
[3] D. J. Manners and D. C. Taylor, *Carbohyd. Res.* **7**, 497 (1968).
[4] J. Schwartz, J. Sloan, and Y. C. Lee, *Arch. Biochem. Biophys.* **137**, 122 (1970).
[5] O. P. Malhotra and P. M. Dey, *Biochem. J.* **103**, 508 (1967).

Almond emulsin[6] (2.5 g) is stirred with 200 ml of 0.01 M sodium acetate buffer (pH 4.8) for 30 minutes at room temperature, and the insoluble residue is removed by centrifugation at 3000 rpm for 10 minutes. The supernatant solution is directly applied to a column of carboxymethyl cellulose CM-52 (2.6 × 30 cm) equilibrated in 0.01 M sodium acetate (pH 4.8) and is eluted first with a linear gradient [1 liter each of (a) 0.01 M sodium acetate buffer (pH 4.8) and (b) 0.01 M sodium acetate buffer (pH 4.8) containing 0.067 M NaCl], then with another gradient [1 liter each of (c) 0.01 M sodium acetate buffer (pH 4.8) containing 0.1 M NaCl and (d) 0.2 M sodium acetate buffer (pH 4.8) containing 0.3 M NaCl]. The elution profile is shown in Fig. 1.

The α-D-mannosidase fractions as marked in Fig. 1 are pooled, dialyzed against 0.01 M sodium acetate buffer (pH 4.8), and concentrated by adsorption on a 1 × 2-cm CM-52 column (or equivalent) and desorption from it with 0.5 M sodium acetate buffer, pH 4.8. The yield of α-D-mannosidase in the combined fraction from the CM-52 column was usually 30–40%, although 60–70% of the total enzyme activity could be accounted for in the effluent. The concentration step usually gave 85–95% recovery. The specific activity of the concentrated α-D-mannosidase was 2–3 units per milligram of protein.

Optimal pH for the α-D-mannopyranosidase activity is between 4 and 5. The K_m values for methyl, phenyl, and p-nitrophenyl α-D-mannopyranosides are 88.7, 23.8, and 4.2 mM, respectively.

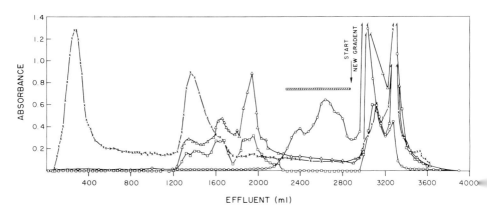

FIG. 1. Chromatography of almond emulsin on CM-cellulose. Elution conditions are described in the text. $A_{289\,nm}$, x—x; α-D-mannoside, ◯—◯; β-D-glucosidase, □—□; β-D-galactosidase, △—△.

[6] Almond emulsin is marketed under the name "β-glucosidase" and is available from Worthington Biochemical Corp., Sigma Chemical Corp., and other suppliers of biochemicals.

The enzyme liberates mannosyl residues from ovalbumin glycopeptide.[7] However, the rate of hydrolysis of mannosyl bond in the glycopeptide is only 0.2% of that of p-nitrophenyl α-D-mannopyranoside. It also releases mannose from glycopeptides prepared from pineapple stem bromelain[8] and α-amylase from *Aspergillus oryzae*.[9] Although the enzyme hydrolyzes mannotetraose, -triose, and -biose[10] completely, it digests yeast mannan, from which the oligosaccharides are derived, only to the extent of 0.1–1.0%. The enzyme is free of endo-mannosidase activity.

The α-D-mannosidase thus isolated is found to be free of β-D-N-acetyl-D-glucosaminidase, β-D-xylosidase, and α-L-fucosidase when p-nitrophenyl glycosides were used as substrates. However, β-D-glucosidase and β-D-galactosidase activities are detectable.

The mannosidase is competitively inhibited by D-glucal and D-mannono-1,5-lactone with K_i of 10 mM and 0.75 mM. D-Glucono-1,5-lactone inhibits the enzyme, but less efficiently than the mannonolactone. Inhibition by D-mannose is not pronounced. Thus when the incubation mixture contains 0.033, 0.133, 0.167, 0.500, and 1.000 M D-mannose, the relative activities toward p-nitrophenyl α-D-mannopyranoside are 72, 53, 42, 22, and 17% of the control in which D-mannose is absent.

$HgCl_2$ exhibits noncompetitive inhibition of the enzyme, although high concentrations are required. Preincubation of the enzyme with $HgCl_2$ for 1 hour greatly enhances the inhibiting effect.

Separation of Glycosidases by Electrofocusing

Assays of β-D-glucopyranosidase and β-D-galactopyranosidase activities are carried out in the same way as for the α-D-mannopyranosidase, except that p-nitrophenyl β-D-glucopyranoside and p-nitrophenyl β-D-galactopyranoside, respectively, are used.

Almond emulsin (0.5 g) is extracted with water (25 ml) and centrifuged to remove insoluble residues. A part of the supernatant (about 60 mg of protein containing about 100–150 units of β-D-glucosidase) can be fractionated by electrofocusing technique in 110 ml of 1% ampholine, pH 3–6. Usually two major β-D-glucosidase peaks (Glu-I at pH 4–6, Glu-II at pH > 6), each containing β-D-galactosidase activity, and a minor glucosidase peak partially overlapping with an α-D-mannosidase peak can be seen. The α-D-mannosidase peak from two such runs can be combined (ca. 45 ml) and reelectrofocused without additional ampholine to remove β-D-glucopyranosidase and β-D-galactopyranosidase

[7] Y. C. Lee and R. Montgomery, *Biochem. J.* **91**, 9c (1964).
[8] J. Scocca and Y. C. Lee, *J. Biol. Chem.* **244**, 4852 (1969).
[9] J. McKelvy and Y. C. Lee, *Arch. Biochem. Biophys.* **132**, 99 (1969).
[10] Y. C. Lee and C. E. Ballou, *Biochemistry* **4**, 257 (1965).

TABLE I

GLYCOSIDASE FRACTIONS FROM ELECTROFOCUSING

Fractions	Activity	K_m (M)	V_{max} (μmoles/min)	Optimal pH
Glu-I	β-D-Glucosidase	0.067	0.072	5.2–6.0
	β-D-Galactosidase	0.182	0.010	4.2–4.8
Glu-II	β-D-Glucosidase	0.080	0.482	5.2–6.0
	β-D-Galactosidase	0.040	0.006	5.7–6.2

TABLE II

INHIBITION OF β-D-GLUCO- AND β-D-GALACTOSIDASES BY HgCl$_2$

	Remaining activity[a]			
	Glu-I		Glu-II	
Inhibitor (HgCl$_2$) (mM)	β-Glucosidase (%)	β-Galactosidase (%)	β-Glucosidase (%)	β-Galactosidase (%)
0	100	100	100	100
0.1	102	98	96	99
1.0	54	78	38	82
10	11	27	Negligible	21

[a] Using p-nitrophenyl glycosides as substrates.

activities. This provides another route for purification of α-D-manno-sidase.

Enzymological parameters of Glu-I and Glu-II are shown in Table I. Inhibition of β-D-glucosidase and β-D-galactosidase activities by HgCl$_2$ are shown in Table II.

[90] α-Mannosidase, β-N-Acetylhexosaminidase, and β-Galactosidase from Jack Bean Meal[1]

By YU-TEH LI and SU-CHEN LI

The utilization of glycosidases to liberate individual carbohydrate units is a specific and convenient method for studying the sequence and

[1] This chapter was written while the authors were under the auspices of grants from the United States Public Health Service NS09626 and the National Science Foundation GB 18019. Y.-T. Li is a recipient of a Research Career Development Award, 1 K04 HD 50280 from the National Institute of Child Health and Human Development, United States Public Health Service.

anomeric configuration of the carbohydrate units of complex saccharide chains. The jack bean contains several exo-glycosidases, including α-mannosidase, β-N-acetylhexosaminidase, and β-galactosidase. These glycosidases have been found to be useful in the study of the sequence and the anomeric configuration of sugar chains of various complex carbohydrates.[2-7] The isolation and properties of α-mannosidase, β-N-acetylhexosaminidase, and β-galactosidase from jack bean meal are described herein. Unless otherwise indicated, all carbohydrates and glycosides referred to below are of the D-configuration and pyranosides, respectively.

Assay Methods

p-Nitrophenyl Glycosides as Substrates

Principle. The enzymes are allowed to react with the appropriate p-nitrophenyl glycoside, and the yellow color of the p-nitrophenol released by enzymatic hydrolysis is estimated at pH 9.8 in a spectrophotometer at 400 nm.

Reagents

Sodium citrate buffer, 0.05 M, pH 3.5, 4.5, and 5.0
p-Nitrophenyl α-mannoside, 2 mM in sodium citrate buffer, pH 4.5
p-Nitrophenyl β-N-acetylglucosaminide, 2 mM in sodium citrate buffer, pH 5.0
p-Nitrophenyl β-galactosides, 2 mM in sodium citrate buffer, pH 3.5
Sodium borate buffer, 0.2 M, pH 9.8

Procedure. One milliliter of p-nitrophenyl glycoside is incubated with 5–100 μl of the enzyme solution for 5–30 minutes at 25°, depending on the activity of the preparation. The reaction is stopped by adding 3 ml of sodium borate buffer, and the intensity of the color produced is estimated at 400 nm, using a Beckman DU-2 spectrophotometer. Under these conditions the molar extinction coefficient of p-nitrophenol is 1.77×10^4 M^{-1} cm^{-1}.

Unit. One unit of enzyme is defined as the amount of enzyme which hydrolyzed 1 μmole of p-nitrophenyl glycoside per minute at 25°.

[2] Y.-T. Li, *J. Biol. Chem.* **241**, 1010 (1966).
[3] Y.-T. Li, *J. Biol. Chem.* **242**, 5474 (1967).
[4] Y.-T. Li, S.-C. Li, and M. R. Shetlar, *J. Biol. Chem.* **243**, 656 (1968).
[5] S.-C. Li and Y.-T. Li, *J. Biol. Chem.* **245**, 5153 (1970).
[6] S. Hakomori, B. Siddiqui, Y.-T. Li, S.-C. Li, and C. G. Hellerqvist, *J. Biol. Chem.* **246**, 2271 (1971).
[7] Y.-T. Li and S.-C. Li, *J. Biol. Chem.* **246**, 3769 (1971).

Oligosaccharides, Glycopeptides, or Glycolipids as Substrates

Principle. The appropriate glycosidase is used to hydrolyze the monosaccharide units in the oligosaccharides, glycopeptides, or glyco-lipids, and the liberated neutral sugar, such as mannose or galactose, is quantitatively determined by an automatic sugar analyzer.[8] The free *N*-acetylhexosamines (*N*-acetylglucosamine or *N*-acetylgalactosamine) liberated in the reaction mixture are determined by Morgan-Elson's re-action.[9] It must be emphasized that the rate of the liberation of a mono-saccharide unit from different glycoproteins by a specific glycosidase varies depending upon the polypeptide moiety and the length and se-quential arrangement of the oligosaccharide chain. Therefore, it is es-sential to check the extent of hydrolysis several times during the course of incubation. When no hydrolysis is obtained, conclusions should be carefully qualified, for the same enzyme obtained from different sources may have different specificities toward different glycopeptides.

Procedure. (a) When α-mannosidase or β-galactosidase is used to hydrolyze complex carbohydrates containing terminal α-mannosyl or β-galactosyl residues, 0.05–0.2 μmole of the complex carbohydrate is dis-solved in 100–200 μl of 0.05 M sodium citrate buffer and incubated with 0.5–5 units (10–50 μl) of the enzyme at 37° for 2–40 hours. At the end of the reaction, an aliquot of the enzymatic digest is directly applied to an automatic sugar analyzer[8] for the quantitative determination of man-nose or galactose.

(b) When β-*N*-acetylhexosaminidase is used to hydrolyze the ter-minal β-*N*-acetylglucosaminide or β-*N*-acetylgalactosaminide in com-plex carbohydrates, 0.05–0.2 μmole of the substrate is dissolved in 100–200 μl of sodium citrate buffer and incubated with 0.5–5 units (10–50 μl) of β-*N*-acetylhexosaminidase at 37° for 2–20 hours. After the reaction 100 μl of the reaction mixture is transferred to a 9 × 7.5 mm test tube, mixed with 100 μl of 0.8 M potassium borate buffer, pH 9.2, and heated at 100° for 5 minutes. The mixture is then cooled, 1 ml of *p*-dimethyl-aminobenzaldehyde reagent[9] is added, and the mixture is incubated at 37° for 15 minutes. Absorbance of the color produced is measured at 585 nm with a Beckman DU-2 spectrophotometer.

Enzyme Purification Procedure

Unless otherwise indicated, all the operations are carried out at 4°. Samples are routinely centrifuged in a Sorvall RC 2-B refrigerated cen-

---

[8] Y. C. Lee, J. F. McKelvy, and D. Lang, *Anal. Biochem.* **27**, 567 (1969). See also this volume [6].

[9] J. L. Reissig, J. L. Strominger, and L. F. Leloir, *J. Biol. Chem.* **217**, 959 (1955).

trifuge for 15 minutes at the following rates: 5000 rpm for the GS-3 large capacity rotor (6 × 500 ml); 7000 rpm for the GSA high speed rotor (6 × 250 ml); 12,000 rpm for the SS-34 superspeed rotor (8 × 50 ml). A 1-kg portion of jack bean meal (Nutritional Biochemical Corporation, Cleveland, Ohio) is suspended in 6 liters of distilled water and stirred for 1 hour at room temperature. The suspension is strained through cheesecloth. The turbid filtrate is adjusted to pH 5.5 at room temperature with 1.5 M sodium citrate buffer, pH 2.7, and centrifuged to obtain about 5.3 liters of clear yellow extract. Solid ammonium sulfate is added to the extract to obtain 30% saturation. After the preparation has stood for 2 hours, the precipitate is removed by centrifugation, and more ammonium sulfate is added to the supernatant to obtain 60% saturation. After standing overnight, the mixture is centrifuged to collect the precipitate, which is then redissolved in 250 ml of 0.1 M sodium phosphate buffer, pH 7.0. This preparation is designated as the crude enzyme fraction.

Alcohol Fractionation. To the crude enzyme fraction, 95% ethanol is added dropwise at room temperature with constant stirring to a concentration of 25%. The mixture is cooled to −10°. The precipitate is collected by centrifugation at −10°, dissolved in 100 ml of 0.1 M sodium phosphate buffer, pH 7.0, and designated as the α-mannosidase-rich fraction. The supernatant is warmed to room temperature, adjusted to pH 4.9 by adding 1.5 M sodium citrate buffer, pH 2.7, again cooled to −10°, and kept at this temperature overnight to precipitate β-N-acetylhexosaminidase. The precipitate is collected as before, dissolved in 60 ml of 0.1 M sodium phosphate buffer, pH 7.0, and designated as the β-N-acetylhexosaminidase-rich fraction. Both the α-mannosidase-rich fraction and the β-N-acetylhexosaminidase-rich fraction are, at this step, contaminated with other glycosidases.

Isolation of α-Mannosidase from the α-Mannosidase-Rich Fraction

Gel Filtration. An upward elution technique is used. The α-mannosidase-rich fraction (30 ml, containing 2.5 g of protein) is applied to a Sephadex G-200 column (5 × 90 cm, Pharmacia K-series with flow adaptor) which has been equilibrated with 0.1 M sodium phosphate buffer, pH 7.0. The column is eluted with the same buffer at a flow rate of 30 ml per hour. The elution pattern is shown in Fig. 1. The fractions in the first protein peak containing α-mannosidase and β-N-acetylhexosaminidase are pooled and dialyzed overnight against saturated ammonium sulfate. The precipitated protein is collected by centrifugation, redissolved in 10 ml of 0.05 M sodium phosphate buffer, pH 7.0, and designated

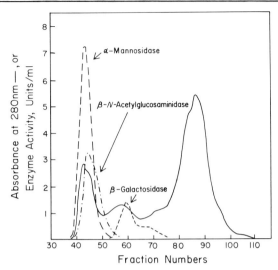

Fɪɢ. 1. Sephadex G-200 filtration of α-mannosidase-rich fraction. An upward flow elution technique is used. The enzyme solution (30 ml) containing 2.5 g of protein is applied to a Sephadex G-200 column (5 × 80 cm) which has been equilibrated with 0.1 M sodium phosphate buffer, pH 7.0. The column is eluted with the same buffer at a flow rate of 30 ml per hour controlled by a peristaltic pump; 25 ml per fraction is collected. ———, absorption at 280 nm; — — —, α-mannosidase activity; ·—·—, β-N-acetylhexosaminidase activity; ----, β-galactosidase activity.

as the partially purified α-mannosidase fraction. The fractions containing β-galactosidase activity are also pooled and precipitated in the same manner by reverse dialysis against saturated ammonium sulfate. The precipitate is dissolved in 10 ml of 0.05 M sodium phosphate buffer, pH 7.0. This preparation is designated as the β-galactosidase-rich fraction obtained from the α-mannosidase-rich fraction and is saved for the further purification. It should be pointed out that some batches of Sephadex G-200 gel absorb α-mannosidase; however, this capacity differs considerably from batch to batch.

DEAE-Sephadex Chromatography. The partially purified α-mannosidase fraction from five Sephadex G-100 filtrations are pooled, dialyzed exhaustively against 0.05 M sodium phosphate buffer, pH 7.0, and applied to a DEAE-Sephadex A-50 column (5 × 30 cm) which has previously been equilibrated with 0.05 M sodium phosphate buffer, pH 7.0. β-Galactosidase is washed off the column with the starting buffer. α-Mannosidase is then eluted with 0.1 M sodium phosphate buffer, pH 7.0, and β-N-acetylhexosaminidase is finally eluted with 0.05 M sodium citrate buffer, pH 6.0, containing 0.05 M NaCl. The elution profile is shown in Fig. 2. The fractions containing α-mannosidase, β-N-acetyl-

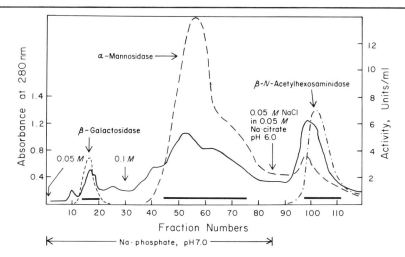

FIG. 2. DEAE-Sephadex A-50 chromatography of the partially purified α-mannosidase fraction. Protein solution (60 ml) containing 2.0 g of protein is applied to a DEAE-Sephadex A-50 column (5 × 30 cm) which has been equilibrated with 0.05 *M* sodium phosphate buffer, pH 7.0; 25 ml per fraction is collected. ———, absorption at 280 nm; - - - -, β-galactosidase activity; — — —, α-mannosidase activity; ·—·—, β-*N*-acetylhexosaminidase activity.

hexosaminidase, and β-galactosidase are separately pooled and precipitated by reverse dialysis against saturated ammonium sulfate. The precipitate is dissolved in 0.05 *M* phosphate buffer, pH 7.0. The α-mannosidase fraction is rechromatographed on the DEAE-Sephadex column (2 × 20 cm) as described above. β-*N*-Acetylhexosaminidase can be further purified by CM-Sephadex chromatography and the β-galactosidase by electrofocusing (see sections: Isolation of β-*N*-Acetylhexosaminidase-Rich Fraction and Isolation of β-Galactosidase).

Pyridine Treatment. In some cases, the α-mannosidase obtained from DEAE-Sephadex chromatography is still contaminated with a trace amount of β-*N*-acetylhexosaminidase activity. This activity can be selectively inactivated by pyridine. Five milliliters of α-mannosidase containing 500 units of enzyme are mixed with 0.5 ml of pyridine and incubated at room temperature for 1 hour. The precipitated protein is removed by centrifugation and the clear supernatant which contains α-mannosidase is dialyzed thoroughly against 0.05 *M* sodium phosphate buffer, pH 7.0, to remove pyridine. The protein is then precipitated by reverse dialysis against saturated ammonium sulfate and redissolved in 1 ml of 0.05 *M* sodium phosphate buffer, pH 7.0, to obtain an α-mannosidase free from β-*N*-acetylhexosaminidase and other contaminating enzymes.

Isolation of β-N-Acetylhexosaminidase from β-N-Acetylhexosaminidase-Rich Fraction

Sephadex G-200 Filtration and DEAE-Sephadex Chromatography. The β-N-acetylhexosaminidase-rich fraction obtained at ethanol fractionation is further purified by Sephadex G-200 filtration and DEAE-Sephadex A-50 chromatography according to the procedure described for the isolation of α-mannosidase from the mannosidase-rich fraction obtained at ethanol fractionation step. The chromatographic patterns are very similar to that of Figs. 1 and 2. The first peak of the Sephadex G-200 filtration containing mainly β-N-acetylhexosaminidase and some α-mannosidase activities (see Fig. 1) are pooled and precipitated by reverse dialysis against saturated ammonium sulfate. The fraction obtained is designated as partially purified β-N-acetylhexosaminidase fraction. The fractions containing β-galactosidase activity (see Fig. 1) are also pooled and precipitated by the same way and is saved for further purification later. When the partially purified β-N-acetylhexosaminidase fraction is subjected to DEAE-Sephadex chromatography, α-mannosidase is eluted by 0.1 M sodium phosphate buffer, pH 7.0 and β-N-acetyl-

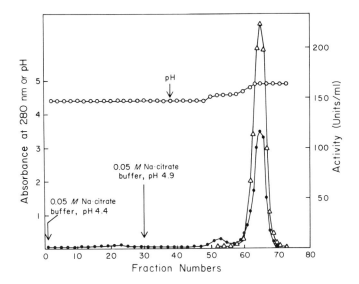

Fɪɢ. 3. CM-Sephadex C-50 chromatography of β-N-acetylhexosaminidase obtained from DEAE-Sephadex chromatography step (see text). Protein solution containing 0.33 g of protein is applied to a CM-Sephadex C-50 column (2 × 30 cm) which has been equilibrated with 0.05 M sodium citrate buffer, pH 4.4; 10 ml per fraction are collected. △—△, Enzyme activity; ○—○, pH; ●—●, E, 280 nm. Reproduced from S.-C. Li and Y.-T. Li, *J. Biol. Chem.* **245**, 5153 (1970).

hexosaminidase is eluted by 0.05 M sodium citrate buffer, pH 6.0 containing 0.05 M NaCl (see Fig. 2). Fractions containing α-mannosidase and β-N-acetylhexosaminidase activities are separately pooled and precipitated by reverse dialysis against saturated ammonium sulfate. The α-mannosidase obtained can be further purified according to the procedure decribed for the purification of α-mannosidase.

CM-Sephadex Chromatography. The β-N-acetylhexosaminidase fraction obtained from DEAE-Sephadex chromatography is dialyzed against 0.05 M sodium citrate buffer, pH 4.4, and applied to a CM-Sephadex C-50 column (2.0 $\times$ 30 cm) previously equilibrated with 0.05 M sodium citrate buffer, pH 4.4. The column is eluted with the same buffer with stepwise increase in pH as shown in Fig. 3. β-N-Acetylhexosaminidase is eluted at its isoelectric point, pH 4.9. The fractions containing β-N-acetylhexosaminidase activity are pooled and precipitated by dialysis against ammonium sulfate. The precipitate is dissolved in 0.05 M sodium phosphate buffer, pH 7.0. As shown in Fig. 4 the β-N-acetylhexosaminidase obtained by this procedure is electrophoretically homogeneous when examined by disc electrophoresis and can be crystallized.[5]

Isolation of β-Galactosidase

The β-galactosidase-rich fraction obtained by Sephadex G-200 filtration of α-mannosidase-rich fraction or of β-N-acetylhexosaminidase-rich fraction are pooled and further purified by DEAE-Sephadex chromatography as described above. With 0.05 M sodium phosphate buffer, pH 7.0, β-galactosidase is not absorbed to the DEAE-Sephadex and is eluted at the breakthrough peak (Fig. 2). The fractions containing β-galactosidase activity are pooled and precipitated by dialysis against saturated ammonium sulfate. The precipitate is dissolved in 0.05 M sodium phosphate buffer, pH 7.0, and is subsequently fractionated by isoelectrofocusing, using ampholine with the pI distributed between pH 7 and 10. β-Galactosidase is focused between pH 7.8 and 8.0. The β-galactosidase fractions are pooled and thoroughly dialyzed against 0.05 M sodium phosphate buffer, pH 7.0, to remove sucrose and ampholine and the enzyme is then precipitated by dialysis against saturated ammonium sulfate.

Properties

α-Mannosidase. The α-mannosidase prepared in this manner can be crystallized in ammonium sulfate solution. After pyridine treatment, the purified α-mannosidase is adjusted to a protein concentration of 3–4%. Solid ammonium sulfate (twice recrystallized and finely ground) is slowly added to the enzyme solution until the solution becomes slightly

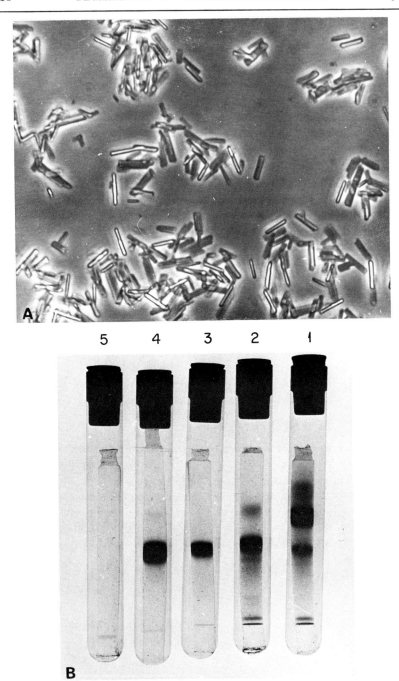

FIG. 4. (A) Crystal of β-N-acetylhexosaminidase. (B) Polyacrylamide gel electrophoresis of β-N-acetylhexosaminidase preparations: 1, preparation obtained by

turbid. The solution is kept at 4° until white needle-shaped crystals (Fig. 5A) appear (approximately 6 weeks). The crystalline α-mannosidase is electrophoretically homogeneous when examined by disc electrophoresis (Fig. 5B).

The molecular weight of α-mannosidase is estimated by gel filtration to be about 190,000. With p-nitrophenyl α-mannoside as substrate, maximal enzyme activity is obtained between pH 4.0 and 5.0. When benzyl- or methyl α-mannoside is used as substrate, maximal activity is found between pH 4.0 and 4.2. The K_m with p-nitrophenyl α-mannoside is $2.5 \times 10^{-3} M$; benzyl α-mannoside, $3.1 \times 10^{-2} M$; methyl α-mannoside, $1.2 \times 10^{-1} M$. The enzyme is strongly inhibited by Ag^+ and Hg^{2+}. Mannono-$(1 \rightarrow 4)$- and $(1 \rightarrow 5)$-lactones act as competitive inhibitors, the latter being the more potent. Jack bean α-mannosidase has been shown to be a zinc enzyme.[10]

In addition to hydrolyzing methyl-, benzyl-, and p-nitrophenyl α-mannosides, α-mannosidase also liberates mannose from various natural α-mannosides, which include α-$(1,2)$-mannobiose,[3] α-$(1,2)$-mannotriose,[3] α-$(1,2)$-mannotetraose,[3] α-$(1,6)$-mannobiose,[3] and mannosylrhamnose.[3] Jack bean α-mannosidase also liberates the mannose unit present in various glycoproteins and glycopeptides such as ovalbumin,[2,3] ovomucoid,[2,3] α_1-acid glycoprotein,[2,3] pineapple stem bromelain,[11] ribonuclease B.[12] Taka amylase[13] and the glycopeptide are isolated from microsomal membrane.[4]

β-N-Acetylhexosaminidase. The molecular weight of β-N-acetylhexosaminidase is estimated by gel filtration to be about 110,000. β-N-Acetylhexosaminidase hydrolyzes both p-nitrophenyl β-N-acetylglucosaminidase and p-nitrophenyl β-N-acetylgalactosaminide. For p-nitrophenyl β-N-acetylglucosaminide, the optimum pH is between 5.0 and 6.0 and K_m is 0.64 mM. For p-nitrophenyl β-N-acetylgalactosaminide, the optimum pH is between 3.5 and 4.0 and K_m is 0.31 mM. Studies of pH and thermal inactivation, mixed substrate analysis, and the K_i values for several competitive inhibitors indicate that the β-N-acetylglucosaminidase and the β-N-acetylgalactosaminidase activities are catalyzed by the same enzyme at the same site. Ag^+, Hg^{2+}, and Fe^{3+} ions are

[10] S. M. Snaith and G. A. Levvy, *Biochem. J.* **114**, 25 (1969).
[11] J. Scocca and Y. C. Lee, *J. Biol. Chem.* **244**, 4852 (1969).
[12] A. Tarentino, T. H. Plummer, Jr., and F. J. Maley, *J. Biol. Chem.* **245**, 4150 (1970).
[13] Y. C. Lee, *Fed. Proc., Fed. Amer. Soc. Exp. Biol.* **30**, 1223 (1971).

Sephadex G-200 chromatography; 2, preparation obtained by DEAE-Sephadex chromatography; 3, preparation obtained by CM-Sephadex chromatography; 4, crystalline β-N-acetylhexosaminidase; 5, blank gel. Reproduced from S.-C. Li and Y.-T. Li, *J. Biol. Chem.* **245**, 5153 (1970).

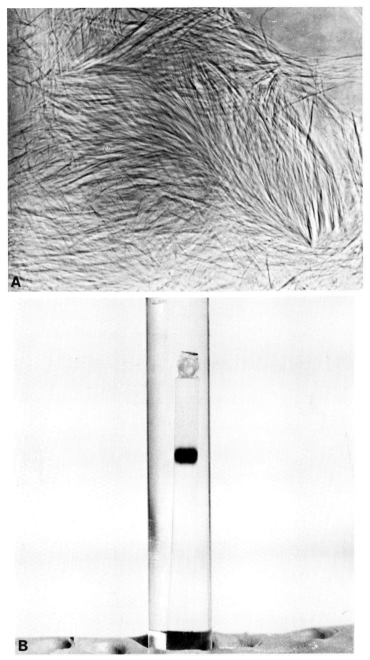

Fig. 5. (A) Crystal of α-mannosidase, $\times 800$. (B) Polyacrylamide gel electrophoresis of crystalline α-mannosidase.

potent inhibitors of this enzyme. β-N-Acetylhexosaminidase liberates terminal β-N-acetylglucosaminyl or β-N-acetylgalactosaminyl residues from the following compounds: ovalbumin and its glycopeptide,[5] ovomucoid,[5] N,N'-diacetylchitobiose,[5] sialic acid and galactose free α_1-acid glycoprotein obtained from human and chimpanzee plasma,[5] cell wall fragments from *Staphylococcus aureus*,[5] globoside isolated from human kidney or red blood cells, asialoderivative of Tay-Sachs ganglioside,[5] and various steroid β-N-acetylglucosaminide.[14]

β-*Galactosidase.* The jack bean β-galactosidase is a basic protein with the pI between pH 7.8 and 8.0[15,16] and an optimum pH between pH 3.5 and 4.5. This enzyme is able to hydrolyze terminal β-galactosyl residues present in various glycoproteins and glycolipids including α_1-acid glycoprotein isolated from human and chimpanzee plasma, ovalbumin-glycopeptide, glycopeptide isolated from rat liver microsome,[4] glycopeptide isolated from the freezing point depression glycoprotein of fish,[17] lacto-N-tetraose and lacto-N-neotetraose,[18] lactosyl ceramide,[7] and monogalactosyl ceramide.[19] The specificity of jack bean β-galactosidase deserves special emphasis. This enzyme can be used to distinguish Gal-β-$(1 \rightarrow 4)$ and Gal-β-$(1 \rightarrow 3)$ linked N-acetyllactosylamine. When 20 μg of Gal-β-$(1 \rightarrow 4)$-GlcNAc in 50 μl of 0.05 M glycine-HCl buffer, pH 3.5, is incubated with 0.1 unit of jack bean β-galactosidase at 37°, 2.8 μg of galactose is liberated after 10 minutes of incubation. Whereas, with Gal-β-$(1 \rightarrow 3)$-GlcNAc as substrate, no liberation of free galactose can be detected after 60 minutes of incubation under the same conditions. Similar results have been obtained for hydrolysis of the galactosyl residue in lacto-N-neotetraose, Gal-β-$(1 \rightarrow 4)$-GlcNAc-β-$(1 \rightarrow 3)$-Gal-β-$(1 \rightarrow 4)$-Glc, and lacto-N-tetraose, Gal-β-$(1 \rightarrow 3)$-GlcNAc-β-$(1 \rightarrow 3)$-Gal-β-$(1 \rightarrow 4)$-Glc.[18]

[14] Y.-T. Li, S.-C. Li, and D. K. Fukushima, *Steroids* **17**, 97 (1971).
[15] Y.-T. Li and S.-C. Li, *J. Biol. Chem.* **243**, 3994 (1968).
[16] Y.-T. Li and S.-C. Li, *in* "Proceedings of the 17th Colloquim on Protides of Biological Fluids" (H. Peeters, ed.), p. 455. Pergamon, Oxford, 1970.
[17] A. L. DeVries, S. K. Komatsu, and R. E. Feeney, *J. Biol. Chem.* **245**, 2901 (1970).
[18] A. Kobata and V. Ginsburg, *J. Biol. Chem.* **247**, 1525 (1972).
[19] Y.-T. Li, S.-C. Li, and G. Dawson, *Biochem. Biophys. Acta* **260**, 88–92 (1972).

[91] α-Galactosidase from Figs[1]

By Yu-Teh Li and Su-Chen Li

Several α-galactosidases have been isolated from various sources,[2,3] and thus far, all which have been studied are active in cleaving the terminal galactose present in the oligosaccharides of the raffinose family. However, for some unknown reason, they have great difficulty in cleaving the terminal α-galactosyl linkages present in sphingoglycolipids and glycoproteins. For example, the crystalline α-galactosidase which we isolated from *Mortierella vinacea*[4] cleaves melibiose and raffinose efficiently, but cannot hydrolyze the α-galactosyl residues present in glycoproteins or glycolipids. On the other hand, fig α-galactosidase can hydrolyze the α-galactosyl linkages present in globoside,[5,8] ceramide trihexoside isolated from various tissues,[5,6] digalactosyl ceramide isolated from the kidney of a patient with Fabry's disease,[7] and a glycopeptide isolated from earthworm cuticle collagen[9] with ease. Therefore, this enzyme is of great value in studying the sequence and anomeric configuration of the α-galactosyl residues in the oligosaccharide chains of various complex carbohydrates. By using fig α-galactosidase we have demonstrated for the first time that ceramide trihexoside contains a terminal α-galactosyl linkage.

Unless otherwise indicated, all carbohydrates and glycosides referred to herein are of the D-configuration and pyranoside, respectively.

Assay Methods

p-Nitrophenyl-α-Galactoside as Substrate

Principle. The enzyme is allowed to react with p-nitrophenyl-α-galactoside. The yellow color of the p-nitrophenol released by enzymatic hydrolysis is estimated at pH 9.8 in a spectrophotometer at 400 nm.

---

[1] This chapter was written while the authors were under the auspices of grants from the United States Public Health Service NS 09626 and the National Science Foundation GB 18019. Y.-T. Li is a recipient of a Research Career Development Award 1 K04 HD 50280 from the National Institute of Child Health and Human Development, United States Public Health Service.

[2] S. Veibel, in "The Enzymes" (J. B. Sumner and K. Myrbäck, eds.), Vol. I, Part I, p. 621. Academic Press, New York, 1950.

[3] K. Wallenfels and O. P. Malhotra, *Advan. Carbohyd. Chem.* **16**, 239 (1961).

[4] H. Suzuki, S.-C. Li, and Y.-T. Li, *J. Biol. Chem.* **245**, 781 (1970).

[5] S. Hakomori, B. Siddiqui, Y.-T. Li, S.-C. Li, and C. G. Hellerqvist, *J. Biol. Chem.* **246**, 2271 (1971).

Reagents

Sodium citrate buffer, 0.05 M, pH 4.0

p-Nitrophenyl-α-galactoside, 2 mM in sodium citrate buffer, pH 4.0

Sodium borate buffer, 0.2 M pH 9.8

Procedure. One milliliter of p-nitrophenyl α-galactoside is incubated with 5–100 μl of the enzyme solution for 5–30 minutes at 37°, depending on the activity of the enzyme preparation. The reaction is stopped by adding 3 ml of sodium borate buffer, and the intensity of the color produced is estimated at 400 nm, using a Beckman DU-2 spectrophotometer.

Unit. One unit of enzyme is defined as the amount of enzyme which hydrolyzes 1 μmole of p-nitrophenyl-α-galactoside per minute at 37°.

Oligosaccharides, Glycopeptides, or Glycolipids as Substrates

Principle. Fig α-galactosidase is used to hydrolyze the terminal α-galactosyl residue present in oligosaccharides, glycopeptides, or glycolipids; the liberated galactose is quantitatively determined by an automatic sugar analyzer.[10]

Procedure. Glycopeptide or glycolipid 0.05–0.2 μmole is dissolved in 100–200 μl of 0.05 M sodium citrate buffer, pH 4.0, and incubated with 0.2–1.0 unit (10–50 μl) of the enzyme at 37° for 2–40 hours. At the end of the reaction, an aliquot of the enzymatic digest is applied directly to an automatic sugar analyzer for the quantitative determination of free galactose.

Procedure for Enzyme Purification

Fresh fig latex contains α- and β-galactosidase, α-mannosidase, and β-N-acetylhexosaminidase. The dried fig latex (ficin), obtained from Nutritional Biochemical Corporation, contains only α-galactosidase. Therefore, ficin was chosen as the source for fig α-galactosidase.

Unless otherwise indicated, all the operations are carried out at 0–5°. One hundred grams of ficin (Nutritional Biochemical Corporation) is dissolved in 1 liter of tap water and dialyzed against tap water for 2 days. The large amount of protein precipitated during dialysis is removed by centrifugation. α-Galactosidase activity in the supernatant is fractionated by the gradual addition of acetone. The precipitate formed

[6] Y.-T. Li and S.-C. Li, *J. Biol. Chem.* **246**, 3769 (1971).

[7] Y.-T. Li, S.-C. Li, and G. Dawson, *Biochim. Biophys. Acta* **260**, 88–92 (1972).

[8] S.-C. Li and Y.-T. Li, *Fed. Proc., Fed. Amer. Soc. Exp. Biol.* **30**, 1118 (1971).

[9] L. Muir and Y. C. Lee, *J. Biol. Chem.* **244**, 2343 (1969).

[10] Y. C. Lee, J. F. McKelvy, and D. Lang, *Anal. Biochem.* **27**, 567 (1969). See also this volume [6].

at 30–45% acetone concentration is collected and dissolved in 30 ml of
0.1 M sodium phosphate buffer, pH 7.0. This material is then passed
through a Sephadex G-100 column (5 × 85 cm). The elution pattern
of fig α-galactosidase from the Sephadex G-100 column is shown in
Fig. 1A. The enzyme is eluted at approximately the position of human
serum albumin. The eluted enzyme is then precipitated by reverse
dialysis against saturated ammonium sulfate. The precipitate is collected
by centrifugation and redissolved in 2 ml of 0.05 M sodium phosphate

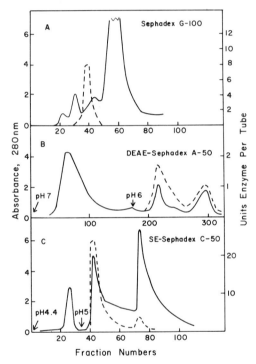

Fig. 1. (A) Sephadex G-100 chromatography of fig α-galactosidase. The column
has been equilibrated with 0.1 M sodium phosphate buffer, pH 7.0, and the enzyme
is eluted with the same buffer. The flow rate of the column is controlled at 30 ml
per hour by a peristaltic pump. ———, absorbance at 280 nm; - - - -, α-galactosidase
activity; 25 ml per fraction is collected. (B) DEAE-Sephadex A-50 chromatography
of fig α-galactosidase. The column is eluted with 0.05 M sodium phosphate buffer,
pH 7.0, and 0.05 M sodium citrate buffer, pH 6.0, containing 0.05 M NaCl as indicated.
———, absorbance at 280 nm; - - - -, α-galactosidase activity; 10 ml per fraction
is collected. (C) SE-Sephadex C-50 chromatography of fig α-galactosidase. The
column has been equilibrated with 0.05 M sodium acetate buffer, pH 4.4. The sample
is applied, then the column is eluted with sodium acetate buffer with various pH as
indicated. ———, absorbance at 280 nm; - - - -, α-galactosidase activity; 5 ml per
fraction is collected.

buffer, pH 7.0, then dialyzed exhaustively against the same buffer. The dialyzed solution is next applied to a DEAE-Sephadex A-50 column (2 × 30 cm) previously equilibrated with 0.05 M sodium phosphate buffer, pH 7.0. The column is washed with the starting buffer, then α-galactosidase is eluted with 0.05 M sodium citrate buffer, pH 6.0, containing 0.05 M NaCl (Fig. 1B). The main α-galactosidase peak is then pooled and precipitated by reverse dialysis against saturated ammonium sulfate. The precipitate is redissolved in 2 ml of 0.05 M sodium acetate buffer, pH 4.4, and dialyzed thoroughly against this buffer. The enzyme solution is applied to a SE-Sephadex C-50 column (2 × 30 cm), previously equilibrated with 0.05 M acetate buffer, pH 4.4. The column is washed with the starting buffer, and the α-galactosidase is eluted with 0.05 M sodium acetate buffer, pH 5.0 (Fig. 1C). The enzyme fraction is concentrated to 1 ml and further purified by passing it through a Sephadex G-100 column (1 × 75 cm). The elution pattern is shown in Fig. 2. Finally, the enzyme fraction obtained from the last Sephadex G-100 column is subjected to electrofocusing using ampholine with the pI distributed between pH 3 and 6. α-Galactosidase activity is focused between pH 4.7 and 4.9. From 100 g of ficin containing about 100 units of enzyme, about 20 units of purified enzyme are recovered. By this procedure the

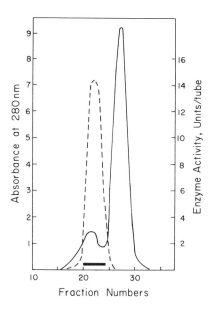

Fig. 2. The second Sephadex G-100 chromatography of fig α-galactosidase obtained from the step using SE-Sephadex C-50. ———, absorbance at 280 nm; - - - -, α-galactosidase activity; 2 ml per fraction is collected.

enzyme is purified approximately 2000 fold and is 80–90% pure as judged by polyacrylamide gel electrophoresis. The purified α-galactosidase is free of contaminating enzymes including glycosidase activity and protease activity.

General Properties

With p-nitrophenyl α-galactoside as substrate, the maximal enzyme activity is between pH 2 and 5.5. With ceramide trihexoside as substrate, a similar optimal pH curve is obtained. The K_m for this enzyme toward p-nitrophenyl α-galactoside is 0.5 mM.

The table summarizes the specificity of fig α-galactosidase. In all cases, 0.2 unit of the enzyme is incubated with the amount of substrate shown in the table for 20 minutes except for methyl-β-galactoside and lactose. The efficiency of the enzyme in cleaving the terminal galactose present in ceramide trihexoside, digalactosyldiglyceride, is fairly close to that of melibiose or raffinose. The glycopeptide isolated from earthworm cuticle collagen contains a mixture of α-(1,2)-galactobiose and α-(1,2)-galactotriose; therefore, the value of the liberated galactose (21.5%) is expressed as the percentage of total galactose rather than the percentage of terminal galactose. The enzyme is not able to hydrolyze lactose and methyl-β-galactoside.

SPECIFICITY OF FIG α-GALACTOSIDASE[a]

Substrates[b]	Substrate concentration (μmoles/0.1 ml)	Time (minutes)	Terminal galactose liberated (%)
Melibiose	0.2	20	70.5
Raffinose	0.2	20	83.0
Gal(1 $\xrightarrow{\alpha}$ 6)Gal	0.15	20	58.6
CH$_3\alpha$-Gal	0.2	20	29.6
Gal-Gal-Glu-Cer	0.17	20	65.1
Gal-Gal-Cer	0.2	20	40.2
Digalactosyl diglyceride	0.17	20	75.6
Earthworm cuticle collagen glycopeptide	0.21	20	21.5
CH$_3\beta$-Gal	0.4	24 (hours)	0
Lactose	0.4	24 (hours)	0

[a] In all cases, 0.2 unit of the enzyme was used. Incubation was carried out at 37°.

[b] Gal = galactose; Cer = ceramide.

Analysis of Anomeric Configuration of Galactosyl Residues
in Ceramide Trihexoside

Ceramide trihexoside is the glycolipid accumulated in abnormally high concentration in tissues of patients with Fabry's disease. The structure of the carbohydrate moiety in this glycolipid has been determined to be galactosyl-$(1 \rightarrow 4)$-galactosyl-$(1 \rightarrow 4)$-glucosyl-$(1 \rightarrow 4)$-ceramide. The anomeric linkages of the glycosides, especially that of the terminal galactose, were not established until 1971.[6,11,12] Although nuclear magnetic resonance (NMR) spectra have been useful in assigning the anomeric configurations in oligosaccharides, it requires rather large quantities of sample and the results are not always easily interpretable. Because the utilization of specific glycosidases for the structural analysis of complex saccharides takes advantage of the strict stereochemical specificity of the glycosidases and requires only very small amounts of substrate, this method is much more convenient and decisive than NMR spectra.

Figure 3 shows the kinetics of the liberation of galactoses from ceramide trihexoside by fig α-galactosidase and jack bean β-galactosidase. Fig α-galactosidase liberates 50% of the galactose whereas jack bean β-galactosidase is not able to liberate any terminal galactose from this

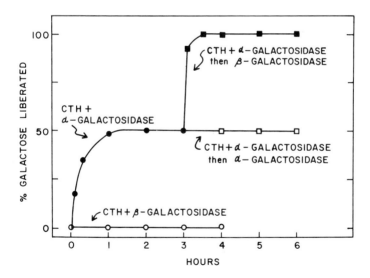

FIG. 3. The kinetics of the liberation of galactose from ceramide trihexoside by fig α-galactosidase and jack bean β-galactosidase. Reproduced from Y.-T. Li and S.-C. Li, *J. Biol. Chem.* **246**, 3769 (1971).

[11] S. Handa, T. Ariga, T. Miyatake, *J. Biochem.* **69**, 625–627 (1971).
[12] J. T. R. Clarke, L. S. Wolfe, and A. S. Perlin, *J. Biol. Chem.* **246**, 5563 (1971).

glycolipid. After the liberation of galactose from ceramide trihexoside reaches its plateau, the remaining 50% of the galactose which is resistant to fig α-galactosidase can be liberated by jack bean β-galactosidase. These results clearly indicate that ceramide trihexoside contains terminal α-galactosyl and penultimate β-galactosyl residues.

[92] α-Galactosidase, β-Galactosidase, β-Glucosidase, β-N-Acetylglucosaminidase, and α-Mannosidase from Pinto Beans (*Phaseolus vulgaris*)

By K. M. L. Agrawal and Om P. Bahl

Phaseolus vulgaris (pinto beans) are readily available and offer a good source for the preparation of several glycosidases suitable for the structure and function studies of complex macromolecules, such as glycoproteins[1] and glycolipids. The germinating seeds are employed for the preparation of the enzymes. The germination is carried out in the dark in sand and in the absence of any external nutrients. During germination an increase of about 20- to 35-fold in the specific activities of the enzymes is observed.[2] This may be due to the net degradation of seed proteins and/or *de novo* synthesis of the glycosidases. The polysaccharides are probably also degraded to provide energy for the embryonic growth of the plant.

The scheme[2] for the simultaneous purification of β-galactosidase, β-N-acetylglucosaminidase, α-mannosidase, and a mixture of α-galactosidase and β-glucosidase is described below. The enzyme preparations are almost free of cross-contamination with the exception of α-galactosidase and β-glucosidase which are obtained as a mixture. The preparation of β-N-acetylglucosaminidase is homogeneous by the criteria of ultracentrifugation and electrophoresis in polyacrylamide gels.

Assay

$$p\text{-Nitrophenyl }\alpha\text{- or }\beta\text{-D-glycopyranoside} \xrightarrow[\text{Enzyme}]{} p\text{-nitrophenol} + \text{monosaccharide}$$

Principle. p-Nitrophenyl α- or β-D-glycopyranoside is hydrolyzed by the appropriate glycosidase liberating p-nitrophenol and the monosaccharide. The activity of the enzyme can be followed simply by the

[1] O. P. Bahl, *J. Biol. Chem.* **244**, 575 (1969).
[2] K. M. L. Agrawal and O. P. Bahl, *J. Biol. Chem.* **243**, 103 (1968).

estimation of p-nitrophenol which forms a yellow chromogen under alkaline conditions with a λ_{max} of 400 nm.[3]

Reagents

p-Nitrophenyl α- and β-D-galactopyranosides, p-nitrophenyl β-D-N-acetylglucosaminide and β-D-N-acetylgalactosaminide, p-nitrophenyl β-D-glucopyranoside, and p-nitrophenyl α-D-mannopyranoside (Pierce Chemical Company, Rockford, Illinois), 25 mM solution[4] of each in 0.05 M sodium citrate buffer, pH 4.6

Sodium citrate buffer, 0.05 M, pH 4.6

Sodium carbonate, 0.2 M

Procedure.[5] A 100-μl aliquot of a 25 mM solution of p-nitrophenyl α- or β-D-glycopyranoside in 0.05 M sodium citrate buffer, pH 4.6, is incubated at 30° with 350 μl of the same buffer for 5 minutes. After the addition of 50 μl of an appropriately diluted enzyme solution, the digest is incubated at 30° for 15 minutes. The reaction is stopped by the addition of 700 μl of 0.2 M sodium carbonate. The yellow color formed is measured at 420 nm by a spectrophotometer.

Definition of Unit. One unit is defined as the amount of enzyme which hydrolyzes 1 μmole of substrate per minute at 30°. Specific activity is expressed as units per milligram of enzyme protein.

Purification Procedure

All steps of extraction and purification are carried out at 4° unless otherwise specified. At each step, the enzyme solution is concentrated either by lyophilization in the presence of 5 mg/ml of ammonium sulfate, by ultrafiltration using commercial ultrafiltration cells (Amicon Corporation, Cambridge, Massachusetts), or by precipitation with 100% ammonium sulfate. The ultrafiltration is convenient and not only concentrates the sample, but also effects partial purification.

Step 1. Extraction. The pinto beans are germinated in the dark in sand flats over a period of 6 days at 26–28°. The cotyledons (1000 g) are washed with water, suspended in 1000 ml of 0.2 M sodium citrate buffer, pH 6.0 and homogenized in a Waring Blendor for three closely spaced 30-second intervals. After the homogenate is allowed to stand for 1 hour, it is centrifuged in a Sorvall RC-2 refrigerated centrifuge

---

[3] J. Findlay, G. A. Levvy, and C. A. Marsh, *Biochem. J.* **69**, 467 (1958).

[4] Some glycosides are difficult to dissolve; 10 mM solution may be substituted and the assay changed accordingly so as to have a net 5 mM substrate concentration in the incubation mixture.

[5] O. P. Bahl and K. M. L. Agrawal, *J. Biol. Chem.* **243**, 98 (1968).

at 14,000 *g*. The clear supernatant is collected and the sediment is discarded.

Step 2. Ammonium Sulfate Fractionation. To the above supernatant, 43 g of solid ammonium sulfate per 100 ml of the supernatant are added over a period of 30–45 minutes with continuous stirring to 65% saturation. The resulting mixture is centrifuged for 1 hour at 37,000 *g*. The brownish precipitate thus formed is dissolved in 400 ml of 0.05 *M* citrate buffer, pH 4.6, and allowed to stand in the cold for 3 hours. The turbid solution is again centrifuged at 37,000 *g* for 1 hour and the crude mixture of the enzyme is precipitated from the supernatant solution by bringing it to 100% saturation with solid ammonium sulfate. The precipitate is redissolved in 400 ml of 0.05 *M* citrate buffer, pH 4.6, and the resulting solution is assayed for various glycosidases.

Step 3. Chromatography on DEAE-Sephadex. To a column (5 × 110 cm) of DEAE-Sephadex, previously equilibrated with 0.04 *M* sodium phosphate buffer, pH 6.5, a 175-ml portion of the enzyme solution (2.7 g of protein) is applied. The column is eluted with 4 liters of a linear salt gradient between 0.04 *M* sodium phosphate buffer, pH 7.5 and 1 *M* sodium chloride in the same buffer. The eluate was collected in 18-ml fractions which are assayed for the various enzyme activities and read at 280 nm for protein content. Fractions 20–70 (D-1) contain α- and β-galactosidases, α-mannosidase and β-glucosidase and fractions 79 through 100 (D-2) only β-N-acetylglucosaminidase. The enzymes are recovered from the two peaks by 100% saturation of the solution with solid ammonium sulfate or alternatively, by dialysis against 0.5% ammonium sulfate and lyophilization of the resulting solution.

Step 4. Chromatography of Fraction D-1 from DEAE-Sephadex on CM-Sephadex. A 5-ml aliquot (60 mg of protein) of D-1 dialyzed against 0.25 *M* sodium acetate buffer, pH 5.5, is applied to a column (2.5 × 70 cm) of CM-Sephadex C-50, previously equilibrated with the same buffer. The column is eluted with a decreasing pH gradient of 400 ml each of 0.25 *M* sodium acetate buffer pH 5.5 and pH 4.0. The eluate is collected in 7-ml fractions. Tubes are pooled into three major fractions, CM-1 (tubes 20–25) contains β-galactosidase, CM-2 (tubes 35–44) α-mannosidase and CM-3 (tubes 45–58) α-galactosidase and β-glucosidase.

Step 5. Further Purification[5] of β-N-Acetylglucosaminidase on DEAE-Sephadex A-50. Fraction D-2, obtained above, is further purified on another DEAE-Sephadex A-50 column (2.5 × 60 cm) previously equilibrated with 0.04 *M* phosphate buffer, pH 6.5. A 5-ml aliquot of the enzyme (19 units) is applied to the column which is eluted with a linear salt gradient of 200 ml of 0.04 *M* phosphate buffer, pH 7.5, and 200 ml of NaCl in the same buffer. The eluate is collected in 4-ml fractions. The tubes (40–44) containing the enzyme activity in a single peak are pooled (fraction D-4) and concentrated.

Step 6. Chromatography of Fraction D-4 on Sephadex G-100. A 7-ml aliquot of fraction D-4 (50.0 mg protein) from the previous step is placed on a column of G-100 (2 × 180 cm), equilibrated with 0.04 M sodium phosphate buffer, pH 7.5. The column is eluted with 500 ml of the same buffer and 4.6-ml fractions are collected. The β-N-acetylglucosaminidase activity appears in tubes 38–44, which are pooled.

Step 7. Stepwise Elution Chromatography of β-N-Acetylglucosaminidase on DEAE-Sephadex A-50. A 150-mg sample of the enzyme, specific activity 7.8, from the G-100 column is applied to a column (2 × 38 cm) of DEAE-Sephadex previously equilibrated with 0.04 M sodium phosphate buffer, pH 6.5. The elution is initiated with 100 ml of 0.04 M sodium phosphate buffer, pH 7.5, which is followed by 100, 100, 150, and 200 ml of the phosphate buffer containing 0.1, 0.15, 0.2, and 0.3 M sodium chloride, respectively. Fractions of 5 ml are collected. The enzyme is eluted with 0.2 M NaCl in tubes 66–81, which are pooled and concentrated.

Step 8. Gel Filtration on Sephadex G-150. A 5-ml aliquot of the enzyme from the preceding step (7 mg protein, specific activity 12.0) is applied to a Sephadex G-150 column (2 × 105 cm) packed under a hydrostatic head of 15 cm. The column is developed with 0.01 M sodium citrate buffer, pH 5.0 containing 0.1 M sodium chloride. Fractions of 1 ml are collected. The enzyme activity appears in tubes 154–166.

Properties of the Enzymes[2]

Homogeneity. The purification scheme described in Table I allows preparation of β-galactosidase, α-mannosidase, and β-N-acetylglucosaminidase almost free of cross-contamination.[2] The enzymes α-galactosidase and β-glucosidase are obtained as a mixture but free of the other three glycosidase activities. The β-N-acetylglucosaminidase preparation is homogeneous as judged by ultracentrifugation and disc electrophoresis in polyacrylamide gel.[6] All other enzyme preparations obtained as above are not homogeneous.

pH Optima and Stability. With the exception of α-galactosidase which has a pH optimum of 6.5–6.7, all others have a pH optimum ranging between 3.8 and 4.8. The enzymes are stable around their pH optima.[2] All the enzyme preparations are unstable to freezing and thawing but can be freeze-dried in the presence of 5 mg/ml of ammonium sulfate with a slight loss of activity. The enzymes can be stored indefinitely

---

[6] K. M. L. Agrawal and O. P. Bahl, manuscript in preparation.

TABLE I

PURIFICATION OF GLYCOSIDASES OF *Phaseolus vulgaris* FROM 1000 G OF GERMINATING SEEDS

Fraction	α-Galactosidase (specific activity) (×10³)	Recovery (units)	β-Galactosidase (specific activity) (×10³)	Recovery (units)	α-Mannosidase (specific activity) (×10³)	Recovery (units)	β-Glucosidase (specific activity) (×10³)	Recovery (units)	β-Acetylglucosaminidase (specific activity)	Recovery (units)
Crude extract (before germination)	2		2		1		1		2	
Crude extract (after germination)	63	401	45	286	35	223	22	140	72	458
Ammonium sulfate precipitation	409	337	270	223	227	198	136	119	475	430
DEAE-Sephadex A-50	2100	260	1700	157	1200	96	690	95	980	343
CM-Sephadex C-50	5022	188	5152	134	3024	58	2150	71	—	—
DEAE-Sephadex A-50									3900	274
Sephadex G-100									8000	238
DEAE-Sephadex A-50									12,000	210
Sephadex G-150									39,000	125

in the solution form in the presence of 50–60% ammonium sulfate. The α-mannosidase is an exception; it loses activity even at 4° under these conditions. It can, however, be stabilized by adding glycerol to 20% concentration.

Effect of Substrate Concentration and Temperature. The K_m values of all the enzymes for p-nitrophenyl substrates range between 0.83 × $10^{-4}\,M$ and 11.7 × $10^{-4}\,M$. Their molar activation energies vary from 9.8–13.6 kcal/mole.

The properties of the enzymes are summarized in Table II.

β-N-Acetylglucosaminidase

Homogeneity, Molecular Size, and Subunit Structure.[6] The β-N-acetylglucosaminidase preparation is free of other glycosidase activities[5] except β-N-acetylgalactosaminidase, which is intrinsic to the enzyme. It is homogeneous by ultracentrifugation and disc electrophoresis in polyacrylamide gels. The molecular weight of the enzyme determined by gel filtration by the method of Whitaker[7] is 112,000 ± 4000 and $s_{20,w}$ value of 8.04 S is obtained by ultracentrifugation. Preliminary evidence indicates that the molecule is composed of two identical subunits of 53,000 ± 3000 molecular weight, which are obtained by dissociation with 6 M guanidine hydrochloride. The dissociated subunits have higher electrophoretic mobility than the native enzyme in polyacrylamide gel at pH 8.3.

Substrate Specificity. The enzyme hydrolyzes p-nitrophenyl-2-acetamido-2-deoxy-β-D-glucopyranoside as well as p-nitrophenyl-2-acetamido-2-deoxy-β-D-galactopyranoside, the rate of hydrolysis of the latter being 20% of the former. The enzyme also hydrolyzes chitobiose

TABLE II
PROPERTIES OF GLYCOSIDASES OF *Phaseolus vulgaris*

Enzyme	pH optimum	pH stability	K_m (M × 10^{-4})	$E_a{}^a$ (kcal/ mole)	Q_{10} 15–25°	Q_{10} 25–35°
α-Galactosidase	6.5–6.7	4.6–8	6.57	13.6	2.5	1.9
β-Galactosidase	3.8–4.0	4.6–8	9.18	12.4	2.7	1.6
α-Mannosidase	3.8–4.0	6–7	11.70	10.1	2.2	1.8
β-Glucosidase	4.8–5.0	4.6–6.8	0.83	11.6	1.7	1.8
β-Acetylglucos-aminidase	4.6–4.8	4.6–8	4.68	9.8	1.9	1.7

^{a} E_a, activation energy.

[7] J. R. Whitaker, *Anal. Chem.* **36**, 1950 (1963).

and the higher oligosaccharides in the series. The rate of hydrolysis is greater in the case of the lower members than that of the higher members in the series. The enzyme is capable of hydrolyzing the nonreducing N-acetylglucosamine residues in desialyzed fetuin, α_1-acid glycoprotein[2] and human chorionic gonadotropin as well as the tryptic glycopeptides derived from them. The enzyme can liberate 40–50% of the nonreducing terminal N-acetylglucosamine residues in the carbohydrate moiety of tryptic glycopeptides of desialyzed fetuin and α_1-acid glycoprotein.

Inhibition of the Enzyme. The various cations, K^+, Zn^{2+}, Cd^{2+}, Mn^{2+}, Cu^{2+}, Mg^{2+}, Ca^{2+}, and iodoacetamide and EDTA do not cause any detectable inhibition of the enzyme activity. However, both β-N-acetylglucosaminidase and β-N-acetylgalactosaminidase activities are markedly inhibited by Ag^+ and Hg^{2+} ions. N-Acetylglucosaminolactone and N-acetylgalactosaminolactone competitively inhibit both activities. The K_i values calculated from Dixon plots are given in Table III.

The studies indicate that both activities probably reside in the same molecule and in the same site. Further evidence is furnished by the fact that N-acetylglucosamine as well as N-acetylgalactosamine cause inhibition of both of the activities.

α-Galactosidase

Substrate Specificity. The enzyme is specific for the α-anomer and has no activity for the β-anomer. It liberates galactose from melibiose (6-O-α-D-galactopyranosyl-D-galactose), raffinose [O-α-D-galactopyranosyl-$(1 \rightarrow 6)$-O-α-D-glucopyranosyl-$(1 \rightarrow 2)$ β-D-fructofuranoside], and stachyose [O-α-D-galactopyranosyl-$(1 \rightarrow 6)$-α-D-galactopyranosyl-$(1 \rightarrow 6)$-α-D-glucopyranosyl-$(1 \rightarrow 2)$-β-D-fructofuranoside]. The enzyme also can use high molecular weight materials, such as locust beans and guar gums, as substrates liberating approximately 30% and 27% of their $1 \rightarrow 6$ linked galactose residues, respectively.

TABLE III

INHIBITION OF β-N-ACETYLGLUCOSAMINIDASE AND β-N-ACETYLGALACTOSAMINIDASE ACTIVITIES OF THE ENZYME BY N-ACETYLGLUCOSAMINOLACTONE AND N-ACETYLGALACTOSAMINOLACTONE

| | K_i | |
| | --- | --- |
Inhibitor	β-N-Acetylglucosaminidase (μM)	β-N-Acetylgalactosaminidase (μM)
N-Acetylglucosaminolactone	0.75	2.0
N-Acetylgalactosaminolactone	37.50	10.0

β-Galactosidase

Substrate Specificity. The β-galactosidase hydrolyzes p-nitrophenyl β-D-galactopyranoside and does not show the α-galactosidase or β-glucosidase activities. The enzyme can hydrolyze both small and large molecular weight substrates. It hydrolyzes lactose, as well as 40–50% of the galactose residues of desialyzed fetuin and orosomucoid. The tryptic glycopeptides of desialyzed fetuin lose 76% of their galactose when digested with the enzyme. The rate and extent of hydrolysis, in general, is greater in the case of glycopeptides than the intact glycoproteins.

β-Glucosidase

Substrate Specificity. The β-glucosidase does not exhibit any α-glucosidase activity. The enzyme as isolated above is still contaminated with α-galactosidase. The enzyme is able to hydrolyze sophorose [O-β-D-glucopyranosyl-$(1 \rightarrow 2)$-D-glucose], cellobiose [O-β-D-glucopyranosyl-$(1 \rightarrow$

TABLE IV

SUBSTRATE SPECIFICITY OF GLYCOSIDASES OF *Phaseolus vulgaris*

Enzyme or substrate	Enzyme (units)	Amount of substrate (mg)	Period of incubation (hours)	Sugar released (%)
α-Galactosidase				
Melibiose	0.58	1.0	1	16.9
Raffinose	0.58	1.0	2	27.0
Stachyose	0.58	2.0	4	9.4
Guar gum	1.16	3.5	24	27.1
Locust bean gum	1.16	3.5	24	30.0
β-Galactosidase				
Lactose	0.43	0.9	1	28.2
Desialyzed fetuin	6.00	5.0	30	40.0
Glycopeptides of desialyzed fetuin	0.76	10.0	60	76.0
α-Mannosidase				
4-O-β-D-Mannopyranosyl-D-mannose	0.02	1.0	1	
Methyl-α-D-mannopyranoside	0.10	0.25	1	7.8
β-Glucosidase				
Cellobiose	0.13	1.0	1	10.2
Gentiobiose	0.13	0.5	1	20.2
Sophorose	0.12	0.250	1	5.0
Methyl-β-D-glucopyranoside	0.07	0.5	1	8.5
β-Acetylglucosaminidase				
Galactose-free glycopeptides of desialyzed fetuin	0.9	4.0	36	15
Galactose-free glycopeptide of desialyzed orosomucoid	0.75	1	9	50

4)-D-glucose], gentiobiose [O-β-D-glucopyranosyl-(1 → 6)-D-glucose], and methyl β-D-glucopyranoside.

α-Mannosidase

Substrate Specificity. The enzyme hydrolyzes p-nitrophenyl α-D-mannopyranoside and methyl α-D-mannopyranoside. It does not hydrolyze yeast mannan or ovalbumin, indicating that the enzyme does not readily attack high molecular weight substrates. It liberates about 20% of the mannose from pronase glycopeptides of ovalbumin.

Some of the substrate specificity studies are summarized in Table IV.

[93] α-Galactosidase, β-Galactosidase, and β-N-Acetylglucosaminidase from *Aspergillus niger*

By Om P. Bahl and K. M. L. Agrawal

Glycosidases occur widely in nature in animals, microorganisms, and plants. *Aspergillus niger* has been reported to contain several exo-glycosidases[1,2] such as α- and β-galactosidases,[3,4] α- and β-glucosidases,[5] β-N-acetylglucosaminidase,[3] α-N-acetylgalactosaminidase,[6] α- and β-mannosidases,[7] β-xylosidase,[8] 1,2-α-L-fucosidase,[9] 1,2-α-D-mannosidase,[10] and 1,2-α-L-rhamnosidase.[11] These enzymes are highly effective in the study of macromolecules containing oligo- or polysaccharide moieties. Since a partially purified enzyme preparation of *A. niger* is available commercially, Rhozyme HP-150 (Rohm and Haas Company, Philadelphia, Pennsylvania), it renders it a highly suitable source for the purification of various glycosidases. The procedures for the isolation of α- and β-galactosidases, β-N-acetylglucosaminidase, α-N-acetylgalactosaminidase, 1,2-

[1] A. Kaji, K. Tagawa, and K. Matsubar, *Agr. Biol. Chem.* **31**, 1023 (1967).

[2] E. Ahlgren, K. Erickson, and O. Vesterberg, *Acta Chem. Scand.* **21**, 937 (1967).

[3] O. P. Bahl and K. M. L. Agrawal, *J. Biol. Chem.* **244**, 2970 (1969).

[4] Y. C. Lee and V. Wacek, *Arch. Biochem. Biophys.* **138**, 264 (1970).

[5] C. R. Krishna Murti and B. A. Stone, *Biochem. J.* **78**, 715 (1961).

[6] M. J. McDonald, M. L. Chawla, and O. P. Bahl, manuscript in preparation.

[7] K. L. Matta and O. P. Bahl, *J. Biol. Chem.* **247**, 1780 (1972).

[8] M. Claeyssens, F. G. Loontiens, H. K. Hilderson, and C. K. DeBruyne, *Enzymologica* **40**, 12 (1970).

[9] O. P. Bahl, *J. Biol. Chem.* **245**, 299 (1970).

[10] N. Swaminathan, K. L. Matta, L. Donoso, and O. P. Bahl, *J. Biol. Chem.* **247**, 1775 (1972).

[11] O. S. Kishi, K. Higashihara, and M. Fukomoto, *J. Agr. Chem. Soc. Jap.* **37**, 84, 146 (1963).

α-L-fucosidase, 1,2-α-D-mannosidase and α-mannosidase are described below.

It is obvious from the following isolation procedures that the three initial steps are common; extraction and ammonium sulfate precipitation, ultrafiltration, and gel filtration on Sephadex G-150. By taking an appropriate fraction of the effluent from the Sephadex G-150 column, it is possible to separate all of the enzymes by chromatography on DEAE-Sephadex and/or on hydroxyapatite. For example, β-galactosidase completely free of α-galactosidase activity can be obtained by chromatography of the appropriate G-150 fraction on hydroxyapatite. Similarly, a highly purified preparation of β-N-acetylglucosaminidase can be obtained by chromatography of the appropriate G-150 fraction successively on DEAE-Sephadex and hydroxyapatite.

$$p\text{-Nitrophenyl }\alpha\text{- or }\beta\text{-glycoside} \xrightarrow[\text{enzyme}]{} p\text{-nitrophenol} + \text{monosaccharide}$$

Enzyme Assay and Units. The reagents for the assay and the procedures are essentially the same as given elsewhere in this volume.[12,13] One unit of the enzyme is the amount which liberates 1 μmole of p-nitrophenol per minute at 37°. The specific activity of the enzyme is defined as the number of units per milligram of protein.

Purification of Enzymes

The enzymes are purified[3] from Rhozyme HP-150 (Rohm and Haas Company, Philadelphia, Pennsylvania). All operations of extraction and purification are performed at 4° unless otherwise specified. During the enzyme purification, the centrifugation is usually performed at 37,000 g for 1 hour in a RC-2 Sorvall refrigerated centrifuge. The enzymes are unstable to freeze-drying as such, but can be freeze-dried in the presence of ammonium sulfate (5 mg/ml). At each step of purification, the enzyme solutions can be concentrated by ultrafiltration using Diaflo XM-50 or PM-30 membranes (Amicon Corporation, Cambridge, Massachusetts). The use of dialysis in cellophane tubing, at least in the initial stages is avoided because the tubing gets ruptured by β-glucosidase or cellulase present in the commercial enzyme preparation.

Step 1. Ammonium Sulfate Precipitation. A suspension of 50 g of the commercial enzyme preparation (Rhozyme HP-150, Celite free) in 150 ml of water, obtained by continuous stirring for 1 hour, is centrifuged to remove any suspended material. The residue is extracted again

[12] O. P. Bahl and K. M. L. Agrawal, *J. Biol. Chem.* **243**, 98 (1968).
[13] K. M. L. Agrawal and O. P. Bahl, *J. Biol. Chem.* **243**, 103 (1963).

with 100 ml of water, and the combined extracts are made up to 250 ml with water.

To the above supernatant solution, 56 g of solid ammonium sulfate per 100 ml is added gradually with stirring to bring about 80% saturation. The resulting precipitate is removed by centrifugation and dissolved in 500 ml of 0.05 M sodium citrate buffer, pH 4.6. Traces of insoluble material, if any, are removed by centrifugation. A second precipitation is carried out at 100% saturation by the addition of 76.7 g of ammonium sulfate per 100 ml of the supernatant. After the removal of the precipitate by centrifugation, it is dissolved in 500 ml of the above buffer.

Step 2. Pressure Dialysis. The enzyme solution is concentrated to a volume of 150 ml in an Amicon ultrafiltration cell as described above. After the addition of another 500 ml of 0.05 M citrate buffer, pH 5.0, the solution is again concentrated to a final volume of 150 ml.

Step 3. Gel Filtration on Sephadex G-150. The above enzyme solution (2.9 g of protein) is applied to a column of Sephadex G-150 (5 × 110 cm) packed under a hydrostatic head of 20 cm and equilibrated with 0.05 M sodium citrate buffer, pH 5.0. The column is eluted with the above buffer and 10-ml fractions are collected. Every other fraction is read at 280 nm for protein content and is assayed for enzymatic activities. The β-galactosidase and β-N-acetylgalactosaminidase activities appear in tubes 80–100 (fraction S-1) while α-galactosidase activity appears in tubes 140–170 (fraction S-2). These fractions are pooled and concentrated. In recent studies two columns of the above size, connected together in series, have been found to yield improved resolution between the β-galactosidase and β-N-acetylglucosaminidase activities.

Step 4. Purification of α-Galactosidase Fraction S-2 by DEAE-Sephadex Chromatography. A 7-ml aliquot of fraction S-2 (600 mg protein) in 0.01 M sodium phosphate buffer, pH 6.5, is applied to a column (2.5 × 38 cm) of DEAE-Sephadex A-50 equilibrated with 0.04 M sodium phosphate buffer, pH 6.5. The column is eluted with a continuous linear gradient of 400 ml of 0.04 M sodium phosphate buffer, pH 7.5, and 1 M sodium chloride in the same buffer. Fractions of 4 ml are collected. The α-galactosidase eluted in tubes 60–70 (fraction D-1) are pooled and concentrated.

Step 5. Purification of β-Galactosidase and β-N-Acetylglucosaminidase from Fraction S-1 by Gel Filtration on Sephadex G-150. The β-galactosidase and β-N-acetylglucosaminidase activties present in fraction S-1 from step 3 are separated by chromatography on Sephadex G-150. A 7-ml aliquot of this fraction (275 mg of protein) is applied to a column of Sephadex G-150 (2 × 105 cm), previously equilibrated with 0.01 sodium citrate buffer, pH 5.0, and 0.1 M sodium chloride. Elution is performed with the same buffer, and 2-ml fractions are collected. Tubes 55 and 59 (fraction S-3) contain β-N-acetylglucosaminidase and tubes 75–88 (fraction S-4), β-galactosidase activity. The fractions are pooled and concentrated.

Step 6 (a and b). Further Purification of β-Galactosidase Fraction S-4 on DEAE-Sephadex. Further purification of β-galactosidase is achieved by column chromatography of the above fraction on DEAE-Sephadex A-50. A 5-ml aliquot (75 mg protein) is applied to the column (2.5 × 38 cm) prepared as described above. The column is eluted with a linear gradient of 300 ml of 0.04 M sodium phosphate buffer, pH 7.5, and 1 M sodium chloride in the same buffer. Fractions of 4 ml are collected. The β-galactosidase activity appears in tubes 56–64 (fraction D-2). The contents of these tubes are pooled and concentrated. Re-chromatography of fraction D-2 is carried out on a DEAE-Sephadex A-50 column (1.2 × 50 cm) with a stepwise elution using 20 ml each of 0.4 M sodium phosphate buffer, pH 7.5, containing 0.2 M, 0.45 M, 0.5 M, and 0.55 M sodium chloride. The fraction D-3 containing β-galactosidase activity still contains some proteolytic activity, which can be removed by chromatography on CM-Sephadex.

Step 7. Chromatography of β-Galactosidase Fraction D-3 on CM-Sephadex C-50. A 2-ml aliquot of the above fraction (5.2 mg of protein) is applied to a column (1 × 20 cm) of CM-Sephadex C-50 equilibrated with 0.25 M sodium acetate buffer, pH 5.8. The column is eluted with a linear pH gradient of 60 ml of 0.25 M sodium acetate buffer, pH 5.8 and pH 4.0. One-milliliter fractions are collected. The β-galactosidase activity appears in a single peak.

Step 8 (a and b). Purification of β-N-Acetylglucosaminidase Fraction S-3 on DEAE-Sephadex A-50. A 5-ml aliquot of the above fraction (90 mg of protein) is applied to a DEAE-Sephadex column (2.5 × 38 cm) prepared as described in step 6. The column is developed with a linear gradient of 200 ml of 0.04 M phosphate buffer, pH 7.5, and 1 M sodium chloride in the same buffer. The eluate is collected in 4-ml fractions. The contents of tubes 40–48, containing β-N-acetylgluco-saminidase activity, are pooled and concentrated to 5 ml. The resulting solution is rechromatographed on a DEAE-Sephadex A-50 column (1.2 × 50 cm) using a stepwise salt gradient[3] ranging from 0.1 M NaCl to 0.5 M NaCl in 0.04 M sodium phosphate buffer, pH 7.5. The enzyme is eluted with 0.5 M NaCl.

The results of purification are summarized in Table I.

TABLE I

PURIFICATION OF GLYCOSIDASES OF *Aspergillus niger* (50 G OF RHOZYME HP-150)

Step and procedure	α-Galactosidase			β-Galactosidase			β-N-Acetylglucosaminidase		
	Total activity (units)	Specific activity (units/mg $\times 10^{-2}$)	Yield (%)	Total activity (units)	Specific activity (units/mg $\times 10^{-2}$)	Yield (%)	Total activity (units)	Specific activity (units/mg $\times 10^{-2}$)	Yield (%)
Crude extract	4400	88	100	2100	42	100	1900	38	100
1. Ammonium sulfate precipitation	4000	130	90	1800	60	85	1800	60	94
2. Pressure dialysis	3600	143	81	1600	66	76	1600	66	84
3. Sephadex G-150	2200	360	50	1080	390	51	1450	550	76
4. DEAE-Sephadex A-50	1200	5530	27	—	—	—	—	—	—
5. Sephadex G-150	—	—	—	850	1130	40	1050	1160	55
6a. DEAE-Sephadex A-50	—	—	—	700	3790	33	—	—	—
6b. DEAE-Sephadex A-50	—	—	—	600	4190	28	—	—	—
7. CM-Sephadex C-50	—	—	—	430	4500	20	—	—	—
8a. DEAE-Sephadex A-50	—	—	—	—	—	—	730	5610	38
8b. DEAE-Sephadex A-50	—	—	—	—	—	—	610	7170	32

Refinement of the Above Procedure

Purification of β-Galactosidase Fraction S-4 on Hydroxyapatite. The procedure described above yields β-galactosidase still containing traces of α-galactosidase activity which can be removed by chromatography on a hydroxyapatite column (2.5 × 6 cm). The column is equilibrated with 5 mM sodium phosphate, pH 5.8. A 5-ml aliquot of the S-4 fraction (50 mg of protein) is applied to the column and eluted stepwise with sodium phosphate buffer of increasing molarity[7] from 0.005 M to 0.6 M, pH 5.8. The β-galactosidase is eluted by 0.6 M sodium phosphate.

Purification of β-N-Acetylglucosaminidase Fraction S-3 on Hydroxyapatite. Further purification of β-N-acetylglucosaminidase can be achieved by chromatography on a hydroxyapatite column carried out under conditions similar to those described above.

Properties of the Enzymes[3]

The enzyme preparations are unstable to freezing and thawing but can be stored at 4° for an indefinite period in 50% ammonium sulfate in 0.05 M sodium citrate, pH 4.6. The enzyme solutions can be freeze-dried in the presence of 5 mg/ml of ammonium sulfate without any loss of activity. The enzyme preparations, thus obtained, do not contain any detectable proteolytic activity. The enzymes, α-galactosidase, β-galactosidase, and β-N-acetylglucosaminidase, have pH optima between 3.8 and 4.6. They are stable around their pH optima and are quite unstable at extreme acidic or alkaline pH levels. The properties of the enzymes are summarized in Table II.

Substrate Specificity of Glycosidases. The enzyme α-galactosidase causes the complete hydrolysis of methyl α-D-galactopyranoside and oligosaccharides, melibiose, raffinose, and stachyose. It also liberates 37–40% of the α-linked galactose residues from large polysaccharides such as galactomannans isolated from guar and locust beans.

The enzyme β-galactosidase is highly specific for nonreducing ter-

TABLE II

PROPERTIES OF GLYCOSIDASES OF *Aspergillus niger*

Enzyme	pH optimum	pH stability	K_m $(M \times 10^{-4})$	V_{max} (μmoles p-nitrophenol/min/mg)
α-Galactosidase	3.8–4.2	3.8–5.8	3.5	58.8
β-Galactosidase	3.2–4.0	2.8–3.9	10.0	55.5
β-N-Acetylglucosaminidase	3.9–4.6	3.5–7.0	6.6	71.4

minal β-galactosidic bonds. It rapidly hydrolyzes methyl β-D-galacto-pyranoside and lactose. It releases 70–75% of the total D-galactose residues from the nonreducing termini of the carbohydrate chains of sialic acid-free fetuin and of sialic acid-free α_1-acid glycoprotein. Similarly, the enzyme also liberates about 80% of the D-galactose residues from desialyzed human chorionic gonadotropin.[14]

Like the above glycosidases, β-N-acetylglucosaminidase is an exo-enzyme capable of hydrolyzing nonreducing β-linked N-acetylglucosaminidic bonds. It also has β-N-acetylgalactosaminidase activity. It liberates N-acetylglucosamine residues from chitin oligosaccharides containing 2–6 residues of N-acetylglucosamine. The rate of hydrolysis of the lower molecular weight members in the series is higher than that of the higher molecular weight members in this series. The enzyme liberates terminal N-acetylglucosamine residues from the glycopeptides derived from ovalbumin, β-galactosidase-treated desialyzed fetuin and α_1-acid glycoprotein. The amount of N-acetylglucosamine released ranges between 30 and 35% of the total N-acetylglucosamine present in these proteins. When β-galactosidase-treated human chorionic gonadotropin is incubated with the enzyme, about 60% of the total N-acetylglucosamine is liberated.[14]

[14] O. P. Bahl, *J. Biol. Chem.* **244**, 575 (1969).

[93a] α-N-Acetylgalactosaminidase from *Aspergillus niger*[1]

By M. J. McDonald and Om P. Bahl

Phenyl α-N-acetylgalactosaminide $\rightarrow$ phenol + N-acetylgalactosamine

α-N-Acetylgalactosaminyl-serine-(desialyzed OSM) $\rightarrow$

N-acetylgalactosamine + protein

Assay Method

Principle. Phenyl α-N-acetylgalactosaminide or desialyzed ovine submaxillary mucin (OSM) can be used as substrates for the enzyme α-N-acetylgalactosaminidase. The hydrolysis of phenyl α-N-acetylgalactosaminide by the enzyme can be followed by measuring phenol with the Folin-Ciocalteau reagent and hydrolysis of OSM by measuring N-acetylgalactosamine by the method of Reissig, Strominger, and Leloir.[2]

[1] M. J. McDonald, M. L. Chawla, and O. P. Bahl, manuscript in preparation.
[2] J. L. Reissig, J. L. Strominger, and L. F. Leloir, *J. Biol. Chem.* **217**, 959 (1955).

Reagents

Phenyl α-N-acetylgalactosaminide, 5 mM in sodium citrate buffer, 0.05 M, pH 4.6

Phenol reagent (Folin-Ciocalteau reagent) 1 N (Fisher Scientific Company)

Sodium carbonate, 0.4 M

Sodium tetraborate, 0.8 M, pH 9.2

Dimethylaminobenzaldehyde (DMBA), 10 g in 100 ml of glacial acetic acid containing 12.5% (volume per volume) 10 N HCl; diluted 1:10 with glacial acetic acid before use

Procedure.[2] PHENYL α-N-ACETYLGALACTOSAMINIDE AS SUBSTRATE. A 50-μl sample of the substrate is preincubated with 100–150 μl of 0.05 M sodium citrate buffer, pH 4.6. After the addition of 25–50 μl of the appropriately diluted enzyme, the digest is incubated for 15 minutes. The enzyme reaction is stopped by the addition of 100 μl of 1 N phenol reagent, and 1 ml of 0.4 M Na_2CO_3 is added immediately. The tubes are allowed to stand at room temperature for 30 minutes and then read at 625 nm. The enzyme and substrate blanks are run concurrently. Standards of phenol, 1–10 μg, are also run with the experiment.

DESIALYZED OVINE SUBMAXILLARY MUCIN AS SUBSTRATE. To a 50-μl solution containing 20 μg of desialyzed OSM is added 50 μl of the enzyme solution in 0.01 M sodium citrate buffer, pH 4.6. After the addition of a drop of toluene, the digest is incubated at 37° for 24 hours. The enzymatic reaction is stopped by the addition of 100 μl of 0.8 M tetraborate buffer, pH 9.2. The tubes are heated for 3 minutes in a vigorously boiling water bath and cooled for 5 minutes in cold water. After the addition of 1.2 ml of DMBA reagent (diluted 1:10 with acetic acid), the tubes are incubated for 30–60 minutes at 37–40° and read at 585 nm. The standard solutions of N-acetylgalactosamine, 0.01–0.04 μmole, are run concurrently.

Units. One unit of enzyme is defined as the amount which releases 1 μmole of phenol from the phenyl glycoside substrate at 37° per minute. The specific activity is expressed as units per milligram of enzyme protein.

Synthesis of Substrate

Phenyl 2-Acetamide-2-deoxy-α-D-galactopyranoside.[3] The synthesis is carried out essentially by the method of Weissmann. A mixture of 0.5 g of α- and β-galactosamine pentaacetate, 1 ml of freshly distilled phenol, and 0.13 g of $ZnCl_2$ in 0.05 ml of acetic acid containing 5% acetic anhy-

[3] B. Weissmann, *J. Org. Chem.* **31**, 2505 (1966).

dride is heated initially at 130° for 30 minutes in a vacuum distillation apparatus under a pressure of 160 nm. The temperature is then raised to 140–145° and maintained for 70 minutes. The reaction mixture is cooled and extracted with 25 ml of chloroform. The chloroform extract is washed successively with 10% sodium sulfate and 2 N sodium hydroxide to remove the excess phenol. The washing is repeated once more and finally the extract is washed with water, decolorized by activated charcoal, and filtered. The solvent is evaporated to dryness to give a syrupy residue, 0.49 g.

Phenyl 2-acetamido-2-deoxy-3,4,6-tri-o-acetyl-α-D-galactopyranoside in the above residue is purified by chromatography on a silica gel column (6.7 g silica gel, 100–200 mesh) previously washed with benzene:chloroform, 50:50. The residue is applied to the column, and the column is eluted with 15 of the above solvent (fraction I, 0.11 g), 60 ml of benzene: chloroform 20:40 (fraction II, 0.16 g), and finally with chloroform (fraction III). Fractions I and II contain the desired product, which can be combined and o-deacetylated with 0.1 M sodium methoxide in methanol. The resulting phenyl 2-acetamido-2-deoxy-α-D-galactopyranoside is crystallized from ethanol, $[\alpha]_D^{23}$, 217.5 (ca. 1, H$_2$O); literature[3]: $[\alpha]_D^{23}$, 258.3 (ca. 0.41, H$_2$O). This shows contamination with the β anomer and can be further purified by repeated crystallizations from absolute ethanol.

Purification Procedure

Step 1. Extraction. Rhozyme HP-150 (Celite free) from *A. niger* (Rohm and Haas Company, Philadelphia, Pennsylvania), 42.6 g, is suspended in 150 ml of 0.1 M sodium citrate buffer, pH 6.0, and is extracted for 4 hours by gentle stirring. The suspension is centrifuged for 1 hour at 16,500 rpm (32,000 g). The residue is extracted again with 150 ml of the buffer. The combined supernatant solutions are filtered through a plug of glass wool to remove any material that does not sediment during centrifugation.

Step 2. Ammonium Sulfate Precipitation. The extract is made up to 300 ml with water and 212 g of ammonium sulfate is added slowly over a period of 2 hours to 100% saturation. The precipitation is carried out overnight and the suspension centrifuged at 16,500 rpm for 1 hour.

Step 3. Ultrafiltration. The ammonium sulfate precipitate is redissolved in 300 ml of 0.05 sodium citrate buffer, pH 4.6, and the solution is concentrated to 100 ml by ultrafiltration through an XM-50 membrane (Amicon Corporation, Cambridge, Massachusetts). Two hundred milliliters of 0.05 M sodium citrate buffer, pH 4.6, is added and the solution is concentrated again to 100 ml. This operation is repeated 3 more times. Finally, the solution is reduced to 95 ml and any insoluble material formed is removed by centrifugation.

Step 4. Column Chromatography on Sephadex G-150. A column of Sephadex G-150 (5 × 193 cm) is packed in $0.05\,M$ sodium citrate buffer, pH 4.6, under a hydrostatic head of 15 cm. A 90-ml aliquot of the above solution (1.9 g protein) is applied and the column is eluted with $0.05\,M$ sodium citrate buffer, pH 4.6. The flow rate of the column is about 8 ml per hour; 5.7-ml fractions are collected. The enzyme α-N-acetylgalactos-aminidase eluted in fractions 335–360 and concentrated to about 43 ml by ultrafiltration as described above. Two columns of 5 × 100 cm each connected in series can be substituted for the single column.

Step 5. Chromatography on DEAE-Sephadex A-50. The enzyme so-lution (193 mg of protein) is applied to a column (2.5 × 93 cm) of DEAE-Sephadex A-50 previously equilibrated with $0.04\,M$ sodium phos-phate buffer, pH 7.4. The column is eluted with 200 ml of the same buffer, followed by a continuous salt gradient between 500 ml of the buffer and 500 ml of the buffer containing $1\,M$ NaCl; 2.6-ml fractions are collected. Fractions 320–345 containing the α-N-acetylgalactosaminidase are pooled and concentrated.

Step 6. Chromatography of DEAE-Sephadex A-50. A 14-ml aliquot of the above enzyme solution (55.7 mg of protein) is applied to a column of DEAE-Sephadex equilibrated with $0.04\,M$ sodium phosphate buffer, pH 6.8. Elution is carried out by a stepwise NaCl gradient in $0.04\,M$ sodium phosphate buffer, pH 6.8. The elution is initiated with 150 ml of the above buffer followed successively by 150 ml of the buffer containing $0.05\,M$, $0.1\,M$, $0.15\,M$ and $0.2\,M$ NaCl.

The results of purification are summarized in Table I.

Table I indicates that the total number of enzyme units in the crude

PURIFICATION OF α-N-ACETYLGALACTOSAMINIDASE FROM *Aspergillus niger*

Step	Protein (mg)	Total units[a]	Specific activity[b] (10^3)	Purification factor
1. Extraction	4260	17.1	4.0	1
2. Ammonium sulfate precipi-tation	2465	22.2	9.0	2.2
3. Ultrafiltration	1787	23.3	13.1	3.2
4. Sephadex G-150	195	19.7	100.9	25.2
5. DEAE-Sephadex	57	12.8	222.0	55.5
6. DEAE-Sephadex	26.7	12.8	481.6	120.4

[a] Micromoles per minute at 37°.

[b] Micromoles per minute per milligram of protein at 37°.

extract is less than in the succeeding steps. This probably is due to an error in the enzyme assay with the crude extract because one has to use a large amount of protein for the assay, and the enzyme blanks are quite high.

Properties of the Enzyme

Stability, pH Optimum, Effect of Temperature and Substrate Concentration. There is a significant loss in the enzyme activity on freeze-drying and also on freezing and thawing. The enzyme is stable in solution in $0.05\,M$ sodium citrate buffer, pH 4.6, in the presence of 50% ammonium sulfate. The enzyme can be kept in the solution indefinitely providing there is no protease activity or bacterial contamination. The pH optimum of the enzyme for the hydrolysis of OSM and the phenyl glycoside in sodium citrate-phosphate buffers of pH levels ranging from 3 to 6.5 is 4.0–4.2. The enzyme shows a maximum activity at 45°. The K_m and V_{max} values calculated from a Lineweaver-Burk plot are $6.99 \times 10^{-3}\,M$ and V_{max} 0.5 μmole per minute per milligram.

Substrate Specificity. The enzyme is highly specific for α-N-acetylgalactosaminide bonds and readily hydrolyzes phenyl 2-acetamido-2-deoxy-α-D-galactopyranoside. It does not split the corresponding α-N-acetylglucosaminidic bond. The *A. niger* β-N-acetylglucosaminidase, on the other hand, possesses specificity for both sugars. In this respect, the nonmammalian enzyme seems to have specificity identical to its mammalian counterpart.[4] It also splits the N-acetylgalactosaminylserine bond in desialyzed ovine submaxillary mucin. It may be noted that the enzyme does not hydrolyze the *o*-glycosidic bond between N-acetylgalactosamine and serine unless the sugar is terminal. The enzyme acts readily on both high and low molecular weight substrates.

[4] B. Weissmann and D. F. Heinrichsen, *Biochemistry* **8**, 2034 (1969).

[93b] 1,2-α-L-Fucosidase from *Aspergillus niger*[1]

By OM P. BAHL

2-O-α-L-Fucopyranosyl-D-galactose $\rightarrow$ L-fucose $+$ D-galactose

Assay Method

Principle. The assay of the enzyme, 1,2-α-L-fucosidase is based on the estimation by gas chromatography of fucose released by the hydrolysis of 2-O-α-L-fucopyranosyl-D-galactose. Fucose has also been estimated by measuring as semicarbazone, by the Conway diffusion technique, the acetaldehyde released on oxidation of fucose with periodic acid.[2]

Reagents

2-*O*-α-L-Fucopyranosyl-D-galactose, 5.5 mM in 0.01 M sodium acetate buffer, pH 4.0

Sodium acetate, 0.01 M, pH 4.0

Dowex 1-X8, HCO_3^- (100–200 mesh)

Dowex 50-X12, H^+ (100–200 mesh)

Dry pyridine

Hexamethyldisilazane

Trimethylchlorosilane

Procedure. To a 100-μl aliquot of the 5.5 mM solution of substrate in 0.01 M sodium acetate buffer, pH 4.0, 10 μl of the enzyme solution is added. The digest is deproteinated and deionized by passage through a small column (a 2.5-ml disposable syringe with a polyethylene plug is suitable for this purpose) of Dowex 1-X8, HCO_3^- (1 × 1 cm) with Dowex 50-X12, H^+ (1 × 1 cm) stacked on the top. The column is eluted with 10 ml of 5% ethanol and the eluate is dried in a Biodryer (The Virtis Company, Inc., Gardiner, New York). The fucose in the residue is transferred to a small glass-stoppered tube and dried again. The resulting residue containing fucose is converted into its trimethylsilyl ether[3] derivative by addition of a 25-μl portion of the silylating agent (prepared by mixing 1 ml of dry pyridine, 0.2 ml of hexamethyldisilazane, and 0.1 ml of trimethylchlorosilane) to the residue and stirring on a Vortex mixer for 15 minutes. A 1-μl sample is used for gas chromatographic analysis.

Units. One unit of enzyme is defined as the amount that liberates 1 μmole of fucose per hour at 37°. The specific activity is expressed as units of enzyme per milligram of protein.

Synthesis of Substrate[4]

2-O-(2,3,4-Tri-O-acetyl-α-L-fucopyranosyl)-1,3,4,6-tetra-O-acetyl-α-D-galactose. To a solution of 1.96 g of mercuric cyanide and 1.69 g of mercuric bromide in 40 ml of freshly distilled (over P_2O_5) acetonitrile

[1] O. P. Bahl, *J. Biol. Chem.* **245**, 299 (1970).

[2] A. K. Bhattacharyya and D. Aminoff, *Anal. Biochem.* **14**, 278 (1966).

[3] C. C. Sweeley, R. Bentley, M. Makita, and W. W. Wells, *J. Amer. Chem. Soc.* **85**, 2497 (1963).

[4] K. L. Matta and O. P. Bahl, manuscript in preparation.

is added 3.25 g of 1,3,4,6-tetra-O-acetyl-α-D-galactopyranose[5] and 3.29 g of 2,3,4-tri-O-acetyl-α-L-fucopyranosyl bromide with stirring until the solution is clear. The contents are kept at room temperature for 3.5 hours. Thereafter, the solvent is evaporated under reduced pressure, and the residue is dissolved in 100 ml of chloroform and filtered. The filtrate is washed five times with 1 N potassium bromide, using 100 ml of the solution at a time, followed by washing with water. The chloroform extract is dried over sodium sulfate, filtered, and concentrated to dryness to give 6 g of a syrupy product. A 4-g aliquot is chromatographed over a silica gel column (130 g of silica gel). Elution with 350 ml of benzene:ether, 9:1 gives a fast-moving material which is discarded. Further elution is carried out with benzene:ether 6:1 and all the fractions having R_f values of 0.59 on thin-layer chromatography (silica gel H) in solvent benzene:ethyl acetate, 6:4, are pooled. Fucose tetraacetate in the same solvent gives an R_f value of 0.73. $[\alpha]_D^{20}$, -35.5 (ca. 1, CHCl₃).

2-O-α-L-Fucopyranosyl-D-galactose. The heptaacetate of the disaccharide obtained above, is O-deacetylated by treating 1.9 g of the material in 150 ml of 1.5% methanolic ammonia at room temperature for 24 hours. The solvent is removed by evaporation, and the residual syrup is crystallized from ethanol:ethyl acetate mixture to give 0.7 g of 2-O-α-L-fucopyranosyl-D-galactose $[\alpha]_D^{20}$, -58 (ca. 1, H₂O); literature[6]: $[\alpha]_D^{20}$, -56.7.

Purification of the Enzyme

All steps of purification are carried out at 4° unless otherwise specified. The enzyme solution at each step of purification is concentrated by ultrafiltration using XM-50 membrane (Amicon Corporation, Cambridge, Massachusetts). Dialysis in cellophane tubing is avoided, at least in the initial stages of purification, since β-glucosidase or cellulase present in the starting material, Rhozyme HP-150 (Rohm and Haas Company, Philadelphia, Pennsylvania) digests the tubing and causes it to rupture.

Step 1. Extraction and Ammonium Sulfate Precipitation. A sample of 250 g of the commercial enzyme product (Rhozyme HP-150 Celite free) is extracted with 2 liters of cold 0.1 M NaCl solution over a period of 2 hours. The extract is centrifuged for 1 hour at 16,500 rpm in a Sorvall refrigerated centrifuge and the residue is extracted twice using 1 liter of 0.1 M NaCl each time. The combined supernatant solution (4

[5] J. O. Deferrari, E. G. Gros, and I. O. Mastronardi, *Carbohyd. Res.* 4, 432 (1967).
[6] A. Levy, H. M. Flowers, and N. Sharon, *Carbohyd. Res.* 4, 305 (1967).

liters, 72 g of protein) is treated gradually with 2808 g of ammonium
sulfate to bring about 100% saturation. After allowing the precipitate
to settle overnight, the solution is centrifuged at 16,500 rpm. The precipitate is dissolved in 1 liter of water and the resulting solution is subjected to another ammonium sulfate treatment (700 g). Finally, the
precipitate is dissolved in 1000 ml of 0.04 M sodium phosphate buffer,
pH 6.8, and concentrated to 400 ml by ultrafiltration. Subsequently, two
more additions of the buffer, 600 ml at a time, are made, and the solution is concentrated to a final volume of 400 ml.

Step 2. Column Chromatography on DEAE-Sephadex A-50. The enzyme solution (17 g of protein) in 0.02 M sodium phosphate buffer, pH
6.8 is applied to a column (10 × 110 cm) of DEAE-Sephadex A-50 previously equilibrated with 0.04 M sodium phosphate buffer, pH 6.8. The
column is eluted with a stepwise salt gradient from 0 to 1 M NaCl in
0.04 M sodium phosphate buffer, pH 6.8. Fractions of 22 ml are collected.
The various enzymes including α- and β-galactosidases, β-N-acetylglu-
cosaminidase and 1,2-α-L-fucosidase are eluted by about 0.3 M NaCl in
fractions 350–450, which are pooled and concentrated by ultrafiltration
to 40 ml. The solution is diluted with 0.25 M sodium acetate buffer to
800 ml and concentrated again to 40 ml.

Step 3. Gel Filtration on Sephadex G-150. The above solution (3.4 g
of protein) is applied to a column of Sephadex G-150 (5 × 110 cm)
packed in 0.25 M sodium acetate buffer, pH 4.6. The column is eluted
with the same buffer and fractions of 4.8 ml are collected. The enzyme
eluted in fractions 250–270 is concentrated by ultrafiltration as described
above.

Step 4. Chromatography on DEAE-Sephadex A-50. A 2-ml aliquot
from the previous step (100 mg of protein) is applied to a column (2.5 ×
40 cm) of DEAE-Sephadex A-50, and the column is eluted with a continuous salt gradient between 200 ml of 0.04 M sodium phosphate buffer,
pH 7.2, and 200 ml of 1 M NaCl in the same buffer. Fractions of 4.5 ml
are collected. Fractions 67–75 containing the enzyme 1,2-α-L-fucosidase
are pooled and concentrated by ultrafiltration to 4 ml. The solution is
diluted again to 25 ml with 0.25 M sodium acetate, pH 4.6, and concentrated to 4 ml.

Step 5. Gel Filtration on Sephadex G-150. A 4-ml sample (70 mg of
protein) from the preceding step is applied to a Sephadex G-150 column
(2 × 150 cm) packed under a hydrostatic head of 40 cm. The column is
eluted with 0.25 M sodium acetate buffer, pH 4.6. Fractions of 2.7 ml
are collected. The enzyme is eluted in fractions 70–78.

The results of purification are summarized in Table I. The preparation of the enzyme thus obtained is free of proteolytic, β-galactosidase,
and β-N-acetylglucosaminidase activities.

TABLE I

PURIFICATION OF *Aspergillus niger* 1,2-α-L-FUCOSIDASE FROM 250 G OF THE
COMMERCIAL ENZYME PREPARATION

Purification procedure	Total protein (mg)	Enzyme (units × 10²)	Specific activity[a] (units/mg × 10²)
Crude enzyme	72,000	50,400	0.7
1. Ammonium sulfate precipitation	17,000	151,000	8.8
2. DEAE-Sephadex A-50	3,400	143,400	42.2
3. Sephadex G-150	570	62,400	109.7
4. DEAE-Sephadex A-50	250	55,500	222.0
5. Sephadex G-150	124	50,000	403.2

[a] Specific activity is defined as micromoles of fucose released from methyl 2-*O*-α-L-fucopyranosyl-β-D-galactopyranoside per hour per milligram of the enzyme at 37°.

Properties of 1,2-α-L-Fucosidase[1]

pH Optimum and K_m. The pH optimum of the enzyme using 0.01 M sodium citrate buffers of pH ranging between 2.4 and 8.0 is 3.8 ± 0.2. The enzyme has a K_m value of $8.3 \times 10^{-3}\ M$ and a V_{max} of 16.0 μmoles/mg per hour for the substrate 2-*O*-α-L-fucopyranosyl-D-galactose.

Substrate Specificity. The enzyme is highly specific for the 1,2-α-L-fucosidic linkage to D-galactose. It hydrolyzes 2-*O*-α-L-fucopyranosyl-D-galactose, methyl 2-*O*-α-L-fucopyranosyl-β-D-galactopyranoside, and 2-*O*-α-fucopyranosyllactose. The enzyme does not hydrolyze *p*-nitrophenyl α-L and β-L-fucopyranosides, 2-*O*-α-L-, 3-*O*-α-L-, or 4-*O*-α-L-fucopyranosylfucoses. Similarly, 3-*O*-β-L- and 6-*O*-β-L-fucopyranosyl-D-galactoses and their corresponding methyl glycosides are not hydrolyzed. The enzyme is not able to cleave the $1 \rightarrow 3$-α-L-fucosidic linkage to either D-glucose or *N*-acetylglucosamine in 3'-*O*-L-fucopyranosyl lactose or lacto-*N*-fucopentaose III, respectively. Finally, the enzyme does not liberate any fucose from lacto-*N*-fucopentaose II, in which fucose is linked by a $1 \rightarrow 4$ α-linkage to *N*-acetylglucosamine.

The enzyme *A. niger* 1,2-α-L-fucosidase hydrolyzes 80–90% of fucose residues from intact as well as desialyzed canine and porcine submaxillary mucins, indicating that the fucose residues are located at the nonreducing termini and are linked to the adjoining galactose residues by 1,2-α-linkages. The enzyme does not hydrolyze fucose residues present in human chorionic gonadotropin or α_1-acid glycoprotein. Finally, the enzyme releases fucose from blood group H substance resulting in a complete loss of blood group activity.

The substrate specificity of the enzyme is summarized in Table II.

TABLE II

SPECIFICITY OF *Aspergillus niger* 1,2-α-L-FUCOSIDASE

Substrates tested	Hydrolysis (%)
1. 2-*O*-α-L-Fucopyranosyl-D-galactose	100
2. Methyl 2-*O*-α-fucopyranosyl-β-D-galactoside	100
3. 2-*O*-α-L-Fucopyranosyllactose	100
4. Lacto-*N*-fucopentaose I	100

$$\text{Gal} \xrightarrow[\beta]{1 \quad 3} \text{GlcNAc} \xrightarrow[\beta]{1 \quad 3} \text{Gal} \xrightarrow[\beta]{1 \quad 4} \text{Glu}$$

$$\begin{array}{c} 2 \\ \Big| \, \alpha \\ 1 \\ \text{Fuc} \end{array}$$

5. Porcine submaxillary mucin	80–90
6. Canine submaxillary mucin	80–90
7. Human chorionic gonadotropin	None
8. Orosomucoid	None
9. Fucan sulfate	None
10. 3¹-*O*-α-L-Fucopyranosyllactose	None

$$\text{Gal} \xrightarrow[\beta]{1 \quad 4} \text{Glu}$$

$$\begin{array}{c} 3 \\ \Big| \, \alpha \\ 1 \\ \text{Fuc} \end{array}$$

11. Lacto-*N*-fucopentaose II	None

$$\text{Gal} \xrightarrow[\beta]{1 \quad 3} \text{GlcNAc} \xrightarrow[\beta]{1 \quad 3} \text{Gal} \xrightarrow[\beta]{1 \quad 4} \text{Glu}$$

$$\begin{array}{c} 4 \\ \Big| \, \alpha \\ 1 \\ \text{Fuc} \end{array}$$

12. Lacto-*N*-fucopentaose III	None

$$\text{Gal} \xrightarrow[\beta]{1 \quad 4} \text{GlcNAc} \xrightarrow[\beta]{1 \quad 3} \text{Gal} \xrightarrow[\beta]{1 \quad 4} \text{Glu}$$

$$\begin{array}{c} 3 \\ \Big| \, \alpha \\ 1 \\ \text{Fuc} \end{array}$$

13. 2-*O*-α-L-Fucopyranosyl-L-fucose	None
14. 3-*O*-α-L-Fucopyranosyl-L-fucose	None
15. 4-*O*-α-L-Fucopyranosyl-L-fucose	None
16. Methyl 3-*O*-β-L-fucopyranosyl-α-D-galactoside	None
17. Methyl 4-*O*-α-L-fucopyranosyl-α-D-galactoside	None
18. 3-*O*-β-L-Fucopyranosyl-D-galactose	None
19. 4-*O*-α-L-Fucopyranosyl-D-galactose	None
20. 6-*O*-β-L-Fucopyranosyl-D-galactose	None
21. Human blood group substance H	100% loss of activity

[94] 1,2-α-D-Mannosidase from *Aspergillus niger*[1]

By N. SWAMINATHAN, K. L. MATTA, and OM P. BAHL

$$\text{2-}O\text{-}\alpha\text{-}D\text{-Mannopyranosyl-}D\text{-mannose} \xrightarrow[\text{enzyme}]{} \text{mannose}$$

Assay Method

Principle. The assay of the enzyme is based on the determination of mannose by the method of Somogyi and Nelson.[2] The disaccharide under the conditions of the assay shows as little as 7% of the reducing power of mannose.

Reagents

    2-O-α-D-mannobiose, 10 mM solution in water
    Sodium acetate buffer, 0.1 M, pH 4.6
    p-Nitrophenyl glycosides (Pierce Chemical Co., Rockford, Illinois)
    Somogyi-Nelson reagents
      1. Alkaline copper reagent
      2. Arsenomolybdate reagent

Assay Procedure. To 50 μl of the 10 mM substrate solution, 2-O-α-D-mannobiose in water, is added 25 μl of 0.1 M sodium acetate, pH 4.6, and 25 μl of the enzyme solution in the same buffer. The digest is incubated for 30 minutes at 30°, and 500 μl of the alkaline copper reagent is added. The tubes are heated for 30 minutes in a boiling water bath, then cooled; 500 μl of the arsenomolybdate reagent is added. The contents are diluted to 5 ml and the absorbance read at 520 nm against suitable substrate and enzyme blanks.

Units. One unit of the enzyme is defined as the amount which liberates 1 μmole of mannose per minute at 37°. Specific activity is expressed as units per milligram of protein.

Assay of Other Glycosidases and Proteases. The enzyme α- and β-galactosidase and β-N-acetylglucosaminidase are assayed as described in this volume [92].[3] Protease activity in the enzyme preparation is measured by using Azocoll as substrate (Calbiochem, Los Angeles, California).

Synthesis of 2-O-α-D-Mannobiose

2-O-α-(2,3,4,6-Tetra-O-acetyl-α-D-mannopyranosyl)-1,3,4,6-tetra-O-acetyl-D-mannopyranose. To a solution of 5.25 g of 1,3,4,6-tetra-O-

---

[1] N. Swaminathan, K. L. Matta, and O. P. Bahl, *J. Biol. Chem.* **247**, 1175 (1972).
[2] R. G. Spiro, Vol. 8 [1].
[3] K. M. L. Agrawal and O. P. Bahl, *J. Biol. Chem.* **243**, 105 (1968).

acetyl-D-mannopyranose,[4] 1.92 g of mercuric cyanide, and 2.7 g of mercuric bromide in 40 ml of freshly distilled acetonitrile (over P_2O_5) is added 8.2 g of acetobromomannose at room temperature. The reaction mixture is stirred for 4.5 hours. After the evaporation of acetonitrile at 30° under reduced pressure, the residue is extracted with 150 ml of anhydrous chloroform. The extract is filtered and washed thrice with 25 ml of 1 M potassium bromide each time and twice with 25 ml of water. The solution is dried over anhydrous calcium chloride and evaporated to dryness. A 3-g sample of the resulting residue is chromatographed over a silica gel column (2 × 55 cm). The column is eluted with benzene:ether, 5:1 (1300 ml) followed by elution with a 1:1 mixture of benzene:ether (2200 ml). The octaacetate of the disaccharide is eluted with the latter solvent.

2-O-α-D-Mannopyranosyl-D-Mannose. The *O*-deacetylation is carried out by dissolving 1 g of the octaacetate of mannobiose, obtained above, in 35 ml of 1% methanolic ammonia and keeping the reaction mixture overnight at room temperature. The methanol is evaporated and the residue containing 2-*O*-α-D-mannobiose is further purified by charcoal chromatography. Yield, 0.38 g $[\alpha]_d^{25}$, 40 (ca. 1, H_2O); literature[5]: $[\alpha]_d^{25}$, 35 (ca. 1, H_2O).

Purification of 1,2-α-D-Mannosidase

All steps of purification are carried out at 4° unless otherwise specified. Dialysis in cellophane tubing and freeze-drying operations are avoided. Fractions from the columns are concentrated by ultrafiltration using XM-50 or PM-30 membranes (Amicon Corporation, Cambridge, Massachusetts).

Step 1. Extraction and Ammonium Sulfate Precipitation. A sample of 100 g of the commercial enzyme product, Rhozyme HP-150 (Rohm and Haas Company, Philadelphia, Pennsylvania), is extracted with 800 ml of cold 0.1 M NaCl for a period of 2 hours. The extract is centrifuged at 37,000 g for 1 hour in Sorvall refrigerated centrifuge and the residue is extracted twice with 200 ml of 0.1 M NaCl each time. To the combined supernatant solution (1200 ml, 18 g of protein) 840 g of ammonium sulfate is added slowly with stirring. After allowing the precipitate to settle overnight, the solution is centrifuged for 1 hour at 37,000 g. The precipitate is dissolved in 2000 ml of 0.05 M sodium phosphate, pH 6.8, and the solution is concentrated to 400 ml by ultrafiltration. Two further additions of the buffer, 600 ml each time, are made, and the solution is concentrated to a final volume of 500 ml.

[4] J. O. Deferrari, E. G. Gros, and I. O. Mastronardi, *Carbohyd. Res.* **4**, 432 (1967).
[5] P. A. J. Gorin and A. S. Perlin, *Can. J. Chem.* **34**, 1796 (1956).

Step 2. Batchwise Treatment with DEAE-Sephadex A-50. The enzyme fraction from the previous step is treated with a thick slurry of DEAE-Sephadex A-50, equilibrated with 0.04 M sodium phosphate, pH 6.8. The suspension is stirred for 1 hour and centrifuged in a refrigerated centrifuge at 1500 rpm for 30 minutes. The residue is extracted twice with 500 ml of 0.04 M sodium phosphate, pH 6.8, to remove any enzymes loosely held on DEAE-Sephadex. Finally, the residue, after centrifugation, is extracted with 200 ml of the buffer with 0.3 M NaCl. Most of the 1,2-α-D-mannosidase is eluted from the DEAE-Sephadex A-50 under these conditions whereas a large amount of protein and pigmented material remains bound. The enzyme solution is diluted to 7 times its volume with water and concentrated by ultrafiltration.

Step 3. Chromatography on DEAE-Sephadex A-50. The enzyme solution, obtained above, is applied to a column (4 × 100 cm) of DEAE-Sephadex A-50, equilibrated with 0.04 M sodium phosphate buffer, pH 6.5. The column is eluted with a linear NaCl gradient between 1500 ml of 0.04 M sodium phosphate buffer, pH 6.8, and 1500 ml of 0.3 M NaCl in the same buffer. Fractions of 20 ml are collected. The enzyme is eluted in fractions 50–70, which are pooled and concentrated by ultrafiltration to 15 ml.

Step 4. Gel Filtration on Sephadex G-150. An aliquot of the enzyme from the preceding step (170 mg in 10 ml) is applied to a column of Sephadex G-150 (2.5 × 110 cm) equilibrated with 0.01 M sodium acetate, pH 4.0, and the column is developed with the same buffer; 4-ml fractions are collected. The enzyme appears in fractions 70–90, which are pooled and concentrated. The bulk of the contaminating β-galactosidase activity is removed in this step.

Step 5. Chromatography on DEAE-Sephadex A-50. In order to remove the last traces of β-galactosidase activity still associated with the 1,2-α-D-mannosidase fraction, the enzyme (25 mg in 5 ml) is applied to a column of DEAE-Sephadex A-50 (2.5 × 40 cm) equilibrated with 0.04 M sodium phosphate buffer, pH 6.8. The column is eluted with a stepwise NaCl gradient using 0.05 M, 0.1 M, and 0.2 M NaCl in 0.04 M sodium phosphate buffer, pH 7.5. The enzyme is eluted with 0.05 M NaCl and is free of β-galactosidase. The results of purification are summarized in Table I.

Properties of 1,2-α-D-Mannosidase

pH Optimum and K_m. The pH optimum of the enzyme is found to be 4.8 using phosphate citrate buffers at pH levels ranging from 2.5 to 8. The enzyme is stable in the pH range of 5.0–8.0. The enzyme has K_m

TABLE I

PURIFICATION OF 1,2-α-MANNOSIDASE FROM 100 G OF THE COMMERCIAL PREPARATION OF *Aspergillus niger*

| Purification procedure | Total protein (mg) | Total enzyme units | | | Specific activity of 1,2-α-mannosidase | % Recovery enzyme units |
		1,2-α-Mannosidase	β-Galactosidase	β-N-Acetyl glucosaminidase		
Crude extract[a]	18,000	240	9600	5250	0.013	100
1. Ammonium sulfate ppt.	10,000	200	8100	4200	0.020	83
2. DEAE-Sephadex A-50	1,870	159	5700	2940	0.120	67
3. DEAE-Sephadex A-50	204	102	91	19	0.500	42
4. Sephadex G-150	77	94	7	—	1.230	40
5. DEAE-Sephadex	70	91	—	—	1.300	33

[a] Specific activity of the crude extract is determined after gel filtration on Sephadex G-25 since the crude extract itself showed some reducing activity.

and V_{max} values of 2 mM and 1.3 μmoles per minute per milligram of protein, respectively, for the substrate 2-O-α-mannobiose.

Inhibition. With the exception of mercury ions which inhibit the activity by about 50% at a 1 mM concentration, no other ion tested, such as Zn^{2+}, Ag^{2+}, Cd^{2+}, Ca^{2+}, and Cu^{+2} shows any inhibitory or stimulatory effect. The various chelating agents, bipyridyl, 1,10-phenanthroline, and EDTA do not show any detectable effect on the enzyme activity. D-Mannonolactone, a potent inhibitor of other mannosidases, is a poor inhibitor here, causing only 50% inhibition at 27 mM concentration.

Substrate Specificity. The enzyme is highly specific for 2-O-α-D-mannosidic linkages and, therefore, hydrolyzes readily 2-O-α-D-mannobiose and 2-O-α-D-mannotriose. It does not hydrolyze 3-O-α-, 4-O-α-, and 6-O-α-D-mannopyranosyl-D-mannoses. A slight hydrolysis of 3-O-α- and 6-O-α-D-mannobiose shown in Table II is probably due to the limitation of the assay, not to any other intrinsic activities. It does not hydrolyze methyl, benzyl, or p-nitrophenyl-α-D-mannopyranosides. It does not liberate mannose from ovalbumin, β-galactosidase-β-N-acetyl-

TABLE II

SPECIFICITY OF *Aspergillus niger* 1,2-α-MANNOSIDASE

Substrate tested	Time of incubation (hours)	Relative rate
1. 2-O-α-D-Mannobiose	0.5	100
2. 2-O-α-D-Mannotriose	0.5	84
3. 3-O-α-D-Mannobiose	0.5	4
4. 4-O-α-D-Mannobiose	0.5	—
5. 6-O-α-D-Mannobiose	0.5	1.4
6. 4-O-α-D-Mannotriose	0.5	—
7. 2-O-α-D-Mannobiotol	0.5	—
8. 2-Acetamido-2-deoxy-3-O-α-D-mannopyranosyl- D-glucose	0.5	—
9. 2-Acetamido-2-deoxy-6-O-α-D-mannopyranosyl- D-glucose	0.5	—
10. Methyl-α-D-mannopyranoside	0.5	—
11. p-Nitrophenyl-α-D-mannopyranoside	0.5	—
12. Yeast mannan	24.0	—
13. β-Galactosidase-β-N-acetylglucosaminidase treated- desialyzed		
(a) Human chorionic gonadotropin and its glyco- peptides	24.0	—
(b) Fetuin and its glycopeptides	24.0	—
(c) α_1-acid glycoprotein and its glycopeptides	24.0	—
14. Ovalbumin and its glycopeptides	24.0	—

glucosaminidase-treated α_1-acid glycoprotein, fetuin, human chorionic gonadotropin, or the glycopeptides derived from them. Yeast mannan is also not hydrolyzed.

A summary of the substrate specificity is given in Table II.

[95] α-Mannosidase from *Aspergillus niger*[1]

By K. L. MATTA and OM P. BAHL

$$p\text{-Nitrophenyl-}\alpha\text{-D-mannopyranoside} \xrightarrow{\text{enzyme}} p\text{-nitrophenol} + \text{mannose}$$

$$\text{Methyl 4-}O\text{-}\alpha\text{-D-mannopyranosyl-}\alpha\text{-D-mannopyranoside} \xrightarrow{\text{enzyme}}$$
$$\text{mannose} + \text{methyl } \alpha\text{-D-mannopyranoside}$$

Assay Methods

Principle. Although the *p*-nitrophenyl glycoside, unlike *A. niger* β-galactosidase or β-*N*-acetylglucosaminidase, is relatively a poor substrate for this enzyme, it can be used for following the purification of the enzyme. The assay of the enzyme using this substrate is based on the determination of *p*-nitrophenol released. When methyl 4-*O*-α-D-mannopyranosyl-α-D-mannopyranoside, which is readily hydrolyzed by the enzyme, is used as substrate, the assay consists in measuring the release of mannose by the Somogyi-Nelson[2] method for reducing sugars.

Reagents

p-Nitrophenyl-α-D-mannopyranoside, 5.5 mM in 0.01 M sodium acetate, pH 4.0

Sodium acetate, 0.125 M, pH 4

Sodium carbonate, 0.2 M

Methyl 4-*O*-α-D-mannopyranosyl-α-D-mannopyranoside, 27.8 mM in water

Somogyi-Nelson reagents

Assay Procedures. METHYL 4-*O*-α-D-MANNOBIOSIDE AS SUBSTRATE. To 40 μl of a 27.8 mM (1%) solution of the substrate in water, 20 μl of 0.125 M sodium acetate, pH 4.0, 20 μl of water and 20 μl of enzyme are added. After incubation of the reaction mixture for 2 hours at 37°, the

[1] K. L. Matta and O. P. Bahl, *J. Biol. Chem.* **247**, 1780 (1972).
[2] R. G. Spiro, Vol. 8 [1].

mannose released is determined by the method of Somogyi-Nelson. Appropriate substrate and enzyme blanks are set up concurrently.

p-NITROPHENYL-α-D-MANNOPYRANOSIDE AS SUBSTRATE. To 100 μl of a 5 mM solution of p-nitrophenyl-α-D-mannopyranoside in 0.01 M sodium acetate, pH 4, 20 μl of the enzyme solution is added. The reaction mixture is incubated for 2 hours at 37°, after which 1.1 ml of 0.2 M sodium carbonate is added. The yellow chromagen formed is measured at 420 nm.

Units. One unit of the enzyme is defined as the amount which liberates 1 μmole of p-nitrophenol or mannose per hour at 37°. The specific activity is expressed as the number of units per milligram of the enzyme.

The methods of assay for other glycosidases are described elsewhere in this volume.

Synthesis of Methyl 4-O-α-D-Mannobioside

Methyl 4-O-α-(2,3,4,6-tetra-O-acetyl-mannopyranosyl)-2,3,6-tri-O-benzoyl-α-D-mannopyranoside.[1] To a solution of 7.5 g of methyl 2,3,6-tri-O-benzoyl-α-D-mannopyranoside in 42 ml of freshly distilled acetonitrile (over P_2O_5) containing 2.56 g of mercuric cyanide and 3.6 g mercuric bromide, is added a solution of 8.9 g of acetobromomannose in 10 ml of acetonitrile. The reaction mixture is stirred at room temperature for 16 hours. After evaporation of the solvent at room temperature *in vacuo*, the residue is extracted with 250 ml of chloroform and the resulting extract is filtered. The filtrate is washed three times with 500 ml of 1 M KBr with subsequent washing with water. The chloroform layer is dried with anhydrous sodium sulfate and evaporated to give 14 g of a syrupy material.

Methyl 4-O-α-D-mannopyranosyl-D-mannopyranoside. The above syrupy material without any further purification is O-deacetylated as follows. A 14-g sample in 300 ml of methanol is treated with 15 ml of barium methoxide (0.6 g) solution in methanol. The reaction is allowed to proceed at 40° for 36 hours. After neutralization with Dowex-50 (H⁺ form), the reaction mixture is filtered and the filtrate is evaporated to dryness. The resulting residue, approximately 6.0 g, when examined by thin-layer chromatography (TLC) in n-propanol ethyl acetate water, 7:5:2, and sprayed with potassium permanganate and sulfuric acid, shows the presence of mannose, methyl α-D-mannopyranoside, and methyl 4-O-α-D-mannopyranosyl-D-mannopyranoside.

The disaccharide is purified on a column of charcoal as described below. The impure material, 6 g of 20 ml of water, is placed on the column. The elution of the column is initiated with 5 liters of water followed by 1 liter of 5% ethanol and the eluate is collected by an automatic fraction collector in 22-ml fractions. Every third fraction is examined by TLC. The disaccharide appears in fractions 40–80, which are pooled and evaporated to dryness by a rotary evaporator. Yield, 0.75 g (13.9%). $[\alpha]_D^{20}$, 97.1 (ca. 1, H_2O). The product gives a single spot when tested by TLC in *n*-propanol:ethyl acetate:water, 7:5:2, and by paper chromatography in butanol:pyridine:water, 6:4:3, the spray reagent being periodate-ammoniacal silver nitrate.

Purification of *A. Niger* α-Mannosidase

All steps of purification are carried out at 4° unless otherwise specified. The enzyme solution at each stage of purification is concentrated by ultrafiltration employing XM-50 membrane (Amicon Corporation, Cambridge, Massachusetts) as previously described. Freeze-drying is innocuous in the initial stages of purification, but the highly purified enzyme is partially inactivated on freeze drying.

Step 1. Extraction and Ammonium Sulfate Precipitation. A sample of 250 g of the crude enzyme preparation, Rhozyme HP-150, (Rohm and Hass Company, Philadelphia, Pennsylvania) is extracted with 750 ml of 0.05 M sodium citrate, pH 4.6, over a period of 4 hours. The extract is centrifuged for 1 hour at 16,500 rpm in a Sorvall refrigerated centrifuge. The residue is extracted again with 600 ml of the same buffer and centrifuged. To the combined supernatant solution is added 1089.0 g of ammonium sulfate slowly with stirring to 100% saturation. The precipitation is allowed to continue overnight and the resulting precipitate is collected after centrifugation for 1 hour at 16,500 rpm. The residue is dissolved in 800 ml of water and dialyzed in acetylated cellophane tubing[3] for 2 days to remove ammonium sulfate. The acetylation is carried out by immersing the dialysis tubing in acetic anhydride and pyridine at room temperature for 6–12 hours followed by thorough washing with water before use. Acetylation of the tubing is necessary to prevent its digestion by β-glucosidase or cellulase present in the crude material. The dialyzate is freeze-dried to yield 18 g of the partially purified enzyme.

Step 2. Gel Filtration on Sephadex G-150. A 4.2-g sample of the above material in 20 ml of 0.05 M sodium citrate, pH 4.6, is applied to a column of Sephadex G-150 (5 × 190 cm), previously equilibrated in the same buffer. The column is eluted with the above buffer and fractions of 10 ml are collected. Fractions 120–159 containing the α-mannosidase activity are concentrated by pressure dialysis. The concentrated solution is diluted with 10 volumes of 0.04 M sodium phosphate, pH 6.8, and concentrated again. This process is repeated twice to ensure equilibration of the enzyme in phosphate buffer for the following step.

[3] L. C. Craig and J. Konigsberg, *J. Phys. Chem.* **65**, 166 (1961).

Step 3. Chromatography on DEAE-Sephadex A-50. To a column (2.5 × 85 cm) of DEAE-Sephadex equilibrated in 0.04 M sodium phosphate, pH 6.8, a 22-ml aliquot of the enzyme solution from the preceding step is applied. The column is eluted with a continuous salt gradient between 1 liter of 0.04 M sodium phosphate, pH 7.6, and 1 liter of 0.3 M NaCl in the same buffer. Fractions of 9.7 ml are collected. The fractions containing the enzyme activity are pooled and concentrated. The concentrated solution is equilibrated with a 5 mM solution of sodium phosphate, pH 5.8, by ultrafiltration.

Step 4. Chromatography on Hydroxyapatite. A column (2.5 × 6 cm) is packed carefully with hydroxyapatite under the manufacturer's directions (Bio-Rad Laboratories, Los Angeles, California) and equilibrated with 5 mM sodium phosphate, pH 5.8. A 7-ml aliquot of the above protein solution (105 mg of protein) is applied and eluted stepwise with phosphate, pH 5.8, of increasing molarities from 0.005, 0.025, 0.050, 0.1, 0.15, 0.2, 0.3, 0.4 to 0.6 M. As a rule the elution with a buffer is continued until no more protein is eluted with that buffer, and then the change to the next buffer is made. It is found in practice, however, that 100 ml of each buffer is enough for elution.

The results of purification are summarized in Table I.

Properties of α-Mannosidase

pH Optimum and pH Stability of the Enzyme. The pH optimum of the enzyme using citrate buffers at pH levels ranging from 3.0 to 8.0 and methyl 4-O-α-D-mannopyranosyl-D-mannopyranoside as substrate is 4.2 ± 0.2. Similar results are obtained when p-nitrophenyl-α-D-mannopyranoside is used as substrate. The enzyme is quite stable between pH 5 and 7 and is unstable at extreme acidic and alkaline pH.

Effect of Temperature. When the activity of the enzyme is measured at various temperatures, it is found to be maximal at 45°. No appreciable loss of activity is noted when the enzyme is kept for 4 hours at 10–45° and assayed at 37° under standard assay conditions. There is a sharp decline in the activity at temperatures higher than 45°, probably due to thermal denaturation of the enzyme.

Effect of Substrate Concentration. The effect of substrate concentration using two substrates, methyl 4-O-α-D-mannopyranosyl-α-D-mannopyranoside and p-nitrophenyl-α-D-mannopyranoside in the concentration range of 0.6 to 15 mM, gives a straight line relationship between $1/v$ and $1/S$. K_m values for the disaccharide and the p-nitrophenyl derivative calculated from Lineweaver-Burk plots are 8.2×10^{-3} and 1.2×10^{-3} M, respectively. The V_{max} values for the above compounds are 41.3 and 12.3 μmoles per milligram per hour respectively.

TABLE I

PURIFICATION OF *Aspergillus niger* α-MANNOSIDASE FROM 250 G OF THE COMMERCIAL ENZYME PREPARATION

Units and specific activities

Step	Total protein (mg)	A. niger α-Mannosidase	β-Mannosidase	α-Galactosidase	β-Galactosidase	β-N-Acetyl-glucosaminidase	β-Xylosidase	α-Glucosidase
Crude extract	9,920	4770[a] (0.3)[b]	4185 (0.4)	14345.0 (1.5)	11160.0 (1.1)	4743 (0.4)	4185 (0.4)	1023 (0.1)
1. Ammonium sulfate precipitation	4,200	2950 (0.6)	4320 (1.0)	9720 (2.3)	7290 (1.7)	3780 (0.9)	1836 (0.4)	456.0 (0.1)
2. Sephadex G-150	704	1584 (2.3)	1782 (2.5)	881.2 (1.2)	4256 (8.0)	1846.8 (3.2)	1504.8 (2.2)	415.8 (0.6)
3. DEAE-Sephadex A-50	210	1316 (6.3)	856 (4.0)	453.6 (2.1)	1701 (8.1)	96.6 (0.4)	16.8 (0.08)	28 (0.6)
4. Hydroxyapatite	22[c]	613.6 (29.8)	(4.1)	18.6 (1.7)		2.86 (0.02)		(0.13)

[a] Units of enzyme.
[b] Specific activity.

Effect of Cations, Chelators, Polyalcohols, and Lactone on Enzyme Activity. Hg^{2+} is found to be a potent inhibitor of the enzyme. Ca^{2+}, Cu^{2+} and Cd^{2+} inhibit the enzyme moderately at 0.1 mM concentration. Ag^+, Zn^+, Mn^{2+}, Mg^{2+}, Fe^{3+} Ba^{2+}, and Sr^{2+} cause about 40–60% inhibition at 1 mM concentration. None of the chelating agents such as EDTA, bipyridal, and 1,9-phenanthroline can be demonstrated to cause any inhibition. Unlike α-galactosidase, myoinositol, mannitol, and erythritol do not inhibit the enzyme activity. D-Manno-(1 → 5)-lactone is found to be a potent inhibitor of the enzyme.

Substrate Specificity. From the specificity studies, the α-mannosidase appears to be specific for 4-*O*-α- and 6-*O*-α-D-mannosidic linkages to the adjoining mannose or *N*-acetylglucosamine residues. Whereas the rates of hydrolysis of 4-*O*-α- and 6-*O*-α-D-mannobioses and 2-acetamido-2-deoxy-4-*O*-α-mannopyranosyl-D-glucose are similar to that of methyl 4-*O*-α-mannopyranosyl-D-mannopyranoside, the rate of hydrolysis of 2-acetamido-2-deoxy-6-*O*-α-D-mannopyranosyl-D-glucose is only 20%. Unlike *A. niger* β-galactosidase or β-*N*-acetylgalactosaminidase which hydrolyze the appropriate *p*-nitrophenyl glycosides readily, α-mannosidase hydrolyzes *p*-nitrophenyl-α-D-mannopyranoside rather slowly. Nevertheless, this *A. niger* α-mannosidase does not have as rigid a requirement for the aglycon as the *A. niger* 1,2-α-D-mannosidase or *A.*

TABLE II

SPECIFICITY OF *Aspergillus niger* α-MANNOSIDASE

Substrate	Relative rate of hydrolysis (2 hour incubation)
1. 4-*O*-α-D-Mannopyranosyl-D-mannose	100.0
2. 2-Acetamido-4-*O*-α-D-mannopyranosyl-2-deoxy-D-glucose	102.0
3. Methyl 4-*O*-α-D-mannopyranosyl-α-D-mannoside	92.9
4. 6-*O*-α-D-Mannopyranosyl-D-mannose	88.8
5. 2-Acetamido-6-*O*-α-D-mannopyranosyl-2-deoxy-D-glucose	20.0
6. 3-*O*-α-D-Mannopyranosyl-D-mannose	0.7
7. Methyl 2-*O*-α-D-mannopyranosyl-D-mannoside	8.5
8. 2-*O*-α-D-Mannopyranosyl-D-mannose	5.1
9. 2-Acetamido-3-*O*-α-D-mannopyranosyl-2-deoxy-D-glucose	0.0
10. Isomaltose	0.0
11. β-Galactosidase-β-*N*-acetylglucosaminidase-treated-desialyzed:	
i. α₁-Acid glycoprotein	40.0[a]
ii. Glycopeptides of fetuin	50.0[a]

[a] Represents the percentage of hydrolysis of the mannose residues exposed by the sequential treatment of the desialyzed glycoprotein or glycopeptide with β-galactosidase and β-*N*-acetylglucosaminidase.

niger 1,2-α-L-fucosidase. It is not surprising, therefore, that the enzyme also hydrolyzes slightly 2-*O*-α- and 3-*O*-α-mannobioses (Table II).

The sequential digestion of desialyzed α_1-acid glycoprotein and pronase glycopeptides of desialyzed fetuin with β-galactosidase and β-*N*-acetylglucosaminidase results in the removal of 70–80% of the galactose and 35–40% of the *N*-acetylglucosamine. When the above β-galactosidase-β-*N*-acetylglucosaminidase-treated α_1-acid glycoprotein or glycopeptides of fetuin are digested with *A. niger* α-mannosidase, 40–50% of the exposed mannose residues are hydrolyzed. Similar results are obtained with the tryptic glycopeptides of human chorionic gonadotropin, suggesting that all these glycoproteins have 1,4-α- or 1,6-α-mannosidic linkages.

[96] β-*N*-Acetylglucosaminidase, α-*N*-Acetylgalactosaminidase, and β-Galactosidase from *Clostridium perfringens*

By EDWARD J. McGUIRE, STEPHEN CHIPOWSKY, and SAUL ROSEMAN

$$\text{GlcNAc} \xrightarrow{\beta} \text{glycoprotein} \xrightarrow{\text{H}_2\text{O}} \text{GlcNAc} + \text{glycoprotein}$$

$$\text{GalNAc} \xrightarrow{\alpha} \text{mucin (polypeptide)} \xrightarrow{\text{H}_2\text{O}} \text{GalNAc} + \text{polypeptide}$$

$$\text{Gal} \xrightarrow{\beta} \text{glycoprotein} \xrightarrow{\text{H}_2\text{O}} \text{Gal} + \text{glycoprotein}$$

Many glycosidases have been described.[1-3] However, most of these preparations are unsatisfactory for use with glycoproteins or other complex carbohydrates. They are frequently contaminated with other glycosidases and/or proteases, and often show little or no activity with substrates of high molecular weight. The latter property is particularly troublesome and has not been adequately stressed in the published literature.

The present study was undertaken to find and purify glycosidases that would hydrolyze high molecular weight substrates, i.e., larger than glycopeptides. These enzymes must be purified free of contaminating proteases and should be free of other glycosidases. The procedures described below are for the purification of two classes of hexosaminidases,

[1] See Vol. 8 and this volume.
[2] K. Nisizawa and Y. Hashimoto, *in* "The Carbohydrates" (W. Pigman and D. Horton, eds.), 2nd ed., Vol. 2A, p. 242. Academic Press, New York, 1970.
[3] See Vol. 8 [98].

β-D-N-acetylglucosaminidases and an α-D-N-acetylgalactosaminidase, and the partial purification of a β-D-galactosidase from the culture filtrate of *Clostridium perfringens* NTCC 3626.

Assay Method

Reagents and Procedures

β-D-N-*Acetylglucosaminidase*. SYNTHETIC LOW MOLECULAR WEIGHT SUBSTRATE, p-NITROPHENYL-β-N-ACETYLGLUCOSAMINIDE (pNpGlcNAc). The enzyme in 0.01 M potassium phosphate buffer, pH 6.0, is added to 0.4 ml of the same buffer containing 1.75 μmoles of pNpGlcNAc and 0.15 mg of bovine serum albumin. The mixture is adjusted to 0.6 ml and incubated for 15 minutes at 37°. The reaction is terminated by the addition of 0.6 ml of 0.5 M sodium carbonate, and the p-nitrophenol released is measured by its absorbance at 400 nm ($E_{cm} = 1.8 \times 10^4$). One unit of enzyme releases 1 μmole of p-nitrophenol in 15 minutes under these conditions (see Vol. 8 [98]).

HIGH MOLECULAR WEIGHT SUBSTRATE, PRETREATED GLYCOPROTEIN. α_1-Acid glycoprotein[4] (AGP) is pretreated with sialidase[5,6] and β-D-galactosidase[7] yielding asialo, agalacto AGP hereafter designated as AGP (-SA,GAL).[8] This treatment exposes a terminal nonreducing β-linked N-acetylglucosamine residue linked to mannose in the oligosaccharide side chain. The incubation mixture contains 135 μg (0.06 μmole of GlcNAc sites) of AGP (-SA, GAL) and enzyme in 0.01 M potassium phosphate buffer, pH 6.0, in a total volume of 0.15 ml. After 1 hour of incubation at 37°, the reaction mixture is cooled in ice and assayed for the release of free GlcNAc by the Morgan–Elson procedure.[9]

α-N-*Acetylgalactosaminidase*. LOW MOLECULAR WEIGHT SUBSTRATE, PHENYL-α-N-ACETYLGALACTOSAMINIDE[10] (PGalNAc). The reaction mixture contain 3 μmoles PGAlm, 0.1 μmole of CaCl₂, and enzyme in 0.01 M potassium phosphate buffer, pH 6.0, in final volume of 0.15 ml. The mixture is incubated for 1 hour at 37°, and the amount of GalNAc released is determined by the Morgan–Elson procedure.

HIGH MOLECULAR WEIGHT SUBSTRATE, PRETREATED OVINE MUCIN. Ovine

---

[4] K. Schmid, *J. Amer. Chem. Soc.* **75**, 60 (1953).

[5] See Vol. 8 [117].

[6] See Vol. 8 [62].

[7] R. C. Hughes and R. W. Jeanloz, *Biochemistry* **3**, 1535 (1964).

[8] H. Schachter, I. Jabal, R. L. Hudgin, L. Pinteric, E. J. McGuire, and S. Roseman, *J. Biol. Chem.* **245**, 1090 (1970).

[9] C. T. Spivak and S. Roseman, *J. Amer. Chem. Soc.* **81**, 2403 (1959).

[10] The α-phenyl-N-acetylgalactosaminide was a generous gift of Dr. B. Weissman [*J. Org. Chem.* **31**, 2505 (1966)].

submaxillary mucin (OSM)[11] is pretreated with sialidase to expose
nonreducing GalNAc, which is linked through an α-glycosidic linkage
to the protein. The sialidase pretreated mucin is designated as OSM
(-SA).[12] The incubation mixture contains 250 μg (0.25 μmole GalNAc
sites) of OSM(-SA), 0.1 μmole of $CaCl_2$, and enzyme in 0.01 M potassium
phosphate buffer, pH 6.0, in a final volume of 0.3 ml. The mixture is
incubated at 37° for 1 hour, the reaction is terminated by immersion
in ice, and the amount of GalNAc liberated is determined by the Morgan–Elson procedure. One unit of enzyme releases 1 μmole of GalNAc in
1 hour under these conditions.

β-D-*Galactosidase*. p-NITROPHENYL-β-GALACTOSIDE (pNGal). The enzyme in 0.01 M potassium phosphate buffer, pH 6.0, is added to 0.4 ml
of the same buffer containing 1.75 μmoles of pNGal. The volume is adjusted to 0.6 ml and incubated for 15 minutes at 37°. The reaction is
stopped by the addition of an equal volume of 0.5 M Na_2CO_3 and the
p-nitrophenol released is measured as above.

Purification Procedure

Growth of Bacteria. Stock cultures of *C. perfringens* NTCC 3626 are
grown in Difco cooked meat medium at 37° and can be stored in this
medium at 4° for several years. The growth medium for preparative
purposes contains, per liter, 35.6 g of Difco Todd-Hewitt broth, 1.0 g
of yeast extract, 2.5 g of NaCl, 1.8 g of K_2HPO_4, 0.2 g of phenyl methyl
sulfonylfluoride, 0.05 g of cysteine-HCl, and 1.5 g of glucose. The
medium is sterilized by separately autoclaving solutions of the broth,
salts, and glucose, cooling to 37°, and combining the three solutions.
The bacteria are grown in standing culture at 37° in 4-liter Erlenmeyer
flasks containing 2 liters of medium. Two milliliter of a culture grown
overnight in the same medium are used as an inoculum, and the cells
are grown for 2–3 days. The cells reach the stationary growth phase in
18–24 hours; however, enzyme is released into the medium for another
48 hours. At this time the cultures are chilled in ice and the cells are
separated from the growth medium by centrifugation. The enzymes are
further purified from the culture fluid, and the cells are autoclaved and
discarded.

Ammonium Sulfate Concentration. Unless otherwise indicated, all
remaining steps are conducted at 4°, and all columns are packed by
gravity filtration. The culture filtrate is adjusted to 80% of saturation
with solid ammonium sulfate and stirred slowly for 12 hours. The

[11] S. Tsuiki, Y. Hashimoto, and W. Pigman, *J. Biol. Chem.* **236**, 2172 (1961).
[12] See Vol. 8 [62] III.

brown, tarry precipitate is collected by centrifugation and dissolved in a minimum volume of 0.01 M Tris·HCl buffer, pH 8.0, containing 0.05 M NaCl and 0.1 mM dithiothreitol (DIT). This fraction contains the bulk of the glycosidase activity and could be stored at $-10°$ for several years with no significant loss of activity.

Gel Filtration. The ammonium sulfate fraction (ca. 18 ml) is applied to a Sephadex G-150 column, 5 × 120 cm, which has been equilibrated with 0.01 M Tris, pH 8.0, containing 0.1 mM DTT. Elution is conducted in the pH 8, DTT buffer; fractions of 15–20 ml are collected. The β-N-acetylglucosaminidase activity appears in three peaks, designated Gm 1, GM 2, and Gm 3, respectively, as shown in Fig. 1. The β-galactosidase and α-N-acetylgalactosaminidase are present in the void volume, Gm 1 peak. The Gm 1, Gm 2, and Gm 3 fractions are pooled separately and can be stored for several months without apparent loss of activity.

DEAE-Cellulose Column Chromatography. Each of the pooled Sephadex peaks is applied to a DEAE-cellulose column 2.5 × 20 gm (0.4 mg of protein per gram of dry DEAE-cellulose). The DEAE-cellulose is prepared for use by suspending Whatman DE-52 in 0.01 M Tris buffer, pH 8.0, containing 0.1 mM DTT (Tris·DTT), and the stirred suspension is intermittently adjusted to pH 8.0 until the pH remains constant. After equilibration, the fines are removed from the suspension, and

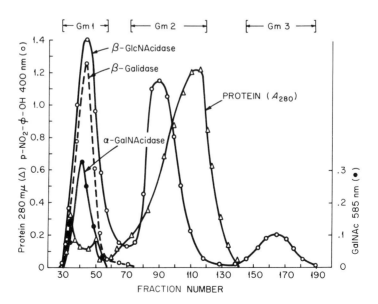

FIG. 1. Sephadex G-150 chromatography of glycosidase (*Clostridium perfringens* growth medium).

the column is packed under pressure using a hand-operated pressure bulb. After the enzyme was applied, the column is washed with 200 ml of Tris·DTT buffer. The enzyme is eluted using a concave salt gradient; the mixing chamber contained 2 liters of Tris·DTT buffer containing 0.15 M KCl, and the reservoir contains 1 liter of 0.4 M KCl in the same buffer. A typical elution pattern obtained with a Gm 1 Sephadex pooled peak is shown in Fig. 2.

The enzymes are unstable at this point unless concentrated. Concentration is effected by dialysis of the pooled fractions against the Tris·DTT buffer and subsequent application to a 1–2 ml DEAE-cellulose column, at pH 8.0. The column is eluted with Tris·DTT buffer containing 1 M KCl, and 2-ml fractions are collected, the major enzyme activity being in tube 2. The concentrated enzymes can be frozen at this stage without loss of activity.

Further Purification of α-N-Acetylgalactosaminidase. The pooled, dialyzed α-N-acetylgalactosaminidase, DE I (fractions 40–60) can be further purified by rechromatography on a DE-52 column (1 × 15 cm) prepared as above. The column is eluted sequentially with 250 ml of the Tris·DTT containing the following KCl concentrations: 0.14 M, 0.16 M, 0.18 M, 0.2 M, and 0.22 M. The major α-N-acetylgalactosaminidase (DE II) is eluted with 0.22 M KCl fraction and can be stored frozen after concentration as described above. A summary of the purification is shown in Table I.

Further Purification of β-N-Acetylglucosaminidases. Further purification of DE-Gm 1, (fractions 25–40) or DE-GM 2, (fractions 80–130) is accomplished by preparative polyacrylamide gel electrophoresis. Typically, a sample volume of 5–10 ml (15–3 mg) is applied to a 4–5 cm

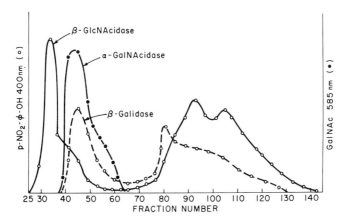

Fig. 2. DEAE-cellulose chromatography of glycosidases.

TABLE I

PURIFICATION OF α-N-ACETYLGALACTOSAMINIDASE FROM *Clostridium perfringens*

Step	Volume (ml)	Total activity (units)	Specific activity (units/mg protein)	Ratio α-GalNAcidase[a] β-GlcNAcidase[b]
Culture fluid	2400	1032	0.39	0.13
(NH$_4$)$_2$SO$_4$, 0–80% saturated	110	1001	0.5	0.13
Sephadex G-150	1467	734	2.5	0.18
DEAE-cellulose I	2640	660	10.4	0.4
DEAE-cellulose II	302	483	26.7	18.2

[a] With OSM(-SA) as substrate.
[b] With pNpGlcNAc as substrate.

7.5% polyacrylamide gel in the Buchler Poly-Prep apparatus. The standard Tris·glycine buffer system is used, and electrophoresis is carried out at 50 mamps for 12 hours at 4°.[13] Electrophoresis of DE-Gm 1 yields a single activity peak, whereas DE-Gm 2 gives two major peaks (PP-Gm 2a and Gm 2c) and occassionally two other peaks (PP-Gm 2b and Gm 2d). The number of peaks obtained upon preparative electrophoresis varies from one batch of enzyme to the next. Upon reelectrophoresis, each peak elutes in the same position and gives only a single activity peak. The enzymes are pooled and concentrated using the DE-52 method above, and can be stored frozen at this stage without apparent loss of activity. The summary of the enzyme purification is shown in Table II.

Properties

Stability. The purified hexosaminidases are stable at $-10°$ for at least 1 year after the concentration steps. They are unstable to freezing in dilute solution (<5 μg/ml). The β-hexosaminidases show differential heat lability at 57°, suggesting that the four major activities are, in fact, different enzymes and not different molecular forms of the enzyme showing the same activity.

pH Optima. The pH optimum for the α-N-acetylgalactosaminidase using PGalm or OSM(-SA) as substrates was found to be pH 5.8 in potassium phosphate buffer. The pH optima for the β-N-acetylglucosaminidases were: Gm 1, pH 6.5; Gm 2a and Gm 2b, pH 5.3; and Gm 3, pH 5.7.

[13] L. Ornstein and B. Davis, *Ann. N.Y. Acad. Sci.* **121**, 321, 404 (1964).

TABLE II

PURIFICATION OF β-N-ACETYLGLUCOSAMINIDASES FROM *Clostridium perfringens*

Step and fraction	Volume (ml)	Total activity (units)	Specific activity (units/mg protein)
Culture fluid	2400	8120	3
(NH$_4$)$_2$SO$_4$, 0–80% saturated	110	7920	4
Sephadex G-150			
Seph-Gm 1	4108	4108	14
Seph-Gm 2	3300	3300	5
Seph-Gm 3	112	112	0.9
DEAE-cellulose			
Of Seph-Gm 1 yielding			
DE-Gm 1	822	493	18
DE-Gm 2	8083	2910	36
Of Seph-Gm 2 yielding DE-Gm 2	6059	3029	63
Of Seph-Gm 3 yielding DE-Gm 3	56	101	180
Preparative electrophoresis			
Of DE-Gm 1 yielding PP-Gm 1	35	295	340
Of DE-Gm 2 yielding			
PP-Gm 2a	116	615	530
PP-Gm 2b	96	192	200
PP-Gm 2c	110	561	510
PP-Gm 2d	137	822	600

Effect of Metals. The α-N-acetylgalactosaminidase is stimulated 1.5- to 2.3-fold, using the above substrates, by 2 mM Ni^{2+}, Co^{2+}, Mn^{2+}, Mg^{2+}, and Ca^{2+}, the most effective being Ca^{2+}, when compared to enzyme incubated in divalent metal ion-free buffer (i.e., NH$_4^+$, Na$^+$, or K$^+$ phosphate buffers). The enzyme is completely inhibited by 2 mM concentrations of Hg^{2+}, Cu^{2+}, Fe^{2+}, and cysteine.

None of the β-N-acetylglucosaminidases, using pNpGlcNAc or AGP (-SA,Gal) as substrates, are stimulated by divalent metal ions and are all inhibited by 2 mM Hg^{2+}. The enzymes show differing susceptibilities toward the other -SH inhibitors. PP-Gm 1 is completely inhibited by 2 mM Cu^{2+}, but none of the PP-Gm 2 fractions are affected. The PP-Gm 2 activities are completely inhibited by 2 mM p-hydroxymercuribenzoate, while PP-Gm 1 is unaffected.

Specificity and Purity. The α-N-acetylgalactosaminidase has been purified free of contaminating proteases and is free of the following glycosidases: α- and β-galactosidase, α- and β-glucosidase, α-L-fucosidase, α-mannosidase, β-glucuronidase, α-N-acetylglucosaminidase, β-N-acetylgalactosaminidase, β-xylosidase, and sialidase. However, the purified, concentrated enzyme still retained low levels of β-N-acetylglucosaminidase activity. Repeated attempts to remove the last traces of β-GlcNAcidase resulted in loss of activity. The α-N-acetylgalactosaminidase acts on both high [OSM(-SA)] and low [α-phenyl-N-acetylgalactos-aminide] molecular weight substrates, releasing free N-acetylgalactos-amine. The asialomucin is the more effective substrate, having a V_{max} 2-fold greater using OSM(-SA) as substrate rather than PGalNAc; and a K_m 10 times lower [K_m: OSM(-SA) = 0.4 mM; PGalm = 3.3 mM].

Each of the purified β-N-acetylglucosaminidases contained no detectable protease activity, and none of the glycosidases listed above. As shown in Table III, the four β-hexosaminidases exhibit different kinetic behavior toward glycoproteins and synthetic glycosides. While all four fractions are more active with the p-nitrophenylglycoside than with the modified glycoprotein, the activity with the latter substrate varied from poor (PP-Gm 1) to substantial (PP-Gm 2a, PP-Gm 2c, PP-Gm 3). In addition to the kinetic differences shown by these preparations, they differ in another potentially important property. The PP-Gm 1 enzyme releases only 50% of the available N-acetylglucosamine from AGP(-SA, -Gal) upon prolonged incubation. The other enzymes (PP-Gm 2a, Gm 2c, or Gm 3) easily hydrolyze 100% of the nonreducing terminal N-acetylglucosamine from AGP(-SA, Gal). This result suggests that the enzymes may differ in their ability to hydrolyze specific β-GlcNAc linkages. This possibility is supported by the finding that the PP-Gm 2a enzyme cleaves. $\beta,1 \rightarrow 3$-N-acetylglucosaminidic linkages at 6 times the rate of PP-Gm 1, but they hydrolyze $\beta,1 \rightarrow 4$ linkages at equal rates.

The β-galactosidase fractions obtained from the DEAE-cellulose chromatography always contained contaminating hexosaminidase activity. Attempts to purify this activity using preparative electrophoresis and isoelectric focusing resulted in the loss of substantial activity. Fractions were obtained which only contained small amounts of β-

TABLE III

K_m AND V_{max} VALUES FOR β-N-ACETYLGLUCOSAMINIDASES

Enzyme	Substrate	K_m (mM)	V_{max} (μmoles/mg/hr)
PP-Gm 1	pNpGlcNAc	1.05	148.0
	AGP(-SA,Gal)	120.0	2.9
PP-Gm 2a	pNpGlcNAc	0.05	167.8
	AGP(-SA,Gal)	9.1	42.8
PP-Gm 2c	pNpGlcNAc	0.07	192.0
	AGP(-SA,Gal)	7.7	28.3
PP-Gm 3	pNpGlcNAc	0.7	108.0
	AGP(-SA,Gal)	7.7	18.2

glucosaminidase and these were tested for the ability to hydrolyze β-galactosidic linkages of high molecular weight substrates [AGP(-SA)]. This β-galactosidase hydrolyzed 100% of this available galactose residues and had a low K_m (10–20 mM) value.

[97] 1,2-α-L-Fucosidase from *Clostridium perfringens*

By DAVID AMINOFF

$$2\text{-}\alpha\text{-L-Fucosyl-lactose} \xrightarrow{\text{H}_2\text{O}} \text{fucose} + \text{lactose}$$

$$2\text{-}\alpha\text{-L-Fucosyl-glycoside} \xrightarrow{\text{H}_2\text{O}} \text{fucose} + \text{aglycon}$$

Assay Method

Principle. Fucosidases are exoglycosidases that cleave fucose-containing glycosides, oligosaccharides, mucins, and glycoproteins to fucose and the corresponding aglycon. Fucosidases have been detected and isolated from a number of sources, including: human,[1] mammalian,[2-5] limpet,[6] liver of abalone,[7,8] and the marine gastropod *Charonia lampas*,[9] *Helix pomatia*,[10] *Trichomonas foetus*,[11] *Aspergillus niger*,[12] gram-negative soil bacterium,[13] oral streptococcus,[14] chemically induced *Rhodopseudomonas palustris*,[15] and phage-induced *Klebsiella aerogenes*.[16]

[1] G. Y. Wiederschain, E. L. Rosenfeld, A. I. Brusilovsky, and L. G. Kolibaba, *Clin. Chim. Acta* **35**, 99 (1971).

[2] G. A. Levvy and A. McAllan, *Biochem. J.* **80**, 435 (1961).

[3] J. R. Esterly, A. C. Standen, and B. Pearson, *J. Histochem. Cytochem.* **15**, 470 (1967).

[4] G. Y. Wiederschain and E. L. Rosenfeld, *Bull. Soc. Chim. Biol.* **51**, 1075 (1969).

[5] G. Y. Wiederschain and E. L. Rosenfeld, *Biochem. Biophys. Res. Commun.* **44**, 1008 (1971).

[6] G. A. Levvy and A. McAllan, *Biochem. J.* **87**, 206 (1963).

[7] N. M. Thannassi and H. I. Nakada, *Arch. Biochem. Biophys.* **118**, 172 (1967).

[8] K. Tanaka, T. Nakano, S. Noguchi, and W. Pigman, *Arch. Biochem. Biophys.* **126**, 624 (1968).

[9] Y. Iijima and F. Egami, *J. Biochem. (Tokyo)* **70**, 75 (1971).

[10] A. Marnay, R. Got, and P. Jarrige, *Experientia* **20**, 441 (1964).

[11] W. M. Watkins, *Biochem. J.* **71**, 261 (1959).

[12] O. P. Bahl, *J. Biol. Chem.* **245**, 299 (1970).

[13] K. Mortensson-Egnund, R. Schöyen, C. Howe, L. T. Lee, and A. Harboe, *J. Bacteriol.* **98**, 924 (1969).

[14] J. K. Pinter, J. A. Hayashi, and A. N. Bahn, *Arch. Oral Biol.* **14**, 735 (1969).

[15] S. A. Barker, G. I. Pardoe, M. Stacey, and J. W. Hopton, *Nature (London)* **197**, 231 (1963).

[16] I. W. Sutherland, *Biochem. J.* **104**, 278 (1967).

In most of these studies nitrophenyl fucoside was used as substrate for ease of following the purification of the enzyme. Few of these enzymes exhibited activity toward naturally occurring oligosaccharides and glycoproteins. This limitation has necessitated the development of a specific assay that will detect and determine the free fucose released in the presence of the bound fucose and other free sugars.[17] Only the free fucose will produce acetaldehyde on oxidation with periodate. Formaldehyde, the oxidation product of other free sugars that may also be present in the incubation mixture, is selectively trapped by the glycine in the oxidation chamber of the Conway unit (Fig. 1). This assay procedure was successfully utilized to screen for fucosidase activity in a number of biological materials including *Clostridium perfringens*, the source of the α-1,2-L-fucosidase to be described.[18-20]

Reagents

Calcium chloride, 0.1 M, adjusted to pH 6.0

Ammonium sulfate, 2.6 M, adjusted to pH 6.0

A$^-$ hog submaxillary mucin,[21] solution to contain 5 μmoles of bound fucose per milliliter. α-L-Fucosyl-1 $\rightarrow$ 2-lactose[22] can also be used as substrate.

NPX-Tergitol in water, 0.025%, a nonionic detergent (Union Carbide Chemical Co.) required for plastic Conway units[17]

Semicarbazide reagent, dissolve the semicarbazide HCl to a concentration of 0.0067 M, in pH 7.0 sodium phosphate buffer (0.06 M NaH$_2$PO$_4$, 0.14 M Na$_2$HPO$_4$)

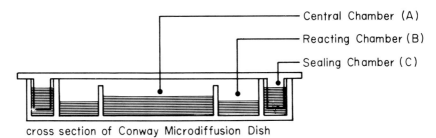

Central Chamber (A)

Reacting Chamber (B)

Sealing Chamber (C)

cross section of Conway Microdiffusion Dish

Fig. 1. Oxidation chamber of Conway unit.

[17] A. K. Bhattacharyya and D. Aminoff, *Anal. Biochem.* **14**, 278 (1966).

[18] D. Aminoff, *Fed. Proc., Fed. Amer. Soc. Exp. Biol.* **25**, 435 (1960).

[19] K. Furukawa and D. Aminoff, *Fed. Proc., Fed. Amer. Soc. Exp. Biol.* **28**, 606 (1969).

[20] D. Aminoff and K. Furukawa, *J. Biol. Chem.* **245**, 1659 (1970).

[21] D. Aminoff and M. Morrow, *FEBS (Fed. Eur. Biochem. Soc.) Lett.* **8**, 353 (1970).

[22] R. Kuhn, *Bull. Soc. Chim. Biol.* **40**, 297 (1958).

Periodate reagent, to be made up fresh just before use, as follows: 0.5 M periodic acid, 10 ml; 0.6 M glycine, 10 ml; 0.75 N NaOH, 10 ml; adjust to pH 7.4–7.5 with 0.75 N NaOH and dilute to 100 ml. Use before $NaIO_4$ begins to precipitate out (2 hours).

L-Fucose, standard, 0.002 M

Procedure. A mixture containing 0.05 ml of $CaCl_2$, 0.05 ml $(NH_4)_2SO_4$, and 0.5 μmole of substrate-bound fucose is incubated with the enzyme at 37° for 15 minutes in a total volume of 0.5 ml and a final pH of 6.0. The reaction is stopped by heating for 1 minute in a boiling water bath. The free fucose released by the enzyme is determined by the Conway unit assay.

Reagents and sample are added to the plastic Conway dish (Fig. 1) in the following order: (1) a drop, 0.05 ml, of NPX in chambers C and B; (2) 2.5 ml of H_2O in chamber C, and evenly distributed to completely cover the area; (3) 0.4 ml of sample, containing 0–0.2 μmole of free fucose as standards, or of incubation mixture, in chamber B adjacent to the drop of NPX; (4) 2.5 ml of semicarbazide HCl *accurately* pipetted into chamber A to completely cover the well (no NPX should be added to this chamber to facilitate wetting, since the NPX solution absorbs in the ultraviolet); (5) 2.0 ml of periodate reagent rapidly added to chamber B (without touching the NPX and sample drops); the Conway unit is *immediately* covered with the plastic lid fitting into the moat, chamber C, to give an air-tight seal. (6) Agitate the unit gently in a circular horizontal motion on surface of bench to completely mix the contents of chamber B. (7) Leave at room temperature for 2 hours without disturbing. (8) Remove an aliquot of the semicarbazone in chamber A and read the absorbance in a spectrophotometer at 224 nm against a reagent bank.

Definition of Unit and Specific Activity. A unit of fucosidase activity is defined as the amount of enzyme that releases 1 μmole of fucose per hour from the A^- hog submaxillary mucin. Specific activity is expressed in terms of micromoles of fucose released per hour per milligram of protein.[23]

Enzyme Preparation

Growth of Cells. Clostridium perfringens, Type 33-4A, is maintained and grown in cooked meat medium (Difco). For the preparation of enzyme the organism is grown in Todd-Hewitt broth medium of the fol-

[23] O. H. Lowry, N. J. Rosebrough, A. L. Farr, and R. J. Randall, *J. Biol. Chem.* **193**, 265 (1951).

lowing composition: Todd-Hewitt broth powder (Difco), 35 g, and $K_2HPO_4 \cdot 3H_2O$, 4.72 g/liter. To maintain adequate anaerobic conditions, 1.5 liters of medium are used in a 2-liter screw-capped Erlenmeyer flask. The medium is autoclaved and cooled just before use. A 0.1% inoculum from the meat broth stock culture is used, and growth is mantained at 37° for 72 hours without agitation and with the flasks tightly closed. The cultures are then rapidly chilled to 4°, and the cells are removed by centrifugation for 20 minutes at 4° and 16,000 g. All subsequent steps in the purification are carried out in the cold room at 4°.

Ammonium Sulfate Fractionation. The culture supernatant solution (94.2 liters) is adjusted to 80% saturation (516 g of solid ammonium sulfate per liter). After 16 hours at 4° the precipitate is collected by centrifugation for 20 minutes at 16,000 g. The sediment is dissolved in 5.5 liters of sodium phosphate buffer.[24] The resulting solution can be lyophilized or stored frozen for at least two years without loss of activity. The crude mixture is refractionated, collecting the precipitates between narrow concentration differences of ammonium sulfate added over a 2–3-hour period with constant stirring. The best fraction is obtained between 52.3% saturation (307 g/liter) and 54.5% (322.4 g/liter) with a fucosidase to sialidase ratio of 15. In the interests of purity, only this fraction is utilized for further purification of the fucosidase despite the resulting lower yield. This represents the best stage for effective removal of most of the contaminating sialidase.

Sephadex G-75 Treatment. The fraction that precipitates between 52.3% and 54.5% ammonium sulfate saturation is dissolved in 200 ml of the phosphate buffer, and 15 ml is applied to a (8.4 × 75 cm) Sephadex G-75 (particle size 40–120 μ) column previously equilibrated with the phosphate buffer.[24] Protein is eluted with the same buffer at a flow rate of 10 ml/hr, and 20-ml fractions are collected. The protein content is determined by the absorption at 280 nm.[25] Fucosidase appears in the first of three major protein peaks. Tubes containing fucosidase activity (650 ml) are pooled for subsequent fractionation.

Alumina C_γ–Celite Chromatography. Alumina C_γ (Sigma) and Celite (Johns-Manville) are both washed with the phosphate buffer,[24] mixed in the ratio 1:4 (v/v), and packed in a 4 × 30 cm column. The fucosidase active eluate from Sephadex G-75 (250 ml) is applied to the column and washed through with 950 ml phosphate buffer, or until the absorption of the effluent at 280 nm is less than 0.025. The column is further washed with 1 liter of 0.75% ammonium sulfate (w/v) to remove extraneous

---

[24] Sodium phosphate buffer, 0.01 M, pH 7.0, and 0.025 M with respect to KCl.
[25] E. A. Peterson and H. A. Sober, Vol. 5 [1].

protein. This is followed by 2.5 liters of 1.00% ammonium sulfate to elute the major fraction of fucosidase.

Treatment with Bentonite. Bentonite (Fisher) is washed with phosphate buffer until the supernatant is completely clear after centrifugation. It is necessary to carry out an initial "titration" to determine the optimum amounts of bentonite required to achieve the maximum of adsorption of extraneous protein with the minimum of fucosidase. In our hands, 5 ml of packed bentonite were required per 100 ml of alumina C_γ eluate. The C_γ eluate is added to the packed bentonite, well mixed for 10 minutes at 4°, and centrifuged at 9000 g for 20 minutes.

Enzyme Concentration. The resulting supernatant may be concentrated by vacuum dialysis in collodion bags (Schleicher and Schuell Co.)[26] against several changes of phosphate buffer in an ice bath at the rate of 25 ml in 8 hours. The enzyme is extremely labile at this stage of the purification.

Properties

Purity. The purified preparation of the enzyme was found to be inactive with the following phenyl-, *o*-nitrophenyl-, or *p*-nitrophenylglycoside substrates: α- and β-D-glucosides, β-D-glucuronide, α- and β-D-galactosides, β-D-xylosides, α-D-mannoside, α- and β-D-*N*-acetylgalactosaminides, and α-D-*N*-acetylglucosaminide. Only a trace of activity was detected with *p*-nitrophenyl-β-D-*N*-acetylglucosaminide after 16 hours of incubation. This was very small compared to the strong reactivity shown by the crude enzyme preparation after 15 minutes of incubation.

PURIFICATION OF α-FUCOSIDASE FROM *Clostridium perfringens*

Fraction	Total units	Specific activity	Purification factor	Recovery (%)	Ratio Fucosidase / Sialidase
I. (NH₄)₂SO₄, 0–80% saturation	182,984	0.68	1	100	1.3
II. (NH₄)₂SO₄, 52.3–54.5% saturation	26,520	4.31	6	15	15.1
III. Sephadex G-75	13,715	4.91	7	8	14.2
IV. Alumina C_γ eluate	3,460	88.6	130	2	1287.5
V. Bentonite	3,170	476	702	2	—

[26] It is necessary to wash the bags prior to use, first with 0.01 N NaOH, then with 0.01 N HCl, followed by distilled water and the phosphate buffer.

No sialidase activity could be demonstrated in the preparation, even after 16 hours of incubation when over 95% of the fucose was liberated from hog H⁺ specific submaxillary glycoprotein (9.4% fucose, 20.8% N-glycolyl neuraminic acid) which was used as the substrate. Likewise, no protease activity was detected in the enzyme preparation.

Stability. The fucosidase is stable to freezing in the phosphate buffer at all stages of purification, but is inactivated on repeated freezing and thawing. Lyophilization results in inactivation of the purified preparation. Vacuum dialysis against phosphate buffer, as indicated above, is the only procedure for effectively concentrating the pure enzyme. Simple dialysis against distilled water results in a rapid loss of activity. Dialysis against buffers at pH below 5.5 results in precipitation of the enzyme with considerable inactivation.

Physical Properties. As isolated, the fucosidase occurs in multiple "isoenzymatic" forms as demonstrated by disc gel electrophoresis.

The purified enzyme has a molecular weight greater than 200,000 daltons as determined by gel filtration on Sephadex G-200.[17]

Kinetic Properties, Metal Requirements, and Inhibitors. In sodium acetate-Veronal buffer the optimum pH is 5.8, whereas in citrate phosphate buffer it is 6.3. The activity is enhanced by $CaCl_2$ $(0.01\,M)$ which also slightly increases the pH optimum of activity in acetate-Veronal to pH 6.0. The activity under these conditions is identical with that of the enzyme in $CaCl_2$ and ammonium sulfate previously adjusted to pH 6.0 (final concentration, 0.01 and $0.26\,M$, respectively). The K_m for the enzyme with hog H⁺ submaxillary mucin as substrate is 0.175 mM. However, this is only a tentative and apparent K_m value on account of the high substrate blank and the fact that the substrate is a high molecular weight polymer with multiple substrate sites.

p-Chloromercuribenzoate, iodoacetamide, and EDTA inhibit the fucosidase activity to a limited extent (50–70%). Fe^{2+}, Fe^{3+}, Hg^{2+}, Zn^{2+}, and Cu^{2+} show marked inhibiting effects on the enzyme.[17]

Substrate Specificity. The fucosidase specificity is restricted to the $\alpha(1 \rightarrow 2)$-L-fucosyl linkage. It is inactive toward simple synthetic aliphatic and aromatic fucosides; α- or β-, -D- or -L-, methyl or p-nitrophenyl pyranosides or furanosides. Of the simple fucosyloligosaccharides found in milk, the most readily susceptible appear to be 2′-fucosyllactose and lacto-N-fucopentaose I. In both cases the L-fucose is bound to galactose by an α-$(1 \rightarrow 2)$ linkage.[22] However, the rate of release from both these oligosaccharides is slower than with the high molecular weight glycoprotein substrate showing H⁺ blood group activity. The rapid but limited release of fucose from A⁺ hog submaxillary mucin[20] is attributed to steric hindrance resulting from the close proximity of the terminal N-acetylgalactosamine, the A-determinant, attached α-(1 → 3) to the same galactose residue to which the fucose is attached α-(1 → 2). The observed 20% of the total fucose released could represent incomplete oligosaccharide chains within the glycoprotein macromolecule. Such chains have been detected and rough quantitative estimates of their proportion[27] indicate that approximately 22% of the total fucose-containing oligosaccharides are oligosaccharides not carrying the nonreducing terminal N-acetylgalactosamine residues.

[27] M. M. Baig and D. Aminoff, unpublished observations (1971).

[98] β-Mannosidase from Snails[1]

By Kazuyuki Sugahara and Ikuo Yamashina

The sources for preparation of β-mannosidase [EC 3.2.1.25] are rather limited. In mammalian tissues, its activity is much lower compared to α-mannosidase. Viscera of gastropods have been used to prepare β-mannosidase.[2] In the viscera of the snails, *Achatina fulica*, collected in the Okinawa Islands, the activity of β-mannosidase is higher than that of α-mannosidase (see the table).

Assay Method

Phenyl β-mannoside, synthesized according to Helferich and Winkler,[3] is used as substrate, and liberated phenol is determined by the method of Asp[4] using a 4-aminoantipyrine reagent. Substrate (5 mM) and enzyme are incubated in a buffer, pH 4.5, prepared from 0.05 M citric acid and 0.025 M K$_2$HPO$_4$ at 37°.

Enzyme Purification

All operations are carried out at 0–4° unless otherwise stated.

Step 1. Preparation of Crude Extract. The acetone powder of snail viscera is used to extract β-mannosidase. A 10-g portion of the powder is homogenized with 200 ml of 0.05 M phosphate buffer, pH 7.24, in a Waring Blendor for 10 minutes. The homogenate is centrifuged at

[1] K. Sugahara, T. Okumura, and I. Yamashina, *Biochim. Biophys. Acta* **268**, 488 (1972).
[2] T. Muramatsu and F. Egami, *J. Biochem. (Tokyo)* **62**, 700 (1967).
[3] B. Helferich and S. Winkler, *Chem. Ber.* **66**, 1556 (1933).
[4] N. G. Asp, *Anal. Biochem.* **40**, 281 (1971).

13,000 g for 20 minutes, then the precipitate is reextracted with 50 ml of the buffer. The combined extract is submitted to freezing with Dry Ice–acetone then thawing at room temperature.

Step 2. Ammonium Sulfate Fractionation. The resulting precipitate formed in the preceding step is removed by centrifugation, then the supernatant is made 32% saturation with $(NH_4)_2SO_4$. The supernatant obtained after centrifugation at 13,000 g for 20 minutes is then brought to 59% $(NH_4)_2SO_4$ saturation. The precipitate collected by centrifugation is dissolved in water to give a protein concentration of 0.3–0.4%.

Step 3. Acetone Fractionation. The solution from the preceding step is made 0.04 M and 0.01 M with $(NH_4)_2SO_4$ and $Zn(CH_3COO)_2$, respectively, then adjusted to pH 6.1 with 1 N NaOH. Cold 50% acetone made up of equal volumes of acetone and 0.01 M $Zn(CH_3COO)_2$–acetic acid buffer, pH 5.9, is added with stirring to give 40% acetone concentration. The resulting precipitate is removed by centrifugation at 13,000 g for 20 minutes, then the supernatant is brought to 70% acetone concentration by adding cold acetone. The precipitate collected by centrifugation is dissolved in 0.01 M phosphate buffer, pH 7.24, then dialyzed against water. The dialyzate is lyophilized after centrifugation to remove insoluble materials.

Step 4. Chromatographic Fractionations. The material from the preceding step (90 mg protein) is dissolved in 6 ml of 0.005 M phosphate buffer, pH 6.8, then applied to a column of hydroxyapatite (prepared according to Siegelman et al.[5]) (1.6 × 16 cm) equilibrated with the same buffer. About 70% of the β-mannosidase activity passes through the column; the rest is eluted with 0.2 M buffer together with β-N-acetylglucosaminidase [EC 3.2.1.30] activity.

Each one third of the effluent (about 10 ml) is applied to a column of DEAE-cellulose (1.6 × 16 cm) equilibrated with 0.005 M phosphate buffer, pH 6.8, then eluted with phosphate buffer of stepwise increasing concentrations; 0.005 M (44 ml), 0.02 M (24 ml) and 0.1 M (92 ml). Flow rate is 15 ml/hour, and 4-ml fractions are collected. Fractions in the peak of β-mannosidase activity eluted with 0.02 M buffer (tubes 16–30) are collected, concentrated and dialyzed. This preparation is designated as purified β-mannosidase. The second peak of β-mannosidase activity is eluted with 0.1 M buffer, which overlaps a small peak of β-N-acetylglucosaminidase activity.

The procedure for preparing the purified β-mannosidase is summarized in the table.

[5] H. W. Siegelman, G. A. Wieczorek, and B. C. Turner, *Anal. Biochem.* **13**, 402 (1965).

PURIFICATION OF β-MANNOSIDASE FROM SNAIL VISCERA

Fraction	Total activity (units)	Specific activity (units/mg protein)	Percent recovery	α-Mannosidase activity (units)	β-N-Acetylglu-cosaminidase activity (units)
Original extract	82	0.041	100	37	366
(NH₄)₂SO₄ fractionation	78	0.103	95	32	266
Acetone fractionation	50	0.54	61	1.0	140
Lyophilization	40	0.43	49	0.4	122
Hydroxyapatite chromatography	27	1.41	33	0.0095	0.368
DEAE-cellulose chromatography	26.5	4.73	32	0.0023	0.096

Properties

1. The purified β-mannosidase is still heterogeneous on discelectrophoresis at pH 9.4,[6] showing several protein bands, but is practically free from α-mannosidase and β-N-acetylglucosaminidase activities. Specific activity of the purified β-mannosidase is 4.73 units per milligram protein under the assay conditions, but is 6.67 on calculation from a Lineweaver-Burk plot.

2. The enzyme has an optimum at pH 4.5, but the peak of activity is broad from pH values 4–5. At the optimum pH, the K_m value with phenyl β-mannoside is 6.5 mM.

3. The enzyme is very stable if its aqueous solution is kept frozen at $-20°$. Heating the enzyme solution at 55° for 5 minutes destroys the activity to various extents depending on pH of the solution. At pH values of 5–7, only 10–20% of the activity is lost, but at pH values of 3–4, more than 80% of the activity is lost.

Hydrolysis of Natural Substrates

1. *Oligomannoside.* A trimannoside, α-Man-(1 → 4)-β-Man-(1 → 4)-Man, from ivory nut mannan[7] is hydrolyzed stepwise by the successive actions of α-mannosidase from hog kidney[8] and the purified β-mannosidase.

2. *Glycopeptides.* Core glycopeptides having a structure, Man-

[6] B. J. Davis, *Ann. N.Y. Acad. Sci.* **121**, 404 (1964).
[7] G. O. Aspinall, R. B. Rashbrook, and G. Kessler, *J. Chem. Soc.* 215 (1958).
[8] T. Okumura and I. Yamashina, *J. Biochem.* (*Tokyo*) **68**, 561 (1970).

(GlcNAc)₂-Asn, obtained from the α-mannosidase digest of the parent glycopeptides which are prepared from ovalbumin or Taka-amylase [EC 3.2.1.1] by pronase digestion are susceptible to the purified β-mannosidase.[9]

[9] K. Sugahara, T. Okumura, and I. Yamashina, *FEBS* (*Fed. Eur. Biochem. Soc.*) *Lett.* **20**, 44 (1972).

[99] β-N-Acetylglucosaminidase from Hen Oviduct

By ANTHONY L. TARENTINO and FRANK MALEY

Although the purification of glycosidases from a wide variety of materials has been presented,[1] the diverse nature and limited availability of some of these enzyme sources have detracted from their utility. Hen oviduct, like most gonadal tissue, has proved to be a fairly rich source of glycosidases, especially β-N-acetylglucosaminidase. As a consequence, this tissue has provided a means of obtaining several glycosidases fairly free of one another. The latter criterion is essential for meaningful results in polysaccharide sequence studies, particularly if lengthy incubation times are required. An indication of the type and quantity of glycosidases in hen oviduct extracts is presented in Table I.

$$\text{RO-}\beta\text{-}N\text{-acetyl-D-glucosaminide} + H_2O \rightarrow \text{ROH} + N\text{-acetyl-D-glucosamine}$$

Assay Method

Principle. The assay is based on the color developed in alkaline solution upon the release of *p*-nitrophenol from *p*-nitrophenly-*N*-acetyl-β-D-glucosaminide. The procedure is essentially that described in detail by Levvy and Conchie,[2] but modified to increase the sensitivity.

Reagents

Sodium citrate, 0.5 *M*, pH 3.0
Sodium citrate, 0.5 *M*, pH 4.5
p-Nitrophenyl-*N*-acetyl-β-D-glucosaminide, 0.012 *M*
p-Nitrophenyl-*N*-acetyl-β-D-galactosaminide, 0.012 *M*
Bovine serum albumin, 1% in 0.9% NaCl

Procedure. The 0.5-ml reaction mixture contains 0.2 ml of citrate buffer (pH 4.5 in the case of β-N-acetyl-D-glucosaminides and pH 3.0 in the case of β-N-acetyl-D-galactosaminides), 0.1 ml of *p*-nitrophenyl-

[1] See Vol. 8 [96] and [117].
[2] See Vol. 8 [98].

TABLE I
GLYCOSIDASES IN HEN OVIDUCT BEFORE AND AFTER PARTIAL PURIFICATION

Enzyme	Enzyme activity	
	Extract[a] (units/mg × 10^{-3})	Partially purified[b] (units/mg × 10^{-3})
β-N-Acetylglucosaminidase	165	29
α-Fucosidase	0.33	0
β-Fucosidase	0.24	0
α-Galactosidase	0.93	2.0
β-Galactosidase	4.2	2.0
α-Glucosidase	1.2	0.74
β-Glucosidase	0.29	0.38
α-Mannosidase	2.34	14.2
β-Mannosidase	2.07	12.5
Glycosylasparaginase	0.29	0.94
Lysozyme	61	2.5

[a] Crude extract of Table II.
[b] P-cellulose (pH 7.1) stage of purification in Table II.

N-acetyl-β-D-hexosaminide, and an aliquot of enzyme diluted in the bovine serum albumin solution, particularly if large dilutions are involved. The reactions are started by the addition of enzyme to the 37° incubation mixture. After 1–5 minutes, the reaction is stopped by the addition of 2.5 ml of 0.1 M Na_2CO_3, and the color developed is determined at 400 nm in a Beckman or comparable spectrophotometer.

Definition of Unit and Specific Activity. A unit is defined as the amount of enzyme that hydrolyzes 1 μmole of substrate per minute under the conditions of assay. Specific activity is based on units per milligram of protein, with the latter determined from the 280 to 260 nm ratio.[3]

Source of Enzyme. The oviducts, obtained from hens whose egg laying capacity was greatly diminished, are washed, extraneous membranes are removed, and the organs are stored in plastic bags at −55°. Alternatively, extracts may be prepared and kept at −10° until used. The glycosidases in the extract are stable for several months at this temperature. All operations are conducted at 0–4°, with twice distilled deionized water used for all solutions. The pH is determined at 25°.

Purification Procedure

Step 1. Crude Extract. Approximately 2 kg of hen oviduct are cut into 3-inch sections and homogenized in a Sorvall omnimixer with 3 liters of 0.01 M potassium phosphate buffer (pH 7.5) containing 0.9% NaCl. The

[3] O. Warburg and W. Christian, *Biochem. Z.* **310**, 384 (1941).

homogenate is centrifuged at 13,000 g for 20 minutes, and the supernatant fluid is filtered through several layers of glass wool. The precipitate is reextracted with 1 liter of the same buffer, centrifuged, and the supernatant fractions are combined. The crude extract is divided into 4 equal portions, each of which is fractionated separately through steps 2 and 3, and then combined as described below.

Step 2. Ammonium Sulfate. The crude extract is diluted with 0.01 M potassium phosphate (pH 7.5) to a protein concentration of 25 mg/ml and solid ammonium sulfate is added with stirring to 0.35 saturation (19.4 g/100 ml extract). After 30 minutes, the suspension is centrifuged at 10,000 g for 20 minutes and the precipitate discarded. The supernatant fraction is increased to 0.65 saturation by the addition of 15.1 g of ammonium sulfate per 100 ml of extract, stirred for 30 minutes, and recentrifuged. The precipitate is dissolved in a minimum volume of 0.01 M potassium phosphate (pH 7.1) and dialyzed against two 6-liter changes of the same buffer.

Step 3. P-Cellulose I. The dialyzed extract from step 2 is applied to a P-cellulose column (7.6 × 16 cm) equilibrated previously with 0.01 M potassium phosphate (pH 7.1). Elution of the column with the same buffer is continued until the absorbance at 280 nm decreases to zero, at which time approximately 75% of the applied protein will have passed through the column. The column is developed with a linear gradient consisting of 1.3 liters of 0.01 M potassium phosphate (pH 7.1) in the mixing chamber and 1.3 liters of 0.3 M potassium phosphate (pH 7.1) in the reservoir. The enzyme is eluted at a buffer concentration of approximately 0.15 M. Fractions with a specific activity of greater than 3 are pooled and the enzyme is precipitated by addition of ammonium sulfate to 0.70 saturation (43.6 g/100 ml). The precipitate is dissolved in a small volume of 0.01 M potassium phosphate (pH 7.5) and dialyzed against two 6-liter changes of the same buffer.

Step 4. DEAE-Cellulose. The enzyme preparations from each P-cellulose column are pooled, concentrated by Amicon ultrafiltration to 76 ml, and applied to a DEAE-cellulose column (3.8 × 26 cm), equilibrated previously with 0.01 M potassium phosphate (pH 7.5). The column is washed successively with 0.01 M and 0.016 M potassium phosphate (pH 7.5), until in each case the absorbance at 280 nm decreases to zero. The β-N-acetylhexosaminidase activity is eluted with 0.10 M phosphate buffer (pH 7.5) and the pooled fractions are concentrated by ultrafiltration to 70 ml. The extract is dialyzed overnight against two 3-liter changes of 0.04 M potassium phosphate (pH 6.0).

Step 5. P-Cellulose II. The dialyzate obtained from step 4 is absorbed to a P-cellulose column (2.5 × 25 cm) equilibrated previously with

0.04 M potassium phosphate (pH 6.0). The column is washed successively with 0.04 M, 0.075 M, 0.10 M, and 0.13 M potassium phosphate (pH 6.0), with absorbance at 280 nm decreasing to zero before the addition of the next higher buffer. The enzyme is eluted with 0.20 M potassium phosphate (pH 6.0), concentrated by ultrafiltration to 30 ml, and dialyzed against two 6-liter changes of 0.1 M ammonium acetate (pH 4.6).

Step 6. CM-Cellulose. The dialyzate from step 5 is applied to a column of CM-cellulose (2 × 15 cm) equilibrated with 0.1 M ammonium acetate (pH 4.6). The column is developed in a stepwise manner with 0.1 M ammonium acetate (pH 4.6) as the starting buffer followed by a series of 0.1 M ammonium acetate buffers of increasing pH. Each buffer differs by 0.4 pH and is changed when the protein eluted reaches a minimum. This procedure is an adaptation of the method of Rhodes *et al.*,[4] who found that egg white proteins are eluted from CM-cellulose at pH levels close to their isoelectric point. The flow rate is 1 ml per minute, and 6-ml fractions are collected. The β-N-acetylhexosaminidase activity elutes sharply from the column at pH 7.1. The enzyme is concentrated by Amicon ultrafiltration and dialyzed against two 3-liter changes of 0.1 M ammonium acetate (pH 4.6). The purification scheme is summarized in Table II.

Properties

Stability. The enzyme is completely stable to storage at 0° for several months and no loss in activity has been encountered on repeated freezing

TABLE II
PURIFICATION OF HEN OVIDUCT β-N-ACETYLHEXOSAMINIDASE

Step	Volume (ml)	Activity (units)	Protein (g)	Specific activity (units/mg)	Recovery (%)
1. Crude extract	5070	64,225	317.7	0.195	100
2. Ammonium sulfate (0.35–0.60)	1510	36,500	61.4	0.509	57
3. P-cellulose I (pH 7.1)	421	19,614	2.85	7.10	31
4. DEAE-cellulose	70	14,371	0.80	17.8	23
5. P-cellulose II (pH 6.0)	30	8,000	0.088	90.0	12
6. CM-cellulose (pH 4.6)	24	3,500	0.0135	252.0	5.4

[4] M. B. Rhodes, P. R. Azari, and R. E. Feeney, *J. Biol. Chem.* **230**, 399 (1958).

and thawing or lyophilization. It may also be kept at 37° for 24 hours at pH 4.6 or 7.1 without loss in activity, but not at higher pH levels.

Isomeric Forms and Molecular Weight. At least two forms of the enzyme are detectable on polyacrylamide gel electrophoresis and isoelectric focusing. The major band has a molecular weight as determined by the electrophoresis procedure of Hedrick and Smith[5] of 118,000 and the minor band, a molecular weight of 158,000. On isoelectric focusing, the former enzyme has an isoelectric point of 6.45 and the latter, 6.86. The purest preparations possess specific activities as high as the most active reported, including the crystalline jack bean meal enzyme.[6]

Specificity. The hydrolytic capacity of the enzyme for the *p*-nitrophenyl-*N*-acetyl-β-D-glucosaminide and galactosaminide does not change throughout the purification procedure. The most active substrate is *p*-nitrophenyl-*N*-acetylglucosaminide (100%), and in the following order: β-methyl-*N*-acetylglucosaminide (9%), di-*N*-acetylchitobiose (4%), tetra-*N*-acetylchitotetrose (3.6%), tri-*N*-acetylchitotriose (2.4%), Asn-(GlcNAc)$_2$ (1.9%), Asn-(GlcNAc)$_2$(Man)$_6$ (GlcNAc)$_2$ (0.1%). The *p*-nitrophenyl-*N*-acetyl-β-D-glucosaminide is hydrolyzed 2.5 times faster than the corresponding *N*-acetylgalactosaminide at their respective pH optima. The hydrolysis of *N*-acetyl-α-D-hexosaminides and the *N*-acetyl-β-D-galactosaminide of Tay-Sachs ganglioside is not detectable.

Optimal pH and Kinetic Constants. The pH optimum for the *p*-nitrophenyl-*N*-acetylgalactosaminides is 3.0 and that for the *N*-acetylglucosaminides, 4.2. The K_m for the latter is 0.56 mM and the V_{max} at 37°, 218 μmoles per minute per milligram protein.

Inhibitors. The hen oviduct β-*N*-acetylhexosaminidase is almost selectively inhibited by Hg^{2+}. Of about 20 metal ions tested, only Ag^+ is inhibitory also, but at concentrations about 100 times greater than Hg^{2+}. Cu^{2+} and Fe^{3+} are ineffective. The inhibition is biphasic with an apparent noncompetitive K_i of $10^{-6}\,M$. Organic mercurials, including mersalyl, *p*-chloromercuribenzoate, CH_3HgI, and $(CH_3)_2Hg$, are only marginally effective in comparison to Hg^{2+}. The inhibition is reversed by thiols.

[5] J. L. Hedrick and A. J. Smith, *Arch. Biochem. Biophys.* **126**, 155 (1968).
[6] S.-C. Li and Y.-T. Li, *J. Biol. Chem.* **245**, 5153 (1970).

[100] α-Mannosidase and β-Mannosidase from Hen Oviduct

By T. Sukeno, A. L. Tarentino, T. H. Plummer, Jr., and F. Maley

$$RO\text{-}D\text{-}Mannoside + H_2O \rightarrow ROH + D\text{-}Mannose$$

Assay Method

Principle. The release of *p*-nitrophenol from the corresponding α- or β-D-mannopyranoside derivatives was measured in a manner identical to that described for β-*N*-acetylglucosaminidase.[1]

Purification Procedure

β-Mannosidase

Step 1. Crude Extract. Hen oviducts (120) stored at $-20°$ are thawed, minced, and homogenized in 4.6 liters of $0.01\,M$ potassium phosphate buffer (pH 7.5), and then centrifuged for 15 minutes at 12,000 *g*. The pellet is washed with 2.0 liters of the same buffer. Supernatant fractions are combined and filtered through glass wool. All the procedures are conducted at 0–4°, unless specified otherwise.

Step 2. Ammonium Sulfate I. The protein concentration of the supernatant fraction is brought to 25 mg/ml with $0.01\,M$ potassium phosphate buffer (pH 7.5), and 19 g of ammonium sulfate per 100 ml of solution is added (0.35 saturation). The resulting suspension is stirred for 20 minutes, then centrifuged at 12,000 *g* for 20 minutes. The supernatant fraction is brought to 0.60 saturation by the addition of 15.1 g of solid ammonium sulfate per 100 ml, stirred for 20 minutes, and centrifuged. The precipitate is dissolved in a minimal volume of $0.01\,M$ potassium phosphate buffer (pH 7.5) and dialyzed against 20 liters of the same buffer with three changes.

Step 3. pH 5.5. The dialyzed extract is adjusted to pH 5.5 with $1\,N$ acetic acid, and the resulting turbidity is removed by centrifugation.

Step 4. Ammonium Sulfate II. The supernatant fraction is brought to 0.70 saturation by adding 43.6 g of solid ammonium sulfate per 100 ml, stirred for 20 minutes, and centrifuged. The resulting pellet is dissolved in a minimal volume of $0.01\,M$ potassium phosphate buffer (pH 7.1) and dialyzed for 24 hours against three 6-liter changes of the same buffer.

Step 5. P-Cellulose (pH 7.1). The dialyzed extract (1600 ml), containing 34.1 g of protein, is applied to a P-cellulose column (7.6 cm × 32 cm),

---

[1] See this volume [99].

which had been equilibrated previously with 0.01 M potassium phosphate buffer (pH 7.1). Elution of the column with the same buffer yields both α- and β-mannosidases as a nonretarded fraction, which can be stored at −10° without loss in activity. For the purification of α-mannosidase, this fraction is placed directly on a P-cellulose (pH 6.1) column, as described in the section on α-mannosidase purification.

Step 6. pH 11.0. The pH of the P-cellulose eluate (1600 ml) is adjusted carefully with 0.1 N NaOH (about 260 ml) until it reaches 11.0. The solution is maintained at this pH for 40 minutes with occasional addition of 0.1 N NaOH, and then brought to pH 7.0 with about 30 ml of 0.5 M sodium citrate buffer (pH 4.5). The resulting turbid solution is centrifuged and the supernatant fraction dialyzed against 0.01 M sodium phosphate buffer (pH 6.1). With this treatment, 95% of the α-mannosidase is inactivated.

Step 7. P-Cellulose (pH 6.1). The dialyzed extract is placed on a P-cellulose column (7.6 cm × 40 cm) equilibrated previously with 0.01 M sodium phosphate (pH 6.1) and washed with the same buffer until the unadsorbed protein is eluted. β-Mannosidase is eluted with a solution of 0.01 M sodium phosphate (pH 6.1), 0.2 M sodium chloride and fractions of 25 ml are collected. The enzyme-containing fractions (50–120) are pooled.

Step 8. Acetone. The pH 6.1 P-cellulose eluate is concentrated to 1200 ml (2.5 mg protein per milliliter) in an Amicon Model 50 ultrafilter containing a PM-10 membrane. Acetone at 4° (514 ml) is added dropwise to 0.30 saturation, and the temperature is maintained at 4°. The resultant suspension is stirred for 30 minutes, and then centrifuged at 0° for 15 minutes at 16,000 g. The supernatant fraction is decanted, brought to 0.60 saturation with 1280 ml of acetone, stirred, and centrifuged. The precipitate is extracted four times with 50 ml volumes of a solution containing 0.05 M sodium citrate buffer (pH 4.6), 0.1 mM zinc acetate, and 0.05 M sodium chloride. The extracts are pooled and concentrated to 13 ml in the Amicon ultrafilter.

Step 9. Sephadex G-200. The concentrated β-mannosidase solution (885 mg protein) is applied to a Sephadex G-200 column (3.7 cm × 173 cm) and eluted with a solution containing 0.05 M sodium citrate (pH 4.6) 0.1 mM zinc acetate, and 0.05 M sodium chloride. Fractions of 8.8 ml are collected. The enzyme in fractions 112–142 is pooled and concentrated to 10 ml in the Amicon ultrafilter. About 96% of the β-N-acetylglucosaminidase from the preceding step is removed by this procedure.

Step 10. Calcium Phosphate Gel. The concentrated β-mannosidase solution (96 mg protein in 8 ml) is dialyzed against 100 volumes of 0.01 M potassium phosphate buffer (pH 7.1) with three changes and applied to a calcium phosphate gel column (2.0 cm × 8.0 cm). The support for the column is prepared by slowly pouring 500 ml of gel (30 mg/ml) into 1 liter of a 10% suspension of Whatman cellulose powder. The column is equilibrated with the same buffer as the dialyzate. Inert protein is removed with 0.05 M phosphate (pH 7.1) and the enzyme-containing fractions eluted with 0.1 M phosphate (pH 7.1) are pooled and concentrated to 1.2 ml in the Amicon ultrafilter. The final purification of β-mannosidase is 10,000-fold over the initial extract, with a recovery of 20%. The purified β-mannosidase from this step is devoid of detectable α-mannosidase, α- and β-galactosidases, α- and β-glucosidases, α- and β-fucosidases, and lysozyme (less than 0.1%), but contains 0.2% glycosyl asparaginase and 0.13% β-N-acetylglucosaminidase. Glycosyl asparaginase is inactivated completely by $2 \times 10^{-3} M$ DONV without any loss of β-mannosidase activity. The purification of β-mannosidase is summarized in Table I.

α-Mannosidase

Step 6a. P-Cellulose (pH 6.1). A portion (700 ml) of the P-cellulose (pH 7.1) elution step is used for the purification of α-mannosidase. This solution, containing a total of 8.5 g of protein, is dialyzed against 15 liters of 0.01 M sodium phosphate buffer (pH 6.1) with three changes. It

TABLE I
PURIFICATION OF HEN OVIDUCT β-MANNOSIDASE[a]

Step	Volume (ml)	Protein[b] (mg)	Total activity (units)	Specific activity (units/mg)	Yield (%)
1. Crude extract	6600	262,000	550	0.0021	100
2. Ammonium sulfate I	2800	72,000	420	0.0058	76
3. pH 5.5	3000	39,000	378	0.0092	69
4. Ammonium sulfate II	1600	34,100	407	0.0119	74
5. P-cellulose (pH 7.1)	2300	28,000	369	0.0125	67
6. pH 11.0	1800	17,400	202	0.0116	53
7. P-cellulose (pH 6.1)	1800	3,240	164	0.0506	43
8. Acetone	200	885	162	0.183	42
9. Sephadex G-200	10	96	150	1.56	39
10. Calcium phosphate gel	1.2	3.2	65	20.6	17

[a] A portion (700 ml) of the eluate from step 5 was taken for the purification of α-mannosidase. To compensate for this removal, the % yield from the pH 11 step on was obtained by multiplying total activity by 23/16 and dividing by 550.

[b] Determined by the method of Lowry et al. [O.H. Lowry, N. J. Rosebrough, A. L. Farr, and R. J. Randall, *J. Biol. Chem.* **193**, 265 (1951)].

is then transferred to a P-cellulose column (3.7 cm × 39 cm), which has been equilibrated with the same buffer as the dialyzate. The column is washed further with the 0.01 M phosphate buffer until most of the unadsorbed protein is removed, followed by elution with a solution of 0.01 M sodium phosphate (pH 6.1), 0.2 M NaCl. Fractions of 20 ml are collected. The enzyme eluted in fractions 27–40 is combined.

Step 7a. Heat Treatment. To the enzyme solution (340 ml), 0.34 ml of 0.1 mM zinc acetate is added. The solution is placed in a water bath at 85° until the temperature of the solution reaches 65° and is maintained at this temperature for 1 hour. The turbid solution is chilled with ice and centrifuged. The precipitate is washed once with a small volume of 0.01 M sodium phosphate buffer (pH 6.1), and the clear supernatant fractions are combined. By this treatment, β-mannosidase, β-N-acetylglucosaminidase, and glycosyl asparaginase are decreased in activity by 94%, 97%, and 91%, respectively, while α-mannosidase is unaffected.

Step 8a. Acetone. A 360-ml solution derived from the above heat step, containing 2.2 mg protein per milliliter, is brought to a 0.30 saturation with 154 ml of cold acetone (4%), stirred for 30 minutes, and centrifuged. Acetone (380 ml) is added to the supernatant fraction to 0.60 saturation. The precipitate is collected by centrifugation at 0° and extracted four times with approximately 50 ml volumes containing 0.05 M sodium citrate buffer (pH 4.6), 0.1 mM zinc acetate, and 0.05 M sodium chloride. The extracts are pooled and concentrated to 6 ml in the Amicon ultrafilter.

Step 9a. Sephadex G-200. The concentrated acetone extract is placed on a Sephadex G-200 column (2.5 cm × 173 cm) equilibrated previously and then eluted with a solution of 0.05 M sodium citrate buffer (pH 4.6), 0.1 mM zinc acetate, and 0.05 M sodium chloride. Fractions 35–45 are combined and concentrated to 5 ml in the Amicon ultrafilter. Each fraction contains 10 ml.

The overall purification of the enzyme from the initial extract is 1290-fold with a recovery of 26%. None of the following enzymes is detectable: β-mannosidase, α- or β-N-acetylglucosaminidase, α- or β-galactosidase, α- or β-fucosidase, α- or β-glucosidase, lysozyme (less than 0.1%), but about 1% glycosyl asparaginase is present. Contamination of the latter enzyme is completely inhibited by $2 \times 10^{-3}\,M$ DONV without any loss of α-mannosidase activity. The purification of α-mannosidase is summarized in Table II.

Properties

Stability. The differential response of the α- and β-mannosidases to heat and pH can be exploited to eliminate a large proportion of the unwanted contaminating glycosidases from each enzyme. Although most

TABLE II
PURIFICATION OF HEN OVIDUCT α-MANNOSIDASE[a]

Step	Volume (ml)	Protein (mg)	Total activity (units)	Specific activity (units/mg)	Yield (%)
6a. P-cellulose (pH 6.1)	340	1840	94	0.0511	43
7a. Heat treatment	360	775	86	0.111	39
8a. Acetone	190	241	76	0.314	35
9a. Sephadex G-200	5	19	58	3.02	27

[a] The purification for α-mannosidase follows almost exactly that described for β-mannosidase in Table I through the P-cellulose (pH 7.1) step. The initial total activity of α-mannosidase in the crude extract of Table I is 718 units.

of the β-N-acetylglucosaminidase is removed by the P-cellulose step,[1] unacceptable levels remain in the mannosidase eluates. In fact, some glycosidases are enriched in comparison to the crude extract.[1] By maintaining the pH of the P-cellulose (pH 7.1) step at 11.0 for 1 hour most of these enzymes, with the exception of β-mannosidase, are greatly reduced. The relative levels of the following enzymes persisting after this treatment are β-mannosidase, 100%; glycosyl asparaginase, 40%; β-N-acetylglucosaminidase, 30%; α-mannosidase, <5%. A similar reduction in activity can be obtained by incubating the P-cellulose (pH 6.1) eluate for 1 hour at 65°, but in this case α-mannosidase is unaffected. The relative activities of the various other enzymes tested in the extract are glycosyl asparaginase, 10%; β-N-acetylglucosaminidase, 5%; β-mannosidase, <1%. Zn^{2+}, as in the case of other α-mannosidases,[2,3] stabilizes the hen oviduct enzyme, but has no apparent effect on β-mannosidase. As anticipated, EDTA promoted the lability of α- but not β-mannosidase.

Inhibitors. Ag^+ at a concentration of 0.6 mM has no effect on β-mannosidase, but impairs α-mannosidase by 90%. Guanidine hydrochloride at a similar concentration impairs both enzymes, but α- is affected twice as much as β-mannosidase. In contrast, Hg^{2+} inhibits both enzymes. Cu^{2+}, Fe^{2+}, Fe^{3+}, and about 15 other metal ions have, at 10–100 times a concentration of Ag^+ that completely inhibited the enzyme, no effect on β-mannosidase. Mannono-$(1 \rightarrow 5)$-lactone is a competitive inhibitor of both enzymes with the K_i for β-mannosidase = $1.7 \times 10^{-5} M$ and for α-mannosidase = $1.1 \times 10^{-4} M$.

Activation. Although Hg^{2+} is an effective inhibitor of the mannosidases, the organic mercurials have little or no influence on the inhibition. However, p-chloromercuribenzoate increases β-mannosidase activity by 65%, provided the substrate concentration is enhanced simultaneously.

[2] S. M. Snaith and G. A. Levvy, *Biochem. J.* **110**, 663 (1968).
[3] T. Okumura and I. Yamashina, *J. Biochem.* (*Tokyo*) **68**, 561 (1970).

Specificity and Kinetics. The p-nitrophenyl-α-D-mannoside is hydrolyzed at 40 times the rate of the corresponding α-methyl derivative, while p-nitrophenyl-β-D-mannoside is hydrolyzed at 14 times the rate of the corresponding β-methyl derivative. No detectable hydrolysis of Asn-(GlcNAc)$_2$(Man)$_1$ could be shown with α-mannosidase, while this compound is hydrolyzed at one-fourth the rate of p-nitrophenyl-β-D-mannosidase by β-mannosidase. (GlcNAc)$_2$(Man)$_1$ is hydrolyzed at one-third the rate of the p-nitrophenyl derivative. At the pH optimum of both enzymes, 4.6, the K_m of β-mannosidase for p-nitrophenyl-β-D-mannoside is $4.5 \times 10^{-3} M$ and that for Asn-(GlcNAc)$_2$(Man)$_1$ is $16.9 \times 10^{-3} M$. However, their V_{max} values are similar. The K_m of α-mannosidase for p-nitrophenyl-α-D-mannoside is $2.9 \times 10^{-3} M$.

Molecular Weights. To estimate the size of each enzyme, a calibrated Sephadex G-200 column is used. The molecular weight for α-mannosidase obtained by this procedure is 250,000, and that for β-mannosidase, 100,000.

[101] Glycosyl Asparaginase from Hen Oviduct

By A. L. TARENTINO and F. MALEY

β-Aspartylglycosylamine + H$_2$O → aspartic acid + 1-amino-N-acetylglucosamine

1-Amino-N-acetylglucosamine + H$_2$O → N-acetylglucosamine + ammonia

Assay Methods

Principle. Glycosyl asparaginase hydrolyzes the synthetic substrate, Asn-GlcNAc,[1] to N-acetylglucosamine, aspartic acid, and ammonia[2-5] and activity is measured usually by assaying for N-acetylglucosamine. However, aspartic acid is also a useful parameter for determining this enzyme.

Method I. The assay is essentially that of Mahadevan and Tappel,[3] but the volume of the reaction mixture is reduced to 0.1 ml. The procedure measures the release of N-acetylglucosamine and is suitable for determining glycosyl asparaginase activity in crude or purified fractions and is convenient for routine purification of the enzyme.

[1] 1-β-Aspartyl-2-acetamido-1,2-dideoxy-D-glucosylamine.

[2] See Vol. 8 [101].

[3] S. Mahadevan and A. L. Tappel, *J. Biol. Chem.* **242**, 4568 (1967).

[4] M. Makino, T. Kojima, T. Ohgushi, and I. Yamashina, *J. Biochem. (Tokyo)* **63**, 186 (1968); see also this volume [102].

[5] A. Tarentino and F. Maley, *Arch. Biochem. Biophys.* **130**, 295 (1969).

Method II. The release of aspartic acid from Asn-GlcNAc is measured spectrophotometrically in the presence of an excess of glutamic–oxalo-acetate transaminase and malate dehydrogenase. The oxidation of NADH is followed continuously at 340 nm, and is directly proportional to the glycosyl asparaginase activity.[5] This procedure can be used only with partially purified enzyme preparations (specific activity at least 0.10), but is more suitable than Method I for studying kinetics and general enzyme properties.

Reagents

> Potassium phosphate buffer, 0.2 M, pH 7.5
> Glutamic–oxaloacetate transaminase (Sigma, specific activity 290 units/mg) diluted with water to 580 units/ml
> Malate dehydrogenase (Sigma, specific activity 125 units/mg) diluted with 2.6 M ammonium sulfate to 120 units/ml
> NADH, 0.012 M, prepared in 1% NaHCO$_3$
> α-Ketoglutarate, 0.1 M, neutralized to pH 7.0
> Asn-GlcNAc, 0.05 M

Procedure. A stock solution containing the following components is prepared daily: phosphate buffer 2.0 ml; NADH, 0.10 ml; malate dehydrogenase, 0.50 ml; glutamic–oxaloacetate transaminase, 0.5 ml; and α-ketoglutarate, 0.10 ml. An aliquot (0.32 ml) of the stock solution is added to cuvettes with a 0.63 ml total capacity (10-mm light path, 2 mm width), followed by Asn-GlcNAc, 0.02 ml; and water and glycosyl asparaginase in a final volume of 0.50 ml. The decrease in absorbance at 340 nm is followed continuously at 30° in a Gilford recording spectro-photometer with a full-scale expansion of 0–0.25 Å. The rate of oxidation of NADH, with enzyme preparations of specific activity at least 0.10, is negligible in the absence of added Asn-GlcNAc.

Definition of Unit and Specific Activity. One unit is defined as the amount of enzyme that releases 1 micromole of *N*-acetylglucosamine from Asn-GlcNAc per minute at 37°. Specific activity is expressed as units per milligram of protein. Protein is measured by the method of Lowry *et al.*[6]

Purification Procedure

The method presented is somewhat simpler than the one described previously.[5] All operations are conducted between 0° and 4° unless speci-fied otherwise. The pH of all buffers is determined at 25°.

[6] O. H. Lowry, N. J. Rosebrough, A. L. Farr, and R. J. Randall, *J. Biol. Chem.* **193**, 265 (1951).

Steps 1–5. These purification steps are identical to those used on the purification of hen oviduct α- and β-mannosidases.[7]

Step 6. P-Cellulose Chromatography. Of the fraction not retained by the P-cellulose (pH 7.1) column in step 5, 20% (460 ml) is dialyzed against three 6-liter changes of 0.01 M potassium phosphate (pH 6.0). The extract is applied to a P-cellulose column (3.6 × 25 cm) previously equilibrated with 0.01 M potassium phosphate (pH 6.0) and the column is washed with the same buffer until the absorbance at 280 nm is negligible. Approximately 70% of the applied protein appears in the nonabsorbed fraction. The column is then developed with a linear gradient consisting of 1.5 liters of 0.01 M potassium phosphate (pH 6.0) in the mixing chamber and 1.5 liters of 0.35 M potassium phosphate (pH 6.0) in the reservoir. The flow rate is 65 ml/hour, and 15-ml fractions are collected.

Glycosyl asparaginase as well as β-mannosidase activity is eluted between tubes 80 and 125. The α-mannosidase activity elutes between tubes 65–110. The enzyme in tubes 90–120 is pooled (425 ml) and concentrated by ultrafiltration to 150 ml. The enzyme is then dialyzed for 24 hours against three 4-liter changes of 0.01 M Tris·HCl (pH 8.0).

Step 7. DEAE-Cellulose Chromatography. The enzyme from step 6 is applied to a column (2 × 25 cm) of DEAE-cellulose that was equilibrated previously with 0.01 M Tris·HCl (pH 8.0). The column is washed with the same buffer until the absorbance at 280 nm decreases to zero and then is developed with a linear gradient consisting of 1 liter of 0.01 M Tris·HCl in the mixing chamber and 1 liter of 0.01 M Tris·HCl containing 0.2 M NaCl in the reservoir. The flow rate is 60 ml/hour, and 12-ml fractions are collected. The enzyme elutes between tubes 100 and 130.[8] The pooled fractions (370 ml) are concentrated by ultrafiltration to 50 ml. The pH of the solution is then adjusted to 4.6 by the slow addition of cold 1 M acetic acid, and dialyzed 24 hours against three 4-liter changes of 0.1 M ammonium acetate buffer (pH 4.6).

Step 8. CM-Cellulose Chromatography. Following dialysis, the enzyme from step 7 is concentrated by ultrafiltration to 10 ml and applied to a column (1.5 × 10 cm) of CM-cellulose that was equilibrated previously with 0.1 M ammonium acetate (pH 4.6). The flow rate is 60 ml/hour and 5-ml fractions are collected. The column is developed stepwise with 0.1 M ammonium acetate buffers of increasing pH (0.3 unit

---

[7] See this volume [100].

[8] β-Mannosidase is very unstable to chromatography on DEAE-cellulose in Tris·HCl (pH 8.0). Less than 0.1% of the total activity is recovered in tubes 100–130.

per buffer change until the protein eluted reaches a minimum).[9] The glycosyl asparaginase activity is eluted sharply from the column at pH 6.0. Although 30% of the enzyme elutes at pH 5.5, this fraction contains the bulk of the applied protein and is discarded. The pooled fractions from the pH 6.0 eluate (20 ml) are concentrated by ultrafiltration to 5.0 ml and dialyzed 24 hours against three 2-liter changes of 0.01 M potassium phosphate (pH 7.1).

Step 9. Calcium Phosphate Gel Chromatography. The enzyme from step 8 (8.3 ml) is applied to a column (1.5 × 5 cm) of calcium phosphate gel[10] that was equilibrated previously with 0.01 M potassium phosphate (pH 7.1). The flow rate is 30 ml/hour, and 5-ml fractions are collected. The column is washed with the starting buffer until the absorbance at 280 nm decreases to zero. The enzyme is eluted sharply by 0.05 M potassium phosphate (pH 7.1). The pooled fractions are concentrated by ultrafiltration to 4.0 ml. (Typical results are summarized in the table.)

Properties

Purity and Molecular Weight. Step 9 enzyme is nonhomogeneous as judged by the presence of 3 bands on disc gel electrophoresis. The preparation still contains 0.05% α-mannosidase, but is devoid of β-mannosidase, β-N-acetylglucosaminidase, β-galactosidase, and L-asparaginase. The molecular weight of the enzyme, calculated from combined

PURIFICATION OF HEN OVIDUCT GLYCOSYL ASPARAGINASE[a]

Step	Volume (ml)	Protein (mg)	Activity (units)	Specific activity (unit/mg)	Yield (%)
1 to 5[b]					
6. P-cellulose (pH 6.0)	395	480	5.5	0.016	18
7. DEAE-cellulose (pH 8.0)	360	116	3.5	0.030	12
8. CM-cellulose (pH 4.6)	20	7.2	1.5	0.208	5
9. Calcium phosphate gel	4	2.6	1.2	0.462	4

[a] These purification steps are identical to those used in the purification of hen oviduct α- and β-mannosidase see this volume [100]. For step 6, 40% of the P-cellulose (pH 7.1) fraction is used.

[b] The specific activity of glycosyl asparaginase in the crude extract (step 1) is 0.00029 unit/mg and at step 5 it is 0.00094 unit/mg.

[9] M. B. Rhodes, P. R. Azari, and R. E. Feeney, *J. Biol. Chem.* **230**, 399 (1958).
[10] The calcium phosphate gel was prepared as described for α- and β-mannosidases in this volume [100].

sucrose density centrifugation and Sephadex G-200 chromatography data, is 101,000.[5]

Stability. The purified enzyme can be stored at 4° for at least 3 months without loss of activity and is stable to freezing and thawing for at least 1 month. Glycosyl asparaginase is completely stable over the pH range 4.5–7.0 at 37° for 1 hour but shows a gradual loss of activity at higher pH values.

pH Optimum, Stoichiometry, and Isoelectric Point. Although maximal activity is at pH 5.1, the enzyme is still 80% active over the pH range 6.0–8.0. Hydrolysis at pH 5.1 yields aspartate, N-acetylglucosamine, and ammonia in a ratio of 1:1:1. 1-Amino-N-acetylglucosamine, an intermediate in the reaction, is nonenzymatically converted to N-acetylglucosamine and ammonia.[2-5] The isoelectric point as determined by isoelectric focusing is 7.2.

Specificity. The enzyme hydrolyzes the amide linkage of Asn-GlcNAc or asparagine-oligosaccharides of ovalbumin, ribonuclease B, and transferrin.[5] A free α-amino and α-carboxyl group on the asparagine moiety is essential for enzyme activity. The 1-N-glycyl-, 1-N-valyl-, and 1-N-seryl-2-acetamido-2-deoxy-β-D-glycopyranosylamines are not substrates for the enzyme.[11]

Kinetic Properties. The K_m value for Asn-GlcNAc is $1 \times 10^{-4} M$ by assay I, and $4.5 \times 10^{-4} M$ by assay II.[5]

Inhibitors. The enzyme is irreversibly inhibited by 5-diazo-4-oxo-L-norvaline (DONV), a compound that also inhibits L-asparaginase.[12] The K_i for DONV inhibition is $9.5 \times 10^{-6} M$.[5] The lactone of N-acetylglucosamine is not inhibitory.

Distribution. Glycosylasparaginase has been isolated from sheep epididymis,[2] rat liver and kidney,[3,11] pig serum,[4] and hen oviduct.[5] The enzymes appear to have the same substrate specificity and kinetic constants and differ primarily in pH optima and molecular weight.

[11] J. Conchie and I. Strachan, *Biochem. J.* **115**, 709 (1969).
[12] R. C. Jackson and R. E. Handschumacher, *Biochemistry* **9**, 3585 (1970).

[102] 4-L-Aspartylglycosylamine Amido Hydrolase (Glycosyl Asparaginase) from Hog Kidney[1]

By MICHIAKI KOHNO and IKUO YAMASHINA

An enzyme capable of hydrolyzing the amide bond of the aspartylglycosylamine which represents the linkage between polypeptide chain

[1] M. Kohno and I. Yamashina, *Biochim. Biophys. Acta* **258**, 600 (1972).

and carbohydrate moiety in many glycoproteins is widely distributed in mammalian tissues.[2] Its lysosomal localization has also been demonstrated.[3,4] The enzyme has been purified partially from various sources, i.e., hog serum,[5,6] rat liver and kidney,[3,7] and hen oviduct.[8] Hog tissue is used to prepare the homogeneous enzyme preparation since this tissue contains the largest amount of the enzyme and the highest specific activity compared to liver, pancreas, and spleen.

Based on the mode of action of the enzyme, it should be named "4-L-aspartylglycosylamine amido hydrolase," but it is referred to as amidase for convenience.

Assay Method

4-L-Aspartylglycosylamine, abbreviated as Asn-GlcNAc, (2-acetamido-1-N-(L-aspart-4-oyl)-2-deoxy-β-D-glucopyranosylamine) is synthesized according either to Yamashina et al.[9] or to Kiyozumi et al.[10] Assay mixture consists of 5 mM Asn-GlcNAc and enzyme preparation in a buffer, pH 5.50, prepared from 0.05 M citric acid and 0.05 M Na$_2$HPO$_4$. The mixture is incubated at 37° for appropriate times, the reaction is stopped by heating at 100° for 5 minutes, and the N-acetylglucosamine released is determined by the Morgan–Elson reaction according to Levvy and McAllan.[11] When the reaction is carried out at alkaline pH values, the reaction is stopped by addition of an equal volume of 10% trichloroacetic acid. The mixture, after being kept for another 60 minutes (shorter period may be used), is then neutralized with 0.6 N NaOH and used for the Morgan–Elson reaction.

The reaction proceeds in two steps (1) and (2), of which step (1) is catalyzed by the enzyme.

$$\text{Asn-GlcNAc} + \text{H}_2\text{O} \rightarrow \text{1-amino-}N\text{-acetylglucosamine} + \text{aspartic acid} \qquad (1)$$

$$\text{1-Amino-}N\text{-acetylglucosamine} + \text{H}_2\text{O} \rightarrow N\text{-acetylglucosamine} + \text{NH}_3 \qquad (2)$$

[2] I. Yamashina, in "Glycoproteins" (A. Gottschalk, ed.). Elsevier, Amsterdam, in press.

[3] S. Mehadevan and A. L. Tappel, J. Biol. Chem. 242, 4568 (1967).

[4] T. Ohgushi and I. Yamashina, Biochim. Biophys. Acta 156, 417 (1968).

[5] M. Makino, T. Kojima, and I. Yamashina, Biochem. Biophys. Res. Commun. 24, 961 (1966).

[6] M. Makino, T. Kojima, T. Ohgushi, and I. Yamashina, J. Biochem. (Tokyo) 63, 186 (1968).

[7] J. Conchie and I. Strachan, Biochem. J. 115, 709 (1969).

[8] A. L. Tarentino and F. Maley, Arch. Biochem. Biophys. 130, 295 (1969); see also this volume [101].

[9] I. Yamashina, M. Makino, K. Ban-I, and T. Kojima, J. Biochem. (Tokyo) 58, 168 (1965).

[10] M. Kiyozumi, K. Kato, T. Komori, A. Yamamoto, T. Kawasaki, and H. Tsukamoto, Carbohyd. Res. 14, 355 (1970).

[11] G. A. Levvy and A. McAllan, Biochem. J. 73, 127 (1959).

Enzyme Purification

Step 1. Extraction and Ammonium Sulfate Fractionation. A 1-kg portion of minced hog kidney is homogenized with 2 liters of 0.05 M citrate buffer, pH 5.0, in a Waring Blendor for 20 minutes, and the homogenate is centrifuged at 8000 g for 20 minutes. The residue is reextracted with 1 liter of the buffer. The extraction can similarly be effected with 0.05 M phosphate buffer, pH 7.5, 0.9% NaCl, or water. Subsequent operations are carried out at 0–4° unless otherwise stated.

Solid $(NH_4)_2SO_4$ is added to the combined supernatant to 35% saturation, and the mixture is centrifuged at 10,000 g for 20 minutes. The supernatant is then brought to 60% $(NH_4)_2SO_4$ saturation by adding solid $(NH_4)_2SO_4$, and the resulting precipitate is collected by centrifugation. The precipitate is dissolved in water and dialyzed. An inactive deposit formed during dialysis is removed by centrifugation.

Step 2. Heat Treatment. The supernatant from the preceding step is diluted with water to give a protein concentration of about 1%, then it is heated at 60° for 15 minutes. After centrifugation to remove insoluble materials, the supernatant is brought to 80% $(NH_4)_2SO_4$ saturation by adding solid $(NH_4)_2SO_4$, and the resulting precipitate is collected by centrifugation. The pellet is dissolved in water, dialyzed, and lyophilized.

Step 3. Acetone Fractionation. Materials from the preceding step is dissolved in a solution containing 0.04 M $(NH_4)_2SO_4$ and 0.01 M $Zn(CH_3COO)_2$ to give about 1% protein concentration, the pH being adjusted to 6.1 with acetic acid. Cold 50% acetone made up of equal volumes of acetone and 0.01 M zinc acetate buffer, pH 5.9, is added with vigorous stirring to give 30% acetone concentration. The mixture is kept at −5 to −10° for 20 minutes, then centrifuged at −10°. The supernatant is dialyzed and concentrated to about one-thirtieth its original volume by ultrafiltration using a membrane filter (Diaflo Ultrafiltration, Model 200, with membrane PM-10, Amicon Corporation, Cambridge, Massachusetts). The concentrate is then dialyzed to equilibrium against 0.02 M potassium phosphate buffer, pH 7.0.

Step 4. DEAE-Cellulose Column Chromatography. The dialyzate from the preceding step, corresponding to about 1.7 g protein, is applied to a column of DEAE-cellulose (4 × 50 cm) equilibrated with 0.02 M potassium phosphate buffer, pH 7.0. After eluting inactive proteins with the same buffer, amidase is eluted with 0.02 M buffer containing 0.1 M NaCl. Fractions containing the peak of amidase activity are pooled and concentrated to about 50 ml by ultrafiltration, then dialyzed to equilibrium against 0.01 M potassium phosphate buffer, pH 7.0.

Step 5. Hydroxyapatite Column Chromatography. Hydroxyapatite is prepared according to Siegelman *et al.*[12] The dialyzate from the preceding step is applied to a column of hydroxyapatite (4.5 × 16 cm) equilibrated with 0.01 M potassium phosphate buffer, pH 7.0. Elution is carried out with potassium phosphate buffer, pH 7.0, of stepwise increasing concentrations, i.e., 0.03 M (640 ml), 0.06 M (1360 ml), and 0.15 M (1170 ml). The flow rate is 35 ml/hour. Two peaks of amidase activity are obtained on elution with 0.06 M and 0.15 M buffer, which are designated as Fr. H-1 and Fr. H-2, respectively. Total recovery of activity is about 70%, of which two-thirds is accounted for by Fr. H-1. Each fraction is further purified.

Step 6. DEAE-Sephadex A-50 Column Chromatography. Fr. H-1 is concentrated to about 50 ml by ultrafiltration, then dialyzed against 0.02 M potassium phosphate buffer, pH 7.0. The dialyzate is applied to a column of DEAE-Sephadex A-50 equilibrated with 0.02 M potassium phosphate buffer, pH 7.0, and eluted with a linear gradient from zero to 0.3 M NaCl in the buffer at a flow rate of 27 ml/hour. Fractions of 9 ml are collected. Amidase activity is eluted in two fractions, i.e., Fr. H-1-A-1 in tubes 133–160 and Fr. H-1-A-2 in tubes 162–176. Fr. H-1-A-1, the major fraction, is pooled, concentrated, and dialyzed against 0.02 M buffer. The dialyzate is rechromatographed on a column of DEAE-Sephadex A-50 under conditions similar to those of the initial chromatography. Fractions with constant specific activity (tubes 107–113) are pooled. This pooled fraction, designated as amidase-1, has a specific activity of 1.55 units per milligram of protein and is homogeneous on disc electrophoresis.

Upon chromatography of Fr. H-2 on a column of DEAE-Sephadex A-50 under conditions similar to those used for Fr. H-1, three peaks of amidase activity are obtained; i.e., Fr. H-2-A-1 in tubes 140–149, Fr. H-2-A-2 in 156–167 and Fr. H-2-A-3 in 175–185. Fr. H-2-A-1 is indistinguishable from Fr. H-1-A-1 in cochromatography on DEAE-Sephadex A-50. Similarly, Fr. H-2-A-2 is indistinguishable from Fr. H-1-A-2, and the combined fraction is designated as amidase-2. Fr. H-2-A-3, which differs from amidases-1 and -2 on DEAE-Sephadex A-50 chromatography, is designated as amidase-3.

The purification procedure for amidase-1 is summarized in the table.

Properties of the Purified Amidases

1. Amidase-1 is homogeneous on disc electrophoresis (at pH 9.4[13]) as well as on that in the presence of sodium dodecyl sulfate (SDS) and

[12] H. W. Siegelman, G. A. Wieczorek, and B. C. Turner, Anal. Biochem. 13, 402 (1965).

[13] B. J. Davis, Ann. N.Y. Acad. Sci. 121, 2, 404 (1964).

PURIFICATION OF GLYCOSYL ASPARAGINASE FROM HOG KIDNEY

Fraction	Total activity (units/10 kg tissue)	Specific activity (units/mg protein)	Percent recovery
Citrate buffer extract, 0.05 M	50.5	0.000081	100
$(NH_4)_2SO_4$ fractionation	46.9	0.000526	92.4
Heat treatment	39.0	0.0018	76.8
Acetone fractionation	40.3[a]	0.0133	79.5[a]
DEAE-cellulose chromatography	32.4	0.0829	63.7
Hydroxyapatite chromatography (Fr. H-1)	15.4	0.161	30.2
DEAE-Sephadex A-50 chromatography (Fr. H-1-A-1)	9.3	1.03	18.3
DEAE-Sephadex A-50 rechromatography	7.0	1.55	13.8

[a] Slight increases in total activity from the previous step may be due to removal of substances interfering with amidase activity.

mercaptoethanol.[14] It has a sedimentation coefficient of 4.60 S. Molecular weight is about 70,000, estimated by SDS-disc electrophoresis and gel filtration. Amidase is a glycoprotein, containing about 3–4 moles of glucosamine per mole.

Amidase-1 shows none of the following glycosidase activities, i.e., α- and β-N-acetylglucosaminidases, α- and β-mannosidases, α- and β-galactosidases, and α-fucosidase, when each substrate (synthetic phenolic glycosides) is incubated with 300 mU/ml of amidase-1 for 5 hours under optimal conditions.

Amidases-2 and -3 are heterogeneous. Isoelectric points, determined by isoelectric focusing, are 5.12, 4.85, and 4.70 for amidases-1 -2 and -3, respectively.

2. The pH-profile of amidase-1 activity varies depending on conditions used to terminate the reaction and also on the assay method. When the enzyme reaction is terminated by adding 10% trichloroacetic acid, a sharp peak is observed at a pH of about 5.5, but activity is still considerable at higher pH values, with a broad peak at around pH 8. However, when the enzyme reaction is terminated by applying ammonia determination[15] directly to the reaction mixture, a single sharp peak of activity is observed at a pH of about 5, due to incomplete release of ammonia from 1-amino-N-acetylglucosamine, the product of the enzymatic reaction (step 1). Note that 1-amino-N-acetylglucosamine is

[14] K. Weber and M. Osborn, J. Biol. Chem. 244, 4406 (1969).
[15] B. Lubochinsky and J. Zalta, Bull. Soc. Chim. Biol. 36, 1363 (1954).

reactive in the Morgan–Elson reaction, the color yield being slightly higher than that of N-acetylglucosamine.

Amidases-2 and -3 are similar to amidase-1 in the pH profile.

The K_m value, determined by a Lineweaver-Burk plot, depends on the enzyme concentration. Using 5 μg/ml of amidase-1, K_m values with Asn-GlcNAc and glycopeptide prepared from ovalbumin[16] are 0.77 mM, but the value decreases as the enzyme concentration increases. It is 0.64 mM when 6 μg/ml of amidase-1 is used. Using the corresponding amounts of amidases-2 and -3, the values are 0.64 mM and 0.40 mM, respectively.

Maximum velocity (specific activity) is also affected by the enzyme concentration, increasing slightly as the enzyme concentration increases.

3. p-Chloromercuribenzoate at 1 mM inhibits amidase 20%. Cu^{2+}, Ni^{2+}, Zn^{2+}, and Mn^{2+} at 1 mM inhibit the enzyme 30–40%, and EDTA at 1 mM about 50%. Of the reaction products, aspartic acid at 1 mM inhibits the enzyme 40%, but neither N-acetylglucosamine nor ammonia at 1 mM inhibit the enzyme.

4. Amidase is invariably stable if stored in 0.02 M phosphate buffer, pH 7.0, at $-20°$. Repeated freezing and thawing do not affect the enzyme activity. The stability depends on the pH of the solution. At pH values above 6, amidase is invariably stable, but inactivation occurs at lower pH values; 50% and 67% of the activities being lost at pH values 5.5 and 5.0, respectively, after standing at 37° for 2 hours. The enzyme is stabilized by its substrate, Asn-GlcNAc. At pH 5.5, addition of the substrate to 5 mM is effective in maintaining the full activity at least for 2 hours at 37°.

Hydrolysis of Various Substrates

In general, amidase requires both free α-amino and α-carboxyl groups of the aspartyl residue of the aspartylglycosylamines for its action. The size and structure of carbohydrate moiety are not important factors for substrates to undergo amidase action. These properties are shared with amidase from hog serum.[6,17]

Synthetic Aspartylglycosylamines.[17] The analogs of the 4-aspartyl-glycosylamine in which N-acetylglucosamine is replaced by galactose, glucose, mannose, or N-acetylgalactosamine, are susceptible to amidase. Under the assay conditions, these analogs are degraded at the following rates: Asn-Gal, 40.2; Asn-Glc, 114; Asn-Man, 134; Asn-GalNAc, 6.4 (figures represent relative activities to Asn-GlcNAc, 100).

The analogs in which carbohydrate is replaced by aniline or cyclo-

[16] I. Yamashina and M. Makino, *J. Biochem.* (*Tokyo*) **51**, 359 (1962).

[17] M. Tanaka, M. Kohno, and I. Yamashina, in preparation.

hexylamine, i.e., Asp-aniline or Asp cyclohexylamine, are not susceptible to amidase. Asparagine is very slowly hydrolyzed by amidase, the rate being less than 1% of that for Asn-GlcNAc. Asn-aniline, Asn-cyclohexylamine and asparagine inhibit amidase in a competitive manner.

Glycopeptides (Asparaginyl-Oligosaccharides). Glycopeptides usually prepared by pronase digestion of glycoproteins should be submitted to enzymatic and/or chemical degradative treatments to remove amino acids other than asparagine linking to carbohydrate moiety prior to the use as substrates for amidase. The products, asparaginyl-oligosaccharides, are susceptible to amidase producing aspartic acid, ammonia, and reducing oligosaccharides. Asparaginyl-oligosaccharides prepared from ovalbumin or Taka-amylase [EC 3.2.1.1] are degraded by amidase with rates and K_m values comparable to Asn-GlcNAc. Asparaginyl-oligosaccharides prepared from α_1-acid glycoprotein of human plasma (orosomucoid) in which oligosaccharides are composed of neutral sugars, amino sugars and sialic acid, are also susceptible to amidase.[18]

[18] T. Yamauchi, M. Makino, and I. Yamashina, *J. Biochem. (Tokyo)* **64**, 683 (1968).

[103] α-Mannosidase from Hog Kidney[1]

By TADAYOSHI OKUMURA and IKUO YAMASHINA

Kidney is used to prepare α-mannosidase since it contains the largest amount of the enzyme and the highest specific activity compared to liver and pancreas.

Assay Method

p-Nitrophenyl α-mannoside (synthesized according to Conchie and Levvy[2]) is used as substrate, and liberated *p*-nitrophenol is determined as described by Conchie *et al.*[3] Substrate (2 mM) and enzyme are incubated in a buffer, pH 4.6, prepared from 0.1 M citric acid and 0.2 M Na_2HPO_4 at 37°. Substrate with concentration higher than 2 mM is inhibitory. When incubation intervals exceed 10 hours, additions of Zn^{2+} (2 mM as zinc acetate) as an enzyme stabilizer and of toluene as an antiseptic are necessary.

[1] T. Okumura and I. Yamashina, *J. Biochem. (Tokyo)* **68**, 561 (1970); and unpublished results. See also previous Volume [98].

[2] J. Conchie and G. A. Levvy, *in* "Methods in Carbohydrate Chemistry" Vol. II (R. L. Whistler and M. L. Wolfrom, eds.), pp. 345–347. Academic Press, New York, 1962.

[3] J. Conchie, J. Findlay, and G. A. Levvy, *Biochem. J.* **71**, 318 (1959).

Enzyme Purification

All operations are carried out at 0–4° unless otherwise stated.

Step 1. Preparation of Crude Extract. Hog kidney (either fresh or frozen) is homogenized with 5 volumes of water for 20 minutes in a Waring Blendor, and the homogenate after standing for 1 hour is centrifuged at 10,000 g for 40 minutes.

The extraction of enzyme can similarly be effected using 0.05 M phosphate buffer, pH 7.5, 0.05 M citrate buffer, pH 4.1 or 0.9% NaCl.

Step 2. Ammonium Sulfate Fractionation. Solid $(NH_4)_2SO_4$ is added to the extract to a final concentration of 38% saturation, and the precipitate is sedimented at 10,000 g. Solid $(NH_4)_2SO_4$ is again added to the supernatant to give a final concentration of 68% saturation. The precipitate is sedimented at 10,000 g.

Step 3. Acetone Fractionation. The sediment from the preceding step (amounting to about 4 g protein from 200 g of the tissue) is dialyzed, and the dialyzate is made 0.04 M and 0.01 M with $(NH_4)_2SO_4$ and $Zn(CH_3COO)_2$, respectively, then adjusted to pH 6.0 with 1 N NaOH, the final volume being 400 ml. Cold 50% acetone, made up of equal volumes of acetone and 0.01 M zinc acetate buffer, pH 5.9, is added with stirring to give 20% acetone concentration. After centrifugation at 10,000 g for 15 minutes, the supernatant is brought to 30% acetone concentration by further addition of 50% acetone. The resulting precipitate is collected by centrifugation.

Step 4. Heat Treatment. The sediment from the preceding step is dissolved in 200 ml of 0.02 M glycine–NaOH buffer, pH 10.5, the solution being adjusted to pH 6.6 with dilute acetic acid. The precipitate formed is removed by centrifugation, and the supernatant is heated at 60° for 15 minutes. The resulting slightly turbid solution is clarified by centrifugation, and the supernatant is dialyzed against distilled water. This enzyme solution can be stored at −20° for several months without any loss of activity. At this step, the preparation is practically free from other major glycosidases contained in the tissue; i.e., α-L-fucosidase, β-*N*-acetylglucosaminidase [EC 3.2.1.30], and β-galactosidase [EC 3.2.1.23]. α-*N*-Acetylglucosaminidase and α-galactosidase [EC 3.2.1.22] are eliminated during ammonium sulfate fractionation.

Step 5. Hydroxyapatite Column Chromatography. Hydroxyapatite is prepared according to Siegelman *et al.*[4] The dialyzate obtained from the preceding step is further dialyzed to equilibrium against 0.05 M phosphate buffer, pH 6.8, and the solution after concentration to about one-tenth its volume by ultrafiltration using a membrane filter (Diaflo Ultrafiltra-

[4] H. W. Siegelman, G. A. Wieczorek, and B. C. Turner, *Anal. Biochem.* **13**, 402 (1965).

tion, Model 200, with membrane PM-10, Amicon Corporation, Cambridge, Massachusetts) is applied to a column of hydroxyapatite (2.4 ✕ 40 cm) equilibrated with the same buffer. After the solution has been adsorbed on the column, inactive proteins are washed out with about 500 ml of 0.11 M buffer at a flow rate of 20 ml/hour, and then the enzyme is eluted with 150 ml of 0.3 M buffer at the same flow rate as above.

Step 6. Sephadex G-200 Column Chromatography. The eluate from the preceding step is concentrated to 5–10 ml using a collodion membrane (Sartorius Membran Filter GmbH, Göttingen, Germany) and applied to a column of Sephadex G-200 (2 ✕ 100 cm) equilibrated with 0.005 M phosphate buffer, pH 6.8. Fractions in the peak of activity are collected, concentrated as above, and rechromatographed. The gel filtration is repeated until specific activity reaches 18 units per milligram of protein.

At this stage, the preparation appears nearly homogeneous on disc electrophoresis. The enzyme solution is very stable, if frozen.

Step 7. Preparative Electrophoresis. Further purification may be obtained by electrophoresis using a block of either starch or Pevikon. Pevikon is superior to starch in obtaining the enzyme preparation for chemical study since the enzyme is a glycoprotein.

The eluate from the Sephadex column is concentrated to 1 ml or less using a collodion membrane and applied to a slit 20 cm from cathode in a block (40 ✕ 1 ✕ 1 cm) prepared with barbiturate buffer, pH 8.6 and $\mu = 0.017$ (for starch) or 0.034 (for Pevikon), containing 0.25 mM Zn(CH$_3$COO)$_2$. To prepare this block, 90 g starch or 110 g Pevikon are required. A voltage of 350 V is applied at 4° for 16 hours. After the electrophoresis, the block is separated into 40 segments, each 1.0 cm wide. Each segment is eluted with 2 ✕ 2 ml of 0.005 M phosphate buffer, pH 6.8. The enzyme activity distributes from fractions 10 to 14 from cathode when starch block is used and from 21 to 25 when Pevikon block is used. By this electrophoresis, specific activity increases about 10%, but the enzyme becomes unstable, losing about 20% of the activity after 24 hours when the enzyme solution after the additions of Zn^{2+} and bovine serum albumin as stabilizers is left standing at 4° or kept frozen at $-20°$.

The purification procedure is summarized in Table I.

Properties

1. The most purified preparation is homogeneous on disc electrophoreses at pH values of 4.0 and 9.4,[5,6] on disc electrophoresis in

[5] B. J. Davis, *Ann. N.Y. Acad. Sci.* **121**, 404 (1964).
[6] R. A. Reisfeld, U. J. Lewis, and D. E. Williams, *Nature (London)* **195**, 281 (1962).

TABLE I

PURIFICATION OF α-MANNOSIDASE FROM HOG KIDNEY

Fraction	Total activity (units/100 g tissue)	Specific activity (units/mg protein)	Percent recovery
Original extract	68	0.011	100
$(NH_4)_2SO_4$ fractionation	53	0.025	79
Acetone fractionation	43	0.057	63
pH Adjustment	25	0.075	37
Heat treatment	17	0.32	25
Hydroxyapatite chromatography	13	3.64	19
Sephadex G-200 chromatography			
1 st elution	11	12.2	16
2nd elution	9.7	18.0	14
Electrophoresis			
Starch block	5.8	20.0	8.6
Pevikon block	6.8	21.1	10

the presence of sodium dodecyl sulfate,[7] and on similar electrophoresis with the use of mercaptoethanol.[7]

Molecular weight of the enzyme is about 100,000, estimated by gel filtration, but is about 42,000 on disc electrophoresis in the presence of sodium dodecyl sulfate. The isoelectric point determined by isoelectric focusing is 5.8. The preparation contains 8.0% neutral sugar, determined by the orcinol–H_2SO_4 method,[8] and 3.3% hexosamine, determined by the Elson–Morgan reaction.[9]

2. The enzyme has a marked optimum at pH 4.6. At this point, the K_m value with p-nitrophenyl-α-mannoside, determined by a Lineweaver-Burk plot, is 0.91 mM. From the plot, the maximum velocity of the most purified preparation having 21.1 units per milligram of protein with the mannoside is estimated to be 33.6 μmoles mg^{-1} min^{-1}.

3. The preparation before step 7 is very stable, but loses its activity on lyophilization. The activity also decreases gradually with time on incubation at pH 4.6 and 37°, but this inactivation is prevented by the addition of 2 mM Zn^{2+}. The stability depends on the protein concentration. At a protein concentration below 0.01%, addition of bovine serum albumin is necessary to maintain the full activity.

Hydrolysis of Various Substrates

Oligomannosides. The K_m values and maximum velocities with oligomannosides are tabulated in Table II.

[7] K. Weber and M. Osborn, *J. Biol. Chem.* **244**, 4406 (1969).

[8] L. F. Hewitt, *Biochem. J.* **31**, 366 (1937).

[9] L. Svennerholm, *Acta Soc. Med. Upsal.* **61**, 287 (1957).

TABLE II
K_m AND V_{max} VALUES WITH OLIGOMANNOSIDES[a]

	Di-		Tri-		Tetra-	
Type of linkage	K_m (mM)	V_{max} (μmoles mg^{-1} min^{-1})	K_m (mM)	V_{max} (μmoles mg^{-1} min^{-1})	K_m (mM)	V_{max} (μmoles mg^{-1} min^{-1})
α-(1 → 2)	71.5	11.8	17.3	3.45	11.8	0.89
α-(1 → 3)					28.0	2.80
α-(1 → 6)	23.8	2.18	11.0	3.70		

[a] The most purified enzyme preparation having 21.1 units per milligram of protein corresponding to V_{max} of 33.6 μmoles mg^{-1} min^{-1} is used.

The terminal mannosidic linkage in a trimannoside, α-Man-(1 → 4)-β-Man-(1 → 4)-Man, is also hydrolyzed.[10]

No mannosyl transferase action is observed.

Glycopeptides. Intact glycoproteins such as ovalbumin or Taka-amylase [EC 3.2.1.1] are not susceptible to this α-mannosidase.

Glycopeptides prepared from ovalbumin, Taka-amylase and stem-bromelain [EC 3.4.4.24] by pronase digestion are hydrolyzed producing certain amounts of mannose. All mannosidic linkages in these glyco-peptides except the innermost one, which is the β-type,[10,11] are susceptible to this α-mannosidase.

[10] K. Sugahara, T. Okumura, and I. Yamashina, *Biochim. Biophys. Acta* **268**, 488 (1972).

[11] K. Sugahara, T. Okumura, and I. Yamashina, *FEBS (Fed. Eur. Biochem. Soc.) Lett.* **20**, 44 (1972).

[104] α-Acetylglucosaminidase from Pig Liver

By BERNARD WEISSMANN

Assay Method

Principle. Enzymatic liberation of aglycon from an aryl N-acetyl-α-D-glucosaminide is measured. Because of the high absorptivity of p-nitrophenolate ion and minor interference in crude extracts, the p-nitrophenyl glycoside is preferred, as in assay of other hydrolases. However, the sometimes more accessible phenyl glycosides may also serve, even with weak enzymatic activities in crude extracts, as for mammalian α-acetylglucosaminidase. It is then important that precaution be taken against bacterial contamination during the longer incubation intervals required and that measurements be arranged to minimize the much larger enzyme blanks, as illustrated in the alternate assay described.

Reagents

Sodium citrate buffer, 0.2 M, pH 4.8.

p-Nitrophenyl N-acetyl-α-D-glucosaminide,[1] 4 mM, in citrate buffer; may be stored at $-18°$

Potassium borate buffer, 1.2 mole H_3BO_3 and 1.0 mole KOH per liter.

Bovine serum albumin, crystalline, 0.2 mg/ml, in 0.002 M sodium citrate buffer, pH 6

p-Nitrophenol standard, 0.025 mM

n-Amyl alcohol–chloroform, 1:5.

Procedure.[2] The extract (suitably diluted with albumin solution), 500 μl, substrate solution, 500 μl, and water, 1.00 ml, are incubated for 2 hours at 37°. The reaction is stopped by addition of 0.50 ml of borate buffer, and the p-nitrophenol liberated is measured at 410 nm. When crude extracts are assayed, solutions may be turbid. In this case, after addition of borate buffer, the digests and blanks are stirred vigorously for 30 seconds with 1.0 ml of amyl alcohol–chloroform (Vortex mixer), and the absorbance of the clear supernatant solutions is read after centrifugation at 3000 rpm.

Definition of Enzyme Unit and Specific Activity. One unit is that amount of enzyme which catalyzes liberation of 1 μmole of p-nitrophenol per minute at 37°. Specific activity is expressed as milliunits per milligram of protein. In measurement of protein,[3] bovine serum albumin is used as a color standard.

Alternate Assay Procedure.[4] Extract, 100 μl, is added to 100 μl of phenyl N-acetyl-α-D-glucosaminide solution (20 mM in 0.1 M sodium citrate buffer, pH 4.8). The digest, covered with 0.3 ml of toluene, is incubated for 16 hours at 37° in a small cork-stoppered tube. Most of the toluene layer is removed with a Pasteur pipette, and the aqueous layer is further extracted with three portions of toluene. The pooled toluene layers are extracted in a small tube with 0.2 ml of 0.05 M NaOH

---

[1] B. Weissmann, *J. Org. Chem.* **31**, 2505 (1966).

[2] B. Weissmann, G. Rowin, J. Marshall, and D. Friederici, *Biochemistry* **6**, 207 (1967).

[3] O. H. Lowry, N. J. Rosebrough, A. L. Farr, and R. J. Randall, *J. Biol. Chem.* **193**, 265 (1951).

[4] B. Weissman and R. Santiago, unpublished results.

and centrifuged, if necessary to separate the layers. After freezing the aqueous layer at $-18°$ or at Dry Ice temperature, the toluene is quickly decanted and the tube is allowed to drain by inversion for 30 minutes over a pad of tissue at $-18°$. Phenol is then estimated in the usual way in the thawed solution by addition of 100 μl of Folin-Ciocalteu phenol reagent and 1.50 ml of 0.4 M sodium carbonate solution, mixing after each addition, and reading the optical density at 650 nm after 30 minutes.

Purification Procedure

Preparative work is done at 0–4°. The purification[2] is summarized in the table.

Extractions and Ammonium Sulfate Fractionation. Fresh pig liver (840 g) is homogenized in portions in a blender with water (2.5 liter). Centrifugation at 20,000 g yields a crude extract, which is subjected to repeated fractionation with solid ammonium sulfate. In each case, the fraction retained is exhaustively dialyzed against 0.01 M sodium citrate buffer of pH 6, and any dialysis precipitate is removed. The limits selected for the three successive fractionations are 0.20 to 0.40 saturation, 0.25 to 0.35 saturation, and 0.00 to 0.31 saturation.

Ion Exchange Chromatography. A column of bed volume 150 ml is formed from 22 g of TEAE-cellulose (0.44 meq/g) which has been equilibrated with 0.006 M Tris·HCl buffer of pH 7.5 containing 0.01 M sodium citrate. The final ammonium sulfate fraction is applied as a load. The column is then washed with starting buffer and developed with starting buffer that is supplemented with NaCl in a linear gradient Fractions of 20 ml are collected. Partially purified enzyme of the highest specific activity, whose properties are summarized, is found in tube 28 of this chromatogram, which is illustrated in Fig. 1.

Comment. Although not economical of activity, the procedure is the only one devised so far that satisfactorily separates the α-acetylgluco-saminidase of the crude pig liver extract from the much more abundant

PURIFICATION OF α-ACETYLGLUCOSAMINIDASE FROM PIG LIVER

Fraction	Total volume (ml)	Total units	Specific activity, (mU/mg of protein)	Recovery (%)
Crude extract	2400	18.7	0.33	(100)
(NH$_4$)$_2$SO$_4$, 0.20–0.40 satd.	505	7.2	1.1	39
(NH$_4$)$_2$SO$_4$, 0.25–0.35 satd.	150	3.33	5.7	18
(NH$_4$)$_2$SO$_4$, 0–0.31 satd.	300	1.42	6.4	8
Column, tube 28	20	0.152	27	1

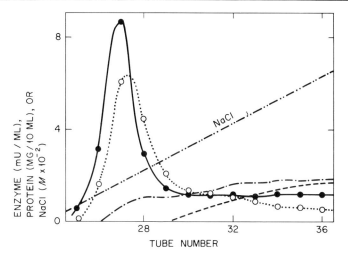

Fig. 1. TEAE-cellulose column developed at pH 7.5 with NaCl in a gradient, as represented, showing the distribution in the effluent fractions of protein (——), α-acetylglucosaminidase (. . . .), β-acetylglucosaminidase (— · —), and α-acetylgalactosaminidase (- - - -).

β-acetylglucosaminidase (1600 U) and α-acetylgalactosaminidase (270 U) activities also present. Because α-acetylglucosaminidase precipitates at unusually low salt concentration, relatively little of the two other acetylhexosaminidases remains after three ammonium sulfate fractionations (β-acetylglucosaminidase, 2.1 U; α-acetylgalactosaminidase, 0.07 U). Most of the residual β-acetylglucosaminidase activity in the final ammonium sulfate fraction apparently resides in isozymes[5] that are not eluted from the TEAE-cellulose column under the conditions used.

Properties

pH Optimum. In 0.05 *M* sodium citrate buffers, activity is optimal at pH 4.7. Activity is somewhat depressed in sodium acetate buffers at comparable values of pH. In absence of substrate, progressively larger proportions of the enzymatic activity become sedimentable below pH 5.0, presumably through aggregation, with essentially complete and irreversible inactivation at pH 3.8.

Purity, Specificity, and Kinetic Properties. The partially purified enzyme shows minor residual enzymatic activity with p-nitrophenyl *N*-acetyl-β-D-glucosaminide (specific activity, 3.7 mU/mg) and with p-nitrophenyl *N*-acetyl-β-D-galactosaminide (1.0 mU/mg), believed to represent contamination with β-acetylglucosaminidase. With phenyl *N*-acetyl-α-

[5] B. Weissmann and D. F. Hinrichsen, *Biochemistry* **8**, 2034 (1969).

D-galactosaminide[5] or with phenyl N-acetyl-α or β-D-mannosaminides, no activity is detectable. A number of other glycosidases abundant in crude extracts are either undetectable or present only in traces in the final preparation.[2] The enzyme catalyzes hydrolysis of p-nitrophenyl N-acetyl-α-D-glucosaminide (K_m, 0.30 mM; relative V_{max} taken as 1.00), o-nitrophenyl N-acetyl-α-D-glucosaminide (K_m 0.08 mM; rel. V_{max}, 2.22), and phenyl N-acetyl-α-D-glucosaminide (K_m 1.4 mM; rel. V_{max} 1.25), but enzymatic hydrolysis of methyl N-acetyl-α-D-glucosaminide is not readily detectable. Little is known at present regarding the natural substrates of this enzyme, which is lysosomal in its occurrence.

Inhibition. Enzymatic activity is irreversibly inhibited by 1 mM $CuSO_4$ (82% inhibition) and by 1 mM $FeCl_3$ (76% inhibition). Competitive inhibitors include: N-acetylglucosamine (K_i 0.68 mM), N-acetylglucosaminolactone[6] (K_i 0.25 mM), methyl N-acetyl-α-D-glucosaminide (K_i 8.2 mM), and N-acetylmannosamine (K_i 0.13 mM). N-Acetylgalactosamine (which inhibits α-acetylgalactosaminidase) and acetamide (which inhibits β-acetylglucosaminidase) are noninhibitory.

Stability. Solutions in dilute sodium citrate buffer of pH 6 may be stored for many months at $-18°$, with repeated thawing, without noticeable loss of activity.

Other Sources. α-Acetylglucosaminidase activity has been shown to occur in a number of gastropods,[6-8] in a variety of mammalian tissues[2,6,9] and in some microorganisms.[9,10] None of the known sources apparently shows vigorous activity with aryl glycoside substrates. A partial purification of the enzyme from extracts of the marine snail *Turbo cornutus* has been reported.[8]

[6] J. Findlay, G. A. Levvy, and C. A. Marsh, *Biochem. J.* **69**, 467 (1958).

[7] L. Zechmeister, G. Toth, and E. Vajda, *Enzymologia* **7**, 170 (1939).

[8] T. Muramatsu, *J. Biochem.* (*Tokyo*) **64**, 521 (1968).

[9] S. Roseman and A. Dorfman, *J. Biol. Chem.* **191**, 607 (1951).

[10] W. M. Watkins, *Biochem. J.* **71**, 261 (1959).

[105] α-Acetylgalactosaminidase from Beef Liver

By BERNARD WEISSMANN

Assay Method[1]

Principle. Enzymatic liberation of phenol from phenyl *N*-acetyl-α-D-galactosaminide[2] is measured. The *p*-nitrophenyl glycoside, a more convenient substrate,[3,4] is unfortunately also less accessible.[5]

Reagents

> Sodium citrate buffer, $0.1 M$, pH 4.7
> Phenyl *N*-acetyl-α-D-galactosaminide, 20 mM, in sodium citrate buffer, $0.1 M$, pH 4.7
> Folin-Ciocalteu phenol reagent, diluted 1:1 with water
> Sodium carbonate, $0.4 M$
> Bovine serum albumin, crystalline, 0.2 mg/ml, in $0.002 M$ sodium citrate buffer, pH 6
> Phenol standard, 1 mM, in $0.05 M$ H_2SO_4

Procedure. The extract (suitably diluted with albumin solution), 100 μl, is incubated for 1 hour at 37° with 100 μl of substrate solution in buffer. The reaction is stopped by addition of 200 μl of diluted phenol reagent, with vigorous mixing. After centrifugation, a 300-μl aliquot of the supernatant fluid is mixed with 1.50 ml of sodium carbonate solution, and the absorbance at 650 nm is read after 30 minutes.

Definition of Enzyme Unit and Specific Activity. One unit is that amount of enzyme which catalyzes liberation of 1 μmole of phenol per minute at 37°, at an enzyme concentration that produces 12 nmoles of phenol per minute per milliliter of digest. (The apparent specific activity of this enzyme is concentration dependent.) Specific activity is expressed as units per milligram of protein. In measurement of protein,[6] bovine-serum albumin is used as a color standard.

[1] B. Weissmann and D. F. Hinrichsen, *Biochemistry* **8**, 2034 (1969).

[2] B. Weissmann, *J. Org. Chem.* **31**, 2505 (1966).

[3] C.-T. Wang and B. Weissmann, *Biochemistry* **10**, 1067 (1971).

[4] B. Weissmann and R. Santiago, unpublished results.

[5] B. Weissmann, *J. Org. Chem.* **35**, 1690 (1970).

[6] O. H. Lowry, N. J. Rosebrough, A. L. Farr, and R. J. Randall, *J. Biol. Chem.* **193**, 265 (1951).

Purification Procedure[1]

Preparative work is done at 4°, and pH is measured at room temperature. Solutions are made in glass-distilled water containing 0.1 mM EDTA. Dialyses are performed against 5 mM sodium citrate buffer of pH 6, using membranes prewashed with EDTA solution, and dialysis precipitates are discarded. The purification is summarized in the table.

Extraction and Ammonium Sulfate Fractionation. Without thawing, frozen fresh beef liver is homogenized in portions in a blender with 3 volumes of water at 25°. Centrifugation at 20,000 g yields the crude extract. This is treated with solid ammonium sulfate. The fraction collected between limits of 0.50 and 0.62 saturation is purified further. (α-Acetylglucosaminidase, now undetectable, and β-acetylglucosaminidase, much decreased in quantity, precipitate at lower salt concentration.)

Ion Exchange Chromatography. The dialyzed ammonium sulfate fraction is applied to a column of DEAE-cellulose (100 g, dry weight, Whatman Type DE-52, 500 ml bed volume), which has been preequilibrated with 5 mM sodium citrate buffer of pH 6.0. After the column is washed with 1 liter of starting buffer and 1.5 liter of 25 mM sodium citrate buffer of pH 6.0, the activity is eluted with 1 liter of 50 mM sodium citrate buffer of pH 6.0, and at once precipitated with solid ammonium sulfate (0.75 saturation).

Gel Filtration. After it has stood 2 days, the ammonium sulfate precipitate is collected and extracted with 10 ml of 50 mM sodium citrate buffer of pH 6. The centrifuged extract is applied to a 5 × 89 cm column of Sephadex G-150, which is prewashed and developed with 50 mM sodium citrate buffer of pH 6. Effluent fractions are assayed at once for protein and enzyme (total activity recovered, 85%) and the most active fractions (815–945 ml) are, without delay, concentrated by ultrafiltration (collodion bags, Schleicher and Schuell Co.) to small volume. The

PURIFICATION OF α-ACETYLGALACTOSAMINIDASE FROM BEEF LIVER

Fraction	Total volume (ml)	Total units	Specific activity (U/mg of protein)	Recovery (%)
Crude extract	2500	955	0.014	100
(NH$_4$)$_2$SO$_4$, 0.50–0.62 satd.	240	412	0.039	43
DEAE-cellulose	1000	298	2.09	31
(NH$_4$)$_2$SO$_4$, 0–0.75 satd.	9.5	248	3.17	26
Sephadex G-150	5	136	7.8	14
(NH$_4$)$_2$SO$_4$, 0.50–0.65 satd.	2	110	12.2	12

concentrate, 5 ml, is treated with neutralized, saturated ammonium sulfate solution. Dialysis of a fraction collected between 0.50 and 0.65 saturation yields the final enzyme preparation.

Comments. Early experience suggested this enzyme to be extremely sensitive to inactivation at low total protein concentration (20–200 μg/ ml), especially in absence of salts. (There were also indications of somewhat improved stability with use of glass-distilled water.) Success of the gel filtration and ion exchange steps apparently depends heavily on the promptness with which the dilute effluent fractions can be concentrated. To obviate necessity for assays during the preparative run, it is recommended that the proper developer concentrations be verified in small pilot columns with the same protein loading and the same conditioned DEAE-cellulose as will be used on larger scale.

The preparation described was carried out twice, with fairly similar results, with a specimen of beef liver having an enzyme content some 3-fold greater than that observed for six other specimens tested since. Electrofocussing experiments[1] with preparations from the beef liver specimen described (not repeated with the others) shows isozyme(s) of unusually low isoelectric point, pI 4.2 or less, to be predominant.[4] This finding tends to explain the excellent purification factor observed in the DEAE-cellulose step.

Limited trials, with minor variations of the procedure described, were made with a less active beef liver.[4] In one run, crude extract of specific activity 0.0058 U/mg was treated with ammonium sulfate; 78% of the enzyme (0.014 U/mg) was recovered in a fraction collected between 0.47 and 0.62 saturation. This was applied to a DEAE-cellulose column (52 mg total protein per milliliter of bed volume), which was washed with 80 mM sodium citrate buffer of pH 6 to remove less active protein. Of the enzyme in the load, 35% appeared in the fraction (0.299 U/mg) now eluted with 0.16 M buffer. At this point β-acetylglucosaminidase (0.013 U/mg) was a minor contaminant and α-acetylglucosaminidase was undetectable.

Properties

pH Optimum. Activity is optimal at pH 4.7 in sodium citrate and sodium acetate buffers.

Specificity.[1] The preparation catalyzes hydrolysis of phenyl, *p*-nitrophenyl, and *o*-nitrophenyl α-glycosides of *N*-acetylgalactosamine, but has no detectable action on *p*-nitrophenyl *N*-acetyl-β-D-galactosaminide, *p*-nitrophenyl *N*-acetyl-α or β-D-glucosaminides, or *p*-nitrophenyl *N*-acetyl-α or β-D-mannosaminides. With considerably lesser efficiency than for aryl *N*-acetyl-α-D-galactosaminides, the enzyme functions as an exoglycosidase in catalyzing release of N-acetylgalactosamine from blood group A substance and from desialized ovine and bovine salivary mucins, incidentally confirming the α-linkage of N-acetylgalactosamine in these polymers. The enzyme is lysosomal in its occurrence and undoubtedly functions in catabolism of glycoproteins.

Kinetic Behavior. With aryl glycosides, progress curves are linear, but the relationship between initial velocity and enzyme concentration may be anomalous, particularly at higher temperatures and at lower substrate concentrations. With 10 mM phenyl glycoside, for example, the relationship is approximated at 37° by: velocity = [enzyme]$^{1.34}$, but at 25° the relationship is linear.[1] Also, a typically allosteric, sigmoid relationship between velocity and substrate concentration is observed at 42°; an essentially normal hyperbolic relationship is observed at 25°; and transitional patterns are seen at intermediate temperatures.[4] Physical studies account for the kinetic abnormalities by revealing existence of rapidly reversible, temperature-dependent association-dissociation phenomena.[3] Oligomeric enzyme species bind substrate or inhibitor tightly and dissociated enzyme species bind these ligands less tightly or not at all.

Quite normal Lineweaver-Burk plots are obtained at higher concentrations of phenyl glycoside, however, even at 37°. From such plots,[4] the apparent K_m for phenyl glycoside is 25 mM at 37°, 9.4 mM at 32°, and 3.1 mM at 25°. For N-acetylgalactosamine, which behaves as a competitive inhibitor, the apparent K_i is 24 mM at 37°, 8.9 mM at 32°, and 2.6 mM at 25°. These marked temperature effects are probably referable largely to changes in dissociation constant of oligomeric enzyme, rather than to changes in intrinsic binding constant for the ligands.

A nonlinear relationship between velocity and enzyme concentration is also observed for α-acetylgalactosaminidase from tissues of the rabbit, frog, and turtle, but not for preparations from the pig, rat, guinea pig, chicken, fish, or earthworm.[3] Kinetic abnormalities are not detected with the bovine enzyme when desialized mucins are used as substrates; enzyme concentrations in such experiments are some 100-fold greater than with aryl glycosides.[1]

Stability. Although the enzyme is conspicuously unstable at low protein concentrations, as already noted, preparations of varying degrees of purity dissolved at 2–5 mg/ml total concentration of protein in dilute sodium citrate buffer of pH 6 have been stored for 3 years at $-18°$ without noticeable loss of activity. Bovine serum albumin and gelatin are both effective stabilizers.

Other Sources of Enzyme. Concurrent studies of the enzyme now recognized as α-acetylgalactosaminidase[1,7] with synthetic glycosides[8] and with desialized salivary mucin[9] as substrates have led to introduction of a separate designation,[9] "seryl acetylgalactosaminide glycosidase," which remains in the literature as a source of confusion. Enzymes have now been partially purified from species including cattle,[1,7] pigs,[1] snails,[10,11] earthworms,[12] and protozoans,[13] which act on α-acetylgalactosaminidic linkages in synthetic glycosides, desialized mucins (carbohydrate-hydroxyamino acid linkages), and blood group A substance (carbohydrate-carbohydrate linkages). An α-acetylgalactosaminidase inducible in *Clostridium perfringens*[14] apparently has greater efficiency with natural polymeric substrates than with synthetic glycosides, a relationship which is the reverse of that found for animal enzymes studied in this regard.

[7] E. Werries, E. Wollek, A. Gottschalk, and E. Buddecke, *Eur. J. Biochem.* **10**, 445 (1969).

[8] B. Weissmann and D. F. Friederici, *Biochim. Biophys. Acta* **117**, 498 (1966).

[9] A. S. Bhargava, E. Buddecke, E. Werries, and A. Gottschalk, *Biochim. Biophys. Acta* **127**, 457 (1966).

[10] H. Tuppy and W. Staudenbauer, *Biochemistry* **5**, 1742 (1966).

[11] A. S. Bhargava and A. Gottschalk, *Biochem. Biophys. Res. Commun.* **24**, 280 (1966).

[12] E. Buddecke, H. Schauer, E. Werries, and A. Gottschalk, *Biochem. Biophys. Res. Commun.* **34**, 517 (1969).

[13] G. J. Harrup and W. M. Watkins, *Biochem. J.* **93**, 9p (1964).

[14] S. Chipowsky and E. J. McGuire, *Fed. Proc., Fed. Amer. Soc. Exp. Biol.* **28**, 606 (1970).

[106] α-Glucosidase from Rat Liver Lysosomes

By DAVID H. BROWN, BARBARA ILLINGWORTH BROWN, and PETER L. JEFFREY

Among the several α-glucosidases which are widely distributed in mammalian tissues one enzyme has its maximal activity at about pH 4, and this has been shown to be present in the lysosomal fraction prepared from rat liver.[1,2] The purified enzyme has both α-1,4-glucosidase and α-1,6-glucosidase activities.[2] Although the enzyme can form glucose readily from oligosaccharides such as maltose and maltotriose, it is probable that the substrate on which it has its physiologically most

[1] N. Lejeune, D. Thinès-Sempoux, and H. G. Hers, *Biochem. J.* **86**, 16 (1963).

[2] P. L. Jeffrey, D. H. Brown, and B. I. Brown, *Biochemistry* **9**, 1403 (1970).

important action is glycogen, since it can catalyze the total hydrolysis of this polysaccharide to glucose,[3] and, in human liver, glycogen accumulates within lysosomes[4] when the α-glucosidase is inactive as it is in type II glycogen storage disease.[5] In the latter circumstance the simultaneous absence of both α-1,4- and α-1,6-glucosidase activities, measured at pH 4, has been shown.[6]

Assay Method

Principle. The assay depends upon measuring the rate of formation of glucose from glycogen. Glucose is measured spectrophotometrically at 340 nm using hexokinase, glucose-6-phosphate dehydrogenase, ATP, and NADP+ in a coupled enzyme system in which NADPH is formed in an amount stoichiometrically equal to that of glucose. The use of glycogen as a substrate is preferred rather than that of maltose, since the enzyme shows marked substrate inhibition by the latter substance[3] (see Properties).

Reagents

> Glycogen, 5% in water. Because of its greater solubility, rabbit liver glycogen should be used, if possible, rather than shellfish glycogen.
>
> Potassium acetate–acetic acid buffer, 0.25 M, containing 1.0 M potassium chloride, pH 4.2
>
> Tris·HCl buffer, 0.05 M containing 1 mM EDTA, pH 8.0
>
> Magnesium chloride, 0.05 M
>
> ATP (disodium salt), 20 mM, adjusted to pH 6.5
>
> Hexokinase (crystalline, from yeast), 15 units/ml. Commercial preparations, which must have negligible contamination by 6-phosphogluconic acid dehydrogenase and by invertase, may be diluted in water or in 5 mM acetate buffer (pH 5.5) without removal of the ammonium sulfate which is usually present.
>
> Glucose 6-phosphate dehydrogenase (crystalline, from yeast) 10 units/ml. See Hexokinase, above, for description of a satisfactory method for diluting the commercial enzyme and for requirements for purity.
>
> NADP+ (monosodium salt), 10 mM, adjusted to pH 6.5
>
> Tris buffer, 5 mM, containing 2 mM EDTA–5 mM 2 mercaptoethanol (ME)–25 mM KCl, adjusted to pH 7.0 with HCl

[3] P. L. Jeffrey, D. H. Brown, and B. I. Brown, *Biochemistry* **9**, 1416 (1970).
[4] P. Baudhuin, H. G. Hers, and H. Loeb, *Lab. Invest.* **13**, 1139 (1964).
[5] H. G. Hers, *Biochem. J.* **86**, 11 (1963).
[6] B. I. Brown, D. H. Brown, and P. L. Jeffrey, *Biochemistry* **9**, 1423 (1970).

Enzyme. Before assay, the α-glucosidase fraction should be diluted in the Tris-EDTA-ME-KCl buffer (pH 7.0) to about 4 mg of protein per milliliter for the lysosomal fraction, or to 500 μg/ml for the ammonium sulfate fraction. The protein concentration of the Sephadex G-100 pool usually makes it possible to assay this fraction without any predilution (see Purification Procedure).

Procedure. A reaction mixture containing 0.1 ml of 0.25 M potassium acetate–1.0 M KCl, pH 4.2, and 0.2 ml of 5% glycogen is warmed to 37°, and 0.2 ml of a suitably diluted α-glucosidase fraction is added to start the reaction. After 20 minutes of incubation at 37°, the solution is heated for 1 minute in a bath of boiling water. After the solution has cooled, it is centrifuged. Glucose is determined in a 0.2-ml aliquot of the opalescent supernatant fluid by measuring the NADPH formed via the action of hexokinase and glucose-6-phosphate dehydrogenase as described below. If there is a possibility that the enzyme fraction to be assayed might itself contain some glucose (as in the case of the cytoplasmic extract which may have glucose derived from blood, or of any fraction containing sucrose from which glucose may be derived easily by chemical hydrolysis), it is necessary to include in the assay a second complete reaction mixture which should be heated in boiling water at zero time and then analyzed, as is the incubated sample. A determination should also be made of the glucose content of a third reaction mixture which should contain all components except the protein fraction to be analyzed and which should be incubated. The latter blank value for the complete assay system should be subtracted from all other analytical results.

Assay for glucose can be conveniently done by adding a 0.2-ml aliquot of each boiled and centrifuged reaction mixture to a spectrophotometer cuvette having a 1-cm light path and a maximum volume of 1.2 ml. Other additions to the cuvette should be made in the following order with mixing: 0.5 ml of 0.05 M Tris-EDTA buffer, pH 8.0; 0.1 ml of 20 mM ATP, pH 6.5; 0.1 ml of 0.05 M MgCl$_2$; 20 μl of 15 units per milliliter of hexokinase; and 40 μl of 10 units per milliliter of glucose 6-phosphate dehydrogenase. After reading the absorbance of this solution at 340 nm against a blank cuvette containing 0.96 ml of water, 40 μl of 10 mM NADP$^+$, pH 6.5, is added to each of the cuvettes and the solutions mixed. The final reaction mixtures (volume, 1.0 ml) are allowed to stand in the spectrophotometer until the absorbance of the sample has reached a maximum value which remains unchanged for several minutes. Ordinarily, formation of NADPH is complete within 10 minutes. The quantity of glucose is calculated from the total increase in absorbance which is due to the formation of NADPH (after correcting numerically for a 4% dilution of the initial absorbance due to the volume of NADP$^+$ solution added). An increase in absorbance of 0.621 corresponds to a glucose content in the cuvette of 0.1 μmole. One unit is defined as the amount of enzyme which catalyzes the formation of 1 μmole of glucose per minute from glycogen.

Purification Procedure

Preparation of Liver Homogenate. Five rats (about 250 g each) are fasted for 24 hours. The animals are sacrificed by intraperitoneal injection of Nembutal, and their livers are removed quickly and put into ice-cold 0.25 M sucrose–2 mM EDTA, pH 7.2. The intact livers should be washed quickly by several successive transfers to fresh sucrose–EDTA (until their surfaces appear to be free of blood). About 30 g of the tissue is cut with scissors into small pieces and *immediately* homogenized in 200 ml of the cold sucrose–EDTA, using a Sorvall Omnimixer operated at 50 V for 30 seconds. The homogenization vessel should be kept in crushed ice at all times. If it is necessary to use any other homogenizing device, the efficacy of the method selected will have to be determined and its results evaluated by using the criteria for the preparation of undamaged lysosomes which have been discussed by Vaes.[7] The homogenate is fractionated centrifugally at 3° according to a procedure which is based on the work of de Duve and collaborators.[8,9] Because of the volumes involved, it is most convenient to use the SS-34 angle-head rotor and a Sorvall RC2-B centrifuge to separate three particulate fractions in successive order as follows: the "nuclear" fraction, sedimented by 4900 g_{av} × minutes (2250 rpm); the "mitochondrial" fraction, sedimented by 18,560 g_{av} × minutes (9500 rpm); the "lysosomal" fraction (L), sedimented by 160,000 g_{av} × minutes (19,500 rpm). In the centrifugal procedure used here each supernatant fluid is separated from its underlying precipitate by careful decantation.[2] In the purification table discussed below, the supernatant fluid remaining after the "nuclear" fraction has been removed as a pellet is referred to as the "cytoplasmic extract." The lysosomal fraction (L) should be washed twice with cold 0.25 M sucrose–2 mM EDTA by suspending it and then recentrifuging each time for 160,000 g_{av} × minutes as was done in its initial separation.

Solubilization of the Lysosomal Fraction. The washed pellet is suspended in 80 ml of 5 mM Tris–2 mM EDTA–5 mM 2 mercaptoethanol

[7] G. Vaes, Vol. 8 [87].

[8] F. Appelmans, R. Wattiaux, and C. de Duve, *Biochem. J.* **59**, 438 (1955).

[9] C. de Duve, B. C. Pressman, R. Gianetto, R. Wattiaux, and F. Appelmans, *Biochem. J.* **60**, 604 (1955).

(pH 7.0) and the suspension is frozen rapidly in a Dry Ice–isopropanol bath and then thawed under cold running water. The freezing and thawing treatment should be repeated four times. All subsequent steps of this procedure are carried out at $3°$–$5°$. The suspension then is centrifuged at 46,000 g for 1 hour, and the supernatant fluid is kept.

Ammonium Sulfate Fractionation. The soluble extract of L is made 42% saturated in $(NH_4)_2SO_4$ by the addition of 243 mg of the solid salt per milliliter of solution. The mixture must be well stirred during addition of the salt and the pH should be kept near 7.0 by the addition of 1 M KOH as required. After standing for 10 minutes, a small precipitate is removed by centrifugation at 46,000 g for 20 minutes. The clear supernatant fluid is made 84% saturated in $(NH_4)_2SO_4$ by the addition of 285 mg of the solid salt per milliliter. After adjusting the pH to 7.0 and allowing the mixture to stand for 10 minutes, the precipitate is collected by centrifugation at 46,000 g as before. It is dissolved in 10 ml of 5 mM Tris–2 mM EDTA–5 mM 2–mercaptoethanol–25 mM KCl, pH 7.0 ("Tris–KCl buffer") and dialyzed overnight against 2 liters of this buffer. The resulting solution can be frozen and kept for at least 1 month without loss of α-glucosidase activity.

Sephadex G-100 Chromatography. Sephadex G-100 should be treated with 6 M urea and then exhaustively washed with water,[10] before a 1.5×70 cm column of it is prepared with previously deaerated gel. The column is equilibrated with Tris-KCl buffer and the flow rate is adjusted to 16 ml per hour. The dialyzed ammonium sulfate fraction (15 ml) is added to the top of the column, and, after the protein fraction has entered the gel bed, the column is connected to a reservoir of Tris-KCl buffer, and fractions (4 ml) are collected at a flow rate of 16 ml per hour. Auricchio and Bruni[11] and Auricchio *et al.*[12] first described the fractionation on Sephadex of a whole homogenate of rat liver in order to separate an α-1,4-glucosidase whose activity was measured at pH 3.6. In the present procedure it is found that a large amount of protein containing a small peak of an α-glucosidase passes through the column without much retardation. As the protein content of the fractions diminishes, a major peak of the lysosomal α-glucosidase is eluted. Because of the low protein concentration in the fractions containing the enzyme, it is necessary to assay for activity against glycogen to provide the information necessary for making a pool. It is usually found that the 100 fractions which are collected after the absorbance at 280 nm has decreased to 0.075 will contain about 50–75% of the total lysosomal glucosi-

[10] F. Auricchio and V. Sica, *J. Chromatogr.* **28**, 26 (1967).
[11] F. Auricchio and C. B. Bruni, *Biochem. J.* **105**, 35 (1967).
[12] F. Auricchio, C. B. Bruni, and V. Sica, *Biochem. J.* **108**, 161 (1968).

dase. It is advisable to exclude in this way the leading edge of the enzyme peak in order to obtain enzyme of highest specific activity in the following step.

DEAE-Cellulose Chromatography. The pool of fractions from the Sephadex G-100 column should be dialyzed overnight against 4 liters of 5 mM Tris–2 mM EDTA–5 mM 2-mercaptoethanol, pH 7.0, in order to reduce the KCl concentration. A 1 × 20 cm column of DEAE-cellulose is prepared and equilibrated with the Tris–EDTA–2-mercaptoethanol buffer. The dialyzed enzyme is allowed to flow through the column at the rate of 40 ml per hour. Under these conditions all the α-glucosidase, but only about 15% of the total protein, is retained by the column. After washing the column with 100 ml of buffer, the enzyme is eluted by buffer containing added 0.25 M KCl. Fractions of about 3 ml should be collected slowly. The α-glucosidase usually is found in fractions 4–7 and must be located by assay as the protein content is too low to be measured accurately especially in view of the ionic composition of the buffer. The four fractions that contain the largest amount of activity are combined and dialyzed overnight against a large volume of Tris–EDTA–2-mercaptoethanol buffer, pH 7.0, containing 25 mM KCl. The resulting enzyme solution should be divided into small aliquots for freezing if it is desired to keep the preparation for several weeks. About 50% of its activity is lost in 1 month at $-20°$. The enzyme may also be lyophilized out of Tris buffer which contains no added KCl, and the dry protein can then be redissolved at a higher concentration in Tris buffer containing added 25 mM KCl. Under these circumstances a preparation which also is stable to freezing is obtained. Upon thawing, all activity is lost if the solution is kept overnight at 4°. The table summarizes results of the purification procedure.

Alternative Preparative Procedures. An enzyme with seemingly similar catalytic properties to that described above has been partially purified by Auricchio and co-workers[11,12] from the supernatant fluid obtained after the centrifugation at 105,000 g of a rat liver homogenate which had been frozen and thawed. The method of purification used differs from the first part of that described in the table chiefly in the inclusion of an early step in which inactive protein is precipitated at pH 4.2. The final enzyme preparation was reported to have a significantly lower molecular weight than that found here for the enzyme obtained by the procedure summarized in the table (see Properties). Bruni *et al.* have described the preparation from bovine liver of an apparently pure α-glucosidase which is active at acid pH.[13] The procedure used was similar in its first steps to that which was developed for the rat liver

[13] C. B. Bruni, F. Auricchio, and I. Covelli, *J. Biol. Chem.* **244**, 4735 (1969).

PURIFICATION OF α-GLUCOSIDASE FROM RAT LIVER LYSOSOMES[a]

Fraction	Protein (mg/ml)	Total protein (mg)	Specific activity (units/mg)	Total activity (units)	Purification
Cytoplasmic extract	18.7	7760	0.052	402	
Lysosomal fraction (L)	10.2	828	0.435	360	1
Soluble extract of L	1.68	133	2.44	325	5.6
(NH$_4$)$_2$SO$_4$ fraction (42–84% saturation)	5.04	75.6	4.63	350	10.6
Pool of Sephadex G-100 fractions	0.037	14.3	19.6	280	45
DEAE-cellulose eluate, dialyzed	0.012	0.16	693	111	1590

[a] The preparation described was from 30 g of liver. Data from P. L. Jeffrey, D. H. Brown, and B. I. Brown, *Biochemistry* **9**, 1403 (1970). Copyright 1970 by the American Chemical Society and reprinted by permission of the copyright owner.

enzyme.[11,12] The procedure of Bruni *et al.* has been modified for the preparation of the corresponding α-glucosidase from rabbit liver by including a step in which the 25,000 *g* supernatant fluid (pH 6.7) is heated at 55° for 5 minutes before carrying out the precipitation of inactive protein at pH 4.2.[14] Subsequently, after one elution of the ammonium sulfate fractionated rabbit liver enzyme from a Sephadex G-100 column, its specific activity is equal to that of the pure bovine liver enzyme.[14] However, at present the only purification of α-glucosidase from lysosomes is that which is described here for the enzyme from rat liver.

Properties

Specificity. The rat liver lysosomal enzyme has both α-1,4-glucosidase and α-1,6-glucosidase activities[2] as shown by its action on maltose and linear oligosaccharides of the maltose series as well as on isomaltose. That both bond types may also be present in the same molecule and be subject to hydrolysis is shown by the conversion to glucose of panose as well as of the series of singly branched compounds isolated as "limit dextrins" from an α-amylase digest of glycogen. In the latter case it has been shown that debranching by α-1,6-glucosidase action is the rate-limiting step in the total hydrolysis of these oligosaccharides. The enzyme is able to catalyze the total hydrolysis of glycogen to glucose.[3] The purified enzyme has no detectable action on D-(+)-cellobiose, D-(+)-melibiose, lactose, D-(+)-turanose, D-(+)-

[14] A. K. Murray and D. H. Brown, unpublished experiments, 1971.

melezitose, sucrose, D-(+)-raffinose, and stachyose. The enzyme can act as a transglucosylase at pH 4 when it is incubated with oligosaccharides such as maltose. In this case, maltotriose and maltotetraose are formed together with a branched trisaccharide of unknown structure which may contain one α-1,3-glucosidic bond, according to the results of preliminary structural studies.[2] The rat liver lysosomal enzyme also catalyzes transglucosylation from maltose to glycogen.[2] The corresponding enzyme in normal human tissues also can catalyze this transglucosylation reaction.[5]

Activators and Inhibitors. Glucose formation from glycogen is stimulated 5-fold at pH 4.0 by 0.2 M KCl. This maximal degree of activation is obtained in both 10 mM acetate and 10 mM histidine–chloride buffers and appears to be the result of a general activating effect of monovalent and divalent cations. This is suggested by the fact that equimolar concentrations of KCl, NaCl, LiCl, and NH$_4$Cl (10–125 mM) and KCl, CaCl$_2$, and MgCl$_2$ (0.1–0.2 M) are equally stimulatory.[2] It has also been shown that this activation is not due to an effect of chloride ion. The enzyme is inhibited in its action on glycogen to the extent of more than 50% by 0.15 M histidine–chloride buffer as compared to 10 mM histidine–chloride at pH values above 4.1. The extent of cation stimulation of glycogen hydrolysis is pH dependent. Above pH 5 there is only a 10% stimulation by 0.2 M KCl, but the extent of activation increases sharply with decreasing pH. The cation concentration has much less effect on maltose or isomaltose hydrolysis than on that of glycogen.[2] It has been stated[15] that the activity of the corresponding enzyme from bovine liver responds in a similar way to changes in cation concentration. Maltose, maltotriose, maltotetraose, and maltopentaose exert pronounced substrate inhibition on the rat liver α-glucosidase when present at concentrations greater than 8 mM.[3] However, neither glycogen nor isomaltose produces substrate inhibition. Maltose and glycogen are mutually competitive substrates. Isomaltose is a competitive inhibitor of maltose hydrolysis.[3] The hydrolysis by the rat liver enzyme of either maltose, glycogen, or isomaltose is inhibited by D-(+)-turanose[1,3,11] (see Kinetic Properties).

pH Optimum and Buffer Effects. In 50 mM acetic acid–potassium acetate buffer containing added KCl (total concentration, 0.2 M in potassium ion), the optimum pH for the activity of the rat liver enzyme at 37° is pH 3.7 for maltose, pH 4.2 for isomaltose, and pH 4.4 for glycogen. The pH optimum for each of these substrates is nearly the same

[15] V. Sica, A. Siani. C. B. Bruni, and F. Auricchio, *Biochim. Biophys. Acta* **242**, 422 (1971).

in citrate or in citrate–phosphate buffers if these have equal cation concentration.[2] In 10 mM acetate buffer containing no added KCl the pH optimum for glycogen is shifted to pH 5.0, but the pH optima for the disaccharide substrates are essentially unchanged (see above).

Kinetic Properties. Although maltose produces strong substrate inhibition, its K_m has been estimated to be 3.8 mM in 40 mM acetate buffer (pH 4.1) containing 0.1 M KCl at 37°.[3] The K_m value for isomaltose is 32.9 mM in pH 3.8 acetate buffer at 37°. The K_m value for rabbit liver glycogen at pH 4.2 depends critically upon the cation concentration of the acetate buffer and upon the particular sample of polysaccharide used, even when calculation of the kinetic data is based upon the analytically determined total outer chain end group content (branch point content) of the glycogen. Values of from 0.7 mM to 6.5 mM have been obtained under various experimental conditions. Glycogen appears to be a competitive inhibitor of maltose hydrolysis (K_i) about 8.6 mM under conditions where the measured K_m for glycogen is 1.1 mM). Maltose also appears to be a competitive inhibitor of glycogen hydrolysis (α-1,4-glucosidase action). A K_i value of 0.6 mM has been found for the reaction in acetate buffer (pH 4.2) containing added 0.2 M KCl. Isomaltose appears to be a competitive inhibitor of maltose hydrolysis (K_i about 61 mM in pH 4.0 acetate buffer).[3] The kinetics of inhibition by turanose are complex, and the inhibition is best described as being of the mixed type with all substrates which have been investigated. With glycogen, turanose inhibition is almost purely noncompetitive, as it is also at relatively high concentrations (5 mM) when isomaltose is the substrate. On the other hand, with maltose as substrate, the kinetic data suggest that inhibition by turanose is of a partially competitive plus a purely noncompetitive type (K_i about 2.8 mM).

Physical Properties. On the assumption that the partial specific volume of the protein is 0.73 ml/g, its molecular weight was found to be 107,000, as determined by equilibrium sedimentation of a solution containing 86 μg of enzyme per milliliter in 5 mM Tris–1 mM EDTA–2 mM 2-mercaptoethanol, pH 7.1, to which 0.1 M KCl had been added. The molecular weight of the enzyme was found to be 114,000 by sucrose density centrifugation in Tris–EDTA–mercaptoethanol buffer containing 25 mM KCl.[2] A molecular weight of about 80,000, as determined by sucrose density centrifugation, has been reported by Auricchio *et al.* for the partially purified rat liver enzyme which they prepared by the alternative procedure described above involving precipitation of inactive protein at pH 4.2 from a homogenate of the tissue.[12] The reason for the discrepancy in the molecular weight values reported for this enzyme is not known.

[107] β-Glucuronidase from Rat Liver Lysosomes

By Philip D. Stahl and Oscar Touster

β-Glucuronidase (EC 3.2.1.31) is a hydrolase found in most animal cells.[1] The enzyme has been partially purified from various sources[1-4]; the liver enzyme has been shown to display multiple forms.[5-7] While the enzyme is largely associated with lysosomes, the subcellular distribution in rat liver is peculiar in that approximately one-third of the enzymatic activity is associated with the endoplasmic reticulum.[8] This unique distribution appears to be true of some other rat tissues as well.[9] Furthermore, the chromatographic properties of rat liver lysosomal and rat liver microsomal β-glucuronidase are different; the respective forms can be separated by anion exchange chromatography.[7] The β-glucuronidase purified by the following procedure is the principal form of the enzyme as found in rat liver lysosomes.[10] The only other animal β-glucuronidase purified to homogeneity is the enzyme from the preputial gland of the female rat.[3]

Assay Method

Principle. β-Glucuronidase is conveniently assayed at pH 5.0 using one of several synthetic β-D-glucosiduronides (glucuronides) which yield chromogenic or fluorescent aglycons. The present method employs phenolphthalein β-D-glucuronide.

Reagents

Sodium acetate, 0.2 M, pH 5.0
Phenolphthalein glucuronide, 0.004 M
Enzyme: about 0.1 unit
Stopping reagent containing 0.133 M glycine, 0.067 M sodium chloride, 0.083 M sodium carbonate

[1] G. A. Levvy and J. Conchie, *in* "Glucuronic Acid" (G. J. Dutton, ed.), p. 301. Academic Press, New York, 1966.
[2] B. V. Plapp and R. D. Cole, *Biochemistry* **6**, 3676 (1967).
[3] K. Ohtsuka and M. Wakabayashi, *Enzymologia* **39**, 109 (1970).
[4] B. U. Musa, R. P. Doe, and U. S. Seal, *J. Biol. Chem.* **240**, 2811 (1965).
[5] B. W. Moore and R. H. Lee, *J. Biol. Chem.* **235**, 1359 (1960).
[6] M. Aoshima and Y. Sakurai, *Gann* **60**, 129 (1969).
[7] R. Sadahiro, S. Takanishi, and M. Kawada, *J. Biochem. (Tokyo)* **58**, 104 (1965).
[8] R. Gianetto and C. de Duve, *Biochem. J.* **59**, 433 (1955).
[9] W. H. Fishman, S. S. Goldman, and R. DeLellis, *Nature (London)* **213**, 457 (1967).
[10] P. D. Stahl and O. Touster, *J. Biol. Chem.* **246**, 5398 (1971).

Procedure. The standard assay used for β-glucuronidase is that of Talalay *et al.*[11] modified as follows; 0.25 ml of 4 mM phenolphthalein glucuronide is added to 0.25 ml of 0.2 M sodium acetate, pH 5.0, together with about 0.1 unit of enzyme in a total volume of 1.0 ml. After incubation at 37° for 15 minutes, the reaction is stopped with 3.0 ml of stopping reagent.[8] The absorbance is measured at 552 nm.

Units. One unit of β-glucuronidase is the activity that catalyzes the release of 1 μmole of phenolphthalein per hour from phenolphthalein glucuronide under standard assay conditions. This unit equals approximately 320 Fishman units.[11] The latter unit is based upon micrograms of product formed per hour.

Purification

Step 1. Preparation of Extract for Chromatography. Sixty fasted male Wistar rats are killed by decapitation. The livers are removed, cooled, and passed through a meat grinder. All subsequent operations are carried out at 5°. From the resulting slurry, a homogenate is prepared in 0.25 M sucrose (3 ml per gram of tissue) with one stroke of a loose fitting Potter-Elvehjem homogenizer. The homogenate is centrifuged at 2500 rpm in a Sorvall GSA rotor for 10 minutes (1×10^4 g min). The pellet is twice suspended in 3 volumes of 0.25 M sucrose, rehomogenized, and recentrifuged. The pooled supernatant fractions from all three centrifugations constitute the cytoplasmic extract. From this extract a mitochondrial-lysosomal (ML) pellet is prepared by centrifugation at 8500 rpm for 40 minutes (4.7×10^5 g min). The ML fraction (containing about 40% of the total β-glucuronidase) is suspended with a loose-fitting Dounce homogenizer in 5 mM Tris-phosphate, pH 7.8 (5 ml per gram of liver), and dialyzed overnight against the same buffer. The dialyzed suspension is centrifuged in a Spinco L-19 rotor at 19,000 rpm for 3 hours (6×10^6 g min). To the supernatant solution ("ML extract"), which should contain about 90% of the enzymatic activity of the ML fraction, is added solid ammonium sulfate to a final concentration of 30% at 5° (the pH of this solution should be 7.0). After 20 minutes, the solution is clarified by centrifugation at 10,000 rpm for 20 minutes (2×10^5 g min). The solution is then adjusted to pH 6.0 with 0.1 N H$_2$SO$_4$ and brought to 70% saturation by adding solid ammonium sulfate over a period of 20 minutes. After a further 20 minutes, the solution is centrifuged to yield a precipitate containing all the enzymatic activity. The pellet is dissolved in a minimum amount of 0.005 M Tris-phosphate, pH 7.8, and dialyzed exhaustively against the same buffer. The yield

[11] P. Talalay, W. H. Fishman, and C. Huggins, *J. Biol. Chem.* **166**, 757 (1946).

from the ammonium sulfate fractionation is usually about 90%. However, occasionally the yield may be as low as 70%.

Step 2. DEAE-Cellulose Chromatography. The dialyzed ammonium sulfate product is applied to a DEAE-cellulose column (Sigma, coarse grade). The column is washed with 1.5 liters of buffer containing 0.1 M NaCl, which elutes a large protein peak and a small but variable amount of activity. The bulk of the enzyme is then eluted from the column with 0.25 M NaCl in buffer. The pooled product, usually 75–99% of the activity applied and representing a 3-fold increase in specific activity, is dialyzed against the starting buffer (5 mM Tris-phosphate, pH 7.8) and applied to a second DEAE-cellulose column (Whatman DE 32, 2.5 × 40 cm), which is eluted with a linear salt gradient (500 ml of starting buffer and 500 ml of 0.4 M NaCl in this buffer) at 50 ml per hour. The product is obtained in a single peak in high yield (80–92%), with a 1.5- to 2-fold increase in specific activity. The enzyme is pooled and dialyzed against 50 mM acetate buffer, pH 5.6. A precipitate, which contained a little enzymatic activity, is formed during dialysis; this precipitate is removed by centrifugation. Dialysis against acetate should be executed as quickly as possible (preferably less than 6 hours), since the dialyzed enzyme is less stable at pH 5.6 than at 7.8.

Step 3. CM-Cellulose Chromatography. The dialyzed product from the DEAE-cellulose step is applied to a column (2.5 × 40 cm) of CM-cellulose (Whatman CM 32), which is equilibrated with 50 mM sodium acetate, pH 5.6. The column is developed with a linear salt gradient (500 ml of the above buffer and 500 ml of 0.3 M NaCl in this buffer) at a rate of 50 ml per hour. The enzyme is eluted, as a single symmetrical peak, at about 40 mM NaCl, between two large protein peaks. The yield of enzyme ranges from 50 to 80%; the specific activity should be increased 6- to 7-fold.

Step 4. Gel Filtration. The pooled fractions from CM-cellulose chromatography are immediately dialyzed against a buffer that is 5 mM in Tris·HCl, pH 8.0, and 5 mM in NaCl. The sample is concentrated to about 5 ml by ultrafiltration and subjected to Sephadex G-200 chromatography. Two columns (2.5 × 90 cm), fitted with flow adapters, are eluted in series with reverse flow at 10 ml per hour. The enzyme is eluted as a symmetrical peak just after the void volume but before the bulk of the protein. This step routinely provides a 2- to 3-fold increase in specific activity with an 80–90% yield.

Step 5. Preparative Electrophoresis. The pooled product from Sephadex G-200 chromatography is concentrated to about 5 ml and further purified with a Canalco "prep-disc" electrophoresis unit. A slight modification is made in the preparation of the gels as suggested by Schenkein et al.[12] The gel is made about 1–2 cm longer than needed and is then cut off to the desired length. This procedure reduces holdup of the eluting proteins at the gel surface. The modification is accomplished by placing a size 00 rubber stopper over the elution tube and a plastic cap (outer diameter 2.5 cm) over the stopper. Saran wrap is then placed around both of these objects and fixed around the column with several rubber bands. The acrylamide monomer is pumped into the column to the desired height filling the space around the rubber stopper. Following polymerization, the Saran wrap is removed and the gel is cut at the base of the column with a fine wire (diameter 0.005 in). The freshly cut gel surface permits a more rapid elution of proteins. The sample is made 10% with respect to sucrose, and 20 μl of 0.04% bromophenol blue is then added. A 5-cm 5% separating gel and 1-cm stacking gel are used in a PD 2/230 column. After the sample is layered onto the stacking gel, a 7- to 9-mA current is applied and elution is begun. The enzyme is routinely eluted in 20–25 hours with 10-minute samples and 30–50 ml per hour elution rate. The pooled product is concentrated and can be subjected to analytical electrophoresis by the method of Takayama et al.[13] as a check for purity. Routinely, the gel, when stained for protein, reveals one major band and several minor bands.

The enzyme is then subjected to electrophoresis again through a 1.5-cm 7% gel (1.0 cm stacking gel) in the manner just described. The product from this step, usually eluting at 20 hours, has a specific activity of approximately 3000. If assayed in the presence of poly-D-lysine (1 mg/ml), a specific activity of about 4400 is usually obtained (with 100 μg of polylysine per milliliter, the specific activity was about 3900). The yield from the 5% and 7% gels varies from 75 to 95%. The yield is much reduced when less pure preparations are subjected to electrophoresis. The entire purification scheme is summarized in the table, which contains the results of a typical preparative experiment. The product obtained by this procedure should migrate as a single band in a number of different electrophoretic buffer systems.[10]

Properties

Molecular Weight and Subunits. The enzyme has a molecular weight of about 280,000[10] as determined by sucrose density gradient centrifuga-

[12] I. Schenkein, M. Levy, and P. Weis, *Anal. Biochem.* **25**, 387 (1968).
[13] K. Takayama, D. H. MacLennan, A. Tzagoloff, and C. D. Stoner, *Arch. Biochem. Biophys.* **114**, 223 (1966).

PURIFICATION OF β-GLUCURONIDASE FROM RAT LIVER LYSOSOMES

Fraction	Protein (mg)	Enzyme activity (units)	Specific activity (units/mg)	Yield (%)	Purification ratio
Homogenate	130,000	46,500	0.36	100	1
Mitochondrial–lysosomal (ML) fraction	16,500	18,200	1.10	39	3
ML extract	9,660	16,000	1.65	34	5
Ammonium sulfate	2,695	10,650	3.95	23	11
Sigma DEAE-cellulose	708	10,550	14.9	23	41
Whatman DE-32	458	9,670	21.1	21	59
Whatman CM-cellulose	40.5	6,000	148	13	410
Sephadex G-200	14.3	4,800	336	10	933
5% gel	1.63	3,940	2400	8	6660
7% gel					
− Polylysine	0.976	2,950	3024	6	8400
+ Polylysine[a]	0.976	4,250	4360		

[a] Assayed with 1 mg per milliliter of poly-D-lysine.

tion,[14] gel electrophoresis,[15] and Sepharose chromatography.[16] When chromatographed on Sephadex G-150 in the presence of 8 M urea or electrophoresed in polyacrylamide gels containing sodium dodecyl sulfate,[17] an apparent subunit molecular weight of about 70,000 is observed. The enzyme is believed to be a tetramer.[10]

pH Optima. β-Glucuronidase from various sources has been shown to display double pH optima. However, it has been shown that, with partially purified lysosomal β-glucuronidase, only a single optimum is obtained in the presence of high salt concentration.[18] Purified lysosomal β-glucuronidase, obtained by the present procedure, also displays dual pH optima when assayed with phenolphthalein glucuronide.[10] The peaks are around pH 4.4 and 4.9 with a trough at 4.6. The dual pH optima are not observed in the presence of high salt concentration, or if the assays are run with lower substrate concentration (10^{-4} M). With this substrate concentration the pH optimum is approximately 4.3. (Preliminary experiments in our laboratory employing a variety of substrates suggest that both the structure and concentration of substrate influence multiplicity of pH optima.)

[14] R. G. Martin and B. N. Ames, *J. Biol. Chem.* **236**, 1372 (1961).

[15] J. L. Hedrick and A. J. Smith, *Arch. Biochem. Biophys.* **126**, 155 (1968).

[16] J. Marrink and M. Gruber, *FEBS* (*Fed. Eur. Biochem. Soc.*) *Lett.* **2**, 242 (1969).

[17] A. L. Shapiro, E. Viñuela, and J. V. Maizel, *Biochem. Biophys. Res. Commun.* **28**, 815 (1967).

[18] E. Delvin and R. Gianetto, *Can. J. Biochem.* **46**, 389 (1968).

Stability. Purified enzyme in a concentration of 90 units/ml is stable for long periods at 5° in 5 mM Tris·HCl, pH 8.0. Enzyme stored at pH 8.0 in more dilute solution slowly loses activity, which can be recovered if a polycation, such as poly-D-lysine or methylated serum albumin, is included in the assay mixture.[10] This phenomenon has been observed with β-glucuronidases from other sources and has been studied by Bernfeld.[19] At pH 5.0 in 50 mM sodium acetate, activity is lost over several days, and this loss of activity cannot be reversed by the addition of polycations to the assay mixture.

Kinetics. With enzyme concentrations up to at least 0.06 μg/ml and under standard assay conditions, phenolphthalein release is directly proportional to the enzyme concentration. With 0.024 μg of enzyme per milliliter, phenolphthalein release is linear for at least 30 minutes. The effect of substrate concentration on enzyme activity at pH 5.0 indicates the reaction obeys normal Michaelis–Menten kinetics; the K_m is $7.05 \times 10^{-4} M$.

Specificity. β-Glucuronidase is an exo-glycosidase with broad specificity for the aglycon.[1] However, although it will hydrolyze oligosaccharides of hyaluronic acid, it does not show activity toward disaccharides.[20,21]

Other Liver β-Glucuronidases. There are several forms of β-glucuronidase in rat liver which differ in their chromatographic properties.[6] Rat liver lysosomes contain two closely related forms.[10] Microsomal β-glucuronidase of rat liver is eluted from DEAE-cellulose columns before the major lysosomal form and has not as yet been extensively purified. Neither microsomal nor lysosomal β-glucuronidase is inactivated by trypsin.[10,22] The chromatographic and electrophoretic properties of microsomal, but not lysosomal, β-glucuronidase are altered by this protease.[10]

[19] P. Bernfeld, *in* "The Amino Sugars" (E. A. Balazs and R. W. Jeanlos, eds.), Vol. 2B, p. 213. Academic Press, New York, 1966.
[20] A. Linker, K. Meyer, and B. Weissman, *J. Biol. Chem.* **213**, 237 (1955).
[21] N. N. Aronson, Jr. and C. de Duve, *J. Biol. Chem.* **243**, 4564 (1968).
[22] E. Delvin and R. Gianetto, *Biochim. Biophys. Acta* **220**, 93 (1970).

[108] β-Galactosidase from Human Liver[1]

By MIRIAM MEISLER

Human liver contains at least two distinct β-galactosidases.[2,3] The enzyme described below is the thermolabile β-galactosidase with optimal activity at pH 3.5. It comprises 85–90% of the total liver activity, measured at pH 5.0 with 5 mM p-nitrophenyl-β-D-galactoside as substrate.

Assay Method

Principle. Purification was monitored with p-nitrophenyl-β-D-galactoside as substrate. The release of galactose from other substrates by the purified enzyme was determined with a coupled galactose dehydrogenase reaction.

Hydrolysis of p-Nitrophenyl-β-D-galactoside

Reagents

1. Sodium chloride, 0.1 M, in 0.1 M sodium acetate buffer, pH 5.0
2. p-Nitrophenyl-β-D-galactoside (PNPG), 0.1 M
3. Trichloracetic acid, 30%
4. 2-Amino-2-methyl-1,3-propanediol (Aldrich), 5.0 M

Procedure. Reaction mixtures containing 0.85 ml of reagent 1, 0.05 ml of reagent 2, and 0.1 ml enzyme preparation are incubated at 37° for 30 minutes. Reactions are terminated by addition of 0.2 ml of reagent 3 and clarified by centrifugation. Supernatants are decanted into 0.2 ml of reagent 4, and the absorption due to p-nitrophenol is measured at 415 nm. An absorbance of 1.0 is produced by 0.1 μmole of p-nitrophenol in a 1-cm light path. Crude homogenates can be directly assayed by this procedure. The reaction rate is constant for at least 3 hours.

Units. One unit of β-galactosidase is defined as the amount of enzyme which catalyzes the hydrolysis of 1 μmole of PNPG per minute under these conditions.

[1] M. H. Meisler, B. J. Fry, and K. Paigen, *Fed. Proc., Fed. Amer. Soc. Exp. Biol.* **30**, 1118 (1971).

[2] M. W. Ho and J. S. O'Brien, *Clin. Chim. Acta* **32**, 443 (1971).

[3] B. Hultberg and P. A. Ockerman, *Scand. J. Clin. Lab. Invest.* **23**, 213 (1969).

Enzymatic Determination of Galactose Release[4]

Reagents

1. Sodium chloride, 0.1 M, in 0.1 M sodium acetate buffer, pH 5.0
2. Tris·HCl buffer, 0.1 M, pH 8.6
3. NAD+, 0.02 M
4. Galactose dehydrogenase, 1 mg/ml, from *Pseudomonas fluorescens* (BMC)

Procedure. β-galactosidase preparation, 0.1 ml, and 0.1 ml of β-galactoside substrate are added to 0.8 ml of reagent 1 and incubated at 37°. The reaction is terminated by chilling in ice water. An aliquot containing 10–50 nmoles of liberated galactose is transferred to the dehydrogenase reaction mixture which consists of 300 µl of reagent 2, 10 µl of reagent 3, and 5 µl of reagent 4. The volume is brought to 0.4 ml with water, and the mixture is incubated at 30° for 30–40 minutes until the oxidation of galactose is complete. Absorption at 340 nm due to NADH is then determined. Controls to which known quantities of galactose are added before and after the 37° incubation should be included. Determinations of PNPG hydrolysis by Method I and Method II are in good agreement.

Purification

Preliminary Steps. β-galactosidase was purified from individual livers obtained at autopsy and frozen immediately. All subsequent operations were performed at 0–5°C. Livers were thawed in 2 volumes of 0.25 M sucrose, 0.05 M Tris·HCl pH 7.5 and homogenized in a Waring Blendor at top speed for 2 minutes. The homogenate was centrifuged at 10,000 g for 45 minutes, and the pelleted material was discarded.

First Ammonium Sulfate Fractionation. To the 10,000 g supernatant, crystalline ammonium sulfate (144 g/liter) was added with stirring for 30 minutes. Centrifugation at 100,000 g for 90 minutes produced a clear supernatant solution. An additional 230 g of ammonium sulfate per liter of supernatant was added by the same procedure. The precipitate was collected by centrifugation at 30,000 g for 30 minutes and dissolved in 0.05 M Tris pH 7.4, and the concentration was adjusted to approximately 60 mg protein per milliliter after determination of protein by the biuret method. From 100 g of starting material approximately 200 ml of this fraction 1 was obtained.

Organic Solvent Precipitation.[5] An equal volume of cold ethanol:

[4] C. A. Mapes, R. L. Anderson, and C. C. Sweeley; *FEBS* (*Fed. Eur. Biochem. Soc.*) *Lett.* **7**, 180 (1970).
[5] B. Fry, unpublished results, 1971.

acetone:ether, 15:4:1, was combined with fraction 1 by gradual and continuous addition of solvent over a period of 1 hour. The mixture was continuously stirred during the addition of solvent and for an additional 15 minutes, and then centrifuged at 14,000 g for 30 minutes. The pellet was discarded. An additional three volumes of solvent were then gradually added to the supernatant by the same procedure. The resulting precipitate was collected by centrifugation at 14,000 g for 30 minutes. The pellet was dissolved in a volume of 0.1 M sodium acetate buffer pH 5.0 equal to the starting volume of fraction 1, which produced a protein concentration of approximately 5 mg/ml. After centrifugation at 100,000 g for 45 minutes to remove insoluble material, the clear supernatant solution was collected (fraction 2). Fraction 2 could be stored at −20° for 6 months without loss of activity. After thawing it was usually necessary to remove additional insoluble material by low-speed centrifugation.

Second Ammonium Sulfate Fractionation. Ammonium sulfate (144 g/liter) was added as described to fraction 2. The precipitate was removed by centrifugation at 30,000 g for 30 minutes and discarded. To the supernatant an additional 230 g/liter of ammonium sulfate was added. The precipitate was collected by centrifugation and dissolved in 0.1 M NaCl, 0.1 M sodium acetate buffer pH 5.0 at approximately 50 mg protein per ml (Fraction 3).

DEAE-Cellulose Chromatography.[6] Prior to chromatography, fraction 3 was dialyzed for 4 hours at 4° against several changes of buffer which had been prepared by titration of 5 mM phosphoric acid with 1 M Tris to pH 6.0. After dialysis, insoluble material was removed by centrifugation at 30,000 g for 10 minutes and the pH was adjusted to 7.3 with 1 M Tris. (Dialysis at pH 7.3 produced considerable loss of activity.) Three hundred milligrams of dialyzed fraction 3 was applied to a column of DEAE-cellulose (2.5 cm × 35 cm) which had been equilibrated with 5 mM phosphate, 50 mM Tris buffer pH 7.3. Enzyme was eluted with a 1 liter gradient from 0.05 to 0.2 M sodium chloride in the equilibration buffer. Twenty-minute fractions (16 ml) were collected into tubes containing 1 ml of 0.5 M sodium acetate, pH 5.0, for immediate adjustment of pH. β-Galactosidase activity was recovered in a single peak in fractions 40–48. This preparation could be stored at −20° after addition of 1/10 volume of glycerol.

G-200 Gel Filtration. The active fractions from DEAE-cellulose chromatography were pooled, concentrated to 7 ml in an ultrafiltration cell (Amicon) and applied to a column (2.5 cm × 45 cm) of Sephadex G-200-120. The enzyme was eluted with 10 mM NaCl in 50 mM sodium phos-

[6] F. B. Jungalwala and E. Robins, *J. Biol. Chem.* **243**, 4258 (1968).

phate buffer, pH 6.0, at a flow rate of 15 ml/hour. A single symmetric peak of enzyme activity was obtained. The active fractions were pooled and concentrated as described, and stored at −20° after the addition of 1/10 volume of glycerol. This preparation was estimated to be 10% pure by polyacrylamide gel electrophoresis.

A typical purification is summarized in the table.

PURIFICATION OF β-GALACTOSIDASE FROM HUMAN LIVER

Fraction	Specific activity (μmoles/min/mg protein)	Total units (μmoles/min)	Total protein (mg)
Homogenate	0.004	300	75,000
Ammonium sulfate, first	0.017	183	10,700
Solvent precipitate	0.15	150	1,000
Ammonium sulfate, second	0.40	125	310
DEAE-cellulose	0.90	33	37
Sephadex G-200	3.6	18	5

Affinity Column Chromatography.[5] Human liver β-galactosidase could be purified with an affinity adsorbent which was originally employed in the purification of galactoside-binding proteins from *Escherichia coli.*[7] Portions of fraction 3 (5 units) were directly applied to a column (1 cm × 13 cm) packed with a mixture of 75 mg of PAPTG-polyBGG[7] plus 1 g of cellulose powder. The enzyme was eluted at room temperature with a gradient of 0–1.0 M KCl in 0.02 M potassium phosphate buffer pH 6.8. A broad peak of activity was obtained, with 9-fold purification of the enzyme.

Properties

Substrate Specificity.[8] Purified enzyme preparations specifically hydrolyze β-galactosides and do not release *p*-nitrophenol from its α-galactoside, β-glucoside, β-glucuronide, or α-mannoside. Our preparations contain minor β-glucosaminidase activity which is probably due to contamination with another enzyme, since it is not inhibited by D-galactonolactone or *p*-chloromercuribenzoate. The K_m for hydrolysis of PNPG is 2×10^{-4} M in the presence of 0.1 M NaCl. Lactose was hydrolyzed at a rate 25% that of PNPG with an apparent $K_m > 5 \times 10^{-2}$ M. Under

---

[7] Polymerized bovine α-globulin containing covalently bound *p*-aminophenylthiogalactoside. The preparation of this material was described by S. Tomino and K. Paigen, *in* "The Lactose Operon" (J. Beckwith and D. Zipser, eds.), p. 233. Cold Spring Harbor Lab. (1970).

[8] Specificity studies were performed on enzyme prepared with affinity columns.

standard reaction conditions the enzyme did not liberate galactose from mixed human gangliosides, human GM-1 ganglioside, desialyated fetuin or its tryptic peptides. With an assay employing radioactive substrates, our purified enzyme preparation has been found to release galactose from lactosylceramide but not from galactosylceramide."[9]

Effects of pH. β-Galactosidase activity exhibits a sharp pH optimum at pH 3.5. However, it is rapidly inactivated at 37° at pH values below pH 4.5. Our enzyme units are defined at pH 5.0, where the rate of reaction is 65% that at pH 3.5. Enzyme stability is also reduced above pH 6.5.

Ionic Strength. β-galactosidase activity is stabilized by 0.1 M concentrations of inorganic salts including NaCl, NH$_4$Cl, CsCl, MgCl$_2$, CaCl$_2$, NaF, MgSO$_4$, and (NH$_4$)$_2$SO$_4$. The major effect of NaCl is to protect the enzyme from thermal inactivation during assay at 37°. Ho and O'Brien[10] have reported that NaCl has the additional effect of reducing the K_m toward 4-methylumbelliferyl-β-D-galactoside.

Inhibitors. The enzyme is completely inhibited by 10^{-5} M p-chloromercuribenzoate. This inhibition can be prevented or reversed by 10^{-3} M dithiothreitol, which suggests the involvement of thiol groups in maintenance of the active conformation of the enzyme. Fifty percent of activity is lost in the presence of 10^{-3} M D-galactono-1,4-lactone. Activity is completely inhibited by 0.1 M sodium nitrite.

Stability. This β-galactosidase is markedly thermolabile: half of the activity is lost during a 10-minute incubation at 37° in 25 mM sodium acetate pH 5.0. Thermal inactivation is prevented by 0.1 M NaCl and is significantly reduced by 5 mM PNPG.

Molecular Weight. By gel filtration on a calibrated column of Sephadex G-200, the molecular weight of the enzyme was estimated to be 80,000.

[9] J. Kanfer, personal communication.
[10] M. W. Ho and J. S. O'Brien, *Clin. Chim. Acta* **30**, 531 (1970).

[109] Neuraminidase (Sialidase) from Rat Heart

By JOHN F. TALLMAN and ROSCOE O. BRADY

Mono-, di-, and trisialogangliosides → desialated derivatives
+ *N*-acetylneuraminic acid

Mucopolysaccharides → desialated mucopolysaccharide + *N*-acetylneuraminic acid

Assay Methods

Ganglioside Substrates

Principle. [³H]*N*-Acetylneuraminic acid is released enzymatically from the specifically labeled ganglioside substrate and is quantified by measuring the trichloroacetic acid-soluble radioactivity which represents the released *N*-acetylneuraminic acid.[1]

Reagents

Potassium acetate buffer, 2 *M*, pH 5.0
Tritiated rat brain gangliosides separated into individual species by preparative thin-layer chromatography,[2] 2 m*M* in distilled water. The specific radioactivity of each *N*-acetylneuraminic acid is 6.0×10^5 cpm/μmole.
Trichloroacetic acid, 100% (w/v)
Human serum albumin, 100 mg/ml H_2O

Procedure. Incubations are carried out in a total volume of 200 μl which include 180 μl of enzyme suspension, 10 μl potassium acetate buffer, and 10 μl of the tritiated ganglioside substrate. After a 3-hour incubation at 37°, 25 μl of human serum albumin, 675 μl of distilled water, and 100 μl of 100% trichloroacetic acid are added; the suspension is mixed and centrifuged in the cold and the supernatant is removed. The pellet is resuspended in 1 ml of 10% trichloroacetic acid and recentrifuged. Both supernatants are combined and the radioactivity in a 1-ml aliquot is determined by liquid scintillation spectrometry. Boiled enzyme controls are incubated simultaneously, and the radioactivity in these controls is subtracted. Activities are calculated from the known specific activity of the substrate.

---

[1] E. Kolodny, J. Kanfer, J. Quirk, and R. Brady, *J. Biol. Chem.* **246**, 1426 (1971).
[2] E. Kolodny, R. Brady, J. Quirk, and J. Kanfer, *J. Lipid Res.* **11**, 144 (1970).

Mucopolysaccharide Substrates

Principle. The free sialic acid released from the glycoprotein is determined by the thiobarbituric acid method for free sialic acid.[3]

Reagents

Sodium metaperiodate, 0.2 M in 9 M phosphoric acid
Sodium arsenite, 10% (w/v) in 0.5 M sodium sulfate in 0.1 N H_2SO_4
Thiobarbituric acid, 0.6% (w/v) in 0.5 M sodium sulfate
Cyclohexanone
Fetuin (GIBCo.), 20 mg/ml

Procedure. Incubations were carried out in a total volume of 200 μl and contained 180 μl of purified enzyme in solution, 10 μl potassium acetate buffer, and 10 μl of an aqueous solution of fetuin (20 μg). After a 1-hour incubation at 37°, 0.1 ml of periodate solution is added and the mixture is kept at room temperature for 25 minutes. One milliliter of arsenite solution is added followed by 3 ml of thiobarbituric acid solution. The mixtures are boiled for 15 minutes and cooled, and 2 ml of cyclohexanone is added. The absorbance of the cyclohexanone layer is read at 549 and 532 nm. Both enzyme and substrate blanks are run simultaneously with the incubation, and their values are subtracted.

Purification Procedure

Solubilization of Enzyme. The highest specific activity of the neuraminidase against ganglioside substrates is found in rat heart muscle; therefore this tissue is employed as the source of the enzyme. Male Sprague-Dawley rats (300 g) were decapitated and their hearts were removed and washed with cold 0.25 M sucrose–1 mM EDTA (pH 7.4). A 10% homogenate of the hearts was prepared in the sucrose EDTA solution and the suspension was centrifuged at 1000 g for 10 minutes. The supernatant was removed and sonicated for 5 × 30 sec (5 mAMP, Branson sonifier) with intermittent cooling at 0°. The mixture was centrifuged at 34,000 g for 40 minutes. The bulk of the enzyme was present in the supernatant fluid. The pH of the supernatant was adjusted to 3.7 and the mixture was stirred in the cold (4°) for 1 hour; and then centrifuged at 10,000 g for 10 minutes. The bulk of the activity was recovered in the pellet which was resuspended in 40 mM potassium phosphate buffer (pH 6.3).

Sephadex G-150 Gel Filtration. Sephadex G-150 was swollen with 40 mM potassium phosphate buffer and poured into a 2.5 × 30 cm

---

[3] L. Warren, *J. Biol. Chem.* **234**, 1971 (1959).

TABLE I

PURIFICATION OF RAT HEART NEURAMINIDASE

Fraction	Total protein (mg)	Total activity $G_{D1a}{}^a$	Specific activity $G_{D1a}{}^b$	Purification factor	% Recovered	Total activity $G_{M2}{}^a$	Specific activity $G_{M2}{}^b$	Purification factor	% Recovered
Homogenate	1327.1	317	0.238	1	100	102.1	0.077	1	100
Supernatant, 34,000 g	381.2	186	0.488	2	58.7	54.9	0.144	2	54
Acid pellet	147.0	100	0.681	2.9	31.6	40.8	0.278	3.6	39.9
Sephadex G-150 (No. 43–48)	9.4	69.4	7.38	31.0	21.9	—	—	—	17.2
Carboxymethyl-sephadex	3.1	61.0	19.6	82.6	19.2	17.6	5.70	74.0	17.2
Isoelectric focusing	—	59.9	—	—	18.9	16.9	—	—	16.5
Sephadex G-150 (No. 16–20)	0.059	47.5	811.9	3412	14.9	3.0	50.9	661	2.9

^{a} Nanomoles per hour.
^{b} Nanomoles per milligram of protein per hour.

column. The excluded volume was determined by blue dextran and the total volume (560 ml) was determined both by calculation and with sucrose. The resuspended pellet was clarified by centrifugation and placed onto the column. The column was eluted with the same buffer, and 10-ml fractions were collected. The enzyme was found in fractions 43–68, which were combined and concentrated to one-half volume using an Amicon concentrator (UM-50 membrane).

Carboxymethyl-Sephadex 50 Chromatography. CM-50 Sephadex was swollen in 40 mM potassium phosphate buffer, and the pH was adjusted to 6.3. The concentrated effluent from the Sephadex column was placed on a 15 × 1 cm column and washed in. A linear gradient between 0 and 0.5 M KCl was applied to the column and 5-ml fractions were collected. The bulk of the neuraminidase was eluted from the column shortly after the salt gradient was applied (fractions 12–20). These fractions were pooled.

Isoelectric Focusing. An isoelectric focusing column was prepared according to the instructions of the manufacturer.[4] The pH range used was 7–10; the effective pH range obtained was slightly more acidic, 6.5–9.0. The fractions from the CM-Sephadex were substituted for the light solution used in preparing the stabilizing sucrose gradient on this column. The enzyme was focused for 48 hours at 4° (500 mV, final current 1.5 mA). Two-milliliter fractions were collected, and the pH, neuraminidase activity, and absorbancy at 280 nm were determined. Active fractions were concentrated between pH 7.0 and 7.3 in a total volume of 10 ml. The pooled active fractions were repurified from ampholine and sucrose on a short (15 × 1 cm) Sephadex G-150 column prepared as above. The elution pattern was identical. The results of a typical purification are shown in Table I.

Properties of the Purified Neuraminidase

Substrate Specificity and Kinetic Parameters. The purified neuraminidase catalyzes the hydrolysis of various gangliosides and fetuin (Table II). It did not hydrolyze neuraminlactose.

Stability. The purified enzyme is not very stable when G_{M2} is used as a substrate. Half the activity is lost in 1 week at 0°; the activity toward the di- and trisialoganglioside and fetuin is retained. Freezing and thawing inactivated the enzyme.

pH Optimum. The pH optimum for the purified enzyme is 5.0.

Effect of Detergents. The reaction was moderately inhibited by sodium taurocholate and Triton X-100; it was severely inhibited by Cutsum.

---

[4] O. Vesterberg and H. Svensson, *Acta Chem. Scand.* **20**, 820 (1966).

TABLE II
K_m AND V_{max} OF VARIOUS GANGLIOSIDES AND PURIFIED NEURAMINIDASE

Ganglioside	K_m (μM)	V_{max} (nmoles/mg/hour)
G_{M2}	43	139
G_{D1a}	113	800
G_{D1b}	100	660
G_{T1}	91	440
Fetuin	—	1430
Neuraminlactose	—	0

Effect of Various Inhibitors. The effect of various substrate and re-action product analogs and inhibitors of the bacterial neuraminidase are shown in Table III. Also shown are the effects of various divalent cations and certain anions.

Nature of Reaction. Neuraminidase catalyzes the hydrolytic cleavage of the α-ketosidic bond of N-acetylneuraminic acid in glycolipids (gangliosides) and glycoprotein (fetuin).[5-7] The activity of this enzyme is present and unchanged in brain and muscle tissue obtained from patients with Tay-Sachs disease and may provide a functioning residual pathway for the catabolism of G_{M2} in the organs and tissues of patients with this disorder.[8]

TABLE III
INHIBITORS OF PURIFIED NEURAMINIDASE[a]

Inhibitor (conc.)	Substrate	% Inhibition
2-Deoxy-2,3-dehydroneuraminic acid (1 mM)	G_{M2}	0
p-Nitrophenyl oxamic acid (1 mM)	G_{D1a}	0
Dihydroisoquinoline derivative (0.5 mM)	G_{D1a}	60
p-Chloromercuribenzoate (0.1 mM)	G_{D1a}	92
Sodium azide (1 mM)	G_{D1a}	9
N-Acetylneuraminic acid	G_{M2}	0
Ceramide trihexoside	G_{M2}	−10
Cu^{2+} (1 mM)	G_{M2}	32
Fe^{3+} (1 mM)	G_{M2}	17
$(NH_4)_2SO_4$ (1 mM)	G_{D1a}	46

[a] Incubations were carried out as described except that 2 μg of purified protein were used in each incubation; the substrate was as indicated.

[5] G. Tettamanti and V. Zambotti, *Enzymologia* **35**, 61 (1968).
[6] Z. Leibovitz and S. Gatt, *Biochim. Biophys. Acta* **152**, 136 (1968).
[7] R. Ohman, A. Rosenberg, and L. Svennerholm, *Biochemistry* **9**, 3774 (1970).
[8] J. Tallman, W. Johnson, and R. Brady, *J. Clin. Invest.* **51**, No. 9 (1972).

[110] Glucocerebrosidase from Bovine Spleen (Cerebroside-β-glucosidase)

By Neal J. Weinreb and Roscoe O. Brady

Glucocerebroside + H₂O → ceramide + glucose

Assay Method

Principle. Enzyme activity effects the liberation of glucose from glucocerebroside. The reaction is monitored by measuring the trichloroacetic acid-soluble radioactivity representing labeled glucose arising from [1-14C]glucose cerebroside.

Reagents

> [1-14C]Glucose cerebroside[1]
> Potassium phosphate buffer, $1.0\,M$, pH 6.0
> Sodium cholate, 50 mg/ml, H₂O
> Human serum albumin, 100 mg/ml, H₂O
> Trichloroacetic acid, 100% (w/v)
> Diethyl ether
> Cutscum, 50 mg/ml, H₂O (Fisher Chemical Co.)
> Phosphonic acid cellulose (Cellex-P; Bio-Rad Laboratories)

Procedure. The assay mixture contains 15 µl of potassium phosphate buffer, pH 6.0, 80 µl of H₂O, 4 µl of Cutscum, and 50 µl of the enzyme preparation. The labeled substrate (7.5 mg/ml) is suspended in a solution of sodium cholate (50 mg/ml) and warmed until clear. A 12-µl aliquot of the water-clear solution is added to the incubation mixture, whose final volume is 0.16 ml. After incubation for 30 minutes in air at 41°, 0.8 ml of cold water, and 0.1 ml of the albumin solution (100 mg/ml) are added, followed by 0.1 ml of 100% trichloroacetic acid. The tubes are chilled in ice, and centrifuged in the cold. The supernatant solution is removed, and the precipitate is washed with 1.0 ml of cold 10% trichloroacetic acid. The pooled trichloroacetic acid solutions are extracted twice with 1.0-ml portions of ether. The radioactivity in an aliquot of the combined aqueous solutions is determined by liquid scintillation spectrometry.

Units. A unit of enzymatic activity is defined as the amount of en-

---

[1] R. O. Brady, J. Kanfer, and D. Shapiro, *J. Biol. Chem.* **240,** 39 (1965). See also R. O. Brady and J. N. Kanfer, Vol. 14 [23].

zyme required to catalyze the hydrolysis of 1 nmole of glucocerebroside per hour. Specific activity is defined as units per milligram of protein. Protein is estimated by the procedure of Lowry.

Purification Procedure

Beef spleen acetone powder was prepared from fresh tissue[2] or purchased from commercial sources. It is finely ground and suspended in 20 volumes of 10 mM potassium phosphate buffer, pH 6.0. After addition of solid sodium cholate to a concentration of 5 mg/ml, the suspension is stirred overnight at 4°, and then centrifuged at 17,300 g for 30 minutes.

TEAE-Cellulose Column Chromatography. From the preceding supernatant, 75 mg of protein are applied to a 2.5 × 4.0 cm column of TEAE-cellulose previously equilibrated with 0.01 M potassium phosphate buffer, pH 6.0, containing 5 mg of sodium cholate per milliliter. The protein is rapidly eluted from the column with twice the bed volume of a solution of 10 mM potassium phosphate, pH 6.0. This entire procedure is repeated a sufficient number of times to obtain the amount of enzyme desired for subsequent steps, and the eluates are pooled.

Butanol Partitioning. To the combined eluates, one-fifth volume of n-butanol at 0° is added dropwise over a period of 15 minutes. The mixture is stirred at 4° for an additional 30 minutes, and then centrifuged at 17,300 g. The upper butanol phase and the precipitate that forms at the interface are discarded. Solid ammonium sulfate is added to the butanol-saturated, aqueous lower phase to a saturation of 60%. After centrifugation, the precipitate is dissolved in 10 mM potassium acetate buffer, pH 4.8, which contains Triton X-100, 0.5%.

Phosphonic Acid Cellulose Chromatography. Protein, 60 mg, from the preceding solution are applied to a 10 × 4 cm column of phosphonic acid cellulose (in a glass Büchner funnel) that has been equilibrated with a solution of 10 mM potassium acetate, pH 4.8-Triton X-100, 0.5%. Stepwise elution is performed with 0.2 M potassium acetate, pH 5.1 and then 1.0 M potassium acetate, pH 5.1, each eluent containing Triton X-100, 0.5%. The 0.2 M eluate is discarded. To the 1.0 M eluate, one-half volume of cold n-butanol is added, and the mixture is stirred for 30 minutes. The butanol-saturated aqueous phase is concentrated by pressure dialysis. Solid ammonium sulfate is added to 60% saturation. After centrifugation, the precipitate is dissolved in 10 mM potassium acetate, pH 4.8-Triton X-100, 0.5%. Of the above solution, 0.06 mg of protein

[2] R. O. Brady, J. S. O'Brien, R. M. Bradley, and A. E. Gal, *Biochim. Biophys. Acta* **210**, 193 (1970).

is applied to a 5×30 mm column of phosphonic acid cellulose equilibrated with the potassium acetate 10 mM, Triton X-100, 0.5%, buffer. Stepwise elution with 1.2 ml of potassium acetate solutions of increasing concentration from 0.1 to 1.0 M in 0.1 M increments is performed. pH is maintained at 5.1, and Triton X-100, 0.5% is in all solutions. The glucocerebrosidase activity is eluted primarily in the 0.4–0.5 M fractions. The enzyme is stabilized in solution by addition of n-butanol to saturation. The final enrichment is approximately 200-fold over the original extract of beef spleen acetone powder. The acetone powder is itself several times enriched over crude, fresh spleen homogenates. The purification procedure is summarized in the Table.

Properties

Homogeneity. On polyacrylamide disc electrophoresis, the bulk of the purified preparation does not enter the gel, suggesting that it probably exists as a large aggregate and is most likely nonhomogeneous.

Stability. The most purified enzyme preparation is stable at 0° for at least 2 weeks, provided that the solution is maintained saturated with respect to n-butanol. After overnight dialysis, approximately 60% of the activity is lost. Activity is partially restored by readdition of n-butanol, but is not restored by addition of lecithin, asolectin, or oleic acid. Enzyme activity was completely lost on dialysis followed by lyophilization and could not be restored by the addition of n-butanol.

Effect of pH. The optimum is pH 6.0 using phosphate and acetate buffers.

PURIFICATION OF GLUCOCEREBROSIDASE FROM BEEF SPLEEN

Fraction	Volume (ml)	Total activity (units)	Protein (mg)	Specific activity (units/mg protein)	% Recovery
I. Cholate extract of acetone powder	150	42,300	1125	38	
II. TEAE-cellulose eluate	550	21,450	231	93	51
III. n-Butanol partitioned TEAE eluate	600	37,200	192	194	88
IV. 60% (NH$_4$)$_2$SO$_4$	250	19,750	60	329	47
V. Cellex-P eluate 0.2–1.0 M KAc	1500	15,750	7.5	2100	37
VI. Cellex-P eluate 0.4–0.6 M eluate	40	5,320	0.72	7400	13

Substrate Affinity. The K_m value for glucocerebroside is $2.5 \times 10^{-5} M$.

Activators and Inhibitors. Activity was not affected by Cu^{2+}, Zn^{2+}, Al^{3+}, ascorbic acid, n-ethylmaleimide, or EDTA. Oleic acid caused a slight stimulation. Activity was unaffected by β-galactal, a potent inhibitor of β-galactosidase. D-glucona-α-lactone, a known inhibitor of plant β-glucosidases, is a competitive inhibitor.

Artificial Substrates. When assayed at pH 5.0, with acetate buffer in the absence of detergents, p-nitrophenyl-β-D-glucoside, and 4-methylumbelliferyl-β-D-glucoside were not hydrolyzed by the purified preparation. On addition of Cutscum to the incubation medium, however, glucose is hydrolyzed from both p-nitrophenyl-β-glucoside, and from 4-methylumbelliferyl-β-glucoside by the purified enzyme preparation. For both artificial substrates, the pH optimum is 5.8–6.0, as with natural glucocerebroside. The specific activity of the reaction with the artificial substrates varies but usually is somewhat greater than with the natural substrate.

Specificity. Galactocerebroside, sphingomyelin, ceramide lactoside, ceramide trihexoside, and globoside were not hydrolyzed. Using Tay-Sachs ganglioside as substrate, no neuraminidase or N-acetyl-β-galactosaminidase activity is detected. Acid phosphatase, measured with p-nitrophenylphosphate, and aryl sulfatase, measured with p-nitrocatechol sulfate are absent. Disaccharidase, using cellobiose and lactose as substrates, is not present. β-Galactosidase, as assayed with p-nitrophenyl-β-galactoside and with 4-methylumbelliferyl-β-galactoside, is absent, even when Cutscum is added to the incubation mixture.

N-Acetyl-β-glucosaminidase is present in the purified preparation, as detected by the hydrolysis of p-nitrophenyl-N-acetyl-β-glucosaminide. The addition of Cutscum, which apparently unmasks activity to artificial glucosides in this preparation, causes total disappearance of N-acetyl-β-glucosaminidase activity in the purified preparation. In crude preparations, similar concentrations of Cutscum do not inhibit N-acetyl-β-glucosaminidase activity, and, indeed, may enhance it.

Nature of the Reaction. Glucocerebrosidase catalyzes the hydrolytic cleavage of the glycosidic bond of glucocerebroside. Attenuation of glucocerebrosidase activity is the biochemical defect responsible for the pathological accumulation of glucocerebroside in patients with Gaucher's disease.[3,4] Assay of glucocerebrosidase activity in a variety of types of

[3] R. O. Brady, J. N. Kanfer, and D. Shapiro, *Biochem. Biophys. Res. Commun.* **18,** 221 (1965).

[4] R. O. Brady, J. N. Kanfer, R. M. Bradley, and D. Shapiro, *J. Clin. Invest.* **45,** 1112 (1966).

cells and tissues permits biochemical confirmation of the diagnosis,[5] documentation of hereditary patterns by identification of carriers,[6] and reliable genetic counseling.[7]

[5] J. P. Kampine, R. O. Brady, J. N. Kanfer, M. Feld, and D. Shapiro, *Science* **155**, 86 (1967).
[6] R. O. Brady, W. G. Johnson, and B. W. Uhlendorf, *Amer. J. Med.* **51**, 423 (1971).
[7] E. L. Schneider, W. G. Ellis, R. O. Brady, J. R. McCulloch, and C. J. Epstein, *J. Pediat.* Submitted for publication, 1972.

[111] Galactosyl Ceramide (Galactocerebroside) β-Galactosidase from Brain

By NORMAN S. RADIN

Galactosyl ceramide + H_2O → galactose + ceramide

Assay Method

Principle. The substrate (galactocerebroside or galactosyl ceramide) is prepared in labeled form, with tritium in the sugar moiety. The labeled sugar released by the enzyme is readily separated from the excess substrate by solvent partitioning and counted by liquid scintillation. Because brain preparations generally contain galactocerebroside, which dilutes the radioactive substrate to an unknown degree, the enzyme must be separated somewhat from the lipid before assay. This is done by solubilization of the enzyme, enzymatic digestion to disintegrate lipids and proteins, and precipitation of the enzyme.

Reagents

Labeled galactosyl ceramide, specific activity about 350 cpm/nmole, in chloroform–methanol (2:1) solution. The preparation of this material is described in this volume [32].

Tween 20, in benzene solution with Myrj 59 (both made by ICI America Inc., Wilmington, Delaware and distributed by Emulsion Engineering Inc., Elk Grove Village, Illinois).

Oleic acid, good quality, in benzene

Sodium taurocholate, crude grade (Mann Research Laboratories), 4 mg/ml together with Tris base, 0.33 mg/ml in water

These solutions are stored in a freezer. To prepare the substrate emulsion, add enough of each solution to a test tube to give 1 mg of cerebroside, 10 mg of Tween, 5 mg of Myrj, and 3.5 mg of oleic acid. Remove the solvents with a stream of nitrogen, with the aid of a warm bath, and add 5 ml of the bile salt–Tris solution. A clear emulsion is obtained by warming the mixture with hot tap water and immersing the tube (capped!) in an ultrasonic bath of the type used for cleaning. (A similar bath can be improvised with a beaker of water and an ultrasonic probe.) The emulsion is stored in the cold room and appears to be stable for at least 1 week.

Sodium citrate buffer, pH 4.5, 1 M
Stopping solution: isopropanol–castor oil–chloroform 6:1:1 (v/v/v)
Carrier galactose: 20 mg/liter in water
Scintillation solvent: 6 g of PPO + 0.2 g of dimethylPOPOP + 1000 ml of toluene + 100 ml of BioSolv BBS-3 (Beckman Instruments, Inc.). We have found this mixture to give higher efficiency than a number of other common mixtures. It is most conveniently dispensed with a valved syringe, such as the Repipet (Labindustries, Berkeley, California).

Enzyme Preparation. Brain is homogenized with 5 volumes of buffer A (50 mM Tris·HCl, pH 7.4 at room temperature, 10 mM MgCl$_2$ 8 mM mercaptoethanol), then sonicated with a dipping probe for about 100 seconds. The suspension is cooled in an ice bath, and the probe power is turned off every 10 seconds to prevent overheating. The homogenate is diluted with 0.5 volume of buffer A containing 6% sodium taurocholate, and "lipase powder" (Mann Research Laboratories) is added in the ratio, 40 mg per gram of brain. This is a crude enzyme preparation from pork pancreas which contains proteolytic enzymes as well as lipolytic enzymes; possibly pancreatin would work as well. The powdered enzyme is suspended in the medium by brief sonication in a bath, and the mixture is then stirred gently in a cold room overnight.

The solubilized galactosidase is removed from debris by centrifugation for 1 hour at 100,000 g. To the supernatant solution is added 1/6 volume of 1 M sodium citrate buffer, pH 3, and the mixture is stirred 30 minutes to complete the precipitation of the enzyme. The galactosidase is collected by centrifugation for 20 minutes at 14,000 g and suspended in 14 ml of water and 7 ml of buffer A per gram of brain. This preparation is the equivalent of about 46 mg of brain per milliliter.

The enzyme preparation can be stored a long time in a freezer, but it seems wise to minimize refreezing by storing it in portions.

Procedure. Into a 20 × 150 mm screw-cap test tube, place 0.1 to 0.4 ml of enzyme preparation, 0.5 ml of substrate emulsion, and 0.1 ml of citrate buffer, pH 4.5; total volume is adjusted with water to 1 ml. Shake the capped tubes at 37° for 3 hours, then add 4 ml of "stopping solution" and mix the contents well (preferably with a vortexing machine). Let the sample sit 15 minutes, then add 5 ml of galactose solution and mix again. Add 6 ml of chloroform, mix again, and centrifuge the tubes briefly to clear the two phases. Transfer 5 ml from the upper phase (7 ml total) to a counting vial and remove the solvent with an air stream, with the aid of a 50–60° water bath.

Redissolve the radioactive galactose in 0.5 ml water. To each vial add 11 ml of scintillation solvent, mix briefly to clear the liquid, and measure the tritium content.

A blank should be run occasionally, with omission of enzyme. This activity, which probably reflects the slight partitioning of the substrate into the upper phase, is approximately 0.15% of the activity in the substrate. The castor oil in the stopping solution serves to overwhelm the micelle-forming ability of the detergents in the substrate emulsion, thereby preventing escape of micellar substrate into the upper phase.

Since the specific activity of the substrate is known (measured in the *same* scintillation mixture), it is a simple matter to calculate the activity of the enzyme preparation. The activity in brain depends somewhat on the age of the animal; in rat it ranged between 0.25 and 1 μmole per hour per gram of brain. Other species seem to exhibit lower specific activities.

Notes on the Procedure. The recovery of galactosidase during the preparation of the partially purified enzyme is not known, as the extraction and digestion process seems to enhance the activity considerably. A small amount of activity is lost in the pellet formed in the first centrifugation and a similar amount is lost during the acid precipitation step. Perhaps a total of 20% of the enzyme is lost this way. In a number of experiments, good linearity, with respect to tissue weight, was obtained even when the acid precipitation was omitted. Thus it is possible that the enzyme preparation procedure can safely be simplified, especially if the amount of enzyme used in the assay is reduced or if the content of endogenous cerebroside is low. The action of the pancreatic "lipase" resulted in considerable stimulation, possibly by breaking down the lysosomes and allowing access of the substrate to the enzyme, so this step should not be omitted if maximal activity is desired. However, it is possible that some tissues contain the galactosidase in particularly fragile lysosomes.

The enzyme preparations are partially insoluble, so it is helpful to stir the suspension magnetically while taking out aliquots, and to take only 2 or 3 aliquots into the pipette. Rehomogenization of stored frozen preparations just before dispensing is helpful.

Duplicate determinations are adequate. The sensitivity of the assay can readily be increased by the use of substrate of higher specific activity and a smaller incubation volume.

Unused substrate can be salvaged from the lower phase.

The above procedure is based on work by Bowen and Arora in this laboratory.[1,2] An earlier version of the procedure has been used successfully by Suzuki and Suzuki, who found a drastic deficiency of the enzyme in children with globoid cell leukodystrophy (see this volume [112]). An alternative method, less sensitive, involves determination of the liberated galactose by an enzymatic dehydrogenation and measurement of the NADH[+] or NADPH[+] thus formed.[3,4] This has the advantage that one need not prepare the substrate in radioactive form. A similar advantage is obtained from another method, which involves hydrolysis of the substrate on a disc of filter paper and assay of the liberated galactose by gas chromatography.[5]

Because cerebroside galactosidase is more concentrated in white matter, and because of the general heterogeneity of the brain, comparisons between samples from different brains will suffer from some uncertainty.

Purification Procedure

The enzyme in rat brain has been purified over 300-fold by conventional methods following the initial solubilization-digestion step.[6] Because the newer affinity columns hold so much promise, it does not seem worthwhile to describe all the details of the procedure.

The best single step in purification was chromatography of the solubilized enzyme with a DEAE-Sephadex A-25 column. The enzyme, in 10 mM Tris pH 7.4 containing 2 mM $MgCl_2$, 8 mM mercaptoethanol, 1% galactose, and 15% glycerol, was adsorbed to the column packing. Elution was carried out with the same buffer containing increasing NaCl concentrations. Most of the galactosidase came out with 0.3 M NaCl. Because the enzyme becomes unstable after this step, we add an equal volume of glycerol to each column fraction and promptly place it into a freezer. The glycerol and galactose act as effective stabilizers, but they must be removed (by dialysis) prior to assay as they inhibit the enzyme.

The solubilization-digestion step also preserves and releases ceramidase and arylsulfatase, which elute from the ion exchange column after the cerebroside galactosidase. Most of the enzymatic activity which acts

[1] N. S. Radin and R. C. Arora, J. Lipid Res. 12, 256 (1971).
[2] D. M. Bowen and N. S. Radin, J. Neurochem. 16, 501 (1969).
[3] S. Gatt, Biochim. Biophys. Acta 218, 173 (1970).
[4] F. B. Jungalwala and E. Robins, J. Biol. Chem. 243, 4258 (1968).
[5] G. Dawson and C. C. Sweeley, J. Lipid Res. 10, 402 (1969).
[6] D. M. Bowen and N. S. Radin, Biochim. Biophys. Acta 152, 587 (1968).

on nitrophenyl β-galactoside is destroyed by the pancreatic digestion step. The purified enzyme still shows appreciable hydrolytic activity toward glucosyl ceramide and lactosyl ceramide, as well as weak activity toward galactosyl sphingosine (psychosine).

Properties[7]

The pH optimum in rat and human brain is 4.5, measured in citrate buffer. The K_m is 22 μM in brains of Sprague-Dawley (derived) rats. The molecular weight is roughly 50,000.

The galactosidase is stimulated greatly by crude taurocholate, less so by free fatty acids and other bile salts. Some bile salts (deoxycholate) actually inhibit the enzyme. The composition of the detergents in the substrate emulsion makes a difference in the observed activity. Stimulation is also achieved, for no obvious reason, by the addition of the decanoyl amide of 2-methyl-2-amino-1-propanol.[8] The rat and human brain enzymes are stimulated as much as 62% by this amide, which faintly resembles the lipoidal product of enzymatic action, ceramide.

The galactosidase is inhibited effectively by galactonolactone, which seems to act against all lysosomal galactosidases. The inhibition is competitive in nature. Galactose is a weak inhibitor, but glucose is inert. Nitrophenyl β-galactoside is a weak inhibitor, possibly because it is also a substrate. Sphingosine is a fairly good inhibitor. Various metal cations are inert: Mg, Mn, Cu, Li, Ca, Fe^{3+}, Zn, and molybdate; also mercaptoethanol and EDTA.

A number of synthetic amides that resemble ceramide were found to be good inhibitors, particularly the decanoyl amide of DL-erythro-3-phenyl-2-amino-1,3-propanediol. Insertion of a DL-hydroxyl group in the fatty acid part increased the inhibition to 69% at 0.3 mM inhibitor concentration. These amides act as noncompetitive inhibitors, which indicates the presence of an effector-sensitive site on the galactosidase molecule.[9]

The enzyme does not appear to distinguish between the two kinds of galactocerebrosides that occur in brain (containing hydroxy and non-hydroxy fatty acids). The enzyme appears to be bound rather firmly to lysosomes, even after strong sonication. It occurs in neurones as well as in glia, and activity is demonstrable in brain well before glial proliferation and myelination take place.

[7] D. M. Bowen and N. S. Radin, *Biochim. Biophys. Acta* **152**, 599 (1968).
[8] R. C. Arora and N. S. Radin, *Lipids* **7**, 56 (1972).
[9] R. C. Arora and N. S. Radin, *J. Lipid Res.* **13**, 86 (1972).

Cerebroside galactosidase appears to be widely distributed in animal tissues and especially high specific activity has been found in small intestine.[10] This enzyme was purified fairly highly and showed two pH maxima.

[10] R. O. Brady, A. E. Gal, J. N. Kanfer, and R. M. Bradley, *J. Biol. Chem.* **240**, 3766 (1965).

[112] Galactocerebroside β-Galactosidase in Krabbe's Globoid Cell Leukodystrophy

By KUNIHIKO SUZUKI

Globoid cell leukodystrophy (Krabbe's disease) is a genetically determined, rapidly progressive, and invariably fatal neurological disorder of infants. The primary enzymatic defect underlying the disease is deficiency of galactocerebroside β-galactosidase.[1-3] The enzyme normally catalyzes the first step of galactocerebroside degradation in which galactocerebroside is cleaved to ceramide and galactose.

$$\text{Ceramide galactose} \xrightarrow[\beta\text{-galactosidase}]{\text{galactocerebroside}} \text{ceramide} + \text{galactose}$$
(galactocerebroside)

Assay Method for Pathological Material

The assay, purification and properties of galactocerebroside β-galactosidase in normal tissues are described in this volume [111]. This section concerns specimens of globoid cell leukodystrophy. The deficiency of galactocerebroside β-galactosidase in globoid cell leukodystrophy has been demonstrated in the brain, liver, spleen, kidney, serum, leukocytes, cultured fibroblasts, and cultured amniotic fluid cells.[1,3-5]

Principle of Assay. The reaction is measured by determining the radioactivity of released galactose from galactocerebroside specifically labeled at C-6 of galactose with tritium. After the reaction, the released

[1] K. Suzuki and Y. Suzuki, *Proc. Nat. Acad. Sci. U.S.* **66**, 302 (1970).
[2] K. Suzuki and Y. Suzuki, *in* "The Metabolic Basis of Inherited Disease" (J. B. Stanbury, J. B. Wyngaaden, and D. S. Frederickson, eds.), p. 760. McGraw-Hill, New York, 1972.
[3] Y. Suzuki and K. Suzuki, *Science* **171**, 73 (1971).
[4] J. Austin, K. Suzuki, D. Armstrong, R. O. Brady, B. K. Bachhawat, J. Schlenker, and D. Stumpf, *Arch. Neurol.* **23**, 502 (1970).
[5] K. Suzuki, E. Schneider, and C. J. Epstein, *Biochem. Biophys. Res. Commun.* **45**, 1363 (1971).

galactose is separated from the unreacted radioactive galactocerebroside by solvent partitioning in a chloroform–methanol–water system.

Reagents

Substrate: Galactocerebroside, labeled and purified according to Radin et al.[6] The specific activity of the working substrate should be at least 1000 cpm/nmole for all pathological specimens, except for serum, for which the specific activity should be at least 5000 cpm/nmole. This can be achieved easily by diluting the original labeled substrate with unlabeled galactocerebroside[7] to the desired specific activity. It can be kept as a solution in chloroform–methanol (2:1, v/v) in the freezer.

Detergents: Tween 20, 1 g, and G-2159 (polyoxyethylene stearate, Myrj 59),[8] 0.5 g, dissolved in 100 ml of benzene

Sodium citrate buffer: 1 M, pH 4.5

Sodium taurocholate: 5 mg/ml in water

Tris base: 1.65 mg/ml in water

Oleic acid: 1 g in 100 ml of hexane

Buffer A[9]: 0.05 M Tris·HCl, 0.01 M $MgCl_2$, and 8 mM mercaptoethanol in water

Galactose: 1 mg/ml in water

Preparation of Tissues. It is essential to always include known normal control specimens in the assay. For assay of the enzyme in solid tissues, postmortem frozen specimens are satisfactory, since the enzyme is stable in organs for many years in the frozen state. Five percent homogenate in water is made using a Dounce homogenizer, sonicated for 1 minute, and then frozen and thawed four times. The homogenate is centrifuged at 1120 g for 10 minutes, and the supernatant is used for the assay.

Serum is used directly for assays without further processing. Plasma is unsatisfactory because commonly used anticoagulants are moderately inhibitory.

Leukocytes can be prepared from anticoagulated blood samples according to any available procedures utilizing differential sedimentation in dextran (e.g., Snyder and Brady[10]). Cultured cells, both skin fibro-

---

[6] N. S. Radin, L. Hof, R. M. Bradley, and R. O. Brady, *Brain Res.* **14**, 497 (1969).

[7] Galactocerebroside satisfactorily pure for this purpose is available from commercial sources (Applied Science Laboratories, State College, Pennsylvania; Supelco, Inc., Bellefonte, Pennsylvania).

[8] Atlas Chemical Industries, Wilmington, Delaware.

[9] D. M. Bowen and N. S. Radin, *J. Neurochem.* **16**, 501 (1969).

[10] R. A. Snyder and R. O. Brady, *Clin. Chim. Acta* **25**, 331 (1969).

blasts and amniotic fluid cells, are harvested from monolayers and centrifuged to obtain cell pellets. Pellets of leukocytes and cultured cells are suspended in water—approximately 1 ml per leukocytes from 10 ml of blood, or per a few million cultured cells. Cells are sonicated for 1 minute, frozen and thawed four times, and centrifuged at 1120 g for 10 minutes. The supernatant is used for the assay.

Procedure. Appropriate amounts of the substrate solution (200 μg per incubation tube) and the detergent solution (1 mg of Tween and 0.5 mg G-2159 per incubation tube) are dried together in a test tube under a stream of nitrogen, and the sodium taurocholate solution is added to make 0.5 mg of labeled substrate per milliliter. The solution is warmed and sonicated in a water-bath-type ultrasonicator to obtain a faintly translucent suspension. The oleic acid solution is similarly dried (0.3 mg per incubation) in a test tube, and the Tris base solution is added to make 3 mg of oleic acid per milliliter. The mixture is sonicated to obtain a uniform suspension.

Incubations are carried out in 10-ml capacity glass-stoppered centrifuge tubes containing 0.2 ml of buffer A diluted with water to 1:2, 0.1 ml citrate buffer, 0.1 ml oleic acid suspension, 0.2 ml enzyme source, and 0.4 ml substrate suspension. Blank tubes contain 0.2 ml of water instead of the enzyme source. The tubes are stoppered and gently shaken in a constant temperature water bath at 37° for 3 hours. At the end of the incubation, 0.1 ml of the unlabeled galactose solution is added as the carrier, and the reaction stopped with the addition of 5 ml of chloroform–methanol (2:1, v/v). The tubes are shaken vigorously with a Vortex stirrer, centrifuged briefly, and the lower chloroform phase removed as completely as possible with Pasteur pipettes. The upper phase is washed twice with the addition of 2 ml of chloroform. When the above procedure is followed exactly, the final upper phase volume is 2.0 ml within the limit of the errors of the entire procedure. An aliquot of this upper phase (conveniently 1.0 ml) is taken into a scintillation vial, dried in an oven at 120°, and the content is dissolved in 0.5 ml of water. Twelve milliliters of a toluene-based scintillation solvent, which contains 10% Bio-Solv III,[11] is added, and the radioactivity is determined. Sample counts are corrected for the blank counts.

Expression of Activity. For solid tissues, for which wet weight can be determined with reasonable accuracy, the activity of galactocerebroside β-galactosidase can be expressed most conveniently as the amount of the substrate hydrolyzed per hour per unit of wet weight. It is impossible to obtain accurate wet weight for fetal tissues, and in these instances

[11] Beckman Instrument, Fullerton, California.

the activity should be expressed on the basis of the protein content of the supernatant enzyme source.[5] The enzyme activities of leukocytes and cultured cells are most reliably expressed on the basis of the supernatant protein. Activities on the basis of cell counts are less reliable because of the inherently larger errors in the determination of cell counts. The serum activity is best expressed simply for a unit volume. The table summarizes the results on galactocerebroside β-galactosidase in globoid cell leukodystrophy, as determined in the author's laboratory by the procedure described above.

Remarks

The standard assay procedure described above incorporates a number of pragmatic considerations which must be kept in mind when the results are to be scrutinized from the more rigorous scientific standpoint.

Brain tissue, particularly white matter, contains large amounts of endogenous galactocerebroside which significantly reduces the specific activity of the added radioactive galactocerebroside. In the standard assay system, the maximum dilution occurs in normal white matter samples in which the exogenous substrate is diluted approximately one to one. In white matter of globoid cell leukodystrophy, however, the amount of galactocerebroside is always less than normal, unlike the case in other sphingolipidoses in which the affected lipids are always present in great excess. Therefore, from the practical viewpoint, the above procedure gives the most conservative estimate of the enzyme deficiency, and the actual degree of the deficiency is always greater.

GALACTOCEREBROSIDE β-GALACTOSIDASE IN GLOBOID CELL LEUKODYSTROPHY[a]

	Normal controls	Globoid cell leukodystrophy
	(nmoles/hour/g wet weight)	
Brain, Gray matter	225 ± 87 ($n = 9$)	10.3 ± 2.3 ($n = 5$)
White matter	366 ± 200 ($n = 9$)	9.3 ± 1.0 ($n = 4$)
Liver	189 ± 126 ($n = 8$)	5.8 ± 4.8 ($n = 4$)
Kidney	460 ± 326 ($n = 9$)	22.7 ± 5.0 ($n = 5$)
Spleen	157 and 186 ($n = 2$)	20.4 ($n = 1$)
	(nmoles/hour/mg protein)	
Fetal brain	5.89 and 6.28 ($n = 2$)	0.028 ($n = 1$)
Fetal liver	2.04 and 2.77 ($n = 2$)	0.025 ($n = 1$)
Leukocytes	2.42 ± 1.13 ($n = 44$)	0.11 ± 0.24 ($n = 7$)
Fibroblasts	3.00 ± 2.38 ($n = 38$)	0.30 ± 0.24 ($n = 7$)
Amniotic fluid cells	1.73 ± 0.47 ($n = 9$)	0.09 ($n = 1$)
	(nmoles/hour/100 ml)	
Serum	23.4 ± 7.83 ($n = 12$)	0.13 ± 0.17 ($n = 4$)

[a] The variations are expressed as standard deviations.

When more precise results are desired from white matter samples, the partially purified enzyme preparation as described in this volume [111] should be used. In cases of gray matter, the maximum isotopic dilution is 10%, and other organs do not contain enough endogenous galacto-cerebroside to cause any detectable dilution.

In normal tissues, the pH optimum of galactocerebroside β-galac-tosidase is 4.5, but there are no significant activity peaks in globoid cell leukodystrophy throughout the pH range of 4.1–8.1.[2] The reaction is linear against time for at least 5 hours. In leukocytes and cultured cells, the reaction is proportional to the amount of protein at least up to 1 mg protein per incubation, except at protein concentrations below 30 μg per incubation, when the measured activity was disproportionately low. See this volume [111] regarding the linearity of the reaction against amounts of solid tissues.

While galactocerebroside β-galactosidase in solid tissues, leukocytes, and culture cells, is stable under frozen conditions, the serum enzyme is relatively unstable (Fig. 1). Therefore, serum samples must be assayed as soon as possible after collection.

With the properly prepared and purified radioactive substrate, the blank counts should be 5% or less of the counts of normal control samples, except in serum samples in which normal control counts are rarely greater than 3 times that of blank tubes.

Leukocytes, skin fibroblasts, and serum of parents of patients ex-hibit galactocerebroside β-galactosidase activities intermediate between

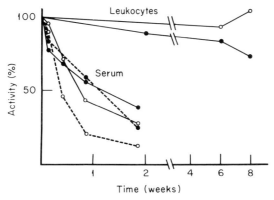

Fig. 1. Effect of storage on the activities of β-galactosidases. Serum and leukocytes were prepared and stored either frozen or refrigerated for designated period of time before enzyme assays. Filled circles, galactocerebroside, β-galacto-sidase; open circles, 4-methylumbelliferyl-β-galactosidase. Solid lines represent samples stored frozen at $-20°$; and dashed lines, samples stored refrigerated at $4°$. Reproduced from Y. Suzuki and K. Suzuki, *Science* **171**, 73 (1971) by permission. Copyright 1971 by the American Association for the Advancement of Science.

those of patients and normal controls.[12] In our experience, freshly obtained serum gives the most reliable results for the purpose of heterozygous carrier detection. There were no overlaps in the enzyme activities among the three groups—normal, patients, and parents. The use of cultured amniotic fluid cells provides the means of prenatal diagnosis of the disorder.[5] Both the deficiency in patients, and the partial deficiency in parents, can be demonstrated only by the use of the natural substrate, galactocerebroside, not by unnatural chromogenic substrates, such as p-nitrophenyl β-galactoside or 4-methylumbelliferyl β-galactoside.

In the author's laboratory, the results on cultured fibroblasts gave a wide variation from one specimen to another. This appears to be the result of wide-ranging different conditions of culture techniques in the different laboratories which contributed samples to the author's laboratory, such as composition of nutrient media, number of passages, the phase of cell growth, and the method of cell harvesting. Whenever possible, efforts should be made to control these factors.

Globoid cell leukodystrophy, clinically and morphologically very similar to the human disease, is known to occur in two strains of terrier dogs (West Highland and Cairn). The same deficiency of galactocerebroside β-galactosidase has been demonstrated in the brain, liver, kidney, and leukocytes,[12,13] utilizing the method described above. However, the canine form of globoid cell leukodystrophy is not identical enzymatically to the human disease. The activities of galactocerebroside β-galactosidase were normal in the serum of affected dogs and of known carrier dogs.[12]

[12] K. Suzuki, Y. Suzuki, and T. F. Fletcher, *in* "Sphingolipids, Sphingolipidoses and Allied Disorders" (B. W. Volk and S. M. Aronson, eds.). Plenum Press, New York, p. 487, 1972.
[13] Y. Suzuki, J. Austin, D. Armstrong, K. Suzuki, J. Schlenker, and T. Fletcher, *Exp. Neurol.* **29**, 65 (1970).

[113] Lactosyl Ceramide β-Galactosidase from Brain

By NORMAN S. RADIN

Lactosyl ceramide + H_2O → galactose + glucosyl ceramide

Assay Method[1,2]

Principle. Lactosyl ceramide (cytolipin H, or ceramide lactoside) is prepared in radioactive form and incubated with the galactosidase.

[1] N. S. Radin, L. Hof, R. M. Bradley, and R. O. Brady, *Brain Res.* **14**, 497 (1969).
[2] S. Gatt and M. M. Rapport, *Biochem. J.* **101**, 680 (1966).

The liberated radioactive galactose is removed from excess substrate by solvent partitioning and counted by liquid scintillation.

Reagents

Labeled substrate: labeled in the galactose moiety as described in this volume [32]. Although it can be labeled by simple exposure to tritium gas, this produces indiscriminate labeling, which may raise the blank value and yield high values when glucosyl ceramide glucosidase is present in the enzyme preparation. A specific activity of about 250 cpm/nmole is adequate. The compound is stored in chloroform–methanol in a freezer.

Triton X-100: in benzene solution, stored in a freezer. The stock Triton (100%) should probably be stored in a freezer too to minimize accumulation of peroxides, since Triton—like other nonionic surfactants—is an ether.

Sodium taurocholate, a crude product from Mann Research Laboratories, 5 mg/ml in water, stored frozen.

Emulsified substrate: prepare in emulsified form by mixing solutions of lactoside (200 nmoles) and Triton (0.5 mg), evaporating to dryness with a stream of nitrogen and a warm bath, and adding 0.4 ml of the taurocholate solution. Warming and sonicating in a cleaning bath yields a clear emulsion. This is enough for one incubation tube. The stability of the emulsion has not been determined but it is probably good for at least a week at 4°.

Sodium acetate buffer, 2 M, pH 5 (measured at room temperature)

Carrier galactose solution: 1 mg/ml

Enzyme Preparation. Rat brain is homogenized in 6 volumes of buffer: 50 mM Tris·HCl (pH 7.4 at room temperature), 10 mM MgCl$_2$, and 8 mM mercaptoethanol. The homogenate is then sonicated in an ice bath with a dipping probe for ten 10-second periods, alternating with 10-second rest periods for cooling. The preparation is diluted 10-fold prior to assay. It may be stored in the freezer prior to dilution.

Procedure. In a screw-cap test tube (20 × 150 mm) place 0.4 ml of the substrate emulsion, 0.2 ml of acetate buffer, 0.2 ml of enzyme suspension (equivalent to 2.9 mg of brain), and enough water to make the total volume 1 ml. Cover the tubes with caps lined with Teflon and incubate the tubes for 2 hours at 37° in a shaker. Stop the reaction with 5 ml of chloroform–methanol 2:1 and 0.1 ml galactose solution. Mix the liquids well and centrifuge the tubes briefly to clear the two phases. Transfer the upper layer to another tube and wash this with 1.5 ml of chloroform–methanol 85:15. Transfer 1.5 ml of the upper phase to a scintillation vial and evaporate off the solvent with a stream of air in a water bath set at 50–60°. The labeled galactose may be counted as described in this volume [111].

A blank run should be carried out without the enzyme or with heated enzyme. Some lactoside probably partitions into the upper layer. In previous runs the blank has amounted to about 0.4 nmole. Rat brain yields approximately 4 nmole of galactose liberated per milligram per hour.

Notes on the Procedure. Since only a portion of the upper phase is used for counting (1.5 ml), a correction must be applied to the observed activity. The calculated volume of the upper phase is about 2.7 ml, so observed values should be multiplied by the factor $2.7/1.5 = 1.8$. However, for more precise calculations one should determine the actual volume of the upper phase at the ambient temperature. The enzyme activity is calculated on the basis of the known specific activity of the galactose moiety, measured in the same scintillation solvent.

It is quite possible that the isolation of the liberated galactose can be carried out more expeditiously with the newer partition system described in this volume [111].

The rate of hydrolysis is constant during the 2-hour incubation, and the rate is proportional to brain weight up to at least 3 mg per tube.

Unhydrolyzed substrate, which is rather expensive, can probably be recovered fairly readily.[3]

The above procedure has been used successfully by Dawson and Stein,[4] who found a case of severe illness in a baby who exhibited a low specific activity of lactoside galactosidase. This galactosidase is evidently distinct from the β-galactosidases in brain which hydrolyze cerebroside and gangliosides.

Purification Procedure

The galactosidase has been purified partially from calf and beef brain.[5] The procedure was developed with nitrophenyl galactoside as the test substrate, so it is not clear that the method is particularly suited to isolation of lactoside galactosidase. This paper also described two additional methods for assaying the enzyme for activity toward lactosyl ceramide. No doubt affinity chromatography will prove very effective in purifying the enzyme and clarifying the question of substrate specificity.

[3] N. S. Radin and R. C. Arora, *J. Lipid Res.* **12**, 256 (1971).
[4] G. Dawson and A. O. Stein, *Science* **170**, 556 (1970).
[5] S. Gatt, see Vol. 14 [29].

Properties[1,2]

Too high a concentration of bile salt is inhibitory; virtually no activity is seen without some bile salt. Inhibition is produced by fatty acid, galactonolactone, nitrophenyl β-D-thiogalactoside (which is not a substrate), sphingosine, and ceramide.

The specific activity of the hydrolase in rat brain changes with age of the animal, reaching a peak at about 24 days. A similar, but not identical, sequence of changes can be seen when the brains are assayed for activity toward nitrophenyl galactoside.[1] The enzyme is probably found in lysosomes, but is readily made "soluble" by sonication with a dipping probe.

A galactosidase in jack beans also hydrolyzes lactosyl ceramide.[6] The K_m of the rat brain enzyme is 22 μM.

[6] Y.-T. Li and S.-C. Li, *J. Biol. Chem.* **246**, 3769, 1971; see also this volume [90].

[114] Lactosylceramide β-Galactosidase in Lactosylceramidosis

By Glyn Dawson

Galactosyl $\beta(1 \rightarrow 4)$ glucosylceramide $\rightarrow$ galactose + glucosylceramide

Principle

Galactosyl $\beta(1 \rightarrow 4)$ glucosylceramide (lactosylceramide; GL-2a) is a catabolite of both globosides (characteristic of erythrocytes, liver, spleen, etc.) and gangliosides (characteristic of gray matter), and any inherited defect in the specific β-galactosidase should manifest itself as a neurovisceral lysosomal storage disease. Such a defect has been described[1,2] in a child who died at the age of 4.2 years, and we have called the disease lactosylceramidosis.

Assay Method

The colorimetric assay system using the synthetic substrate β-nitrophenyl-β-D-galactoside does not measure lactosylceramide β-galactosidase activity (see also [111], galactosylceramide galactosidase). Thus in G_{M1}-gangliosidosis there is a total absence of reactivity toward p-nitrophenyl-β-D-galactoside [118], but GL-2a metabolism is normal. This

[1] G. Dawson and A. O. Stein, *Science* **170**, 556 (1970).
[2] G. Dawson, R. Matalon, and A. O. Stein, *J. Pediat.* **79**, 423 (1971).

has subsequently been confirmed[3] by direct demonstration of elevated (350% over normal) lactosylceramide β-galactosidase activity in G_{M1}-gangliosidosis tissue. Hence it was necessary to use the assay of Radin[4] [113] in which GL-2a is specifically labeled with ^{3}H at C-6 of galactose [111]. After incubation with cell-free tissue extracts, the liberated [^{3}H]galactose is removed from the excess substrate by chloroform–methanol–water partition and counted by liquid scintillation [113]. The procedures for preparation and labeling of substrate are as described in [32] and [113].

Enzyme Preparation

Human liver (obtained at autopsy or biopsy) and cultured human skin fibroblasts (grown to monolayer confluency; stationary phase) were sonicated in 5 vol of buffer (0.1 M sodium acetate, pH 4.5, containing 0.15 M NaCl) at 10,000 cycles per second for 10 × 12 seconds. The temperature should not exceed 4°. The sonicated cells, in approximately 1.0 ml of buffer, were centrifuged at 600 g to remove cell debris. Aliquots (0.1 ml) of the supernatant were incubated with the [^{3}H]-labeled GL-2a in buffer containing sodium taurocholate (2 mg/ml) or Triton X-100 (1 mg/ml) as described by Radin [113]. This supernatant may be stored at −20° for several months without appreciable loss of activity. For assays of urinary lactosylceramide β-galactosidase the urine was either used directly, lyophilized, or concentrated by ultrafiltration. In our hands, dialysis or precipitation with ammonium sulfate led to loss of activity. Tissue from the patient with lactosylceramidosis contained about one-tenth of normal activity (see the table).

Procedure

In addition to using the radioactive assay of Radin [113], the galactose dehydrogenase assay of Mapes et al.[5] was used; it is sensitive enough to detect activity in urine.

Lactosylceramide β-Galactosidase Activity in Normals and in the Patient with Lactosylceramidosis

The enzyme isolated from both normal and pathological liver and skin fibroblasts had a pH optimum of 4.5; other properties were essentially the same as for the rat brain preparation described by Radin

[3] R. O. Brady, J. S. O'Brien, R. M. Bradley, and A. E. Gal, *Biochim. Biophys. Acta* **210**, 193 (1970).
[4] N. S. Radin, L. Hof, R. M. Bradley, and R. O. Brady, *Brain Res.* **14**, 497 (1969).
[5] C. A. Mapes, R. L. Anderson, and C. C. Sweeley, *FEBS (Fed. Eur. Biochem. Soc.) Lett.* **7**, 180 (1970).

Tissue	Lactosylceramide β-galactosidase (nmoles/mg protein/hour)	β-galactosidase (μmoles 4-methylumbelliferone liberated/mg protein/hour)
Normal fibroblasts	1.10	0.63
Fabry hemizygote	0.85	0.60
Fabry heterozygote	0.94	0.65
Hurler homozygote	1.00	0.50
Lactosylceramidosis patient	0.11	0.55
Mother of patient	0.44	0.60
Father of patient	0.48	0.57
Patient + normal (mixed)	0.62	—
Normal liver	0.28	0.69
Lactosylceramidosis liver	0.04	0.24

[113]. The residual enzyme in tissue from the patient with lactosyl-ceramidosis had the same pH optimum (4.5) as normal, but detailed heat-inactivation and kinetic studies have not been carried out. Activity in the patient's liver and cultured fibroblasts was 10–15% of normal (see the table). Mixing experiments gave no evidence for the presence of in-hibitors or activators in the patient's tissue. Lactosylceramide β-galac-tosidase was also present in normal human brain,[3] leukocytes, and urine and deficient in the patient. Both parents of the child showed enzymatic activity midway between normal levels and levels in patients (see the table) suggesting that they were heterozygotes for an autosomal reces-sive enzyme defect.

[115] Ceramide Trihexosidase from Human Placenta

By WILLIAM G. JOHNSON and ROSCOE O. BRADY

Galactosylgalactosylglucosyl ceramide + H_2O → galactosylglucosyl ceramide

+ galactose

Assay Methods

Principle. Enzyme activity causing the release of the terminal α-linked [³H]galactose from tritiated ceramide trihexoside is designated ceramide trihexosidase and is quantified by determination of perchloric acid-soluble radioactivity remaining in the upper phase of Folch partition. Enzyme activity causing release of strongly fluorescent 4-methylumbelliferone from its nonfluorescent α-galactoside is designated α-galactosidase and is quantified by fluorometry at basic pH.

Reagents

CERAMIDE TRIHEXOSIDASE (CTHASE)

O-α-D-galactopyranosyl-$(1 \rightarrow 4)$-O-β-D-galactopyranosyl-$(1 \rightarrow 4)$-O-β-D-glucopyranosyl ceramide-$[^3H]$, $([^3H]CTH)$ [1]
Sodium acetate buffer, 1.0 M, pH 4.4
Sodium cholate purified, 50 mg/ml in H_2O (Mann Biochemicals)
Human serum albumin, 100 mg/ml in H_2O (HSA, Pentex Corporation)
Perchloric acid, 2% (w/v)
Potassium hydroxide, 4% (w/v)
Chloroform–methanol, 2:1 (v/v)
Theoretical upper phase (chloroform:methanol:H_2::3:48:47) [2]
Naphthalene (recrystallized, Eastman Organic Chemicals)
PPO (2,3-diphenyloxazole, New England Nuclear)
POPOP (p-bis[2-(5-phenyloxazolyl)]benzene, New England Nuclear)
1,4-Dioxane (Baker)

α-GALACTOSIDASE

4-Methylumbelliferyl-α-D-galactopyranoside (4-MU-α-galactoside) (Koch-Light Company)
Citrate-phosphate buffer, "0.15 M" (McIlvain's buffer, undiluted) [3]
4-Methylumbelliferone (4 MU, K&K Laboratories, Inc.)
Glycine, 0.2 M in H_2O
Sodium hydroxide, 0.2 M in H_2O

Procedure

CERAMIDE TRIHEXOSIDASE. A solution of the $[^3H]CTH$ (5 mg/ml), in sodium cholate solution (50 mg/ml) is prepared without warming. In addition to the enzyme extract, the incubation mixture contains 30 μg (0.027 μmole) of labeled $[^3H]CTH$, (277 μCi/mmole), 300 μg of sodium cholate, 15 μmoles of sodium acetate buffer (pH 4.4), and water to a final volume of 0.13 ml. The sample tube and an identical control tube (in which the enzyme extract has been boiled 5 minutes) are incubated

[1] R. O. Brady, A. E. Gal, R. M. Bradley, and E. Mårtensson, *J. Biol. Chem.* **242**, 1021 (1967).
[2] J. Folch, M. Lees, and G. H. Sloan Stanley, *Biol. Chem.* **226**, 497 (1957).
[3] G. Gomori, see Vol. 1 [16].

for 1 hour at 37°. The reaction is stopped by adding 10 μg of HSA, 100 mg/ml, followed by 0.15 ml of perchloric acid, 2%. The precipitate is removed by centrifugation at 2000 g for 5 minutes. The supernatant is removed with a capillary pipette and neutralized with 0.09 ml of KOH 4% (w/v). Further precipitate is removed by recentrifugation, and the supernatant is added to 5 ml of chloroform–methanol, 2:1, and 0.5 ml of H$_2$O. After thorough mixing (Vortex mixer) and centrifugation, the aqueous phase is removed and heated at 50° to dryness. The residue is dissolved in 10 ml of counting solution (600 g of naphthalene, 24 g of PPO, 600 mg of POPOP, dioxane to 2000 ml) and counted for ^{3}H.

α-GALACTOSIDASE. A substrate solution of 4-MU-α-galactoside 5 mM, (16.9 mg in 10 ml of H$_2$O) is prepared. An incubation mixture contains up to 50 μl of enzyme extract, 100 μl of citrate-phosphate buffer "150 mM" (made by mixing 22.2 ml of 0.2 M disodium phosphate and 27.8 ml of 0.1 M citric acid solution[3]), and 50 μl of substrate solution to a total volume of 0.2 ml. The blank, containing the same amounts of buffer and substrate, but water instead of enzyme extract, is treated identically to the sample. An enzyme blank is usually unnecessary in the fluoro- metric assay. After incubation for 1 hour at 37°, the reaction is stopped by adding 0.8 ml of glycine–sodium hydroxide buffer, 0.1 M, pH 10.7 (made by mixing equal volumes of NaOH 0.2 M and glycine 0.2 M). The fluorescence is quantitated using a fluorometer with a liquid standard (quinine alkaloid, 10 μg/ml in 0.1 N H$_2$SO$_4$) and the result compared with the fluorescence of known amounts of 4 MU. In this laboratory an Eppendorf fluorometer was used with primary filter transmitting the Hg line, 366 nm, and a double barrier secondary filter transmitting of 430–470 nm. Quartz cuvettes were used throughout.

Purification of Placental α-Galactosidases

Reagents

    Monosodium phosphate
    Disodium phosphate
    Ammonium sulfate
    Sephadex G-200 (Pharmacia)
    DEAE-Sephadex A-25 (Pharmacia)
    Sodium chloride
    Citric acid
    Sodium hydroxide
    CM Sephadex C-25 (Pharmacia)

Procedure. Fresh human placentas were placed on ice immediately after the birth and were processed within 12 hours after delivery. All subsequent procedures were carried out at cold room temperature (3°C). Placentas were perfused with normal (0.9% NaCl in H_2O) saline to remove clots; 2500 g was dissected free from fibrous tissue and homogenized in portions in a Waring Blendor (two, 1-minute runs) in 5 volumes of sodium phosphate buffer 25 mM pH 6.0.[4]

Ammonium Sulfate Precipitation. Ammonium sulfate fractionation of the solution was performed (using 70.5 g $(NH_4)_2SO_4$ per 100 ml of solution as 100% saturation), and the fraction that precipitated between 21% and 50% is resuspended in sodium phosphate buffer to a total volume of 8000 ml (the bulk of the α-galactosidase is precipitated by a 21% to 40% cut, however, since other lysosomal hydrolases were being studied concomitantly, the larger fraction was taken). This highly concentrated protein solution was dialyzed against sodium phosphate buffer, 25 mM, pH 6.0, for 48 hours.

Sephadex G-200. The retentate was centrifuged at 50,000 g for 1 hour, and the supernatant was applied to an upward flowing column of Sephadex G-200 (Pharmacia column, 10 × 100 cm) equilibrated with the same phosphate buffer. Fractions were collected (25 ml) and analyzed for α-galactosidase by the fluorometric assay. A single α-galactosidase peak was found; fractions containing 90% of the total assayed α-galactosidase were pooled and retained. Pooled α-galactosidase fractions from two such runs were combined, concentrated together to a volume of 300 ml (Amicon system with PM-10 ultrafilters), and centrifuged at 30,000 g for 20 minutes; the supernatant was reapplied to a freshly packed column of the Sephadex G-200 in identical fashion. Fractions containing 90% of the total assayed α-galactosidase were collected and concentrated in the same way to a volume of 300 ml.

DEAE-Sephadex. Sodium phosphate buffer (8 liters) 25 mM, pH 6.0, was titrated to pH 6.00 ± 0.02 at 25°. DEAE-Sephadex A-25 (about 150 g) was swollen in 2 liters of this buffer and titrated to pH 6.00 ± 0.02 (25°). The gel suspension was washed two times with fresh buffer and retitrated each time. A Pharmacia column (5 × 60 cm) was packed with this gel and washed with 2 liters of the phosphate buffer. Combined fractions from two runs of the previous step (300 ml) were titrated to pH 6.00 ± 0.02 at 25° and applied slowly (downward flow) to the column; the column was washed with 500 ml of the phosphate buffer and eluted with a linear NaCl gradient, 0–600 mM, total volume 2 liters (Kontes gradient maker). Fractions were collected (25 ml) and

[4] G. Gomori, Vol. 1 [16].

two incompletely separated peaks of α-galactosidase were distinguished, well separated from β-hexosaminidase and other lysosomal hydrolases. The two α-galactosidase peaks were pooled, concentrated to 30–50 ml (Amicon system, PM-10 membranes), and dialyzed overnight against the sodium phosphate buffer 25 mM, pH 6.0.

CM-Sephadex. Citrate phosphate buffer, 30 mM, pH 5.0, was prepared (by mixing 243 ml citric acid, 0.1 M, and 257 ml of sodium diphosphate 0.2 M diluting 5 fold),[3] and titrated to pH 5.00 ± 0.02 (25°). Sephadex C-25 was swollen in this buffer, titrated to pH 5.00 ± 0.02, washed twice, retitrated, and packed into a chromatography column (0.9 × 20 cm, Pharmacia). The column was washed with the same buffer (50 ml). The protein sample (30 ml) was titrated to pH 5.00 ± 0.02 (25°) applied slowly (downward flow) to the column; the column was washed with 50 ml of the citrate phosphate buffer and eluted with a linear NaCl gradient 0–400 mM, total volume 200 ml. Fractions of 3.5 ml were collected. Two well separated peaks of α-galactosides were distinguished on analysis with the fluorescent substrate. The initial peak, consisting of protein which did not stick to the gel, contains two α-galactosidases, P1 and P2, with low CTHase activity. The second peak, eluted with a salt gradient contains α-galactosidase P3, which has high CTHase activity and is designated placental CTHase. All three enzymes can be separated from each other by performing the final step (CM-Sephadex chromatography) at pH 4.80 ± 0.02 (25°), but the yield of enzyme P3 is then much lower, and it is incompletely separated from P2. It is important that fractions be analyzed and peak fractions pooled and titrated to pH 5.8 as quickly as possible since rapid denaturation of P5 occurs at more acid pH.

A typical purification is summarized in Table I.

Properties of the Purified Enzymes

Yield and Purity. The overall yield of α-galactosidase was 28%. The fractional contribution of each α-galactosidase isozyme to the crude extract is not known. The yield (84%) and purification factor (21,800-fold) for P3 CTHase are undoubtedly artifactually high, probably because of the inability to measure accurately the very low level of CTHase in the very large volumes of crude extract. However, the presence of an inhibitor (or another substrate) in the early stages of purification cannot be ruled out. It is not felt that any of the three isozymes of placental α-galactosidase has been purified to homogeneity.

Stability. The pH of maximum stability for P3 and P1 is pH 5.8. P3 is rapidly denatured below pH 5.2 and P1 below pH 4.0. P3 is heat

TABLE I

PURIFICATION OF α-GALACTOSIDASE AND OF CERAMIDE TRIHEXOSIDASE (CTHase) FROM HUMAN PLACENTA[a]

Fraction	Total protein (g)	Total activity[b] α-galacto-sidase (μmoles/hour)	Total activity[c] CTHase (nmoles/hour)	Specific activity α-galacto-sidase (nmoles/mg protein/hour)	Specific activity CTHase (nmoles/mg protein/hour)	Purifi-cation factor α-galac-tosidase	Purifi-cation factor CTHase	Yield α-galacto-sidase (%)	Yield CTHase (%)
Fresh placental tissue, 10 kg	—	—	—	—	—	—	—	—	—
Crude homogenate	582	2100	(29120)	3.6	(.05)	1	1	100	100
Ammonium sulfate, 21–50%	192	1900	(0)	9.9	—	2.7	—	90	—
Sephadex G-200, 2 passes	6.0	1350	(4000)	225	(0.67)	62	(13.4)	64	(14)
DEAE-Sephadex, pH 6.00	0.108	690	28600	6390	265	1770	5300	33	98
CM Sephadex P1 + P2	0.031	385	1020	12400	33	3400	660	18	3.5
P3	0.022	215	24000	9800	1090	2700	21800	10	82
Peak tube, P3	0.000081	2.46	207	30400	2560	8400	51200	0.1	0.7

[a] The values reported are average values from several preparative runs.

[b] All assays for α-galactosidase used 4-methylumbelliferyl-α-D-galactopyranoside as substrate at 1.25 mM concentration.

[c] All CTHase assays used the standard conditions described (0.2 mM CTH). The CTHase could not be accurately assayed until after the DEAE-Sephadex step either because of the low concentration of enzyme present or because of the presence of an in-hibitor substance.

labile (90% of the activity is lost in 15 minutes at 50° at pH 4.4) and P1 heat stable (50% of the activity is lost in 2 hours at 50° at pH 4.4).

Substrate Specificity. Although each of the three α-galactosidases show activity to ceramide trihexoside, P3 contains over 95% of the total CTHase activity and is the only true CTHase of the three α-galactosidase isozymes. The ratio of ceramide trihexosidase to α-galactosidase is 0.004 for P1, 0.01 for P2, and 0.15 for P3, when the assays are performed as described. p-Nitrophenyl-α-D-galactopyranoside and o-nitrophenyl-α-D-galactopyranoside as well as 4-MU-α-galactoside are cleaved by all three α-galactosidases. Glucocerebroside, galactocerebroside, and ceramide lactoside as well as 4-MU-β-galactoside and paranitrophenyl-β-D-galactopyranoside are not cleaved. Table II shows K_m and V_{max} for P3.

Inhibitors. P3 α-galactosidase is inhibited by glucocerebroside, but not by CTH, lactosylceramide, or galactose, as in the case of intestinal CTHase of the rat.[1] However, unlike that enzyme, P3 α-galactosidase is not inhibited by glucosylsphingosine. Table III shows inhibitors of P3.

pH Optimum. P3 has pH optimum of 4.4 for both α-galactosidase and CTHase. P1 has pH optimum of 4.8 for α-galactosidase and CTHase.

Effect of Detergents. P3 α-galactosidase shows inhibition by ionic detergents, but not by nonionic detergents (Table III). P3 CTHase is inhibited by high concentrations of both types of detergents.

Nature of the Reaction. Ceramide trihexosidase catalyzes the hydrolytic cleave of the terminal α-linked galactose of CTH. This activity is present in a variety of human tissues including intestinal mucosa, white blood cells, kidney, and placenta. Attenuation of CTHase activity, a genetic defect carried as a sex-linked recessive trait in humans, is associated with accumulation in tissues of CTH and severe renal disease

TABLE II
K_m AND V_{max} OF CERAMIDE TRIHEXOSIDASE, P3

Substrate	K_m (mM)	V_{max} (nmoles/mg protein/hour)
CTH[a]	0.84	13,400
4-MU-α-galactoside	1.9	80,000

[a] The CTH substrate was used in sodium cholate solution, 50 mg/ml, which was necessary for solution of the CTH. Cholate inhibited both the CTHase and 4-MU-α-galactosidase activity of P3. Therefore, the true V_{max} is no doubt higher than that reported here, and the true K_m may be lower. Though Triton X-100 does not inhibit P3 4-MU-α-galactosidase, CTH in Triton gave lower P3 CTHase activities than CTH in cholate.

TABLE III

INHIBITORS OF 4 MU-α-GALACTOSIDASE, P3 SUBSTRATE: 4-MU-α-GALACTOSIDE, 1.25 nM

Compound	Concentration	Inhibition (%)
FeCl$_3$	2.5 mM in H$_2$O	47
HgCl$_2$	2.5 mM in H$_2$O	17
p-Nitrophenol	2.5 mM in H$_2$O	96
p-Chloromercuryl sulfonate	2.5 mM in H$_2$O	24
Sodium cholate	5 mg/ml in H$_2$O	80
Sodium taurocholate	5 mg/ml in H$_2$O	58
Triton X-100	5 mg/ml in H$_2$O	0
Ca^{2+}, Mg^{2+}, Sr^{2+}	2.5 mM in H$_2$O	0
Galactose	2.5 mM in H$_2$O	0
Melibiose	2.5 mM in H$_2$O	0
Raffinose	2.5 mM in H$_2$O	0
Stachyose	2.5 mM in H$_2$O	0
Lactose	2.5 mM in H$_2$O	0
Tay-Sachs ganglioside	5 mg/ml in H$_2$O	0
Ceramide lactoside	5 mg/ml in water solution with 5 mg/ml Triton X-100	0
Ceramide trihexoside	5 mg/ml in water solution with 5 mg/ml Triton X-100	0
Asialo Tay-Sachs ganglioside	5 mg/ml in water solution with 5 mg/ml Triton X-100	0
Glucosylsphingosine	5 mg/ml in water solution with 5 mg/ml Triton X-100	0
Glucocerebroside	5 mg/ml in water solution with 5 mg/ml Triton X-100	12

in patients with Fabry's disease.[5] Assay of this enzyme permits accurate diagnosis of patients,[5] detection of carriers,[6] and reliable genetic counseling[7] for this disorder.

[5] R. O. Brady et al., N. Engl. J. Med. 276, 1163 (1967).
[6] R. O. Brady, W. G. Johnson, and B. W. Uhlendorf, Amer. J. Med. 51, 423 (1971).
[7] R. O. Brady, B. W. Uhlendorf, and C. B. Jacobson, Science 172, 174 (1971).

[116] β-Hexosaminidase A from Human Placenta

By WILLIAM G. JOHNSON, GEORGE MOOK, and ROSCOE O. BRADY

R-N-Acetylhexosaminide + $H_2O \rightarrow$ R-OH + N-Acetylhexosamine

Assay Methods

Principle. Enzyme activity causing the release of strongly fluorescent 4-methylumbelliferone from its nonfluorescent β-hexosaminide is designated β-hexosaminidase and is quantified by fluorometry at basic pH. Both the β-glucosaminide and the β-galactosaminide derivatives of 4-methylumbelliferone are cleaved by hexosaminidase A. The β-glucosaminide has been used more frequently as a substrate and is used here; however, the β-galactosaminide is probably more convenient for work with tissue.

Reagents

4-Methylumbelliferyl-2-acetamido-2-deoxy-β-D-glucopyranoside (4 MU-β-N-acetylglucosaminide) (Koch-Light Laboratories)
Citrate–phosphate buffer, "0.15 M," pH 4.4[1]
4-Methylumbelliferone (4-MU, K + K Laboratories, Inc.)
Glycine, 0.2 M in H_2O
Sodium hydroxide, 0.2 M in H_2O

Procedure. A substrate solution of 4 MU-β-glucosaminide, 5 mM (19 mg in 10 ml H_2O) is prepared. An incubation mixture contains 10 μl of enzyme extract, 150 μl of citrate–phosphate buffer "150 mM," pH 4.4 (made by mixing 22.2 ml of 0.2 M disodium phosphate and 27.8 ml of 0.1 M citric acid solution[1]), and 50 μl of substrate solution to a total volume of 0.21 ml. The blank, containing the same amounts of buffer and substrate but water instead of enzyme extract, is treated identically to the sample. An enzyme blank is usually unnecessary in the fluorometric assay. After incubation at 37° for up to 15 minutes, the reaction is stopped by adding 0.8 ml glycine–sodium hydroxide buffer, 0.1 M, pH 10.7 (made by mixing equal volumes of NaOH 0.2 M and glycine 0.2 M). Tenfold dilution is performed, if necessary by mixing 0.1 ml of samples and blanks with 0.9 ml of the pH 10.7 buffer. Further dilution should not be attempted. Instead very active enzyme extracts should be diluted with the citrate–phosphate buffer pH 4.4 before addition to the incubation mixture. The fluorescence is measured and the

---

[1] G. Gomori, Vol. 1 [16].

result compared with the fluorescence of known amounts of 4-MU. In this laboratory an Eppendorf fluorometer was used with a primary filter transmitting the Hg line, 366 nm, and a double barrier secondary filter transmitting 430–470 nm. A liquid standard (quinine alkaloid, 10 μg/ml in 0.1 N H$_2$SO$_4$) and quartz cuvettes were used throughout.

Purification of Placental Hexosaminidase A

Reagents

Monosodium phosphate
Disodium phosphate
Ammonium sulfate
Sephadex G-200 (Pharmacia)
DEAE Sephadex A-25 (Pharmacia)
Sodium chloride
Citric acid
Sodium hydroxide
CM Sephadex C-25 (Pharmacia)

Extraction. Fresh human placentas are placed on ice immediately after birth. They are processed within 12 hours after delivery. All subsequent procedures are carried out at cold room temperature (3°). Placentas are perfused with normal saline (0.9% NaCl in H$_2$O) to remove clots. Then 2500 g is dissected free from fibrous tissue and homogenized in portions in a Waring blendor (two 1-minute runs) in five volumes of sodium phosphate buffer 25 mM, pH 6.[2] The homogenate is centrifuged at 10,000 g for 20 minutes, and the supernatant is retained.

Ammonium Sulfate Precipitation. Ammonium sulfate fractionation of the solution is performed (using 70.5 g of (NH$_4$)$_2$SO$_4$ per 100 ml of solution as 100% saturation), and the fraction which precipitated between 25% and 55% is collected by centrifugation at 10,000 g for 30 minutes and resuspended in sodium phosphate buffer 25 mM pH 6 to a total volume of 800 ml. This highly concentrated protein solution is dialyzed against sodium phosphate buffer 25 mM, pH 6, for 48 hours.

Sephadex G-200. The retentate is centrifuged at 50,000 g for 1 hour, and the supernatant is applied to an upward flowing column of Sephadex G-200 (Pharmacia column, 10 × 100 cm) equilibrated with the same phosphate buffer. Fractions are collected (25 ml) and analyzed for β-hexosaminidase. A single β-hexosaminidase peak is found; fractions con-

[2] G. Gomori. Vol. 1 [16].

taining 90% of the total assayed activity are pooled and retained. Pooled fractions from two such runs are combined, concentrated together to a volume of 300 ml (Amicon system with PM-10 ultrafilters), and centrifuged at 30,000 g for 20 minutes, the supernatant is reapplied to a freshly packed column of Sephadex G-200 in identical fashion. Fractions containing 90% of the total assayed β-hexosaminidase are collected and concentrated in the same way to a volume of 300 ml.

DEAE-Sephadex. Sodium phosphate buffer (8 liters) 25 mM, pH 6, is titrated to pH 6.00 ± 0.02 at 25°. DEAE Sephadex A-25 (about 150 g) is swollen in 2 liters of this buffer and titrated to pH 6.00 ± 0.02 (25°). The gel suspension is washed twice with fresh buffer and retitrated each time. A Pharmacia column (5 × 60 cm) is packed with this gel and washed with 2 liters of the buffer. Combined fractions from two runs of the preceding step (300 ml) are titrated to pH 6.00 ± 0.02 at 25° and applied slowly (downward flow) to the column; the column is washed with 500 ml of the phosphate buffer and eluted with a linear NaCl gradient, 0–600 mM (in the same phosphate buffer), total volume 2 liters (Kontes gradient maker). Fractions are collected (25 ml), and two well-separated peaks of β-hexosaminidase are found on assay. The first peak, in the material which does not stick to the column, corresponds to β-hexosaminidase B.[3] The second peak is eluted with the salt gradient and corresponds to β-hexosaminidase A. Fractions containing 80% of the β-hexosaminidase A are pooled, concentrated to 30–50 ml (Amicon system, PM-10 membranes), and dialyzed overnight against the sodium phosphate buffer 25 mM pH 6.

CM-Sephadex. Citrate–phosphate buffer, "30 mM," pH 4.8 is prepared by mixing 24.8 ml disodium phosphate 0.2 M and 252 ml citric acid 0.1 M, diluted 5-fold, and titrated to pH 4.80 ± 0.02 (25°) with citric acid 1 M. CM Sephadex C-25 is swollen in this buffer, titrated to pH 4.80 ± 0.02 (25°), washed twice with fresh buffer, retitrated, and packed into a column (0.9 × 20 cm, Pharmacia). The column is washed with the same buffer (50 ml). The protein sample of β-hexosaminidase A (30 ml) is titrated to pH 4.80 ± 0.02 (25°) and applied slowly (downward flow) to the column; the column is washed with 50 ml of the citrate–phosphate buffer and eluted with a linear NaCl gradient 0–600 mM (prepared in the same citrate–phosphate buffer), total volume 200 ml. Fractions of 3.5 ml are collected. At this pH the β-hexosaminidase is bound to the column and eluted by the salt gradient in a single peak. The peak fractions are pooled, immediately retitrated to pH 6.0, and dialyzed against sodium phosphate buffer 25 mM pH 6.

[3] D. Robinson and J. L. Stirling, *Biochem. J.* **107**, 321 (1968).

TABLE I
PURIFICATION OF β-HEXOSAMINIDASE FROM HUMAN PLACENTA[a]

Fraction	Total protein (g)	Total activity[b] (β-hexos- aminidase mmoles/hour)	Activity (β-hexos- aminidase μmoles/mg protein/hour)	Purification factor (-fold)	Yield (%)
Fresh placental tissue, 10 kg	—	—	—	—	—
Ammonium sulfate concentrate	73.6	260	3.5	1	100
Sephadex G-200, 2 passes	2.6	112	43	12.3	43
DEAE-Sephadex[c]	0.18	55	306	158	38
CM-Sephadex	0.0056	39	6960	3590	27

[a] The values reported are average values from several preparative runs.

[b] Fractions (2) and (3) contain a mixture of β-hexosaminidase isozymes A and B, and specific activities, purification factors, and yields of enzyme activities, listed are totals for the mixture. Fractions (4) and (5) contain β-hexosaminidase A only; enzyme activities, specific activities purification factors, and yields listed are for this enzyme only. All β-hexosaminidase assays used 4-MU-β-glucosaminide at 1.25 mM concentration.

[c] Hexosaminidase A is separated from hexosaminidase B by this step and constitutes 55% of the total β-hexosaminidase recovered at this stage.

A typical purification is summarized in Table I. Substrate K_m and V_{max} are compared in Table II.

Properties of the Purified Enzyme

pH Optimum. The pH-activity curve of β-hexosaminidase A is very broad peaking at in the region of pH 4.2–4.4.

Effect of Detergents. β-hexosaminidase A shows marked inhibition by anionic detergents but not by neutral or cationic detergents (Table III). Some stimulation of activity is seen with cationic detergents.

Inhibitors. Table III shows inhibitors of β-hexosaminidase A.

Nature of the Reaction. Hexosaminidase A catalyzes the hydrolytic

TABLE II
K_m AND V_{max} OF PLACENTAL β-HEXOSAMINIDASE A

Substrate	K_m (mM)	V_{max} (mmoles/mg protein/hour)
4-MU-β-N-acetylglucosaminide	1.4	84
4-MU-β-N-acetylgalactosaminide	0.88	3.8

TABLE III

INHIBITORS OF HEXOSAMINIDASE A; SUBSTRATE: 4-MU-β-GLUCOSAMINIDE
1.25 mM

Compound	Concentration	Effect (%)
N-Acetylgalactosamine	2.5 mM	−75
N-Acetylglucosamine	2.5 mM	−35
Galactosamine	2.5 mM	−25
Glucosamine	2.5 mM	0
Galacturonic acid	2.5 mM	−44
p-Chloromercurisulfonate	2.5 mM	−95
$HgCl_2$	2.5 mM	−97
p-Nitrophenol	2.5 mM	−99
Sodium cholate	5 mg/ml	−69
Sodium taurocholate	5 mg/ml	−66
Triton X-100	5 mg/ml	0
Cetylpyridium chloride	5 mM	−23
Lecithin	5 mM	+14
Cetyltrimethylammonium bromide	5 mM	+16

cleavage of terminally linked β-N-acetylhexosamine. The enzyme is present in a wide variety of human tissues including liver, spleen, kidney, serum, white blood cells, brain, and placenta. Absence of hexosaminidase A, a human genetic defect carried as an autosomal recessive trait, is associated with accumulation of ganglioside GM_2 (with smaller amounts of asialo-GM_2) and lethal brain disease in Tay-Sachs disease.[4,5] Assay of this enzyme permits accurate diagnosis of patients,[4] detection of carriers,[6] and reliable genetic counseling for this disorder.

[4] S. Okada and J. O'Brien, Science 165, 698 (1969).
[5] E. H. Kolodny, R. O. Brady, and B. W. Volk, Biochem. Biophys. Res. Commun. 37, 526 (1969).
[6] J. O'Brien, S. Okada, A. Chen, and D. Fillerup, N. Engl. J. Med. 283, 15 (1970).

[117] Thermal Fractionation of Serum Hexosaminidases: Applications to Heterozygote Detection and Diagnosis of Tay-Sach's Disease

By Michael M. Kaback

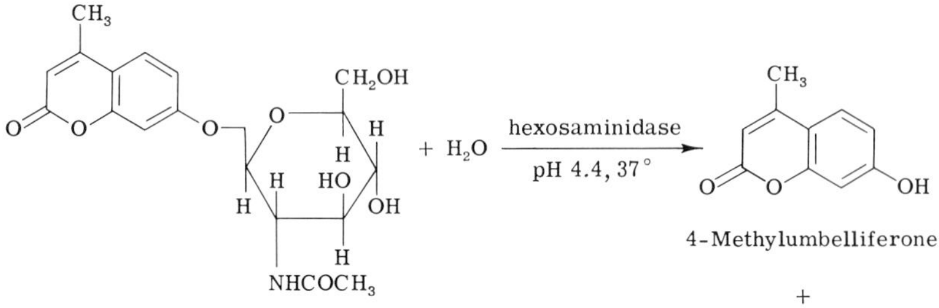

4-Methylumbelliferyl-*N*-acetyl-β-D-glucosaminide

4-Methylumbelliferone

N-acetylglucosamine

With *p*-nitrophenyl or 4-methylumbelliferyl-β-D-*N*-acetyl pyranoside substrates, at least two hexosaminidases can be defined in human tissues by DEAE chromatography, by starch gel, acrylamide, or cellulose acetate electrophoresis, and by isoelectric focusing techniques.[1–4] These isoenzymes, termed hexosaminidase A and B, lack specificity for the C-4 position of the *N*-acetyl hexosaminide moiety of the synthetic substrates. Each has approximately 8-fold higher activity with the *N*-acetyl glucosaminide-substituted substrate, but the *N*-acetylgalactosaminide glycoside is also readily hydrolyzed by both enzymes.

Activity of hexosaminidase A (Hex A), the anionic isoenzyme (at neutral pH in starch gel), is totally deficient in various tissues from children with Tay-Sach's disease (TSD), while hexosaminidase B (Hex B) is present in usual or increased amounts.[2] Serum, leukocytes, and cultivated skin fibroblasts from obligate heterozygotes for this autosomal recessive condition (parents of affected children) demonstrate a partial reduction of Hex A activity when compared with similar tissues from noncarrier individuals.[5,6]

[1] D. Robinson and J. L. Stirling, *Biochem. J.* **107**, 321 (1968).

[2] S. Okada and J. S. O'Brien, *Science* **165**, 698 (1969).

[3] K. Sandhoff, *FEBS* (*Fed. Eur. Biochem. Soc.*) *Lett.* **4**, 351 (1969).

[4] J. Freedland, L. Schneck, A. Saifer, M. Pourfar, and B. W. Volk, *Clin. Chim. Acta* **28**, 397 (1970).

[5] J. S. O'Brien, S. Okada, A. Chen, and D. L. Fillerup, *N. Engl. J. Med.* **283**, 15 (1970).

[6] S. Okada, M. L. Veath, J. Leroy, and J. S. O'Brien, *Amer. J. Hum. Genet.* **23**, 55 (1971).

The deficiency of Hex A activity in TSD, as defined with synthetic chromogenic or fluorogenic substrates, correlates with the inability of muscle extracts from children with TSD to cleave the terminal N-acetylgalactosamine from G_{M2} ganglioside.[7] This is the specific glycosphingolipid which accumulates in excessive amounts in the tissues (particularly within neuronal lysosomes) of infants affected with this fatal condition (see Fig. 1).

In addition to chromatographic and electrophoretic distinctions, differential heat stability of these isoenzymes has been described.[8] Enzymatic activity of Hex A is labile when exposed to 50–52° for defined periods while Hex B activity is essentially unchanged by such treatment. Exact quantitative correlation is not evident between proportions of Hex A and B as defined by electrophoresis versus that ascertained by differential thermal lability studies. Within the context of this chapter, however, Hex A is equated with heat-sensitive hexosaminidase activity and Hex B with the heat-stable enzyme (or enzymes).

Extension of the heat-inactivation method to human serum provides a simple and quantitative method for Hex A determination from a readily accessible human source. Comparison of serum hexosaminidase activity before and after heat exposure permits specific quantification of the heat labile (Hex A) fraction.

Assay Methods

Principle. Hexosaminidase activity in human serum is determined with 4-methylumbelliferyl-β-D-N-acetyl glucosaminide as substrate.[9] The reaction is quantified by fluorometric measurement of the free 4-methylumbelliferone produced in this hydrolysis. Comparison of hexosaminidase activity before and after 2- and 3-hour incubations at 52° permits

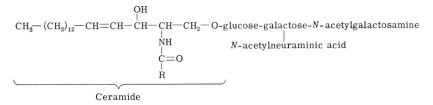

FIG. 1. G_{M2} ganglioside; "Tay-Sachs ganglioside."

[7] E. H. Kolodny, R. O. Brady, and B. W. Volk, *Biochem. Biophys. Res. Commun.* **37**, 526 (1969).

[8] N. Dance, R. G. Price, D. Robinson, and J. L. Stirling, *Clin. Chim. Acta* **24**, 189 (1969).

[9] D. H. Leaback and P. G. Walker, *Biochem. J.* **78**, 151 (1961).

specific quantification of the heat-labile fraction (Hex A). Sera from children with TSD show less than a 5% decrease in hexosaminidase activity after heat treatment. Hex A in sera from heterozygous individuals is 36.1 ± 5.1% as compared with 56.2 ± 4.1% in noncarrier sera.

Reagents

Phosphate–citrate buffer (P-C buffer): Na_2HPO_4 20 mM, citric acid 12 mM; adjust pH to 4.4 with NaOH

4-Methylumbelliferyl-β-D-N-acetyl glucosaminide (Pierce Chemical Co.), MW = 379. Prepare 5 mM solution in P-C buffer.

Glycine–carbonate buffer: glycine 0.17 M, Na_2CO_3 0.17 M; adjust pH to 10.0 with NaOH

4-Methylumbelliferone (Pierce Chemical Co.), MW = 176. Prepare 0.2, 0.5, 1.0, and 2.0 nanomoles/ml concentrations in glycine–carbonate buffer (stable in dark up to 14 days at 25°)

Procedures

Heat Inactivation. Serum, 0.2 ml, is diluted with 1.8 ml of P-C buffer and mixed thoroughly; 50-μl aliquots of the diluted serum is distributed into 7 disposable glass culture tubes (Corning glass, catalog No. 99445). All tubes are corked and distributed into three separate incubation racks. Three tubes are placed in the first rack (A) and maintained at 4°. Of these, two serve as duplicates for the unheated total hexosaminidase determination, and the third as a reaction blank. Duplicates of the remaining aliquots are placed in racks B and C, which are then immersed in a constant temperature water bath at 52° ± 0.2° with constant agitation to ensure uniform heat distribution. The water bath is covered during the period of heat inactivation. At 2 hours, rack B is removed and the duplicates are cooled to 4°. Rack C is removed from the bath after 3 hours, and samples are immediately cooled.

Hexosaminidase Assay. The 0-, 2-, and 3-hours heat-treated duplicates may be assayed immediately or the samples can be frozen (−20°) without change in enzymatic activity for as long as 4 months. At assay, 100 μl of 5 mM substrate (freshly prepared in P-C buffer each day) is added to each sample except the diluted serum blank. An additional 100 μl substrate aliquot is placed in a separate tube without serum to be used in the zero-time reaction blank.

In a final volume of 0.15 ml the reaction mixture contains: Na_2HPO_4, 20 mM; citric acid, 12 mM; 4-methylumbelliferyl-β-D-N-acetyl gluco-

saminide, 3.4 mM; serum equivalent, 5 μl. The reactants are mixed thoroughly, covered, and placed in a constant temperature bath at 37° ± 0.2° with constant agitation.

After 30 minutes, all samples are removed, cooled, and 4.85 ml of 0.17 M glycine-carbonate buffer, pH 10.0, is immediately added to all reacted samples. The nonreacted serum and substrate aliquots are then mixed and glycine–carbonate buffer is added simultaneously. This sample serves as the zero-time reaction blank.

All tubes are mixed, wiped, and then read directly in the fluorometer. (The Corning tubes described have negligible background fluorescence and can be read directly in the Turner fluorometer.)

Fluorometry. Hexosaminidase activity is quantitated by the amount of free 4-methylumbelliferone generated in the reaction mixture. 4-Methylumbelliferone, under these conditions, has an absorption maximum (λ_{max}) at 365 nm and an excitation maximum of 450 nm. In view of these specifications, fluorescence is determined (in a Turner fluorometer with standard door attachment) with a 7-60 Wrattan primary filter and with 48 and 2A secondary filters. The blank is read directly under these conditions and routinely yields minimal fluorescence. The reacted 0-, 2-, and 3-hour samples, however, require the introduction of a 0.1 reduction filter (10% transmission) in conjunction with the 48 and 2A secondary filters to provide on-scale fluorescence levels. Duplicate samples do not vary by more than 5%, and there should be less than 5% variation between the 2- and 3-hour inactivated samples. (Inactivation under these conditions is essentially complete by 2 hours; the 3-hour sample, therefore, serves primarily as an internal control on the heat inactivation process.)

Total hexosaminidase activity, in nanomoles of 4-methylumbelliferone produced, is determined from the amount of fluorescence in the nonheated samples compared with a standard curve developed from known concentrations of 4-methylumbelliferone prepared in 0.17 glycine–carbonate buffer pH 10.0. 4-Methylumbelliferone standards (0.2, 0.5, and 1.0 nanomole/ml) are included with each assay.

It is recommended that appropriate "internal standards" be included with each series of experiments. Sera from a proven TSD patient, from one or two obligate heterozygotes, and from one or more known "controls" are evaluated each time assays are performed for genotype designation in an "unknown" individual. These safeguards serve to delineate rapidly any source of technical or mechanical artifact in the method.

Definition of Unit and Specific Activity. Serum hexosaminidase specific activity and units are expressed in nanomoles of 4-methylumbelliferone produced per hour per milliliter of serum. Hex A is expressed as the percentage of the total hexosaminidase activity inactivated by 2–3 hours of treatment at 52° under the defined conditions.

Properties

Enzyme Stability. Both the total hexosaminidase activity and the percent Hex A (defined by heat inactivation) remain unchanged at −20° for up to 6 months. Reduction in the percentage of heat-labile Hex A can be observed after 24 hours at room temperature.

Assay Conditions. Limited solubility of the umbelliferyl-glycoside substrate does not permit absolute substrate saturation in this assay. However, at 3.4 mM substrate (final assay concentration), saturation is closely approximated. Under the conditions described, the reaction is linear up to 40 minutes at 37° and over a range of serum activities from 0 to 2500 units.

Both the total and heat-stable enzyme activities are affected by ionic strength and pH. With proportionate increases of ionic strength to 100 mM phosphate and 60 mM citrate (over a pH range of 4.0–5.5), a 40% reduction in enzyme activity is observed. Similarly, a reduction of ionic strength to 5 mM phosphate and 3 mM citrate over the same pH range results in a 10–20% reduction in activity.

The actual pH in the reaction mixture under the conditions given is 4.6–5.0. This is slightly above the absolute pH optimum (4.4) for total activity, but because of relative instability of the heat-labile isoenzyme at lower pH, a slightly higher pH is used in the assay.

Conditions of Thermal Inactivation. The quantification of the A isoenzyme by its relative heat instability involves complex physicochemical interactions. In addition to the temperature at which inactivation is conducted, the pH, ionic strength, and protein concentration affect the degree of inactivation observed. Increased loss of enzymatic activity at 52° is observed with: protein concentrations less than 0.1%, both high and low ionic strengths (as discussed previously under assay conditions), pH <4.4 and >5.4, and at temperatures above 54°.

Kinetic Properties. With 4-methylumbelliferyl-β-D-N-acetyl glucosaminide, both Hex A and Hex B in human serum have an apparent K_m of $5.4 \times 10^{-4} M$. This compares with K_m's of $9 \times 10^{-4} M$ in human kidney[8] and 6.7×10^{-4} in spleen.[1]

Applications. This method can be accurately applied to the diagnosis of TSD and to the detection of heterozygotes for this lethal recessive gene (see the table).

In certain individuals the percentage of serum Hex A may incorrectly indicate heterozygous levels. This may be due to an absolute increase in serum Hex B (or some other heat-stable hexosaminidase)

SERUM HEXOSAMINIDASE ACTIVITY

Subjects	Total activity[a]			% Hex A (52°)		
	Mean	SD	Range	Mean	SD	Range
Infants with TSD (13)[b]	588.4	252.0	227.4–926.4	1.7	1.1	0–5.6
Obligate heterozygotes (35)	655.0	221.4	279.8–1109.7	36.1	5.1	27.0–44.2
"Noncarriers" (100)	842.3	166.7	440.4–1312.4	56.2	4.1	50.0–71.3

[a] Nanomoles 4-methylumbelliferone produced per milliliter of serum per hour.
[b] Number of individuals tested is given in parentheses.

and thereby a relative decrease in heat-labile activity. Such individuals include: pregnant females, certain women on contraceptive medications, individuals with diabetes mellitus and possibly other systemic illnesses, and last, a small percentage (3–4%) of people whose serum Hex A level falls into an inconclusive area (arbitrarily defined to prevent false negatives) between known heterozygotes and "noncarrier" individuals. In each of these instances, leukocyte and/or cultured skin fibroblast Hex A levels provide the critical "backup" method by which genotype designation can usually be made.

Total hexosaminidase levels and Hex A by heat inactivation in leukocytes or fibroblasts are determined with sonicates of these cell types prepared at 4°. The addition of bovine serum albumin (hexosaminidase-free) in P-C buffer to a final concentration of 0.5% is required during heat inactivation. The methods for inactivation or assay are otherwise identical to those described for serum. Since these materials are likely to represent a more "primary enzyme source" than serum (hexosaminidase is believed to be lysosomal in origin), they seem less subject to those, as yet, undefined variables that may affect the level of serum enzymes. For purposes of TSD carrier-detection, serum evaluations appear accurate in over 95% of the individuals tested. Leukocytes or fibroblasts are probably greater than 99% accurate in making this distinction.

[118] Ganglioside G_{M1} β-Galactosidase

By HOWARD R. SLOAN

Ganglioside G_{M1} β-Galactosidase from Human Liver

Ganglioside G_{M1} + H_2O → ganglioside G_{M2} + galactose

p-Nitrophenyl-β-D-galactoside + H_2O → p-nitrophenol + galactose

Assay Method

Principle. Enzyme activity can be determined by measuring the conversion of ganglioside G_{M1}, galactosyl-$(1 \to 3)$-N-acetylgalactosaminyl-$(1 \to 4)$-$[(2 \to 3)$-N-acetylneuraminyl]galactosyl-$(1 \to 4$-$)$glucosyl-$(1 \to 1)$-$[2$-N-acyl]sphingosine to ganglioside G_{M2}, N-acetylgalactosaminyl $(1 \to 4)$-$[(2 \to 3)$-N-acetylneuraminyl]galactosyl-$(1 \to 4$-$)$glucosyl-$(1 \to 1)$-$[2$-N-acyl]sphingosine, and galactose. The reaction is quantified by measuring the amount of [³H]galactose released from ganglioside G_{M1} ([³H]galactose). The activity may also be assayed by measuring the conversion of p-nitrophenyl-β-D-galactoside to p-nitrophenol.

Method I

Reagents

Ganglioside G_{M1} ([³H]galactose), 45 μCi/μmole
Sodium acetate buffer, 1.0 M, pH 5.1
Cutscum solution in H_2O, 0.5% (w/v) (Fisher Chemical Co.)
Sodium deoxytaurocholate, 10% (w/v)
Chloroform–methanol solution 2:1 (v/v)
Thin-layer plates of silica gel G (0.25 mm thick)
Scintillation counting solution: 2664 ml toluene, 1332 ml Triton X-100, 22.0 g PPO, and 600 mg dimethyl POPOP[1]

Procedure. The assay mixture contains 25 μmoles of sodium acetate buffer, pH 5.1, 5 nmoles of ganglioside [³H]G_{M1} ([³H]G_{M1}), 600 μg of sodium deoxytaurocholate, 200 μg of Cutscum, an aliquot of the enzyme preparation and water in a final volume of 0.2 ml. The labeled substrate in 1.0 ml of chloroform–methanol, 2:1 is taken to dryness under a stream of nitrogen, dissolved in 2.0 ml of water and sonicated for 1 minute with a Heat Systems-Ultrasonics sonifier (Model W-1850) equipped with a

[1] R. H. Benson, *Anal. Chem.* **38**, 1353 (1966).

micro tip. To the incubation mixture, 5–10 μl of the water-clear solution of ganglioside G_{M1} is added. After incubation at 37°, the reaction is terminated by heating for 3 minutes at 100°. The galactose liberated from G_{M1} is separated from the other reaction product, ganglioside G_{M2}, and from excess G_{M1} by thin-layer chromatography. Chromatography is performed on thin-layer plates of silica gel G with chloroform–methanol–2.5 M ammonium hydroxide (60:35:8) as solvent. The area corresponding to galactose is scraped from the plate with a razor blade and counted in a liquid scintillation counter in 10 ml of the counting solution.

Units. A unit of enzyme activity is defined as the amount of enzyme required to catalyze the hydrolysis of 1 nmole of G_{M1} per hour. Protein is estimated by the procedure of Lowry.[2]

Preparation of Labeled Ganglioside G_{M1}. This procedure describes the selective tritiation of the 6-carbon atom of the terminal galactose moiety of ganglioside G_{M1}. Total gangliosides are extracted from human brain by the method of Kanfer *et al.*[3] G_{M1} is separated from other gangliosides by column chromatography on Anasil S (Analabs. Inc.).[4] Purified G_{M1} (100 mg) is added to 20 mg of galactose oxidase (Worthington Biochemical Co.), 1.5 nmoles sodium phosphate buffer, pH 7.0, 15 ml spectroanalyzed tetrahydrofuran (Fisher Chemical Co.), and 25 ml of water. The resulting suspension is incubated, with gentle stirring, at 37° for 40 hours, and the reaction is terminated by the addition of 200 ml of chloroform–methanol, 1:2 (v/v). The precipitated galactose oxidase is removed by centrifugation in the cold at 1000 g. The supernatant (containing the aldehydic form of ganglioside G_{M1}) is concentrated to dryness *in vacuo*, dissolved in 50 ml of water, and the pH of the resulting solution adjusted to 7.6 by the dropwise addition of 0.5 M Na_2HPO_4. Eight milligrams of tritiated sodium borohydride with a specific activity of 500 mCi/mmole is dissolved in 1.0 ml of 0.01 M NaOH. The tritiated borohydride solution is promptly added dropwise to a stirred solution of the aldehydic form of G_{M1} at 25°; addition is completed within 10 minutes. Stirring is continued for 2 hours. To ensure complete reduction of the aldehydic form of G_{M1}, 200 mg of nonradioactive $NaBH_4$ is added during the next 20 minutes. After four additional hours of stirring, the entire reaction mixture is placed in dialysis sacs. The solution is dialyzed

[2] O. H. Lowry, N. J. Rosebrough, A. L. Farr, and R. J. Randall, *J. Biol. Chem.* **193**, 265 (1951).

[3] J. N. Kanfer, R. S. Blacklow, L. Warren, and R. O. Brady, *Biochem. Biophys. Res. Commun.* **14**, 287 (1964).

[4] R. J. Penick, M. H. Meisler, and R. H. McCluer, *Biochim. Biophys. Acta* **116**, 279 (1966).

at 4° against 1000 volumes of 0.1 M sodium acetate buffer, pH 5.2, to dissociate sugar-borate esters; two changes of this buffer are employed. Dialysis is then continued against four changes of distilled water. The retentate is lyophilized and the ganglioside $[^3H]G_{M1}$ purified on columns of Anasil S.[4]

The purified radioactive G_{M1} migrates as a single band on silica gel G plates developed in either chloroform–methanol–2.5 M NH$_4$OH (60:35:8) or n-propanol–water (7:3) and has the same R_f as authentic G_{M1}. Autoradiography demonstrates that the location of the radioactivity coincides precisely with that of authentic G_{M1}. The yield of the reaction is approximately 60% and the specific radioactivity of the product is 45 μCi per micromole. Treatment of the pure ganglioside $[^3H]G_{M1}$ with rat brain β-galactosidase yields ganglioside G_{M2} and galactose. More than 95% of the radioactivity in G_{M1} is present in the terminal galactose moiety.

Method II

Principle. o- or p-Nitrophenyl-β-D-galactopyranoside is employed as substrate. The intensity of the color of p-nitrophenol released by the enzymatic hydrolysis is measured at pH 11 in a spectrophotometer.

Reagents

Sodium acetate buffer, 1.0 M, pH 5.1
o- or p-Nitrophenyl-β-D-galactopyranoside, 0.8% (w/v)
Cutscum solution in H$_2$O, 0.5% (w/v) (Fischer Chemical Co.)
Human serum albumin solution in H$_2$O, 10% (w/v)
Trichloroacetic acid, 4% (w/v)
Sodium carbonate, 1 M

Procedure. The assay mixture contains 25 μmoles of sodium acetate buffer, pH 5.1, 400 μg of o- or p-nitrophenyl-β-D-galactopyranoside, 200 μg of Cutscum, an aliquot of the enzyme preparation, and water in a final volume of 0.2 ml. After incubation for 90 minutes at 37°, the reaction is terminated by the addition of 0.1 ml of the 10% human serum albumin solution and 0.5 ml of 4% trichloroacetic acid. Of the clear supernatant solution obtained by low speed centrifugation (600 g), 0.6 ml is added to 0.4 ml of 1 M sodium carbonate. The amount of nitrophenol liberated is quantified by measuring the optical density at 415 nm.

Units. A unit of enzyme activity is defined as the amount of enzyme required to catalyze the hydrolysis of 1 nmole of o- or p-nitrophenyl-β-D-galactopyranoside per hour. Protein is estimated by the procedure of Lowry.[2]

Purification Procedure

Fresh human liver is obtained at autopsy and homogenized in 9 volumes (w/v) of cold 0.25 M sucrose in 10 mM sodium citrate, pH 6.0 with the use of a Potter-Elvehjem homogenizer with a loosely fitting Teflon pestle. The homogenate is centrifuged in the cold at 600 g for 15 minutes and the residue is discarded. The 600 g supernatant solution is then centrifuged at 10,000 g for 30 minutes. The supernatant solution is discarded; the pellet is resuspended in approximately 25 volumes of 0.5% (w/v) Triton X-100 in 10 mM sodium citrate, pH 6.0. The resulting suspension is stirred overnight at 4° and then centrifuged at 100,000 g for 7 hours. The 100,000 g supernatant is dialyzed exhaustively against 10 mM sodium citrate, pH 6.0. The retentate is centrifuged at 20,000 g for 60 minutes; the supernatant is decanted carefully, and the sediment is discarded.

Heat and Acid Precipitation. The pH of the 20,000 g supernatant from the preceding step is adjusted to 4.2 by the dropwise addition of 1.0 M sodium formate, pH 4.0. The mixture is then heated at 45° for 30 minutes, and the insoluble protein is removed by centrifugation and discarded.

Ammonium Sulfate Fractionation. Solid ammonium sulfate (24.3 g/100 ml) is added to the supernatant solution from the previous step in order to raise the concentration to 40% of saturation. The precipitate is removed by centrifugation at 10,000 g for 15 minutes, and sufficient ammonium sulfate (9.7 g/100 ml) is added to the supernatant solution to adjust the concentration to 55%. The precipitate is collected by centrifugation, dissolved in a small amount of 10 mM sodium citrate, pH 6.0, and dialyzed at 4° against three changes of this buffer. Protein that comes out of solution during dialysis is removed by centrifugation.

Acetone Fractionation. This fractionation is performed at −20°. Acetone (50 ml/100 ml) is slowly added to the supernatant solution from the previous step. The precipitate that forms is removed by centrifugation at 10,000 g for 30 minutes. The supernatant is decanted, and to it additional acetone is added (100 ml/100 ml). A second precipitate forms and the mixture is recentrifuged. The supernatant is discarded and the precipitate is suspended in 10 mM sodium citrate, pH 6.0. The final enrichment is approximately 75-fold over the original particulate suspension. The purification procedure is summarized in the table.

Properties

Stability. The most highly purified preparation is stable at −20° for at least 2 months.

PURIFICATION OF GANGLIOSIDE G_{M1}-β-GALACTOSIDASE FROM HUMAN LIVER

Fraction	Total activity (units)	Protein (mg)	Specific activity (units/mg protein)
I. 10,000 g particles	187	260	0.72
II. Triton X-100 supernatant	361	42	8.6
III. Heat and acid precipitation	275	28	9.8
IV. Ammonium sulfate, 40–55%	245	7	35
V. Acetone, 33–67%	108	2	54

Activators and Inhibitors. Ganglioside G_{M1} is hydrolyzed at a very slow rate in the absence of detergent. The hydrolysis is stimulated 10- to 20-fold by the addition of sodium deoxytaurocholate (3 mg per milliliter, of reaction mixture) and Cutscum (1 mg/ml). With *o*- or *p*-nitrophenylgalactopyranosides as substrate, hydrolysis is stimulated 1.3- to 1.8-fold by the addition of detergent. Hydrolysis of both ganglioside G_{M1} and the nitrophenylgalactosides is competitively inhibited by galactose and γ-galactonolactone.

Effect of pH. The optimum pH is between 4.6 and 5.2 with both the authentic and the artificial substrate.

Specificity. The most highly purified preparations hydrolyze both the nitrophenylgalactosides and ganglioside G_{M1}. Whole homogenates of liver cleave the artificial substrate approximately 100 times as rapidly as the authentic substrate. This difference in the rate of hydrolysis of these two substrates does not change significantly during the course of enzyme purification. The purest preparations cleave galactosyl ceramide and galactosyl sphingosine at a very slow rate.

Nature of the Reaction. The purified enzyme catalyzes the hydrolytic cleavage of the terminal galactose moiety from ganglioside G_{M1}. Ganglioside G_{M2} and galactose are the products of the reaction. The activity of this enzyme is markedly reduced in various tissues from patients with G_{M1} gangliosidosis.[5] The deficiency of ganglioside G_{M1}-β-galactosidase activity probably accounts for the accumulation of ganglioside G_{M1} in this disease.

Histochemical Assay of β-Galactosidase Activity

5-Bromo-4-chloro-3-indolyl-β-D-galactopyranoside + H_2O →

dibromodichloro-indigo + galactose

Assay Method

Principle. Enzyme activity can be detected by demonstrating the conversion of the colorless 5-bromo-4-chloro-3-indolyl-β-D-galactopyran-

[5] J. O'Brien, *J. Pediat.* **75**, 167 (1969).

oside to the insoluble, intensely blue, dibromodichloro-indigo. The marked insolubility of the product permits precise subcellular localization of enzymatic activity.[6]

Reagents

Glutaraldehyde, 1.25% (w/v) in Dulbecco's saline without calcium and magnesium[7]

5-Bromo-4-chloro-3-indolyl-β-D-galactopyranoside (Cyclo Chemical Co.) (BCI-gal)

Dimethyl formamide

Sodium acetate buffer, 0.1 M, pH 5.4

Sodium chloride in water, 0.085% (w/v)

Potassium ferricyanide, 50 mM

Potassium ferrocyanide, 50 mM

Human serum albumin in water, 10% (w/v)

Procedure

Preparation of Incubation Solution. BCI-gal, 10 mg, is dissolved in 2.5 ml of dimethyl formamide. To this solution, 155 ml of 0.1 M sodium acetate buffer, pH 5.4, 2.5 ml of 0.085% sodium chloride, 7.5 ml of 50 mM potassium ferricyanide and 7.5 ml of 50 mM potassium ferrocyanide are added. The resulting incubation solution is stable at -20 for at least 6 months. It may be reused repeatedly if particles of indigo pigment are removed by centrifugation at 10,000 g.

Preparation of Tissue. 6 μ thick sections of liver, spleen, lymph node, kidney, or brain are cut on a Linderstrøm-Lang cryostat at -20 and are transferred carefully to glass cover slips. Peripheral blood leukocytes are cultured in suspension[8] and harvested by low speed centrifugation (600 g). The cell pellet is dispersed gently in 1.0 to 2.0 ml of 10% human serum albumin and smeared onto cover slips. Fibroblasts derived from human skin or bone marrow[9] may be grown directly on cover slips. The cover slips are placed in a carriage (Arthur H. Thomas) and incubated at 4° for 12 minutes in the 1.25% glutaraldehyde solution. The carriage is then dipped serially into three beakers containing Dulbecco's saline without calcium and magnesium.[7] The carriages are then transferred to small jars (fitted with ground glass tops) that contain the in-

[6] B. Pearson, P. L. Wolf, and J. Vazquez, *Lab. Invest.* **12**, 1249 (1963).

[7] R. Dulbecco and M. Vogt, *J. Exp. Med.* **99**, 167 (1954).

[8] J. Paul, "Cell and Tissue Culture," 4th ed., p. 261. Williams and Wilkins, Baltimore, Maryland, 1970.

[9] H. R. Sloan, B. W. Uhlendorf, C. B. Jacobson, and D. S. Frederickson, *Pediat. Res.* **3**, 532 (1969).

cubation solution. After 24–36 hours of incubation at 37°, the reaction is terminated by transferring the carriage into a beaker of methanol. The slides may be counterstained with hematoxylin and eosin[10] and then viewed under a light microscope. The presence of dense blue granules indicates β-galactosidase activity.

Inhibitors. The reaction is completely inhibited by 50 mM solutions of galactose or γ-galactonolactone. Glucose, fucose and mannose are not inhibitors. All tissues from normal individuals, including cultured peripheral leukocytes and fibroblasts, show intense blue granular staining with the BCI-gal technique; the same tissues from patients with the disease G_{M1} gangliosidosis show a complete absence of the indigo granules.[11]

[10] A. G. E. Pearse, "Histochemistry-Theoretical and Applied." Little Brown, Boston, Massachusetts, 1960.
[11] H. R. Sloan, *Chem. Phys. Lipids* **5**, 250 (1970).

[119] Sphingomyelinase from Human Liver (Sphingomyelin Cholinephosphohydrolase)

By Howard R. Sloan

$$\text{Sphingomyelin} + H_2O \rightarrow \text{ceramide} + \text{phosphorylcholine}$$

Assay Method

Principle. Enzymatic activity can be assayed by measuring the conversion of sphingomyelin to ceramide and phosphorylcholine. The reaction is conveniently quantified by measuring the amount of trichloroacetic acid-soluble product, [14C]phosphorylcholine released from the acid-insoluble [methyl-14C]sphingomyelin substrate.[1] Several alternative methods for measuring the hydrolysis of sphingomyelin have been described[2-4]; the following method is, however, the most convenient.

Reagents

[14C]Sphingomyelin ([methyl-14C]sphingomyelin), 0.6 μCi/μmole
Sodium acetate buffer, 1.0 M, pH 5.1
Sodium cholate, 10 mg per milliliter of H_2O

[1] J. N. Kanfer, O. M. Young, D. Shapiro, and R. O. Brady, *J. Biol. Chem.* **241**, 1081 (1966).
[2] S. Gatt and Y. Barenholz, Vol. 14 [26].
[3] D. Rachmilewitz, S. Eisenberg, Y. Stein, and O. Stein, *Biochim. Biophys. Acta* **144**, 624 (1967).
[4] P. B. Schneider and E. P. Kennedy, *J. Lipid Res.* **8**, 202 (1967).

Human serum albumin solution in H_2O, 10% (w/v)

Trichloroacetic acid, 100% (w/v)

Cutscum solution in H_2O, 0.5% (w/v) (Fisher Chemical Co.)

Chloroform–methanol solution, 2:1 (v/v)

Scintillation counting solution: 2664 ml of toluene, 1332 ml of Triton X-100, 22.0 g of PPO, and 600 mg of dimethyl POPOP[5]

Procedure. The assay mixture contains 50 μmoles of sodium acetate buffer, pH 5.1, 75 nmoles of [^{14}C]sphingomyelin, 100 μg of sodium cholate, 250 μg of Cutscum, an aliquot of the enzyme preparation, and water in a final volume of 0.2 ml. The labeled substrate in 1.0 ml of chloroform–methanol, 2:1 is taken to dryness under a stream of nitrogen, suspended in a solution of sodium cholate (10 mg/ml) and sonicated for 1 minute with a Heat Systems-Ultransonics sonifier (Model W-1850) equipped with a micro tip. A 10-μl aliquot of the water-clear suspension of [^{14}C]sphingomyelin is added to the incubation mixture. After incubation for 2 hours at 37°, the reaction is terminated by the addition of 0.1 ml of the 10% human serum albumin solution, 0.1 ml of 100% trichloroacetic acid, and 0.8 ml of cold water. The tubes are placed in an ice bath for 5 minutes and centrifuged for 10 minutes at 600 g in a refrigerated centrifuge. The supernatant solution is decanted from the precipitate which is washed once with 1 ml of cold 10% tricholoroacetic acid. A 1.0-ml aliquot of the combined supernatant solution is added to 10 ml of the scintillation counting solution and the amount of radioactivity determined by liquid scintillation spectrometry. Corrections for quenching by the trichloroacetic acid may be made by the automatic external standard technique.

Units. A unit of enzymatic activity is defined as the amount of enzyme required to catalyze the hydrolysis of 1 nmole of sphingomyelin per hour. Protein is estimated by the procedure of Lowry.[6]

Preparation of Labeled Sphingomyelin

Sphingomyelin labeled with ^{14}C in the methyl carbon atoms of the choline portion of the molecule may be chemically synthesized according to the procedure described by Shapiro and Flowers.[7] A similar product may be more conveniently prepared biosynthetically.

Method I. Young (80 g) albino rats are placed on a diet that is markedly deficient in methionine, choline, and vitamin B_{12}. The diet has

---

[5] R. H. Benson, *Anal. Chem.* **38**, 1353 (1966).

[6] O. H. Lowry, N. J. Rosebrough, A. L. Farr, and R. J. Randall, *J. Biol. Chem.* **193**, 265 (1951).

[7] D. Shapiro and H. M. Flowers, *J. Amer. Chem. Soc.* **84**, 1047 (1962).

the following composition: hydrogenated vegetable oil, 300 g; liquid corn oil, 50 g; complete vitamin mixture, 40 g (without choline and vitamin B$_{12}$)[8]; salt mixture W,[8] 40 g; Alphacel, hydrolyzed,[8] 50 g; sucrose, 410 g; cystine or cysteine, 5 g; assay protein C-1,[9] 100 g.

The hydrogenated vegetable oil and liquid corn oil are heated to 50° and placed in the bowl of a Hobart mixer. The solid constitutents are added slowly, and the semisoft mixture is stirred for 30 minutes. It is advisable that four rats be placed on this diet because 1 or 2 of the animals may not survive the 4 weeks of the special diet. After 4 weeks the animals appear to be severely malnourished and are markedly deficient in methionine and choline. 200 μCi of [methyl-^{14}C]choline chloride (specific activity, 30–50 μCi/μmole) and 200 μCi of L-[methyl-^{14}C]-methionine (specific activity, 40–55 μCi/μmole) in saline are injected once daily into the peritoneal cavity of each animal for the next 5 days. Twenty-four hours after the last injection, the animals are killed with ether, and the liver and spleen are removed. Lipids are extracted in 20 volumes of chloroform–methanol 2:1[10] and then portioned and washed as described by Folch, Lees and Sloane Stanley.[11] Sphingomyelin is purified from the lipid extract by standard techniques.[10,12] Treatment of the pure sphingomyelin with phospholipase C derived from *Clostridium perfringens*[13] yields ceramide and phosphorylcholine. Less than 0.1% of the radioactivity of the purified sphingomyelin is present in the ceramide portion of the molecule. The specific activity of different preparations of sphingomyelin ranged between 1 and 3 μCi/μmole.[10]

Method II. This biosynthetic method is a modification of the procedure described by Rachmilewitz *et al.*[14] Pregnant mice are maintained on the diet deficient in methionine, choline, and vitamin B$_{12}$ for the last 7 days of pregnancy and through the first 10 days postpartum. 100 μCi of [methyl-^{14}C]choline chloride (specific activity, 30–50 μCi/μmole) and 100 μCi of L-[methyl-^{14}C]methionine (specific activity, 40–55 μCi/μmole) in saline are injected into the peritoneal cavity of the mother on the sixth postpartum day. The injection is repeated once daily for the next 4 days. Twenty-four hours after the last injection the mother is killed with ether and the liver is removed. Sphingomyelin is purified from

[8] Nutritional Biochemical Co.

[9] Skidmore Enterprises, Cincinnati, Ohio.

[10] P. O. Kwiterovich, Jr., H. R. Sloan, and D. S. Fredrickson, *J. Lipid Res.* **11**, 322 (1970).

[11] J. Folch, M. Lees, and F. H. Sloane Stanley, *J. Biol. Chem.* **226**, 497 (1957).

[12] C. C. Sweeley, *J. Lipid Res.* **4**, 402 (1963).

[13] M. Kates, *in* "Lipid Metabolism" (K. Block, ed.), p. 219. Wiley, New York, 1960.

[14] D. Rachmilewitz, S. Eisenberg, Y. Stein, and O. Stein, *Biochim. Biophys. Acta* **144**, 624 (1967).

the lipid extract as described in Method I. The sphingomyelin obtained by Method II is almost identical with that obtained by Method I; the specific activities are, however, somewhat greater (2–4 μCi/μmole).

Method III. Fibroblasts derived from human skin are grown in Eagle's "minimum essential medium" containing the nonessential amino acids, 10% (v/v) fetal bovine serum and neomycin (50 μg/ml).[15] Twenty-five percent of the methionine and choline of the culture medium is replaced with L-[methyl-¹⁴C]methionine (specific activity, 40–55 μCi/ μmole) and [methyl-¹⁴C]choline chloride (specific activity, 30–50 μCi/μmole), respectively. The cells are maintained in culture until they form a confluent monolayer. The fibroblasts are removed from the walls of the glass bottles by treatment with a dilute trypsin solution.[15] Lipids are extracted from the cells and the sphingomyelin purified as described in Method I above. More than 99% of the ¹⁴C in the sphingomyelin is localized in the choline portion of the molecule and specific activities as great as 12 μCi/μmole may be achieved.

A useful by-product of each of these three biosynthetic methods is [methyl-¹⁴C]phosphatidylcholine with more than 99% of the ¹⁴C in the choline portion of the molecule.[10–13] In each of the methods described in this section [¹⁴C-methyl]methionine and [¹⁴C-methyl]choline may be replaced by the appropriate tritium-labeled compound. Higher specific activities are obtained. The softer radiation of tritium is more susceptible to quenching by tricholoroacetic acid and therefore tritium-labeled sphingomyelin is less useful than the ¹⁴C labeled compound in the determination of sphingomyelinase activity. Recently Stoffel, LeKim, and Tschung[16] have described a simplified chemical method for labeling phosphatidylcholine and sphingomyelin in the choline moiety.

Purification Procedure

Frozen human liver is homogenized in 9 volumes (w/v) of cold 0.25 M sucrose, 10 mM sodium cacodylate buffer, pH 6.0, with the use of a Potter-Elvehjem homogenizer with a tightly fitting Teflon pestle. The homogenate is centrifuged in the cold at 600 g for 15 minutes and the residue is discarded. The 600 g supernatant solution is then centrifuged at 10,000 g for 30 minutes. The 10,000 g pellet is resuspended in approximately 10 volumes of 0.5% (w/v) Triton X-100 in 10 mM sodium cacodylate buffer, pH 6.0. The resulting suspension is stirred overnight at 4° and then centrifuged at 100,00 g for 2 hours. The 100,000 g supernatant solution is dialyzed exhaustively against 6 changes of 10 mM

[15] H. R. Sloan, B. W. Uhlendorf, C. B. Jacobson, and D. S. Fredrickson, *Pediat. Res.* 3, 532 (1969).

[16] W. Stoffel, D. LeKim, and T. S. Tschung, *Hoppe-Seyler's Z. Physiol. Chem.* 352, 1058 (1971).

sodium cacodylate, pH 6.0. The retentate is centrifuged at 20,000 g for 60 minutes; the supernatant is decanted carefully and the sediment is discarded.

Ammonium Sulfate Fractionation. Solid ammonium sulfate (24.3 g/100 ml) is added to the 20,000 g supernatant in order to raise the concentration to 40% of saturation. The precipitate is removed by centrifugation at 10,000 g for 15 minutes and sufficient ammonium sulfate (9.7 g/100 ml) is added to the supernatant solution to adjust the concentration to 55%. The precipitate is collected by centrifugation, dissolved in a small amount of 10 mM sodium cacodylate, pH 6.0, and dialyzed at 4° against three changes of this buffer. Protein that comes out of solution during dialysis is removed by centrifugation.

Butanol Extraction. Over a 30-minute period BuOH (20 ml/100 ml) is added dropwise to the supernatant solution from the preceding step. The mixture is stirred at 25° for 1 hour and then centrifuged at 12,000 g for 1 hour. Following centrifugation, there are four distinct zones in the glass centrifuge tube. The upper (butanol) phase is intensely yellow and contains most of the lipid of the original solution. The lower (aqueous) phase contains essentially all the sphingomyelinase activity. Between the upper and lower layer there is a brown interface that contains a large amount of protein. At the bottom of the tube there is a dense brown button. The yellow supernatant solution is removed and discarded; the aqueous phase is carefully aspirated with a syringe and needle, diluted with an equal volume of water and lyophilized to dryness.

Affinity Chromotography. Sphingosine phosphorylcholine is prepared as described by Kaller[17] and coupled to agarose (Sepharose 4B Pharmacia) that has been converted to the succinylated derivative.[18] Approximately 50 ml of succinylated Sepharose 4B is suspended in 100 ml of 0.1 M sodium citrate, pH 6.0; 500 mg of sphingosine phosphorylcholine and 4 g of 1-ethyl-3-(3-dimethylaminopropyl) carbodiimide are added and the suspension is stirred for 12 hours at room temperature. A portion of the affinity labeled Sepharose 4B is thoroughly washed with 0.15 M sodium chloride followed by 0.1 M sodium cacodylate, pH 6.1, and then packed into a 5 × 100 mm column. The lyophilized aqueous phase from the butanol extraction step is dissolved in a small volume of 0.1 M sodium cacodylate, pH 6.1, and applied to the column. Thorough washing with 0.1 M sodium cacodylate, pH 6.1, does not elute any sphingomyelinase activity. The enzyme is completely eluted from the column by a small volume (10–15 ml) of 0.1% Triton X-100 (w/v) in 0.1 M sodium cacodylate, pH 6.1.

[17] H. Kaller, *Biochem. Z.* **334**, 451 (1961).
[18] P. Cuatrecasas, *J. Biol. Chem.* **245**, 3059 (1970).

The final enrichment of enzyme is approximately 120-fold over the total liver homogenate. The purification procedure is summarized in the table.

Properties

Stability. The most highly purified preparation is stable for at least 2 months at $-20°$.

Effect of Detergent. Enzymatic activity is stimulated approximately 20% by the addition to the incubation mixture of 2–10 mg/ml of Triton X-100, Cutscum, or sodium cholate. Sodium dodecyl sulfate is a potent inhibitor.

Inhibitors. Phosphate ion is a weak inhibitor as is phosphotidylcholine and sphingosinephosphorylcholine. Phosphorylcholine and glycerylphosphorylcholine are not inhibitors.

Effect of pH. The optimum pH is between 4.8 and 5.2.

Specificity. The most highly purified preparations appear to be specific for sphingomyelin and dehydrosphingomyelin. Phosphatidylcholine, phosphatidylethanolamine, and sphingosinephosphorylcholine are not hydrolyzed.

Substrate Affinity. The apparent K_m value for sphingomyelin is $2.1 \times 10^{-5} M$.

Nature of the Reaction. The purified enzyme catalyzes the hydrolytic cleavage of the phosphodiester bond of sphingomyelin. The products of the reaction are ceramide (*N*-acylsphingosine) and phosphorylcholine. The activity of this enzyme is markedly reduced in various tissues from patients with Niemann-Pick disease, types A and B.[19] The deficiency of sphingomyelinase activity probably accounts for the accumulation of sphingomyelin in many tissues of patients with these disorders.

PURIFICATION OF SPHINGOMYELINASE FROM HUMAN LIVER

Fraction	Total activity (units)	Protein (mg)	Specific activity (units/mg protein)
I. 10,000 *g* particles	1690	310	5.5
II. Triton X-100 supernatant	3740	84	45
III. Ammonium sulfate, 40–55%	3500	28	125
IV. Butanol extraction	2800	13.7	205
V. Affinity chromatography	2700	4.1	655

[19] D. S. Fredrickson and H. R. Sloan, "The Metabolic Basis of Inherited Disease" (J. B. Stanbury, J. B. Wyngaarden, and D. S. Fredrickson, eds.), 3rd ed., p. 783. McGraw-Hill, New York, 1972.

[120] Cerebroside Sulfate Sulfatase (Arylsulfatase A) from Human Urine

By J. L. BRESLOW and H. R. SLOAN

Three enzymes with arylsulfatase activity (aryl sulfate sulfohydrolases, EC 3.1.6.1) designated A, B, and C have been detected in mammalian tissues and in lower organisms.[1] It is generally accepted that arylsulfatase A and cerebroside sulfate sulfatase are the same enzyme.

Assay Methods

The activity of the enzyme can be assayed by measuring either the conversion of *p*-nitrocatechol sulfate to *p*-nitrocatechol or the rate of hydrolysis of cerebroside sulfate [galactosyl (3-sulfate) ceramide] to galactosyl ceramide and inorganic sulfate.

Method 1

Principle. Human urine contains both arylsulfatase A and B, and each can be measured individually in a mixture of the two.[2] *p*-Nitrocatechol sulfate (PNCS) is employed as substrate. The intensity of the color of *p*-nitrocatechol (PNC) released by the enzymatic hydrolysis is measured at pH 12 or greater in a spectrophotometer.

Baum's Reagent for Arylsulfatase A (Baum's Reagent A)

Reagent	Final concentration
PNCS	0.01 *M*
Sodium acetate buffer, pH 5	0.5 *M*
Na$_4$P$_2$O$_7$	0.0005 *M*
NaCl	1.71 *M*

Procedure. The enzyme preparation (0.2 ml) is preincubated for 5 minutes at 37°; 0.1 ml of Baum's reagent, previously warmed to 37°, is then added. The reaction is terminated after 5 minutes of incubation at 37° by the addition of 1.0 ml of 2.5 *M* NaOH. The amount of *p*-nitrocatechol liberated is quantified by measuring the optical density at 515 μmoles.

Units. The molar extinction coefficient of PNC is 12,400. A unit of enzymatic activity is defined as the amount of enzyme required to cata-

[1] T. G. Flynn, K. S. Dodgson, G. M. Powell, and F. A. Rose, *Biochem. J.* **105**, 1003 (1967).

[2] H. Baum, K. S. Dodgson, and B. Spencer, *Clin. Chim. Acta* **4**, 453 (1959).

lyze the hydrolysis of 1 μmole of PNCS per minute. Specific activity is defined as units per milligram of protein. Protein is estimated by the method of Lowry.[3]

Method 2

Principle. Cerebroside sulfate-^{35}S is biosynthesized as described by Mehl and Jatzkewitz[4] and purified by column chromatography.[5] The [35S]sulfate liberated from cerebroside [35S]sulfate by the enzymatic reaction is conveniently quantified by measuring the amount of radioactivity in the upper phase of a standard Folch extraction.[6]

Reagents

Cerebroside [35S]sulfate, 1.2 μCi/μmole
Sodium acetate buffer, 1.0 M, pH 4.5
Sodium deoxytaurocholate, 2% (w/v)
Chloroform–methanol solution, 2:1 (v/v)
Potassium chloride solution, 0.1 M
Scintillation counting solution: 2664 ml of toluene, 1332 ml of Triton X-100, 22.0 g of PPO, and 600 mg of dimethyl POPOP

Procedure. The assay mixture contains 5 μmoles of sodium acetate buffer, pH 5.0, 0.5 mg of sodium deoxytaurocholate, 30 nmoles of cerebroside [35S]sulfate, an aliquot of the enzyme preparation and water in a final volume of 0.2 ml. The labeled substrate in 1.0 ml of chloroform–methanol, 2:1 is taken to dryness under a stream of nitrogen and suspended in a solution of sodium deoxytaurocholate (2%) and sonicated for 1 minute with a Heat-Systems Ultrasonic sonifier (Model W-1850) equipped with a micro tip; a 25-μl aliquot of the water-clear suspension of cerebroside [35S]sulfate is added to the incubation mixture. After incubation for 2 hours at 37°, the reaction is terminated by the addition of 4 ml of chloroform–methanol, 2:1. The tube is shaken vigorously for 3 minutes, and 0.5 ml of 0.1 M potassium chloride is added. The tubes are shaken for an additional 2 minutes and then centrifuged at 600 g for 5 minutes; 0.4 ml of the upper (aqueous) phase is carefully removed and added to 20 ml of the scintillation counting solution, and the amount of radioactivity is determined by liquid scintillation spectrometry. Alternatively, 1.0 ml of the upper phase may be taken to dryness in a

---

[3] O. H. Lowry, N. J. Rosebrough, A. L. Farr, and R. J. Randall, *J. Biol. Chem.* **193**, 265 (1951).

[4] E. Mehl and H. Jatzkewitz, *Hoppe-Seyler's Z. Physiol. Chem.* **339**, 260 (1964).

[5] P. O. Kwiterovich, Jr., H. R. Sloan, and D. S. Fredrickson, *J. Lipid Res.* **11**, 322 (1970).

[6] J. Folch, M. Lees, and G. H. Sloane Stanley, *J. Biol. Chem.* **226**, 497 (1957).

scintillation vial under a stream of nitrogen; 1 ml of water and 10 ml of the counting solution are added to the dry residue, and the radioactivity is quantified.

Units. A unit of enzymatic activity is defined as the amount of enzyme required to catalyze the hydrolysis of 1 nmole of cerebroside sulfate per hour. Specific activity is defined as units per milligram of protein. Protein is estimated by the procedure of Lowry.[3]

Purification Procedure

Polyacrylamide gel electrophoresis is employed as the criterion for assessing protein purity. Electrophoresis is performed as described by Davis[7]; a 1.5 cm stacking gel (2.5% acrylamide) and a 4-cm resolving gel (5% acrylamide) are used. The gels are routinely run at pH 8.9. The purity of the product of step 5 was also verified by gel electrophoresis at pH 2.9 in the presence of 8 M urea.

Twenty-four-hour urine collections are obtained from normal individuals and stored at 4° until processed.

Step 1. Ammonium Sulfate Precipitation, 0–50%. Step 1 is performed at 4°. Urine is collected over a 24-hour period and then adjusted to pH 5.7 by adding 100 ml of 1.0 M NaAc, pH 5.7, to 900 ml of urine. This solution is filtered through glass fiber filter paper (Reeve Angel grade 934 AH) to remove cellular and other debris; 313 grams of enzyme grade ammonium sulfate (Schwarz-Mann) is added slowly to each liter of the filtrate. A copious precipitate forms during the next 2 hours, and the mixture is then filtered through glass fiber filter paper. The precipitate is dialyzed exhaustively against 0.1 M NaAc buffer pH 5.7, and the retentate is centrifuged; the sediment is discarded, and the supernatant is used in step 2.

Step 2. Acetone Fraction, 33–76%. This fractionation was carried out at −20°. Acetone (A.C.S. analytical reagent, Mallinckrodt) is slowly stirred into the soluble material from step 1 in the ratio of 1:2 (v:v). The precipitate that forms is removed by centrifugation at 10,000 g for 30 minutes. The supernatant is decanted and to it additional acetone added, 1:1 (v:v). A second precipitate forms and the mixture is recentrifuged. The supernatant is discarded, and the precipitate is resuspended in 0.1 M NaAc, pH 5.7.

Acetone fractionation reduces neutral and polar lipid concentrations below the limit of detection of thin-layer chromatography, decreases

---

[7] B. J. Davis, *Ann. N. Y. Acad. Sci.* **121**, 404 (1964).

the amount of phosphorus associated with the protein to $<3\%$ and is necessary for further purification of the enzyme.[8]

Step 3. Affinity Chromatography. Psychosine sulfate (Pierce Chemical Company) is coupled to agarose as previously described.[8] The affinity labeled Sepharose 4B is thoroughly washed with $0.1\,M$ sodium acetate, pH 5.7, before use and packed into a 9×250 mm column to which the final solution from step 2 is applied. The column is washed thoroughly with $0.1\,M$ sodium acetate, pH 5.7, which does not remove any enzyme activity. The enzyme activity is eluted with 0.1% Triton X-100 (w:v) in $0.1\,M$ NaAc, pH 5.7, buffer. Fractions with enzyme activity are pooled and dialyzed vs. $0.1\,M$ NaAc, pH 5.7, buffer in order to reduce the concentration of Triton X-100. The retentate is then concentrated to 1.5 ml by filtration through a UM-2 Diaflo membrane (Amicon).

Step 4. Sephadex G-200 Chromatography. A Sephadex G-200 column (10×1500 mm) is equilibrated with $0.1\,M$ sodium acetate, pH 5.7, at $4°$. The final solution from step 3 is applied to the column and the flow rate is maintained at 3 ml per hour; 3-ml fractions are collected. Those fractions with the greatest enzyme activity are pooled and concentrated to 0.5 ml by filtration through a UM-2 Diaflo membrane for use in step 5.

Step 5. Preparative Polyacrylamide Gel. The concentrate from step 4 is then subjected to preparative polyacrylamide gel electrophoresis. The standard Ornstein-Davis alkaline buffers are used.[7,9] A 2.5% stacking gel (1 cm) and a 5% resolving gel (2 cm) are employed. The material from step 4 is applied to the stacking gel, and electrophoresis is performed at 3 mA; the system is maintained at $4°$ throughout the procedure. Eluting buffer (a 1:3 dilution of lower gel buffer) is pumped at a rate of 50 ml per hour, and 5-ml fractions are collected.

Alternate Step 5. It is also possible to rechromatograph the concentrate from step 4 on columns of Sephadex G-200. Pure arylsulfatase A is obtained in the ascending portion of the peak of enzyme activity.

The final enrichment is 175-fold over the original urine specimens. The purification procedure is summarized in the table.

Properties

The purest preparations from either step 5 or alternate step 5 migrate as a single band on analytical polyacrylamide gels at pH 8.9 and 2.9.

[8] J. L. Breslow and H. R. Sloan, *Biochem. Biophys. Res. Commun.* **46**, 919 (1972).
[9] L. Ornstein, *Ann. N. Y. Acad. Sci.* **121**, 404 (1964).

PURIFICATION OF ARYLSULFATASE A FROM HUMAN URINE

Step	Total activity (units)	Protein (mg)	Specific activity (units per mg protein)
Urine	12.6	704	.02
Step 1. Ammonium sulfate precipitate, 0–50%	15.1	53	.29
Step 2. Acetone, 33–67%	14.5	35	.41
Step 3. Affinity chromatography	10.1	11	.90
Step 4. G-200 chromatography	8.1	2.8	2.9
Step 5. Preparative polyacrylamide gel electrophoresis	1.6	.45	3.5

Physical Properties. Human urinary arylsulfatase A elutes from Sephadex G-200 with a K_{av} of 0.30 at both pH 5.7 and pH 8.0. This is compatible with a molecular weight of approximately 110,000.

The amino acid composition does not reveal any unusual features except for a relatively high proline content. This has also been observed for ox liver arylsulfatase A.[10] The pH optimum of the pure enzyme is 4.75.

Nature of the Reaction. The enzyme catalyzes the hydrolytic cleavage of the sulfate ester bond of cerebroside [^{35}S]sulfate and p-nitrocatechol sulfate. In addition to the sulfate ion, the products are galactosyl ceramide and p-nitrocatechol, respectively. The activity of this enzyme is markedly reduced in various tissues from patients with metachromatic leukodystrophy.[11] The deficiency of this enzyme probably accounts for the accumulation of galactosyl (3-sulfate) ceramide in metachromatic leukodystrophy.

[10] L. W. Nichol and A. B. Roy, *Biochemistry* **4**, 386 (1965).
[11] H. Jatzkewitz and E. Mehl, *J. Neurochem.* **16**, 19 (1969).

[121] Corrective Factors for Inborn Errors of Mucopolysaccharide Metabolism

By MICHAEL CANTZ, HANS KRESSE, ROBERT W. BARTON, and ELIZABETH F. NEUFELD

There exist several genetically distinct defects of dermatan sulfate and heparan sulfate catabolism that are expressed in cultured fibroblasts: the Hurler, Hunter, Sanfilippo, Scheie, and Maroteaux-Lamy syndromes (mucopolysaccharidoses I, II, III, V, and VI, respectively[1]). In each case, there is increased accumulation and prolonged turnover time of one or both of these sulfated mucopolysaccharides, readily measured with isotopic methods.[2] The abnormal metabolism is the consequence of a deficit of a protein (different in the various genetic conditions) required for mucopolysaccharide degradation. When the appropriate protein is added exogenously, it causes a normalization of the kinetics of mucopolysaccharide turnover in the recipient cells. The proteins have been designated "corrective factors" and qualified by the name of the disease of which they correct the metabolic defect—e.g., the Hunter corrective factor specifically normalizes mucopolysaccharide metabolism in fibroblasts derived from patients with the Hunter syndrome.[3] Probably all the corrective factors are catabolic enzymes that assist the recipient cells in degrading mucopolysaccharide; the enzymatic activity of several factors is already known. Conventional assays of the appropriate enzymes may soon supersede the bioassay for corrective factors described in this chapter. However, the bioassay will remain useful when the enzymatic nature of the factor remains to be elucidated or the substrate is difficult to obtain. In addition, it is possible that the ability of an enzyme to function as a corrective factor may require some specific structural feature that enables it to be taken up by the cells, so that assays for enzymatic activity and for corrective factor activity may not always measure precisely the same entity.

Culture of Fibroblasts

We assume that the reader is familiar with the basic techniques of cell culture,[4,5] and shall therefore limit ourselves to those aspects specific to the study of sulfated mucopolysaccharide metabolism.

Eagle's Minimal Essential Medium in Earle salts is modified by (a) replacement of $MgSO_4$ by an equivalent amount of $MgCl_2$; (b) reduction of $NaHCO_3$ to 1.6 g/liter; (c) addition of "nonessential amino acids." Fetal calf serum (*not* heat-inactivated) 100 ml, is added to 900 ml of the modified Eagle-Earle medium. Antibiotics and glutamine are added shortly before use. We routinely add penicillin and streptomycin (100 units of each per milliliter) and nystatin (25 units per milliliter)

[1] V. A. McKusick, "Heritable Disorders of Connective Tissue." Mosby, St. Louis, Missouri, 1966.
[2] J. C. Fratantoni, C. W. Hall, and E. F. Neufeld, *Proc. Nat. Acad. Sci. U. S.* **60**, 699 (1968).
[3] E. F. Neufeld and M. J. Cantz, *Ann. N. Y. Acad. Sci.* **179**, 580 (1971).
[4] R. G. Ham and T. T. Puck, see Vol. 5 [9].
[5] J. A. Boyle and J. E. Seegmiller, see Vol. 22 [17].

but recommend that the latter be omitted unless the operator encounters frequent yeast infections; because of its limited solubility, nystatin precipitates on the surface of the cells, thereby spoiling them for studies by microscopy. Addition of Aureomycin (50 mg/ml) protects against mycoplasma infections. The sulfate concentration of this medium is ca. 0.4 mM, mostly from the streptomycin solution, with a small contribution from the fetal calf serum.

The use of medium buffered with bicarbonate requires that the cells be kept in an atmosphere of 95% air–5% CO_2. It is critical that during experiments the pH be kept no higher than 7.0 (orange color of the phenol red indicator); at pH 7.4 and higher, there is a significant inhibition of mucopolysaccharide catabolism in *normal* cells.[6] A pH somewhat lower than 7.0 is not detrimental. In the absence of a CO_2 incubator we have found it convenient to perform experiments in 100-mm plastic petri dishes (Falcon), placed into suitcase-like plexiglass boxes with a rubber gasket and two sealable outlets for gassing.

Measurement of $^{35}SO_4$-Labeled Mucopolysaccharide

Principle. Sulfated mucopolysaccharides are the only macromolecules into which $^{35}SO_4$ is incorporated by cultured fibroblasts.[2] Therefore, separation of sulfate-labeled mucopolysaccharides, [^{35}S]MPS, by any procedure that would separate inorganic sulfate from polymeric sulfate is adequate (e.g., extraction with hot 80% ethanol or dialysis). This contrasts with the need for extensive purification of the mucopolysaccharide when ^3H or ^{14}C precursors are used to label the carbohydrate chains.

Reagents

> Trypsin, commercial preparation for tissue culture, 2.5%, filtered; freshly diluted with 4 parts of 0.9% NaCl
> NaCl, 0.9% (w/v)
> Ethanol, 80% (v/v)
> NaOH, 10% (w/v)
> Scintillation fluid: 0.2% 2,5-diphenyloxazole and 0.0025% 1,4-bis-2'(5-phenyloxazolyl)benzene in toluene-ethylene glycol monomethyl ether 1:1, v/v
> Acetic acid, 2 N
> Reagents for protein determination

Procedure. After cells have been incubated in medium containing

---

[6] S. O. Lie, V. A. McKusick, and E. F. Neufeld, *Proc. Nat. Acad. Sci. U. S.* **69**, 2361 (1972).

$^{35}SO_4$ for the desired length of time, the medium is removed and the cells, still attached to the dish, are rinsed with 5 ml of 0.9% NaCl. They are detached from the plate with trypsin (10 minutes at 37° usually suffices, but the time must be adjusted for variations in the adhesiveness of cells or potency of the trypsin). The cell suspension is transferred to a conical 12-ml centrifuge tube, and centrifuged at room temperature at ca. 1000 g for 4 minutes. The supernatant fluid is aspirated, and the tube is rinsed with 2 ml of 0.9% NaCl (do not stir cells at this time). After centrifugation and removal of the saline supernatant, 2 ml of 80% ethanol is added to each tube, and the cell clump is broken up with a sealed Pasteur pipette. The tubes are topped with a tapered glass bead or marble and placed into a gently boiling water bath. When the ethanol begins to bubble, each tube is removed from the bath. After all the tubes have come to the boiling point, they are replaced into the boiling water and again brought to boiling. The suspensions are cooled to room temperature and centrifuged; the ethanol supernatant is discarded. Again 2 ml of 80% ethanol is added and extraction of the pellet is repeated.

After cooling, centrifugation, and removal of the supernatant fluid, 1 ml of 10% NaOH is added and the suspensions placed in the boiling water bath (no stirring at this time). After the precipitate has dissolved, the tubes are tilted and rotated so that any remaining bits of tissue adhering to the glass can be washed into the alkali.

After cooling, an aliquot (0.5 ml) of the solution is mixed with 9.5 ml of scintillation fluid for measurement of radioactivity, whereas another aliquot, (0.1–0.2 ml) is neutralized with 1.5 volumes of 2 N acetic acid and used for protein determination by the method of Lowry *et al.*[7] [^{35}S]MPS accumulation is expressed as counts per minute per milligram of cell protein.

For diagnostic work, and for the purification of corrective factors, measurement of this intracellular [^{35}S]MPS suffices. Should it be desirable, however, to measure [^{35}S]MPS secreted into the medium or attached to the cell periphery, extensive dialysis of the medium or trypsin supernatant is recommended. After dialysis for 5 hours against 0.1 M $(NH_4)_2SO_4$, followed by 20 hours against running tap water, an aliquot (0.2 ml) is mixed with 0.3 ml water and 9.5 ml scintillation fluid for determination of radioactivity.

Assay of Corrective Factor Activity

Principle. Corrective factors reduce the accumulation of [^{35}S]MPS in fibroblasts derived from mucopolysaccharidosis patients. Under rigor-

[7] E. Layne, see Vol. 3 [73].

ously standardized conditions, the reduction (i.e., correction) can be reproducibly related to the amount of factor applied.

Procedure. Fibroblasts are used the day after transplantation. At that time, the dishes should contain cells at or close to confluence (ca. 0.5–1.0 ml cell protein or 1 to 2 million cells, per 100 mm Falcon plate).

Radioactive medium (5 ml, containing about 20 μCi $H_2{}^{35}SO_4$, or about 5×10^6 cpm/ml) is added to each plate, with 1.0 ml of suitably diluted factor preparation or buffer control. Solutions applied to cells must be sterilized by filtration through a cellulose ester membrane (Millipore, 0.45 μm, type HA, available in disposable sterile form as "Swinnex"). Some factors are adsorbed to the membrane unless protected by foreign proteins; therefore medium, which contains proteins of fetal calf serum, is premixed with factor before filtration.

The cells are incubated for 40–48 hours at 37°, in 95% air–5% CO_2. Intracellular [^{35}S]MPS is then measured as described above. Correction is the difference between [^{35}S]MPS (cpm/mg cell protein) in the sample with corrective factor and that in the buffer control.

Calculation of Units. Increasing amounts of corrective factor cause increasing correction, asymptotically approaching a maximum. The factors, some of which are known and others presumed to be enzymes, behave kinetically as substrates in Michaelis-Menten kinetics. A double reciprocal plot, of 1/correction vs. 1/protein concentration, gives a line of which the y intercept is the reciprocal of maximal correction (C_{max}).[8] A unit is defined as that activity which gives half-maximal correction, and can be obtained graphically from the double reciprocal plot in a manner analogous to K_m. Alternatively, the following equation can be used:

$$C_{max}/c - 1 = 1/unit$$

where c is the correction given by the sample under consideration. A convenient alternative to determining C_{max} graphically is to perform one assay with so much factor (over 30 units) that the correction is within experimental error of C_{max}.

Some Precautions. The need for maintaining the pH of the medium at or below neutrality has already been stressed; in addition, solutions to be tested for corrective factor activity must be completely free of inorganic sulfate (which would reduce the specific activity of $^{35}SO_4$ and thus simulate correction), of toxic substances, and of inhibitors of protein or mucopolysaccharide synthesis. One milliliter of a hypotonic solution added to 5 ml of medium is well tolerated, but the same amount of hypertonic solution is not. Thorough dialysis against 0.01 M sodium

[8] R. W. Barton and E. F. Neufeld, *J. Biol. Chem.* **246**, 7773 (1971).

phosphate, pH 6.0, in 0.9% NaCl is recommended to remove interfering substances of low molecular weight.

Should it be necessary to test for correction in the presence of substances interfering with mucopolysaccharide biosynthesis, a "chase" experiment is recommended. Fibroblasts are prelabeled for 48 hours in 6 ml of radioactive medium; the medium is removed, the cells are rinsed with 5 ml of sterile 0.9% NaCl, and 5 ml of unlabeled medium with factor preparation or buffer control is added. [^{35}S]MPS remaining after 24 hours is determined.

Because there are significant differences between cell lines in responsiveness to corrective factor, it is desirable to use one line throughout a purification procedure.

Hurler Corrective Factor[8]

Purification

General Comments. The starting material is urine from normal adult young men. Although human urine is in some ways an excellent source (it is available in large quantities, and all the factors are present at high specific activity), it has the serious disadvantage of unpredictability. Age and sex of the donors may affect the protein composition and, hence, the results of purification; diet is no doubt an important variable, since the initial ammonium sulfate step depends on coprecipitation of small amounts of protein with a copious quantity of urates. Because of the potentially troublesome variability of human urine, the investigator is cautioned to monitor his yields carefully, and, if necessary, to test fractions other than the ones recommended.

Specific activity is expressed as corrective factor units per absorbance unit at 280 nm. In the first step, it may be necessary to verify protein concentration by the Lowry method.[7] Enzyme grade ammonium sulfate is used throughout. All procedures are carried out at 4°, except that the initial addition of ammonium sulfate may be made before the urine has been chilled.

Step 1. Ammonium Sulfate Precipitation. Morning urine, 120 liters, is made 70% saturated with ammonium sulfate (450 g/liter) within 3 hours after voiding. The precipitate is allowed to settle for at least 24 hours; as much of the supernatant fluid as possible is removed by siphoning or by gentle suction, and the remainder by centrifugation at 8000 g for 20 minutes. The precipitate is suspended in 1.5 liters of 30% saturated (164 g/liter) ammonium sulfate solution, and the pH adjusted to 6.0 by addition of 0.5 M Na$_2$HPO$_4$. After stirring for 1 hour, the suspension is centrifuged at 8000 g and the supernatant solution made

70% saturated in ammonium sulfate. The suspension is centrifuged once more, and the supernatant solution discarded.

Step 2. Sephadex G-200 Chromatography. The above precipitate is dissolved in approximately 150 ml of 0.5 M NaCl containing 0.01 M sodium phosphate, pH 6.0 (Buffer A) and dialyzed for 6 hours against 2 changes of 6 liters each of this buffer. The dialyzed sample, 230 ml, is loaded onto a Sephadex G-200 column (90 × 10 cm), previously equilibrated with Buffer A, and eluted with the same buffer in 20-ml fractions at a flow rate of 100 ml/hr. Hurler factor is eluted as a single peak. On this and subsequent column fractionations, the following procedure must be followed to prevent inactivation of the factor: after a small aliquot has been removed for assay of corrective factor activity, the fractions are made approximately 80% saturated by addition of solid ammonium sulfate and stored at 4°.

Step 3. Carboxymethyl Cellulose Chromatography. The most active fractions from the gel filtration step (stored in 80% ammonium sulfate) are pooled and centrifuged at 10,000 g; the precipitate is dissolved in 50 ml of 0.1 M NaCl containing 0.1 M sodium acetate, pH 4.0 (Buffer B). The sample is equilibrated with Buffer B by dialysis for about 10 hours, using three 6-liter changes. Thoroughness of dialysis should be ascertained by measurement of conductivity. The dialyzed solution is loaded onto a column of carboxymethyl cellulose (20 × 4 cm, previously equilibrated with Buffer B). The column is washed with 200 ml of Buffer B and then eluted in 17-ml fractions with a linear gradient formed by 2 liters of Buffer B and 2 liters of 0.5 M NaCl in 0.1 M sodium acetate, pH 4.0. Flow rate is 170 ml per hour. Fractions are handled as described above, with the additional precaution of adjusting the fractions to pH 6.0 with 0.5 M Na$_2$HPO$_4$ before addition of ammonium sulfate. Hurler factor is eluted as a single peak.

Step 4. Hydroxylapatite Chromatography. The ammonium sulfate precipitate from the pooled fractions of the carboxymethyl cellulose peak is dissolved in a minimal amount of Buffer A and equilibrated against this buffer by dialysis. The sample, 13 ml, is applied to a hydroxylapatite column (21 × 1.5 cm, equilibrated with Buffer A); after washing with 100 ml of the same buffer, a linear gradient of phosphate ion is applied (500 ml of Buffer A and 500 ml of 0.25 M sodium phosphate, pH 6.0, in 0.5 M NaCl). Fractions containing 10 ml are collected at a flow rate of 45 ml per hour. Hurler corrective factor emerges as two major peaks, preceded at times by a minor peak. The most active fractions from the two major peaks are about 1000-fold purified over the starting material. These may be stored at 4° in 80% ammonium sulfate, but because of the low protein concentration, carrier protein

TABLE I
SUMMARY OF THE PURIFICATION OF HURLER CORRECTIVE FACTOR[a]

Purification step	Total protein (A_{280})	Total activity (units)	Specific activity (units/A_{280})	Purification (fold)	Recovery (%)
1. Ammonium sulfate, 30–70%	12,800	69,600	5.4	—	—
2. Sephadex G-200	1,000	46,800	46.8	8.7	67
3. Carboxymethyl cellulose	114	61,900	542	100	89
4. Hydroxylapatite					
Total	10	46,000	4,600	850	66
Most active fraction of peak 1	0.65	3,800	5,800	1,070	—
Most active fraction of peak 2	0.70	4,000	5,700	1,050	—

[a] R. W. Barton and E. F. Neufeld, *J. Biol. Chem.* **246**, 7773 (1971).

(bovine serum albumin, 1 mg/ml) must be added prior to further manipulations.

The purification procedure is summarized in Table I.

Properties

Physicochemical Characteristics. Hurler corrective factor has an apparent molecular weight of 87,000, as determined by gel filtration on Sephadex G-200. The isoelectric point is between 4.6 and 6.3.

Function. Hurler corrective factor is an α-L-iduronidase[9]; it is presumed to be an exoglycosidase, which removes L-iduronic residues from chains of dermatan sulfate and heparan sulfate.

Specificity. Hurler corrective factor accelerates mucopolysaccharide degradation only in those cells that are deficient in α-L-iduronidase— cells from Hurler patients, from patients with the Scheie syndrome, and from some patients with a phenotype intermediate between Hurler and Scheie (it is believed that these diseases represent different combinations of allelic mutations). It has no effect on the mucopolysaccharide metabolism of fibroblasts from patients with any other mucopolysaccharidosis, nor from normal individuals.

Stability. The Hurler corrective factor is maximally stable at pH 6.0; inactivation is rapid above pH 7.0 and below pH 4.0. Ionic strength is a critical parameter; at low ionic strength, the factor aggregates with

[9] G. Bach, R. Friedman, B. Weissmann, and E. F. Neufeld, *Proc. Nat. Acad. Sci. U. S.* **69**, 2048 (1972).

an eventually irreversible loss of activity. The factor can be stored frozen for several weeks, but repeated freezing and thawing destroy the activity. In 80% ammonium sulfate at 4°, even the purest fractions retained 20% of their initial activity after one year.

Purity. The best fractions obtained so far are estimated to be 10% pure, because of the 6 or 7 protein bands in polyacrylamide gel electrophoresis, only one corresponded to Hurler factor activity. N-Acetyl-β-glucosaminidase and acid phosphatase are particularly strong contaminants of peaks 1 and 2 eluted from hydroxylapatite, respectively.

Hunter Corrective Factor[10]

Purification

Step 1. Ammonium Sulfate Fractionation. A 70% ammonium sulfate precipitate obtained from 110 liters of urine as described for the Hurler factor, is suspended in 500 ml of H_2O and stirred for 1 hour (see General Comments and Step 1, under Hurler corrective factor). After centrifugation, the supernatant fluid (630 ml) is made nominally 50% saturated in ammonium sulfate (on the assumption that the supernatant fluid was zero saturated), the pH being held at 6.0 by periodic additions of $0.5\,M$ Na_2HPO_4. After standing for 1 hour, the precipitate is removed by centrifugation and the ammonium sulfate concentration in the supernatant fluid is raised to 80% saturation, the pH being controlled as above. The precipitate is collected by centrifugation, redissolved in a minimal amount of $0.5\,M$ NaCl–$0.01\,M$ sodium phosphate pH 6.0 (Buffer A), and dialyzed overnight against four 6-liter changes of this buffer.

Step 2. Sephadex G-200 Chromatography. The dialyzed solution from step 1 (110 ml) is chromatographed on Sephadex G-200 (90 × 10 cm) previously equilibrated with Buffer A, at a flow rate of 100 ml/hr. Hunter factor is eluted as a single peak, located on the ascending limb at a large protein peak which contains primarily albumin. The fractions with the highest specific activity are pooled and concentrated to a volume of about 60 ml by ultrafiltration on a Diaflo PM 30 membrane (Amicon 400 ultrafiltration cell).

This material is rechromatographed on the same column; when pooling the active fractions, a sharp cut is made on the retarded side of the peak, to exclude as much albumin as possible. The pooled fractions are concentrated by ultrafiltration to about 25 ml.

Step 3. Affinity Chromatography on Antialbumin Sepharose. This step was devised to separate albumin from Hunter corrective factor, since conventional procedures failed to resolve these two proteins. The anti-

---

[10] M. Cantz, A. Chrambach, G. Bach, and E. F. Neufeld, *J. Biol. Chem.* **247**, 5456 (1972).

body-Sepharose is prepared by linking Sepharose 4B (140 ml after settling) with immunoglobulin-G from a goat immunized with human serum albumin (140 ml of solution containing 5 g of purified IgG, of which 800 mg is specific antibody, dialyzed against 0.2 M sodium citrate buffer, pH 6.0) using the cyanogen bromide procedure of Cuatrecasas.[11] The antialbumin Sepharose, which has a binding capacity of about 0.7 mg human serum albumin per milliliter, is packed in a 2 × 33.5 cm column and equilibrated with 0.1 M sodium phosphate buffer, pH 7.0. Hunter corrective factor from step 2, 26 ml (containing 2 mg of albumin per milliliter) is dialyzed against the same buffer and passed through the column at room temperature (30 ml per hour, 5.4-ml fractions). The effluent fractions with A_{280} greater than 0.1 are pooled and concentrated by ultrafiltration to 6 ml.

Step 4. Preparative Polyacrylamide Gel Electrophoresis. A Buchler "Polyprep 100" preparative electrophoresis apparatus is operated at about 0°C by means of circulating liquid from a cooling bath. The basic methodology is that of Rodbard and Chrambach,[12] with multiphasic buffer system 1958.8,[13] the composition of which follows. Upper Buffer: N-tris(hydroxymethyl)methyl-2-aminoethane sulfonic acid (TES), 0.04 M; 4-picoline, 0.044 M. Lower Buffer: 4-picoline, 0.315 M; HCl, 0.25 M. Concentration Gel Buffer: phosphoric acid, 0.049 M; 4-picoline, 0.054 M. Separation Gel Buffer: HCl, 0.0594 M; 4-picoline, 0.858 M. Elution Buffer: 0.132 M in 4-picoline, titrated to pH 6.8 with HCl, and 25% (w/v) in sucrose. The operative pH of the separation phase is 7.95.

Gel concentration in the separation gel is $T = 5\%$, cross-linking $C = 2\%$; the volume of the gel is 40 ml. For the concentration gel, $T = 3.12\%$, $C = 20\%$ are employed; the volume is 12 ml. The gel solutions are deoxygenated prior to polymerization by bubbling with argon at a rate of 20 ml per minute for 5 minutes. Polymerization is initiated by 0.55 mM potassium persulfate and 0.014 mM riboflavin; N,N,N',N'-tetramethylethylenediamine (TEMED) in the separation gel is 3.13 mM, and 6.25 mM in the concentration gel. Photopolymerization is done for 60 minutes.

The concentrate of step 3 is dialyzed against Upper Buffer; solid sucrose to a concentration of 25% and 0.1 ml of a 0.1% bromophenol blue solution are added. The sample is layered on top of the concentration gel and the electrophoresis column is operated at a constant cur-

[11] P. Cuatrecasas, *J. Biol. Chem.* **245**, 3059 (1970).
[12] D. Rodbard and A. Chrambach, *Anal. Biochem.* **40**, 95 (1971).
[13] T. M. Jovin, M. L. Dante, and A. Chrambach, "Multiphasic Buffer System Output," PB 196085–196091, National Technical Information Service, Springfield, Virginia 22151 (magnetic tapes) (1970).

rent of 25 mA. The column is eluted at 1 ml per minute; 2-ml fractions are collected into a cooled fraction collector.

Hunter corrective factor emerges from the column shortly after the bromophenol blue front has been eluted. Two peaks of activity are obtained, with R_f's of 0.91 (peak I) and 0.76 (peak II), respectively.

The active fractions are combined on the basis of specific activity and protein patterns obtained upon reelectrophoresis in polyacrylamide gel (analytical scale, $T = 7.5\%$; $C = 2\%$, same buffer system as above).

The purification procedure is summarized in Table II.

Properties

Physicochemical Characteristics. Hunter corrective factor has an apparent molecular weight of 65,000, determined by polyacrylamide gel electrophoresis, and 114,000, determined by gel filtration on Sephadex G-200. Its isoelectric point is between pH 3.0 and 4.9.

Function. When supplied to fibroblasts from Hunter patients, the Hunter corrective factor increases the rate of degradation of dermatan sulfate. The precise mechanism by which it produces this acceleration of catabolism is not yet elucidated.

Specificity. Purified Hunter corrective factor (pools 4/4 and 4/5) have no effect on the mucopolysaccharide metabolism of fibroblasts from normal individuals or from patients with the Hurler, Sanfilippo, Scheie, or Maroteaux-Lamy syndromes.

TABLE II

SUMMARY OF THE PURIFICATION OF HUNTER CORRECTIVE FACTOR[a]

Purification step	Total protein (A_{280})	Total activity (units)	Specific activity (units/A_{280})	Purification (fold)	Recovery (%)
1. Ammonium sulfate					
0–70%	12,300	540,000	44	—	—
50–80%	3,500	512,000	146	3	95
2. Sephadex G-200					
Column 1	274	393,000	1,440	33	73
Column 2	166	322,000	1,950	44	60
3. Antialbumin Sepharose	102	260,000	2,550	58	48
4. Polyacrylamide gel electrophoresis					
Pool 4/1	7.4	13,200	2,320	53	2.4
Pool 4/2	8.0	20,440	3,330	76	3.8
Pool 4/4	2.9	11,810	5,250	119	2.2
Pool 4/5	8.2	29,200	4,620	105	5.4

[a] M. Cantz, A. Chrambach, G. Bach, and E. F. Neufeld, *J. Biol. Chem.* **247**, 5456 (1972).

Purity. Although the Hunter corrective factor of pool 4/4 manifests but one protein band in polyacrylamide gel electrophoresis, it is not considered homogeneous, since several protein bands can be resolved by isoelectric focusing. The preparation is free of the common lysosomal glycosidases and of α-L-iduronidase.

Stability. The Hunter corrective factor is most stable between pH 5.0 and 8.0. Below pH 5 it is poorly soluble, with irreversible loss of activity; above pH 8.0, inactivation occurs in a few hours. It can be stored frozen for many months; some loss of activity is noted upon repeated freezing and thawing of purified material. In contrast to Hurler and Sanfilippo corrective factors, the Hunter corrective factor is not inactivated in solution of low ionic strength and does not have to be stored in ammonium sulfate.

Sanfilippo A Corrective Factor[14]

Purification

Step 1. Ammonium Sulfate Precipitation. The 70% ammonium sulfate precipitate from 120 liters of urine is collected as described for the Hurler corrective factor and extracted for at least 1 hour with 1.5 liters of 0.03 M sodium phosphate pH 7.0, 0.15 M in NaCl. The suspension is centrifuged at 10,000 g for 30 minutes and the supernatant is made 70% saturated with ammonium sulfate (see General comments and Step 1 under Hurler corrective factor).

Step 2. Sephadex G-200 Chromatography. This step is carried out precisely as described for the Hurler corrective factor. In this and subsequent fractions the same precautions are taken against inactivation: an aliquot of selected fractions is taken for assay, and the remainder is immediately made 70% saturated with ammonium sulfate, and stored at 4°.

Step 3. Carboxymethyl Cellulose Chromatography. The ammonium sulfate precipitate derived from the most active fractions of step 2, is collected by centrifugation and suspended in 50 ml of 0.1 M sodium acetate buffer, pH 4.2 (Buffer C), and dialyzed for 6 hours against six 6-liter changes of the same buffer. Thoroughness of dialysis is monitored by conductivity. The solution is loaded onto a column of carboxymethyl cellulose, 4 × 22 cm, preequilibrated with Buffer C. A linear gradient is applied by mixing 2 liters of 0.1 M sodium acetate, pH 4.2, with 2 liters of 0.4 M NaCl in 0.1 M sodium acetate, pH 4.2. The column is eluted at a flow rate of 90 ml per hour; 17-ml fractions are collected. The Sanfilippo A corrective factor emerges in one sharp peak.

[14] H. Kresse and E. F. Neufeld, *J. Biol. Chem.* **247**, 2164 (1972).

Step 4. Hydroxylapatite Chromatography. The ammonium sulfate precipitate derived from the active fractions of the preceding step is collected, dissolved in 12 ml of $0.005 M$ sodium phosphate, pH 6.8, containing $0.15 M$ NaCl (Buffer D). The sample is dialyzed for 20 hours against four 6-liter changes of Buffer D, and chromatographed on a hydroxylapatite column, 1.5×36 cm, previously equilibrated with the same buffer. A linear gradient is applied by mixing 750 ml of Buffer D with 750 ml of $0.1 M$ sodium phosphate, pH 6.8, in $0.15 M$ NaCl. Fractions of 6 ml are collected at a flow rate of 24 ml per hour. The Sanfilippo A corrective factor emerges in one minor peak (peak 1) followed by two major peaks of activity (peaks 2 and 3 in Table III).

Step 5. Polyacrylamide Gel Electrophoresis. This procedure is performed as described by Rodbard and Chrambach.[12] The multiphasic buffer system (2243.3) has the following composition[13]—Upper Buffer: N-2-hydroxymethylpiperazine-N'-2-ethane sulfonic acid (HEPES), $0.04 M$; Tris, $0.005 M$. Lower Buffer: HCl, $0.05 M$; Tris, $0.062 M$. Concentration Gel Buffer: acetic acid, $0.065 M$; Tris, $0.029 M$. Separation Gel Buffer: HCl, $0.036 M$; Tris, $0.04 M$. Polymerization is initiated for the concentration gel by 0.55 mM potassium persulfate, 0.014 mM riboflavin, and 6.55 mM N,N,N',N'-tetramethylethylenediamine (TEMED) and for the separation gel by 0.046 mM potassium persulfate, 0.014 mM riboflavin, and 1.25 mM TEMED. The gel concentration in the separation gel is $T = 6\%$, and cross-linking, $C = 2\%$.

The ammonium sulfate precipitate obtained from fractions of peak 3 of the preceding step is collected, dissolved in 2 ml of Upper Buffer,

TABLE III

SUMMARY OF THE PURIFICATION OF SANFILIPPO A CORRECTIVE FACTOR[a]

Purification step	Total protein (A_{280})	Total activity (units)	Specific activity (units/A_{280})	Purification (fold)	Recovery (%)
1. Ammonium sulfate, 0–70%	10,700	419,000	39	—	—
2. Sephadex G-200	989	309,000	312	8	74
3. Carboxymethyl cellulose	75	125,000	1,660	42	30
4. Hydroxylapatite					
Peak 2	9.6	33,700	3,520	90	8
Peak 3	11.3	47,500	4,190	107	11
5. Polyacrylamide gel electrophoresis of peak 3	0.1	3,480	33,100	850	0.8

[a] H. Kresse and E. F. Neufeld, *J. Biol. Chem.* **247**, 2164 (1972).

dialyzed for 16 hours against four 6-liter changes of the same buffer, and applied to tubes of 17-mm diameter containing 9.5 ml of separation gel solution and 3 ml of concentration gel solution. The amount of protein used is 4 mg per tube. After electrophoresis, gels are frozen and cut transversely with a razor blade into 2-mm slices, from which the factor is recovered by homogenization in 3 ml of $0.1\,M$ Tris buffer, pH 7.0, containing $0.15\,M$ NaCl, followed by centrifugation.

The purification procedure is summarized in Table III.

Properties

Physicochemical Characteristics. The Sanfilippo A corrective factor has an apparent molecular weight estimated at 114,000 by Sephadex G-200 filtration and 186,000 by polyacrylamide gel electrophoresis. The isoelectric point is between pH 5.6 and 6.9.

Function. The Sanfilippo A corrective factor hydrolyses sulfate groups from the heparan sulfate accumulated in fibroblasts from Sanfilippo A patients. It is not clear whether the linkages hydrolyzed are O-sulfate or N-sulfate.

Specificity. The most purified preparation of Sanfilippo A corrective factor (step 5) has no effect on the mucopolysaccharide metabolism of fibroblasts from normal individuals or from patients with other mucopolysaccharidoses.

Purity. The most purified fraction (step 5) is estimated to contain about 40% Sanfilippo A corrective factor. It contains N-acetyl-β-glucosaminidase as major and aryl sulfatase A as minor contaminants.

Stability. The Sanfilippo A corrective factor is unstable at low ionic strength ($0.01\,M$ NaCl), but stable in $0.15\,M$ NaCl from pH 4 to 7. It may be stored for many months at $4°$ in 70% saturated ammonium sulfate. Heparan sulfate (0.5 mg/ml) acts as stabilizer even at low ionic strength.

[122] Purification of Neuraminidases (Sialidases) by Affinity Chromatography

By PEDRO CUATRECASAS

Principle

Neuraminidases from various sources can be purified rapidly and effectively[1] by chromatography on columns containing selective adsorb-

[1] P. Cuatrecasas and G. Illiano, *Biochem. Biophys. Res. Commun.* **44**, 178 (1971).

ents prepared by the covalent attachment of a specific enzyme inhibitor to an insoluble matrix.[2,3] A selective neuraminidase adsorbent is prepared by attaching through azo linkage an inhibitor of this enzyme, N-(p-aminophenyl)oxamic acid, to agarose beads which contain the tripeptide, glycylglycyltyrosine (Scheme 1). The enzymatic activity

present in extracts of *Clostridium perfringens*, *Vibrio cholerae*, and influenza virus is extracted by columns containing this adsorbent, and nearly quantitative elution is achieved by modifying the pH and ionic strength of the buffer. The advantages of these procedures over the conventional methods of purification[4] for these enzymes include greater rapidity and yield, and at least in some cases a higher degree of purity. The affinity columns can be used to purify or concentrate neuraminidases, and to remove denatured or inactivated forms of the enzyme in already purified preparations.

Preparation of Selective Adsorbent[1]

Preparation of Tyrosine–Agarose. Fifty milliliters of packed Sepharose 4B (Pharmacia) is suspended in distilled water to achieve a total volume of 100 ml. Twelve grams of crushed, solid CNBr (Eastman) is added[5] at once to the suspension, which is vigorously mixed by magnetic stirring in a well ventilated hood equipped with a pH meter. The pH is adjusted and maintained at 11 by continuous titration with $4 M$ NaOH. When the rate of change in pH decreases substantially (about 10 minutes), the suspension is rapidly cooled and washed over a Buchler funnel with 1 liter of ice-cold $0.2 M$ NaHCO₃ buffer, pH 9.5. The activated moist agarose cake is suspended in 100 ml of the same buffer containing 2 g of glycylglycyltyrosine (Fox Chemical Co., Los Angeles,

---

[2] P. Cuatrecasas, M. Wilchek, and C. B. Anfinsen, *Proc. Nat. Acad. Sci. U.S.* **61**, 636 (1968); P. Cuatrecasas and C. B. Anfinsen, *Annu. Rev. Biochem.* **40**, 259 (1971); P. Cuatrecasas and C. B. Anfinsen, Vol. 22 [31].

[3] P. Cuatrecasas, *Advan. Enzymol.* **36**, 29 (1972).

[4] J. T. Cassidy, G. W. Jourdian, and S. Roseman, *J. Biol. Chem.* **240**, 3501 (1965); G. L. Ada, E. L. French, and P. E. Lind, *J. Gen. Microbiol.* **24**, 409 (1961); M. E. Rafelson, Jr., M. Schneir, and V. S. Wilson, *Arch. Biochem. Biophys.* **103**, 424 (1963); M. E. Rafelson, Jr., S. Gold, and I. Priede, Vol. 8 [116]; J. T. Cassidy, G. W. Jourdian, and S. Roseman, Vol. 8 [117].

California), and the slurry is gently mixed at 4° for about 16 hours.[5] The modified agarose is washed at room temperature with 8 liters of 0.1 M NaCl during an 8-hour period. This washed derivative, which by amino acid analysis contains 6–10 μmoles of peptide per milliliter of packed agarose, can be stored at 4° for at least one year. However, if the derivative is stored for more than 1 week it should be washed again before use since small quantities of free peptide may be discharged into the medium during this time.

Preparation of Enzyme Inhibitor. The inhibitor, N-(p-aminophenyl) oxamic acid is prepared by catalytic hydrogenation of N-(p-nitrophenyl) oxamic acid (K and K). Five grams of the compound are dissolved in 30 ml of dimethyl formamide, and 0.5 g of 10% palladium on charcoal is added. Reduction is performed for 4 hours at room temperature in a Parr at 40 mm of pressure. The reduced compound, after separation from the charcoal by filtration, is evaporated to dryness, resuspended in a small volume of dimethyl formamide, and precipitated with ether. The compound is stored at $-20°$ in a desiccator under vacuum and protected from light.

Preparation of Adsorbent. Eighty milliliters of the inhibitor is dissolved in 30 ml of ice-cold 0.4 M HCl, and 150 mg of sodium nitrite (in 1 ml of cold water) is added over a 1-minute period. After 5 minutes an 80-ml slurry of 0.5 M NaHCO$_3$ buffer, pH 8.9, containing 30 ml of packed agarose-Gly-Gly-Tyr is added at 24° and the pH of the suspension is adjusted immediately to 8.8. After stirring gently for 8 hours at 24° the adsorbent is washed with about 12 liters of 0.1 M NaCl over a 6- to 8-hour period. The derivative can be stored for at least one year at 4°, although it must be washed again briefly before use.

The selective adsorbent can be used repeatedly; after each use, however, it is advisable to unpack the column and to gently stir the agarose beads in suspension for about 1 hour before packing a new column. Adsorbents prepared by attaching N-(p-aminophenyl) oxamic acid directly to unsubstituted, cyanogen bromide-activated agarose[2] are ineffective for enzyme purification, in conformity with the frequently observed requirement in affinity chromatography for spatial separation (tripeptide in Scheme 1) between ligand and matrix backbone.[2,3,5]

Conditions for Chromatography

Conditions for Adsorption of Enzyme to the Column. The crude mixture containing the enzyme to be purified must be thoroughly dialyzed

[5] P. Cuatrecasas, *J. Biol. Chem.* **245**, 3059 (1970); P. Cuatrecasas, *Nature (London)* **228**, 1327 (1970).

for about 16 hours at 4° against 50 mM sodium acetate buffer, pH 5.5, containing 2 mM CaCl$_2$ and 0.2 M sodium EDTA. This is the same buffer which is used to equilibrate the column. The solution, which is centrifuged at 20,000 g for 40 minutes to remove all particulate material, must not contain more than 30 mg of total protein per milliliter since such highly concentrated solutions interfere with the extraction of the enzyme by the column.

For reasons described in detail elsewhere[3] a batchwise type of purification with these adsorbents is ineffective and column procedures must therefore be used. The size of the column used will depend on the total amount of enzyme to be purified, the total sample volume, and the specific neuraminidase being purified. Although the optimal column size must be empirically determined in each case, generally 10% of the theoretical capacity of a column can be experimentally achieved. Thus, a column having dimensions of 0.5 × 15 cm, which can theoretically adsorb more than 200 mg of enzyme, can safely be used to purify at least 20 mg of enzyme.

The column is equilibrated at room temperature with 20 volumes of 50 mM sodium acetate buffer, pH 5.5, containing 2 mM CaCl$_2$, and 0.2 mM sodium EDTA. These conditions are optimal for adsorption of neuraminidase from *Vibrio cholerae, Clostridium perfringens,* and influenza virus. The pH of this buffer is quite critical; for example, the columns are not effective if the equilibrating buffer has a pH of 6.5. The crude protein mixture is applied on the column, and this is followed by 10–30 column volumes of buffer or until absorbance at 280 nm is negligible.

Elution of Enzyme from the Column. The enzyme is eluted with 0.1 M NaHCO$_3$ buffer, pH 9.1. The progress of the eluting buffer through the column can be followed conveniently by the intensification of the color of the absorbent which results upon ionization of the azo moiety by the higher pH of the buffer. To rapidly lower the pH to which the purified enzyme is exposed, the effluent fractions are collected in tubes which contain 0.5 M sodium acetate buffer, pH 5, in a volume equal to 10% of the desired fraction volume.

The enzyme emerges sharply in very concentrated form with the front of the eluting buffer. The pH of the fraction containing the enzyme varies from 7.5 to 8. However, attempts to elute the enzyme by using directly buffers having these lower pH values are not successful, for reasons described elsewhere.[3] The elution fractions which immediately follow the early protein peak frequently have a slight brownish color which should not be mistaken as protein. This represents trace amounts of ligand which are removed from the column and are of little consequence. Some of this material, however, may also be present in the fraction containing the eluted enzyme so that care must be taken not to estimate the total protein content in that fraction by absorbancy (280 nm) measurements alone. To determine protein content, it is necessary to use colorimetric procedures or to first subject the enzyme to dialysis. Most neuraminidases withstand dialysis against 50 mM NH_4HCO_3 buffer, pH 7.5, followed by lyophilization.

Specific Examples. Figure 1 describes a representative chromatographic experiment for purifying neuraminidase from *Clostridium per-*

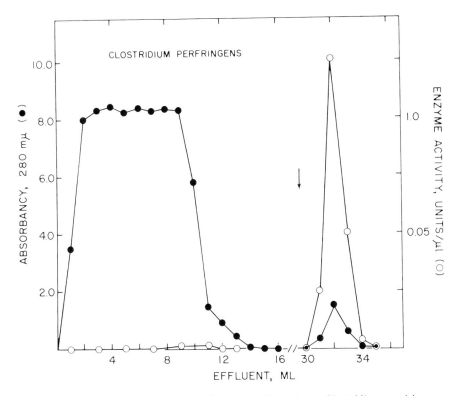

Fig. 1. Affinity chromatography of neuraminidase from *Clostridium perfringens* on the agarose derivative shown in Scheme 1. Seven milliliters of a commercially (Worthington) "purified" enzyme containing 73 mg of protein were dialyzed for 16 hours at 4° against 4 liters of 50 mM sodium acetate buffer, pH 5.5, containing 2 mM $CaCl_2$ and 0.2 mM EDTA. This material was applied to a 0.4 × 4 cm column containing the adsorbent which had been equilibrated with the same buffer. Elution was achieved with 0.1 M $NaHCO_3$ buffer, pH 9.1 (arrow); the pH of the eluted sample was immediately lowered to 6.0 with 1 N NaOH. The purification data of this experiment are summarized in the table. From P. Cuatrecasas and G. Illiano, *Biochem. Biophys. Res. Commun.* **44**, 178 (1971).

AFFINITY CHROMATOGRAPHY OF NEURAMINIDASES FROM VARIOUS SOURCES

	Source of enzyme		
Purification step	Clostridium perfringens[a]	Vibrio cholerae[b]	Influenza virus[c]
Sample applied on column			
Protein (mg)[d]	73	92	1.2[e]
Volume (ml)	7	12	5
Specific activity[f]	3.6	0.071	0.86
Fraction eluted from column			
Protein (mg)[e]	1.1	0.2	0.5[e]
Specific activity	264	30	1.9
Yield of activity (%)	100	92	91
Purification	73	420	2.2

[a] Summarizes the experiment described in Fig. 1.

[b] A crude filtrate (140 mg of protein) from Vibrio cholerae (Sigma) was suspended in 12 ml of 50 mM sodium acetate buffer, pH 5.5, 2 mM CaCl$_2$, 0.2 mM EDTA, centrifuged at 2000 rpm for 20 minutes, and dialyzed against the same buffer for 16 hours (4°). Chromatography was as described in Fig. 1.

[c] Purified monovalent inactivated influenza virus vaccine strain A2/Aichi/2/68, containing 400 CCA units/ml.

[d] Determined by absorbancy at 210 nm [M. P. Tombs, F. Souter, and N. F. Maclagan, Biochem. J. **73**, 167 (1959)].

[e] For convenience, this was determined by ultraviolet spectroscopy, using $E_{280}^{0.1\%} = 3.0$.

[f] Micromoles of N-acetylneuraminic acid formed per minute per milligram of protein. Data from P. Cuatrecasas and G. Illiano, Biochem. Biophys. Res. Commun. **44**, 178 (1971).

fringens. The data of this experiment is summarized in the table, which also presents the purification of the Vibrio cholerae and influenza virus enzymes on the same column described in Fig. 1. The specific activities of the purified enzymes are at least as high as those previously reported for the most highly purified enzymes.[4]

[123] Heparinase and Heparitinase from Flavobacteria

By ALFRED LINKER and PETER HOVINGH

A fairly large number of enzymes obtained from microorganisms have the unusual property of degrading acidic polysaccharides by an elimination mechanism rather than by the normal hydrolytic pathway. This includes enzymes acting on hyaluronic acid, the chondroitin sulfates, heparin, heparitin sulfate, pectins, and alginates.

Eliminases isolated from induced and uninduced flavobacteria can degrade several acid mucopolysaccharides.[1] Purification of these enzymes has made it possible to degrade specific substrates.[2,3] Recently,[4] the enzyme complex of heparin induced flavobacteria was fractionated more extensively than previously. An eliminase acting specifically on heparin and an eliminase acting specifically on heparitin sulfate were obtained.

Assay

Principle. The degradative mechanism via elimination results in the formation of α,β unsaturated uronides which have a strong absorption maximum at 232 nm. Changes in absorption at this wavelength can, therefore, be used very readily for enzyme assay.

In a crude system containing large amounts of protein or nucleic acid, which would interfere with this absorption, a procedure, based on increase in reducing end groups or a paper chromatographic assay will have to be used.

Reagents

Sodium acetate, 0.1 M, adjusted to pH 7.0 with acetic acid
Hydrochloric acid, 0.03 M
Calcium acetate, 0.01 M

Substrates at 50 mg/ml in distilled water:

Heparin (commercially available, most preparations with good anticoagulant activity should be suitable)
Heparitin (heparan) sulfate. This can be prepared by the method of Cifonelli.[5,6] This polysaccharide is not a single species, but a family of related compounds. Enzyme activity varies somewhat with the sulfate content[4] of a given species. Therefore, a fraction with intermediate sulfate content should be chosen for assay purposes.[4]

[1] A. Linker, P. Hoffman, K. Meyer, P. Sampson, and E. D. Korn, *J. Biol. Chem.* **235**, 3061 (1960).
[2] A. Linker and P. Hovingh, *J. Biol. Chem.* **240**, 3724 (1965).
[3] T. Yamagata, H. Saito, O. Habuchi, and S. Suzuki, *J. Biol. Chem.* **243**, 1523 (1968).
[4] P. Hovingh and A. Linker, *J. Biol. Chem.* **245**, 6170 (1970).
[5] J. A. Cifonelli and A. Dorfman, *J. Biol. Chem.* **235**, 3283 (1960).
[6] J. A. Cifonelli, *in* "Chemistry and Molecular Biology of the Intercellular Matrix" (E. A. Balazs, ed.), Vol. 2, p. 961. Academic Press, New York, 1970.

Chondroitin sulfate. Either 4- or 6-sulfate can be used. These are commercially available.

Procedure. Assay for following the chromatographic purification of the enzymes. To 0.02 ml of substrate solution (1 mg), 0.1 ml of a given column eluate and 0.02 ml of the 10 mM calcium acetate reagent are added. The solution is incubated for 1–3 hours at 30° or 43°, 2.9 ml of 30 mM HCl are added at the end of the incubation time to stop the reaction and the absorbance at 232 nm is determined.

Assay of enzyme preparations. The enzyme is dissolved in a given volume of 0.1 M sodium acetate, adjusted to pH 7.0 (or brought to 0.1 M in sodium acetate if in solution). A suitable portion of this is added to 0.04 ml (2 mg) of substrate solution, 0.03 ml of 10 mM calcium acetate reagent are added and the final volume made up to 0.3 ml with 0.1 M sodium acetate reagent. The solution is incubated at 30° or 43° and at 0, 1, 2, and 4 hours an aliquot of 0.05 ml is withdrawn and added to 2.9 ml of 30 mM HCl. Absorbance at 232 nm is determined. For good accuracy a change in absorbance of 0.2 per hour is desirable. Activity is expressed as the change of absorbance at 232 nm per hour at 30° or 43°. Specific activity is the change of absorption at 232 nm per hour per milligram of protein.

If the enzymes are used as a tool to determine the presence or absence of heparin or heparitin sulfate in a crude system where interference with absorbance at 232 nm may be present, a different assay procedure will have to be used. Increase in reducing end groups with time can be measured. However, as oligosaccharides obtained from heparin, in particular, have very low reducing values, this assay is only about one-tenth as sensitive as the procedure using ultraviolet absorption. The method has been described in detail previously.[7] To be used in the present context the incubation of enzyme and substrate can be carried out as described above but larger total volumes and larger aliquots for assay will have to be used (the addition to the HCl solution is omitted; the aliquot is added to the reducing sugar reagent instead).

An alternate qualitative assay can be carried out by paper chromatography. This again would be most suitable for the determination of presence of a particular substrate in crude systems but can also be used to follow the chromatographic purification of the enzymes. In order to obtain sufficient material for visualization on chromatograms without having to use large volumes for spotting, the substrate concentration for the incubation mixture should be increased to a final value of 10 mg/ml. At appropriate time intervals aliquots of 20 μl (at most 10 μl at one time

[7] A. Linker, Vol. 8 [112].

to avoid large spots) are spotted on Whatman No. 1 paper. Controls of known substrate and enzyme digests should be included. The paper is then developed with butanol:acetic acid:water (50:15:35) for 24–48 hours by downward irrigation. The spots are visualized with an alkaline silver nitrate reagent[8] or by UV absorption using a short wavelength mineral light. Products obtained can be compared with controls. As these digests contain several oligosaccharides[2,9] quantitation which is possible for some eliminases[10] would be difficult here. However, reliable qualitative data can be obtained.

Preparation of Crude Enzyme

Flavobacteria were maintained in stock culture with heparin as the only carbon source. Culture medium: 1 liter of pH 7.4, 0.1 M sodium phosphate buffer containing 1 g of ammonium sulfate, 1 g of heparin (approximately 150 units/mg), and trace amounts of calcium chloride, ferrous sulfate, magnesium sulfate, manganous chloride, and phosphomolybdic acid. The organisms are grown at 25° and subcultured every 3 weeks.

For the preparation of enzyme, 2 ml of stock culture are seeded into 2 liters of trypticase soy broth (6 g of B.B.L. media per liter), also containing 20 mg of heparin per liter (activity about 50 units/mg). After growth for 72 hours at room temperature, the cells are collected by centrifugation and reinoculated into 22 liters of the same broth (also containing 20 mg/liter of heparin). After the culture is grown for an additional 24 hours with intermittent shaking at room temperature, the cells are collected by centrifugation and washed with 25 mM sodium phosphate buffer, pH 8.0. Cells grown on this low heparin concentration are unable to degrade heparin, but can easily be adapted to grow on heparin as follows:

The cells are suspended in 3 liters of 25 mM sodium phosphate buffer, pH 8.0, containing 1.5 g of casein hydrolyzate (Difco) and 3 g of the same heparin as above. The suspension is aerated vigorously for 17 hours, the cells are collected by centrifugation and washed with 25 mM sodium phosphate buffer, pH 8.0. The following procedures are carried out at 4°. The centrifuged organisms are suspended in about 70 ml of 0.7 M sodium acetate, pH 7.0, and sonicated for 25 minutes in an ice–alcohol bath at full power on a Bronwill Biosonik model BPI sonifier. The sonicate is then centrifuged at 25,000 g for 1.5 hours. The supernatant is

[8] W. E. Trevelyan, D. P. Procter, and J. S. Harrison, *Nature* (*London*) **166**, 444 (1950).

[9] A. Linker and P. Hovingh, *Biochemistry* **11**, 563 (1972).

[10] H. Saito, T. Yamagata, and S. Suzuki, *J. Biol. Chem.* **243**, 1536 (1968).

recentrifuged at 25,000 g for 1.5 hours and this supernatant dialyzed overnight against 100 volumes of distilled water. The dialyzate is either lyophilized, frozen, or used immediately for further purification (usual batch size is 80 ml of solution or 600–800 mg of lyophilized material).

This crude preparation contains sulfatases, glycuronidases, chondroitinases, heparinase, and heparitinase. For a better yield of enzymes, the residue from sonication can be resonicated under the same conditions, and the sonicate centrifuged and dialyzed. The dialyzate can be combined with the main preparation or worked up separately.

Chromatographic Purification of Enzymes. Chromatographic materials are prepared as follows: 5 g of hydroxyapatite (BioGel HTP) or cellulose phosphate (Whatman P11) is suspended in 5.0 ml of 1.0 M potassium phosphate buffer, pH 6.8. The suspension is diluted to 100 ml and allowed to settle; the fines are removed by washing 4 times with 20 volumes of distilled water. The final suspension is then poured into a column of appropriate size. In general, 6–7 ml fractions are collected during elution. All following procedures are carried out at 4°. Hydroxyapatite columns are eluted with a linear gradient formed by addition of 50 mM potassium phosphate buffer, pH 6.8, from one vessel to a mixing vessel containing 0.5 M sodium chloride in the same buffer. The cellulose phosphate columns are eluted in a stepwise fashion using sodium chloride at concentrations of 25 mM, 50 mM, 0.10 M, and 0.15 M all in 50 mM potassium phosphate buffer, pH 6.8.

Eluate fractions are dialyzed when necessary against 50 volumes of distilled water with stirring overnight.

1. When the final dialyzate from the sonication step is used directly (usually 80 ml from one batch of organisms), it is brought to 0.1 M in sodium acetate by addition of the proper amount of this salt (check pH at this point, it should not be below 6.5). To this solution 75 mg of protamine sulfate in 3 ml of water are added. After 1 hour the precipitate is removed by centrifugation at 15,000 g for 15 minutes, and the supernatant, containing about 350 mg of protein, is diluted to 400 ml with distilled water and added to a column (6.0 × 2.1 cm) containing 7.5 g of hydroxyapatite.

2. For lyophilized crude enzyme, about 1 g (equivalent to one batch of organisms and consisting of approximately 400 mg of actual protein) is dissolved in 25 ml of 0.5 M sodium acetate overnight. The pH at this point should be above 6.5. As not all the crude material redissolves, the suspension is centrifuged at 15,000 g for 15 minutes. The precipitate is washed with 10 ml of 0.5 M sodium acetate and centrifuged again at 15,000 g for 15 minutes. The supernatant and wash fluid are combined and 25 mg of protamine sulfate in 1 ml of water, are added. After 1 hour

the suspension is centrifuged at 15,000 *g* for 15 minutes, the supernatant diluted to 400 ml with distilled water and this is then added to a hydroxyapatite column as described above.

Elution. After the addition of enzyme solution, the column is first washed with 200 ml of 50 m*M* potassium phosphate buffer, pH 6.8, and then a linear gradient of sodium chloride in phosphate buffer is applied (see above); 300 ml is used in each vessel. All eluates are checked for activity as described under assay procedures with the 3 substrates indicated. Three major peaks should be obtained (see Fig. 1). Peak I contains heparinase; peak II, heparitinase; and peak III, chondroitinases. As can be seen, considerable overlap occurs between heparinase and heparitinase activities. Therefore, further fractionation is necessary.

Heparinase Purification. Fractions in peak I indicated by crosshatching are combined, dialyzed overnight and adjusted to 10 m*M* potassium phosphate, pH 6.8, by using an 1.0 *M* solution of this buffer. This solution is then applied to a cellulose phosphate column (7 cm × 1.1 cm). The column is first eluted with 100 ml of 50 m*M* potassium phosphate, pH 6.8, followed by 100-ml fractions of each sodium chloride concentration mentioned above. Eluates are checked using heparin as substrate (Fig. 2). Fractions indicated by the cross-hatched area are

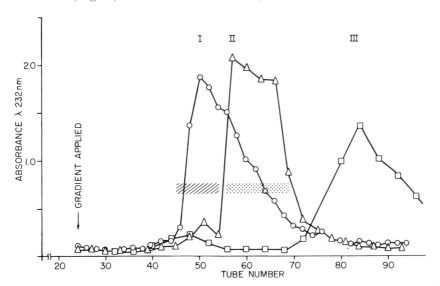

Fig. 1. Hydroxyapatite column (6.0 × 2.1 cm). Conditions for elution are described in the text. Heparinase activity was assayed for 2 hours at 30° using heparin as substrate, ○——○. Heparitinase activity was assayed for 1 hour at 43° using heparitin sulfate as substrate, △——△. Chondroitinase activity was assayed for 2 hours at 30° using chondroitin 6-sulfate as substrate, □——□.

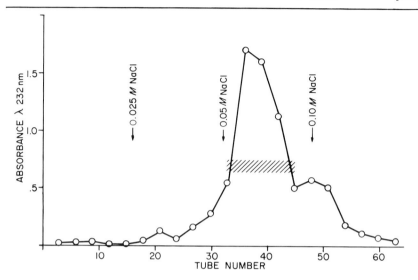

FIG. 2. Cellulose phosphate column (7.0 × 1.1 cm). Conditions for elution are described in the text. All the salt solutions indicated also contain 50 mM potassium phosphate buffer pH 6.8. Heparinase activity was assayed for 2 hours at 30° with heparin as substrate.

combined, dialyzed overnight, and lyophilized. Final yield of heparinase is usually about 1 mg of protein containing 15,500 units. (Owing to the presence of salt, the total amount of material is usually 10 mg; extensive dialysis may lead to some inactivation of enzyme.) This preparation contains no heparitinase. (It should be remembered that heparinase will have some activity on heparitin sulfate due to the heparinlike portion present in this polysaccharide.[4] However, when assayed at 43°, the temperature optimum of heparitinase, no activity can be detected.) Glycuronidase[11] chondroitinases and sulfatase (as measured on glucosamine disulfate)[11] are absent.

Heparitinase Purification. Fractions in peak II (Fig. 1) indicated by stippling are combined, dialyzed, and adjusted to 10 mM potassium phosphate as above. This solution is then applied to a cellulose phosphate column (4.5 cm × 1.1 cm). Elution is carried out as above for the heparinase. Results are shown in Fig. 3. The fractions in peak II indicated by stippling are combined, dialyzed against two changes of distilled water and lyophilized. A yield of about 0.2 mg of protein containing 4600 units (i.e., specific activity of 23,000, assayed at 43°) can be expected. Due

[11] C. T. Warnick and A. Linker, *Biochemistry* 11, 568 (1972).

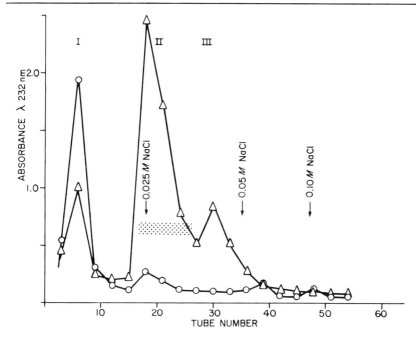

FIG. 3. Cellulose phosphate column (4.5 × 1.1 cm). Conditions for elution are described in the text. All salt solutions indicated also contain 50 mM potassium phosphate buffer, pH 6.8. Heparinase activity was assayed for 2 hours at 30° with heparin as substrate, ○——○. Heparitinase activity was assayed for 2 hours at 43° with heparitin sulfate as substrate, △——△.

to the presence of salt the total amount of material is usually 2–4 mg. This heparitinase preparation is free of heparinase, chondroitinase, glycuronidase[11] or sulfatase (as measured on glucosamine disulfate).

The purification steps are summarized in the table. The heparinase prepared here has a ten times higher specific activity than that reported previously,[4] and the heparitinase, three times higher.

Chondroitinases are obtained as a side product of this procedure (see Fig. 1).

Comments. It should be emphasized that column chromatography is not a completely reproducible procedure. This is particularly true for the hydroxyapatite columns. Therefore, the purification of each batch has to be followed carefully. It may be necessary at times to reapply fractions to another hydroxyapatite column if poor separation was achieved the first time.

It is also very important to use the dimensions of the cellulose phosphate columns suggested here, as excessive amounts of this support may inactivate the heparitinase.

CHROMATOGRAPHIC PURIFICATION OF THE ENZYMES

Fraction	Total protein[a] (mg)	Heparin as substrate		Heparitin sulfate as substrate	
		Total activity 30°	Specific activity	Total activity 43°	Specific activity
Crude enzyme	425.0				
Protamine sulfate, supernatant fluid	346.0				
Heparinase					
Hydroxyapatite column, peak I	10.3	17500	1700	0	
Cellulose phosphate column	0.7	15400	22100	0	
Heparitinase					
Hydroxyapatite column, peak II	12.1	16300	1350	23810	1970
Cellulose phosphate column	0.2	120	600	4600	23100

[a] O. H. Lowry, N. J. Rosebrough, A. L. Farr, and R. J. Randall, *J. Biol. Chem.* **193**, 265 (1951).

Properties

Specificity. The heparinase acts on heparin and, to a minor degree, on heparitin sulfate. As heparitin sulfate appears to contain some heparin-like segments,[12,13] this is not too surprising. No activity on heparitin sulfate is obtained at 43°, the temperature optimum of heparitinase, indicating that the activity at 30° is due to heparinase itself, not to contaminating heparitinase. Other mucopolysaccharides are not degraded by the heparinase, while modified heparins are poor substrates.[4] Sulfatase acting on glucosamine disulfate is absent in this preparation. However, the possible presence of a sulfatase acting on iduronic acid sulfate, a postulated component of heparin,[14] cannot be excluded.

The heparitinase acts on heparitin sulfate only, but a very small contamination of heparinase may be present in this preparation. As above, the known sulfatase is absent.

pH Optimum. Both enzymes show quite broad optima at pH 7.0 depending somewhat on the buffer used.[4]

Effect of Temperature. The heparinase has an optimum at 30°, and

[12] A. Linker and P. Hovingh, *Biochim. Biophys. Acta* **165**, 89 (1968).
[13] J. A. Cifonelli, *Carbohyd. Res.* **8**, 233 (1968).
[14] U. Lindahl and O. Axelsson. *J. Biol. Chem.* **246**, 74 (1971).

essentially no activity above 40°. The heparitinase has a fairly sharp optimum at 43°.[4] This difference in optima is very useful for assay purposes to test contamination of one enzyme by the other.

Inhibitors and Activators. Heparinase is inhibited by Hg^{2+}, Cu^{2+}, and Fe^{3+}. The heparitinase is inhibited by Cu^{2+}, Hg^{2+}, Zn^{2+}, and Cd^{2+}.

Ca^{2+} appears to activate the enzyme, or at least much more reproducible kinetics are obtained in its presence, particularly when heparin preparations from various sources are used as substrates. Calcium ions, therefore, should always be included in assays.

General. Enzyme solutions or lyophilized preparations are quite stable when kept at $-20°$.

[124] Chondroitinases from *Proteus vulgaris* and *Flavobacterium heparinum*[1]

By SAKARU SUZUKI

Reactions Catalyzed by Chondroitinase-AC[2]

$$\xrightarrow{\beta 1,4} (\text{GlcUA} \xrightarrow{\beta 1,3} \text{GalNAc})_n \rightarrow n\Delta\text{Di-OS} \tag{1}$$

Chondroitin

$$\xrightarrow{\beta 1,4} (\text{GlcUA} \xrightarrow{\beta 1,3} \underset{\underset{4S}{|}}{\text{GalNAc}})_n \rightarrow n\Delta\text{Di-4S} \tag{2}$$

Chondroitin 4-sulfate

$$\xrightarrow{\beta 1,4} (\text{GlcUA} \xrightarrow{\beta 1,3} \underset{\underset{6S}{|}}{\text{GalNAc}})_n \rightarrow n\Delta\text{Di-6S} \tag{3}$$

Chondroitin 6-sulfate

[1] This article is based on work supported by grants from the Ministry of Education, Japan, the Toray Science Foundation, and the Takeda Science Foundation.

[2] The enzymes catalyze several other reactions which are not shown here (see section on specificities). The abbreviations: GlcUA, D-glucuronic acid; IdUA, L-iduronic acid; GalNAc, GalNAc-4S, and GalNAc-6S *N*-acetylgalactosamine and its 4-sulfate and 6-sulfate, respectively; ΔDi-OS, 3-*O*-(β-D-Δ4,5-glucuronopyranosyl)-*N*-acetylgalactosamine; ΔDi-4S and ΔDi-6S, 3-*O*-(β-D-Δ4,5-glucuronopyranosyl)-*N*-acetylgalactosamine 4- and 6-sulfate, respectively.

Reactions Catalyzed by Chondroitinase-ABC.[2] Reactions (1), (2), (3), and

$$\xrightarrow{\beta1,4} (\text{IdUA} \xrightarrow{\alpha1,3} \text{GalNAc})_n \rightarrow n\Delta\text{Di-4S}$$
$$\underset{\text{4S}}{|} \tag{4}$$

Dermatan sulfate

where n indicates the number of repeating units.

The term "chondroitinase" was introduced by Dodgson and Lloyd[3] to describe an enzyme from *Proteus vulgaris* NCTC 4636 that converted chondroitin sulfate to sulfated disaccharide. The disaccharide product was originally characterized by these same authors as a saturated sulfated disaccharide, i.e., *N*-acetylchondrosine sulfate. However, Linker *et al.*,[4] Suzuki,[5] and Nakada and Wolfe[6] have pointed out that the strain of *P. vulgaris* used in their own work yields Δ4,5-unsaturated disaccharides. A somewhat similar enzyme also has been found in *Flavobacterium heparinum* ATCC 13125.[4] More recently, Yamagata *et al.*[7] have purified the preparations from *P. vulgaris* and *F. heparinum*. The resulting purification indicated that extracts from *F. heparinum* contain two distinct chondroitinases showing different substrate specificities while extracts from *P. vulgaris* contain only one chondroitinase whose specificity is similar to one of the *F. heparium* chondroitinases. The method of assay and the method of purification of the two distinct chondroitinases (termed chondroitinase-ABC and chondroitinase-AC) are presented below.

Assay Method

Principle. The assay depends upon measurement of formation of ethanol-soluble products from ^{14}C-labeled chondroitin 4- and (or) 6-sulfates. The methods for preparing ^{35}S-labeled chondroitin sulfates have been described in detail elsewhere.[7]

Reagents

Tris·HCl, pH 8.0, 0.5 M
Sodium acetate, 0.6 M
Bovine serum albumin, 0.1% (w/v)

[3] K. S. Dodgson and A. G. Lloyd, *Biochem. J.* **68**, 88 (1958).
[4] A. Linker, P. Hoffman, K. Meyer, P. Sampson, and E. D. Korn, *J. Biol. Chem.* **235**, 3061 (1960).
[5] S. Suzuki, *J. Biol. Chem.* **235**, 3580 (1960).
[6] H. I. Nakada and J. B. Wolfe, *Arch. Biochem. Biophys.* **94**, 244 (1961).
[7] T. Yamagata, H. Saito, O. Habuchi, and S. Suzuki, *J. Biol. Chem.* **243**, 1523 (1968).

A mixture of [^{14}C]chondroitin 4- and 6-sulfate with specific activity of about 10^6 cpm/μmole of glucuronic acid, 2 mM
Enzyme

Procedure. To microtubes containing 5 μl each of Tris buffer, sodium acetate, bovine serum albumin, and [^{14}C]chondroitin sulfate is added 0.0001–0.0005 unit of enzyme and sufficient water to give a final volume of 50 μl. Controls contain heat-inactivated enzyme. After incubation for 10 minutes at 37°, the reaction mixtures are heated at 100° for 1 minute. Unreacted [^{14}C]chondroitin sulfates are precipitated by sequential addition of 50 μl of a carrier solution containing 3.5 mg each of chondroitin 4- and 6-sulfate per milliliter, and 0.2 ml of ethanol containing 1% potassium acetate. The mixtures are allowed to stand at 0° for 30 minutes, and the precipitates are removed by centrifugation. Aliquots of the supernatant fluids are applied on disks (2.4 cm in diameter) of filter paper, which are then dried in an oven at 60°. The radioactivities of the disks are determined in a liquid scintillation spectrometer.

Definition of Units and Specific Activity. The reaction rate is proportional to the amount of enzyme added under the conditions described above. One unit is defined as the amount of enzyme that catalyzes the release of 1 μmole (as unsaturated disaccharide) of soluble products per minute. The specific activity is expressed as units per milligram of protein. Protein is determined by the method of Lowry *et al.*[8]

Alternative Assay Method. By the use as substrate of commercially available chondroitin 6-sulfate (unlabeled), the chondroitinase reaction can be measured by the liberation of $\triangle$Di-6S which is active in the Morgan–Elson reaction or has a marked absorption in the ultraviolet region.[7]

Purification Procedure[9]

Chondroitinase-ABC

Preparation of Crude Extract. *Proteus vulgaris,* strain NCTC 4636, is grown at 30° with aeration in a medium containing 15 g of peptone, 4.5 g of meat extract, 1.5 g of NaCl, and 1 g of crude chondroitin 6-sulfate (Seikagaku Kogyo Co., Tokyo) per liter. The cells are harvested by centrifugation at 3° at the end of the logarithmic phase (usually between 8 and 10 hours), and are washed with cold water. Approximately 350 g of cells, wet weight, are obtained from a 50-liter cultivation.

---

[8] O. H. Lowry, N. J. Rosebrough, A. L. Farr, and R. J. Randall, Vol. 3 [73].
[9] Specific and purified chondroitinases (types ABC and AC) and the three reference disaccharides ($\triangle$Di-OS, $\triangle$Di-4S, and $\triangle$Di-6S) are available from Miles Laboratories, Inc., Elkhart, Indiana 46514, or from Seikagaku Kogyo Company, Nihonbashi-Honcho 2-9, Tokyo 103, Japan.

All operations described below are conducted between 0° and 4°. All centrifugations are at 17,500 g for 30 minutes.

The washed cells, 1 volume, are suspended in 3 volumes of 20 mM Tris·HCl, pH 7.2, and disrupted with the use of the French press. The resulting suspension is centrifuged, and the supernatant fluid is collected.

Streptomycin Treatment. To 500 ml of the crude extract is added, with stirring, 125 ml of 5% streptomycin sulfate in 20 mM Tris·HCl, pH 7.2. After standing for 1 hour, the suspension is centrifuged. The supernatant is allowed to stand for 12 hours, and the resulting precipitate is removed by centrifugation.

Ammonium Sulfate Fractionation. To the streptomycin-treated supernatant (573 ml) is added, with stirring, 140 g of solid ammonium sulfate. The suspension is allowed to stand for 1 hour, and then it is centrifuged. To the supernatant fluid is added, with stirring, 84 g of ammonium sulfate. The suspension is allowed to stand and is centrifuged as before. The resulting precipitate is dissolved in 50 ml of 20 mM Tris·HCl, pH 7.2, and dialyzed for 24 hours against five 2-liter changes of the same buffer.

DEAE-Cellulose Chromatography. A column (4.5 × 40 cm) of DEAE-cellulose (Brown Company) is prepared and equilibrated with 6 liters of 20 mM Tris·HCl, pH 7.2. The dialyzed ammonium sulfate fraction (65 ml) is applied to the column at the rate of 60 ml per hour. The adsorbent is then washed with 1 liter of 0.02 M Tris·HCl, pH 7.2. The bulk of the activity applied to the column is recovered in the wash.

Phosphocellulose Chromatography. A column (2.1 × 15 cm) of phosphocellulose (Serva) is equilibrated with 0.5 liter of 20 mM Tris·HCl, pH 7.2. The DEAE-cellulose wash, about 500 ml, is applied to the column at the rate of 30 ml per hour, and the adsorbent is washed with 100 ml of 0.1 M NaCl in 20 mM Tris·HCl, pH 7.2. A linear gradient of elution is applied with 0.1 M and 0.6 M NaCl in the same buffer as limiting concentrations. The total volume of the gradient is 600 ml. The flow rate is 30 ml per hour, and 5-ml fractions are collected. The enzyme appears as a single peak in the effluent after 100 ml of buffer have passed through the column. The fractions containing the major portion of this peak are pooled, and concentrated to 6 ml by pressure dialysis against 3 liters of 20 mM Tris·HCl, pH 7.2.

This procedure yields a chondroitinase-ABC preparation purified about 134-fold over the crude extract (see the table).

Chondroitinase-AC

Preparation of Crude Extract. Flavobacterium heparinum, strain ATCC 13125, is grown at 30° with aeration in a medium containing 7 g

PURIFICATION OF CHONDROITINASE-ABC FROM *Proteus vulgaris* AND CHONDROITINASE-AC FROM *Flavobacterium heparinum*

	Chondroitinase-ABC		Chondroitinase-AC	
Fraction	Total units	Specific activity	Total units	Specific activity
Crude extract	6000	0.28	1440	0.15
Streptomycin	6720	0.45	970	0.40
Ammonium sulfate	3800	0.80	1220	0.75
DEAE-cellulose	2500	11.0	—	—
Phosphocellulose	990	37.5	944	139
Sephadex G-200	—	—	170	43

of Trypticase, 6 g of Phytone (both from Baltimore Biological Laboratory), 1 g of glucose, 3 g of NaCl, 1 g of K_2HPO_4, and 1 g of crude chondroitin 6-sulfate per liter. The medium is adjusted to pH 7 with HCl. The cells are harvested by centrifugation at 3° at the end of the logarithmic phase (usually between 10 and 12 hours), and are washed with cold water. Approximately 410 g (wet weight) of cells are obtained from a 50-liter cultivation.

All operations described below are conducted between 0° and 4°. All centrifugations are at 17,500 g for 30 minutes.

The cells are suspended in 3 volumes of cold 0.02 M Tris·HCl, pH 7.2, and are treated in a 10-kc sonic oscillator for 5 minutes. The resulting suspension is centrifuged, and the supernatant fluid is collected.

Streptomycin Treatment. To 500 ml of the crude extract is added, with stirring, 125 ml of 5% streptomycin sulfate in 0.02 M Tris·HCl, pH 7.2. After standing for 4 hours, the resulting precipitate is removed by centrifugation.

Ammonium Sulfate Fractionation. To the streptomycin-treated supernatant (580 ml) is added, with stirring, 181 g of solid ammonium sulfate. The suspension is allowed to stand for 1 hour, and then it is centrifuged. To the supernatant fluid is added, with stirring, 145 g of ammonium sulfate. The suspension is allowed to stand and is centrifuged as before, The resulting precipitate is dissolved in 50 ml of 20 mM Tris·HCl, pH 7.2, and dialyzed for 24 hours against five 2-liter changes of the same buffer.

Phosphocellulose Chromatography. A column (2.5 × 30 cm) of phosphocellulose is equilibrated with 2 liters of 20 mM Tris·HCl, pH 7.2. The dialyzed ammonium sulfate fraction is applied to the column at the rate of 60 ml per hour, and the adsorbent is washed with 400 ml of 20 mM Tris·HCl, pH 7.2. The column is developed by linear gradient elution with 400 ml of 20 mM Tris·HCl, pH 7.2, in the mixing flask and 400 ml of 0.5 M NaCl in the same buffer in the reservoir. The flow rate is 60 ml per hour, and fractions of 8 ml are collected. The chondroitinase activities are separated into two components; the early peak (an enzyme with the specificity of chondroitinase-ABC) appears in the fractions from tubes 30–50 and the late peak (chondroitinase-AC) from tubes 55–80. The late peak fractions are pooled and concentrated to 2 ml by pressure dialysis against 2 liters of 0.02 M Tris·HCl, pH 7.2.

Sephadex G-200 Chromatography. A 1-ml portion of the phosphocellulose fraction is applied to the top of a Sephadex G-200 column (1.8 × 63 cm) equilibrated with 0.1 M NaCl in 20 mM Tris·HCl, pH 7.2. Elution with the same salt solution is carried out at the rate of 10 ml per hour with collection of 2-ml fractions. The enzyme appears as a single peak in the fractions from tubes 38–58. The tailing portion (tubes 48–58) of this peak is contaminated with a glucuronidase which hydrolyzes the β-glucuronidic bond of nonsulfated and 6-sulfated unsaturated disaccharide (for the assay of this glucuronidase, see Yamagata *et al.*[7]). The glucuronidase-free fractions (tubes 38–47) are pooled and concentrated to 2.5 ml by pressure dialysis against 2 liters of 200 mM Tris·HCl, pH 7.2.

This procedure yields a chondroitinase-AC preparation purified about 287-fold over the crude extract (see the table).

Properties

Physical Characteristics. The purified chondroitinase-ABC and chondroitinase-AC contain no protein impurity, as judged by the electrophoresis on acrylamide gel.

The purified chondroitinase-ABC and chondroitinase-AC are stable for at least 2 months when stored in an ice bath. Repeated freezing and thawing results in destruction of the activities.

Specificities. Chondroitinase-ABC acts with almost equal facility on chondroitin 4- and 6-sulfate as well as dermatan sulfate, and at a reduced rate on hyaluronic acid. The specificity of the enzyme does not involve recognition of a particular size or a particular sulfate content of substrate molecule. Thus, the pretreatment of chondroitin 4- and 6-sulfates with testicular hyaluronidase has no significant effect on the rate of chondroitinase reaction. Chondroitin sulfate "D" from shark cartilage and chondroitin sulfate "E" from squid cartilage, which contain 2(or 3)-O-sulfated glucuronic acid residue and 4,6-di-O-sulfated hexosamine residue, respectively, are also degraded to give rise to the corresponding disulfated disaccharides.[10] Likewise, dermatan sulfate "H"

[10] S. Suzuki, H. Saito, T. Yamagata, K. Anno. N. Seno, Y. Kawai, and T. Furuhashi, *J. Biol. Chem.* 243, 1543 (1968).

from hag fish notochord consisting mainly of alternating α-1,3-L-iduronic acid and β-1,4-N-acetylgalactosamine-4,6-disulfate units is attacked by chondroitinase-ABC.[11]

Chondroitinase-AC differs from chondroitinase-ABC in that it does not attack dermatan sulfate with alternating α-1,3-L-iduronic acid and β-1,4-N-acetylgalactosamine-4-sulfate units. The D-glucuronic acid-containing sections of hybrid dermatan sulfates serve as substrates for chondroitinase-AC. Subsequent characterization of the fragments produced by chondroitinase-AC digestion serves to examine the hybrid structure of a given dermatan sulfate sample.[12]

Since the chondroitinases attack the bond between O and the C-4 of uronic acid in an elimination reaction, either saturated monosaccharide (if the terminal group is N-acetylgalactosamine or its sulfate ester) or saturated disaccharide (if the terminal group is uronic acid) is released from the nonreducing end of substrate after exhaustive digestion. Polysaccharides with different chain lengths, therefore, vary in their ratios of the saturated products to the unsaturated products. Such ratio gives a measure of the degree of polymerization.

[11] K. Anno, N. Seno, M. B. Mathews, T. Yamagata, and S. Suzuki, *Biochim. Biophys. Acta* **237**, 173 (1971).

[12] L.-Å. Fransson and B. Havsmark, *J. Biol. Chem.* **245**, 4770 (1970).

[125] Chondrosulfatases from *Proteus vulgaris*[1]

By SAKARU SUZUKI

Chondro-4-sulfatase[2]

$$\Delta\text{Di-4S} \rightarrow \Delta\text{Di-OS} + \text{SO}_4{}^{2-}$$

N-Acetylchondrosine 4-sulfate → N-acetylchondrosine + $\text{SO}_4{}^{2-}$

Chondro-6-sulfatase[2]

$$\Delta\text{Di-6S} \rightarrow \Delta\text{Di-OS} + \text{SO}_4{}^{2-}$$

N-Acetylchondrosine 6-sulfate → N-acetylchondrosine + $\text{SO}_4{}^{2-}$

The name "chondrosulfatase" was originally given to the enzyme

[1] This article is based on work supported by grants from the Ministry of Education, Japan, the Toray Science Foundation, and the Takeda Science Foundation.

[2] The enzymes catalyze several other reactions which are not shown here (see section on specificities). The abbreviations: ΔDi-OS, 3-O-(β-D-Δ4,5-glucuronopyranosyl)-N-acetylgalactosamine; ΔDi-4S and ΔDi-6S, 3-O-(β-D-Δ4,5-glucuronopyranosyl)-N-acetylgalactosamine 4- and 6-sulfate, respectively.

[3] C. Neuberg and E. Hofmann, *Biochem. Z.* **234**, 345 (1931).

which liberates SO_4^{2-} from a family of chondroitin sulfates.[3] However, Dodgson and Lloyd[4] made the important observation that the true substrate of the chondrosulfatase of *Proteus vulgaris* (NCTC 4636) is not intact chondroitin sulfate but the oligosaccharides produced therefrom through the action of either testicular hyaluronidase or the chondroitinase which is present in the same microorganism.[5] More recently, Yamagata *et al.*[6] have demonstrated the separation of two sulfatases (termed chondro-4-sulfatase and chondro-6-sulfatase) from extracts of *P. vulgaris*. In this work, it was found that the 4- and 6-sulfatase catalyze the conversion of ΔDi-4S (or *N*-acetylchondrosine 4-sulfate) and ΔDi-6S (or *N*-acetylchondrosine 6-sulfate), respectively, to ΔDi-OS (or *N*-acetylchondrosine) and inorganic sulfate. The methods of assay and purification and the properties of the bacterial enzymes are presented below. Also added is a brief note on the occurrence in animals of distinct chondrosulfatases which hydrolyze intact chondroitin sulfate or relatively high molecular weight oligosaccharides in the hyaluronidase digest.

Assay Method

Principle. The assay measures the conversion of ^{35}S-labeled ΔDi-4S or ΔDi-6S to $^{35}SO_4^{2-}$ by electrophoretic separation. The methods for preparing the ^{35}S-labeled disaccharides have been described in detail elsewhere.[6]

Reagents

Tris·acetate, pH 7.5, 0.5 M
Bovine serum albumin, 0.1% (w/v)
ΔDi-4^{35}S or ΔDi-6^{35}S containing about 5×10^5 cpm/μmole, 2 mM
Enzyme
$BaCl_2$, 0.02% (w/v) in 75% aqueous methanol
Potassium rhodizonate,[7] 12 mg in 15 ml of water to which are added 10 ml of concentrated ammonium hydroxide and 25 ml of ethanol

Procedure. To microtubes containing 5 μl each of Tris·acetate, bovine serum albumin and ΔDi-4^{35}S or ΔDi-6^{35}S is added 0.0001–0.0005 unit of enzyme and sufficient water to give a final volume of 50 μl. This is incubated for 10 minutes at 37°. Control tubes containing heat-

[4] K. S. Dodgson and A. G. Lloyd, *Biochem. J.* **66**, 532 (1957).
[5] S. Suzuki, this volume [124].
[6] T. Yamagata, H. Saito, O. Habuchi, and S. Suzuki, *J. Biol. Chem.* **243**, 1523 (1968).
[7] J. J. Schneider and M. L. Lewbart, *J. Biol. Chem.* **222**, 787 (1956).

inactivated enzyme are also incubated in each series of assays. The reaction is terminated by placing the tubes in a boiling water bath for 1 minute and centrifuged. A 10-μl portion of the supernatant is applied to Whatman No. 1 filter paper (60 cm long) together with 0.03 μmole of unlabeled Na_2SO_4 as an internal marker. Electrophoresis,[8] is carried out in 50 mM ammonium acetate, pH 5.0, at 30 V/cm for 20 minutes. After drying, the paper is sprayed with the $BaCl_2$ reagent, then dried in an air stream for 10 minutes, and finally sprayed with the freshly prepared rhodizonate reagent. The area corresponding to SO_4^{2-} appearing as a light yellow spot against a pink background is cut out, and counted in a liquid scintillation spectrometer.

Definition of Unit and Specific Activity. The reaction rate is proportional to the amount of enzyme added under the conditions described above. One unit is defined as the amount of enzyme that catalyzes the release of 1 μmole of SO_4^{2-} from either ΔDi-4S or ΔDi-6S in 1 minute. The specific activity is expressed as units per milligram of protein. Protein is determined by the method of Lowry *et al.*[9]

Purification Procedure[10]

Preparation of Crude Extract, Streptomycin Treatment, and Ammonium Sulfate Fractionation. With 500 ml of the crude extract of *Proteus vulgaris* (NCTC 4636), a preliminary concentration of the sulfatases as far as the stage for absorption onto DEAE-cellulose (see Chondroitinases[5]) is done.

DEAE-Cellulose Chromatography. The DEAE-cellulose column (4.5 × 40 cm) is washed with 1 liter of 20 mM Tris·HCl, pH 7.2, then the adsorbent is eluted first with 3 liters of 70 mM NaCl in 20 mM Tris·HCl, pH 7.2, and then with 3 liters of 0.15 M NaCl in the same buffer. The flow rate is 60 ml per hour, and 20-ml fractions are collected. Most of the activity for ΔDi-4³⁵S appears in the 70 mM NaCl part, and that for ΔDi-6³⁵S in the 0.15 M NaCl part. The fractions (tubes 95–170) containing the major portion of the 4-sulfatase activity and free from the 6-sulfatase activity are pooled and concentrated to 10 ml by pressure dialysis against 10 liters of 20 mM Tris·HCl, pH 7.2 (chondro-4-sulfatase).

The peak for the 6-sulfatase (tubes 230–270) is usually overlapped with the tailing portion of the 4-sulfatase peak. To remove the 4-

[8] R. Markham and J. D. Smith, *Biochem. J.* **52**, 552 (1952).
[9] O. H. Lowry, N. J. Rosebrough, A. L. Farr, and R. J. Randall, Vol. 3 [73].
[10] Specific and purified chondrosulfatases are available from Miles Laboratories, Inc., Elkhart, Indiana 46514, or from Seikagaku Kogyo Company, Nihonbashi-Honcho 2-9, Tokyo 103, Japan.

sulfatase, the 6-sulfatase fractions are pooled, concentrated as above, and subjected to a second chromatography on a column (2.1 × 24 cm) of DEAE-cellulose equilibrated with 0.02 M Tris·HCl, pH 7.2. The column is washed with 650 ml of 70 mM NaCl in the same buffer and then eluted with 1200 ml of 0.15 M NaCl in the same buffer. The fractions (tubes 100–150) containing the major portion of the 6-sulfatase activity and free from the 4-sulfatase activity are pooled and concentrated to 10 ml by pressure dialysis against 10 liters of 20 mM Tris·HCl, pH 7.2 (chondro-6-sulfatase).

A summary of the purification procedure is shown in the table.

PURIFICATION OF CHONDROSULFATASES

Fraction	Chondro-4-sulfatase		Chondro-6-sulfatase	
	Total units	Specific activity	Total units	Specific activity
Crude extract	4100	0.19	1700	0.08
Streptomycin	4460	0.29	1660	0.11
Ammonium sulfate	1640	0.34	630	0.13
DEAE-cellulose				
Early peak	650	4.75	0	0
Late peak	24	0.03	386	0.5
Second DEAE-cellulose				
Late peak	0	0	220	1.7

Properties

Specificity. Chondro-4-sulfatase catalyzes the conversion of ΔDi-4S and *N*-acetylchondrosine 4-sulfate to the corresponding nonsulfated disaccharides and SO_4^{2-}, but does not attack ΔDi-6S and *N*-acetyl-chondrosine 6-sulfate. In contrast, chondro-6-sulfatase carries out the desulfation of the disaccharide 6-sulfates but it does not attack the disaccharide 4-sulfate isomers. No $^{35}SO_4^{2-}$ is released from ^{35}S-labeled chondroitin 4- and 6-sulfate, from *N*-acetylgalactosamine 4- and 6-[^{35}S]-sulfate, and from hexa-, penta-, tetra-, and trisaccharides prepared from ^{35}S-labeled chondroitin 4- and 6-sulfates, when they are incubated with a 1:1 mixture of chondro-4-sulfatase and chondro-6-sulfatase under the condition described above. Chondro-6-sulfatase catalyzes the selective hydrolysis of the sulfate ester at position 6 of *N*-acetylgalactosamine 4,6-disulfate, 3-*O*-(β-D-Δ4,5-glucuronopyranosyl)-*N*-acetylgalactosamine 4,6-disulfate, and 3-*O*-(2- or 3-*O*-sulfo-β-D-Δ4,5-glucuronopyranosyl)-*N*-acetylgalactosamine 6-sulfate. Chondro-4-sulfatase, on the other hand, does not attack the monosaccharide disulfate but hydrolyzes the sulfate at position 4 of the disaccharide 4,6-disulfate as well as the sulfate at position 4 of 3-O-(2- or 3-O-sulfo-β-D-Δ4,5-glucuronopyranosyl)-N-acetylgalactosamine 4-sulfate.

Stability. Both chondro-4-sulfatase and chondro-6-sulfatase can be stored in an ice bath for at least 2 months without significant loss of activity.

Comments on Chondrosulfatases from Animal Sources

It has long been known that animals can desulfate chondroitin sulfates *in vivo,* but until recently all attempts to prepare from animal tissues a chondrosulfatase of defined specificity have been unsuccessful. Quite recently, several investigators have reported the purification from animal tissues of sulfatases capable of attacking intact chondroitin sulfates or partially depolymerized chondroitin sulfates. Held and Buddecke[11] have purified a sulfatase from ox aorta. This enzyme can liberate $SO_4{}^{2-}$ from intact chondroitin 4-sulfate but not from intact chondroitin 6-sulfate. A sulfatase preparation of similar specificity has been obtained from squid liver.[12] With the squid enzyme, it has been shown that prior depolymerization of chondroitin 4-sulfate with testicular hyaluronidase results in a marked decrease in the rate of sulfatase reaction. A sulfatase of somewhat different specificity has been found in the lysosome of rat liver.[13] Specificity studies indicated that the enzyme is active only toward relatively high molecular weight oligosaccharides. Intact chondroitin 4- and 6-sulfates and tetra- and hexasaccharides are not attacked by this enzyme.

[11] E. Held and E. Buddecke, *Hoppe-Seyler's Z. Physiol. Chem.* 348, 1047 (1967).

[12] T. Yamagata, N. Kato, A. Hagishita, T. Naiki, and S. Suzuki, unpublished experiments, 1971.

[13] N. Tudball and E. A. Davidson, *Biochim. Biophys. Acta* 171, 113 (1969).

[126] Phosphomannanase from *Bacillus circulans*

By WILLIAM L. McLELLAN

Phosphomannanase is an extracellular enzyme of *Bacillus circulans* that cleaves mannan from yeast cell walls or intact yeast,[1] and depolymerizes phosphomannans produced by the *Hansenula* species.[2,3] The

[1] W. L. McLellan, L. E. McDaniel, and J. O. Lampen, *J. Bacteriol.* 102, 261 (1970).

[2] M. E. Slodki, *Biochim. Biophys. Acta* 57, 525 (1962).

[3] M. E. Slodki, *Biochim. Biophys. Acta* 69, 96 (1963).

enzyme specifically cleaves a glycosyl bond adjacent to a phosphodiester-linked mannan.[4] The product of cleavage of the cell wall of yeast is a mannan terminating $(man)_n$-man^6-P-^{1}man. An intact diester bond is required for the action of the enzyme. Other possible substrates for the enzyme have not been investigated.

Assay Method

Principle. A suspension of yeast cell walls is incubated with enzyme, the walls are removed by centrifugation, and the released mannan is determined colorimetrically.

Preparation of Yeast Cell Walls.[5] Yeast cells are grown in Wicker-ham medium[6] and harvested in the logarithmic phase of growth. They are washed with distilled water, and 10 g (wet wt) of cells are suspended in 20 ml of ice cold buffer (27.5 mM, KH_2PO_4, 115 mM Na_2HPO_4, 0.55 M mannitol, 1 g of sodium dodecyl sulfate per liter); prewashed glass beads (0.25 mm, Will Corp., Rochester, New York) are added (40 g), and the cells are disrupted by shaking for 3 minutes in a Braun cell homoge-nizer, maintaining the temperature below 6° with a stream of liquid CO_2. The glass beads are removed by centrifugation for 5 minutes at 500 g, washed with buffer, and recentrifuged to remove cellular material. The suspension is then centrifuged for 10 minutes at 1000 g, and the precipi-tate is resuspended in buffer by shaking for 2 minutes on a Vortex mixer; the procedure is repeated twice, the supernatant being discarded. The particulate material is suspended in 40 ml of buffer, and equal volume of ether is added; the suspension is mixed for 5 minutes with a Vortex mixer and then centrifuged for 10 minutes at 2000 g. The cell walls sedi-ment, and the debris and whole cells collect at the interface. The ether extraction is repeated 3 times. The walls are then washed three times by suspension in and centrifugation from 0.9% NaCl and finally washed with distilled water three times. They are stored in suspension at 4° for as long as a week.

Procedure. The reaction consists of 10 μmoles of tris(hydroxymethyl)-aminomethane–succinate buffer (pH 6.0), suspended yeast walls (2 mg of carbohydrate), and enzyme (0.005–0.01 unit) in a total volume of 0.25 ml. The mixture is incubated at 30° for 10 minutes, chilled in ice to stop the reaction, and centrifuged at 2° for 5 minutes at 10,000 g. An aliquot of the supernatant is assayed for carbohydrate by the phenol–

[4] W. L. McLellan and J. O. Lampen, *J. Bacteriol.* **95**, 967 (1968).

[5] P. J. Mill, *J. Gen. Microbiol.* **44**, 329 (1966).

[6] L. J. Wickerham, *U.S. Dep. Agr. Tech. Bull.* **1029**, (1951).

sulfuric acid method of Dubois *et al.*[7] A blank is incubated without enzyme. The assay is not linear with time, but under the conditions used carbohydrate released is proportional to enzyme concentration. Glucanases in the enzyme preparation do not interfere with the assay.[4]

Definition of Unit and Specific Activity. One unit of phosphomannanase is defined as the amount of enzyme that will catalyze the release of 1 μmole of carbohydrate per minute under the conditions of the assay. Mannose is the standard reference carbohydrate. Specific activity is expressed as units per milligram of protein. The method of protein determination is that of Lowry *et al.*[8]

Purification Procedure

Growth of Bacillus circulans.[4] An inoculum of *B. circulans* strain 63-7[9] is grown for 24 hours with shaking at 37° in a medium containing per liter: 0.145 g of yeast nitrogen base (Difco, without amino acids and ammonium sulfate), 1 g of $(NH_4)_2SO_4$, 5 g of enzymatically hydrolyzed casein (Calbiochem), 12.6 g of KH_2PO_4, and 20 g of glucose (autoclaved separately). A 5% inoculum is made in a fermentor containing the same medium except KH_2PO_4 is 1.26 g and glucose 8 g/liter. Polyglycol P-2000 (Dow Chemical Co.), 0.02 ml/liter, is added to prevent foaming, and the cells are grown with good mixing and aeration. The pH is maintained at 7 by metering in 5 N KOH with a pH control unit (New Brunswick Scientific Co.). Under these conditions of growth, glucose is exhausted at 14 hours and maximum enzyme levels appear at about 17 hours.

Concentration of Medium. The culture is centrifuged with a Sharples super centrifuge and the supernatant fraction concentrated with a Rodney-Hunt turba-film evaporator, maintaining a temperature of less than 35°. (Smaller batches can be concentrated with a Rinco flash evaporator.) The concentrated medium is dialyzed against cold running tap water for 15 hours and then reconcentrated. It is dialyzed until the salt concentration is less than 0.01 M (checked with a conductivity meter using phosphate buffer pH as a standard).

Protamine Sulfate and Ammonium Sulfate Treatment. The concen-

[7] M. K. Dubois, K. A. Gilles, J. K. Hamilton, P. A. Rebers, and F. Smith. *Anal. Chem.* **28**, 350 (1956).

[8] O. H. Lowry, N. J. Rosebrough, A. L. Farr, and R. J. Randall, *J. Biol. Chem.* **193**, 265 (1951).

[9] This strain was obtained from K. Arima, University of Tokyo, and does not require the presence of yeast cell walls in the medium for production of the enzyme. K. Horikoshi, H. Koffler, and K. Arima, *Biochim. Biophys. Acta* **73**, 268 (1963).

trated medium is adjusted to pH 6.0 and a 25% solution of protamine sulfate (Calbiochem) is added dropwise with stirring to give a
final concentration of 0.7%. The solution is left at 4° overnight and then
centrifuged to remove the inactive precipitate. This step removes much
nonprotein contamination as well as some inactive protein. The supernatant is brought to 47% saturation by the addition of solid ammonium
sulfate (30.4 g/100 ml). The resulting precipitate, which contains the
enzyme, is dissolved in 10 mM Tris-succinate buffer (pH 6) and dialyzed
against the same buffer for 24 hours. An enzymatically inactive precipitate which forms on dialysis is removed by centrifugation.

Gel Filtration on P-100. A column of P-100 (Biorad Laboratories)
25 × 85 cm is prepared and equilibrated with 0.1 M Tris–succinate pH 6.
An aliquot (20 ml) of the dialyzed ammonium sulfate fraction is applied,
and the column is developed with the same buffer. Fractions are collected and assayed for protein and phosphomannanase. After elution of
a large band of inactive protein, the enzymatic activity appears.

Ammonium Sulfate Fractionation. The peak fractions (60 ml) are
pooled, placed in a dialysis bag, and concentrated with Aquacide 1
(Calbiochem). The protein concentration is adjusted to 10 mg/ml and
solid ammonium sulfate is added to achieve 50% saturation (29.1 g/100
ml). The precipitate is removed by centrifugation and the supernatant
brought to 70% saturation ammonium sulfate (15.6 g/100 ml). The
precipitate containing the enzyme is collected by centrifugation, suspended in 10 mM Tris–succinate buffer (pH 6.0) and dialyzed overnight against the same buffer.

Ampholine Electrophoresis. If further purification of the enzyme
is desired, aliquot (16 mg protein) may be treated by density gradient
electrophoresis with ampholine. A gradient of 5–25% glycerol containing
2% ampholine pH 4–8 is prepared in a 110-ml column (LKB Instruments, Inc.). The sample is introduced, and electrophoresis is carried
out for 62 hours at 700 V. The fractions are collected and assayed. The
peak of activity is pooled, concentrated to 2 ml with Aquacide 1; ampholine is removed by passage through a P-10 column (1.2 × 40 cm).
The specific activity of the enzyme is double by this procedure.[4]

A typical purification is summarized in the table.

Properties

pH Optimum. The enzyme has a pH optimum of 6. It exhibits 80%
activity at pH 5 or 7.

Molecular Weight and Isoelectric Point. The molecular weight estimated by gel filtration is about 47,000. Ampholine electrofocusing indicates an isoelectric point of 6.8.

PURIFICATION OF PHOSPHOMANNANASE FROM *Bacillus circulans*

Fraction	Volume (ml)	Protein (mg/ml)	Enzyme activity (units/ml)	Specific activity (units/mg)
1. Medium	100,000	1.7	5.6	3.3
2. Concentrate	900	50	164	3.3
3. Protamine supernatant	1,325	13	53	4.1
4. 47% (NH$_4$)$_2$SO$_4$	40	42	1500	35.6
5. P-100 pool	120[a]	2.3	163	71
6. 50–70% (NH$_4$)$_2$SO$_4$	28[a]	5.0	610	122

[a] This value is the summation of two runs on P-100.

Stability. The purified enzyme is stable to freezing and thawing. It is quite stable at 30°. At 55° the half-life of activity is 15 minutes.

Glucanase Activity. As prepared, the enzyme has β-1 → 3-glucanase activity. It will hydrolyze soluble laminarin but will not hydrolyze yeast glucan or remove measurable amounts of glucan from yeast walls or living yeast. The glucananase cannot be separated from phosphomannanase on the basis of size, charge, or heat inactivation.

Other Sources of Enzyme. Small gut juice contains small amounts of phosphomannanase activity (it can depolymerize *Hansenula* phosphomannan). The enzyme has also been purified from *Flavobacterium dormitator* var. *glucanolyticae* grown on a medium containing baker's yeast.[10] The enzyme purified from this source also has β-1 → 3-glucanase activity that cannot be separated on basis of size or charge.

Usefulness of Phosphomannanase. The enzyme can be used in combination with enzymes from small gut juice[4,11] for production of yeast protoplasts. The enzyme is present in snail gut juice but its activity is rate limiting. The enzyme has been useful for the study of the organization of cell wall polymers in yeast wall and may prove to be useful in preparation of a primer to study mannan biosynthesis.

[10] S. Yamamoto and S. Nagasaki, *J. Ferment. Technol.* **50**, 127 (1972).
[11] S. Nagasaki, N. P. Neuman, P. Arnow, L. D. Schnable, and J. O. Lampen, *Biochem. Biophys. Res. Commun.* **25**, 158 (1966).

[127] Amylase from *Pseudomonas stutzeri*

By JOHN F. ROBYT and ROSALIE J. ACKERMAN

One of the intriguing properties of enzymes is their high degree of substrate and product specificity. The use of enzymes as catalysts for the preparation of specific compounds is superior to the usual methods employed by the organic chemist. Enzymes are required in very low concentrations, and they produce high yields of the compound with a minimum or no contamination from side reactions. Of the enzymes that degrade starch, those with an exo mechanism (i.e., those enzymes that have a specificity for degrading the starch chains from the nonreducing end) are found to form single products in high yields.

The use of fungal glucoamylase to produce high-dextrose syrups and the use of plant β-amylase to produce commercial maltose are classical examples. The search for other enzymes that form specific products from starch has resulted in the discovery that an amylase from *Pseudomonas stutzeri* NRRL B-3389 forms high yields of maltotetraose.[1]

Maltodextrins (maltose through maltooctadecaose) have also been prepared by the use of either acid[2,3] or endo enzymes,[4,5] i.e., enzymes that hydrolyze the linkages of a polymer located in the inner regions to produce more than one product. These methods of preparation require chromatographic techniques to separate the component products. Cellulose column chromatography[2] has been used to separate the products, but the yields are low, frequently less than 50 mg per compound. Higher yields (100–500 mg) have been obtained by charcoal column chromatography.[3-5] However, the relatively tedious character of column preparations, with their long periods of time required for resolution and with their low yields, make this a less desirable technique than the use of exoenzymes that form a single specific product.

Preparation of *Pseudomonas stutzeri* Amylase

Culturing the Organism

Pseudomonas stutzeri NRRL B-3389 is cultured in a liquid medium consisting of 1% trypticase (BBL), 0.5% yeast extract (Difco), 0.28% anhydrous potassium monohydrogen phosphate, 0.1% anhydrous potassium dihydrogen phosphate, and 1% soluble potato starch (Baker). Stock cultures are maintained on slants of the liquid medium plus 3% agar (Difco) and are stored at 5°. A loopful from a stock slant is used to inoculate 2 ml of liquid medium; after 24 hours at 30°, a 2% inoculum is added to 50 ml of liquid medium which is incubated at 30° with reciprocal shaking for 14–16 hours. Volumes up to 1 liter of medium

[1] J. F. Robyt and R. J. Ackerman, *Arch. Biochem. Biophys.* **145**, 105 (1971).
[2] J. A. Thoma, H. B. Wright, and D. French, *Arch. Biochem. Biophys.* **85**, 452 (1959).
[3] P. M. Taylor and W. J. Whelan, *Chem. Ind. (London) 1962*, p. 44 (1962).
[4] D. French, J. F. Robyt, M. Weintraub, and P. Knock, *J. Chromatogr.* **24**, 68 (1966).
[5] J. H. Pazur, *in* "Methods in Carbohydrate Chemistry," Vol. I, (R. L. Whistler and M. L. Wolfrom eds.), p. 337. Academic Press, New York, 1962.

in 2-liter shake flasks can be prepared. Larger quantities (10 liters or more) are prepared in fermentors. A 10-liter culture is prepared by adding 250 ml of inoculum to 10 liters of sterile medium. The culture is incubated 3 days at 30°, with an aeration rate of 0.5 liter of air per liter of culture per minute, and a stirring rate of 200 rpm; 0.5 g of Antifoam A (Dow-Corning) is used to control foaming. The enzyme is excreted into the medium with maximum amounts (10–15 units/ml) occurring after 3 days.

Purification of the Enzyme

The cells from 20 liters of culture are removed by continuous centrifugation. Several drops of 1-octanol are added to the centrifugate to prevent foaming. The clear supernatant is concentrated to 2 liters in a rotating drum humidifier. The temperature remains between 15° and 17°, and the concentration takes about 18 hours. The concentrate is collected, and the reservoir and filter of the humidifier is rinsed with small volumes of 20 mM sodium glycerophosphate buffer (pH 7) containing 5 mM calcium chloride.

The concentrate is filtered and cooled to 4° by immersion into an ethanol-ice bath. Solid ammonium sulfate is added slowly with stirring to a concentration of 25% (w/v). The temperature is maintained at 4° and the precipitate is allowed to settle several hours; it is then centrifuged and dissolved in 200 ml of 20 mM sodium glycerophosphate buffer (pH 7). The solution is again immersed into an ethanol-ice bath and the temperature allowed to drop to 2°; 2 volumes of acetone ($-10°$) is added dropwise with stirring, and the temperature is maintained between 2° and 5°. The solution is centrifuged, and the precipitate is dissolved in 20 ml of pH 7 buffer. At this stage, the enzyme should have been purified approximately 1000-fold. A summary of the purification steps and assay data is given in the table. If the enzyme is to be kept long periods, it is precipitated with 50% ammonium sulfate and the precipitate is stored at 4° as an ammonium sulfate suspension. The enzyme has maintained its activity, stored in this way, for over two years.

Assay and Units

The amylase is assayed by measuring the increase in reducing value either by alkaline copper[6] or by alkaline ferricyanide[7] after the enzyme is incubated with soluble starch. One milliliter of properly diluted enzyme is added to 1 ml of starch (10 mg/ml buffered with 20 mM sodium glyc-

---

[6] J. F. Robyt and W. J. Whelan, *in* "Starch and Its Derivatives" (J. A. Radley, ed.), 4th ed., p. 431. Chapman & Hall, London, 1968.
[7] J. F. Robyt, R. J. Ackerman, and J. G. Keng, *Anal. Biochem.* **45**, 517 (1972).

PURIFICATION SCHEME FOR *Pseudomonas stutzeri* AMYLASE[a]

Purification steps	Activity (U/ml)	Protein (mg/ml)	Specific activity (U/mg)	Purification factor
1. Culture supernatant (20 liters)	9.4	3.7	2.5	1.0
2. Tenfold concentration (2 liters)	96.6	48.5	2.0	0.8
3. 25% ammonium sulfate precipitate from step 2 dissolved in pH 7 buffer (200 ml)	345.7	1.13	306	122
4. 70% acetone precipitate from step 3 dissolved in pH 7 buffer (20 ml)	8426	3.25	2599	1036

[a] Data taken from J. F. Robyt and R. J. Ackerman [*Arch. Biochem. Biophys.* **145,** 105 (1971)] by permission.

erophosphate, pH 7), and the digest is incubated for 10 minutes at 37°. The reaction is stopped by adding 0.1 ml of 5 M trichloroacetic acid. After 5 minutes, the solution is diluted to 20 ml, and the reducing value is determined using a series of maltose concentrations as standards.

The unit of activity is defined as the number of micromoles of α-1 → 4 glycosidic bonds hydrolyzed per minute and may be calculated from the reducing value as follows:

$$\text{unit} = \frac{(\mu g \text{ of maltose equivalents produced/min/ml of digest)} \times 2}{342}$$

The factor 2 is the volume of the digest, and 342 is the molecular weight of maltose, which gives the total number of micromoles of α-1 → 4 bonds cleaved per minute.

Properties of the Amylase

The enzyme displays a broad pH-activity curve in which more than 80% of the activity is retained between pH 6.5 and 10.5. The pH optimum is 8. Incubations of the enzyme at various temperatures between 10° and 55° for 1 hour showed that it was relatively stable between 10° and 40° but that its stability rapidly decreased above 40° and was completely lost at 55°.[1] Initial velocity measurements showed that the optimum temperature was 47°. But because of the instability of the enzyme at this temperature, digests are usually run between 35° and 40°.

Large-Scale Preparation of Maltotetraose

Five hundred to 100 g quantities of maltotetraose are prepared by a 35% enzymatic conversion of Waxy-maize starch (Amioca, National Starch Corp., Plainfield, New Jersey). A 50 mg/ml slurry of starch is prepared and autoclaved to obtain an uniform solution. After cooling to 40°, 10 units of enzyme per gram of starch are added with stirring, and the digest is incubated at 40° for 40 minutes. The reaction is stopped instantly by adding two volumes of ethanol. The precipitate is removed and discarded, and the clear supernatant is reduced to one-sixth its volume by flash evaporation at 50° under vacuum. The concentrated maltotetraose solution is lyophilized, and the residual water is removed from the maltotetraose by treating the lyophyl powder with anhydrous acetone. This step is repeated twice and is followed by a treatment with anhydrous ethanol. The maltotetraose is then vacuum dried at 40° for 18–24 hours.

A system for the continuous enzymatic degradation of starch has been suggested.[8] In this procedure a starch solution is continuously added from a reservoir to an ultrafiltration cell by means of positive pressure from a compressed nitrogen tank (see Fig. 1). The enzyme is added to the ultrafiltration cell (C) where the product passes through a permeable membrane into a collecting vessel (E). The membrane is of sufficiently low porosity that the enzyme and undegraded starch are retained in the cell.

A 2% Waxy-maize starch solution is placed in the reservoir (B) and the reaction cell (C). The reaction is initiated by the addition of 8 units of enzyme to the 400-ml reaction cell. An Amicon PM-10 membrane gives a flow rate of 50–100 ml/hr when the nitrogen pressure is 55 psi. This rate could be maintained for 24 hours.

Paper chromatographic analysis of the product filtrate (E) showed that the only product was maltotetraose. This continuous system has a number of advantages over the static system. The amount of enzyme for

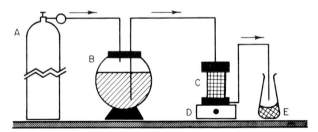

FIG. 1. Diagrammatic scheme for the continuous enzymatic conversion of starch to maltotetraose by *Pseudomonas stutzeri* amylase. A, compressed nitrogen gas; B, 3 liters of 2% starch reservoir; C, 400 ml reaction vessel, ultrafiltration cell with a PM-10 Amicon membrane filter; D, magnetic stirrer; E, product collection vessel. Redrawn from J. J. Marshall and W. J. Whelan, *Chem. Ind.* (*London*) 1971, **701** (1971).

[8] J. J. Marshall and W. J. Whelan, *Chem. Ind.* (*London*) 1971, **701** (1971).

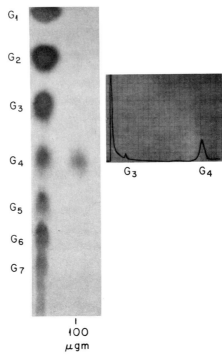

Fig. 2. Chromatographic analysis of the maltotetraose product. *Left:* paper chromatography on Whatman 3 MM using water:95% ethanol:nitromethane (18:46:35 v/v) for 2 ascents at 65° [J. F. Robyt and D. French, *Arch. Biochem. Biophys.* **138**, 662 (1970)]. *Right:* gas–liquid chromatography of the trimethylsilylated maltotetraose.

equivalent amounts of substrate is considerably reduced. Because the product is continuously separated from the reaction site by passage through the membrane into the product vessel, the yield is increased to approximately 50%, and the product does not undergo secondary degradation by the enzyme. The limit dextrin or undigested starch do not have to be separated from the product by the addition of ethanol. Thus, several steps are eliminated, and the yield and purity of the product are increased.

Properties of the Maltotetraose

The paper and gas–liquid chromatographic analysis (Fig. 2) of the product showed it to be maltotetraose of high purity. It was readily degraded to maltose by sweet potato β-amylase. The specific rotation (sodium D line, 22°) of a sample containing 1.0 g per 100 ml of water was $+172°$.

[128] Glucoamylase from *Aspergillus niger*

By JOHN H. PAZUR

$$\alpha\text{-}(1 \rightarrow 4, 1 \rightarrow 6)\text{-Glucan} + H_2O \rightarrow \text{D-glucose}$$

Assay Method

Principle. A rapid micro procedure for assaying for glucoamylase is employed for following enzymatic activity during synthesis and purification of glucoamylase.[1] The procedure is based on the use of glucose oxidase and peroxidase for measuring the D-glucose liberated from amylose by the glucoamylase. A chromogen (*o*-dianisidine) is used in the reaction mixture, and the ultraviolet absorption of the reduced chromogen is monitored continuously in a Gilford recording spectrophotometer at 400 nm. Other assay procedures for glucoamylase are based on the measurement of the D-glucose produced by starch by reducing sugar methods.[2] In the latter procedures, it may be necessary to isolate the D-glucose by suitable chromatographic procedures prior to the final step of the assay.

Reagents

Amylose, 0.1% in 0.1 M sodium acetate buffer of pH 5

Peroxidase and glucose oxidase, 0.1% of partially purified preparations (free of D-glucose) in 0.1 M sodium acetate buffer of pH 5

o-dianisidine, 0.5% in 50% ethanol

Procedure. Samples of 200 µl of the amylose peroxidase-glucose oxidase solution and 30 µl of the *o*-dianisidine were added to a silica cuvette with a 1-cm light path in a Gilford recording spectrophotometer. A sample of 10 µl of the enzyme solution was injected into the cuvette and the change in color is recorded as a function of time. The change in absorbance in a 10-minute period is calculated. A unit of enzyme activity is defined as that amount of enzyme which gives a change in absorbance of 1 in the 10-minute interval.

Purification Procedure

Step 1. Preparation of Crude Enzyme Solution. Aspergillus niger (strain NRRL 330) was maintained on agar slants in a media consisting of: D-glucose, 30.0 (all values in g/liter); nutrient broth, 8.0;

---

[1] J. H. Pazur, A. Cepure, and H. R. Knull, *Carbohyd. Res.* **21**, 1 (1971).

[2] J. H. Pazur and T. Ando, *J. Biol. Chem.* **234**, 1966 (1959).

$NaNO_3$, 3.0; K_2HPO_4, 1.0; $MgSO_4 \cdot 7H_2O$, 0.5; KCl, 0.5; and 0.1 ml of a trace element solution of (values in g/liter): $CuSO_4 \cdot 5H_2O$, 1.25; $FeSO_4 \cdot 7H_2O$, 8.0; $ZnSO_4 \cdot 7H_2O$, 6.2; $Na_2MoO_4 \cdot 2H_2O$, 0.31; $MnCl_2 \cdot 4H_2O$, 0.52; and $GaCl_2$, 0.025. The spores from the slants were used to inoculate 50-ml starter cultures which were used in turn to inoculate 1 liter cultures. For maximal growth the cultures were agitated on a laboratory shaker at room temperature. After an 85-hour period, culture contents were pressed through cheese cloth to remove the mycelium and the filtrate was used for the isolation of the glucoamylase. The glucoamylase activity and the protein content[3] of the filtrate were determined and specific activities were calculated (Table I). Since many strains of *Aspergillus*[2,4,5] produce isoenzymatic forms of glucoamylase, the specific activity of crude preparations represents the sum of the activities of all forms of the enzyme. The isoenzymatic forms of glucoamylase are separable by ion-exchange chromatography[2] and the final purified product consists of one isoenzyme.

Step 2. Precipitation with Alcohol. The *A. niger* filtrate from step 1 was stirred with four volumes of cold 95% ethyl alcohol and maintained in the cold for 1 hr. The precipitate which formed settled to the bottom of the container and was collected by decantation and centrifugation. The precipitate from 3.5 liters of culture filtrate was dissolved in 200 ml of McIlvain citrate–phosphate buffer of pH 8.0 (30 ml of $0.1 M$ citric acid and 970 ml of $0.2 M$ disodium hydrogen phosphate) and analyzed for glucoamylase activity and protein content.

TABLE I

GLUCOAMYLASE ACTIVITIES OF FRACTIONS OBTAINED AT VARIOUS STAGES OF PURIFICATION

Fraction	Volume	Units	Protein	Specific activity	Percentage yield
Filtrate	3500	58,000	9,500	6	—
After ethanol precipitation	200	52,000	186	280	90
After DEAE-cellulose chromatography	50	37,300	17	2,180	64
After ethanol precipitation	5	30,100	15	2,070	52
After second DEAE-cellulose chromatography	30	18,700	8	2,390	32
After ethanol precipitation	2	18,300	8	2,310	31

[3] O. H. Lowry, N. J. Rosebrough, L. A. Farr, and R. J. Randall, *J. Biol. Chem.* **193**, 265 (1951).

[4] Y. Morita, K. Shimizu, M. Ohga, and T. Korenaga, *Agr. Biol. Chem.* **30**, 114 (1966).

[5] D. R. Lineback and W. E. Baumann, *Carbohyd. Res.* **14**, 341 (1970).

Step 3. Chromatography on DEAE-Cellulose. The solution of gluco-amylase from step 2 was introduced from a separating funnel slowly onto a glass column (450 mm × 35 mm) packed with 30 g of DEAE-cellulose (0.86 meq/g capacity, medium mesh, Sigma Chemical Co., St. Louis, Missouri) and washed thoroughly with the McIlvain citrate–phosphate buffer of pH 8.0. After adsorption of the sample on the DEAE-cellulose was completed, the column was washed with 500 ml of McIlvain citrate–phosphate buffer of pH 8.0 (diluted 1:1 with distilled water) followed by 500 ml of citrate–phosphate buffer of pH 6.0 (93 ml of 0.1 M citric acid and 157 ml of 0.2 M disodium hydrogen phosphate diluted to 500 ml) and finally with 500 ml of citrate–phosphate buffer of pH 4.0 (154 ml of 0.1 M citric acid and 96 ml of 0.2 M disodium hydrogen phosphate diluted to 500 ml). Fractions of 10–15 ml of the eluate were collected by means of a fraction collector. Activity and pH measurements showed that the glucoamylase was present in fractions of pH 5.5 to 6.0 and in fractions of pH 4.5 to 5.0. Since the isoenzyme which was eluted at pH 4.5 was present in considerably greater amounts in the culture filtrate this isoenzyme has been purified further as the enzyme representative of the glucoamylases.

Step 4. Ethanol Precipitation. The fractions at pH 4.5–5.0 containing the glucoamylase were combined (50 ml) and stirred with 4 volumes of ethyl alcohol. The precipitate which formed settled on refrigeration of the sample, and it was collected by decantation and centrifugation methods. This precipitate was dissolved in 5 ml of citrate–phosphate buffer at pH 6.0 and assayed for enzyme activity and protein content.

Step 5. Rechromatography on DEAE-Cellulose. The sample of gluco-amylase was rechromatographed on a second column of DEAE-cellulose which had been washed with McIlvain citrate buffer at pH 6.0. The adsorption of the glucoamylase onto the DEAE-cellulose was effected at pH 6.0 and the elution was effected by the citrate–phosphate buffer of pH 4.0. Again the glucoamylase was located in fractions of pH 4.5–5.0. These fractions were combined (30 ml) and the glucoamylase was precipitated with 4 volumes of ethyl alcohol. The precipitate was dissolved in 2 ml of water and assayed for activity. The enzyme can be maintained in solution in the cold or in the dry state after lyophilization. The enzyme is extremely stable and can be maintained for long periods with little loss of activity.

Properties

The glucoamylase prepared by the above procedure was homogeneous when examined by centrifugation, electrophoresis, and column chromatography. Although many different crystallization procedures have been

TABLE II

RELATIVE RATES OF HYDROLYSIS OF α-D-GLUCOSIDES BY PURIFIED AMYLOGLUCOSIDASE

α-D-Glucoside	Structure	Relative rate
Maltose	4-α-D-Glucopyranosyl-D-glucose	100
Maltobionic acid	4-α-D-Glucopyranosyl-D-gluconic acid	34
Nigerose	3-α-D-Glucopyranosyl-D-glucose	6.6
Isomaltose	6-α-D-Glucopyranosyl-D-glucose	3.6
Arabinosyl glucoside	3-α-D-Glucopyranosyl-D-arabinose	1.66
Maltulose	4-α-D-Glucopyranosyl-D-fructose	1.39
Phenyl glucoside	Phenyl α-D-glucoside	0.87
Glycerol glucoside	1-α-D-Glucopyranosyl-D-glycerol	0.87
Turanose	3-α-D-Glucopyranosyl-D-fructose	0.67
Methyl glucoside	Methyl α-D-glucoside	0.072

attempted, the enzyme has not been obtained in crystalline form. The molecular weight of the glucoamylase is 100,000 as determined from sedimentation and diffusion data[6] and verified by density-gradient centrifugation data; the pH optimum for enzyme activity is 5.0, and the isoelectric point is 4.2. Glucoamylase contains carbohydrate residues, D-mannose, D-glucose, and D-galactose as integral structural units of the enzyme molecule.[1] The function of these residues appears to be as stabilizers of the tridimensional structure of the molecule.[7]

Glucoamylase is an α-glucosidase with a very broad substrate specificity. Table II contains data on the relative rates of action of the enzyme on a variety of compounds containing α-linked glucosyl residues.[6] In view of its broad substrate specificity, the enzyme has been widely used in structural studies of various types of glycans,[8,9] glucosides,[10] and glycoproteins.[11]

[6] J. H. Pazur and K. Kleppe, J. Biol. Chem. 237, 1002 (1962).

[7] J. H. Pazur, H. R. Knull, and D. L. Simpson, Biochem. Biophys. Res. Commun. 40, 110 (1970).

[8] M. H. Saier, Jr., and C. E. Ballou, J. Biol. Chem. 243, 4319 (1968).

[9] W. J. Whelan, Biochem. J. 122, 609 (1971).

[10] J. H. Pazur and A. Cepure, Carbohyd. Res. 5, 359 (1967).

[11] R. Spiro, J. Biol. Chem. 242, 4813 (1967).

[129] Sucrose Phosphorylase from *Pseudomonas saccharophila*

By JOHN J. MIEYAL

Sucrose + $P_i \rightleftharpoons \alpha$-D-Glucose-1-P + D-Fructose

Above is the reversible reaction for which sucrose phosphorylase (disaccharide glucosyl transferase, EC 2.4.1.7) is named. The mechanism of this and of other glucosyl transfer reactions which the enzyme catalyzes has been reviewed recently.[1]

Assay Method

Principle. The formation of the Glc-1-P is monitored spectrophotometrically by coupling the phosphorolysis of sucrose with pyridine nucleotide reduction in the presence of NADP, phosphoglucomutase, and Glc-6-P dehydrogenase. Enzyme activity is directly related to the rate of change of absorbance at 340 nm due to NADPH production.

Several alternative procedures for the assay of this and of the other reactions which the enzyme catalyzes have been described. These procedures include spectrophotometric determination of glucose and fructose in coupled assay systems requiring ATP, Mg^{2+}, NADP, hexokinase, phosphoglucoseisomerase, phosphoglucomutase, and Glc-6-P dehydrogenase.[2,3] Glucose has also been determined colorimetrically with the Glucostat reagent (Worthington Biochemical Corporation). Phosphate has been measured by the method of Lowry and Lopez.[2,4]

Reagents

> Buffer, potassium phosphate, 0.1 M, pH 7.0
> Mg sulfate, 6 mM
> Sucrose, 0.4 M
> Mercaptoethanol, neat liquid
> NADP, 4 mg/ml ($\sim$5 mM)
> Phosphoglucomutase, 10 mg/ml (Boehringer Mannheim Corporation)

[1] J. J. Mieyal and R. H. Abeles, *in* "The Enzymes" (P. D. Boyer, ed.), 3rd ed., Vol. 7. Academic Press, New York, in press.
[2] R. Silverstein, J. Voet, D. Reed, and R. H. Abeles, *J. Biol. Chem.* **242**, 1338 (1967).
[3] J. J. Mieyal, M. Simon, and R. H. Abeles, *J. Biol. Chem.* **247**, 532 (1972).
[4] Vol. 3 [115].

Glc-6-P dehydrogenase, 5 mg/ml (Boehringer Mannheim Corporation)

Procedure. The first four reagent solutions are combined in the proportions 33, 22, 11, and 0.16 volumes, respectively, and diluted with H_2O to 100 volumes to give a convenient, stable, stock solution of assay mixture. For each assay, 0.9 ml of this mixture is combined in a 1-cm path, 1.5-ml cuvette with 0.1 ml of the NADP solution, 5 μl of the phosphoglucomutase, and 1 μl of the Glc-6-P dehydrogenase. The appropriate microliter-quantity of a solution of sucrose phosphorylase is then added to initiate the reaction. The composition of each assay solution ($\sim$1 ml) therefore is: 30 mM K phosphate, pH 7.0, 0.6 mM Mg sulfate, 80 mM sucrose, 20 mM mercaptoethanol, 0.5 mM NADP, 50 μg phosphoglucomutase, 5 μg Glc-6-P dehydrogenase, and 0.005–0.01 unit of sucrose phosphorylase. The assay is carried out at 30°. Under these conditions the rate of reduction of NADP (ΔA_{340nm}) is linear with sucrose phosphorylase concentration.

Definition of Unit and Specific Activity. One unit of enzyme is the amount which catalyzes the formation of 1 μmole of Glc-1-P (i.e., measured as 1 μmole of NADPH) per minute per milliliter at 30°. Specific activity is expressed as units of enzyme per milligram of protein. Protein concentrations are determined by the method of Lowry *et al.*[5]

Purification Procedure[6]

Growth and Maintenance of P. saccharophila (ATCC 9144). The bacteria are grown according to a method adapted from that of Doudoroff.[7] The aqueous medium is composed of 33 mM KH_2PO_4–Na_2HPO_4 buffer, pH 6.5, 0.1% NH_4Cl, 0.05% $MgSO_4 \cdot 7H_2O$, 0.005% $FeNH_4(SO_4)_2 \cdot 12H_2O$ (or $FeCl_3$), 0.001% $CaCl_2$, and 0.5% sucrose.

The following is the procedure for large-scale production. A dry salts medium is premixed and contains 1456 g of KH_2PO_4, 1134 g of Na_2HPO_4, 560 g of NH_4Cl, and 280 g of $MgSO_4 \cdot 7H_2O$. A trace salts solution is prepared by dissolving 16 g of $FeNH_4(SO_4)_2$ (or $FeCl_3$) and 3.2 g of $CaCl_2$ in 1.6 liters of distilled water. To each of six 5-gallon carboys are added 85.8 g of the dry salts mix, 70 ml of the trace salts solution, and 14 liters of distilled H_2O. Each bottle is fitted with a two-hole stopper containing a bubbler tube and a glass plug in order to permit aeration and sample taking and addition. All are then sterilized. Two 50% solutions of sucrose

---

[5] Vol. 3 [73].
[6] J. G. Voet, Ph.D. Thesis, Brandeis University, Waltham, Massachusetts, 1969.
[7] Vol. 1 [28].

(80 g of sucrose in 160 ml of aqueous solution) are prepared for each carboy and sterilized separately. To initiate growth, one portion of sucrose, a 2-liter inoculum (grown previously), and approximately 20 ml of a sterilized, silicon antifoaming agent are added to each carboy using sterile technique, and vigorous aeration is begun. Aliquots are withdrawn periodically to check the pH of the medium and the progress of the growth. The medium is maintained throughout the growth at pH 6.2–7.0 by periodic additions of concentrated NH_4OH. When the culture in each bottle reaches a density which gives a reading of 200–300 Klett units (No. 54 filter), the second portion of sucrose is added. The growth rate is dependent on aeration and temperature (25–30°), increasing as each is increased, but the bacteria do not survive temperatures much above 30°. When the turbidity of the culture gives a reading of 400–600 Klett units, the cells are harvested in a Sharples (or De La Val) continuous flow centrifuge. The usual yield is 100–200 g of cell paste per bottle. The cell paste is stored frozen and may be kept for more than 6 months without significant loss of sucrose phorphorylase activity.

The stock culture of *P. saccharophila* is maintained by alternate transfer from liquid to solid medium (agar plate) and back again to liquid. Once a month an agar plate is struck from the most recent liquid culture. It is checked after 3 days for growth and lack of contamination. After a week, a liquid culture (50 ml of medium in a 250-ml flask) is inoculated from the successful plate, preferably from a single colony. For agar plates, 20 g of 2% Difco Bacto-agar, 5 g of dry salts mix, and 5 ml of trace salts solution are diluted to 960 ml, and the solution is sterilized. 2.5 g of sucrose is dissolved in 40 ml of water, the solution is sterilized then combined (while both are still hot) with the 960-ml solution to make 1 liter of agar plate medium. Approximately 25 ml is transferred to make each plate. Note that in the agar plates the sucrose concentration is 2.5%. The medium for the small liquid cultures is the same as that for the large liquid cultures except that sucrose concentration is 2.5%. These small cultures are used as inocula for the 2-liter cultures which are used to inoculate the 16-liter cultures.

Step 1. Preparation of Crude Extract. Frozen bacterial paste, 1000–1500 g, is homogenized in a Waring blendor in small portions with 0.1 M K phosphate, pH 7.0, and recombined to give three ∼ 1-liter batches of suspended cells. Each batch is sonically disrupted with a Bronson Sonic Power sonic oscillator for 30 minutes. The temperature of the cell suspension during sonication is maintained between 5° and 20° by means of a methanol-ice bath. All other operations described are performed at 0–5°. After sonication, the suspensions are centrifuged for 1 hour at 16,000 g (10,000 rpm), and the cell debris is discarded. The average yield of enzyme activity in the supernatant is 15 units per gram of initial bacterial paste.

Step 2. Protamine Sulfate Fractionation. To this supernatant fluid is added, dropwise, sufficient protamine sulfate solution (10% in 0.1 M K phosphate buffer, pH 7) to give 0.38 mg of protamine sulfate per milligram of protein. The resulting suspension is stirred for 1 hour (or longer). An aliquot is withdrawn and centrifuged in a clinical centrifuge; a dilution of the supernatant liquid is measured for A_{260nm} and A_{280nm}. If the ratio A_{280}/A_{260} is <0.7, more protamine sulfate solution is added in volumes containing 1 g of protamine sulfate and the above procedure is repeated until $A_{280}/A_{260} \geq 0.7$. The entire suspension is then centrifuged for 45 minutes at 16,000 g, and the precipitate is discarded.

Step 3. Ammonium Sulfate Fractionation. The supernatant solution (3–4 liters) from the previous step is adjusted to pH 7.0 with 40% KOH and is then brought to 50% saturation with respect to ammonium sulfate by the slow addition of granulated solid ammonium sulfate (313 g per liter initial volume of solution). During addition of ammonium sulfate, the solution is maintained at pH 7.0 by periodic addition of 40% KOH. The resulting suspension is stirred for at least 30 minutes (it may be left at this stage overnight or for several days without significant loss of enzymatic activity). Centrifugation at 16,000 g for 45 minutes precipitates essentially all the enzyme. The supernatant solution is discarded. The protein pellets are washed from the bottles with a minimum amount of 5 mM K citrate buffer, pH 5.7, and recombined to give a final volume of ~300 ml. This solution is dialyzed against two 12-liter changes of 5 mM K citrate, pH 5.7. Any resulting precipitate in the dialyzate is removed by centrifugation at 16,000 g for 45 minutes. The supernatant solution contains the enzymatic activity.

Step 4. DEAE-Cellulose Chromatography. The protein solution from the previous step is applied to a (10 × 60 cm) DEAE-cellulose column that has been equilibrated with 5 mM K citrate, pH 5.7. The column is then washed with the same buffer until the A_{280nm} of the effluent is <0.2. Usually 8–10 liters of effluent are collected at the rate of ~10 ml/minute. A convex exponential gradient is then applied to the column. The lower reservoir which is connected directly to the column originally contains 8 liters of 5 mM K citrate, pH 5.7. The solution in this reservoir is stirred mechanically throughout the operation. The upper reservoir, which feeds into the lower, originally contains 8 liters of 0.3 M KCl in 5 mM K citrate, pH 5.7. The effluent is collected in 250-ml fractions at the rate of ~4 ml/minute. The fractions are monitored for protein concentration and enzymatic activity. Enzymatic activity typically begins to elute just after the fraction containing the maximum concentration of protein, and is generally collected within fractions 20–50. The fractions of approximately the same specific activity and containing at least 90% of the eluted activity are pooled (4–5 liters) and adjusted to pH 7.0.

Step 5. Ammonium Sulfate Fractionation. The enzyme is again precipitated from solution by the addition of solid ammonium sulfate (313 g per liter initial volume of solution), and the solution is maintained at pH 7.0 with 40% KOH. The resulting suspension is allowed to stand for at least 1 hour (this period can be extended to days without significant loss of enzymatic activity), and is then centrifuged at 16,000 g for 45 minutes. The supernatant is discarded, and the precipitated protein is recombined in a minimum volume (40–50 ml) of 5 mM K phosphate pH 6.5, then dialyzed against two 8-liter changes of that buffer.

Step 6. Calcium Phosphate-Cellulose Chromatography. A calcium phosphate column is prepared according to the method of Massey[8] as follows. Whatman ashless cellulose powder, 100 g, and 500 ml of a calcium phosphate gel suspension,[9] containing 70 mg of solid per milliliter, are added to 1 liter of 5 mM K phosphate, pH 6.5. The resulting slurry is degassed under vacuum and then poured into a 4.4 cm-diameter column with a sintered glass disc, containing at the bottom a ~¼-inch layer of wet cellulose powder which had been suspended in 5 mM K phosphate, pH 6.5. A reservoir containing 4 liters of 5 mM K phosphate, pH 6.5, is connected to the top of the column, and the buffer is allowed to flow through the bed as it packs. For best results, all these operations should be done in the cold. 2 liters of the buffer are passed through the column to equilibrate it, before the dialyzed concentration enzyme solution is applied. The column is then washed with an additional 500 ml of the 5 mM K phosphate. The enzyme is eluted from the column with 12 mM K phosphate, pH 6.5. Fractions (20 ml) are collected at the rate of 1 ml/minute and analyzed for protein concentration and enzymatic activity. The fractions are pooled which have similar specific activity and contain at least 90% of the eluted enzymatic activity.

Step 7. Concentration with DEAE-Cellulose. Sufficient packed, wet DEAE-cellulose (in the phosphate form and equilibrated with 12 mM K phosphate, pH 6.5) is added to the enzyme solution to bind all the sucrose phosphorylase. Small portions of the DEAE-cellulose are added until the supernatant solution of a centrifuged aliquot of the suspension shows insignificant enzymatic activity. Usually 3–5 ml of packed DEAE-cellulose are required per 100 mg of protein in the enzyme solution. The suspension is then filtered on a Büchner funnel under vacuum in order

[8] V. Massey, *Biochim. Biophys. Acta* **37**, 310 (1960).
[9] S. M. Swingle and A. Tiselius, *Biochem. J.* **48**, 171 (1951).

to diminish the volume of the suspended DEAE-cellulose, but the cellulose is kept wet throughout the filtration. The resulting concentrated slurry is poured into a 1×50 cm column and is allowed to pack. The excess buffer is eluted from the column until it just enters the column bed at the top. The enzyme is then eluted as a sharp band with 0.5 M KCl in 12 mM K phosphate, pH 6.5. In most cases the protein can be seen on the column as a yellow band. Fractions of 3 ml volume are collected at the rate of ~ 1 ml/minute, and those fractions of similar specific activity are pooled and then dialyzed against 1600 ml of 7 mM Tris·HCl, pH 7.5, containing 0.05 ml of N,N,N',N'-tetramethylethylenediamine (TEMED). This buffer is solution A' (see below) for preparative polyacrylamide gel electrophoresis, diluted 80-fold.

Step 8. Preparative Polyacrylamide Gel Electrophoresis.[10] After dialysis the protein solution is concentrated to at least 7 mg/ml using a Diaflo ultrafiltrator under 40 psi pressure of N_2. The electrophoresis is carried out using a modification of the Canalco "Prep-Disc" system. The procedure involves placing the protein solution on top of a photopolymerized spacer gel and electrophoresing the enzyme toward the anode. The sample passes through the spacer gel into the separating gel and is eluted from the bottom of the separating gel. The system is set up as follows. The bottom of the PD 2/320 upper column is covered with parafilm, and 14 ml is added of separating gel mixture containing 2 ml of solution A', 1 ml of CN-5%, 2 ml of CN-10%, 2 ml of water, and 8 ml of AP (the composition of the stock solutions is given in Table I). Water is gently layered over the surface of the mixture in order to prevent miniscus formation on the top of the gel. The solution is allowed to polymerize for 45 minutes. The water layer is removed and replaced with 14 ml of spacer gel mixture containing 2 ml of solution B', 4 ml of solution DN, 2 ml of solution E, and 8 ml of water. A layer of water is placed on top of the solution, which is photopolymerized for 1 hour using a fluorescent lamp approximately 4 inches from the gel column. After the gels are polymerized, the parafilm at the bottom of the column is removed and the elution tubing is cleared of gel, if necessary. Cheese cloth is placed over the bottom of the column in order to keep the gels in place. The lower column of the apparatus is covered with a single thickness of dialysis tubing and filled with electrode buffer containing 5.52 g of diethylbarbituric acid (Veronal) and 1.0 g of Tris base per liter of solution, pH 7.0. The upper and lower columns are then fitted into the elution jacket and secured. The elution system is set up using solution A' diluted 1:8, i.e. 200 ml to 1600 ml, as the elution buffer. Fourteen milli-

[10] D. E. Williams and R. A. Reisfeld. *Ann. N. Y. Acad. Sci.* **121**, 373 (1964).

TABLE I

COMPOSITION OF STOCK SOLUTIONS FOR PREPARATIVE GEL ELECTROPHORESIS[a]

A': 240 ml 1 M HCl
   34.25 g Tris-base
   2.3 ml TEMED (N,N,N',N'-tetra-
    methylethylenediamine)
   Water to 500 ml
   pH 7.5

B': 39 ml 1 M H$_3$PO$_4$
   4.95 g Tris·HCl
   0.46 ml TEMED
   Water to 100 ml
   pH 5.5

CN—5%: 20 g acrylamide
   0.12 g Bis
   Water to 100 ml

CN—10%: 40 g acrylamide
   0.12 g Bis (N,N'-methylene-
    bisacrylamide monomer)
   Water to 100 ml

AP: 0.14 g ammonium persulfate
   100 ml water
   Made fresh each time

DN: 14 g acrylamide
   0.25 g Bis
   Water to 100 ml

E: 4 mg riboflavin
   Water to 100 ml

[a] J. G. Voet, Ph.D. Thesis, Brandeis University, Waltham, Massachusetts, 1969.

liters of concentrated enzyme solution (7 mg/ml),[11] to which has been added 2.1 g of electrophoresis-grade acrylamide and 0.5 ml of 0.005% bromophenol blue, is placed on top of the spacer gel, and electrode buffer is layered on top of it. The electrodes are connected, and the sample is electrophoresed at 2.5 mA until the blue band is below the spacer gel. The current is then adjusted to 12.5 mA. Fractions of 20 ml volume are collected at the rate of ~1 ml/minute. Sucrose phosphorylase routinely begins to elute 2 hours after the blue dye has eluted under these conditions. The fractions containing enzyme of similar high specific activity are pooled, and the solution is dialyzed against two 1-liter changes of 10 mM K maleate, pH 7.0. The dialyzate is then concentrated to ~10 mg/ml on a Diaflo ultrafiltrator. A very good purification is shown in Table II. Generally, a 15% recovery of total activity can be expected and a specific activity of approximately 55 units/mg.

Properties

Physical Characteristics.[2,7] The purified protein appeared homogeneous upon ultracentrifugation, as well as starch gel and acrylamide gel electrophoresis. Molecular weight determinations by Sephadex chromatography and from ultracentrifugation data yielded a molecular weight

---

[11] Using a more concentrated solution of enzyme or a larger volume than 14 ml of the same concentration has led to poorer recoveries of total enzymatic activity at this step.

TABLE II

PURIFICATION OF SUCROSE PHOSPHORYLASE FROM *Pseudomonas saccharophila*[a]

Fraction	Protein (mg/ml)	Activity (units/ml)	Total activity (units)	Specific activity (units/mg protein)
Crude sonic extract	48	7.9	22,500	0.17
Protamine sulfate, ammonium sulfate, and dialysis	89	42	21,800	0.47
DEAE-cellulose chromatography	0.72	1.9	13,800	2.6
Ammonium sulfate and dialysis	28	76	13,700	2.7
Calcium phosphate-cellulose chromatography	0.66	19.2	10,700	29
DEAE-cellulose concentration and dialysis	—	202	8,800	—
Polyacrylamide gel electrophoresis, dialysis, and concentration	7.6	508	5,260	67

[a] J. J. Mieyal, unpublished results, 1971.

of 80,000–100,000. Molecular weight determination by SDS-acrylamide gel electrophoresis[12] indicated a molecular weight of 50,000 suggesting that the enzyme is composed of two identical subunits.

Specificity. Sucrose phosphorylase is apparently absolutely specific for the glucosyl group which is transferred.[13] The glucosyl acceptor specificity, however, is quite broad. Table III lists the values of apparent K_m and V_{max} which have been determined for the various acceptors. There seems to be an inverse relationship between the values of K_m for the acceptors and the observed maximal velocities.[1,3]

Inhibitors. In general, the glucosyl acceptors at high concentrations will act as competitive inhibitors of the reaction, but their binding constants are low,[2] e.g., K_i values for P_i and sorbose are 250 and 67 mM, respectively. On the other hand, glucose and gluconolactone are much more effective competitive inhibitors,[1,2] having K_i values of 0.46 and 0.62 mM, respectively.

Stability, pH Optima. The enzyme is stable for many months when stored as a frozen concentrated solution. Repeated freezing and thawing of such solutions does not markedly affect the enzymatic activity. The enzyme loses all activity instantly when heated to 80° or above. The enzyme can be lyophilized to dryness and redissolved several times with-

[12] J. Henkin, unpublished results, 1970.
[13] M. Doudoroff, *in* "The Enzymes" (P. D. Boyer, ed.), 2nd ed., Vol. 5, p. 229. Academic Press, New York, 1961.

TABLE III
RELATIVE REACTIVITIES OF GLUCOSYL ACCEPTORS

Acceptor	Apparent K_m (M)	Apparent V_{max} (μmoles/ min/unit)	V_{max}/K_m	Buffer[a]
P_i[b]	0.002	1.40	700	TM
Fructose[b]	0.013	0.91	70	TM
Sorbose[b]	0.130	0.62	4.8	TM
Ethylene glycol[c]	0.87	0.053	0.061	TM
	2.7	0.046	0.017	MES
trans-1,2-Cyclohexanediol[c]	0.40	0.023	0.057	TM
	0.27	0.014	0.052	MES
Methanol[c]	2.3	0.38	0.017	TM
	2.1	0.011	0.005	MES
cis-1,2-Cyclohexanediol[c]	0.39	0.0046	0.015	MES
Ethanol[c,d]	"2.1"	0.004	0.002	TM
Water[c,e]	"20"	0.023	0.001	TM
	"20"	0.005	0.00025	MES

[a] TM = tris maleate; MES = sodium morpholinoethanesulfonate.

[b] Values taken from: R. Silverstein, J. Voet, D. Reed, and R. H. Abeles, *J. Biol. Chem.* **242**, 1338 (1967).

[c] Values taken from: J. J. Mieyal, M. Simon, and R. H. Abeles, *J. Biol. Chem.* **247**, 532 (1972).

[d] Single experiment: K_m is simply taken as one-half the concentration of ethanol used.

[e] The upper limit of the K_m for water as 20 M was estimated as follows: Since ethylene glycol–water mixtures of up to 25% ethylene glycol in most cases do not significantly affect the rate of glucose production, it is assumed that 75% water is still a "saturating" concentration for the enzyme; hence, one-half of this value may represent an upper limit for "K_m^{water}," i.e., $0.75 \times 55\ M$ (pure water) $\div 2 \approx 20\ M$.

out loss of activity. Chloroethanol and mercaptoethanol at concentrations as low as 7% by volume completely and irreversibly inactivate the enzyme. The pH optimum[2] for phosphorolysis is 7.0, while that for hydrolysis is 6.3.

[130] Cellobiose Phosphorylase from *Clostridium thermocellum*

By JAMES K. ALEXANDER

Cellobiose + $P_i \rightleftharpoons$ α-D-glucose 1-phosphate + D-glucose

Assay Method

Principle. The liberation of P_i is measured in a reaction in which glucose 1-phosphate is the glucosyl donor and D-xylose is the glucosyl acceptor.

Reagents

> α-D-Glucose 1-phosphate, sodium salt, 0.2 M, pH 6.4
> D-Xylose, 1.67 M
> L-Cysteine, 80 mM, pH 6.4 (freshly prepared)
> EDTA, 4 mM, pH 6.4
> Acetic acid, 0.1 M

Procedure. Before the assay, the enzyme is incubated in 20 mM cysteine and 1.0 mM EDTA, pH 6.4, for 5 minutes at 37°. The reaction mixture in a total volume of 1.0 ml contains 0.2 ml of glucose 1-phosphate, 0.2 ml of D-xylose, 0.2 ml cysteine, 0.2 ml EDTA, and 0.2 ml of preincubated enzyme (<5 units of enzyme). Four samples, each containing 0.2 ml of the reaction mixture, are transferred to separate tubes for incubation at 37°. In one sample the reaction is terminated immediately by the addition of 7.0 ml of acetic acid. In the remaining samples, the reaction is terminated at 5 minute intervals by the addition of acetic acid. The amount of P_i liberated is determined by the Fiske-SubbaRow method.[1]

Definition of Unit and Specific Activity. One unit of cellobiose phosphorylase is defined as the amount which catalyzes the liberation of 1 μmole of P_i per 15 minutes under the conditions described. Specific activity is expressed as units per milligram of protein. Protein is determined by the procedure of Lowry *et al.*[2] or spectrophotometrically.[3]

[1] C. H. Fiske and Y. SubbaRow. *J. Biol. Chem.* **66**, 375 (1925).

[2] O. H. Lowry, N. J. Rosebrough, A. L. Farr, and R. J. Randle, *J. Biol. Chem.* **193**, 265 (1951).

[3] E. Layne, Vol. 3 [73].

Purification Procedure[4]

Step 1. Growth of Bacteria. Clostridium thermocellum (strain 651), a strict anaerobe, is cultivated by the technique described by Hungate.[5] Stock cultures are stored at 5° in cellulose broth[6] which contains the following: 0.3% NaCl, 0.1% $(NH_4)_2SO_4$, 0.05% KH_2PO_4, 0.05% K_2HPO_4, 0.01% $MgSO_4 \cdot 7H_2O$, 0.001% $CaCl_2 \cdot 2H_2O$, 0.02% sodium thioglycolate, 0.1% yeast extract, 0.25% cellulose,[7] and 0.5% $NaHCO_3$. Solid sodium bicarbonate is sterilized in a hot air oven and added aseptically to the autoclaved media. After inoculation, cultures are flushed thoroughly with CO_2 gas to obtain anaerobic conditions and to complete the bicarbonate buffer system. For the purification described here, *C. thermocellum* is grown at 50–55° in a carboy containing about 20 liters of cellulose broth.[8] The temperature is maintained by a heating jacket connected to a Thermo-o-watch Laboratory Controller Model L-6. Uniform cellulose suspensions are maintained with a magnetic stirrer. Carbon dioxide is flushed through the media until gas production by the culture can be seen. Incubation is continued until nearly all the cellulose has been utilized and gas production has diminished. The inoculum is 1 liter of a 3- or 4-day-old culture grown in cellulose broth. The cells are harvested by centrifugation in a Sorvall Continuous Flow System Type KSB-4. The yield is usually from 0.5 to 1.0 g (wet weight) of cells per liter.

Step 2. Preparation of Extracts. The cells are suspended in 10 ml of 20 mM PP$_i$ and 1 mM EDTA, pH 7.0, per gram (wet weight) of cells and disrupted by a 60 second treatment with a Branson Sonifier. In case the pH of the cell suspension becomes lower than 7, it should be neutralized with NaOH before disruption. The debris is removed by centrifugation at 27,000 g for 1 hour, and the supernatant fraction is used as the cell extract. Except for the column separations, which are carried out at room temperature, the fractionations described below are performed at 0–4°.

---

[4] J. K. Alexander, *J. Biol. Chem.* **243**, 2899 (1968).

[5] R. E. Hungate, *Bacteriol. Rev.* **14**, 1 (1950).

[6] R. H. McBee, *J. Bacteriol.* **56**, 653 (1948).

[7] Nonhydrolyzed Alphacel, obtained from Nutritional Biochemicals Corporation, is used as a source of cellulose. Cellulose is washed thoroughly with water before use.

[8] Cells are grown in cellulose medium because the activity of cellobiose phosphorylase in cellulose-grown cells is greater than in cellobiose grown cells (N. J. Patni and J. K. Alexander, unpublished data, 1971). Both cellobiose phosphorylase and cellodextrin phosphorylase are inducible enzymes; no activity of either enzyme is present in glucose-grown cells [N. J. Patni and J. K. Alexander, *Bacteriol. Proc.* p. 152 (1971)].

Step 3. Protamine Treatment. To the cell extract, 0.07 mg of prot-amine sulfate per milligram of protein is added, and, after standing 1 hour, an additional 0.04 mg of protamine sulfate per milligram is added. After standing for 30 minutes, the preparation is centrifuged at 27,000 g for 60 minutes.

Step 4. Ammonium Sulfate Fractionation. To the supernatant fraction from the protamine treatment, solid ammonium sulfate (enzyme grade from Mann Research Laboratories) is added until the solution is 50% saturated. After standing for 30 minutes, the mixture is centrifuged at 12,000 g for 30 minutes. The concentration of ammonium sulfate in the supernatant fraction is increased to 65% saturation. After standing for 30 minutes, the mixture is centrifuged at 12,000 g for 30 minutes. The precipitate is dissolved in 4 ml of 10 mM triethanolamine–2.5 mM EDTA–2.5 mM mercaptoethanol buffer at pH 7.5 (unless indicated otherwise, this buffer is used throughout the purification procedure).

Step 5. Bio-Gel P-200 Chromatography. The above fraction is placed on a column (2.5 × 50 cm) of Bio-Gel P-200 which has been equilibrated with buffer. The enzyme is eluted in buffer and the samples with the highest specific activity are combined.

Step 6. Alumina C$_\gamma$ Gel. The above fraction is absorbed by two suc-cessive additions of alumina C$_\gamma$ gel (Sigma Chemical Co.). In the first absorption, 0.5 mg of alumina per milligram of protein is added, the preparation is stirred for 15 minutes and then centrifuged at 12,000 g for 10 minutes. The supernatant fraction is absorbed by adding the same ratio of alumina to protein and is treated in the same way as in the first absorption. The alumina precipitates are eluted several times with buffer containing 1 mM PP$_i$. For each elution, 1 ml of buffer per milligram of alumina is added. The alumina eluants with the highest specific activity are combined.

Step 7. DEAE-Cellulose Chromatography. The combined alumina-treated fraction is placed on a column (1.2 × 24 cm) of Whatman DEAE-cellulose (DE 52). The column, previously equilibrated with buffer, is eluted with buffer containing increasing amounts of PP$_i$. After the addition of the enzyme, the column is rinsed with 30 ml of buffer containing 10 mM PP$_i$. The column is eluted with a linear gradient in which the mixing chamber initially contains 100 ml of buffer with 10 mM PP$_i$ and the reservoir contains 100 ml of buffer with 30 mM PP$_i$. Most of the enzyme is eluted with 15–20 mM PP$_i$. The fractions with the highest specific activity are combined resulting in a final purification of about 130-fold. The purified preparation is free of all known inter-fering enzymes except traces of cellodextrin phosphorylase.

The purification procedure is summarized in the table.

PURIFICATION OF CELLOBIOSE PHOSPHORYLASE FROM *Clostridium thermocellum*[a]

Fraction	Volume (ml)	Total activity (units)	Total protein (mg)	Specific activity (units/mg)	Yield (%)
Crude extract	218	4905	1310	3.75	100
Protamine sulfate supernatant	223	5207	1137	4.58	106
Ammonium sulfate, 50–65%	6.0	3318	220	15.1	66
Bio-Gel, P-200	15.5	2480	58.6	42.3	51
Alumina C$_\gamma$ gel	18	1555	13.5	115	32
DEAE-cellulose	33.6	807	1.61	500	16

[a] J. K. Alexander, *J. Biol. Chem.* **243**, 2899 (1968).

Properties[4]

Enzyme Stability. The crude enzyme is stable at $-5°$ for more than one year. Purified preparations can be stored at $4°$ for a month or longer with less than 10% loss of activity. Occasionally large losses in activity occur during purification.

pH Optimum. The pH optimum is 6.5 in triethanolamine–citrate buffer. The enzyme is active over a range of pH 4.5–9.0.

Specificity. Like most phosphorylase reactions, an inversion occurs in the cellobiose phosphorylase reaction. Thus, the phosphorolysis of cellobiose results in the formation of α-D-glucose 1-phosphate. The enzyme is inactive on β-D-glucose 1-phosphate.[9] In addition to cellobiose, the enzyme catalyzes the phosphorolysis of celtrobiose (4-O-β-D-glucopyranosyl-D-altrose). Cellobiose phosphorylase catalyzes the arsenolysis of cellobiose.[9]

The enzyme is active on 10 different glucosyl acceptors, namely, D-glucose, 2-deoxyglucose, 6-deoxyglucose, D-glucosamine, D-mannose, D-altrose, L-galactose, L-fucose, D-arabinose, and D-xylose. In the reactions with glucose, 2-deoxyglucose, glucosamine, mannose, arabinose, and xylose as acceptors, the resulting disaccharides all contain β-1,4-glucosidic linkages.[10] It seems reasonable to expect that the remaining disaccharides, likewise, contain this linkage.

No activity was observed when the following compounds were examined as acceptors: L-arabinose, D-lyxose, D-ribose, 2-deoxyribose, L-xylose, D-allose, D-fructose, D-glucitol, i-inositol, D-gluconate, 3-O-methyl-D-glucose, α-methyl-D-glucose, β-methyl-D-glucose, or N-acetyl-D-

[9] J. K. Alexander, *J. Bacteriol.* **81**, 903 (1961).
[10] J. K. Alexander, *Arch. Biochem. Biophys.* **123**, 240 (1968).

glucosamine. The slight amount of activity with cellobiose as an acceptor appears to be due to traces of cellodextrin phosphorylase in the enzyme preparation.

Cellobiose phosphorylase presumably is active on D-glucose, 2-deoxyglucose, 6-deoxyglucose, D-glucosamine, D-mannose, and D-xylose as they exist in the C-1 conformation. However, from specificity and kinetic considerations, it has been proposed that enzyme is active on the 1-C conformers of D-altrose, D-arabinose, L-galactose, and L-fucose.

Kinetics. The relative V_{max} and apparent K_m values are 202 and 9.2 mM for 6-deoxyglucose, 100 and 35 mM for D-xylose, 53 and 73 mM for 2-deoxyglucose, 31 and 9.5 mM for D-glucosamine, 22 and 85 mM for D-mannose, 9 and 240 mM for D-arabinose and 8 and 160 mM for L-fucose. The apparent K_m for glucose 1-phosphate is 2.1 mM. The K_i for D-glucose is 1.2 mM. The apparent K_m value is 7.3 mM for cellobiose and 2.9 mM for P_i.

[131] Cellodextrin Phosphorylase from *Clostridium thermocellum*

By JAMES K. ALEXANDER

$$(\text{Glucose})_n + P_i \rightleftharpoons \alpha\text{-D-Glucose 1-phosphate} + (\text{Glucose})_{n-1}$$

Assay Method

Principle. The liberation of P_i is measured in a reaction in which glucose 1-phosphate is the glucosyl donor and cellobiose is the glucosyl acceptor.

Reagents

α-D-Glucose 1-phosphate, sodium salt, 0.4 M, pH 7.5
Cellobiose, 0.1 M
Dithiothreitol, 0.16 M
Tris, 0.16 M, pH 7.5
EDTA, 0.16 M, pH 7.5
Acetic acid, 0.1 M

Procedure. Before the assay, cellodextrin phosphorylase is incubated with 40 mM dithiothreitol, 20 mM Tris, and 20 mM EDTA at 37° for 30 minutes. The reaction is started by the addition of 0.8 ml of a mixture containing 0.2 ml of glucose 1-phosphate, 0.2 ml of cellobiose, 0.2 ml of dithiothreitol, 0.0375 ml of Tris, 0.0375 ml of EDTA, and 0.125 ml of water to 0.2 ml of the preincubated enzyme solution. The reaction is terminated immediately in one sample by the addition of 0.2 ml of the reaction mixture to 7 ml of acetic acid. After incubation at 37° for periods of 15, 30, and 60 minutes, the reaction is terminated in a similar manner in three other 0.2-ml samples. The amount of P_i liberated is determined by the Fiske-SubbaRow method.[1]

Definition of Unit and Specific Activity. One unit of cellodextrin phosphorylase is defined as the amount which catalyzes the liberation of 1 μmole of P_i per 15 minutes under the conditions described. Specific activity is expressed as units per milligram of protein. Protein is determined by the procedure of Lowry *et al.*[2]

Purification[3]

Step 1. Growth of Bacteria. *Clostridium thermocellum* (strain 651) is the source of the enzyme. The method of growing this organism and preparing cell extracts has been described.[4] In the present case, the cells are suspended in 10 mM Tris·maleate buffer, pH 7.0, before disruption.

Step 2. Protamine Treatment.[5] To the crude enzyme 1 mg of protamine sulfate per 8 milligrams of protein is added slowly with constant stirring. After 15 minutes the mixture is centrifuged at 27,000 g for 60 minutes and the precipitate is discarded.

Step 3. First Ammonium Sulfate Fractionation. To the supernatant fraction, solid ammonium sulfate (enzyme grade, Mann Research Laboratories) is added slowly with constant stirring until the solution is 60% saturated. After 15 minutes, the preparation is centrifuged at 8000 g for 60 minutes, and the precipitate is discarded. Additional ammonium sulfate is added to the supernatant fraction to increase the concentration to 80% saturation and, after standing 30 minutes, it is centrifuged at 8000 g for 60 minutes. The supernatant fraction is discarded and the precipitate is dissolved in 15 ml of 10 mM Tris, pH 7.0.

Step 4. First Alumina C_γ Treatment. To the 60–80% ammonium sulfate fraction, 0.4 mg of alumina C_γ gel (Sigma Chemical Co.) per milligram of protein is added. After stirring for 15 minutes, the gel is removed by centrifugation at 12,000 g for 10 minutes and discarded. The supernatant fraction is dialyzed against 1 liter of 10 mM Tris–1 mM

[1] C. H. Fiske and Y. SubbaRow, *J. Biol. Chem.* **66**, 375 (1925).
[2] O. H. Lowry, N. J. Rosebrough, A. L. Farr, and R. J. Randall, *J. Biol. Chem.* **193**, 265 (1951).
[3] K. Sheth and J. K. Alexander, *J. Biol. Chem.* **244**, 457 (1969).
[4] J. K. Alexander, this volume [130].
[5] With the exception of the chromatographic separations, which were carried out at room temperature, all steps in the purification procedure were carried out at 0–5°.

EDTA–0.5 mM dithiothreitol buffer, pH 7.5, for 16 hours with one change of buffer (unless indicated otherwise, this buffer is used throughout the purification procedure).

Step 5. First DEAE-Cellulose Chromatography. The dialyzed fraction is placed on a column (2 × 18 cm) of Whatman DEAE-cellulose (DE52) which previously has been equilibrated with buffer. The unabsorbed protein is removed with 100 ml of buffer and the elution is carried out with 60 ml of buffer containing 50 mM KCl, 100 ml of buffer containing 100 mM KCl, and 100 ml of buffer containing 125 mM KCl. The enzyme is eluted at a concentration of 125 mM KCl. The fractions with the highest specific activity are combined and used in the next purification step.

Step 6. Second Alumina C_γ Treatment. To the combined fractions from the preceding step, 0.2 mg of alumina C_γ gel per milligram of protein is added. The preparation is stirred for 15 minutes, then the gel is removed by centrifugation and discarded. The supernatant fraction is dialyzed for 6 hours against 500 ml of buffer with two changes of buffer.

Step 7. Second DEAE-Cellulose Chromatography. The dialyzed fraction from the previous step is applied to a column (1.2 × 14 cm) of DEAE-cellulose (Schleicher and Schuell, type 70) which previously has been equilibrated with buffer. The unabsorbed protein is removed with 40 ml of buffer and the elution is carried out with 30 ml of buffer containing 5 mM PP$_i$, 40 ml of buffer containing 10 mM PP$_i$, and 60 ml of buffer containing 12.5 mM PP$_i$. The enzyme is eluted at a concentration of 12.5 mM PP$_i$. The fractions with the highest specific activity are combined and used in the next purification step.

Step 8. Second Ammonium Sulfate Fractionation. To the combined fractions, solid ammonium sulfate is added until the concentration is 62% of saturation. After the preparation has stood for 15 minutes, the precipitate is removed by centrifugation and discarded. To the supernatant fraction, additional ammonium sulfate is added to increase the concentration to 80% of saturation. After standing for 30 minutes, the mixture is centrifuged and the supernatant portion is discarded. The precipitate is dissolved in 8 ml of 10 mM Tris, pH 7.0 and used in the following step.

Step 9. Third Alumina C_γ Treatment. To the above fraction, 0.3 mg of alumina C_γ gel per milligram of protein is added. The preparation is stirred for 30 minutes, the suspension is centrifuged, and the supernatant fraction is discarded. The gel is washed three times in 10 ml of 10 mM Tris, pH 7.0, and once with 10 ml of 10 mM Tris containing 2 mM PP$_i$ at pH 7.0. The enzyme is recovered from the gel by three elutions with 5 ml of 10 mM Tris containing 5 mM PP$_i$ at pH 7.0.

Step 10. Third Ammonium Sulfate Fractionation. To the combined fractions recovered from the gel, solid ammonium sulfate is added until the solution is 63% saturated. After standing for 15 minutes, the mixture is centrifuged and the precipitate is discarded. The concentration of ammonium sulfate in the supernatant fraction is increased to 80% saturation and, after standing for 30 minutes, the mixture is centrifuged. The supernatant fraction is discarded and the precipitate is dissolved in 10 ml of 10 mM Tris, pH 7.0.

The purification procedure is summarized in the table. This procedure results in a 450-fold purification of the enzyme. None of the following enzymes, which are present in the crude extracts, can be detected in the purified preparations: cellobiose phosphorylase, β-glucosidase, phosphoglucomutase, or phosphatase.

Properties[3]

Enzyme Stability. Certain preparations of the purified enzyme have been stored for over 4 months at $-5°$ without significant loss in activity, however, some preparations have been unstable under these conditions. The enzyme is stable to repeated freezing and thawing. Crude enzyme preparations can be stored at $-5°$ for at least 6 months without significant loss in activity.

pH Optimum. In Tris·acetate buffer, the enzyme is active over a pH range of 5.5–9.0 with an optimum at about pH 7.5.

Activators and Inhibitors. The purified enzyme exhibits an absolute requirement for reducing compounds such as cysteine, dithiothreitol,

PURIFICATION OF CELLODEXTRIN PHOSPHORYLASE FROM *Clostridium thermocellum*[a]

Fraction	Volume (ml)	Total activity (units)	Total protein (mg)	Specific activity (units/mg)	Yield (%)
Crude extract	120	655	1715	0.38	
Protamine sulfate supernatant	138	635	1170	0.54	97
Ammonium sulfate I (60–80%)	16	477	301	1.6	73
Alumina C_γ gel I supernatant	24	462	154	3.0	70
DEAE-cellulose I	29.5	366	25.2	14.5	56
Alumina C_γ gel II supernatant	35	347	16.6	20.9	53
DEAE-cellulose II	20	234	2.80	83.6	36
Ammonium sulfate II (62–80%)	8	200	1.67	120	31
Alumina C_γ gel III eluate	15	117	0.75	156	18
Ammonium sulfate III (63–80%)	10	68	0.40	170	10

[a] K. Sheth and J. K. Alexander, *J. Biol. Chem.* **244,** 457 (1969).

2-mercaptoethanol, reduced glutathione, or sodium sulfite. Crude enzyme preparations are partially active in the absence of added reducing substances. No activation is obtained with non-sulfur reducing compounds, such as sodium borohydride or ascorbic acid.

No activation or inhibition is found with AMP, cyclic AMP, ATP, glucose 6-phosphate, glucose 1,6-diphosphate, pyridoxal 5-phosphate, or 0.1% yeast extract. No inhibition is observed with 100 mM P$_i$ (in the direction of phosphorolysis), 50 mM cellobiose, or 10 mM glucose. No loss in activity is observed from extensive dialysis, an indication that the enzyme contains no readily dissociable cofactors.

Specificity. The phosphorolysis of β-1,4-oligoglucans by cellodextrin phosphorylase proceeds with an inversion which results in the formation of α-D-glucose 1-phosphate. The enzyme catalyzes the phosphorolysis of cellohexaose, cellopentaose, cellotetraose, and cellotriose; however, it is unable to catalyze the phosphorolysis of cellulose, cellobiose, laminaritriose, melezitose, or raffinose. Celloheptaose apparently can be phosphorolyzed, since cellohexaose is active as a glucosyl acceptor. The arsenolysis of cellotriose is evidence that arsenate can substitute for phosphate in the reaction.

The enzyme is active on the following glucosyl acceptors: cellobiose, cellotriose, cellotetraose, cellopentaose, cellohexaose, cellobiitol, 4-O-β-D-glucosyl-D-altrose, 4-O-β-D-glucosyl-2-deoxy-D-glucose, 4-O-β-D-glucosyl-D-mannose, 4-O-β-D-glucosyl-D-xylose, and laminaritriose. The glucosyl units added in these reactions presumably are linked to the acceptor through a β-1,4-glucosidic bond. This is supported by evidence that in at least one case in which cellobiose as an acceptor a β-1,4-glucan, cellotriose, is formed in the reaction. Further, since the phosphorolysis of laminaritriose does not occur, the enzyme would not be expected to synthesize β-1,3-glucans.

The following compounds were inactive as glucosyl acceptors: raffinose, melezitose, trehalose, turanose, melibiose, sucrose, maltose, lactose, gentiobiose, methyl-α-D-glucopyranoside, methyl-β-D-glucopyranoside, methyl-α-D-mannopyranoside, methyl-α-D-xylopyranoside, methyl-β-D-xylopyranoside, salicin, 3-O-methyl-D-glucose, D-glucosamine, D-mannitol, D-mannose, D-fructose, D-galactose, L-rhamnose, 2-deoxyribose, D-xylose, or D-glucose.

The Use of Phosphorylases for the Synthesis of Oligosaccharides. One of the most interesting features of microbial phosphorylases is their ability to synthesize oligosaccharides. In the case of cellodextrin phosphorylase, at least five analogs of cellobiose can serve as glucosyl acceptors, thus making it possible to synthesize a series of oligosaccharides, each with a different reducing portion. In addition, the activity with laminaritriose as an acceptor makes it possible to synthesize oligosaccharides containing both β-1,3- and β-1,4-linkages.

Similarly, the β-1,3-glucan phosphorylases are active on cellobiose and other β-1,4-glucans as acceptors,[6-9] providing another means of synthesizing oligosaccharides with mixed β-1,3- and β-1,4-linkages. In view of the broad specificity of the β-1,3-glucan phosphorylases, a large number of oligosaccharides undoubtedly can be synthesized with these enzymes.

It should be possible to utilize the β-glucan phosphorylases to synthesize oligosaccharides with any combination of β-1,3- or β-1,4-linkages. In combination with the suitable acceptors available commercially and the analogs of cellobiose that can be synthesized with cellobiose phosphorylase,[10] the possibilities for synthesis become even greater. These enzymes appear to provide the most versatile means available for oligosaccharide synthesis.

Kinetics. The apparent K_m value for the glucosyl acceptor, cellobiose, is 1.2 mM and for the glucosyl donor, glucose 1-phosphate, it is 4.7 mM. In the direction of phosphorolysis, the apparent K_m value is 1.0 mM for cellotriose, cellotetraose, and cellopentaose, and 0.37 mM for cellohexaose. The $V_{\max}$ is nearly the same for cellopentaose, cellotetraose, and cellotriose. Both the K_m value and the $V_{\max}$ are lower for cellohexaose than for the other cellodextrins. The apparent K_m value for P_i is 0.13 mM, 0.19 mM, 0.24 mM, and 0.26 mM for the phosphorolysis of cellotriose, cellotetraose, cellopentaose, and cellohexaose, respectively.

[6] S. H. Goldemberg, L. R. Maréchal, and B. C. De Souza, *J. Biol. Chem.* **241**, 45 (1966).
[7] L. R. Maréchal, *Biochim. Biophys. Acta* **146**, 417 (1967).
[8] D. J. Manners and D. C. Taylor, *Arch. Biochem. Biophys.* **121**, 443 (1967).
[9] H. Kaus and C. Kriebitzsch, *Biochem. Biophys. Res. Commun.* **35**, 962 (1969).
[10] J. K. Alexander, *J. Biol. Chem.* **243**, 2899 (1968).

[132] Laminaribiose Phosphorylase and β-1,3-Oligoglucan Phosphorylase from *Euglena gracilis*

By Sara H. Goldemberg and Luis R. Maréchal

$$\text{Laminaribiose} + P_i \rightleftharpoons \text{glucose} + \text{glucose 1-phosphate} \tag{1}$$

$$\text{Laminaritriose} + P_i \rightleftharpoons \text{laminaribiose} + \text{glucose 1-phosphate} \tag{2}$$

$$\text{Laminaritetraose} + P_i \rightleftharpoons \text{laminaritriose} + \text{glucose 1-phosphate} \tag{3}$$

$$\text{Laminaripentaose} + P_i \rightleftharpoons \text{laminaritetraose} + \text{glucose 1-phosphate} \tag{4}$$

$$\text{Laminarihexaose} + P_i \rightleftharpoons \text{laminaripentaose} + \text{glucose 1-phosphate} \qquad (5)$$

$$\text{Laminariheptaose} + P_i \rightleftharpoons \text{laminarihexaose} + \text{glucose 1-phosphate} \qquad (6)$$

$$\text{Laminarioctaose} + P_i \rightleftharpoons \text{laminariheptaose} + \text{glucose 1-phosphate} \qquad (7)$$

These reactions are catalyzed by two enzymes found in cell-free extracts from *Euglena gracilis*. They have been named laminaribiose phosphorylase and β-1,3-oligoglucan phosphorylase, and differ in some properties: they can be separated by column chromatography and calcium phosphate gel, have different requirements for -SH groups, and act on β-1,3-oligoglucans with different initial reaction rates.

Laminaribiose Phosphorylase[1]

Assay Method

Principle. The assay is based on the estimation of inorganic phosphate released from α-D-glucose 1-phosphate, using glucose as acceptor.

Reagents

Imidazol·HCl buffer, $0.2\,M$, pH 6.5, containing $0.02\,M$ EDTA.
α-Glucose 1-phosphate, dipotassium salt, $0.1\,M$
α-Glucose, $0.5\,M$

Procedure. The reaction mixture contains 10 μl of imidazole–HCl–EDTA buffer, 10 μl of glucose 1-phosphate, 10 μl of glucose, and 0.01–0.04 unit of enzyme, in a final volume of 50 μl. After incubation for 10 minutes at 37°, the reaction is stopped by heating, and the P_i liberated is determined.[2] Controls without glucose or glucose 1-phosphate are also included.

Definition of Unit of Enzyme Activity. A unit of enzyme is the amount which catalyzes the formation of 1 μmole of P_i per minute under the conditions of the assay.

Enzyme Preparation

Growth of Euglena Cells. Euglena gracilis strain z is grown in a medium of the following composition: peptone (Difco), 5 g; yeast extract (Difco), 2 g; glucose, 15 g; vitamin B_{12}, 10 μg; and distilled water, 1 liter. A 5-ml inoculum prepared from an agar slant is incubated at room temperature for a week and added to 200 ml of medium in Roux bottles. These are kept at 28–30° for 6–8 days, the cells are centrifuged at 10,000 g for 5 minutes, and washed twice with distilled water. About

---

[1] S. H. Goldemberg, L. R. Maréchal, and B. C. de Souza, *J. Biol. Chem.* **241**, 45 (1966).
[2] See Vol. 3 [115].

10 g of wet cells per liter of medium are obtained. All subsequent operations are carried out at 0–4°.

Step 1. Preparation of Crude Extract. The cells obtained from 1.5
liters of culture are suspended in 2 volumes of 2.2% $(NH_4)_2HPO_4$ and
disrupted through a French press at 10,000 psi. After centrifugation at
10,000 g for 10 minutes, the precipitate is washed twice with the phosphate solution. The original supernatant fluid and the washings are
pooled and passed through wet cotton to retain fat pellicles.

Step 2. Protamine Sulfate Precipitation. To the crude extract (70 ml),
4.2 ml of a freshly prepared 5% protamine sulfate solution is added.
After the preparation has stood for 5 minutes, the suspension is centrifuged for 10 minutes at 10,000 g and the precipitate is discarded. The
supernatant solution is much clearer and generally more active than the
crude extract.

Step 3. Ammonium Sulfate Fractionation. A solution of saturated ammonium sulfate, pH 7 (35 ml) is slowly added to the protamine sulfate
supernatant (70 ml). After standing for 20 minutes, the precipitate is
removed by centrifugation at 15,000 g for 10 minutes. The supernatant
fluid is again precipitated with 70 ml of saturated ammonium sulfate
solution. The precipitate is collected by centrifugation and dissolved in
7 ml of water. After overnight dialysis against water, the inactive protein which generally sediments under these conditions is centrifuged off.

Step 4. Adsorption on Calcium Phosphate Gel. The 33–66% ammonium
sulfate fraction (11.6 ml) is treated with calcium phosphate gel (0.37
mg of calcium phosphate per milligram of protein). After 10 minutes, the
suspension is centrifuged and the gel is discarded. Although specific activity usually increases about 2-fold, occasionally no purification is obtained
with this treatment. The active supernatant fluid is treated with 5 mg of
calcium phosphate gel per milligram of residual protein. The enzyme is
eluted with 5 mM sodium pyrophosphate, pH 8.6, in fractions of 4, 3,
and 2 ml. They are dialyzed overnight separately against 1 liter of 10
mM Tris–1 mM EDTA buffer, pH 7.2 (with one change). Generally the
first or second eluate or both are used, according to their specific activity.

Step 5. DEAE-Cellulose Chromatography. A 1 × 14 cm column of
DEAE-cellulose is equilibrated with 10 mM Tris–1 mM EDTA buffer,
pH 7.2, to chromatograph the calcium phosphate gel eluate. About 3 mg
of protein are applied in each run. The column is washed with 20–30 ml
of the buffer solution, and then elution is carried out with a linear gradient of NaCl, going from 0–0.3 M and keeping a constant concentration
of 10 mM Tris and 1 mM EDTA (total volume of the gradient is 50 ml).
Fractions of 1 ml are collected. The enzyme is eluted between 0.15 and
0.25 M NaCl concentration. Generally only the fractions between 0.18

and 0.22 M NaCl are pooled. They are concentrated in a rotatory evaporator to 1–2 ml at about 15°, and dialyzed overnight against 1 liter of distilled water.

The purification is summarized in Table I.

Properties

Stability. The enzyme eluted from DEAE-cellulose and the calcium phosphate gel fraction maintain 90–100% of their activity when stored at −14° for 3 months, and repeatedly thawed and frozen.

Reaction Products. On incubation with 0.1 M glucose and 20 mM glucose 1-phosphate for 10–20 minutes, laminaribiose and P_i are formed. When the incubation time is much longer, both at 0.1 M and 10 mM glucose, laminaritriose, laminaritetraose, and trace amounts of laminaripentaose are also formed. It seems that only one enzyme catalyzes the reactions (Eqs. 1–4).

Equilibrium of the Reaction. The average values for K'_1, K'_2, and K'_3 (Eqs. 1–3) are 0.3, 0.3, and 0.4, respectively.

pH Optimum. Maximum activity lies between pH 6.3 and 6.9.

Kinetics. The K_m for glucose is 19 mM; inhibition is observed with concentrations higher than 0.1 M. The K_m for glucose 1-phosphate is 2.1 mM; for P_i it is 2.5 mM, and 5 and 6 mM for laminaribiose (Eq. 1) and laminaritriose (Eq. 2), respectively.

Specificity. Only α-D-glucose 1-phosphate can serve as donor, β-D-glucose 1-phosphate, D-glucose 6-phosphate, D-galactose 6-phosphate, D-fructose 1-phosphate, D-galactose 1-phosphate, or D-mannose 1-phosphate do not substitute for it. Glucose, on the contrary, at 10 mM concentra-

TABLE I

LAMINARIBIOSE PHOSPHORYLASE PURIFICATION

Fraction	Volume (ml)	Activity (units/ml)	Protein (mg/ml)	Specific activity (units/mg)	Yield (%)
Crude extract	70	1.4[a]	12	0.1	100
Protamine sulfate supernatant	70	1.6[a]	7.4	0.2	112
Ammonium sulfate, 33–66%	11.6	6.5	13	0.5	77
Calcium phosphate gel supernatant	11.6	5.0	10.5	0.5	60
Calcium phosphate gel eluate	3	6.0	0.93	6.5	18
DEAE-eluate	1.7	4.8	0.2	24.0	8.4

[a] Activity determined after dialysis against water.

tion, may be replaced by a number of β-glucosyl derivatives: β-phenylglucoside, arbutin, and salicin are very good acceptors, and cellobiose, laminaribiose, laminaritriose, laminaritetraose, β-methylglucoside, and 2-deoxyglucose are less efficient.

Inhibitors. Slight inhibition is observed with 1 mM phlorizin, $HgCl_2$, NADP, pyrophosphate (pH 7), fructose 6-phosphate, 0.4 M imidazole; 40% and 80% inhibition is detected with 0.2 mM and 2 mM p-hydroxymercuribenzoate, respectively.

Other Sources. Laminaribiose phosphorylase has also been found in other protists (*Astasia ocellata*).[3]

β-1,3-Oligoglucan Phosphorylase[4]

Assay Method

Principle. The assay is based on the estimation of P_i liberated from α-D-glucose 1-phosphate, using laminaribiose as acceptor.

Reagents

Imidazole–HCl buffer, pH 7.2, 0.2 M, containing 0.02 M EDTA
Mercaptoethanol, 0.1 M
Laminaribiose, 0.05 M
α-D-Glucose 1-phosphate, dipotassium salt, 0.1 M

Procedure. (i) PREINCUBATION. The following components are mixed: 10 μl of imidazole–HCl–EDTA buffer, 10 μl of mercaptoethanol and 0.01–0.04 unit of enzyme in a total volume of 30 μl. They are preincubated for 30 minutes at 30°.

(ii) INCUBATION. Laminaribiose (10 μl) and 10 μl of α-D-glucose 1-phosphate are added to the preincubated mixture; the final volume is 50 μl. After 10 minutes at 37°, the reaction is stopped by adding 2 ml of water and the released P_i is determined.[2] Controls without laminaribiose are also included.

Definition of Unit of Enzyme Activity. A unit of enzyme is the amount which catalyzes the formation of 1 μmole of P_i per minute under the conditions of the assay.

Enzyme Preparation

The same procedure as for laminaribiose phosphorylase (Section I) is followed, up to the ammonium sulfate step.

Adsorption and Elution from the Calcium Phosphate Gel. The pre-

[3] D. J. Manners and D. C. Taylor, *Arch. Biochem. Biophys.* **121**, 443 (1967).
[4] L. R. Maréchal, *Biochim. Biophys. Acta* **146**, 417, 431 (1967).

cipitate of 21 ml of a suspension of calcium phosphate gel (40 mg solids per milliliter) is used to adsorb 12 ml of the enzyme fraction (ammonium sulfate step, 14 mg of protein per milliliter; ratio mg phosphate gel/mg protein: 5). After 10 minutes in the cold, with occasional stirring, the suspension is centrifuged and the supernatant is discarded. The enzyme is eluted from the gel with sodium pyrophosphate solutions, pH 8.6, as follows: 5, 4, and 4 ml of a 5 mM solution, and 4 and 4 ml of a 10 mM solution. The eluates are separately dialyzed overnight against 1 liter of 10 mM Tris·HCl–1 mM EDTA buffer, pH 7.2, with one change. Fractions I and II are enriched with laminaribiose phosphorylase (see Table II), while β-1,3-oligoglucan phosphorylase is eluted with 10 mM sodium pyrophosphate. However, occasionally the latter enzyme appears after the third elution with 5 mM sodium pyrophosphate.

The purification is summarized in Table II.

Comments. As shown in Table II, fractions I and II (laminaribiose phosphorylase) catalyze the release of P_i from glucose 1-phosphate using laminaribiose instead of glucose at a rate 10 to 20-fold lower. Fractions III–V (β-1,3-oligoglucan phosphorylase), on the other hand, act with similar rates on both substrates.

These ratios of activities are consistently observed in different preparations.

Properties

Stability. About 50% of the original activity is observed in 8-week-old preparations kept at $-14°$, and repeatedly thawed and frozen.

Reaction Products. β-1,3-Oligoglucan phosphorylase catalyzes also the formation of several β-1,3-oligoglucans, especially the higher members of the series (up to about laminarioctaose), using both 10 mM glucose or laminaribiose as acceptors (Eqs. 1–7).

Laminaribiose phosphorylase (see above) acts on laminaribiose 3–4 times faster than on higher oligosaccharides (Eq. 1); the inverse is true for β-1,3-oligoglucan phosphorylase.

pH Optimum. The optimal pH when glucose or laminaribiose are used as substrates is 7–7.5.

Kinetics. The K_m for laminaribiose is 4 mM, and for laminaritriose, 4.5 mM; for P_i it is 2 mM, and for α-D-glucose 1-phosphate, 1.8 mM (Eq. 2). For glucose the K_m is 40 mM (Eq. 1); no inhibition is observed with high concentrations of the monosaccharide.

Specificity. The enzyme is quite specific for α-D-glucose 1-phosphate. α-D-Mannose 1-phosphate, α-D-galactose 1-phosphate, β-D-glucose 1-phosphate, and fructose 1,6-diphosphate are ineffective as glucosyl

TABLE II

PURIFICATION OF β-1,3-OLIGOGLUCAN PHOSPHORYLASE FROM Euglena gracilis[a]

Step	Volume (ml)	Protein (mg/ml)	Activity (units/ml)		Specific activity (units/mg)		Yield (%)		Ratio glucose/lamina-ribiose
			Lamina-ribiose	Glucose[b]	Lamina-ribiose	Glucose[b]	Lamina-ribiose	Glucose[b]	
A. Crude extract[c]	100	7.0	1.5	2.8	0.2	0.4	100	100	1.9
B. Protamine sulfate supernatant[c]	100	3.0	1.5	2.8	0.5	0.9	100	100	1.9
C. Ammonium sulfate 33–66%[c]	12	14.0	8.5	16	0.6	1.1	68	69	1.9
D. Calcium phosphate gel eluate									
I. 5 mM sodium pyrophosphate	4.4	3.1	0.7	15	0.2	4.8	1.9	23.5	22
II. 5 mM sodium pyrophosphate	2.2	3.5	0.6	8	0.2	2.3	0.8	6.0	14
III. 5 mM sodium pyrophosphate	2.2	4.0	5.6	6	1.6	1.5	8	4.7	1.0
IV. 10 mM sodium pyrophosphate	3.6	5.2	9.2	9.5	1.8	1.8	22	11.8	1.0
V. 10 mM sodium pyrophosphate	3.0	4.8	3.3	3.3	0.7	0.7	6.6	3.5	1.0

[a] The enzyme activity was determined as described under Assay Method.

[b] Activity measured as described in Assay Method, but substituting 5 μmoles of glucose for laminaribiose. Under these conditions, both phosphorylases were actually measured in A, B, and C. Enzyme units defined as described in the text but with the mentioned change.

[c] Activity measured after dialysis against water.

donors. On the other hand, 10 mM laminaribiose may be replaced with equal efficiency by laminaritriose, laminaribiosyl β-1,4-glucose, laminaritriosyl-saligenin, laminaritriosyl-p-hydroquinone, arbutin, salicin, and laminaritetraose; β-phenyl- and β-methylglucoside, cellobiose and laminaripentaose are also good acceptors. Laminariheptaose, glucose, and gentiobiose are less efficient.

Activators. Sulfhydryl groups seem to be essential for β-1,3-oligoglucan phosphorylase activity. This requirement is not so evident in fresh enzyme preparations, but it is quite marked in aged extracts. Optimal activity is obtained preincubating for 20–30 minutes at 30° with 0.02 M mercaptoethanol; cysteine or glutathione have similar effects.

[133] Glycogen Phosphorylases *a* and *b* from Yeast [1]

By Michel Fosset, Larry W. Muir, Larry Nielson,
and Edmond H. Fischer

$$\text{Glucose}_n + P_i \rightleftharpoons \text{glucose}_{n-1} + \text{glucose-1-P}$$

Yeast phosphorylase is routinely assayed by a minor modification of the colorimetric method for determining the release of inorganic phosphate (see also this volume [132]).

Assay Method

Reagents

Dilution buffer: 0.1 M sodium succinate, 0.1% bovine serum albumin, pH 5.8

Substrate: 0.1 M sodium succinate, 0.10 M glucose 1-phosphate, 2% shellfish glycogen, pH 5.8

Stopping reagent: 14 ml of 36 N H_2SO_4, 2.5 g of ammonium molybdate in 900 ml of water

Reducing reagent: 2.5 g of aminonaphthylsulfonic acid, 148.7 g of NaHSO$_3$, 5.0 g of Na$_2$SO$_3$ per liter of water

Procedure.[2] Phosphorylase is diluted into the succinate buffer and preincubated at 30° for 5 minutes in 0.1-ml aliquots; the reaction is started by adding 0.1 ml of the substrate solution. After 5 minutes,

[1] M. Fosset, L. W. Muir, L. Nielson, and E. H. Fischer, *Biochemistry* **10**, 4105 (1971).
[2] J. L. Hedrick and E. H. Fischer, *Biochemistry* **4**, 1337 (1965).

2.6 ml of the stopping reagent and 0.2 ml of the reducing reagent are added. After 5 minutes, the optical density is measured in a colorimeter or spectrophotometer. Since there is a slow hydrolysis of glucose-1-P during color development, blanks should be included.

Units. One unit of enzyme catalyzes the release of 1 μmole of P_i per minute under these conditions. Protein can be determined by any standard procedure. With highly purified yeast phosphorylase, the absorbancy index, $A_{280}^{1\%}$, of 14.9 can be used to determine the protein concentration.

Purification Procedure

All steps are carried out at 0–4°. Either diisopropyl phosphorofluoridate or phenylmethane sulfonylfluoride, or both, are added to a final concentration of $10^{-4} M$ at each step of the purification to inhibit some of the strong proteases that are present in the yeast extract.

Commercial baker's yeast (Fleischmann) is obtained in two-pound pressed cakes. Each cake is crumbled into 1 liter of water. When the yeast is fully suspended (approximately 30 minutes), 1200 ml of washed glass beads (120 μ diameter, 3 M Co.) are added, and the mixture is ground for 30 minutes in an Eppenbach colloid mill (Gifford Wood, Inc., Hudson, New York, Model MV-6-3), which is cooled to 4° by circulating a water–alcohol mixture from a cooling bath. The following procedure is described for 8 lb (3.65 kg) of yeast and, therefore, requires four separate grinding steps.

The pooled extract is decanted from the glass beads and centrifuged at 12,000 g for 30 minutes to remove unbroken cells and debris, and the supernatant is passed through glass wool in a funnel to remove lipids. To the filtrate (crude extract) streptomycin sulfate is added to a final concentration of 0.7%. The solution is stirred for a few minutes, then centrifuged at 12,000 g for 1 hour and the pellet is discarded.

The streptomycin sulfate supernatant is adjusted to pH 5.8 with 2 N NaOH and passed through a large Büchner funnel containing 100 g of DEAE-Sephadex A-50 (Pharmacia) equilibrated in 0.1 M sodium succinate, 1 mM EDTA, pH 5.8. The breakthrough fraction (which contains the yeast phosphorylase kinase)[3] is discarded and the ion-exchanger is washed with approximately 20 liters of the above buffer; phosphorylase activity is then eluted with 4 liters of 0.5 M sodium succinate, 1 mM EDTA buffer, pH 5.8. This eluate is brought to 55% saturation with solid ammonium sulfate and centrifuged for 30 minutes

[3] L. Muir and E. H. Fischer, manuscript in preparation.

at 12,000 g; the pellet obtained is resuspended in 55% saturated ammonium sulfate and recentrifuged to wash off contaminating proteases.

The ammonium sulfate pellet is suspended in ca. 25 ml of 0.1 M sodium succinate, 1 mM EDTA, pH 5.8, and dialyzed against several changes of this same buffer. The dialyzate is centrifuged for 1 hour at 100,000 g and the clear solution obtained is applied to a 2.5 × 30 cm column of DEAE-Sephadex A-50 equilibrated with the same buffer. The breakthrough fractions are discarded and phosphorylase activity is eluted in a single peak with a linear gradient from 0.1 to 0.5 M sodium succinate, 1 mM EDTA, pH 5.8. The active fractions are pooled and concentrated to 4 ml by vacuum dialysis.

The concentrated solution is then applied to a 2.5 × 120 cm column of Sephadex G-200 swollen in 0.1 M sodium succinate, 1 mM EDTA, pH 5.8. This column is eluted with the above buffer and phosphorylase activity emerges in a peak skewed toward the leading edge. The first active fractions contain a mixture of phosphorylase b and a whereas the trailing fractions contain primarily phosphorylase a. The active fractions are pooled and concentrated to approximately 5 ml by vacuum dialysis.

This concentrated solution is dialyzed against 0.13 M sodium succinate, 1 mM EDTA, pH 5.8, then applied to a 1.2 × 80 cm column of DEAE-Sephadex A-50 equilibrated against the same buffer. The column is eluted in a stepwise fashion with 0.13 M, 0.14 M, 0.15 M, 0.16 M, and 0.17 M sodium succinate. At each step the column is eluted until no protein can be detected in the effluent. After elimination of a breakthrough protein peak, phosphorylase a first emerges at 0.15 M succinate, then phosphorylase b at 0.17 M; these two active fractions are collected separately and concentrated by vacuum dialysis. The final solution is stable at 4° for several weeks and can be lyophilized. A rapid and irreversible loss of activity occurs at pH values greater than 7 which can be prevented by inclusion of 10 mM Mg^{2+}.

The purification is summarized in the table. The overall procedure takes 1–2 weeks; the ratio of phosphorylase b to a in the final material is usually 60 to 40.

Properties

Physical, Chemical, and Enzymatic Properties. Both forms of the enzyme are pure and homogeneous by the criteria of electrophoresis in polyacrylamide gels in the presence of sodium dodecyl sulfate, isoelectric focusing, and sedimentation in the ultracentrifuge. Both have a similar subunit molecular weight of 103,000 ± 3000, amino acid composition and contain 1 molecule of pyridoxal 5′-P per subunit. Phosphorylase b exists

PURIFICATION OF YEAST GLYCOGEN PHOSPHORYLASE

Step	Specific activity (U/mg)	Purification (-fold)	Recovery (%)
Crude extract	0.15	—	100
Streptomycin sulfate supernatant	0.20	1.3	94
DEAE-Sephadex eluate	1.8	12.0	55
Ammonium sulfate pellet, 55% sat.	2.2	14.7	47
First DEAE-Sephadex column	16	106	40
G-200 Sephadex column	45	300	31
Second DEAE-Sephadex column			
Total activity	63	420	27
First peak (a form)	135	—	—
Second peak (b form)	25	—	—

predominantly as a tetramer with a molecular weight of 390,000 and phosphorylase a as a dimer with a molecular weight of 250,000.

Kinetic Properties. The optimal pH of activity is 5.8. Neither form of the enzyme is affected by 2′-, 3′-, or 5′-AMP. Phosphorylase a shows some inhibition at glucose-1-P concentrations greater than 0.1 M. The specific activities of the a and b forms are given in the table; they have similar Michaelis constants of 2.9 and 2.2 mM for glucose-1-P and 0.51 and 0.65 mg/ml for glycogen, respectively. They differ, however, in their affinities for a single substrate in the absence of the second, phosphorylase a having higher affinity in both cases. Glucose-6-P noncompetitively inhibits both forms of the enzyme with K_i values of 11 and 1 mM for a and b, respectively.

The two forms of the yeast phosphorylase are not interconverted by rabbit skeletal muscle phosphorylase kinase or phosphatase. A protein kinase has been purified from baker's yeast[3] which catalyzes the yeast phosphorylase $b \rightarrow a$ reaction by transferring one molecule of phosphate from the γ-position of ATP to each subunit of the enzyme.

[134] Phosphorylase from Dogfish Skeletal Muscle[1]

By PHILIP COHEN and EDMOND H. FISCHER

$$\text{Glycogen}_n + \text{P}_i \rightleftharpoons \text{glycogen}_{n-1} + \text{glucose-1-P}$$

The purification and properties of glycogen phosphorylases (EC 2.4.1.1) from a number of sources has already been described in previous

[1] P. Cohen, T. Duewer, and E. H. Fischer, *Biochemistry* **10**, 2683 (1971).

volumes. These include enzymes from rabbit muscle (Vol. 1, p. 200; Vol. 5, p. 369), rabbit liver (Vol. 1, p. 215), and potato (Vol. 4, p. 504). The present article describes the isolation of two enzymes from sources that could be of interest relative to comparative biochemistry or evolution, namely, the Pacific dogfish, a primitive vertebrate, which has separated from the main line of evolution leading to mammals ca. 450 million years ago, and yeast, a simple unicellular eukaryotic organism which lends itself to genetic manipulation.

Assay Method

Glycogen phosphorylase can be assayed in the direction of glycogen synthesis by determining the release of P_i[2,3] or by the incorporation of [14C]glucose into glycogen from [14C]glucose-1-P.[4] The enzyme can also be assayed in the direction of glycogen breakdown by measuring the incorporation of $^{32}P_i$ into glucose-1-P[3,5] or the reduction of NADP in a coupled system containing phosphoglucomutase and glucose-6-P dehydrogenase.[6] Routinely, the standard assay described for rabbit muscle phosphorylase is used since dogfish phosphorylase has similar K_m's for glucose-1-P and AMP but an 8-fold higher K_m for glycogen as primer.

Reagents

Dilution buffer: $0.1 M$ sodium maleate, $0.04 M$ 2-mercaptoethanol, 0.1% bovine serum albumin, pH 6.5

Substrate: $0.1 M$ sodium maleate, $0.15 M$ glucose 1-phosphate, 2% shellfish glycogen, 2 mM AMP (when added), pH 6.5

Stopping reagent: 14 ml 36 N H_2SO_4, 2.5 g of ammonium molybdate, in 900 ml water

Reducing reagent: 2.5 g aminonaphthyl sulfonic acid, 148.7 g $NaHSO_3$, and 5.0 g Na_2SO_3 per liter of water.

Procedure and Units of Activity. Phosphorylase (0.2 ml) appropriately diluted in the dilution buffer is allowed to react with 0.2 ml of substrate at 30°. After 5 minutes, the reaction is stopped by addition of 9.1 ml of stopping reagent; 0.5 ml of reducing reagent is then added, and the blue color that develops is measured in a colorimeter (e.g., Klett with a red, No. 66 filter) or a spectrophotometer at 660 nm.

Units. One unit of activity represents that amount of enzyme which

[2] J. L. Hedrick and E. H. Fischer, *Biochemistry* 4, 1337 (1965).

[3] A. M. Gold, R. M. Johnson, and J. K. Tseng, *J. Biol. Chem.* 245, 2564 (1970).

[4] D. Shepherd and I. H. Segel, *Arch. Biochem. Biophys.* 131, 609 (1969).

[5] D. Shepherd, S. Rosenthal, G. T. Lundblad, and I. H. Segel, *Arch. Biochem. Biophys.* 135, 334 (1969).

[6] O. H. Lowry, D. W. Schulz, and J. V. Passonneau, *J. Biol. Chem.* 239, 1947 (1964).

releases 1 μmole of P_i per minute under these conditions. Proteins are measured either according to Lowry *et al.* (1951)[7] or, on purified solutions, by using an absorbancy index $A_{280}^{1\%}$ of 12.9.

Preparation

Muscle is excised from the back and tail regions of Pacific dogfish (*Squalus sucklii*), netted in Puget Sound. The material is rapidly frozen and generally is used within 3 months, since the activity declines with time, usually by about 5% per month. The enzyme is isolated in a form totally dependent on AMP for activity and therefore designated as phosphorylase *b* by analogy with the rabbit muscle enzyme. The purification is carried out at 4°.

Muscle (1000 g) is thawed for 1 hour under 20° running tap water, minced in a coarse meat grinder, extracted with 1000 ml of water for 15 minutes at room temperature, and the solution is collected by straining through four layers of cheese cloth and squeezing. The muscle is extracted twice, first with 1000 ml of water for 10 minutes, then with 500 ml for 5 minutes. The combined extracts (step 1) are pooled, adjusted from pH 6.0 to pH 5.4 with 1 M acetic acid, and centrifuged for 20 minutes at 10,000 g. The supernatant is filtered through fluted filter paper to remove lipid and titrated to pH 6.8 with solid potassium bicarbonate (step 2). To each liter of protein solution, 1250 ml of 3.6 M ammonium sulfate (475 g/liter), containing 0.6 ml of conc. NH_4OH per liter is added making the final concentration 2.0 M; 2-mercaptoethanol (7.0 mM) and EDTA (0.5 mM) are included and also added to all buffers after this stage. The preparation is left at 4° overnight, then the precipitate is collected by centrifugation at 10,000 g for 20 minutes, resuspended in 5 mM glycerophosphate, pH 70, and dialyzed for 30 hours against two changes of 50 volumes of the same buffer. The solution is then centrifuged for 1 hour at 80,000 g and filtered through glass wool to remove a fine precipitate of denatured protein and traces of lipid still remaining (step 3), and layered on a 30 × 4 cm DEAE-cellulose column equilibrated with the dialysis buffer. The column is washed extensively with this buffer and the enzyme eluted with 20 mM glycerophosphate, pH 7.0. At this salt concentration dogfish phosphorylase is essentially the only protein eluted, and begins to emerge after 4–5 void volumes. The active fractions are pooled (step 4) and the solution taken to 40% saturation in ammonium sulfate. Slight turbidity is removed by centrifugation and the ammonium sulfate increased to 50%. After standing for 2 hours at 4°,

[7] O. H. Lowry, N. J. Rosebrough, A. L. Farr, and R. J. Randall, *J. Biol. Chem.* **193**, 265 (1951).

TABLE I

A Typical Purification of Dogfish Skeletal Muscle Glycogen Phosphorylase[a]

Step	Volume (ml)	Protein (mg)	Units	Specific activity	Puri-fication (-fold)	Yield (%)
1. Extract	5750	46,000	153,000	3.3	1.0	100
2. Acid (pH 5.4) supernatant	5350	40,000	149,000	3.7	1.1	97
3. 50% $(NH_4)_2SO_4$ ppt. after dialysis and centrifugation	240	8,160	100,000	12.3	3.7	67
4. DEAE cellulose chromatography (pH 7.0)	300	1,080	65,000	60.0	18.2	43
5. 50% $(NH_4)_2SO_4$	30	1,000	62,000	62.0	18.8	41

[a] 2000 g of muscle was used in this preparation.

the precipitate is collected and dissolved in 50 mM glycerophosphate, 1.0 mM EDTA and 7 mM mercaptoethanol, pH 7.0.

A summary of a typical purification is shown in Table I. It can be seen that there is approximately 1.25 g of phosphorylase per 1000 g of dogfish muscle, i.e., slightly less than half the amount present in rabbit skeletal muscle but as much as found in human muscle. Routinely 2000 g of muscle are used yielding approximately 1 g of purified enzyme (40% yield) in 3–4 days.

TABLE II

Properties of Dogfish Phosphorylase b and a

Property		b Form	a Form
1. 260/280 nm absorbance ratio		0.56	—
2. $A^{1\%}_{280}$		12.9	—
3. Specific activities (units/mg)	$-$AMP	<1	48
	$+$AMP	62	62
4. K_m glucose 1-phosphate (mM)		24	—
K_m AMP (mM)		0.15	—
K_m Glycogen (%)		0.12	—
5. Apparent specific volume		0.746	—
6. Subunit molecular weight		99,000	99,000
7. Subunit structure	—	Dimer	20% associated from dimer to tetramer
8. $s_{20,w}$ (S)	±	8.9	10.0
9. PLP per subunit		1	1
10. Moles of phosphate incorporated per subunit ($b \rightarrow a$ conversion)		0	1
11. Rapidly reacting SH groups		2	—

Properties

The purified enzyme is stable for at least 4 months when frozen in 20% glycerol at pH 7.0 in the presence of 1.0 mM AMP, EDTA, and mercaptoethanol. The preparation is homogeneous by the criteria of gel electrophoresis in the presence and in the absence of sodium dodecyl sulfate, and by sedimentation velocity and equilibrium in the ultracentrifuge. Dogfish phosphorylase b may be converted to the a form using rabbit phosphorylase kinase, and reconverted to the b form with rabbit muscle or liver phosphorylase phosphatase. Unlike rabbit muscle phosphorylase, neither the b nor the a form of dogfish phosphorylase will crystallize in the presence or absence of Mg^{2+} and AMP. Some further properties are summarized in Table II. Preliminary sequence information which includes the regions around the phosphoserine, phosphopyridoxyl lysine, and amino terminus has indicated an identity between dogfish and rabbit muscle phosphorylases in close to 90% of the residues.

[135] N-Acetylgalactosamine Deacetylase from Clostridium tertium[1]

By Donald M. Marcus

Assay Method

Principle. Deacetylation of N-acetylgalactosamine, the terminal non-reducing residue of the blood group A determinant,[2] destroys the serological activity of blood group A substance. Blood group A activity of the enzyme-treated and control substances is measured by hemagglutination inhibition.

Reagents and Apparatus

Tris·HCl, 10 mM, pH 7.5, containing 0.9% NaCl, 5 mM $MnCl_2$, and 10 mM mercaptoethanol

Blood group A substance: hog gastric mucin (Wilson Laboratories, Chicago), a mixture of A and H substances, or saliva from a group A secretor, boiled 5 minutes, or a glycoprotein purified from gastric or ovarian cyst mucin[3]

[1] D. M. Marcus, E. A. Kabat, and G. Schiffman, *Biochemistry* **3**, 437 (1964).
[2] D. M. Marcus, *N. Engl. J. Med.* **280**, 994 (1969).
[3] E. A. Kabat, "Experimental Immunochemistry," 2nd ed. Thomas, Springfield, Illinois, 1961.

2% suspension (v/v) of washed human erythrocytes of blood group A

Microtitrator (Cooke Engineering Co.) apparatus with loops and pipettes that deliver 0.025 ml

Procedure. Of a solution containing 400 μg of mucin, 0.2 ml is incubated at 37° for 16 hours; a drop of toluene is added to the surface of the reaction mixture to prevent bacterial growth. A solution containing the same quantity of substrate without enzyme, or with boiled enzyme, is incubated at the same time as a control. After incubation the reaction mixtures are boiled for 5 minutes, and serial 2-fold dilutions of each reaction mixture are prepared with the microtitrator. A 0.025 ml aliquot of anti-A antibody, containing four agglutinating units, is added to each well, the reactants are mixed and incubated at room temperature for 30 minutes, and 0.025 ml of a 2% suspension of group A cells is added to each well. After incubation at room temperature for 1 hour the agglutination patterns are read. The lowest concentration of blood group substance that inhibits hemagglutination completely is considered to be the end point.

Units. The amount of enzyme required to inactivate 100 μg of blood group substance under these conditions is defined as 1 unit.[1]

Cultivation of Clostridium tertium. A strain of *Clostridium tertium* isolated originally by Iseki[4] was obtained from Dr. C. Howe. A modification of the medium used by Howe[5] provided maximum induction of the deacetylase; 1 liter of medium contained the following ingredients: 5 g of Sheffield N-Z amine type A (an enzymatic hydrolyzate of casein), 1 g of NaCl, 5 g of K_2HPO_4, 1 g of D-glucosamine hydrochloride, 20 ml of ferrous sulfate (120 mg in 100 ml of 0.1 N HCl), and 20 ml of a mixture of vitamins containing 20 mg each of calcium pantothenate, nicotinic acid, pimelic acid, pyridoxine, and thiamine, and 2 mg of riboflavin in 100 ml of water. For initiation of growth and serial passage, 10-ml cultures were grown in an anaerobic jar and transferred every 48 hours. For bulk production of enzymes, 6-liter lots of culture medium were inoculated with 100 ml of a 24-hour anaerobic culture and incubated aerobically for 24 hours. The bacteria were sedimented by centrifugation at 1000 g for 1 hour at 4°, and the supernatant fluid was concentrated to 200 ml by dialysis against Carbowax 20,000 (Union Carbide Chemicals Co.).

[4] S. Iseki and S. Okada, *Proc. Jap. Acad.* **27**, 455 (1951).
[5] C. Howe, G. Schiffman, A. E. Bezer, and E. A. Kabat, *J. Bacteriol.* **74**, 365 (1957).

Purification Procedure

All the purification steps were carried out at room temperature.

Step 1. DEAE-Cellulose. The concentrated culture medium was dialyzed against a buffer containing 10 mM Tris·HCl, pH 7.5, 0.9% NaCl, and 1 mM MnCl$_2$, and passed through a 20 g DEAE-cellulose column, 3.4 × 20 cm, equilibrated with the same buffer. The deacetylase is not retained on the column under these conditions, several glycosidases and a peptidase are bound to the column and can be eluted with NaCl.[6] Recovery of the enzyme was usually 90–100%.

Step 2. Hydroxyapatite. Hydroxyapatite suitable for column chromatography was prepared by two methods.[7,8] Hydroxyapatite is now sold by several vendors, but I have not used these materials for this purification. A slurry of 17 g of hydroxyapatite and 17 g of Celite in 1 mM phosphate buffer, pH 7.0, was packed in a column 3–5 × 16 cm; the flow rate of narrower columns was too slow. The initial effluent from the DEAE-cellulose column was concentrated to about 70 ml by ultra-filtration and applied to the hydroxyapatite column in the same Tris buffer used in step 1. Most of the protein applied to the column is washed through with the 1 mM phosphate buffer, and the deacetylase was eluted with a phosphate gradient constructed with an external reservoir containing 0.15 M phosphate, pH 6.8, and a mixing chamber containing 300 ml of the 1 mM buffer. The flow rate was 80–90 ml per hour, and 12-ml fractions were collected. The deacetylase was eluted in two peaks which were similar in substrate specificity, specific activity, sensitivity to activation by Mn^{2+} and mercaptoethanol and inhibition by EDTA. The recovery of enzyme in this step averaged about 50%; the degree of purification varied from 10- to 125-fold, depending on the quantity of non-specific protein in the culture medium.

Step 3. Carboxymethyl (CM) Cellulose. The first deacetylase peak eluted from the hydroxyapatite column is contaminated with a neuraminidase which can be removed by absorption to CM cellulose. The enzyme mixture is dialyzed against a buffer containing 50 mM sodium acetate, pH 6.0, and is passed through a 1-g column of CM cellulose equilibrated with the same buffer. The deactylase is not retained by the column and is usually recovered completely in the initial effluent.

Properties

Stability. The enzyme has been stable at 4° over 5 years, and it is stable at room temperature for 1–2 weeks.

[6] D. M. Marcus, *J. Exp. Med.* **118**, 175 (1963).
[7] A. Tiselius, S. Hjertén, and O. Levin, *Arch. Biochem. Biophys.* **65**, 132 (1956).
[8] W. R. Anacker and V. Stoy, *Biochem. Z.* **330**, 141 (1958).

Contaminating Enzymes. No peptidase, protease, or glycosidase activities were detected in the purified enzyme preparation. The glycosidase activities tested, with *p*-nitrophenylglycosides as substrates, included α- and β-galactosidase, β-hexosaminidase, and β-glucosidase. No dialyzable sugars are released from blood group A substance by the action of this enzyme.

Specificity. The only known substrates for this enzyme are blood group A glycoproteins isolated from human and porcine mucins, and human group A erythrocytes. There was no detectable deacetylation of blood group B or H substances, *N*-acetylgalactosamine, *N*-acetylglucosamine, and two oligosaccharides with terminal nonreducing *N*-acetylgalactosamine residues obtained by hydrolysis of blood group A substance.[1]

Activators and Inhibitors. The deacetylase is inhibited by EDTA and activated by Mn^{2+} or Co^{2+} in the range of 4×10^{-5} to $5 \times 10^{-3} M$ concentrations.[1,5] The enzyme is also activated by 1×10^{-3} to $1 \times 10^{-2} M$ mercaptoethanol.

pH Optimum. The enzyme is most active between pH 7 and 8, and is inactive below pH 6 and above pH 8.5.[5]

Use of the Enzyme. The loss of blood group activity caused by deacylation of the terminal *N*-acetylgalactosamine residue is reversible by *N*-acetylation with acetic anhydride in sodium bicarbonate.[1] Blood group A substance can be labeled in this manner with $[^3H]$ or $[^{14}C]$ acetic anhydride. The deacylated galactosamine residue is easily hydrolyzed from the glycoprotein by treatment with nitrous acid,[1] the amino sugar is deaminated and converted into a 2,5-anhydrosugar which is readily hydrolyzed by dilute acid. The protonated amino group stabilizes the glycosidic bond between galactosamine and galactose to acid hydrolysis, increasing the yield of the terminal disaccharide obtained by this procedure.[9]

[9] M. E. Etzler, B. Anderson, S. Beychok, F. Gruezo, K. O. Lloyd, N. G. Richardson, and E. A. Kabat, *Arch. Biochem. Biophys.* **141**, 588 (1970).

[136] Periplasmic Enzymes That Hydrolyze Nucleoside Diphosphate Sugars

By LUIS GLASER and JOHN MAUCK

Most gram-negative bacteria and a few gram-positive bacteria contain enzymes on their surface that can hydrolyze nucleoside diphosphate sugars. The term periplasmic[1] is meant to indicate that these proteins are released from the cell surface during spheroplast formation or by the osmotic shock procedure. These proteins are probably loosely attached to the cell surface[2] rather than free in the periplasmic space, between cell wall and cell membrane, and in the case of *Escherichia coli* may be specifically localized at the ends of the cell.[3,4]

The preparation of several partially purified proteins with different substrate specificities will be described in this section.

UDP-Sugar Hydrolase from *E. coli*, and Its Protein Inhibitor

This enzyme catalyzes the reactions:

$$\text{UDP-sugar} \rightarrow \text{uridine} + P_i + \text{sugar-1-P}$$

$$\text{Nucleoside 5'-monophosphate} \rightarrow \text{nucleoside} + P_i$$

This same enzyme has been purified as a 5'-nucleotidase by Neu.[5] The purification method of Glaser *et al.*[6] for both the enzyme and protein inhibitor will be described in this section.

Assay Method

The assay is based on the liberation of α-D-glucose-1-P from UDP-D-glucose.

Reagents

Tris·HCl, 50 mM, pH 8.0–10 mM MgCl$_2$–1 mM EDTA
UDP-Glucose, 10 mM, pH 7.0
α-D-Glucose-1-6 diP, 20 μM
TPN, 10 mM, pH 7.0
ATP, 10 mM pH 7.0

The following auxiliary enzymes obtained from commercial sources are used. Their activity is expressed in micromoles of product formed per minute.

Phosphoglucomutase
Hexokinase

[1] L. Heppel, *Science* **156**, 1451 (1967) and *in* "Structure and Function of Biological Membranes," (ed. L. Rothfield) p. 224. Academic Press, New York, 1971.

[2] J. B. Ward and L. Glaser, *Biochem. Biophys. Res. Commun.* **31**, 671 (1968); *Arch. Biochem. Biophys.* **134**, 612 (1969).

[3] B. K. Wetzel, S. S. Spicer, H. F. Dvorak, and L. A. Heppel, *J. Bacteriol.* **104**, 529 (1970).

[4] H. F. Dvorak, B. K. Wetzel, and L. A. Heppel, *J. Bacteriol.* **104**, 543 (1970).

[5] H. C. Neu, *J. Biol. Chem.* **242**, 3896, 3905 (1967).

[6] L. Glaser, A. Melo, and R. Paul, *J. Biol. Chem.* **242**, 1944 (1967).

Glucose-6-P dehydrogenase

E. coli alkaline phosphatase chromatographically purified. Commercial samples should be assayed to ensure that they do not degrade UDP-D-glucose, assay A (below) is used for this purpose.

Assay A. This assay is used in crude preparations which contain phosphatases which hydrolyze α-D-glucose 1-P to glucose. The reaction mixtures contain in a final volume of 0.6 ml, 0.3 ml of Tris buffer, 0.05 ml of UDP-glucose, 50 μg of *E. coli* alkaline phosphatase and the enzyme to be assayed. After incubation at 37° the reaction is stopped by heating at 100° for 60 seconds. Controls containing no enzyme, and enzyme but no substrate are also included. Several levels of enzyme should be used to ensure proportionality. Crude extracts of many bacteria contain an inhibitor of this enzyme which obscures the presence of the enzyme (see below).

The heated reaction mixture is centrifuged to remove denatured protein, and a suitable aliquot of the supernatant fluid is used for determination of glucose (0.1–0.3 ml).

The reaction mixture for glucose determination contains 0.5 ml of Tris buffer, 0.1 ml of TPN, 0.5 ml of ATP, and 0.2 unit of hexokinase and glucose-6-P dehydrogenase, in a final volume of 1 ml. The quantity of glucose is determined by the formation of stoichiometric quantities of TPNH at 340 nm.

Assay B. This is a coupled assay and can be used when other phosphatases are not present in the enzyme preparation. The reaction mixture contains, in a volume of 1 ml, 0.5 ml of Tris buffer, 0.1 ml of α-D-glucose-1-6-diP, 0.05 ml of TPN, 0.05 ml of UDP-glucose, and 0.2 unit each of phosphoglucomutase and glucose 6-P dehydrogenase. The rate of the reaction is determined from the rate of TPNH formation at 340 nm reduction in a thermostatted spectrophotometer: 1 unit of the UDP-sugar hydrolase is the quantity of enzyme that will form 1 μmole of glucose-1-P per hour under these assay conditions.

Other assays using radioactive substrates have been described[6,7] for special situations but will not be described here.

Purification of *E. coli* Sugar Hydrolase

The enzyme is purified initially as the enzyme inhibitor complex; advantage is taken of the change in molecular weight on removal of the inhibitor, to purify the enzyme. In order to assay the enzyme in the presence of the inhibitor, the enzyme is heated at 56° for 15 minutes in the presence of $10^{-4} M$ CoCl$_2$ before assay.

[7] J. Mauck and L. Glaser, *Biochemistry* **9**, 1140 (1970).

Escherichia coli ATCC 12793 is grown to stationary phase in rotary shaker at 37° in Difco Antibiotic medium 3.

Fifteen liters of culture are harvested by centrifugation in a Lourdes continuous flow centrifuge and washed twice by centrifugation with 500–ml of $0.05\,M$ Tris·HCl (pH 8.0)–10 mM $MgCl_2$–1 mM EDTA. The cells are suspended in the same buffer (final volume, 200 ml), and 70-ml aliquots are disrupted by sonic oscillation in a 10-kc Raytheon magneto-striction oscillator for 10 minutes, followed by centrifugation at 78,000 g for 1 hour.

Salt Fractionation. To the supernatant fluid (150 ml) are added 75 ml of freshly prepared 5% streptomycin sulfate, and after 15 minutes the precipitate is removed by centrifugation at 14,000 g for 10 minutes. The supernatant fluid should give no further precipitation on the addition of streptomycin sulfate. To the supernatant fluid is added with stirring neutral saturated ammonium sulfate to 43% saturation. After 30 minutes at 3°, the precipitate is collected by centrifugation (12,000 g for 10 minutes) and discarded. The supernatant fluid is brought to 60% saturation, and after 30 minutes the precipitate is collected by centrifugation as above and dissolved in 30 ml of 50 mM Tris·HCl (pH 8.0), 10 mM $MgCl_2$, and 1 mM EDTA and dialyzed with stirring for 12 hours against four 2-liter changes of the same buffer.

DEAE-Cellulose Chromatography. The dialyzed solution is put on a column, 2.5×40 cm, of DEAE-cellulose equilibrated with 50 mM Tris·HCl (pH 8.0), 10 mM $MgCl_2$, and 1 mM EDTA eluted with linear gradient of 200 ml of the same buffer in the mixing flask and 200 ml of 50 mM Tris·HCl (pH 8.0), 0.5 M KCl, 10 mM $MgCl_2$, and 1 mM EDTA in the reservoir. Six-milliliter fractions are collected at a rate of 9 ml per hour and fractions 56–64, containing the enzyme–inhibitor complex, are pooled and lyophilized to a final volume of 12 ml. The enzyme–inhibitor complex after DEAE-cellulose chromatography is free of α-D-sugar-1-P-phosphatase and of the ADP-D-sugar pyrophosphatase also present in this organism.[8] The ADP-D-sugar pyrophosphatase can be recovered from the DEAE-cellulose column by further elution with 50 mM Tris·HCl (pH 8.0), 0.5 M KCl, 10 mM $MgCl_2$, and 1 mM EDTA. Although the ADP-sugar pyrophosphatase is also present in crude extracts as an inactive enzyme–inhibitor complex, the inhibitor is very labile[8] and the free enzyme is recovered from the DEAE-cellulose column. (Its properties are briefly described below.)

The concentrated UDP-sugar hydrolase inhibitor complex is further purified by chromatography on a column, 2.5×100 cm, of Sephadex

[8] A. Melo and L. Glaser, *Biochem. Biophys. Res. Commun.* **22**, 524 (1966).

G-100 equilibrated with 0.05 M Tris·HCl (pH 8.0)–10 mM MgCl$_2$–1 mM EDTA. Three 3-ml fractions are collected per hour. The enzyme–inhibitor complex (fractions 36–38) is pooled and heated at 56° for 18 minutes to destroy inhibitor and release active enzyme. The time required to release free enzyme completely from the enzyme–inhibitor complex has varied with different enzyme preparations, and a pilot run and a small scale should be carried out with each fresh enzyme preparation. After cooling to 3°, denatured protein is removed by centrifugation (20,000 g for 15 minutes), and the supernatant fluid is concentrated by lyophilization to a volume of 5.6 ml and again chromatographed on a Sephadex G-100 column. The fractions containing free enzyme appear after the main protein peak. They are pooled and will remain fully active for at least 1 month if kept frozen at −20°. A summary of a typical purification is shown in Table I.

Properties of the UDP-Sugar Hydrolase

The enzyme hydrolyzes all uridine diphosphate sugars at approximately equal rates. The ratio of 5′-nucleotidase to UDP-sugar hydrolase is constant through purification, and both activities are equally affected by heat denaturation, or by the protein inhibitor described below. Free UMP has been shown to not be an intermediate in the hydrolysis of UDP-sugar. The enzyme is protected against heat denaturation by Co^{2+}, Mg^{2+}, Mn^{2+}, Ca^{2+}, and Zn^{2+}. The first three metals are cofactors in the reaction at neutral pH. At pH 5.0 the enzyme is active with Co^{2+} and

TABLE I

PURIFICATION OF UDP-SUGAR HYDROLASE FROM *Escherichia coli*[a]

Fraction	Volume (ml)	Protein (mg/ml)	Activity (units)	Specific activity (units/mg protein)	Ratio of 5′-nucleotidase to UDP-sugar hydrolase
Streptomycin supernatant	225	25	3300[b]	0.60	0.77[c]
Ammonium sulfate, 43–60%	70	27.2	2394[b]	1.3	1.2
DEAE-cellulose	12	40.8	1752[b]	3.6	—
First Sephadex G-100	5.6	131	1430[b]	11.7	—
Second Sephadex G-100	42	0.027	870[d]	755	1.5

[a] Reproduced with permission from L. Glaser, A. Melo, and R. Paul, *J. Biol. Chem.* **242**, 1944 (1967).

[b] Assays of the enzyme were carried out after heating a 0.2-ml aliquot of the fraction at 56° for 15 minutes to destroy the enzyme–inhibitor complex.

[c] Assays in crude extracts are only approximate.

[d] These were recovered as active enzyme and the remainder as enzyme–inhibitor complex.

Mn^{2+} as cofactors but not with Mg^{2+}.[6] The K_m for UDP-glucose is 2 × 10^{-4} M. The same K_m is obtained with CDP-glucose, GDP-glucose, ADP-glucose, and dTDP-D-glucose, but the maximal velocity with these nucleotides is only 1 or 2% of that with UDP-D-glucose. The K_m for all the 5′-mononucleotides is 1 to 2 × 10^{-6} M.[6] The protein as isolated has been shown to contain bound Zn^{2+}.[9] The enzyme has a molecular weight of 52,000, and its amino acid composition has been determined.[5]

Properties of ADP-Glucose Pyrophosphatase

This enzyme is obtained as a by-product of the DEAE-cellulose chromatography step (see above). This enzyme catalyzes the conversion of ADP-glucose to 5′-AMP and α-D-glucose 1-P. The assay system is the same as described above using ADP-glucose as a substrate. The enzyme has also been partially purified from *Salmonella typhimurium*.[6] Although its properties have not been investigated in detail, the enzyme appears highly specific for ADP-sugar, and has a K_m of 1 × 10^{-4} M for ADP-glucose. No measurable activity can be detected with other sugar nucleotides. The pH optimum is 8. The enzyme from *Salmonella typhimurium* will cleave guanosine diphosphate sugars, but with a K_m of 4 × 10^{-3} M and at a third of the maximal velocity of ADP-glucose.

UDP-Sugar Hydrolase Inhibitor

A large number of *E. coli* strains contain an inhibitor of the UDP-sugar hydrolase. This inhibitor is a protein, which is located intracellularly, and is present in about 5-fold excess over the quantity of UDP-sugar hydrolase in the same cells.

Assay. The assay is based on the inhibition of activity of purified UDP-sugar hydrolase using assay A or B above. Since the reaction of inhibitor with enzyme is time dependent,[6] the inhibitor is incubated with enzyme for 10 minutes at 37°, and then residual enzyme activity is assayed. At least 2 levels of inhibitor have to be used to ensure that the assay is proportional to inhibitor concentration. One unit of inhibitor is equivalent to one unit of enzyme activity.

Purification of UDP-Sugar Hydrolase Inhibitor

Preparation of Spheroplasts. Stationary cells of *E. coli* (see above) harvested by centrifugation are washed twice with cold 10 mM Tris·HCl pH 8.0, and are suspended in 10 mM Tris·HCl, pH 8.0–20% sucrose at 25°, in a ratio of 1 g of cells to 80 ml of medium; 1 ml of 0.1 M EDTA is added followed by 1 ml of 10 mg/ml lysozyme in the same buffer. The

[9] H. F. Dvorak and L. Heppel, *J. Biol. Chem.* **243**, 2647 (1968).

cells are incubated at 37°, with gentle agitation; aliquots are withdrawn at intervals and diluted 10-fold in H_2O, and their optical density at 600 nm is determined. Spheroplasting is indicated by a drop in optical density due to cell lysis in hypotonic medium. After complete spheroplasting, the optical density is about 20% of the original value.[10] Spheroplasting is usually complete in 10 minutes. Alternatively, spheroplasting can be followed by examining the cells by phase microscopy. Spheroplasting serves to release most of the UDP-sugar hydrolase into the spheroplasting medium.

Extraction of Spheroplasts. The spheroplasts are collected by centrifugation (15,000 g for 20 minutes at 3°) and suspended in 80 ml of 50 mM Tris·HCl (pH 8.0)–10 mM $MgCl_2$–1 mM EDTA and stored frozen. All subsequent operations are carried out at 3°.

To 400 ml of lysed spheroplast suspension is added 18 mg of DNase, and the suspension is incubated at 25° for 20 minutes cooled to 0°, and further disrupted in 70-ml aliquots in a 10 kc Raytheon sonic oscillator for 5-minute periods followed by centrifugation at 78,000 g for 1 hour.

Salt Fractionation. To the supernatant fluid is added with stirring enough neutral saturated ammonium sulfate to bring the final concentration to 39% saturation. After 30 minutes at 3°, the precipitate is removed by centrifugation (12,000 g for 10 minutes) and discarded. The supernatant fluid is brought to 60% saturation, and after 30 minutes the precipitate is collected by centrifugation as above and dissolved in 25 ml of 50 mM Tris·HCl (pH 8.0)–10 mM $MgCl_2$–1 mM EDTA and dialyzed for 4 hours against four 1-liter changes of the same buffer.

To the dialyzed ammonium sulfate fraction is added with stirring 6.5 ml of 2% protamine sulfate, and after 15 minutes the precipitate is removed by centrifugation. The supernatant fluid is concentrated in a dialysis bag packed in solid sucrose for 4 hours and dialyzed for 16 hours against two 2-liter changes of 0.5 M Tris·HCl (pH 8.0)–10 mM $MgCl_2$–1 mM EDTA. A precipitate, formed during dialysis, is removed by centrifugation (18,000 g for 10 minutes) and discarded.

DEAE-Sephadex Chromatography. The dialyzed solution is passed through a column, 2.5 × 30 cm of DEAE-Sephadex equilibrated with 50 mM Tris·HCl (pH 8.0)–10 mM $MgCl_2$–1 mM EDTA. The inhibitor is not adsorbed under these conditions, and the column serves to remove inert protein.

CM-Sephadex Chromatography. The DEAE-Sephadex eluate is dialyzed against three 1-liter portions of 10 mM histidine, pH 6.0, for 16 hours and placed on a column, 2 × 30 cm, of carboxymethyl Sepha-

---

[10] H. C. Neu and L. Heppel, *J. Biol. Chem.* **239**, 3893 (1964).

dex G-50 equilibrated with 10 mM histidine, pH 6.0. The column is eluted with a linear gradient; the mixing flask contains 200 ml of 10 mM histidine, pH 6.0, and the reservoir 200 ml of 10 mM histidine (pH 6.0)–0.3 M KCl. Three fractions of 4.8 ml each are collected per hour. Fractions 85–95 contain the inhibitor. A summary of a typical purification is shown in Table II. The inhibitor is somewhat unstable and will lose 20–30% of its activity per month at $-20°$.

Properties of Inhibitor

Reaction of the protein inhibitor with the enzyme is time (minutes) dependent, and appears to involve a conformational change in the inhibitor. Heating of the enzyme inhibitor complex releases free enzyme, but no conditions have been found to release free inhibitor from the enzyme inhibitor complex. The molecular weight of the inhibitor is approximately 60,000.[5]

Nucleoside Diphosphate Sugar Hydrolase from *B. subtilis*

Bacillus subtilis W-23 under conditions of phosphate limitation produces a periplasmic nucleoside diphosphate sugar hydrolase similar to the enzyme from *E. coli*, but with much broader substrate specificity.

Assay

The assay is the same as that used for the *E. coli* enzyme.

Purification Procedure

Phosphate-limited cultures of *B. subtilis* W-23 are grown in a medium of the following composition in grams per liter: NH$_4$Cl, 1.5;

TABLE II
PURIFICATION OF UDP-SUGAR HYDROLASE INHIBITOR FROM *Escherichia coli*[a]

Fraction	Volume (ml)	Protein (mg/ml)	Activity (units)	Specific activity (units/mg)	Ratio of inhibition of 5'-nucleotidase to UDP-sugar hydrolase
Ammonium sulfate, 39–59%	96.5	50	8500	1.75	—
Protamine supernatant	54.5	5.5	5620	18.7	—
DEAE-Sephadex	40.5	3.2	5450	42.5	1.48
Carboxymethyl Sephadex[b]	50	0.10	4100	528.0	1.28

[a] Reproduced with permission from L. Glaser, A. Melo, and R. Paul, *J. Biol. Chem.* **242**, 1944 (1967).

[b] Only 4100 out of 5450 units of inhibitor were chromatographed on the carboxymethyl Sephadex column.

KCl, 0.45; Tris base, 12.1; $MgSO_4 \cdot 7H_2O$, 0.05; $FeSO_4 \cdot 7H_2O$, 0.005; $MnCl_2 \cdot 4H_2O$, 0.005; and K_2HPO_4, 0.035; and adjusted to pH 7.6. Glucose is autoclaved separately and added to a final concentration of 6 g/liter. Cells are grown in a rotary shaker at 37° in 2-liter flasks containing 600 ml of medium.

Purification of Nucleoside Diphosphate Sugar Hydrolase

The following procedure is used to obtain a partially purified nucleoside diphosphate sugar hydrolase. *B. subtilis* W-23 is grown in phosphate-limited medium to an optical density of 0.65 at 650 nm. The cells from 4.8 liters of medium are harvested by centrifugation and washed twice with 25 ml of 0.1 M Tris·HCl (pH 8.0)–10 mM $MgCl_2$–1 mM EDTA and then suspended in 100 ml of 0.1 M HEPES[11] (pH 7.0)–10 mM $MgCl_2$–10 μM $CaCl_2$–20% w/v sucrose at 25°, containing 0.2 mg/ml of lysozyme and incubated at 25° until most of the cells have been converted into spheroplasts as judged by phase microscopy (usually 25–30 minutes). All subsequent steps in the purification were carried out at 3°. The spheroplasts are removed by centrifugation at 15,000 g for 15 minutes and small particles by centrifugation at 105,000 g for 2 hours. The supernatant fluid is dialyzed against two 8-l changes of 0.1 M Tris·HCl (pH 8.0)–10 mM $MgCl_2$–1 mM EDTA for 24 hours and concentrated by pressure filtration using an Amicon XM-50 membrane.

The enzyme is placed on a 2.5 × 30 cm column of DEAE-cellulose equilibrated with 0.1 M Tris·HCl (pH 8.0)–10 mM $MgCl_2$–1 mM EDTA and eluted with a linear gradient. The mixing flask contains 500 ml of this buffer and the reservoir 500 ml of 0.5 M KCl in the same buffer. The enzyme is eluted between 140 and 170 ml. The peak tubes are concentrated by pressure filtration as described above to a volume of 6 ml; 1-ml aliquots are further purified by chromatography on a 1 × 60 cm column of Sephadex G-200, equilibrated with 0.1 M Tris·HCl (pH 8.0)–10 mM $MgCl_2$–1 mM EDTA–0.2 M KCl. Fractions of 1 ml are collected. The purified enzyme can be kept frozen in the buffer used in the gel filtration step for at least 1 month with only minor loss in activity. The purification is summarized in Table III.

Properties of the Enzyme

The enzyme is extremely similar to the *E. coli* enzyme, but of much broader specificity. It hydrolyzes a large variety of sugar nucleotides and is a 5' nucleotidase. It also hydrolyzes p-nitrophenyl phosphate, and bis-p-nitrophenyl phosphate. Typical K_m values are UDP-glucose 5 × $10^{-6} M$, CDP-ribitol 4 × $10^{-7} M$. The enzyme has no metal requirement

---

[11] HEPES is N-2-hydroxyethylpiperazine-N'-2-ethanesulfonic acid.

TABLE III

PURIFICATION OF NUCLEOSIDE DIPHOSPHATE SUGAR HYDROLASE FROM
Bacillus subtilis W-23[a]

Fraction	Volume (ml)	Activity (units)	Specific activity (units/mg of protein)	Ratio of 5'-Nucleotidase: nucleoside diphosphate sugar hydrolase
Spheroplast supernatant	152	6.1	0.31	—
DEAE-cellulose eluate	6	3.3	1.2	1.3
Sephadex G-200[b]	3	0.22	2.3	1.2

[a] For details see text. Reproduced with permission from J. Mauck and L. Glaser, *Biochemistry* **9**, 1140 (1970).

[b] Only 1 ml of the DEAE-cellulose eluate was chromatographed on Sephadex G-200.

and a pH optimum of 8.0. Free 5'-nucleotides are not obligatory intermediates in the hydrolysis of nucleoside diphosphate sugars.

Other Periplasmic Nucleoside Diphosphate Sugar Hydrolases

Salmonella typhimurium contains one or more nucleoside diphosphate sugar hydrolases, of broad specificity firmly attached to the cell membrane. Like typical periplasmic enzymes, these enzymes are accessible to external substrate, but so far have not been purified.[6]

[137] GDP-Glucose Glucohydrolase from Yeast[1]

By ENRICO CABIB, SIMONETTA SONNINO, and HÉCTOR CARMINATTI

$$\text{GDP-glucose} + H_2O \rightarrow \text{GDP} + \text{glucose}$$

Assay Method

Principle. GDP-[^{14}C]glucose is used as substrate. After incubation, the unreacted nucleotide is adsorbed on an ion-exchange resin, and the radioactive glucose in the filtrate is counted.

Reagents

GDP-[^{14}C]glucose, 10 mM, specific activity 1×10^5 cpm/μmole. The labeled nucleotide can be synthesized[2] or purchased com-

[1] S. Sonnino, H. Carminatti, and E. Cabib, *J. Biol. Chem.* **241**, 1009 (1966). S. Sonnino, H. Carminatti, and E. Cabib, *Arch. Biochem. Biophys.* **116**, 26 (1966).

mercially (International Chemical and Nuclear Corporation, California).[2]

Sodium glycerophosphate, 1 M, pH 7.2

EDTA, 0.1 M, pH 7

Dowex-1 (acetate form) suspension in water containing 50 mg of dry resin per milliliter

Procedure. The reaction mixture contains 3 μl of GDP-[^{14}C]glucose, 2.5 μl of sodium glycerophosphate, 2.5 μl of EDTA, and enzyme, in a total volume of 33 μl. Incubation is for 1 hour at 30°. The reaction is stopped by freezing the tubes in an ice–salt mixture. Then 0.5 ml of ice-cold water is added, followed by 0.2 ml of the Dowex 1 suspension. The tubes are brought to room temperature and vigorously agitated on a reciprocating shaker for 10 minutes. The suspension is filtered under vacuum into another test tube, and two washings are made with 0.2-ml portions of water. The combined filtrates are poured on an aluminum planchet and are followed by a 0.5-ml washing of the test tube. After drying on a boiling water bath, the samples are counted.

Definition of Unit. One unit is defined as the amount of enzyme which catalyzes the liberation of 1 μmole of glucose per minute under the conditions of assay. Specific activity is expressed as units per milligram protein.

Purification Procedure

Starch-free commercial baker's yeast, lyophilized and kept at −20°, is used as source of enzyme. Operations are performed at 0–2° and centrifugations at 25,000 g for 5 minutes, unless otherwise stated.

Step 1. Extraction and Sephadex Filtration. Lyophilized yeast (4 g) is ground vigorously for 20 minutes in a precooled mortar, placed in a freezer at −10°. To prevent condensation of water on the yeast, the mortar is enclosed in a plastic bag, with a hole for the pestle. The mass of broken cells is blended in a VirTis homogenizer, with 16 ml of 50 mM Tris·maleate buffer, pH 7.5, containing 1 mM EDTA. After centrifugation for 15 minutes, the pellet is reextracted for 1 hour with 8 ml of buffer, and again centrifuged off. The very turbid extracts are combined (19.2 ml) and centrifuged for 3 hours at 100,000 g. The supernatant fluid (13.3 ml) is sucked off, without disturbing the loose layer of particles that overlies the translucent pellet. It is then passed through a Sephadex G-25 column (2.5 × 31 cm), previously equilibrated with

[2] A. D. Elbein, Vol. 8 [16].

10 mM NaCl, containing 1 mM EDTA. The fraction corresponding to the void volume is discarded and the turbid fraction that follows is collected (26.5 ml, see Table, p. 982).

Step 2. Ammonium Sulfate Fractionation. To the Sephadex filtrate, 1.3 ml of 1 M MnSO₄ is added, and 20 minutes later the flocculent precipitate is centrifuged off and discarded. To the supernatant fluid (26.5 ml) an equal volume of ammonium sulfate solution (486 g/liter, brought to pH 7.5 with ammonia and containing 1 mM EDTA) is added. After 20 minutes the suspension is centrifuged; the supernatant fluid is mixed with one half-volume (26 ml) of the ammonium sulfate solution and centrifuged again 20 minutes later. The precipitate is redissolved in 8.2 ml of 50 mM sodium glycerophosphate buffer, pH 7.2, containing 1 mM EDTA and dialyzed 90 minutes against 1500 ml of the same buffer and another 90 minutes against 1500 ml of 2-fold diluted buffer.

Step 3. DEAE-Cellulose Chromatography. The dialyzate (10.3 ml) is applied to a 1.8 × 4.5 cm DEAE-cellulose column, previously washed with 200 ml of 25 mM sodium glycerophosphate buffer, pH 7.2, containing 1 mM EDTA. After filtering the sample, the column is washed with 20 ml of the same buffer. The enzyme is eluted with a stepwise NaCl gradient obtained in the following way. For each step, an 8-ml portion of eluent is fed to the column and recovered in a single fraction. Each fraction contains sodium glycerophosphate, pH 7.2 and EDTA at constant concentrations of 25 mM and 1 mM, respectively. The concentration of NaCl for each successive fraction is as follows (expressed in molarity): 0.02, 0.04, 0.06, 0.09, 0.12, 0.15, 0.18, 0.21, 0.24, 0.27, and 0.30. The enzyme is usually found in fractions 3–8. The five best fractions are pooled.

Step 4. Calcium Phosphate Gel Adsorption. The DEAE-cellulose eluate is dialyzed for 5 hours against 3 changes (2 liters each) of 25 mM sodium glycerophosphate, pH 7.2, containing 1 mM EDTA. Calcium phosphate gel (8.4 ml, containing 20 mg of solids per milliliter) is centrifuged, and the pellet is resuspended in 0.5 ml of water. The dialyzate (36.5 ml) is added to the gel and 10 minutes later the suspension is centrifuged and the sediment is discarded. The supernatant liquid is added to the pellet from 15.3 ml of gel suspension, previously taken up in 0.9 ml of water. After 10 minutes the suspension is centrifuged and the supernatant liquid is discarded. The gel is extracted twice, for 10 minutes each time, with 50 mM potassium phosphate, pH 7, containing 1 mM EDTA; 7.3 ml is used in the first extraction and 3.6 ml in the second.

Step 5. Sephadex G-100 Chromatography. The combined extracts (10.7 ml) are dialyzed 3.5 hours against 2 liters of 5 mM sodium glycerophosphate at pH 7.2, containing 0.1 mM EDTA. The dialyzate is concentrated to 0.4–0.5 ml in a rotary evaporator at 10°. The concentrated enzyme is applied to the top of a Sephadex G-100 column (0.8 × 80 cm), previously equilibrated with 25 mM sodium glycerophosphate at pH 7.2, containing 1 mM EDTA, 0.125 M NaCl, and 10 mM mercaptoethanol. Elution is carried out with the same buffer. After the first 12 ml have passed through, 1.2-ml fractions are collected. The activity usually emerges in fractions 5–10. The four best fractions are combined.

Note. In different batches, the amounts of DEAE-cellulose and calcium phosphate gel are kept proportional to the amount of protein in the corresponding fractions. In different preparations the yield has varied between 10 and 17% and the purification between 110- and 190-fold. A typical purification is summarized in the table.

Properties

Stability. The purified enzyme is fairly stable when kept at −20°. A loss of activity of about 50% has been observed after 6 months.

Contaminating Activities. The enzyme preparation is free from sugar nucleotide phosphorylase.[3] A slight activity of GDP phosphatase is observed only in the absence of EDTA.

Specificity. The enzyme is specific for GDP-glucose. ADP-glucose, UDP-glucose, TDP-glucose, CDP-glucose, GDP-mannose, GDP-galactose, and glucose 1-phosphate are not substrates.

Kinetics. The enzyme shows a very broad pH-optimum, between 5 and 9. The Michaelis constant for GDP-glucose at pH 7.2 and 30° is 0.23 mM.

Effectors. No change in the enzymatic activity is observed by omitting EDTA or adding divalent cations, except as noted below. Tris·HCl,

PURIFICATION OF GDP-GLUCOSE HYDROLASE FROM YEAST

Fraction	Volume (ml)	Total activity (munits)	Total protein (mg)	Specific activity (munits/mg)	Recovery (%)
Sephadex filtrate	26.5	758	493	1.54	100
Ammonium sulfate	10.8	509	139	3.66	67
DEAE-cellulose eluate	40	417	28.4	14.7	55
Calcium phosphate gel eluate	10.7	213	3.36	63	28
Dialyzed and concentrated enzyme	0.42	192	—	—	25
Sephadex G-100 eluate	4.8	133	0.64	208	17.5

[3] See Vol. 8 [34].

75 mM, pH 7.5, inhibits 50–60%. GTP and GDP are also inhibitors. The inhibition by GDP is competitive, with a K_i of 80 μM. The addition of excess Mg^{2+} partially relieves this inhibition.

[138] CMP-Sialic Acid Hydrolase

By EDWARD L. KEAN

CMP-sialic acid → sialic acid + (CMP)

CMP-Sialic acid has not as yet been detected in mammalian tissue. A knowledge of the mechanisms that regulate the level of this sugar nucleotide within the cell may be of considerable aid in understanding the processes involved in the biosynthesis of the sialic acid-containing heteropolymers. Among the factors that contribute to the failure of this compound to accumulate in the cell, such as its chemical lability and its utilization in synthetic reactions by the sialyl transferases, is its enzymatic cleavage.[1,2]

Assay Methods

Principle. Two assay methods are used to measure the activity of CMP-sialic acid (CMP-NAN) hydrolase. Both methods involve the separation of product from substrate and measurement of the amounts of each by radioactive procedures. The first method uses ion exchange column chromatography and the second, high voltage paper electrophoresis. The disappearance of the substrate, CMP-[1-^{14}C]NAN,[3] and the appearance of [1-^{14}C]NAN are measured by each of these techniques.

An evaluation of enzymatic activity that is based upon measuring

---

[1] This presentation is based upon work presented previously in preliminary form: E. L. Kean and S. R. Coghill, *Fed. Proc., Fed. Amer. Soc. Exp. Biol.* **30**, 1117 (1971), and a manuscript in preparation.

[2] M. Shoyab and B. K. Bachhawat, *Biochem. J.* **102**, 13C (1967); M. Shoyab and B. K. Bachhawat, *Indian J. Biochem.* **4**, 142 (1967); M. Shoyab and B. K. Bachhawat, *Indian J. Biochem.* **6**, 56 (1969).

[3] CMP-[1-^{14}C]NAN is prepared by using CMP-NAN synthetase of purified nuclei from rat liver [E. L. Kean, *J. Biol. Chem.* **245**, 2301 (1970)]. See also this volume [51]. The compound is isolated from the incubation mixture and purified by ion exchange and paper chromatographic techniques. See Vol. 8 [31] and E. L. Kean and W. E. Bruner, *Exp. Cell. Res.* **69**, 384 (1971). [1-^{14}C]NAN, used as the substrate in this reaction, is prepared by the NAN-aldolase procedure [P. Brunetti, G. W. Jourdian, and S. Roseman, *J. Biol. Chem.* **237**, 2447 (1962)] using [1-^{14}C]-pyruvate, *N*-acetyl D-mannosamine and bacterial NAN-aldolase (type III from Sigma Chemical Corp.). See also Vol. 8 [31].

only the disappearance of substrate, CMP-NAN, measures change in an initially large value and may lack reliability when dealing with small changes.[4] Methods that measure product formation are preferred. The sensitivity of the present methods are such that both the disappearance of substrate and the appearance of product are measured with each determination. These procedures thus provide also an evaluation of the stoichiometry of this relationship with each incubation, a relationship that has proved to be 1:1.

Method 1. Ion Exchange Column Assay

Reagents

CMP-[1-^{14}C]NAN, 12 mM, specific activity, 0.39 $\times$ 10^6 cpm/μmole
Tris·HCl, 1.0 M, pH 9.0
NH$_4$HCO$_3$, 0.1 M
EDTA, 1.0 M, pH 9.0
KCl, 1.0 M
Anion exchange resin AG 1-X8, 200–400 mesh, HCO$_3$-(BioRad). This form of the resin is prepared by treating the chloride form of the resin with 2 N HCl, 1 N NaOH, and 1 M NaHCO$_3$, sequentially, accompanied by extensive washing with water between each change. The resin is stored as a slurry in the cold.
Radioactive counting system composed of: 667 ml of toluene, 333 ml of Triton-X-100, 5.5 g of 2,5-diphenyloxazole (PPO), and 0.1 g of 1,4-bis[2-(5-phenyloxazoyl)]benzene (POPOP).

Procedure. INCUBATION. Into test tubes, 75 mm $\times$ 10 mm, add 0.015 ml of Tris buffer, 0.025 ml of CMP-[1-^{14}C]NAN (112,000 cpm), enzyme (0.04–0.12 unit), and water to a total volume of 0.1 ml. Incubation is carried out at 37° for up to 20 minutes, after which the reaction is stopped by the addition of 0.02 ml of EDTA.[5] The mixture is frozen

---

[4] Colorimetric methods, such as the borohydride-stable thiobarbituric acid assay (used in previous reports[2]) were tried and found not acceptable for these investigations. The high concentration of CMP-NAN that is required to achieve substrate saturation in the present work magnifies greatly the difficulties described above when one measures only the decrease in yield of borohydride-stable color. The apparent K_m for CMP-NAN observed in these studies is 6.6-fold greater than that reported previously for the enzyme in sheep liver.[2]

[5] Although no metal requirement could be demonstrated for this enzyme, high concentrations (0.17 M) of EDTA were effective in stopping the reaction. Exposure to high temperature (used previously[2]) will also destroy enzymatic activity. However, CMP-NAN is very labile to this treatment. Thus, when a solution containing 0.033 μmole of CMP-NAN in Tris buffers at either pH 7.2 or 8.5 is placed in a boiling water bath, about 28% is destroyed in 30 seconds and 50% in 1 minute (E. L. Kean, unpublished observations, 1969).

in a Dry Ice–methyl Cellosolve bath and stored frozen until analyzed. A control is run for each complete incubation tube and contains all the components of the complete incubation except enzyme. After incubation of the control, EDTA is added, followed by the enzyme, and the mixture is then frozen.

ISOLATION. A small column of resin is used for each sample. The barrel of a 5-ml disposable plastic syringe (Becton, Dickinson & Co., Rutherford, New Jersey) to which is attached a 22-gauge disposable needle serves conveniently for this purpose. The syringe is plugged with glass wool, and 2 ml of resin (gravity packed) is added. [¹⁴C]NAN and CMP-[¹⁴C]NAN are separated by batchwise elution of the resin. The contents of the incubation tube are added to the resin with a Pasteur pipette, and the tube is washed with water. The combined initial eluate and water washes are collected in a graduated tube to a total of 5 ml. [1-¹⁴C]NAN is eluted with 9.8 ml of 0.1 M NH$_4$HCO$_3$, which is collected in a 10-ml volumetric flask, and diluted to volume. CMP-[1-¹⁴C]NAN is then eluted with 1.0 M KCl. The effluent is collected in 2 batches of 10 ml each although over 92% of the counts are collected in the first 10 ml. Aliquots of the fractions are counted as follows: 1.0 ml of the combined initial eluate and water wash, and 2.0 ml each of the NH$_4$HCO$_3$ and KCl fractions are pipetted into counting vials. Ten milliliters of counting solution is added per milliliter of sample. Cloudy solutions that may result, especially in the first KCl fraction, clarify in the cold. In preliminary studies, no differences in counting rates were observed between samples of ¹⁴C in water, NH$_4$HCO$_3$, or KCl. In addition, even the persistence of cloudy solutions did not decrease the counting rate. Over 95% recovery of the radioactivity added to the column is obtained by this procedure.

Calculations. The difference in total counts of NAN and CMP-NAN between the complete incubation and its control is obtained. The preparations of CPM-[1-¹⁴C]NAN contain about 3% free [1-¹⁴C]NAN.[6] The control incubation corrects for this and any nonenzymatic hydrolysis of CMP-NAN.

Recoveries of radioactivity from the complete and control incubations are within 2% of each other. In the presence of active enzyme (e.g., where there is the production of about 10,000 cpm of NAN over the control of about 2500 cpm) there is less than 10% difference between the production of NAN and the decrease in CMP-NAN. Where there is little enzymatic activity, the relatively small change in the initially high concentration of CMP-NAN may be masked by the differences in re-

---

[6] It is the experience of this laboratory that small amounts of free sialic acid are present in all preparations of CMP-sialic acid.

covery from the columns. Under these circumstances, the 1 to 1 reciprocal relationship between substrate and product may not be apparent. This difficulty is resolved, however, if the counts are normalized to the recovery of either the control or complete tubes. Normalization is not necessary in the measurement of the production of NAN alone since one is dealing with relatively large changes superimposed over small values. Less than 2% difference in the incremental production of counts in NAN is observed between normalized and nonnormalized data.

Method 2. Paper Electrophoresis

This method involves separating NAN and CMP-NAN by means of high voltage paper electrophoresis, locating the radioactive areas by scanning the electrophoretogram, and finally measuring the amount of radioactivity by liquid scintillation spectrometry. Electrophoresis is performed on 10 inch-wide, 22.5 inch-long sheets of Whatman 3 MM paper in 1% sodium tetraborate buffer, pH 9.4. Temperature regulation of the system is important in obtaining reproducible separations. A Gilson high-voltage electrophorator, Model D, used in these studies was attached to a Precision Scientific circulating refrigerated bath, Model 154. Electrophoresis, performed at 4500 V for 40 minutes, is started when the temperature of the reservoir is at $0°$. At the end of this period, the temperature of the reservoir rises to about $+10°$. These conditions were optimal for obtaining reproducible, distinct separations of substrate and product.

Reagents

CMP-[1-^{14}C]NAN, 60 mM, specific activity, 0.27×10^6 cpm/μmole (counted by drying a sample on borate-impregnated filter paper).
Tris·HCl, 1.0 M, pH 9.0
EDTA, 1.0 M, pH 9.0
Sodium tetraborate buffer, 1%, pH 9.4
Toluene counting system, composed of a solution of 5 g of PPO and 0.3 g of POPOP per liter of toluene

Procedure. INCUBATION. Enzyme (0.02–0.1 unit) is incubated with 0.005 ml of CMP-[1-^{14}C]NAN, 0.015 ml of Tris·HCl buffer, and water in a final volume of 0.10 ml at $37°$ for up to 20 minutes. The reaction is terminated by the addition of 0.02 ml of EDTA. Each incubation is accompanied by a control treated in an identical manner, except that enzyme is added after EDTA. The samples are frozen until analyzed.

ELECTROPHORESIS. A sheet of filter paper is soaked in borate buffer and blotted to remove excess buffer; 0.05 ml of the incubation mixture is applied as a streak no longer than 1.5 cm in length. Two samples and their controls are applied to a sheet. After electrophoresis, the sheets are dried and inspected under ultraviolet light to reveal the location of UV quenching zones. The sheets are cut lengthwise into 1.5 inch-wide strips that encompass the electrophoretic pattern of the samples, and they are scanned to locate the areas of radioactivity. The strip is cut into 0.5-inch sections which are placed into counting vials. The amount of radioactivity is measured in a scintillation spectrometer, using 20 ml of the toluene system per vial.

Over 95% of the radioactivity that is applied to the sheets for electrophoresis is recovered. Differences in recovery of about 1–2% are encountered between the complete incubations and their controls. As discussed previously, the data are normalized to adjust for unequal recovery between the complete and control incubations. The increase in radioactivity in the NAN area and decrease in the CMP-NAN area, relative to the control, are obtained. The one-to-one stoichiometry between the appearance of NAN and the disappearance of CMP-NAN when calculated by these means is greater than 95%.

Table I shows the results of analyses of identical samples of complete and control incubations by both the column and electrophoretic techniques. Each of the methods give similar values for the appearance of NAN and the disappearance of CMP-NAN. In addition, the production of NAN as measured by these two methods is similar to the value obtained when the highly specific NAN-aldolase procedure[7] is used.

TABLE I
CMP-NAN HYDROLASE ASSAYS

	nmoles per incubation tube[a]	
Assay method	Increase in NAN	Decrease in CMP-NAN
Ion exchange column	16.9	16.9
Paper electrophoresis	18.3	18.2
NAN-aldolase[b]	18.4	—

[a] Multiple, identical incubations were performed as indicated in the text and were analyzed for the decrease of CMP-NAN and/or the production of NAN by the method indicated.

[b] P. Brunetti, A. Swanson, and S. Roseman, Vol. 6 [68].

[7] P. Brunetti, A. Swanson, and S. Roseman, Vol. 6 [68]. The applicability of this method for the routine assay of CMP-sialic acid hydrolase is being examined.

Definition of Unit and Specific Activity. One unit of CMP-NAN hydrolase is defined as that quantity of enzyme cleaving 1 μmole of CMP-sialic acid, or producing 1 μmole of sialic acid, per hour when incubated at 37°. Specific activity is defined as units per milligram of protein.

Crude Homogenate. Rat livers are homogenized in 0.25 M sucrose or 0.25 M sucrose-TKM buffer.[8] With a ratio in the homogenate of about 1 g of rat liver per 10 ml of homogenate, 0.01–0.02 ml incubated for 10–20 minutes at 37° will be within the linear range of the assay. Enzymatic activity in crude homogenates varied from about 40 to 70 units of enzyme per gram wet weight of liver.

Subcellular Distribution. As shown in Table II, when homogenates of rat liver are fractionated by the procedures of de Duve *et al.*[9] and of Schneider and Hogeboom,[10] most of the CMP-NAN hydrolase activity is distributed between the nuclear and microsomal fractions.[11]

TABLE II
SUBCELLULAR DISTRIBUTION OF CMP-NAN HYDROLASE

| | Fractionation procedure | | | |
| | de Duve *et al.*[a] | | Schneider and Hogeboom[b] | |
Fraction	Enzyme units/g liver	Relative[c] yield (%)	Enzyme units/g liver	Relative yield (%)
Crude homogenate	71	—	41	—
Cytoplasmic extract	31	—	—	—
Nuclear	65	66	15	33
Mitochondria, heavy	0.46	0.47	3.3	7
Mitochondria, light	0.94	0.95		
Microsomes	27	27	27	59
Supernatant	5.8	5.9	0.28	1
Recovery (%)[d]	103	—	111	—

[a] C. de Duve, B.C. Pressman, R. Gianetto, R. Wattiaux, and F. Appelmans, *Biochem. J.* **60**, 604 (1955).

[b] W. C. Schneider and G. H. Hogeboom, *J. Biol. Chem.* **183**, 123 (1950).

[c] Relative yield is based upon the total recovery of enzymatic activity.

[d] Recovery (%) in the procedure of de Duve *et al.*[a] is based upon the summation of the activities in the cytoplasmic extract plus nuclear fractions, taken as 100%.

[8] G. Blobel and V. R. Potter, *Science* **154**, 1662 (1966). The TKM buffer used by these workers is composed of 50 mM Tris·HCl, pH 7.5, 25 mM KCl, and 5 mM MgCl$_2$.

[9] C. de Duve, B. S. Pressman, R. Gianetto, R. Wattiaux, and F. Appelmans, *Biochem. J.* **60**, 604 (1955).

Purified Nuclei. The enzymatic activity detected in the nuclear fraction probably reflects its nonhomogeneity since nuclei purified by the procedure of Blobel and Potter[8] are essentially inactive. Thus, purified nuclei which retained 26% of the CMP-NAN synthetase activity of the crude homogenate, had only 0.3% of the CMP-NAN hydrolase activity. This is observed with either intact nuclei, nuclei that are disrupted by sonication or homogenization, or nuclei that are dialyzed against 10 mM Tris·HCl buffer, pH 7.5.

Plasma Membranes[11]

Plasma membranes were prepared from rat liver by the procedure of Touster *et al.*[11] The yields (percent of the activities that are present in the crude homogenate) of the marker enzymes, 5'-nucleotidase and phosphodiesterase I, in the plasma membranes were 31% and 33%, respectively. CMP-NAN hydrolase was recovered in the plasma membranes in 21% yield, accompanied by a 12-fold increase in specific activity.

Properties

pH Optimum and Effect of Metals. When measured in Tris·HCl buffers the pH optimum is at about pH 9.0. Although magnesium was used in early studies, as suggested by previous reports,[2] a requirement for metal ions could not be demonstrated.

K_m *for CMP-NAN.* The apparent K_m of the enzyme in the microsomal fraction is 0.59 mM.

Stability. The enzyme in the crude extract and microsomal fraction of rat liver retains full activity for several months when stored at $-20°$. When kept in an ice bath, enzymatic activity declined steadily with about 40% of the original activity remaining after 1 month. Enzymatic activity is inhibited by sulfhydryl compounds, such as mercaptoethanol and glutathione. Thus, the dialysis medium that is used in the preparation

---

[10] W. C. Schneider and G. H. Hogeboom, *J. Biol. Chem.* **183**, 123 (1950).

[11] The subcellular distribution of CMP-NAN hydrolase observed in this study, i.e., present in crude nuclear and microsomal fractions but absent from purified nuclei, is similar to observations made by O. Touster, N. N. Aronson, Jr., J. T. Dulaney, and H. Hendrickson [*J. Cell Biol.* **47**, 604 (1970)] on the subcellular distribution of nucleotide pyrophosphatase and phosphodiesterase I of rat liver. Touster *et al.* concluded that these enzymes are derived from plasma membranes that distribute themselves between the crude nuclear and microsomal fractions. A similar situation appears to hold for CMP-sialic acid hydrolase. (E. L. Kean and K. J. Bighouse, manuscript in preparation, 1972.)

of CMP-NAN synthetase (10 mM Tris·HCl, pH 7.5, containing 1% (v/v) mercaptoethanol)[3] completely inactivates the hydrolase.

Stability of CMP-NAN to Other Hydrolytic Enzymes. CMP-sialic acid is resistant to the action of NAN-aldolase, purified snake venom 5'-nucleotidase, venom phosphodiesterase, alkaline phosphatase, 5'-nucleotidase of bull semen, and neuraminidase.[3,12]

[12] S. Roseman, *Proc. Nat. Acad. Sci. U.S.* **48**, 437 (1962); L. Warren and R. S. Blacklow, *J. Biol. Chem.* **237**, 3527 (1962); D. G. Comb, D. R. Watson, and S. Roseman, *J. Biol. Chem.* **241**, 5637 (1966).

[139] Bacteriophage-Induced Capsular Polysaccharide Depolymerase[1]

By EDWARD C. YUREWICZ, MOHAMMED ALI GHALAMBOR, and EDWARD C. HEATH

A common feature of phage infection of encapsulated bacteria is the induction of enzymes that degrade the capsular (or slime) polysaccharide of the host. Such enzymes commonly termed polysaccharide depolymerases, have been examined from a number of phage–host cell systems.[2-8] In all cases where the mechanism of polysaccharide degradation has been studied, the phage-induced polymerases have been shown to be glycanohydrolases that exhibit a high degree of specificity for the specific extracellular polysaccharide produced by the host organism. The availability of specific glycanohydrolases of this type have been of extraordinary utility in determining the structural characteristics of complex heteropolysaccharides. On the other hand, because of the specificity of these enzymes for the particular polysaccharide produced by any given host organism, it appears essential to isolate a phage that will induce the production of a capsular polysaccharide depolymerase in the specific organism that elaborates the polysaccharide of interest. This report describes the isolation of a virulent bacteriophage for an encapsulated

[1] E. C. Yurewicz, M. A. Ghalambor, D. H. Duckworth, and E. C. Heath, *J. Biol. Chem.* **246**, 5607 (1971).
[2] B. H. Park, *Virology* **2**, 711 (1956).
[3] M. H. Adams and B. H. Park, *Virology* **2**, 719 (1956).
[4] C. Eklund and O. Wyss, *J. Bacteriol.* **84**, 1209 (1962).
[5] I. W. Sutherland and J. F. Wilkinson, *J. Gen. Microbiol.* **39**, 373 (1965).
[6] P. F. Bartell, T. E. Orr, and G. K. H. Lam, *J. Bacteriol.* **92**, 56 (1966).
[7] A. M. Chakrabarty, J. F. Niblack, and I. C. Gunsalus, *Virology* **32**, 532 (1967).
[8] P. F. Bartell, G. K. H. Lam, and T. E. Orr, *J. Biol. Chem.* **243**, 2077 (1968).

strain of *Aerobacter aerogenes* as well as the purification and properties of a glycanohydrolase which specifically degrades the capsular polysaccharide of the host organism.

Host Organism and Growth Conditions

The strain of *Aerobacter aerogenes* used in these studies, when grown on solid media, yields colonies which are highly mucoid due to the production of large amounts of capsular polysaccharide. When the organism is grown in liquid media, particularly in a synthetic medium containing low nitrogen content, large amounts of extracellular polysaccharide and slime may be isolated from the viscous culture filtrate. Detailed structural analysis of this polysaccharide[9] indicated that it is a high molecular weight polymer composed of repeating sequences of the following tetrasaccharide:

$$\overset{3}{\longrightarrow} \text{D-galactose} \overset{1\alpha3}{\longrightarrow} \text{D-mannose} \overset{1\alpha3}{\longrightarrow} \text{D-galactose} \overset{1\alpha}{\longrightarrow}$$

$$\beta\underset{1}{\overset{2}{\uparrow}}$$

$$\text{D-glucuronic acid}$$

Moreover, these studies also indicated that the phage-induced polysaccharide depolymerase described below is a glycanohydrolase which specifically cleaves the galactosyl-$1 \overset{\alpha}{\to} 3$-galactose linkages in the capsular polysaccharide.

For production of polysaccharide depolymerase, the host organism is grown in a chemically defined medium of the following composition (grams per liter); $(NH_2)_2SO_4$, 10; Na_2HPO_4, 10; KH_2PO_4, 3; K_2SO_4, 1; NaCl, 1; $MgSO_4 \cdot 7H_2O$, 0.2; $CaCl_2 \cdot 2H_2O$, 0.014; $FeSO_4 \cdot 7H_2O$, 0.001; casamino acids, 0.5; and glucose, 10. Glucose is prepared as a 50% solution, sterilized separately, and aseptically added to the sterilized salts medium. Organisms are grown at 37° in New Brunswick fermentors with mechanical agitation and forced aeration.

Stock cultures of *A. aerogenes* are maintained at $-15°$ in Trypticase soy broth containing 15% glycerol; vials containing 1 ml of bacterial suspension are grown in Trypticase soy broth overnight and plated on the same medium containing 2% agar. Single mucoid colonies are isolated and used to inoculate media for further studies.

Isolation of Bacteriophage

A virulent bacteriophage for *A. aerogenes* may be isolated as follows: 25 ml of raw sewage is clarified by passage through a bacterial filter

[9] E. C. Yurewicz, M. A. Ghalambor, and E. C. Heath, *J. Biol. Chem.* **246**, 5596 (1971).

and then added to a flask containing 5 ml of 5-fold concentrated Trypticase soy broth (15 g/100 ml) and inoculated with 1 ml of a log phase culture of the organism. After 5 hours of incubation at 37° on a rotary shaker, cells are removed by centrifugation. A 10-ml aliquot of the supernatant fluid is diluted to 20 ml with 6% sterile Trypticase soy broth and inoculated with 1 ml of a log phase culture of the organism. Complete lysis occurs after incubation for 2 hours at 37°. Appropriate dilutions of the crude lysate are plated with *A. aerogenes* with a soft agar layering technique according to Adams.[10] Several types of plaques are observed although the predominant morphology characteristically exhibits a translucent, depressed halo surrounding the clear zone of lysis. This type of plaque morphology is formed characteristically by bacteriophage that are capable of inducing the synthesis of a capsular polysaccharide depolymerase in infected host cells. A bacteriophage, designated as K2 in the remainder of this paper, may be purified by repeated single plaque isolation as described by Adams.[10]

Cell lysates used in the preparation of phage K2 or polysaccharide depolymerase are prepared routinely as follows: 9000 ml of synthetic medium is inoculated with 1000 ml of an overnight culture of *A. aerogenes*. After incubation for 75–85 minutes at 37°, phage K2 is added at a multiplicity of infection of 1/25 and incubation is continued; lysis is usually complete within 2 hours. The lysate is then chilled immediately to 4° and cell debris is removed by centrifugation.

Purified phage suspensions are prepared as follows: solid ammonium sulfate (50 g/100 ml) is added to a cell lysate; the resultant precipitate is dissolved in T2 buffer (see below) and centrifuged at 900 g for 30 minutes to remove insoluble material. The supernatant fluid from the previous step is centrifuged at 78,800 g, the supernatant is discarded, and the pellet is resuspended in T2 buffer. To 4.5 ml of crude phage suspension are added 2.4 g of CsCl, and the suspension is centrifuged to equilibrium (36 hours) at 39,000 rpm in a Beckman SW 39 rotor. The phage are concentrated in a narrow band near the bottom of the tube and are collected by puncturing the bottom of the tube and allowing the contents to flow dropwise from the tube. Phage suspensions are stored at 4° in T2 buffer containing the following components (grams per liter): Na_2HPO_4, 3.0; KH_2PO_4, 1.5; NaCl, 4.0; K_2SO_4, 5.0; $CaCl_2 \cdot 6H_2O$, 0.02; and $MgSO_4 \cdot 7H_2O$, 0.2. The phage titer is stable for at least one year if the suspension contains more than 10^{10} phage per milliliter.

[10] M. H. Adams, *in* "Methods in Medical Research" (J. H. Comroe, Jr., ed.), Vol. 2, p. 1. Year Book Publishers, Chicago, 1950.

Purification and Properties of Phage-Induced Depolymerase

Assay Method

Principle. The most sensitive and reliable assay for depolymerase is based on the determination of the rate of formation of reducing sugar equivalents resulting from the endogalactosidase activity of the depolymerase on capsular polysaccharide. The most sensitive method of determining reducing sugar equivalents is based on the method of Park and Johnson[11] with galactose as a standard.

Reagents

Sodium acetate buffer, 0.5 M, pH 5.2

Purified capsular polysaccharide, 1.2 mg/ml

Potassium ferricyanide, 0.5 g in 1 liter of water (stored in brown bottle)

Carbonate-cyanide reagent consisting of 5.3 g of sodium carbonate and 0.6 g of KCN per liter of solution

Ferric iron reagent containing 1.5 g of ferric ammonium sulfate and 1 g of sodium monolauryl sulfate (D-duponol) in liter of 0.05 N H_2SO_4

Procedure. For routine assay of depolymerase, incubation mixtures contain the following components in a total volume of 0.25 ml; purified capsular polysaccharide, 0.12 mg; sodium acetate buffer, 25 μmoles; and appropriate dilutions of the enzyme. Incubation is conducted at 42° and the reaction is followed by removing 50-μl aliquots at appropriate intervals. Each aliquot is transferred directly to a tube containing 0.2 ml of water and 0.2 ml each of the ferricyanide solution and of the carbonate-cyanide reagent. After each desired aliquot has been treated in this manner, the tubes are heated in a boiling water bath for 15 minutes, cooled, and 1 ml of ferric iron reagent is added to each. The mixtures are shaken, allowed to stand for 15 minutes and the resultant blue color is quantitated at 620 nm. The assay is linear for a period of at least 10 minutes and over a concentration range of at least 10-fold.

Definition of a Unit. One unit of enzyme activity is defined as that amount which catalyzes the formation of 1 μmole of reducing sugar (galactose equivalents) per minute under the conditions specified above, and specific activity is defined as units per milligram of protein. Protein

[11] J. T. Park and M. J. Johnson, *J. Biol. Chem.* **181**, 149 (1949).

was estimated by the ultraviolet-spectrophotometric method of Waddel,[12] using bovine serum albumin as a standard.

Purification Procedures

Three 10-liter cultures of log phase *A. aerogenes* are infected with phage K2, and lysis is allowed to proceed to completion. The lysate is chilled to 4° and all subsequent purification procedures are conducted at 0–4° unless otherwise indicated. Because of interfering reactions, it is not possible to accurately determine enzyme activity in the crude lysate.

Step 1. Ammonium Sulfate Precipitation. After removal of cell debris by centrifugation, ammonium sulfate (43.6 g/100 ml) is added to the lysate and the resultant precipitate is allowed to settle overnight; the major portion of the supernatant is siphoned off and discarded. The remaining suspension is centrifuged at 15,000 *g* for 25 minutes, and the pellet is retained.

Step 2. Extraction of Ammonium Sulfate Pellet. The pellet is suspended in 148 ml of 10 m*M* Tris buffer, pH 7.5, stirred for 1 hour, and then centrifuged at 7800 *g* for 1 hour; the resultant supernatant is discarded since sufficient residual ammonium sulfate was present in the pellet to prevent extraction of depolymerase by the indicated amount of buffer. Most of the depolymerase activity is then extracted from the pellet by the following procedure: the pellet is resuspended in sufficient 10 m*M* Tris buffer, pH 7, containing 1% ammonium sulfate, to give a final volume of 310 ml. The resultant suspension is dialyzed against three changes (12 liters each) of the same buffer over a period of 22 hours. The suspension is then centrifuged at 12,000 *g* for 10 minutes and the pellet is discarded. Phage are removed from the supernatant fluid by centrifugation at 78,800 *g* for 90 minutes.

Step 3. DEAE-Cellulose Chromatography. The supernatant fluid is diluted to 1400 ml with 10 m*M* Tris buffer, pH 7, and quantitatively adsorbed to DEAE-cellulose by batchwise addition of adsorbent. The depolymerase-containing resin is packed into a column (4 × 26 cm), washed with buffer, and then eluted with a linear gradient of KCl in the same buffer. Depolymerase activity elutes as a single peak at approximately 0.18 *M* KCl. Prior to the final step of purification, the majority of the KCl is removed from the depolymerase solution by concentrating the solution to 52 ml by pressure dialysis and then diluting to 500 ml with 10 m*M* Tris buffer, pH 7.

Step 4. CM-Cellulose Chromatography. The depolymerase solution is

[12] W. J. Waddell, *J. Lab. Clin. Med.* **48**, 311 (1956).

passed through a column of CM-cellulose (4 × 40 cm) previously equilibrated with 10 mM Tris buffer, pH 7. Due to a rather low affinity of depolymerase for the resin only 66% of the total units applied are adsorbed. The saturated resin is washed with buffer and the column eluted with a linear gradient of KCl. The specific activity of depolymerase in each of the fractions is constant across the peak, suggesting homogeneity. Fractions containing depolymerase activity are pooled, concentrated to 3.6 ml by pressure dialysis, and dialyzed against 10 mM Tris buffer, pH 7.5, containing 1% ammonium sulfate.

The purification of depolymerase is summarized in the table.

Properties of Depolymerase

Criteria of Purity. Disc gel electrophoresis of purified depolymerase at pH 4 reveals a single protein component which migrates toward the cathode. Elution of slices of an unstained control gel permits demonstration of depolymerase activity coincident with the protein band. Electrophoresis of depolymerase at pH 8.3 reveals a protein component which barely entered the gel, migrating toward the anode. Again, enzymatic activity is found to be coincident with the protein band.

Purified depolymerase may be subjected to molecular sieve chromatography on a Sephadex G-200 column. When this is done all the protein emerges as a single symmetrical peak just within the void volume of the column. Depolymerase activity is coincident with the protein peak and exhibits a constant specific activity in all fractions, again suggesting that purified depolymerase is homogeneous.

The molecular weight is estimated to be approximately 380,000. Studies on the subunit structure of depolymerase indicate that the enzyme is composed of four each of two nonidentical subunits of 36,000 and 63,000 molecular weights, respectively.

PURIFICATION OF DEPOLYMERASE[a]

Fraction	Units	Protein (mg)	Specific activity (units/mg)	Purification (-fold)
I. Ammonium sulfate pellet	5870	3410	1.7	1
II. Second extract of I	4970	750	6.6	4
III. DEAE-cellulose	4160	48.5	86	51
IV. CM-cellulose	1150	4.9	235	140

[a] Depolymerase was purified from 30 liters of crude cell lysate as described in the text.

Endohydrolase. The action of depolymerase on capsular polysaccharide is characterized by (a) a rapid reduction in the viscosity of the substrate and (b) a relatively slow release of reducing groups. These results strongly suggest an endohydrolytic mode of action. Under conditions of limited hydrolysis of the capsular polysaccharide, the release of reducing groups is linear with respect to both time and enzyme concentration and is the basis of the enzyme assay.

Effect of pH, Buffer Concentration, and Temperature. Maximum depolymerase activity is observed at pH 5.2 in 100 mM acetate buffer. The buffer concentration is important; lowering of the acetate concentration below 40 mM results in significant loss of depolymerase activity. Depolymerase activity increases with increasing temperature up to 55°, indicating a relatively heat-stable protein. Heating of the enzyme at 68° for 10 minutes results in 52% inactivation.

Inhibitors. Depolymerase activity does not require the addition of metal ions and is not inhibited by a variety of chelating agents, including: EDTA, potassium cyanide, *o*-phenanthroline, and 8-hydroxyquinoline. Of the various metal ions tested for inhibition, only Fe^{3+} markedly effects depolymerase activity; the presence of $5 \times 10^{-5} M$ $FeCl_3$ results in 50% inhibition. Dithiothreitol and *p*-chloromercuribenzoate have no effect on either depolymerase activity or stability of the enzyme.

Specificity. As indicated above, the phage-induced depolymerase is highly specific for the capsular polysaccharide of the host. Thus, the glycanohydrolase exhibits no detectable activity on a variety of synthetic galactosides or other galactose-containing heteropolysaccharides from various sources.

[140] Trehalase from *Streptomyces hygroscopicus* and *Mycobacterium smegmatis*[1]

By Betty W. Patterson, Ann Hey Ferguson, Mike Matula, and Alan D. Elbein

Trehalase has been isolated and purified from both *Streptomyces hygroscopicus*[2] and *Mycobacterium smegmatis*. In *S. hygroscopicus*, the

[1] This work was supported by grants from the Robert A. Welch Foundation and the National Institute of Allergy and Infectious Diseases.
[2] A. E. Hey and A. D. Elbein, *J. Bacteriol.* 96, 105 (1968).

enzyme is isolated from the soluble fraction whereas in *M. smegmatis* the enzyme is membrane-bound and is solubilized by butanol treatment.

$$\text{Trehalose} + \text{H}_2\text{O} \rightarrow 2 \text{ glucose}$$

Assay Method

Principle. The activity of the enzyme is measured either by the formation of reducing sugar using the Nelson method[3] or by the formation of glucose with glucose oxidase.

Reagents

α,α-Trehalose, 0.1 M
Sodium cacodylate buffer, 0.1 M, pH 6.5, or sodium phosphate buffer, 0.1 M, pH 7.5
Nelsons reducing sugar reagents or glucostat

Procedure. For routine assay of the streptomyces enzyme, incubation mixtures contained 20 μmoles of sodium cacodylate buffer, 3.0 μmoles of trehalose and an appropriate amount of enzyme (10–50 μl) in a final volume of 0.1 ml. For assay of the mycobacterial enzyme, incubation mixtures contained 100 μmoles phosphate buffer, 5 μmoles trehalose, and 10–100 μl of enzyme in a final volume of 0.2 ml. Mixtures were incubated for 15–30 minutes at 37°. The reaction was stopped by placing tubes in a boiling water bath for 5 minutes. The volume was then adjusted to 1 ml, and insoluble protein was removed by centrifugation. The supernatant liquid was then tested by the Nelson method for reducing sugar or by glucostat for free glucose. Protein was measured by the method of Sutherland *et al.*[4]

Definition of Enzyme Activity. One unit of enzyme is defined as that amount which hydrolyzes 1 μmole of trehalose in 1 minute at 37°.

Trehalase from S. hygroscopicus

Purification

Growth of Bacteria and Preparation of Crude Extract. S. hygroscopicus is grown at room temperature with shaking in a medium of the following composition[5] (g/100 ml): $(\text{NH}_4)_2\text{SO}_4$, 0.1; NH_4NO_3, 0.5; $\text{MgCl}_2 \cdot 6\text{H}_2\text{O}$, 0.5; K_2HPO_4, 0.3; CaCO_3, 0.1; glucose, 8.0 and mineral

[3] N. Nelson, *J. Biol. Chem.* **153**, 375 (1944).

[4] E. W. Sutherland, C. F. Cori, R. Haynes, and N. S. Olsen, *J. Biol. Chem.* **180**, 825 (1949).

[5] A. D. Elbein, R. L. Mann, H. E. Renis, W. M. Stark, H. Koffler, and H. R. Garner, *J. Biol. Chem.* **236**, 289 (1961).

mixture C, 1.0. Mineral mixture C has the following composition (g/100 ml); FeSO$_4$·7H$_2$O, 0.28; ferric ammonium citrate, 0.27; CuSO$_4$·5H$_2$O, 0.0125; MnSO$_4$·H$_2$O, 0.1 and CoCl$_2$·6H$_2$O, 0.01. Cells are harvested by filtration, washed well with distilled water and stored in the freezer. Sixty grams of cell paste is suspended in 180 ml of 0.1 M phosphate buffer, pH 7.5 containing 90 mg EDTA (0.5 mg/ml) and 180 mg of egg white lysozyme (1 mg/ml). After incubation for 1 hour, the cell debris is removed by centrifugation. All operations are conducted at 0°C.

Manganese Precipitation and First Ammonium Sulfate Fractionation. To 180 ml of crude extract, 7.2 ml of 1 M MnCl$_2$ is added slowly with stirring. The precipitate is removed by centrifugation, and the supernatant liquid is brought to 30% saturation by the addition of 31.5 g of solid ammonium sulfate. The precipitate is removed by centrifugation, and an additional 38.7 g of solid ammonium sulfate is added to the supernatant liquid (to 60% saturation). The precipitate is removed by centrifugation, dissolved in 30 ml of distilled water, and dialyzed overnight against distilled water.

Calcium Phosphate Gel Treatment. To 33 ml of the ammonium sulfate fraction, 49.5 ml of calcium phosphate gel (15 mg/ml) is added. The mixture is allowed to stand for 15 minutes and is then centrifuged. The supernatant fluid is discarded, and the gel is suspended in 50 ml of 40 mM phosphate buffer, pH 6.5. The gel suspension is stirred mechanically for 30 minutes and then centrifuged. Enzymatic activity is recovered in the supernatant fluid.

Second Ammonium Sulfate Fractionation. To 50 ml of supernatant liquid from the calcium phosphate gel elution, 12.1 g of solid ammonium sulfate (40% saturation) is added. The precipitate is removed by centrifugation, discarded, and an additional 5.1 g of solid ammonium sulfate is added to the supernatant fluid (55% saturation). The precipitate is removed by centrifugation, dissolved in 10 ml of distilled water and dialyzed overnight against distilled water.

Hydroxyapatite Chromatography. A 10-ml amount of the second ammonium sulfate fraction is applied to a column (1.5 × 6 cm) of hydroxyapatite which was previously equilibrated with 1 mM phosphate buffer, pH 6.5. The column is washed with 50 ml of 1 mM phosphate buffer and then with 50-ml portions of 10 mM, 20 mM, and 30 mM phosphate buffer. Enzymatic activity is removed at 30 mM phosphate buffer.

The results of the purification procedure are summarized in Table I.

Properties

Stability. The enzyme is stable to freezing at all stages of purification. The purified enzyme shows no loss in activity after storage for 4 weeks

TABLE I

PURIFICATION OF *Streptomyces hygroscopicus* TREHALASE

Fraction	Total units	Specific activity (units/mg of protein) $\times 10^{-2}$
Crude extract	35.0	2.2
MnCl$_2$ and ammonium sulfate I	30.0	7.2
Calcium phosphate gel	28.3	14.8
Ammonium sulfate II	15.8	21.8
Hydroxyapatite	3.3	166.7

at $-20°$ and can be frozen and thawed several times with no loss of activity.

Specificity and Kinetics. The enzyme is highly specific for trehalose. None of the other glucose disaccharides (kojibiose, sophorose, nigerose, laminaribiose, maltose, cellobiose, isomaltose, gentiobiose) are hydrolyzed by this enzyme preparation. The enzyme exhibits some activity (13% the rate of trehalose) on trehalosamine (α-D-glucose-α-D-glucosamine). The K_m for trehalose is $1.8 \times 10^{-2} M$.

Effect of pH. The pH optimum in cacodylate buffer is about 6.5. High concentrations of phosphate $(0.2 M)$ cause a 20% inhibition of activity whereas Tris buffer at this concentration completely inhibits the activity.

Trehalase from *Mycobacterium* smegmatis

Purification

Growth of Bacteria and Preparation of Crude Extract. *M. smegmatis* is grown with shaking in Trypticase Soy Broth at 37° for 72 hours. Cells are harvested by centrifugation, washed with distilled water and stored at $-20°$. One hundred grams of cells are suspended in 500 ml of 10 mM phosphate buffer, pH 7.5, and subjected to sonic oscillation for 6–8 minutes. The extract is then centrifuged at 20,000 g for 15 minutes. The supernatant liquid is further centrifuged at 150,000 g for 14–18 hours. The supernatant liquid is discarded and the pellet is resuspended in about 30 ml of phosphate buffer and homogenized.

Solubilization of Trehalase. Cold butanol is slowly added to the above suspension until 10% saturation is reached. The mixture is allowed to stir at 0° for 30 minutes and then centrifuged at 150,000 g for 14–18 hours. The supernatant liquid is removed and lyophilized. Alternatively, the butanol-treated mixture can be lyophilized without centrifuging. In this case, the resuspended powder is centrifuged at 20,000 g for 1 hour to remove debris.

TABLE II

PURIFICATION OF *Mycobacterium smegmatis* TREHALASE

Fraction	Total units	Specific activity (units/mg of protein) $\times 10^{-3}$
Crude	108	1.3
Pellet	108	7.8
Butanol + lyophilization	45	5.3
Ammonium sulfate	35.8	9.8
DEAE 0.4 M KCl	3.6	13
DEAE 0.6 M KCl	3.3	44

Ammonium Sulfate Fractionation. The lyophilized powder is suspended in half-volume of distilled water, allowed to stand for 30 minutes with occasional stirring, and centrifuged at 20,000 g for 1 hour. The supernatant liquid is then brought to 35% saturation by the addition of solid ammonium sulfate (20.9 g/100 ml). After the preparation has stood for 30 minutes, the precipitate is isolated by centrifugation, resuspended in 10 mM phosphate at pH 7.5, and dialyzed against the same buffer.

DEAE-Cellulose Chromatography. The above fraction is applied to DEAE cellulose (2.5 $\times$ 30 cm) and the column is washed with 10 mM phosphate buffer. The column is then eluted with 200 ml of each of the following molar concentrations of KCl in 10 mM phosphate buffer: 0.1, 0.2, 0.3, 0.4, and 0.6. Trehalase activity is eluted in both the 0.4 M and 0.6 M KCl elutions. Although the specific activity is higher in the 0.6 M fraction, this fraction contains a greater amount of maltase activity, whereas the 0.4 M fraction has only slight maltase activity.

The purification procedure is outlined in Table II.

Properties

Stability. All preparations through the ammonium sulfate fraction can be stored at $-20°C$ for at least 2 months without any loss in activity.

Metal Ion Requirement. The enzyme exhibits a requirement for Mg^{2+}, which is most pronounced in the ammonium sulfate and DEAE-cellulose fractions. The optimum concentration of Mg^{2+} is 2.5×10^{-3} M.

Specificity. The enzyme fraction shows slight activity on maltose but other glucose disaccharides are not hydrolyzed. If activity on trehalose is taken as 100, the activity on maltose in the 0.4 M fraction is 13.

Effect of pH. The optimum pH is 7.5 in phosphate buffer.

Author Index

Numbers in parentheses are reference numbers and indicate that an author's work is referred to, although his name is not cited in the text.

Chizhov, O. S., 188
Chrambach, A., 892, 893, 894, 896(12, 13)
Chrambach, R. A., 411
Christian, W., 773
Cifonelli, J. A., 77, 79, 88, 96, 125, 126, 313, 682, 903, 910
Claeyssens, M., 728
Clamp, J. R., 20, 148, 149, 161
Clarke, J. T. R., 719
Clauser, H., 36, 46
Cleland, R. L., 130, 134
Cline, M. J., 314
Coffin, A., 328
Coghill, S. R., 416, 983
Cohen, A., 551, 552, 553
Cohen, P., 963
Cohn, M., 391
Cole, R. D., 814
Colowick, S. P., 469
Colvin, B., 500, 501, 503(3), 506(3, 5), 507, 508(2)
Comb, D. G., 990
Conchie, J., 786, 787, 792, 814, 819(1)
Conrad, H. E., 180, 281, 544, 682
Constantopoulos, G., 137
Copeland, P. C., 240, 241, 242, 244
Corey, E. J., 180
Cori, C. F., 997
Corneil, I., 216
Costantino-Ceccarini, E., 492
Courtois, J. E., 523
Covelli, I., 810
Craig, L. C., 751
Cranzer, E., 154
Cree, G. M., 180
Cremer, N. E., 333
Crimmin, W. R. C., 44
Croon, I., 192
Cruickshank, C. N. D., 94
Cuatrecasas, P., 206, 208(5), 354, 392, 393, 878, 893, 897, 898(1), 899(2, 3), 900(3), 902
Cunningham, L. W., 16
Cynkin, M. A., 260, 584, 589(8), 591(8), 599, 601(24)

D

Dalferes, E. R., 114, 117(44), 118(44)
Dance, N., 863, 866(8)
Danker, A. K., 54

Dankert, M., 583, 584(2), 594, 602, 606, 623(2)
Dante, M. L., 893, 896(13)
Davidson, E., 82, 83(19)
Davidson, E. A., 70, 673, 921
Davies, M., 77
Davis, B. J., 335, 410, 412(2), 447, 543, 760, 771, 789, 794, 882, 883(7)
Davis, R. V., 235
Dawson, G., 149, 152, 159, 160(5), 162, 713, 715, 837, 846, 847
Dawson, R. M. C., 567
DeBelder, A. N., 194
DeBernardo, S. L., 20
DeBruyne, C. K., 728
Decker, L., 135
DeDuve, C., 420, 808, 814, 815(8), 819, 988
Defendi, V., 232
Deferrari, J. O., 740, 745
DeJongh, D. C., 159, 160(5)
Dekaban, A. S., 137
DeLellis, R., 814
Delvin, E., 818, 819
Denborough, M. A., 328
Denton, W. L., 509
Desai, P. R., 383, 384(8, 11), 385(8, 9), 386(4, 11), 387(3, 11)
De Souza, B. C., 953, 954
DeTyssonsk, E. R., 90
DeVries, A. L., 713
Dey, P. M., 699
Dietzler, D. N., 692
DiJeso, F., 447
Dintzis, H. M., 231
Dische, Z., 96, 113, 119(40), 196, 241, 285, 290, 455, 456, 672
Distler, J. J., 44, 276, 277(3), 297, 482, 484(1), 561, 653, 655(20)
Dittmer, J. C., 148, 151(15), 606
Dodge, J. T., 245, 253
Dodgson, K. S., 122, 880, 912, 918
Doe, R. P., 814
Doi, A. K., 281
Doi, K., 281
Donald, A. S. R., 12
Donoso, L., 728
Dorfman, A., 77, 88, 92, 96, 130, 641, 645, 655, 657(15, 21), 800, 903
Doty, M., 225

Hill, D. L., 493
Hill, E. A., 296, 297(1)
Hill, R. L., 500, 503(4), 505, 506(4)
Himmelspach, K., 222, 224(7), 225(7), 229(7), 231(7)
Hinrichsen, D. F., 20, 799, 800(5), 801. 802(1), 803(1), 804(1), 805(1)
Hirs, C. H. W., 15, 49, 60, 694
Hirsch, E. I., 678, 679(8), 681(8), 683(8)
Hirst, J. W., 391
Hjertén, S., 364, 969
Ho, M. W., 820, 824
Hodge, J., 218
Högman, C., 382
Höök, M., 681
Hof, L., 300, 303(1), 304(1), 840, 844, 847(1), 848
Hoffman, P., 35, 78, 82, 83(19), 671, 673 (40), 903, 912
Hofmann, E., 917
Hofreiter, B., 218
Hogeboom, G. H., 420, 988, 989
Hollerman, C. E., 313, 314(3)
Hopfer, U., 567
Hopton, J. W., 763
Horecker, B. L., 254, 256(5), 561, 584. 585(5), 586(5), 587(5), 588(5), 589 (5), 591(5), 602, 623(3)
Horikoshi, K., 923
Horton, D., 28
Horton, R. E., 385
Horwitz, A. L., 645, 657(15)
Hough, L., 149
Hovingh, P., 903, 905(2), 908(4), 909(4), 910(4)
Howard, A. N., 212
Howard, I. K., 332, 333(2, 4), 334(2, 4). 337(2, 4, 5, 6), 339(6)
Howe, C., 236, 237(5, 7), 239(7), 240, 241, 242, 243(7), 244(7), 246, 248(3). 250. 763, 968, 970(5)
Howell, S. F., 313, 314(2)
Hribar, J. D., 159, 160(5)
Huang, C.-C., 16
Hudgin, R. L., 416, 756
Huggins, C., 815
Hughes, R. C., 20, 756
Hultberg, B., 820
Hulyalkar, R. K., 174
Hungate, R. E., 945

Hunkler, F. L., 532, 538(6)
Hunter, W. M., 347
Hurwitz, J., 49
Huskins, R. H., 195

I

Iijima, Y., 763
Illiano, G., 206, 208(5), 897, 898(1), 902
Inbar, M., 314
Inman, J. K., 231
Inoue, K., 232, 233
Inouye, M., 124
Irvine, R. A., 207, 208(8)
Isbell, H. S., 366
Iseki, S., 968
Ishihara, H., 285, 399, 403
Ishiyama, I., 369
Ito, E., 692
Iyer, R. N., 46, 216
Izaki, K., 689

J

Jabbal, I., 285, 399, 416, 756
Jackson, R. C., 786
Jackson, R. L., 15, 54, 55(7), 59(7), 61 (7, 8), 254
Jacobson, C. B., 856, 873, 877
Jaffé, W. G., 314
Jamieson, G. A., 20, 341, 343(9)
Jan, K., 254, 257(2), 258(2)
Janis, R., 232, 233
Jarrige, P., 763
Jatzkewitz, H., 881, 884
Jayme, G., 581
Jeanes, A., 8, 151, 159, 166(3), 167, 241
Jeanloz, R. W., 20, 76, 210, 233, 756
Jeffrey, P. L., 114, 805, 806, 808(2), 811 (2, 3), 812(2, 3), 813(2, 3)
Jencks, W. P., 252
Jett, M., 20
Jirgensons, B., 387
Jocius, I. B., 391, 392
Jörnvall, H., 372
Joffe, S., 233
John, C. E., 435
Johnson, C. H., 250
Johnson, G. S., 64, 69(2)
Johnson, M. J., 359, 993

Subject Index